# Fundamentals of Physical Geography

SECOND CANADIAN EDITION

# Fundamentals of Physical Geography

SECOND CANADIAN EDITION


David Briggs
University of Huddersfield

Peter Smithson
University of Sheffield

Timothy Ball
University of Winnipeg

Peter Johnson
University of Ottawa

Peter Kershaw
University of Alberta

Antoni Lewkowicz
University of Toronto

Copp Clark Pitman Ltd.
A Longman Company
Toronto

Cover photo: Spring thaw along Hudson Bay coast of Quebec. The long dark swirls are the low sandy Belcher Islands, a remnant of the continental glacial age. The photo is from *Challenger* 7, 51b, 29 April 1985—6 May 1985 (Courtesy NASA).

**ISBN** 0-7730-5173-2

Executive editor: Jeff Miller
Editors: John Meyers, Melanie Sherwood
Photo research: Walter Wagnleithner
Cover design: Avril Orloff
Design: Steve MacEachern/Smart Work
Illustrations: Allan Moon, Dave McKay, Valentino Sanna
Maps: Martyn Lengden
Typesetting: Marnie Morrissey

**Canadian Cataloguing in Publication Data**
Fundamentals of physical geography
2nd Canadian ed.

Includes bibliographical references and index.
ISBN 0-7730-5173-2

1. Physical geography. I. Briggs, David J. (David John), 1948– .

GB55.F86 1993 910'.02 C92-095646-7

Copp Clark Pitman Ltd.
2775 Matheson Blvd. East
Mississauga, Ontario
L4W 4P7

Associated companies:
Longman Group Ltd., London
Longman Inc., New York
Longman Cheshire Pty., Melbourne
Longman Paul Pty., Auckland

Printed and bound in Canada

2 3 4 5 5173-2 97 96 95 94

# Overview of Contents

# Contents

# Preface

The need for a more complete awareness of the earth and the way it works is growing and geography as the original interdisciplinary discipline can help. The world is not a series of separate units, but is integrated and interdependent. We have learned how to dissect the world into discrete units, each studied by specialists. We have to put the units together to understand how they interrelate if we are to make proper decisions about the problems facing the earth.

This philosophy was basic to the work of David Briggs and Peter Smithson in their original text, *Fundamentals of Physical Geography*, published in the U.K. in 1985. Its appeal was the impetus for taking the decision to adapt the book for Canadian classrooms. Building on the excellent foundation laid by Briggs and Smithson, the first Canadian edition of *Fundamentals of Physical Geography*, prepared by Tim Ball, sought to bring a Canadian perspective to the table, introducing major changes to the sections on climate, soils, periglacial and glacial environments, and including throughout examples, terms, and idioms more appropriate to North America.

The enthusiastic reception that greeted the first Canadian edition brought with it the clamour for more Canadian material. Four years have passed since publication and in that time there have been advances in the discipline. As well, there have been changes in our physical world brought on by both natural and human activity. Together these factors led to the decision that the time was right to bring out a new edition.

The preparation of the second edition takes the process begun by Tim Ball a major step forward. It is a collaborative effort of four authors, each of whom has brought their particular expertise to the project, but all of whom share a common pedagogical goal. Tim Ball is principally responsible for updating and rewriting the first fourteen chapters. Peter Johnson is the principal author for the geomorphology chapters (fifteen through twenty-five) with the exception of the chapter on periglacial systems. Antoni Lewkowicz has updated the periglacial chapter which he wrote for the first edition. The final six chapters on ecology, soils, and biogeography have been thoroughly revised by Peter Kershaw.

Many colleagues read and commented on all or parts of this text. In particular we would like to acknowledge Geoff Scott of the University of Winnipeg, Wayne Rouse of McMaster University, Fes de Scally of Okanagan College, and Ian Campbell of the University of Alberta for their thorough reviews of the earlier edition of the text.

One of the improvements made in the second Canadian edition is the replacement of all photographs. We are extremely grateful to the following for their assistance in locating or providing many of the photographs for this book: Dr. D.R. Grant, Geological Survey of Canada; Dr. Robert Schemenauer and his colleagues, Atmospheric Environment Service; Dr. J. Shaw, Geography Department, University of Alberta; Dr. Denis A. St-Onge, Polar Continental Shelf Project, Energy, Mines and Resources Canada, and Ms. Debbie Dodds, NASA Johnson Space Center.

John Meyers edited the text and, in that capacity, exhibited considerable skill, patience, and good humour. The art and desktop publishing departments at Copp Clark are responsible for the visual attractiveness of this volume. Together, the publisher's editorial and production staff have produced a first-rate publication.

Finally, we would like to acknowledge our families. To Marty Ball for her continued understanding, assistance, and support; to Eileen Johnson and family for enduring the long summers away from home; and to Linda Kershaw whose careful editorial work immeasurably improved the text, we dedicate this book.

# Preface to the original edition

The basic aim of physical geography has always been to understand how the world works. In the past, this aim was pursued through the careful description and classification of features, and the search for patterns in their distribution across the world. Explanation was then achieved by relating these patterns to what were believed to be the causative phenomena—often in a non-quantitative and largely subjective fashion. Thus, theories of landscape development were proposed which stressed the role of time (or stage) and saw the convergence of the landscape towards relatively flat surfaces (peneplains). Similarly, patterns of soil and vegetation were explained in terms of the overriding influence of climate, and much attention was given to the definition and classification of climatic regions, weathering regions, and morphogenetic regions.

Without doubt these approaches had their value, and there is still scope for some of these ideas. But in the last twenty years or so physical geographers have begun to adopt different approaches and to ask different questions. Increasingly, emphasis has been placed on monitoring and understanding processes and predicting change in the physical world. At the same time, we have started to consider the world in terms of physical systems—dynamic, and closely interrelated sets of objects which can only be understood when viewed as a whole. To a great extent, these changes have occurred in association with technical developments, such as the production and adoption of powerful computers, the evolution of sophisticated statistical methods, advances in satellite imagery, and improvements in the design of laboratory and field equipment. The consequences, though, are far-reaching and fundamental. Physical geography today is more quantitative, more analytical, and more rigorously scientific than ever before.

There are other consequences, one of them being that the subject has become more specialized. We now have more information to deal with, and this information is much more technical and complex. We cannot hope to master it all, so we have to concentrate our energies on selected parts. As a result, the discipline tends to have split more and more into all its different '-ologies': climatology, meteorology, hydrology, various types of geomorphology (fluvial, glacial, etc.), pedology, ecology, and so on. A glance at the range of specialist books on the average student reading list, or on the shelves of the typical academic bookshop, shows this clearly. In many ways this is no bad thing. It encourages us to become experts rather than generalists. But it also has a number of adverse implications. In the first place, specialism may narrow our breadth of knowledge in physical geography. Breadth of understanding, however, has a value; it gives us one of our unique skills, the ability to step back from the details and see the wider context and significance. Second, it runs counter to that very search for integration and wholeness which a systems approach requires. Third, it can make physical geography a more exclusive subject, barred to all but the experts.

It was, therefore, these considerations which motivated us to write this book. We wished to begin to bring the '-ologies' back together; to build a foundation from which those who are interested could step more easily into their selected specialisms, yet which would also serve for those who required only the fundamentals of the subject as a whole. We wished also to emphasize the links within the environment, and thus we have adopted a systems approach. At the same time we wanted to show how important humankind is in the physical world and how physical processes in turn affect our daily life; consequently we have tried to integrate the human and applied issues. Additionally, it is our belief that physical geography lies outside our own front door, and we have therefore stressed the everyday examples and occurrences of the features and processes we discuss. Finally, we needed to build the whole thing in a logical and rational manner, piece by piece, without losing sight of the broader 'wholeness' of the physical world. For this reason, we begin by presenting an overview—a framework—of the earth and its subsystems. Then we examine each of these subsystems in turn: the atmosphere, the hydrosphere, the lithosphere, and the biosphere. In the process we consider each systematically, showing how the various components are related and recalling the links with other parts of the world; showing, too, how humankind fits into the scheme. By the end, we hope that the reader will not only have a grasp of the fundamentals of physical geography, but will also be aware of the excitement and importance of the subject, and thus be stimulated to take it further.

Many years ago, we were given this stimulation by our own teachers of geography—Gordon Roberts at Lewes Grammar School and Allan Griffiths at West Leeds Boys' High School. To them we owe a great debt. If we can pass on their sense of excitement and interest to some of our readers, we will consider this book a success.

# Physical geography: a starting point

Along the Manitoba Escarpment there is a river valley. It is not a spectacular valley like those of the Colorado, Rhine, or Fraser rivers. It is like many thousands of others throughout the world. Wilson Creek, which rises a little way to the west in Riding Mountain National Park, tumbles amid the till and shale of the second Prairie level before flowing into Dauphin Lake. It is possible to stand beside the stream at the base of the escarpment and look up the valley to the rolling terrain of Riding Mountain. What we see is a distinctive notch that the stream has cut in the escarpment over the last 8000 years.

It is a scene that can tell us a great deal, and it contains the key to understanding the physical world around us. If we can start to understand this valley—if we begin to appreciate how the individual features we see are linked together, how they function—then we have opened a door on our physical environment. For this small valley, like almost any other we might have chosen, demonstrates many of the fundamental processes and factors that make up the environment. If we can in some way recognize and comprehend these things, we will have a basis for answering a vital question: how does our natural world work?

Let us look a little more closely at the view. The river, we can see, splashes over the rocks and rapids that line its channel. It is bounded in front of us by a fairly flat area—the flat plain of glacial Lake Agassiz. On the right, the valley side rises to the bush- and tree-covered flanks of the hills; on the left, there is a small scar where the shale bedrock is exposed in the hillside, and above that extensive woodland. Here and there, boulders and outcrops of bedrock protrude through the vegetation. It is a scene typical of countless parts of upland Canada and it is similar to those in upland regions throughout North America and the world. How can we start to learn from it?

## Processes in the Wilson Creek valley

It becomes clear if we inspect this scene that a great deal is happening. Some of the events are obvious: the air is constantly moving, bending the grass, loosening leaves from the trees, sweeping clouds across the sky. At times rain or snow falls and storms unleash their fury on the land. All the time, too, water is flowing through the stream channel, carrying with it logs and twigs washed from the banks and woods, pebbles and boulders that have slid or rolled down the valley side. The boulders can be seen in periods of flood, bouncing and shuffling along the channel floor; they come to rest in shoals and bars at the margins of, or within, the channel, and are moved on again as the river rises. Cattle are moving around on the floodplain, grazing on the vegetation. Occasionally a hiker walks along the creek.

Other events are less apparent. Beneath the soil, the rocks are being weathered and broken down. On the faces of the cliffs, ice freezes in crevices and loosens small fragments of rock. Most of this material, called talus, is carried downslope by gravity. Plant roots extend into larger fissures and open up the cracks. The soil is also invisibly active. Within the soil, rock fragments are being broken down further. Leaves and twigs are being chewed up and decomposed by soil organisms. The soil itself is moving, pushed and prodded downslope by the development of ice crystals during the winter, carried by the wash of water over and through the soil, splashed by the impact of raindrops. Water and dissolved substances are being drawn from the soil by plants; the water is being released to the atmosphere from the leaves. Slowly, the vegetation is changing, advancing in some areas, retreating in others. Unseen, energy is being passed through the atmosphere to the land and vegetation, and returned to the air and space.

### A MODEL OF THE WILSON CREEK VALLEY

These processes are the lifeblood of the Wilson Creek valley. They are the means by which the valley is being shaped and developed. They show the pathways by which material and energy is being moved through the valley. They thus provide us with a way of understanding the scene. We can start to build up a picture of the way the valley is functioning; we can start to account for the features we see in the valley now, and even to predict some of the changes that may occur in the future.

We might go even further, for we can represent this picture in diagrammatic form. We might use the technique employed in car manuals, for example, to show the workings of the valley as we have interpreted it. Just as the car designer explains the position and operation of all the components of the engine by drawing exploded diagrams of the vehicle, so we can construct a similar picture of the Wilson Creek valley. We can show the various features of the valley, use arrows and lines to indicate the main processes and movements affecting them, and label the diagram to show what is happening (Figure 1.2). What we end with is a picture—a model—of the valley.

This model, however, is a little limited. It helps us to see what is happening within the Wilson Creek valley, but it does not tell us much about the rest of the world. The model is too specific; it retains too much of the detail of this particular valley. What we need is a more general, a more abstract model. How can we produce it?

### FROM WILSON CREEK TO THE WORLD!

The first thing we must do is omit some of the unnecessary detail. After all, it matters little when we are trying to understand the broad aspects of our environment that the river in this little valley twists in this way here, or that the slope has a little undulation in it there. We can dispose of much of this

FIGURE 1.1 ▼ Wilson Creek cuts a youthful valley through the shale of the Manitoba Escarpment 400 km northwest of Winnipeg. The valley has formed in the 10 000 years following the retreat of glacial ice (photo: J. Thomlinson, courtesy Manitoba Department of Natural Resources)

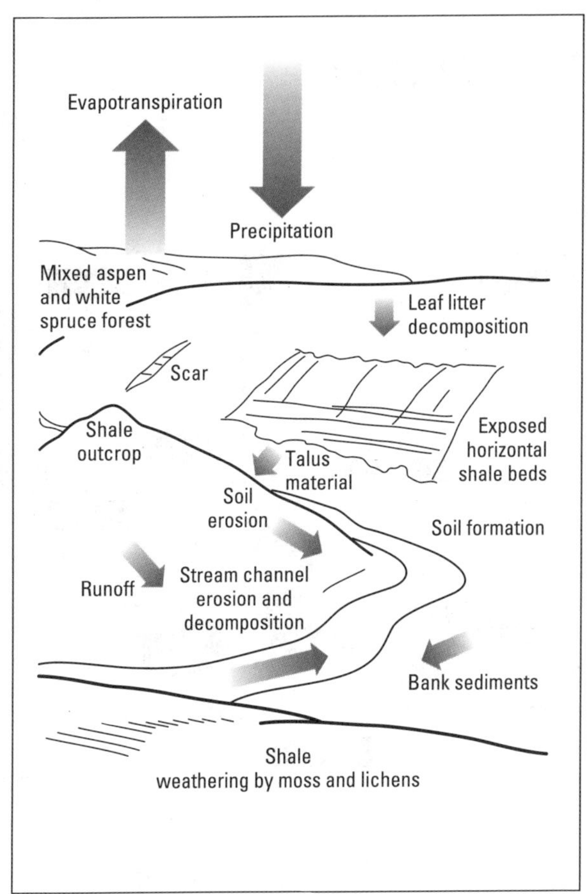


FIGURE 1.2 ▼ The features and processes in the Wilson Creek valley make up an environmental system

local detail and concentrate instead on the most important features of the area. We can show the clouds, the stream, the floodplain, the hillsides, the woods and bush—all the features that we feel characterize the valley. We can represent all these features as boxes and we can show the processes operating within and between them by arrows (Figure 1.3). As a consequence, we obtain a more abstract, yet simpler and clearer, picture of the Wilson Creek valley.

We now have a model that is less specific to the particular view before us. Nevertheless, we still cannot use it to understand the world as a whole, for not every valley or area we might be interested in contains the features we have recognized here. To progress farther, to get to the basic structure and character of the scene, we need to take a rather different approach. We need to ask two even more fundamental questions. What does this valley (or indeed any other part of the natural world) consist of? And what is it that links these components together?

The atmosphere, the water, the rocks and landforms, the soil, animals, and vegetation—these are the basic building blocks of our environment. And the mortar between them is provided by the flow of energy and matter: the movement of heat and other forms of energy from the atmosphere to the ground, through the soil, vegetation and landscape and ultimately back to space; the movement of water from the atmosphere to the oceans and back again; the movement of rock debris from the rocks through the landscape; the flow of nutrients from the soil to the vegetation and back to the soil. If we describe the Wilson Creek valley in this way, as we have done in Figure 1.4, we have a model that is relevant not just to this area but anywhere in the world.

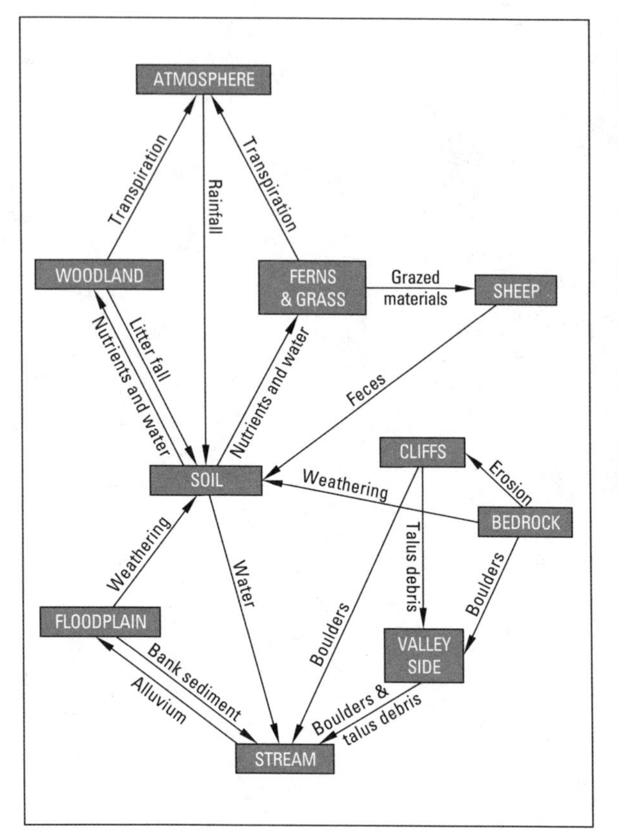


FIGURE 1.3 ▼ A preliminary model of the Wilson Creek valley, showing general relationships between the main features of the environment

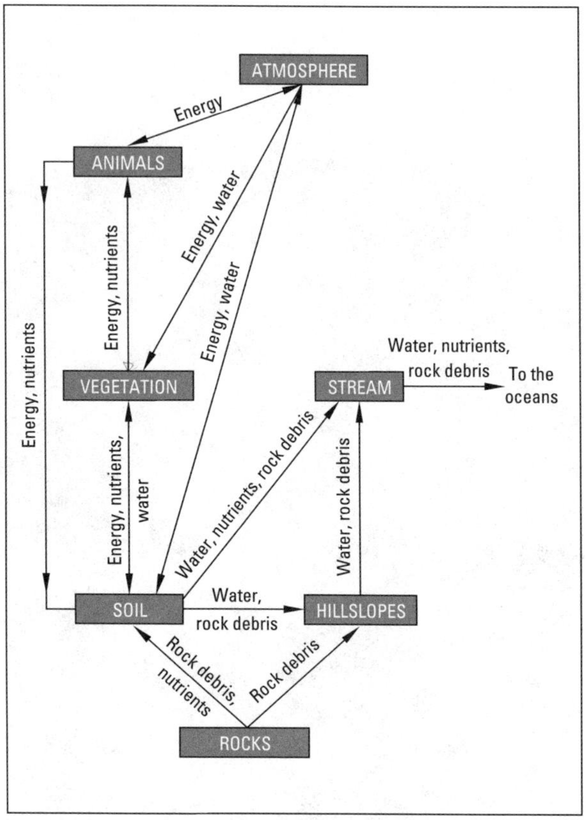

FIGURE 1.4 ▼ A generalized model of the Wilson Creek valley showing the major flows of energy, water, rock debris, and nutrients

## WE HAVE DESCRIBED A SYSTEM

We have made some real progress, then. We have discovered what we set out to find: a key to understanding our environment. And we have done more than that, for we have recognized and described a system.

Let us define what we mean. A system is simply a set of objects and the relationships between them. Almost anything, at any scale, can be seen as a system, from a droplet of water on a table top to the whole universe. But the concept of a system gives us a way of looking at those objects. It is an approach that helps us focus on the way the objects interact. This is important because the natural world is highly interactive, and if we are to understand—or, even more critically, to manage—the world effectively, we need to be aware of these interactions.

The other important step we have made is to represent this system in terms of a model. What we have done is to simplify the reality into a more manageable form. The need for this is all too apparent. The world is an extremely complex place, and we cannot hope to comprehend all the details of its complexity. We need, instead, to produce models that reflect the most important components and linkages of the system, but allow us to ignore (at least for a while) the less important details. Subsequently we might wish to extend our models, and bring them closer to reality, though doing so can be complicated and defeat the purpose of simplification.

There are, in fact, many different sorts of models we can construct: **hardware** or **physical models** in which we actually build a physical likeness of the system we are concerned with; or **mathematical models** in which we represent components of the system by mathematical symbols and the relationships between them by equations, as in Figure 1.5; and more general, conceptual models like the one we have built of the Wilson Creek valley.

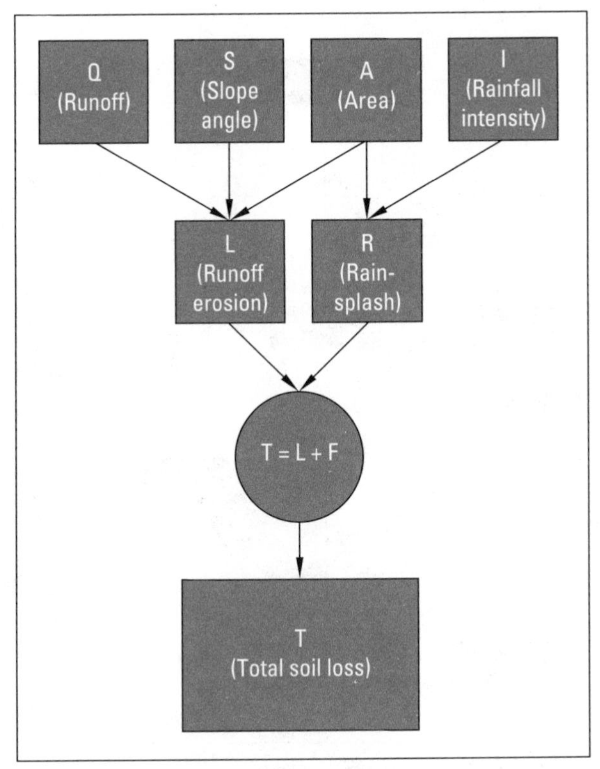


FIGURE 1.5 ▼ A mathematical model: soil erosion by water (based on Meyer and Wischmeier, 1969)

# A systems approach

## THE STRUCTURE OF SYSTEMS

Our example of a system in Figure 1.2 leaves something to be desired. It is not, in truth, a very clearly defined system; its boundaries are determined simply by the limits of our view. In most cases we are dealing with rather more distinctive systems, bounded by more easily definable features. We may, for example, take a whole river valley, the edges of which are marked by the line of the highest ground that encircles the valley, and treat this as a system. Or we may take an area covered by a particular vegetation type, a field or farm, a glacier or cloud.

Although the boundaries of these systems may appear obvious, they are rarely impermeable. The systems are not self-contained. Instead, matter and energy flow into and out of the system, across its boundaries as inputs and outputs (Figure 1.6). We can see this in our example: inputs of water, sediment and other materials enter the system from upstream and leave it downstream. This flow of inputs and outputs is typical of all natural systems except, perhaps, at the scale of the whole earth or universe. The earth, being a closed system, has inputs and outputs of energy, but virtually no inputs and outputs of matter. An open system would have inputs and outputs of both.

Sometimes that is all we know about a system—its boundaries and the inputs and outputs. We can detect a relationship between the inputs and outputs, but we do not understand what goes on inside. In this case we are looking at the system as a **black box** (Figure 1.7). As our knowledge progresses, however, we may be able to discover what lies inside the system. What we find is a series of smaller subsystems, each linked by a series of flows of energy and matter. We then have a view of the system as a **grey box**. But delve a little deeper and we may be able to see the whole internal working of the system: its individual components, the pathways by

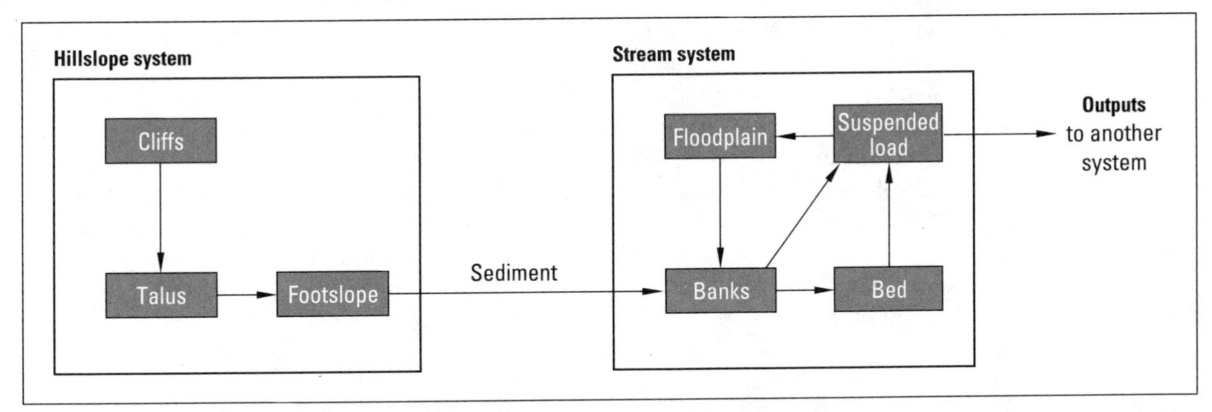


FIGURE 1.6 ▼ The relationship between two connected systems. Sediment moves through the hillslope system and into the stream system; there it is transferred through the banks, bed, suspended load and floodplain deposits before being lost as outputs. Many other examples fit this general structure

which energy and matter flow between them, the storages where energy and matter may be held for certain periods of time. We now have a view of the system as a **white box** (Figure 1.7).

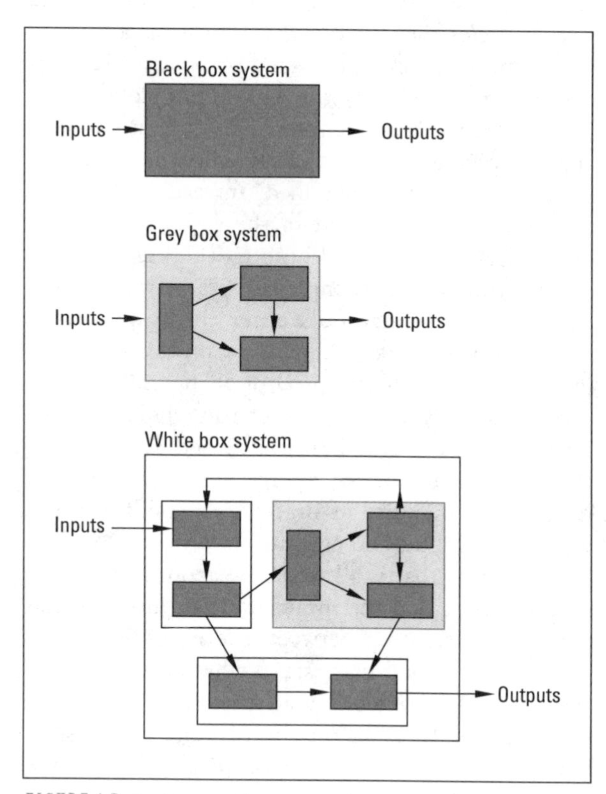


FIGURE 1.7 ▼ Types of systems

## TYPES OF SYSTEMS

A system, therefore, is a set of components linked by flows of energy and matter. But what does that mean? If we take our example of the Wilson Creek valley (Figure 1.1) we could, for example, measure all the main features and attributes of the valley and look at the relationships between them. We might see how the size of the trees relates to soil depth, how soil depth relates to slope angle, how slope angle relates to rock structure, and so on (Figure 1.8). Is that a system?

The answer is yes. It is referred to as a **morphological system**. In looking at the system in this way we are concerned not with the dynamics of the interactions and flows, but merely their morphological expression. Another way of looking at the system, however, could be to focus attention on the flows of matter and energy through the valley. We might, for example, represent the movement of sed-

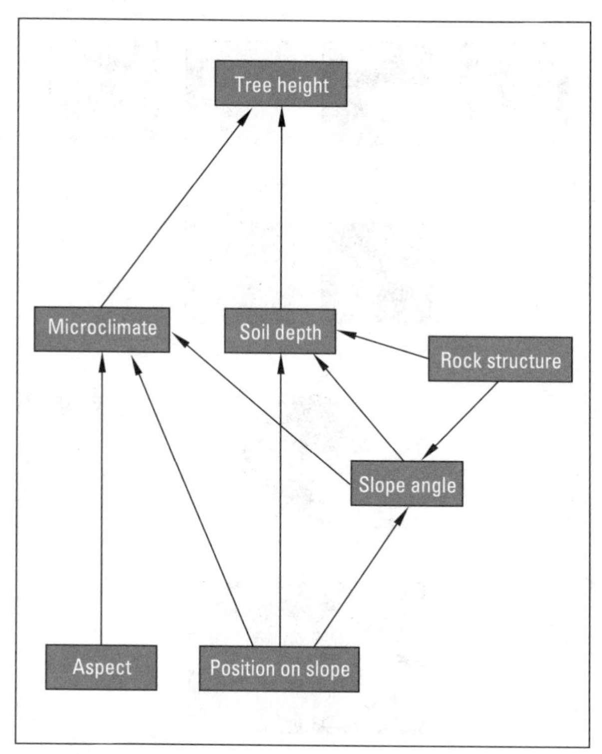


FIGURE 1.8 ▼ Factors affecting rate of tree growth in Wilson Creek valley: an example of a morphological system

iment through the system as in Figure 1.9. In this case we have described the valley as a **cascading system**, one in which there is a cascade of energy and matter through the environment from one component to another. It is a particularly useful way of dealing with systems and we will use it to look at the flow of solar energy through the atmosphere (in Chapter 3) and the cascade of sediment through the landscape (in Chapters 16–19).

There are other ways of looking at environmental systems. We can combine morphological and cascading systems to define what are called **process-response systems**. Figure 1.10 depicts part of the Wilson Creek valley in these terms. As this shows, the morphology of the system is related to the flows of energy and matter. In other words, the form of the system is a function of the processes operating within it. This is a vital concept and one we will use repeatedly throughout this book, for it helps us to see the ways the environment develops and is maintained.

Finally, we can also define what are referred to as **ecosystems**. These are concerned with the biological relationships within the environment—the interactions between plants and animals and their

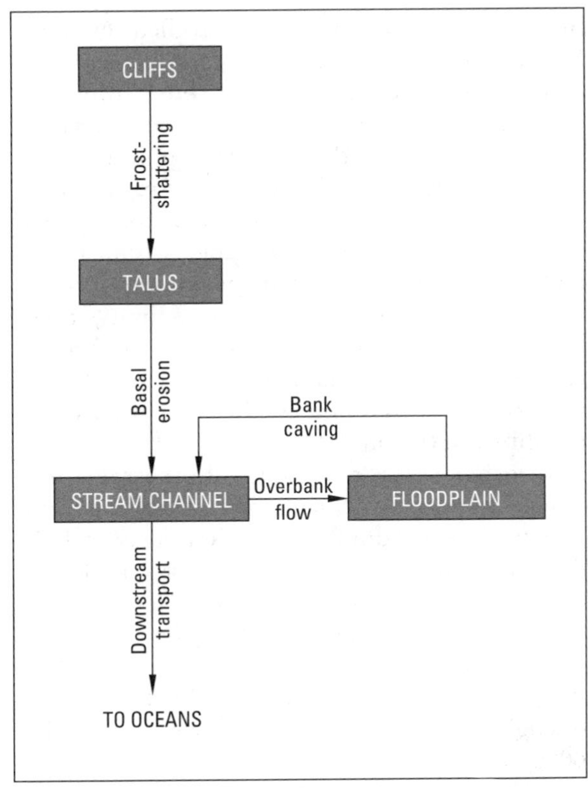


FIGURE 1.9 ▼ Sediment movement through the Wilson Creek valley: an example of a cascading system

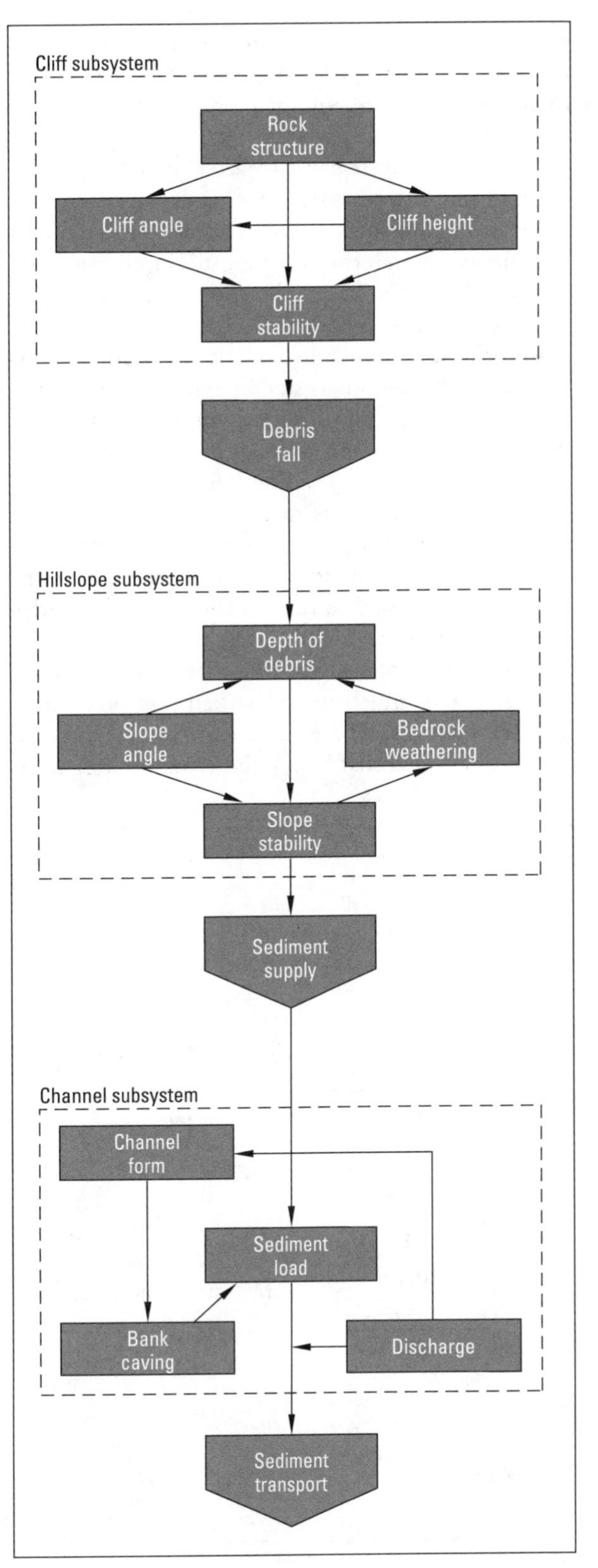


FIGURE 1.10 ▼ The cliff-hillslope-channel system of the Wilson Creek valley: an example of a process-response system

physical surroundings. It is an approach we will adopt in the last section of this book when we look at soil–vegetation–animal relationships.

## SYSTEM CHANGE AND SYSTEM STABILITY

The environment is dynamic. It is constantly changing. What causes this change? The immediate answer is energy. As we will see in the next chapter, the earth obtains most of its energy from the sun (as **radiant energy**). Energy within the environment tends to be unevenly distributed, however, and so it tends to flow through the environment in an attempt to produce a more equitable distribution. In the process it is itself changed, but it also carries out work. It alters the environment.

Over almost twenty years beginning in the 1960s, Dr. Robert Newbury carried out an extensive and classic study of the entire Wilson Creek drainage basin. He set up climate stations at several sites within the valley and placed flow monitoring stations at key points in the stream. The study showed that changes were occurring constantly within the entire basin. Much of this change would often appear haphazard and random. Changes in one direction at one time might be reversed at the next observation.

Students were required to account for energy in each section of the system using data collected at the flow monitoring stations. They accounted for all the energy, though the variables might change from one section to the next. The reason for this is that although the system is active, it is also in a state of equilibrium; over time it tends to maintain its general structure and character in sympathy with the processes acting upon it. Thus, although the channel position varies, on average it follows the same route down the valley. Although the extent of the woodland fluctuates, its mean area and position are constant. Though the weather varies, the longterm average climate is more or less consistent. Systems such as this are said to be in a condition of **steady-state equilibrium** (Figure 1.11).

Not all systems behave like this, however. Some are subject to much more marked and often irreversible changes. These changes may be triggered by certain events that knock the system out of equilibrium. Hillslopes provide an excellent example. Within certain limits they remain stable and their form does not change. But if those limits are exceeded—if the ground becomes too wet or the pressures exerted on the slope are too great—slope failure may occur and a new slope form is created. The same is true of many atmospheric systems. The air near the ground may be stable, within limits, but if the system is disturbed (for example, if the air is overheated) so that it exceeds these limits, it may become unstable: the air may start to rise until it reaches a new equilibrium position. Systems of this type are said to be in a condition of **metastable equilibrium** (Figure 1.11).

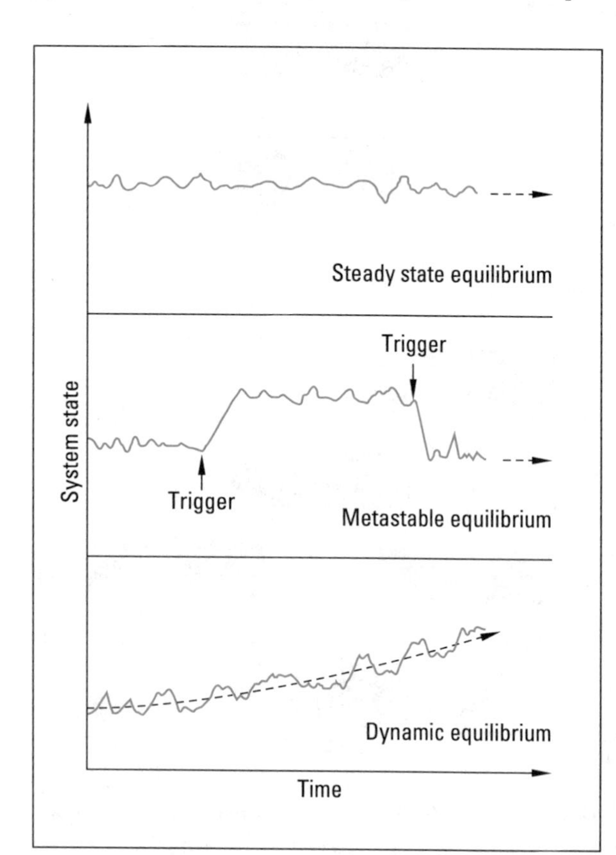


FIGURE 1.11 ▼ Types of equilibrium

In other cases, environmental systems may change more gradually and progressively over time. Short-term, random fluctuations may still be detectable due to the effects of minor variations in environmental conditions. But in the longer term a distinctive and consistent trend may be visible. This is characteristic of climatic change. As we will see in Chapter 9, the climate is gradually changing (though with minor fluctuations), and over the last 6000 years average summer temperatures in northwest Europe have fallen by about 2°C. Under these circumstances, the system is said to be in a state of **dynamic equilibrium**.

All these types of equilibrium are illustrated in Figure 1.11. Although they vary in detail, they all reflect a fundamental principle: that the form of any system adjusts toward a state of equilibrium with the processes and conditions to which it is subject. However, these examples also illustrate a further property of environmental systems—the operation within them of feedback processes. **Feedback** refers to the ability of a system to modify earlier links in the chain of interrelationships so that an initial change in the system is either amplified (**positive feedback**) or damped down (**negative feedback**).

We can see how these feedback processes operate with an example. Suppose the Manitoba Escarpment were affected by a prolonged drought (or even a more permanent change in climate), which resulted in the area receiving much less rainfall than it presently does. What would be the effect on Wilson Creek?

In the short term the consequences are fairly obvious. Less rainfall would mean that there was less water available for runoff. Less water would therefore enter the stream channel, and stream discharge would decline. At the same time, less sediment would be brought into the stream by water flowing down the hillside, so the sediment load of the stream would fall. In addition, the stream itself would be unable to transport its load as effectively, and would deposit much of its material on the channel bed. Reduction of the amount of water and sediment load of the stream would reduce the ability of the stream to erode its banks. This, too, would reduce the amount of sediment entering, or passing down, the channel. Thus, the initial decline in the amount of rainfall would trigger a cycle of positive feedback resulting in progressively less sediment being transported by the stream (Figure 1.12).

But the story does not end there, for in the longer run other changes would take place. The increased aridity might lead to changes in the vegetation cover, for the trees and shrubs would start to experience moisture stress. In time, much of the vegetation might wilt and die (we saw this in Western Canada in the drought of 1988). This would reduce the amount of evapotranspiration and would leave

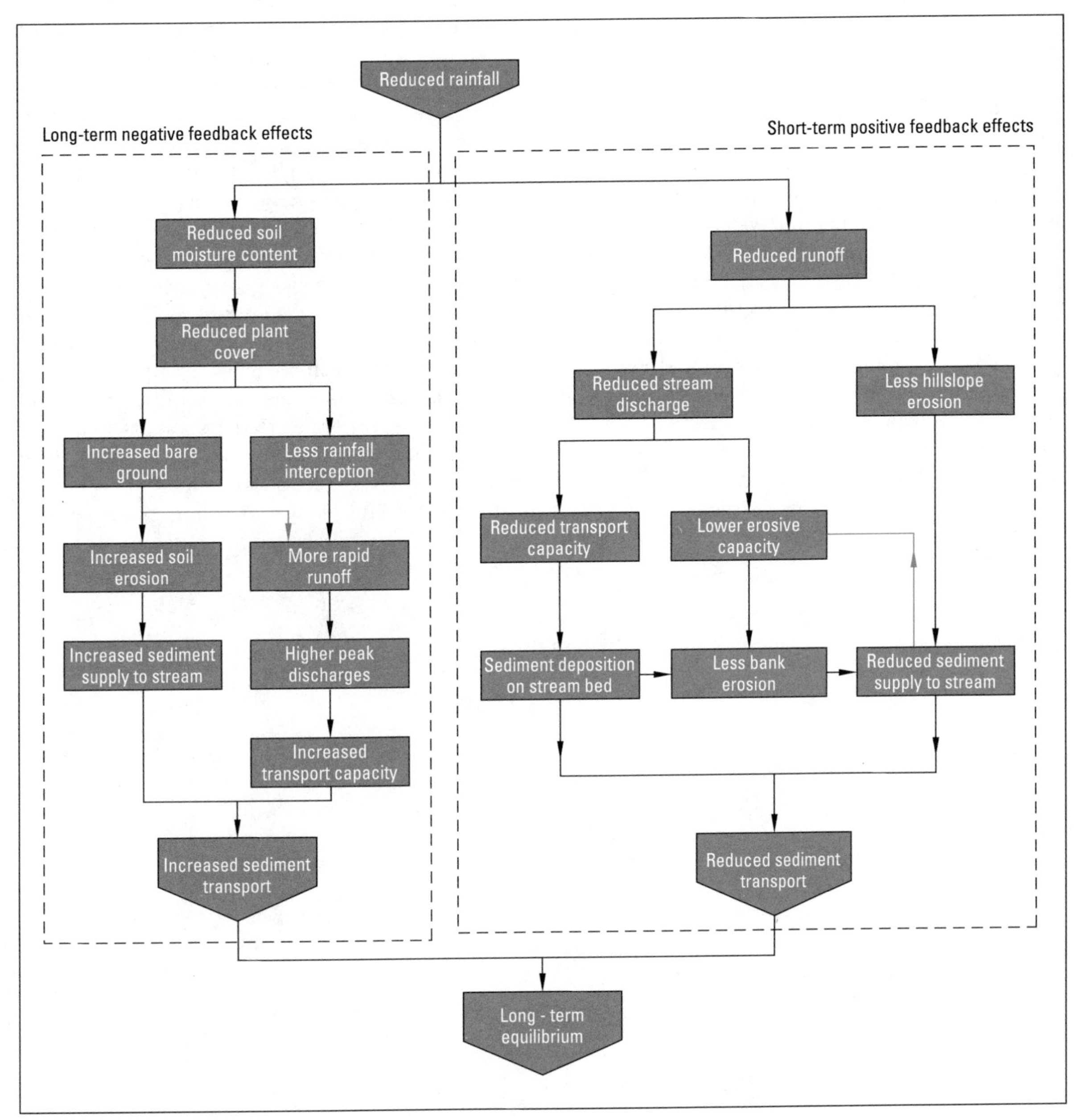


FIGURE 1.12 ▼ Potential effects of reduced rainfall on the Wilson Creek valley, showing the operation of feedback relationships

areas of bare soil. The rain that did fall, therefore, would not be trapped or taken up by vegetation, but would be free to run off across the unprotected surface. This would ultimately increase the quantity of sediment being washed into the stream. Moreover, because there would be less vegetation to intercept or slow down the water after rainfall, much of the water would reach the stream channel relatively swiftly, so that the stream would respond more quickly to storm events. At the peak of its flow, it would be able to carry large amounts of material and would have the capacity once more to erode its bed and banks. Thus, a longer-term cycle is operating to restabilize the system, to halt the progressive reduction in the rate of sediment movement. This is a process of negative feedback (Figure 1.12).

This example is simple but it illustrates a number of important principles that are worth emphasizing. First, environmental systems are dynamic and intimately interrelated. Changes in one part of the system work through to affect other parts. Second, change tends to be perpetuated by processes of positive feedback, inhibited by negative feedback. Third, over time, systems may undergo alternating periods of positive and negative feedback, as they change in response to some external impulse, then restabilize (this is characteristic of systems that are in a state of metastable equilibrium).

In addition, we may draw two further inferences about change and stability in environmental systems. The first is that because conditions change, and because systems adjust to change, many of the features we see in the environment at any moment might, in part, be inherited from previous times when conditions and processes were different. We can certainly see this in the case of the Wilson Creek valley. Several features there are due not to any present-day process but to the effects of glacial conditions several thousand years ago. The glaciers have gone, but the features do not simply disappear. They remain and are slowly modified by new processes. Their present form is thus a product of several superimposed periods of development, during which different processes have been at work.

The final point is that, because of their complex interrelationships and their potential for change, environmental systems are difficult to manage. All too easily people may trigger changes in these systems, more far-reaching and destructive perhaps than they intend, simply because they do not appreciate all the ramifications of their actions. For example, construction of a road across the natural slope can alter the drainage over a large area even if culverts are used. We will see many examples of this in the rest of this book. In almost every case the lesson will be the same. We need to understand how environmental systems work, how they change, how they respond to external conditions, if we are to manage them effectively. For this reason a systems approach to physical geography is useful.

# The global energy system

## Introduction

The sun is the primary source of energy that fuels earth's natural systems. A basic understanding of how energy moves through this system is essential to understanding global weather and climate. In this chapter, we examine the nature of solar energy and how it is emitted by the sun. Variations in sun–earth relationships create changes in the pattern and distribution of energy at the top of the atmosphere. Finally, we study the mechanisms that carry this energy to all parts of the global system.

## The nature of inputs and outputs of energy

Imagine the earth from 300 000 km into space. An isolated sphere; blue, patched with brown and green and wreathed in white. A world of water, dotted with land, partly clothed in swirling cloud. This is a view of the global system. Into this system pour the inputs of solar energy; from it come reflected and reradiated energy, which are its outputs. From our privileged vantage point, we could measure the inputs to the earth and, with suitable equipment, monitor the global outputs. We could therefore draw up a simple model of the globe as an energy system (Figure 2.1), showing the inputs and outputs, but this gives us no idea of what happens inside. It is a picture of the globe as a black box system. It is the simplest view of the system we can obtain, but it tells us nothing about the internal components or subsystems, nor the relationships between them, only what enters and leaves the globe.

Let us start by looking at those energy flows we *can* examine. Without doubt, the main input of energy to the global system comes from the sun. Compared with the solar contribution, all other inputs are almost negligible. The gravitational effect of the moon and the sun, and reflected and radiant energy from the moon provide some energy, with even smaller inputs from the impact of objects such as meteorites and comets. Many of these extraterrestrial objects burn upon entering the atmosphere and do not reach the surface, although they do act as a minute input of energy to the atmosphere.

Just as the input of energy is dominated by radiant energy from the sun, so the output of energy from the earth is almost entirely radiant energy, although this time with somewhat different properties. Much of the energy has been radiated or emitted by the earth and its atmosphere, being modified in the process, but some of the output is represented by solar radiation reflected from clouds or from the earth's surface without any major modification. As the overall energy level of the earth is not changing, we can assume that there must be a balance between the energy input to and energy output from the globe as a whole.

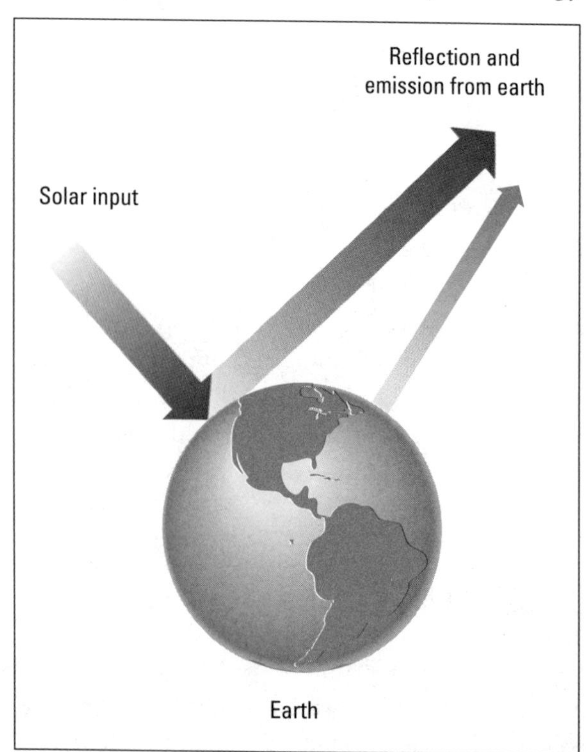


FIGURE 2.1 ▼ A black box model of the earth's energy system

## Concepts of energy

Before discussing the quantities of energy received by the earth, it is necessary to consider, briefly, the nature of energy and the ways we measure it.

Energy exists in a variety of forms. We are familiar with electrical energy in the home and increasingly with nuclear (or atomic) energy. Neither of these has any great significance with regard to environmental processes. More important as far as environmental processes are concerned are **radiant**, **thermal**, **kinetic**, **chemical**, and **potential energy**.

**Radiant energy** is the most relevant to our discussion here, for it is in this form that the sun's energy is transmitted to the earth. The heat from the sun excites or disturbs electric and magnetic fields, setting up a wave-like activity in space, known as electromagnetic radiation. The length of these waves—that is, their distance apart (Figure 2.2)—varies considerably, so that solar radiation comprises a wide range of electromagnetic wavelengths (Figure 2.3). Only a very small portion of these are visible to the human eye, reaching us as light, but all transmit energy from the sun to the

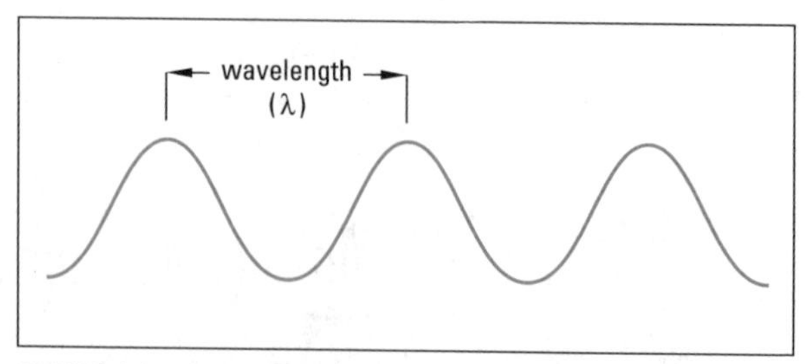


FIGURE 2.2 ▼ Electromagnetic radiation: the distance from one crest to one crest or from one trough to one trough is known as the wavelength ($\lambda$). It is an important indicator of the properties of the electromagnetic radiation

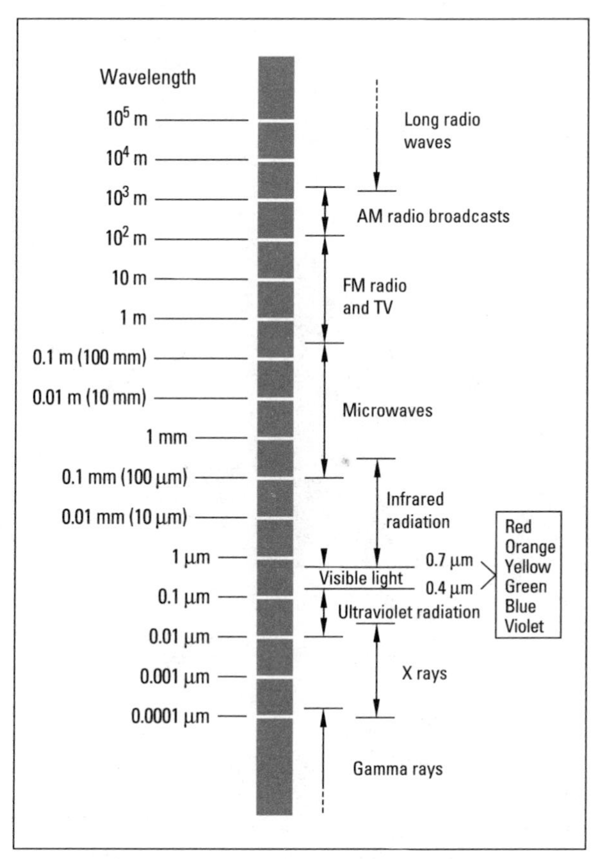


FIGURE 2.3 ▼ The electromagnetic spectrum (after Neiburger *et al*, 1982)

earth. Assuming a mean distance from sun to earth of about $15 \times 10^7$ km, it takes this energy about 8.3 minutes to reach us.

On passing through the atmosphere that surrounds the earth, some of this radiant energy is reflected or absorbed. Because of this interception, not all the radiant energy finds its way to the earth's surface. That which does, and that which is absorbed by the atmosphere, is converted from radiant to other forms of energy. Much is altered to **thermal (heat) energy**. It warms the earth's surface and the atmosphere by exciting the molecules of which they are composed. In simple terms, the radiant energy (which involves disturbance of magnetic and electric fields) is transmitted into the molecules making up the earth and atmosphere with a resulting change in the type of energy.

Thermal energy can therefore be considered as energy involved in the motion of extremely small components of matter. The energy of motion is referred to as **kinetic energy** (and thus thermal energy is sometimes described as the kinetic energy of molecules). Any moving object possesses kinetic energy, and it is through the use of this energy that, for example, a stone thrown into a lake can disturb the water and produce waves. It is also through the exploitation of kinetic energy that turbines and engines are able to produce heat, light, and so on.

**Chemical energy** represents a form of electrical energy bound up within the chemical structure of any substance. It is released in the form of thermal or kinetic energy when the substance breaks down. Coal, when it is burnt, releases heat. Food, when it is digested, is used to provide the body with heat and movement.

**Potential energy** is related to gravity. Because of the apparent pull that the earth exerts upon objects within its gravitational field, material is drawn toward the earth's centre. Thus, objects lying at greater distances from the earth's centre (for example, rocks on a hillside, water at the top of a waterfall or the air near a mountain summit) possess more potential energy. This energy is converted to kinetic energy when the rock, the water or the air descends to lower levels; some energy is converted to heat through friction.

Thermal, kinetic, chemical, and potential energy are important to the earth's system but operate internally and so cannot be observed directly from space. To understand the results of these different flows of energy, we must look more closely at them, concentrating on the forms of energy that have significance for the physical geography of the earth.

## TRANSFERS OF ENERGY

The inputs and outputs of energy we have considered so far do not have a uniform distribution over the globe. Both the earth and the air experience major inequalities in energy receipts and emissions. As a result of these differences, spatial transfers of energy take place, for energy is redistributed to minimize the inequalities, to maintain (or to achieve) an equilibrium.

To understand how energy is transferred we need to consider a little further the principles of energy transformation and modification. We have seen already that energy can exist in a number of forms, and as a general principle energy will be transferred from areas of high energy status to areas of low energy status in an attempt to eradicate the differences. Thus, energy differences expressed by the level of temperature in two bodies, such as the air and the soil, tend to be reduced over time as heat is transferred from the hotter to the cooler body. In this way the soil is heated during the day when the air is warm and it loses heat energy back to the air at night when the atmosphere is cool (Figure 2.4).

heat is released. We shall return to these mechanisms in more detail in Chapters 4 and 5.

Transfers also occur between other forms of energy. If two objects with different kinetic energies are brought together, a transfer takes place between

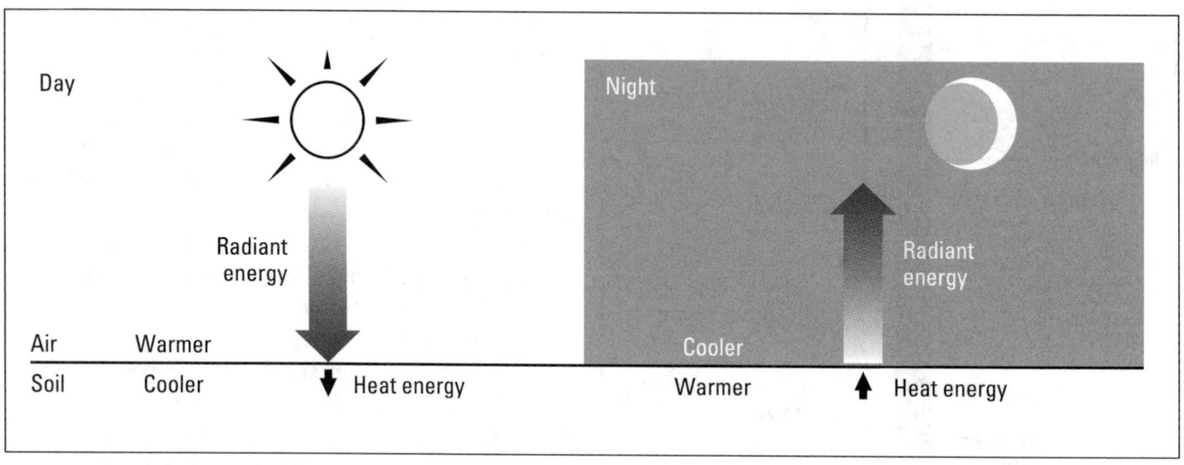


FIGURE 2.4 ▼ Energy exchange at the soil surface

the two that tends to equalize the energy levels. For example, a rapidly flowing stream (high kinetic energy) that comes into contact with a static boulder (no kinetic energy) tends to push the boulder into motion. In doing so, the stream loses energy by friction but imparts some of this energy to the boulder in the form of motion (kinetic energy). Similar principles apply to chemical energy.

In the case of thermal energy, three main methods of transfer can be identified: radiation, convection, and conduction. **Radiation** is the process by which energy is transmitted through space, mainly in the form of electromagnetic waves. **Convection** involves the physical movement of substances containing heat, such as water or air, and is not possible in a solid. **Conduction** is the transfer of heat through a medium from molecule to molecule.

These three processes of transfer are often closely related. Thus, energy may be conducted through the soil to the surface and then radiated or convected into the atmosphere. Similarly, in the air, convection currents may raise warm air masses to higher levels, and then conduction to surrounding cooler atmosphere may occur (see Chapter 4), while condensation of water vapour releases latent heat. Convection is very important as an energy transfer mechanism because it transfers energy in two forms. The first is the **sensible heat** content of the air, which is transferred directly by the rising and mixing of warmed air. It can also be transferred by conduction. The other form of energy transfer by convection is less obvious as there is no temperature change involved, hence its name, **latent heat**. The evaporation of water into vapour or the melting of ice into water involves a supply of heat to allow the change to take place. When the reverse process operates, from vapour to liquid, or liquid to ice, this

One way of looking at these transfers of energy is to regard them as movements down an energy gradient. It is easier to see this principle in the example of heat energy, for we can all appreciate that heat moves from hotter to colder areas. Heat the end of a metal bar in a fire and the heat will move along the metal until it burns your fingers! Heat energy in this case moves down the energy gradient in the bar. The same general processes operate with other forms of energy (Figure 2.5).

## ENERGY TRANSFORMATIONS

During these transfers of energy it is clear that the nature of the energy often changes, although the total quantity of energy involved remains constant. Radiant energy heats the objects it meets; it is converted from radiant energy to heat energy. Kinetic energy may similarly be converted to heat energy; the friction of a moving body against another liberates heat, as we can demonstrate by filing or sawing a piece of metal—or even by rubbing a finger over a slightly rough surface.

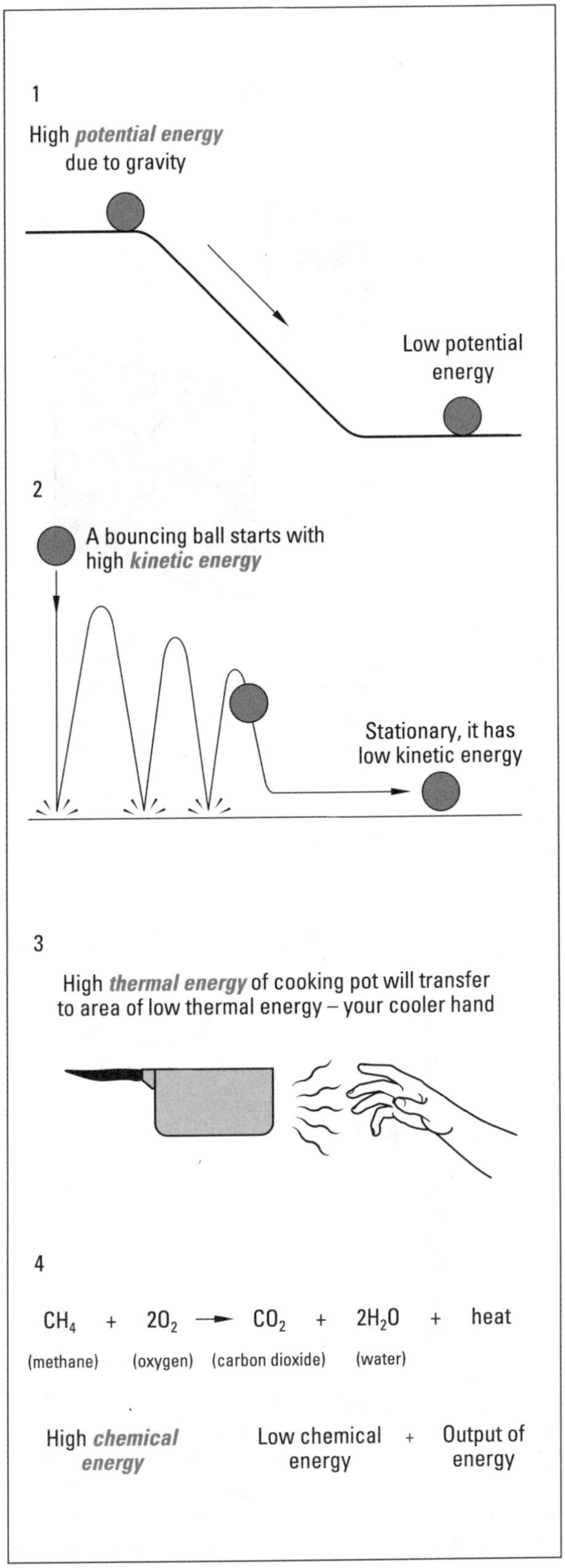

Under natural conditions the range of probable transformations is fairly limited. That is, the various forms of energy are normally able to be converted to all other forms, but follow relatively well-defined pathways (Figure 2.6) toward the lowest level of energy—that of heat.

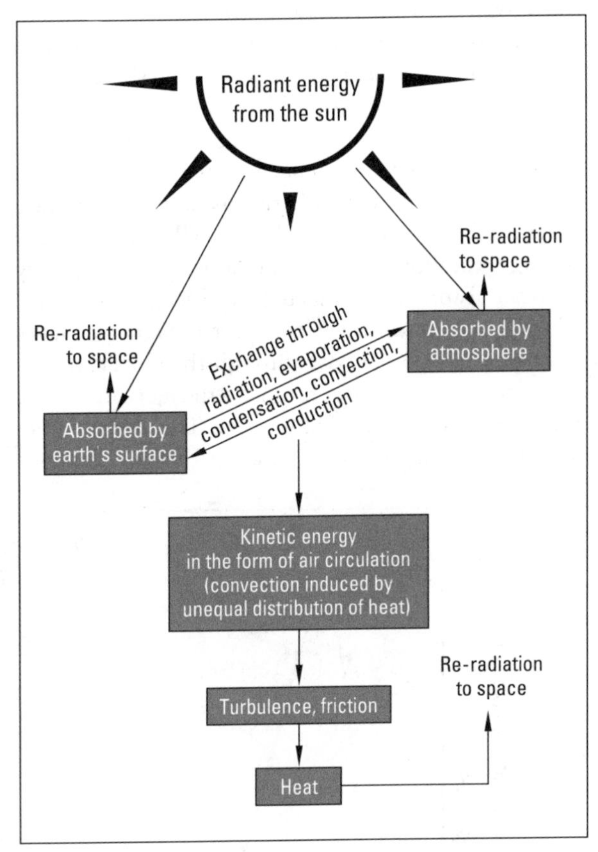


FIGURE 2.6 ▼ Energy transformations in the atmosphere (after Miller, 1966)

# General patterns and principles of electromagnetic radiation

**Solar energy** is transmitted to the earth in the form of radiant energy. How does this energy reach us? Why do we receive the amount we do? Why does the energy have the particular properties it has? To answer these questions we need to examine some of the principles of radiation.

1

High *potential energy* due to gravity

Low potential energy

2

A bouncing ball starts with high *kinetic energy*

Stationary, it has low kinetic energy

3

High *thermal energy* of cooking pot will transfer to area of low thermal energy – your cooler hand

4

$CH_4$ (methane) + $2O_2$ (oxygen) → $CO_2$ (carbon dioxide) + $2H_2O$ (water) + heat

High *chemical energy* → Low chemical energy + Output of energy

FIGURE 2.5 ▼ Examples of energy gradients

## PRINCIPLES OF RADIATION

Radiant energy consists of electromagnetic waves of varying lengths. Any object whose temperature is above **absolute zero** (0°K or –273°C) emits radiant energy. The intensity and the character of this radiation depend upon the temperature of the emitting object. As the temperature rises, the radiant energy increases in intensity, but its wavelength decreases; as the temperature falls the intensity decreases and the wavelength increases (Figure 2.7). In addition, the amount of radiation reaching any object is inversely proportional to the square of the distance from the source (Figure 2.8). This distance decay factor accounts for the difference in solar inputs to the various planets in our solar system.

To a certain extent radiation is able to penetrate matter, as for example X rays which can pass through the human body, but most radiant energy is either absorbed or reflected by objects in its path. Absorption occurs when the electromagnetic waves penetrate but do not pass through the object; reflection involves the diversion or deflection of the waves from the surface of objects. The ability of an object to absorb or reflect radiant energy depends

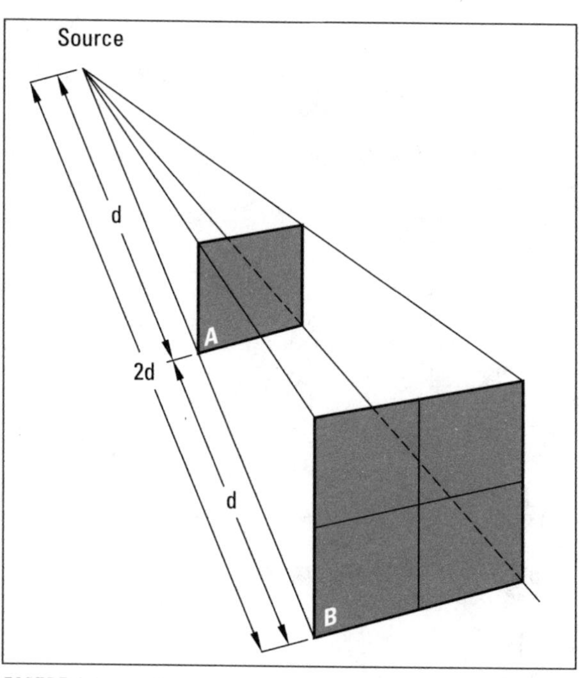


FIGURE 2.8 ▼ The inverse square law. Intercept area at distance d from the source is just one-fourth that at distance 2d; energy passing through area A is spread over an area four times as large at B

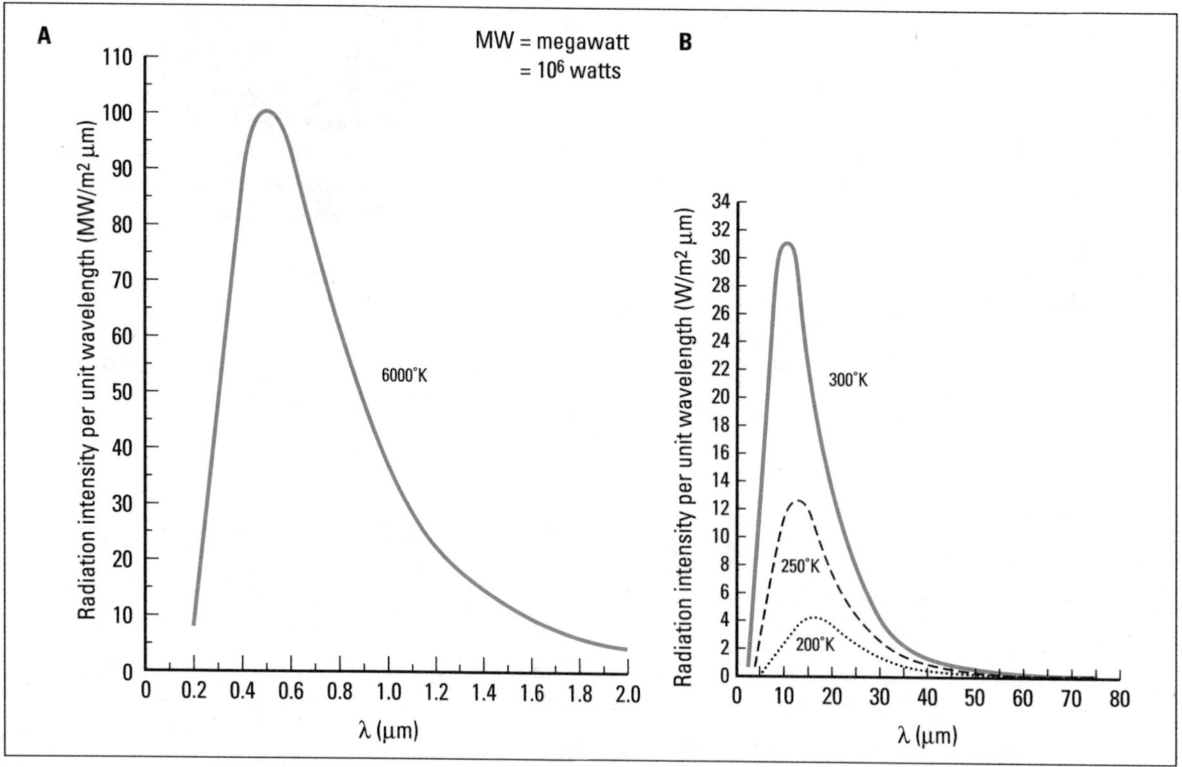


FIGURE 2.7 ▼ Variation of intensity of black-body radiation with wavelength: (A) T = 7000°K (approximately the emission temperature of the sun); (B) T = 200°K, 250°K, and 300°K (range of earth emission temperatures). Note the differences in scales (after Neiburger *et al*, 1982)

upon a number of factors, including the detailed physical structure of the material, its colour and surface roughness, the angle of the incident radiation and the wavelength of the radiant energy.

An object that is able to absorb all the incoming radiation is referred to as a **black body**. Although this has conceptual value, a perfect black body does not exist in reality. All objects absorb a proportion of incoming energy and reflect the remainder. The amount of shortwave radiation reflected from a surface is called the **albedo**. It is calculated by dividing the amount reflected by the total amount arriving at a surface, and is expressed as a percentage. The colour of the surface determines the amount reflected. Solar collection panels are flat black to ensure that the maximum amount of shortwave energy is absorbed and converted to sensible heat. (The albedo values of some common surfaces are given in Table 3.2). Differences also occur according to the wavelength of the energy. Thus snow and sand both absorb **longwave radiation** (5–50 μm) quite efficiently, but they reflect relatively large proportions of **shortwave radiation** (0.4–0.8 μm). Indeed, under constant conditions, it is possible to define the wavelengths that specific materials selectively absorb, and this knowledge can be used to characterize or identify materials in remote sensing. It is frequently used in astronomy to determine the gases in stellar atmospheres. Whereas solid substances usually absorb most wavelengths of radiation, gases tend to be very selective in their absorption and therefore emission wavelengths (Figure 2.9). This property is very important to the earth, as it means that the atmosphere only absorbs and emits in certain wavelengths. At other wavelengths, radiation is able to pass right through the atmosphere without modification. The atmosphere is composed of gas molecules, particles of matter such as dust, water droplets, and ice crystals. Light waves striking these obstacles are scattered in all directions, so that radiant energy is scattered back to space as well as down to the surface. There is no change of wavelength in this process, known as **scattering**, simply a change of direction for some of the radiant energy.

The gas molecules are most effective at scattering light in the blue wavelength. Since gas molecules compose much of the atmosphere, we see the sky as blue whether we view it from the ground or from space. When the sun is setting or rising, the radiant energy passes at a lower angle through the larger particles of dust in the lower atmosphere. The result

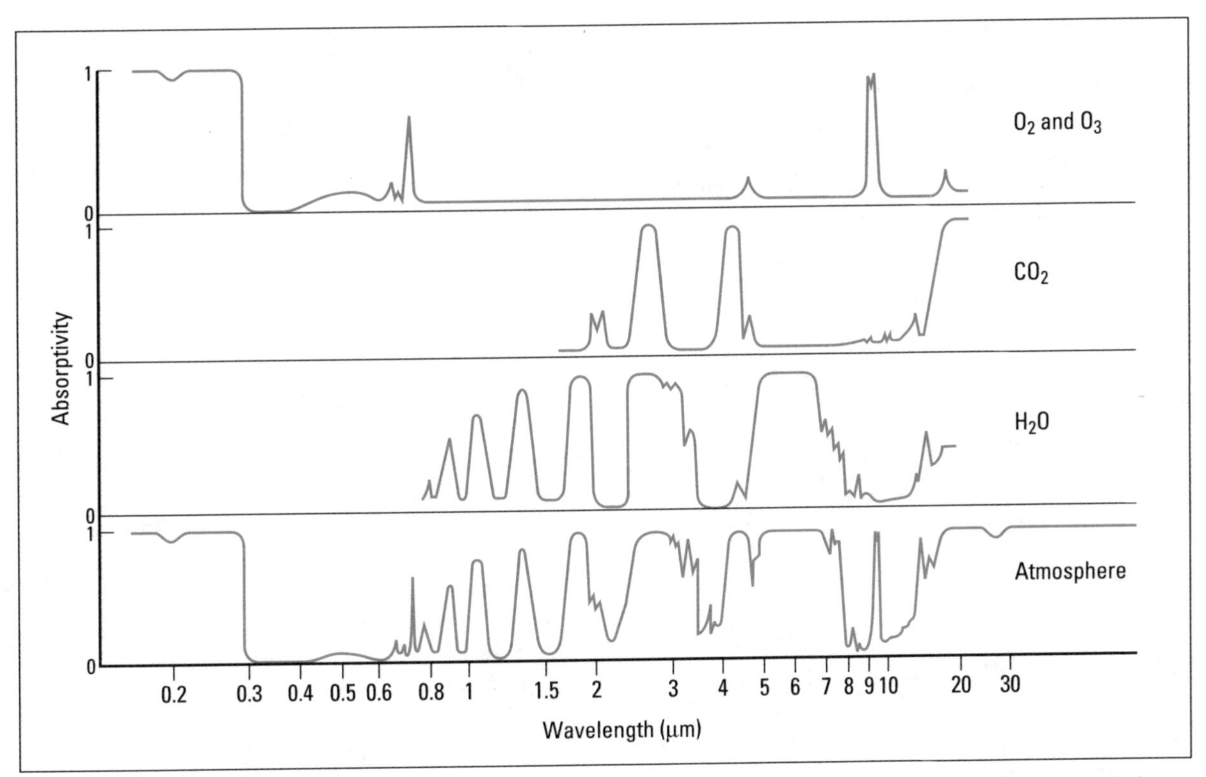


FIGURE 2.9 ▼ Absorptivity at different wavelengths by constituents of the atmosphere and by the atmosphere as a whole (after Fleagle and Businger, 1963)

is that more of the red wavelength is scattered, producing the colourful skies at the beginning and ending of the day.

**Absorption** of radiant energy has more far-reaching consequences than reflection or scattering. As an object absorbs energy its temperature rises, because the radiant energy is converted to heat (thermal energy). Reradiation of this energy tends to occur at a temperature different to that of the initial, radiating object, and thus the radiation emitted is at a different wavelength. The earth, for example, is considerably cooler than the sun; thus the energy it emits is characteristically of longer wavelengths than the initial solar inputs.

We can summarize the radiation laws as follows:

1. All substances emit radiation when their temperature is above absolute zero (–273°C or 0°K).
2. Some substances absorb and emit radiation at certain wavelengths only. This is mainly true of gases.
3. If the substance is an ideal emitter (a black body), the amount of radiation given off is proportional to the fourth power of its absolute temperature. This is known as the Stefan-Boltzmann law and can be represented as $E = \sigma T^4$, where E equals the maximum rate of radiation emitted by each square centimetre of the surface of the object, $\sigma$ is a constant (the Stefan-Boltzmann constant) with a value of $5.67 \times 10^{-8}$ W $m^{-2}$ $K^{-4}$, and T is the absolute temperature.
4. As substances get hotter, the wavelength at which radiation is emitted will become shorter (Figure 2.4). This is called Wien's displacement law, which can be represented as $\lambda_m = \alpha/T$, where $\lambda_m$ is the wavelength at which the peak occurs in the spectrum, $\alpha$ is a constant with a value of 2898 if $\lambda_m$ is expressed in microns, and T is the absolute temperature of the body.
5. The amount of radiation passing through a particular unit area is inversely proportional to the square of the distance of that area from the source ($\lambda/d^2$), as shown in Figure 2.5.

# Solar radiation input

Now that the principles of radiation have been outlined, we can look at the details of solar radiation input to the earth in a more meaningful manner.

Because we know the mean distance of the earth from the sun we can work out, from law 5 above, how much radiation the earth should receive. This amount is the solar constant and has a value of about 1370 W $m^{-2}$ at the top of the atmosphere. Recent work from satellites shows that the solar constant decreased about 0.1% between 1978 and 1985, with an increase from 1985 to 1988. A 1% increase would be adequate to cause an increase of 0.5°C in global temperature. By measuring how much radiation reaches the top of the atmosphere and knowing the size of the sun, as well as the earth's mean distance, the emission temperature of the sun can be determined from law 3. For the photosphere or visible light surface of the sun, this value works out to about 6000°K. This figure then enables us to determine at what wavelength most radiation will be emitted from the sun from law 4, that is,

$$\lambda_m = 2898/6000$$
$$= 0.48\ \mu m$$

From Figure 2.3, we can see that this value is in the middle of the visible part of the spectrum. Note that it is the wavelength of blue light.

From the radiation laws it has been possible to determine how much radiation the earth ought to receive, as well as the amount and properties of solar radiation. Similar calculations can be made for the earth when we are considering outputs.

The input of energy to the earth at its mean distance from the sun is only an average value, for changes are taking place all the time. For example, the earth is rotating on its axis once in 24 hours, the earth is orbiting the sun once in 365 days and, as its axis of rotation is at an angle of about 23.5° to the vertical, the distribution of radiation at the top of the atmosphere is constantly changing. Over even longer periods of time, the nature of the earth's orbit and the angle of tilt of the earth also change, thus affecting the amount and distribution of radiation over the earth. These, however, are only important on a time scale of thousands of years and will be discussed more fully in Chapter 9.

The sun also emits energy in what is called corpuscular radiation (sometimes referred to as the **solar wind**), which is comprised primarily of ionized particles and magnetic fields. There is a connection between variations in the strength of the solar wind and activity on the surface of the sun. This activity is most clearly seen in the form of sunspots and solar flares or prominences. We will discuss these as well in Chapter 9. The solar wind interacts with the magnetosphere, the magnetic field that surrounds the earth, and this interaction is most brilliantly visible in the **aurora borealis** in the northern hemisphere and **aurora australis** in the southern hemisphere. Some research suggests that there is a connection between solar disturbances and various patterns of climate on earth, but the mechanisms have not been established. One of the most significant connections suggests that there is a correlation between a 22-year sunspot cycle and drought across the Great Plains of North America. If it can be shown that the event does recur regularly, prediction and alleviation of the consequences could be of great social and economic benefit.

Let us look in more detail now at the diurnal and seasonal effects.

## DIURNAL VARIATION

As the earth rotates on its axis, a different portion of the top of the atmosphere will be exposed to the incoming solar radiation (often abbreviated to **insolation**). At dawn, the sun will be low in the sky and the amount of radiation passing through a unit area normal to the line from the sun will be spread over a large area (Figure 2.10). As the sun rises in the sky the surface area decreases, and so intensity increases. If our surface is eventually at right angles to the solar beam it will receive the maximum intensity of radiation—the surface area is at its smallest. As well as the angle between the sun's rays and the top of the atmosphere and the earth's surface, the length of daylight will also affect the amount of radiation received. At the equator, day-length remains at approximately 12 hours throughout the year. At the poles it varies between 0 and 24 hours depending upon the time of the year.

## SEASONAL VARIATION

Seasonal variations in insolation arise from the changing axial tilt of the earth throughout the year (Figure 2.11) and the eccentricity of the earth's orbit. The orbit is an ellipse, not a circle, so that the earth is slightly nearer the sun (147 000 000 km) on 4 January and at its farthest distance (152 000 000 km) on 4 July. The variation in distance means that the amount of energy received also varies. The variation in energy received is ±3.5%, which does make a measurable difference in total insolation received in the two hemispheres (Figure 2.12). Being nearer the sun means that the radiation input will be slightly higher. The earth is closest to the sun (that is, at perihelion) in the northern hemisphere winter and farthest away (aphelion) in the southern hemisphere winter at the present time. Because of changes in the elliptical orbit of the earth, to be discussed in Chapter 9, these relationships are constantly changing.

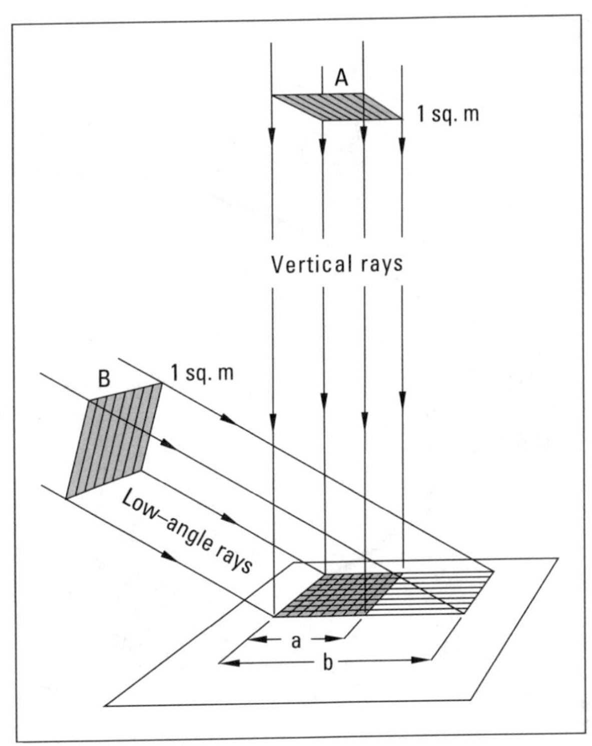


FIGURE 2.10 ▼ Energy distribution on an intercepting surface depends upon the angle of the incoming energy rays. Energy distribution is more concentrated on a perpendicular surface (a) than on a surface at a lower angle (b)

As the earth orbits the sun with its axis of rotation pointing in a constant direction, the area that is illuminated by the sun and the angle between the sun's rays and the top of the atmosphere will change. At the June solstice, the sun is above the horizon throughout the 24 hours for all latitudes north of the Arctic Circle, while south of the Antarctic Circle the sun would not be visible. Between the autumn equinox (22 September) and the winter solstice (22 December), the latitude at which the midday sun is overhead gradually moves southward from the equator to the Tropic of Capricorn (23.5°S). By 22 December insolation will

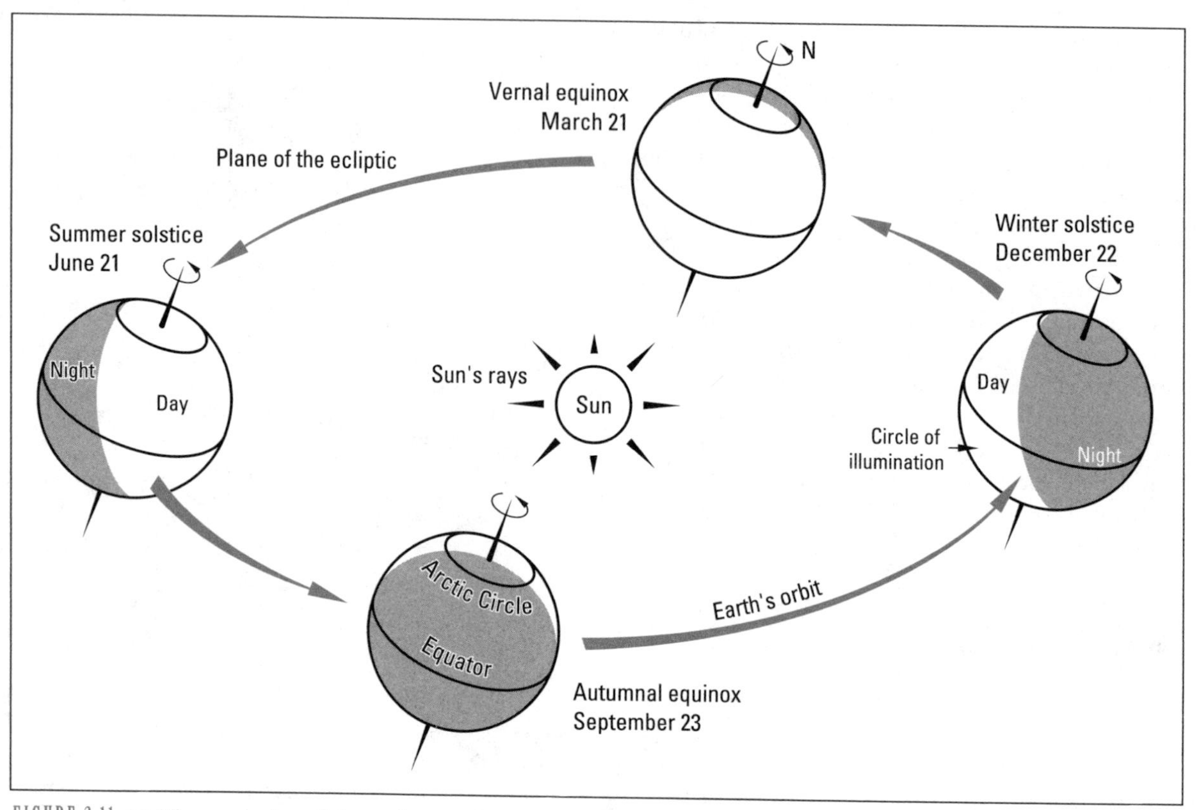


FIGURE 2.11 ▼ The revolution of the earth around the sun. The seasons result because the earth's tilted axis remains in a constant orientation in space as the earth revolves around the sun

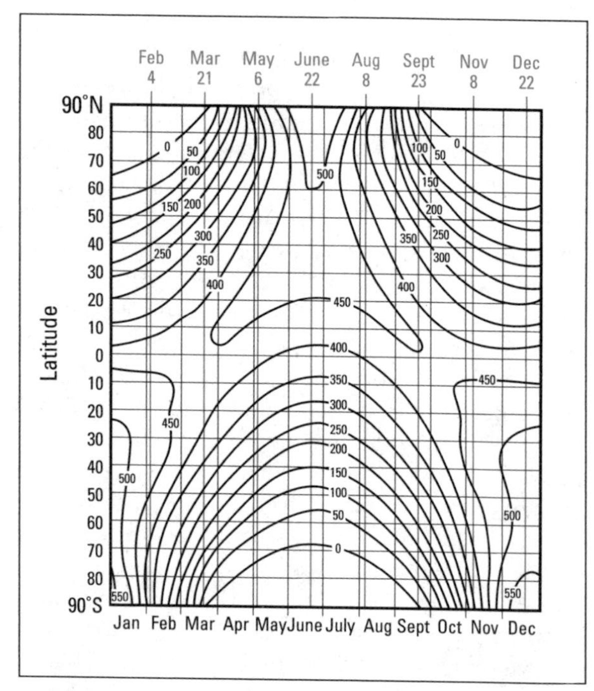


FIGURE 2.12 ▼ Solar radiation (W $m^{-2}$) falling on a horizontal surface at the top of the atmosphere (after Neiburger *et al*, 1982)

be at a maximum at that latitude, and zero north of the Arctic Circle. Between 22 December and 21 March, the sequence is reversed, and in the period leading up to the summer solstice, the latitude of the overhead sun moves northward from the equator to the Tropic of Cancer, insolation increases in the northern hemisphere, and the South Pole is thrown progressively into shadow (Figure 2.13).

If you stand with your back to the North Pole, the **altitude** of the sun is the angle between the horizon and the sun at noon. A navigator uses a sextant to measure this angle. Latitude can be calculated with the following formula:

$$\text{Altitude} = 90° - \text{Latitude} \pm \text{Declination}$$

Declination is the latitude at which the sun's rays are vertical at noon. You add declination if you are in the same hemisphere as the sun, subtract if the sun is in the opposite hemisphere.

For example, the altitude of the sun at Halifax (latitude about 45° N) would be:

September 23 (equinox),
altitude = $90° - 45° \pm 0°$
= 45°

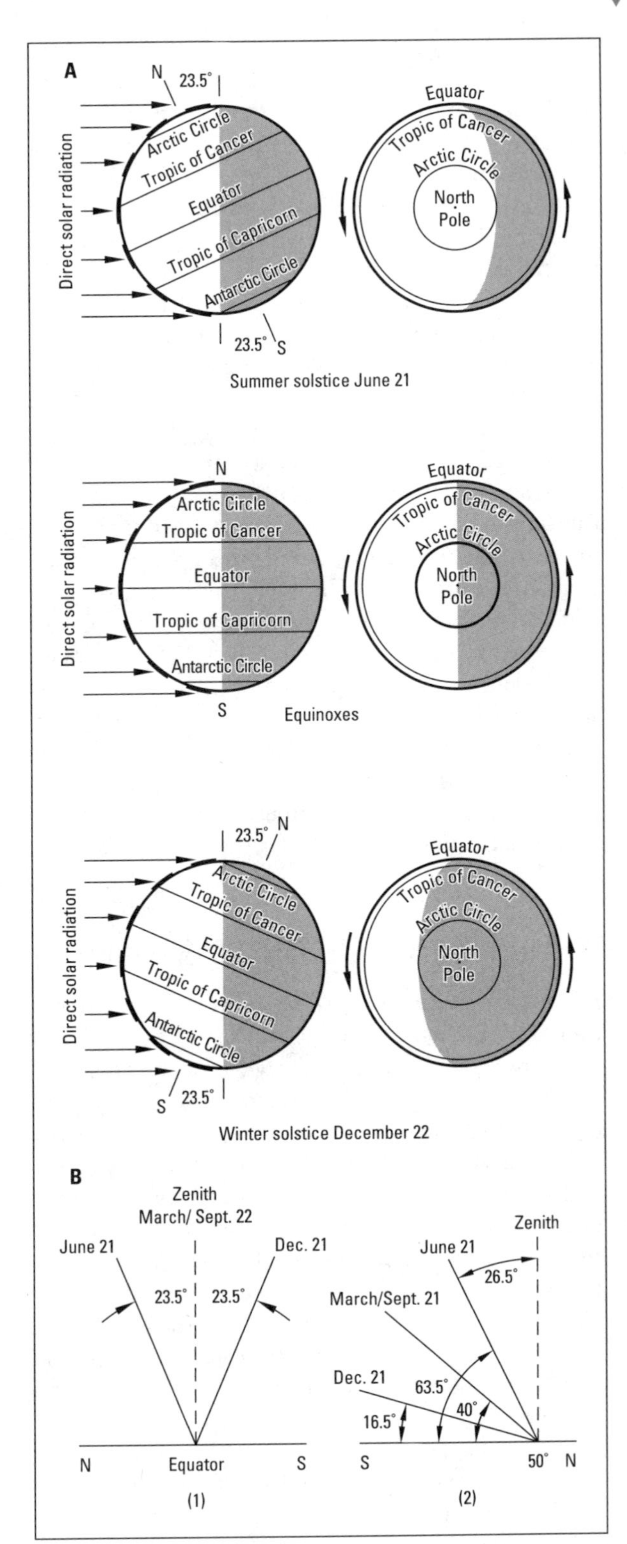


FIGURE 2.13 ▼ (A) Exposure of the earth to the sun's radiation at the solstices and the equinoxes; (B) position of the midday sun at the equator (1) and at 50° N (2) at the solstices and the equinoxes

December 22 (winter solstice),

$$\text{altitude} = 90° - 45° - 23.5° = 45° - 23.5° = 21.5°$$

June 21 (summer solstice),

$$\text{altitude} = 90° - 45° + 23.5° = 45° + 23.5° = 68.5°$$

Calculate the angle of the sun for your community or for the North Pole. These calculations seem unnecessary and simple today. However, similar (but more sophisticated) calculations are essential for effective development of solar collectors.

The presence of the earth's atmosphere has a dramatic effect on the input of radiant energy, but we can illustrate the essentially astronomic controls of the input either at the top of the atmosphere or by assuming there was no atmosphere; the result is the same. It is clear from Figure 2.12 that, taking an annual figure, the tropics would receive the most radiation as the input never falls to low values, unlike the situation at the poles, where 24 hours of daylight in summer becomes 24 hours of darkness in winter.

## Endogenetic energy

Energy supplied by the sun is known as **exogenetic**, in that it is derived from outside the earth (Greek *exo* = outside + *genos* = creation). It can be argued that, indirectly, almost all the earth's energy is exogenetic and all but a small amount is derived from the sun; the remainder comes from other cosmic bodies, including the attractive force of the moon, which causes tidal activity (a form of kinetic energy). In addition, a minute proportion of the total energy comes from within the earth and is thus **endogenetic** (Greek *endon* = within + *genos*).

The most obvious source of endogenetic energy is from the hot interior of the earth. The outer core of the globe consists of molten materials at immense pressures, and at temperatures up to 2600°C. There is an almost immeasurably small and continuous conduction of this heat to the earth's surface that adds to the energy inputs acting on the landscape. This can be detected in deep mines and caves. Locally, the decay of radioactive minerals can provide energy to the surface. More dramatic leakages of this endogenetic energy are seen in the form of volcanoes, hot springs, and various other tectonic activities. Taken together, all sources of endogenetic energy appear to contribute no more than 0.0001 percent of the total energy supply averaged over the earth's total surface.

## ENERGY OUTPUTS OF THE GLOBE

The output of energy from the earth is in radiant form, but it is not identical to the input of radiant energy from the sun. The earth has modified the input by a variety of processes. Some of the original solar energy input is reflected by clouds or the ground surface and returned to space with little change in its radiative properties; it is still shortwave radiation. As insolation passes through the atmosphere it is scattered, much of it toward the earth, but a small proportion back to space as an output of shortwave energy.

Of much greater importance is the emission of radiant energy from the earth itself. As a result of the absorption of solar energy in the atmosphere and at the surface, the earth will have gained energy that will be converted into heat. In turn, the earth and its atmosphere emit radiation because of their lower emission temperatures, following Wien's law. The average temperature of the earth is about 290°K, while that of the atmosphere is a chilling 250°K. Consequently the energy emission will reach a maximum at a wavelength of 2898/290 or 2892/250, which is 9.99 or 11.59 μm; and overall emission is entirely within the infrared range (Figure 2.3).

In this form the energy is susceptible to absorption by the atmosphere, so very little escapes directly to space; most is repeatedly reabsorbed and re-emitted before it is able to leave the system. The ability of the globe to trap energy in this way helps to keep the temperature of the earth and atmosphere higher than it would otherwise be. In other words, it promotes energy storage within the system.

At a global scale these processes lead to energy outputs of which about 36 percent are in the short wavelengths derived from reflected insolation, and about 64 percent in the long wavelengths, largely from emission by the atmosphere.

So far we have discussed the amount of energy received at the top of the atmosphere, and it is apparent that the angle of the sun's rays, the latitude, and the time of year are the three most important controlling factors. The distinctive latitudinal distribution of energy at the top of the atmosphere is also generally repeated in the amount received at the bottom of the atmosphere (see Figure 3.15). The energy given back to space must eventually equal that coming in or the mean temperature of the earth will change. The difference between the incoming radiation and the outgoing radiation is the earth's **net radiation budget**. On average there is a surplus of energy in the tropical and subtropical regions and a deficit in the middle latitudes and polar regions. The point of balance is at approximately 38°N in the northern hemisphere and 40°S in the southern hemisphere. The difference is primarily due to the different ratios of land and water in the two hemispheres. The latitude of zero balance is reflected in the temperature gradient. If the mean temperature for each latitude is calculated, the greatest change from one latitude to the next (moving from the equator to the pole) is coincident with the point of zero energy balance (see Figure 3.17). The point of zero balance migrates poleward with the sun in the summer (reaching approximately 60° of latitude), and equatorward in the winter (to about 30° of latitude). This point of balance is very important in the general pattern of world climate, particularly in the middle latitudes, as we will discuss in later chapters.

## SPATIAL AND TEMPORAL VARIATIONS IN OUTPUTS

Radiation outputs from the globe vary considerably over time and across the global surface. Spatial fluctuations depend upon a number of factors, including the character of the atmosphere (e.g., its temperature and the degree of cloudiness) and the nature of the earth's surface (e.g., vegetation cover and topography). From the polar regions an output of about 140 W $m^{-2}$ compares with 250 W $m^{-2}$ from equatorial areas—a ratio of about 2:1—whereas the ratio for the short wavelength input is about 6:1. These aspects will be covered in Chapter 3.

Over the long term the fluctuations in global energy outputs possibly relate to outside influences; a change in input may lead to an adjustment in the output. The ways in which these adjustments take place are complex, and involve interactions called feedback relationships (Figure 2.14). Vegetation cover, atmospheric conditions (including moisture content and cloud cover), the extent of polar and mountain snow cover, the area of the sea surface, and even soil cover and roughness may change in response to alterations in energy inputs. Through such changes, the earth is able to adjust its energy outputs in the event of any longterm variation in inputs by altering the balance between absorption, retention, emission and reflection of energy.

The question, however, is whether longterm variations of this kind occur.

Certainly over geological time quite marked fluctuations in climate have taken place, as attested by the evidence of ice ages and tropical conditions contained in, for example, the rocks of Arctic Canada and many other areas. Some of these changes are due to movement of the continental plates but some may be related to alterations in energy inputs and, if so, it is clear that outputs, too, must have changed. As the snow cover was extended during the ice ages, reflection must have increased, while absorption (and hence reradiation) must have been reduced. We will examine some of the possible consequences of this process in Chapter 9. Ultimately, however, a new equilibrium seems to be established as energy outputs decline to match the new, lower level of inputs.

It is an intriguing question, also, to ask whether changes in global conditions could arise due to adjustments in the outputs independently of change in energy inputs. Any event that significantly alters the reflectivity of the earth's surface might trigger such changes. An increase in the extent of the oceans relative to land due, perhaps, to major earth movements; increased snow cover as a result of mountain-building; changes in vegetation cover due to these events (or even due to human activity); or changes in the atmosphere brought about by massive volcanic eruptions—all could lead to significant changes in the global climate and hence in energy outputs. The implications for the world's climate are very important.

What is certain is that marked variations in global energy outputs do occur in the long term. Many of these variations are probably cyclical, related to changes in solar inputs such as those resulting from differences in the tilt and orbit of the

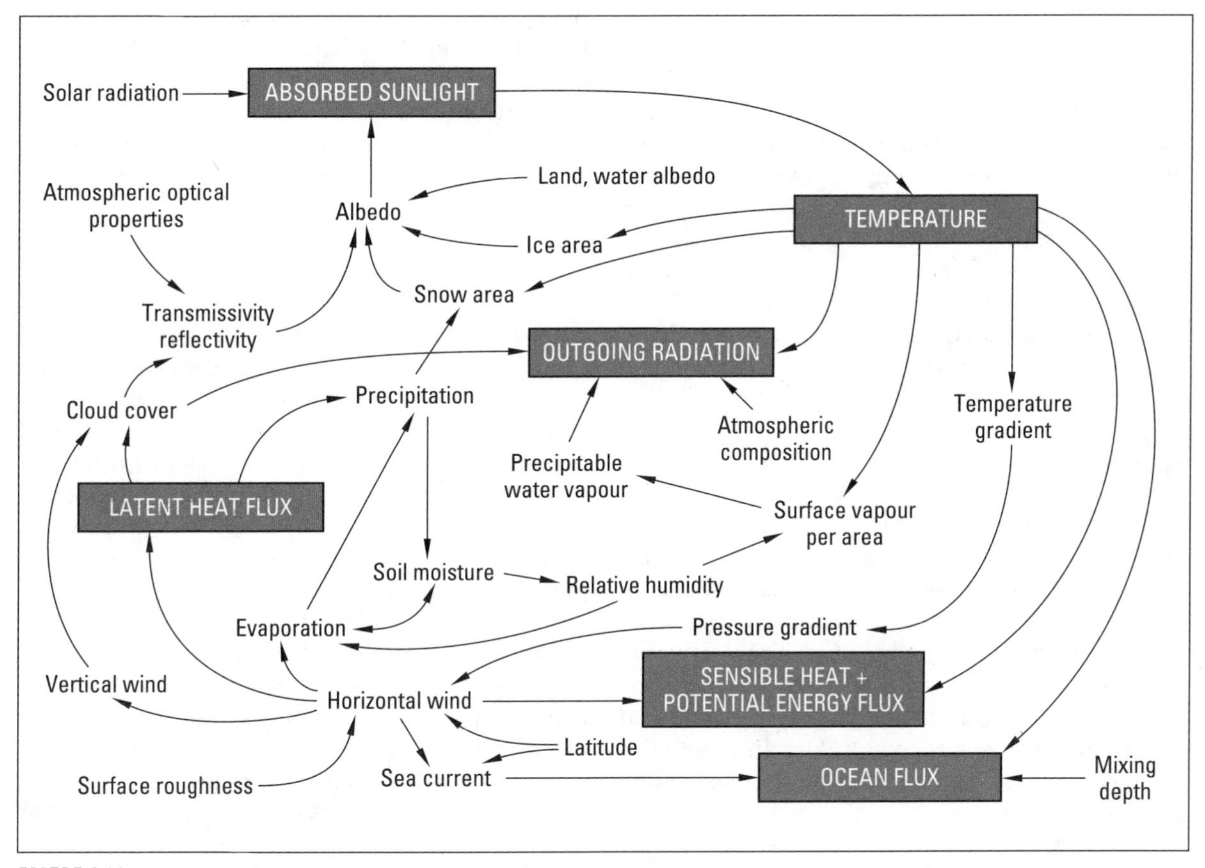


FIGURE 2.14 ▼ The complexity of the interactions in the global energy system are partially revealed here. This indicates why it is so difficult to model global climate (after Kellogg and Schneider, 1974)

earth. It is also apparent that such variation in outputs are critical if the earth is to adjust to alterations in the energy inputs that are known to occur, and thereby maintain steady-state equilibrium. An unanswered question is, to what extent can humans change these outputs and upset the equilibrium?

# Thermodynamics

The basic principles of energy are embodied in the laws of thermodynamics. These were initially developed in 1843 by Prescott Joule to explain processes seen in steam engines. Since then it has been appreciated that they have far wider significance, and they now represent basic precepts of science.

The first two laws of thermodynamics state that:

1. Energy can be transformed but not destroyed.
2. Heat can never pass spontaneously from a colder to a hotter body; a temperature change can never occur spontaneously in a body at uniform temperature.

The first law therefore defines the conservation of energy. The second law leads to the principle that energy transfers are a result of inequalities in energy distribution and that energy is always transferred from areas of high energy status to areas of low energy status, that is, down the energy gradient.

The third law of thermodynamics is less easy to understand. In very general terms, it says that systems tend toward equilibrium, that is, a random distribution of energy over time.

## ENERGY AND WORK

The transfer and conversion of energy are associated with the performance of work. The sun performs work in heating the earth through its provision of radiant energy. A river uses kinetic energy to perform the work of moving boulders. The weathering of rocks or the decomposition of plant debris involves work carried out largely by chemical energy. Indeed, it is the work done in these ways that characterizes the myriad of processes operating in the environment.

When this work is carried out, therefore, energy is transferred from one body to another, and in some cases it is also converted from one form to another. In the process, the total energy content remains the same, it is only changed in form. When a river or glacier cuts a valley, the energy they use is not destroyed but transferred or converted to other forms—some to heat energy, some to potential energy, some remaining as kinetic energy. When a plant grows, it takes in energy from the sun, from the air, and from the soil and stores it; the energy is not lost, merely transferred and transformed.

## SUMMARY

Energy is the driving force for all the processes operating in the global system. It performs the work in processes such as moving rocks, cutting valleys, lifting mountains, making water flow, the wind blow and plants grow. This work is performed through the transfer and transformation of energy. These transformations tend to follow well-defined routes.

The work is carried out because of differences in the energy status of different objects or conditions. Inequalities in the distribution of available energy (that which is capable of performing work) lead to energy transfers; in the process of these transfers work is done. The energy involved in these transfers is not destroyed; it merely changes form.

## GLOBAL ENERGY TRANSFERS

Every feature and every part of the globe is at some stage or another involved in energy transfers and transformations, and, as conditions change, so the nature of the transfers and conversions operating at any one place also change. We cannot, therefore, describe the processes operating throughout the entire global system in any detail. We can, however, try to identify the dominant transfers operating at a global scale and indicate, within this general pattern, the roles played by the various subsystems.

We have already noted that the balance between incoming and outgoing radiation is such that marked disparities occur between the energy status of different parts of the globe. The most obvious effect of this is the range in temperature we find when travelling from pole to equator, a range of 30° to 60°C depending upon the time of year and the hemisphere. In very simple terms, it is these differences that drive the global energy circulation. In order to achieve equilibrium, energy is transferred from the warmer parts of the globe to the cooler. If someone turned off the sun these transfers would result eventually in a more or less uniform distribution of energy across the globe; the fact that the sun continues to supply this unequal distribution of energy, however, maintains the imbalance and makes the attempt to achieve uniformity a losing battle. On the other hand, if this battle were not fought, the fact that the equatorial areas are constantly gaining more energy than they lose, while the polar areas are losing more than they gain, would result in a massive accumulation of heat in lower latitudes and indescribable cold in higher latitudes.

Thus, there exists a net poleward transfer of energy, and this transfer maintains the existing pattern of energy distribution; it feeds the higher latitudes and drains the lower latitudes (Figure 2.15). This transfer is performed by a variety of processes. Undoubtedly, the main transfers occur in the atmosphere. Winds carry warm air and water vapour away from the tropics. The warm air thus transfers heat (thermal energy) to the cooler latitudes. The water vapour carries energy in the form of latent heat. When the water vapour condenses, this energy is released as heat and warms the surrounding atmosphere. The oceans, too, transfer significant amounts of energy poleward. Heating of the sea in equatorial areas creates a temperature gradient between the lower and higher latitudes. Ocean currents carry the warmer waters down this gradient by a process of **lateral convection**, that is, the force created by the difference in temperature between one part of the ocean and another. Equally important, surface winds move this water poleward, a reflection of the close interaction between atmosphere and ocean.

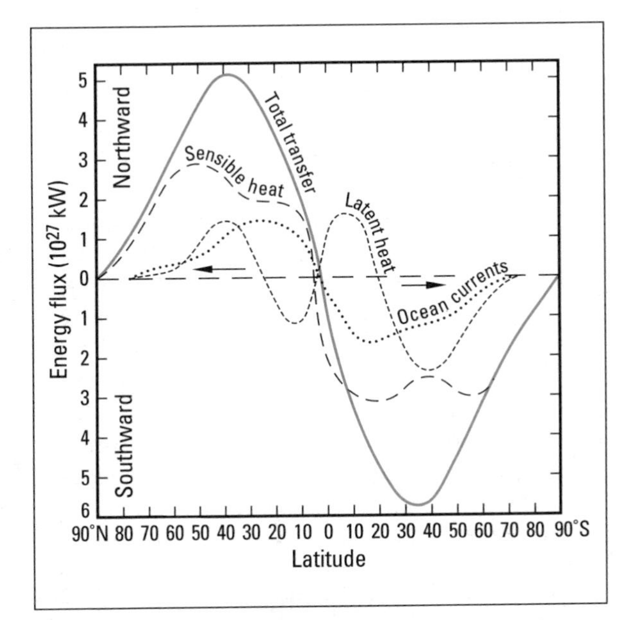


FIGURE 2.15 ▼ The average annual latitudinal distribution of the components of the poleward energy transfer in the earth–atmosphere system (after Sellers, 1965)

In both air and sea, however, the transfers are not one-way. If this were the case, we would be faced with a buildup of air and water in the higher latitudes and a slow emptying of the tropical areas. Clearly this does not happen; the warm air and ocean currents that flow toward the poles are replaced by a counter-flow of cooler air and water moving from the poles. In the case of the sea, the flow tends to occur at depth, for the cool water sinks. In the air, the pattern is a little more complex. The transfer of latent heat (that is, energy tied up within water vapour) occurs mainly in the lowest 2 or 3 km of the atmosphere. It is closely related to the surface wind network, which we will discuss in Chapter 4. The sensible heat transfer (that is, of warm air masses) occurs both close to the earth's surface and also at high altitudes (around 10 km). Both flows, however, are reflected by counter-flows of cooler air from higher latitudes. We will examine these processes closely in Chapter 3.

Thus the three main processes of energy transfer at a global scale are:

1. The horizontal transfer of sensible heat by warm air masses.
2. The transfer of latent heat in the form of atmospheric moisture.
3. The horizontal convection of sensible heat by ocean currents.

Of these three processes, the first is the most important, accounting for about 50 percent of the total annual energy flow. The other two processes account for about 20 percent and 30 percent, respectively.

We will consider the detailed processes involved in these transfers in Chapter 4, and will see there the factors that lead to the spatial distribution of these transfer mechanisms, but it is worth noting here that marked latitudinal variations in the three processes occur. Sensible heat transfers by the atmosphere, for example, are at a maximum between 50° and 60° north and south of the equator, and again at 10° to 30° north and south (Figure 2.15). This pattern reflects the two types of transfers referred to earlier; the higher level transfers are dominant in the subtropical zone while surface transfers are most active in middle latitudes.

The transfer of latent heat also shows a complex pattern, related to the distribution of water vapour in the atmosphere and the dominant, lower-air wind patterns. Thus its main effects are seen between 20° and 50° north and south of the equator, where winds blowing outward from the subtropics carry moist air poleward. Nearer the equator the pattern is reversed. Winds created by equatorial low pressures carry this air into the lower latitudes. As we will see in Chapter 4, this pattern is closely related to the global wind system.

Oceanic transfers of energy are most important either side of the equator, reflecting the outward movement of warm water from the tropical region. We will discuss these processes fully in Chapter 14.

In total, the processes of energy transfer maintain a steady-state equilibrium within the global system; they replenish energy losses in areas where outputs exceed inputs (the higher latitudes) and they remove energy from areas where inputs are in excess (the lower latitudes).

# Local and regional energy transfers

While these atmospheric and oceanic processes account for the spatial redistribution of energy at a global scale, they are not the only means of energy transfer in the global system. At a more local level, numerous other transfers are taking place.

## ATMOSPHERIC TRANSFERS

Within the atmosphere, local and regional winds, convection currents, and air masses carry energy as sensible heat and as latent heat. The uplift of air and the water contained within it transforms some of this energy into potential energy, which is released when the air sinks or the water falls as rain. Small, local transfers of energy to the earth's surface occur due to friction, while the kinetic energy of the wind is transmitted to soil and rock particles as these are picked up and blown along. Heat energy from the atmosphere is also transferred to soils and plants through conduction and radiation (Figure 2.16).

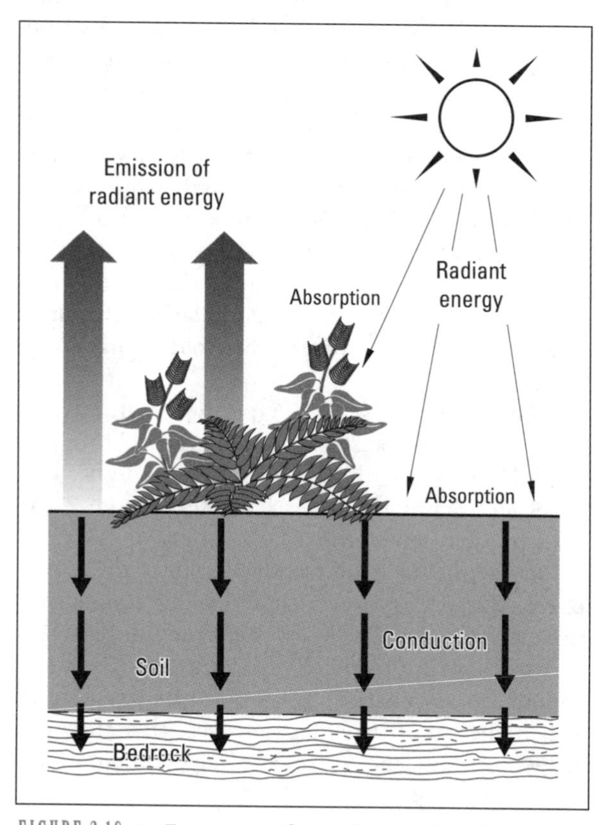


FIGURE 2.16 ▼ Energy transfers at the earth's surface

## HYDROLOGICAL TRANSFERS

Water similarly takes part in a variety of transfer processes (Figure 2.17). Water condensing in the atmosphere releases latent heat; this warms the surrounding atmosphere. Potential energy derived from the initial uplift of water vapour into the atmosphere is transformed into kinetic energy as raindrops fall, and some of this kinetic energy is transmitted to the earth's surface as rock and soil particles are splashed into motion. Further potential energy is expended and converted to kinetic energy as the water percolates through the soil, runs into streams and flows to the sea. The flowing water again imparts some of its kinetic energy to material that it catches up and carries along.

The water also takes part in chemical processes of weathering and thus chemical energy is transferred to heat energy, given off during the chemical reac-

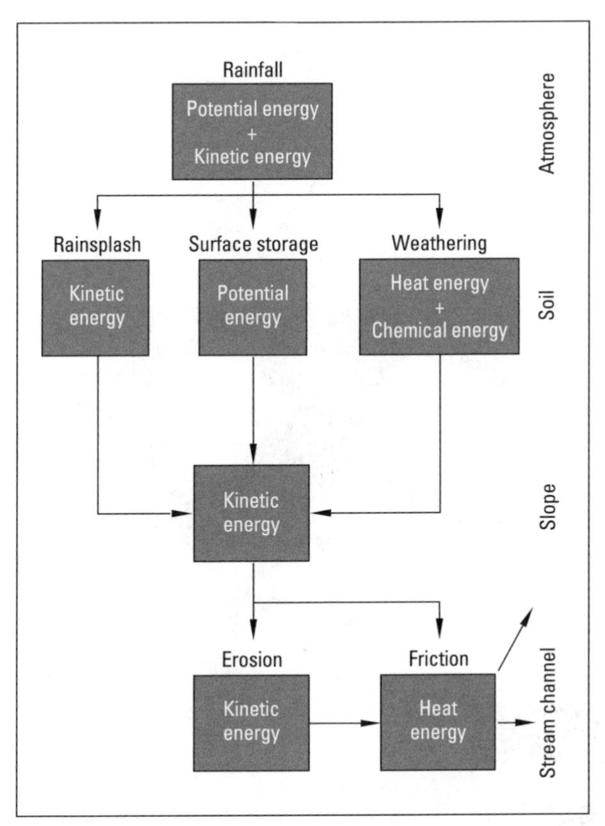


FIGURE 2.17 ▼ Energy transfer processes involving water

tions. In the sea, the currents transfer energy laterally, while the upwelling and sinking of water masses leads to vertical transfers. Finally, the evaporation of water from the sea, from rivers and lakes, and from the soil involves the conversion of thermal or radiant energy to kinetic and potential energy as the water is again raised from its original position and carried to higher levels in the atmosphere.

## LANDSCAPE TRANSFERS

Many of these transfers influence landscape processes, for the movement of water through the landscape is one of the main ways the earth's surface is altered and moulded. The potential energy possessed, for example, by boulders on a slope is a product of the erosion of the valley by the water and ice. Potential energy is also derived from earth movements, for mountain-building lifts the rock to leave it higher than the surrounding earth surface. Since these mountain-building processes are powered by heat energy within the earth, they represent the transformation of heat energy to kinetic and, ultimately, potential energy. The potential energy is subsequently converted to kinetic energy as the rock particles tumble, sludge or wash downslope. Friction with the surface and between the particles releases further energy in the form of heat.

## ECOLOGICAL TRANSFERS

On land, the formation of soil, the growth of plants, and the support given by this vegetation to animals all reflect further energy transfers and conversions.

In the case of terrestrial ecosystems (Figure 2.18), the development of a soil cover involves weathering, which in turn reflects the transfer of chemical energy from rocks to soil. Plants take up substances from the soil and store the chemical energy in their tissue. They also use radiant energy from the sun, and chemical and heat energy from the atmosphere, all three forms being converted to chemical energy by the plant. As the vegetation dies, or animals devour the plant materials, this energy is cycled through the environment. Animals convert the chemical energy to heat for bodily warmth and to kinetic energy for motion. They return some energy to the soil and the atmosphere as chemical energy.

Similar processes operate in aquatic ecosystems, although in this case much of the initial input of chemical energy is derived from organic matter washed into the waters from the land.

On a global scale it is impossible to quantify precisely the effects of all these processes. What is clear is that energy transfers create a fabric of relationships that bind the global system together, and provide the motive power for the processes that operate within our world, and which are the very foundation of our existence.

# Conclusion

The energy transfers that operate in the global system derive from inequalities in inputs and outputs across the world. At a global scale, they involve the movement of energy as sensible heat and latent heat by the atmosphere and as sensible heat by the oceans.

At a more detailed level, these transfers permeate every part of the global system. They involve transfers from rocks to soil, soil to plants, plants to

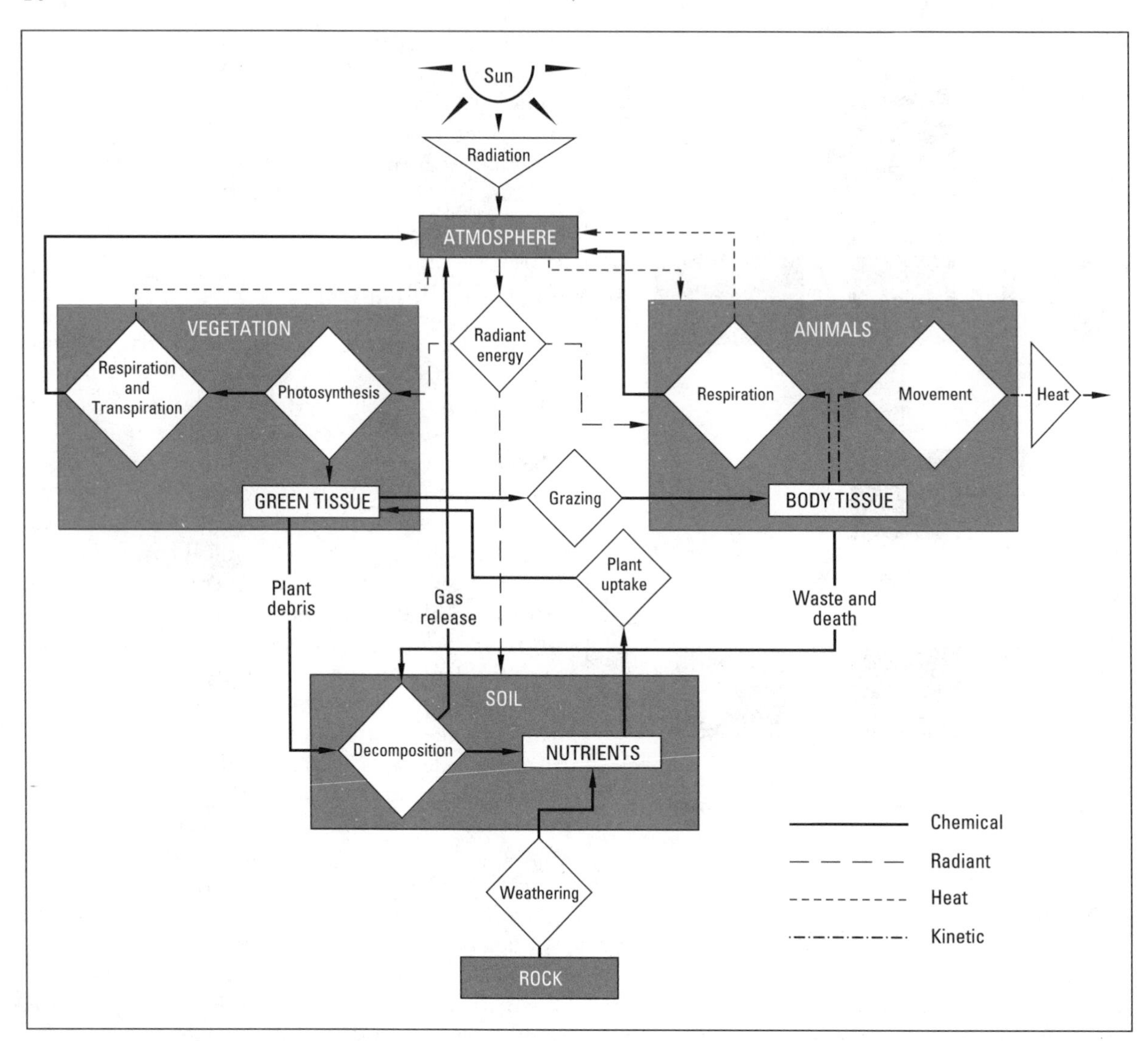


FIGURE 2.18 ▼ Simplified depiction of energy flows and transformations in terrestrial ecosystems

atmosphere; in fact all the components of the world are interconnected by these transfers. They also involve transformations of energy from one state to another.

Together these transfers and transformations provide the power for all the processes operating in our environment. They bind the global system into a unified whole. They are the lifeblood of our planet.

# Energy balance of the atmosphere

In the opening chapters we have seen the ways that energy reaches the earth from the sun and how a balance is achieved between the inputs and outputs of energy. In the case of the earth's energy system, we can take measurements of what goes on within the atmosphere and at the surface. In this chapter, we will examine the role of these energy flows in generating atmospheric movement, examine the nature and constitution of the atmosphere itself and what occurs within it, and explore how these help produce and maintain our climates.

## The solar energy cascade

What happens to the energy that pours into the atmosphere from the sun? We can start at the top of the atmosphere with the input of solar radiation (Figure 3.1). Following this input of energy, the first important effect is **absorption** by the atmosphere. Energy that is not absorbed proceeds to the next step—

reflection. If reflection takes place, either from clouds or from **scattering** by gas molecules, then the shortwave radiation returns to space as part of the earth's output. The remaining radiation will reach the ground surface where some of it is again reflected by the earth. The proportion of radiation that is not reflected is absorbed by the ground in the earth subsystem; this is calculated as (Q + q)(I – a).

Over a long time period, we can assume that the net gain to the ground is zero. The amount of energy received here must eventually be lost in order to maintain a balance. It is lost partly as **longwave radiation** and partly during **evaporation** and **convection** (the transfer of sensible heat). The energy from evaporation and **sensible heat**, together with longwave radiation, is absorbed by the atmosphere, which in turn radiates some of this energy back to space to complete the system.

The main components of this system are the atmosphere itself and the earth. These serve two functions: they trap and store energy and they divert or reflect it. They act as both stores and regulators of the energy that flows between them.

This model of heat energy transfers in the atmosphere (Figure 3.1) is highly simplified; it is no more than a summary of what we call the energy cascade. We have not yet allowed for the energy that takes other forms in the atmosphere. There is energy in the movement of air—**kinetic energy**; some air has energy depending upon its position above the earth's surface—**potential energy**. There are even small amounts of **chemical energy** that could be included in a larger, more realistic model of the earth–atmosphere system.

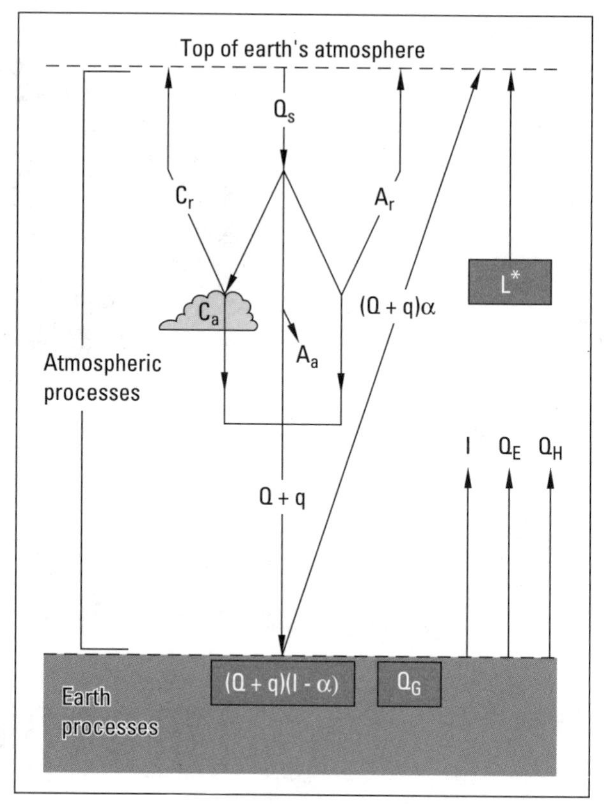


FIGURE 3.1 ▼ The solar energy cascade. $Q_S$ is incoming shortwave radiation; $C_r$ is shortware radiation reflected from clouds; $A_r$ is shortwave radiation reflected and scattered by the atmosphere; $C_a$ is shortwave radiation absorbed by the clouds; $A_a$ is shortwave radiation absorbed by the atmosphere; $\alpha$ is the surface albedo; Q is shortwave radiation arriving directly at the surface; q is shortwave radiation that has been reflected off another surface and arrives indirectly at the surface and $(Q + q)\alpha$ is the amount reflected; $Q_G$ is the net heat flow into the earth; $Q_E$ is the heat transfer from surface to atmosphere by latent heat of vaporization; $Q_H$ is the sensible heat transfer from surface to atmosphere; I is longwave emission from the surface; $L^*$ is the net longwave radiation from the atmosphere

## THE SUBSYSTEMS

The interface between the atmosphere and the earth plays a vital role in energy exchanges. It is a very diverse zone with surfaces ranging from simple forms like smooth water or ice to complex ones such as forests or urban areas. Each surface affects the way energy is apportioned (Figure 3.2). Some surfaces, such as water, use the majority of available energy for evaporation; at the other extreme we have urban surfaces and deserts where most of the energy is used for heating (Figure 3.3). The influence exerted by the surface will depend upon its areal extent. A small pool of water surrounded by an extensive grassland will have little effect on the local climate—it is too small. Lakes the size of Ontario or Superior are sufficient to produce major modifications of energy use and therefore of climate.

Water, ice, and desert surfaces may be thought of as simple interfaces between air and earth, but for the majority of land surfaces we are dealing with a zone, a three-dimensional layer from canopy top to the roots, where a myriad of complicated subsystems may exist. In Figure 3.4 we can see possible

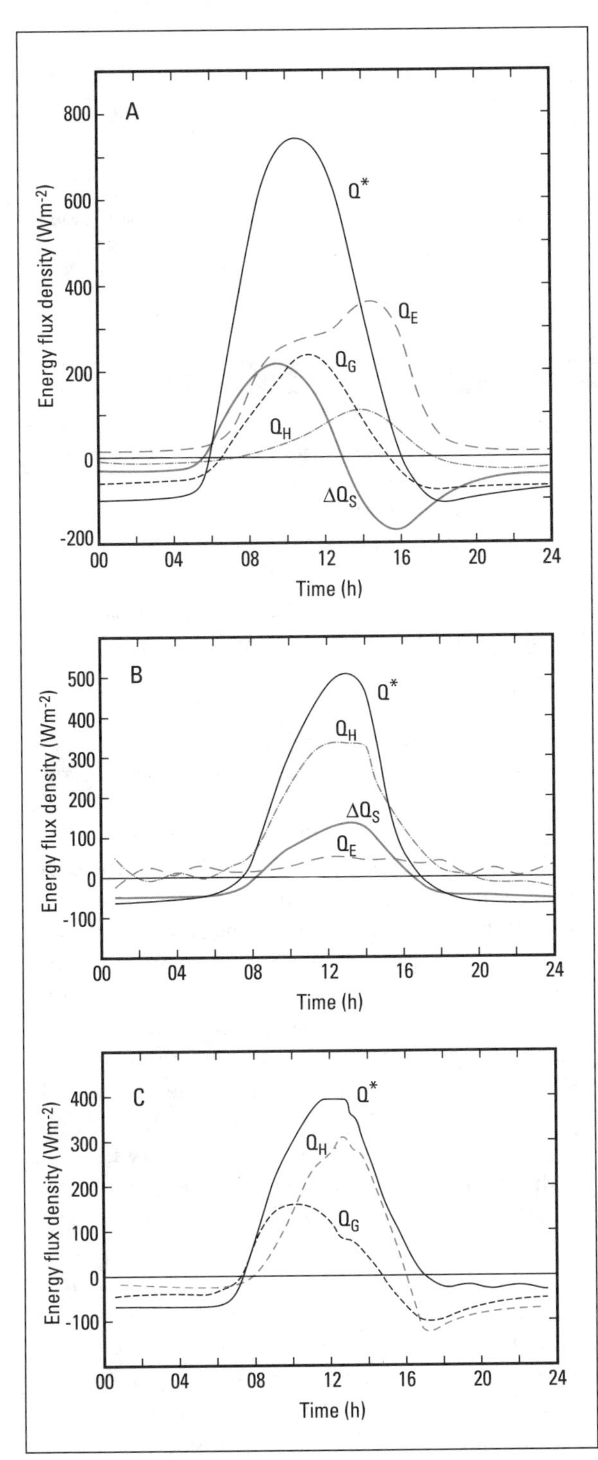


FIGURE 3.2 ▼ Diurnal variation of the energy balance components for (A) a shallow water layer in September; (B) the top of an urban surface, and (C) a desert. $Q^*$ is available energy (net radiation); $\Delta Q_S$ is net energy storage ( (A) and (B) after Oke, 1978, (C) after Stearns, 1969)

FIGURE 3.3 ▼ Meteosat image for 4 September 1982, infrared waveband. Light areas are relatively cool and dark areas are relatively warm. The strong heating over the Sahara Desert is evident by its dark appearance in this waveband

flows of energy within a forest, where the canopy top acts as a major interface between air and ground. As well as the canopy top we also have to consider the trunk space, where further energy transfers take place: the two-way transfer of infrared radiation to the soil and the canopy, the effects of latent and sensible heat losses and the influence of transpiration. The artificial surface of the city presents an even more complicated situation (Figure 3.5). Buildings vary in height and surface material, while heat is released in vast quantities but with an irregular distribution, so that radiation exchanges are difficult to determine or predict. Yet it is this type of surface on which an increasing proportion of the world's population are dependent.

## SPATIAL AND TEMPORAL VARIATIONS

Clearly our system is beginning to get more complex. However, the diagrams we have seen so far show the system at a single point assuming average conditions. All the individual inputs and outputs of the system, and the character of each of the subsystems, vary according to a range of conditions—time of day, time of year, regional climate, state of the

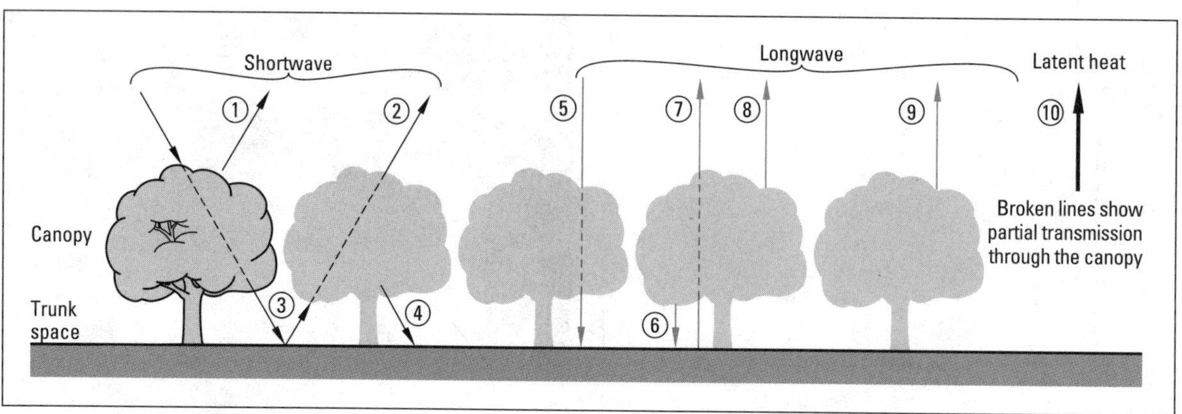


FIGURE 3.4 ▼ Radiation exchanges above and within a forest. 1: Shortwave energy reflected back to space from the trees. 2: Shortwave energy reflected back to space from the surface. 3: Shortwave energy received at the surface (direct). 4: Shortwave energy reflected back to the surface from the trees (diffuse). 5: Longwave energy counterradiated from the atmosphere. 6: Longwave energy counterradiated from the trees. 7: Longwave energy leaving the surface and being trapped in the atmosphere. 8: Longwave radiation leaving the trees and being trapped in the atmosphere. 9: Longwave energy escaping from the trees directly back to space. 10: Latent heat escaping to the atmosphere from moisture evaporated in the forest. Broken lines indicate partial transmission through the leaf canopy

ground, and so on. To understand the role and function of the atmosphere and its reactions with the surface more fully, we need to move away from the individual sites to consider the global interactions.

In Figure 3.2 we can see that different surfaces may possess common features. On a daily basis, available energy changes from being slightly negative at night to strongly positive values by day, before returning to negative amounts in the evening. The intensity of the daytime surplus will depend upon such factors as latitude and amount of cloud. The available energy can then be converted into latent heat, sensible heat or heat exchange with the soil. Over water or ocean surfaces the situation may differ (Figure 3.2A). The vast energy store of deep water means that latent and sensible heat flows can be maintained throughout the day and night, even when more energy is being lost than gained.

Extending our time period to one year, the detailed experience of the energy balance components will depend upon the nature of the climate. In equatorial areas (Figure 3.6), available energy ($Q^*$) is high throughout the year. The two maxima are associated with the passage of the overhead sun at the equinoxes (see Figure 2.12). The soil is permanently moistened by abundant precipitation so that most of the surplus energy is used in evapotranspiration ($Q_E$, the combination of evaporated water and water transpired through plants). Values for sensible heat ($Q_H$) are low apart from a small increase toward the end of the drier season. Few parts of the world are as well watered as São Gabriel, Brazil. Yuma, Arizona is one of the driest parts of the United States. The clear desert air gives high values of energy input, but there is little moisture to evaporate so the curve for sensible heat ($Q_H$) closely follows that for available energy ($Q^*$), as shown in Figure 3.7. Moving poleward, the period of surplus radiant energy becomes more concentrated into the summer period (Figure 3.8), and more strongly negative in winter. The subdivision into latent and sensible heat fluxes will depend upon the moistness of the climate. On the polar icecaps we find yet another change (Figure 3.9). First, the energy balance components are small in absolute terms throughout the entire year. Second, the annual cycle of the latent heat flow ($Q_E$) is similar to that of available energy ($Q^*$), but the energy use for evaporation is greater than that from radiation. Third, throughout the year, the flow of sensible heat is from the atmosphere to the surface because of the semi-permanent surface temperature inversion; this inversion occurs because the ice is usually colder than the air (see Chapter 7).

The variety of surface energy balances is important because it affects the overlying atmosphere to a marked degree. Across the globe, the atmosphere experiences a variation of energy input from the ground surface to produce an almost infinite mosaic of microclimates. These are very important for any particular location but their effects would be

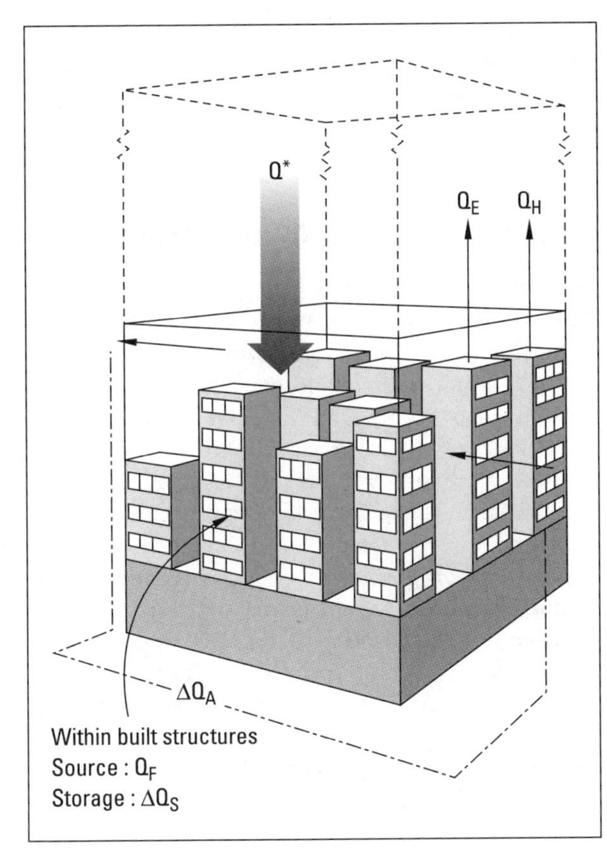


FIGURE 3.5 ▼ The energy balance of an urban air volume. $Q^*$ is available energy (net radiation); $Q_H$ is sensible heat; $Q_E$ is latent heat; $Q_F$ is anthropogenic heat; $\Delta Q_S$ is the net energy storage by buildings; $\Delta Q_A$ is the net energy (sensible and latent) advected into and out of the city (after Oke, 1978)

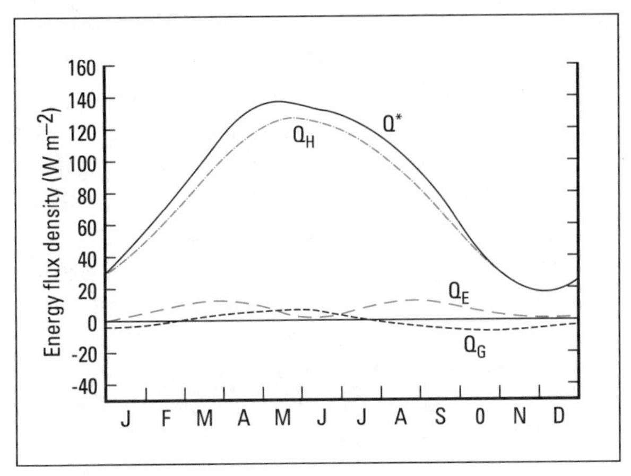


FIGURE 3.7 ▼ The surface energy budget for Yuma, Arizona, a desert climate (after Sellers, 1965)

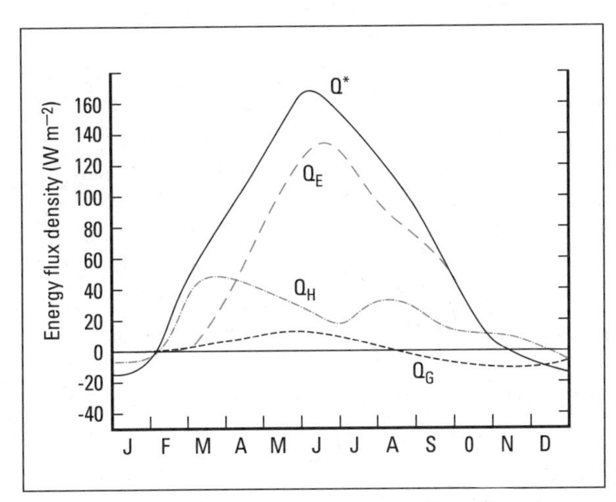


FIGURE 3.8 ▼ The surface energy budget for Madison, Wisconsin, a humid midlatitude climate (after Sellers, 1965)

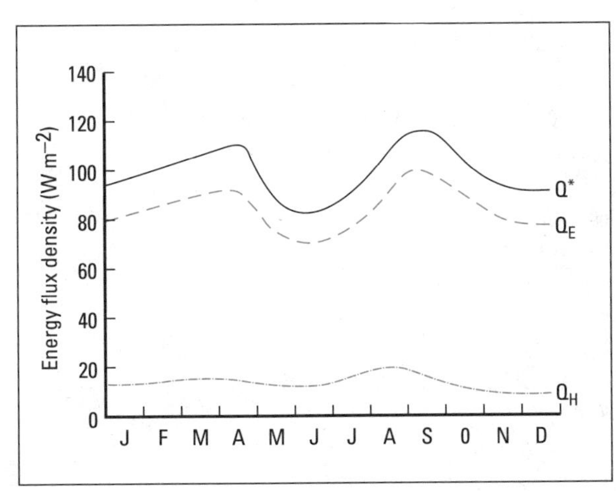


FIGURE 3.6 ▼ The surface energy budget for São Gabriel, Brazil, an equatorial continental climate

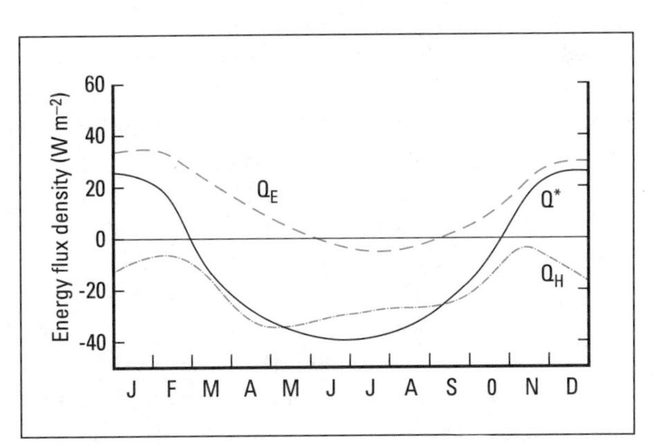


FIGURE 3.9 ▼ The surface energy budget for Mirny, Antarctica (66°33'S), a polar climate

insignificant at the global scale. However, a consistent pattern can be observed. More energy is received in tropical areas than in temperate and polar regions, an energy difference that generates horizontal temperature gradients between tropics and poles. In turn, the temperature gradients drive the major airflows and the weather systems that develop within them. Together, these flows of air help to overcome the imbalance of energy, and they represent a basic element of our global climate.

Of course, the atmosphere does not act in isolation. It is influenced by—and in turn influences—the other main components of the global system: the **hydrosphere** (water), **cryosphere** (ice), **lithosphere** (rocks), and **biosphere** (life). Changes in any one of these components may lead to changes in atmospheric—and therefore climatic—conditions. Some effects may be persistent because the oceans respond much more slowly than the atmosphere; once a temperature anomaly has been produced, it takes a long time for it to be eradicated. Moreover, relatively small changes in the hydrosphere may have drastic implications for atmospheric conditions. For example, when a dam is built a lake is created, and almost regardless of its size it will alter the hydrosphere and consequently the atmosphere.

Similarly, rapid changes in the biosphere affect the atmosphere. The nature and extent of the vegetation control the amount of energy reflected by the earth. In this context, the human impact upon vegetation, through forest clearance and agriculture, may be particularly important.

# The atmosphere

A thin shell of gases surrounds the earth. This is the atmosphere. It plays a vital role in allowing life to survive. It shields us from harmful radiation. It helps to stabilize the differences in temperature between tropical and polar regions. It reduces losses of heat from the earth's surface. It provides water to sustain life. We only have to contrast the earth with the moon to see what it would be like without an atmosphere, or with Venus or Mars if we had a different type or amount of atmosphere.

The gravitational force of the earth is sufficient to retain the present atmosphere, but its density is not uniform. As we move away from the earth's surface, the density becomes progressively lower until we reach the edge of space. What is the nature of this atmosphere? How does it achieve the functions outlined above? We shall be examining these points throughout this section, but first the general properties of the atmosphere must be considered.

## CHEMICAL COMPOSITION OF THE ATMOSPHERE

The chemical composition of the atmosphere is remarkably constant in the lowest 16 km, with nitrogen and oxygen being dominant, but the composition is not static. Volcanic activity, biological and chemical reactions, radioactive decay, gravitational losses and human activity all act together to produce, maintain, and alter the atmosphere (Table 3.1). There are two climatologically important gases that vary in proportion. The first of these is **water vapour**, which gets into the atmosphere through evaporation, mainly from oceans and land surfaces. In arid areas, there is little water to be evaporated so the atmosphere is also dry, perhaps containing only 0.2 to 1 percent of water vapour. In very humid areas this figure may rise to 3 percent or even 4 percent. Although it is the most variable of the greenhouse gases, it is by far the most abundant and

Table 3.1

*Composition of the earth's atmosphere*

| *Constituent* | *Formula* | *Percent by volume* |
|---|---|---|
| Nitrogen | $N_2$ | 78.08 |
| Oxygen | $O_2$ | 20.95 |
| Argon | Ar | 0.93 |
| Carbon dioxide | $CO_2$ | 0.035 |
| Neon | Ne | 0.0018 |
| Helium | He | 0.0005 |
| Methane | $CH_4$ | 0.0001 |
| Krypton | Kr | 0.0001 |
| Hydrogen* | H | |
| Nitrous oxide* | $NO_2$ | |
| Xenon* | Xe | |
| Carbon monoxide* | CO | |
| Ozone* | $O_3$ | |
| Water vapour | $H_2O$ | Variable (0–4) |

*Negligible percentage

important in the amount of heat energy it traps and reradiates.

The second variable gas is **carbon dioxide** ($CO_2$). It is a by-product of combustion and photosynthesis. Increases in fuel consumption have led to global increases in the carbon dioxide content of the atmosphere to above 350 parts per million or 0.035 percent (Figure 3.10). In addition, plants use carbon dioxide for growth, and plants cover a large proportion of the (northern) land hemisphere. In summer, when the plants are growing most rapidly, they consume more carbon dioxide than they release. So the atmospheric levels of carbon dioxide fall. In winter, plant growth is greatly reduced in the northern hemisphere, with no corresponding increase in the southern hemisphere where so much of the surface is sea, so levels of carbon dioxide show a seasonal change.

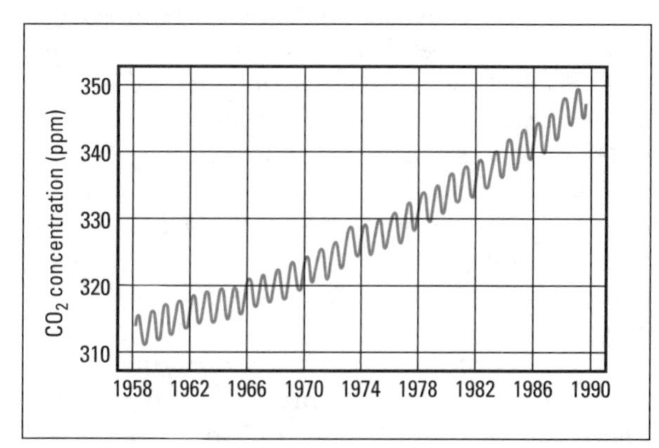


FIGURE 3.10 ▼ Measurements of atmospheric carbon dioxide taken at Mauna Loa, Hawaii. Fluctuations are seasonal variations as summer plant growth removes $CO_2$ from the atmosphere

At higher levels in the atmosphere, water vapour and carbon dioxide decrease in importance, but between 20 and 55 km **ozone** ($O_3$) becomes more common. Ozone is an enriched form of oxygen containing three atoms. It is formed by photochemical interaction between the ultraviolet (UV) portion of the solar spectrum and molecular oxygen. As the ultraviolet passes through the upper atmosphere, it is absorbed by the molecular oxygen, breaking the bond and producing two oxygen atoms. These can combine with a free oxygen atom, producing ozone. After formation, the ozone easily photodissociates or is broken down by other processes so that overall there is a balance between formation and destruction. The amount of UV absorbed within the ozone layer varies depending on the chemical constituency and temperature. Maximum concentrations occur between 15 and 30 km. Approximately 5 percent of the ultraviolet radiation reaches the surface of the earth.

Ozone amounts vary horizontally as well as vertically; they also vary seasonally as the sun changes its angle in the sky. The amount of ozone above any region can vary considerably in a matter of hours as currents swirl in the upper atmosphere. Globally, the maximum amount of ozone occurs in the middle latitudes while minimums are found over the poles. Amounts at the equator remain relatively constant year round and show no decrease in the record since 1960.

At the poles it is a different story. Here, in winter, within what is called the polar vortex (Chapter 5), extremely cold air is trapped. Ozone that is normally transported from the tropics to the poles is blocked from entering, and without sunlight ozone cannot be formed locally. The vortex is stronger over the Antarctic than the Arctic because of land/water differences. If temperatures become very low, as they often do in the Antarctic in winter, clouds called polar stratospheric clouds (PSC) form from frozen nitric acid and other particles. Some of the particles are from human sources; however volcanic eruptions, such as Mount Pinatubo in 1991, are still the largest contributor.

The so-called "ozone hole" is not a hole, but an area of thinning purportedly caused by manufactured gases, particularly chlorofluorocarbons (CFC). It is believed that CFCs are carried upward from the surface to the stratosphere where photodissociation frees reactive chlorine, which in turn can destroy ozone. It must be understood that the full chemistry and function of the atmosphere at these levels is not known and monitoring at these levels is very recent. The thinning over Antarctica was first reported in 1978, and since then the "hole" has been larger or smaller from year to year.

In Canada, there were only a few monitoring stations until the issue caught public attention and concern in 1992. At these sites a spectrophotometer measured the difference in the amount of light received over the ultraviolet segment of the solar spectrum. Differences are attributed to reduction by ozone absorption, and a figure of the amount of ozone is given in Dobson units (DU). The global average is roughly 350 DU. The lowest reading over Antarctica in 1991 was less than 150 DU.

## AEROSOLS

The term **aerosol** is used for particles in the atmosphere, and the most common of these are dust and smoke. There have been many attempts to estimate the general volume of aerosols in the atmosphere, but it is a very difficult problem. One estimate puts the total natural volume of small particulate matter in the atmosphere at 1250 million tonnes. The amounts vary from year to year and from region to region, and much depends on the location, surface configuration, time of year, and other factors. For example, there will be a great deal of dust over a desert region during a dust storm, whereas the volumes will tend to be low over the ice of the high arctic, especially in winter.

Surprisingly, there is a high level of aerosols over the oceans because of sea salt particles created from spray and evaporation. It is estimated that sea salt makes up the single largest volume of atmospheric particulate matter. These salt particles and much of the fine dust in the atmosphere are extremely important in the formation of precipitation, discussed in Chapter 4. It has been argued that they also act as important blockers of energy in the atmosphere, but have two opposing tendencies. The scattering of incoming radiation back to space tends to cause a cooling of the lower atmosphere, while the blocking of outgoing radiation acts like a blanket and tends to warm the lower atmosphere.

There are many natural sources of aerosols, but one of the most interesting is volcanoes, which can inject very large quantities of dust into the troposphere and stratosphere. The eruptions of Mount St. Helens in Washington state (1980), El Chicon in Mexico (1982), and Mount Pinatubo (1991) heightened the interest in volcanoes and their potential effect on climate. The importance of volcanic eruptions as causes of climate change will be discussed later.

A single volcanic eruption can have considerable effect but it is usually relatively short-lived. The addition of aerosols by human activities is different because it is increasing, and is potentially more devastating. It was estimated as early as 1968 that 280 million tonnes per year, a higher volume than from windblown dust, was being produced by human activities. It has been suggested that dust veils over some areas could become so severe that they would dramatically alter the climate in a region. The continuing degradation of soils and expansion of agricultural activity into more arid regions are likely to continue this increase in aerosols. Evidence of the increase of dust levels since the beginning of the Industrial Revolution is found in glacier ice from different parts of the world.

## TEMPERATURE STRUCTURE

Every day, hundreds of meteorological balloons are released into the atmosphere to take observations of temperature, wind velocity and humidity. Each ascent shows some differences in the temperature structure of the air, mainly related to local weather. In general they all show a decrease of temperature with height up to about 9 to 15 km above the surface. Then temperatures stop falling and even begin to rise (Figure 3.11) from the lowest values of about –60°C. At 30 km we are approaching the limit of most standard radiosonde ascents and additional information about the atmospheric temperature structure has to be obtained from rockets or satellites. These show that there is a progressive increase in temperature up to about 50 km, where values reach about freezing point, followed by a further decrease to the lowest atmospheric temperatures of about –100°C at about 80 km. Beyond this level, there is a steep rise of temperature in a physical sense, but as the air is so rarified, it holds little heat.

Each layer of the atmosphere is identified by its temperature structure (Figure 3.11). The **troposphere** and to a lesser extent the **stratosphere** are where most climate occurs.

Within the atmosphere there are three zones of heat production separated by two cooler zones. The earth's surface, at the base of the troposphere, is where most heat is produced by conversion of solar energy. Ozone reacts with ultraviolet light to produce heat in the stratosphere. In the **thermosphere**, particles emitted by the sun give high temperatures, but a low heat content.

Figure 3.11 shows the tropopause at a fixed height. Actually the troposphere is oval, being twice as deep over the equator (17–18 km) as at the poles (7–10 km). The difference is created partly by rotation of the earth but mostly by temperature. The low figure is for winter and the high for summer. Equatorial temperature is more consistent, therefore the seasonal difference is less than that for the poles. This difference in thickness affects the distance solar energy must travel passing through the atmosphere.

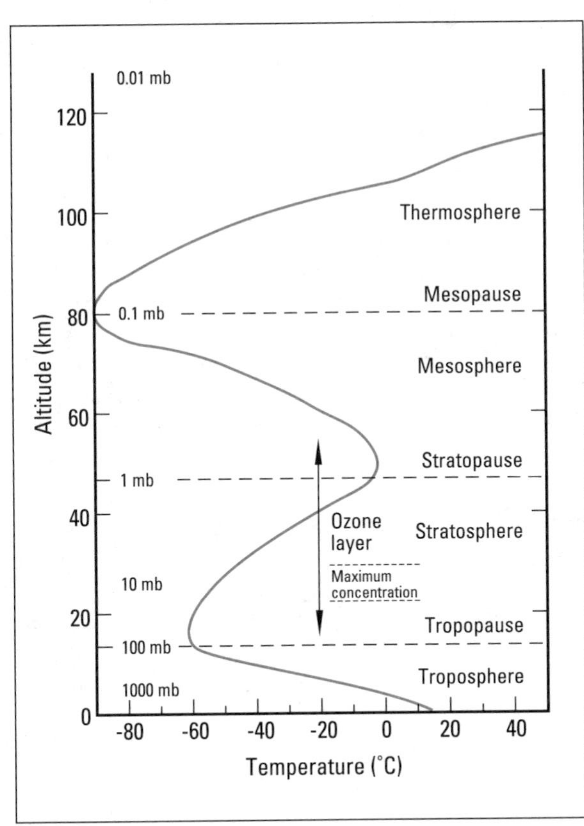


FIGURE 3.11 ▼ Layers of the atmosphere in relation to the mean temperature and pressure profile

# Energy in the atmosphere

## THE ENERGY SYSTEM

The sun's energy represents the prime source for our climatic system. Energy flows from the sun and interacts with the earth's atmosphere and surface in a complex manner to give rise eventually to the climatic zones of the earth. As mentioned in Chapter 2, heat storage and transport are also involved, and the best way to understand the processes is to examine them in turn. The laws of radiation have been outlined in Chapter 2, so here we look first in detail at the internal mechanisms of the radiation flow, then consider the spatial variability of the flows that give rise to different climates.

Perhaps the best way to explain what is happening is to follow the path of sunlight from the top of the atmosphere and describe what affects it on its journey to the earth's surface. Longwave exchanges can then be described.

## SHORTWAVE RADIATION IN THE ATMOSPHERE

As our beam of sunlight enters the atmosphere it begins the first of many interactions with the gases and particles. One of the first is the interaction between ultraviolet light and molecular oxygen that results in the higher temperature of the thermosphere. After this it passes through the **mesosphere** with little change.

In the stratosphere, the density of atmospheric gases increases. There is more oxygen available, which reacts with the shortest or ultraviolet wavelengths and effectively removes them, warming the atmosphere in the process. Already about 2 percent of the original beam strength has been lost.

It is in the troposphere that most effects take place. In the upper troposphere, the atmosphere is relatively dense, with a pressure about 20 percent of that at the surface. The size of the gas molecules of the air is such that they interact with the **insolation**, causing some of it to be scattered in many directions. This process depends on wavelength. The shorter waves are scattered more than the longer waves and so we see these scattered waves as blue sky. If the reverse were true the sky would be permanently red, and if there were no atmosphere, as on the moon, the sky would be black. Dust and haze in the atmosphere produce further scattering; some of the scattered radiation is returned to space, but much is directed down toward the surface as down-scatter or diffuse radiation. This is also the type of radiation we experience during cloudy conditions with no direct sunlight when the solar beam is 'diffused' by the water droplets or ice particles. Without diffuse radiation, everything we see would be either very bright, when in direct sunlight, or almost black, when in shadow. Lunar photographs are a good example of this.

Another type of shortwave energy loss is absorption. The gases in the atmosphere absorb some wavelengths (see Figure 2.9), as do clouds. In this way, we have a warming of the atmosphere though the amounts involved are small. The most important loss of shortwave radiation in its path through the atmosphere is by reflection. The water droplets or ice crystals in clouds are very effective in reflecting insolation. Satellite evidence shows that a mean figure of 19 percent of the original insolation is reflected by clouds. The degree of reflection is usually called the albedo. **Albedo** is expressed as the ratio of the amount of reflected radiation to the incoming radiation. If multiplied by 100, this can be expressed as a percentage. The lowest and thickest clouds tend to reflect most while the thin, high-level ice clouds have an albedo of only about 30 percent.

Now the beam reaches the ground surface, carrying about 50 percent of its original energy. Even now, not all of this is absorbed, as the surface itself has a reflecting effect. The global average albedo represents some 6 percent of the radiation *at the top of the atmosphere*, so the loss is not great. However, the figure may seem large when expressed as a percentage of the radiation actually reaching the surface. For example, the albedo of freshly fallen snow may reach as high as 90 percent (Table 3.2). the greatest variability is over water. When the sun is high in the sky, water has a very low albedo. This is the reason the oceans appear dark on satellite photographs. At low angles of the sun, such as at sunset, at sunrise or in midwinter in temperate and subpolar latitudes, the albedo may reach nearly 80 percent.

Table 3.2

*Albedos for the shortwave part of the electromagnetic spectrum*

| | *Percent* |
|---|---|
| Water (sun angles above 40°) | 2–4 |
| Water (sun angles less than 40°) | 6–80 |
| Fresh snow | 75–90 |
| Old snow | 40–70 |
| Dry sand | 35–45 |
| Dark, wet soil | 5–15 |
| Dry concrete | 17–27 |
| Black road surface | 5–10 |
| Grass | 10–20 |
| Deciduous forest | 10–20 |
| Coniferous forest | 5–15 |
| Crops | 15–25 |
| Tundra | 15–20 |

The sunlight reaching the earth's surface that is not reflected by the earth is absorbed and converted into heat energy. The distribution of energy received at the surface is shown in Figure 3.12. Thus, incoming radiation can be absorbed (in the atmosphere and at the surface), scattered (in the atmosphere), or reflected (by clouds and at the surface). When reflected, the radiation is returned to space in the shortwave form and becomes part of the outflow of energy from the earth (see Chapter 2). Similarly, some of the scattered radiation is returned to space, to give a shortwave albedo for our planet of 28 percent. The modifications of the solar beam by the atmosphere are shown in Figure 3.13.

The expansion of agriculture, especially in North America in the last 100 years, has meant that more and more natural vegetation has been removed. Usually this means that forests are removed and replaced with agricultural crops and in most instances this results in a shift from a darker surface to a lighter one. Compare the values for forests with those for crops in Table 3.2.

Although the general trend is to higher albedo surfaces, this is not always the case, and often one trend is counteracted by another. For example, expansion of deserts would increase albedo, while expansion of irrigated areas would decrease it. Regardless of the net effect there is no doubt that a change in the local and regional climates will occur. This has been well documented in urban areas, where surface changes are probably most dramatic. There are calculations of the amount of change in the global albedo that have already occurred or that may occur, but these are only crude estimates and the global impact has yet to be specifically determined.

## Longwave radiation

All substances emit longwave radiation in proportion to their absolute temperatures. Despite all the losses, the earth's surface receives the most shortwave radiation and therefore has the highest temperatures. It follows that most longwave emission will be from the ground surface. The atmosphere is much more absorbent to longwave radiation than to shortwave radiation. Carbon dioxide and water vapour are very effective absorbers of much of the longer part of the spectrum (see Figure 2.9), except between 8 and 12 mm. This gap is known as a window. As water vapour is concentrated in the lowest layers of the atmosphere this is where most absorption will take place. Clouds are also very effective at absorbing longwave radiation and hence their tem-

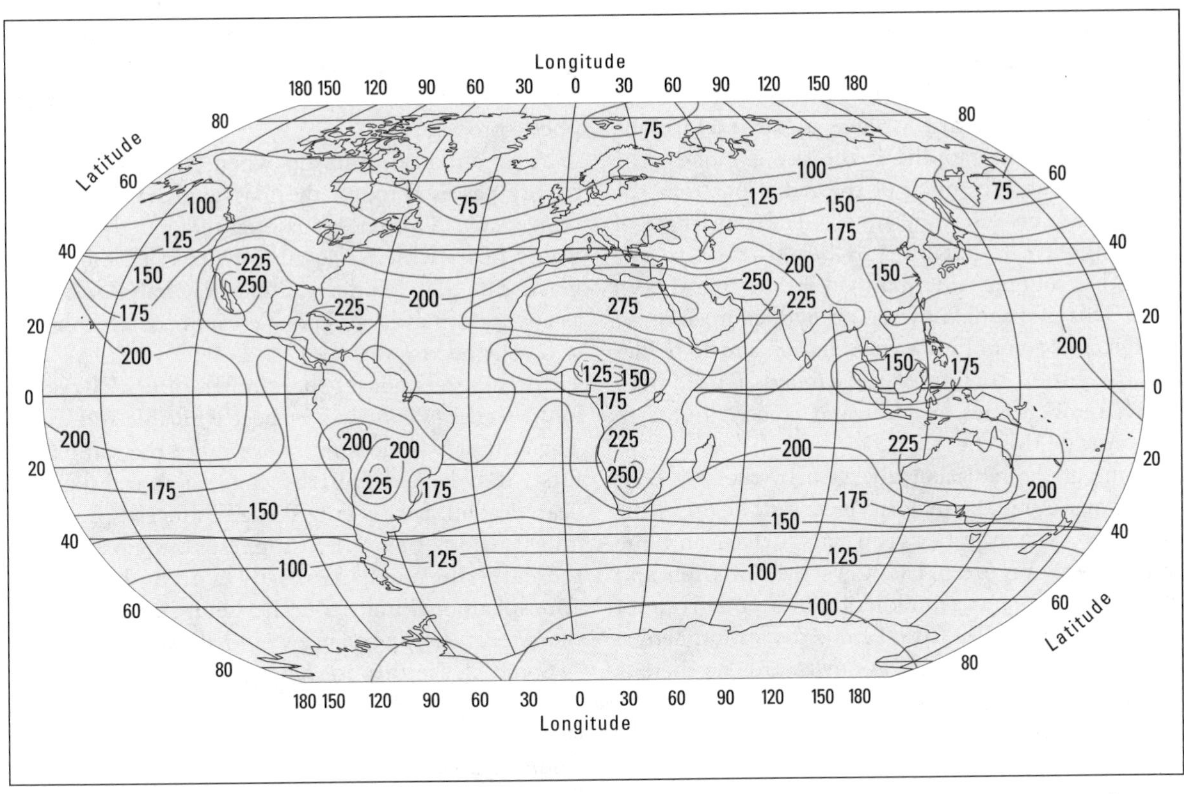


FIGURE 3.12 ▼ The average annual solar radiation received on a horizontal surface at the ground. The units are W m$^{-2}$ (after Sellers, 1965)

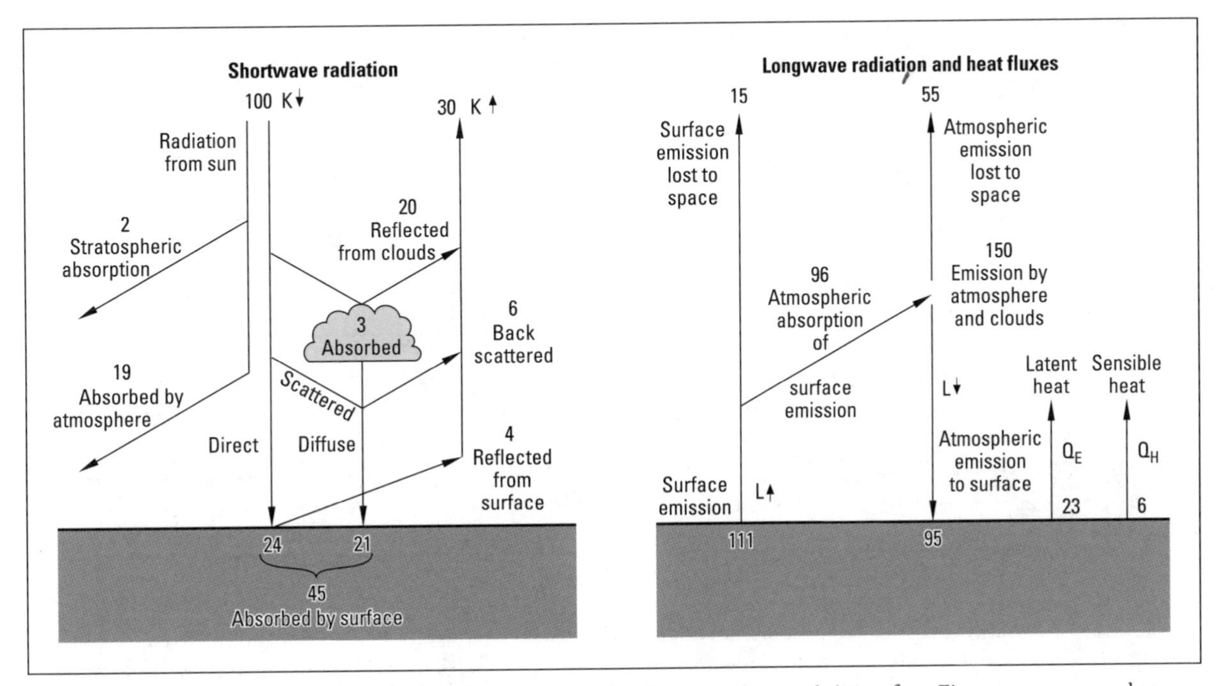


FIGURE 3.13 ▼ Modification of short and longwave radiation by the atmosphere and the surface. Figures are expressed as a percentage of incoming shortwave radiation at the top of the atmosphere, based on a global mean

perature will be higher than otherwise. This cloud effect is most noticeable at night. With clear skies, radiation is emitted by the surface but little is received from the atmosphere and therefore the temperature falls rapidly. If the sky is cloudy, the clouds will absorb much of the radiation from the surface and, because they are also emitters, more of the longwave radiation will be returned to the ground as counter-radiation than if the sky had been clear. This is absorbed by the ground, compensating for the emission of longwave radiation and so reducing the rate of cooling of the ground. Figure 3.14 shows temperatures on clear and cloudy nights to demonstrate this effect.

Some of the radiation given off by the surface is lost to space but the majority gets caught up in the two-way exchange between the surface and the atmosphere. Figure 3.13 shows the emission and absorption of longwave radiation as a proportion of incoming energy. Radiation from the atmosphere is emitted spaceward as well as downward. As there is less water vapour at higher levels, absorption by the atmosphere is less and proportionally more is lost to space.

## THE GREENHOUSE EFFECT

Table 3.1 lists the gases that constitute the earth's atmosphere. Water vapour, carbon dioxide, and methane are usually referred to as the **greenhouse gases**. It is argued that the gases act like the glass in a greenhouse.

The analogy is based on the properties of these gases outlined in the preceding section. It was originally developed to help students understand atmospheric processes.

A greenhouse heats up when (shortwave) solar energy passes through the glass and heats the surfaces inside. This energy is then reradiated as (longwave) heat, which is unable to pass through the glass. The glass works as a one-way valve. As long as energy comes in and none escapes the greenhouse will continue to warm.

In the greenhouse, the temperature will rise as long as energy comes in or heat is unable to escape. The atmosphere does not work quite the same way. First, it is heated more by convective transport of heat. Second, heat is only slowed in its escape. Many other analogies have been suggested, none of which recreate the mechanisms by which the earth's atmosphere maintains a higher temperature than its short-term energy balance would allow.

Some of the disputes about global temperatures and climate mechanisms are evident in the calculations of the effect of greenhouse gases. How cold would the earth be without the greenhouse gases? Estimates range from –18°C to –22°C. The average annual temperature for the world is approximately 15°C. Estimates of the total temperature difference because of the greenhouse gases range from 33°C to 37°C. A one-degree change in the annual average temperature is very significant, and would alter regional climates considerably. The greenhouse effect is discussed further in Chapter 9.

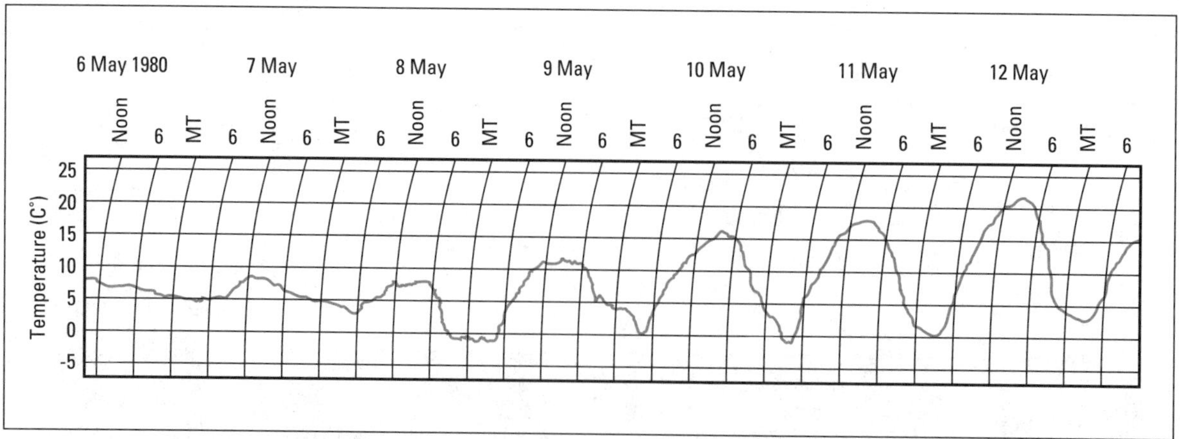


FIGURE 3.14 ▼ Contrasting diurnal temperature variations on cloudy and clear days. Cloudy weather prevailed for the first three days, giving a small diurnal temperature range. As the skies cleared later in the week, daytime temperatures increased but nighttime temperatures were lower. There is some indication of a slight, progressive rise of both day and night temperatures as a result of the storage of solar energy in the ground

## GLOBAL RADIATION BALANCE

Figure 3.12 shows the amount of solar radiation received at the surface of the earth after depletion losses in the atmosphere. Note that the highest values are not in the tropics but in the subtropics; this is because of cloud cover and is a clear indication of the effects of the atmosphere. We are interested now in the amount of energy available at the surface to create the dynamics of climate; this amount is referred to as net radiation or the radiation balance, and can be shown as follows:

$$Q^* = (Q + q)(I - \alpha) + L^*$$

where $Q^*$ is net radiation
$Q$ is direct insolation
$q$ is diffuse insolation
$(I - \alpha)$ is albedo losses
$L^* = I\downarrow - I\uparrow$
$L^*$ is the net longwave radiation
$I\downarrow$ is longwave energy returned to earth by the atmosphere (known as counter-radiation)
$I\uparrow$ is longwave energy lost from the surface

The general values for the earth's surface are shown in Figure 3.15.

More shortwave radiation appears to be absorbed by the earth than leaves it by a mixture of short and longwave radiation. The surface seems to be gaining heat. Similarly, the atmosphere seems to be losing heat. If radiation were the only process operating, the earth's surface should be getting warmer and the atmosphere cooler. They do not do so because, in addition to radiation, there are thermal energy transfers in the form of convective heat exchanges. Much of this takes place through evaporation and is discussed later in this chapter.

## SPATIAL VARIABILITY OF RADIATION EXCHANGES

The earth is a spheroidal body which spins on an axis tilted at about 23.5° to the vertical and has an elliptical orbit around the sun. These factors have a considerable influence on how radiation is distributed at the earth's surface.

In Chapter 2, we described the input of solar energy at the top of the atmosphere and how it was determined by these astronomic controls. Figure 2.12 shows how the radiation would be distributed at the top of the atmosphere, or at the earth's surface if we had no atmosphere. However, if we look at a

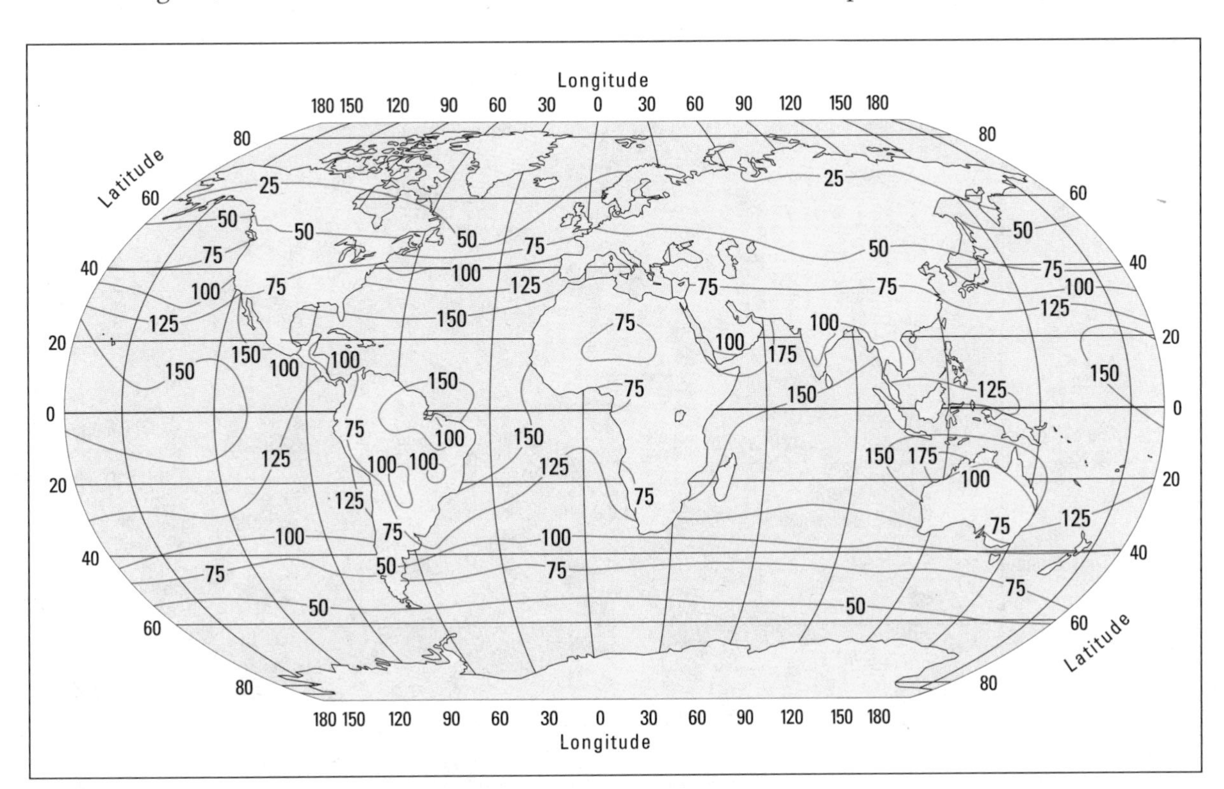


FIGURE 3.15 ▼ Global distribution of mean annual net radiation. Units are $W m^{-2}$ (after Budyko *et al*, 1962)

map of the average annual shortwave radiation reaching the ground, it is appreciably different (Figure 3.12). The general impression from the map is of a decrease of energy input toward the poles, with local anomalies. Most of these are caused by the distribution of clouds. High values are found over the Sahara, Australian, Asian, and American deserts. The lowest values occur in regions of high cloudiness such as Iceland, the Aleutian Islands in the North Pacific, the Congo Basin, and parts of West Africa.

This map may be a little misleading. By showing the radiation reaching the ground surface we are omitting an important factor. The surface albedo influences how much radiation is absorbed and this is not indicated on the map. Oceanic areas absorb a similar total to that shown on the map, but for ice-covered surfaces, such as Greenland, and areas with light-coloured, dry soils, such as the Sahara, the total radiation absorbed may be significantly lower (Figure 3.16).

What is it that produces this spatial pattern of radiation? Obviously, the astronomic factors have a great effect, giving rise to the overall decline poleward. But this decrease is far greater than one would expect from the distribution at the top of the atmosphere.

## THE ANGLE OF THE SUN'S RAYS

Thus, we have to look for other reasons. One of the most important is the angle between the sun's rays and the earth's surface, known as the **angle of incidence**. The input is greatest whenever the surface is at right angles to the sun's rays. If the sun is overhead, a horizontal surface will receive the highest intensity of radiation. When the sun is low in the sky horizontal surfaces receive the lowest intensity radiation. However, a steep slope facing the sun will receive higher values because the angle of incidence is increased. As the earth is a sphere at a great distance from the sun, the sun's rays are essentially parallel and hit the surface at different angles (Figure 2.13).

A secondary effect that further decreases radiation intensity is the longer path through the atmosphere at higher latitudes. Scattering and absorption

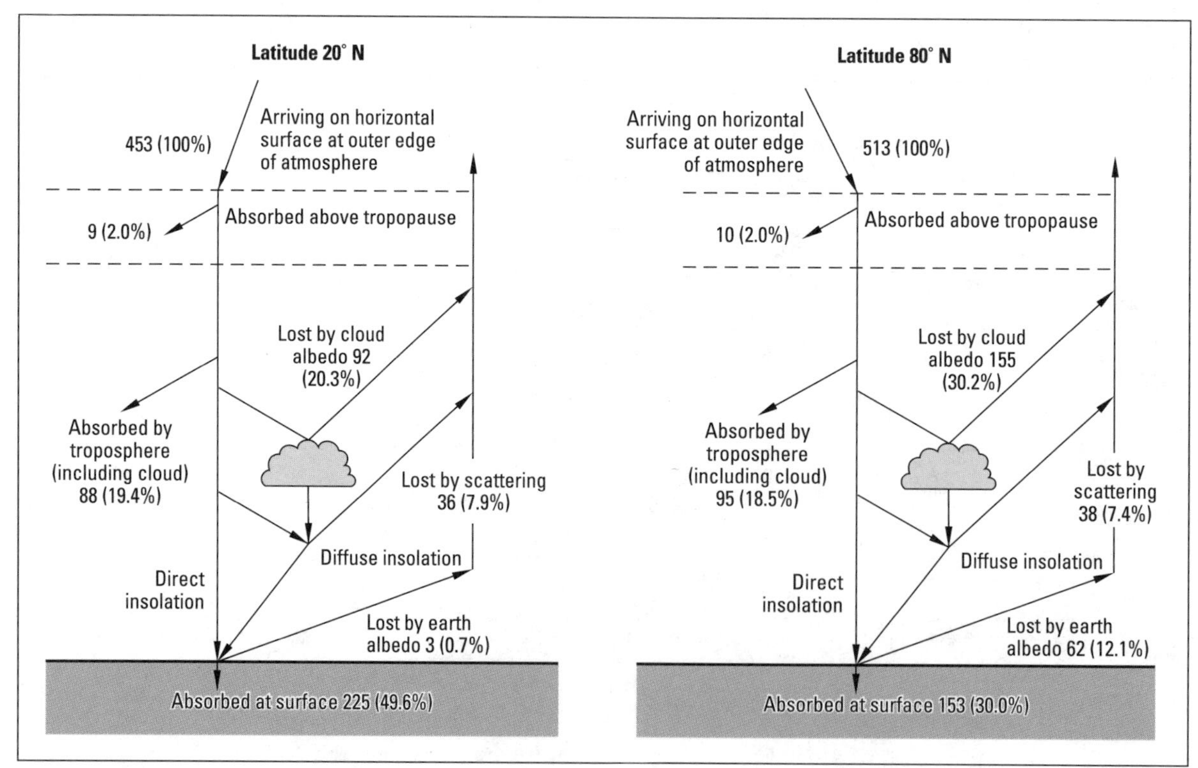


FIGURE 3.16 ▼ Contrasts in the disposition of insolation at summer solstice on a sand surface at 20°N and on ice surface at 80°N. Units are W $m^{-2}$. "Absorption" implies the conversion of the radiation into some other form of energy. "Lost" indicates a loss to the earth–atmosphere system (after Jackson, 1963)

will be higher, and this increases diffuse radiation at the expense of direct radiation. The amount of scattering and absorption vary depending upon the degree of haziness of the atmosphere. Where the atmosphere is very dusty, as in semi-arid or desert areas, more radiation will be absorbed and scattered, preventing it from reaching the ground surface. As the dust particles are much larger than gas molecules, scattering is not dependent upon wavelength and the sky has a whitish hue rather than the deep blue of a clear atmosphere. This effect is also noticeable over urban areas, where pollution produces the same effect.

# Latitudinal radiation balance

To see how much radiant energy we have available at any location we must know how much radiation is being lost as well as how much is reaching that location. Longwave radiation emission is proportional to the absolute temperature of the surface. It is far less variable than the input of solar radiation, ranging from about 350 W $m^{-2}$ in the Sahara to about 100 W $m^{-2}$ in Antarctica. For any particular latitude, we can sum the surface and atmospheric radiation balances to find out which areas of the earth have a radiation surplus and which areas have a deficit. The results are shown in Figure 3.17A, which shows the annual average balance; note that there is a surplus of energy in the northern hemisphere from the equator to 38°N and a deficit from there to the pole. In the southern hemisphere the surplus occurs between the equator and 40°S. The difference is due to the difference in the land-to-water ratio between the northern and southern hemispheres. The latitudes 38°N and 40°S represent the **points of zero energy balance** and are important in understanding the whole pattern of global climate.

Figure 3.17B shows the difference in the position of the point of energy balance for summer and winter in the northern hemisphere. As the vertical position of the sun migrates north and south of the equator the point of zero energy balance also migrates. In the northern hemisphere it moves between approximately 65°N in the summer and 30°N in the winter. There is a similar pattern in the southern hemisphere, although again the latitudes are slightly different, reflecting the surface configuration. Remember that these are only approximations representing average locations, but they do correspond with the average latitudes of the **snowline**. If there is an energy surplus the snow will melt, while if there is a deficit the snow will remain. Variations in the energy balance will be reflected in shifts in the points of balance and changes in the climate of these regions. When we understand this point of balance and the way it moves, we can see why the climate of the middle latitudes is so variable from season to season and from decade to decade, and we can understand other major elements of the atmospheric system, such as the Rossby waves and jet streams, which will be discussed in detail later.

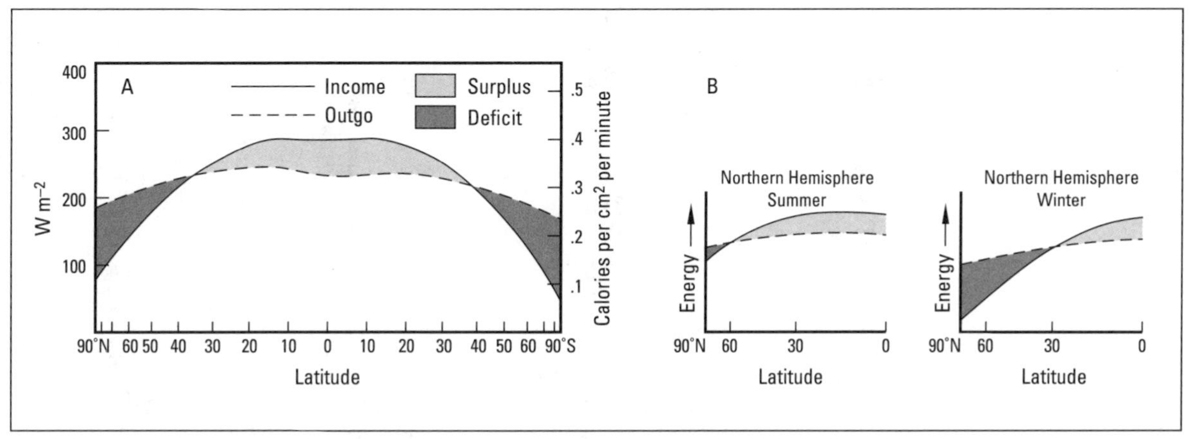


FIGURE 3.17 ▼ (A) Mean latitudinal distribution of the earth's radiation. Note the points of balance where the lines cross, that is, where incoming and outgoing radiation are equal. This is known as the point of zero balance. (B) The point of zero energy balance for summer and winter in the northern hemisphere. The point moves with the seasonal migration of the sun and is generally coincident with the Polar Front

The point of zero balance is reflected in the latitudinal temperature distribution shown in Table 3.3. The change of temperature from one latitude belt to the next is known as the **temperature gradient**. Note that the greatest change in the gradient occurs in the belts coincident with 38°N and 40°S. Since wind speed tends to be greater when the temperature gradient is higher we find that the point of zero energy balance is the normal latitude for the strongest winds. Of course, there are other factors as we will see later, but the temperature gradient remains the major driving force.

Over a given period of time the amount of energy entering the earth's atmospheric system from the sun should equal the amount leaving, as discussed earlier. However, in Figure 3.17A there is an imbalance. It shows a surplus of energy between 38°N and 40°S and a deficit toward the poles. Naturally, the magnitude of the surplus should be equal to the deficit, but it does mean that there must be a steady transfer of energy from the tropics poleward, otherwise the tropics would get hotter and polar regions cooler. It is the winds and ocean currents of the world that bring about the necessary heat transfer.

## Variability through time

As well as changing spatially, the components of the radiation balance at any point will be changing through time. For example, we could measure the values for a particular day. The results we obtain would depend upon the time of year and the nature of the weather conditions on that day. Atmospheric transmission, cloudiness, moisture levels, surface albedo, and even the moisture content of the soil (affecting longwave emission) would cause changes in the components. Figure 3.18 shows two extreme examples with clear and cloudy conditions.

### Table 3.3

*The average temperature and percentage of cloud cover for each 10° of latitude**

| | *Lateral zone* | *Temperature (°C)* | *Albedo (%)* | *Cloud cover (%)* |
|---|---|---|---|---|
| 0–90°N | 80–90 | –23.6 | 61 | 62 |
| | 70–80 | –15.9 | 46 | 66 |
| | 60–70 | – 7.2 | 24 | 65 |
| | 50–60 | 0.5 | 14 | 60 |
| | 40–50 | 7.5 | 12 | 53 |
| | 30–40 | 14.0 | 10 | 46 |
| | 20–30 | 20.4 | 10 | 43 |
| | 10–20 | 25.4 | 9 | 47 |
| | 0–10 | 25.5 | 8 | 52 |
| 0–90°S | 0–10 | 24.8 | 7 | 52 |
| | 10–20 | 23.3 | 8 | 48 |
| | 20–30 | 18.8 | 8 | 48 |
| | 30–40 | 13.5 | 8 | 54 |
| | 40–50 | 8.7 | 7 | 66 |
| | 50–60 | 1.2 | 17 | 72 |
| | 60–70 | –11.0 | 21 | 76 |
| | 70–80 | –29.5 | 63 | 65 |
| | 80–90 | –47.9 | 84 | 54 |

*The difference between each zone is known as the temperature gradient.

Data: S. Ladochy *et al*, *Climate and Landforms, a laboratory manual in physical geography* (1987), Kendall Hunt.

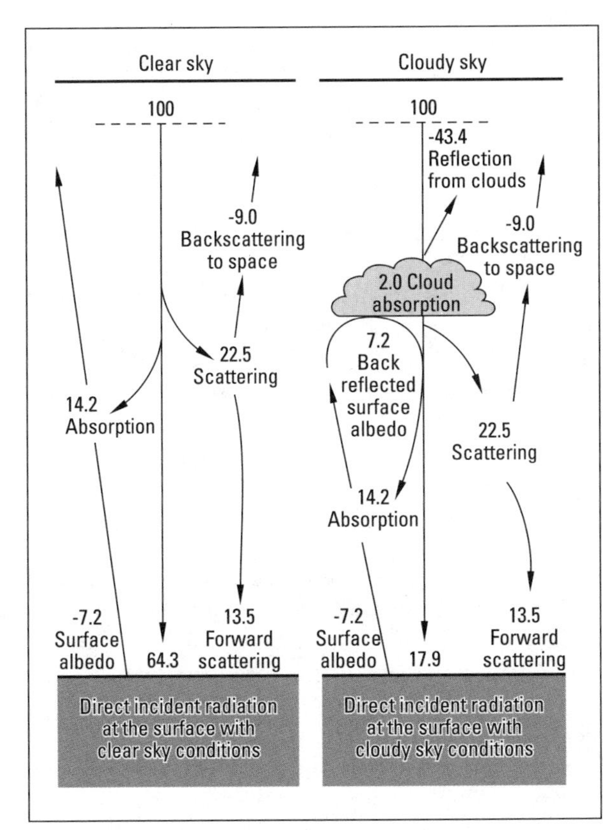


FIGURE 3.18 ▼ The effects of cloud on the incoming shortwave radiation balance. Figures are expressed as a percentage of insolation at the top of the atmosphere (after Bridgman, 1969)

On an annual basis, fluctuations in weather and the changing astronomic input of solar radiation lead to further alterations of the radiation balance components (Table 3.4). In turn, these variations help to explain the seasonal changes in heat absorption at the surface and hence the climatic regimes. For longer time periods, cyclic variations in the earth's orbit and its axis of rotation produce even more changes in the input of solar radiation. This subject will be covered in more detail in Chapter 9.

All components of radiation show a seasonal variation. In many cases it is the net effect or difference between incoming and outgoing radiation that is important. Downward shortwave radiation (K↓) is direct and diffuse radiation from the sun; some of this radiation is reflected at the surface, especially in winter, to leave us with net shortwave radiation (K*). The earth emits longwave radiation as well as receiving it from the atmosphere; the difference between emission and absorption is the net longwave radiation (L*). As the values are determined by the absolute temperature of the emitting body, seasonal differences are not pronounced. The value in Table 3.4 is negative because more longwave radiation leaves the surface than reaches it. The difference between all incoming and all outgoing radiation, irrespective of wavelength, is called net all-wave radiation (Q*) and determines how much energy is available at the surface.

## Table 3.4

*Radiative fluxes for Copenhagen**

| | *Range of monthly means* | | |
|---|---|---|---|
| | *Low* | *High* | *Yearly mean or total* |
| Downward shortwave radiation (K↓) | 1.8 | 28.1 | 3456 |
| Albedo (α) | 19% | 50% | 22% |
| Net shortwave radiation (K*) | 1.4 | 22.3 | 2682 |
| Net longwave radiation (L*) | −2.6 | −7.0 | −1649 |
| Net all-wave radiation (Q*) | −1.8 | 16.1 | 1033 |

*All values are MJ $m^{-2}$, except as indicated.

After D.H. Miller, *Advances in Geophysics*, 11 (1965), pp. 175–302, and D.H. Miller, personal communication (1985).

# Heat balance at the earth's surface

## USES OF AVAILABLE ENERGY

In the previous section, we showed how the earth's surface normally receives a surplus of radiation, which leads to a warming of the surface. This situation cannot last indefinitely, as the temperature gradient in the air and the soil would become enormous. Energy tends to flow down a gradient and as radiation is absorbed by the surface, so heat is transmitted into the soil and into the air. This takes place in proportion to the amount of energy originally absorbed:

$$Q^* = Q_H + Q_G$$

where $Q^*$ is the net radiation
$Q_H$ is the sensible heat transfer into the air
$Q_G$ is the heat flow in the ground

If the surface is damp some of the energy will be used in evaporation. Over the oceans most of the net radiation energy is used for evaporation. Therefore:

$$Q^* = Q_H + Q_G + Q_E$$

where $Q_E$ is the heat used for evaporation

This is a simplification, as changes in heat storage can take place and a small amount of energy is used in plant growth.

The energy transfer into the atmosphere is the final component of the radiation imbalance between surface and atmosphere. The net radiational loss in the atmosphere is counteracted by this heat transfer from the surface. So, over a long period, the atmosphere gains as much energy as it loses.

## SENSIBLE HEAT

How do these processes that use the net radiation energy take place? One way is by sensible heat transfer.

Sensible heat transfer is the exchange of warm air down the temperature gradient. By day, this will normally be upward, but at night there may be a weak transfer of sensible heat down to the cooler ground surface. It takes place because the air in contact with the surface becomes warmer through conduction. Being warmer, the air will be less dense than its surroundings and, like a cork in water, it will tend to rise until it has the same density (temperature) as its surroundings. Occasionally, this process is visible. If the ground is being warmed intensely, the rate of sensible heat transfer is high. The rising air can then be seen as a shimmering of the air layer near the ground due to the variable refractive indices of light through the air of different temperatures. Replacing the rising warm air are pockets of cooler air descending toward the ground.

The significance of sensible heat in the local heat budget depends upon the frequency and intensity of surface heating. Where the surface is usually hotter than the air, values maybe high, but where there is little temperature difference, sensible heat transfer from the ground to the air will be low (Figure 3.19).

## LATENT HEAT

The concept of latent heat can best be understood by conducting a small experiment. Start with a large block of ice out of a freezer and measure its temperature; perhaps it may be –10°C. Then place it in a heat resistant glass beaker and heat the beaker at a constant rate, monitoring the temperature of the ice continuously. Keep heating the beaker until all the ice has melted into water; eventually it will reach boiling point and vaporize as steam. If the temperature values are then plotted against time, we find a steady increase of temperature (representing heat input from the heater and some heat flow from the air, which will be warmer than the ice) until melting starts. Despite the steady addition of heat, there is then no increase in temperature until the ice melts completely (Figure 3.20). A similar effect is found on vaporization. Where has the heat gone? It was not being used to raise the temperature during melting or vaporization but to change the physical state of water, either from solid to liquid or liquid to vapour. As the heat appears to be hidden, it is known as latent heat.

## HEAT CONSUMPTION

Change of state involves a considerable use of energy to change from solid to liquid, or from liquid to vapour. In the first case, we need $3.33 \times 10^5$ J kg$^{-1}$; this quantity of heat is called the **latent heat of fusion**. In the second, much more energy is needed. At 10°C the **latent heat of vaporization** is $2.48 \times 10^6$ J kg$^{-1}$, but it falls slightly with increasing temperature. To get a better idea of this large quantity of energy needed for evaporation, the amount con-

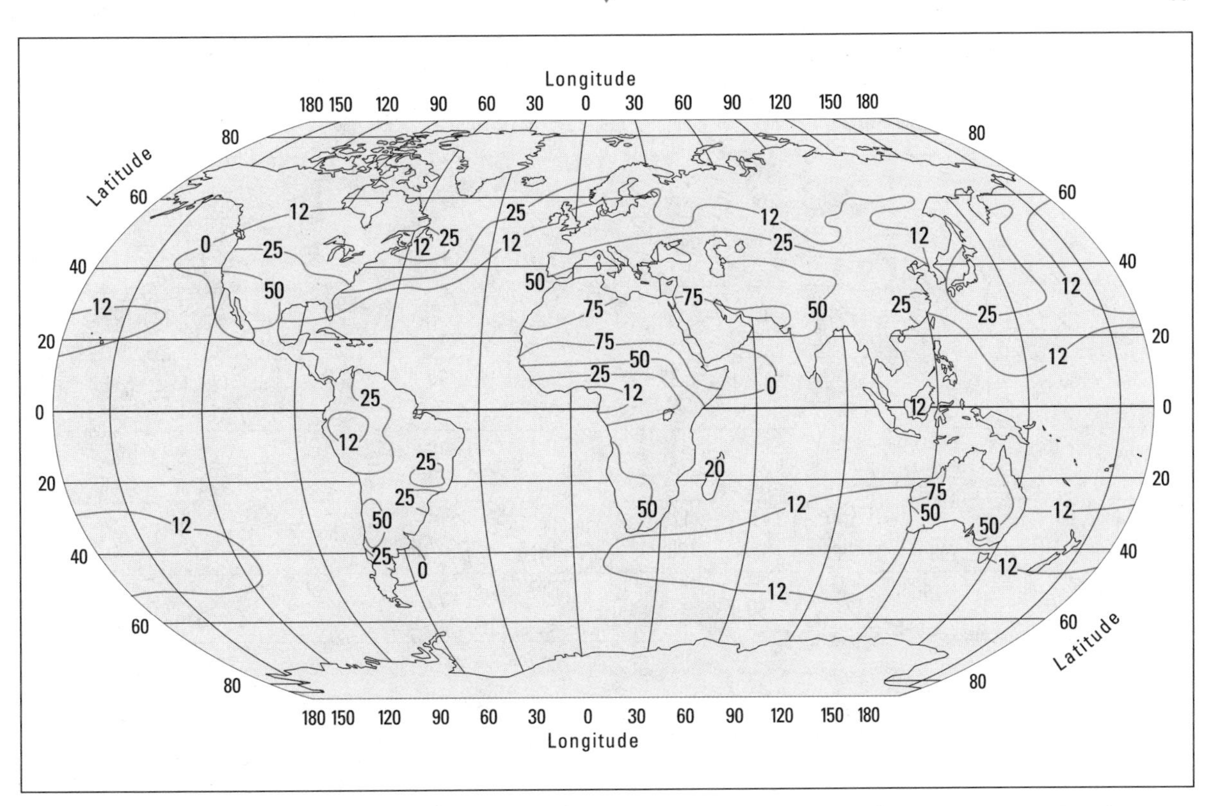


FIGURE 3.19 ▼ Global distribution of the vertical transfer of sensible heat. Units are W $m^{-2}$ (after Budyko *et al*, 1962)

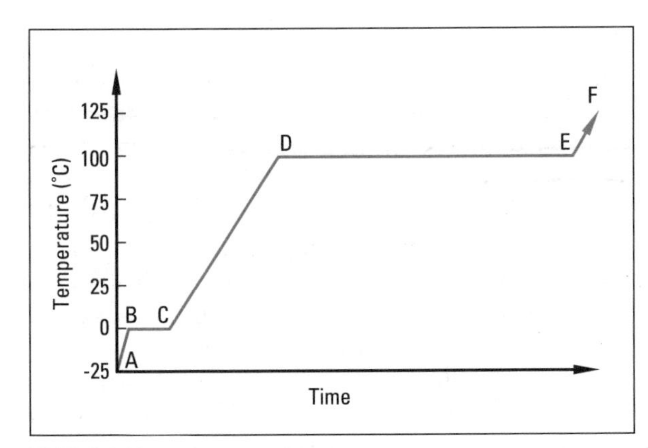


FIGURE 3.20 ▼ The pattern of temperature and phase changes for water. The temperature remains constant during each phase change as long as pressure remains constant. Specific heat is the amount of energy needed to raise the temperature of a given volume of material

sumed in evaporating only 10 g of water is about the same as that needed to raise the temperature of 60 g of water from 0°C to boiling point (100°C).

We tend to be most aware of evaporational cooling after swimming. The effect of evaporation leads to an extraction of heat from the skin surface; sweating works in a similar way.

Where the state of water changes to a lower energy level (i.e., from vapour to liquid or liquid to solid), it will release the same quantity of energy that was originally used. This is especially important in our atmospheric heat balance. Water that is evaporated from the surface will extract energy from that surface, where there is usually a surplus anyway. Eventually the vapour will condense in the atmosphere, probably as a cloud droplet, releasing latent heat originally extracted from the surface and so helping to warm the atmosphere. This can also take place poleward of the original evaporation point, so evaporation can transfer heat energy both into the atmosphere and poleward.

Much of the earth's surface is covered by oceans, where evaporation takes place continuously. Even a large proportion of the land surface is moist much of the time. Consequently the role of latent heat in balancing the heat budget of the earth is vital. Latent heat transfer by convection carries about one-

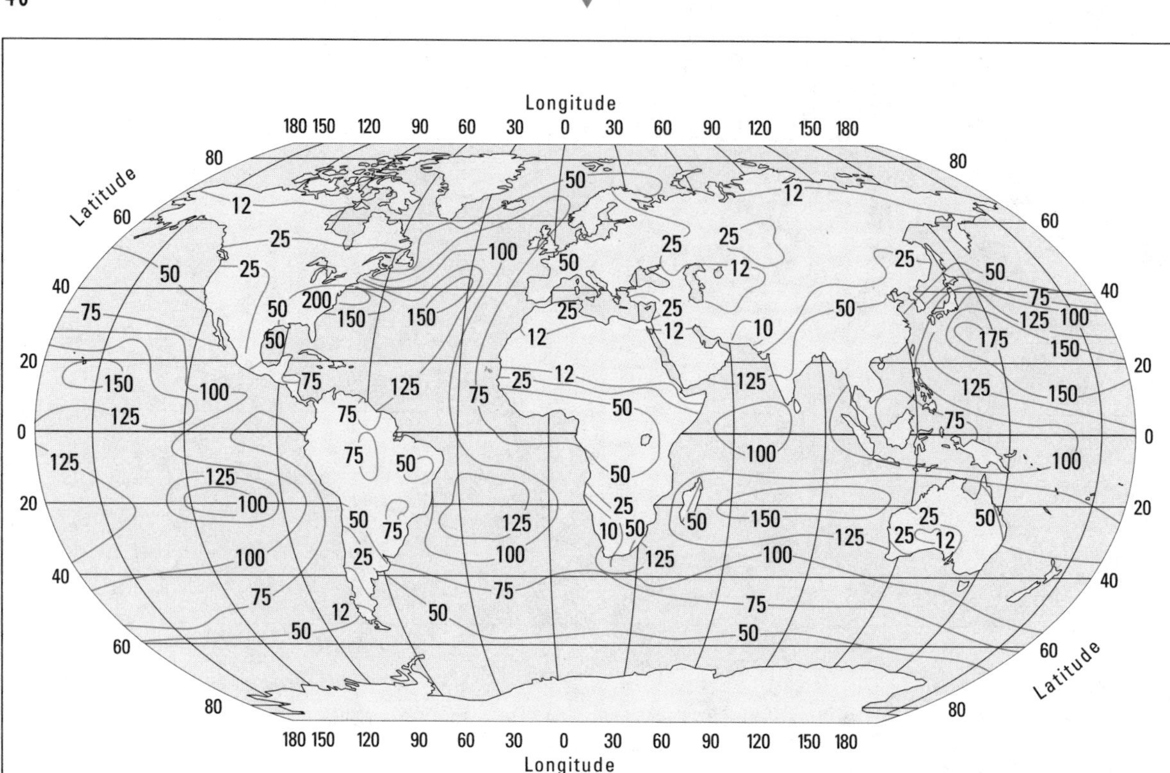


FIGURE 3.21 ▼ Global distribution of the vertical transfer of latent heat. Units are W $m^{-2}$ (after Budyko *et al*, 1962)

fifth of the energy of incoming solar radiation back to the atmosphere (Figure 3.21).

The heat used for evaporation over land areas depends upon the availability of moisture and of energy. In polar regions it is small, but increases equatorward, reaching a maximum in the moist equatorial forests of South America, central Africa and Indonesia. Over the desert areas, there is little moisture available and evaporation is insignificant.

# The atmospheric energy system

The details of the atmospheric energy system may appear complicated, but the system is very important as the driving force of all climates. To recap what happens: energy enters from the sun; it is reflected, absorbed, and reradiated within the system but does not form a uniform distribution; some areas receive more energy than they lose, in other areas the reverse occurs. If this situation were to continue the areas with an energy surplus would get hotter and those with a deficit would get cooler. This does not happen because the temperature differences help to drive the wind and ocean currents of the world. They carry heat with them, either in the sensible or latent forms, and counteract the radiation imbalance. Winds from the tropics are therefore normally warm, carrying excess heat with them. Polar winds are blowing from areas with a deficit of heat and so are cold.

The actual flows of energy can be shown more simply if we consider them as part of a large system in which we distinguish the inputs and outputs with feedback between the different subsystems (Figure 3.22).

While the general principles of flow are known, the figures quoted are, in most cases, best estimates. Measurements have been made at a number of places, but in insufficient quantities to give a reliable global figure. It is little use giving a global average based on a few clustered observations. Satellite observations have helped (and led to appreciable changes in estimates of the earth's albedo), but there are still numerous flows that are imper-

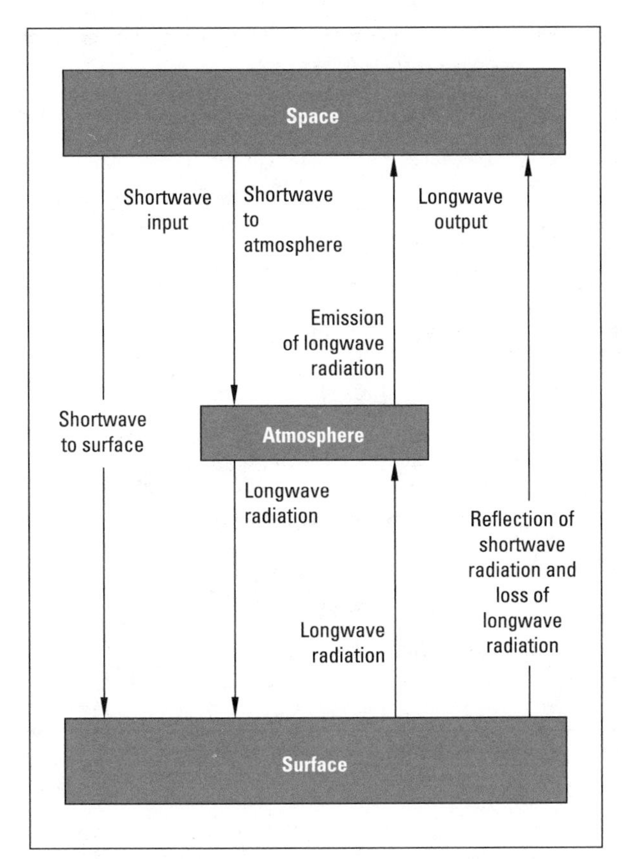


FIGURE 3.22 ▼ The earth–atmosphere radiant energy system

fectly measured. Longwave emission by the atmosphere, the separation of shortwave radiation into direct and diffuse radiation, and sensible and latent heat transfer are the main problems as conditions vary quickly, and, until measurements are more comprehensive, some of the figures are little more than intelligent guesses. The actual value of the flow will depend, in part, on the nature of the assumptions made about them. But we can be sure that what comes into the earth–atmosphere system must eventually leave.

## FEEDBACKS

Changes in energy absorption at the earth's surface may produce a response in the atmosphere. In turn, further responses by the atmosphere may be initiated to accentuate or suppress the initial change. Positive feedback has a 'snowball' effect that could produce a potentially damaging situation, whereas negative feedback tends to restore conditions to their original state. A simple example of negative feedback is shown by the effect of an increase in the mean temperature of the lower atmosphere. If this occurred, it would be accompanied by an increase in emission of longwave radiation to space, which would tend to cool the air. A positive feedback system is shown in Figure 3.23. These feedback processes will be discussed further in Chapter 9, where their importance in longterm changes of climate is outlined.

Some feedback mechanisms in the atmosphere, like the ones described above, are quite common and simple to understand. Problems arise when we try to anticipate their consequences. A feedback is likely to affect other parts of the atmospheric system. We have to appreciate this point when attempting to understand and model our atmospheric circulation. While the essential controls of the global atmosphere are all identified in general terms, some of the feedback interactions are so difficult to determine that we do not even know whether they are positive or negative (Table 3.5). For example, an increase in the amount of cloud will decrease the input of shortwave radiation to the surface and reduce the loss of longwave radiation to space; the former should produce a decrease in surface temperature and the latter an increase. Hence the net effect will depend upon the precise changes in shortwave input and longwave output, which in turn will depend upon the type, height, thickness, and global distribution of cloud. We cannot easily tell which way the change will be.

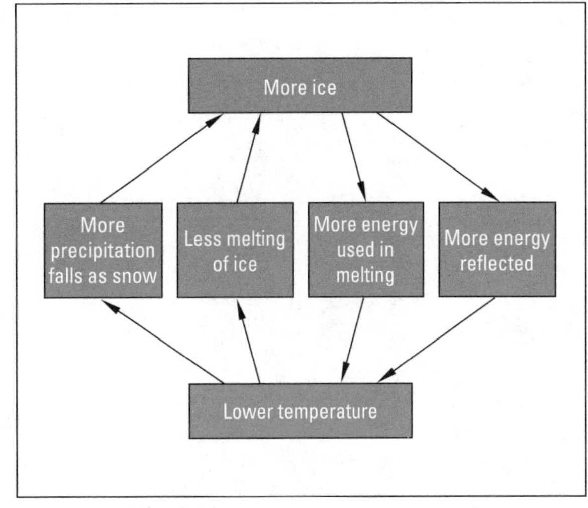


FIGURE 3.23 ▼ Positive feedback system initiated by a decrease in temperature

## Table 3.5

*Climate feedback mechanisms*

| *Mechanisms* | *Linkages* | *Feedback* |
|---|---|---|
| Temperature—radiation | Increased $T_{a,s}$ → Increased IR loss → Cooling | – |
| Water vapour—greenhouse | Increased $T_a$ → Increased IR trapping → Warming (at constant RH) | + |
| Snow/ice albedo—temperature | Decreased $T_s$ → Increased albedo → Cooling | + |
| Cloudiness—temperature[1] | Change of cloud amount → Radiative fluxes → Temperature change | ?0 or – |
| | Decreased $T_s$ → Decreased cloud altitude → Less radiation gets through → Cooling | + |
| Lapse rate—surface temperature | Local convective adjustment → Large-scale baroclinic effects | ? |
| Sea temperature anomalies — atmospheric circulation[2] | Wind field forcing of surface water drift and divergence Heat fluxes | ? |

*Notes*

T = equilibrium temperature of atmosphere ($T_a$) or surface ($T_s$)
IR = infrared radiation, longwave energy
RH = relative humidity

[1] Cloud effects are complex due to several factors: cloud amount, height of cloud tops, thickness, and optical properties. The responses will vary with season and latitude.

[2] Air-sea interactions probably differ according to time scale. In the short term, sea temperature anomalies appear to be forced by the atmospheric circulation.

After R.G. Barry, *Progress in Physical Geography*, 3 no. 1 (1979), pp. 119–31.

# Heat and moisture in the atmosphere

Water is the most important element on earth. Indeed, it is more appropriate to call it planet "Water" rather than Earth. We have already seen the importance of water in trapping heat to keep the earth at a livable temperature. In this chapter, we will examine how water moves through the atmosphere. The water is derived from the earth—evaporated from the oceans, rivers, and soil or transpired by the vegetation. Once in the atmosphere, cooling converts water vapour to water droplets. By various processes the droplets combine to form clouds, some of which produce life-sustaining precipitation.

## The sun, insolation, and temperature

### THE DAILY MARCH OF THE SUN

Let us start by considering the most persistent and fundamental cause of these processes: the daily progress of the sun from east to west across the heavens. As we know, this apparent movement of the sun occurs because of the rotation of the earth as it circles the sun, a rotation that takes fractionally under 24 hours to complete. As the sun moves from the horizon to its midday position high in the sky and back to the horizon, the angle its rays make with the earth varies. We saw in the last chapter that when the sun is low in the sky the rays strike the earth obliquely and the radiation is spread out over a wider area. In addition, relatively more incoming radiation is intercepted by vapour and dust and gases in the atmosphere. Consequently the average insolation per unit area is relatively low. In contrast, when the sun is high in the sky its rays strike the ground almost perpendicularly; they are concentrated in a small area and they experience relatively little interception as they pass through the atmosphere. Average insolation is at a maximum.

The effect upon surface temperatures is obvious. If we consider a clear day at solstice (June 21) in, say, Regina, sunrise is at about 5 A.M. Temperatures are low, for during the night the earth has been losing heat by radiational cooling. But, slowly, as the sun rises the earth warms and, in turn, the air in contact with the surface is heated too (Figure 4.1). By about 2 P.M. the earth and air are at their warmest, the maximum temperature at the surface coming earlier than that in the air because that is where the conversion of solar energy to thermal energy takes place. From now on, as the sun gradually sinks, the ground surface and the overlying air will cool. The sun sets at about 9 P.M.; cooling continues throughout the night until minimum temperatures are reached just before dawn.

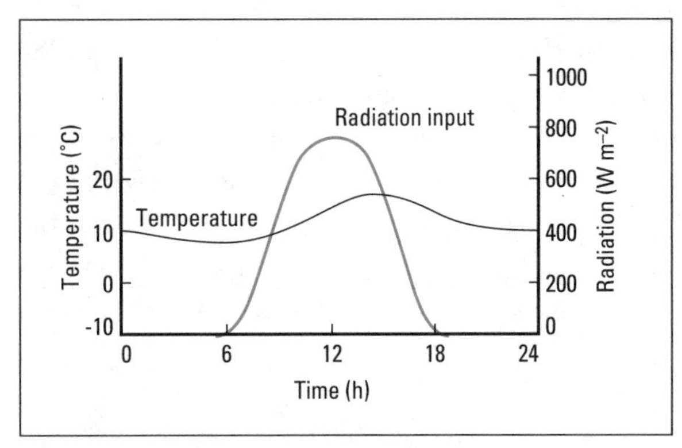


FIGURE 4.1 ▼ Diurnal changes in shortwave radiation input and temperatures on a clear day

This daily variation in insolation and temperature is one of the most basic components of our weather. It is so obvious and so regular that we take it for granted. And yet quite marked differences in atmospheric conditions occur in response to the daily progress of the sun. As we will see later, the associated changes in temperature may lead to significant changes in humidity, and they often spark major atmospheric processes such as vertical movements of air and even heavy storms.

Before we examine the seasonal a daily pattern of insolation and temperature it is necessary to understand how the properties of land and water affect radiation and energy. If the earth was a smooth sphere with a single surface, land or water, the isotherms would parallel the lines of latitude and temperature would decrease uniformly from the equator to the poles. This is not the case. Variations in the pattern are due primarily to the distribution and different properties of land and water.

Four factors create the difference. Water has a higher specific heat than soil. It takes more energy to raise the temperature of a unit volume of air one centigrade degree than a similar volume of soil. As a result, water heats more slowly than soil, but it also cools down more slowly. Shortwave energy from the sun can penetrate much further into water, as much as 50 m in the tropics, than it can into soil. A thicker layer of the water is heated than soil. The mobility of water means that heat from this thicker surface layer can be mixed throughout the entire ocean system. Soil does not mix, although a high water content in a porous soil can transfer heat energy. The heat is transferred more easily because of the different properties of water and the movement of water through the soil. Finally, the unlimited supply of water from the ocean creates a greater cooling through evaporation than over the land.

It is also apparent that the daily pattern of insolation and temperature change itself varies according to atmospheric conditions. The effects are most obvious when the air is clear and still, for then heating and cooling proceed uninterrupted. If the sky is cloudy or very hazy, however, the daily pattern of temperatures is much more subdued (Figure 3.14). Similarly, the pattern varies spatially. It is less marked over the sea, for much more of the incoming energy is used to heat and evaporate the water, and less is returned directly to heat the atmosphere. During the night, the sea cools slowly, with the result that temperatures do not fall so much as on land—one reason why coastal areas are less prone to nighttime frosts (Figure 4.2). The pattern is most apparent in areas with dry air and ground surfaces, such as deserts. Here, incoming radiation is large and little energy is used for evaporation so temperatures are high, while radiational cooling at night is intense, giving rise at times to extremely low air temperatures.

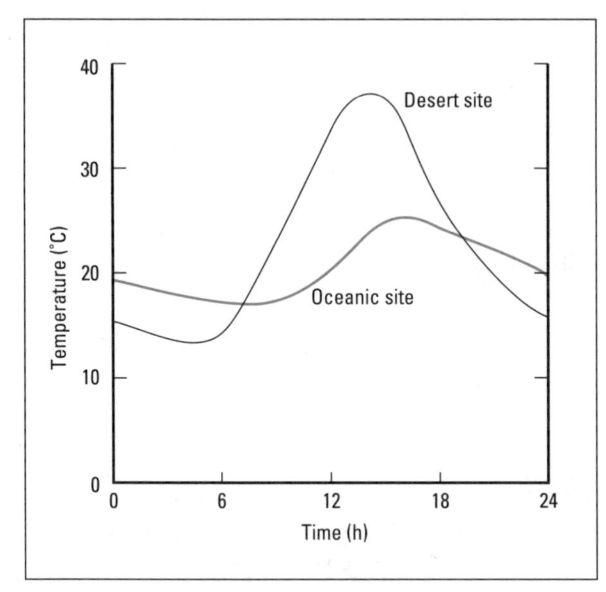


FIGURE 4.2 ▼ The effect of water on diurnal temperature ranges. Oceanic sites have a small diurnal temperature range, whereas in a desert the range is high

## THE SEASONAL PATTERN

A very similar pattern of variation takes place on a seasonal scale. The cause in this case is not the earth's rotation, but its changing relationship with the sun, the variation within its orbit that produces the apparent seasonal progress of the sun from the Tropic of Cancer to Capricorn and back.

This change in the position of the sun leads to changes both in the angle of the incoming rays and in the duration of daylight. Both factors influence the amount of insolation received by the earth and, therefore, the degree of atmospheric heating. Considering again our area in Regina we would find that in the winter the maximum elevation of the sun, at midday, was about 16°, for the sun stands approximately over the Tropic of Capricorn. Thus the rays of the sun still strike the earth at a relatively low angle and the degree of midday heating is limited.

As the sun moves northward to the equator and thence to the Tropic of Cancer its midday position rises and the rays strike the earth less obliquely. Moreover, days become longer and nights shorter. Maximum temperatures increase until, in about July, they reach their highest values, slightly after the maximum radiation in late June. From then until mid-December the sun returns south, its midday position in the sky declines, the quantity of insolation received at the surface is reduced, and temperatures fall.

It is apparent that, in Regina, the winter months represent a period when incoming radiation is low. Outputs of energy from the earth continue, however, so the area experiences a **net radiation deficit**. During the spring, as the overhead sun moves north of the equator, radiation inputs rise to match outputs, but the degree of atmospheric warming is restricted because much of the excess energy is used to reheat the earth and oceans. By August the earth has warmed up; during autumn the sun returns to its position over the equator but now the earth still retains much of its heat gained during the summer. The air, therefore, remains relatively warm compared with spring even though the sun is at the same midday zenith angle.

This effect is most marked in coastal areas. Water heats more slowly than the ground (it has a higher heat capacity) and, by the same token, retains its heat longer. Regina and Vancouver are at approximately the same latitude but the difference between their temperatures is apparent. Vancouver has cooler summer temperatures and warmer winter temperatures than Regina. This difference in temperature range is referred to as an index of **continentality**; that is, the temperature range, from summer maximums to winter minimums, will increase the closer a station is to the centre of a continent and away from the modifying influence of the oceans (Figure 4.3).

The seasonal pattern of radiation and associated temperature conditions varies latitudinally. In polar areas, the sun never gets high in the sky, but daylength varies markedly, so that during summer months these areas experience perpetual daylight. Conversely, in the winter months they are in continuous darkness. The seasonal radiation balance is therefore very variable. At the North Pole, for example, from April to September there is a potential continuous radiation surplus, for the sun will shine for 24 hours a day if the sky is cloud-free, so there is less nighttime cooling. In contrast, for the rest of the year a radiation deficit occurs. No insolation is experienced for six months, so radiational cooling continues, interrupted only by the transfer

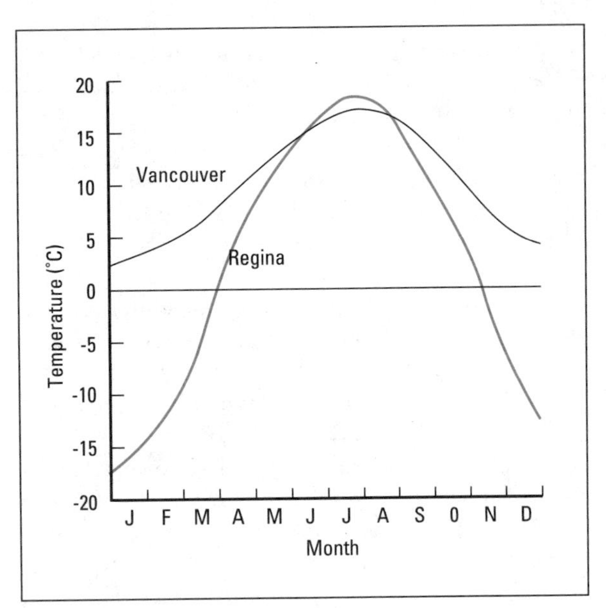


FIGURE 4.3 ▼ The effects of water on seasonal temperatures. Regina is hotter in summer and cooler in winter than Vancouver

of air from warmer latitudes and heat transferred through the ice from the warmer water beneath (Figure 4.4).

The pattern in the tropics is very different. Here the sun never strays far from its overhead position, so seasonal variations in radiation are limited. The daily pattern of temperature and insolation is marked, but it varies little throughout the year (Figure 4.5).

# The effects of heating and cooling of the atmosphere

## GENERAL EFFECTS

Changes in insolation over time affect other aspects of climate as well as temperature. The atmosphere is a highly complex system, and the effects of changes in any single property tend to be transmitted to many other properties. Thus heating and cooling of

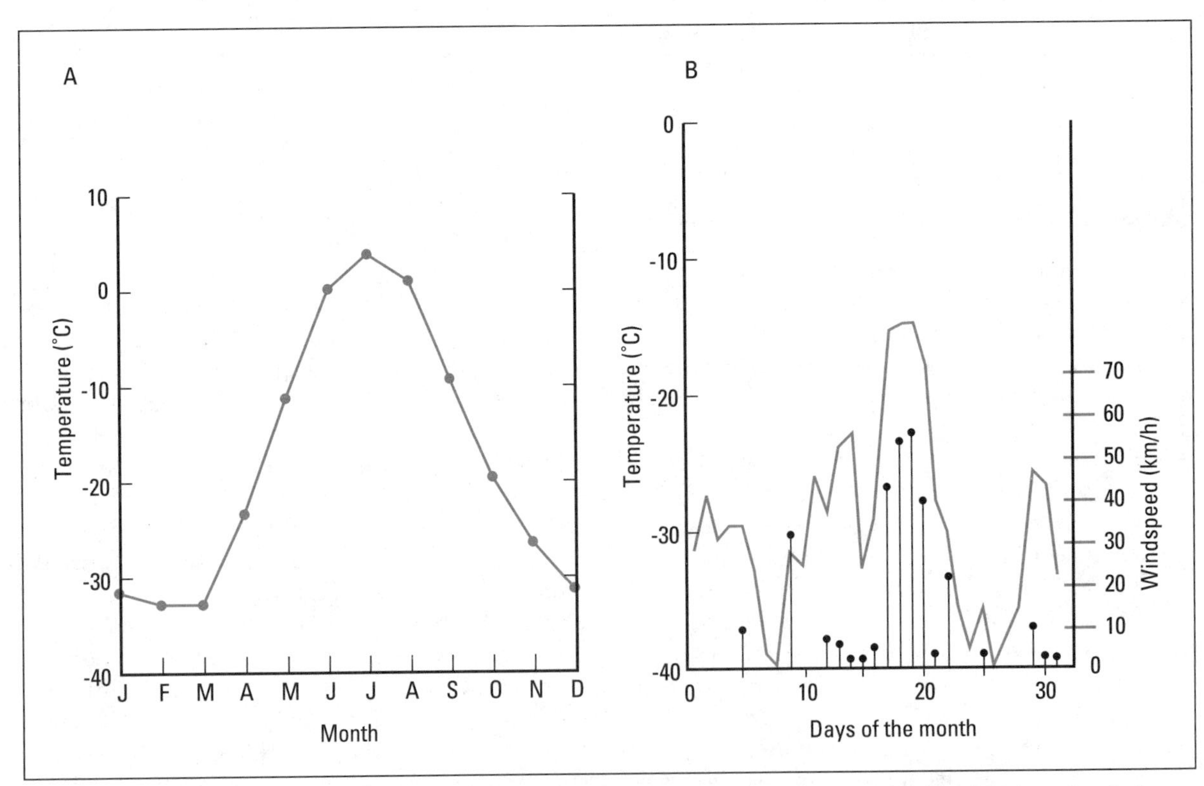


FIGURE 4.4 ▼ (A) Mean daily temperature at Alert (82.5°N). (B) Temperatures (solid curve) at Alert in winter in relation to wind speed (vertical lines). Under calm conditions, temperatures are very low but stronger winds disturb the surface inversion and bring warmer air from more southerly latitudes

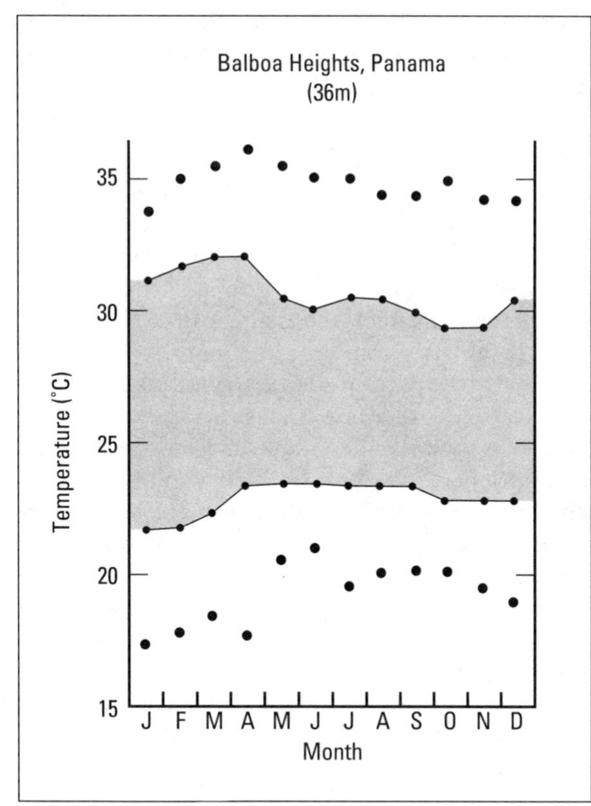


FIGURE 4.5 ▼ Mean monthly temperatures for Balboa Heights, Panama. Symbols from top to bottom indicate highest maximum and mean maximum, mean minimum and lowest minimum respectively for each month

the air cause adjustments in humidity and air stability; they may cause condensation and evaporation, cloud formation, and the development of storms.

What happens, then, when air is heated? To simplify the problem we will consider a parcel of air in contact with a warm ground. Like any other substance its temperature rises and it expands. Gases expand on heating more than either liquids or solids, so this effect is quite marked. Moreover, as the air expands its density falls; the same mass of air now occupies a larger volume. As its density falls so it becomes lighter than the surrounding air and it tends to rise like a bubble. Reverse the process, cool the parcel of air, and the opposite occurs. It contracts, its density increases and it sinks.

One effect of heating and cooling of the atmosphere is therefore to cause vertical movements of air. But there are other effects. As the air becomes cooler its ability to hold moisture in the form of water vapour is reduced. If it cools to the point where it can no longer hold the water in vapour form, condensation occurs and water droplets form. As the air is heated, these droplets tend to evaporate and become water vapour once more. Thus, heating and cooling are intimately linked to the processes of evaporation, condensation, and precipitation formation.

Let us consider these main effects in turn.

## VERTICAL MOVEMENTS

The rising of warm air is a process we can see at work on a hot day by the shimmering effect of air near the surface; we can see it too if we watch the beautiful and immense towers of a convectional cloud forming. Such vertical movements usually develop if localized heating of the atmosphere takes place, so that individual parcels of air become warmer and lighter than the air around them.

Localized heating occurs for a number of reasons. Variation in the wetness or albedo of the surface may cause differences in atmospheric heating; the air above dark coloured or dry surfaces heats up more rapidly than that above light or wet surfaces. Differences in slope angle may have the same effect. But there is another factor that plays an important role in these vertical air movements. This is the vertical change in air temperature away from the ground surface.

## ENVIRONMENTAL LAPSE RATE

If we measured air temperatures in the troposphere at different heights under stable, cloudless atmospheric conditions, we would find that temperature usually falls with height. The reason is quite simple. The incoming radiation heats the ground. A small proportion of this heat is transmitted downward into the soil by conduction, but the majority is returned as either sensible heat or longwave radiation to heat the atmosphere. Heating is greatest close to the ground surface and declines with height.

The rate at which the temperature falls with increasing altitude is, on average, 6.4 C°/1000 m, and this pattern continues as far as the tropopause. This is the **environmental lapse rate**. It is not constant, however, for it is affected by atmospheric and surface conditions. When the air is turbulent, or is being mixed by strong winds, the environmental lapse rate, at least in the lower layers of the atmosphere, is low. With strong surface heating, it is steep, meaning air temperature declines rapidly with height. Under some conditions, temperatures may even rise with height for short distances; this is known as an **inversion**. Whatever its value, it is this environmental lapse rate that greatly influences vertical air movements.

## STABILITY AND INSTABILITY

We can start to understand the importance of the environmental lapse rate by considering a simple example. Imagine localized heating of the air above an island in the sea. The island, because it converts sunlight to heat more effectively than the surrounding water, will act as a thermal source. Above this thermal source the air will become warmer, its density will decrease, its surface pressure will fall and the air will rise. Typically, after this bubble of air has risen a distance equal to about one or two times its own diameter it sinks back. New and larger bubbles form in its wake, however, and each rises a little higher. The bubble of air is called an **adiabat**, a word of Greek origin that loosely means *without exchange*. Air has conservative properties, which means that it takes some time before the bubble dissipates and mixes back into the surrounding air. It is reasonable to compare the bubble to a hot air balloon.

What controls this movement? The answer, simply, is temperature. If the bubble of air is warmer than its surroundings it will continue to rise; if it is cooler, it will sink. We know already that the general temperature of the air declines upward—this is the environmental lapse rate. We might imagine, therefore, that once the bubble starts to rise it will continue to do so indefinitely, for the air around it is becoming progressively cooler with height. This does not happen, however, and the reason is that as the air bubble rises it also cools. The critical factor that determines the height to which the bubble rises is the relative rate of cooling of the bubble and the surrounding air.

The next question, then, is why does the bubble get cooler? As the air bubble rises, it comes into contact with less dense surrounding air. The pressure confining the bubble is reduced and it expands. As the bubble expands, there is less contact between the molecules and this results in cooling. This is in accord with the **Gas Laws** which state that:

$$PV/T = K(\text{constant})$$

In other words, the pressure (P), volume (V), and temperature (T) of a gas are interdependent. A change in any one of these properties tends to cause changes in the other.

The rate at which the rising parcel of air cools with height is a constant, at about 10C° for each 1000 m. It is known as the **dry adiabatic lapse rate** (DALR). Adiabatic means that there is no heat exchange between the bubble and its surroundings and, as long as the bubble of air rises rapidly, this condition applies. It is called dry, not because the air does not contain any moisture, but because no condensation has taken place.

We have, therefore, a framework for determining how far the bubble will rise. As long as no condensation occurs it will cool at the dry adiabatic lapse rate. The surrounding air cools at the environmental lapse rate. The bubble rises until its temperature

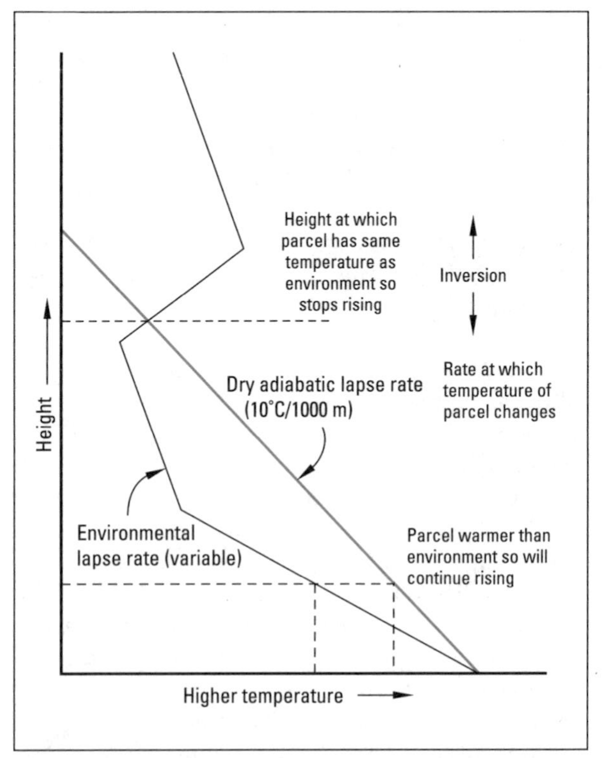


FIGURE 4.6 ▼ Thermal buoyancy of an air parcel. The parcel will continue to rise as long as it is warmer than the surrounding air

(and therefore its density) is equal to that of the surrounding air. This is shown in Figure 4.6.

Clearly the dry adiabatic lapse rate is a constant; the two variables in this relationship are the environmental lapse rate and the initial temperature of the air bubble. The bubble will only rise if it is warmed sufficiently to overcome the confining effect of the environmental lapse rate. If the bubble cannot rise it is said to be **stable**. If the temperature of the bubble is raised enough, however, or if the environmental lapse rate is great enough, the air bubble can rise a considerable distance. It is in these circumstances that part of the troposphere is said to be **unstable**.

## CONDENSATION

'As long as no condensation occurs. . . .' That was the proviso we established when considering the dry adiabatic lapse rate. But all air, even the driest desert atmosphere, contains some moisture, and if it is cooled sufficiently it will experience condensation. When this happens the processes of vertical uplift are modified. When air is cooled, by whatever process, it can reach a temperature at which water vapour (gas) condenses to water (liquid). This is known as the **dew point** temperature.

The ability of the air to hold moisture is dependent upon the temperature. As the temperature of the air increases, so its moisture-holding capacity also rises; to put it another way, more moisture must be added to reach saturation at a higher temperature.

The amount of moisture that air holds may be assessed in a number of ways. **Relative humidity** is the most frequently used measure. It is the ratio of the amount of moisture the air contains to the amount of moisture the air could hold when saturated at that air temperature, expressed as a percentage. Relative humidity may be measured indirectly from wet-bulb and dry-bulb temperature readings using psychrometric tables (see Appendix I). Evaporation of moisture from the wet bulb leads to a cooling that is inversely proportional to the relative humidity of the air. If the air is saturated, there will be no evaporation, no cooling, and so no difference in temperature between the dry and wet bulbs. Although frequently used, relative humidity does have the disadvantage of being temperature dependent. For example, as air temperature rises, relative humidity will fall even though the moisture content of the air has remained constant, because the air is able to hold more moisture. An absolute method of measuring moisture content is to determine the **vapour pressure**, which is that part of the total atmospheric pressure exerted by water vapour. Again it can be obtained indirectly from the wet- and dry-bulb thermometer using tables. The **humidity mixing ratio**, another absolute measure, is the ratio of the mass of water vapour to the mass of dry air with which the water vapour is associated. It is usually expressed in g $kg^{-1}$. The relationship between temperature and the moisture content at saturation is indicated by the **saturation vapour pressure** curve (Figure 4.7). Thus, as a rising air bubble cools, it approaches the temperature at which condensation occurs. When the air bubble reaches that temperature, it becomes saturated and condensation takes place.

If condensation was the only thing that happened on saturation, then the effect on the air bubble would be small apart from the extra weight of the droplets. There is, however, another effect. As water changes from its vapour state to a liquid, it releases latent heat. This heat acts to warm the air and thereby counteracts the cooling resulting from expansion.

We can readily see the implications for our air bubble. Now, instead of cooling at 10C°/1000 m (its dry adiabatic lapse rate), it cools more slowly as it rises. This new, lower rate of cooling is known as the **saturated adiabatic lapse rate** (SALR) (also referred to as the wet [WALR] or moist [MALR] in some literature). Unlike the dry rate this is not a constant, for as we can imagine it depends upon the amount of heat released by condensation, and this, in turn, depends upon the moisture content and therefore the temperature of the air. Warm air is able to hold a lot of moisture and thus, on cooling, it releases a lot of latent heat; cold air is able to hold far less moisture, so the heat production during condensation is much less. This is one reason why some of the world's most severe storms, the tropical cyclones, are found in warmer climatic regions.

Let us illustrate the effect of condensation by considering a specific example. Figure 4.8 shows the path curve for the bubble. Its initial temperature is 20°C and it will cool at 10C°/1000 m until saturation point is reached. It is at this level that we first see the visible evidence of our bubble—a small cloud will form. Above condensation level, the rate of cooling slows to the saturated adiabatic lapse rate

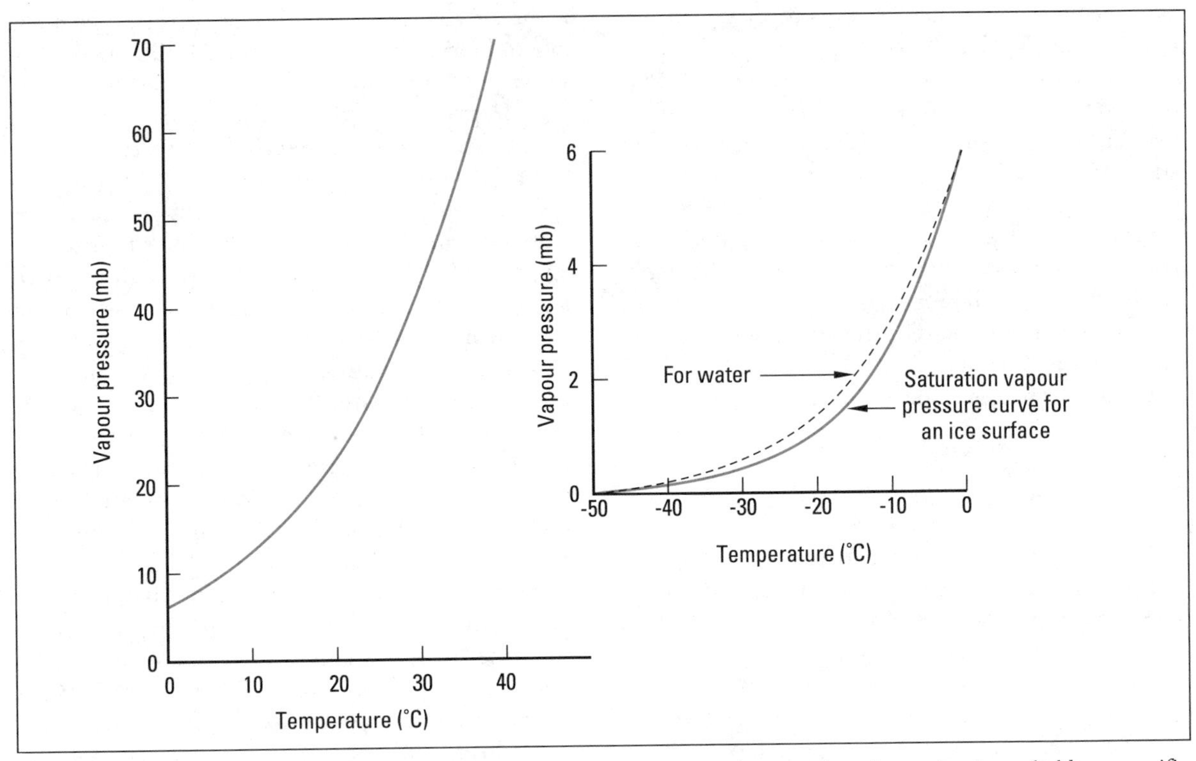


FIGURE 4.7 ▼ The saturation vapour pressure curve. The curve demonstrates how much moisture the air can hold at a specific temperature. Below 0°C, the curve is slightly different for an ice surface than for a supercooled water droplet

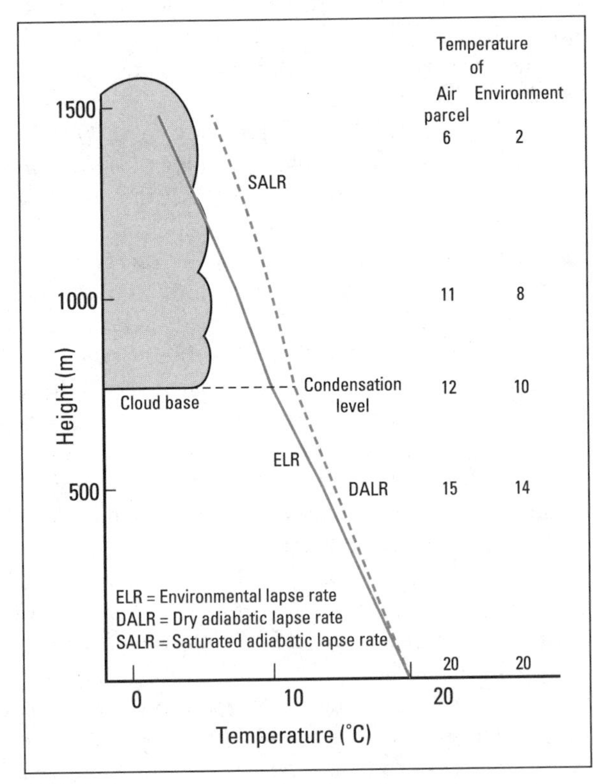


FIGURE 4.8 ▼ The effect of condensation on the rate of cooling of an air parcel

as long as the bubble's temperature is still higher than that of the environment. If it is, we get large convectional clouds building up, and these will probably give rain.

Whether the atmosphere is still stable or not will depend upon the relative rates of cooling of the dry bubbles, the saturated bubbles, and the environment. We can summarize this in Figure 4.9. If the environmental lapse rate is less than the dry adiabatic lapse rate, we have **absolute instability** as bubbles of air, even if they cool at their maximum rate (the DALR), will be warmer than the environment. If the environmental lapse rate is cooling more slowly than the saturated adiabatic lapse rate, we have **absolute stability**. If the environmental lapse rate is between the DALR and the SALR we have **conditional instability**; in other words, instability depends upon the air reaching saturation point.

Stability has a considerable effect upon the degree to which convective activity will take place. If the air is unstable it will rise and may produce clouds, whereas if it is stable, convection will be reduced. If the air beneath an inversion is fairly moist, a layer of cloud may develop. Moist air will have been brought to the inversion by convection

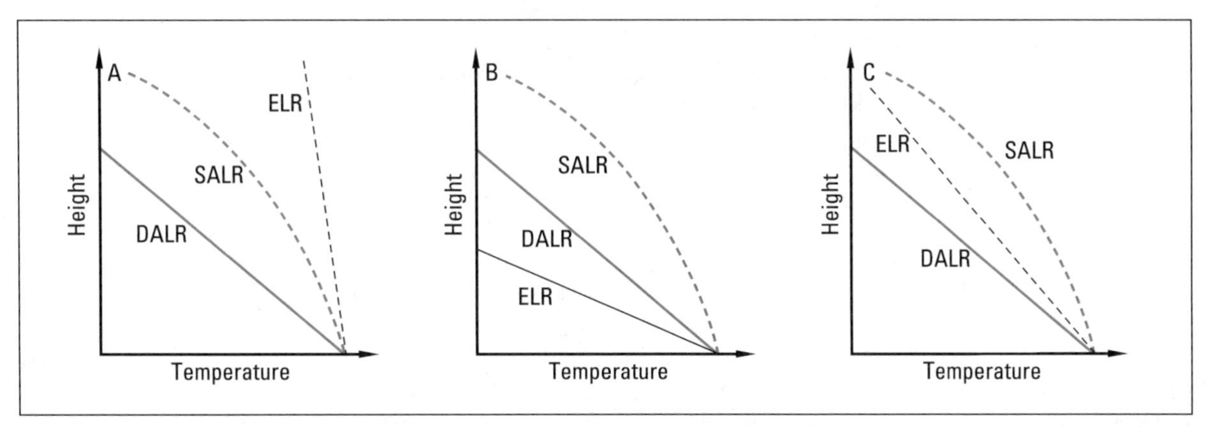


FIGURE 4.9 ▼ (A) Absolute stability: the parcel is always cooler than the surrounding air. (B) Absolute instability: the parcel is always warmer than the surrounding air. (C) Conditional instability: before condensation occurs the parcel is cooler than the surrounding air, afterward it is warmer

and, as it cannot rise further, it spreads out beneath the inversion to give a dense layer of stratus cloud (Figure 4.10).

Absolute instability in the atmosphere is infrequent except very close to the ground—the convection it initiates helps to transfer heat upward and so reduces the environmental lapse rate. What is much more common is for the environmental lapse rate to lie between the dry adiabatic lapse rate and the saturated adiabatic lapse rate. In this situation of conditional instability, the atmosphere is stable for air that has not reached saturation point, but is unstable for saturated air. If the air can be forced to reach the condensation level, either by ascent over hills or mountains, or by convergence associated with a depression, then the air will become unstable and assist vertical motion. The former process is one of the mechanisms that leads to higher rainfall over mountains.

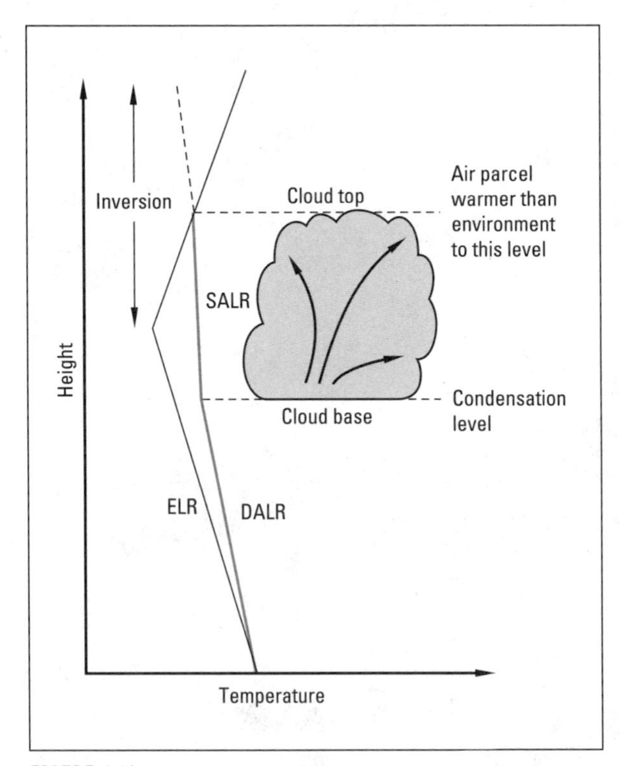


FIGURE 4.10 ▼ The effect of an inversion on cloud development with moist air: stratus cloud formation

# The effects of condensation

## CAUSES OF CONDENSATION

Clouds are one of the most interesting aspects of the sky. They are constantly changing their shape and form to reflect the processes of formation and the environment in which they are developing. To produce clouds, we need the air to reach saturation point. It is clear that saturation can be reached either by adding water to air or by cooling the air (Figure 4.7). The first process occurs over warm water surfaces such as the Great Lakes in the autumn, or over the Arctic Ocean, where water will evaporate from the relatively warm sea surface and rapidly condense into the cold air above to give **arctic sea smoke**. Vertical air motion sufficient to produce clouds can be caused by the following:

- by orographic lifting, that is, air rising over hills or mountains
- by convergent uplift, that is, air moving into a low pressure area and rising, or within air streams
- or by convection

Radiational cooling or contact cooling at a cold ground surface may also be sufficient to produce saturation, but as these are ground-based processes the resulting condensation is known as **fog**. It is like cloud because it is composed of myriads of water droplets, but the detailed mechanisms of formation are different.

## FOG

Fogs are a common feature of the climate of some parts of the world. For example, they are frequent on the North Sea coast of Britain in summer and on the Grand Banks of Newfoundland. The Grand Banks has the most days of fog each year of any region in the world, over 200, as the cold waters of the Labrador Current flow down Davis Strait and meet the very warm waters of the Gulf Stream. Although not as dramatic, the coast of British Columbia also experiences frequent fogs as cool and warm waters mix.

There are two weather situations that can form fog when the ground surface is cooler than the air in contact with it. First, when the ground loses heat at night by longwave radiational cooling, usually with the clear skies of an anticyclone; second, when warm air flows from a warm region to cover a cold surface, particularly a melting snow surface with lots of moisture about. The first type is called **radiation fog**, and the second **advection fog**.

The fog consists of microscopic droplets of water, between 1 and 20 μm in diameter. Visibility in a fog will depend upon the sizes and concentration of droplets in it. When they are small and numerous, visibility is poor, perhaps as little as 5 m. If pollution adds suitable nuclei, condensation of water vapour is favoured. When the droplets are large or sparse, visibility is less affected. The actual formation of radiation fog represents a delicate balance between radiational cooling, air movement, and condensation. It only forms when cooling occurs faster than the rate at which latent heat is added by condensation. Because vapour is converted into water droplets, the moisture content and saturation temperature fall, so further cooling is necessary to give saturation. For this reason, fog is more frequent during the long nights of autumn and spring than

FIGURE 4.11 ▼ Typical towering cumulus clouds over southern Ontario; the clouds have sharply defined tops indicating the presence of liquid droplets. The bases are normally flat and uniform (photo: Robert S. Schemenauer)

in summer. If winds are strong, saturated air near the ground will mix with drier air above and prevent fog forming.

By reducing visibility, fog can be a major environmental hazard. Airports may be closed and road transport is hazardous. Economic losses can result from these delays, but the potential for artificial fog clearance seems low; too much energy would be needed to warm or dry the air to prevent condensation.

## CLOUDS

Clouds and fog are the result of air being cooled below the dew point temperature. We also need **condensation nuclei**, that is, particles in the air, mostly dust or aerosols and salt, onto which water vapour can condense. Clouds are composed of a mass of water droplets or ice crystals of almost microscopic size. The number of droplets per unit volume of cloud varies considerably depending upon its origins; lower concentrations of larger droplets occur in clouds formed in the middle of the oceans, while high concentrations of smaller droplets are found in continental regions. Clearly this is a consequence of the greater availability of nuclei over the dusty continental interiors, but polluted industrial areas may have a similar effect. Studies of these condensation nuclei have shown that there are two broad classes: those with an affinity for water, called **hygroscopic particles**, like salt; and **non-hygroscopic particles** that require relative humidity above 100 percent before they can act as centres for condensation. There are always insufficient nuclei for the amount of water vapour in a cloud, particularly in dry regions. Many researchers have attempted to show that this can be corrected by seeding the cloud with artificial nuclei, the most common type being silver iodide.

We can find out much about what is happening in the atmosphere by looking at the type of clouds, and especially their shape. There is an almost infinite variety of forms: the basic division is between clouds that are layered, known as **stratiform**, and those in which the vertical extent of the cloud is more important. These are known as **cumuliform** (Figure 4.11).

In stratiform clouds, the rate of upward motion is slow, but it may take place over many hundreds or even thousands of square kilometres. At low levels, they are composed of water droplets, but at higher levels (2000 to 6000 m) we see a mixture of water droplets and ice crystals. Above about 6000 m, stratiform clouds are composed mainly of ice crystals (Figure 4.12). Some clouds may show signs of convection, even if it is weak. These types have 'cumulus' incorporated into their names, such as stratocumulus or altocumulus.

Low- and medium-level stratiform types are the main rain-bearing clouds of temperate latitudes. Around the centre of a low pressure area we often see a characteristic sequence of clouds as the warm air associated with the depression advances over the retreating cold air (Figure 4.13). As the depression approaches, the first signs are the high cirrus clouds that slowly expand to become cirrostratus. This cloud sheet thickens, therefore the base is closer to the ground, producing altostratus, followed by nimbostratus and precipitation. By this final stage,

FIGURE 4.12 ▼ Cirrostratus can be a thin veil of high, ice-crystal clouds, or it can be a layer with striations or streaks. A 22° halo is sometimes seen in cirrostratus and is partly visible in this photo from Mont Tremblant, Quebec (photo: Robert S. Schemenauer)

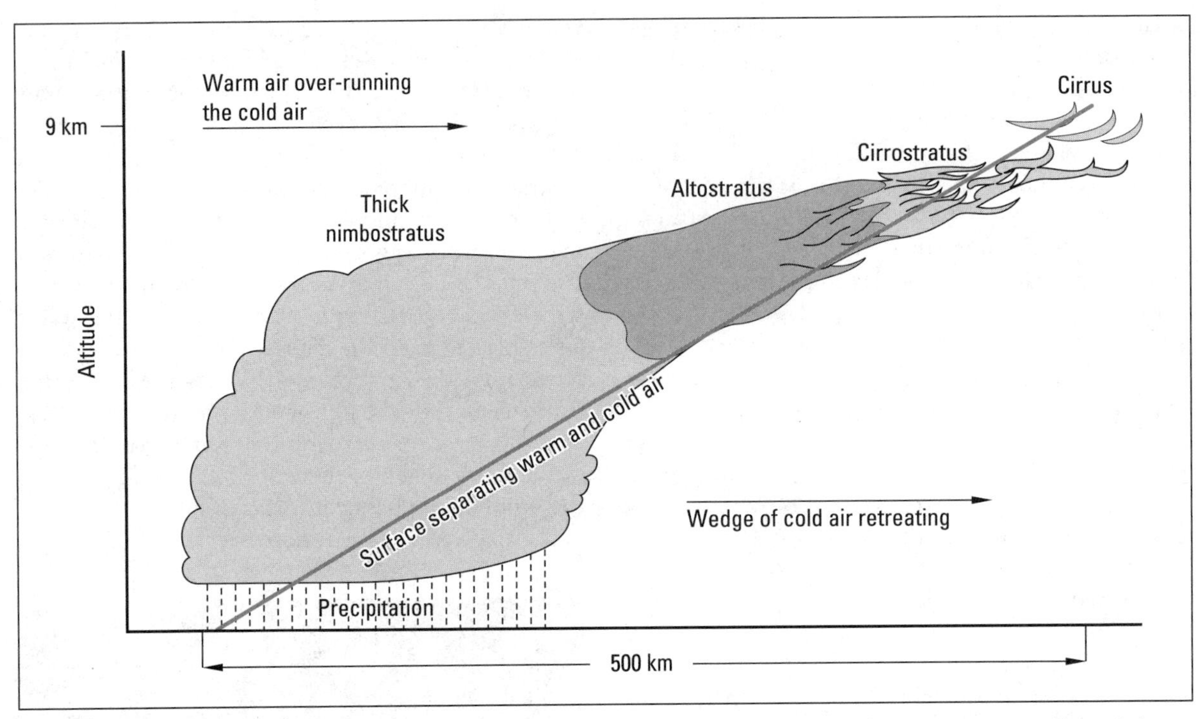


FIGURE 4.13 ▼ The characteristic sequence of clouds associated with a warm front ahead of a depression or midlatitude low pressure area

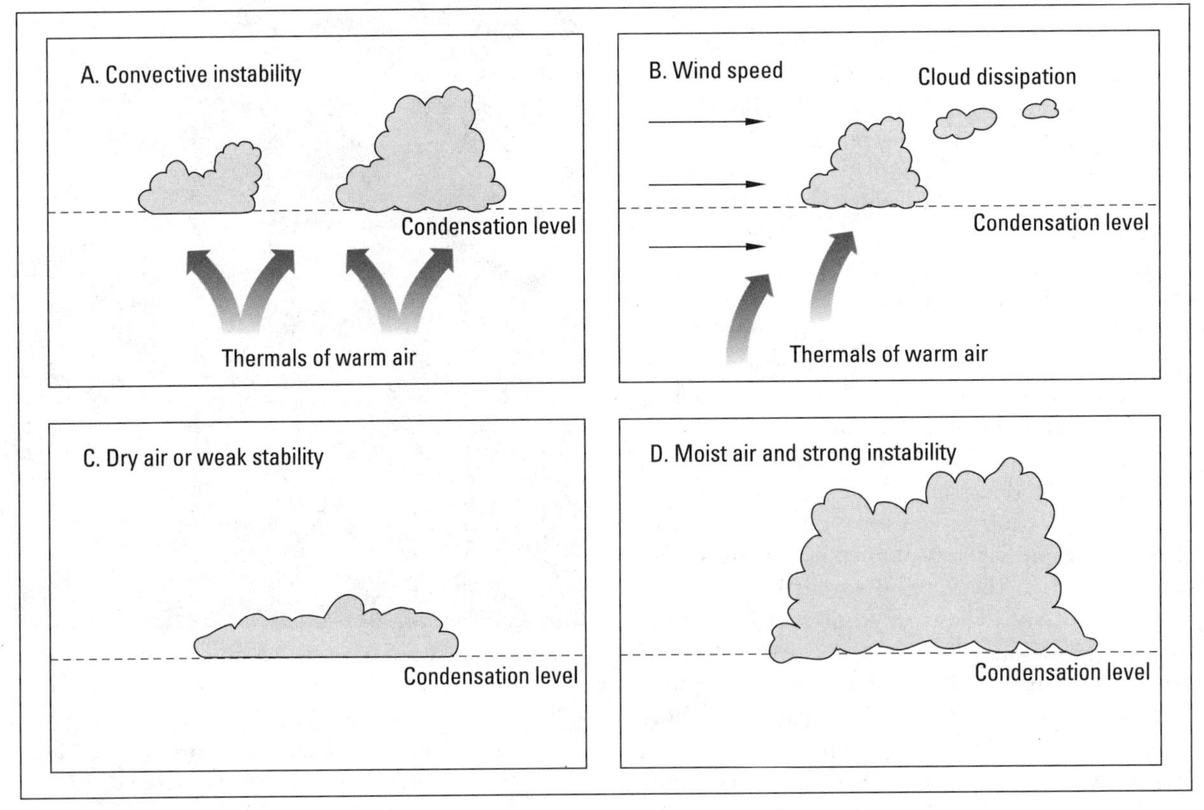


FIGURE 4.14 ▼ The controls affecting the form of cumulus clouds

there is a whole complex of cloud sheets that are difficult to differentiate from the ground or even from space.

The other main group of cloud, cumuliform, is the result of localized convection or instability. Bubbles of warm air, rising beyond the condensation level (if the air is unstable), are seen as cumulus clouds. The precise shape of the cloud will depend upon the degree of instability, the water vapour content of the air, and the strength of the horizontal wind (Figure 4.14). There are many different types of cumulus cloud subdivided on the basis of their appearance. If an inversion of temperature exists, as often happens with anticyclones, the bubbles will rise to the inversion and then start to level out or descend, evaporating the cloud droplets as they do so. This type of cloud is known as **fair weather cumulus**, as it rarely grows sufficiently to give rain (Figure 4.15).

FIGURE 4.15 ▼ Fair weather cumulus cloud over Brantford, Ontario. Fair weather cumulus have limited vertical development and are commonly seen in periods of fine weather, in the absence of frontal activity (photo: Robert S. Schemenauer)

Some cumulus clouds may grow larger and taller. The sharp and clear outlines of the cauliflower-like cumulus become more diffuse and ragged as the upper part of the cloud becomes a fibrous mass of ice crystals. The **cumulonimbus** stage has now been reached. At this stage precipitation is usually occurring, sometimes accompanied by lightning and thunder. As the mass of ice crystals develops, it is often blown downwind by the strong winds of the high atmosphere to give an anvil shape (Figure 4.16) characteristic of cumulonimbus clouds. Convection may initiate other clouds nearby or on the flanks of the parent cloud as it gradually decays and evaporates. In this way a cloud mass may travel as a single entity for several hours.

FIGURE 4.16 ▼ Cumulonimbus clouds can rise to altitudes of 10 000 m; the tops, composed mostly of ice crystals, spread out in the well-known anvil shape. Left: a thunderstorm forming over Saskatchewan (photo: Robert S. Schemenauer); right: cumulonimbus over the Gulf of Mexico was photographed by a space shuttle crew in August, 1984 (photo courtesy of NASA)

For rain to fall, we need clouds, but many clouds survive for hours without giving rain. What special circumstances enable some clouds to produce rain whereas others give none? We will examine this problem in the next section.

# Precipitation

## FORMATION OF PRECIPITATION

The average annual precipitation on the summit of Mt. Waialeale, Kauai, Hawaii, is 11 684 mm. Without doubt the processes producing precipitation can be very effective when conditions are favourable. But how do these minute cloud droplets (Figure 4.17) get large enough to fall as rain within about twenty minutes of the cloudy air reaching saturation?

To answer this question, we must delve inside a cloud and see what is happening there. In a cloud made up entirely of water droplets there will be a variety of droplet sizes. The air will be rising within the cloud, perhaps at a rate of 10 to 20 cm per second, though much more rapidly in cumulonimbus clouds. As it rises, so the drops get larger through **collision** and **coalescence**; some will reach drizzle size. When the uplift is stronger, say 50 cm per second, the downward movement of the drops will be reduced, so there will be more time for them to grow. If the cloud is about 1 km deep, small raindrops of 700 μm in diameter may be formed.

When the temperature of a water droplet falls below the point when it should freeze yet it remains liquid, it is said to be **supercooled**. The temperature at which the water droplet becomes an ice crystal depends upon the type of nucleus. Water droplets tend to remain liquid to about –9°C, with many liquid still at –20°C, and a few at –40°C. With cooling to –10°C, ice crystals may start to develop among the water droplets. This mixture of water and ice would not be particularly important but for a peculiar property of water. The saturation vapour pressure curve for ice (Figure 4.7) is slightly different from that of water. The air can be saturated over ice when it is not saturated over water. Thus, at –10°C, air saturated with respect to liquid water is **supersaturated** relative to ice by 10 percent, and at –20°C by 21 percent. As a result, the ice crystals in the cloud tend to grow and become heavier at the expense of the water droplets.

FIGURE 4.17 ▼ Cloud droplets range from 1 to 40 μm in diameter. Their sizes and concentrations can be measured by laser-optical probes mounted on aircraft (photo: Environment Canada)

As the ice crystals sink into lower layers of the cloud where temperatures are only just below freezing, they have a tendency to stick together to form snowflakes. This is brought about because the supercooled droplets of water in the cloud act as an adhesive. After the snowflakes have melted the resulting drops may grow further by collision with cloud droplets before they reach the ground as rain. This method of producing rain drops is known as the **Bergeron-Findeisen process**, after the developers of the theory. Beneath the base of the cloud, however, evaporation will take place in the drier air and if the drop is small it may be evaporated completely. Often, rain falling into very hot dry air over the Prairies in summer evaporates. This is called **virga**.

Precipitation formation both by collision and coalescence and by the Bergeron-Findeisen process undoubtedly occurs in the atmosphere, though clearly the Bergeron-Findeisen process can only operate when cloud temperatures are well below freezing. The rate at which vapour is converted into water droplets and precipitation depends upon three main factors:

- the rate of coalescence and ice-crystal growth
- cloud thickness
- the strength of updrafts in the cloud

The total amount of rain will be determined by the lifespan of the cloud, the height of the cloud above the ground, the moisture content of the air mass, and how long these processes operate. Cloud thickness and updraft speed are largely dependent upon instability and vertical motion in the atmosphere. Precipitation has been classified in terms of the factor that gives rise to the upward movement, so let us look at this in more detail.

## Convectional precipitation

The spontaneous rising of moist air due to instability is known as convection. We have seen that upward-growing clouds are associated with convection. Since the updrafts are usually strong, cooling of the air is rapid and lots of water can be condensed quickly. Collisions and coalescence are likely to be frequent, so the larger droplets rapidly increase in size. Eventually, growing larger and heavier, the droplets overcome the lift provided by the updraft, and they fall through the cloud into the clear air beneath. As the volume of water in these big droplets is large relative to their surface area, little evaporation takes place in the nonsaturated air below the cloud. At the ground there will be a burst of heavy rain as the cloud passes.

Unstable air that favours convectional rain is most frequently found in warm and humid areas. In most areas in Canada and the United States, 90 percent of the annual rainfall is by convection. Often this is the result of very cold air moving over a warmer ground surface, or differential heating of the ground, to give the steep lapse rates characteristic of instability and convection.

### THUNDERSTORMS

When the atmosphere is very unstable, cumulonimbus clouds develop, sometimes accompanied by lightning and thunder. At night an intense thunderstorm can provide one of the most spectacular displays of the atmosphere. Flashes of lightning shoot from cloud to earth or within the clouds (Figure 4.18), accompanied by great crashes of thunder. How can such dramatic manifestations of energy build up in a cloud? This has puzzled meteorologists for many years. In some way electrical charges in the cloud are separated; Figure 4.19 shows regions of different electrical charge in the cloud and on the ground. Electrical discharge is from the positive to the negative and the diagram indicates the possibilities. About 80 percent of the discharges are between positive and negative areas of the cloud, thus most **lightning** occurs within the cloud. Some 20 percent of lightning is between cloud and ground. When the lightning passes through the cloud it heats the air along its path to 30 000°C, causing the air to expand explosively to produce **thunder**.

FIGURE 4.18 ▼ Lightning can move in vertical strokes from cloud to ground, or pass horizontally from cloud to cloud; thunderstorm over Toronto, Ontario (photo: Robert S. Schemenauer)

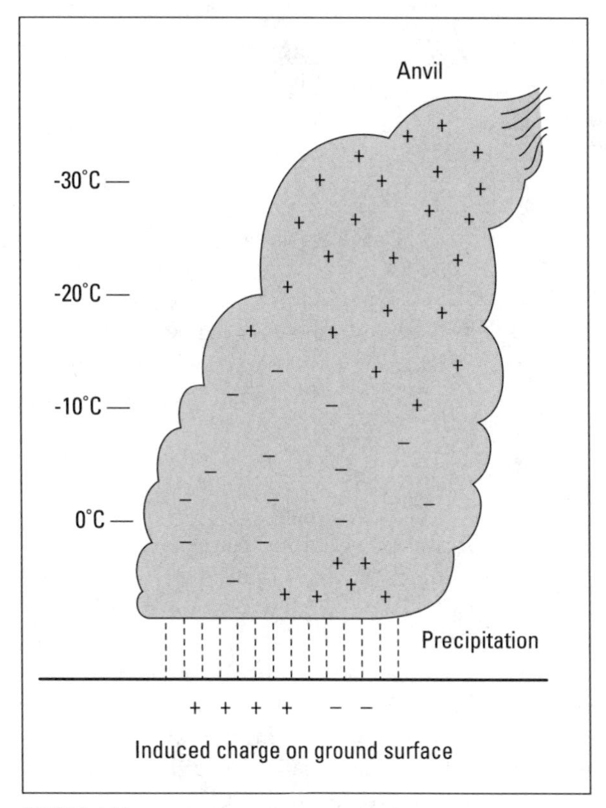



FIGURE 4.19 ▼ Electrical charge separation in a thunder cloud

As well as electricity, the thunderstorm is often accompanied by squalls of cold wind blowing away from the cloud. They usually originate as downdrafts of air near the main burst of rain (Figure 4.20). These are the microbursts identified as a cause of some recent aircraft crashes. Hail may fall near the centre of the storm, sometimes causing great damage.

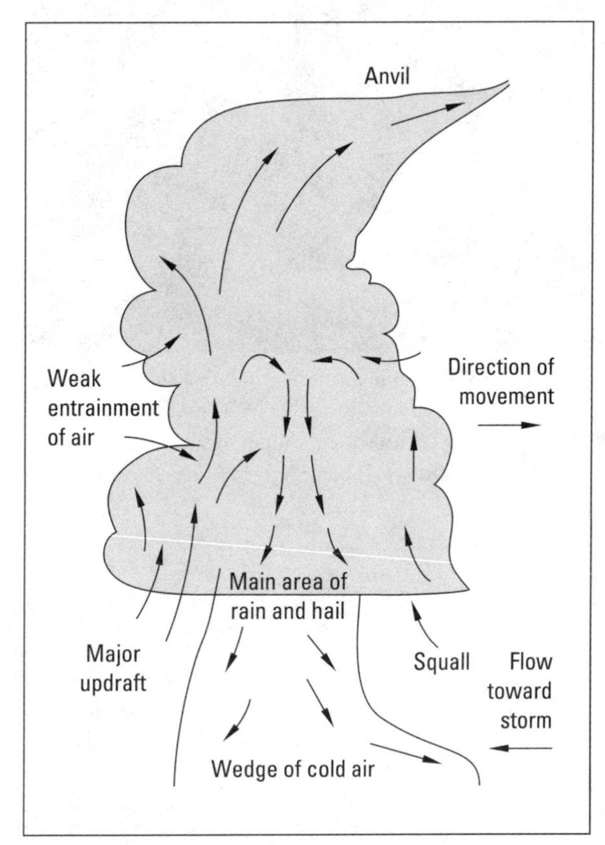


FIGURE 4.20 ▼ General air movement and areas of precipitation in and around a thunderstorm

FIGURE 4.21 ▼ Hailstones from a summer thunderstorm can badly damage crops and property (photo courtesy of NASA)

FIGURE 4.22 ▼ Cross section through a hailstone collected near Ledoux, Saskatchewan, in August, 1973; it is 16 cm in diameter. Regions of clear ice appear black, and opaque ice is white (photo courtesy of E.P. Lozowski, University of Alberta)

## HAIL

Hail is a form of precipitation composed of spheres or irregular lumps of ice (Figure 4.21). It falls in narrow bands associated with cumulonimbus clouds. However, the destruction it can produce is dramatic. Crops can be torn to shreds, greenhouses ruined, and cars dented by the weight of half a kilogram or more of ice falling from the sky.

Splitting open a large hailstone will show that it is composed of alternating layers of clear and opaque ice (Figure 4.22). It appears that the stone is involved in complex movements within the cloud, being swept up to the higher, colder parts of the cloud several times. Any moisture condensing on the stone will freeze instantly, producing the opaque ice visible in Figure 4.22. At lower levels in the cloud it takes condensed water a longer time to freeze. Air bubbles can then escape, leaving a layer of clear ice. The alternating layers of clear and opaque ice indicate the number of times the hailstone has been swept up by the cloud updrafts.

## Cyclonic precipitation

In temperate and sub-polar latitudes, some precipitation comes from **midlatitude cyclones**. The cyclone is characterized by areas of rising air. A satellite photograph of a cyclone shows the extensive areas of cloud resulting from this slow but widespread ascent of the air (Figure 4.23).

There are a number of differences from convectional precipitation. The areal extent of rising air associated with a cyclone is much larger, and the rate of upward movement and the rate of condensation in the clouds are much less. Because of this, the droplets grow more slowly and fall out of the cloud sooner. Being small, they can be greatly affected by evaporation in the drier air beneath the cloud base. For example, in an atmosphere with a relative humidity of 90 percent, a droplet of radius 10 mm will only fall 3 cm before evaporating; drops of 100 mm and 1 mm would fall 150 m and 40 km, respectively. For a particular rainbelt, it may take several hours of steady rain before the system has passed, giving a total fall of perhaps 5 to 10 mm.

In the deep and widespread clouds associated with a cyclone, it is quite common for ice crystal clouds at higher levels to act as a source of supply to the mixed clouds of ice and water droplets at lower levels. The addition of extra ice crystals speeds up the precipitation process and leads to more intense rainfall.

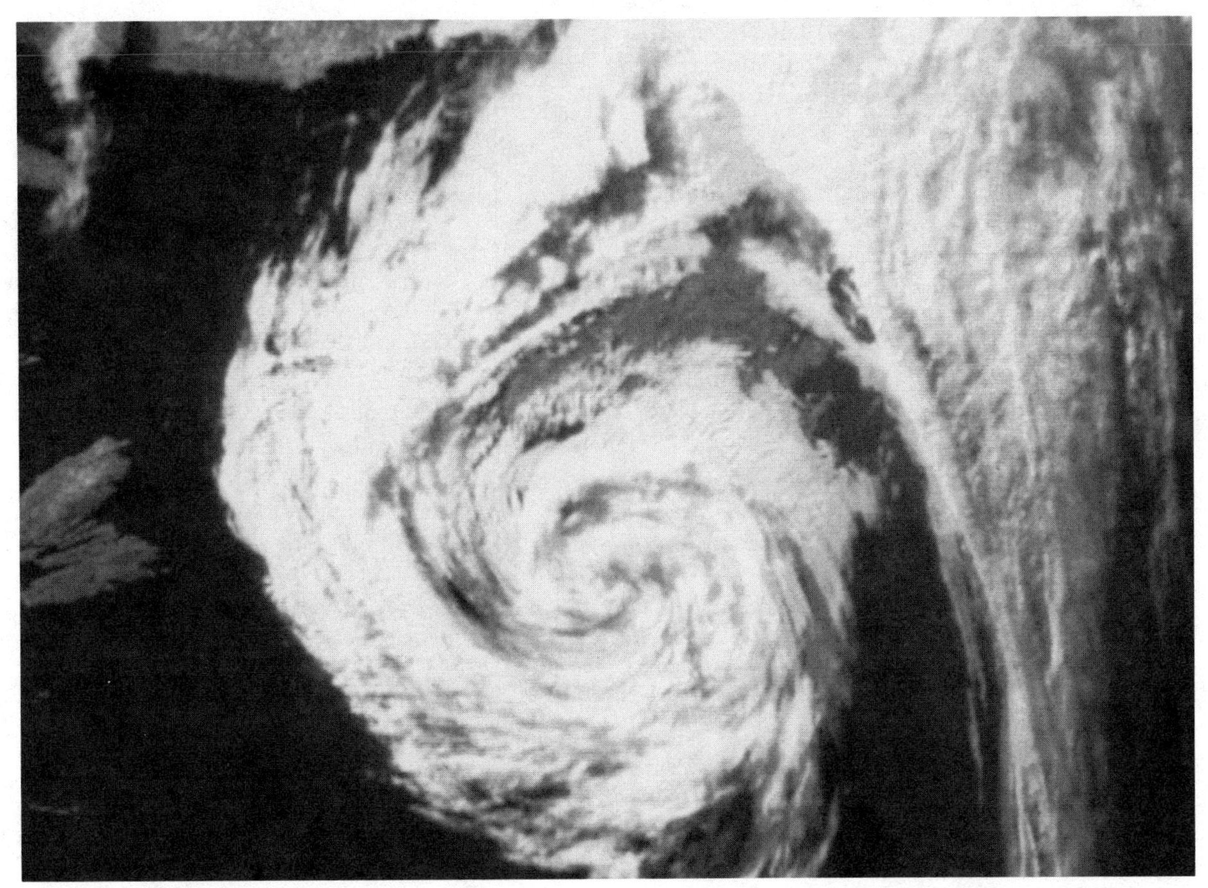

FIGURE 4.23 ▼ A well-developed cyclone south of Newfoundland (Cape Breton visible to the left). The extensive areas of cloud formed by widespread ascent can be seen clearly; the shadows indicate that the clouds are multi-layered. The north–south band of clouds to the right is a cold front (photo: Atmospheric Environment Service, Environment Canada)

# Orographic precipitation

Almost all mountain areas are wetter than the surrounding lowlands. To take two examples, annual rainfall on the west coast mountains of Vancouver Island frequently exceeds 3000 mm (Henderson Lake recorded 8123 mm in 1931). The station at the top of Mt. Waialeale on Kauai in Hawaii has an annual average of 11 684 mm. Even the Ahaggar and Tibesti mountains in the centre of the Sahara receive more rain than do the surrounding lowlands—Asekrem, at 2700 m, has an annual average of about 125 mm compared with only 13 mm at Silet, 720 m above sea level. Why should this be so?

Where air meets an extensive barrier it is forced to rise. Rising, as we know, leads to cooling of the air, and cooling encourages condensation. On mountain slopes and above the mountain summits, the clouds start to pile up, reflecting the forced ascent of air (Figure 4.24). Often they reach thickness sufficient to give drizzle and rain. Notice that orographic rain can occur in coincidence with changes in the stability of the air. Air that is initially stable may be forced to rise up the side of a mountain.

If the air is very moist near the ground but much drier above, as it rises the rates of cooling between the top and bottom of the layer will be different (Figure 4.25). The upper part will cool more quickly and so become relatively colder, leading to less stable air. The cloud development associated with instability will increase and rain may fall over the mountains. This situation is known as convective or potential instability.

Hills as well as mountains act as favourable areas for convectional showers. The slopes facing the sun will be warmed more rapidly than flatter areas, because of the higher angle of incidence. The resulting cloud may produce rainfall that is restricted to the upland area.

The orographic effect is most pronounced when it is already raining upwind of the hills or mountains. Where air is rising, associated with a cyclone for example, the rate of uplift is increased by the extra forced ascent provided by the hills. This leads to a greater rate of condensation on the windward side, larger drops of rain being formed, and heavier rainfall at the surface. Coupled with the slowing of the rainbelt as it passes due to increased friction, the net effect is considerably greater rainfall (Figure 4.26).

On the leeward side of the hills, subsidence or descending air begins to dominate; the cloud sheet thins or even dissipates and rainfall declines. As the air descends, it gets warmer because of compression. The higher temperature results in lower relative humidity, the capacity of the air to hold moisture is increased, and the probability of precipitation is reduced. This results in a **rain shadow effect** on the leeward slope of the mountains. The Okanagan Valley in the interior of British Columbia is semi-arid because of this effect. Here, rainfall is far less than on the upslope side, and sunshine amounts and daytime temperatures are normally higher. It is this

FIGURE 4.24 ▼ Air ascending over hills or mountains cools and orographic clouds are formed—these are altocumulus lenticularis near Clearwater, B.C. (photo: Robert S. Schemenauer)

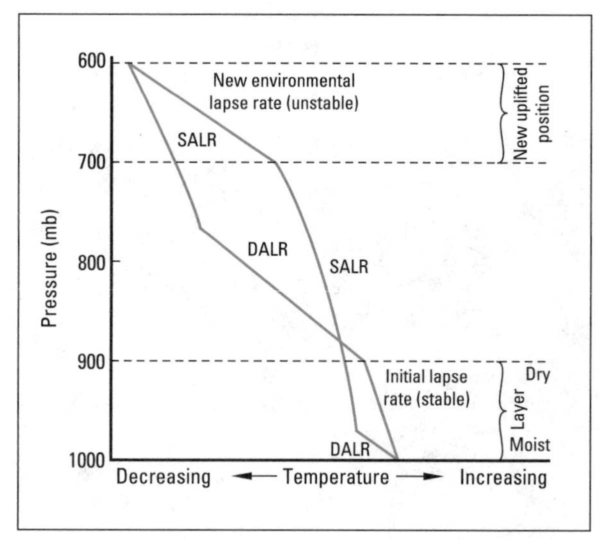


FIGURE 4.25 ▼ Destabilization of the lower atmosphere through uplift of a layer of air 100 mb in thickness. Between 1000 mb and 900 mb the air is initially stable. It is moist near the base and drier aloft. As a result of uplift the new environmental lapse rate indicates instability

FIGURE 4.26 ▼ Production of greater rainfall over hills is the result of forced ascent of moisture-laden air. Compare this with the pattern of rainfall in Figure 11.11

mechanism that also explains the Chinook effect—the warm winds that blow down the east side of the Canadian Rockies.

These processes ensure that rainfall normally increases with altitude, at least up to about 2500 m; above this level, the lower water content of the atmosphere becomes more significant. In Hawaii, rainfall on windward slopes at 1000 to 1300 m is about 7500 mm $yr^{-1}$, but on the summit of Mauna Loa at 4000 m it is only 380 mm $yr^{-1}$.

As much of the precipitation in mountains is due to an intensification of existing rain, it would be wrong to think of orographic precipitation as a truly separate category. It can occur as drizzle or by convective instability, but much more frequently it will depend upon cyclonic or convection processes already operating. These two types can occur together in cyclones, so perhaps we should identify convectional, cyclonic, and orographic precipitation as interrelated mechanisms of rainfall rather than classifying them into these types.

# Conclusion

In this chapter we have tried to follow and explain the exchange and movements of heat and moisture between the earth's surface and the atmosphere. The process of evaporation provides the supply of moisture in the lower atmosphere. The prevailing winds then circulate the moisture and mix it with drier air elsewhere. Only if we have a dry surface in areas well away from the oceans and where dry subsiding air is dominant will moisture levels be low.

Water vapour is only the first stage of the precipitation chain; we must convert the vapour into liquid form. This is usually achieved by cooling as a result of uplift, either rapidly as in convection, or slowly as in cyclonic storms; mountains will also cause uplift but the rate will depend upon their shape and the direction of the wind. Even this is insufficient as we can tell from the large number of clouds in the sky that never give precipitation. To produce precipitation, cloud droplets must become large enough to reach the ground without evaporating. The cloud must possess the right microphysical properties for growth of the droplets to take place. It must have ice crystals if the Bergeron-Findeisen process is to operate, or a wide spectrum of drop sizes with plenty of moisture condensing for the collision-coalescence system to work. Even these suitable conditions may be unrewarding if the cloud does not last long enough for the growth to take place. Clearly precipitation development represents a delicate balance of counteracting forces, some leading to droplet growth, others to droplet destruction. Nevertheless, where conditions are basically favourable—where air can rise high enough to produce large vertical developments of cloud—copious amounts of precipitation occur.

The effect upon the climate is obvious, but precipitation is also important in other ways. As we will see in Chapters 11 and 13, it is a major component of the hydrological cycle. Rainfall also takes part in many of the processes that build our landscape. And plants and animals are highly dependent upon precipitation. Therefore, in Chapter 11 we will be returning to the question of precipitation and following its progress to earth in more detail.

# The atmosphere in motion

## Atmospheric circulation and winds

The earth's atmosphere is in perpetual motion, movement that is striving to eradicate the constant differences in pressure and temperature between different parts of the globe. This motion produces the winds and storms we are all familiar with. It is this circulation that plays a basic part in maintaining a steady state in the atmosphere and generating the climatic zones that characterize the earth.

Nature does not like imbalance or inequity. Variations in temperature create variations in pressure—as a result air moves from areas of high pressure to low to restore balance. We sense this movement as wind. More correctly, it is the advection or horizontal movement of air.

The wind in turn transports heat energy and moisture, influencing the global climate. The jet stream is a major wind mechanism, created by the differences in temperature between the polar and tropical regions. Great waves that develop in the upper atmosphere in the middle latitudes are a major influence on energy and the pattern of climate in those middle latitudes and therefore in Canada.

## ATMOSPHERIC PRESSURE

Earlier we saw that atmospheric pressure decreases as we move away from the earth's surface. Now, to learn how the atmosphere moves we need a better understanding of vertical and horizontal pressure.

The earth is surrounded by a gaseous envelope that we call the atmosphere. It is held to the earth by gravity, consequently heavier gases occur in the lower atmosphere, and as we move upward the types of gas and the number of molecules for each decrease. For example, oxygen decreases with altitude so that above 3000 m most humans are noticeably affected and cannot carry out normal activities.

In 1643, the Italian scientist Torricelli placed a tube, closed at one end, upside down in a dish of mercury. He found that the weight of the atmosphere on the mercury in the dish pushed the mercury 76 cm (29.92 inches) up into the tube. The mercury rose higher when pressure increased and lowered when pressure fell.

Imagine a column of air 1 metre square extending from the surface of the earth to the top of the atmosphere. The average number of molecules in the column exert a pressure on the surface at the bottom of the column. This pressure is measured in newtons, which is the force that would accelerate 1 kilogram of mass 1 metre per second squared. In Canada, we are accustomed to measuring the pressure of air in kilopascals. A **pascal** is equal to 1 newton per square metre (N $m^2$). **Standard atmospheric pressure** at sea level produces a force of 101 325 N $m^2$ or 101.325 kilopascals (kPa). Countries not using metric, such as the United States, convert newtons to millibars. One hundred newtons equals 1 millibar (mb), therefore standard atmospheric pressure is 1013.25 millibars.

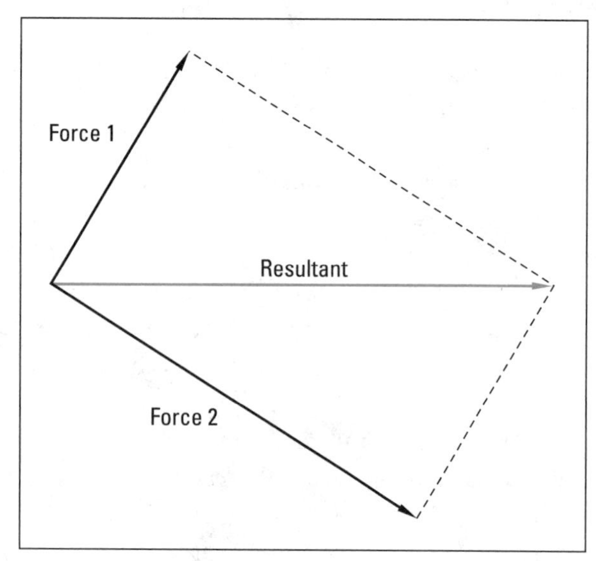


FIGURE 5.1 ▼ The resultant of two forces acting in different directions. The length of a force arrow is proportional to the strength of the force

The number of molecules in our column of air varies with temperature because the column is not a confined space. If the temperature rises, the molecules move more rapidly and extensively, thus reducing the number of molecules in the column. Fewer molecules means less weight and therefore lower atmospheric pressure on the surface. Similarly, lower temperatures slow the molecules, increasing the number within the column, and this increases pressure. In summary, warm air is generally associated with lower pressure and cold air with higher pressure.

## CAUSES OF AIR MOVEMENT

Why do we have winds at all? Certainly the explanation is not unique to the earth, for similar patterns of atmospheric circulation have been identified on other planets.

To answer this question it is useful to consider some of the basic principles of motion. Our understanding of these is due in large degree to Isaac Newton, who first formulated laws of motion.

There are two main laws:

1. A particle will remain at rest or in uniform motion unless acted upon by another force.
2. The action of a single force upon a particle causes it to accelerate in the direction of the force. If there is more than one force the particle is accelerated in the direction of the resultant (Figure 5.1).

These forces are particularly important for movement in the atmosphere because forces are continuously acting on particles of air, causing them to accelerate or decelerate and change their direction.

## FORCES ACTING UPON THE AIR

*Pressure gradient force* Let us imagine a small parcel of air some distance above the ground. What forces will act upon it? The most obvious is the force of gravity, which tends to attract all mass toward the earth's centre. In addition, we have the pressure exerted by the air surrounding the parcel (Figure 5.2). If this pressure was the same on all sides of the parcel then its effects would cancel out. But this is not so. Pressure decreases upward in our atmosphere, as we saw in Chapter 4. The force pushing the parcel of air upward is greater than the downward force from the overlying atmosphere; there is a potential upward acceleration of the parcel. Luckily, this vertical force is almost exactly balanced by the force of gravity, otherwise we would have lost our atmosphere long ago. Most air movements that we observe are horizontal. Forces are also operating in this dimension. Where the atmosphere is dense, the lateral pressure on the parcel of air is great; where the atmosphere has a low density, the lateral pressure is less. Variations in the density of the atmosphere from one part of the globe to another result in an imbalance of forces and lateral movement of the air (Figure 5.3A). The air is pushed from areas of high pressure to areas of low pressure.

This, in fact, is the basic force affecting the earth's atmospheric motion. It is called the **pressure gradient force**. As we have seen, a pressure

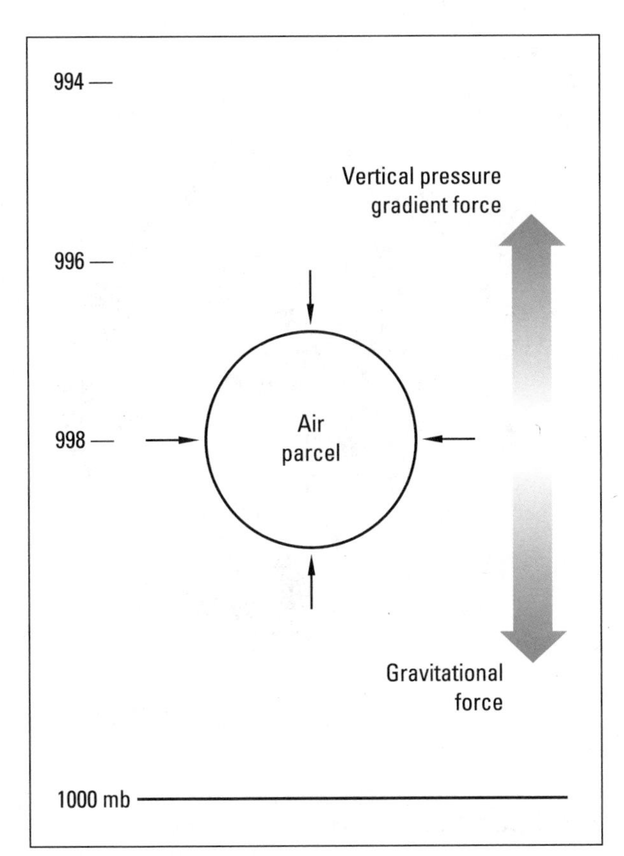


FIGURE 5.2 ▼ Pressure forces acting on a parcel of air

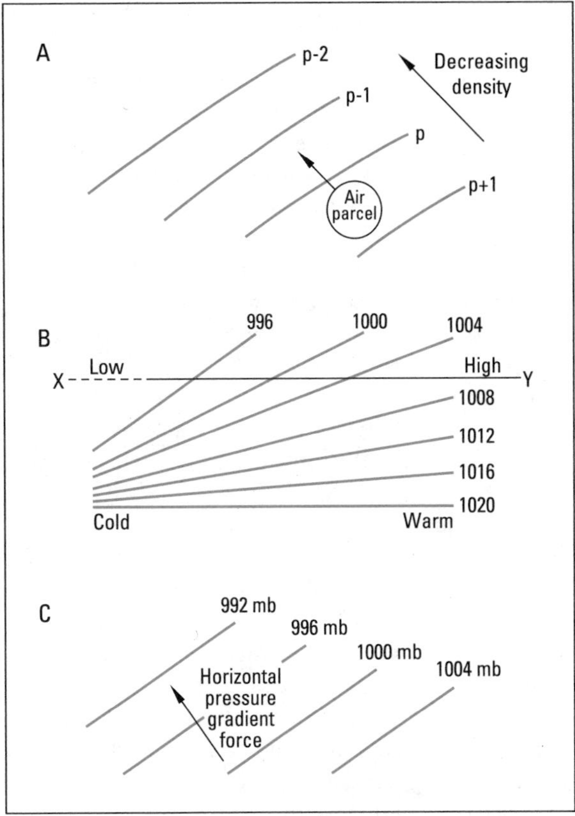


FIGURE 5.3 ▼ (A) Force exerted on an air parcel produced by density differences. (B) Change of pressure with height above areas of warm and cold air. At X, above the cold air, pressure is lower than at the same height, Y, above the warm air. (C) Horizontal pressure gradient force acting at right angles to the isobars

gradient exists both vertically and horizontally. Pressure decreases vertically because, as we move upward through the atmosphere, the weight of overlying air diminishes. It varies laterally because of differences in the intensity of solar heating of the atmosphere. Where solar radiation is intense the air warms and expands, and its density declines. Air pressure falls. Where cooling occurs, the air contracts, its density increases, and air pressure becomes greater.

A corollary of this principle is that the pattern of air pressure close to the surface is reversed in the upper atmosphere. Because cold air contracts, the upward decline in pressure is rapid and at any constant height above a zone of cool air the pressure is relatively low. Conversely, warm air expands and rises, so that the vertical pressure gradient is less steep. Above areas of warm air, therefore, the pressure tends to be relatively high (Figure 5.3B). The effect upon atmospheric motion is clear. At the surface the air will move from cold to warm zones; at higher altitudes the flow will be from warm to cold.

Differences in air pressure may be mapped by defining lines of equal pressure (usually measured in millibars). These are known as **isobars**. Air movement occurs at right angles to the isobars, down the pressure gradient; that is, from areas of high pressure to areas of low pressure (Figure 5.3C). The magnitude of the force causing movement (the pressure gradient force) and thus the speed of the wind, is inversely proportional to the distance between isobars. Thus, the closer the isobars are together, and the more rapidly pressure falls with distance, the stronger is the wind.

Mathematically, the relationship can be written:

$$F = -\frac{1}{p} \times \frac{p_2 - p_1}{n}$$

where pressure values at points 2 and 1 are $p_2$ and $p_1$, $n$ is the distance separating 2 and 1, $p$ is air density, and $F$ is the resulting acceleration. We can use formula to indicate how quickly the parcel ought to accelerate. The standard isobaric interval on a pressure chart is 4 mb and air density is 1.29 kg $m^{-3}$. Suppose the isobars are 300 km apart on a sea level chart. What will the acceleration down the pressure gradient be? In uniform units, the formula will become

$$F = \frac{1}{1.29} \times \frac{4 \times 10^2}{300 \times 10^3} = 0.00103 \text{ m s}^{-1}$$

If this rate is kept up for 1 hour (3600 seconds) we would have a value of 3.72 m $s^{-1}$ after 1 hour. As pressure gradients of this size can last for days, we might expect very high wind speeds to develop unless other forces interfere. There are two main forces that prevent this happening. One is friction and the other is a result of the earth's rotation.

If we look at the wind field on a weather map, it would be immediately apparent that air does not flow down the pressure gradient toward areas of low pressure. If it did, the low pressure areas would fill and the wind movement would stop. Instead we find that the wind is blowing parallel (or almost) to the isobars, rather than across them.

This is the effect of the earth's rotation.

*Coriolis force* Although we are not aware of it, the earth is rotating from west to east at 15° longitude per hour. Reference to Newton's laws shows that if we have a parcel of air moving southward and there are no forces acting upon it, it will continue to move in the same absolute direction (that is, a straight path as viewed from space). However, the earth is gradually turning and so, relative to the ground surface, this parcel will appear to have followed a curved track to the right (Figure 5.4). To explain this apparent deflection in Newton's terms, we have to introduce a force to account for the movement as observed from the ground; it is called the **Coriolis force**, after the French mathematician who formalized the concept. The value of the Coriolis force changes with the angle of latitude and the speed of the air. Mathematically, for a unit mass of air, it is CF = $2\omega v \sin \theta$ (where $\omega$ is angular momentum, which varies because the earth's rotation varies with latitude, v is the velocity of anything moving across earth's surface, and $\theta$ is the latitude). It is greatest at the poles, where the earth's surface is at right angles to the axis of rotation, but it diminishes progressively toward the equator, where it reaches zero. The reason for this is

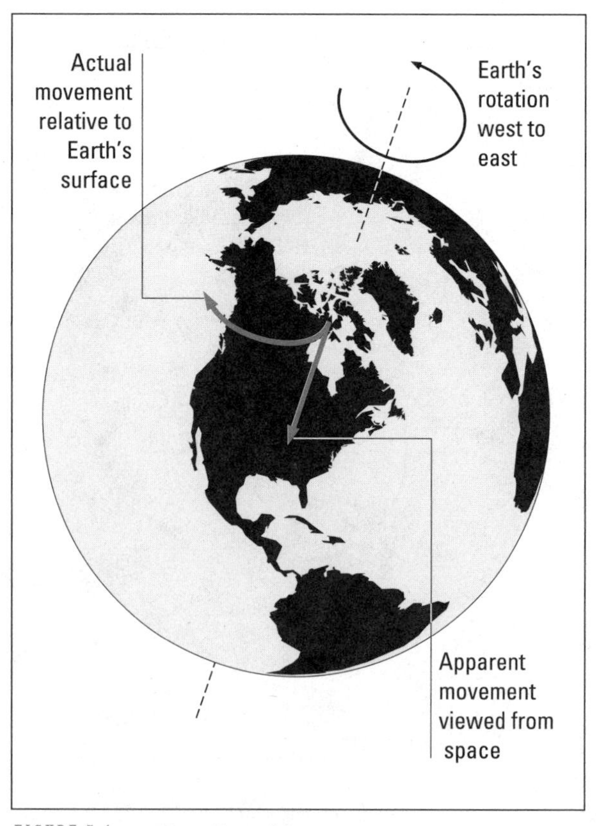


FIGURE 5.4 ▼ The effect of the earth's rotation on air movement

shown in Figure 5.5. As one moves toward the equator, so the earth's surface eventually becomes parallel to the axis of rotation. Note that in the formula, when the velocity (v) of the object is zero, Coriolis force is zero. Similarly, the sine of $\theta$ (latitude) is zero at latitude 0° and 1 at 90°, therefore Coriolis force is zero at the equator and maximum at the poles.

*Geostrophic wind* Let us return to our parcel of air experiencing a pressure gradient force on the rotating earth. Initially the parcel will move down the pressure gradient, but as soon as it begins to move, it will start to be affected by Coriolis force. It will be deflected toward the right in the northern hemisphere, toward the left in the southern hemisphere (Figure 5.6). As the wind accelerates, its speed will increase and, because Coriolis force is related to speed ($CF = 2\omega v \sin \theta$), so the two forces pulling together eventually produce an equilibrium. This will occur when the two forces are equal and opposite, the resultant wind blowing parallel to the isobars; it is known as the **geostrophic wind**. Its velocity will be determined primarily by the pressure gradient force, but because Coriolis force increases toward the pole it takes a stronger pressure gradient to achieve the same geostrophic wind velocity.

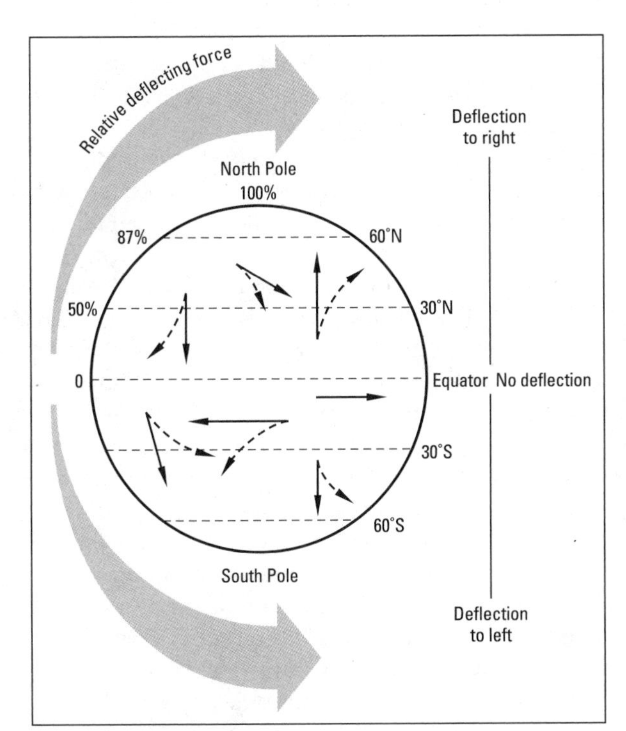


FIGURE 5.5 ▼ The changing magnitude of Coriolis force with latitude

Although we have only considered two of the forces acting upon the air parcels, nevertheless the geostrophic wind is a useful concept. It approximates closely the winds observed in the atmosphere above the friction zone (that is, in the upper atmosphere). Strictly, it operates only when the isobars are straight—a rare event. Normally isobars are curved and winds are subject to another force, termed centripetal acceleration, which acts toward the centre of rotation. When this rotational component is included, the resultant wind is called the **gradient wind**, which is closer to observed flow in the upper atmosphere (Figure 5.7).

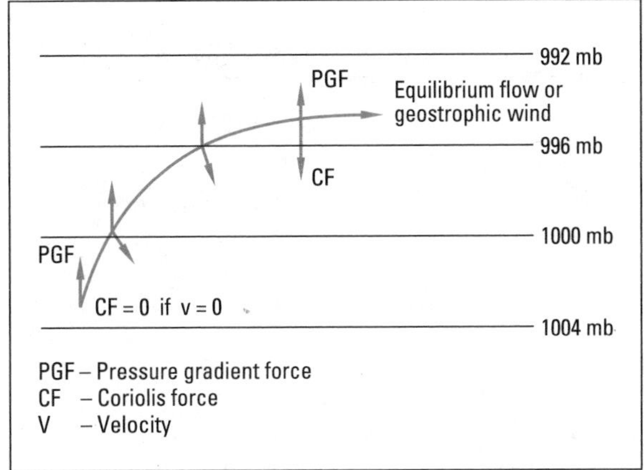


FIGURE 5.6 ▼ Balance of forces for the geostrophic wind in the northern hemisphere

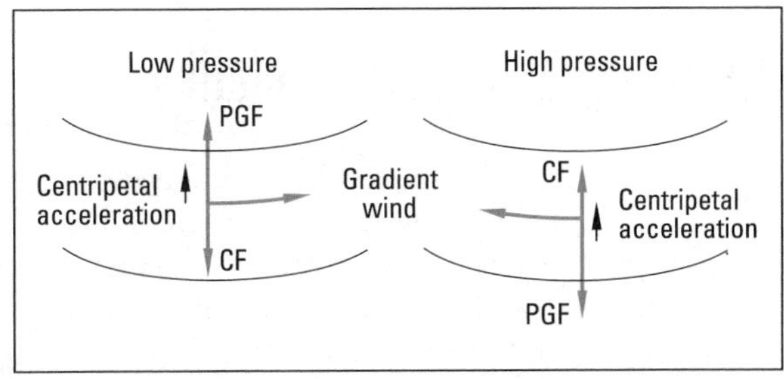


FIGURE 5.7 ▼ Balance of forces for the gradient wind where curvature of the isobars is important

*Friction* Inspection of a surface weather map will show that, at ground level, the wind does not blow parallel to the isobars. It blows across the isobars toward the area of lower pressure. The more observant may notice that this angle between the wind flow and the isobars is greater over land areas than over seas. This may give a clue to the reasons for the change. Land surfaces are rougher than seas; they tend to slow the wind through friction more effectively. Friction acts as a force pulling against the direction of flow. We can now arrange our balance of forces to include friction. To achieve balance, the flow will now be across the isobars because the Coriolis pull to the right (northern hemisphere) decreases as the air velocity falls (Figure 5.8).

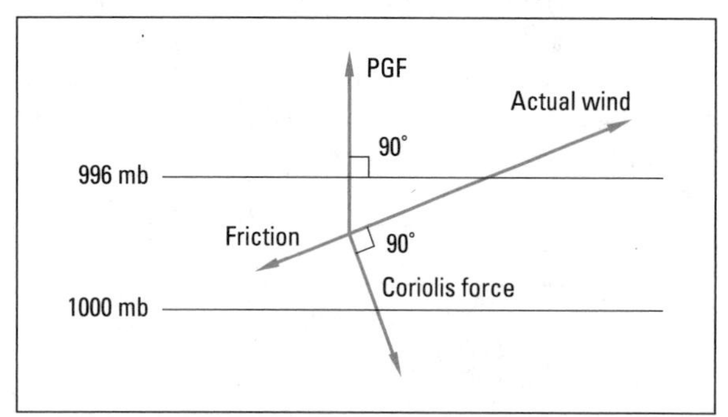


FIGURE 5.8 ▼ The effect of friction on the geostrophic wind. Coriolis force is always at right angles to the actual wind. It is smaller than the pressure gradient force because friction has reduced the speed of the wind

From these forces we can now explain horizontal flows of air. They are initiated by pressure differences, then modified by friction and the effects of the earth's rotation.

# The global pattern of circulation

With these principles in mind we can try to build up a picture of the global pattern of circulation in the earth's atmosphere. We start by considering a highly simplified model of the atmospheric system: a uniform, non-rotating, smooth earth.

Friction is greatest at the surface, therefore it slows the layer of air directly in contact with the ground the most. As we move away from the surface the frictional effect decreases and wind speed increases. Since wind speed determines the amount of Coriolis force, the balance of forces illustrated in Figure 5.8 changes. The result is a different wind speed and direction at different altitudes. When we plot this on a vertical axis we see that speed increases and direction changes in a distinctive effect known as the **Ekman spiral**.

Wind blowing over the ocean is reduced in speed by friction, but this energy is partially used to move the surface layer of water. This is the mechanism that creates ocean currents, as we see in Figure 5.9. Since water moves in layers like the air, the next layer down moves with diminished speed. Eventually a depth is reached at which there is no movement due to surface winds. In the ocean, similarly to the pattern of the wind, Coriolis force deflects moving water to the right in the northern hemisphere and to the left in the southern hemisphere. Each successive layer of water moves more slowly as you go deeper and is deflected to the right of the layer above. This is also called an Ekman spiral.

As we have seen, the basic force causing atmospheric motion is the pressure gradient; this gradient arises from the unequal heating of the atmosphere by solar radiation. At the equator—the 'firebox' of the circulation, as it has been called—solar radiation is converted into heat. The air expands and rises and flows out toward the poles. Cool, dense air from the poles returns to replace it. We can readily demonstrate the pattern of circulation by heating a dish of water at its centre. Hot water bubbles up above the heat source and flows across the surface to the cold 'polar' areas. At depth the flow is reversed. So long as this unequal heating is continued, the cellular flow is maintained.

In reality, however, the pattern is more complex for, instead of flowing directly to the poles at high altitudes, the warm air from the equator gradually undergoes radiational cooling and sinks. Most of it reaches the surface between about 20° and 30° latitude, and this subsiding air gives rise to zones of high pressure at the tropics. These used to be known as the horse latitudes because sailing ships were often becalmed and ran short of food and water. The unfortunate horses were then eaten or thrown overboard. Now we use the less picturesque term subtropical high pressure belts.

As the descending air reaches the surface it diverges, some returning toward the equator to complete the cellular circulation of the tropics, the remainder flowing poleward (Figure 5.10).

Various other factors disrupt this pattern further, for the earth is not at rest nor uniform as we have so far assumed. It rotates. Its surface is highly variable; it has oceans and continents; it consists of a mosaic of mountains and plains. Moreover, the inputs of

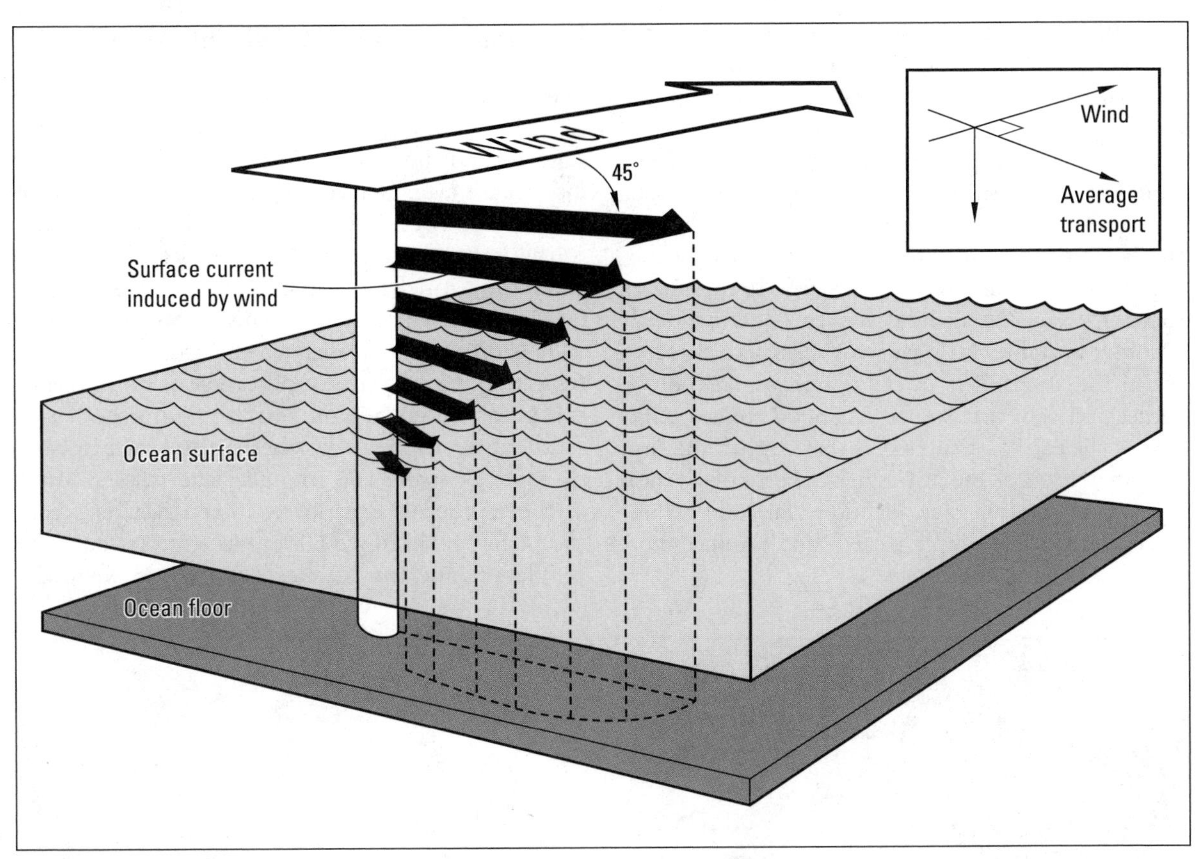


FIGURE 5.9 ▼ The Ekman spiral. Wind is deflected by Coriolis force—clockwise in the northern hemisphere; each layer closer to the surface moves more slowly and to the right of the layer above it. Moving air at the surface of a lake or ocean will induce a current in the water, and the Ekman effect will continue down through layers of water, each successive layer moving more slowly and to the right of the one above. Inset: the average transport of surface water in the Ekman layer is at right angles to the prevailing winds. Similar effects in the southern hemisphere are deflected counterclockwise

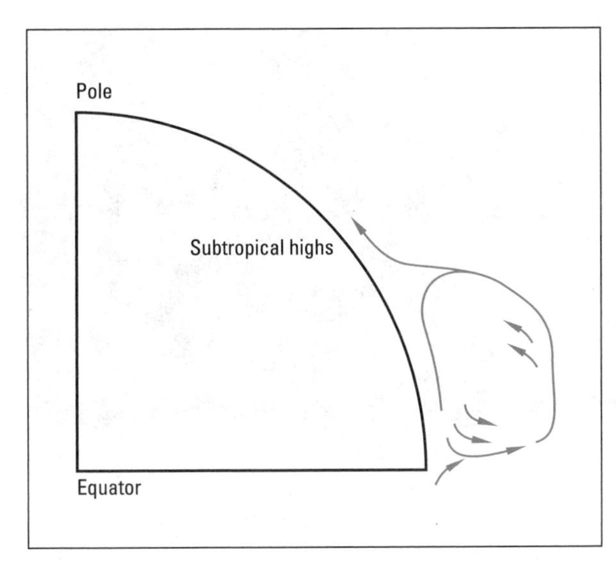


FIGURE 5.10 ▼ The meridional air circulation in the tropics

solar radiation vary considerably both on a seasonal and a daily basis.

## THE EFFECT OF THE EARTH'S ROTATION

The rotation of the earth causes the winds to be deflected from the simple pattern we have distinguished in Figure 5.10. The deflection is toward the right in the northern hemisphere and toward the left in the southern hemisphere. Instead of the direct meridional flow, therefore, the so-called Coriolis force produces a surface flow similar to that shown in Figure 5.1.

This is not the only effect of the earth's rotation. Air moving toward the poles from the tropics forms a series of irregular eddies, embedded within the generally westerly flow. These can be seen on the satellite photographs as spiralling cloud patterns, similar to the patterns we can see in a turbulent river (Figure 5.11).

Again we can understand the cause of these eddies with the help of a simple experiment. A pan of water is heated at the rim and cooled at the centre. If the pan is slowly rotated it is seen that a simple thermal circulation is produced. If the rate of rotation is increased, however, the flow suddenly becomes unstable. New patterns form like those we see in the atmosphere of the temperate latitudes—eddies and waves. It seems that rapid rotation, like that of the earth, sets up forces that disturb the simple circulation of the atmosphere. Particularly near the axis of rotation (i.e., in higher latitudes), these forces destroy the simple pattern and produce more complex circulations (Figure 5.12).

## THE EFFECT OF SURFACE CONFIGURATION

Even now, our picture of atmospheric circulation is far from complete. As we have noted, the earth's surface is not uniform, and the variations in surface form cause ever more disruptions to the pattern of circulation. Friction bends the winds perpendicular to the isobars, reducing the effect of Coriolis force, and locally it deflects the surface flow of air to produce highly complicated systems of movement.

It is difficult to model the effects of surface configuration, but a general indication of its influence can be obtained by comparing the northern and southern hemispheres. In the northern hemisphere, there are extensive and irregular land masses. Much of the southern hemisphere, in contrast, is ocean, except for the high ice plateau of Antarctica, where very low temperatures are experienced. As we might

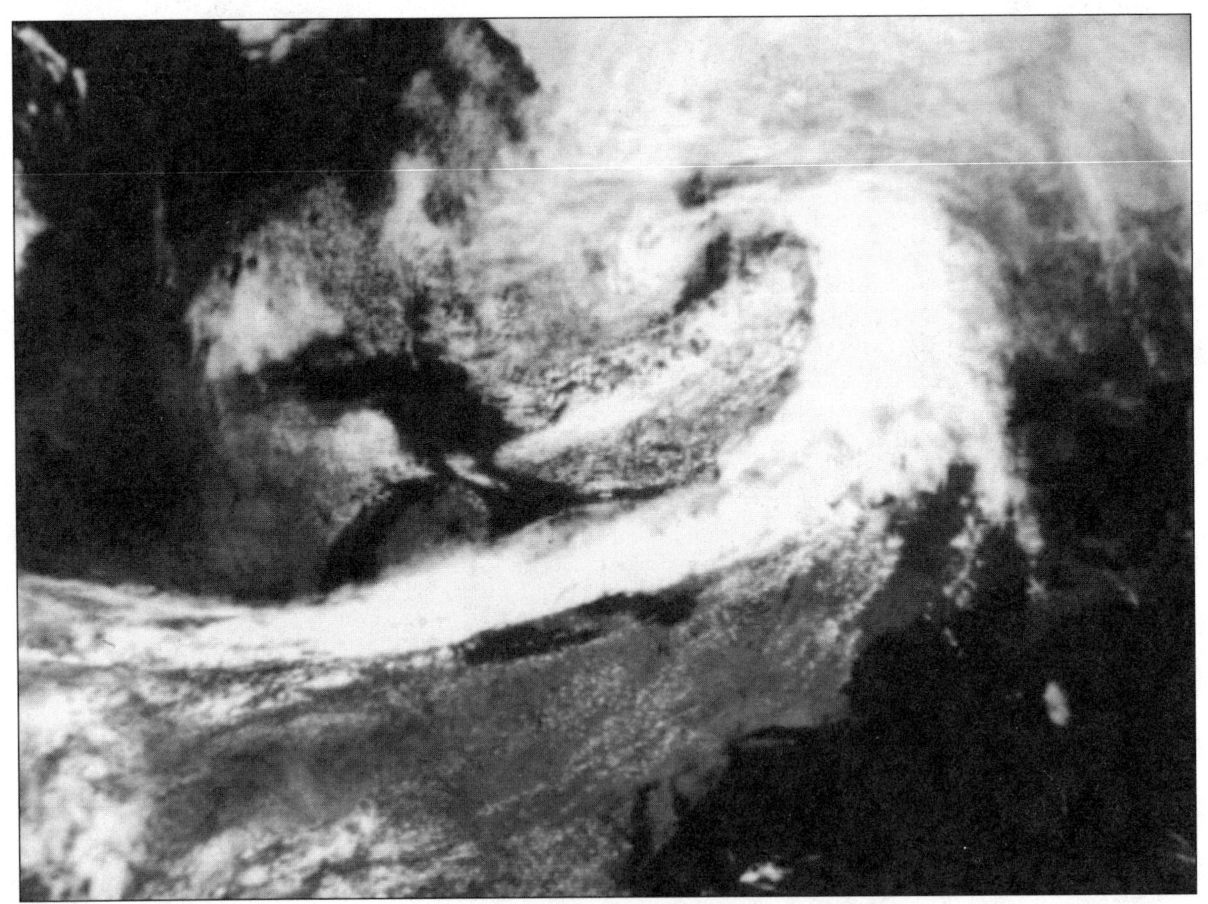

FIGURE 5.11 ▼ Spiral cloud around a mid-latitude depression. A spring storm has formed in central Quebec east of James Bay, with a cold front stretching south to Montreal and southwest to Lake Michigan; there is considerable convection producing cumulus cloud north of the front (photo: Atmospheric Environment Service, Environment Canada)

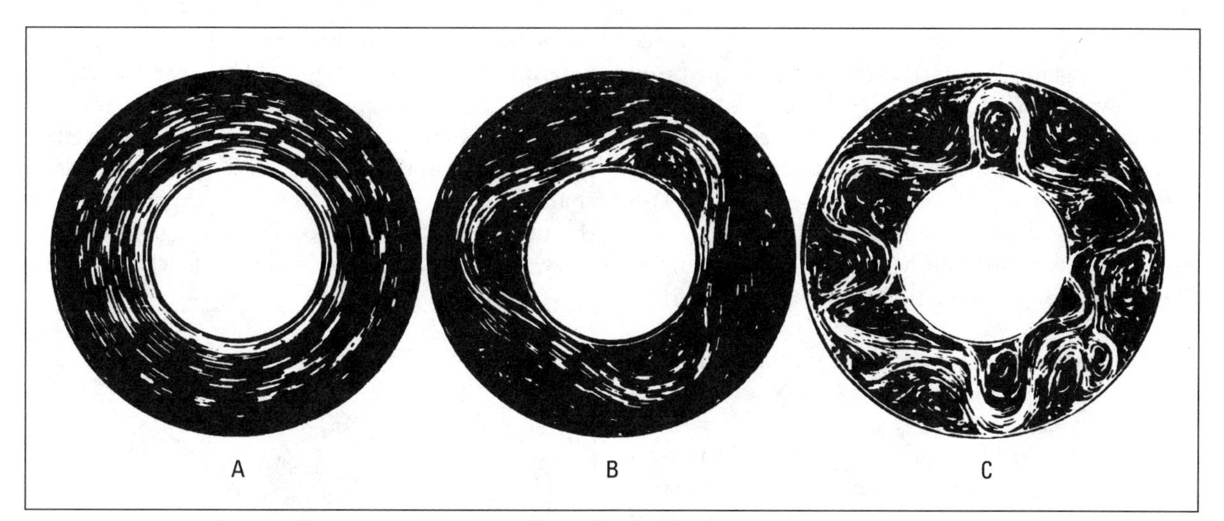


FIGURE 5.12 ▼ Streak photographs giving an example of the top-surface flow of patterns in a rotating fluid subject to heating at the outer sidewall and cooling at the inner sidewall. At low rates of rotation (A) the flow is symmetric about the axis of rotation. As the rotation rate increases (B) the flow develops jet streams and waves. At higher rates (C) the flow is highly irregular, resembling the cyclonic and anticyclonic eddies found in westerly air circulation (photograph by courtesy of Dr. R. Hide)

expect, the pattern is much simpler in the southern hemisphere. A strong westerly flow of cool polar air occurs even in the southern summer. Conversely, in the northern hemisphere, the flow is more irregular, the circulation weaker. The temperature differences between pole and equator are less marked (about 30°C compared with 60°C in the southern hemisphere) and so the driving force for the winds—the pressure gradient—is reduced.

# Energy transfers and the global circulation

## FORMS OF ENERGY IN THE ATMOSPHERE

The general circulation of the atmosphere is powered by solar radiation. It operates in response to the disparities in energy inputs between different parts of the world. It acts to reduce these disparities; it is unsuccessful only because the energy differences are maintained by constant inputs of solar radiation.

The circulation, however, represents one of the main processes of energy transfer through the global system and by this process it preserves a steady state within the system, preventing the accumulation of heat in the tropics or the perpetual cooling of the poles. Energy is transferred in a variety of forms, however, and during these transfers it undergoes numerous transformations.

Four main forms of energy exist in atmospheric circulation: latent heat, sensible heat, potential energy, and kinetic energy. The total energy of a unit mass of air can therefore be described as follows:

$$E_t = L_q + C_pT + gz + V^2 2$$

where $L_q$ is latent heat content
$C_pT$ is sensible heat content
$gz$ is potential energy
$V^2 2$ is kinetic energy

We have discussed these forms of energy in previous chapters. As a reminder, potential energy represents the energy derived from the position of matter and kinetic energy is the energy of motion. The latter is readily understood as a function of wind velocity.

Latent heat is more difficult to visualize. It is the quantity of heat released or absorbed, without any change in temperature, during the transformation of substance from one state to another (e.g., from solid to liquid). The main source of latent heat in the atmosphere is water vapour; when this condenses to water it releases heat, when it is formed by the evaporation of water it absorbs heat.

The sensible heat can be thought of as the temperature of the atmosphere. More specifically it is the temperature ($T$) times the specific heat ($C_p$) of the air at a constant pressure. Sensible heat is gained from the ground surface after the absorption of shortwave radiation, or by the release of latent heat through condensation. (The specific heat of a substance is the amount of heat required to raise the temperature of 1 g of that substance by 1 C°. This is defined at a constant pressure because adding heat normally alters the volume/pressure relationship of the substance. The specific heat of still air at 10°C is 1.010 J $g^{-1}$ $°C^{-1}$).

The potential energy of the atmosphere is essentially a function of its height above the ground surface ($z$); gravity ($g$) is a constant. As air moves in the atmosphere it tends to change its height and alter its energy content. If the air sinks slowly, the potential energy decreases. Normally it is converted to sensible heat, and the air becomes warmer as it subsides. If the air rises, the temperature tends to decline but the potential energy increases.

Kinetic energy is proportional to the square of the velocity of the wind ($V^2/2$). Therefore strong winds have more kinetic energy than gentle winds, as the damage they cause indicates. On a global scale, hurricanes and other strong winds are relatively rare, so the quantity of energy in the form of kinetic energy is limited. Even in the regions of strongest winds it probably reaches no more than 0.5 percent of the total energy content of the atmosphere.

## ENERGY TRANSFER IN THE ATMOSPHERE

The pattern of energy transfer in the atmosphere is complex, and we can only consider here some of the general components of the pattern. As a starting point let us look at a simplified model of what happens in the tropics (Figure 5.13).

The circulation between the tropics consists of two cells, known as **Hadley cells**. Air blows in toward the low pressure belt of the equator (the equatorial trough) across the subtropical seas. As it does so evaporation of water from the ocean uses vast quantities of energy so that the sensible heat transfer to the atmosphere is often small. In contrast, over the desert land masses very little evaporation occurs, energy loss is limited, and the incoming radiation heats the ground surface which then heats the atmosphere. Thus much more of the energy is in the form of sensible heat. During the night this energy is reradiated back to space, for the dry air is unable to intercept much outgoing longwave radiation. Anyone who has camped in desert areas will know the effect is to produce great ranges in temperature from day to night. More specifically, the net surplus of radiation is fairly small. As we have seen, winds approaching the equator rise as they meet the equatorial trough, creating a cloudy zone that can often be seen on satellite photographs (Figure 5.14). The ascent of this air is not a continuous, widespread phenomenon, but occurs mainly

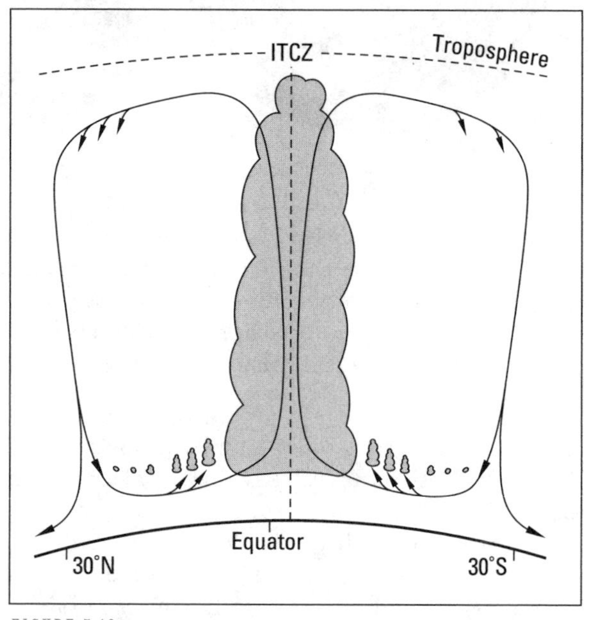


FIGURE 5.13 ▼ Details of the meridional air circulation or Hadley cell in the intertropical zone

FIGURE 5.14 ▼ Linear cloud development along the equatorial trough or Intertropical Convergence Zone across the Atlantic; a view from the *Apollo 15* mission en route to the moon, July, 1971 (photo courtesy of NASA)

in association with localized, often intense and short-lived updrafts such as in thunderstorms. As the air rises and cools, the water vapour condenses and releases latent heat. The increased height of the air also represents increased potential energy. The frequency and consistency of this mechanism is reflected in the fact that stations in Java in south-east Asia have recorded thunderstorms on as many as 322 days of the year, and one station has recorded rainfall on 355 days of the year.

The equatorial air then diverges and flows poleward, so the potential energy is exported to higher latitudes. The cycle is completed as radiational cooling causes subsidence of the air. In the process the air dries and warms as the potential energy is converted to sensible heat. It also checks the rise of convection currents in these subtropical areas, producing clear, cloudless skies.

In temperate and polar areas the processes of energy transfer are more complex. There is no general, cellular circulation of air as in the tropics, but instead a complicated pattern in which individual rotating storms play an important part. Within these storms warm air masses rise, releasing latent heat and gaining potential energy. They then become intermixed with descending cold air and gain sensible heat (Figure 5.15). The rotating storms are moving, so the position of this intermixing changes constantly, although there is a tendency for a concentration in certain zones in the northern hemisphere. Labrador, Newfoundland, and Greenland are associated with these areas of activity, experiencing cool, southward-moving flows of air. Britain and Scandinavia, in contrast, tend to be influenced far more by warm northerly-moving air, a phenomenon that greatly improves their climate.

All these transfers of energy through the atmosphere are highly variable, and major differences in the intensity and character of transfers occur over time. Thus, the flows of energy represent net increments, often produced by individual, temporary processes. For this reason it is difficult to detect the nature of energy transfer directly from the pattern of general circulation.

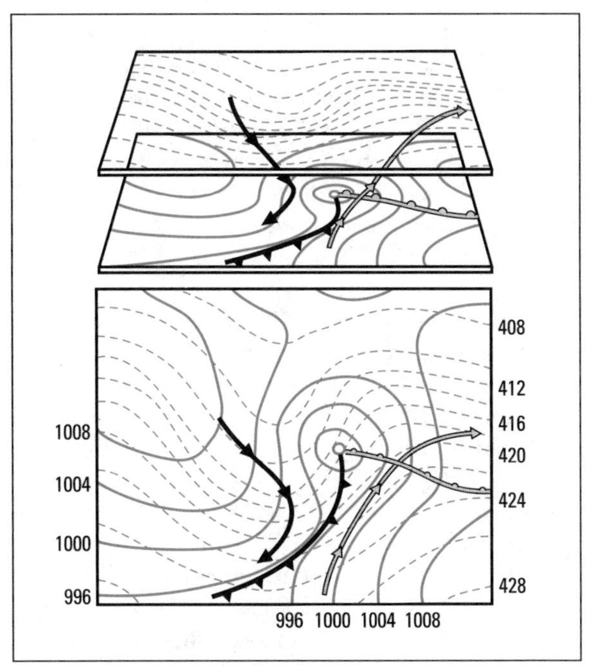


FIGURE 5.15 ▼ Energy exchanges in a northern hemisphere midlatitude depression. The lower diagram shows 1000 mb and 600 mb contours; upper diagram shows a perspective view of the same with three-dimensional trajectories of air parcels originating in the central part of the cold air and in lower levels in the warm sector. The three portions of each trajectory are for approximately equal time intervals (after Palmen and Newton, 1969)

# Wind patterns

The general circulation of the atmosphere reflects the operation of the atmospheric system as a whole. It is clear, however, that the system is composed of many important subsystems and it is these—the main wind belts of the globe—that provide much of the climatic variation and consistency in the world. We have already indicated that the westerly winds dominate the climate of the temperate latitudes; similarly, the equatorward movement of air in the regular easterly trade winds have a prevailing influence on tropical climate.

## SURFACE WINDS

Three main surface wind belts can be distinguished in each hemisphere. Around the equator, in the low pressure equatorial trough, there occurs a zone of convergence where the northeasterly trade winds blowing from the Tropic of Cancer meet the southeasterly trade winds blowing from the Tropic of Capricorn. Either side of the equatorial trough these winds dominate, giving the trade wind belt.

Poleward of the tropics, in the temperate latitudes, we find a zone of prevailing westerlies, while around the poles occur a belt of easterlies (Figure 5.16). We will examine each of these zones separately, but as we do so it is important to remember that, in reality, these wind belts do not operate in isolation. They are closely interrelated.

*The equatorial trough* The equatorial trough or **intertropical convergence zone** (ITCZ) is a shallow trough of low pressure generally situated near the equator (Figure 5.17). Over the oceans it is fairly static, because seasonal temperature changes are small. In the Pacific, for example, its average position varies by no more than 5° of latitude within the course of a single year. The situation is very different over the continents, however. During summer in continental areas the trough sweeps poleward, reaching 30° or even 40° latitude over eastern China. Behind the trough the winds are predominantly westerly and are the main rain-bearing winds to these areas (Figures 5.16, 5.18). Where they reach into higher latitudes, they are called **monsoons** (an Arabic word meaning season) and they show an almost complete reversal of direction from summer to winter, a change that tends to occur with uncanny regularity about the same dates each year.

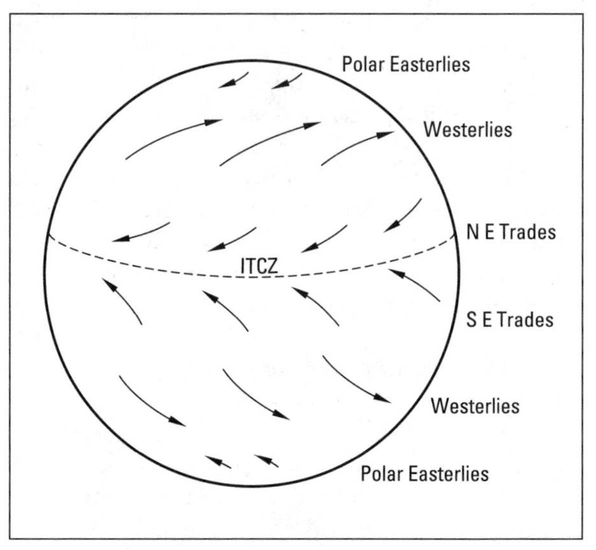


FIGURE 5.16 ▼ The main surface wind belts. The equatorial trough is the convergence zone between the trade wind belts; it is relatively narrow

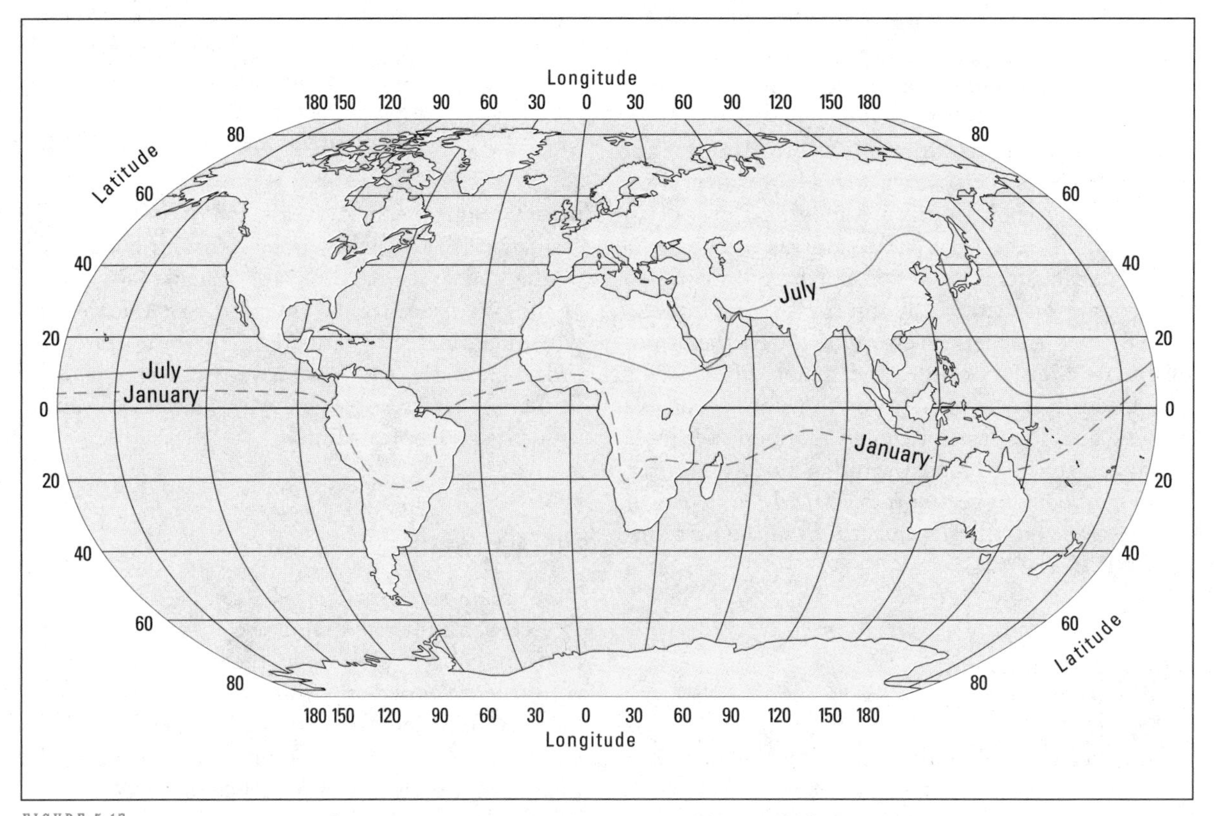


FIGURE 5.17 ▼ Mean positions of the equatorial trough—the ITCZ—at the times of the extreme seasons

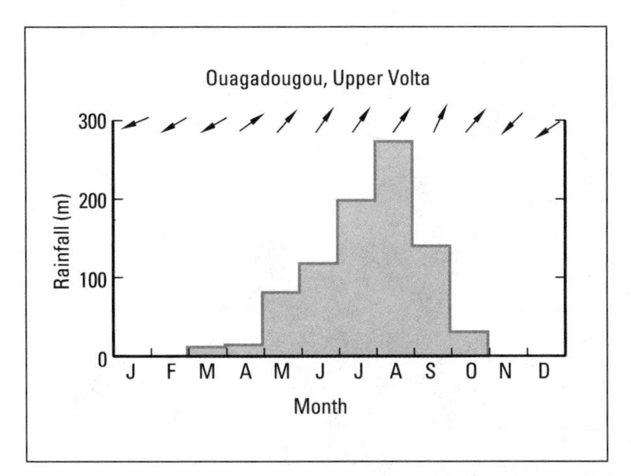


FIGURE 5.18 ▼ Mean monthly precipitation and wind direction at Ouagadougou, Upper Volta. Arrows indicate mean wind direction. Arrow pointing upward represents a southerly wind

With the exception of the monsoon, the winds in the equatorial trough tend to be light and variable and, because sailors often found themselves becalmed there, this area became known as the Doldrums (Figure 5.19).

*The trade winds* The **trade wind** belts lie between the equatorial trough and the subtropical highs at 20° to 30° of latitude. This zone occupies nearly half the globe, much of it ocean, and within this area the steady easterly trades provide a stable and constant climate (Figure 5.20). At the surface the winds have a component toward the equator, being from the northeast in the northern hemisphere and the southeast in the southern. The wind blows out of the northeast 96 percent of the year at Mt. Waialeale in Hawaii (Figure 5.21). Above the surface friction layer, the winds become more easterly.

Viewed from the air, the oceanic trade winds contain innumerable uniform small clouds, all with a similar base and depth (see Figure 6.17). These are the visual expression of the transfer of latent heat from the sea surface, through evaporation, before condensing at cooler levels in the atmosphere. These are the visual expression of the transfer of latent heat from the sea surface, through evaporation, and then condensation at cooler levels in the atmosphere.

As we have noted, the seasonal movement of the equatorial trough is slight over the oceans, so the oceanic tropical areas are dominated by the trades.

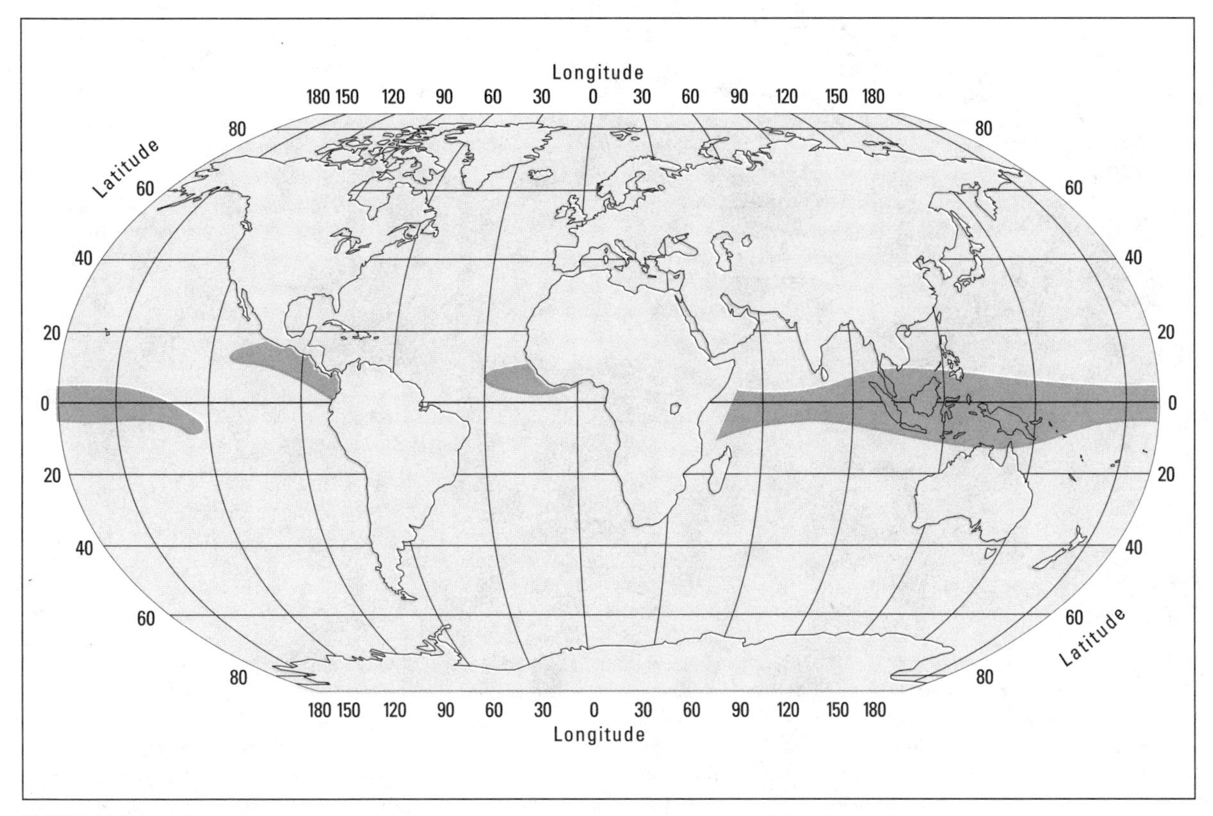


FIGURE 5.19 ▼ The principal areas of the equatorial trough zone or Doldrums (after Crowe, 1951)

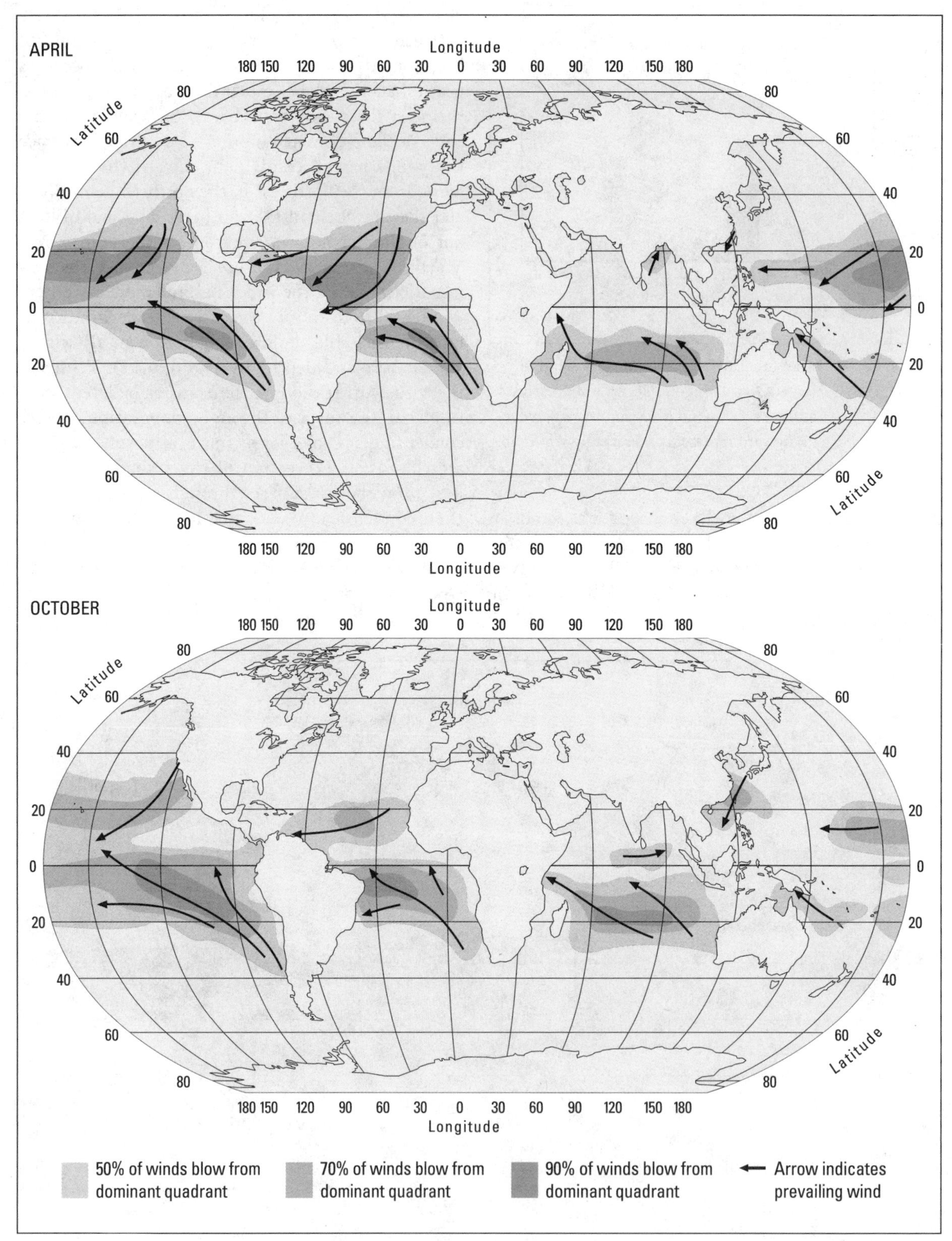


FIGURE 5.20 ▼ The trade wind systems of the world in April and October. The isopleths (areas of winds of equal strength) are in terms of relative constancy of wind direction and enclose shaded areas where 50, 70, and 90 percent of all winds blow from the predominant quadrant with speeds above 3.3 m $s^{-1}$ (after Crowe, 1971)

FIGURE 5.21 ▼ In the Pacific Ocean, the trade winds propel clouds from east to west. When the air current is intercepted by a high land mass, such as the Big Island of Hawaii, the stable cloud pattern is divided to bypass the island in a wide arc forming an 'island wake.' This photograph, taken from a space shuttle mission in June, shows how the prevailing wind direction dictates that the north and northeast of the islands are wetter than the western side and frequently under cloud (photo courtesy of NASA)

This was very critical in the days of sailing vessels plying trade between Asia and Europe, giving rise to the name. On the continents the trades are far more restricted in extent, and the equatorial westerlies and monsoons are more important. The two belts interact closely; it is the convergence of moisture in the trade winds that feeds the equatorial trough. The shift in the position of the trough thus determines the relative extent of the easterlies and westerlies. When the trough is farther north, with the overhead sun in July, the trades are restricted in the northern hemisphere, particularly over the land. In January the trough is at its most southerly position and the trades extend to the equator. In the southern hemisphere less marked variations occur, for the predominance of ocean means that the southern limit of the trough remains close to the equator (Figure 5.17).

*The westerlies* In comparison to the winds of the tropics, the westerlies of the midlatitudes seem unreliable and fickle. They are westerlies only on average.

Poleward of the subtropical anticyclones, rotating storms are the main mechanism of energy transfer. Unlike hurricanes, these systems cover vast areas and can clearly be seen from space, identified by their characteristic spiral of clouds (Figure 5.11). They tend to move northeastward, although directions vary from north to southeast. Typically, they follow an evolutionary pattern that we will be examining more closely in the next chapter. The storms are initiated in areas of strong temperature gradients, such as off Newfoundland, where the cold Labrador Current and warm Gulf Stream are in close proximity, forming almost circular air masses with low pressure at the centre and a rotational movement of winds around the periphery (Figure 5.22). They are known as lows, cyclones, or depressions. As they evolve they become initially more intense, the central pressure falling to as little as 960 mb. Air moves into the low, gradually raising the pressure. On average the location of maximum intensity is in the areas of Iceland in the Atlantic and the Aleutian Islands in the Pacific. Thus meteo-

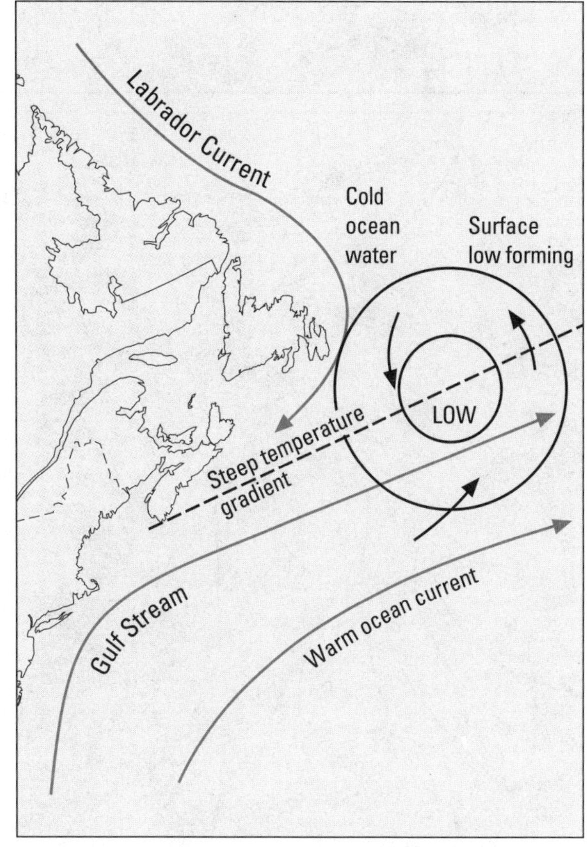


FIGURE 5.22 ▼ Generation of low pressure systems along the steep temperature gradient at the juxtaposition of the Gulf Stream and Labrador Current

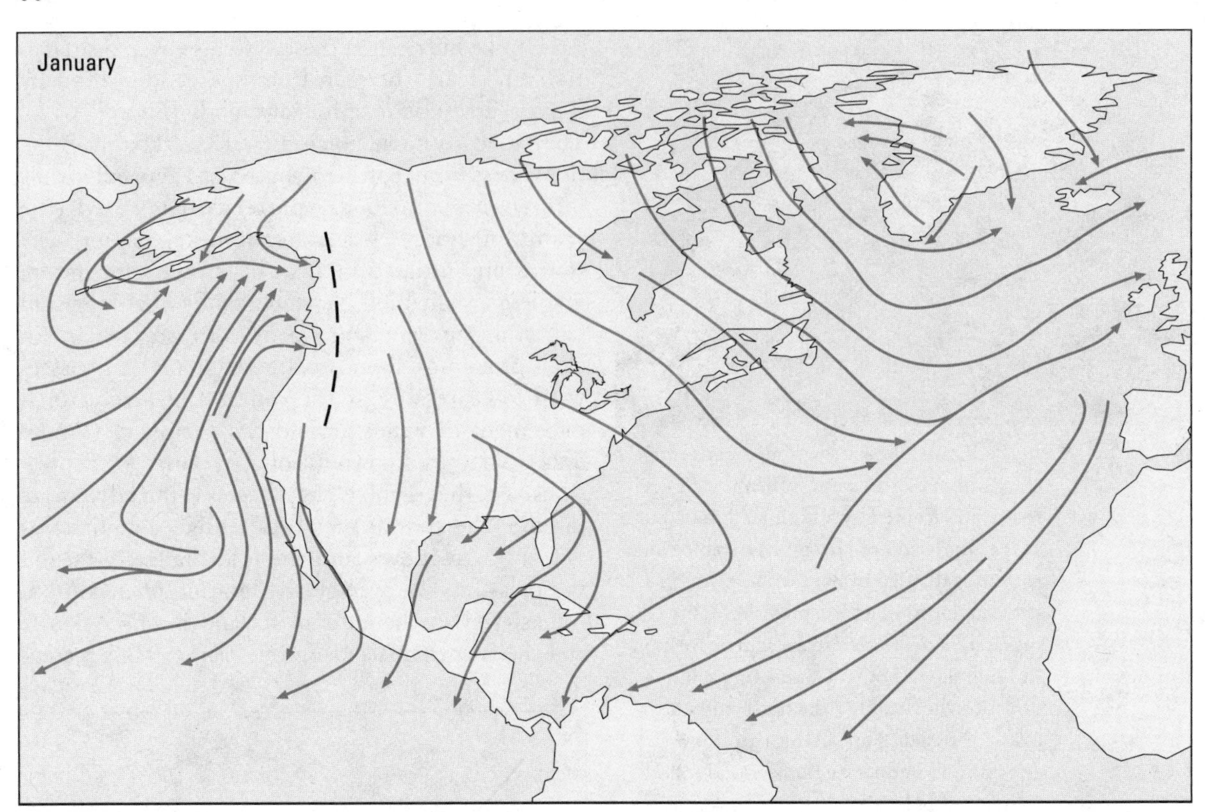


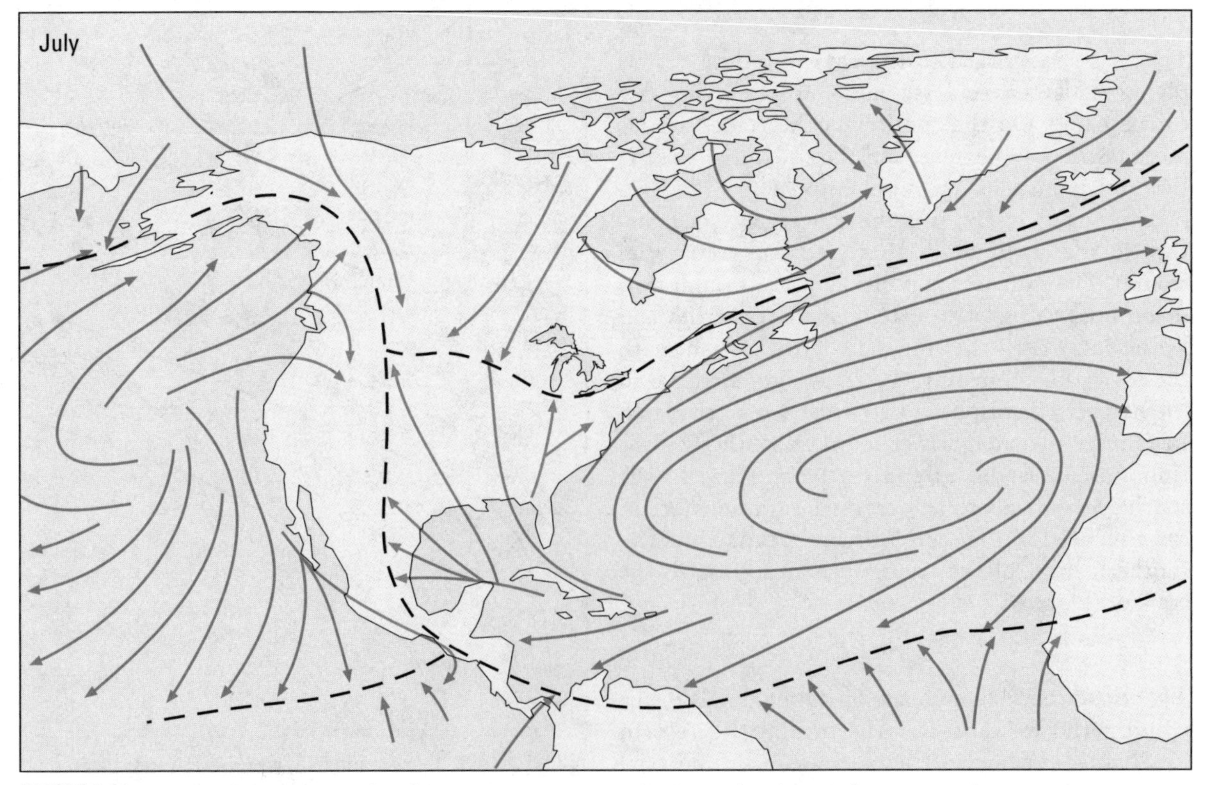


FIGURE 5.23 ▼ Mean flow patterns of surface winds in January and July. Broken lines indicate general contact zones between the airflows

rologists speak of the Iceland Low and the Aleutian Low. In the southern hemisphere there are no distinct areas for the genesis of storms, so lows form throughout a wide belt.

As a low approaches, winds increase in strength, initially from a southerly direction, then become westerly, and finally, as the low moves away, they veer to northwesterly or even northerly. The tracks of the lows reach farther poleward in summer than in winter, so the area affected by the storms varies seasonally. Nowhere is this seasonal pattern more clearly seen than over the Mediterranean basin and in California. In winter, when the cyclones follow a more southerly path, they bring rain to more southerly latitudes. In summer, the cyclones move away to be replaced by the subtropical anticyclones and dry, hot weather.

The regular march of cyclones and anticyclones through the temperate latitudes produces a majority of winds between northwest and south in the northern hemisphere, and southwest and north in the southern hemisphere. This pattern is far from invariable, however, and depending upon the precise tracks taken by the lows, winds from any direction are possible (Figure 5.23). The prevailing westerlies, therefore, are anything but prevailing. Moreover, the strong north–south component of winds in these areas allows a more active transfer of energy between the tropics and polar regions.

*Polar easterlies* Around the poles, beyond the main westerly belt, there is some evidence of prevailing easterlies. The winds are variable and linked to the shallow polar anticyclones. In the northern hemisphere, they are often influenced by the circulation around the northern edge of cyclones. As a result, they change direction according to the local weather and topography.

In the southern hemisphere the vast Antarctic icecap controls the atmospheric circulation around the pole. Anticyclones develop frequently over eastern Antarctica, and strong southeasterly winds develop around the margins of the ice plateau with consistencies similar to those of the trades, and occasionally of great strength. The hostility of the climate in these regions can only be imagined. At Byrd (85°S), for example, the mean wind speed in August (midwinter) is 10.2 m $s^{-1}$ with a constancy of 85 percent. The daily mean temperature in this month is –37.0°C.

# Upper winds

## THE NATURE OF THE UPPER WINDS

When we look up at high clouds on a clear day it is not unusual to find that their direction of movement is different from that of the surface winds. As this implies, winds in the upper atmosphere can be affected by forces operating in a different direction to those at the surface, and may appear to be part of a different system of circulation to the surface winds. If we were to make an ascent by balloon into these upper wind systems we would find that the change from surface to upper atmosphere conditions was not abrupt but transitional. With increasing height, we would discover that the winds tend to follow a gradually more distinct zonal (west-to-east) direction and that they become stronger. The main reason for this change is the disappearance of the frictional influence of the ground surface upon the winds. In other words, the flow more nearly approximates to the geostrophic winds that, you will remember, result from the interaction of the pressure gradient and Coriolis forces (Figure 5.24).

The zonal flow of the upper winds can be shown on average as a cross section from north to south (Figure 5.25). In fact, variations around this average picture are slight, except in the monsoon areas of Asia. At each season the same basic pattern exists. Between about 30°N and 30°S we have a zone of high-level easterly winds that are relatively weak, reaching a maximum speed of 4 to 5 m $s^{-1}$ (about 17 km $hr^{-1}$) at about 3 km above the surface. On either side of this belt occurs a ring or vortex of much stronger westerly winds.

*The upper westerlies* These high-altitude westerly winds are a major feature of our atmosphere. They reach their maximum speed at approximately 12 km above the surface between 30° and 40° latitude. The mean speed is as much as 34 m $s^{-1}$ (125 km $hr^{-1}$) and maximum speeds of several hundred kilometres per hour are not uncommon. It is not surprising that aircraft can travel from North America to Europe more quickly than on the return journey.

Although these wind patterns are generally constant, seasonal variations do take place, especially in the northern hemisphere. The upper westerlies are strongest in the winter when the temperature differ-

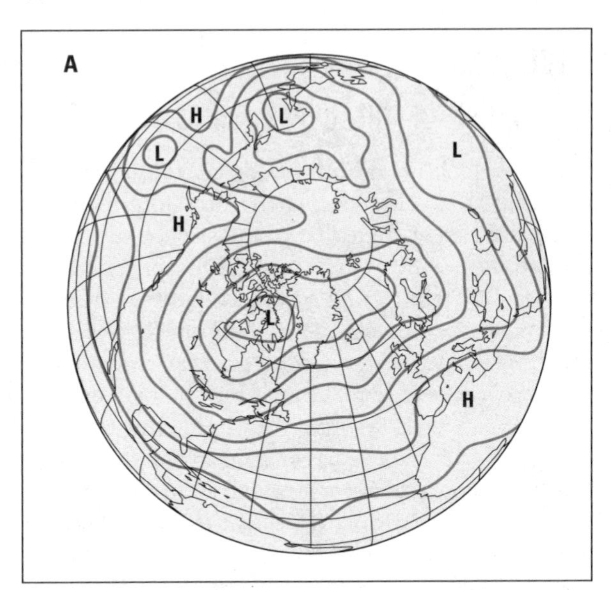


FIGURE 5.24A ▼ The upper westerlies at 500 mb on 9 February, 1981. The map shows the height of the 500 mb pressure level above a fixed datum near sea level. Winds blow parallel to the contours at a speed proportional to the gradient. Although flow is dominantly westerly, well-marked troughs (eastern North America) and ridges (Alaska) can be seen. Isolines are normally drawn every 8 mb; here they are every 24 mb to indicate the pattern more clearly

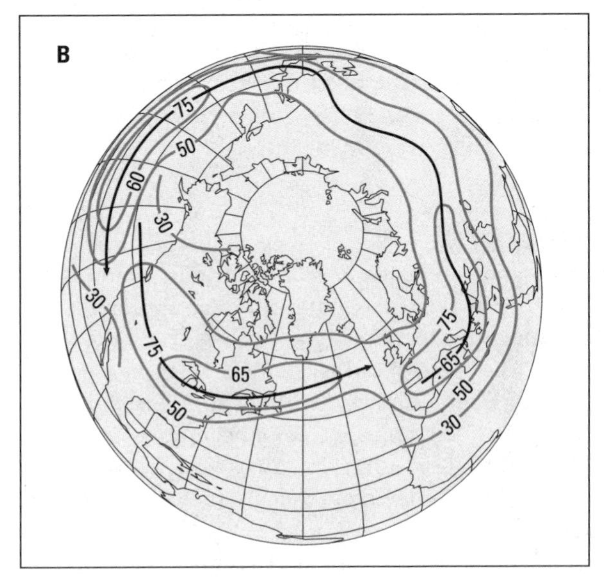


FIGURE 5.24B ▼ A typical summer northern hemisphere circumpolar vortex is shown separate from the normal isolines. Isotachs (wind speed) are shown in kilometres per hour

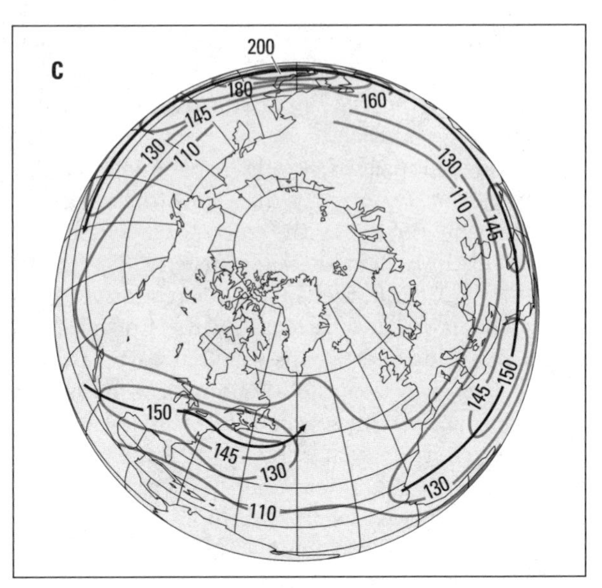


FIGURE 5.24C ▼ A typical winter northern hemisphere circumpolar vortex. Note the more southerly position and the higher windspeeds

ences between the tropics and temperate latitudes are at their greatest. From June to August temperatures in the northern hemisphere are relatively warm, even in polar regions, so the pressure gradient is reduced and the upper westerlies decline to speeds as low as 15 m $s^{-1}$ (55 km $hr^{-1}$).

As ever, changes in the southern hemisphere are less pronounced, largely due to the greater thermal stability there. The vast areas of ocean absorb large quantities of heat without any significant increase in temperature. The ice plateau of Antarctica also stays very cool, so the temperature gradients do not change very much from winter to summer.

The position of the boundary between the westerlies and easterlies (of both the upper and surface winds) varies throughout the year. In December to February, the polar vortex of the winter (northern) hemisphere expands, pushing the belts southward so that, at the surface, the boundaries are at about 30°N and 35°S. As the year progresses, the other polar vortex begins to expand as winter sets in over the southern hemisphere. The boundaries eventually reach about 35°N and 30°S by June to August. The separation between the two systems is not vertical. As a result, we have some parts of the tropics with easterlies in the lower atmosphere and westerlies above. Only over a small area of the globe do easterlies occur at all levels, whereas westerlies extend throughout the atmosphere over a large proportion of the earth.

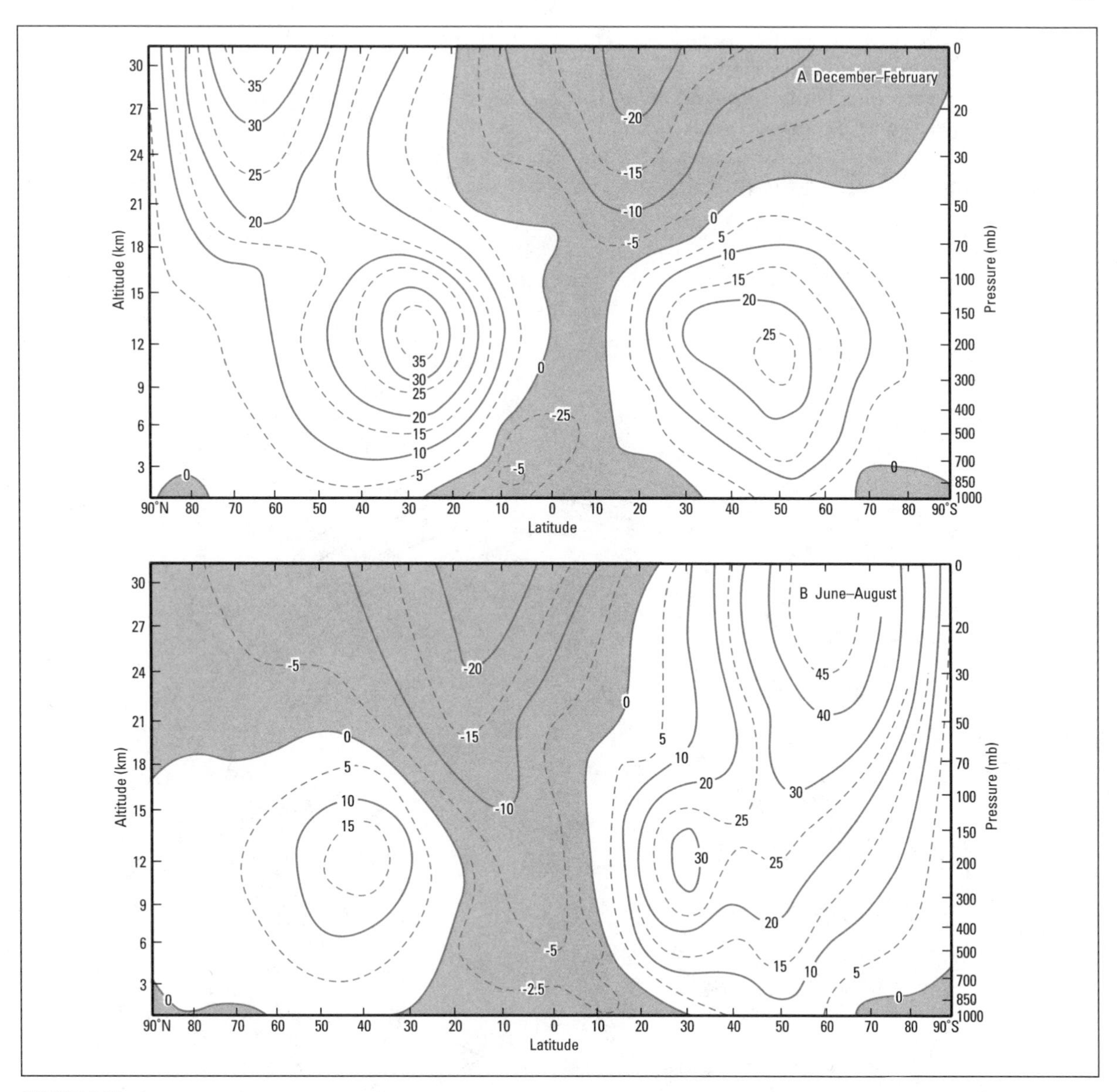


FIGURE 5.25 ▼ Mean zonal wind for (A) December–February and (B) June–August. Units are m $s^{-1}$. Shaded areas (negative values) denote easterly winds (after Newell *et al*, 1969)

*Rossby waves and jet streams* The pattern of easterlies and westerlies in the upper atmosphere is only part of the total picture. In addition to the marked zonal flows, there are less apparent but nonetheless important meridional flows. In the circumpolar areas, for example, there occur wave-like patterns of flow called **Rossby waves** (after C.G. Rossby, the Swedish meteorologist) that play a vital role in the energy exchange between the temperate and polar areas.

It is not easy to detect these meridional flows within the pattern of strong zonal circulation by normal methods of depicting winds. These normally show average conditions, so that processes that balance each other, flowing northward for six months, perhaps, then southward for the next six months, are lost. Yet this is exactly what happens in the case of the Rossby waves. At a particular location southerly flows may last for a few days to be followed by more northerly winds as the wave progresses eastward.

In order to see these waves it is necessary to use a rather different technique of presenting atmospheric circulation. Instead of mapping the actual wind directions or speeds, we can plot the height at which a particular pressure surface is reached. This may seem a strange way of depicting winds, but as we know, the geostrophic winds blow parallel to the isobars, at a speed inversely proportional to the distance between the isobars. Similarly the winds blow parallel to the contours of the pressure surface. Where the contours are close together, the winds are rapid. Irregularities in the pressure surface indicate local patterns of wind movement.

Figure 5.26 shows a pressure surface (500 mb) map for January. The projection of the map may make it difficult to appreciate the direction of flow

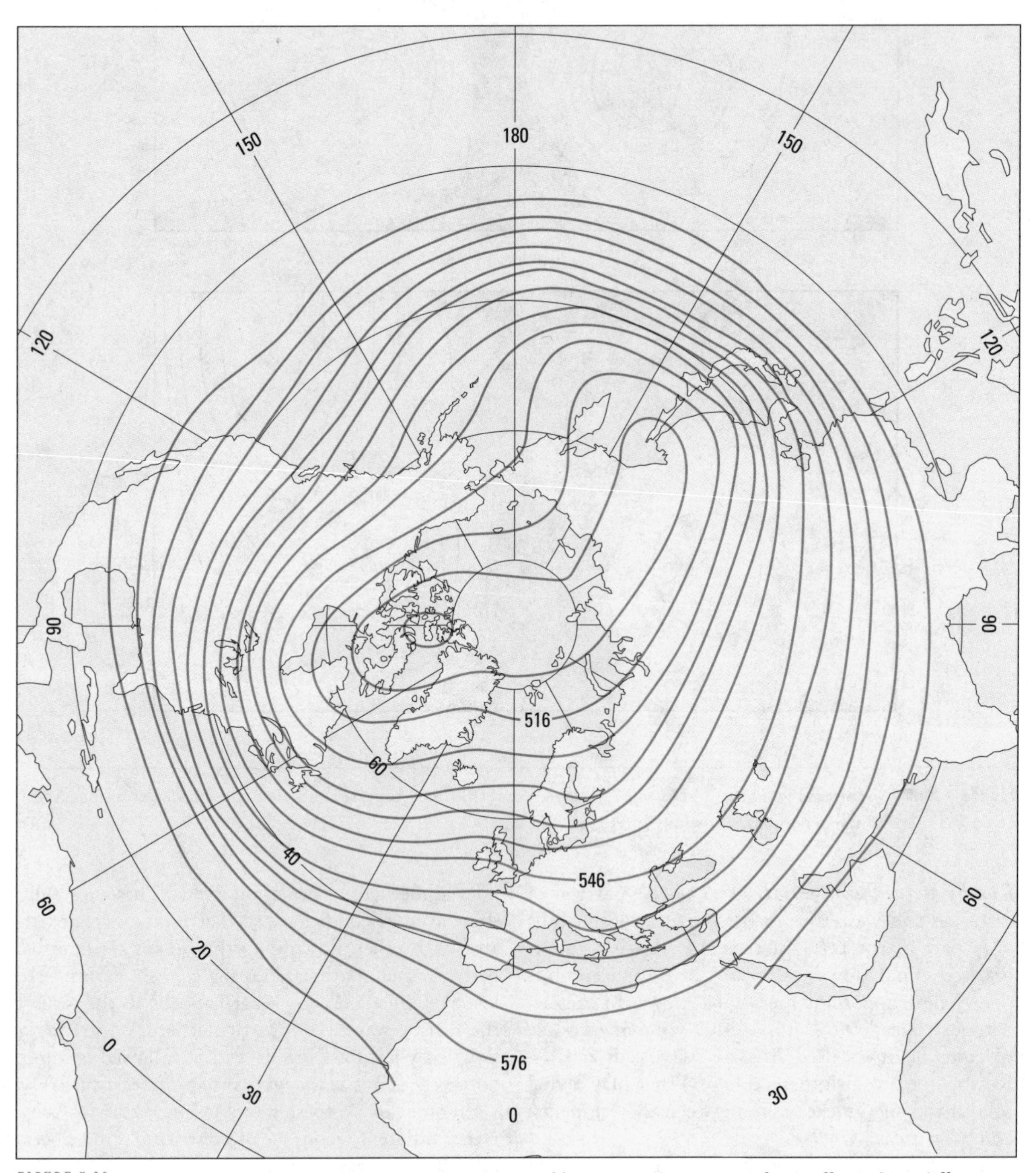


FIGURE 5.26 ▼ Monthly mean 500 mb contours for January 1951–66. Units are decametres (after Moffitt and Ratcliffe, 1972)

immediately. What is clear is that the flow is not perfectly circular around the North Pole. Areas occur, even on this monthly chart, where the mean flow is northward, and in other areas it is southward. Effectively the air is flowing in a series of waves around the pole, carrying warmer air northward on parts of its track and cold air southward elsewhere. These are the Rossby waves. In January the most prominent features of these waves are the pronounced troughs in the pressure surface near 80°W and 140°E, with a weaker trough between 10°E and 60°E. In July (Figure 5.27), the circula-

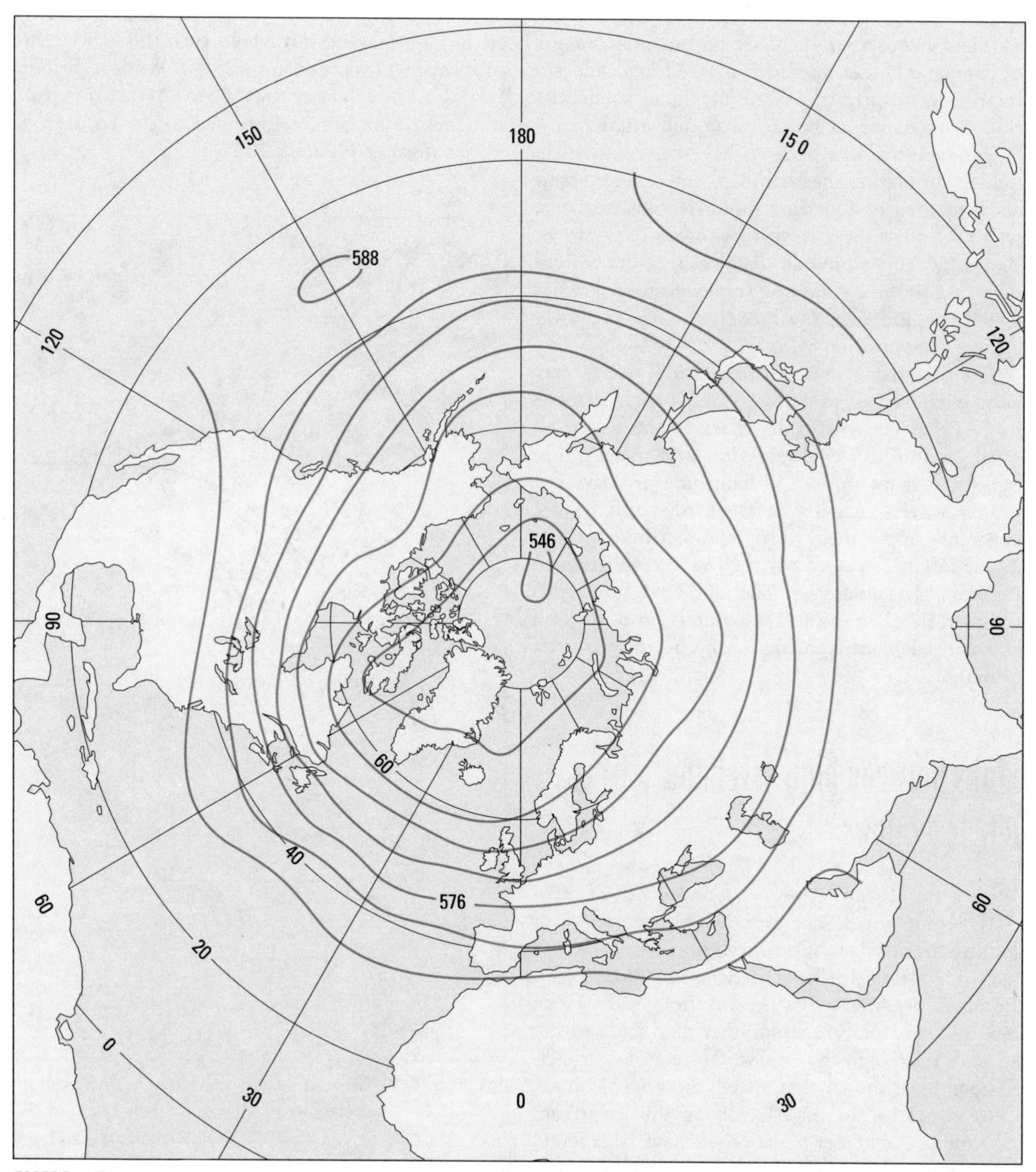


FIGURE 5.27 ▼ Monthly mean 500 mb contours for July 1951–66. Units are decametres (after Moffitt and Ratcliffe, 1972)

tion is less intense and the troughs less well-marked.

Many experiments have been conducted to determine the reasons for this pattern. Clearly surface features play an important part, even at this height. The presence of the Rocky Mountains and the Himalayas is believed to 'lock' the troughs at 80°W and 140°E, respectively. The distribution of land and sea is also thought to be of importance. In the southern hemisphere there are no mountain ranges of comparable size, nor such marked land and sea temperature contrasts. As a result, the mean circulation is much more symmetrical around Antarctica.

Similar wave-like patterns have been formed in models simulating the earth. If a temperature gradient is applied in a rotating fluid, the heat exchange will take place in a wave-like form. This can be shown by using aluminum flakes as markers of fluid flow (Figure 5.12). Because surface features are not usually included in the model, the waves slowly change their position.

Even during a shorter time period, waves may exist in the upper westerlies, though their shape is less regular. The smaller waves tend to be associated with an individual depression and move more rapidly, perhaps up to 15° longitude per day. The longer waves—usually between four and six are apparent—move more slowly and are linked to the major circulation features such as the subtropical highs and Icelandic lows. The long wave flow tends to 'steer' the shorter waves, moving them northward when ahead of a trough and southward to the rear of a trough.

## Links between high-level and low-level flows

Within the Rossby waves are found bands of especially strong winds. The existence of these winds or **jet streams** was not appreciated until the increased use of aircraft during the Second World War. Bombers heading across the Pacific toward Japan reported headwinds so strong that they could hardly advance relative to the ground. More recent investigations have shown that speeds up to 135 m $s^{-1}$ (490 km $hr^{-1}$) can exist locally in the jet stream maximum. A number of major jets have been found in the troposphere—the polar-front jet, the subtropical jet, and the tropical easterly jet—but others exist in the stratosphere.

What is a jet? Basically it is a very narrow current of air travelling at great speeds through the lower westerly flow. Jets are formed in regions of rapid temperature gradient and they can lead to intense accelerations (and decelerations) of air in their vicinity. As we shall see later, when air is forced to change its rate of flow, tropospheric vertical motion may be started. In turn this may influence weather events at lower levels. Work in Britain and the USA has shown how surface rainfall amounts could be closely linked to the position of the jet streams (Figure 5.28).

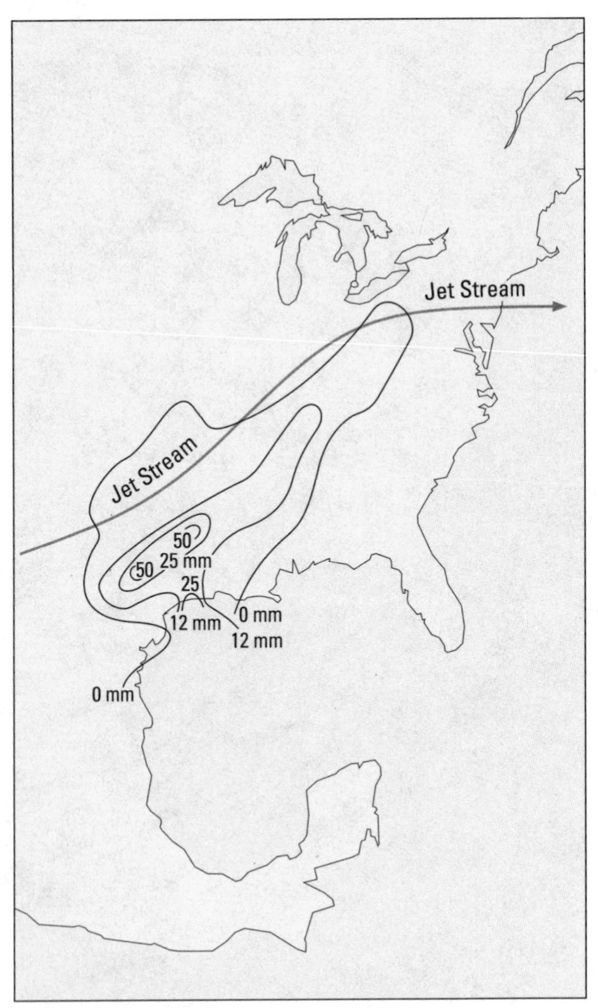


FIGURE 5.28 ▼ The relationship between jet stream maxima and surface precipitation (shown in mm), 20 September 1958 (after Richter and Dahl, 1958)

## The zonal index

The atmosphere has, in simple terms, two distinct air masses; these consist of a dome of cold air over each polar region, with the remainder being warmer tropical air (Figure 5.29). Where these two meet, generally referred to as the Arctic Front (or Polar Front) in the northern hemisphere, the temperature gradient is greatest with the result that the strongest winds occur. The atmosphere behaves according to the rules of fluid dynamics, therefore the river of air develops sinuosity or meandering, which we have discussed as the Rossby waves. The number of waves and their amplitude varies from season to season, year to year and apparently from decade to decade.

The polar vortex expands and contracts on a seasonal basis as the sun migrates, but it also expands and contracts over much longer periods, apparently in conjunction with variations in solar activity as manifested by the sunspots. Generally, strong circulation corresponds to a tight vortex and relatively consistent patterns of weather in the vast expanse of the middle latitudes. The flow is predominantly from west to east and is the result of a high temperature and therefore high pressure gradient in the middle latitudes. (Measurements usually compare the pressure difference between 35° and 55° of latitude.) In this case there is said to be a high **zonal index**.

Weak circulation of the vortex corresponds to a more erratic pattern of the jet stream and the Arctic Front, with the result that the large amplitude waves extend much farther north and south. In this case the pressure gradient is weak latitudinally, and the flow is more meridional, that is, north and south. Meteorologically it is called a low zonal index. This pattern results in extremely variable weather and the possible development of sluggish, or blocking, high pressure regions that can result in droughts in western North America. It was this type of pattern that existed over the Canadian Prairies in the summers of 1961, 1967, and 1988, creating clear skies, high temperatures, and dry conditions.

The **index cycle** is the name given to the average sequence of transitions from a high zonal index to a low zonal index over a period of about four to six weeks, as shown in Figure 5.30. Note how the pattern can create situations of temperature extreme between the west and east side of North America.

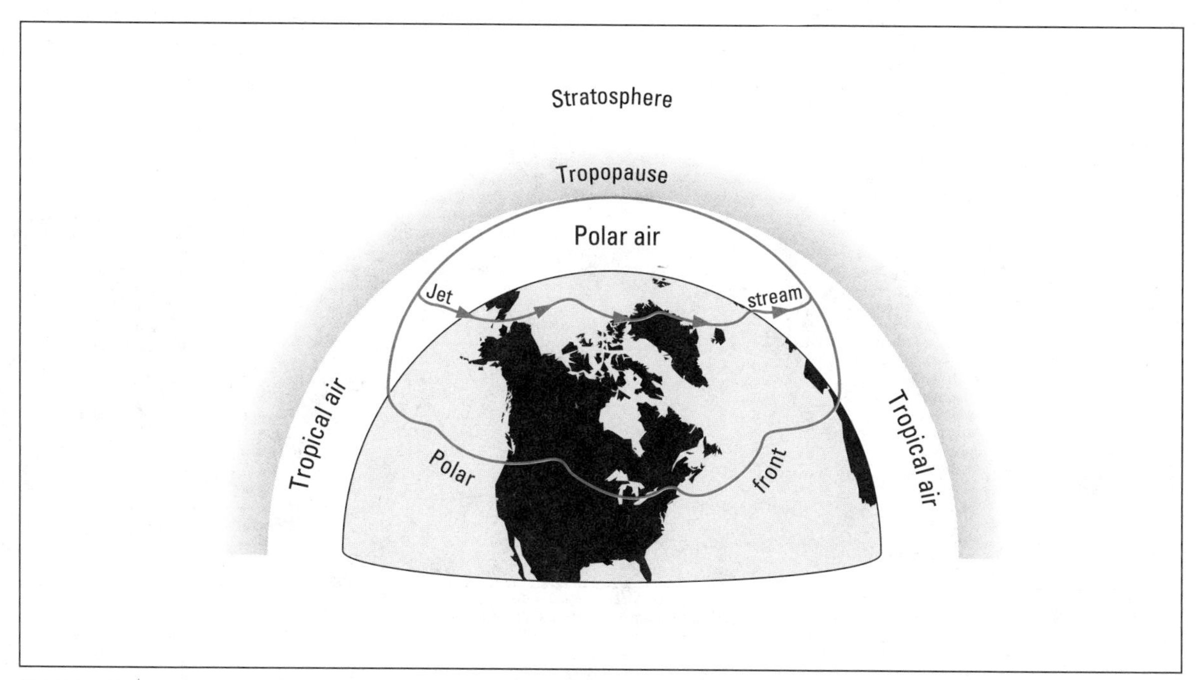


FIGURE 5.29 ▼ Two major air mass systems. Note the dome of Arctic air and the overrunning tropical air along the Polar Front, and the position of the jet stream. The correlation between the waves in the jet stream and the Polar Front is clearly visible

The winter of 1986–87 had warm air extending north in the western part of the continent, giving one of the warmest winters on record, while cold air dominated the east, giving a very cold winter.

## Conclusion

Atmospheric movements, together with oceanic circulation, provide the main processes by which energy is transferred through the global system. They act to maintain a steady state in the system by transporting excess energy from areas that receive high inputs of solar radiation to areas where inputs are small. These movements involve two general patterns of flow; the dominantly zonal flow of air within the main wind belts and the less apparent but equally important meridional transfers. Both circulations are controlled by the pressure gradient force, which acts as the driving force for atmospheric motion. The earth's rotation—the so-called Coriolis force—and friction modify the simple pattern of circulation initiated by the pressure gradient force to give the complex systems we find in the earth's atmosphere.

These atmospheric movements are vital for a number of reasons. Many of the features of the world's climates are dependent upon the character of atmospheric circulation, as we shall see in Chapter 7. Seasonal and daily variations in the circulation affect our life directly, and extreme events may have dramatic implications for humans, topics that will be covered in Chapter 6.

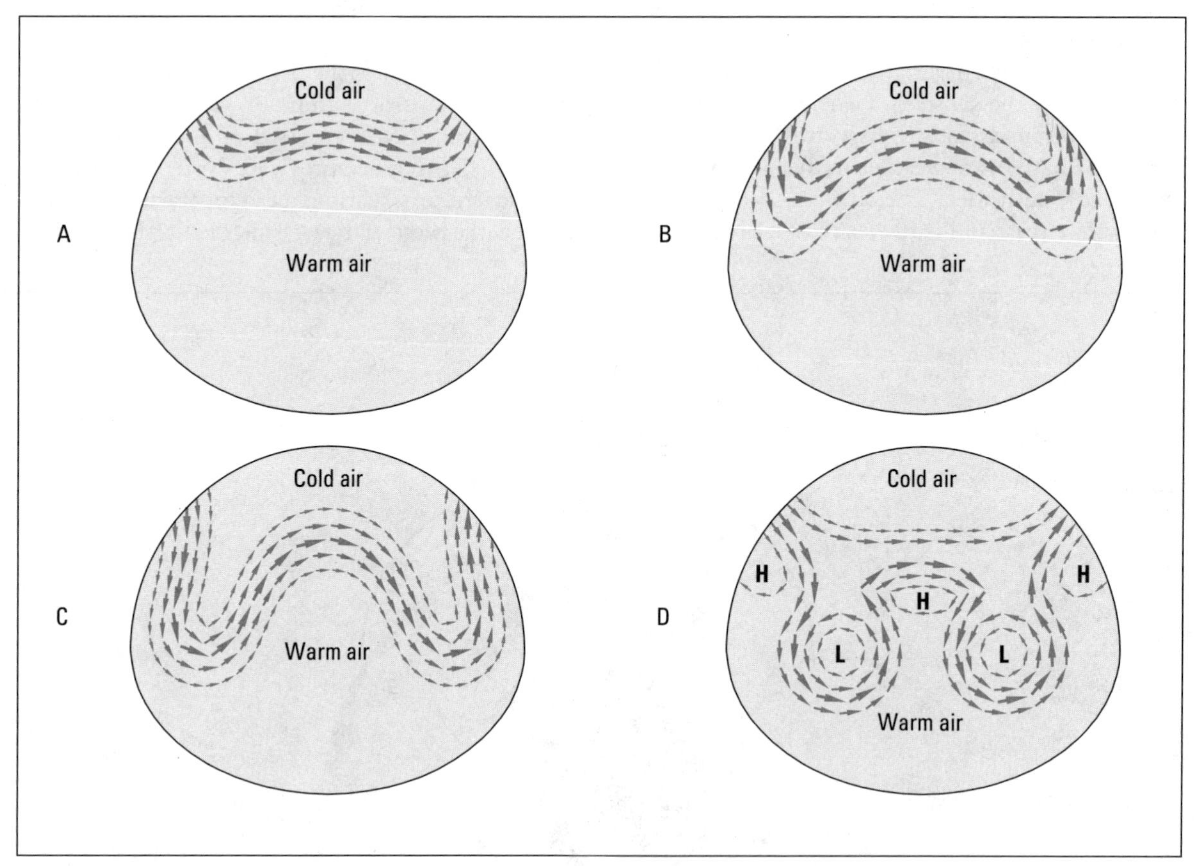


FIGURE 5.30 ▼ The sequence of the index cycle. Note the gradual deepening of the waves and the return to a west-to-east pattern after a normal cycle

# Weather-forming systems

## Weather and weather-forming systems

Not a year goes by without weather events somewhere in the world causing damage or loss of life. Floods, blizzards, tornadoes, hurricanes, and even heat waves can create problems and generate much economic stress over the affected areas. To be prepared for such events, it is vital for us to understand the weather, to be able to predict with accuracy, and preferably well in advance, events such as these. It is important too that we understand the vagaries not only of day-to-day weather conditions, but also of longer-term trends. How useful it would be for the farmer to know what the weather over the next few weeks or even the whole growing season will be; then sowing or ploughing or harvesting could be planned far more successfully. How useful it would be to have a clear idea of the weather in the year ahead so that wheat harvests could be predicted, plans for winter snow could be made and contingency plans could be made to deal with droughts. Any such detailed understanding is a long way away. It will come only as we gather far more knowledge about the medium-term processes operating within the atmosphere, and about the myriad factors that influence these processes.

FIGURE 6.1 ▼ Cloud formations over Africa and the Atlantic and Indian oceans; note the clear skies of the Sahara and southern Africa. View from an *Apollo* mission, December, 1972 (photo courtesy of NASA)

The key to understanding and predicting the weather lies in understanding what we call weather-forming systems. If we look at a satellite photograph showing half the globe it is clear that the distribution of clouds is not random (Figure 6.1). In some areas clouds are abundant, sometimes showing certain patterns that make it possible to identify their means of formation. Many areas are devoid of cloud altogether and surface features can be seen. By comparing this photograph with a map of surface pressure we would see that the large spirals of cloud are associated with cyclones in the middle latitudes, and the main cloud-free areas with the large anticyclones of the subtropics. Between these areas, the cloud patterns are less clear, though over the South Atlantic Ocean the trade winds have produced some interesting forms, and over the cold Benguela current off southwest Africa there are extensive layers of low cloud. By viewing this instantaneous picture, we can see the way different areas of the atmosphere interact, and by using the surface pressure information we can relate these cloud patterns to the weather systems that produce them.

## Air masses

An air mass is a large, uniform body of air with no major horizontal gradients of temperature, wind, or humidity. In the anticyclonic areas of the world, where air movement is slight, the air is in contact

### Table 6.1

*Average properties of air masses*

| *Air mass* | *Symbol* | *Properties* | *Temperature (°C)* | *Specific humidity ($g\ kg^{-1}$)* |
|---|---|---|---|---|
| Continental arctic | cA | Very cold, dry | –40 | 0.1 |
| Continental polar | cP | Cold, dry (winter) | –10 | 1.4 |
| Maritime polar | mP | Cool, moist | 4 | 4.4 |
| Continental tropical | cT | Warm, dry | 24 | 11 |
| Maritime tropical | mT | Warm, moist | 24 | 17 |
| Maritime equatorial | mE | Warm, very moist | 27 | 19 |

After A.N. Strahler and A.H. Strahler, *Elements of Physical Geography* (1979), Wiley.

with the ground surface and gradually acquires the thermal and moisture properties of the ground. We find that the air then has relatively uniform distributions of temperatures and humidity over large areas—for example, the Canadian Arctic in winter. Whether or not the air will fully reach equilibrium with the surface characteristics will depend upon how long it remains in the source region.

The character of an air mass is dependent upon conditions in the area in which it forms. Because of this it is possible to classify air masses on the basis of their source area. Four main types are recognized: Arctic (or Antarctic), Polar, Tropical, and Equatorial, and these are further subdivided into continental (for those forming over large land masses) and maritime (for those forming over the oceans) (Table 6.1, Figure 6.2).

Figure 6.3 shows the source region and general pattern of movement of air masses for North America. Examine the map in conjunction with the seasonal positions of the **Arctic Front**. The cold dense arctic air dictates the pattern. It moves into an area and pushes the warm air out of the way, or it retreats allowing warm air to move in behind.

The frequency with which different air masses invade a region is determined by the season and the latitude. Northern Canada is usually dominated by arctic or polar air masses while southern Canada only experiences these in winter. Note that some regions of southern Canada are affected by as many as five of the types listed in Table 6.1.

The Arctic Front was illustrated in Figure 5.28. It marks the boundary between the two major global air types, tropical and Arctic. Arctic and Antarctic air sit over the polar regions, forming domes of cold, dry air. The junction where the two air types meet is referred to as the Arctic Front. It marks a major boundary in the atmosphere, and coincides with the point of zero energy balance and the snow line, as discussed in Chapter 3. With migration of the sun the domes of cold air expand in the winter and shrink in the summer.

The mean summer position of the Front is about 65° and the mean winter position about 38° of latitude. These seasonal positions are generally illustrated by the cP and midlatitude air mass positions in Figure 6.2. Notice that the mean positions are not parallel with the lines of latitude in the northern hemisphere. They dip toward the equator over the land in January and toward the pole in July. The pattern is created by the different heating and cooling capacities of land and water. This is confirmed by the uniform pattern in the southern hemisphere.

Temperature contrast across the Front can be very dramatic. For example, in January, 1943, the temperature rose by a full 27C° in 2 minutes as the Front moved through Spearfish, South Dakota. Changes in the air occur as it moves over new surfaces, but it is the formation of cyclones that are necessary for mixing the surplus energy from the tropics with the deficit energy from the polar latitudes.

As the air mass moves away from its source area, its character changes due to the influence of the underlying surface. Air moving toward the poles generally comes into contact with cooler surfaces. This causes it to be cooled from below, so that it may become saturated, with the result that low clouds are formed. In addition, the air is made more stable, so rainfall is less likely (Figure 6.4). Conversely, air moving toward the equator becomes warmer as it meets warmer surfaces. As we saw in Chapter 4, warming of the lower layers of the air steepens the lapse rate, making the air less stable and convectional showers more likely.

Changes in air masses in these ways are particularly marked in the midlatitudes. Here, cyclones draw in air from several sources; the air is modified by the new surfaces it encounters, and is gradually mixed as it rises around the cyclone centre.

## Weather-forming systems of temperate latitudes

In 1918, J. Bjerknes, a Norwegian meteorologist, published an article entitled "On the Structure of Moving Cyclones." Weather forecasting developed from the demands of pilots flying during the First World War. Many of the terms, such as "outbreaks of air" or "fronts," reflect this origin. The Bjerknes theory became known as the polar front theory because it was along the front that these swirling vortices formed. They act to mix the air and transport heat energy. They also bring extensive and sometimes damaging storms to the middle latitudes. Arctic Canada rarely sees these systems. The subarctic experiences them in the summer months, while central and southern Canada are affected mostly in the spring and fall as the Arctic Front migrates between its mean summer and winter positions.

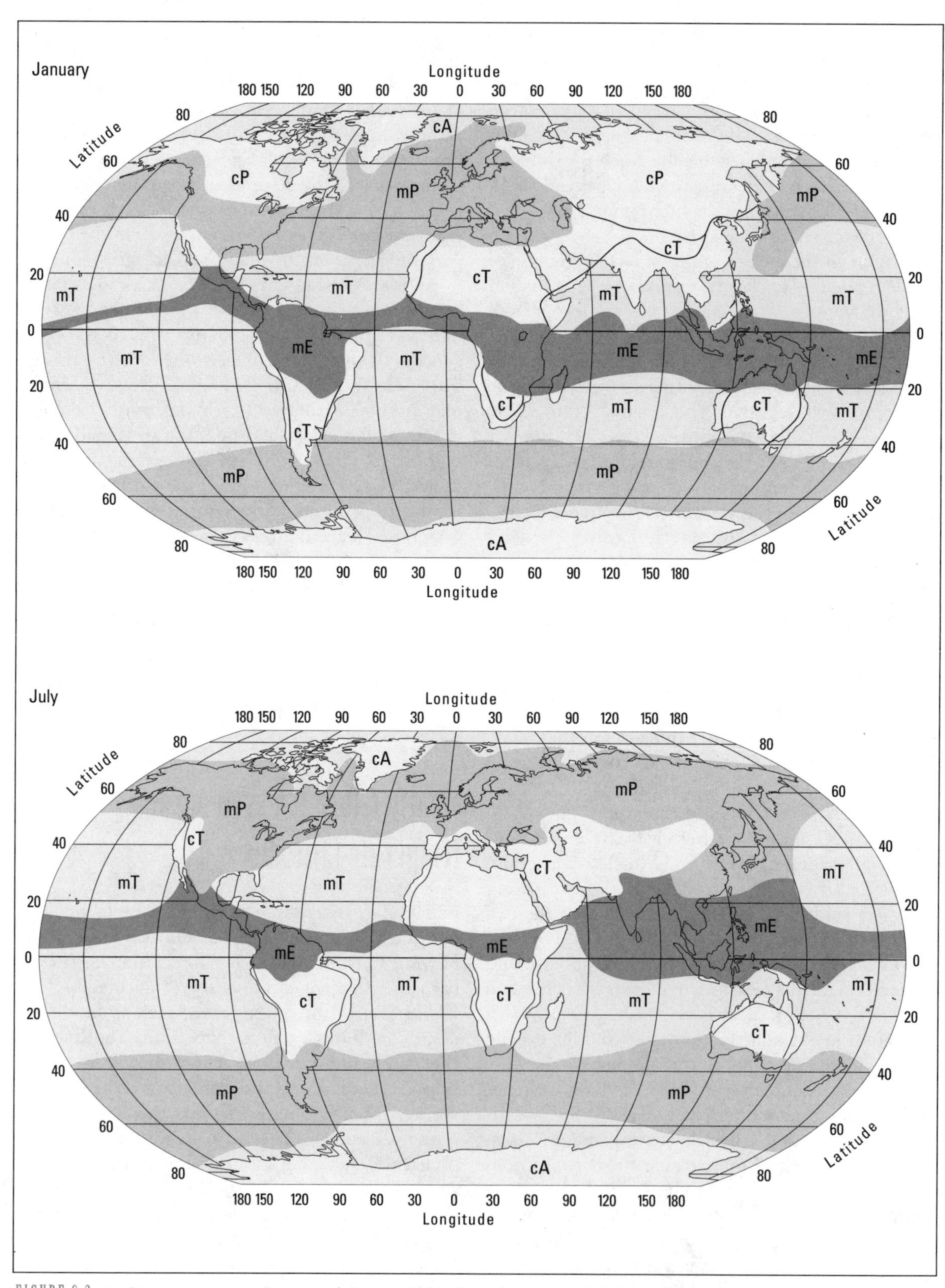


FIGURE 6.2 ▼ Air mass source regions; symbols as in Table 6.1 (after Crowe, 1971)

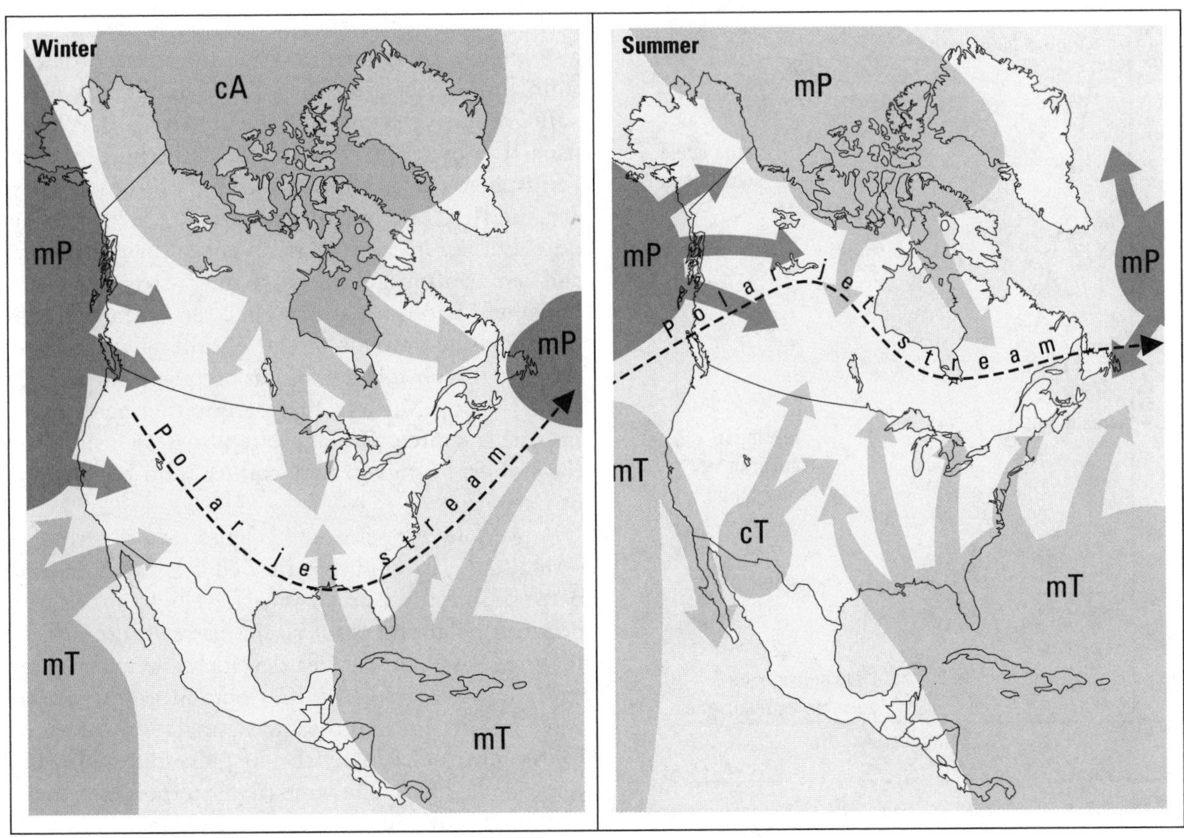


FIGURE 6.3 ▼ The source region and direction of movement of air masses in North America. Note that the Prairie region is the only area that experiences all types from tropical to arctic; symbols as in Table 6.1

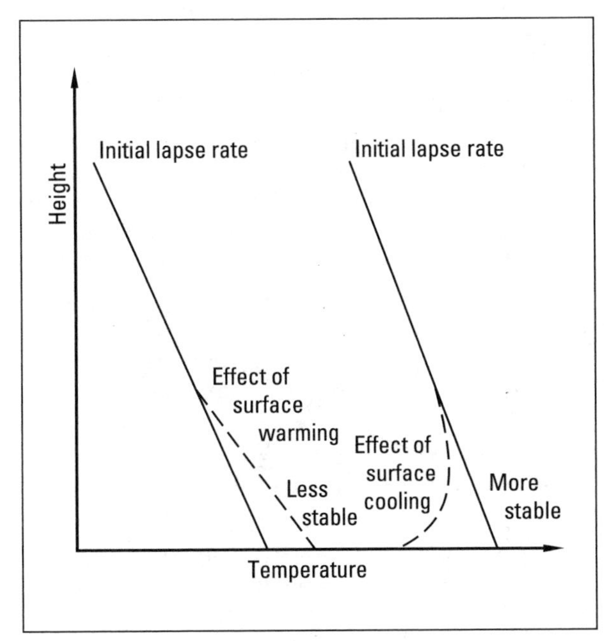


FIGURE 6.4 ▼ The effects of surface warming and cooling on lapse rates

## ANTICYCLONES

The **anticyclone** is a mass of relatively high pressure within which the air is subsiding. The major anticyclonic belts are in the subtropics, centred about 30° from the equator. They represent the descending arm of the Hadley cell circulation of the trôpics. As air descends it gets warmer and drier (Figure 6.5), but in these regions its descent is restricted by the layer of cool oceanic air below. This produces a semi-permanent inversion. This combination of circumstances results in very stable atmospheric conditions, reducing the possibility of precipitation. These anticyclonic belts are therefore associated with the main desert areas of the world.

In the middle latitudes, anticyclones often develop as a result of convergence in the upper westerlies, particularly where the waves in those winds are large. The anticyclones then build up within the depression tracks, diverting the cyclones from their usual routes and giving rise to exceptional patterns of weather. **Blocking anticyclones,**

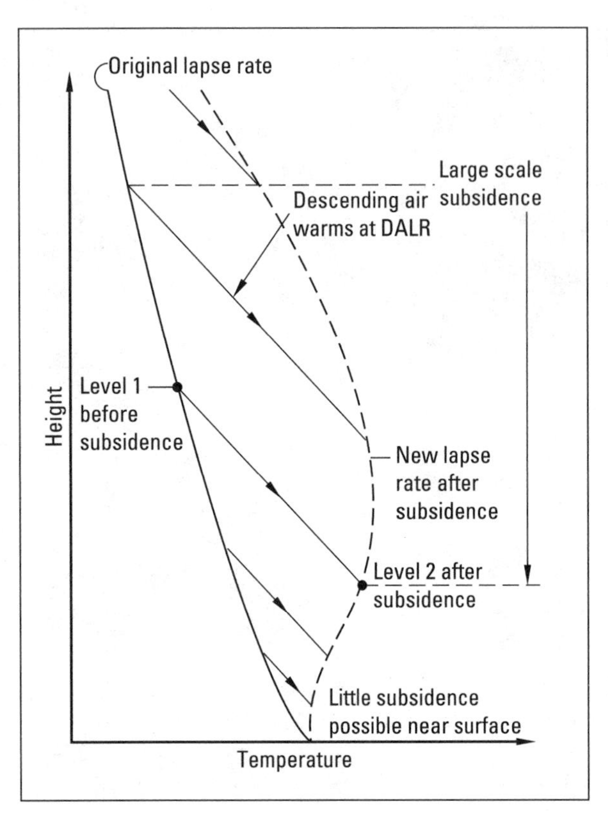


FIGURE 6.5 ▼ The effect of large-scale subsidence on lapse rates

as they are called, are most frequent over northwest Europe and the north Pacific. Blocking in the Atlantic caused the drought in northwest Europe in 1975–76, the severe winter of 1978–79, and the droughts of 1961 and 1967 in western North America. Unfortunately we do not yet know enough about the causes of blocking anticyclones to predict their occurrence or behaviour.

Anticyclones are normally associated with dry, calm weather. Clear skies or extensive cloud and very warm or very cold conditions generally occur. The combination we get depends upon the time of year, the degree of moistness, the source of the air, and the intensity of the anticyclone. In North America in summer, anticyclones often bring hot dry weather, but in winter, cold weather is usual with the amount of cloud and position of the anticyclone determining the intensity of the cold.

## CYCLONES

Wind speed rises, pressure falls and the clouds get thicker—a common sequence of events in the mid-latitudes, heralding the approach of yet another cyclone. The **cyclone**, or depression or low as it is also known, brings with it conditions very different from those associated with anticyclones. Air pressure is relatively low and the air rising. Cyclones usually move relatively quickly; in the northern hemisphere this is often toward the northeast. They are smaller, but within them air is rising more quickly than it descends in an anticyclone. Pressure and temperature gradients are much steeper so that horizontal winds are strong. In essence, they are the main mobile systems of the middle latitudes and they are responsible for the characteristic climates of these regions. Much of the precipitation here comes from this source. Cyclones are also responsible for the sudden swings in temperature from hot to cold or vice versa.

If we follow a series of cyclones over a period of several days, we find that they conform to a general pattern. Initially, a small wave develops in the front between polar and tropical air masses (Figure 6.6). In some cases, no further development takes place and the wave gradually dies out. More often, the wave begins to amplify and a small low pressure centre forms. Gradually the air pressure within this centre falls, but at the same time, to the observer on the ground, it becomes smaller. Eventually the system starts to fill as air moves in and the cyclone disappears.

What we see at the surface is only part of the story, however, for the cyclone also extends up into the atmosphere. The low pressure centre represents a column of rising air—one that is often visible on satellite photographs (Figure 6.7). To understand the cyclone more completely, we need to ascend to the top of this column, to the upper atmosphere where we find the ridges and troughs in the upper westerlies. The flow around these waves is not always in equilibrium with the pressure gradients. Where air moves out of a trough it accelerates; as it approaches a trough it slows down (Figure 6.8). The air moving away from the trough draws air from the lower atmosphere, causing a reduction in pressure at the surface. Thus, air is seen to converge at the ground within the cyclone, rise upward into the upper atmosphere, and there diverge as it flows away from the trough (Figure 6.9). The relative rates of surface convergence and upper-air divergence control the development of the cyclone. If divergence exceeds convergence, the cyclone intensifies as air is drawn out of the system. At this stage we find air pressure at the ground falling. If convergence exceeds divergence, the cyclone fills and air

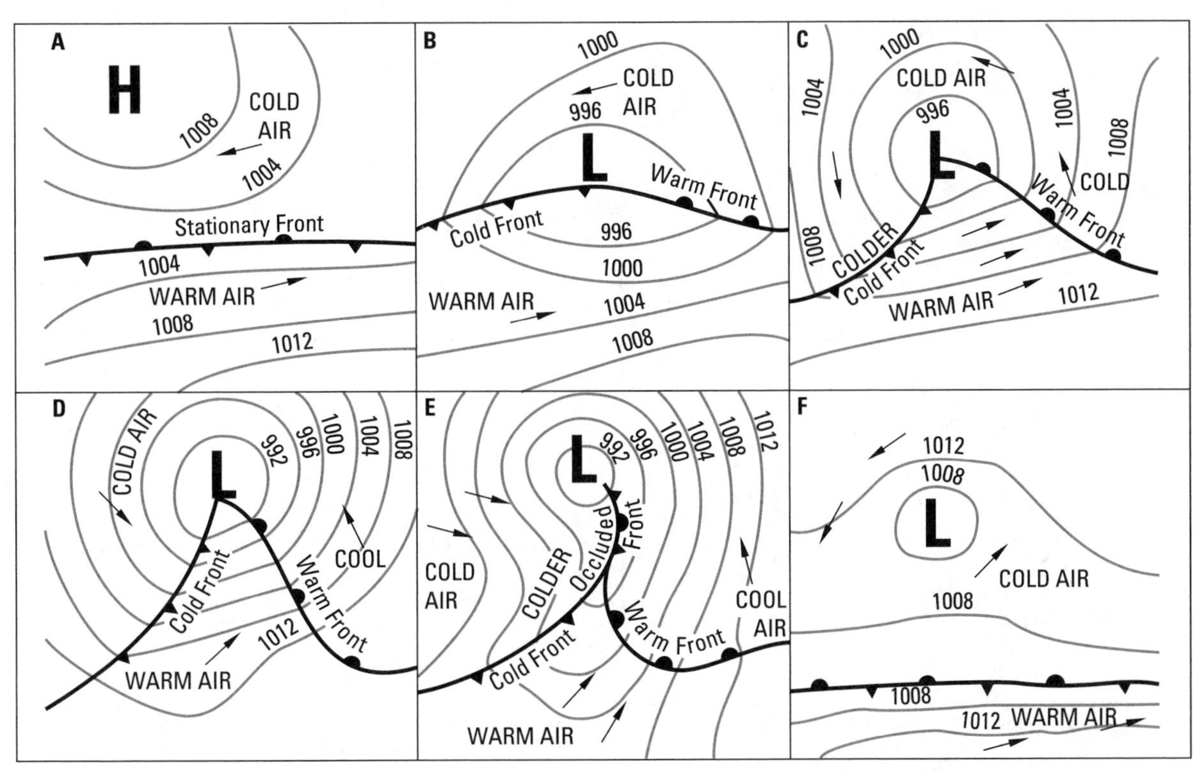


FIGURE 6.6 ▼ Six stages in the formation (frontogenesis) and demise (frontolysis) of a midlatitude cyclone in the northern hemisphere. D is the mature stage and E the occluded stage. F shows frontolysis, with the pool of warm air (L) trapped above the surface

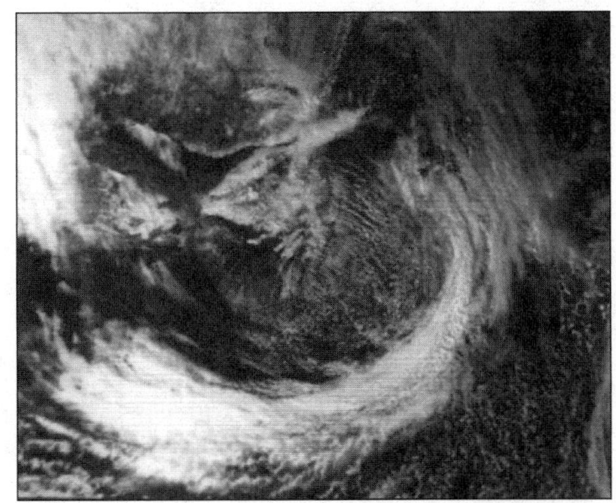

FIGURE 6.7 ▼ Spiral cloud system associated with a midlatitude cyclone southeast of Newfoundland. Convective clouds on the warm side of the front are in a characteristic open cellular pattern; on the cold side of the front the convective clouds are in bands, which usually align parallel with the wind. There is ice in the Gulf of St. Lawrence (photo: Atmospheric Environment Service, Environment Canada)

pressure at the surface rises. This is what happens in the final stages of the cyclone.

In the northern hemisphere, the troughs and ridges of the upper westerlies tend to favour certain locations. There is normally a ridge near the Western Cordillera and a trough near the east coast of the USA. This means that the area most favoured for cyclone formation is south of Newfoundland. The depressions grow more pronounced, reach their maximum intensity, then decay near Iceland. They reach their lowest average pressure along this track (Figure 6.8).

Because the cyclones are areas of rising air, they are almost always accompanied by extensive cloud and precipitation. The steep pressure gradients and rapid falls of pressure that occur can cause weather problems in the affected areas. Details of cloud location and thickness will depend upon the nature of the upper atmospheric divergence and temperatures, on the time of year and sources of the air. If we look at the paths of cyclones (Figure 6.10), we can see that they cross the continent from west to east with the general circulation of the westerlies; most of these cyclones cross in conjunction with the major

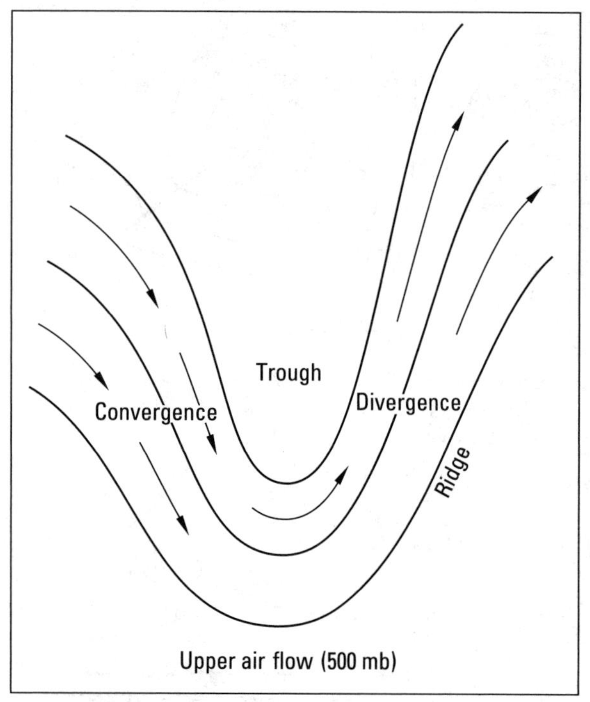


FIGURE 6.8 ▼ Air flow around a trough in the upper atmosphere. As the air approaches the trough strong convergence builds up. On leaving the trough, the air diverges, as shown by the streamlines spreading out

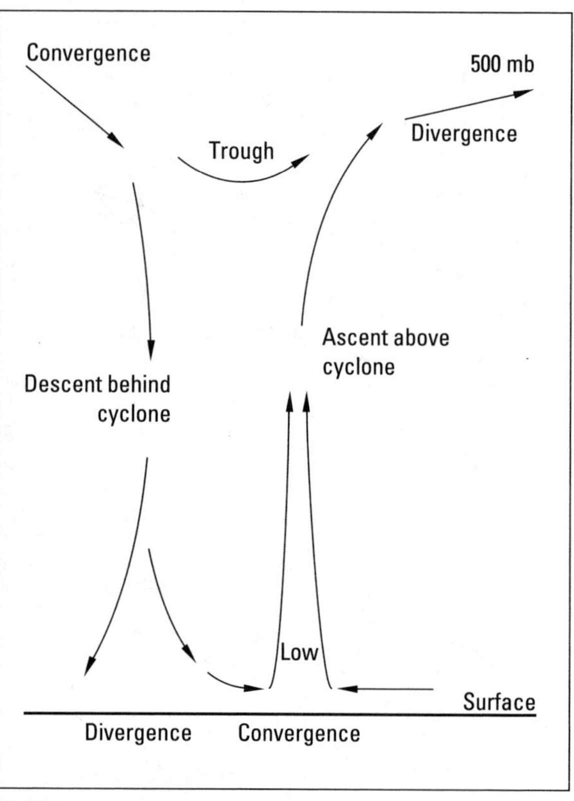


FIGURE 6.9 ▼ Interaction between surface and upper atmospheric flow near an upper trough

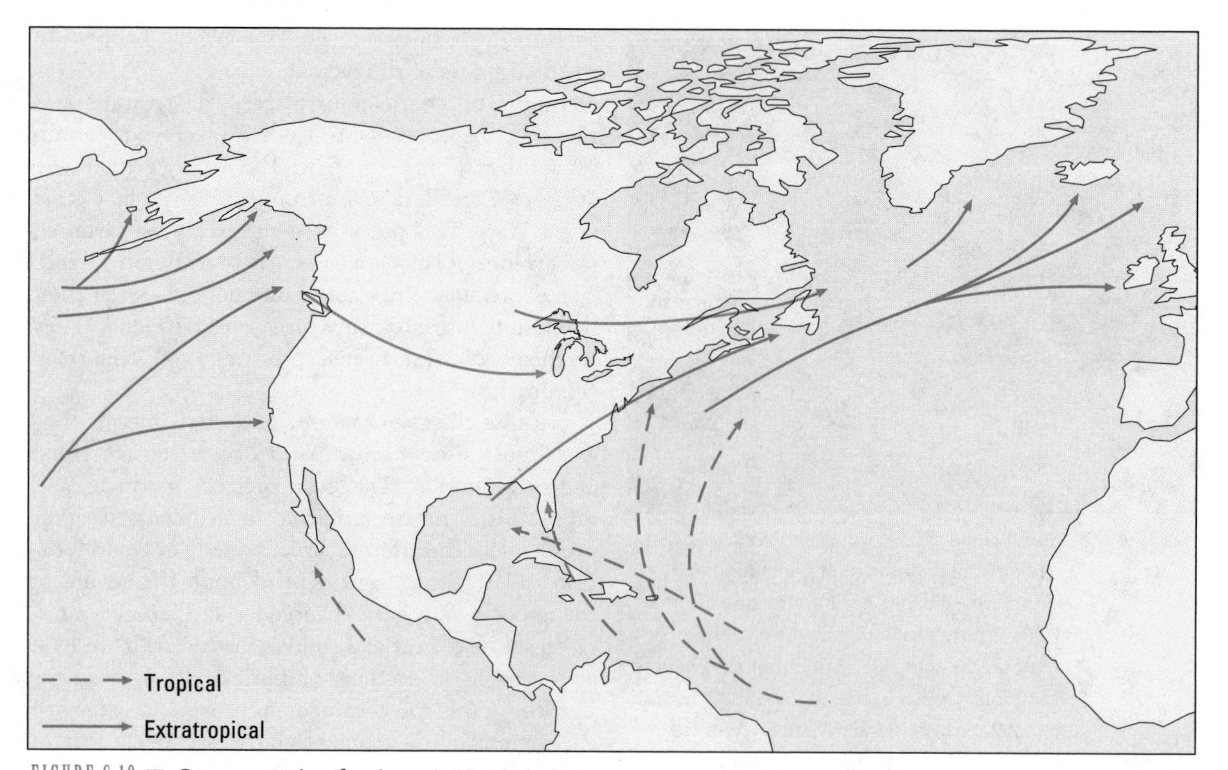


FIGURE 6.10 ▼ Common paths of cyclones in North America

air mass fronts. Canadian cyclonic precipitation occurs predominantly in the spring and fall as the Arctic Front makes its seasonal migration.

## FRONTS

In many cyclones, we would find that there is not a gradual change of temperature as the systems pass, but several sudden changes. Figure 6.11 shows the trace of a thermograph during the passage of a cyclone. If it has been cold before the storm approaches, temperatures may rise slightly. This is due to heat being trapped by cloud cover and wind mixing up the cold air. If it has been warm, temperatures may fall, because the sun will no longer be shining. Suddenly the temperature starts to rise, perhaps by several centigrade degrees within a few hours. Temperatures will then remain fairly stable until the cold air in the rear of the cyclone reaches us. The fall of temperature is usually even more sudden than the earlier rise; falls of 10C° within a few minutes are not unknown. The most rapid change ever recorded was the rise that took place at Spearfish, South Dakota, on 22 January, 1943, when the temperature jumped 27C° in two minutes, from –20°C to +7°C.

The sudden change of temperature clearly indicates a change of air mass. The separation surface between air of different origins is called a **front**. Where warm air is replacing cold air, we have a warm front, and where cold air is replacing warm air we have a cold front. The clouds show the main areas of rising air produced by divergence in the upper atmosphere. This is the reason the clouds do not follow the frontal surface as closely as one might expect.

The warm front slopes at a low gradient of about 1 in 300, which means that the first clouds associated with the front can be seen long before the surface front is near. Cirrus clouds are the first indicators of the approach of the front, followed by

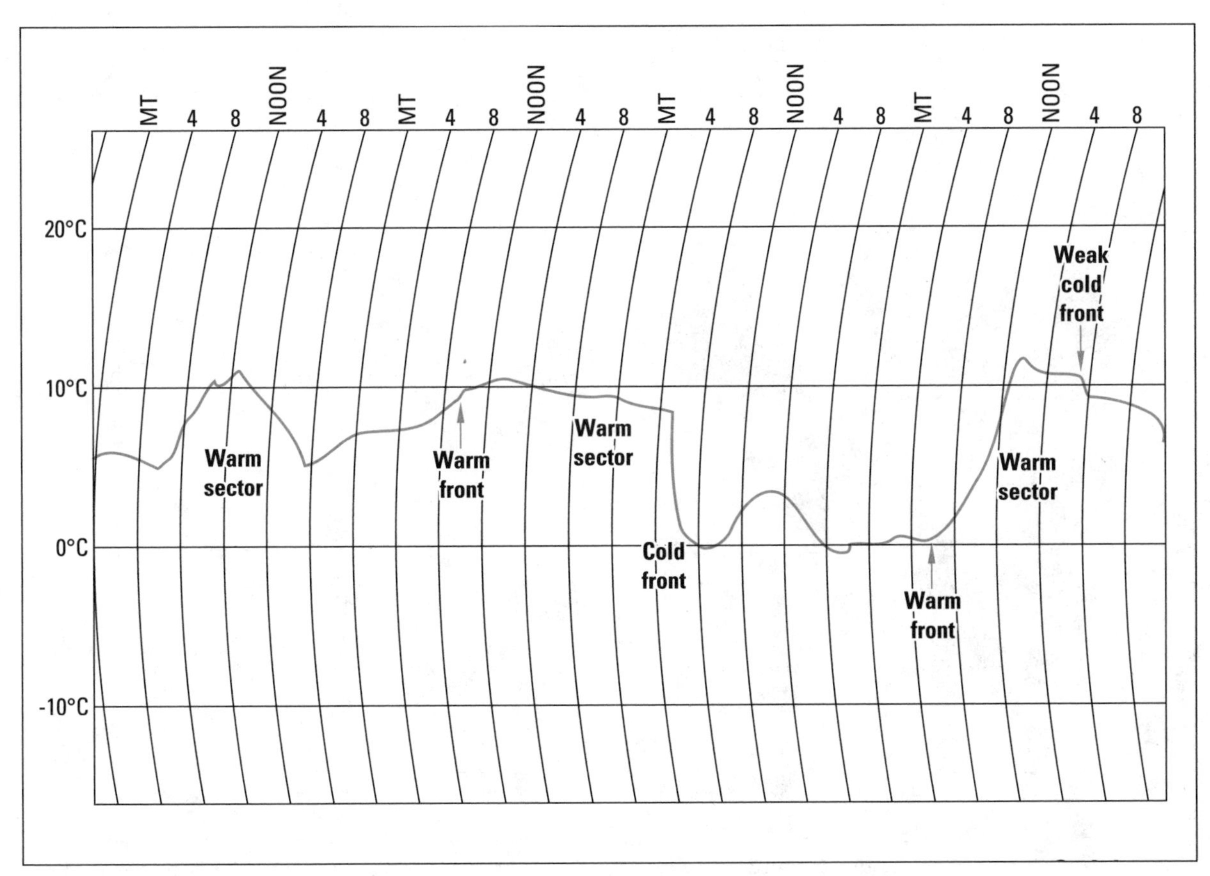


FIGURE 6.11 ▼ Thermograph trace recorded in Sheffield, England, 20–23 March, 1981, illustrating the effects on temperature of the passage of two cyclones. The fall of temperature at the cold front on 22 March was unusually large for the UK

a sequence of a gradually thickening and lowering cloud base. Cirrostratus clouds are followed by altostratus, then nimbostratus clouds, by which time rain will be falling. In general, the atmosphere is fairly stable at a warm front, but some convection does occur in the middle levels, producing areas of heavier precipitation. Figure 6.12 shows an example of the rainfall patterns associated with a warm front.

The slope of the cold front is much steeper, at about 1 in 50. Weather activity at the cold front is often much more intense than at a warm front. If the warm air is unstable, the effect of uplift at the front generates thunderstorms and even tornadoes. The line of deep cloud may be seen on satellite photographs (Figure 6.13) as a very distinct band. The cold air descending with the heavy rain can intensify the effect of the fall in temperature.

Where the air in the warm sector between the fronts is rising, cloud development near the fronts follows the pattern described above; this is known

FIGURE 6.13 ▼ A cold front in the Atlantic off Newfoundland associated with a cyclone to the northeast (photo: Atmospheric Environment Service, Environment Canada)

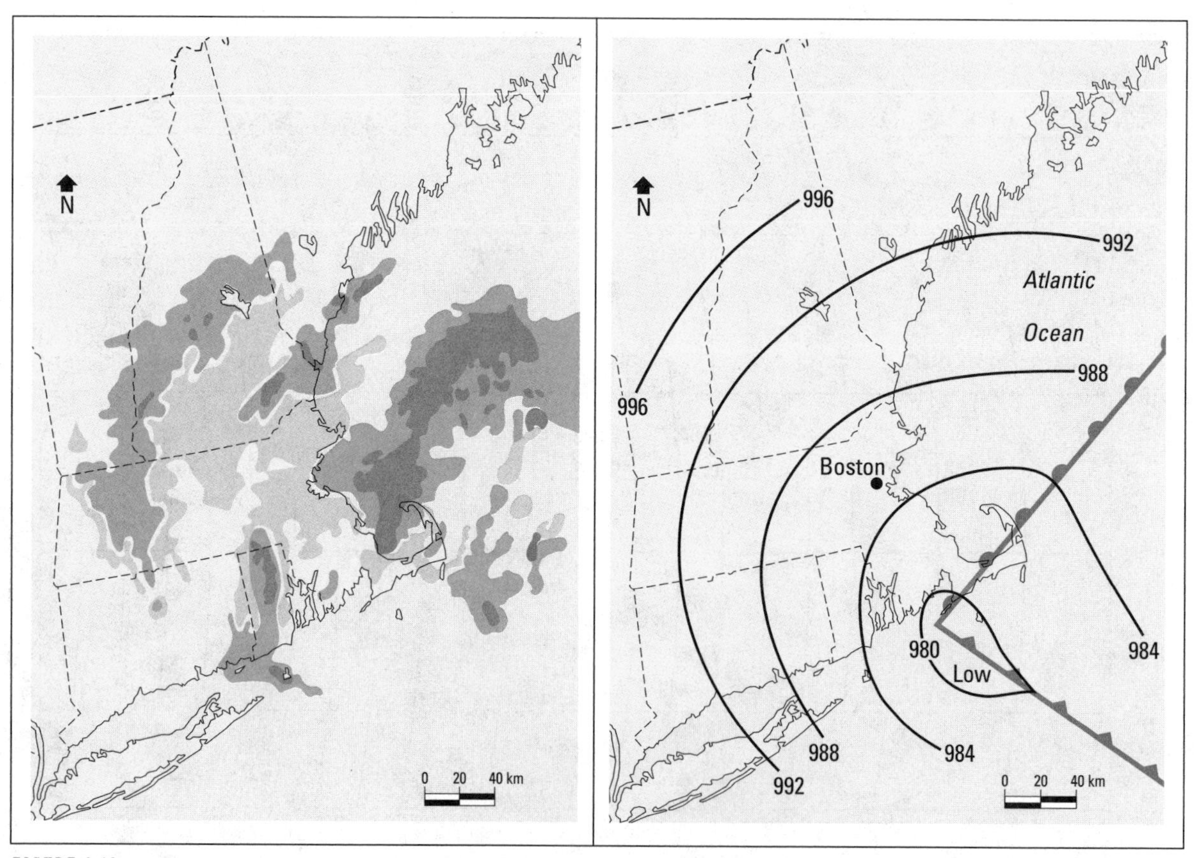


FIGURE 6.12 ▼ Precipitation patterns associated with a cyclone off Massachusetts; data obtained by radar (after Austin and Houze, 1972). Density of shading represents areas of precipitation; note the heaviest precipitation corresponds to the warm front shown on the map at right

as an **ana-front** (from the Greek word meaning up). However, farther away from the cyclone, the intensity of uplift decreases and cloud may grow gradually thinner as the front dies out. In this stage of only weakly rising air, the front is termed a **kata-front** and the transition zone of temperature is fairly broad (Figure 6.14). Rainfall is slight from kata-fronts as the clouds are not deep enough and the updrafts are weak.

In most cyclones, the cold front moves more rapidly than the warm front. The air of the warm sector is raised above the ground surface as the cold front catches up with the warm front. this is known as the stage of **occlusion**, or the **occluded front**. The nature of the front will now depend upon the relative temperatures of the two cold air masses (Figure 6.15). Where the air behind is colder than that ahead, we will have a structure rather like a cold front. If it is warmer than the air ahead, the structure will resemble a warm front.

The detailed air movements and cloud distribution at an occluded front are complex. As it represents a mixing of air of different origins, humidities, temperatures, and stabilities it is not surprising that great variation can occur between fronts or even along the same front.

At one time it was believed that it was the air rising along the frontal surface that caused the development of a cyclone. However, the role of divergence in the upper atmosphere is now believed to be the most important factor, with the fronts being a consequence of the rotation of air around the cyclone's centre. From being a cause of the depression, the front has been relegated to a conse-

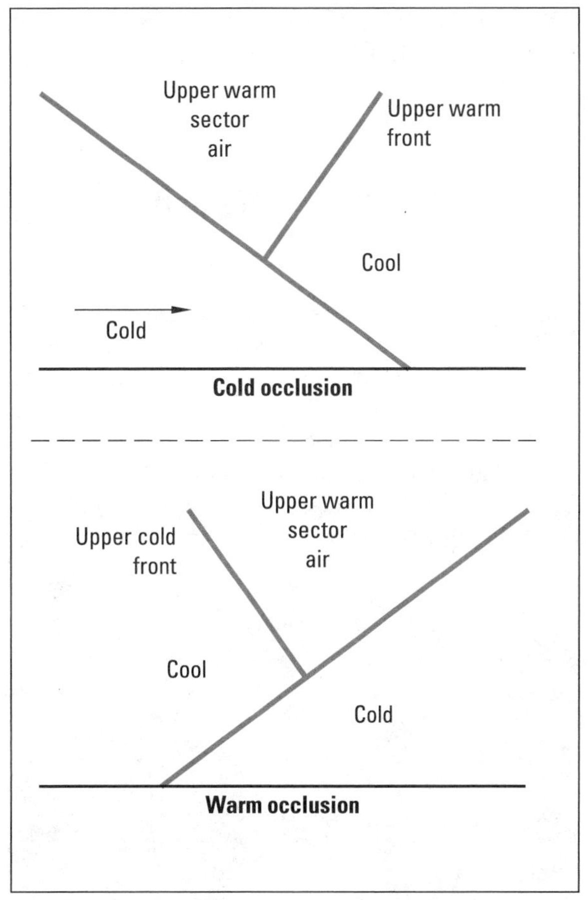


FIGURE 6.15 ▼ Simplified cross sections through cold and warm occluded fronts

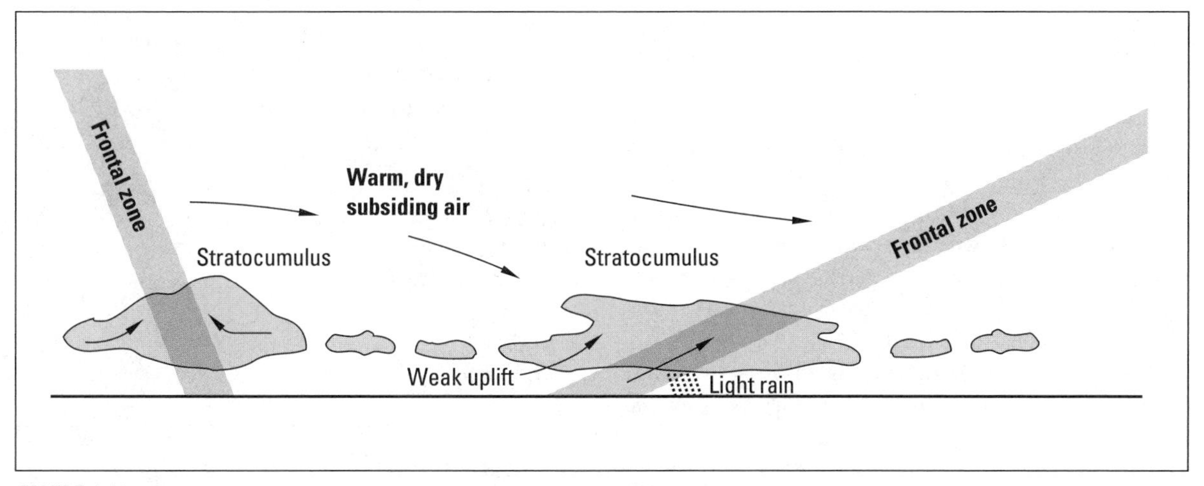


FIGURE 6.14 ▼ Cloud structure at a kata-cold and a kata-warm front

quence. Nevertheless the weather activity associated with fronts is still a very important aspect of the cyclone.

# Weather-forming systems of the tropics

## EASTERLY WAVES

The weather of the trade wind zone normally shows little variety. It is characterized by small convectional clouds drifting across the sky in response to the prevailing winds (Figure 6.16). Showers may develop in the afternoon and they are likely to be heavier and more frequent in the summer season, but otherwise the weather remains remarkably constant throughout the year.

FIGURE 6.16 ▼ Trade wind cumulus clouds over St. Lucia, in the Caribbean, provide visible evidence of continuous evaporation from the warm tropical seas (photo: Robert S. Schemenauer)

Occasionally disturbances arise to upset this quiet regime. On a dramatic scale there is the tropical cyclone, which is discussed in the next section, but on a smaller scale there is the easterly wave. As its name implies, this represents weather-forming systems related to wave-like structures in the easterly flow of air. They reach their maximum intensity at about the 700 mb level.

The wave does not necessarily move at the same speed as the easterly flow, and it may even exceed the average wind speed. Preceding the wave, convectional cloud dies down due to surface divergence and subsidence of the air, while the wind backs toward the northeast (Figure 6.17). (A wind that changes direction in a counterclockwise direction on the compass is said to back; for example, in this case it shifts from east to northeast. A wind is said to veer if it changes direction in a clockwise direction, for example, from northeast to east.) As the main axis of the wave approaches, convergence becomes dominant, causing ascent of the air, cloud formation, and precipitation just ahead of the low pressure trough. The wind veers suddenly as the wave passes, to be followed fairly quickly by a clearance of the cloud and a return to undisturbed trade wind flow.

The passage of the wave is not dramatic, therefore, but in areas where the weather hardly changes it does at least provide a little variety. Moreover, in areas such as the Caribbean, where the waves are frequent, they are responsible for a significant proportion of the annual precipitation.

## TROPICAL CYCLONES

Throughout the tropics one main feature of the weather is the tropical cyclone. Unlike its counterparts in the middle latitudes, this is not large; instead it consists of a small, intense, revolving storm. It goes under a variety of names: the hurricane in the Caribbean and North America, typhoon in the Pacific, or cyclone in the Bay of Bengal. To qualify as a hurricane, the storm must contain winds over 63 knots (32 m $s^{-1}$). Less intense storms are called tropical cyclones or tropical storms.

### Table 6.2

*Average number of tropical cyclones per year by ocean basins (1958–77)*

| | |
|---|---|
| Northwest Atlantic | 8.8 |
| Northeast Pacific | 13.4 |
| Northwest Pacific | 26.3 |
| North Indian | 6.4 |
| South Indian | 8.4 |
| Australian | 10.3 |
| South Pacific | 5.9 |

After W.M. Gray, in D.B. Shaw (ed.), *Meteorology of the Tropical Oceans* (1978), Royal Meteorological Society.

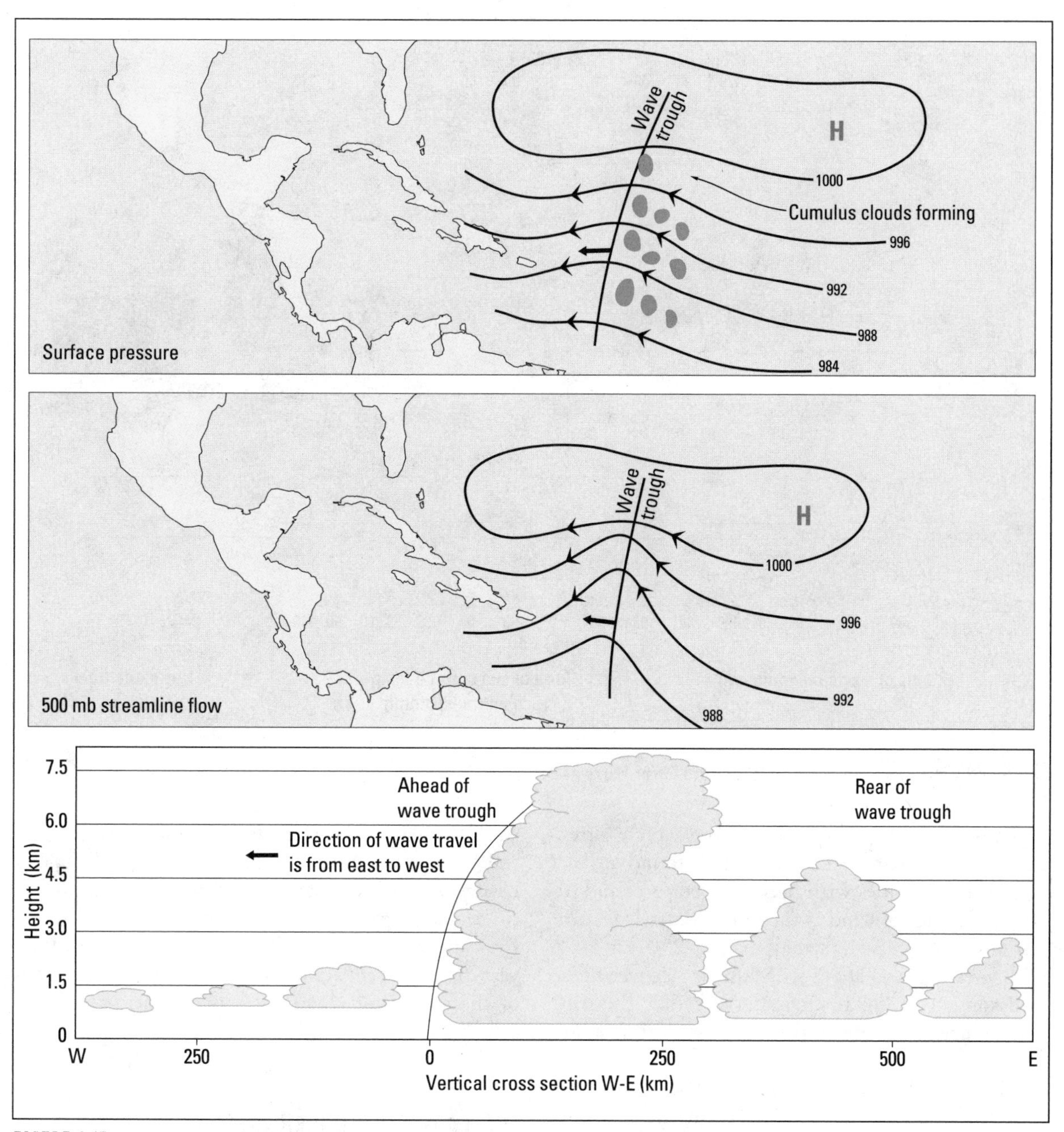


FIGURE 6.17 ▼ Surface pressure, 500 mb streamline flow, and vertical structure of an easterly wave (after Malkus, 1958)

If we look at the parts of the globe affected by these cyclones, it is apparent that they only develop over the warmer parts of the seas (Figure 6.18). For each hemisphere, it is the summer and autumn periods when cyclones are most likely to strike.

The Pacific Ocean has the most hurricanes (Table 6.2). Some affect California, but the majority form farther west and cross the Philippines before curving northward toward Hong Kong, China, or Japan. Other areas have fewer storms, but they can still cause serious damage in Australia, India, Madagascar, and even Arabia.

## EFFECTS OF HURRICANES

To people who have never experienced a hurricane the devastation is unbelievable. The storm consists

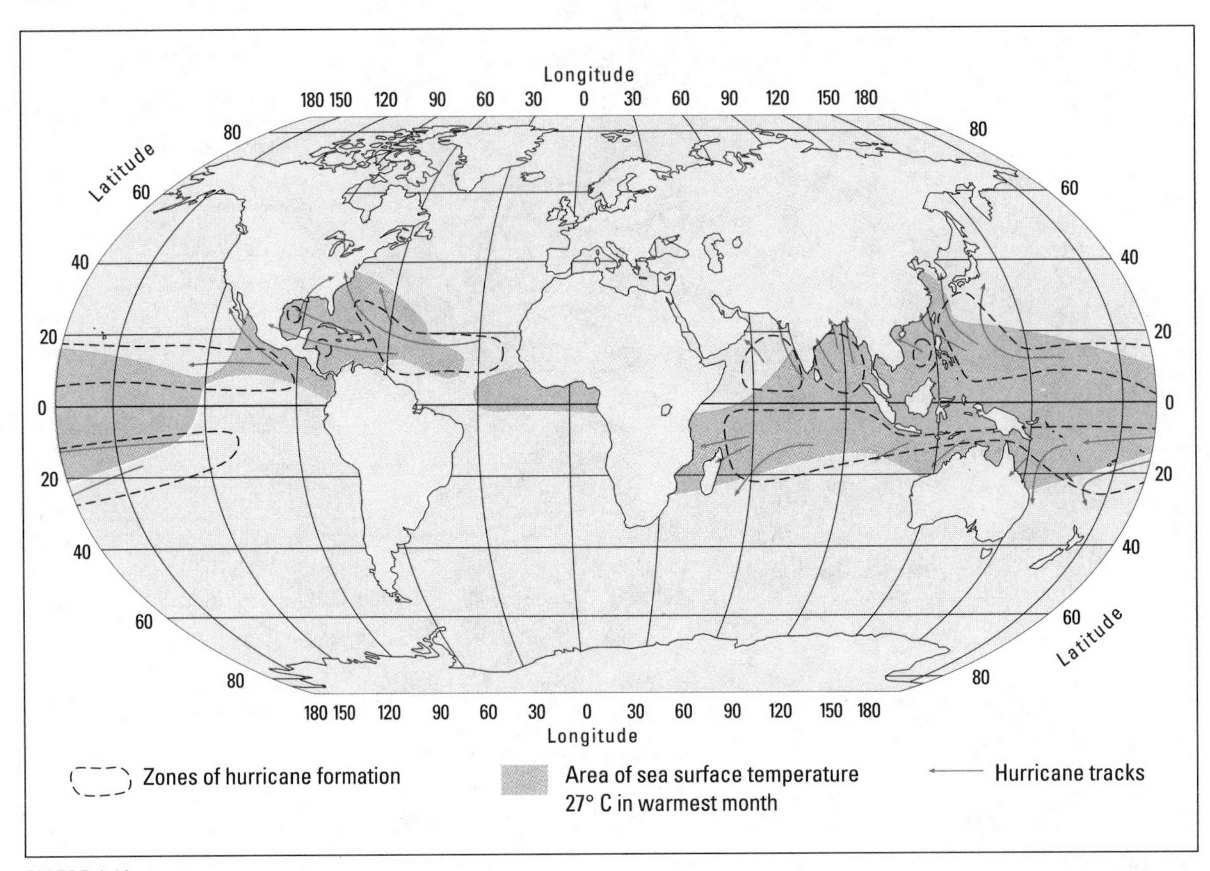


FIGURE 6.18 ▼ Areas of hurricane formation (after Barry and Chorley, 1982)

of a spiral of thick cloud bands from which torrential rain falls (Figure 6.19), centred around an 'eye' of almost clear skies with high temperatures and little wind. But the wind is the most dramatic feature. Around the eye winds may exceed 250 km $hr^{-1}$, with mean speeds above 175 km $hr^{-1}$, covering an area some 150 km in diameter. Beyond this ring speeds gradually decrease, but gale-force winds (above 60 km $hr^{-1}$) can extend over an area 600 km in diameter. Ahead of the storm, winds blow from one direction, usually northerly in the northern hemisphere. After the eye has passed, the wind resumes with equal ferocity from the opposite direction (Figure 6.20).

When the storms pass onto land they weaken. Winds decrease and the structure of the storm becomes less clear, but heavy rain may still fall. It has been estimated that hurricane rainfall accounts for nearly one-quarter of the annual precipitation in the southeastern USA. Even in the northeast, heavy falls may occur. In June, 1972, Hurricane Agnes gave more than 30 cm of rain over parts of Pennsylvania in a single day during the decaying stage. It is from heavy rain and storm surges along the coast that most hurricane deaths occur. Today's better weather forecasts enable preventive measures to be taken more quickly than before, and fewer lives are now lost, though storm damage can still be vast. Satellites constantly monitor the skies for signs of the spiralling cloud pattern indicating a hurricane.

## THE CAUSES OF HURRICANES

Despite the danger and damage of hurricanes, we know surprisingly little about their origins, except that they all form over the tropical seas where temperatures are above 27°C, and that they do not form within about 5° latitude of the equator. Once developed, they move toward the west within the trade winds, gradually increasing in intensity. Before dying out the storm begins to swing poleward. A few manage to maintain their identity but they gradually decay and acquire the characteristics of a midlatitude depression. Many September storms

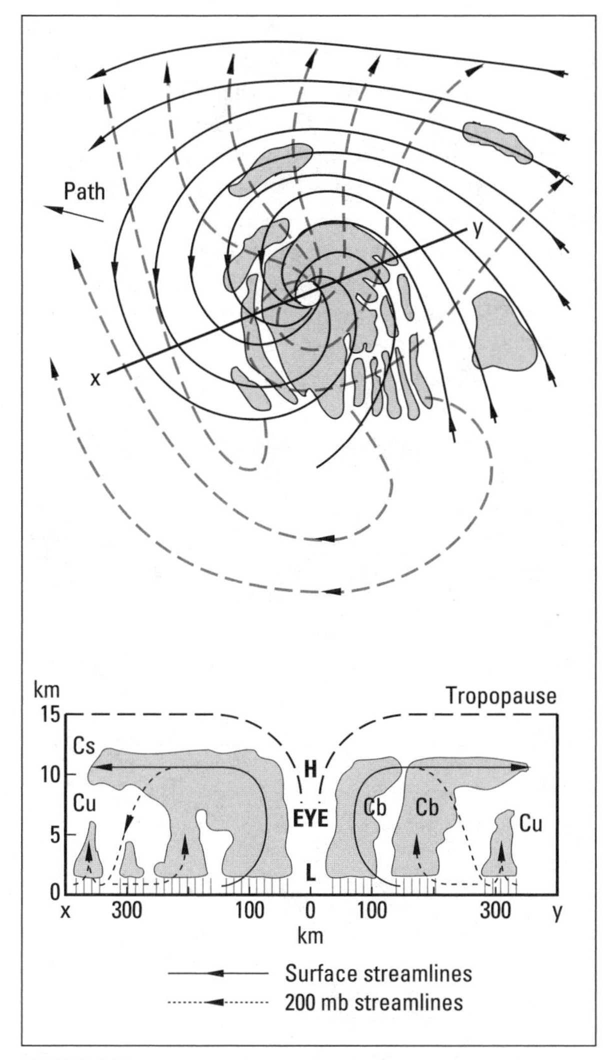


FIGURE 6.19 ▼ A model of the areal (above) and vertical (below) structure of a hurricane (after Barry and Chorley, 1982)

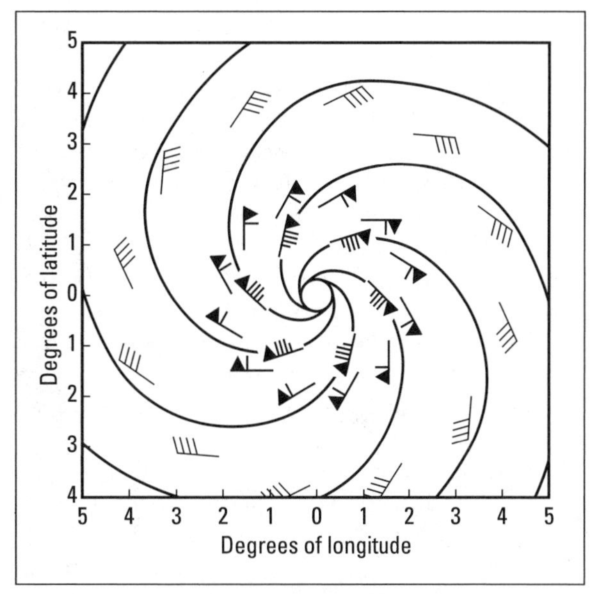


FIGURE 6.20 ▼ Mean vector wind field and streamlines. A solid triangle represents 50 knots (25.8 m $s^{-1}$), a single feather 10 knots (5.1 m $s^{-1}$), and a half feather 5 knots (2.6 m $s^{-1}$) for wind speed (after Riehl, 1979)

and floods in northwest Europe can be traced back to Caribbean hurricanes.

Once started, the development of the hurricane is fairly predictable. But what starts it off? In order for the cyclonic wind circulation to develop, we must have air converging, which requires some form of initial disturbance. We do find a variety of small disturbances within the tropics where vertical movements and rotation can be started. As they are small in size and found over the seas, their location is difficult to predict. Once the air begins to rise, it cools and, on reaching saturation, large quantities of latent heat are released. It is this process that is believed to be responsible for giving so much energy to the storm. Once the storm moves over land, the main source of energy is lost and so the storm decays.

## TORNADOES

Tornado! The very word brings alarm in areas such as the Midwest and Mississippi Valley of the USA, but they also occur in all Canadian provinces. A darkening sky and the appearance of a pale cloud may be a warning before the familiar and frightening tornado funnel appears. The funnel may descend from the cloud base, getting larger and darker, until it virtually touches the ground, accompanied by a tremendous roaring wind that everyone describes as sounding like the rumbling of a freight train. Debris is caught in the funnel, and as it moves across the countryside it leaves devastation in its wake. Contact with the ground causes loss of energy, so contact is not maintained for long. The spinning motion continues, however, and contact is reestablished. This results in the skipping path of the tornado as it travels. The frequency is higher in the USA, about 700 per year compared to 70 in Canada, because of greater confrontation between warm tropical air and cold arctic air. Tornadoes can occur in all months of the year in the southern states, while in Canada they are exclusively a summer phenomenon.

Water acts as a moderating influence creating more stable conditions, and as a result there are fewer tornadoes on the coasts. Similarly, the Great Lakes reduce the number in Ontario; southwestern Ontario, however, does have the highest number of recorded tornadoes in Canada. The intrusion of warm, moist, unstable, tropical air necessary for tornadoes is usually restricted to the southern regions.

The frequency of tornadoes in Canada is shown in Figure 6.21. They have been sighted in every province and usually between April and October. Despite this, a tornado was reported in January, 1954, in Nova Scotia.

The tornado is normally narrow, about 0.5 km wide, and seldom does it move more than 20 km. But exceptions do occur, with some being up to 1.5 km wide and travelling 500 km. How fast it moves we cannot tell; no recorder has survived the passage of a tornado. But speeds of over 400 km $hr^{-1}$ are believed to occur.

Tornadoes occur in most parts of the world, but they achieve their greatest strength and frequency over the continental plains of the USA. The reason for this concentration is the frequent juxtaposition of layers of air with great contrasts in air temperature, moisture, and wind. Air ahead of the cold front may be drawn in from the Gulf of Mexico. Behind it, cold air may be sweeping southward from the Canadian Arctic. Such a situation is ideal for the development of cumulonimbus clouds needed to spawn tornadoes.

Like hurricanes, the precise mechanism whereby a tornado funnel forms is not understood. It is probable that tornadoes are produced by thermal and mechanical effects acting in the cloud. But why some clouds generate tornadoes and others do not is a mystery. Nevertheless, favourable conditions can be recognized and tornado warnings are issued by local weather services.

The necessary ingredients are very warm, moist, and unstable air in which well-developed cumulonimbus can form. A cold front pushing in pro-

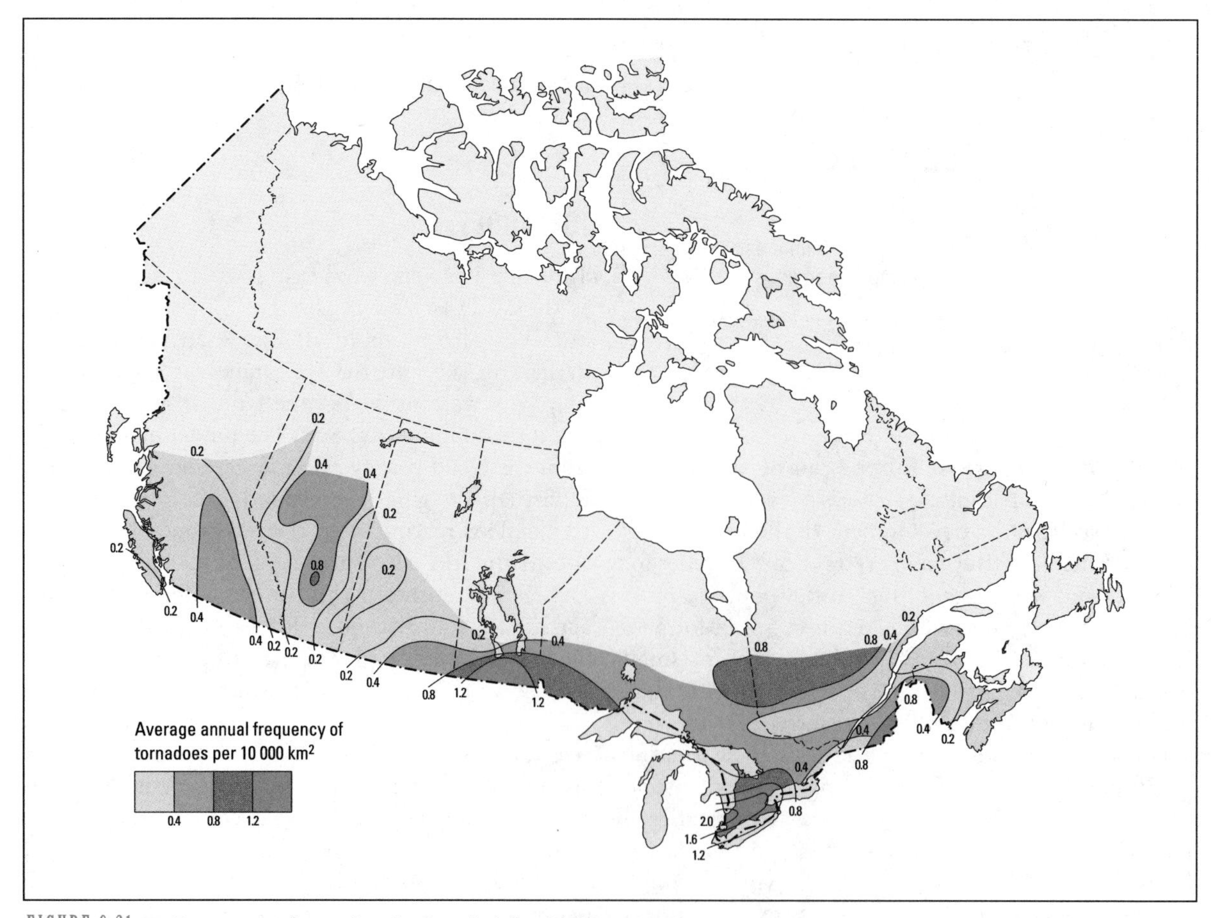


FIGURE 6.21 ▼ Frequency of tornadoes in Canada (after Phillips, 1990)

duces extreme instability—and very strong updrafts. These interact with winds in the lower troposphere, producing a shearing effect.

Rotation of the tornado is counterclockwise in the northern hemisphere and clockwise in the southern hemisphere. Wind speeds vary but often exceed 180 km $h^{-1}$. Tornadoes with very high winds often have smaller whirls called **suction vortices** within the main funnel. These are known as **multi-vortex tornadoes**. A good example of such a tornado occurred in Edmonton on 31 July, 1987, taking 27 lives (Figure 6.22).

Over the sea, tornado funnels are termed waterspouts. As convection over the sea tends to be less intense than over land, the waterspout is much weaker than the tornado, but may damage small boats or buildings if it passes over land.

FIGURE 6.22 ▼ A tornado strikes Edmonton, Alberta, July, 1987. The funnel becomes visible when moist air moves into the area of sharply lowered pressure and condenses (photo courtesy of the Alberta Public Safety Service)

# Weather prediction

We all know from experience how much daily and seasonal variations in weather influence our lives. Clearly it is useful to have an idea of the weather that is in store for us. But how is it possible to foretell the weather? In the past we relied heavily on folklore. 'Red sky at night, shepherd's delight; red sky in the morning, shepherd's warning,' says one country adage. 'When there's sheep-backs [cumulus clouds] in the sky, not long wet, not long dry,' goes another. Clearly these sayings sometimes contain a grain of truth—that is presumably why they have survived.

As we have started to understand atmospheric processes in more detail, methods of forecasting have become more sophisticated. There are a number of different approaches. One of the main methods is extrapolation. Using experience gained from years of analyzing weather patterns, the forecaster predicts the behaviour of the weather system from a knowledge of its recent history. This approach depends upon the principle that weather, like history, repeats itself; thus weather analogies can be drawn from the past to use as a basis for understanding the near future.

The trouble with this approach is that the weather does not always follow the expected course; similar conditions may result in very different weather. Moreover, it is difficult to obtain all the necessary information. The range of predictability is also limited by the extent of the record. In most places this is very limited indeed. A second approach is therefore used more widely. This involves understanding the basic processes of weather information and using these physical laws to predict events. Unfortunately, although we know many of the basic physical laws, and can express them mathematically, the equations we end up with are difficult to solve. Only recently, with the development of computers, has it been possible to tackle this mind-stretching task; the first attempt to forecast weather in this way, without computers, in 1921, took several months! Today, however, numerical approaches to forecasting are widely used. They involve integrating information on many different atmospheric properties.

To forecast values of pressure and wind, the northern hemisphere is subdivided into a grid consisting of about 4000 squares, each with an area of 90 000 $km^2$. For each point, the upper atmosphere is subdivided into fifteen levels and values of the critical atmospheric properties determined, mainly from satellite information. The physical equations of motion, continuity, and thermodynamics are applied to each grid point at each level to predict the new value a short time ahead. The new data set then provides the starting point for the next set of predictions and so on, until the 24-hour or 48-hour forecast is available.

We might get the impression from this technique that we can forecast the weather for the distant future, but this is not true. It appears that small deviations can seriously affect the development of weather-forming systems. New predictions have to be made on a daily basis to incorporate small-scale changes that could become very important. The problem is that we do not have enough information (nor large enough computers) to solve the equations accurately.

## WEATHER PREDICTIONS AND HAZARDS

It is possible to predict many of the weather hazards discussed earlier, but we have to distinguish between large-scale and small-scale hazards. Tropical and temperate latitude cyclones can be predicted reasonably well, so that we know approximately the areas they are likely to affect with their strong winds and heavy precipitation. At the smaller scale, the warnings of tornado formation are announced for a large area, but precisely where the funnel clouds will touch down on the surface is not known. It is probably impossible to forecast such conditions for more than a few minutes ahead. Flash floods from a single thunderstorm are in a similar category. We have to accept them as one of the microscale features of our atmosphere that occasionally may cause devastation over a small area. The chance of any one site being affected by them is very small.

# Climates of the world

On a July evening in Ottawa the residents are sweltering. Temperatures have touched 29°C during the day, a little above average, but relative humidity is over 60 percent and this makes it feel very uncomfortable. Meanwhile, at about the same latitude on the east coast, Halifax is much cooler at 18°C. Even though the levels of humidity are slightly higher than in Ottawa the air feels more pleasant; discomfort is less. Why are these climates so different at the same distance from the equator?

If we take an example of a difference in latitude rather than longitude, the contrast becomes even more pronounced. In Northwest Territories, there is Resolute, on the southern tip of Cornwallis Island. The average July daytime maximum temperature is now only 4.1°C and nighttime temperatures often fall below freezing. If we head south to Maracaibo on the Venezuelan coast at the same longitude as Ottawa, but at latitude 10°N, we find average temperatures in July to be 34°C, a large jump from Resolute, or even Ottawa.

Repeating the experience in January would not find these communities in the same order of warmth (Table 7.1). At Maracaibo, daytime maximum temperatures at 32°C would be only slightly cooler than in July. Halifax comes

Table 7.1

*Latitudinal temperature transect of the Americas*

| *City* | *July (mean)* | *January (mean)* |
|---|---|---|
| Maracaibo (10°N) | 34°C | 32°C |
| Halifax (44°53′N) | 18.2°C | –6°C |
| Ottawa (45°N) | 20.9°C | –10.2°C |
| Resolute (74°43′N) | 4.1°C | –32.1°C |

second warmest at –6°C; Ottawa would be shivering at about –10.2°C but at Resolute, north of the Arctic Circle, there would be no daylight and with average *maximum* temperatures of –28.4°C it would be decidedly chilly.

Almost all the figures we have quoted have been averages. Clearly, if we measured temperatures on any single day, we might find a very different pattern, due to the influence of different weather systems, or winds from different directions. What we are dealing with here is not the weather—the short-term, transient conditions of the atmosphere—but the regional climate, the average, longterm pattern.

What we get is a static picture, a statistical summary of the climate, described by graphs and tables. It is a useful picture, for it helps us to compare different areas or to define briefly the character of a region. It allows us to see the way general atmospheric conditions vary across the globe. Later in this chapter we will look at the nature of this regional variation in climate, and in Chapter 29 we will discover how these climatic patterns relate to patterns of vegetation and soil and landscape. First, however, we need to ask two more fundamental questions. What are the characteristics that make up what we call 'climate'? And what are the factors that influence climate at a regional scale?

# The elements of climate

Our first question refers to what we generally call the elements of climate, the properties that we need to measure if we are to describe the climate. The main elements are the temperature, the rainfall, the nature of the winds, and the degree of humidity. But each of these elements can be measured in a variety of different ways.

We can determine the mean annual temperature, for example, and this gives us some indication of average conditions throughout the year. Yet other aspects of temperature are also important. The monthly pattern of temperature is useful, because it shows the seasonal variations. The temperature range (the difference between the temperature of the hottest and coldest month) is also helpful. All three may be used to describe and classify the regional climate.

Similarly, in measuring precipitation, it is important to know not just the annual average, but also the monthly or seasonal pattern. In some cases the reliability or variability of the rainfall is significant, since this reflects the likelihood of excessive precipitation or drought.

These two elements—temperature and rainfall—possibly illustrate the main characteristics of the climate. But if we are to get a more complete picture we also need information on wind speed and direction and frequency, on average humidity, and perhaps on the rates of evaporation throughout the year. As we will see in later sections, all these elements may be used in climatic classifications.

## FACTORS AFFECTING THE CLIMATE

The climatic elements are controlled on a day-to-day basis by the passage of the sun, by the nature of the weather systems, and by local atmospheric factors, such as local winds and air movements. In the longer term, climate is determined by the relationship of the area to the sun and by its position relative to major atmospheric features such as the permanent centres of high or low pressure, or the main components of the circulation. Thus it is possible to distinguish seven main factors that control climate:

1. latitude
2. altitude
3. the distribution of land and sea
4. the nature of ocean currents
5. the distribution of mountain barriers
6. the pattern of prevailing winds
7. the location of the main centres of high or low pressure

Latitude affects the seasonal pattern of temperature (Figure 7.1), for as we have seen, the inputs of solar radiation decline markedly away from the

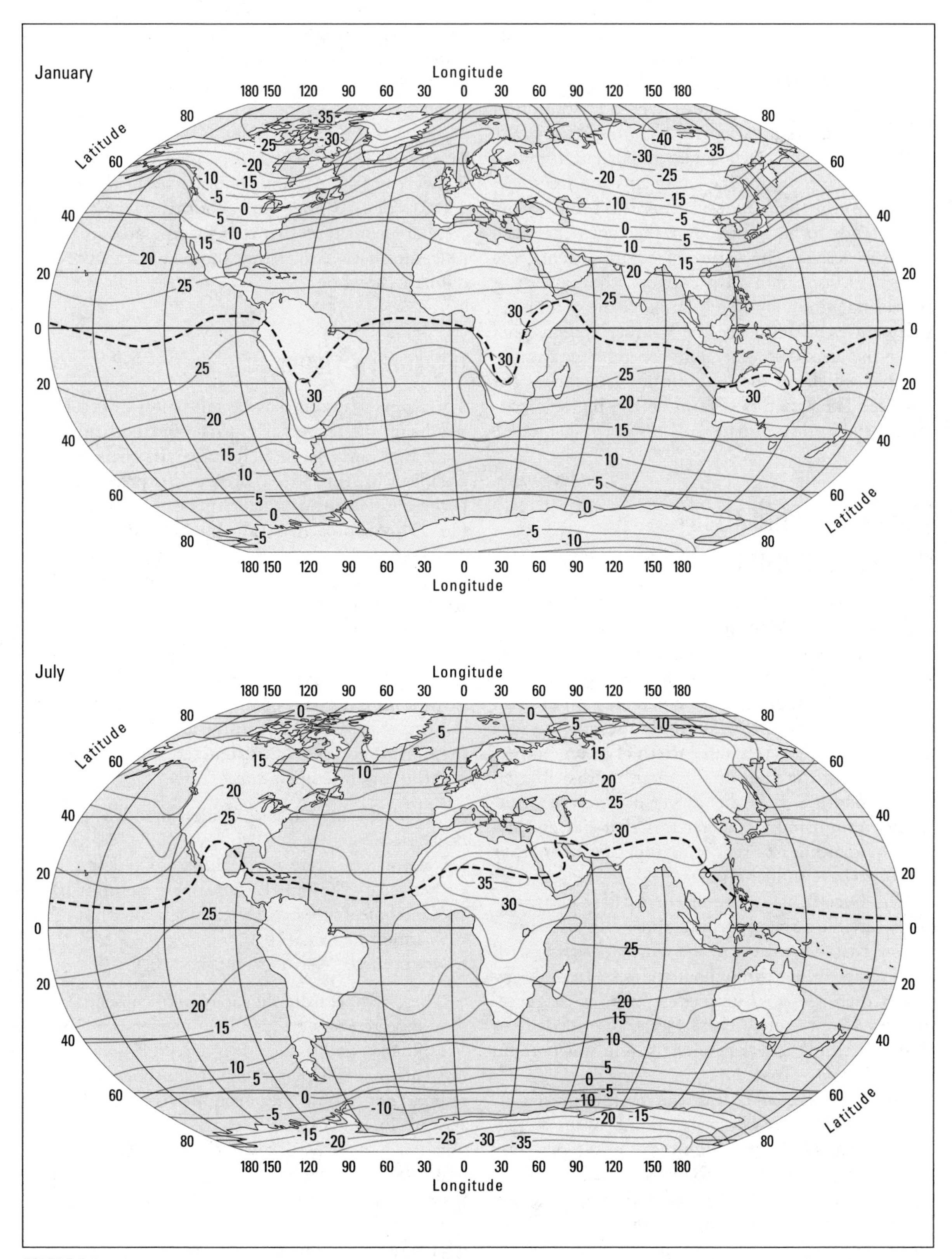


FIGURE 7.1 ▼ Mean sea level temperatures in January and July (°C). The approximate positions of the thermal equator are shown by the heavy broken line (after Barry and Chorley, 1982)

equator. Altitude also affects temperature, for the atmosphere becomes cooler with height at an average rate of about 6.4C° for every 1000 m (the average environmental lapse rate). The distribution of land and sea has a variety of complex effects. We will mention one of the more important implications in Chapter 8, for the land gains and loses heat more rapidly than the sea; thus the temperature range tends to be greater over the continents than over the oceans. The nature of ocean currents, the distribution of mountain barriers, the pattern of prevailing winds, and the location of the main pressure centres all influence the weather systems that affect the area. They therefore control the 'average' weather which, in the long term, characterizes the climate. These factors determine the regional pattern of the world's climate.

# Climatic classification

## CLASSIFYING CLIMATES

If we compared the climatic record of all the observing sites in the world, it is highly unlikely that we should find any two that were identical. Each location possesses an individuality that is unique. On the other hand some of these differences are so small as to be insignificant. Many similarities may be seen between, say, the levels and pattern of mean temperature during the year, or the monthly distribution of precipitation. It is this aspect of similarity running through individual records that has encouraged climatologists to group together areas that appear to have similar climates.

The Greeks made one of the earliest classifications, dividing the earth into three climatic zones. In low latitudes there was the winterless zone, with high temperatures throughout the year. In high latitudes there was the summerless zone, with generally low temperatures. There was an intermediate or middle latitude zone where one season was hot and the other cool or cold.

Such a scheme has at least the merit of great simplicity. The system works if you only want to distinguish areas of the world in terms of temperature. But it fails to take into account other aspects of climate such as precipitation, nor does it provide any information about the range of values or their variability.

After the Greeks, few attempts were made to classify climates until our century, when interest revived. Numerous classifications have been proposed based upon, for example, vegetation, water needs, atmospheric circulation, and human physiological responses. Some of the best known are by Miller (1951), de Martonne (1948), Köppen (pronounced Kerp'n) (1918) and a derivation by Trewartha (1954), and Thornthwaite (1933 and 1948). The Köppen and Thornthwaite schemes were noteworthy in that they incorporated quantitative information to define the boundaries between climatic categories.

## THE KÖPPEN SYSTEM

The Köppen system has become most popular as a teaching aid, especially with the modifications made by Trewartha or Geiger. Details of the scheme are given in Appendix II. Köppen is presented here because it was the original climate classification system. This approach is valuable because it helps us understand the objectives of climate classification. Despite criticisms, Köppen holds up remarkably well. At the time it was developed, very few weather stations existed to establish climate patterns. The lack of data forced Köppen to approach the problem from another direction. Briefly, the classification is based on annual and monthly means of temperature and precipitation. Five of the six major categories are recognized on the basis of monthly mean temperature:

A. Tropical rainy climate. Coldest month > 18°C.
B. Dry climate.
C. Warm temperate rainy climate. Coldest month between –3°C and 18°C; warmest month > 10°C.
D. Cold boreal forest climate. Coldest month < –3°C; warmest month > 10°C.
E. Ice climate. Warmest month < 10°C.

These temperature divisions were chosen because they appeared to correspond with vegetation boundaries. Plants were assumed to provide a good overall indication of climatic conditions because they respond not only to mean conditions of temperature and precipitation but also to the variability and seasonality. Not surprisingly, a close correspondence is found between many climatic regions and vegetation. (Compare the maps in Appendix II and Chapter 29.) It must be stressed that the Köppen system deals with natural vegetation regions, but in many parts of the world the vegetation pattern has been dramatically changed by human activities. The major change has been in the removal of forests, which began some 10 000 years ago and has increased in rate and extent ever since. In Europe, deforestation has gone on for over 2000 years, with the most intense period beginning around A.D. 1500. Temperate North American deciduous forests have been dramatically reduced in the last 200 years, from 170 million hectares to 10 million hectares. Tropical rain forests are experiencing the same onslaught. To what extent does this change the climate? The Köppen system assumes that climate creates vegetative patterns under natural conditions. The evidence is increasing that removal of the forests in turn changes the climate.

If these had been the only zones identified, the classification would have been very limited. In addition, further descriptions are provided to yield more information about precipitation and temperature conditions. From these combinations, twelve major climatic types were designated. These are summarized here in Table 7.2, and detailed descriptions are given in Appendix II.

Table 7.2

*The Köppen-Geiger climate classification system, revised 1953*

| | |
|---|---|
| Af | Tropical rain forest climate |
| Am | Monsoon variety of Af |
| Aw | Tropical savanna climate |
| BS | Steppe climate |
| BW | Desert climate |
| Cf | Mild humid, no dry season |
| Cw | Mild humid, dry winter |
| Cs | Mild humid, dry summer |
| Df | Snowy forest climate, moist winter |
| Dw | Snowy forest climate, dry winter |
| ET | Tundra |
| EF | Perpetual frost |

### THORNTHWAITE CLASSIFICATION

A development away from vegetationally derived classifications came with the Thornthwaite scheme in 1948. Thornthwaite based his work on the concept of potential evapotranspiration (PE) and the soil moisture budget. Potential evapotranspiration is defined as 'the water loss estimated to occur from a full vegetative cover growing on a soil that is always saturated.' Potential evapotranspiration is essentially a function of climate alone and so can be used as a classification system. The most important influences are solar radiation and temperature, atmospheric moisture, and wind. In general, PE increases as sunshine, temperature, and wind increase, and as the humidity decreases.

In this classification, boundaries are drawn by comparing precipitation with potential evapotranspiration. If we compare the monthly variation of precipitation and PE in graphic form (Figure 7.2), we can easily see whether or not precipitation exceeds PE. The graph for Seattle illustrates the classic case of rainy winters and dry summers. When the rainfall is greater than PE, the excess is assumed to be absorbed by the soil until an equivalent of 100 mm of precipitation has been added. As rainfall increases rapidly in the autumn, this recharge of soil moisture is achieved by November, and until PE becomes greater than rainfall in late April we have a water surplus amounting to 391 mm. Most of this will reach the river systems as runoff. During early summer, the 100 mm of moisture in the soil is used, primarily by plant growth and evaporation, then further evapotranspiration leads to a water deficit amounting to 208 mm, which lasts until October. Distinct break-points in the curves can be seen.

Thornthwaite's system has a certain scientific rigour but it is difficult to use in large areas of the world because of the absence of the necessary climatic data. Köppen had to deal with the problem of lack of data as well. The problem still exists for most regions. Either there is no data being collected or records have not been kept long enough to be usable.

Thornthwaite's system is more valuable because it underlines the importance of water for vegetation, soils, and climate. A basic exercise using the Thornthwaite water balance approach is in Appendix II.

## Regional climates

Whatever might be argued for or against different methods of climatic classification, the main application for climatic data is descriptive; our interest is in temperature levels and rainfall totals, their seasonal-

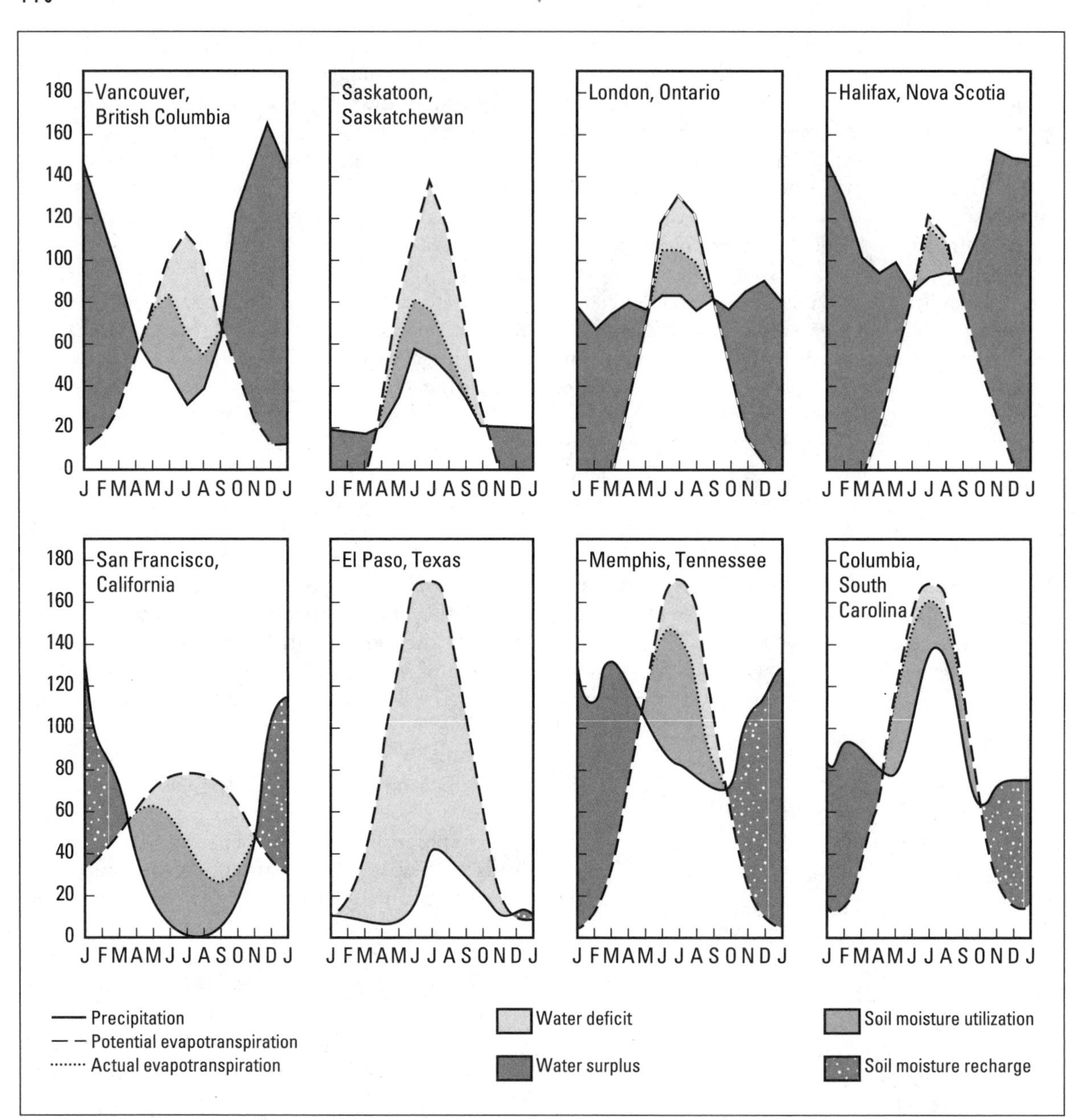


FIGURE 7.2 ▼ Average monthly water budgets at selected stations in North America

ity and extremes. If we also gain some understanding of why the climates are as they are, so much the better. It is clearly impossible (and unnecessary) to describe all individual examples of climate, so some scheme must be adopted to identify groups. In this book, we shall make a basic subdivision into tropical, temperate, and polar climates. First we shall examine the main features of each atmospheric circulation system, and then look at examples of different climatic regions within each area.

# Tropical climates

The tropics have been described as the firebox of our atmospheric engine. Most of the sun's energy is absorbed here, energy that is transferred eventually into the cooler, energy-poor latitudes. This transfer is brought about by wind flows and ocean currents.

A simple approach to climate in the tropics is to distinguish four main areas, as shown in Figure 7.3:

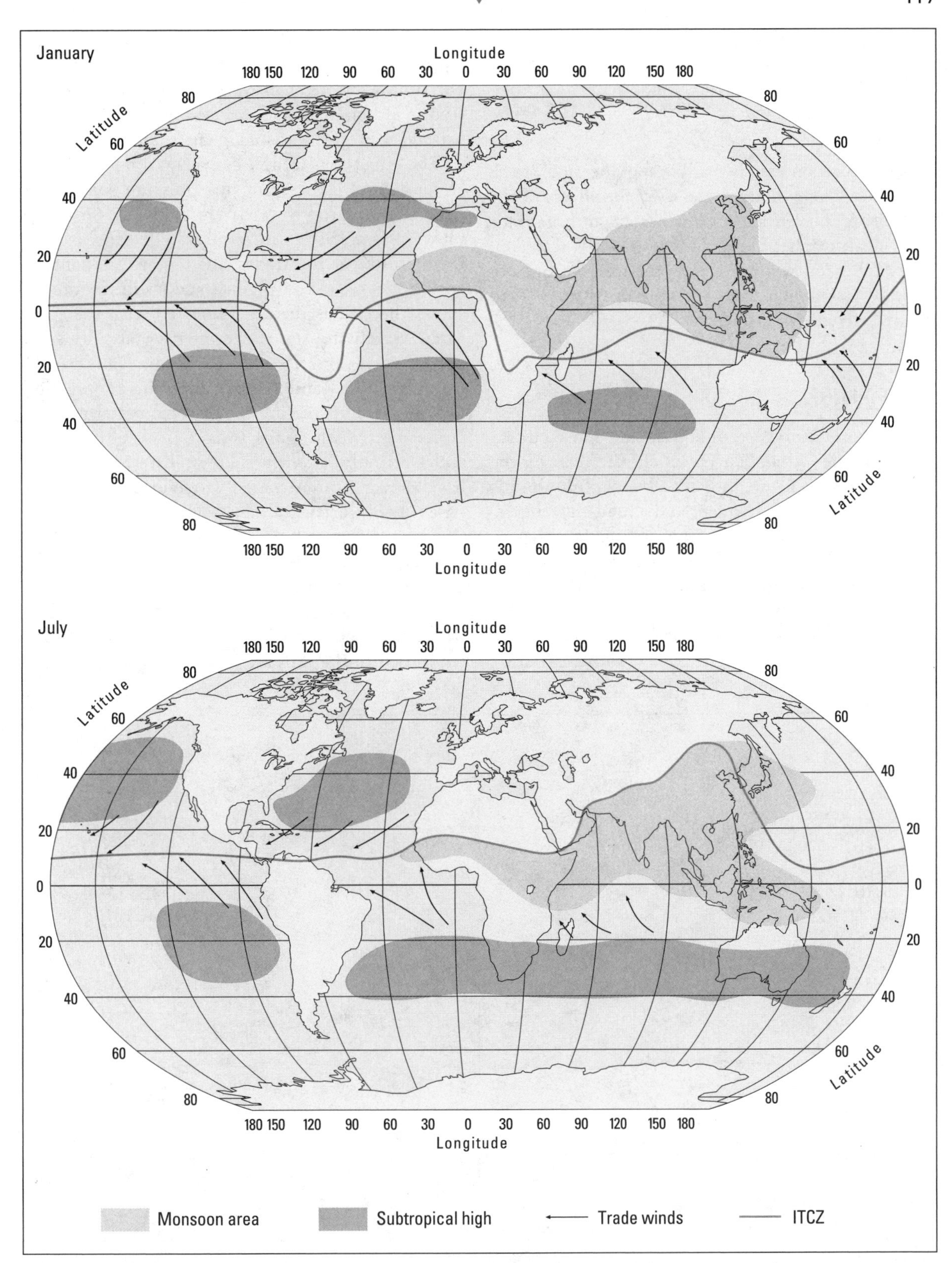


FIGURE 7.3 ▼ Areas affected by the four main climatic zones in the tropics in January and July. The boundaries are really transition zones and may show considerable shifts in position from year to year

- the equatorial trough zone or Intertropical Convergence Zone (ITCZ)
- the subtropical highs
- the trade wind areas
- the monsoons

The monsoon area is really a modification of the trade wind zone brought about by the effects of the continents. Nevertheless, as it is such an important part of atmospheric circulation, it will be treated separately. You will recall that the main features of tropical circulation were outlined in Chapter 5, to which reference may be made.

## THE EQUATORIAL TROUGH

It is the equatorial trough area that most closely meets people's idea of a tropical climate. During the day, clouds build up into massive cumulonimbus displays. Rainfall is frequent and abundant, temperatures and humidity are high, acting together to give us the tropical rain forests. The structure of the atmosphere, though, is not so simple as this model may suggest. The multitude of names that have been used for the area give some idea of its variety—the doldrums, the intertropical front, intertropical convergence zone, intertropical trough, equatorial trough, or intertropical confluence zone. For simplicity we shall refer to it as the equatorial trough although it does extend toward the subtropics, and it is quite variable in character.

The equatorial trough has many different forms. It represents the area of low pressure somewhere near the equator toward which the trade winds blow. The precise form it takes will depend upon the stability of the trades, their moisture content, and the degree of convergence and uplift. Figure 7.4 shows the basic structure associated with the equatorial trough and the Hadley cell; we also see some of the variations that can develop. The trough migrates north and south, following the vertical rays of the sun, creating the seasonal pattern of precipitation. Much of the trough is over the oceans and it is only recently that satellite photographs have shown us more detail of cloud forms and the structure of the equatorial trough.

The structure of the trough is variable (Figure 7.5), and careful scrutiny can show that a hierarchy of cloud is present. The larger element is a cloud cluster, perhaps 100 km to 1000 km in length. Within the cluster there are convective cells and embedded in the cells are individual convective ele-

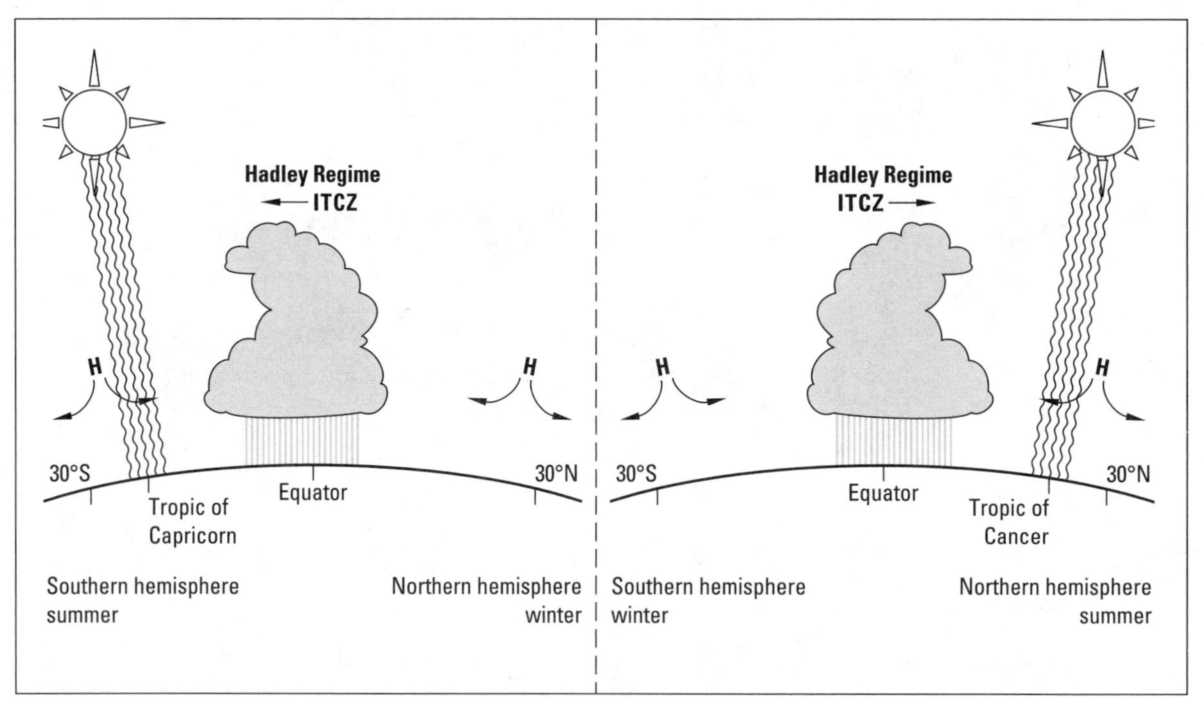


FIGURE 7.4 ▼ Generalized north–south cross sections through the equatorial trough zone (ITCZ). Note the relationship between the seasonal movement of this zone and the wet/dry seasons

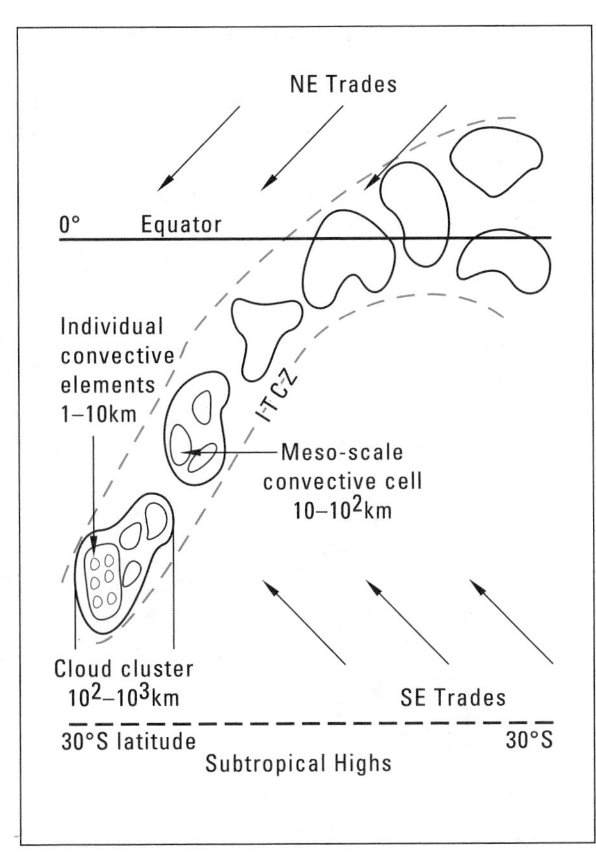


FIGURE 7.5 ▼ The Intertropical Convergence Zone (ITCZ) or equatorial trough, showing the organization of associated convective cloud systems. Plan view (after Mason, 1970)

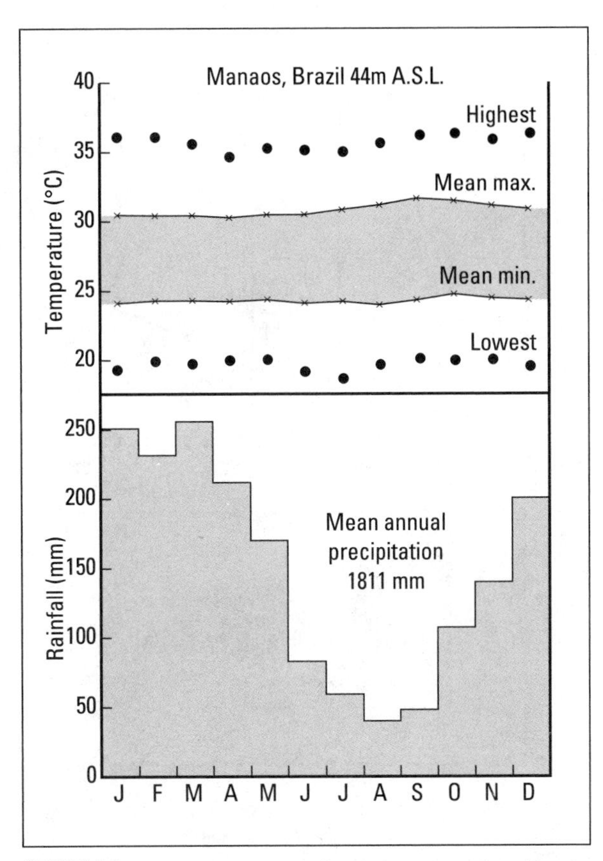


FIGURE 7.6 ▼ Climatic graphs for Manaos, Brazil, in the equatorial trough zone

ments that can give the heavy rain characteristic of the equatorial trough.

What is the climate of the equatorial trough like? Figure 7.6 gives an example of monthly rainfall and temperatures at Manaos in Brazilian Amazonia. The mean monthly maximum temperature varies by 2.8C° over the year, and the mean monthly minimum by only 0.6C°. Extremes are rare and insignificant by temperate latitude standards. Mean annual rainfall is high, with 1811 mm, though even in this zone there is a drier season when rain days are fewer. This is true of most of the equatorial trough zone, though the intensity and duration of the dry season vary. Only a few areas have no dry season, as in Indonesia, where Padang in Sumatra (7 m above sea level) receives an average rainfall of 4427 mm and only one month has less than 250 mm. The driest season occurs when the trough moves poleward in response to continental heating in the summer hemisphere. As we move further away from the equatorial trough zone, so the dry season lengthens and we reach the monsoon or trade wind areas.

## THE SUBTROPICAL HIGHS

The **subtropical high pressure zones** act as the meteorological boundary between the tropical and temperate latitudes. The dominant air movement is away from the highs; circulation is maintained by the subsiding air from the Hadley cell. Because the air is subsiding it tends to be warm and dry. An inversion develops in the lower atmosphere (Figure 7.7) and so these subtropical highs are generally cloud-free and deficient in rain. Where the highs remain fairly constant in position we find the main desert areas of the world—the Sahara, the Kalahari, the Great Australian Desert, and the Great American Desert.

If we look at a map of surface pressure, the high pressure centres that we would expect over the

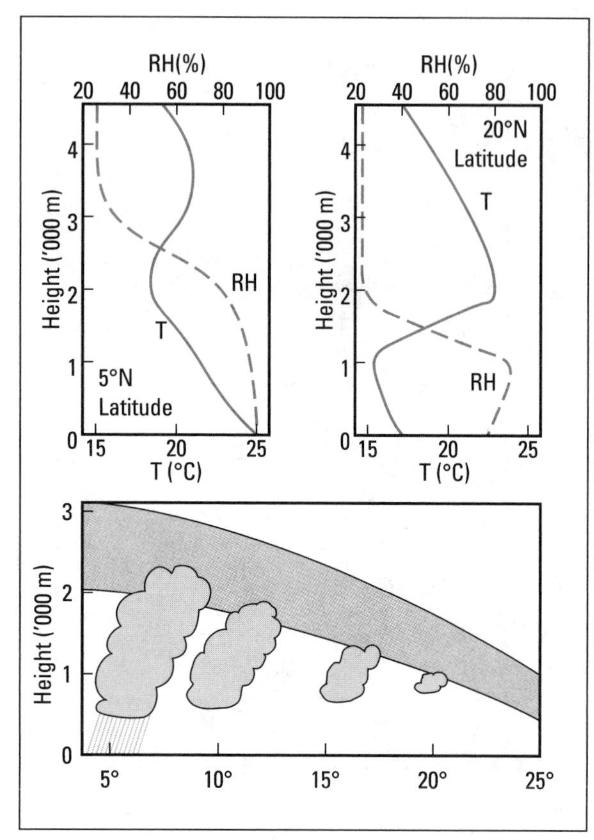


FIGURE 7.7 ▼ A generalized meridional cross section in the trade wind zone and (above) vertical temperature and humidity profiles at 5°N and 20°N (after Nieuwolt, 1977)

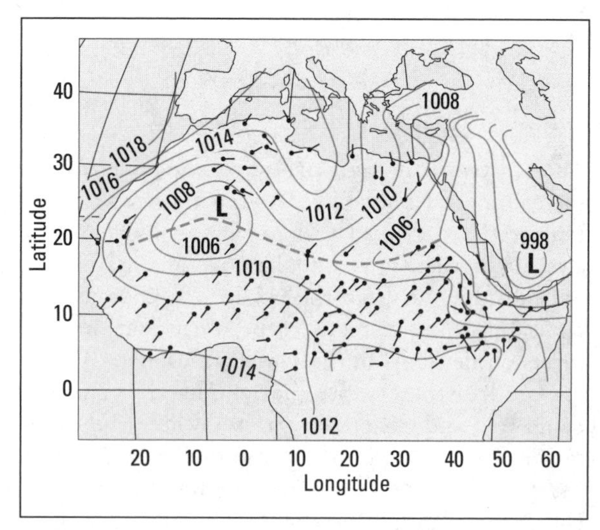


FIGURE 7.8 ▼ Mean surface pressure and wind direction over North Africa in July. The broken line indicates the mean position of the equatorial trough (after Thompson, 1965)

desert areas may be absent (Figure 7.8). On the contrary, there is often a weak low pressure area. These lows are the result of intense heating of the ground surface during the cloudless days, taking temperatures above 40°C in summer. The air becomes less dense and thermal lows form. As they are a consequence of surface heating they tend to be fairly shallow and are replaced by high pressure by the 850 mb level (Figure 7.9).

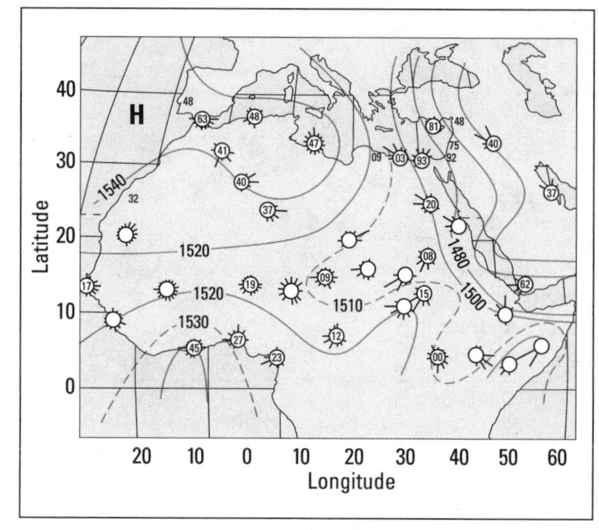


FIGURE 7.9 ▼ Mean 850 mb pressure surface and wind rose over North Africa in July. The figures inside the wind rose denote the mean height of the pressure surface, i.e., 37 indicates 1537 metres (after Thompson, 1965)

We would expect the climate of these zones to be characterized by little rain and extremes of temperature. The data for Atbara (Figure 7.10) confirm this point. In midsummer, mean maximum temperatures are 42°C, but in winter the mean minimum temperatures are only 8°C and frost can occur occasionally. The very dry atmosphere helps by allowing longwave radiation from the ground to escape to space, with little counter-radiation from water vapour or clouds. Even on the coast of Africa, at Bahrain, nighttime temperatures are cool in winter, though frost is very rare. Precipitation is negligible. Rain falls on about ten days per year giving a total of about 75 mm. Most of this falls in winter and spring, when temperature latitude depressions extend their effects far south and do give occasional rain.

Desert rainfall is notoriously variable. No rain may fall for several years, to be followed by several

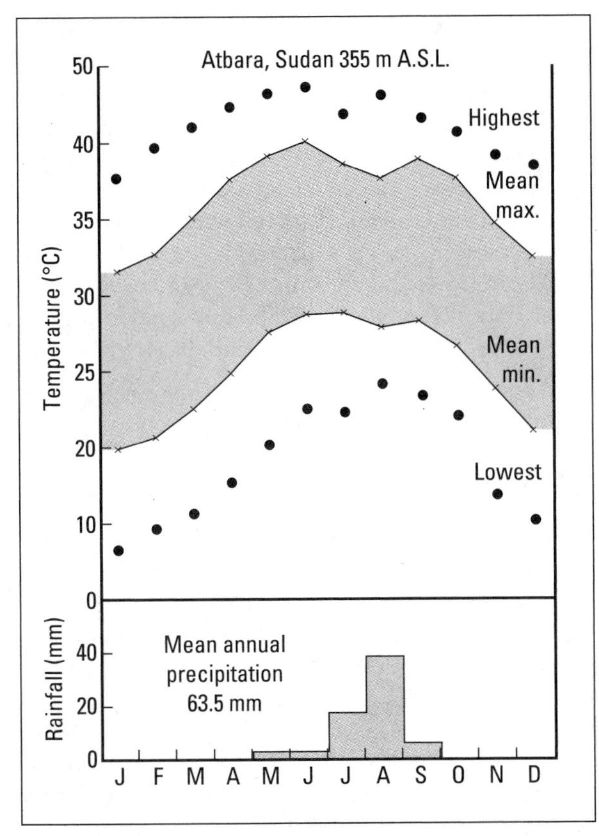


FIGURE 7.10 ▼ Climatic graphs for Atbara, Sudan, dominated by the subtropical anticyclones

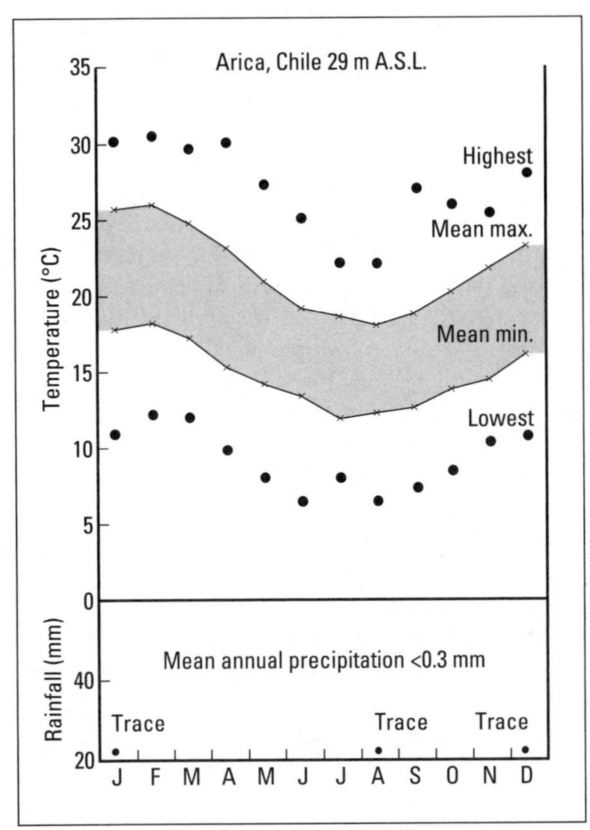


FIGURE 7.11 ▼ Climatic graphs for Arica, Chile, in the subtropical high pressure belt. Global and local factors ensure that there are few occasions favouring precipitation

heavy showers giving a few centimetres. It is this variability that makes the average rainfall figures for desert areas almost meaningless. Annual rainfall totals at Phoenix, Arizona, for a 100-year period, for example, show a longterm mean rainfall of 186.9 mm, but most years (55 percent) have less than this, and a few years have very much more. It has been said that for deserts average rainfall is the total that never falls.

In some of the subtropical high pressure belts additional factors reduce the likelihood of rain. On the west coast of the Sahara, the Kalahari, the Atacama, and Lower California deserts, cold ocean currents flow just offshore. They cool the air and make it even more stable. Mist and fog may be frequent but rain is rare. One of the driest places in the world is Arica in the Atacama Desert of Chile (Figure 7.11). Years have elapsed between rainstorms, and even then only a few millimetres may fall. Conditions here are similar to other coastal deserts near a cold ocean current, but in addition the prevailing winds blow from the southeast. To reach the Chilean coast they must cross the main mountain barrier of the Andes, some 5000 m, which further emphasizes stability and dryness. The result of all these factors, acting against the mechanisms of rainfall generation, is to produce one of the driest parts of the earth.

## THE TRADE WINDS

Blowing away from the subtropical anticyclones of each hemisphere are the **trade winds**—northeasterlies in the northern hemisphere and southeasterlies in the southern hemisphere. The trades can be some of the most constant and reliable winds of the world. If we measure constancy by the persistence of winds from the same general direction (±45° of the mean direction), then over 80 percent of the time the winds will be from this direction. Hawaii is a good example, with winds on the north coast blowing from the northeast 96 percent of the year (Figure 5.20).

Around the tropics the trade winds are very stable, being greatly affected by subsidence so the moist layer near the surface is thin (Figure 7.7). The sudden rise of temperature and drying of the air at about the 900 mb pressure surface is known as the trade wind inversion. In the northeast of the Atlantic and Pacific oceans, it may be only a few hundred metres about the surface, effectively preventing rainfall over the oceans. When islands, such as the Canary Islands, rise through the inversion, the lower windward slopes may be moist due to cloud and some rain, but above the mean level of the inversion and on the leeward slopes we have deserts. Figure 7.12 shows the climatic conditions at Santa Cruz de Tenerife in the Canary Islands. The cool Canary current helps to keep temperatures moderate in summer.

The trade winds gradually pick up moisture as they blow away from their source areas—the anticyclones. We can see this in the shape of the trade wind cumulus clouds. They are the visible sign that moisture is being evaporated from the seas and partly condensing as clouds. With more moisture being added and the influence of the anticyclones weakening, the intensity of the trade wind inversion weakens and it gets higher. Rainfall is now more likely to occur, and if we look at the situation on the western side of the Atlantic Ocean we have a much moister climate, although with a distinct wet and dry season. In the Caribbean, the wet season is usually summer and autumn when rainfall from the moist trades is supplemented by torrential rain from tropical storms. Their effects will be discussed in more detail later.

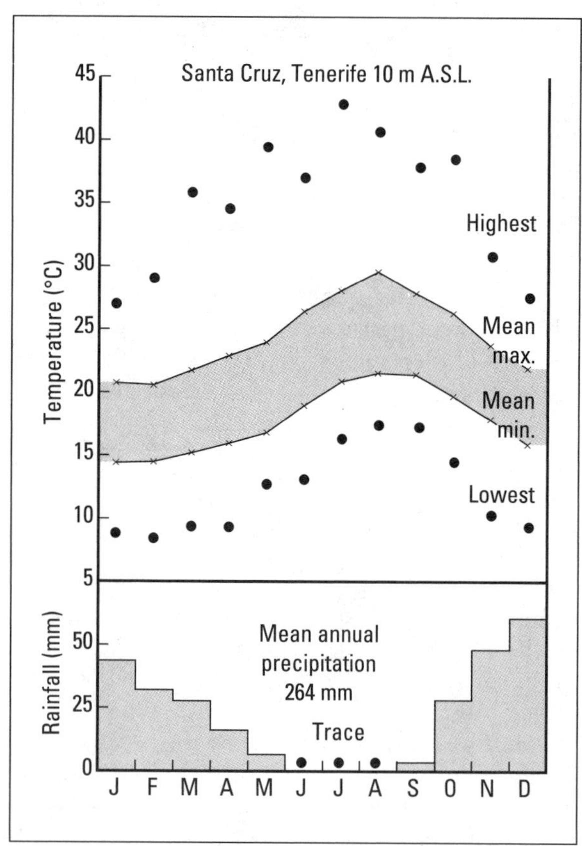


FIGURE 7.12 ▼ Climatic graphs for Santa Cruz de Tenerife, Canary Islands, displaying a maritime trade wind climate

## THE MONSOONS

In some parts of the world, the wind systems appear to experience a seasonal reversal. In one season they may be blowing from the southwest; in the other season they are from the northeast. There has been much discussion among learned authorities about the precise limits of the **monsoon**, but Figure 7.3 summarizes these ideas. Without a doubt, a large area of the tropics is affected by a seasonal reversal in areas where we might expect the trade winds to be dominant.

The reason for the reversal is the positions of the continents in the northern hemisphere. During the northern hemisphere summer, surface heating of the continental land masses is intense. A shallow surface low pressure centre forms over the Sahara, over India, and over central Asia. The equatorial trough moves northward, allowing an inblowing of moist southwest to south winds to give the wet season in West Africa, India, China, and most of southeast Asia. In winter the continents cool down, high pressure becomes established at the surface, and winds between northeast and north predominate. This is the cool dry season for the monsoon areas of the northern hemisphere.

In the southern hemisphere, land masses are smaller and only Australia develops the semblance of a monsoon. Even then its influence does not extend very far inland. Over East Africa, set astride the equator, a seasonal reversal does occur, but the winds tend to be blowing parallel to the coast. As a result, the rainy season is between the main monsoon flows, rather than during one of them, as in most of the other regions. Nairobi, Kenya, has its rainy season from March to May and again in November and December (Figure 7.13), in the lulls between the main northeast and southeast monsoon.

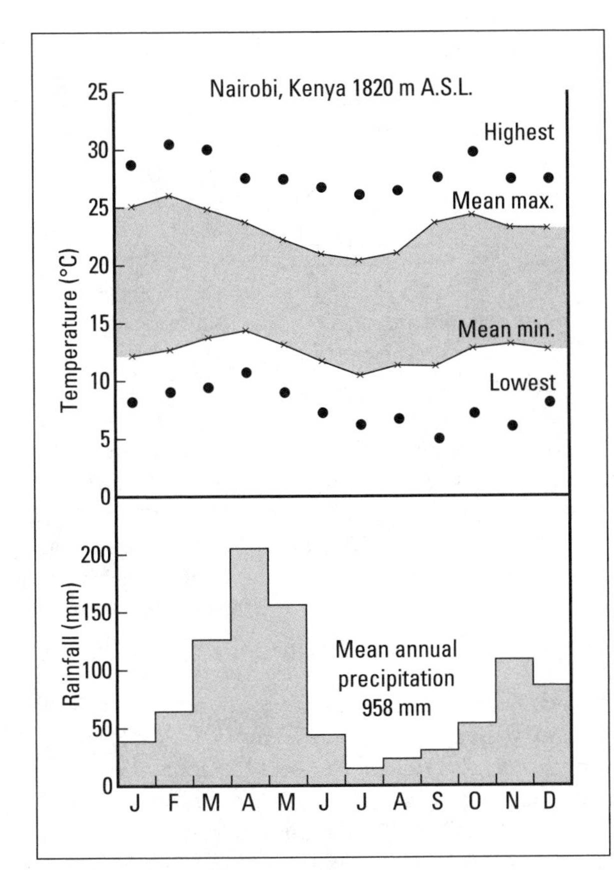


FIGURE 7.13 ▼ Climatic graphs for Nairobi, Kenya. There is a bimodal rainfall distribution related to lulls in the main monsoonal flows, unlike that of other monsoon areas where the heaviest rains are associated with the strongest monsoon flows (see Figure 7.14)

Freetown, Sierra Leone, illustrates the classic monsoon climate of West Africa. Temperatures are high throughout the year; there is no source of cold air to affect the area in winter, though away from the influence of the sea, temperatures fall as low as 15°C. From January to March, rainfall is unusual. In April and May, thundery activity begins to develop, often associated with westward-moving storms or disturbance lines in the northeasterlies, or the Harmattan, as they are known locally. Toward the end of May, the monsoon rains start and rain falls on most days. Large totals accumulate; water is flowing everywhere and crops are growing rapidly. The air is very humid even during the day. During October, the steady rains cease, to be replaced briefly by thundery activity of the disturbance lines again. By December the dry season has returned and the land dries out.

This sequence of events—dry season, disturbance line thunderstorms, monsoon rain—has been used to distinguished the climates of West Africa. The zones are aligned parallel to the equatorial trough and move northward and southward with the season. To the north of the area called the Sahel, the monsoon rains may not penetrate in some years and rainfall will be from the disturbance lines only. When the trough is at its northernmost extent, a short dry season appears in the southernmost parts of the region on the coast. The effects of the South Atlantic subtropical high are believed to cause this midsummer dry season.

The best known area of the monsoon is India. Variability does exist across the vast subcontinent, but for much of the area there is a summer wet season and a winter dry season. New Delhi (Figure 7.14) illustrates typical conditions. The wet season is shorter and not so rainy as Freetown; most of the rain falls in the three months July to September

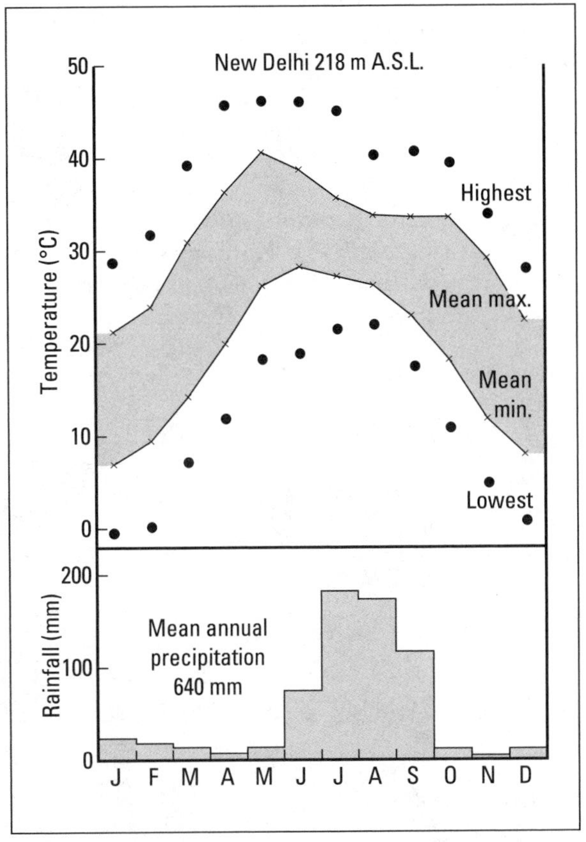


FIGURE 7.14 ▼ Climatic graphs for New Delhi, India, showing the typical single precipitation peak for most monsoon areas. Note the summer temperature peak occurs before the onset of the main rains

with little at other times. The cloudiness associated with the rain reduces sunshine and tempers the heat. As a result, May and June are the hottest months of the year, with mean maximum temperatures well above 40°C. The monsoonal cloud clears in late September and October, so the temperatures remain high, the lack of cloud compensating for the reduced intensity of the sun. In winter, when the winds blow from the continental interiors of Kashmir, temperatures fall and frost may occur. Occasional midlatitude depressions enter from the northwest and give light rain to the northwestern areas of India including Delhi. Farther northwest into northern Pakistan, the summer monsoon rainfall is weak and the winter depressions make December and January the main rainy season.

In detail, the onset, retreat, and origins of the monsoon are more complex. They depend upon an interaction between the massive land and sea breeze system generated by heating and cooling of the continents, the winds of the upper atmosphere and, in Asia, by a modification by the Tibetan Plateau, whose mean altitude reaches 3660 m—almost halfway through the atmosphere. Normally the southwesterly monsoons are overlain by easterly winds. Waves may develop in the upper easterly flow and, when they favour ascending motion by divergence, rainfall develops in the moist monsoonal air (Figure 7.15). Northward the layer of moist air gets thinner and thus rain frequency and intensity decrease, as shown in West Africa.

## TROPICAL CYCLONES

The above sections describe the main flows of air of the tropics. Within these airflows, disturbances can develop that alter their characteristics and are often the source of extensive precipitation.

The major disturbance of tropical latitudes is the cyclone. The main features have already been mentioned in Chapter 6, so here we summarize their regional distribution and effects on climates.

If we look at the parts of the globe affected by tropical cyclones, it is apparent that they only develop over the warmer parts of the oceans (Figure 6.18). In each hemisphere, cyclones are most likely to strike in the summer and autumn. Along the Atlantic and Gulf coasts of the USA, the normal hurricane season is from June to November. Early in the season, storms develop in the Gulf of Mexico or the Caribbean but there is a progressive eastward movement of their starting points until September, when they may start as far east as the Cape Verde Islands off West Africa. After September the area of origin shifts back toward the Gulf of Mexico. Another seasonal change affects the zone of recurvature. Most storms initially move westward but at some stage many begin a curving track toward the north and then the northeast (Figure 7.16). The average latitude of recurvature is at its northernmost position in August and farthest south in November, linking with the change in position of the upper westerlies that help to steer the hurricanes.

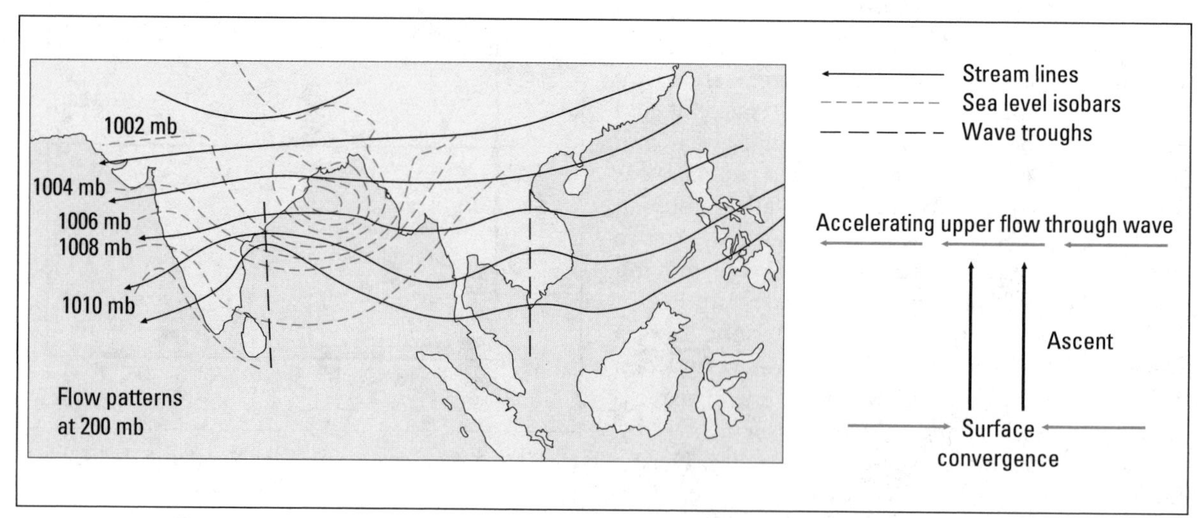


FIGURE 7.15 ▼ Relationship between surface and upper atmospheric flow during the Indian summer monsoon. The map shows the flow patterns at 200 mb on 26 August, 1955 (after Koteswaram and George, 1958)

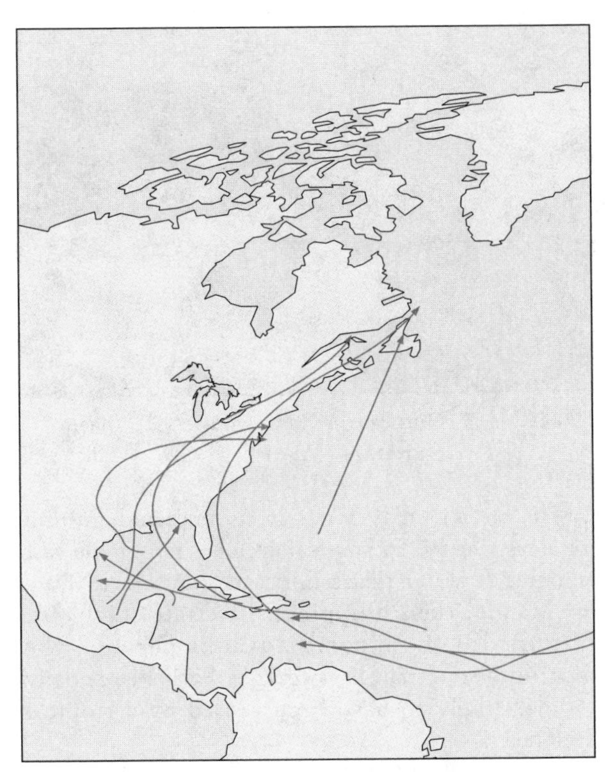

FIGURE 7.16 ▼ Tracks of hurricanes and tropical storms during the 1979 Atlantic hurricane season. Not all tropical storms reach hurricane intensity (> 33 m $s^{-1}$) but exhibit similar properties (after Hebert, 1980)

The Pacific Ocean has the most hurricanes but the figures are difficult to compare (they are called typhoons in southeast Asia and the western Pacific, and tropical cyclones in the Indian Ocean). Until satellite photography became available, many storms could reach hurricane intensity and decay without being recorded by the global observing network. It was only in the Atlantic that shipping was sufficiently dense to record most of the storms. As a result, in all areas except the Atlantic, the number of reported hurricanes has increased markedly, especially in the eastern Pacific and the south Indian Ocean. Changes in the frequencies of hurricanes do occur. Figure 7.17 shows these variations in the Atlantic, where records are most reliable. The number of storms is closely linked to the surface temperatures of the sea and the degree of cyclonic curvature in the trade winds; warmer seas and more cyclonicity mean more storms.

Hurricanes affecting the USA have been classified according to their intensity (Tables 7.3 and 7.4). Few recent storms have been category 5, though Camille (1969), David and Frederick (1979), and Gilbert (1988) reached this level. Hurricane Andrew (1992) was only classified as a class 4 storm, yet it gives some measure of the potential for damage of a class 5 storm (Figure 7.18). There has been a trend toward fewer deaths due to storms but an increased cost of damage. Improved forecasts of storm tracks have enabled preparations to be made to reduce risk to life but with greater material prosperity plus inflation, the cost of a storm inevitably rises, especially if it crosses an urban area.

In coastal regions a major cause of damage is the storm surge. The very low atmospheric pressures mean that the weight of air at the surface is reduced. This allows the water surface to bulge up as much as 6 m, a bulge that moves along with the hurricane. When it moves on-shore, it acts like a very long wave, causing extensive flooding inland. If the surge arrives at high tide with an onshore wind the damage can be even greater.

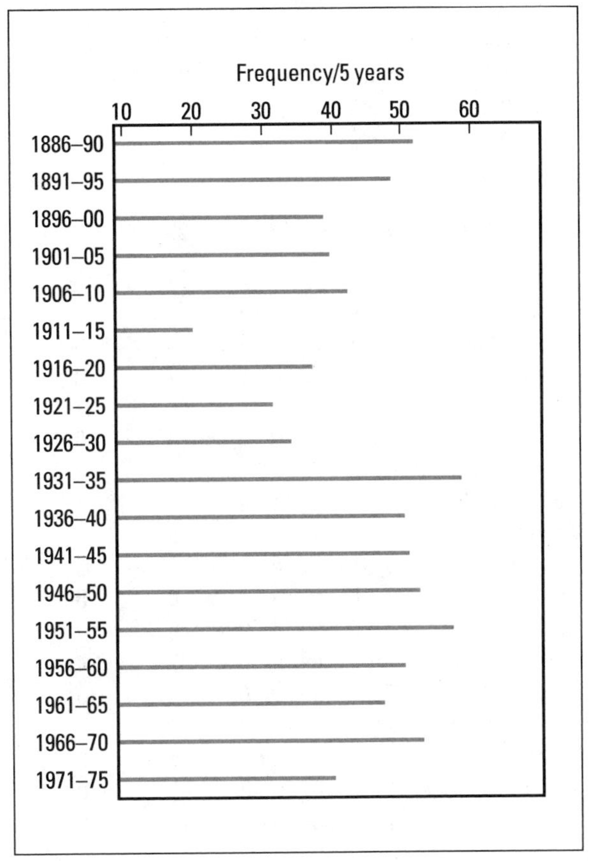


FIGURE 7.17 ▼ Atlantic hurricane frequency 1886–1975 (after Riehl, 1979)

## Table 7.3

*Hurricane scale of intensity*

| *Scale number* | *Central pressure (mb)* | *Winds ($m\ s^{-1}$)* | *Surge (m)* | *Damage* |
|---|---|---|---|---|
| 1 | 980 | 33–42 | 1–1.5 | Minimal |
| 2 | 965–979 | 43–49 | 1.5–2.5 | Moderate |
| 3 | 945–964 | 50–58 | 2.5–3.7 | Extensive |
| 4 | 920–944 | 59–69 | 3.7–5.5 | Extreme |
| 5 | <920 | >69 | >5.5 | Catastrophic |

After P.J. Hebert and G. Taylor, 'Hurricanes', *Weatherwise*, 32 (1979), pp. 100–7.

## Table 7.4

*Number of hurricanes (direct hits) affecting United States and individual states, 1900–78*

| *Area* | *Categories* 1 | 2 | 3 | 4 | 5 | *All* | *Major hurricanes* |
|---|---|---|---|---|---|---|---|
| United States | 47 | 29 | 38 | 13 | 2 | 129 | 53 |
| Texas | 9 | 9 | 7 | 6 | 0 | 31 | 13 |
| Louisiana | 4 | 6 | 6 | 3 | 1 | 20 | 10 |
| Mississippi | 1 | 1 | 2 | 0 | 1 | 5 | 3 |
| Alabama | 3 | 1 | 3 | 0 | 0 | 7 | 3 |
| Florida | 18 | 11 | 15 | 5 | 1 | 50 | 21 |
| Georgia | 1 | 3 | 0 | 0 | 0 | 4 | 0 |
| South Carolina | 4 | 3 | 2 | 1 | 0 | 10 | 3 |
| North Carolina | 9 | 3 | 6 | 1 | 0 | 19 | 7 |
| Virginia | 1 | 1 | 1 | 0 | 0 | 3 | 1 |
| Maryland | 0 | 1 | 0 | 0 | 0 | 1 | 0 |
| Delaware | 0 | 0 | 0 | 0 | 0 | 0 | 0 |
| New Jersey | 1 | 0 | 0 | 0 | 0 | 0 | 0 |
| New York | 3 | 0 | 4 | 0 | 0 | 7 | 4 |
| Connecticut | 2 | 1 | 3 | 0 | 0 | 6 | 3 |
| Rhode Island | 0 | 1 | 3 | 0 | 0 | 4 | 3 |
| Massachusetts | 2 | 1 | 2 | 0 | 0 | 5 | 2 |
| New Hampshire | 1 | 0 | 0 | 0 | 0 | 1 | 0 |
| Maine | 4 | 0 | 0 | 0 | 0 | 4 | 0 |

After P.J. Hebert and G. Taylor, 'Hurricanes', *Weatherwise*, 32 (1979), pp. 100–7

FIGURE 7.18 ▼ Hurricane Andrew, Florida, 1992 (photo courtesy of NASA)

The mean rainfall totals in summer and autumn for areas affected by tropical cyclones reflect the vast amounts of water that a hurricane can release. Data for Manila, the Philippines, illustrate this point (Figure 7.19). In some 24-hour periods, the monthly average fall has occurred. Such a large daily total is likely to have been caused by a tropical cyclone.

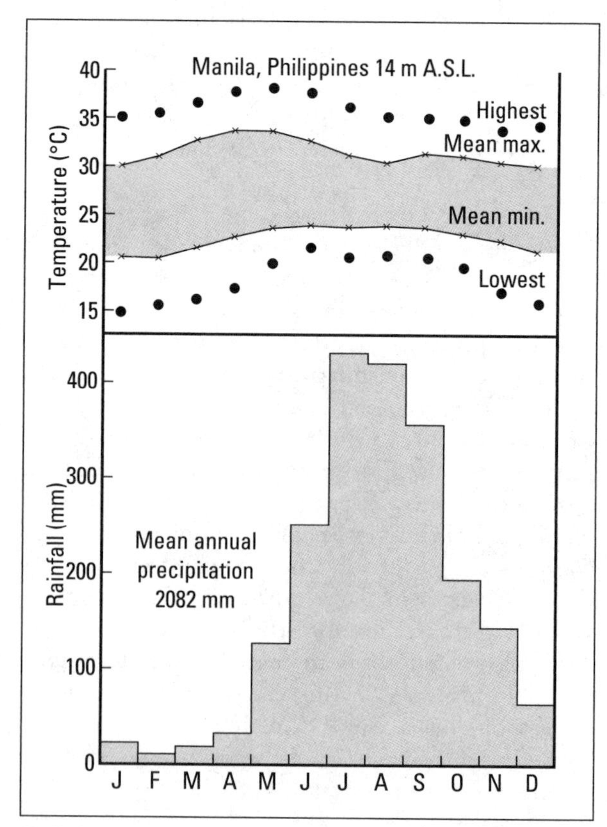


FIGURE 7.19 ▼ Climatic graphs for Manila, Philippines, a maritime tropical climate affected by typhoons

Not all tropical areas are affected by tropical storms or easterly waves, but other, less organized disturbances give appreciable precipitation, as in Mozambique and parts of Brazil. In India, monsoon depressions develop in the moist southeasterly flow up the Ganges valley and bring extensive rains. A few areas miss major disturbances altogether. Anomalous dry zones occur in northeast Brazil, where annual rainfall of less than 500 mm is found. In Somalia in Africa, mean annual precipitation is below 250 mm and aridity prevails (Figure 7.20), although the area is only just north of the equator.

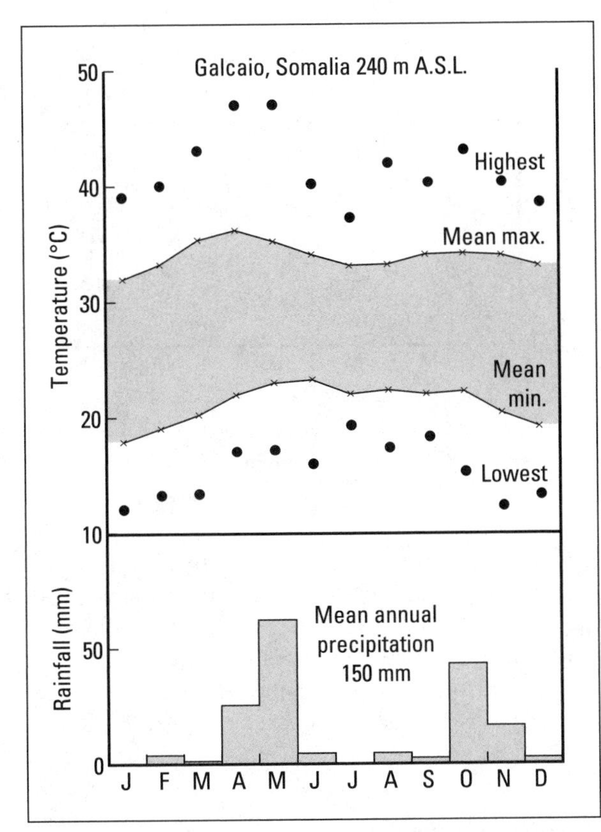


FIGURE 7.20 ▼ Climatic graphs for Galcaio, Somalia, in the equatorial dry zone of Africa

# Temperate climates

## OCEANIC AREAS

Temperate latitudes are dominated by westerly winds. At the surface, winds can be very variable depending upon the movement of depressions and anticyclones. In the upper atmosphere the westerlies really become dominant, forming a series of waves (Figure 5.26) that affect the movement and development of the main surface pressure systems. There is a clear seasonal movement of the main centres of activity. In the summer, the depressions tend to be less intense and follow more poleward tracks; the anticyclone centres also have a more northerly location. During winter the main centres of the subtropical highs move southward and depressions become more frequent in lower latitudes. This type of climate, with mild wet winters and hot dry summers (when anticyclones are dominant), has been called a Mediterranean climate.

As winter is the only rainy season, the frequency and intensity of the actual depression tracks will determine how much rain falls. Consequently, variability is often great with very dry years interspersed with sudden flooding or prolonged heavy rain. Eureka, California, has a classic Mediterranean climate (Figure 7.21) with the distinctly dry summer and wet winter. Perth, in Western Australia, has a similar type of climate (Figure 7.22), although winter in the southern hemisphere is May to September.

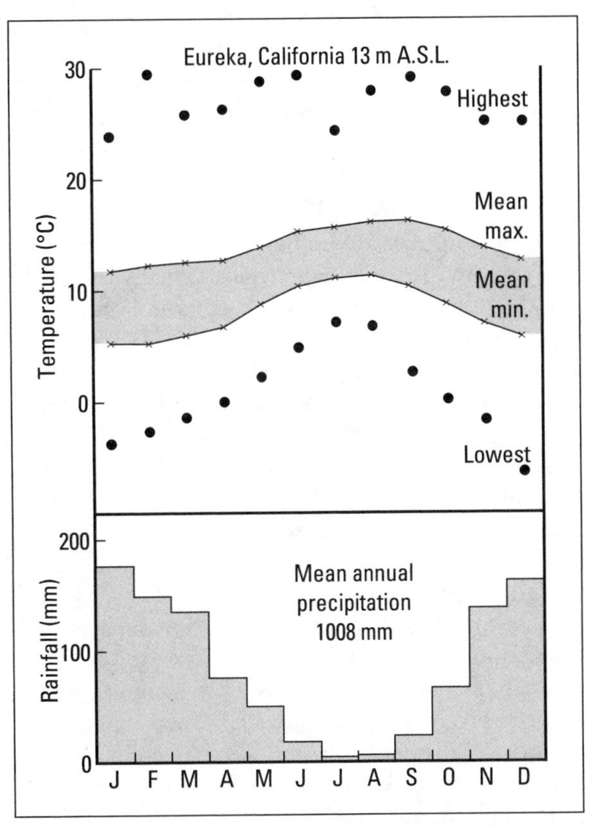


FIGURE 7.21 ▼ Climatic graphs for Eureka, California. The rainfall regime is similar to Perth (Figure 7.22) in the southern hemisphere

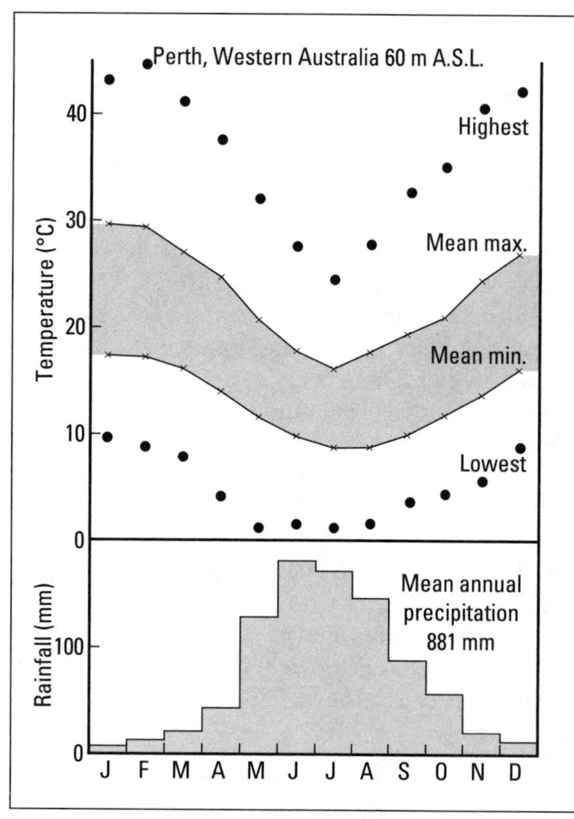


FIGURE 7.22 ▼ Climatic graphs for Perth, Western Australia, showing a Mediterranean climate; in the southern hemisphere, the May to September rainy season is during winter

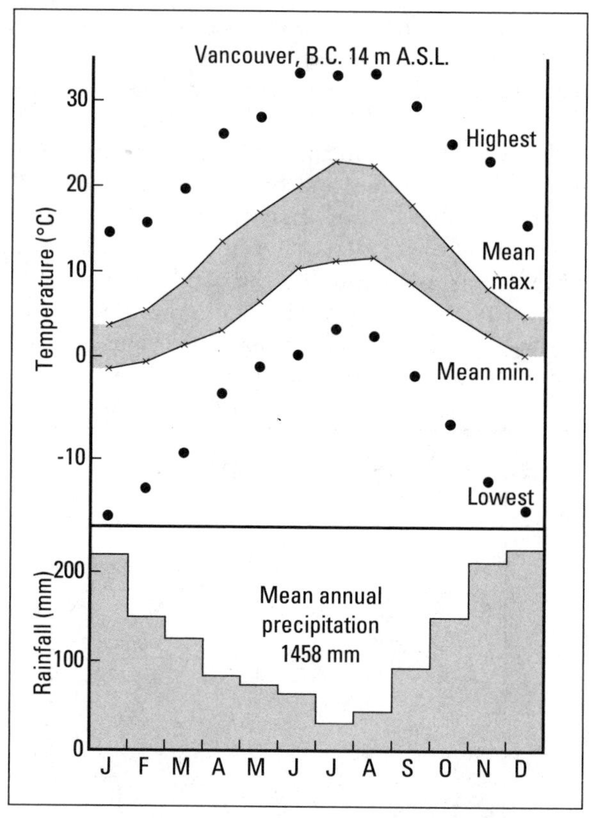


FIGURE 7.23 ▼ Climatic graphs for Vancouver, British Columbia, with a moist, temperate climate

In the higher latitudes beyond the Mediterranean zone, the influence of depressions extends throughout the year and so the rainfall is more evenly distributed in time as well as space. Depressions normally move from west to east, with the consequence that, on the western side of the continents, winds blow onshore. The sea is cooler than the land in summer, so air blowing off the sea tends to temper the continental effect, reducing summer temperatures. In winter, the sea is warmer, maintaining higher temperatures in the cool season. Vancouver (Figure 7.23) is an excellent example of this type of climate, with abundant rainfall at all seasons and its minimum in summer. Temperatures are equable throughout the year, uncomfortable extremes being a rare event. Much of northwest Europe experiences a similar climate and, because there is not a comparable range of mountains to the Rockies to prevent mild air penetrating the continent, this climatic type extends much farther into the continent of Europe than it does in America. The effect of prevailing westerly winds across the Pacific and Atlantic oceans drives warm water northeastward in the Kuroshio and Gulf Stream-North Atlantic Drift. Temperatures are maintained at a higher level than would be expected on the western side of the continents by this oceanic energy transport.

## CONTINENTAL AREAS

On the eastern side of the continents (e.g., northeastern USA, eastern China, eastern Siberia) the prevailing airflow is offshore, so oceanic influence is reduced. In summer, air will be blowing off the heated land mass, but in winter the land cools and even on the coast very low temperatures can be found. New York is a good example of this climatic region (Figure 7.24) with hot summers, cool winters, and precipitation evenly distributed throughout the year. Comparison with Eureka (Figure 7.21) at an identical latitude, stresses this contrast between the windward and leeward coasts. The cool

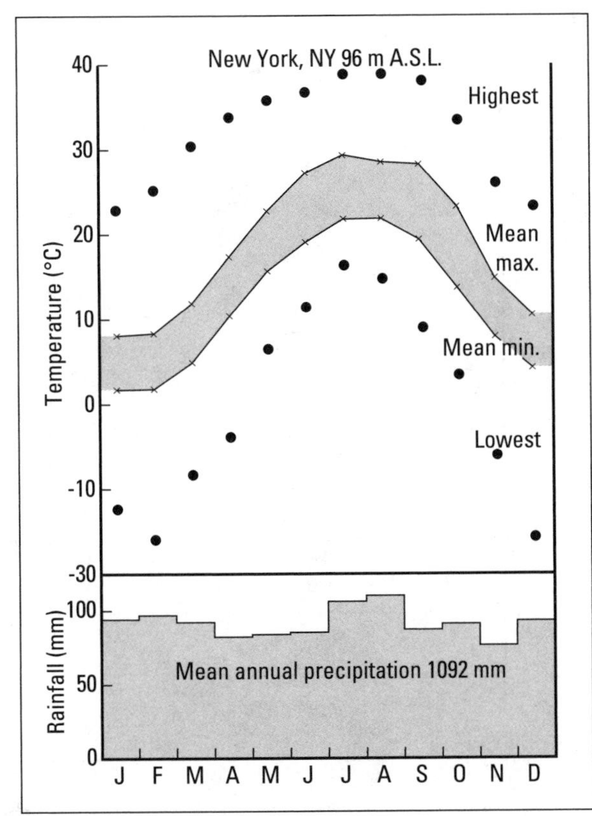


FIGURE 7.24 ▼ Climatic graphs for New York City. Note the contrast with Eureka (Figure 7.21) on the west coast of the continent at the same latitude

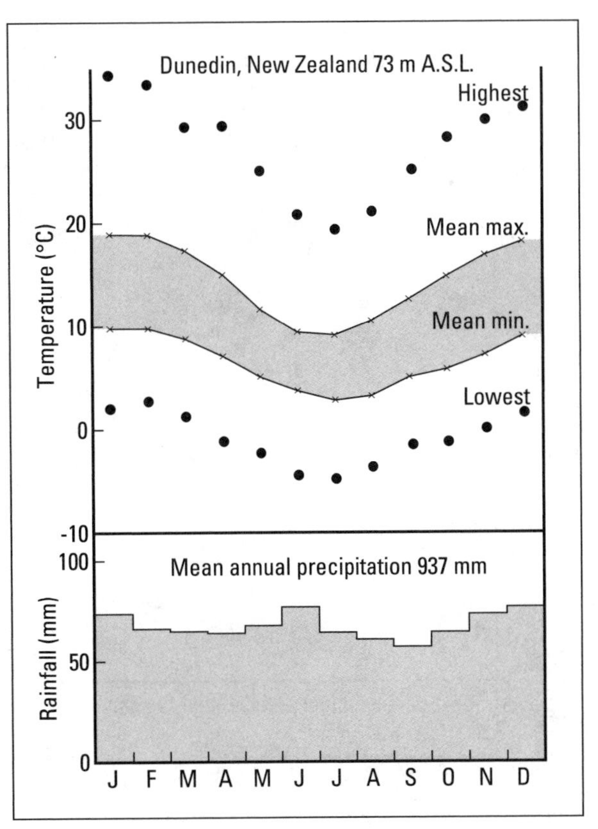


FIGURE 7.25 ▼ Climatic graphs for Dunedin, New Zealand, in a maritime temperate climate

California Current makes summer temperatures particularly low in coastal northern California.

In the southern hemisphere, only South America and New Zealand extend so far southward, but continental areas at these latitudes are small. Hence the capacity for major heating or cooling of the air is low. The record for Dunedin (Figure 7.25) shows this, with a small annual temperature range. The frequency of depressions means that rainfall is abundant (161 days with rain per year), evenly distributed, and of low intensity, giving a mean annual total of 937 mm.

Away from the coasts, rainfall gradually decreases, and temperatures become more extreme, with cold winters and hot summers. For example, at Winnipeg, Manitoba, conditions are extreme. In winter, beyond all warming influences and with the sun low in the sky at midday, average daily maximum temperatures in January are –13°C. Average minimum temperatures drop to as low as –48°C. Once more radiation starts to be received from the sun, temperatures start to rise rapidly. Highest temperatures occur in July soon after the radiation maximum, which is characteristic of continental areas, and temperatures up to 42°C have been observed. The moisture content of the atmosphere in winter is very small. Summer is the rainy season, convectional showers giving short spells of heavy rain, although depressions do give some light snowfall in winter. Mean annual totals remain small.

In the centre of the Eurasian landmass, where depressions penetrate less frequently and moisture sources are distant, rainfall can fall to very low values even in the westerly belt. Kazalinsk, Kazakhstan (Figure 7.26), has a mere 124 mm on average which, with the hot summers, gives a desert climate. Even near the coast, mountain barriers can dry out the air. When this dry air descends into intermontane basins or on the leeward side of the barrier, the likelihood of rain is small.

In the central areas of the USA, the climate is renowned for its extremes. Although this area is still within the main westerly zone, a spell of prolonged northerly or southerly winds can draw air

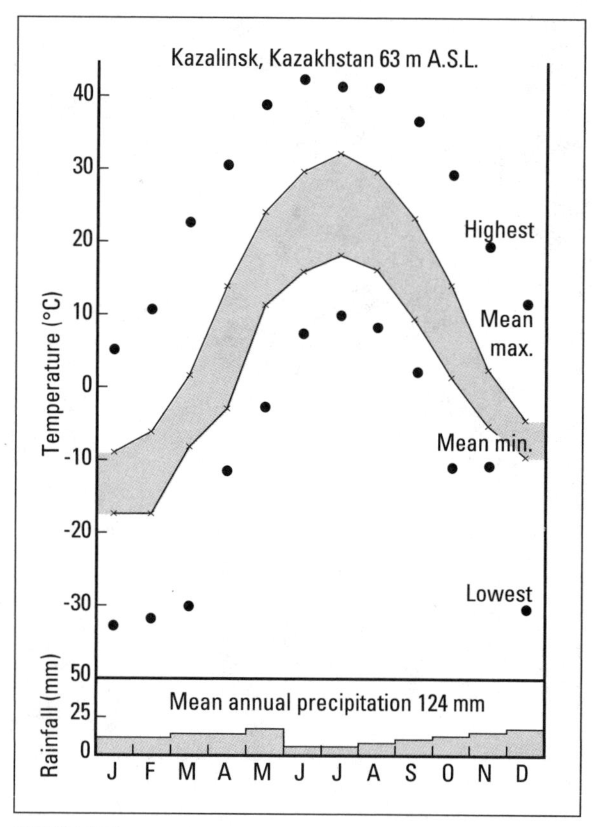


FIGURE 7.26 ▼ Climatic graphs for Kazalinsk, Kazakhstan (45.7°N, 61°E), with a continental climatic regime

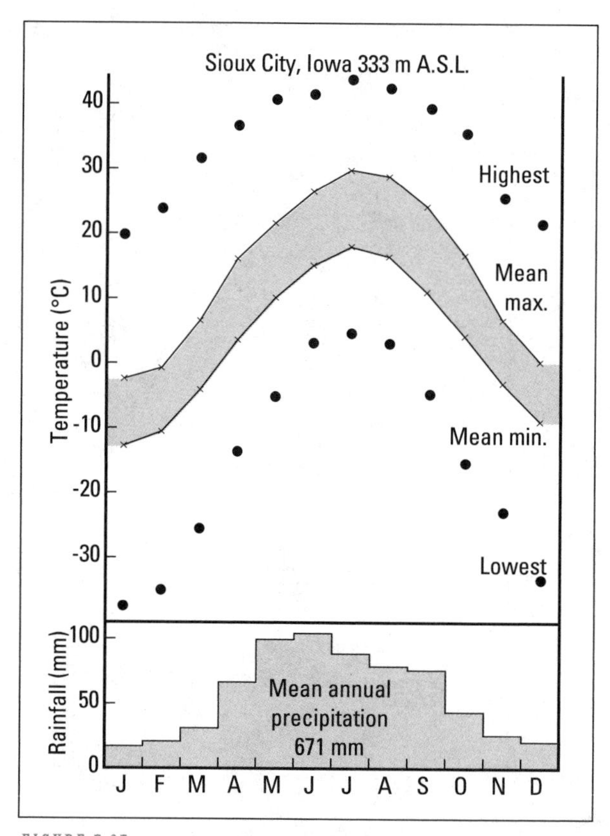


FIGURE 7.27 ▼ Climatic graphs for Sioux City, Iowa. The city is well away from the moderating influences of the oceans and so has an extreme climate with cold winters and hot summers, but is much wetter than Kazalinsk (Figure 7.26), at a similar latitude in the continental interior of Asia

from the Arctic or the Gulf with few relief features to provide shelter. Thus at Sioux City, Iowa (Figure 7.27), the mean temperature in January is –7°C, but the average of the highest temperature is 11°C with 20°C being recorded once; the mean lowest temperature is –26°C with an extreme of –37°C. So the mean figure of –7°C is a little misleading unless this is borne in mind. In July, extremes from 44°C to 5°C have occurred. Some of the waves of cold air from the north sweep southward and occasionally reach the Gulf coast, injuring frost-sensitive crops. These waves usually occur to the rear of a cold front when high pressure builds up to the west, strengthening the west–east pressure gradient.

## THE NORTHERN EXTREMES

In more northern parts of the oceanic temperate latitudes, the effects of proximity to the pole become apparent. These areas are directly in line with depression centres, or even to the north of them. As a result, they experience cold easterly winds when the depression centres pass to the south, rather than the mild air of the warm sector (Chapter 6). Because of this, the climate is cool, moist, and cloudy with frequent gales. Reykjavik, Iceland, shows the extreme oceanicity of many of these areas, being exceptionally mild for its latitude (64°N) in winter but cool in summer (Figure 7.28). Highest precipitation is in winter, when some will be in the form of snowfall, and least in late spring and early summer. This area is surrounded by oceans, so extremes are rare. Nearer the continents, colder or warmer air may be drawn in occasionally. At Tromsö, Norway, at latitude 70°N, the same as Point Barrow in Alaska, the mean annual temperature is 2.5°C, compared with –12°C at Barrow. Extremes range from 28°C to –18°C. The reason for this tongue of warmth that extends so far poleward is the direction

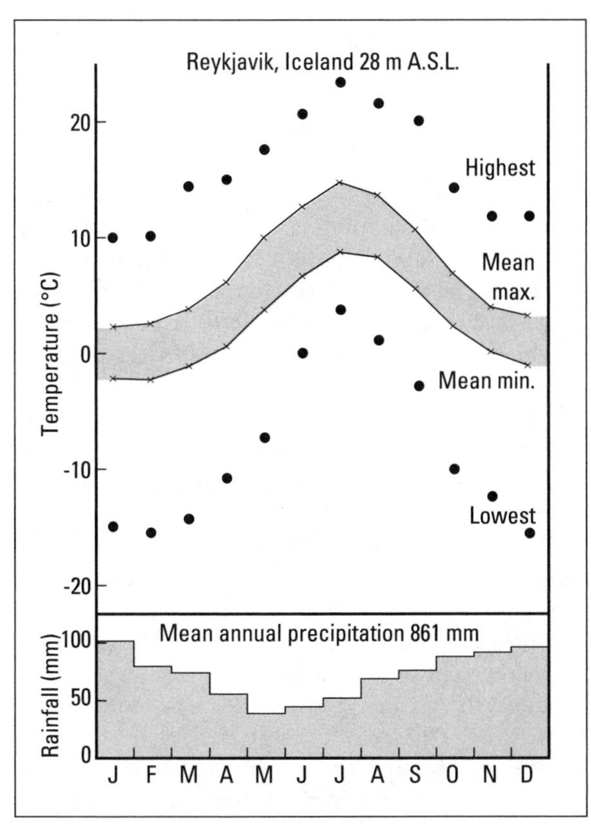


FIGURE 7.28 ▼ Climatic graphs for Reykjavik, Iceland. Despite being 20 degrees of latitude north of Sioux City, the oceanic location of Reykjavik ensures much warmer winter temperatures

of the depression tracks in the North Atlantic. The majority start in the Newfoundland area, and reach their greatest intensity near Iceland, where most decay. Some, however, carry through to northern Norway, bringing with them air that originated in more southerly latitudes. Ocean currents follow a similar trend to maintain ice-free waters so far north.

# Polar climates

The polar regions have low energy inputs. Even though the sun may be above the horizon for 24 hours in midsummer the intensity of its radiation is low. Also, so much energy is reflected from the ground or used in melting ice that the potential for heating is limited. As a result, temperatures are low. Only when warm air is rapidly advected poleward can even relatively warm temperatures result.

Under polar climates we can distinguish three regimes:

- In North America and Eurasia we have the tundra zone. The winter snows do melt in summer to give a brief growing season but temperatures soon fall and winters are bitter.
- The Arctic Ocean, covered by ice throughout the year, experiences summer melting, creating water on the ice surface. With the moisture available it becomes a very cloudy area. In addition, a great deal of heat energy passes from the relatively warm water through the ice to raise air temperatures, one of the factors that makes the Arctic warmer than the Antarctic.
- Antarctica, which again is an ice-covered surface, is separated from any source of warm air by the cold southern oceans, and has a mean altitude over 3000 m. As a result little melting occurs, except on the coast, and summer days are much sunnier than in the Arctic.

Let us look at these three areas in more detail.

## THE TUNDRA ZONE

The tundra zone is based on a vegetation type rather than a climatic region. It is the area of lichens, mosses, sedges, grasses, and a few birch trees. Much of the ground is permanently frozen, though the surface layer thaws in the summer and is known as the **active layer**. This presents great problems for human development. If the permafrost thaws, perhaps as a result of an influx of heat from a building, the meltwater will flow away from the site and subsidence takes place. Exploration for oil in the Arctic has experienced many problems because of this. We are dealing with a very sensitive environment where small changes to the system can generate major consequences.

Depressions, with their fronts and warm air, rarely penetrate into the tundra zone so the climate is determined by more local energy inputs and outputs. In central Canada the 10°C summer isotherm coincides with the mean summer position of the Arctic Front, which also coincides with the northern limit of trees. The isotherm is also generally coincident with the southern limit of discontinuous permafrost (Figure 23.3). The tundra climatic region lies north of the treeline and is under the influence of cP or cA air masses for most of the year. In the northern hemisphere it occupies a relatively narrow zone along the coastlines of North America and Eurasia, extending farther south on the east side of these land masses where there is more continentalism. The climate only occurs on a few island and small fringe areas of Antarctica in the southern hemisphere. Because of the cold and lack of depressions, winter precipitation is scanty. It is effectively a cold desert. In winter there is little energy input and temperatures fall dramatically, as shown by Coppermine, Northwest Territories (Figure 7.29).

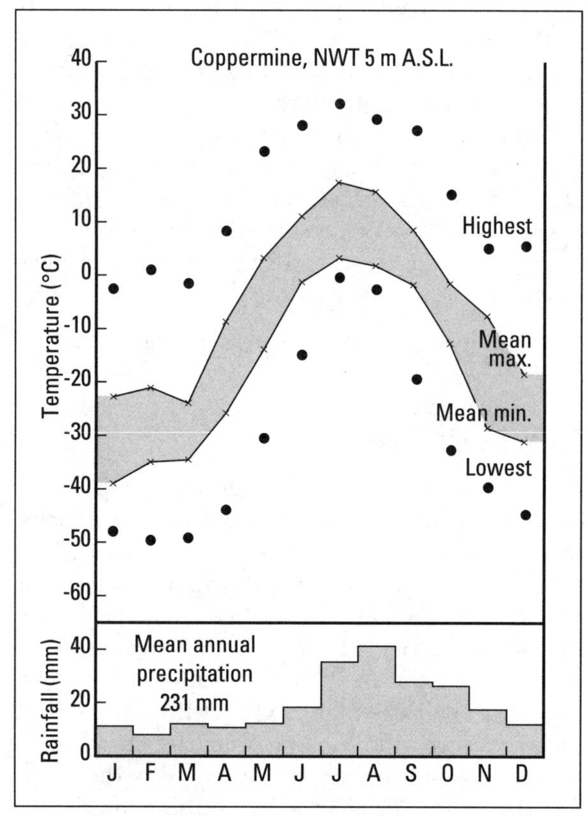


FIGURE 7.29 ▼ Climatic graphs for Coppermine, Northwest Territories (67°49'N, 115°05'W), in the tundra zone, with brief warm summers and intensely cold winters

As the sun gets higher in the sky during spring, the land responds quickly to heating once the snow cover melts. Because the snow is thin, this does not take long. Temperatures increase rapidly so that by midsummer mean temperatures reach 17°C, and even temperatures over 30°C are not unknown. The heating can generate convectional showers so summer is the wetter season. Even then amounts are small, and if it were not for the low evaporation these areas would be like deserts. The mean annual total of 231 mm for Coppermine is very similar to Pheonix, underlining the concept of hot and cold deserts.

## ARCTIC OCEAN BASIN

The Arctic icecap covers a large area, varying from 11.7 million km$^2$ in winter to 7.8 million km$^2$ in summer. The ice is constantly rotating from east to west, pushed by the polar easterly winds shown in Figure 5.16. This has created a political problem because research stations established on the ice in the sovereign territory of one nation have drifted into the territory of another. For example, in the 1970s, the Soviet scientific ice station designated T4 drifted into Canadian territory.

The ice varies in thickness up to 5 m, with the oldest ice usually being the thickest. The Arctic Basin has a net loss of energy of about 60 percent. In order to maintain a balance there is a great deal of energy transported in by air and water currents. The North Atlantic Drift is a major contributor. Surprisingly, there is a great deal of heat energy transferred through the ice when the air temperature drops below the temperature of the underlying salt water (–1.6°C).

Climatic stations are nonexistent on the icecap and very sparse on lands surrounding the ocean basin. Most have been in existence for only a brief period, therefore their records are short. Most of our knowledge of climate conditions of the High Arctic in Canada comes from the work of the Polar Shelf project, which is primarily interested in the glaciers of these high latitudes.

In winter the shallow polar anticyclone dominates the scene, though its centres are often over Siberia or Canada. Midlatitude depressions approach the margins of the ocean, especially between Norway and Spitzbergen, and between Greenland and Baffin Island. With few depressions, skies are fairly clear, with a mean value of cloudiness of about 40 percent. Precipitation is very slight and temperatures fall to a daily mean value of about –35°C. In summer, pressure gradients are weak over the polar ocean, with a low pressure centre near the North Pole and a high near Alaska. The warmer temperatures associated with constant daylight result in the surface ice melting. The process of melting and evaporation both require energy. Almost all available energy during summer goes into melting or evaporation, leaving virtually nothing for heating the air. Away from the land surfaces temperatures rarely rise above 5°C.

With so much moisture around from melting ice and relatively low temperatures the air soon reaches saturation and cloud is abundant. The mean cloudiness for much of the Arctic in July is 90 percent so, in this part of the Land of the Midnight Sun, the sun is rarely seen. With conditions favouring rising air being infrequent, precipitation tends to be small. Absolute amounts are difficult to measure because of drifting snow, but 100–200 mm per year is a realistic estimate.

## ANTARCTICA

Conditions in the Antarctic continent are very different. It is isolated from other continents by the vast expanse of the southern oceans. More than 55 percent of its surface lies at an elevation of more than 2000 m and about 25 percent at more than 3000 m above sea level. The effects of latitude, altitude, and insolation result in the vast majority of the continent being permanently ice-covered. Only

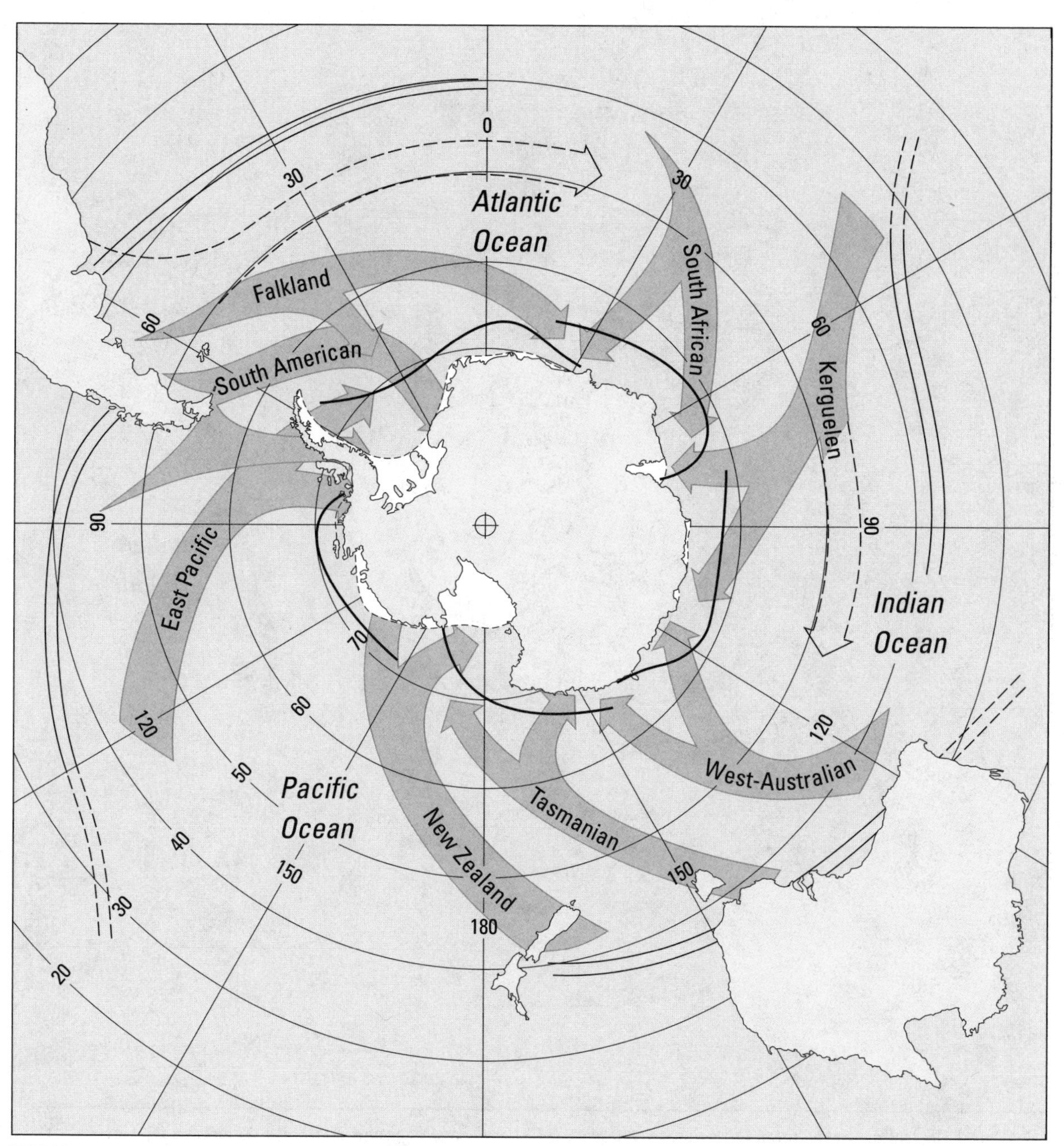


FIGURE 7.30 ▼ Paths of cyclones during winter in Antarctica (after Rusin, 1964)

about 3 percent of its area is believed to be free of a permanent icecap. This occurs in the coastal zone or in some of the arid valleys near McMurdo Sound.

Recording sea level pressure is difficult when much of the continent is so high. The lowest standard pressure surface that is above land level is the 500 mb surface. In both winter and summer low pressure prevails, but this is due to the cold, dense air rather than the area being a centre of active depression formation. In fact, depressions rarely penetrate onto the icecap, especially the higher ice plateau of eastern Antarctica (Figure 7.30). We can visualize the continent as an area of relative calm surrounded by the midlatitude storms that carry clouds and blizzards. As sources of warm air are so distant, even warm sectors of the depressions are

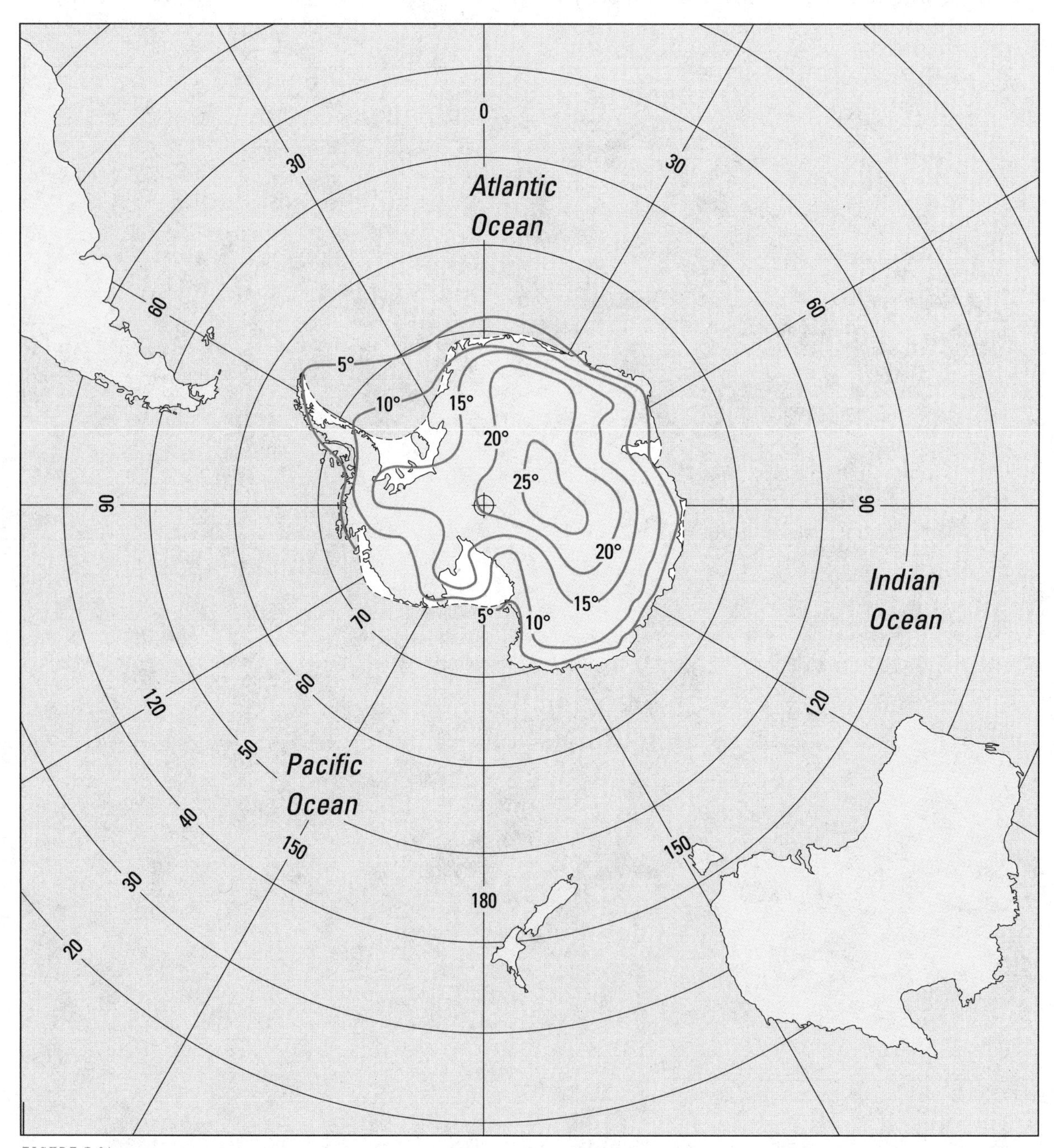


FIGURE 7.31 ▼ Isolines of the mean temperature difference (°C) between surface and 20 m over Antarctica in winter (June–August) (after Phillpot and Zillman, 1969)

cool and warm fronts much less significant in the climatology of the Antarctic than in the northern hemisphere.

The small number of depressions over much of the continent means that cloudiness is low. This affects the radiation exchanges. In summer radiation input is high because there is little cloud to reflect the insolation. In winter the absence of cloud and sunshine means extensive longwave radiation losses and a large temperature drop. From this we might expect hot summers and cold winters, but much of the surface is ice with a high albedo (>80 percent). Hence most incoming radiation is reflected and lost to the surface. The high emissivity of snow and ice assists the loss of longwave radiation, and with little counter-radiation the surface has a large radiational deficit. Coupled with the high altitude, therefore, this means that temperatures rarely rise above freezing, and melting, such a feature of Arctic summers, does not take place.

As most of the longwave losses are from the surface and advection is limited, so the surface usually has the lowest temperatures, producing a marked inversion. This is a conspicuous feature of the Antarctic climate. Except for the two summer months it is ever-present over the high ice plateau and frequent elsewhere. Figure 7.31 shows the average intensity of the inversion in winter. The presence of the inversion means that surface temperatures for Antarctica tend to be much lower than those in the air above. Nevertheless, it is at this level that human activity takes place. Figure 7.32 illustrates some features of the climate. At the South Pole there is an annual temperature range of 30C° but temperatures never rise above freezing. Precipitation is impossible to measure because of drifting snow and so is omitted from the graph. On the coast at McMurdo Sound, daytime summer temperatures do rise above freezing but the warmth is short-lived and night-time temperatures fall below freezing on average.

Another distinctive feature of the Antarctic climate is the wind. Figure 7.33 shows the average pattern of surface wind flow. In essence it resembles a vast katabatic wind (one blowing down a slope) flowing off the icecap. Because of Coriolis force the usual direction is southeasterly rather than southerly. Average speeds are strong: 6.5 m $s^{-1}$ at McMurdo and 6.2 m $s^{-1}$ at the pole. Near the coast, the winds can be funnelled by glacier valleys descending steeply to the coast to give much stronger than average flows. They are strongest when the katabatic wind is supplemented by the

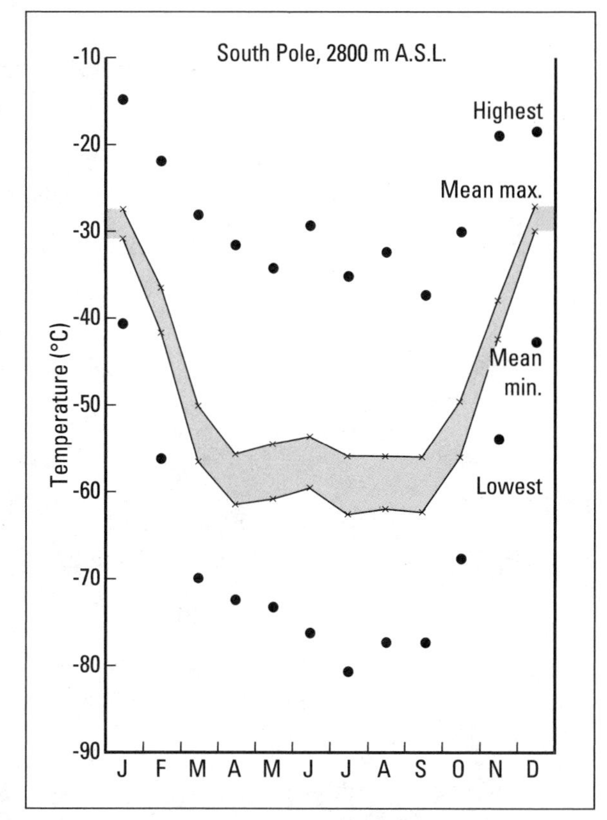


FIGURE 7.32 ▼ Climatic graphs for the South Pole. Altitude greatly affects the temperature levels. Precipitation is believed to be low but is very difficult to measure because of drifting snow

pressure gradient wind. At one particularly severe site, Cape Denison, the annual mean wind speed is 19.4 m $s^{-1}$, with the highest monthly mean speed being 24.9 m $s^{-1}$. The combination of strong winds, low temperatures, and driving snow can produce the most extreme weather conditions on earth.

# Conclusion

The regional variation of climate across the earth's surface may appear complex at first, but it is controlled by a few dominant factors: the distribution of solar radiation, the global circulation of the atmosphere, and the general pattern of continents and oceans. What we experience as we move from one climatic zone to another are changes in all the interacting components of the climate in response to variations in these factors.

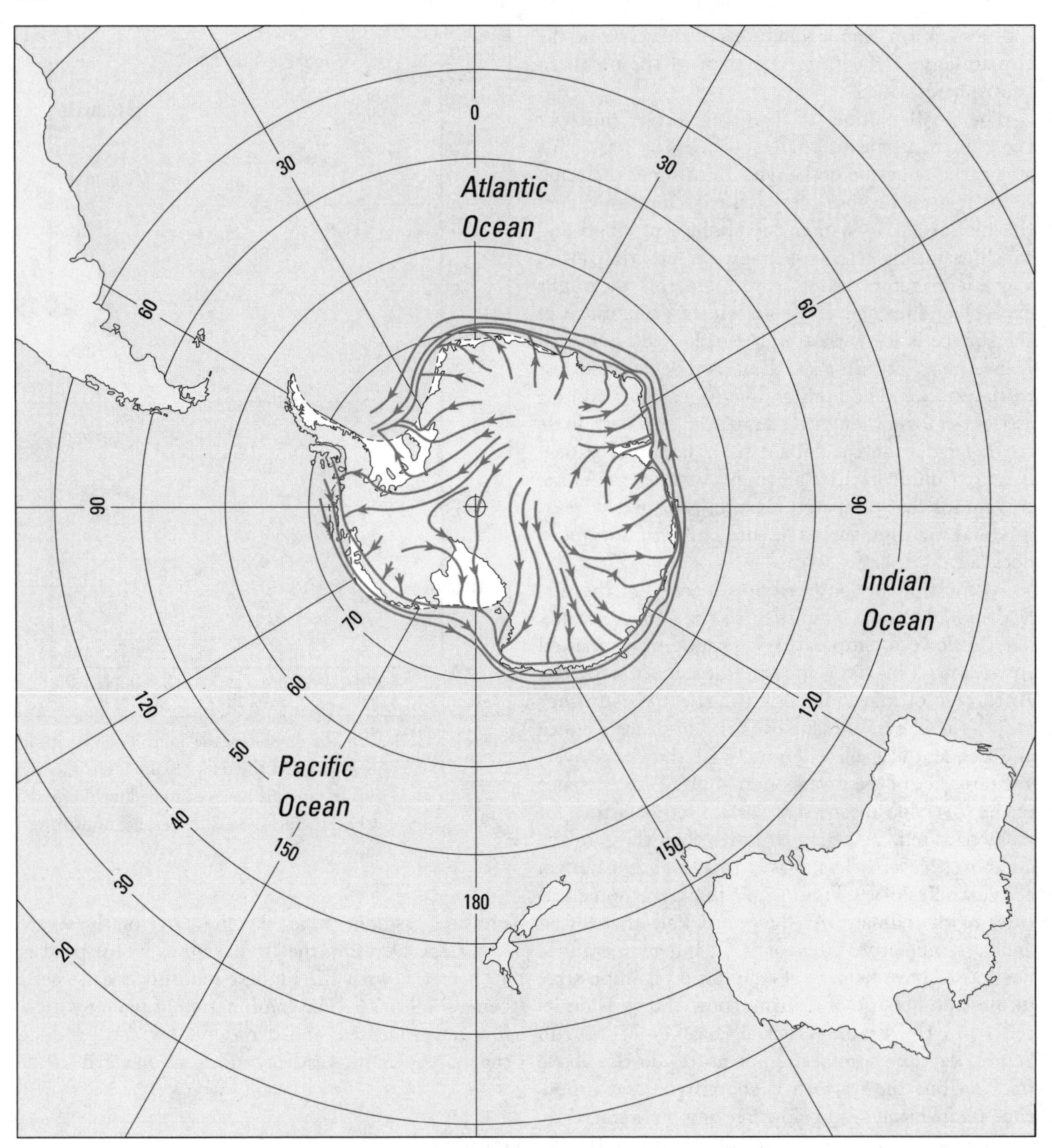


FIGURE 7.33 ▼ Average pattern of the surface wind flow, inferred from predominant wind frequencies at stations and from traverse records (after Mather and Miller, 1967)

The pattern of the world's climates is not only important in relation to atmospheric processes; many other aspects of the global system show a similar variation, for it is the climate that governs processes of landscape formation, soil development, and vegetation evolution. All three vary in relation to climate. Thus, knowledge of the climatic regions of the world will provide an important starting point for looking at other aspects of the environment.

# Micro and local climates

## The climate near the ground

Living as we do in the lowest few metres of the atmosphere, we should have a special interest in the diverse and complicated climate of this zone. Climatic differences equivalent to a change in latitude of several degrees can occur in a matter of a few metres. In 1965, Geiger's classic study *The Climate Near the Ground* was published. More recently this layer has been referred to as the **boundary layer**. More specifically it is called the planetary or atmospheric boundary layer and represents the interface between the atmosphere and the earth. It is comparable in function and importance to the human skin.

We have said that significant differences can occur within this layer. When the sun is shining, for instance, the ground may become too hot to walk on barefooted, as on a dry, sunny beach in midsummer or a black road surface. At the ground surface the temperature may reach 65°C although at head height it may only be 30°C, and in the shade, where most temperature observations are made, it might be as low as 20°C. Similar variations can be found in wind and humidity. So, we may ask, what is it about this boundary layer that produces such major differences; differences that are not repeated anywhere in the free atmosphere?

The reason for this variability is that we are dealing with the main exchange or activity zone between the ground surface and the atmosphere. Energy is reaching this zone from both the sun and, to a much lesser extent, the atmosphere. It is absorbed and then returned to the atmosphere in a different form, or is stored in the soil as heat. This absorption process is very sensitive to the nature of the ground surface. Conditions such as surface colour, wetness, vegetation, topography, and aspect all affect the interaction between the ground and the atmosphere. We can see these effects clearly in snowy weather. Clean snow reflects solar radiation; the surface remains cool and the snow fails to melt. But where the snow is dirty, it absorbs the radiation, heats up, and melts. Vegetation, too, may protect the snow from the heat of the sun, while even late in the spring snow may be preserved in sheltered hollows or on hillslopes facing away from the sun. Let us look at the causes of these differences in more detail.

# Microclimate over different surfaces

## THE NATURE OF ENERGY TRANSFER OVER A SURFACE

As the ground surface gets hotter through absorbing the sun's energy, the layer of air in contact with the ground becomes warm by conduction. If this was the only mechanism of heat transfer, it would take a very long time before even the lowest 1 m of air was warmed. The daytime maximum temperature at this height would not occur until about 9 P.M. Clearly this cannot be the only process transferring heat, although it is the most important in the lowest few millimetres, where temperature gradients are extreme. Above this level, the effect of heating the air causes it to become less dense than its surroundings and so it rises, carrying heat with it. Cooler air then moves in to take its place. This air is heated in turn. Consequently we have **convection currents** rapidly transferring heat to the cooler layers of the lower atmosphere. If there is a strong wind blowing, mixing of the heat is encouraged and the temperature profile in the lower atmosphere becomes less steep (Figure 8.1).

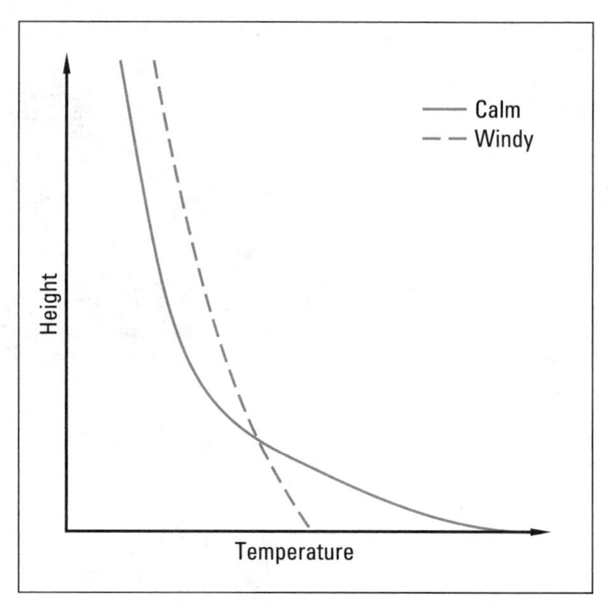


FIGURE 8.1 ▼ Temperature profiles above a heated ground surface on calm and windy days

## THE RADIATION BALANCE AT NIGHT

At night the radiation balance at the ground surface is negative (Figure 8.2). More longwave radiation is lost from the ground than is returned as counter-radiation from the atmosphere. This is especially true with clear skies and dry air, which allow the radiation to escape to space more easily. In cloudy or humid conditions heat loss is less effective, for water vapour readily absorbs longwave radiation. During the night, therefore, the surface gets cooler although heat may flow up from lower levels in the soil to maintain surface temperatures. In a sandy soil, with its large air spaces, this process is limited so the surface becomes particularly cool, as anyone who has slept outdoors on beach sand will know. The air in contact with the ground also gets cooler, making the air denser and preventing any thermals of warmer air rising and mixing with the air above.

The nighttime profile of temperature during calm conditions is shown in Figure 8.3, and this illustrates the major cooling at the surface. If it is windy, the cooler air will be mixed with the warmer air above to give a smaller increase of temperature with height. Clouds are efficient emitters or radiators of longwave radiation, so low clouds encourage counter-radiation to the surface and the net loss of energy from the surface is reduced. The factors most favouring low surface temperatures at night are consequently clear skies, dry air, no wind, and sandy or

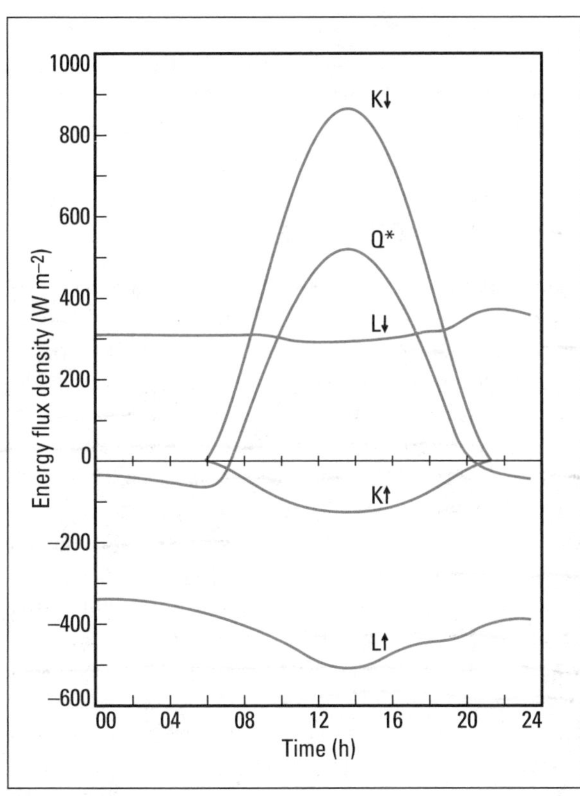


FIGURE 8.2 ▼ Radiation budget components for 30 July, 1971 at Matador, Saskatchewan (50°N), over a 0.2 m stand of native grass. Inputs to the surface have been plotted as positive and outputs negative to assist interpretation. K is shortwave radiation, L is longwave radiation, and $Q^*$ is net radiation (after Oke, 1978)

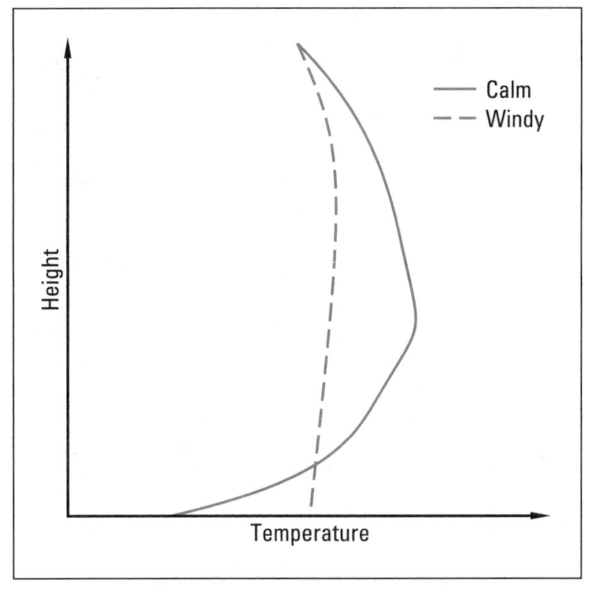


FIGURE 8.3 ▼ Nighttime temperature profiles on calm and windy nights with clear skies

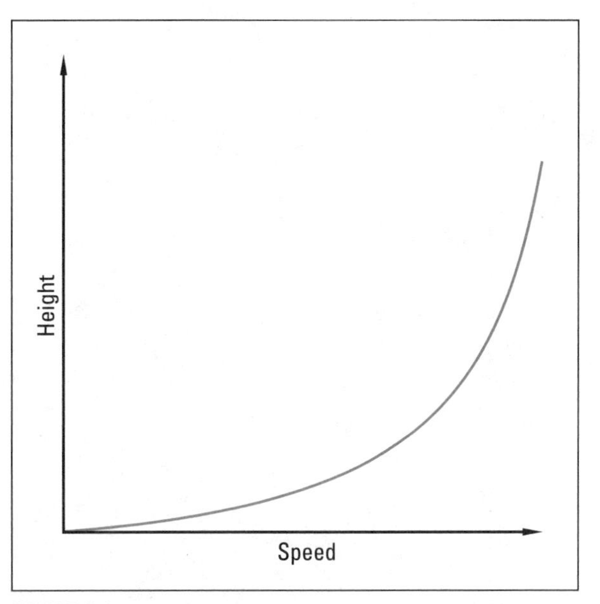


FIGURE 8.4 ▼ Wind speed profile near the ground. The precise form of the profile will depend upon the roughness of the surface as well as upon any buoyancy

peaty soil. If these conditions occur at the beginning of the growing season in most temperate latitudes then frost-sensitive crops are likely to be damaged.

## WIND NEAR THE GROUND

The profile of wind speed near the ground is somewhat similar to that of temperature (Figure 8.4). As we approach the ground, wind speed decreases very rapidly to almost zero in contact with the soil surface. This is largely due to the frictional drag exerted on the air by the underlying rigid surface; the rougher the surface, the more it slows the air (see Chapter 24). Over a soil surface, the effect on the wind is fairly simple, but when we are dealing with a vegetation layer or an urban area, interference is much greater, as we shall see later.

This microclimate at a soil surface represents the simplest case of modification of the energy balance by ground conditions. Both the inputs and outputs of radiation are changed and this alters the way the energy is used in terms of sensible and latent heat, and heat flow or storage into the soil. This is illus-

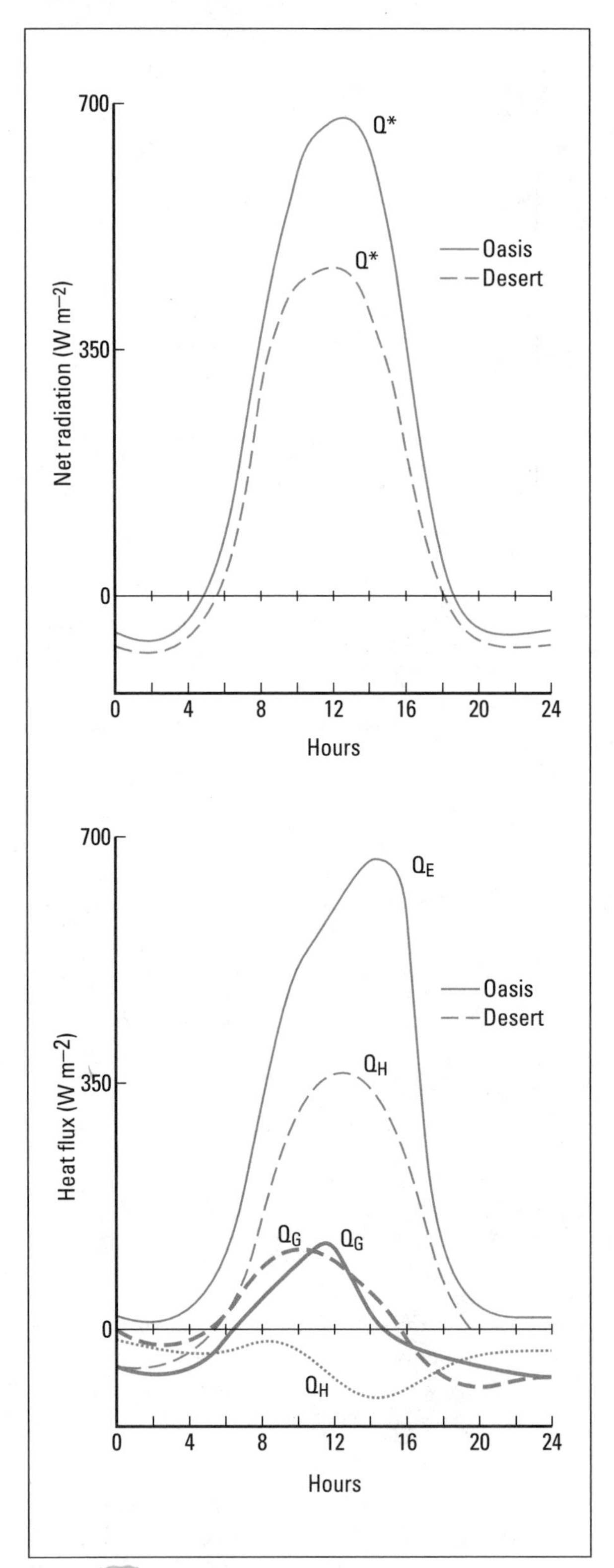


FIGURE 8.5 ▼ Net radiation and the energy budget in an oasis and in a desert. $Q_H$ is sensible heat flux, $Q_E$ is latent heat flux, and $Q_G$ is soil heat flux

trated in Figure 8.5, where the energy balances over a wet and dry soil are contrasted, and in Table 8.1, where the thermal differences above the surfaces are demonstrated.

## MICROCLIMATE OVER A BARE SOIL

Many different properties of the soil influence conditions in the thin layer of atmosphere above it. Soils vary in colour, and darker soils, such as those rich in organic matter, absorb radiant energy more efficiently than do light-coloured soils. Moisture in the soil is also important. Wet soils are normally dark, but water has a large **heat capacity**, that is, it requires a great deal of energy to raise its temperature. A moist soil, therefore, warms more slowly than a dry one (Figure 8.6).

A further complication in heat transfer into soils is that air is a poor conductor of heat. If there is a large amount of air between the soil particles, heat transfer into the soil is slow. This means that heat is trapped in the upper layers, so the surface layers warm more rapidly. Because of this, dry sandy soils get very hot on sunny days. Water conducts heat more easily than air, so soils that contain some moisture are able to transmit warmth away from the surface more easily than dry soils. However, if the soil contains a lot of water, the large heat capacity of the water will prevent the soil warming despite heat being conducted from the surface. For most agricultural crops a balance is needed so that soils warm up fairly quickly at depth and are neither too wet nor too dry. This is achieved when the moisture content of the soil is about 20 percent.

## MICROCLIMATE ABOVE A VEGETATED SURFACE

The nature of microclimatic conditions and processes becomes far more complex when a vegetation cover is present, for not all the energy is absorbed at a single surface. Some is absorbed by the top of the vegetation, some penetrates into the plants, and some may even reach the soil surface. The amount that gets through the soil depends upon the height of the crop, the density of the leaves, and the angle of the sun's rays. As the size of the plants increases, so does the degree of microclimate modification.

Let us look at some of the detailed effects of plants on the microclimate by considering condi-

## Table 8.1

*Twenty-four hour diurnal air temperature variation in July*

| *Height above surface (m)* | *1* | *5* | *7* | *9* | *11* | *13* | *15* | *17* | *19* | *21* |
|---|---|---|---|---|---|---|---|---|---|---|
| | | | | | *Irrigated oasis* | | | | | |
| 2 | 21.4 | 18.9 | 20.7 | 25.4 | 30.5 | 33.2 | 33.9 | 33.7 | 30.0 | 26.4 |
| 25 | 23.8 | 20.8 | 21.8 | 25.3 | 30.2 | 33.0 | 33.9 | 34.3 | 30.9 | 29.7 |
| 50 | 26.2 | 22.6 | 22.5 | 25.5 | 30.0 | 33.1 | 33.5 | 34.5 | 31.6 | 32.7 |
| 100 | 28.6 | 25.9 | 23.8 | 25.9 | 29.9 | 33.0 | 33.3 | 34.0 | 31.9 | 34.2 |
| | | | | | *Semidesert* | | | | | |
| 2 | 23.0 | 19.9 | 23.1 | 28.4 | 33.5 | 37.0 | 36.7 | 37.8 | 33.9 | 29.4 |
| 25 | 24.5 | 21.4 | 23.5 | 28.1 | 33.4 | 35.6 | 35.3 | 36.5 | 33.5 | 32.2 |
| 50 | 26.2 | 22.6 | 23.9 | 28.2 | 33.4 | 35.0 | 34.7 | 36.3 | 33.0 | 32.9 |
| 100 | 28.6 | 23.9 | 25.1 | 27.8 | 32.9 | 34.6 | 33.9 | 35.8 | 32.8 | 33.1 |

(Column headings 1–21 are *Hours of day*.)

After I.A. Goltsberg, *Microclimates of the U.S.S.R.* (1969), Israel Program for Scientific Translation.

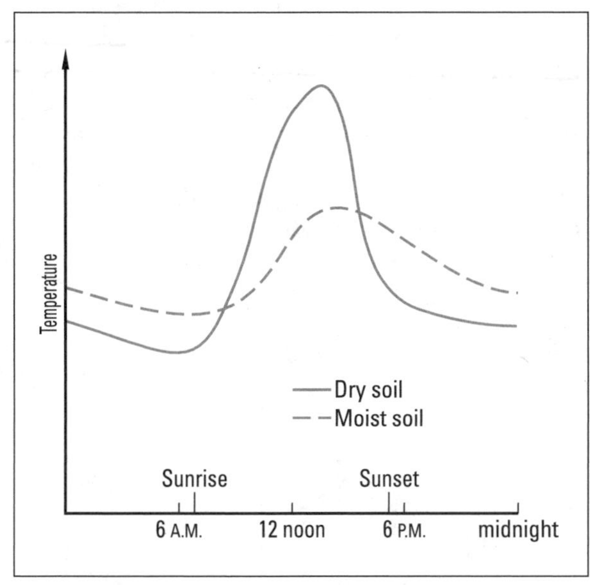


FIGURE 8.6 ▼ Diurnal temperature changes over a dry soil and a moist soil surface

tions around a single leaf. The amount of shortwave radiation absorbed by a leaf depends upon the quantity of radiation reaching its upper surface, the angle between the leaf and the sun's rays, and the colour of the leaf. Through absorption, the temperature of the leaf rises and, consequently, the amount of longwave radiation is transferred downward toward the soil, and some flows upward. With a large number of leaves, the sun's rays are increasingly obstructed so the amount of sunlight reaching the ground may be small. The actual quantity depends upon the type and amount of leaves (or **leaf area index**) and the crop height.

Because of its agricultural importance there have been numerous studies of the climate within crops. Agronomists and plant physiologists use this information to increase yields from plants best suited to the micro- and macroclimate in which they grow. It is now possible to determine the type of plants growing and to check their health by aircraft or satellite photography. The nature of the radiation reflected and emitted from leaves varies from one species to another and from healthy to unhealthy plants due to alterations in the distribution of pigments in the leaves.

*Temperatures in the vegetated layer* If we look at mean profiles of wind speed, temperature, and humidity within a plant crop, there is some similarity with those found above a bare soil surface (Figure 8.7). In this instance, the main heat exchange zone is found slightly below the canopy top rather than at the soil surface. As a result, daytime temperatures reach their maximum values within the canopy. The actual location represents a balance between the reduction in sunlight intensity as it penetrates the crop and the decrease in wind speed and turbulence that would help to remove the heated air. At night under clear skies, longwave radiation continues to flow from the leaf surfaces, but only that from the upper leaves is able to escape from the plant system. At lower levels in the crop, radiation is trapped and re-emitted, maintaining warmer temperatures. Thus the temperature profile has a minimum value just below the canopy top and gets warmer both into the atmosphere and down to the soil surface. If the crop has a low density, with large gaps between plants, the air cooled by contact with the radiating leaves becomes denser, and sinks to the ground to give the minimum temperatures there. In the soil, temperature changes are much smaller because the surface heating and cooling are greatly reduced through shading by the leaves.

*Wind in the vegetated layer* The wind speed profile is also more complex due to the presence of the crop. Its precise form depends upon the nature of the crop and the prevailing wind speeds. By day, there is normally a progressive decrease in speed as far as the middle canopy. Below this level, most crops have fewer leaves, enabling the wind to blow through the crop. So we get a slightly windier zone before the final decrease toward the soil surface. This effect can be felt behind a hedgerow or windbreak where the stems are not so effective at reducing wind speed as the leafier branches at higher levels.

*Moisture in the vegetated layer* Daytime humidity levels usually show a progressive decrease from the soil surface, through the crop, into the atmosphere. Moisture is evaporated from the soil and transpired by the plant leaves, so that the main moisture sources are within the crop. As wind speeds are low, much of the moisture remains within the vegetation, but that in the upper layers may be carried away by convection to mix with the drier air above. At night the humidity profile is more complicated. Cooling may give rise to dewfall on the upper leaves, producing an inverted profile for a short distance, but normally humidity differences are relatively weak throughout the crop.

Within a plant canopy, moisture exchanges are extensive and of vital importance to the well-being of the crop. In reality these processes are highly complex, but we can get an idea of the exchanges by constructing a simple model of the water balance.

Figure 8.8 shows the inputs and outputs of moisture we might expect for an ideal crop. The major input for most climatic regimes is precipitation, either in the form of snow or rain, but hail, dewfall, frost, and fog can add small amounts. Some of this moisture is intercepted by the leaves. Depending upon the intensity and duration of the precipitation and the nature of the leaf, the water may drip off

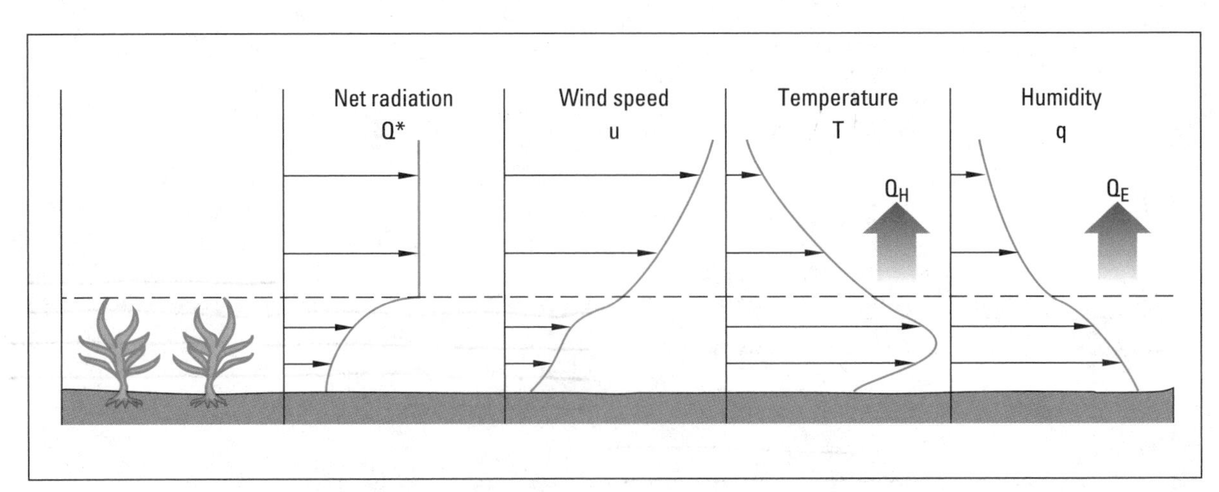


FIGURE 8.7 ▼ Typical profiles of net radiation, wind speed, temperature and humidity above and within a plant canopy (after Saugier, 1977)

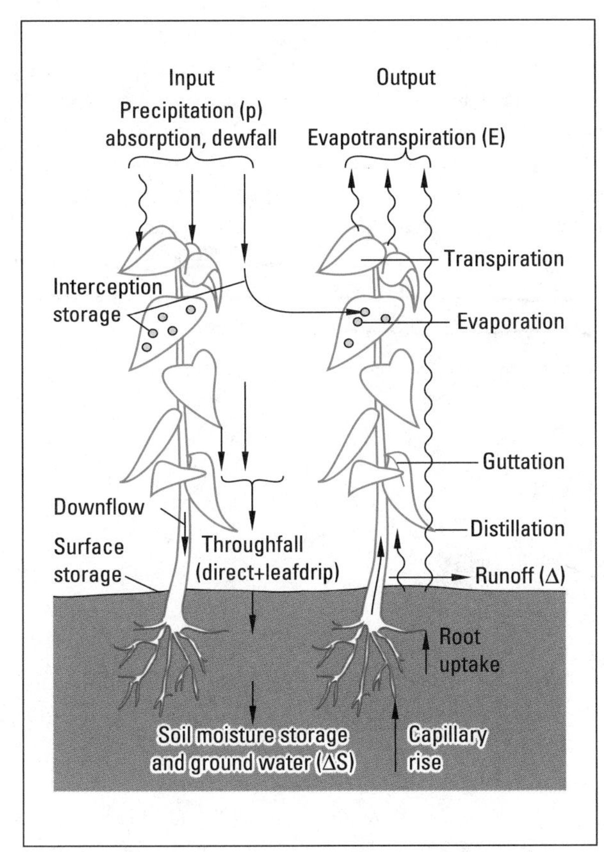


FIGURE 8.8 ▼ The hydrological cascade in a soil–plant–atmosphere system (after Oke, 1978)

the leaves or be directly evaporated without ever reaching the gound surface. This effect is greatest when rainfall is light and the leaf density high—it is much more sensible to shelter from the rain under a beech or white spruce than a tree of low leaf density such as the ash or jack pine. Small quantities of moisture may flow down the stems of the plants, but with heavy or prolonged rain some droplets will fall right through the crop to moisten the soil surface, eventually reaching the plant roots.

The output from the system is primarily through transpiration from the leaves and evaporation from both soil and leaves. Moisture is extracted from the soil to maintain transpiration, but if it becomes too dry during droughts, the plants may wilt or even die. During periods of rain, input is usually far higher than evapotranspiration alone. This surplus then goes to recharge moisture in the soil or it becomes runoff—the horizontal flow of moisture on the soil—which eventually forms part of the river system. We will examine these processes in more detail in Chapter 13.

# The microclimate of woodland

So far we have been dealing with the microclimate on, or very close to, the ground surface. From this, we have been able to illustrate the processes controlling climate at that level. As the crop or vegetation gets larger, so the degree of modification increases and the active zone extends from the higher canopy down to the soil surface. The extreme example of this effect is seen in the mature forest. So much has been written about the microclimate within a forest that the term forest climate is frequently used to indicate the wide variety of conditions.

On a hot summer's day it is noticeable that temperatures in a forest are much lower than outside, providing a respite from the glare and heat of the sun. Air movement is weak. It feels humid, and the impression is of an entirely different climate. This affects plant and animal life as well as humans. Quite different ecosystems develop because of the climatic environment produced by the forest. Because of the differences in scale, microclimates within a forest are more distinct than in grasslands or with low crops.

## RADIATION EXCHANGES IN WOODLAND

It is apparent on entering a forest that the forest canopy cuts out much of the incoming radiation. Most of the energy is absorbed by the tree canopy. A significant proportion is reflected; about 5 to 15 percent on average, although in some cases reflection may reach 20 percent. Only a very small proportion reaches the ground directly, and this is normally in the form of small patches of light called sun-flecks. The remainder penetrates the vegetation indirectly; it is reflected, absorbed, and re-emitted and arrives as diffuse radiation. During the progress of radiation though the forest vegetation, considerable changes in spectral composition take place, as specific wavelengths are filtered out or scattered by the canopy. The shorter wavelengths (i.e., blue light) are preferentially removed by the leaves, while amounts of longer-wave red and infrared radiation increase. This change in the composition of the light is responsible for the characteristic colours that we encounter in woodlands. It also makes the light less suitable for plant growth. As a result, the range of plants that can survive on the forest floor is limited.

The woodland affects not only the input of radiation; it similarly affects outputs. The manner of this modification is far more complex, for outgoing longwave radiation comes from a wide range of sources—from the atmosphere, the top of the canopy, from the leaves and branches of the trees, from the undergrowth, and from the soil surface. There is inevitably a great deal of interception, absorption, and re-emission of the longwave radiation, so that little escapes directly to space.

## VARIATIONS OVER TIME

These patterns of microclimate are only averages. Considerable variations occur over time due to changes in the inputs of solar radiation and to changes in the woodland itself. If we measured shortwave inputs of radiation throughout the day we would find that levels remained low, with the exception of brief periods associated with the development of sun-flecks; the peak intensity would occur about midday (Figure 8.9). During the night, cooling is slow, for the vegetation traps and returns much of the outgoing longwave radiation.

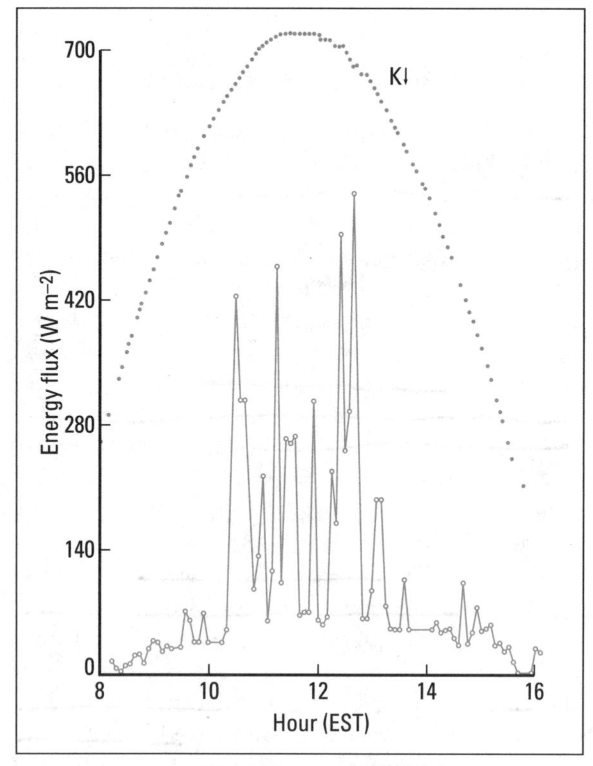


FIGURE 8.9 ▼ Global solar radiation (K↓) above a pine plantation and at one point on the floor (after Gay *et al*, 1971)

This pattern also changes seasonally. In winter the inputs of radiation are low and the effect of the forest on the microclimate diminishes. Moreover, in deciduous woodlands, the trees lose their leaves so that there is much less interception and absorption. If we compare woodland temperatures with those in open land, therefore, we find a much smaller difference in winter. The effect of the woodland is at a maximum when the trees are in full leaf and radiation inputs are high (Figure 8.10).

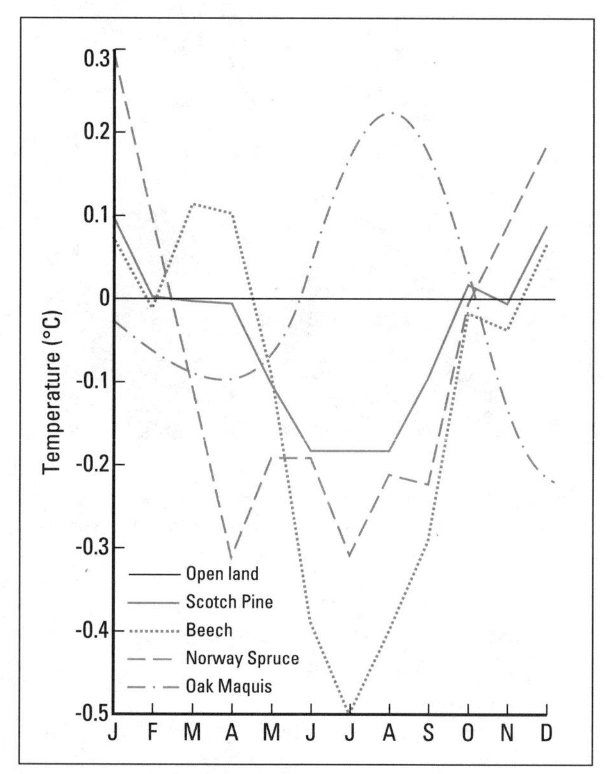


FIGURE 8.10 ▼ Mean monthly forest temperatures compared with thermal conditions in the open. The woodlands are normally cooler in summer and slightly warmer in winter. Anomalies are beech in summer because of the late opening of the leaves, and the very dense oak maquis, which has little transpiration in summer (after Smith, 1975)

## THE EFFECTS OF WOODLAND TYPE

The microclimate of woodland depends very much upon the type of woodland we are dealing with.

Deciduous trees show a much more marked seasonal change than conifers, which continue to intercept radiation throughout the winter. But considerable variations occur between different species of deciduous trees. Birch leaves, for example, are smaller and have a lower density than beech or oak, so that even when they are in full leaf, birch trees allow more light to reach the ground. As a result more plants grow on the woodland floor. Similarly, pine trees give a less dense canopy than do spruce; the dark, unvegetated floor of plantations of Sitka spruce contrasts with the much lighter conditions in the pine woodland.

It is not only the amount of radiation absorbed by the trees that varies with tree species. So too does the amount of reflection from the canopy. Coniferous trees have albedos of 5 to 15 percent, while deciduous trees may reflect as much as 20 percent (although the average is about 12 percent; see Table 8.2). In addition, of course, the nature of the understorey is important. An open canopy allows the development of one or more layers of understorey plants and these, too, intercept both incoming and outgoing radiation. The extreme example is shown by the tropical rain forest. Although radiation inputs are high, the successive layers of trees, bushes, and shrubs intercept so much radiation that only about 1 percent reaches a height 2 m above the ground. Less than 0.1 percent may reach the forest floor.

### Table 8.2

*Tree albedos (percent)*

| | | | |
|---|---|---|---|
| Aleppo pine | 17 | Eucalyptus | 19 |
| Monterey pine | 10 | Sitka spruce | 12 |
| Loblolly pine | 11 | Norway spruce | 12 |
| Lodgepole pine | 9 | Birch and aspen | |
| Scotch pine | 9 | late winter | 25 |
| Oak | | Orange trees | 32 |
| summer | 15 | Tropical rain forest | 13 |
| spring | 12 | Cocoa | 16 |

## WINDS IN WOODLAND

Patterns of wind in woodland are similar to those in grassland, although the zone of modified flow extends to a much greater height. Above the canopy, wind speed may be increased slightly, but, as the canopy is approached, velocity falls rapidly. Lowest wind speeds are often found within the leafy canopy, and where the undergrowth is also dense, velocities may remain low. In most cases, however, the main trunk zone is more open, so there is less interference with airflow and wind speeds increase again. Near the ground, friction and the effect of low-growing plants cause velocity to fall once more. Complex patterns of flow often develop in the forest, with local funnelling and deflection of the wind. We can often see the nature of these flow patterns in the distribution of dead leaves on the woodland floor. Sheltered areas trap deep layers of leaves, while more exposed zones are swept clear by the wind.

## MOISTURE IN WOODLANDS

In general, vapour pressure is slightly higher in a forest or woodland than outside it. This is mainly due to the large amount of leaf area in a forest, transpiring moisture into the atmosphere; this moisture is not easily dispersed because of the lighter winds. On the other hand, interception of moisture by vegetation reduces the amount of water available at the forest floor, so the net effect on humidity levels is small. As daytime temperatures are cooler than those outside, the relative humidity of the air should also be greater even if the forest atmosphere contained the same absolute amount of water vapour. Experiments suggest values about 5 percent above those outside, though the precise differences depend upon the type of woodland as well as the time of year and weather conditions (Table 8.3).

### Table 8.3

*Difference of relative humidity (percent) between the inside and outside of a forest*

| *Forest* | *January* | *April* | *July* | *October* | *Year* |
|---|---|---|---|---|---|
| Deciduous broad-leaf | 3.4 | 3.2 | –0.8 | 1.1 | 2.2 |
| Evergreen needle-leaf | 4.8 | 4.8 | 6.5 | 9.5 | 6.8 |
| Japanese cedar | 1.6 | –1.1 | 1.5 | 0.5 | 0.8 |

Positive values indicate that inside, the forest is more humid.

After M.M. Yoshino, *Climate in a Small Area* (1975), University of Tokyo Press.

### SNOW SURFACES

In Canada, microclimate studies with relevance to the potential effects of global warming have been carried out on the tundra. Permafrost covers large areas of the world, especially in North America and Eurasia. In winter the surface is covered with snow, which has one set of microclimate conditions. In the summer the upper layer of the permafrost, known as the **active layer**, thaws creating unique conditions.

Temperature profiles in a deep snow pack are different than those of soil (Figure 8.11). Note that temperatures just below the surface are slightly warmer because shortwave radiation is transferred through the snow more easily than longwave. Also, during the day radiative heat transfer is greater than heat conduction in the upper layer. Energy is lost by reflection of shortwave radiation and escape of small amounts of longwave energy.

At night, there is no shortwave and longwave loss to the air if the air is colder than the snow surface. As a result, changes in energy are active right at the surface. Once the sun rises, the shortwave penetrates and overcomes the longwave loss, and the area of maximum temperature moves down below the surface by conduction.

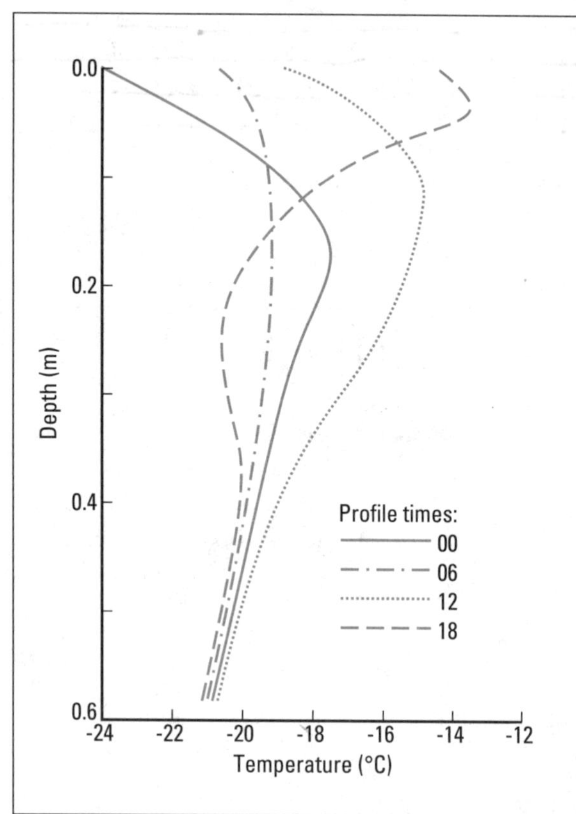


FIGURE 8.11 ▼ Diurnal snow temperature profiles from Devon Island icecap (after Oke, 1978; Holmgren, 1971)

The depth of the snow is very important in energy transfer. If the snow depth is less than 0.15 m, energy will be transferred down to raise the temperature of the underlying layer. This, in turn, will cause melting of the snow from below.

The amount of melting from below is primarily a function of the depth of snow and the ground material. However, on a short-term basis even a very thin snow cover is a good insulator preventing the downward diffusion of heat energy. For many plants and animals, the ability of snow to prevent ground temperatures falling too rapidly can be critical to their survival. Too much snow is a problem, especially for animals, but too little snow can be devastating.

Farmers are usually interested in maintaining adequate snow cover. It prevents frost penetrating too deeply. Meltwater is valuable to replenish soil moisture in the spring, and early growth of some seedlings can be protected by the insulating effects of the snow. Rouse, in studies done at Churchill, Manitoba, has shown how critical this snow cover is to the survival and annual growth of trees, especially in the first year. In the grain-growing region of Canada, crops such as fall rye and winter wheat can only be successful if there is adequate snow cover throughout the winter. Winter wheat can be an important crop for several reasons. The farmer is independent of spring seeding conditions. The crop is growing the minute the soil temperature is adequate, thus reducing the problems of a short growing season due to early fall frosts. Perhaps more important in today's world, it allows the wheat to develop ahead of weeds and thus reduces the need for chemical weed control.

## Local climates

### URBAN CLIMATES

The climate modifications found in woodlands are small compared to what humans achieve in cities. Instead of a mixture of soil and vegetation, we have a mosaic of concrete, glass, brick, asphalt, and stone surfaces ranging to heights of several hundred metres. Among this, we scatter grass surfaces and a few trees to variegate the 'concrete jungle.' The building materials have vastly different physical properties from soil and plants. For example, the warmth of concrete in the evening is due to its high heat capacity. This means that if large quantities of heat are added to the material while the sun is shining it is slowly released during the night, adding warmth to the urban atmosphere. In this way, city temperatures are kept relatively high. We notice the effect most in the evening when we travel from the cool of the open countryside to the heat of the city (Figure 8.12). It is an effect called the **urban heat island**. Early blooming of flowers and decreased snowfall and frost are both indicators of this effect.

*The urban heat island* We can illustrate the different responses of the city and rural areas by comparing their heat budgets, as shown in Figure 8.13. It is the change of the heat budget by the urban surface that helps to produce the distinctive urban climate, so let us look in more detail at the way changes are produced. By day, both rural and urban areas experience a radiation surplus. Smoky urban atmospheres may reduce the size of this surplus slightly, but as the quality of urban air improves because of pollution controls, the differences in inputs have become slight.

At a smaller scale, differences are more significant. Trees and crops allow a certain amount of radiation to pass through to the ground surface. They transpire moisture and have a low heat capacity. As we saw earlier, this results in cooler temperatures beneath the canopy. In the city, the building materials of concrete, brick, and stone all have high heat capacities, enabling them to store large amounts of heat. Shadowing can be important, but there are still numerous surfaces exposing large, dry areas to the sun's rays. When the angle between the receptive surface and the sun's rays approaches 90° the heat input will reach its maximum. This effect is likely to occur much more frequently in an urban area, with its vertical walls, than in a rural area.

Of the energy that is available as net radiation, some is used to heat the air, some in evaporation, and the remainder is absorbed by the soil or buildings and other artificial surfaces. Here the main contrasts arise. In a city, sewers and drainage sys-

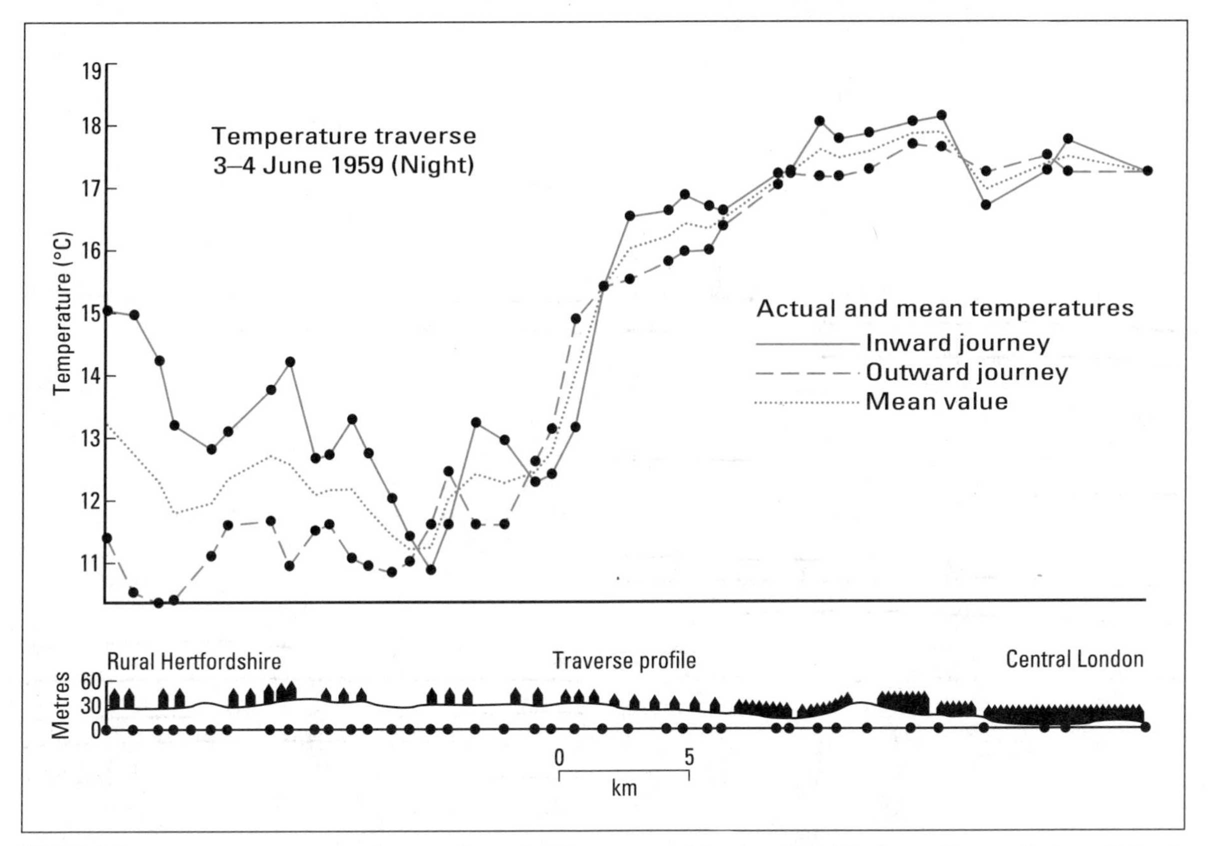


FIGURE 8.12 ▼ Temperature traverse from rural Hertfordshire to central London, U.K. The lower diagram depicts relief and building density (after Chandler, 1965)

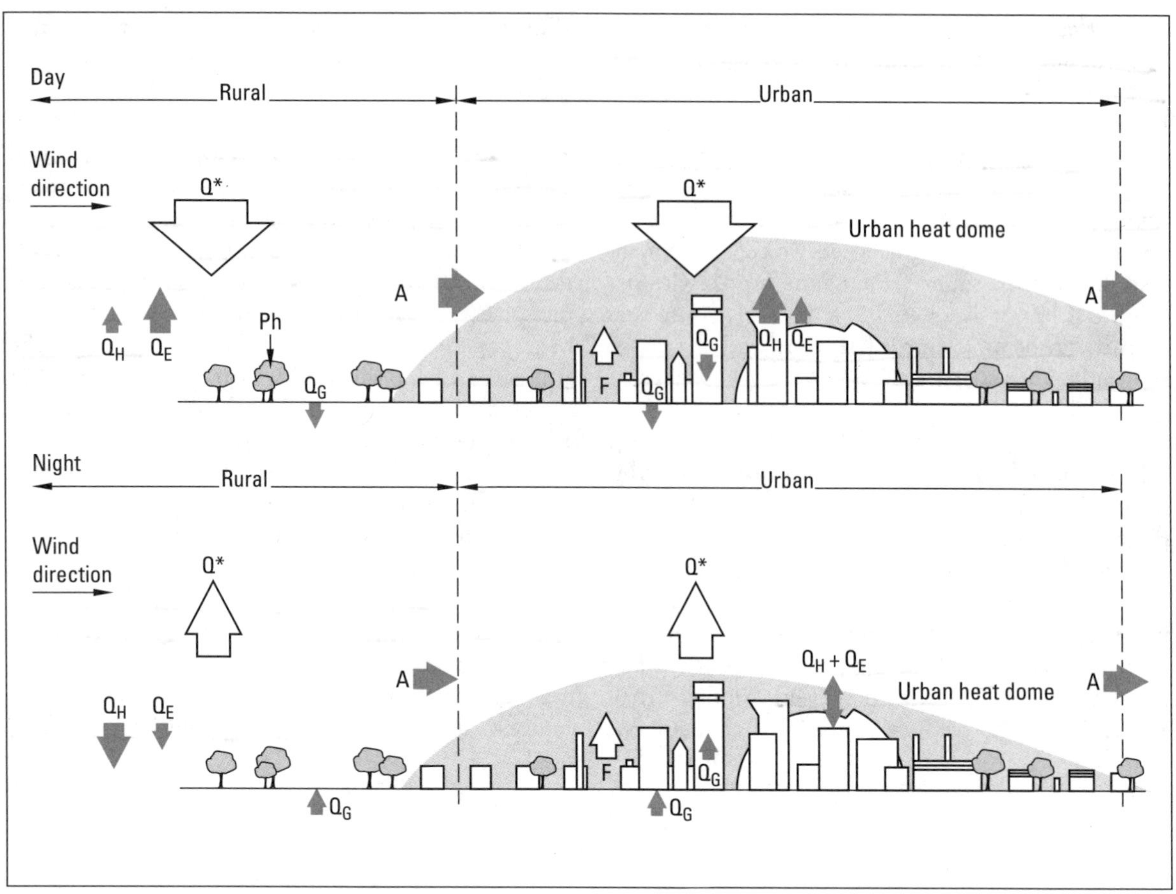


FIGURE 8.13 ▼ Heat balances of urban and rural surfaces by day (top) and by night (bottom). Ph is energy used in photosynthesis and A is net advected energy. See Figure 8.5 for meaning of other symbols (after Fuggle and Oke, 1970)

tems lead to a rapid removal of water. Surfaces soon become dry once rain has stopped, so the use of energy for evaporation is small. This means that more is available for heating the air and buildings rather than for evaporation, which is 'non-productive' in terms of heating. Conversely, when rain falls in a rural area, leaves intercept some and the soil soaks up much of the rest so that there is plenty of moisture around for evaporation. The energy used in evaporation cannot be available for heating the air or soil and so temperatures stay cooler. A final fact is significant in the city. Large amounts of fuel are used to heat buildings, in industrial processes, and in car travel. Even human activity generates appreciable amounts of heat where population density is high, and all this heat is eventually released into the urban atmosphere (F in Figure 8.13). On Manhattan Island, research has shown that in the average January, the amount of heat produced from combustion alone is greater than the amount of energy from the sun by a factor of 2.5. In summer this ratio is only about 0.15.

At night, the ground surface loses energy, resulting in cooling. In rural areas, the ground becomes cooler than the air above giving an inversion of temperature. There is then a weak transfer of heat to the surface from the soil and from the atmosphere, but these additions do not compensate for the radiational losses and so temperatures fall. In a hot summer this may feel refreshing compared with the humid warmth of the city. There the buildings continue to give off heat they absorbed and stored during the day ($Q_G$ in Figure 8.13). This is added to the heat of combustion from furnaces used in the building, shown as F in Figure 8.13, to reduce the rate of cooling. We can see this in Figure 8.14, where the cooling rates in Montreal are compared with a nearby rural site.

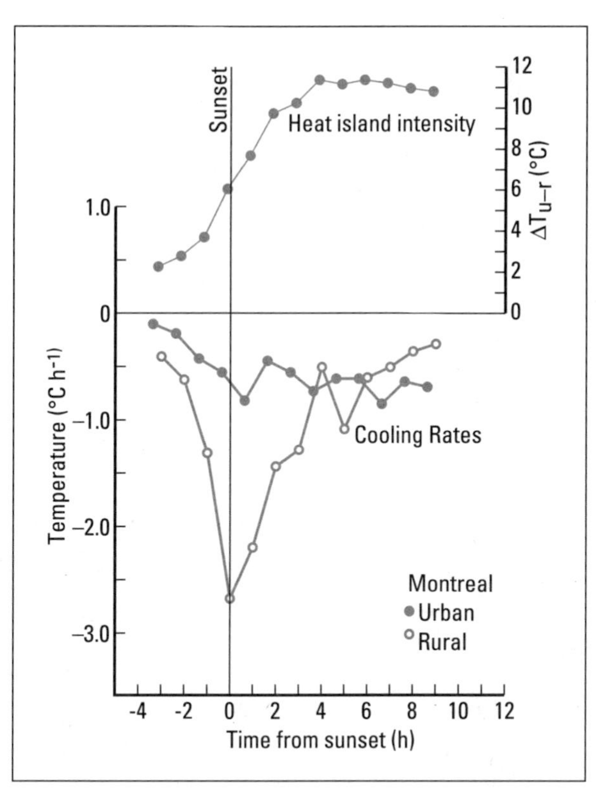


FIGURE 8.14 ▼ Mean hourly urban-rural cooling rates and heat island intensities based on data from summer nights with clear and calm conditions in Montreal, Quebec (after Oke and Maxwell, 1975)

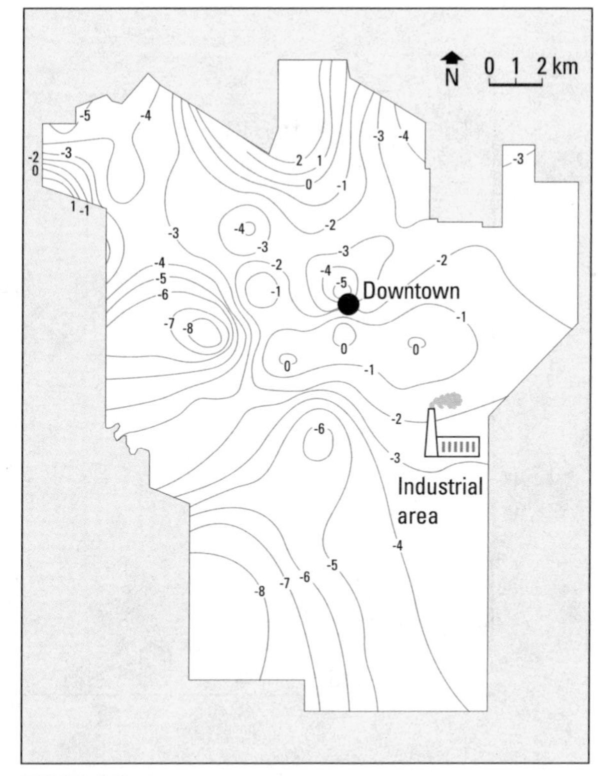


FIGURE 8.15 ▼ Nighttime isotherms (in °C) for Calgary, Alberta, showing the urban heat island in December 1975 (after Nkemdirim, 1976)

This relative warmth prevents the development of an inversion so heat transfer and evaporation still take place. Dewfall or condensation is much less frequent than in rural areas. It is this urban heat that many city dwellers find uncomfortable in the summer.

## THE EFFECT OF WINDS

If winds were strong, all this surplus heat would be rapidly removed from the city to be mixed with cooler surrounding air, and there would be no such thing as the urban climate. When there is little wind and clear skies, we find the greatest differences between urban and rural areas (Figure 8.15). The pattern of nighttime minimum temperatures usually shows highest values near the city centre, fairly uniform levels in the low-density suburbs, and then a sharp boundary into the cooler rural areas. This is seen most clearly in cities with fewer relief features. Valleys, hills, and parkland within the urban area can produce major changes. Slopes can give rise to areas of greater heating by day but, by night, cold air may move downslope and accumulate in the valley bottom, counteracting any urban effect. The parklands have different heat capacities, albedos, and emission temperatures from surrounding buildings, giving slightly lower day and nighttime temperatures. We see that the advantages of these 'urban lungs' extend well beyond their aesthetic appeal.

Even when winds are not light, the presence of the urban structure tends to slow air movement. Wind records from city centre sites show lower average speeds than suburban or rural locations nearby, although the degree of gustiness may be higher, especially in summer. As the air flows over the very irregular surface of a city, friction with the buildings retards the wind in the lowest layers (Figure 8.16). The presence of skyscrapers, however, produces eddies (Figure 8.17) that can cause strong local winds. At street level these can become quite

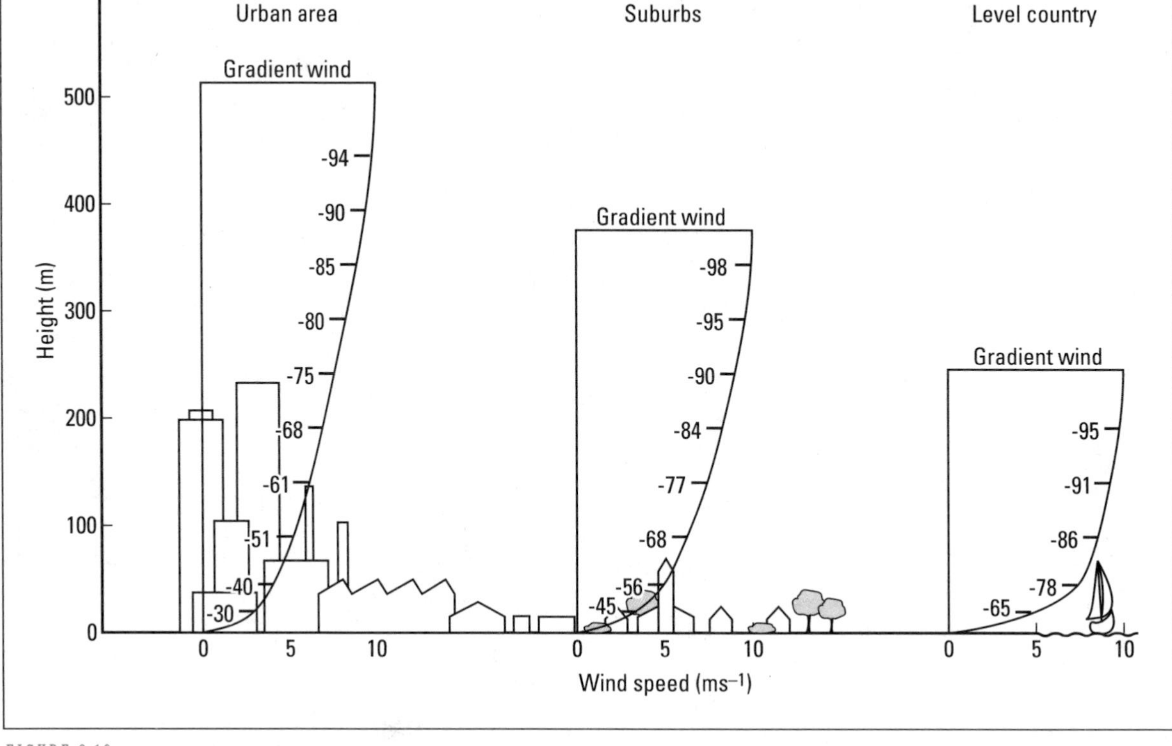


FIGURE 8.16 ▼ The effects of terrain roughness on the wind speed profile. With decreasing roughness, the depth of the affected layer becomes shallower and the profile steeper (after Davenport, 1965)

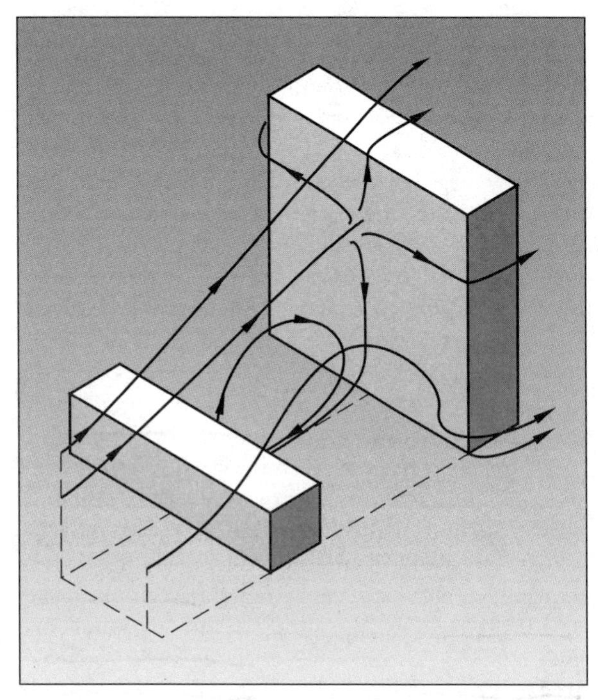

FIGURE 8.17 ▼ Typical flow pattern on the windward face of a building (after Wise, 1971)

unpleasant, raising dust and rubbish, and making walking difficult. Quite a few shopping precincts have been unpopular with shoppers until the architects realized that such winds could be a problem and took measures to prevent them.

## CLOUD AND PRECIPITATION IN CITIES

Most climatic changes brought about by urbanization have been well documented. These are summarized in Table 8.4. Some of the changes are appreciable, though a global decrease in use of coal as a fuel and energy source could lead to smaller modifications in insolation, contaminants, and fogs, as has been demonstrated in the United Kingdom. The increase in cloud and precipitation over cities was the one aspect that took some time to prove. It is only recent work, particularly in St. Louis, Hamilton, Winnipeg, and Vancouver, that has confirmed conclusively the urban effect. There appear to be multiple causes for the increases in cloud cover and precipitation. Added heating by air crossing the city, increases in contaminants, the mechanical effects on airflow, and altered moisture all appear to have a role.

Table 8.4

*Effects of urbanization on climate. Average urban climatic differences expressed as a percentage of rural conditions (except for temperature)*

| | *Annual* | *Cold season* | *Warm season* |
|---|---|---|---|
| Pollution | +1000 | +2000 | +500 |
| Solar radiation | –22 | –34 | –20 |
| Temperature (°C) | +2 | +3 | +1 |
| Humidity | –6 | –2 | –8 |
| Visibility | –26 | –34 | –17 |
| Fog | +60 | +100 | +30 |
| Wind speed | –25 | –20 | –30 |
| Cloudiness | +8 | +5 | +10 |
| Rainfall | +14 | +13 | +15 |
| Thunderstorms | +15 | +5 | +30 |

Source S.A. Changnon, 'Inadvertent weather modification,' *Water Resources Bulletin*, 12 (1976), pp. 695–718.

The confluence zones induced by these urban effects may even lead to preferential development of clouds and rain. Which factor becomes dominant in a particular storm varies depending upon the nature of air circulation over the city on that day. As the effects are less noticeable in winter than in summer, it follows that it is the natural, not artificial, heating effects that are most important. Observations have shown that in some cities precipitation is higher from Monday to Friday when industrial activity and pollution levels are greater than on the weekend. However, contradictory results have also been found, so the role of industrial activity on precipitation is still under investigation.

As the degree of urbanization has increased, so an ever greater number of people are affected by an urban climate. Apart from the more obvious effects of pollution, wind, and summer heat, few people may realize that their urban area has changed other aspects of the climate. The nature of the urban area represents an extreme example of the way humans can change the climate near the ground.

## The microclimate of slopes

So far, all examples have assumed that the ground surface is almost flat. In reality there are few areas of the world that are so level that the effect of topography can be ignored. The reason we need to know more about the topography is that slopes modify how much shortwave radiation reaches the surface. We saw earlier that the maximum intensity or radiation is received when the angle between the surface and the sun's rays is 90°. If a horizontal surface is tilted so it is at right angles to the rays, then the amount of radiation received increases. This factor is exploited by sunbathers who can tilt the angle of their reclining seats to achieve a maximum heat input. If this was the only factor, calculating the new input for a slope would be easy. However, while the slope remains constant in its angle and direction of slope, the sun is continuously changing its position in the sky throughout the day and throughout the year. Slopes, unlike sunbathers, cannot adjust their position. Consequently a slope that receives maximum intensity at one time on a certain day of the year may be in shadow at other times.

### EFFECTS ON THE RADIATION BALANCE

As the movement of the sun across the sky is known, it is possible to calculate the intensity of shortwave radiation falling on a slope of any combination of gradient and orientation (**azimuth**) for clear skies (Figure 8.18). More frequently we are interested in the total radiation rather than the intensity, but even this problem has been overcome using computers. A computer program can be devised to calculate the intensity of radiation on the surface for any particular time and slope. So, for the start of the program, radiation intensity is calculated for sunrise. Then the computer determines the sun's position in the sky, say ten minutes later, works out the new radiation intensity, and then adds this value to the previous total. This is continued until sunset (Figure 8.19). We then have the daily total of shortwave radiation based on intensity values every ten minutes. The contribution from diffuse radiation is assumed to be constant throughout the day, and so does not add to the spatial variability of solar receipt at the surface. Nonetheless it is vitally important for slopes with a northerly aspect that would otherwise receive very little shortwave radiation. Moon explor-

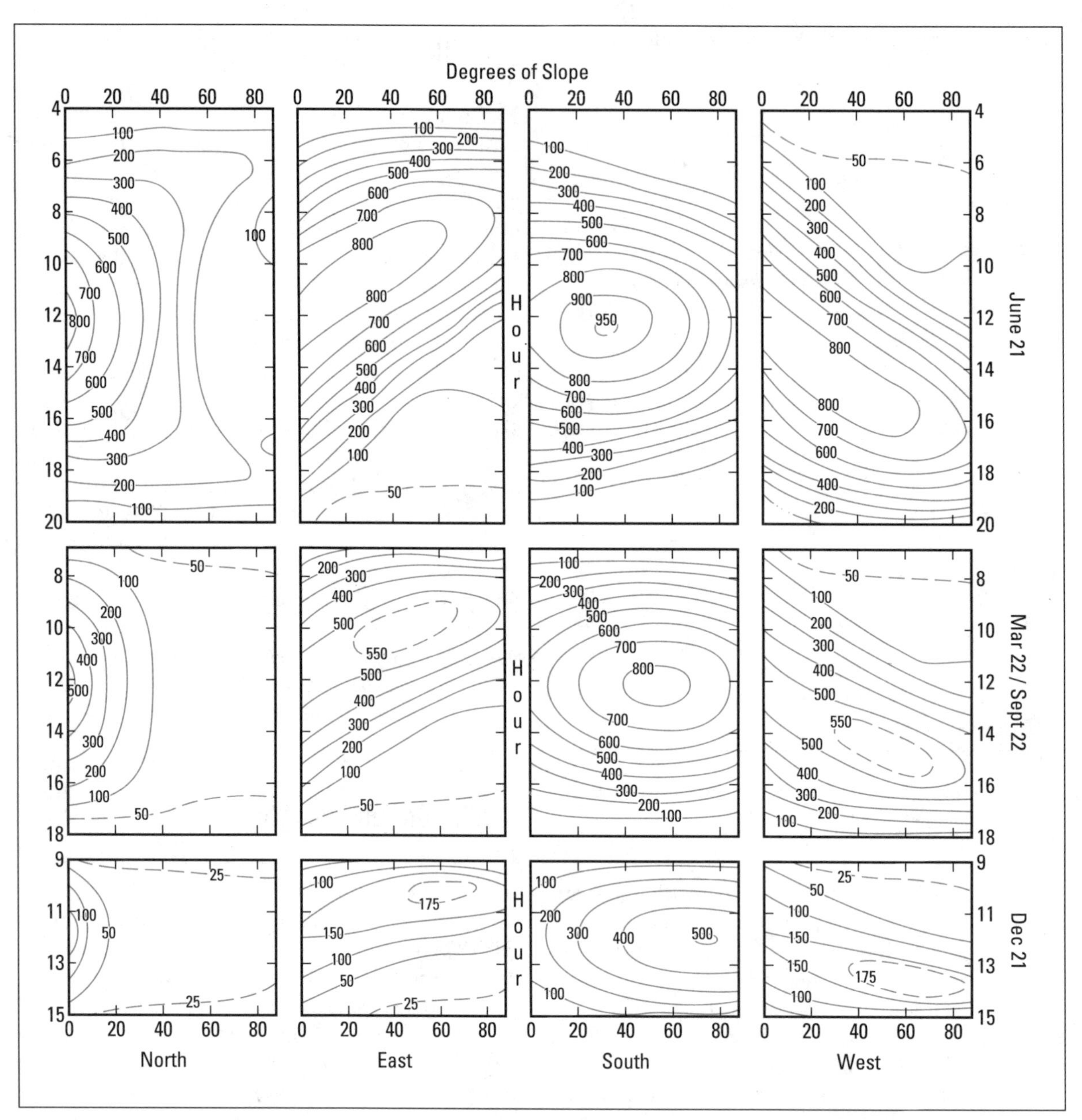


FIGURE 8.18 ▼ Direct and diffuse shortwave radiation on slopes of different gradients and orientation at the solstices and the equinoxes under clear skies at 53°N. Units are $Wm^{-2}$. Apart from minor differences due to atmospheric moisture content, the values will be the same for 22 March and 22 September. Based on a model developed by the Department of Building Science, University of Sheffield, U.K.

ers are able to see this, for with no atmosphere there is no diffuse radiation and any surface that is not directly in sunlight appears almost black.

These effects of slopes upon radiation inputs mean that the radiation balance varies locally with topography. In the northern hemisphere, slopes with a southerly aspect receive a greater input of radiation than northerly ones, resulting in larger exchanges in sensible heat and higher temperatures (Table 8.5). In high latitudes this additional energy may be an advantage in sunshine-starved areas, but in more arid countries the increased radiation will evaporate moisture more quickly and may produce moisture stresses in cultivated plants.

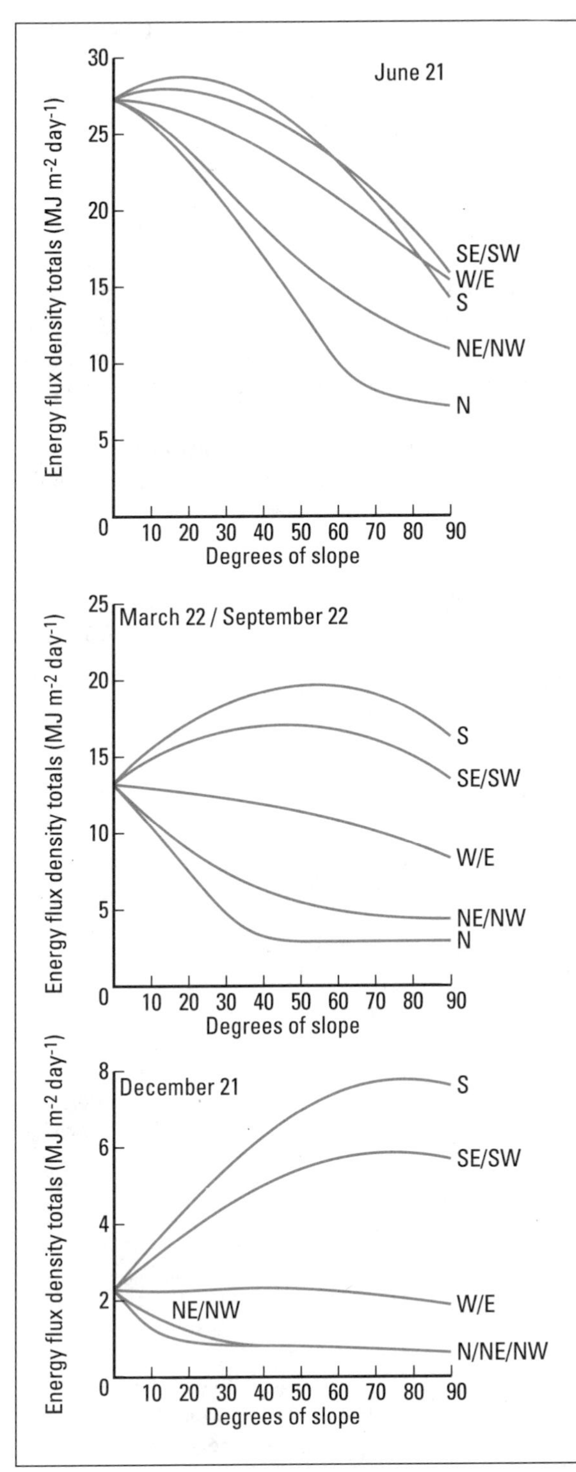


FIGURE 8.19 ▼ Total daily direct and diffuse solar radiation incident upon slopes of differing angle and aspect at 53°N. Note different scale for December. Based on a model developed by the Department of Building Science, University of Sheffield, U.K.

## Table 8.5

*Influence of slope orientation on microclimate*

| | | *After five dry days* | *After two days with rain* |
|---|---|---|---|
| Maximum temperature (°C) | N | –1.9* | –1.5 |
| | E | –1.3 | 0.0 |
| | S | 2.6 | 1.4 |
| | W | 0.5 | 0.2 |
| Minimum temperature (°C) | N | –0.3 | –0.4 |
| | E | –0.1 | –0.4 |
| | S | 0.4 | 0.3 |
| | W | 0.0 | 0.5 |
| Daily mean temperature (°C) | N | –0.9 | –0.4 |
| | E | 0.1 | –0.3 |
| | S | 1.1 | 0.6 |
| | W | –0.4 | 0.2 |
| Relative humidity at 1:00 P.M. | N | 8% | 1% |
| | E | 3% | 5% |
| | S | –13% | –3% |
| | W | 6% | –4% |

*Figures are relative to a horizontal surface nearby.

Translated from Fuh Baw-Puh, 'Influence of slope orientation on the microclimate,' *Acta Meteorologica Sinica*, 32 (1962), pp. 71–86.

## SLOPES AT NIGHT

At night, when there is no input of shortwave radiation, the effect of a sloping ground surface on the energy budget is less pronounced. Figure 8.20 shows the exchanges taking place. For slopes between 0° and 30°, emission of longwave radiation follows the cosine law ($E_{sl} = E_{horiz} \cos \theta$); at higher angles more radiation is emitted than would be predicted. The only effect of slope direction is in influencing surface heating during the day which, through heat storage, may affect nighttime temperatures and hence emission rates. If the sky is obstructed by trees, other valley slopes, or even buildings, much of the longwave emission is absorbed and reradiated back to the ground. This reduces the rate of cooling from the ground and is one of the factors important in keeping city centre

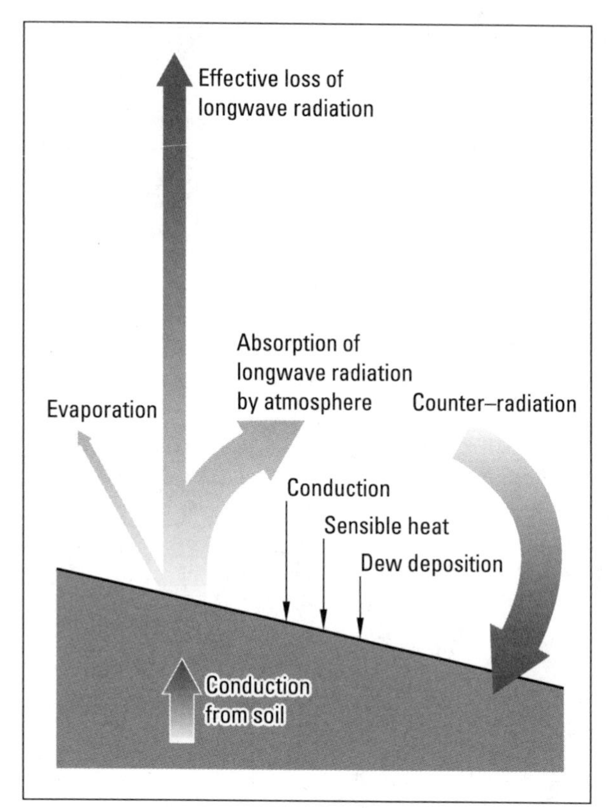


FIGURE 8.20 ▼ Nighttime energy exchanges. The magnitude of the components will vary greatly depending upon weather conditions such as cloud amounts, wind speed, and humidity

temperatures high at night. It can sometimes be seen in frosty weather when open grassy surfaces are white, but, beneath trees or near buildings where counter-radiation has been greater, there is no sign of frost on the ground.

Of much greater importance at night is what happens to the air as it cools through contact with the ground surface. As the air becomes cooler, it gets denser. If the surface is flat, the cold air remains at ground level. However, on a slope, cool air may move downslope as a **katabatic wind**, increasing in strength and volume until it meets a physical barrier, such as a fence, wall, or embankment, or until it is no longer colder than the surrounding air. Once the cold air stops moving it continues to cool through radiation emission and may eventually reach very low temperatures. This microclimatological effect can be very pronounced on clear, calm nights that allow radiational cooling to continue at a high rate.

One result of this process is the formation of **frost hollows**. Farmers should take care that frost-sensitive crops are not grown where cold air is likely to accumulate and give ground or even air frosts. It is for this reason that in frost-susceptible areas, fruit orchards are cultivated on valley slopes, allowing the cold air to drain through the trees without accumulating. A classic example of a frost hollow was found in the European Alps. A limestone sinkhole with a steep backwall facing northeast allowed cold air to become stagnant. Figure 8.21 shows temperatures at different levels on one particular night. Towards the WSW the sinkhole is intersected by a col that allows the stagnant cold air to remain in the lowest 50 m of the hollow. Temperatures as low as –51°C have been recorded when the ground was snow-covered. Even coastal Antarctica is usually much warmer than this!

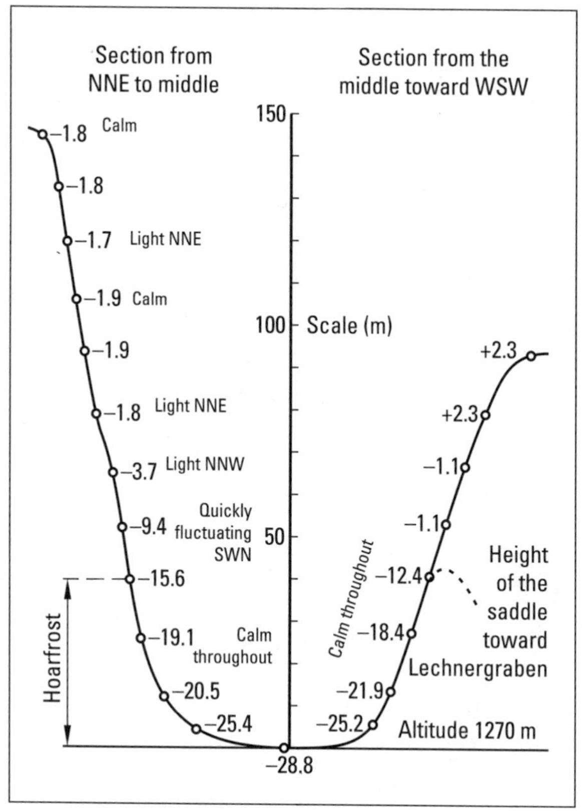


FIGURE 8.21 ▼ Temperature distribution in the Gstettneralm sinkhole near Lunz, Austria, 21 January 1930 (after Schmidt, 1930)

## VALLEY BREEZE SYSTEMS

If the katabatic winds, described above, are not prevented from flowing, then we find they begin to form an organized system of cold air drainage downslope and down-valley. Speeds are low, perhaps 1 m per second or less, and the movement tends to pulsate with intermittent surges—like that seen in water running off a sloping road surface. The downslope flows eventually combine into a down-valley flow, known as a mountain wind, where they emerge onto the lowlands.

By day, this **cold air drainage** does not occur, except where snow and ice surfaces maintain cooling. Instead, it is replaced by **anabatic winds** upslope. These are produced by heat on the slope, which forces the warm air to rise upslope. Cool air from the valley floor flows in to replace the warm air and a valley breeze is generated (Figure 8.22). These valley breeze systems could not last long unless a continuity of the flow was maintained. This is usually found as a counter-wind at higher levels. If the pressure gradient wind is strong this increases local mixing so that major temperature differences are prevented. No cold air is available to sink downslope nor warm air to rise upslope, so the formation of the breeze is stopped. Like so many microclimatological phenomena, valley and mountain breezes require clear skies and light winds for their operation.

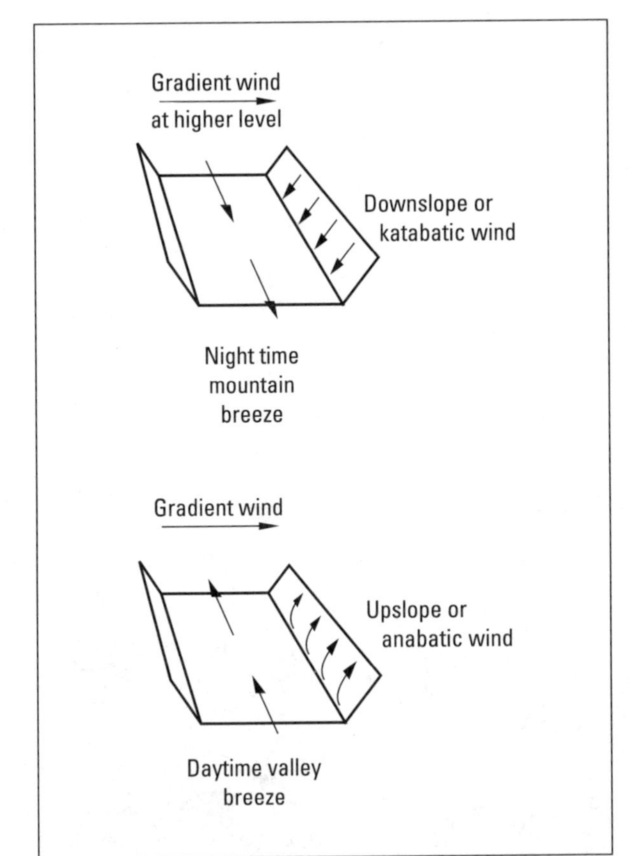


FIGURE 8.22 ▼ Katabatic (left) and anabatic (right) airflow in a valley–mountain breeze system

## SEA BREEZES

The driving force for the valley and mountain breezes is a temperature gradient. Temperature contrasts develop between slopes and valley floors, between uplands and lowlands, so that the nature and strength of the wind depends upon the precise form of the gradient. This thermal control of wind occurs at all scales from the general circulation of the atmosphere (Chapter 5) right down to the smallest eddy of heat rising from the ground. In Chapter 7 we discussed the mechanism of the monsoons as a regional climate that had great importance for many parts of the world, but particularly for the Indian subcontinent. The more rapid heating and cooling of the land mass of Asia results in low pressure over the land in the heat of summer, and high pressure with the cold of winter. The result is that winds tend to blow from the water to the land in summer, and from the land to the water in winter. The word *monsoon* means 'season' in Arabic, a reference to the seasonal change in wind direction. Monsoon winds from the Indian Ocean bring rains to a land dessicated by dry winter winds from central Asia, which are heated by compression as they descend from the Tibetan plateau. The same pattern of wind reversal often occurs along the coasts of oceans and large lakes on a daily basis. Here it is referred to as a **sea breeze**, when blowing from water to land, and a **land breeze** when blowing from land to water.

Sea breezes are formed by the different responses to heating of water and land. If we have a bright, sunny morning with little wind, the ground surface warms rapidly as it absorbs shortwave radiation. Most of this heat is retained at the surface although some will be transferred through the soil at a rate dependent upon the amount of air and water in the soil. As a result, the temperature of the ground surface increases and some of this heat warms the air above. When the sun sets, the surface starts to cool rapidly as there is little store of heat in the soil. Thus we find that land surfaces are characterized by high day (and summer) temperatures and low night (and winter) temperatures.

The response of the sea is very different. First, sunshine can penetrate the water to about 30 m, as any skindiver knows. Second, water has a large heat capacity, so a lot of sunshine has to be absorbed to raise its temperature. In addition, the warming surface water will be mixed with cooler, deeper water through wave action and convection. Instead of a thin active layer, which we have in a soil, the top 20 m or so of water provides the active layer; consequently temperature changes are slow. Slight warming occurs during the day and slight cooling at night. This means that the sea is normally cooler than the land by day and warmer by night. (On a longer time scale, the sea is cooler relative to the land in summer and warmer in winter unless there are unusual currents offshore.)

The higher temperature over the land by day generates a weak low pressure area. As this intensifies during daytime heating, a flow of air from the cool sea develops, gradually changing in strength and direction during the day. At night the reverse circulation evolves, with a flow of air from the cooler land to the warmer sea, though as the temperature difference is usually less and the atmosphere stable, the land breeze is weak. At higher levels, we find a flow in the opposite direction (Figure 8.23) compensating for the surface land or sea breeze. Even large lakes show a breeze system of this nature (Figure 8.24), though in winter the lakes may freeze, so that temperature differences do not develop.

In tropical areas, the strength and reliability of the sea breeze brings a welcome freshness to the area along the coast, and its effect can extend up to 150 kilometres inland.

# Conclusion

The atmospheric processes of radiation, convection, evaporation, and advection interact with the variety of surfaces at the ground level to produce a mosaic of microclimates. Distinctive effects can be found at a wide variety of scales in increasing size from the microclimate of a single leaf through crops, forest, valley slopes, urban areas, and sea–land breezes. In most cases, there is not a firm boundary between scales, the micro and local climates form part of a continuum from smallest to largest. Certainly within the larger scales like urban climates there would be innumerable microclimates resulting from surface modifications. This diversity makes their investigation fascinating. Equally, it presents prob-

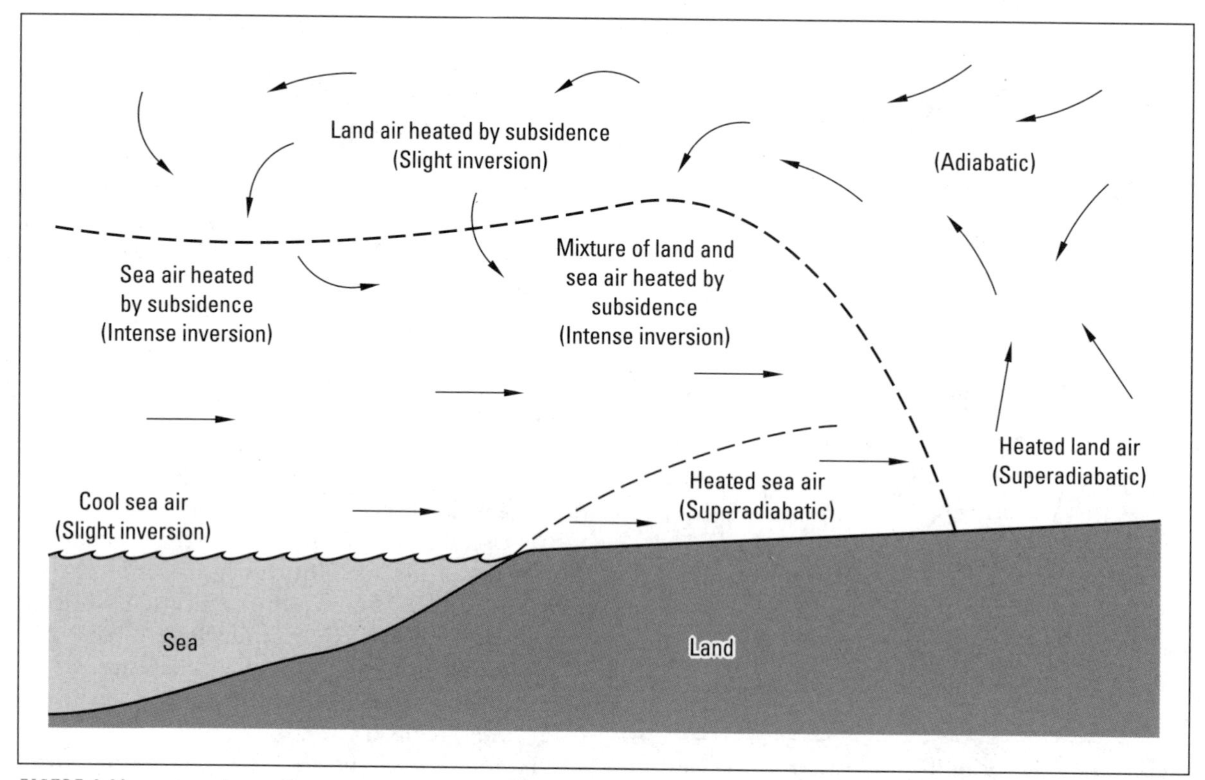


FIGURE 8.23 ▼ A sea breeze with a light geostrophic wind (after Munn, 1966)

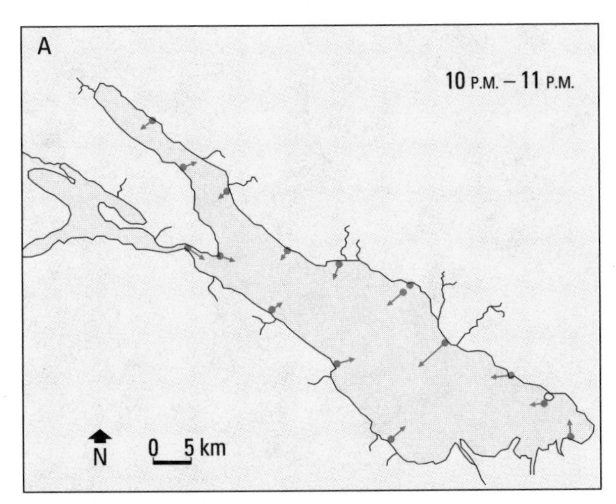


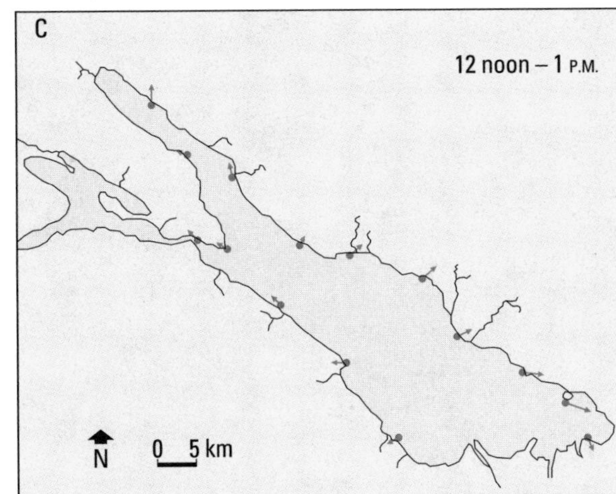


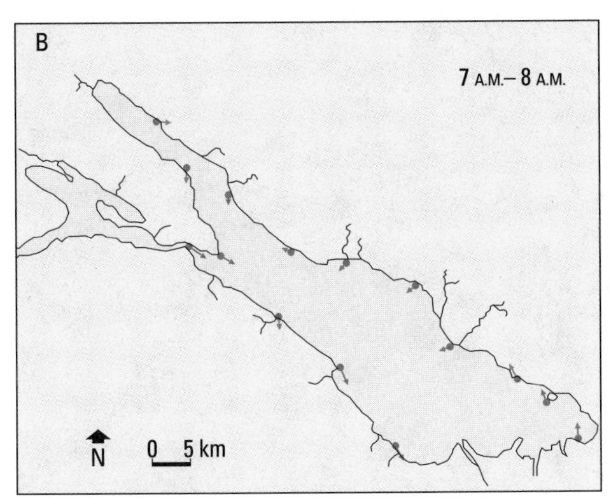


FIGURE 8.24 ▼ The land-lake breeze over Lake Constance, Germany. A: 10–11 P.M., land breeze phase; B: 7–8 A.M., intermediate phase; C: 12 noon–1 P.M., lake breeze phase (after Huss and Stranz, 1970)

lems of explanation and interpretation, as it is physically impossible to measure the wide variety of possible microclimates, and it is easy for so-called understanding to degenerate into a series of case studies. A final understanding (if there is such a thing!) will only come when we appreciate the interactions and links between the myriad of atmospheric processes and surface conditions.

The importance of microclimate modifications goes far beyond the study of climate, however. It is at this scale that we can see the relationships between climatic processes, landscape and ecosystems. Landforms and vegetation modify the microclimate; the microclimate in turn controls many of the processes involved in landscape, soil development and plant growth.

# Climatic change

## Introduction

Ironically, it was a change in global climate starting about 1976 that made people aware that the climate could change. A few scientists had been studying and revealing evidence of the extent and speed with which the climate had varied. In the late 1970s, the world became gripped by predictions of doom that the world was cooling and another Ice Age was pending. Lowell Ponte published a book titled *The Cooling* in 1976. He wrote, "It is cold fact; *the global cooling presents humankind with the most important social, political, and adaptive challenge we have had to deal with for ten thousand years.*" The author provided the italics to emphasize the threat. Just ten years later, the same type of statements were being made about global warming.

By the late 1980s a majority of scientists accepted that global warming had been occurring for the last one hundred years. There was debate about the amount of warming, but most agreed that it had occurred. A greater debate raged over the cause. Some argued that it was carbon dioxide from anthropogenic sources creating an enhanced greenhouse effect. Others, such as Richard Lindzen, said, "The consensus has been reached before the research has even begun." Who is right? What do we know?

To answer these questions we need to examine three major components.

- How the evidence of climate change is gathered and measured.
- The mechanisms of climate and how they interact to cause change. In this portion we will also examine the role of computers in simulating regional and global climates.
- Finally, we will present the pattern of climate change that has been determined so far.

# Longterm changes

We must consider how our system changes over longer periods of time. It is clear that local surface conditions vary considerably from year to year, season to season, and even from day to day. In addition, we have stated that the major driving force of atmospheric circulation—the input of solar radiation—changes when a sufficiently long time period is considered. The balance between the processes operating in the atmosphere, therefore, is not constant; the details of the system shown in Figure 9.1 vary over time. If we had been around 10 000 years ago, at the end of the last Ice Age, we would have seen just how different these conditions could be. A few million years earlier, in the period geologists call the Tertiary, the system would have been different again. Thus the climatologist must understand what causes changes in the atmospheric system, and what are the likely effects of these changes.

## HISTORY OF CLIMATOLOGY

When climatology was in its infancy in the nineteenth century, it was a common belief that climate was an unchanging feature of our environment. This was consistent with the scientific philosophy of uniformitarianism, which was based upon the Darwinian concept that things changed only very gradually over long periods of time. It was assumed that by averaging climatic data over a sufficiently

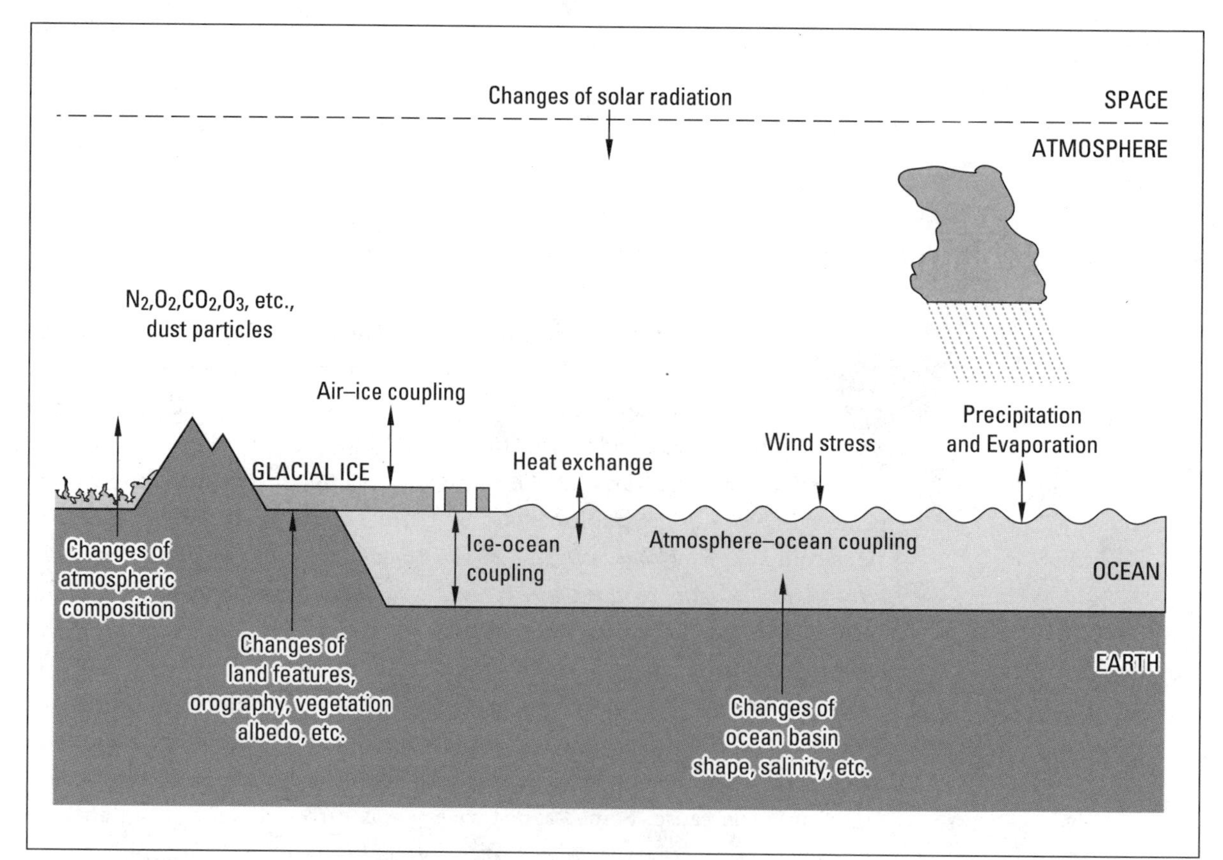


FIGURE 9.1 ▼ The physical process and properties that govern the global climate and its changes (after US National Academy Report *Understanding Climate Change*, 1975)

long period, perhaps thirty or fifty years, the true climate would be determined.

As more was learned about former climates, it became apparent that climates fluctuate all the time, and much more widely than previously imagined. The extreme climates of the Ice Ages were discovered first. The work of scientists such as Louis Agassiz determined that huge icecaps had formed in the last two to three million years, events that could only have occurred under dramatically different climatic conditions.

More recently, the classic book *Climate Through the Ages*, written by H. Brooks in the 1930s, and Hubert Lamb's work *Climate: Past, Present and Future*, published in 1972, have clearly demonstrated the extent of change and variability within historic times. The evidence of these climatic fluctuations is well documented from a variety of sources.

For example, for the eighteenth century in Canada, weather diaries and instrumental records show that conditions were colder and more extreme than at present. In Europe the records show similar patterns, with colder conditions and more frequent harvest failures from the seventeenth to the nineteenth centuries. Before that, weather diaries, tree ring records, and other evidence show a much warmer climate in a three-hundred-year span from about A.D. 900 to 1200. Switch the clock back another 12 000 years and we find most of Canada still covered with the huge icecap of the Wisconsinan glaciation. The site of Churchill, Manitoba, lay buried and the land depressed under the weight of almost 2000 metres of ice. Remarkably, and still not fully explained, sufficient surplus heat energy occurred to melt that vast ice sheet in a mere 6000 years.

Since the ice melted, the climate has fluctuated a great deal and vegetation, animals, and humans have had to react. For example, Nichols has shown that the boreal forest extended into the High Arctic in the period now known as the Climatic Optimum some 4000 to 6000 years ago.

Thus we have the evidence of historical records and a variety of geological and biological materials demonstrating frequent, and often drastic, variations in climate. The subject of former climates and the evidence used to interpret climatic variations is now so broad that this chapter must be only a brief and relatively simple introduction. Let us start by examining what types of evidence are used and how they can help us to interpret the climatic record.

## LARGE-SCALE CHANGES

Previously, we saw how short-term changes in the hydrosphere and biosphere could affect atmospheric circulation. At a longer time scale, the four major components of the global system have potential climatic consequences. Changes in any one of these components—the hydrosphere, lithosphere, cryosphere, and biosphere—may lead to changes in atmospheric, and therefore climatic, conditions. Let us look at a few examples of changes that are likely to have occurred. We know that during the Quaternary Ice Ages (Table 9.1, see also Table 15.1), sea levels fell by up to 150 m, relative to the present. In the areas of wide continental shelves, this would have exposed large areas of dry land, which would have had a very different albedo, and especially a changed heat storage capacity, from the water it had replaced. The lowered sea level could also change the positions of the ocean currents, tongues of warm water being deflected from their previous patterns, though currents would also be affected by the atmospheric circulation.

Expansion or contraction of the icecaps, even by quite small amounts, may lead to prolonged changes in climate over a wide area. Concern has been expressed recently about the apparent increase of winter snow cover in the northern hemisphere because of these potential changes, but the trend appears to have been short-lived. The effects of changes in the lithosphere may be similarly far-reaching. As we shall discover in Chapter 17, the earth's crust is highly mobile; it is constantly being raised into mountain chains, and moved across the surface of the globe. In the process, the character and distribution of the continents change. As a result, the patterns of both atmospheric and oceanic circulations are disturbed and the climate, too, is modified. It is interesting to speculate what would be the effect on the climate of the northern hemisphere today with much less north–south movement of air; the southwest to northeast flow of air across the Atlantic would be greatly reduced, together with the speed of the North Atlantic Drift. Locally, rainfall would penetrate more easily into the Great Plains, the Chinook would disappear, and the wide swings of temperature that can affect the Mississippi Valley would be reduced due to fewer depressions moving southeastward or northeastward. However, because of all the complex interactions involved, much of this is speculation. We cannot be sure of the precise consequences, and so

Table 9.1

*World Quaternary nomenclature*

| | *North America* | *Britain* | *North European Plain* | *Alps* |
|---|---|---|---|---|
| | | Flandrian | Flandrian | |
| Lower | WISCONSINAN | DEVENSIAN | WEICHSELIAN | WÜRM |
| | Sangamonan | Ipswichian | Eemian | Riss-Würm |
| | ILLINOIAN | WOLSTONIAN | SAALIAN | RISS |
| Middle | Yarmouthian | Hoxnian | Holsteinian | Mindel-Riss |
| | KANSAN | ANGLIAN | ELSTERIAN | MINDEL |
| | Aftonian | Cromerian | Cromerian | Gunz-Mindel |
| Upper | | BEESTONIAN | | |
| | | Pastonian | | |
| | NEBRASKAN | BAVENTIAN | MENAPIAN | GUNZ |
| | | Antian | Waalian | Donau-Gunz |
| | | THURNIAN | EBURONIAN | DONAU |
| | | Ludhamian | Tiglian | Biber-Donau |
| | | WALTONIAN | PRAETIGLIAN | BIBER |

Note: Glacial periods are printed in capitals; interglacials in lower case.

proposals for widespread weather modification must be treated with great caution.

## SURFACE CHANGES

The surface of the earth is the interface between the atmosphere and the lithosphere, and, like our skin, is an extremely important area that is misunderstood, neglected, and taken for granted. It is the site of great interaction as the surface affects the atmosphere and the atmosphere affects the surface. The biosphere, or organic zone, is a product of the interaction between these great spheres. Changes in climate produce changes in the lithosphere, which in turn can cause changes in the climate. Humans are a natural agent of change at the surface, as are all species. Recently, however, the extent and potential implication of the human-induced changes have become a cause for concern. We have changed the surface dramatically in our cities, through agriculture, forestry, and many other activities.

The reason we need to know about these changes in surface conditions depends upon the nature of our atmospheric system. As we have seen, the primary driving force is radiant energy from the sun. Some radiation is absorbed in the atmosphere, some is absorbed at the surface, but in both cases energy is reradiated back to space. The difference in absorption between surface and atmosphere produces vertical temperature gradients that result in vertical flows of energy.

Similarly, the surplus of energy that builds up in tropical areas and the deficit in polar regions produces horizontal temperature gradients across the earth's surface. These drive the major horizontal airflows. The temperature gradients give rise to pressure gradients which in turn result in the wind systems and the kinetic energy of atmospheric motion. Kinetic energy can operate over a wide time range, but eventually it is dissipated through turbulence and friction, ending as heat. To complete the energy cycle, this heat is returned to space as infrared radiation emitted from the atmosphere.

This sequence represents the basic controls of the atmosphere. What is clear, though, is that in detail the atmospheric system is highly dynamic. Changes outside the globe (e.g., solar radiation amounts), within the atmosphere, and within the rest of the global system may all cause complex reactions. And because atmospheric and climatic conditions have such a forceful impact on our existence, it is essential that we come to understand the system thoroughly.

## CAUSES OF CLIMATIC CHANGE

One question must be asked. Why does the climate vary so much? It is a question that may not have only one answer. In looking for possible causes we can think of influences external to our planet, of purely internal factors between the atmosphere and the surface, or a combination of both. Let us look at these in turn.

## VARIATIONS IN ENERGY INPUTS

As the sun is the main source of energy, is there any indication that its output has changed, with a consequent effect on our climate? Is it just coincidence that from 1645 to 1715, the coldest period of the Little Ice Age, there were virtually no sunspots? The period is known as the **Maunder Minimum**. Increasingly, evidence from data collected outside the earth's atmosphere by such sources as Skylab indicate that the sun's energy output varies much more than we had previously considered. Many attempts have been made to correlate sunspots with variations in climate. Some people claimed success, others did not reach any conclusions, and still others claimed that there were no correlations. It is interesting to note that people studying longer continuous records such as tree rings tend to find correlations, while those studying shorter modern records report little success. There is no doubt that there are cycles of sunspot activity, but the difficulty is finding correlations with climate. One problem is the lack of usable longterm records; another is a lack of understanding of the mechanisms of climate as they relate to the solar wind and the earth's magnetic field. Finally, there is the problem that the cycles, if they exist, are being masked by other factors such as volcanic dust. It seems logical that there would be cycles, because we recognize the phenomenon in many other aspects of nature.

An indisputable cause of variation in solar energy input to the earth's surface is the astronomical relationship between the sun's rays and the earth. The orbit of the earth around the sun is roughly elliptical. The nearest point of this orbit to the sun is known as the **perihelion** (Greek *peri* = near + *helios* = sun), and is about $14.71 \times 10^7$ km from the sun. The farthest point is known as the **aphelion** (Greek *ap* = far + *helios*) and is about $15.2 \times 10^7$ km from the sun. At present the perihelion occurs on 3 January while the aphelion is on 4 July, but over time the dates change gradually. The eccentricity produces a minor fluctuation in the input of solar radiation, with a maximum of 1400 W $m^{-2}$ at the aphelion. Thus, solar inputs at the extremes of the orbit vary by about 7 percent. The time of year at which the earth is nearest the sun changes through time in a cycle that takes about 21 000 years. Its effect is referred to as the **precession of the equinoxes** (Figure 9.2).

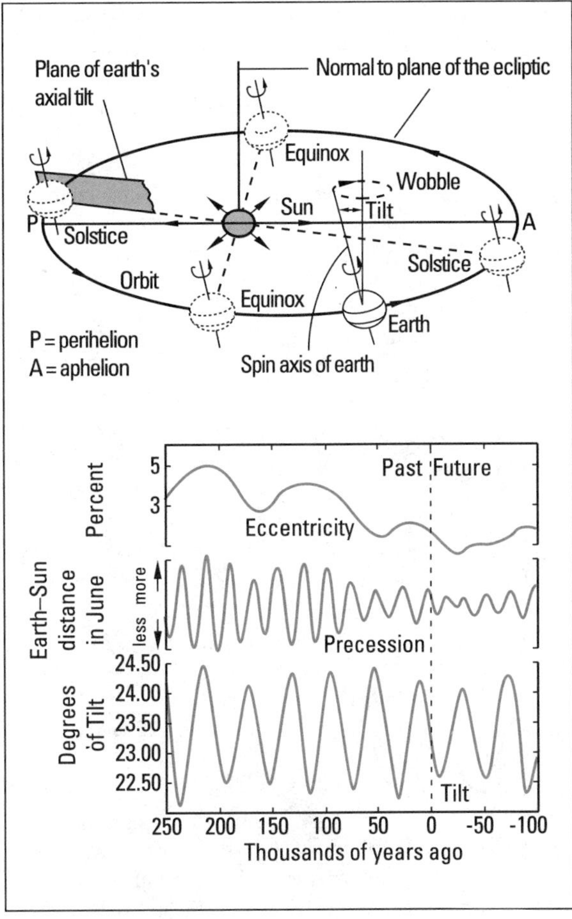


FIGURE 9.2 (Above) Geometry of the sun–earth system showing factors causing variation in radiation receipt by the earth (after Pittock *et al*, 1976); (below) changes in eccentricity, tilt, and precession for the last 250 000 years and the next 100 000 years (after Imbrie and Imbrie, 1979)

An even longer cycle of variation occurs due to slight changes in the shape of the orbit. Over a period of about 90 000 years this varies from almost circular to strongly elliptical, and as a consequence the seasonal variation in solar inputs shows a marked fluctuation (Figure 9.2). Its effect is referred to as the **eccentricity of the orbit**.

The final source of variation in the distribution of solar inputs occurs due to the changes in the tilt of the earth's axis of rotation. Although, at present, the tilt is about 23.5°, the limits vary from about 22° to 24.4°. This means, in essence, that the Tropics of Cancer and Capricorn shift slightly. When the axis has a greater tilt, the position of the overhead sun at midday at the solstices is farther poleward by about 2.5° than when the tilt is at its least. Again the change is regular, with the full cycle taking about 40 000 years. The variation is sometimes referred to as the **obliquity of the ecliptic** or more simply as the variation in tilt (Figure 9.2).

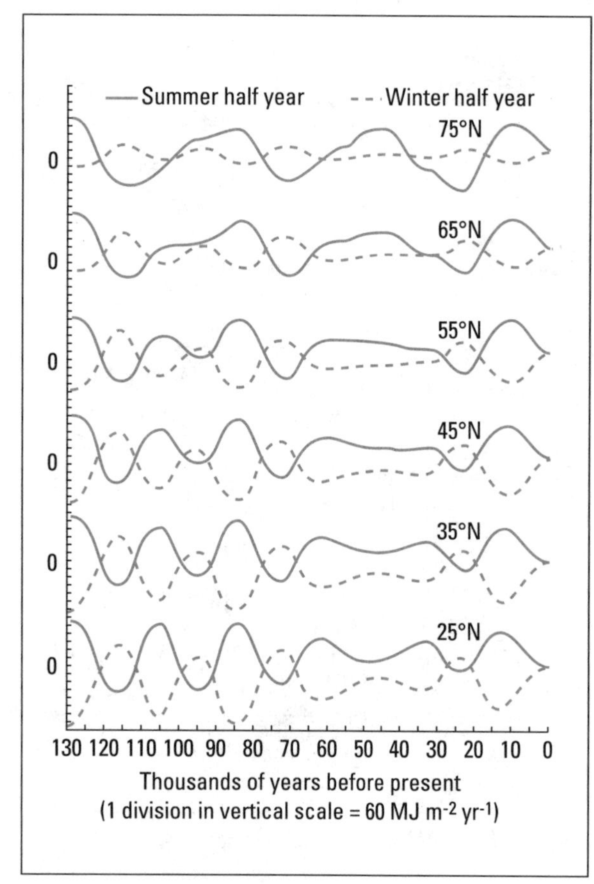


FIGURE 9.3 ▼ The variation over the last 130 000 years of the radiative flux between latitudes 25°N and 75°N in the summer and winter half-years based on calculations by Milankovitch (after Manley, 1976)

The variations in solar radiation at different latitudes of the earth's surface due to these orbital changes were first calculated by **Milankovitch** in 1930. Apart from a few modifications, the basic pattern has remained unchanged and is shown in Figure 9.3. In high latitudes it is the 40 000-year cycle that dominates, but at lower latitudes the 21 000-year cycle is more important.

Calculations have been made of the amount of heat that would be available to the different latitudes based on the Milankovitch variations, with allowances being made for the amounts of ice existing at each period of time. These show that the orbital variations did have the correct timing and size to start the succession of major advances and retreats of the icecaps during the last 300 000 years. This is seen most clearly in some of the ocean cores where undisturbed sediments have accumulated over many thousands of years. Fluctuations in temperature as determined from their fossil and carbonate contents do tie in closely with the Milankovitch cycles (Figure 9.4).

Figure 9.4 shows a spectrum analysis of isotope records from the Indian Ocean. The ratio of oxygen-16 to oxygen-18 isotopes varies as the temperature of the atmosphere varies, and these variations are then recorded in ocean sediments. A spectrum analysis is a statistical technique to determine if there are cycles or repetitive sequences in a long record. When such an analysis was run on the isotope record, it indicated cycles of the length shown in Figure 9.4. These were remarkably similar to those of the Milankovitch theory. Evidence such as this, from a completely independent source, is especially powerful in support of a theory.

Orbital changes take place only slowly. However, sudden changes in climate can occur. For example, in one deposit near Birmingham, UK, a typical northern assemblage of beetles was found and dated to 10 025 ± 100 years B.P. As indicated 10 cm higher in that deposit, no Arctic fauna survived at an age of 9970 ± 110 years B.P. Conversely, the rapid cooling at about 10 900 B.P. brought a catastrophic readvance of the ice, which destroyed fully grown forests, and caused desiccation in Colombia and a marked cooling in Antarctica within a time span of only 200 to 300 years. It seems highly unlikely that orbital variations could be responsible for such sharp climatic fluctuations as these. For these changes we must look to other mechanisms.

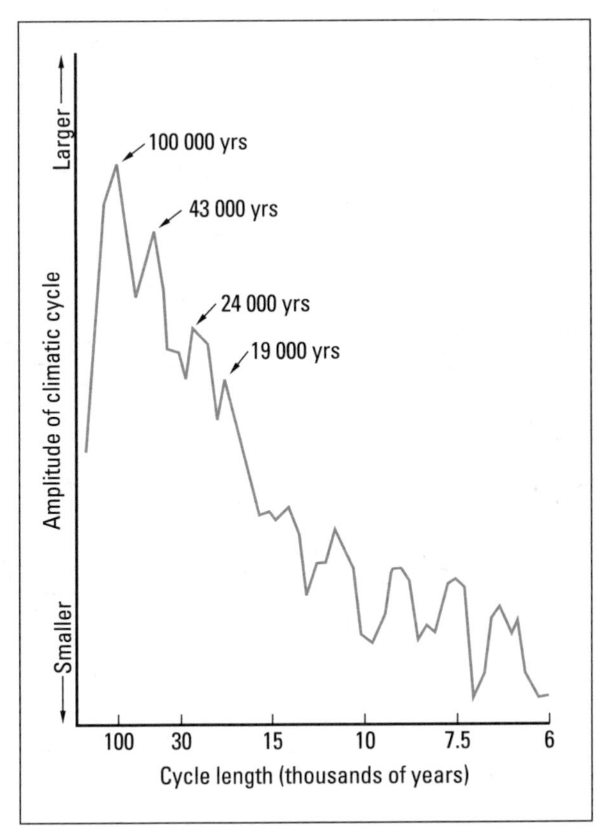


FIGURE 9.4 ▼ Spectrum of climatic variation over the past half million years. This graph—showing the relative importance of different climatic cycles in the isotope record of two Indian Ocean cores—confirmed many predictions of the Milankovitch theory (after Imbrie and Imbrie, 1979)

## CHANGES IN SOLAR OUTPUT

Our sun is described as a medium size, variable star. It is a nuclear fusion reactor of hydrogen, emitting all the energy and particles associated with such a reaction. At present we only consider the electromagnetic energy, as discussed in Chapter 2. We assume that this energy coming from the sun does not vary, a fact reflected in the term **solar constant**. The term **irradiance** is also used to refer to this energy.

The particles of matter emitted by the sun are collectively referred to as **corpuscular energy**, after the Latin word *corpus*, meaning body. These particles have mass and stream through space toward the earth in what is known as the **solar wind**. This wind compresses the earth's magnetic field on the sun side and stretches it thousands of kilometres into space on the opposite side, creating the **magnetotail** (Figure 9.5). Variations in the solar wind create changes in the magnetosphere that result in magnetic storms in the earth's magnetic field at ground level. Such storms can seriously affect power systems, overloading power grids and causing shutdowns. They also affect communication systems and satellites.

The interaction between the highly energized solar particles trapped in the magnetic field and the atmosphere also results in the **aurora borealis** in the northern hemisphere and **aurora australis** in the southern hemisphere. The aurora are evidence of varying activity on the sun. From approximately A.D. 1645 to A.D. 1715 very few aurora were witnessed because there was little activity on the sun. This period is known as the **Maunder Minimum** and coincidentally matches the nadir of a cool climate period known as the Little Ice Age. Generally, it appears that when the sun is quiet the earth is cooler and when it is active the world is warmer. There have been many studies showing correlations between climate and sunspots. The problem is that these are only correlations, and nobody has discovered a mechanism to explain the relationship.

Sunspots are darker areas that appear on the **photosphere** (light surface) of the sun. Each spot is made up of two areas, a dark inner area called the **umbra** and a lighter surrounding area called the **penumbra**. The average temperature of the photosphere is 6000°K but the sunspots are cooler. The umbra is roughly 4000°K while the penumbra is about 5600°K.

Sunspots appear in groups and only occur within 40° of the solar equator. At the beginning of a cycle they appear between 30° and 40° north and south of the equator, but appear at lower latitudes as the cycle progresses. The number and size of spots and the number of groups all vary through time. Information is standardized by the Zurich or Wolf relative sunspot number, $R$, obtained by the formula

$$R = k(f + 10G)$$

where $f$ is the number of spots
$G$ is the number of groups
$k$ is a correction factor for the individual observation site

The system was introduced in 1848 by Rudolf Wolf, who used earlier records to extend the data back to 1610, when Galileo first observed sunspots with a telescope.

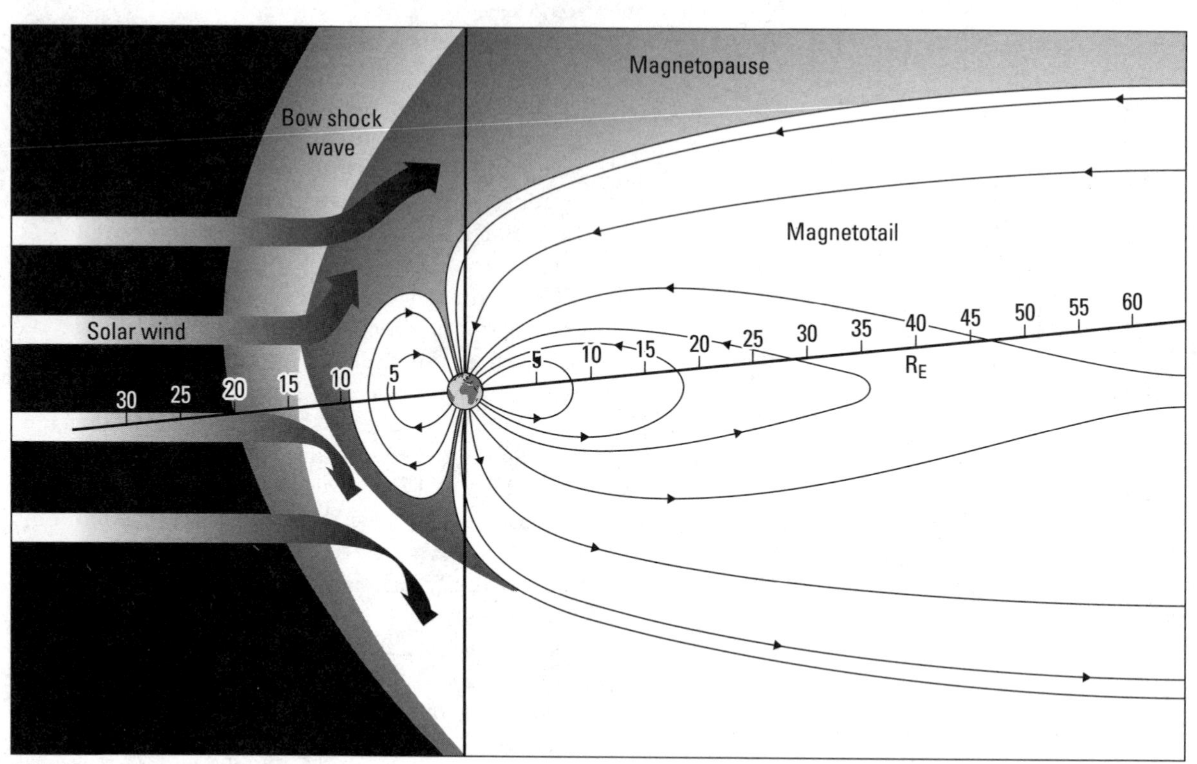


FIGURE 9.5 ▼ The magnetosphere of the earth showing the impact of solar wind. The scale $R_E$ is earth radii. Note the deflection of the solar wind by the magnetic field (after Moore and Hunt, 1987)

A plot of the relative number of sunspots from A.D. 1700 to 1987 is shown in Figure 9.6. Note the reduced number of spots of the Maunder Minimum in the first two decades. However, also note the low number in the period 1800 to 1820, another period of colder temperatures. This will be discussed later in the section on the climate change record.

Fluctuations in the number of sunspots are very evident in Figure 9.6. From maximum to maximum the longterm average is roughly 11 years, although this varies from 7 to 17 years. The cycle is named the Schwabe cycle, after the scientist who first noted it in 1843. In the first 11 years the sunspots in each hemisphere of the sun have different magnetic

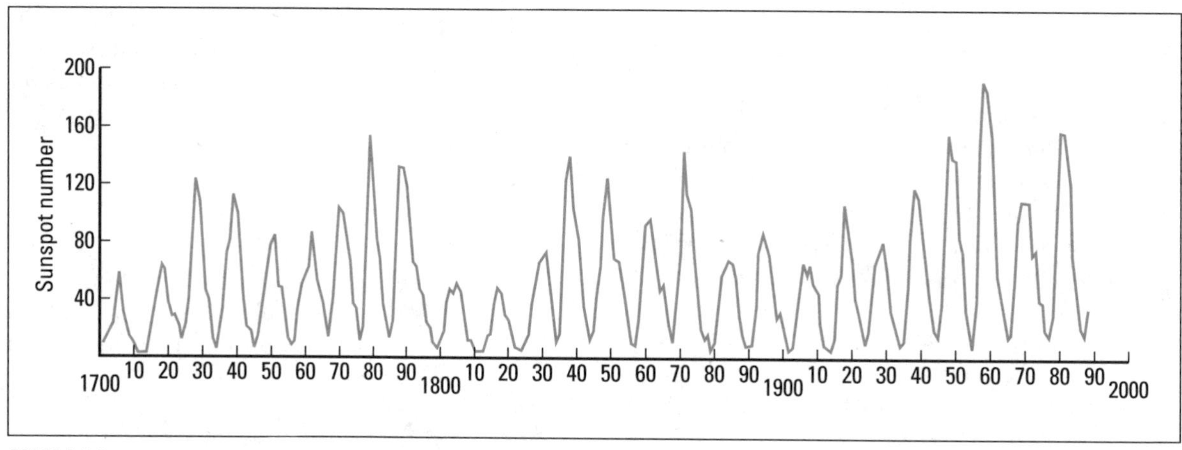


FIGURE 9.6 ▼ Yearly mean sunspot numbers, 1700 to 1987 (from D.C. Wilkinson, World Data Center for Solar Terrestrial Physics, Boulder, Colorado)

polarities which then reverse in the second 11 years, creating the 22-year or Hale cycle.

A third cycle of 80 to 100 years is marked by a gradual increase and then a rapid fall in the amplitude of the cycles. This is called the Gleissberg cycle after its discoverer. Longer fluctuations of 180, 1300, and 2600 years are claimed by some studies. Others observe the size or area of the sunspots and correlate these with variations in solar energy output. An increasing number of studies correlate sunspot activity with climate events. The pattern of drought on the Great Plains of North America was one of the first. More recently, the occurrence of El Niño off the coast of South America has been related to changes in the Quasi-biennial oscillation. This fluctuation of winds in the upper atmosphere has, in turn, been related to sunspot activity. The difficulty with all the correlations is that they lack a mechanism. It is not known how solar changes result in climate changes on earth, and research continues. One problem is that we have only a limited record of solar variability beyond the atmosphere. The space vehicle Skylab provided about 14 years of data and showed small variations, but greater than expected. In the late 1980s the Solar Maximum Mission satellite was launched to study the sun during a period of greater activity, and provided additional information. Despite the small amount of variation measured, it must be remembered that models show that known ranges of temperature variation throughout history can be created by a variation in solar energy of less than 10 percent.

# Variations in internal conditions

## FEEDBACK PROCESSES

Our climatic system consists of several subsystems, such as the atmosphere, the oceans, the icecaps, and the land surfaces. These are all closely related and changes in one may affect the others. Moreover, changes within one of these components may act as positive or negative feedbacks, ultimately influencing inputs of solar radiation to the earth. These feedback mechanisms may be responsible for many of the more rapid fluctuations in climate that have occurred throughout the earth's history.

Positive feedback leads to more dramatic and far-reaching changes. The initial effect is magnified, so that quite small changes in the environment produce major adjustments in the system. Perhaps this is why the climate sometimes changes abruptly without any evidence of a clear change in external conditions. Figure 9.7 shows such an effect, which has been proposed as a cause of the Ice Ages. A quite small cooling of temperatures at the poles of only 1C° or 2C° delays the summer melting of the Arctic icecap. Because the ice survives longer, the albedo of the surface is high for longer. More incoming shortwave radiation is reflected back to space; heating of the surface is therefore reduced, allowing the icecaps to survive even longer, which increases reflection further, which lowers temperatures further. . . . The cycle is self-perpetuating. Once they have been initiated, positive feedback processes magnify the effect of the initial change and cause major adjustments in the system, possibly even an Ice Age.

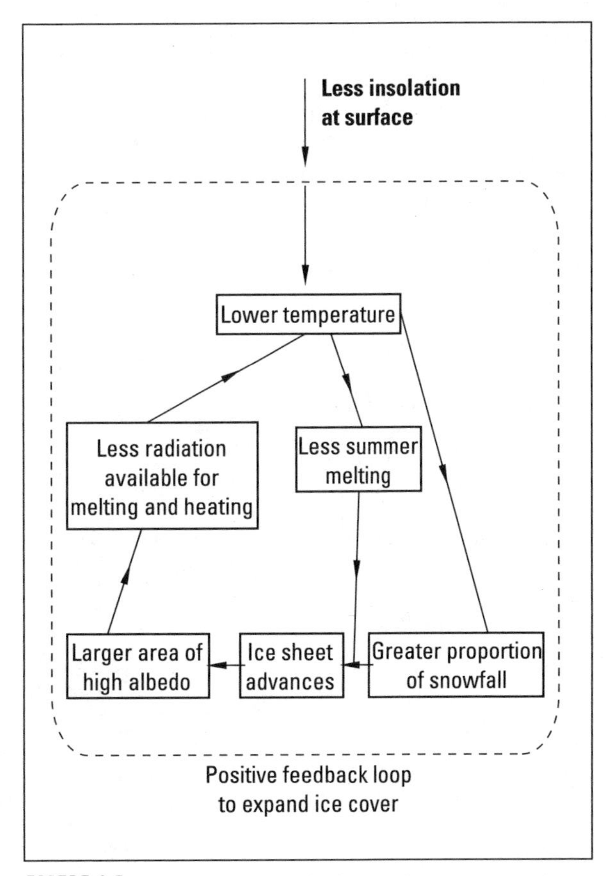


FIGURE 9.7 ▼ A positive feedback loop demonstrating how a decrease in insolation and lower surface temperatures may generate further cooling and perhaps even an Ice Age

# Mountain-building and plate tectonics

Mountain-building can be responsible for climate change, although the uplift is generally too slow to account for the more rapid variations in climate. Its effect can be significant as the mountain builds and rises gradually through the treeline and the permanent snowline (Figure 9.8). Imagine the changes in climate created as the Rocky Mountains grew. The extent of snow-covered land increases, altering the earth's albedo. They form a barrier across the westerly upper level winds, causing changes in climate for a great distance downwind. This barrier creates what is referred to as a planetary wave in the lee of the mountains and influences the pattern of climate in western Canada today.

Plate tectonics refers to the slow movement of the individual plates that make up the earth's crust. These mechanisms are described in Chapter 17. Today the pattern of land and water affects the circulation of the oceans and alters the basic latitudinal distribution of climate. For example, consider how different the climate of Europe would be if Central America did not exist and the Gulf Stream flowed into the Pacific instead of transporting its heat energy across the Atlantic. How different would the climate of the Indian subcontinent be if the Himalayas were not being pushed up by tectonic forces?

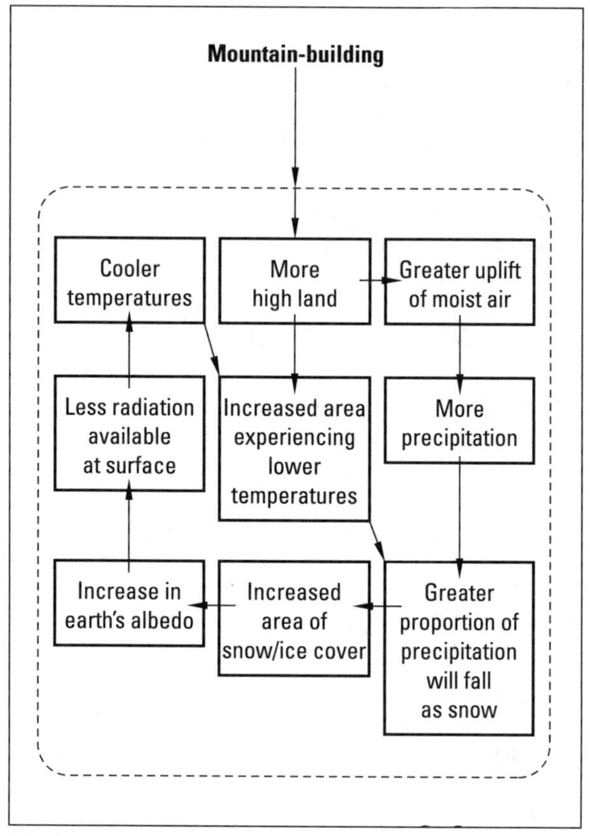


FIGURE 9.8 ▼ A positive feedback loop showing some of the effects of mountain-building on surface climate

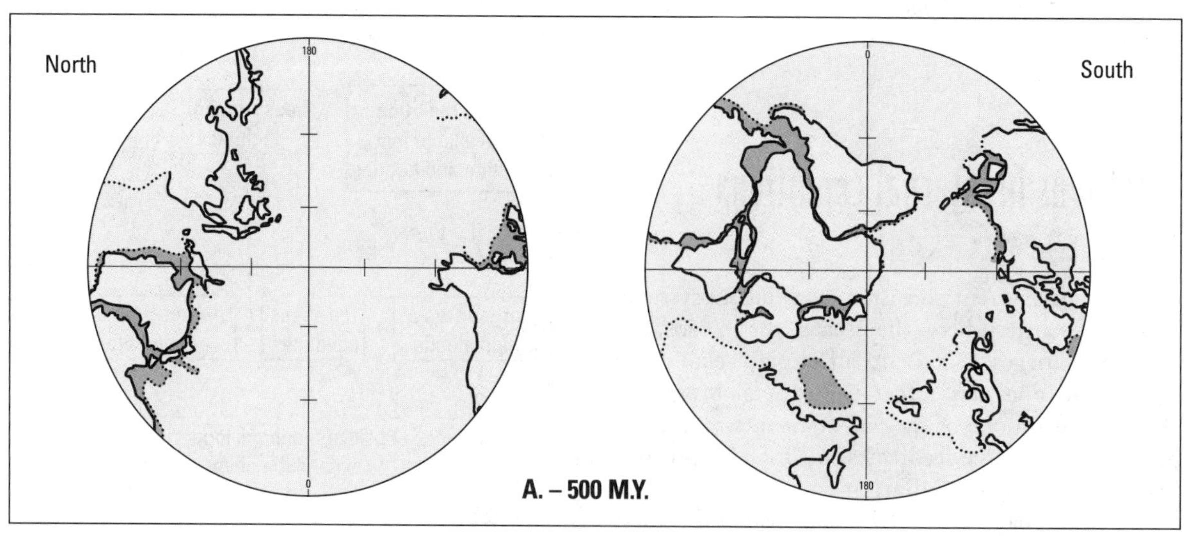


FIGURE 9.9 ▼ Positions derived from paleomagnetic data showing the continent blocks (A) about –500 M.Y., in late Cambrian–early Ordovician time; (B) about –340 M.Y., early in the Carboniferous era; (C) about –250 M.Y., in the Permian era; (D) about –50 M.Y., in the early Tertiary. We can only speculate on the shape of land masses in these times, as indicated by broken lines. The circumference of each map is the equator. Latitudes 30° and 60° are marked; the longitudes are arbitrary (from Lamb, 1977)

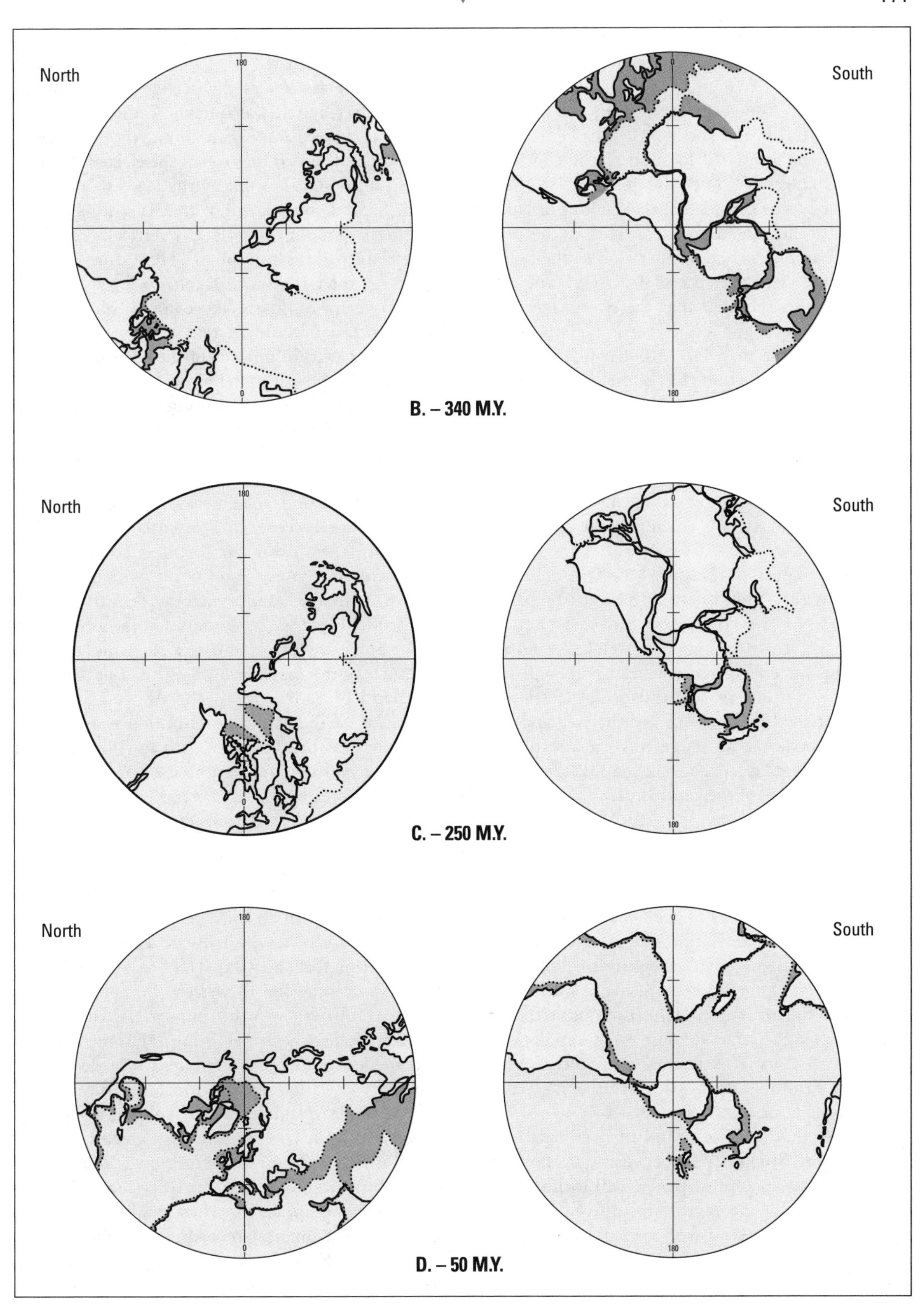
North
South
180
0
B. – 340 M.Y.
North
South
C. – 250 M.Y.
North
South
D. – 50 M.Y.

Throughout geologic time the crust that makes up the continents has constantly moved and changed in area. At present we have about 30 percent land and 70 percent ocean with the predominant land mass in the northern hemisphere. Figure 9.9 shows the estimated location of the continents at various stages of the earth's history. Two major factors need to be considered, the amount of land at the equator, and the amount of land at the poles. Some climatologists believe that land at the poles is essential for the occurrence of Ice Ages. The difficulty with this theory is that the Arctic Basin was in existence throughout the recent Pleistocene Ice Age. Indeed, the map for –50 M.Y. (million years ago) shows a small but clearly established Arctic Ocean. Figure 9.9B shows the configuration for –340 M.Y., when the single continent of Pangaea existed. Notice how the continents essentially extend uninterrupted from the North to the South Pole. Japan is surprisingly stable close to the North Pole, while Antarctica is equally stable close to the South Pole.

None of these diagrams show the mountain ranges, but they would have influenced the pattern of climate of the time, especially if they ran north–south across the prevailing global wind patterns. The patterns of erosion, transport, and sedimentation continuing through geologic time are only understood if the climate of the area and time are understood. It is an interesting exercise to compare the periods of coal formation and deposition with the positions of the continents.

## VOLCANIC ACTIVITY

If we attempt to classify climate change by extraterrestrial, geologic, and atmospheric influences, we run into problems with categorizing volcanoes. They are geologic events that have a much more immediate impact. Generally, it is thought that the effect of a single, large volcanic event can continue for 10 to 15 years. This assumes that volcanic eruptions are random and that clusters of eruptions could have a more significant impact.

The idea that volcanoes could affect climate is relatively recent. Studies by Hubert Lamb in the 1970s introduced the concept of a **dust veil index** (DVI). According to this theory, the amount of dust in the upper atmosphere varies with volcanic activity; sufficient dust can block the sun, causing cooling. The eruption of Mount St. Helens in 1980 and El Chichón in 1982 stimulated interest in the effect of such events on climate.

Initially, it was assumed that all volcanoes would have an effect, but it was quickly learned that three basic factors were important. First, the latitude of the volcano; if it is in the tropics the ejected material can be transported upward by the Hadley cell circulation. Second, the nature of the volcano; a stratovolcano that erupts violently and vertically can eject material into the stratosphere. Here, fine dust, of radius less than 10 μm, can remain for at least 12 years, creating dramatic sky colours. Most of the heavier material would be removed by gravity and precipitation within months of the eruption. Third, the nature of the ash material determines the effect on optical properties. For example, sulphur and sulphates produce greater scattering because they operate in the yellow portion of the visible spectrum.

Comparison of the temperature record with the dust veil index and volcanic events indicates that cooling occurs, however, it is essential to be careful. For example, volcanic dust acts as condensation nuclei, creating more cloud cover, which can also alter the climate. Many volcanic eruptions have been studied, primarily because of the impact on history. Some are cataclysmic to a particular civilization, such as the eruption in 1470 B.C. of Thera on the island of Santorini in the Aegean Sea. The destruction of the Minoan civilization is attributed to this explosion. Others, such as Tambora, an Indonesian volcano that erupted in 1815, created harrowing conditions that forced adaptations and change. It is estimated that Tambora ejected 150 $km^3$ of volcanic dust into the atmosphere, believed to be the most in some 10 000 years.

It was initially assumed that Tambora caused the infamous "year with no summer" in 1816. A conference held in Ottawa in 1989 to examine global information for the year 1816 disclosed that extreme cold occurred in western Europe and eastern North America. More important, the global temperature had been declining for some years, a phenomenon that Eddy attributed to a decline in solar activity called the Dalton Minimum or the Little Maunder Minimum. The volcano apparently pushed an already cooling earth to a greater cold.

Perhaps the best-known eruption is that of the Indonesian volcano Krakatoa in 1883. It is the first large volcanic eruption for which we have relatively extensive instrumental records. For example, the barometer at Kew Gardens in London reportedly recorded the atmospheric shock wave five times as it sped around the earth. Solar phyrheliometer measurements indicate that radiation levels were lower for some years after the event.

In 1991, the eruption of the Philippine volcano Mount Pinatubo added large volumes of dust and gases to the atmosphere. This threw off predictions of increased global warming as the sunlight was blocked. It also added gases that interfered with understanding of the human impact upon the ozone layer. It is just one more of the difficulties of separating human contributions to climate change from natural variability.

## THE SURFACE

You will examine the surface of the earth and the forces that shape it in the second half of this book. One of the forces is climate, as heating and cooling, freezing and thawing, wetting and drying work to break down the rock. The materials and colour of the surface are critical because they determine how the energy from the sun is absorbed. Changes to the surface would result in changes to the energy balance and ultimately a change in climate.

The large-scale changes to the landscape are mostly the result of agriculture, although, as we saw in the section on cities, we have created unique islands of climates. Plowing the land and resculpturing the surface does occur over time, but it is the alteration of vegetation that causes the most extensive change.

Outside urban areas, the human effect on climate has been less marked, but evidence is growing that these changes are detectable on a larger scale. Forest clearance, tillage of grassland, and desertification due to soil erosion all increase the albedo of the surface and alter the radiation balance. Little work has been done to determine the impact of these changes on global climate, however, estimates by Budyko

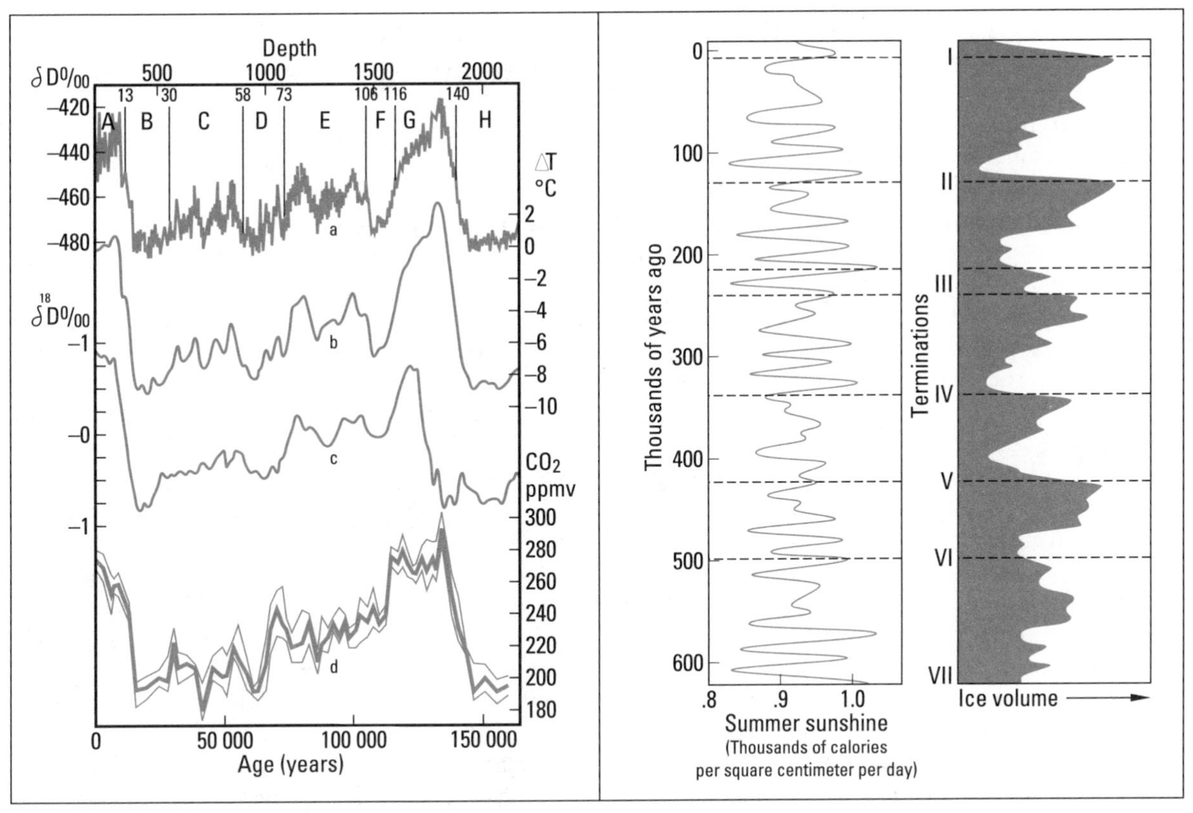


FIGURE 9.10 ▼ (Left) From the top, the diagram indicates (a) the amount of deuterium, (b) the estimated atmospheric temperature, (c) the ratio of oxygen isotopes 16 and 18, and (d) the level of carbon dioxide. The temperature curve is derived from the oxygen isotope data. Note how the curves show a steady increase followed by a sudden collapse in a 'sawtooth' pattern (from Lorius *et al*, 1985). (Right) A comparison of summer sunshine variability with variations in ice volume. Note the sawtooth pattern as the ice volume increases and then collapses (Source *Scientific American*)

and others indicate that changes should occur. Compare the albedos of different surfaces in Table 3.2 with the changes in heat budget after conversion from forest to agriculture shown in Table 9.4.

Two major periods of forest clearing have occurred. The first was the deforestation coincident with the Little Ice Age, the cool period from A.D. 1450 to 1850. Clearing was extensive in western Europe, the Indian subcontinent, and Asia. The second clearing is still in progress and includes the tropical rainforest areas of South and Central America, and southeast Asia.

### THE ATMOSPHERE

Volcanoes are geologic events, but they change the composition of the atmosphere, which in turn changes the climate. In the early chapters of this book we examined the climate as if the atmosphere were a fixed, unchanging envelope. Of course, this is not the case. The problem is to determine how much variability occurs independent of human activity.

Until recently it was difficult to even estimate the composition of the atmosphere in the past. Flora and fauna could be used as secondary indicators of the changes, but this assumes that they responded the same way in the past as they do today. Ice cores had been used, like tree rings, as a chronologic record, and they did show layers of ash associated with volcanic eruptions. Then techniques were developed that allowed the extraction of gas from bubbles in the ice. These apparently provide an encapsulated sample of past atmospheres. Ice cores from Antarctica have been drilled to a depth of 2000 m and are reputed to reveal 150 000 years of atmospheric change.

Figure 9.10 (left) shows the pattern of carbon dioxide, deuterium, oxygen-16/oxygen-18 isotopes, and deduced temperatures from these cores. Notice that the curves appear to be coincident. It has been proposed that the temperature and carbon dioxide curves are evidence that changes in carbon dioxide result in changes in temperature. The problem is that the curves are not detailed enough for us to determine which is causing which. Some investigators suggest that temperature increase results in a carbon dioxide increase. One of the problems is that the age of the air differs from the age of the ice by as much as 4000 years. On this basis it was determined that carbon dioxide increase occurs approximately 1000 years after, or at least at the same time, as the increase in temperature.

At the bottom of the core the ice is so compressed that roughly 8 metres is needed to produce enough gas to obtain a single reading. This probably represents some 10 000 years of ice accumulation, almost enough to cover an interglacial, the period between two glacial eras. The cores help us understand more recent variations, but as with most measures of climate the accuracy decreases as you go back in time.

Notice the sawtooth pattern of the curves and compare them with the pattern of variation shown in Figure 9.10 (right). There appears to be a gradual increase to a peak, followed by a relatively rapid collapse. It appears that the Milankovitch orbital changes are the major mechanism at work. What are the other mechanisms? If carbon dioxide causes climate change, what causes the variation in carbon dioxide?

## The evidence for climate change

### GEOLOGICAL AND BIOLOGICAL EVIDENCE

A wide range of evidence will reveal clues about climate. By comparing certain landforms and sediments with the type of environment in which they are forming today, it may be possible to infer the climate during their formation. The relationship between landforms and climate is not simple, but some landforms yield information about former climates (Table 9.2). Permafrost features such as pingos only develop where mean annual temperatures are below about –5°C.

Similarly, **cirques** can be used to determine former snowlines. In arid areas, we find that sand dunes form only where precipitation totals fall below 200 mm; the prevailing wind direction during their formation can be found from the alignment of the dunes. In the Orinoco Basin of South America, fossil sand dunes indicate that at some time in the past dry, wind-blown material was available in sufficient quantities to form dunes in an area that is now quite wet. Some glacial and fluvial sediments are sufficiently distinctive and related to climate to be useful as indicators of former climate. It was from the recognition of fossil tills, called **tillites**, that evidence of early glaciations in southern Africa was first accepted. Another characteristic sediment is **varve clay**, a layered deposit of clay and

## Table 9.2

*Some geomorphological indicators of climate*

| *Landform* | *Type of climatic information* |
|---|---|
| Pingos, palsas, ice wedge casts, giant polygons | Permafrost: mean annual temperatures below –5°C |
| Cirques | Temperature, through their relationship to snowlines |
| Closed lake basins | Precipitation levels associated with ancient shoreline formation |
| Fossil dunes of continental interiors | Former wind directions and precipitation levels |
| Angular talus | Frost action with the presence of some moisture |
| Misfit valley meanders | Higher discharge levels can be determined from meander geometry |
| Aeolian-fluted bedrock | Aridity and wind direction |
| Tufa mounds | Higher groundwater levels, wetter conditions |

After A.S. Goudie, *Environmental Change* (1983), Oxford University Press.

silt that accumulates in still, lacustrine waters. Varves are a series of alternating coarse and fine layers which are believed to represent an annual input. The relative thickness of the layers is related to the amount of sediment entering the lake, and therefore to the rainfall and ice melt during that year. During the winter, the lake freezes over so that the finest clay sediment is deposited and no coarser material is brought in. The cycle begins again in spring when the ice melts. Over time, the changing frequency of thick and thin layers can give some idea of climatic fluctuations.

Frequently, it is biological material contained in the sediments that is more useful in making interpretations about climate. Particularly useful are the remains of pollen from plants. The study of pollen grains and the reconstruction of pollen sequences is called palynology. Changes in the percentages of pollen grains reflect changes in the vegetation, which is a reflection of climate change. Seed-bearing plants release large quantities of pollen into the atmosphere during their flowering period. The pollen returns to the earth's surface by settling or washout in rain, and may be preserved if trapped in an anaerobic environment such as lake muds or peat. If we take a series of samples through the deposit, extract the pollen by chemical methods and identify them under a microscope, we can build up a picture of the vegetation of the area at the time of deposition (Figure 9.11). The different composition of the preserved pollen for different time periods at our site indicates something about the vegetational history of the area. On the assumption that plants producing the pollen responded to climatic conditions in the past as they do now, we can draw conclusions about the nature of the climate.

The method is not without difficulties. Not all plants produce the same amount of pollen, so some species will be over-represented relative to others even if they occupy a similar proportion of the land surface. Not all the pollen will be locally derived; some may have been blown in from other locations where different plant species grow. Suitable sites for preservation are not uniformly distributed, so vegetation growing around marshes or lakes is more likely to be preserved. We are also making the assumption that the vegetation is in equilibrium with the prevailing climate, which may not have been true for all occasions. Nevertheless, despite its limitations, pollen analysis can tell us about broad changes in vegetation and, indirectly, in climate as long as the limitations are recognized and understood.

Numerous other fossil remains provide similar evidence. The fragmentary remnants of beetles, snails, plants (including seeds, nuts, leaves, and logs), marine and freshwater molluscs such as foraminifera, ostracods, and vertebrates all provide information about former climatic conditions.

## TREE RINGS

One of the most widely used sources of information on past climates is **tree rings**. Most temperate latitude trees have an annual growth ring (Figure

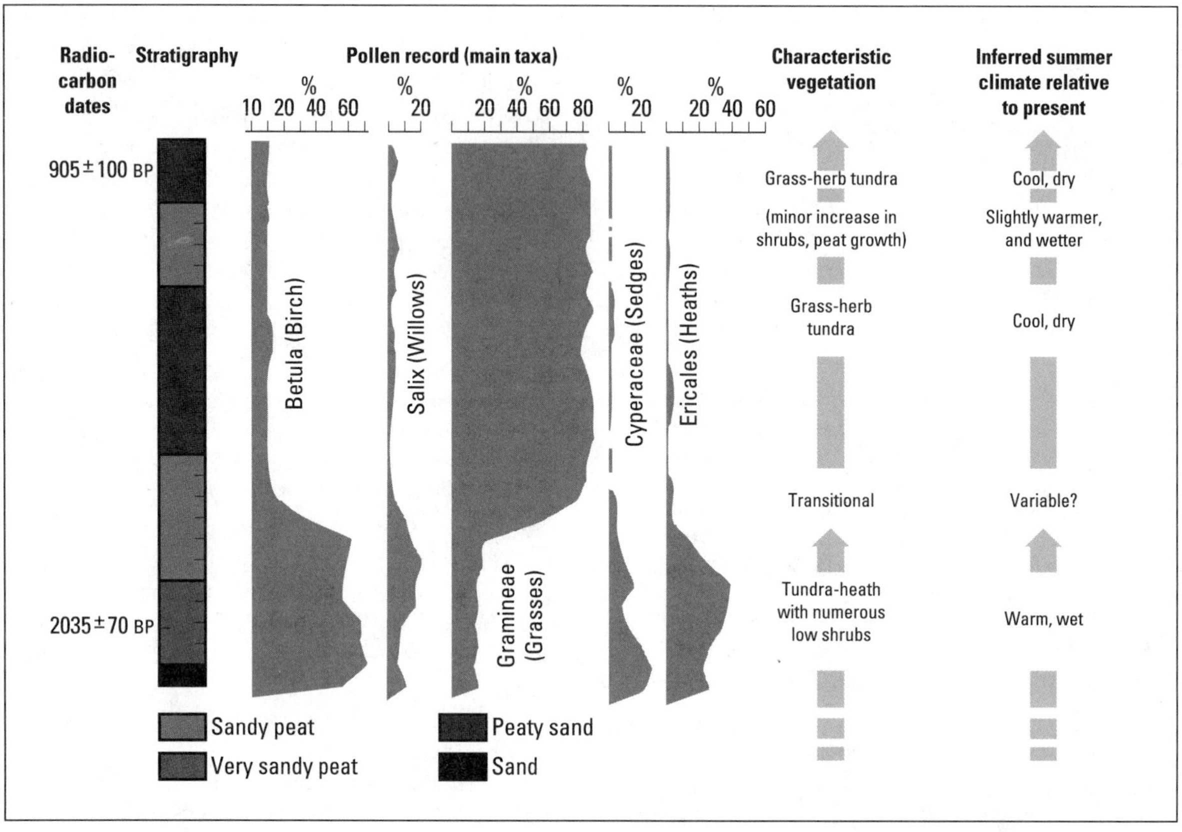


FIGURE 9.11 ▼ Pollen diagram from an arctic environment, Burton Bay, near Frobisher Bay. This shows the sequence of relatively simple plant assemblages. Sequences of vegetation, soils, and climate are clearly correlated; dates are years before present (B.P.) (after Short and Jacobs, 1982)

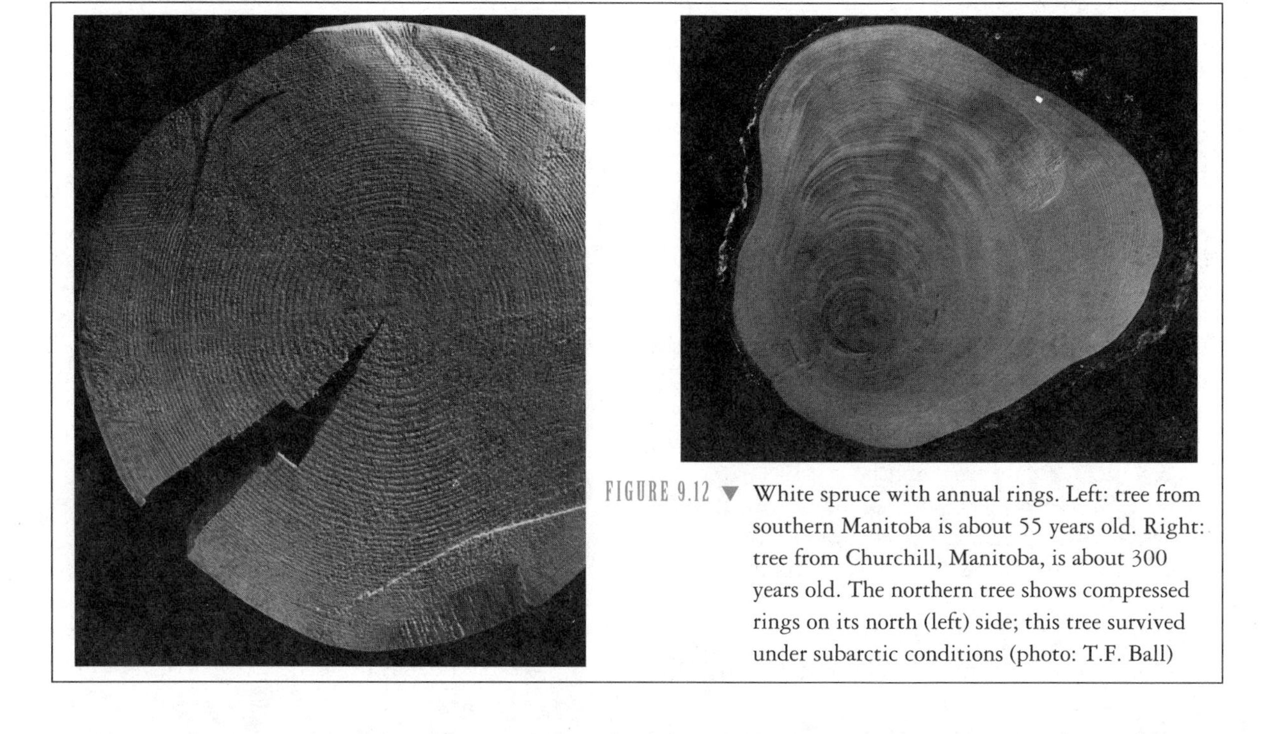

FIGURE 9.12 ▼ White spruce with annual rings. Left: tree from southern Manitoba is about 55 years old. Right: tree from Churchill, Manitoba, is about 300 years old. The northern tree shows compressed rings on its north (left) side; this tree survived under subarctic conditions (photo: T.F. Ball)

9.12), the width of which is influenced by weather conditions during the growing season. By comparing the pattern of rings with climatic data from periods for which instrumental records are available, we can determine the relationship between ring width and climate. This relationship can then be used to ascertain the general nature of climatic conditions for earlier periods.

Tree rings respond to different climatic factors in different areas, of course. In arid areas moisture is often limited, so growth relates to rainfall conditions. In other cases, such as in the mountains of California, temperature is more important. More often, the ring widths depend upon a combination of temperature and precipitation, but sophisticated statistical techniques enable us to reconstruct former climates, for example in maritime Europe, on the basis of the changes in ring widths of oak trees.

## The dating problem

One of the problems of using geological evidence of past climatic conditions is that we cannot always date the materials accurately. This is a severe limitation, for there is little use in knowing that, at one site, the climate was once cooler than at present if we cannot say when, or compare it with other areas. It is sometimes possible to date events relative to each other without any reference to their absolute age; for example, where organic material is found above and below a distinctive inorganic horizon, it can be assumed that the material above the layer is younger than that below, but there is no indication of the time period involved. This is called relative dating. For example, archeologists have recognized different cultural levels by their artifacts and given them a relative age through the evolution of styles, but this is not very satisfactory or reliable.

Of much greater use is absolute dating, where we can give a definite age to material. Two techniques already mentioned have provided methods of absolute dating. The alternation of a coarse and a fine layer in a varve is believed to represent an annual input. Some years produce particularly thick bands, perhaps as a result of more meltwater or rainfall than usual. Varve chronology depends upon the correlation of these distinctive bands in different areas so that older and older bands can be recognized and counted. A similar technique has been used with tree rings. The science of **dendrochronology** involves comparing the pattern of rings from trees of different ages to reveal an overlapping sequence that takes us farther into the past (Figure 9.13). In this way, using bristlecone pine trees from California, the oldest living things on earth, we have been able to trace and date climatic conditions back more than 8000 years. Recent techniques provide increasingly accurate measures of the amount of growth in the summer season. Since this is a direct function of the temperature and precipitation, it provides a picture of the climate conditions from one year to the next. Recent work on trees at Churchill, Manitoba, involved intensive study of slices cut from the tree every 15 cm; this method, though time-consuming, has provided a great deal more detail about the pattern of growth and therefore about the climate during its entire life.

Radioisotope methods now provide the main method of dating. Radioactive substances decay at

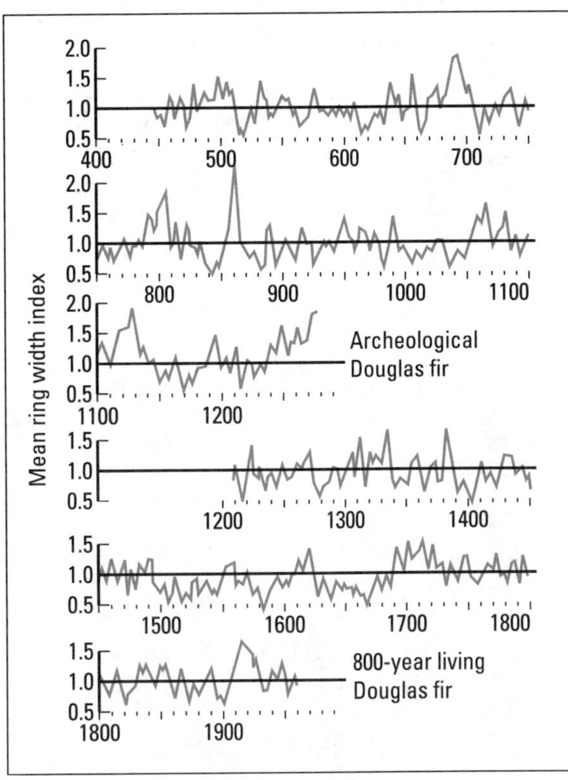


FIGURE 9.13 ▼ Ring width index series from Douglas fir trees on the southwest Colorado plateau, 2000 to 2500 m above sea level, from A.D. 440 to 1964: five-year means plotted at the middle year. Ring widths are directly related to precipitation and inversely related to temperature over about thirteen months from June to June (after Fritts *et al*, 1965)

fixed and known rates. For example, carbon-14, an isotope of carbon, decays slowly, so that in 5730 years half the radioactive carbon atoms will have broken down into carbon-12. This method of measuring the amount of radioactive carbon can be used on substances containing suitable carbon atoms such as charcoal, plant material, or shells for ages less than about 60 000 years. For older samples, there is too little radioactivity left to be measured.

A number of other radioisotopes can be used based on the uranium series of elements, on potassium, or on rubidium. The reversals of the earth's magnetic field also provide a reference series for the last four million years. Table 9.3 shows the known reversals dated by the potassium method. This catalogue is useful, as the reversals are a global event. Oceanic sediments from different places will have had different rates of deposition but their identification is helped by the additional detail provided by the reversals (Figure 9.14).

## Table 9.3

*Major paleomagnetic reversals dated by the potassium-argon method*

| | |
|---|---|
| Brunhes normal epoch | Present to 690 000 years B.P. |
| Matuyama reversed epoch | 690 000 to 2.43 million years B.P. |
| Gauss normal epoch | 2.43 to 3.32 million years B.P. |
| Gilbert reversed epoch | 3.32 to over 4.55 million years B.P. |

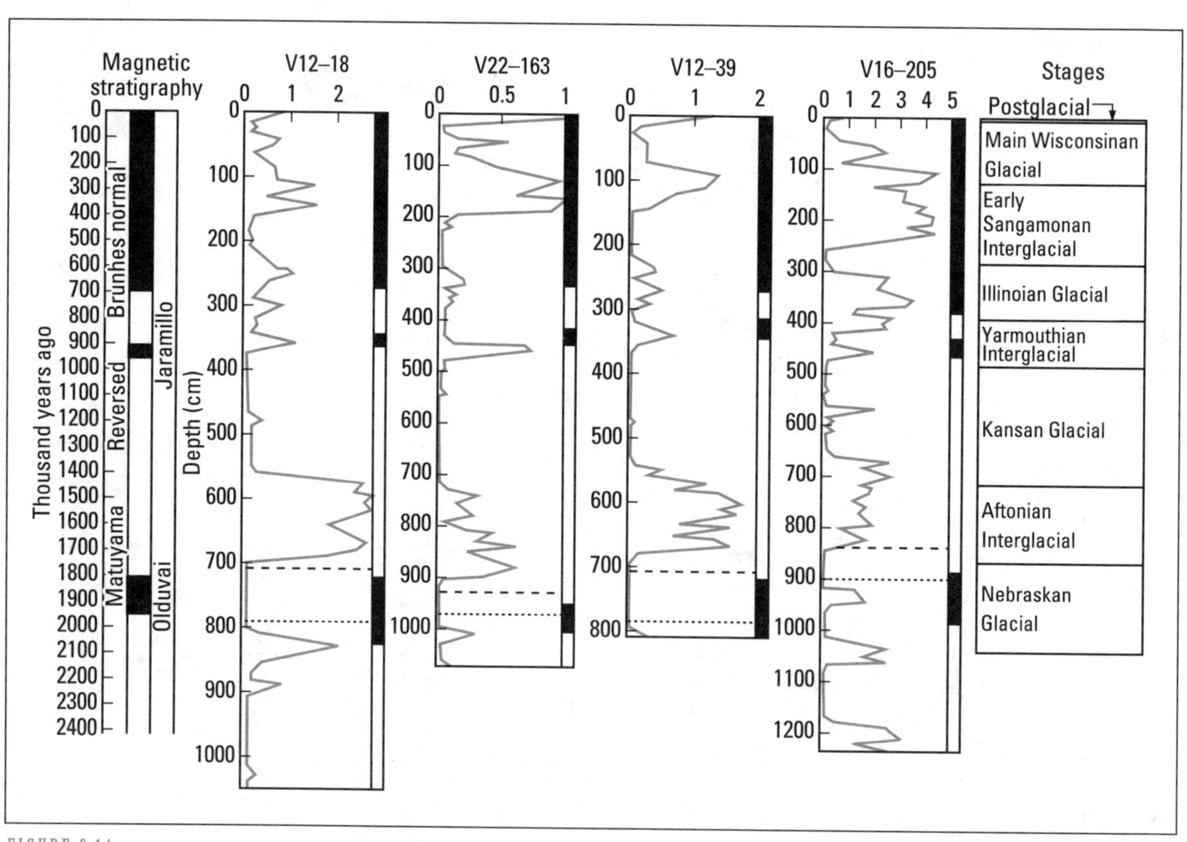


FIGURE 9.14 ▼ Frequency curves of *Globorotalia menardii* species in relation to magnetic and stratigraphic time zones for four cores in the tropical South Atlantic off South America. Scales at the top of the column are ratios of the number of shells of the *Globorotalia menardii* complex to the total number of foraminifera in the sample. Broken and dotted lines indicate the levels at which certain species become extinct in each core (after Ericson and Wollin, 1968)

## DOCUMENTARY AND HISTORICAL EVIDENCE

Weather often makes the headlines in our heavily-populated world, perhaps as a result of excessive rains, severe cold, droughts, or heatwaves in some part of the globe. Similar events occurred in the past, but our knowledge of them often depends upon the chance factor of documentary records. Many documents have survived from historic times, perhaps the best-known being the Bible, and some of these provide contemporary reports of the weather and climate. Unfortunately, many records give a distorted view of what really happened and it is very difficult to discern objective reality for any length of time. Some attempts have been made, such as the analysis of wheat prices in England and wine harvest records in France, but weather is not the only factor determining price or harvest times. Lamb, the well-known climatologist, who has spent much time devoted to this problem, believes that there is only limited scope for deriving climatic information from human history.

The most recent period of our climatic history has the advantage of introducing instruments that provide a direct method of measuring climate. By A.D. 1700, most of the standard meteorological instruments had been invented, though the quality of manufacture and the siting of the equipment left a lot to be desired. Much time has been spent trying to improve the quality of early data as rarity enhances their value. As a result of this work we now have average temperatures for central England by season and for the whole year going back to A.D. 1659 (Figure 9.15). Precipitation measurements dating to 1677 have survived in a form suitable for analysis, but it was not until 1725 that sufficient data became available to derive monthly values of rainfall that are reasonably accurate, and even then it was for only a few parts of the country. Over the years, instruments and their siting have become standardized, at least within a particular country, so now it is possible to see directly how climate has varied. Unfortunately, our understanding of the causes of these variations is far less complete.

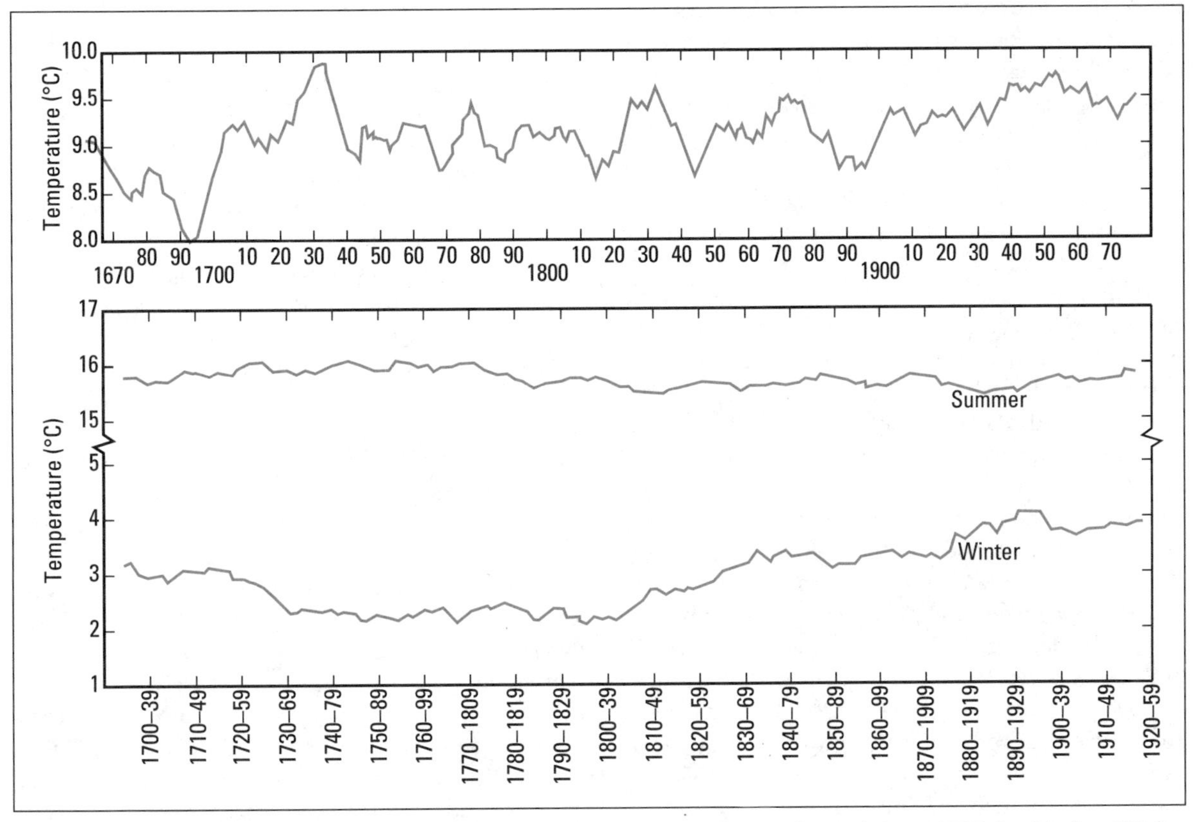


FIGURE 9.15 ▼ (Above) Ten-year running means of Central England temperatures from 1650 to 1975 (after Manley, 1976); (Below) forty-year running means for summer and winter temperatures of Central England (after Lamb, 1977)

# Nature of changes during the Quaternary

We have indicated how former climates may be interpreted from a wide range of different types of evidence. In some instances there is abundant information about the type of climate that must have existed and for other periods it is rather sparse, but wherever possible all lines of evidence should be used to reinforce the conclusions. Occasionally, evidence may be contradictory, but with the large amounts of information about climatic history it is now possible to outline the changes that have taken place.

During the last three million years, large parts of the globe have experienced much colder conditions than at present. At times, icecaps have advanced equatorward over areas that now have a temperate climate. This period of cooling is known as the **Pleistocene Epoch**, and together with the recent period since the last glaciation (the **Holocene**) it forms the Quaternary Period. Recent studies have indicated that our original ideas on the glaciations of the Pleistocene Epoch were simplistic. For a long time it has been argued that there were four distinct glacial episodes in this period, separated by long, warmer interglacial stages. The names given to these episodes in North America and their assumed correlations with similar events in other parts of the world are shown in Table 9.1.

It now seems certain that these events were far more complex than this implies. During the glacial phases there occurred relatively warm periods (**interstadials**) when the ice retreated, and cold periods (**stadials**) when it advanced. The interglacials were not long, uninterrupted periods of warmth, but undoubtedly contained cooler phases (Figure 9.16). It is probable that there were many more than four periods in North America deserving the name glacial, and more than three interglacials. But the evidence of the Pleistocene is far from complete. The deposits laid down by the various icecaps and glaciers have been eroded, rearranged, often removed; as a result they are difficult to interpret. In addition, deposits from interglacial periods are rare.

Some of the best evidence for events during the Pleistocene comes from deep sea sediments. These show that the early Pleistocene probably involved a long period of irregular cooling, during which the

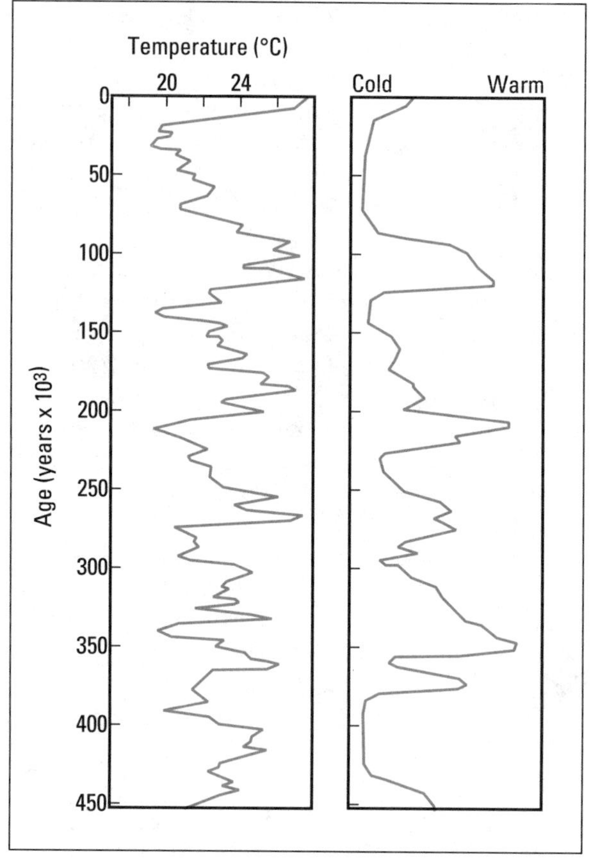


FIGURE 9.16 ▼ (Left) A temperature curve derived from oxygen isotope ratios in the Caribbean showing fluctuations in temperature during the interglacials; (right) a foraminifera curve from the Caribbean showing less detail of the temperature fluctuations and poorer correspondence with the isotope method for older sediments (after Pittock *et al*, 1978)

polar and mountain icecaps expanded. As they did so, water was taken out of the hydrological cycle; less reached the sea and sea level fell. Eventually, perhaps about 1 to 1.5 million years ago, the icecaps had grown large enough to affect the climate significantly, and more rapid expansion occurred as the ice triggered further cooling. Interestingly, it seems that as the icecaps grew, a gradual westward migration of the main centres of ice accumulation took place, as snow was concentrated on the windward side of the icecaps.

Quite suddenly (in geological terms) this glacial advance turned into a retreat, and the ice melted; sea levels rose; land that had been depressed by the weight of the ice rebounded through isostatic recovery. The world entered an interglacial period. But this, too, was temporary. The climate cooled again, and the processes were repeated.

Exactly how many times this happened we cannot tell. At their maximum extent the icecaps covered some 34 percent of the globe, in North America reaching the present Ohio and Missouri rivers, and extending across most of northern Europe as far south as southern Britain. Ice also spread out from the mountain ranges, from the Alps, from the Himalayas, from the Andes, and even from the mountain areas of Australia and Africa. The exact relationship between these events in the northern and southern hemispheres remains unclear. It is not certain, although it seems likely, that the glacial advances coincided in both hemispheres. As can be seen, there is still much work to be done in deciphering the glacial history of the Pleistocene.

## GLACIALS

During glacial phases, we can usually tell how far the icecaps extended by moraines or outwash material. Few organic remains are preserved near the ice front to indicate temperature levels, where the close proximity of ice would ensure that summer air temperatures would not rise much above freezing and winter temperatures would be very low. Farther away from the ice, the glacial climate would remain cold but some plants grew, to give a vegetation resembling that of the Arctic tundra today. On a global scale, there appears to have been an equatorward shift of the vegetation zones, though compression of these zones also occurred as most of the present desert areas were very arid during the last glacial phase (see Figure 28.10).

Temperatures may be assessed by five main lines of evidence: the nature of floral and faunal remains, the extent of permafrost, the limits of frost-affected sediments, the level of cirques and the snowline, and isotopic measurements. Although some inconsistencies arise, all indicators confirm a major drop in temperature during the glacials of approximately 5C°, with the change being even greater in ice-covered areas. A much clearer record is now available for the oceans, where fossils preserved in ocean sediment cores have enabled scientists to reconstruct the probable sea surface temperatures during the maximum of the last glaciation.

Not all the glacial periods consisted of massive ice advances. During the interstadials, ice-covered areas were considerably less extensive, although the climate retained tundra characteristics in much of North America. At times, warm summers may have lasted for a long enough period to affect the flora and fauna. For example, about 60 000 years ago, boreal forest existed in the English Midlands and about 40 000 years ago beetle remains suggest summer temperatures at least 5C° higher than those in the following glaciation. It is possible that average July temperatures became slightly warmer than those of today, though winter temperatures were much colder. The climate would appear to have been less affected by oceanic influences than today.

## INTERGLACIALS

During the interglacial periods, temperatures were similar to those of today and in some cases slightly higher. The change from glacial to interglacial appears to have been sudden in most parts of the world, so sudden that vegetational changes do not always accord with the temperature levels indicated by beetles, which are able to respond rapidly to climatic changes. During the warm interglacials, vegetational succession in Europe saw a change from tundra through birch and coniferous woodland to mixed oak forest, though each interglacial differed in detail over the timing of when species appeared and which species became dominant. In more northerly latitudes the climate did not always become warm enough for mixed oak forest to grow and coniferous forest or even birch woodland may represent the climax vegetation. In earlier interglacials, the oak forest eventually gave way to coniferous forest before the gradual return to tundra and the glacial conditions.

Rainfall is much more difficult to determine except on the very general scale of dry, moist, and wet. In temperate latitudes rainfall does not seem to have differed greatly from that of today, but in the tropics periods of greater moisture have alternated with drier times. Most of the evidence is relatively recent and little is known about rainfall or lake levels in previous interglacials.

In the past, the duration of full interglacial temperatures has been from 10 000 to 15 000 years interspersed with more prolonged cold periods, during which the icecaps expanded. We should not forget that, as our present interglacial has lasted already for about 10 000 years, we may be close to its natural end.

## PLEISTOCENE CLIMATIC HISTORY

In the space available it would be impossible to cover this topic in detail, so in this section an outline is given of how the North American climate has changed during the Pleistocene, with emphasis on the recent period. The main feature of the Pleistocene has been the oscillation of cooler and warmer climates, identified as glacials and interglacials in Table 9.1. Many minor fluctuations were superimposed on this basic larger cycle of glacial and interglacial. Originally, it was believed that there were only the four major glaciations listed in Table 9.1, but increasing evidence, especially from the 1950s onward, shows a much more complex picture. It became clear that although there had been four major periods of continental icecap formation, identified by the southern limits of their expansion (Figure 9.17), advances and recessions of the ice had not been synchronous. There was a lack of correlation of events between continents and even on the same continent. For example, there is considerable disagreement over where and when the Sangamonan interglacial occurred in North America. These disputes will continue as long as there is only a limited amount of evidence available to build a complete picture. Part of the difficulty is that the evidence of the first icecaps is obliterated by the advance of the subsequent sheets. The distinctive evidence of four major events tends to be confined to the southern margins, where the greatest advances are separated (Figure 9.17).

The terminal moraines that mark the most southerly position of the Wisconsinan glaciation contain tree trunks and vegetative material that have provided radiocarbon dates of about 12 000 years. From that time forward the glaciers retreated steadily but quite rapidly as the earth went through a warming toward the present interglacial. The maximum advance and the beginning of the retreat marks the end of the Pleistocene Epoch and the transition to the Holocene. Radiocarbon dating is reasonably accurate for materials 60 000 years old or less, therefore it is only useful in the latter stages of the Pleistocene, that is, during the Wisconsinan glaciation and up to the present. The extreme extent of the icecap and various positions during the retreat are shown in Figure 9.18.

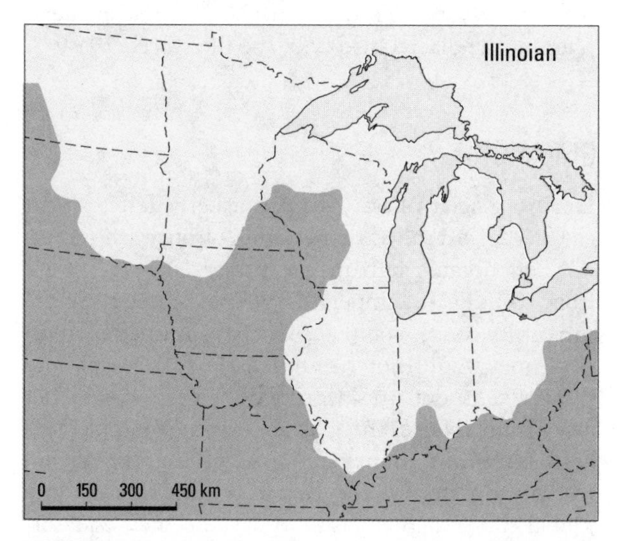


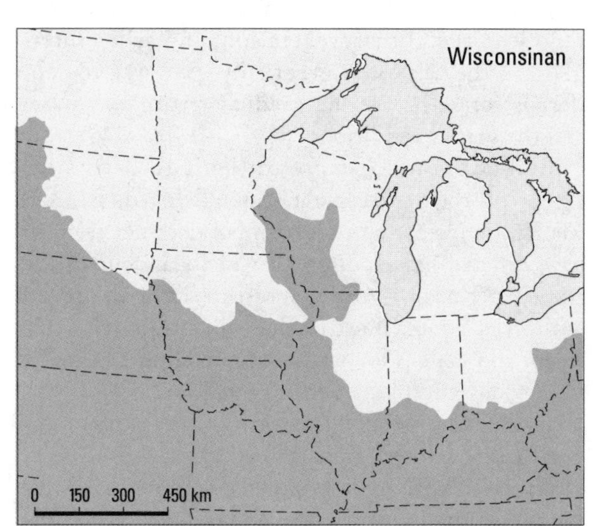


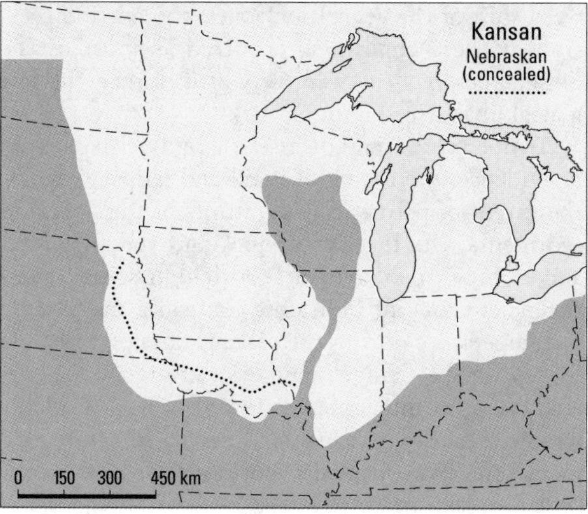


FIGURE 9.17 ▼ The maximum extent of the four major ice advances during the Pleistocene. Note that the Nebraskan was obliterated by the Kansan and only fragmentary evidence of it remains. Note the 'driftless' area of southwestern Wisconsin (after Flint, 1957)

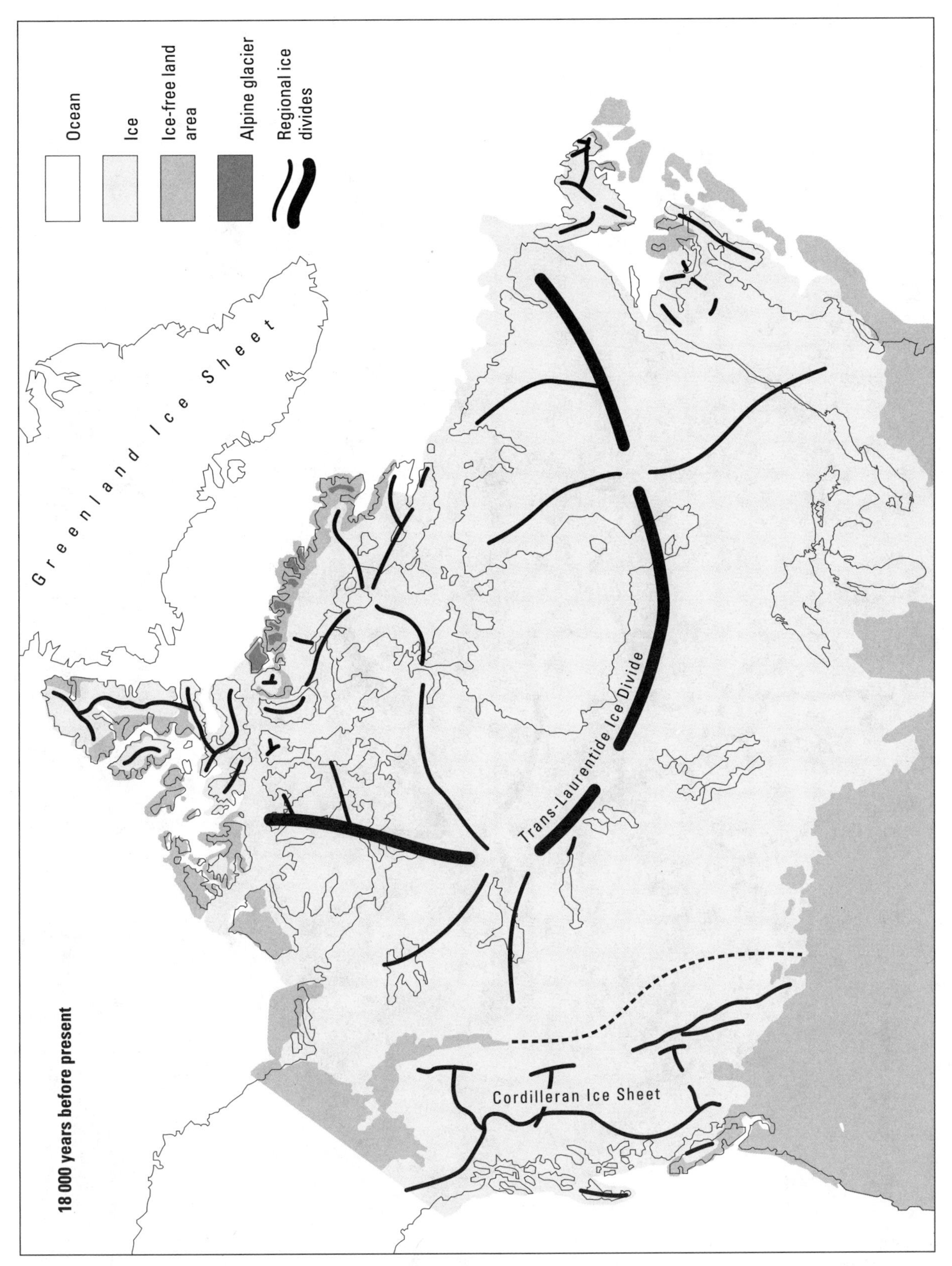


FIGURE 9.18A ▼ The Laurentide Ice Sheet at maximum extent during the Wisconsinan glaciation about 18 000 years ago. Note the separate Cordilleran Ice Sheet on the west side (after Dyke and Prest, 1987)

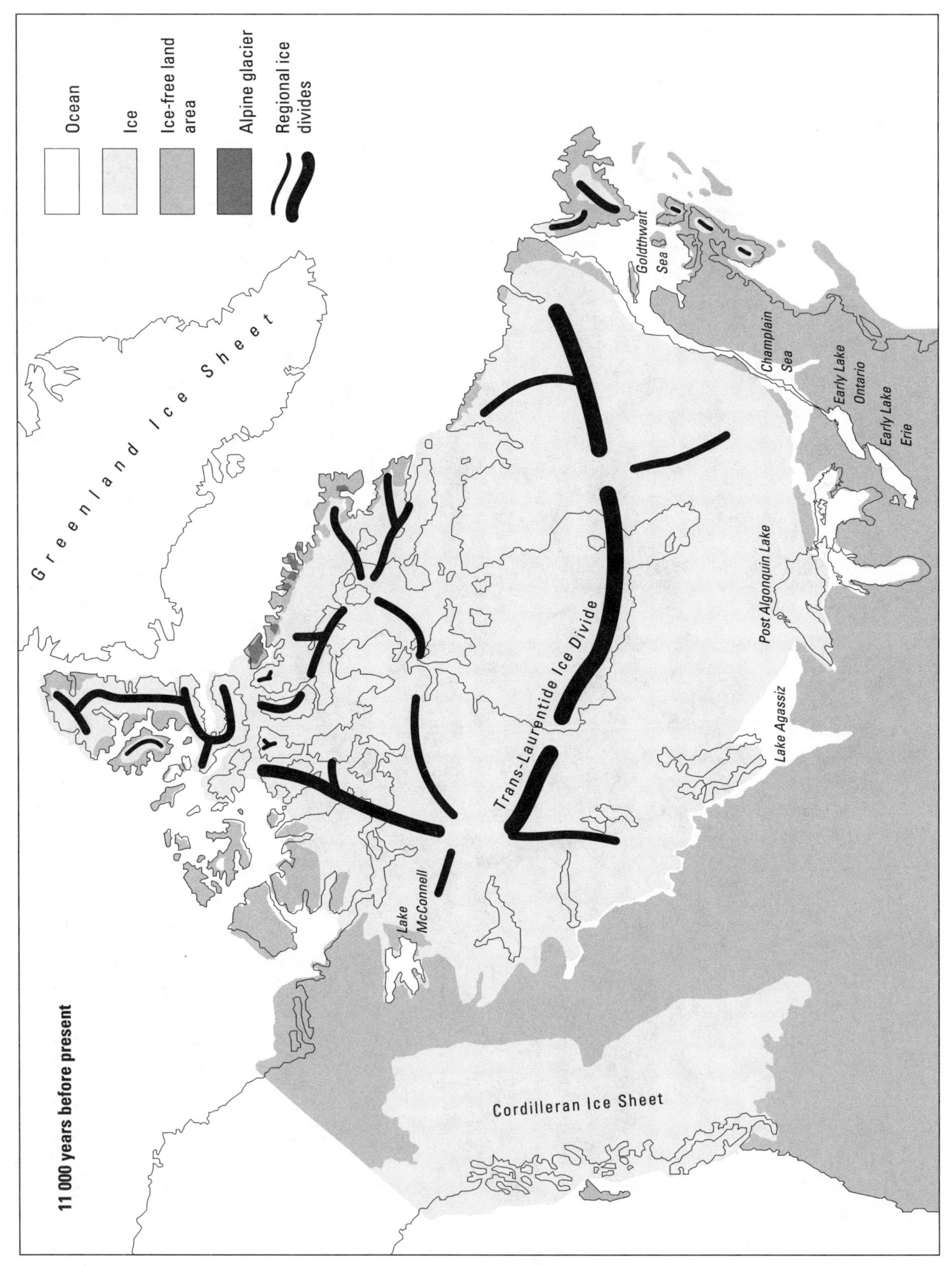


FIGURE 9.18B ▼ The ice sheet in full retreat 11 000 years ago. Note the proglacial lakes created along the edge of the retreating ice (after Dyke and Prest, 1987)

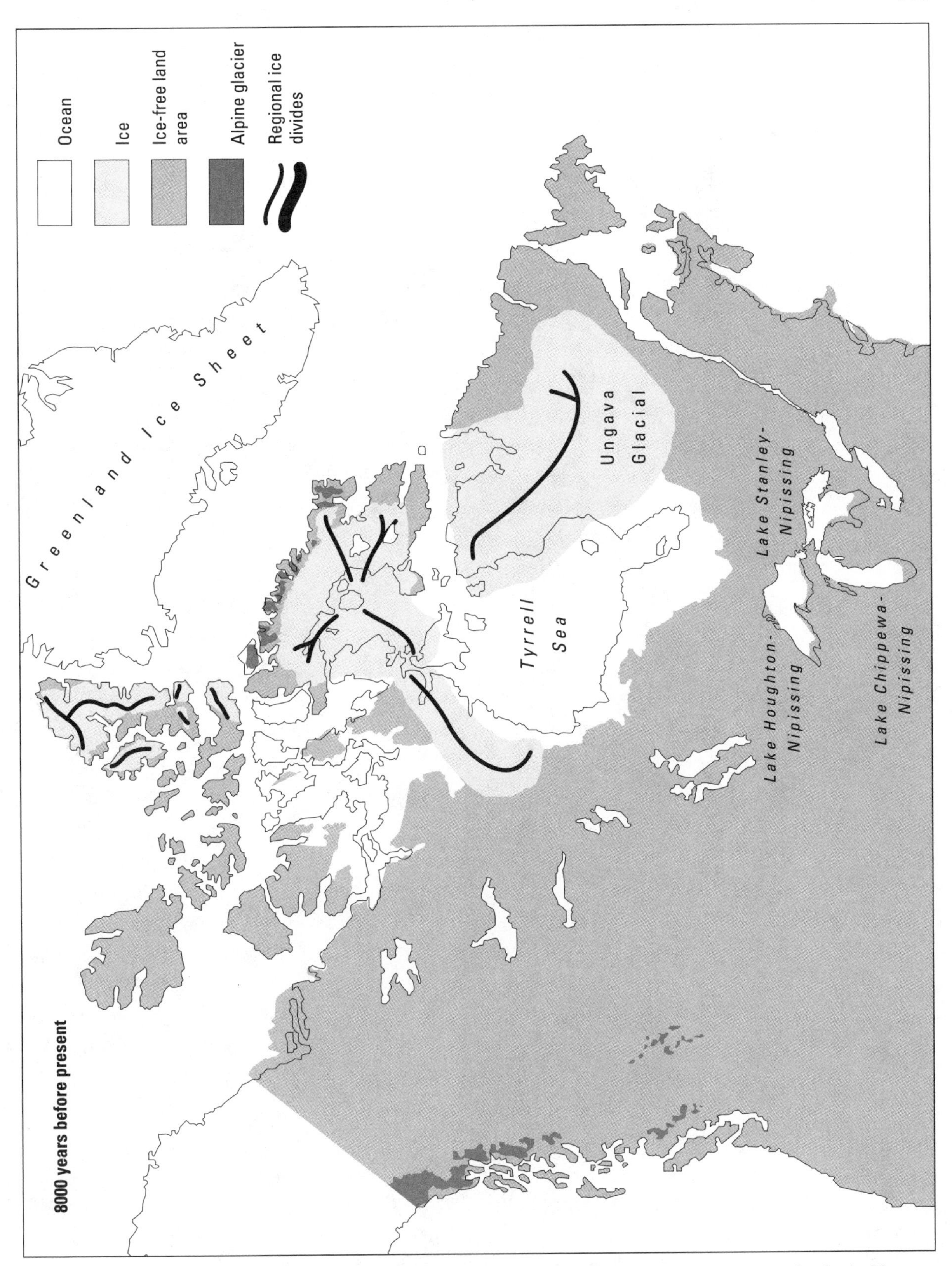


FIGURE 9.18C ▼ Global warming has reached into the High Arctic, Hudson Bay is open to the ocean, and only the Ungava region is heavily glaciated (after Dyke and Prest, 1987)

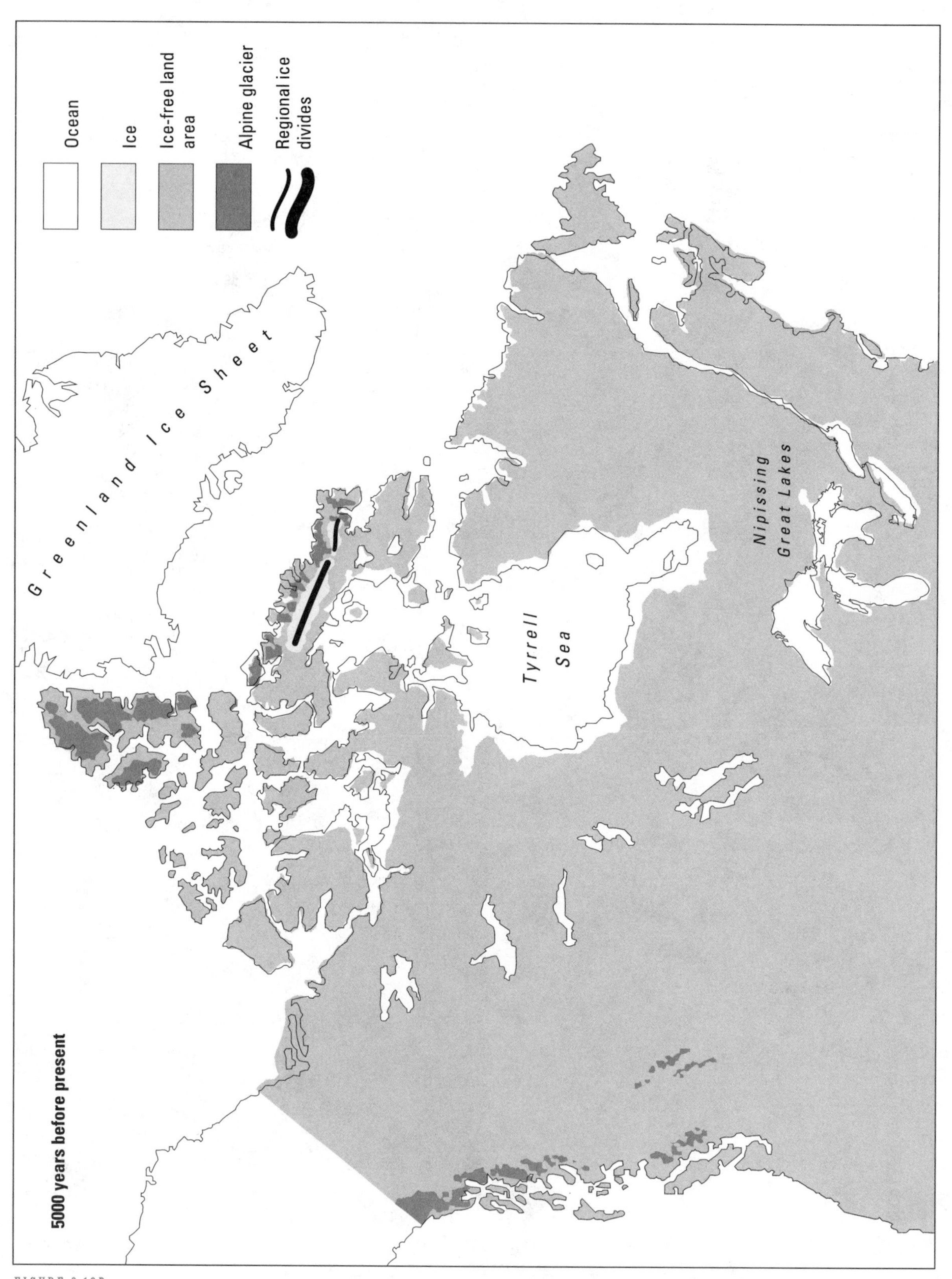


FIGURE 9.18D ▼ Canada at the peak of warming associated with the Climatic Optimum. Conditions were warmer than today and the treeline extended to the High Arctic (after Dyke and Prest, 1987)

The extreme southerly extent of the ice and the colder global conditions meant that the latitudinal climate zones were different than those of today. In North America the periglacial region occupied the northern portions of the Unites States, while the central regions experienced higher levels of precipitation and lower evaporation. This resulted in the formation of many lakes, referred to as pluvial lakes. The larger of these occupied the valleys and intermontane basins of the western United States. The largest was Lake Bonneville, with a maximum area of some 50 000 km$^2$; it covered most of western Utah (Figure 25.3). Similar lakes formed in other parts of the world, for example, Lake Chad in Africa and Lake Eyre in Australia (Figure 25.5).

Most of Canada, northern New York State, New England, and much of the Great Lakes region was still under the Laurentide icecap 12 000 years ago. Retreat had begun in Europe, but Scandinavia remained covered. Gradually, the ice front retreated, filling the low areas with meltwater. Sedimentary rocks along the edge of the Canadian Shield offered less resistance than the igneous rock and there was more erosion. The large lakes that formed in these areas are shown in Figure 9.18B as a great sweep from Great Bear Lake down through the Great Lakes. In several areas meltwater was trapped between the moraines and the retreating ice sheet, forming large proglacial lakes. The largest of these was glacial Lake Agassiz, and at its greatest extent it occupied a triangular region as shown in Figure 9.18B. Lakes Winnipeg and Manitoba are the remnants of this lake, which gradually diminished as various outflow channels opened.

The retreating ice fronts left behind their debris as mounds and vast plains of **till** that was often reworked by the meltwater. The result of this and the clay deposits of the glacial lakes make up many of the landforms we see today in the area that was covered by ice. The hollows of the exposed rocks of the Shield where filled with water in an irregular drainage pattern that is the familiar landscape of so much of northeastern North America. Generally, a landscape with many lakes is one that is very recent in geologic time, because lakes are caused by obstruction in the drainage system. Thus the northern half of North America has many lakes while the southern half has very few.

As the ice retreated, the vegetation moved north, colonizing the gradually thawing ground, a sequence that has been partly reconstructed by palynology. The postglacial period has not been fully reconstructed, however a general pattern of climatic fluctuations can be identified. Britain experienced a warmer period for some 1000 years, commencing about 12 000 years ago. This was followed by a cooler period that saw tundra reappear in northern Britain and across the south German plain until about 10 000 years ago, after which there was a sustained warming.

In North America, it appears that this sequence was delayed, probably because the sheer size of the Laurentide Ice Sheet absorbed much of the increased energy in melting. A slight readvance of the ice about 8700 years ago was followed by an overall warming trend and a retreat of the ice. Now the boreal forest followed the retreating ice, so that by 6500 years ago it was north of its present position (Figure 9.19). During this period oak became a major species on the eastern seaboard and in the midwestern United States. Farther north, the boreal forest was advancing and Nichols has estimated, from pollen samples taken primarily in the Ennadai Lake region north of Manitoba, that July temperatures were about 4C° warmer than at present. This period of maximum warmth has been termed the **Climatic Optimum**, but it is more often referred to now as either the Hypsithermal or the Altithermal. It appears to have been a worldwide event. Trees extended farther poleward and to higher altitudes than now. Some wetter and drier periods have been identified by changes in the character of the vegetation.

Following the Climatic Optimum, temperatures in general have decreased. Nichols found that summer temperatures were below today's levels around 2500 years ago and again 400 years ago, but 1200 and 800 years ago they were about 1C° warmer. These changes have been confirmed from other sources, including historical evidence. About 1000 years ago the climate was in a warm, dry phase, with few winter storms in the North Atlantic. This warmer period is called the Little Climatic Optimum or the Medieval Optimum. The Vikings took advantage of this quieter and warmer period to colonize Iceland, Greenland, and North America, as the settlement at L'Anse aux Meadows on the northern tip of Newfoundland testifies. Hubert Lamb has estimated that temperatures were between 1C° and 2C° warmer than at present. By A.D. 1200, cooling set in, with increased storminess. Large parts of Friesland in north Holland were flooded after storm damage in 1216. At the same time, drought was starting to affect Indian settlements in Iowa and

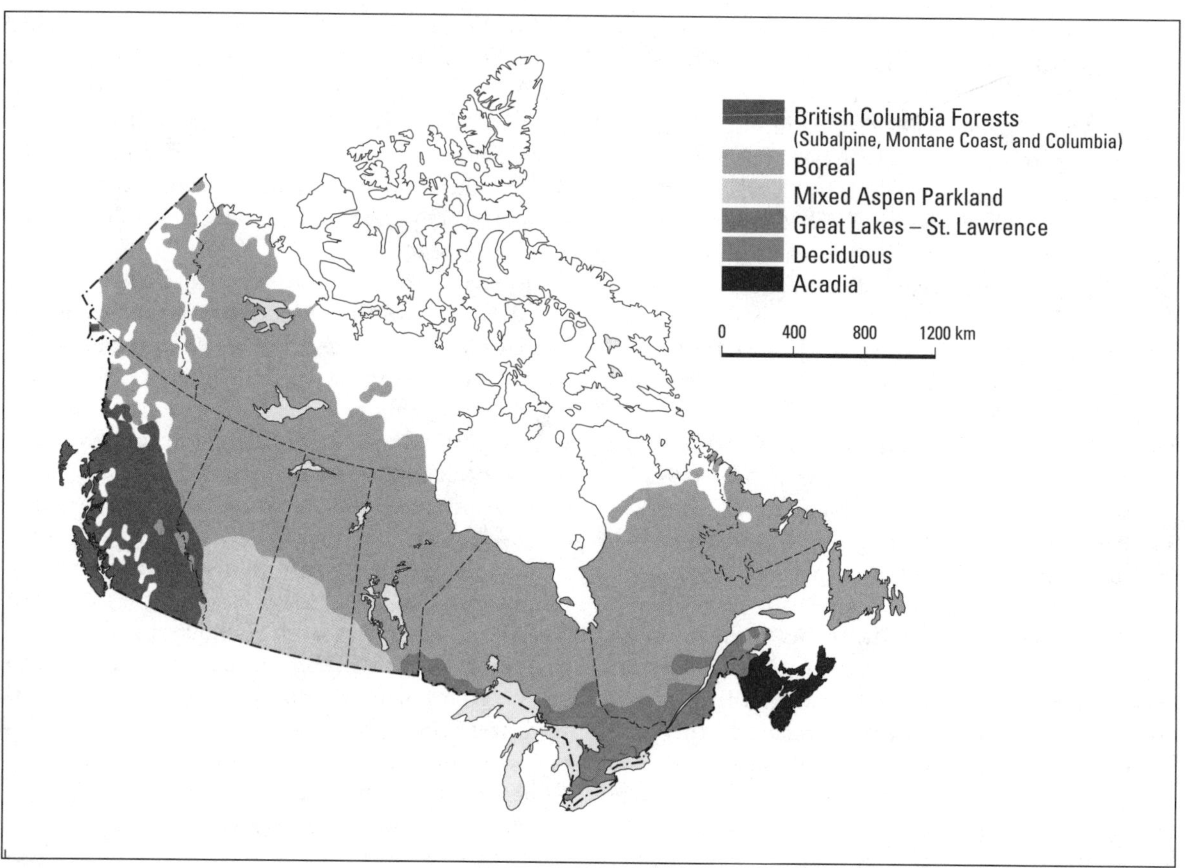


FIGURE 9.19 ▼ Major forest regions in Canada. Note dominance and extent of the Boreal Forest; Canada has approximately 10 percent of the world's productive forest area (after G.A.J. Scott)

South Dakota, and particularly the Anasazi peoples of the southwest.

# The Little Ice Age

After a partial return to more favourable conditions from 1400 to 1550, the climate grew colder again and for 300 years Europe experienced a very distinct cold spell. This was known as the **Little Ice Age**. In central England, the mean annual temperature for the 1690s was only 8.1°C—about 1.5C° cooler than at present. Lamb has put the main phase between 1550 and 1700, but in North America it seems to have extended to at least 1750. Ball has shown that there was a dramatic shift in climate around 1750, according to records maintained by the Hudson's Bay Company. The company carried on the fur trade in the northern regions of the continent, and its records indicate considerably different conditions than today, similar to reports in European diaries and records. Ice conditions noticeably more severe than at present are reported for the North Atlantic, Hudson Strait, and Hudson Bay. The southerly advance of cold air pushes the pack ice farther south and the number of severe storms tracking across the Atlantic increases.

Severe winters are reported for eastern North America by early settlers, with 1607–1608 being one of the earliest and the worst. Samuel Champlain, the great French explorer, reported that there was ice thick enough to bear the weight of a man on the edges of Lake Superior in June, 1608. Tree ring studies indicate a decrease in westerly flow and an increase in north winds through this period. Hudson's Bay Company records indicate similar changes in wind patterns at least until 1760. The severe weather conditions created problems for George Washington and his troops at Valley Forge. Increased precipitation and cold hampered many of the campaigns. Glaciers advanced noticeably in many places as more snow fell and survived the cooler summers. Food production was seriously hampered, leading to famine and hardship in many parts of the world.

# Toward the present

By the middle of the nineteenth century, the Little Ice Age was waning and the steady warming trend that reached its peak in the 1930s and 1940s began. During this warm phase, westerly winds were much more frequent in the middle latitudes and the strength of the northern hemisphere circumpolar westerly winds was high. This pattern of winds did not bring benefits everywhere; it also coincided with drought in the Great Plains of North America. After this peak, there was a decline in mean northern hemisphere temperatures to the early 1980s, followed by a warming in the later part of the decade. The effect of a hemisphere mean temperature change of a few tenths of a degree may seem very small, but change is not uniform. Some areas experienced more significant increases or decreases of temperature. Similarly, rainfall patterns are notoriously variable. For example, in the Sahel area of Africa, we can see from the rainfall record that at certain times there have been sequences of higher than average rainfall followed by periods with lower than average rainfall (Figure 9.20). These changes may have significant effects for humans. The most recent drought has had serious consequences for the population, which had been increasing rapidly during the previous wet spell. An estimated 100 000 people died, largely from starvation, as a result of the drought, and it continued in the 1980s.

## CLIMATE MODELS

Some idea of the complexity of climate is seen in the schematic in Figure 2.14. The problems of measuring, storing, and manipulating the variables and the interrelationships was impossible before the advent of the computer. For example, Milankovitch spent most of his life calculating the net changes in energy created by changes in the sun–earth relationship. The computer offers some opportunities to simulate or model global climate, but it will be some years before computer capacity is sufficient to handle even a few of the variables and interactions shown in Figure 2.14.

Climate models are actually mathematical models in which climate is represented by equations. They occur in a wide range of sophistication and scale from a simple energy balance model (EBM) to the general circulation model (GCM) that simulates the entire global climate. We will only discuss the GCM here.

Each GCM is a mathematical formula for the atmosphere comprised of formulas for several major discrete segments of the earth–atmosphere system. Four equations are always included; they represent

- conservation of energy
- conservation of momentum
- conservation of mass
- an ideal gas law: pressure × volume = absolute temperature × gas constant

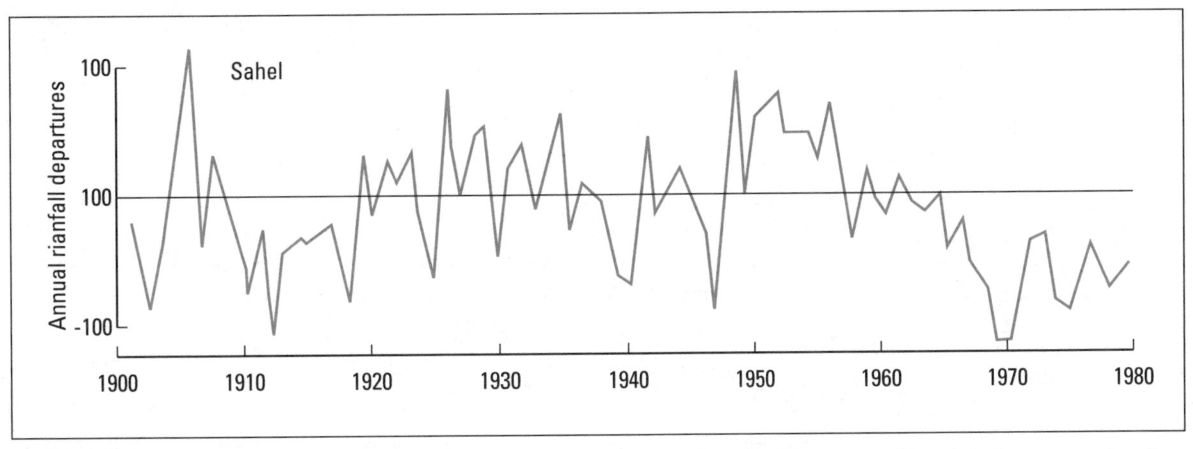


FIGURE 9.20 ▼ Annual rainfall departures for the Sahel, 1900–80. The plotted values are roughly equivalent to a regionally averaged 'percent of standard departure' (after Nicholson, 1983)

These are fundamental laws of physics which should yield accurate predictions—but this is not the case.

The same is true of daily weather forecasts. The atmosphere functions by the laws of physics, yet predictability beyond a certain time period is increasingly difficult. Some observers believe that the atmosphere acts chaotically and therefore is not predictable. Our knowledge of the interactions and causes of climate and climate change are inadequate to confirm this. This is also the reason that climate models are limited in their predictive value. Their original purpose was not prediction, although that has become their role. GCM should be a tool for learning how the global climate functions and reacts under probable conditions.

The mathematical formulas of a GCM are applied to the data for each point on a three dimensional grid, as illustrated in Figure 9.21. The world is divided into a grid, usually with a 3° by 5° latitude and longitude spacing (some are 4° by 6°). Look at an atlas and determine how large an area in your region is covered by a 3° by 5° rectangle. Move the rectangle around and determine how different the climate is within the area covered. As Figure 9.21 shows, the grids extend up from the surface to the top of the atmosphere. The number of vertical divisions used by different models varies between six and fifteen. The average climate is determined for each rectangle, but in most we have virtually no longterm data. The climate for each square of the grid is determined by what is called **parameterization**. This requires that variables be simplified or neglected, and as this is done the model moves farther from reality.

Imagine the complexity of calculations because of an open atmosphere with horizontal and vertical variation and movement in all directions. There is one great limiting factor in any climate model, beyond the lack of capacity of computers, and that is the lack of data. The model is derived from data and is only as good as the amount and length of record. Similarly, the results are limited by the data available. Recently, satellite information has provided wider coverage in space and time. The oceans cover 70 percent of the earth and we have virtually

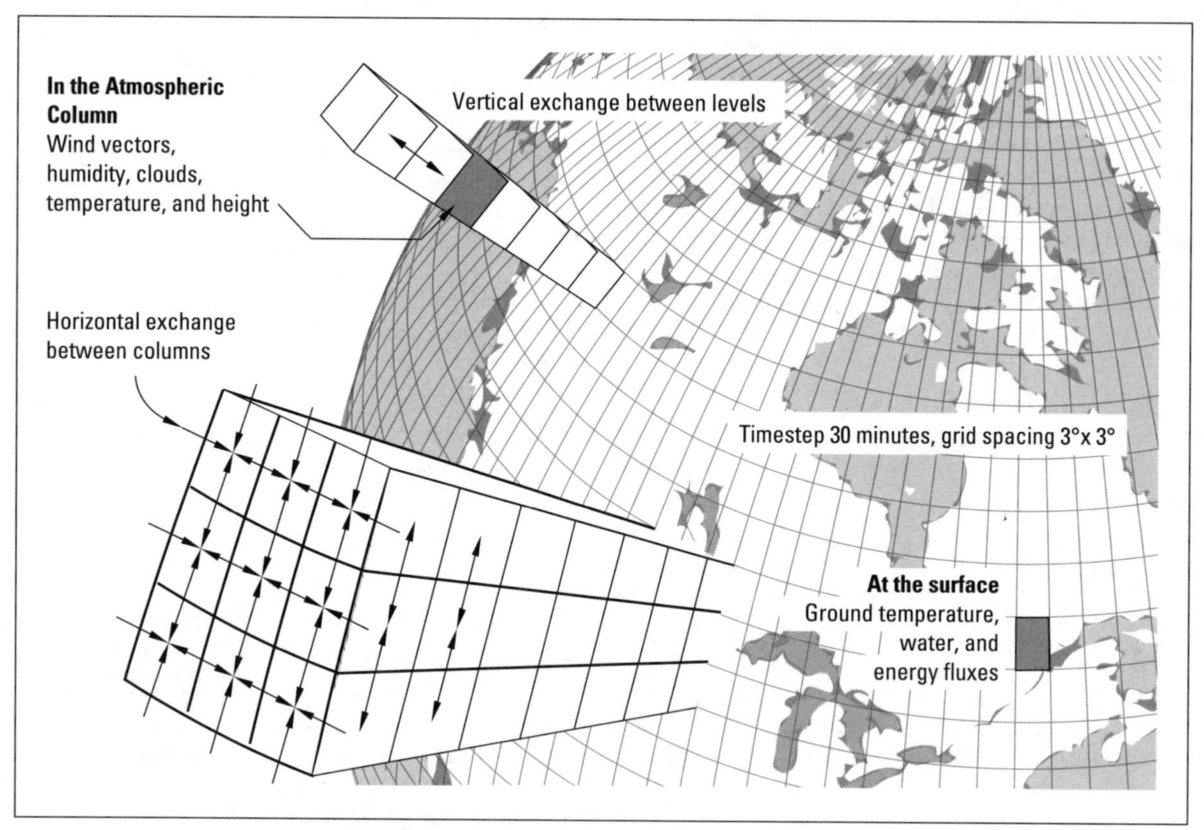


FIGURE 9.21 ▼ The basic characteristics of a general circulation model of climate. The model is a grid of interacting columns spread across the surface of the globe

no historic data for them. Most of the land surface has no record. In Canada we have very limited coverage of the subarctic and Arctic yet these are crucial areas that influence climate. Surface data is limited, but upper air data is even more scarce.

A major interconnection in climate is between the atmosphere and the oceans. The role of the ocean in climate change is critical, but again we have little knowledge of the three dimensional ocean. There are ocean models and atmospheric models, but none that effectively interconnect the two. The more sophisticated GCM can cope with a very shallow ocean, but lack of understanding of the interactions—and limits of computer capacity—prevent any meaningful simulation of reality.

Five GCM incorporate a shallow ocean layer. The Geophysical Fluid Dynamics Laboratory (GFDL) model was the first scale GCM; the Goddard Institute for Space Studies model (GISS) is perhaps the best-known because it uses the most sophisticated computer and its output is used by more people in other areas of planning and research. The National Center for Atmospheric Research (NCAR), Ohio State University (OSU), and the United Kingdom Meteorological Office (UKMO) models are less well known, but still valuable. In 1991, a Canadian GCM was completed by researchers at the Canadian Climate Centre in Toronto. It has a slightly smaller grid and better parameterization of atmospheric components, but also lacks an interactive connection with a deep and circulating ocean.

The GCM have been used to recreate glacial climates and other periods in history, but their major role has been in estimating the greenhouse effect. All the predictions of global warming are based on computer generated effects of doubling carbon dioxide. All models show that a doubling would result in a warmer earth, but this could be because the models are designed to have that result. None of the models are able to simulate natural feedback mechanisms, and even if they could we do not know how they would work.

A study by Schlesinger showed that when the same data is put into each of the five major GCM they yield different results. When the same data is put into the *same* model it yields different results. When the estimated increase in carbon dioxide for the last century is put into the models, the results show that the global temperature should be approximately 1C° warmer than at present.

GCM are very simplistic simulations of the complexity that is the atmosphere, nonetheless they provide the only means we have to try to understand how global climate works. The ability to correlate and manipulate the volume of data makes the computer an invaluable tool. Thus the models do not predict, but create 'what if' scenarios.

# The effects of humans

## AIR POLLUTION

The period for which accurate instrumental records are available is the period within which human impact upon climate could have been most marked. Superimposed upon the natural changes of the atmosphere, therefore, are the effects of human activity. Air pollution in particular may have caused variations in temperature. Smoke and carbon dioxide are two of the most widespread pollutants. Less obvious pollutants are sulphur dioxide, nitrogen oxides, and hydrocarbons, from sources such as electric power stations and vehicle exhausts.

The effect of these pollutants upon the local climate is often straightforward. The particles of smoke and dust act as nuclei for condensation, encouraging the formation of fog. In Los Angeles, photochemical fogs form as a result of the interaction between strong sunlight and hydrocarbons emitted by automobiles. Instead of being dispersed by winds and mixed with cleaner air, the choking pollutants hang in the still, sheltered air, causing problems for transport and health. Similar dense fogs used to be common in London, due to coal smoke. In 1952, the deaths of 5000 people were blamed on a particularly bad fog. Following this, legislation was introduced to restrict the use of coal, and since then the problem has almost disappeared.

Solid atmospheric pollutants also influence the receipt of solar radiation by the surface. Direct radiation is reflected and scattered, so the amount of energy reaching the earth is reduced. We might expect that this would cause a decline in surface temperatures, but in reality the opposite seems to happen. The reason is apparently that the pollutants intercept, absorb, and reradiate outgoing longwave radiation, so that more energy is trapped within the lower layers of the atmosphere.

It is not only solids in the atmosphere that affect climate. Gaseous pollutants may also interfere with the inputs of shortwave solar radiation and the outputs of longwave radiation. Carbon dioxide intercepts little shortwave radiation. This again traps energy in the lower atmosphere and raises temperatures. In the period from 1880 to 1940, there was a progressive rise in the mean temperature of the northern hemisphere, and this was thought to be a result of the increase in levels of carbon dioxide. Since then, the trend has reversed twice, with decreasing temperatures until the early 1980s, with a subsequent rapid increase. Clearly, the explanation is more complex than we once believed.

Direct warming of the atmosphere by waste heat also raises temperatures. Estimates of the global energy production have indicated that $8 \times 10^6$ MW are generated annually, most of it in densely populated urban and industrial areas. Long-period temperature records at city-centre sites usually show an increase of temperature through time because of this effect, coupled with storage of heat by buildings.

### CHANGES IN THE SURFACE VEGETATION

Outside urban areas the human effect on climate has been less marked, but extensive changes in vegetation and land use will have an effect at the microscale, and may be important at larger scales. Forest clearance, tillage of grassland, and desertification due to soil erosion all increase the albedo of the surface and may alter the earth's radiation balance slightly (Table 9.4). A change in land use from forest to farmland is believed to lead to an increase in the carbon dioxide content of the atmosphere, which again would affect the radiation balance. In tropical areas, where most rainfall is the result of convection, the reduced radiation absorbed at the surface would result in less heating, and so less convection, which in turn means less rainfall. We are in a positive feedback situation. Recent work on the prolonged drought in the Sahel region of West Africa has suggested that changes in surface vegetation through overgrazing and land mismanagement are probably crucial factors in the major decrease of rainfall that has occurred (Figure 9.20).

## Conclusion

It is clear that in the medium to long term, over a time scale of tens to thousands of years, our climate varies, not randomly, but systematically. Broad, consistent fluctuations occur, giving periods of relative warmth and periods of cold, years of aridity and years of wetness. The reasons for these fluctuations are not clear; variations in the earth's orbit and rotation, changes in solar output, internal adjustments in vegetation, topography, and atmosphere may all be contributory factors. The time scale of possible causative factors is shown in Figure 9.22, but we must not forget that most factors are interactive; we cannot isolate a single process and describe its consequences with much confidence. In recent centuries human activity, as well, has had an impact on climate.

Two questions remain. What is the effect of these climatic fluctuations? And where are we going now? Some of the effects are all too apparent to us. In areas that are marginal for agriculture and human habitation, minor changes in the climate may have an appalling consequence, bringing crop failure, soil erosion, and famine. Some of the effects are more

**Table 9.4**

*Changes of heat budget after conversion of forest to agricultural use*

| | *Albedo* | *Bowen ratio ($Q_H/Q_E$)* | *$Q^*$ ($W\ m^{-2}$)* | *$Q_H$ ($W\ m^{-2}$)* | *$Q_E$ ($W\ m^{-2}$)* | *Evapotranspiration (mm month$^{-1}$)* |
|---|---|---|---|---|---|---|
| Coniferous forest | 0.12 | 0.50 | 60 | 20 | 40 | 41 |
| Deciduous forest | 0.18 | 0.33 | 53 | 13 | 39 | 40 |
| Arable land, wet | 0.20 | 0.19 | 50 | 8 | 42 | 43 |
| Arable land, dry | 0.20 | 0.41 | 50 | 15 | 35 | 36 |
| Grassland | 0.20 | 0.67 | 50 | 20 | 30 | 31 |

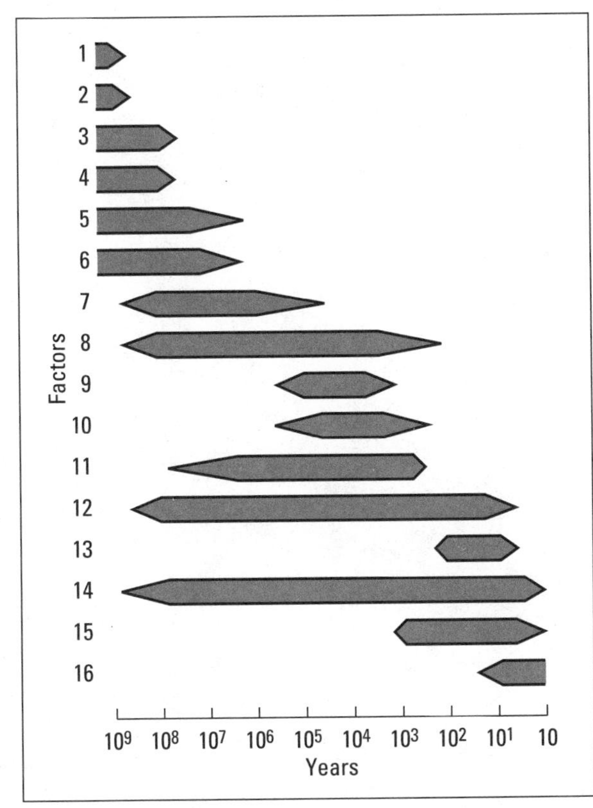


FIGURE 9.22 ▼ Potential causative factors in climatic change and the probable range of time scale of change attributable to each (after Mitchell, 1968). Factors: (1) evolution of the sun; (2) gravitational waves in the universe; (3) galactic dust; (4) mass and composition of the air (except $CO_2$, $H_2O$, and $O_3$); (5) polar wandering; (6) plate tectonic movement; (7) orogenic and continental uplift; (8) $CO_2$ in air; (9) earth-orbital variations; (10) air–sea–icecap feedback; (11) abyssal oceanic circulation; (12) solar variability; (13) $CO_2$ additions by fossil fuel combustion; (14) volcanic dust in the stratosphere; (15) ocean–atmosphere autovariation; (16) atmospheric autovariation

subtle, but nonetheless significant for humans. As the pattern of climate changes, people tend to move; new areas become favourable for habitation, others become unfavourable. It has been suggested that the stimulus for the Viking invasions and settlement of Iceland, Greenland, and Britain was climatic deterioration in Scandinavia. Nomadic tribes today respond to similar stimuli.

The effects of climatic change are not confined to agriculture and other human activities. As we will see in later chapters, these fluctuations also influence landscape processes. Throughout the temperate regions of the world, the imprint of past climatic change is clear within the landscape. Glacial landforms lie hundreds of kilometres beyond the limits of the present icecaps; lakes that were at one stage huge inland seas are now small pools by comparison; river valleys that once carried vast torrents of water are now occupied by small, placid streams. Effects of a similar magnitude can be detected in the vegetation. In many areas the range of plants that we find today is a result of migration and mixing of vegetation in response to climatic change. The global system, as we have noted before, is intricately interrelated. Changes in one part affect others, and the effect is nowhere more apparent than in the influence of climatic change.

So to our last question. What will be the climate of the future? Numerous predictions have been made. It has been suggested that the temperate latitudes are entering a phase of cooler climate, possibly another Little Ice Age. It is also argued that we are at present within an interglacial; at some stage in the future we will move into another glacial period. The problem with such predictions is that at the time scale of our measurements, changes in climate are almost imperceptible. Moreover, climatic change involves numerous different trends, superimposed upon one another and each operating on a different periodicity. Consequently, it is almost impossible to tell how long any trend we identify will persist.

# The global water balance

## The hydrological cycle

In some parts of the world raindrops splattering against the window panes or drumming on the roof are such a common occurrence that people give them little thought. But what happens to this water descending from the skies, and how did it get there in the first place? Some regions experience several centimetres in a day, but can this amount of moisture be held in the atmosphere or does it have to be constantly replenished from elsewhere? Do the vast quantities of water that flow down our rivers each day come directly from precipitation alone, or is there some underground source that provides a steady flow of water? Questions like this make us think more about the movement of water between the atmosphere, the lithosphere, and the oceans. Clearly they must be interlinked, but how does this movement take place, and why? In this chapter we shall describe the main features of the hydrological cycle and outline at a general level its main processes and components.

The whole system of water movement has been termed the world **hydrological cycle**. In Figure 10.1 we can see the main components of the system,

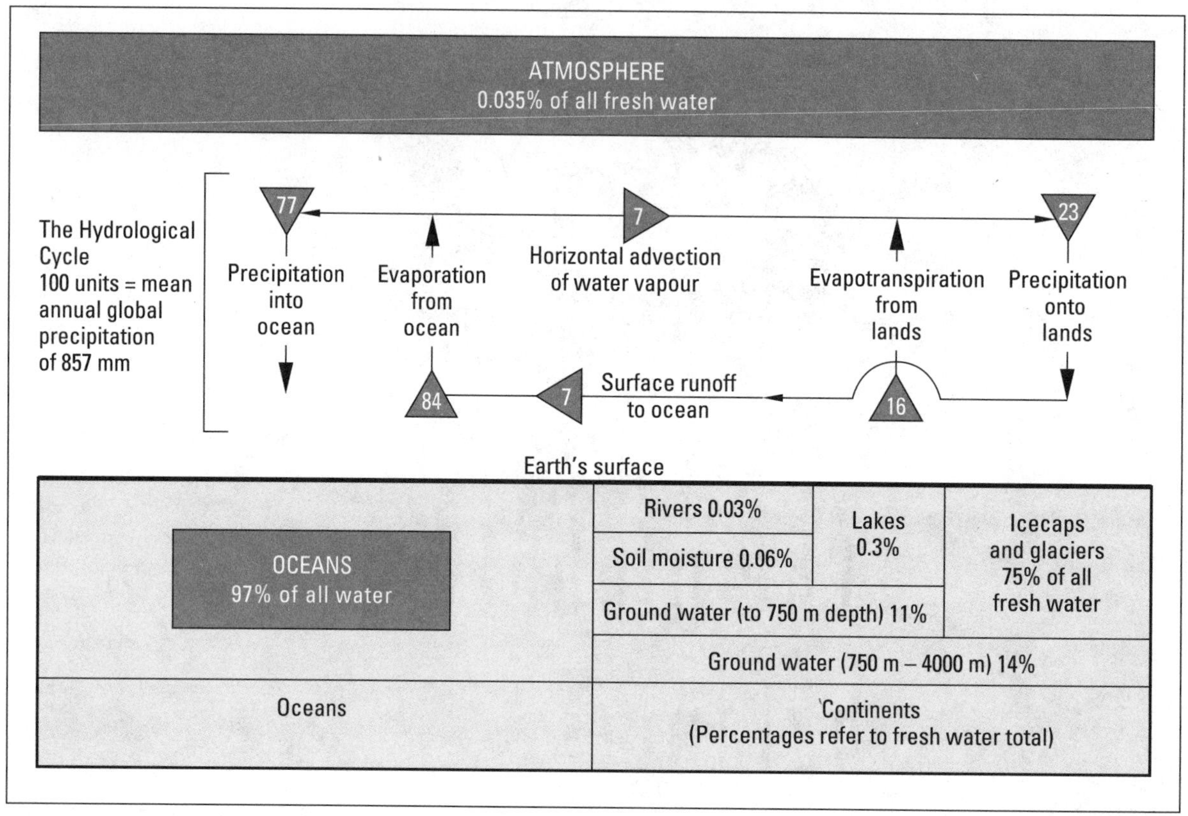


FIGURE 10.1 ▼ The global hydrological cycle (after More, 1967)

rather like the solar energy cascade we looked at in Chapter 3. The main differences here are that the storage components in the system are much larger relative to the inputs, outputs, and flows than they were for solar energy. The heavy line represents the surface of the earth, the interface between the lithosphere and the atmosphere. Water is evaporated from this surface using energy from the solar cascade. It is returned to the surface in a variety of forms of precipitation, but primarily rain and snow. The left side of the diagram shows that 97 percent of water is in the oceans. Most of the evaporation and precipitation occurs over this surface, which comprises 70 percent of the earth's surface. The right side is the remaining 30 percent that is land. Total evaporation from these surfaces is 84 units (from the ocean) plus 16 units (from the land), or 100 units. This equals the input of 100 units provided by the global mean annual precipitation of 857 mm. However, only 77 units fall on the oceans and 23 units over the land. More water evaporates from the oceans than falls back as precipitation—84 units and 77 units, respectively. There is a net horizontal transfer of the surplus 7 units to the land in the form of water vapour. As a result, more precipitation falls over the land than is provided by evaporation—23 and 16 units, respectively. The net surplus here is returned to the ocean as surface runoff, mostly in rivers. The vast amount of water stored in the oceans represents an enormous stabilizing influence on the cycle, as do the rather smaller volumes stored as ice in the icecaps of Antarctica, Greenland, and smaller mountain glaciers, as well as the sea ice of the Arctic Ocean.

Over time, the hydrological cycle is more variable than the solar energy cascade, for changes can take place in the amounts of water stored in the ice and oceans, as the Ice Ages of the recent geological past have indicated. However, on a shorter time scale, there is a rough balance between the various components, and no sudden, major changes in ocean content, ice cover, or groundwater storage take place.

Figure 10.1 also shows the distribution of water on the continents. Note that icecaps and glaciers plus groundwater contain almost 100 percent of all fresh water. Lakes, rivers, and soil moisture have about 0.39 percent. Canada is estimated to have between one-third and one-quarter of all the world's fresh water. Most is locked in to glacier ice, permafrost, and muskeg. Lakes cover our landscape. You can remove water from lakes, but if you take out more than is replaced, the lake will be reduced. It can only be refilled by blocking the flow, thus denying water to downstream uses. Groundwater is a potential source, but over half of this is below 750 m depth, and shallow groundwater is being polluted in many regions. The only short-term renewable portion of this is the annual flow in the rivers. Fresh water is a very scarce resource in many parts of the world, and is of increasing importance as demand grows and supply is diminished.

# Inputs to the global hydrological cycle

## PRECIPITATION

Precipitation is effectively the only input at the surface. (It is assumed that the amount of water in the global system is constant. In fact, water in the form of water vapour is added to the atmosphere every time a volcano erupts, but the amount has never been calculated.) The distribution of this input across the world shows a marked relationship to the distribution of factors influencing precipitation, in particular, the incidence of storms, the atmospheric moisture content, and the oceans. Unfortunately, it is difficult to be precise about the amount of precipitation occurring in different parts of the world, for data on rainfall are not available everywhere. Over the continents, there is a satisfactory network of recording stations only in the developed countries (Figure 10.2). Over the oceans, measurements are sparse. No rain gauges are maintained at sea, so precipitation has to be estimated by indirect methods. Formerly, ships' reports were used to give an approximate measure of precipitation, but more recently estimates have been made from satellite images of the clouds. As yet, these results have not been incorporated into global figures, so maps such as Figure 10.3 are little more than intelligent guesses for many parts of the world.

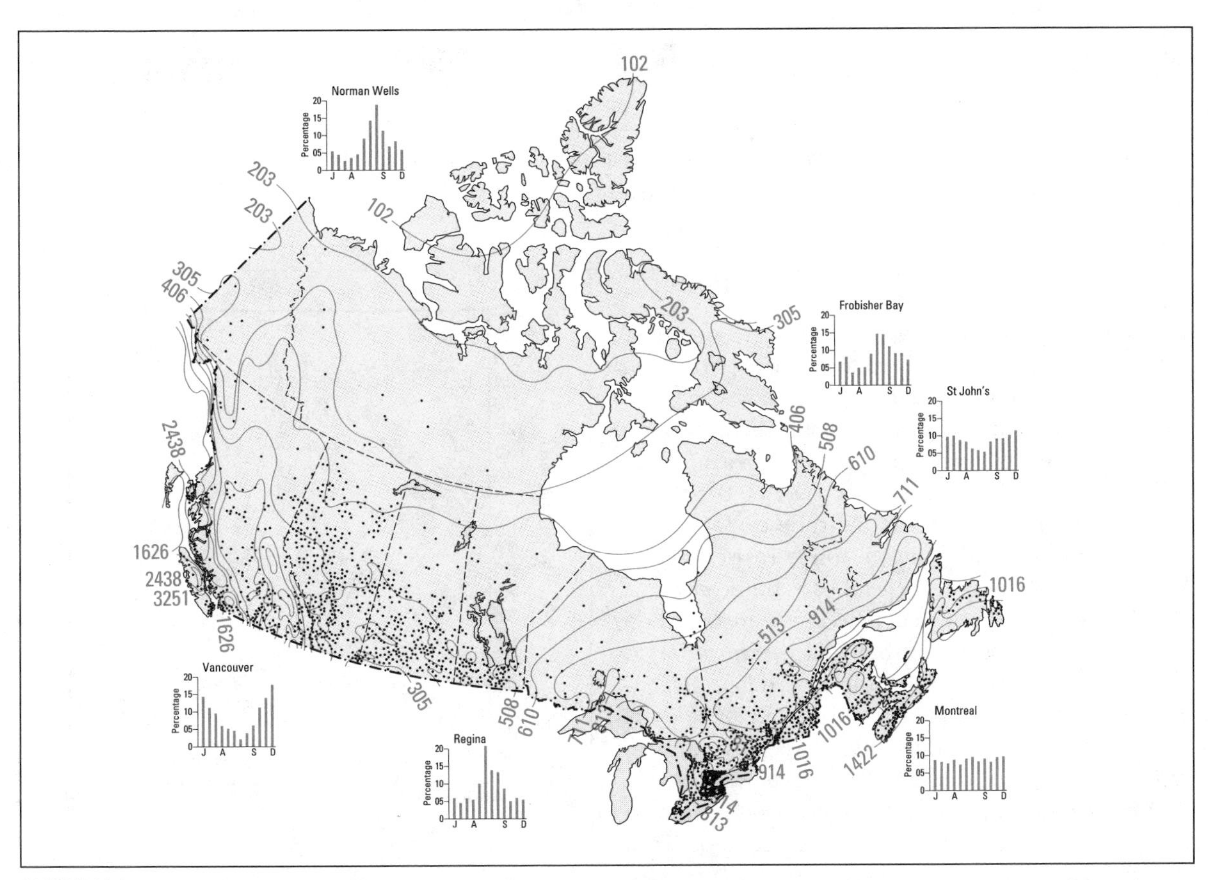


FIGURE 10.2 ▼ Mean annual precipitation for Canada (millimetres), 1931–60. Inset graphs show percentages of annual precipitation in each month for six stations (after Hare and Thomas, 1979)

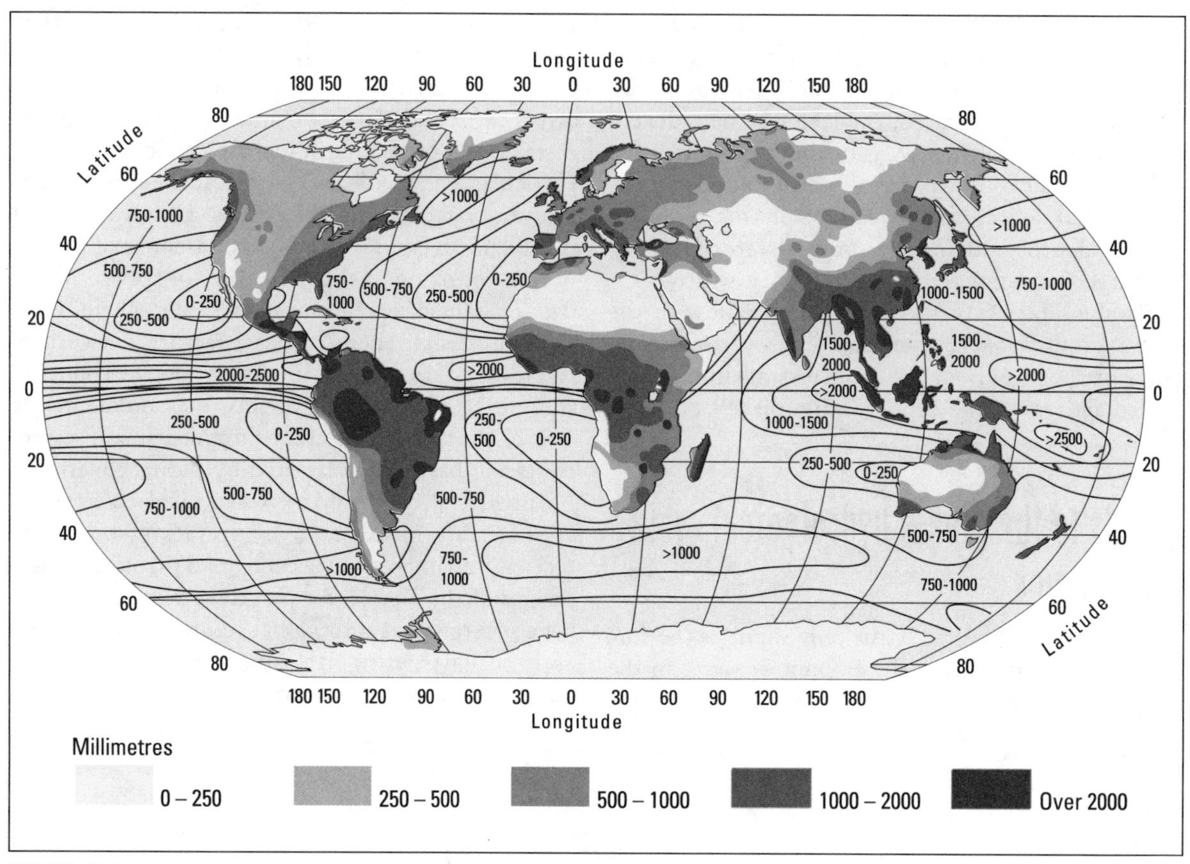


FIGURE 10.3 ▼ Mean annual precipitation over the earth. Values over the oceans are approximations, as no rain gauges exist apart from island sites (after Critchfield, 1983)

## LATITUDINAL DISTRIBUTION OF PRECIPITATION

Putting the continental and oceanic data together, we can obtain a general assessment of precipitation inputs across the world. Figure 10.4 shows the latitudinal distribution of precipitation. The curves are roughly symmetrical, with peaks of rainfall at 40°–50°N and S, and a further peak at 0°–10°N. These correspond to the midlatitude cyclone belts and the equatorial trough zone, respectively. Superimposed upon this general pattern, however, is a smaller-scale pattern that reflects the distribution of oceans in the world. Prolonged precipitation is only possible if moisture in the atmosphere is constantly replenished, and the main source of renewal is the oceans. Thus, rainfall tends to be greater in association with areas of extensive ocean. We can see this by comparing the patterns of precipitation in the two hemispheres; both in the dry zone between 20° and 30°, and in the wet zone from 40° to 50°, the southern hemisphere receives significantly higher amounts of precipitation. Toward the poles,

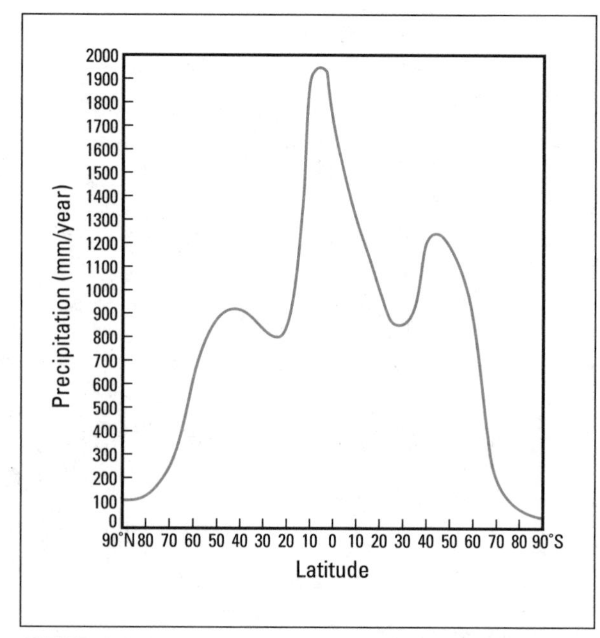


FIGURE 10.4 ▼ Average annual latitudinal distribution of precipitation (after Sellers, 1965)

precipitation declines dramatically in both hemispheres, totals becoming almost zero over the South Pole.

This description indicates only the average latitudinal pattern of precipitation inputs. What we must not forget is that precipitation is highly variable in both quantity and character. Especially within the subtropics, the inputs of rainfall are very unreliable, and long dry periods may result in extensive drought, interspersed by periods when excessive rainfall occurs, causing flooding and soil erosion. In contrast, many parts of the middle and high latitudes, which receive much less rainfall in total, are blessed with very consistent and reliable patterns of rainfall. From both the human and the geomorphological point of view, variability may be as important as quantity, as we will see in Chapter 11.

Precipitation occurs in a variety of forms. Throughout most of the world, the major input is in the form of rainfall, but in the high latitudes, and in many mountain areas, much of it falls as snow. Moreover, significant inputs may occur in the form of fog, dew, or rime (frozen fog). Again, the character of the precipitation has an important influence on what happens to the water after it has reached the ground. So, too, does the intensity of the precipitation. We will see in the next chapter that rainfall may take place as gentle, prolonged drizzle or as sudden, intense storms; snow may fall as light flurries or as blizzards. The effects upon the hydrological cycle, upon geomorphological processes and, above all, upon humans are almost always greater when precipitation is intense.

## INFILTRATION, THROUGHFLOW, AND PERCOLATION

We only have to watch what happens during a rainstorm to appreciate that a considerable proportion of precipitation does not flow across the surface as runoff, but soaks into the ground. This process is known as infiltration, and in humid temperate areas of the world as much as 95 percent of the precipitation may be absorbed by the soil in this way. Once in the soil, the water comes under the influence of a variety of forces, including gravity and the suction forces applied by small soil particles. Under the influence of gravity, water drains downward into the soil and downslope. Downslope movement carries

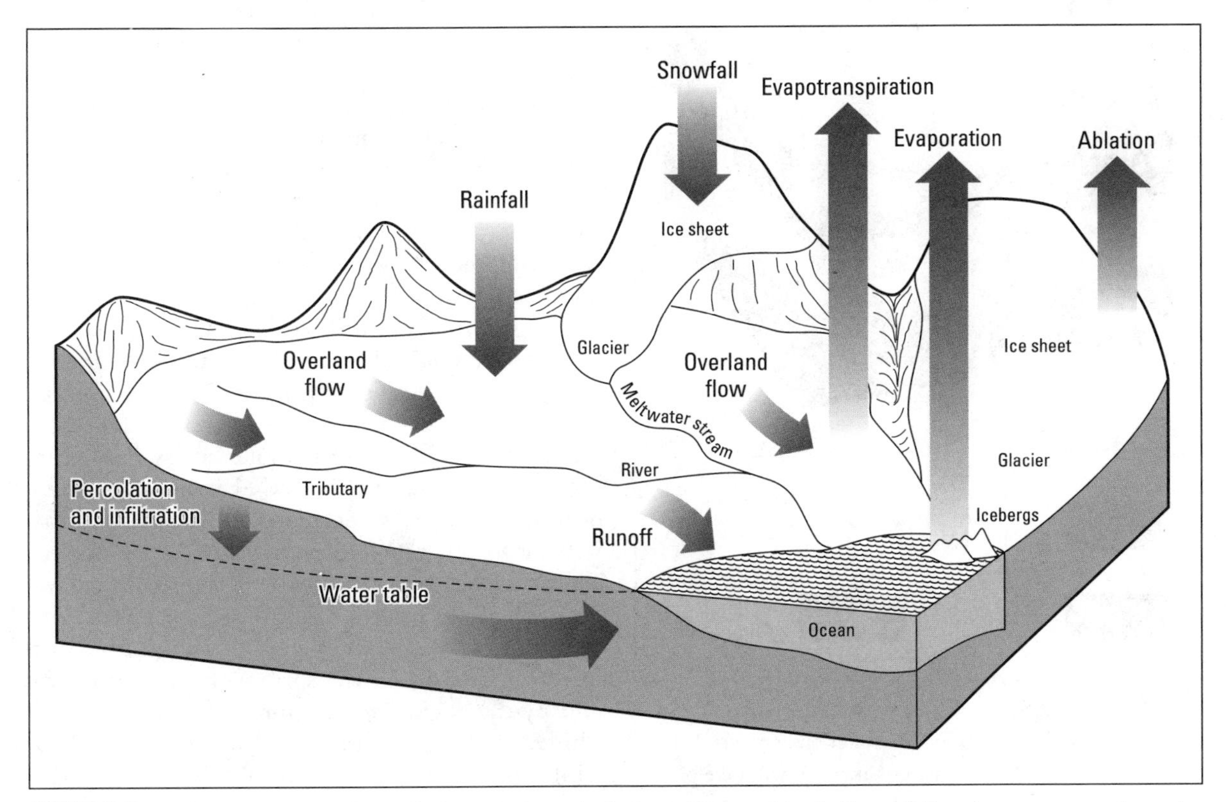


FIGURE 10.5 ▼ The main components of a basin drainage hydrology. The combined effects of all such systems constitute the global hydrological cycle

the water through the soil into river valleys, a process referred to as **throughflow** or **interflow**. In the valley, the water may emerge to contribute to the stream runoff (Figure 10.5). Drainage through the soil is called **percolation**, and much of the percolating water makes its way to the bedrock where it meets a zone of almost permanent saturation. The top of this saturated layer is marked by the **water table**; beneath the water table lies the groundwater (Figure 10.6). This is often almost immobile, and it may lie dormant in the bedrocks for thousands of years. In some cases, as in well-fissured limestone rocks, the groundwater is more active, however, and it flows through narrow cracks, tunnels, and caves to reemerge as major springs, often only a few hours after it entered the ground. Again, once it is at the surface, the water joins with the channel runoff to flow to the sea. Release of groundwater is slow and almost independent of rainfall events, so the flow of water from springs and seepage contributes a very steady and reliable supply to streams. It is largely because of this supply that most streams go on flowing even during long periods without rain.

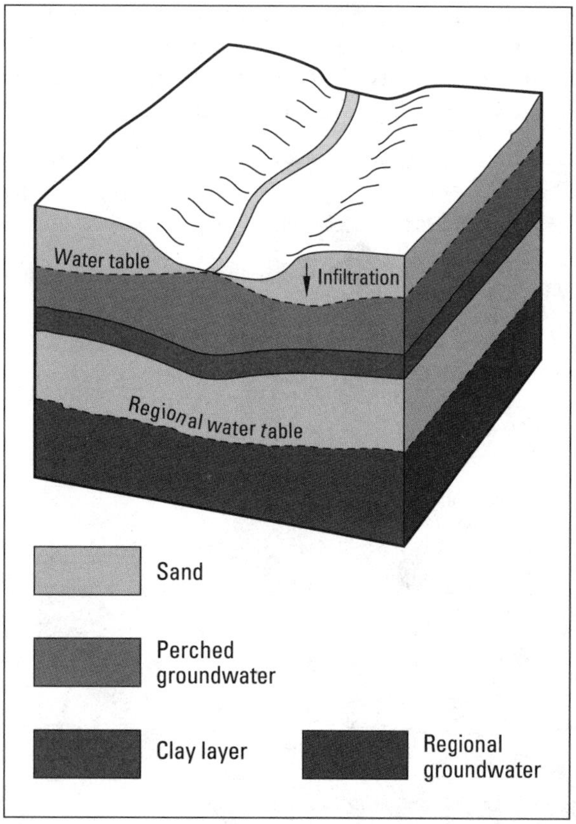

FIGURE 10.6 ▼ The relationship between topography, lithology, and the water table. No recharge of the regional groundwater will take place in this situation because the clay layer will prevent percolation to the lower bed

## STORAGE AND MOVEMENT IN THE CRYOSPHERE

Another important store of fresh water is the cryosphere—the vast ice-covered regions of the world. This contains about 26 million cubic kilometres of water, about 75 percent of the world's fresh water, or enough to maintain the Mackenzie River, NWT, for about 240 000 years! The water enters this huge store largely as snowfall and makes its way to the sea at imperceptibly slow rates, often only a metre or less per year. Thus, snow falling on the central parts of the Greenland or Antarctic icecaps may be trapped there for many thousands of years; dating of ice in Antarctica, for example, has shown that some of it collected over 100 000 years ago. Even in the fastest moving glaciers, ice may take several centuries to flow from the source area to the snout.

Many Canadian rivers are fed by glacial meltwater, the Bow River that flows through Calgary being an important example. The flow regime of these rivers is directly influenced by the seasonal and diurnal pattern of temperature. In winter, cold temperatures reduce the melting, and so reduce the flow. In the spring and early summer, melting is greatly increased and rivers rise significantly. Similarly, melting and flow increase during the day and decrease at night as temperatures rise and fall. This is especially noticeable in rivers along the Rocky Mountains where water levels can increase dramatically in just a few hours.

The meltwater from glaciers has often been in storage for considerable periods. Water is also stored cryogenically for shorter periods. Snow accumulations in the winter months are important sources of water to refill lakes and potholes across Canada in the spring. Because the ground is usually frozen beneath the snow when melt begins, most of the water runs off into the rivers, lakes, and potholes. This seasonal pattern creates a very distinctive annual flow sequence for most Canadian rivers. The flow is low in the later part of the winter but increases strongly in the spring; this causes flooding. By the beginning of summer, flows are diminishing, and by late summer flows approach their winter averages. This pattern of surplus in the spring and deficit in the late summer and winter is an obstacle to agencies such as hydroelectric facilities, urban water facilities—and beavers, which require more consistent supplies.

## OCEANIC FLOWS AND STORAGE

It is apparent if we examine the volumes of water entering the oceans from surface runoff (Table 10.1) that the oceans do not receive equal amounts of runoff. The Atlantic Ocean collects vast quantities of fresh water from the Amazon, the Mississippi, and the Congo. The Pacific is supplied by relatively few major rivers, mainly because of the mountain ranges near the coasts of the American continents, from the Andes right through to Alaska. Only from China does the Pacific get an appreciable contribution. Despite its size, the Arctic Ocean receives the greatest inflow, not from the icepack but from the northward flowing rivers, the Ob, Yenisei and Lena in Russia, and the Mackenzie in Canada. The vast proportion of this inflow occurs during the summer months, for in winter most of these rivers are frozen and flow is reduced.

### Table 10.1

*Continental runoff (mm yr$^{-1}$)*

| | |
|---|---|
| Europe | 300 |
| Asia | 286 |
| Africa | 139 |
| North and Central America | 265 |
| South America | 445 |
| Australia, New Zealand, and New Guinea | 218 |
| Antarctica and Greenland | 164 |

After M.I. Lvovitch, *Symposium on World Water Balance* (IASH–UNESCO, 1972).

If we consider also the input of water through precipitation over the oceans (Figure 10.3) and the losses by evaporation (Figure 10.8), it is clear that some oceans receive far more water than they return to the atmosphere. As the oceans are all interlinked and must keep the same mean level, it is clear that a balance is maintained by the lateral flow of water from areas of excess to areas of deficit. We cannot be precise about the magnitude of these flows, for measurements of ocean water movement are sparse, but as Table 10.2 indicates, large quantities of water must be involved.

### Table 10.2

*Oceanic flows of water*

| | |
|---|---|
| Atlantic Ocean | $+0.14 \times 10^6$ m$^3$ s$^{-1}$ |
| Indian Ocean | $+0.70 \times 10^6$ m$^3$ s$^{-1}$ |
| Pacific Ocean | $-0.68 \times 10^6$ m$^3$ s$^{-1}$ |
| Arctic Ocean | $-0.16 \times 10^6$ m$^3$ s$^{-1}$ |

*Note* Positive values indicate an inflow. Negative values indicate an outflow. For comparison, the average Gulf Stream total transport is approximately $70 \times 10^6$ m$^3$ s$^{-1}$.

Both the Atlantic and Indian oceans, for example, are 'dry' oceans, in that they lose more water by evaporation than they gain by precipitation and runoff. Conversely, the Pacific and Arctic oceans have a moisture surplus, precipitation and runoff exceeding evaporation. To maintain a balance, water therefore flows from the Arctic to the Atlantic and from the Pacific to the Indian Ocean. The net flow into the Atlantic is estimated to be about $0.14 \times 10^6$ m$^3$ s$^{-1}$ (about 100 times the flow of the Mississippi). Of course, these are not only net flows. In reality, water is constantly moving from one ocean to another. For example, there is a considerable surface flow across the Bering Strait and near Iceland into the Arctic Ocean, but this is more than compensated by deeper flows back into the Atlantic and Pacific. Similarly, a two-way flow occurs between the Pacific and Indian oceans. In the long term, however, these flows maintain equilibrium in the oceanic water budget.

This flow of water within the oceans cannot conceal that the world's ocean basins represent a vast trap for water in the hydrological cycle. The oceans contain about 1 350 400 000 km$^3$ of water, or about 97 percent of the total global supply. Not a lot is known about how long the water stays in these immense reservoirs, but it is clear that much of it remains there for thousands of years before being evaporated. It is also clear that the size of this store may vary markedly over geological time. During glacial periods, for example, release of water from the continents in the form of runoff was slowed down as more and more water became locked up in the expanding icecaps. The water budget became imbalanced, therefore, and the oceans lost more water than they gained. Sea levels throughout the world fell; at their extreme they were possibly 100 metres lower than they are today. Imagine the effects of this. The Bering Strait would have become an isthmus; Newfoundland would no longer have been an island. The British Isles would have become linked to the mainland of Europe as the North Sea dried up. Similarly, Borneo, Sumatra, and Java would have been joined by land to Malaysia. The whole oceanic circulation would have changed as a consequence. Indeed, the world's hydrological system must have been very different from what it is now.

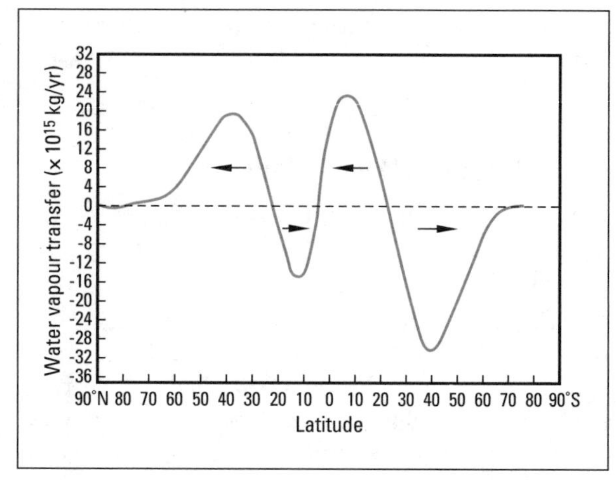


FIGURE 10.7 ▼ Latitudinal distribution of the average annual meridional transfer of water vapour in the atmosphere (after Sellers, 1965)

### ATMOSPHERIC MOVEMENT AND STORAGE

We started by talking of precipitation as the main input of moisture to the hydrological system, and this might imply that the atmosphere contains large quantities of moisture waiting to be released. This is far from the truth, however; it has been estimated that no more than 0.035 percent of the world's fresh water—a meaninglessly small proportion of the total water volume of the world—is contained in the atmosphere. Precipitation, and the continued operation of the hydrological system, is only possible if the atmospheric moisture is constantly being replenished by evaporation.

The inputs of moisture to the atmosphere are not everywhere in balance with outputs by precipitation. Over the oceans, for example, evaporation is high and precipitation relatively low; the atmosphere gains more moisture than it loses. Over the continents, even Australia, evaporation is less than precipitation, and the atmospheric moisture budget is negative. As with the oceans, therefore, horizontal flows of moisture (advection) must occur to maintain equilibrium.

We can get some idea of these flows (Figure 10.7) by comparing evaporation and precipitation levels throughout the world. Unlike other diagrams of heat and water budgets, this does not show maximum values over the equator. Nor does it show flows in the same direction throughout each hemisphere. What we find is that in the northern hemisphere there is a net transfer of moisture toward the equator in the tropical and subtropical areas. To the north of the subtropical high, flow is toward the pole. A similar pattern exists in the southern hemisphere.

The subtropics, therefore, constitute the main source of moisture for the atmosphere; as we saw earlier, it is in this zone that evaporation is at a maximum. The water from this area is carried both equatorward and poleward. In the midlatitudes, at about 35°N and S, the quantity of advected moisture is at its greatest. This moisture is then carried by the midlatitude cyclones into the temperate areas of the world, where it falls as cyclonic precipitation.

## Outputs from the surface hydrological system

### EVAPORATION

Evaporation is a much less dramatic process than precipitation. The only visible evidence we have of it is the gradual drying-up of the surface—rapidly in the case of roads, or the fog rising from warm water into cool air. Despite this, evaporation is one of the major outputs from the earth's surface (Figure 10.8).

The rate of evaporation depends upon a number of factors. Most important is the supply of energy, for evaporation involves the conversion of water to water vapour, and this requires considerable inputs of energy. Another factor is the availability of moisture at the surface. As the surface dries out and moisture becomes less available, rates of evaporation tend to decline. Thus, all else being equal, wet sur-

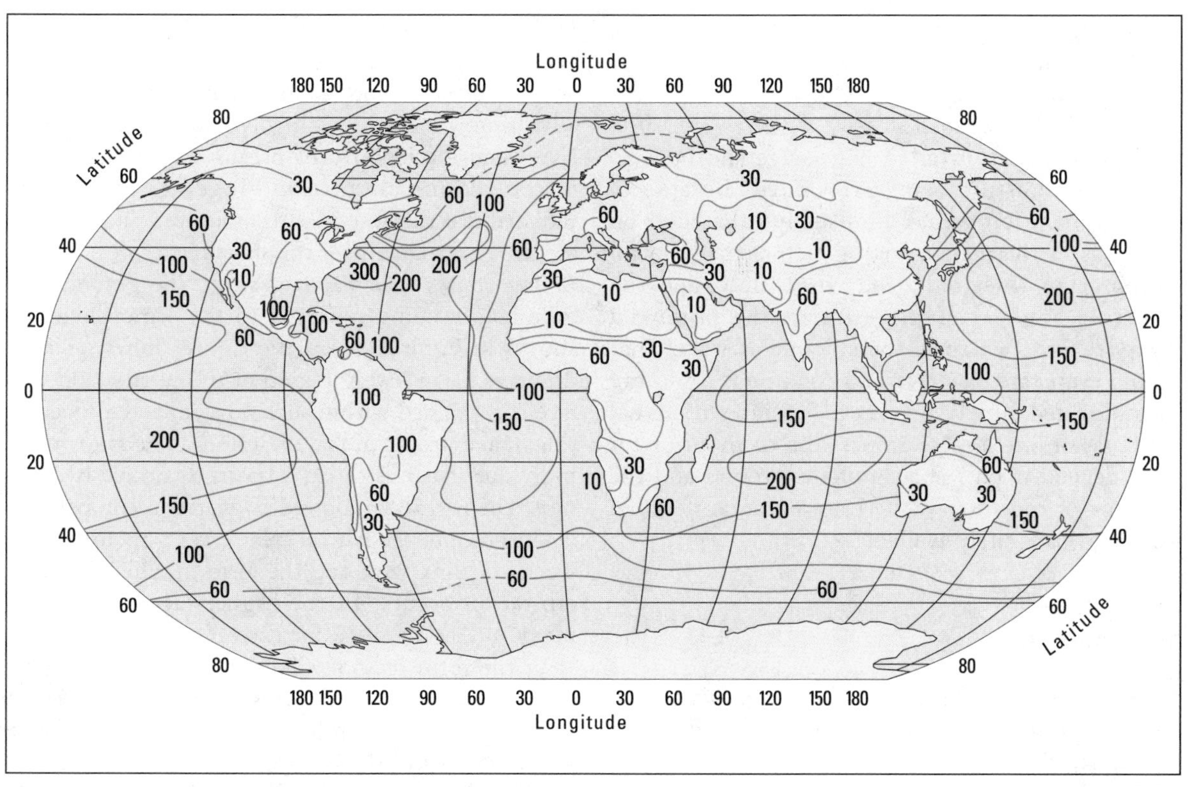


FIGURE 10.8 ▼ Mean annual evaporation (cm) (after Barry, 1969b)

faces, such as oceans or lakes, result in higher rates of evaporation than relatively dry surfaces. In addition, evaporation is favoured by a moisture gradient between the surface and the air above, and thus rates decline when the atmosphere is moist. Finally, wind plays an important part by removing the moist air and maintaining a moisture gradient.

The global pattern of evaporation reflects the interaction of all these factors. As we have seen, energy inputs are greatest in the tropical and equatorial areas, and we might expect that these would be the zones of greatest evaporation. But it is in these areas that the atmosphere is often most humid; clouds build up and intercept the incoming radiation, restricting the amount of evaporation. In many of the subtropical continental areas, energy inputs are also high and humidity is low, but so is rainfall, with the result that there is insufficient moisture to maintain high levels of evaporation. As a consequence, we find the greatest rates of evaporation in the subtropical ocean areas; here there is an open water surface, plenty of sunshine because of the low cloudiness, and the trade winds to remove the moisture from near the sea surface. Some of the moisture evaporated in these areas can be seen in the characteristic trade wind cumulus clouds, but much of the remainder is not visible. It stays in the form of vapour until it reaches the equatorial trough zone where it contributes to the cloudiness and precipitation that typifies the humid tropics.

We might expect that negligible amounts of evaporation occur from the final component of the hydrological system—the cryosphere—yet this is not entirely true. One of the processes by which the icecaps decay is direct evaporation of ice, a process known as **sublimation**. This takes place mainly at the margins of the iceaps, where high seasonal inputs of solar radiation supply sufficient energy to convert the ice crystals to water vapour. Today, this process accounts for no more than 2 percent of the moisture return to the atmosphere, but in the past it was considerably more important. During the latter part of the Ice Age, when the extensive icecaps that had invaded the temperate latitudes were stagnating and decaying over an area of thousands of square kilometres, sublimation must have been much greater.

The details of the precise distribution of evaporation across the world remain obscure. The data from which the precipitation maps were drawn may have been unreliable over the oceans, but in the case of evaporation they are no more than intelligent guesses almost everywhere. Unlike precipitation, evaporation is not something we can easily measure; we cannot use an inverted rain gauge and hope to catch the moisture being evaporated! Because of this, evaporation values are usually calculated on the basis of the meteorological factors that control evaporation—net radiation, wind flow, and the moisture gradient. These estimate the **potential evaporation**, which is the amount of water that would evaporate under various conditions assuming an unlimited supply of water. The rates of actual evaporation are almost impossible to monitor for they depend upon the state of the ground and the amount of moisture in it. Data on such detailed conditions are rarely available.

## TRANSPIRATION

Over vegetated land, much of the moisture is returned to the atmosphere not by evaporation from the ground surface but by **transpiration** from plants. In reality it is often difficult to separate the two processes; they are both similar in mechanism and they often act simultaneously. By transpiration we mean evaporation from the leaves of plants and, where the surface is fully covered by vegetation, such as in a forest or a well-grassed pasture, this is the main process by which water is lost. We can sometimes see the effects of the process. On cold spring or autumn mornings, temperatures may fall low enough for the air in immediate contact with the leaves to be close to saturation. As moisture is liberated from the plant by transpiration, it is not evaporated, for the air is too humid. Instead, it forms a small bead of water on the leaf, called a **guttation drop**. Similarly, on humid days, we may see clouds of water vapour rising from trees, as moisture released by transpiration condenses in the moist air.

Nevertheless, it is often more convenient to consider evaporation and transpiration as a single process. We then combine them to give the composite term **evapotranspiration**.

## INTERNAL TRANSFERS

Between the input of water in the form of precipitation and its output in the form of evaporation, transpiration and runoff, a great deal may happen. Extensive movements of water take place, in the atmosphere, on land, and in the oceans. In the atmosphere these horizontal transfers are known as advection. On land they involve the flow of water in rivers and glaciers. In the oceans they occur as part of the oceanic circulation. Together they act to redistribute the water, so that inputs and outputs are kept in balance over the short term.

The routes that water reaching the ground may take before being returned to the atmosphere are shown in Figure 10.5. This figure shows a single drainage basin. Every region of the earth's land surface is contained within such a basin, and altogether they make up the global hydrology. The study of an individual basin is called a **basin drainage hydrology**. The pathway followed by water in any particular instance depends to a great extent upon the area we are dealing with and the form in which the precipitation occurs. In the higher latitudes and in many mountain areas, for example, the main input is in the form of snow. This may accumulate over time, slowly being compressed into ice. As ice, it may slowly move under gravitational forces as a glacier or part of the icecap. Eventually the water is released by melting, to form streams and rivers which then flow to the sea. In addition, many large glaciers in the Arctic and Antarctic terminate in the sea and release their water directly into the oceans as huge icebergs (Figure 10.9).

In temperate areas, the route taken by the water is more complex, but the processes of transfer are much more rapid. Most precipitation occurs as rainfall, and this either collects on the ground surface or soaks into the soil. The water that remains on the surface tends to flow over the land as runoff. This rapidly reaches the streams and is then carried to the sea. On the whole, channel flow is rapid. It takes water about three days to travel from the source to the mouth of the Thames in England, a distance of 350 km. It probably takes about eight weeks for water to pass the 6300 km length of the Mississippi River. Thus streams represent one of the most dynamic, important—and also sensitive—routes for these transfers of water from land to the oceans.

The water that soaks into the ground travels more slowly. Some of it flows gradually through the soil and ultimately emerges as seepage water in springs or in stream banks. Some of it drains downward into the bedrock and there enters the almost immobile groundwaters, where it may be stored for many years before emerging again at the surface.

FIGURE 10.9 ▼ Iceberg, Disko Bay, western Greenland. Waterlines are clearly visible that show how it has tilted to different positions in its lifetime (photo: John Dunn)

*Runoff* The most obvious process of water movement in the hydrological system is surface runoff: the flow of water over the ground in distinct channels. The process is easy to follow during a period of heavy and prolonged rainfall. Initially, the soil may be dry and most of the rain may soak into the soil, but as the surface becomes wetter, or where it is impermeable, signs of running water start to appear. Small depressions fill with water, spilling over as the water flows downslope. Tiny channels called **rills** are cut in the soil, and these are gradually enlarged as they merge to form more permanent gullies. If we follow the gullies, we find them entering stream channels, and if we follow the channels we see them joined by other streams, forming rivers that flow to the sea.

Not all rivers reach the sea, however. On the way, water is evaporated, and in drier parts of the world streams may dry up before they get to the sea. Often they gather in the wetter mountain areas and spread out on the flat arid plains at the mountain edge, where much of the water soaks into the ground. In other cases, the waters flow into large inland depressions where they are trapped, producing permanent inland lakes such as Lake Eyre in Australia, Lake Chad in Africa, or the Caspian Sea in Asia. A few are diverted by human intervention before they reach the sea. The Colorado River is a dramatic and disturbing example.

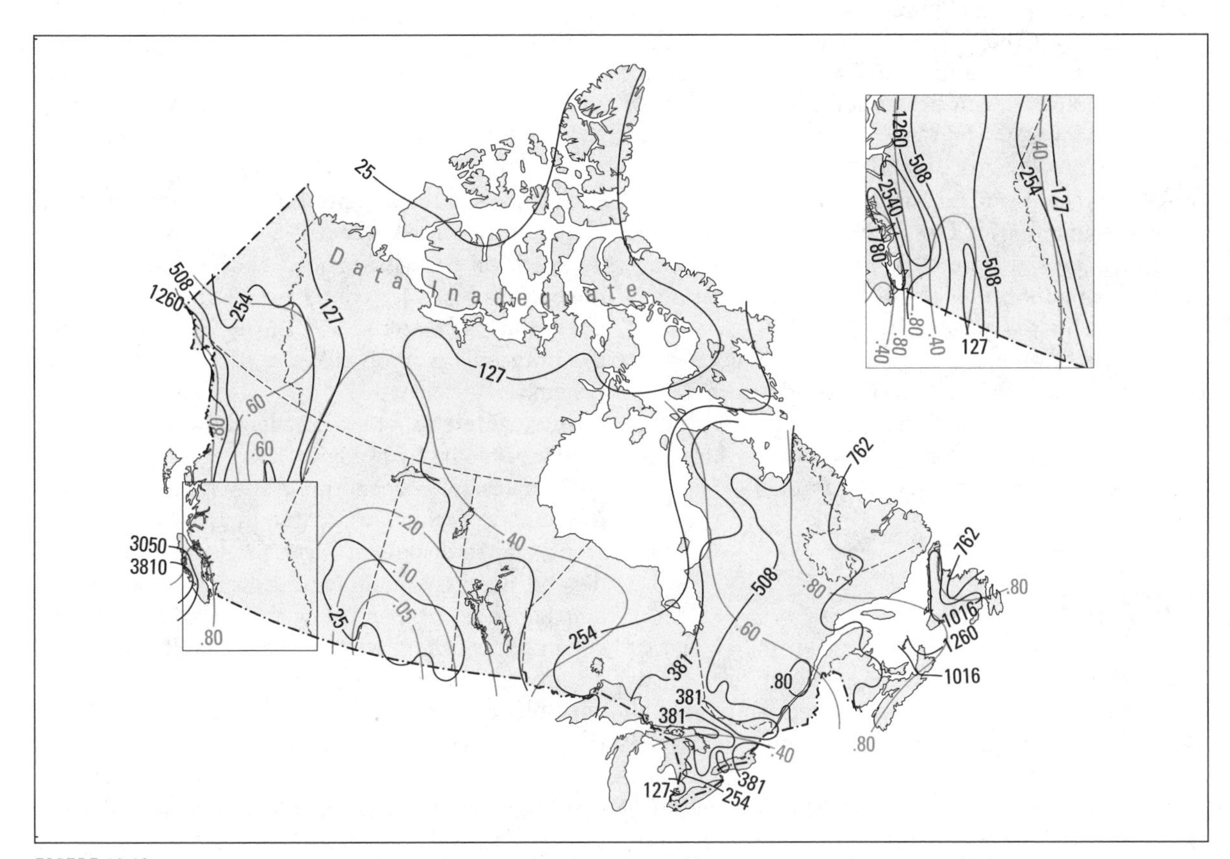


FIGURE 10.10 ▼ Mean annual runoff in millimetres (black) and runoff ratio (colour) (after Hare and Thomas, 1979)

*The global distribution of runoff* Seen on a global scale, the distribution of runoff is very unequal (see Table 10.1). Some of the continents receive large amounts of rainfall and experience plenty of runoff; others are relatively dry and surface runoff is rare. The areas having most runoff are those with high rates of precipitation and low rates of evaporation. Thus, South America has the largest value, mainly because much of the continent lies in the humid tropics. The temperate latitude continents of Europe, Asia, and North America all have similar amounts of runoff. Africa has a low ratio of runoff to evaporation because of the large area of desert across the Sahara and the smaller proportion of the surface with tropical rain forest. The driest continent is Australia, where large permanent rivers are scarce. Antarctica is rather an unknown quantity. Evaporation must occur, but is very small. The snow that falls and accumulates on the continent is eventually returned to the oceans through the slow process of glacial flow.

## Table 10.3

*Mean annual stream discharge to the oceans for selected Canadian rivers*

| *River* | *Area ($km^2$)* | *Discharge ($m^3 s^{-1}$)* |
|---|---|---|
| Saguenay | 90 100 | 1 820 |
| St. Lawrence | 1 026 000 | 9 860 |
| Churchill | 281 300 | 1 200 |
| Nelson | 722 600 | 2 370 |
| Albany | 133 900 | 1 400 |
| Koksoak | 133 400 | 2 550 |
| Yukon (at Alaska border) | 297 300 | 2 320 |
| Fraser | 219 600 | 3 540 |
| Columbia (at Washington border) | 154 600 | 2 800 |
| Mackenzie | 984 195 | 10 800 |

After Hare and Thomas (1979).

*The Canadian distribution of runoff* Figure 10.10 shows the mean annual runoff for Canada as determined by the Canadian Committee for the International Hydrological Decade. The mean annual runoff is shown by the solid lines. Note the very high values on the west coast and the very low values on the Prairies and in Arctic Canada. The broken lines and vertical bars for individual stations show the runoff ratio. This is calculated by dividing the runoff by the amount of precipitation. It indicates the amount of moisture that goes to other outputs in the hydrological cycle.

The size of drainage basin and mean annual discharge for selected Canadian rivers is given in Table 10.3. Compare the size of the drainage area with the amount of discharge. Notice how the size alone is not a determinant of the amount of runoff.

# Conclusion

The movement of water through the hydrological cycle is one of the vital processes operating within the global system. A major component of this cycle is the oceans, which cover some 70 percent of the earth's surface and contain 97 percent of the world's water. Because the oceans store so much of the total water volume, they act as a major regulator upon the flow of water through the system. They particularly influence evaporation and the return of water to the atmosphere, and thereby control to a great extent flows of moisture in the atmosphere and rates of precipitation. Another major control upon the hydrological cycle is provided by the cryosphere. As the events of past Ice Ages have testified, changes in the volume of water stored in the icecaps have far-reaching effects upon the rest of the hydrological system.

Increasingly in recent decades, however, a further regulator of the hydrological system is emerging—we humans. We depend upon water for life; we use it not only for drinking, but in industry, to feed crops, and to dispose of waste products. Because we depend upon the hydrological cycle, we also try to manage and control it. How successfully we do this depends upon how well we understand the system we are dealing with, and especially the interactions with other natural systems such as the atmosphere, the lithosphere, and the biosphere. For our own sake, therefore, it is vital that we begin to comprehend the delicate processes and interrelationships of the hydrological system.

# Precipitation

## The nature of precipitation

For several months during the summer of 1988 much of western North America, including the Canadian Prairies, experienced prolonged drought. The effects were widespread. Crops and vegetation suffered, many trees died, extensive forest fires devastated vast tracts in northern Manitoba, streams no longer flowed. Lakes, like Old Wives Lake in southern Saskatchewan and Whitewater Lake in southern Manitoba, dried up for the first time in memory including the Dust Bowl times of the 1930s. Potholes and other small bodies of water disappeared, causing a dramatic drop in the waterfowl population. Reservoirs fell and water use was restricted in many areas (Figure 11.1). In places, drying of the soil led to shrinkage of the ground, and buildings and bridges were weakened and cracked.

For people who live in the humid regions of the world, precipitation is normally so common that it is taken for granted. During times of drought, the importance of precipitation and its role in feeding the hydrological system becomes all too apparent. For precipitation represents the input of water to the global hydrological system. It is the nature of this input—the character and distribution of precipitation—that we will consider in this chapter.

FIGURE 11.1 ▼ Effects of drought at the Indian Head reservoir, part of Diefenbaker Lake, north of Regina, Saskatchewan (photo courtesy of Prairie Farm Rehabilitation Administration, Agriculture Canada)

## TYPES OF PRECIPITATION

To most people, three types of precipitation come immediately to mind: rain, snow, and hail. As we would see if we intercepted these and looked at the raindrops, snowflakes, or hailstones more closely, the distinction between them is not always clear, and the terms mean different things in different areas. Moreover, they are not the only forms in which moisture inputs occur. Dew, fog-drip, and rime all transfer water from the atmosphere to the ground (Table 11.1). Their contribution, however, is often small.

*Rain* We have already discussed the main processes of rainfall generation in Chapter 4; we are concerned here with the nature of the rainfall after it has fallen from the cloud.

Typically, rainfall consists of water drops that vary in size. Where the rain is produced by thin, low-level stratus clouds, droplets tend to be small, with a majority 0.2 to 0.5 mm in diameter. Where the clouds are thicker, strong updrafts hold droplets in the atmosphere longer, so the number of collisions increases and the rain is composed of larger droplets, often several millimetres in diameter. Droplet diameter is also affected by events during its fall through the atmosphere. Further collision and coalescence cause them to grow, while evaporation makes them smaller, and may cause the total loss of very tiny droplets. In general, however, the droplets reaching the ground show a logarithmic

**Table 11.1**

*Types of precipitation*

| *Type* | *Characteristics* | *Typical amounts* |
|---|---|---|
| Dew | Deposited on surfaces, especially vegetation; hoar frost when frozen | 0.1 to 1.0 mm per night |
| Fog-drip | Deposited on vegetation and other obstacles from fog; rime when frozen | Up to 4 mm per night |
| Drizzle | Droplets <0.5 mm in diameter | 0.1–0.5 mm $hr^{-1}$ |
| Rain | Drops >0.5 mm in diameter, usually 1–2 mm | Light < 2 mm $hr^{-1}$<br>Heavy >7 mm $hr^{-1}$ |
| Hail | Roughly spherical lumps of ice 5– > 50 mm in diameter, often showing a layered structure of opaque and clear ice in cross section | Highly variable |
| Snowflakes | Clusters of ice crystals up to several cm across | |
| Granular snow | Very small, flat opaque grains of ice; solid equivalent of drizzle | |
| Snow pellets (graupel or soft hail) | Opaque pellets of ice 2–5 mm in diameter falling in showers | |
| Ice pellets | Clear ice encasing a snowflake or snow pellet | |
| Sleet (UK) | Mixture of partly melted snow and rain | |
| (USA) | Frozen rain or drizzle drops | |

size distribution, with a large number of small droplets and a much smaller number of large drops. The details of size distribution vary according to the rainfall intensity, as Figure 11.2 indicates.

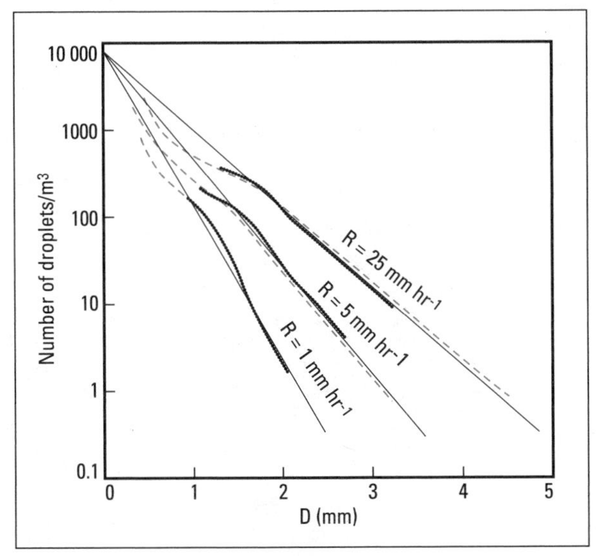


FIGURE 11.2 ▼ Measured drop-size distributions (broken lines) compared with best-fit exponential curves (straight lines) (after Marshall and Palmer, 1948)

The size of the droplets has considerable significance, for, together with the strength of updrafts in the air, it controls the fall velocity of the rain. In still air, the fall speed of a droplet 0.2 mm in diameter is about 70 cm $s^{-1}$; for a drop of 2 mm diameter it is about 650 cm $s^{-1}$. The momentum of the droplet when it reaches the ground is known as the **terminal velocity**, and, with the mass of the drop, determines its kinetic energy.

The total kinetic energy of a storm depends upon the number of raindrops reaching the ground. This is a measure of the rainfall intensity and in Figure 11.3 the relationship between the intensity and total kinetic energy of rainfall is shown. Rainfall intensity varies considerably both within an individual storm and between storms. Rainfall from thick cumulus-type clouds is particularly variable due to spatial differences in cloud thickness and updraft strength, but intensities may be as high as 200 mm $hr^{-1}$ or more. Precipitation from stratiform clouds is less variable and intensities are often low—less than 10 mm $hr^{-1}$. The average annual rain and snow for Canada is shown in Figure 11.4. Compare the different pattern of the isolines with those for snowfall alone in Figure 11.5.

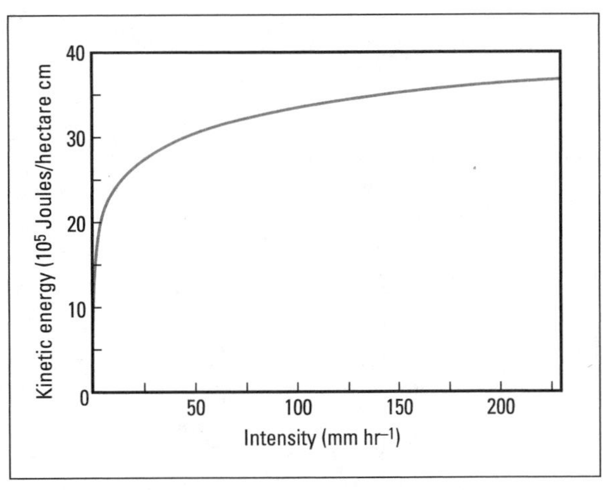


FIGURE 11.3 ▼ Rainfall intensity measured against the kinetic energy released by rainstorms (after Bennett and Chorley, 1971)

*Snow* In most areas of the world, rainfall is by far the most important input to the surface hydrological system. Snow occurs mainly in winter or at high altitude year round, even in the tropics, and despite its thickness and persistence, the quantities of moisture involved are relatively small (Table 11.2). In general, 12 cm of freshly fallen snow produces only

## Table 11.2

*Water equivalents of some Canadian snowfall records*

| *Location* | *Date* | *Amount (cm)* | *Moisture equivalent (mm)* |
|---|---|---|---|
| *24-hour record* | | | |
| Lakelse Lake, B.C., 17 Jan. 1974 | | 118.1 | 99 |
| *Average January* | | | |
| Halifax, N.S. | 1951–80 | 63.1 | 52 |
| Montréal, Qué. | 1951–80 | 52.7 | 44 |
| Toronto, Ont. | 1951–80 | 33.4 | 28 |
| Winnipeg, Man. | 1951–80 | 23.7 | 20 |
| Regina, Sask. | 1951–80 | 20 | 16 |
| Edmonton, Alta. | 1951–80 | 27.2 | 23 |
| Vancouver, B.C. | 1951–80 | 25.7 | 21 |
| Whitehorse, Yukon | 1951–80 | 21.3 | 18 |

Source Environment Canada

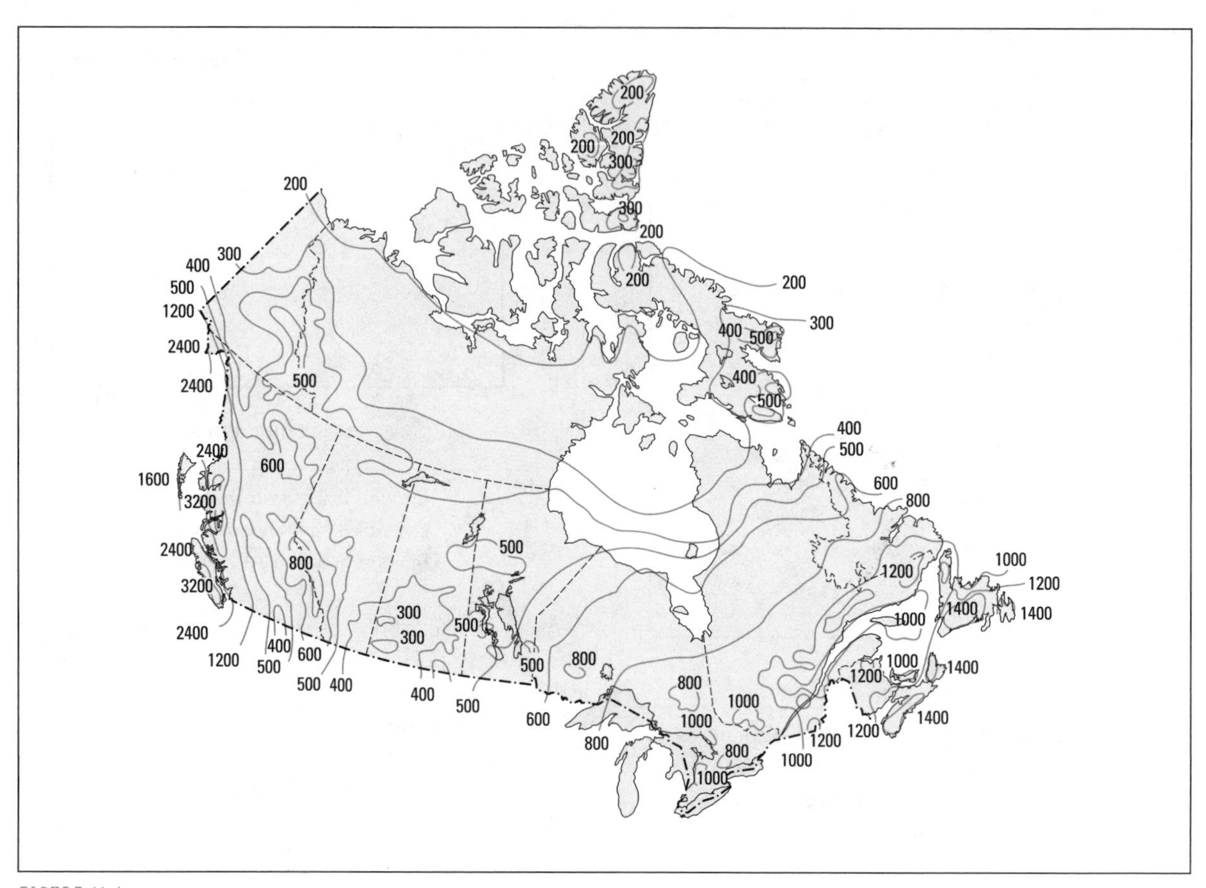


FIGURE 11.4 ▼ Average annual rain and snow for Canada (mm) (after Phillips, 1990)

about 1 cm of water. Where snow is formed in very cold, dry air the moisture equivalent is even smaller, and it may take as much as a metre of snow to produce 1 cm of water. In high mountain and polar regions where temperatures are low throughout the year, the majority of precipitation falls as snow. Even so, because of the low temperatures preventing the atmosphere holding much moisture, many of these areas are, in fact, quite arid. There are no substantial data to provide accurate estimates, but it seems likely that on a world basis no more than 1 percent of the total annual precipitation occurs as snow.

Despite this, snow is an important source of moisture in many middle- to high-latitude regions. Dr. Wayne Rouse, in a study of microclimates in the Churchill region of Manitoba, showed how snowmelt was essential to the growth performance of trees in the region. Lakes, rivers, and groundwater are all essentially dependent upon snowmelt to replenish them after the high evaporation of the summer months. Although snow melts down to yield relatively small amounts of water, it is effective because it usually accumulates and melts in a relatively short period of time.

It can be shown that the drought of 1988 was aggravated by the lack of snowfall in the preceding winter. Little snow meant that surface water supplies were significantly reduced. These are the source of much of the precipitation that falls during the summer. As moisture is evaporated from the surface, it forms convective clouds which produce showers and thundershowers, the major form of rainfall in the summer months. Therefore, unless moisture is transported in from the oceans, the drought is self-perpetuating. Snowcover is also important to protect vegetation and prevent frost penetrating deeply into the ground. Animals and birds, such as lemmings and ptarmigan, are dependent upon adequate snowfall for winter habitat.

Snowfall usually starts in the atmosphere as tiny ice crystals produced at temperatures well below freezing. As the crystals fall they tend to aggregate, particularly where there is sufficient moisture in the air to bind the crystals together. This mainly occurs where temperatures are close to the freezing point,

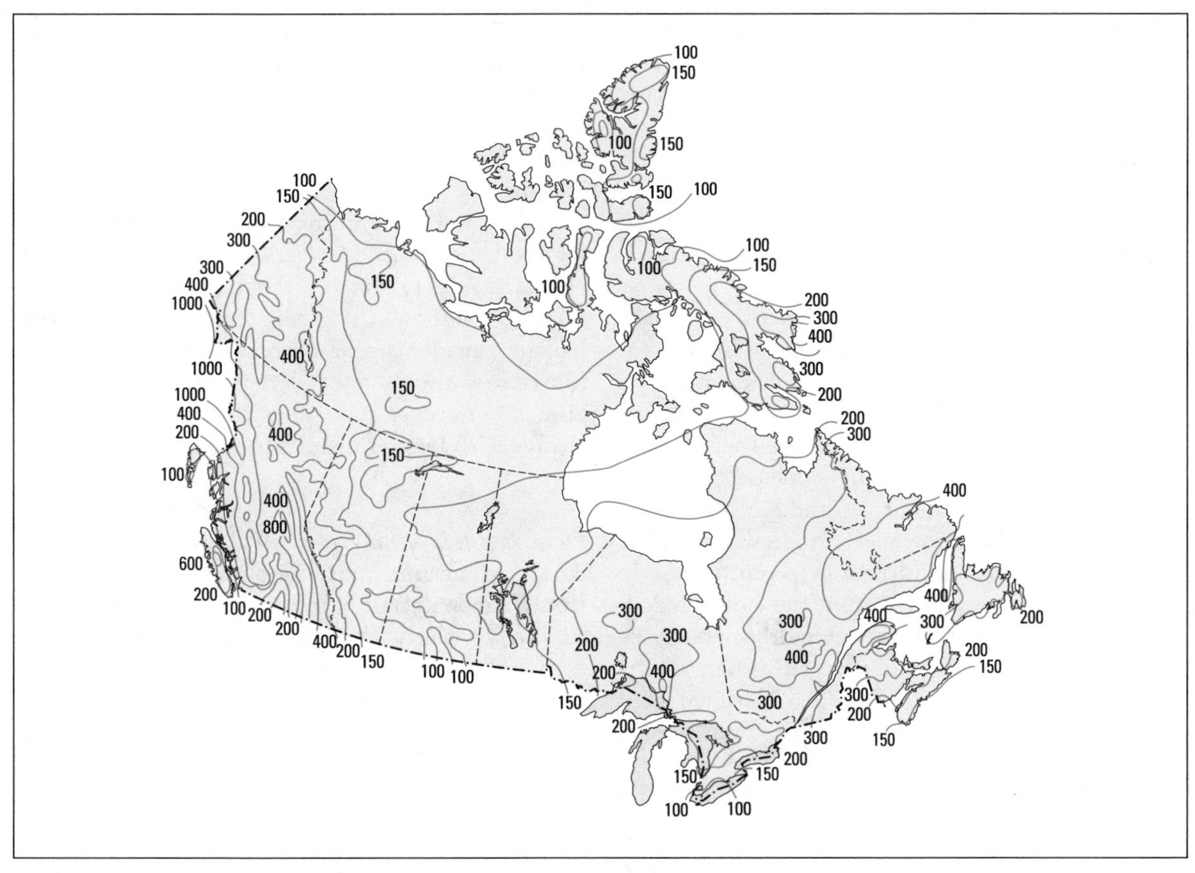


FIGURE 11.5 ▼ Average annual snowfall for Canada (cm) (after Phillips, 1990)

and in these conditions large snowflakes may be formed. At lower temperatures, however, moisture is lacking and the crystals do not aggregate.

As with rain, the fall velocity of the snowflakes depends on size and, all else being equal, large flakes fall more rapidly than small ones, with maximum speeds of about 100 cm $s^{-1}$. As we all know, however, snowflakes vary considerably in shape and this too may influence fall speeds. Moreover, the density of snowflakes is very low; large flakes often have a density of as little as 100 kg $m^{-3}$ (compared with approximately 1000 kg $m^{-3}$ for raindrops). Consequently, for their size, snowflakes are light and they are readily blown by the wind. For this reason the distribution of snowfall during a storm is greatly influenced by surface wind conditions, and even after reaching the ground the snow may be redistributed to form deep drifts and snow-free areas.

Another important feature of moisture inputs in the form of snow is that it is often many weeks or, in polar areas, even years before the water is actually released. Thus, in mountain areas, winter snowfall may survive into the spring, and the snow represents a temporary store of water which is released only by melting. In Arctic and Antarctic regions, the snow accumulates for centuries, then moves with imperceptible slowness in the icecaps and glaciers before melting perhaps thousands of years later. Unlike rainfall, therefore, snow is not always a direct input to the hydrological system.

*Hail* The word hail can strike fear into the heart of farmers in many parts of the world. Damage to crops can be severe, though normally the devastation is localized. Hailstorms usually produce a swath of stones as the parent cloud moves across the country. Because of the limited areal extent of the storms and their relative infrequence, their contribution to water inputs is generally small.

Hailstone sizes vary considerably, but are usually less than 1 cm in diameter. Stones of this size can cause damage, but it is the larger stones, possessing considerable kinetic energy, that produce spectacular effects, such as damage to cars, greenhouses, and vegetation.

Data on the frequency of hailstorms are not entirely reliable. Standard statistics probably underestimate the true frequency of hail because many storms pass between observing stations. For example, in South Africa, where hailstorms are prevalent, the standard network of recording stations gave an average of five storms per year. When the network was increased to one observer per 10 $km^2$, eighty days with hail were recorded.

In North America, hail causes extensive damage to crops and in some years this results in great financial loss. Some attempts have been made to alleviate the problem by seeding potential thunderstorm clouds with silver iodide before they develop hail. Most clouds lack sufficient condensation nuclei to convert all water vapour to liquid. Silver iodide particles act as nuclei to convert more water vapour into water droplets and ultimately precipitation. In this way, energy is removed from the cloud, reducing convective forces that carry raindrops above the freezing level to create hail. Generally these efforts have not been very successful. The National Hail Research Program in the United States was unable to replicate reported Soviet successes after three years and it was abandoned in 1975. The research program at Penhold, Alberta, lasted longer but did not have much more success and was essentially abandoned in 1986. Hail damage continues to be a serious problem on the Great Plains of North America, and some farmers rely on crop insurance for protection. Most farmers know the probability of hail in their region or can obtain the information. Figure 11.6 shows the average number of days with hail in Canada, but as noted above the data are a function of the number of reporting stations. It is also important to remember that the pattern will change with longterm shifts in the climate.

*Dew, fog-drip, and rime* Walking through fields on a cold autumn morning after a clear night, we would almost certainly be conscious that the ground surface was wet with dew. Similarly, in a dense forest with lots of mist or cloud we might see water dripping from the leaves as fog-drip. In these two cases, we are dealing with some of the smallest con-

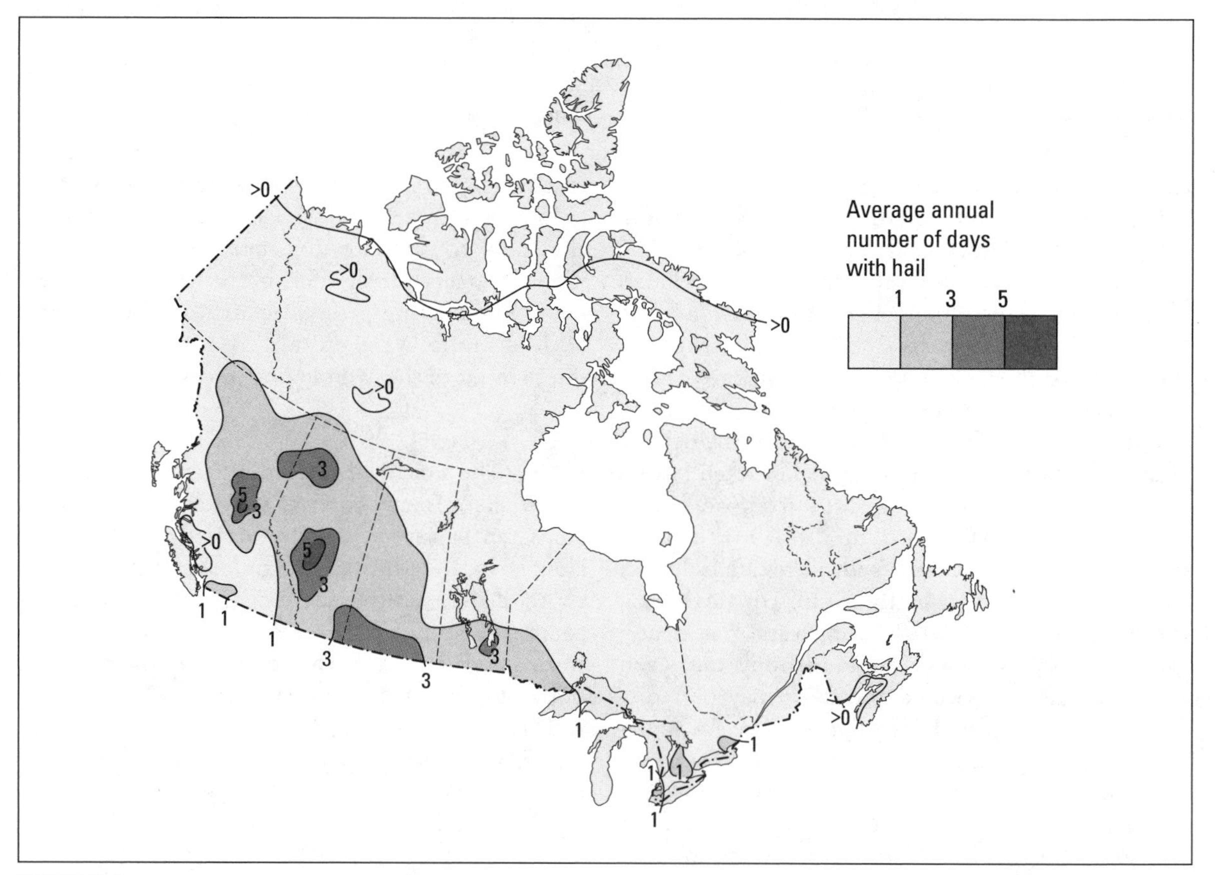


FIGURE 11.6 ▼ Average number of days with hail in Canada (Environment Canada)

tributors to the precipitation input, although they may have some importance. Dew forms on cold surfaces at night when the air is close to saturation. Under these conditions, of course, the air can hold little moisture and, as the atmosphere loses moisture to the ground in the form of dew, it dries out further. Consequently, the total amount of dewfall that can occur in a single night is normally limited, rarely more than 0.6 mm. As evaporation rates are high once the sun rises, such small quantities of moisture are soon returned to the atmosphere, so the contribution of dew to the local water budget is likely to be small.

Where cloud droplets are blowing continuously across a rough surface, such as a forest, we get fog-drip. The process results from the deposition of small water droplets moving horizontally by contact with the vegetative surface. Eventually the droplets combine to form larger drops which fall to the ground. The effect is accentuated if the trees increase the turbulent motion of the air as it moves over the canopy. The vertical motion brings cloud droplets downward and onto the leaves.

Fog-drip is most important in areas where forested mountain ridges extend into persistent cloud sheets, for example, in California, Hawaii, the Canary Islands, Japan, and Germany. Studies in all these areas indicate that there is a significant increase in moisture input at the ground, which would not otherwise occur. At Berkeley, California, as much as 200–300 kg of water per square meter of surface is found during the summer when little rain falls. On Hawaii, trees planted at 800 m altitude catch trade wind cloud droplets at a rate of about 4 mm per day; over the year, this represents an input of about 750 mm to the island's water budget. Without this additional input it is unlikely that the forests would be able to survive.

If temperatures at the ground are below freezing point, then the drifting cloud droplets freeze on the vegetation to form rime (Figure 11.7). The weight of rime can be damaging to trees; the contribution to water input and to tree damage has been investigated in Germany, where coniferous tree growth can be severely hampered by damage to the growth points.

FIGURE 11.7 ▼ Rime ice formed on these trees when droplets in early morning fog touched the frozen branches (photo: Robert S. Schemenauer)

# Measurement of precipitation

## RAIN

As every farmer knows, it is easy to measure rainfall. Any watertight container sited well away from buildings and trees will act as a rain gauge. How much it collects will depend not only upon the amount of rain, but also on the gauge diameter and its height above the ground (Table 11.3). Because of this, rain gauges in the UK have a standard diameter of 12.7 cm and are set a fixed distance of 30 cm above the ground. Unfortunately, the standard varies from country to country, so that in

Table 11.3

*Variation of rainfall catch with height*

| Height of gauge mouth above ground (cm) | 5 | 10 | 15 | 20 | 30 | 46 | 76 | 152 | 610 |
|---|---|---|---|---|---|---|---|---|---|
| Catch as percentage of that at 30 cm | 105 | 103 | 102 | 101 | 100 | 99.2 | 97.7 | 95.0 | 90.0 |

After J.P. Bruce and R.H. Clarke, *Introduction to Hydrometeorology* (1966), Pergamon Press.

Canada the diameter is 9 cm at a height of 30 cm above the ground, while in the USA the gauges are 20 cm wide and 78 cm high (Figure 11.8). Comparisons of rainfall totals between countries are therefore more difficult than might be expected.

The main reason for these differences in gauge height is snow. Snowfall is difficult to measure because of its lightness and tendency to drift. In the USA the same gauge is used to measure both snowfall and rain, so it has to be well above the level of drifting snow. In Canada, separate gauges are used, while in Britain snowfall is a relatively small component of the annual precipitation.

In normal operation, the amount of rainfall collected in a gauge is measured once a day. A standard rain gauge will only record the total rain that has fallen between readings; in many cases it is important to know when the rain fell and at what intensity. For this purpose, recording rain gauges are used.

## SNOW

In some countries, the water equivalent of snowfall is found by melting the snow that has accumulated in the gauge. Another method of determining water equivalence required pushing a hollow pole vertically into the snow. The pole has a balancing point that slides so that the pole can be made to balance like a scale. A reading of the water equivalency can then be made directly from markings on the pole opposite the balance point. Both methods provide an estimate of the amount of water in the snowfall, which is useful for flood forecasting and other purposes. They do not provide a measure of how much snow has fallen. Often this is done by collecting snowfall in a cylinder like rainfall. Indeed, in the United States the same gauges are used for snow and rain. In Canada separate gauges are used. This is not usually very accurate because even minor wind eddies around the cylinder can reduce the amount of

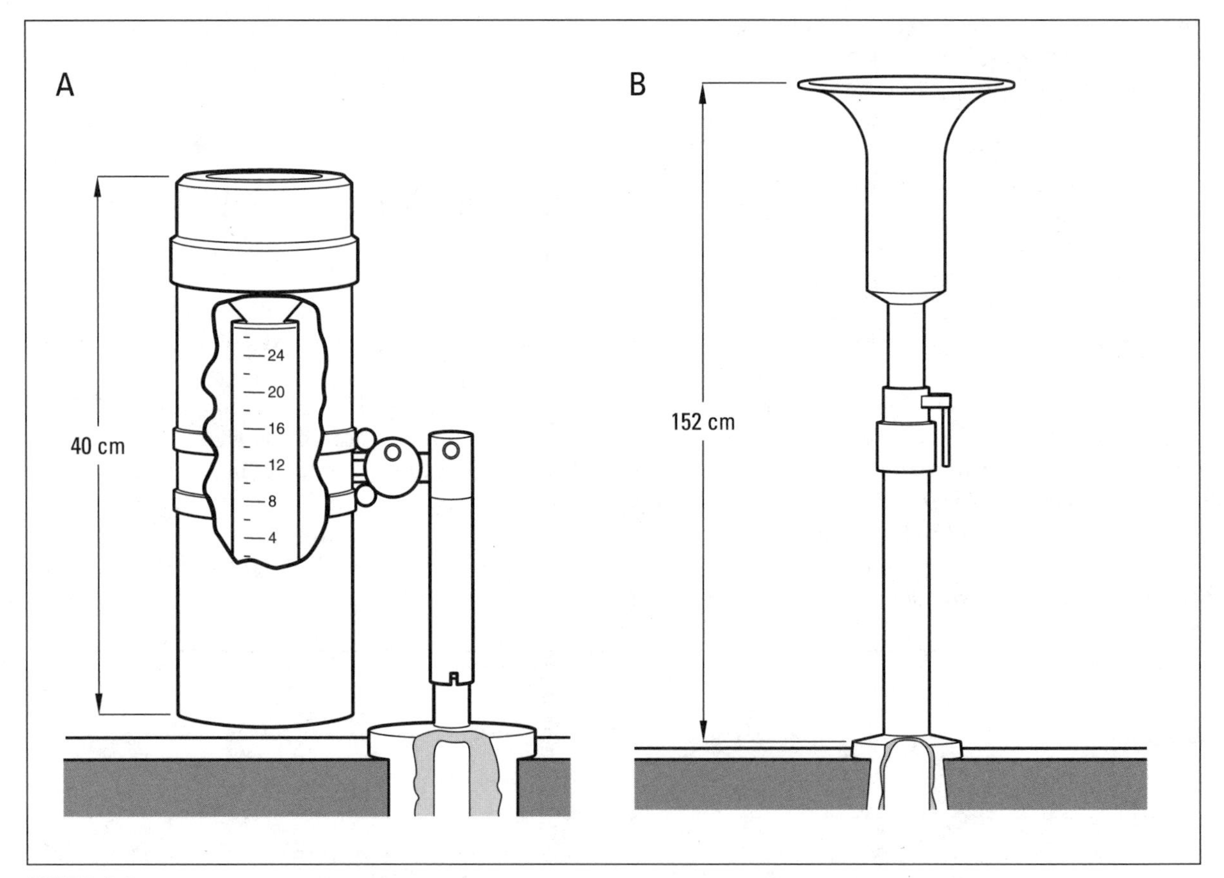


FIGURE 11.8 ▼ Precipitation gauges used in Canada. (A) The type B rain gauge is used to measure daily and 6-hourly rainfall amounts at 2500 stations; (B) the Nipher snow gauge is used to determine the water equivalent of snowfall at 350 observing stations (Source Atmospheric Environment Service, Environment Canada)

snow caught, thus under-measuring the actual snowfall. Various deflectors have been designed and work reasonably well, but most investigators acknowledge the limitations of accurately determining snowfall.

A low gauge can be completely covered in a heavy snowfall. This problem is resolved in the USA where a tall gauge is used, however, the gauge tends to underestimate the amount of snow reaching the ground. Many countries also use a graduated pole pushed into the snow in two or more representative areas to determine the depth of fall. The subjectivity of such a method is obvious. Recently, experiments have also been made to measure snow depth photogrammetrically, either with aerial or satellite photography. Where the snowfall is substantial, the depths can be measured fairly accurately, but without ground observations the water equivalent of the snow is unknown.

By any method, measurements of snowfall and water equivalent of snowfall always present problems and probable inaccuracies. Apart from a few areas of intensive observations, we do not know the precise input of water to the surface by snow. Figure 11.5 shows the average annual snowfall for Canada. The map is compiled from individual station records, which we have conceded are not truly representative. The number of stations is limited considering the size of the country, and as a result the isolines are very crude approximations of the actual average snowfall.

### HAIL

Hail measurement is even more difficult. Hailstones possess considerable kinetic energy and many will bounce out of a conventional gauge, causing underestimation of the total fall. The size distribution of hailstones can be obtained from a hail pad that measures the degree of impact by the stones. If pads are left out for known times, the amount of ice and the water equivalent can be found. As we have seen, hail is insignificant as a precipitation input to the hydrological cycle, so it is normally recorded separately in terms of the number of days with hail.

### FOG-DRIP AND DEWFALL

The water content of fog-drip and dew is small, so special measurement techniques have to be used. Fog-drip falls to the surface after contact with leaves or trees, so trough-shaped rain gauges have been designed to increase the sampling area and make measurements more accurate. In principle they work like an ordinary gauge.

The most commonly used instrument for dewfall is an accurate weighing device. The dewdrops collect on hygroscopic plates that are attached to a balancing system to weigh the collected water. All methods suffer from the basic uncertainty of how accurately the gauges collect dew compared to natural surfaces. However, these water quantities are minute, so that even large errors are insignificant in relation to the total precipitation input.

## Temporal variations of precipitation

### VARIABILITY IN THE SHORT TERM

The variations of rainfall over time are of vital importance to hydrologists. Decisions about bridge size, storm sewer construction, culvert dimensions, and even flood protection measures must be made on the basis of the expected inputs of precipitation. For this type of decision, a single total is not very informative. We need to know not only how much rain is likely to fall, but over what period of time. In a single day, 25 mm of rainfall may not be significant, but if that amount fell in an hour, or even less, there could be drastic consequences. Surface runoff may occur, soil erosion might be initiated, streams might start to swell, and flooding might result. Clearly, precipitation intensity is extremely important.

If we monitored a storm, we would normally find that precipitation intensity—that is, the amount of rainfall per unit time—varied considerably. Heavy bursts of rain are normally seen to alternate with relatively quiet periods. All types of rainfall show these variations; there is rarely such a thing as steady rain. In convectional storms, variations are often associated with the passage of the main convection zones across the land. Where the updrafts are strong, raindrops are held in the cloud and prevented from falling, but as the updrafts weaken the drops fall more easily to the ground, giving periods of higher intensity (Figure 11.9). Cyclonic rain, too, shows considerable variability, often associated with temporary zones of instability in the cyclone (Figure 11.10).

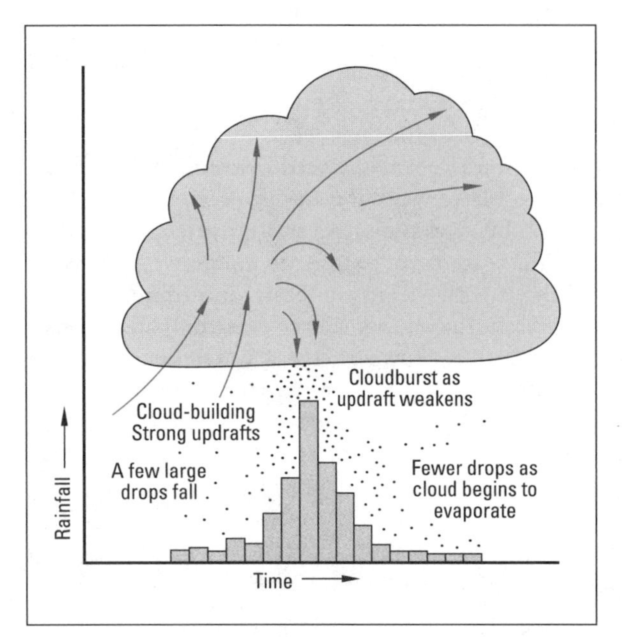


FIGURE 11.9 ▼ Surface rainfall rates relative to cloud development. In the early stages of cloud buildup, strong updrafts prevent most drops reaching the ground. Later, when the cloud drifts downwind, or as the updraft weakens, a period of much heavier rain may ensue

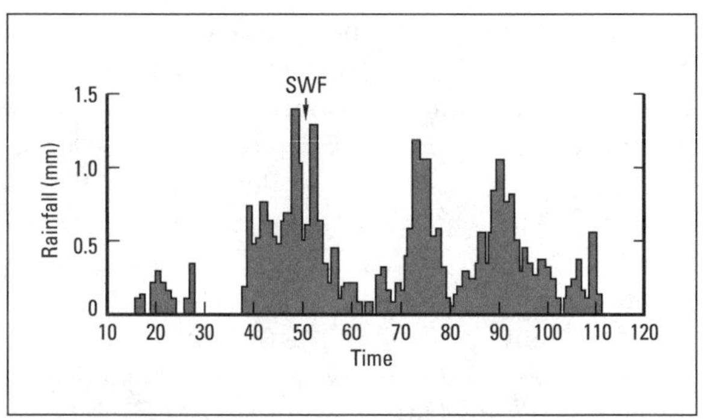


FIGURE 11.10 ▼ Variations in rates of precipitation ahead of, during, and after the passage of a surface warm front (SWF), Formby, England. Numbers on the horizontal axis represent intervals of time after an arbitrary starting point. Each interval is about 15 minutes (after Atkinson and Smithson, 1972)

Clouds form and cover large areas of the earth; however, the amount of rain that falls is surprisingly low on average. Often the cloud-forming mechanisms exist but the precipitation-forming mechanisms do not. Similarly, in some regions precipitation-forming conditions are always available. A principle mechanism required for cloud and precipitation formation is uplift of moist air. One of the most common situations in which this occurs is where moist air is blowing over a mountain barrier. The air is forced to rise and cool, producing relatively high volumes of precipitation. In the lee of the mountain the air is now drier because moisture has been removed. It holds this moisture more effectively, especially since the air is now heated by compression as it descends the mountain. This creates the distinctive **rainshadow effect**, illustrated in the cross section through the Western Cordillera of Canada in Figure 11.11.

The short-term variability of rainfall differs greatly from one area to another. It tends to be greatest in the tropics, and at Djakarta, Indonesia, for example, the annual rainfall of 1800 mm falls in only 360 hours on average. By contrast, the average

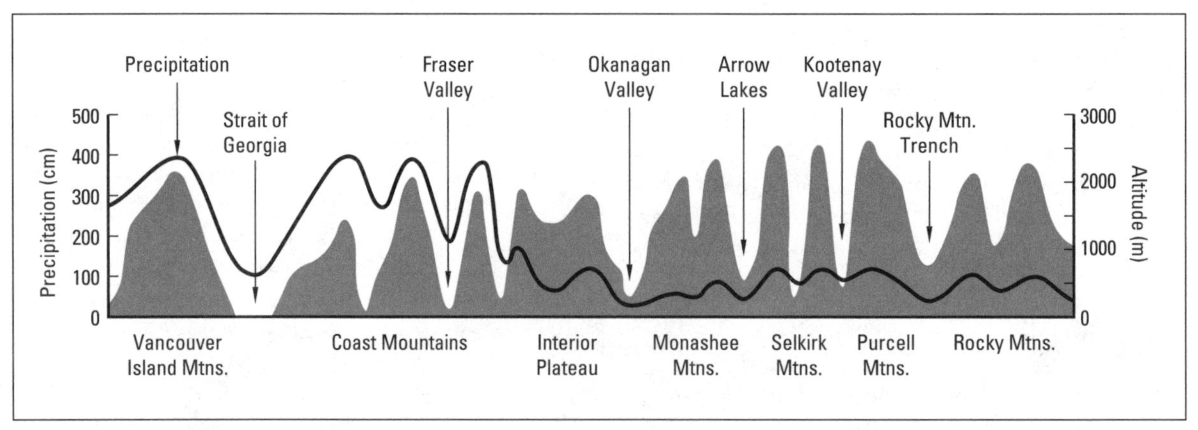


FIGURE 11.11 ▼ Schematic cross section through the Western Cordillera showing the variation of precipitation with varying topography (after Scott, Ladochy, and Sweet, 1987)

rainfall in London, England, is only 600 mm, yet this takes 500 hours to fall. Variability in precipitation is often most important, however, in the more arid parts of the world, for here even quite small storms may be a rare event (Table 11.4); channels that have been dry for months or even years may fill with water, and the baked clay used to make houses may crumble and be washed away. Within a matter of hours the rainfall may have ceased and the water almost vanished; within weeks the vegetation will have disappeared again.

## SEASONAL VARIABILITY

In many climates there is a predictable and consistent cycle of rainfall during the course of the year related to the latitudinal migration of the wind and pressure systems. Precipitation areas associated with areas of convergence and uplift tend to shift poleward in summer and equatorward in winter. Some areas, like the British Isles, remain within the same pressure system throughout the year and so seasonal variations are subdued. This is also true in the equa-

### Table 11.4

*Rain days and amounts at Timimoun, Algeria, 29°50'N, 0°30'E, 1926–50*

| | | | | | | | | | | | | | | |
|---|---|---|---|---|---|---|---|---|---|---|---|---|---|---|
| Days with no rain | 9007 | (98.8 percent) | | | | | | | | | | | | |
| Days with rain | 118 | (1.2 percent) | | | | | | | | | | | | |
| Rain amount (mm) | >1 | >2 | >3 | >4 | >5 | >6 | >7 | >8 | >9 | >10 | >15 | >20 | >25 | >30 |
| Number of days in 25 years | 63 | 40 | 30 | 18 | 13 | 12 | 9 | 9 | 7 | 5 | 3 | 2 | 1 | 1 |

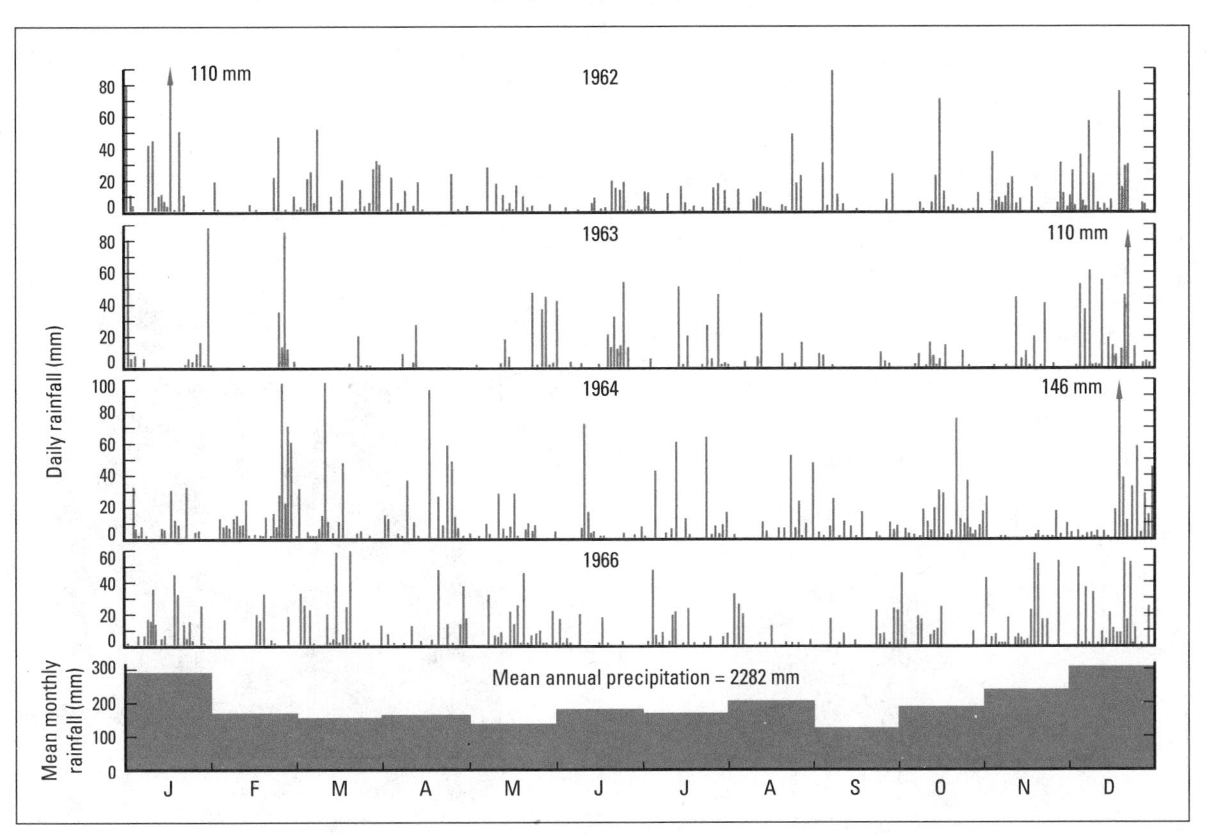


FIGURE 11.12 ▼ Daily rainfall over a four-year period at Singapore, 1962–66 (after Nieuwolt, 1968), compared with mean monthly rainfall for one year

torial trough zone, where rainfall can occur at any time throughout the year (Figure 11.12), and in deserts, where rainfall is almost negligible. The brief, rare storms that do occur can come at any time, so monthly rainfall, averaged over the long term, shows little variation (Figure 11.13). Even within the same pressure system, some seasonal pattern may be evident. In the midlatitudes, where rainfall is associated with the activity of the rain-bearing cyclones, the winter and autumn are relatively wet, for it is at these periods that the westerlies bring the most intense storms.

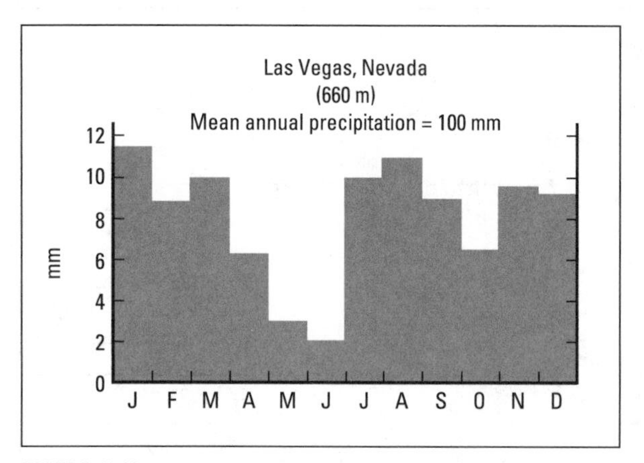


FIGURE 11.13 ▼ Mean monthly precipitation at Las Vegas, Nevada

In the tropics and subtropics, where convectional rainfall is more important, precipitation tends to be more abundant during the summer months (Figure 11.14). The magnitude of these seasonal variations is even more marked in the monsoonal areas of the world where the year can be subdivided into a wet and a dry season. At Tilembeya in the Sahel region of Mali, for example, rainfall during August is over 200 mm; from December to February it is almost zero. Similarly, in the monsoon areas of Asia and northern Australia (Figure 7.15), seasonal differences are great, so that hydrological conditions vary considerably throughout the year. During the dry season there is practically no surface runoff, followed by a wet season when runoff is extensive. Vegetation, geomorphological processes, and humans all respond to these changes.

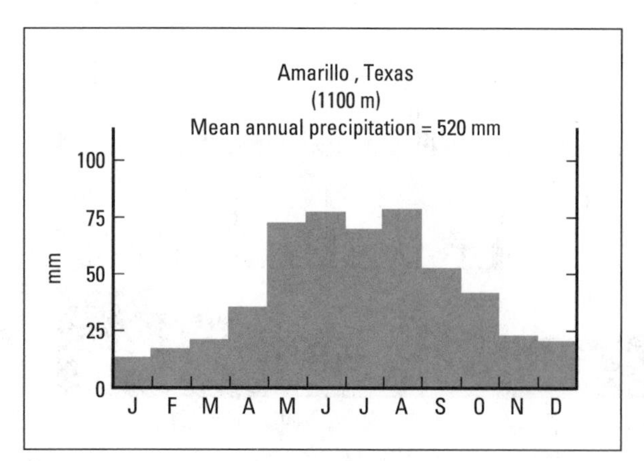


FIGURE 11.14 ▼ Mean monthly precipitation at Amarillo, Texas

Precipitation is more abundant during winter months in those parts of the world that experience

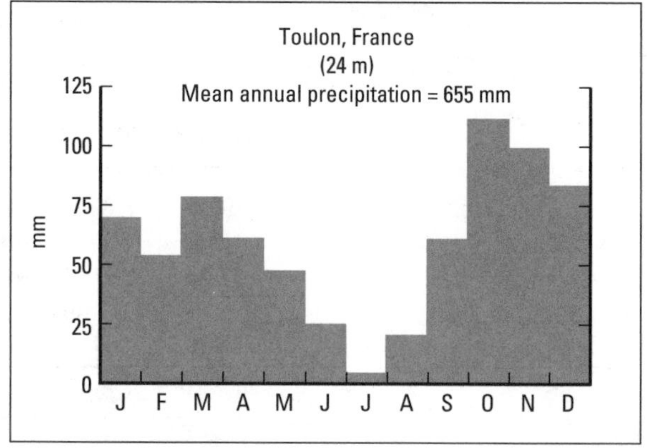


FIGURE 11.15 ▼ Mean monthly precipitation at Toulon, France

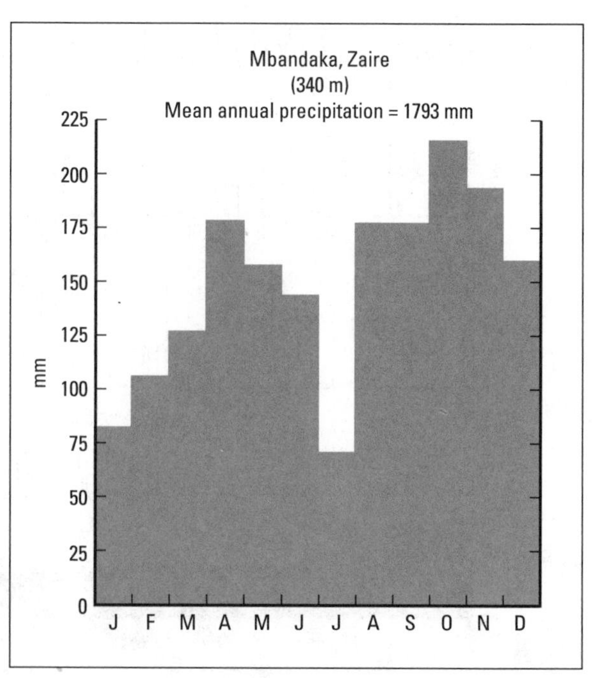


FIGURE 11.16 ▼ Mean monthly precipitation at Mbandaka, Zaire

the midlatitude depressions during winter only. The Mediterranean area is the best example of this type of precipitation regime, although other areas do experience similar rainfall patterns. In some areas, the seasonal patterns may be more complex. There may be more than one peak in rainfall totals, as in many areas of supposedly Mediterranean climate (Figure 11.15), and in the tropics where the seasonal migration of the equatorial trough produces two maxima (Figure 11.16). In part of East Africa, a similar pattern occurs (Figure 11.17), but not as a result of the passage of the trough; the rainfall occurs when the monsoonal flow is at its weakest. There are still two peaks of precipitation but there is a delay between the passage of the equatorial trough, the influx of the monsoonal flow, and the onset of rain. The rain occurs when daytime heating is sufficient to convect the new moisture to produce thunderstorms by late afternoon.

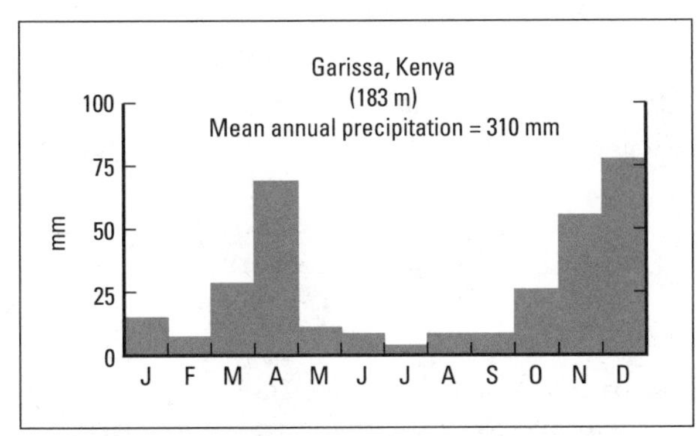


FIGURE 11.17 ▼ Mean monthly precipitation at Garissa, Kenya

## RAINFALL FREQUENCIES

In view of the important consequences of extreme variations in rainfall, it is useful to have some measure of the reliability of precipitation. This may be expressed in a number of different ways. One of the most common is to plot graphs of what are called rainfall **recurrence intervals**. Using data from a long time period, say fifty years, it is possible to estimate the frequency with which storms of a particular size or intensity are exceeded. In general, small storms occur most commonly, and very heavy storms only rarely. Thus, a graph like that in Figure 11.18 is obtained. From this, it is possible to tell how frequently a storm giving, for example, 50 mm or less in a day will occur; or how many years it will be on average between storms of 100 mm or more

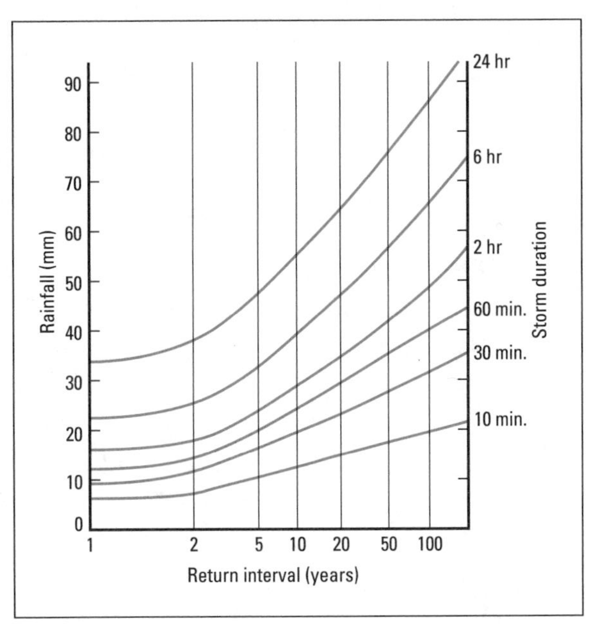


FIGURE 11.18 ▼ The return period for precipitation totals at Sheffield, England (based on methods in Flood Studies Report, NERC, 1976)

per day. Such information can be very useful in planning bridges, drains, or canals, when the aim is normally to produce something that will cope with all but the most extreme events. It is important to remember, however, that the figures are only probabilities, derived from average conditions. It is quite possible for two storms with an average recurrence interval of fifty years to occur on successive days!

Another way of expressing information on rainfall variation is to plot annual rainfall totals on similar graphs. Thus, in Figure 11.19 the frequencies of annual rainfall for Sheffield, England, and Timimoun, Algeria, are shown. We can see that there is a 50 percent probability of at least 760 mm of rainfall occurring in any year at Sheffield, while at Timimoun the equivalent total is 14 mm. Notice that the range of rainfalls is from a low of about 500 mm to 1000 mm for Sheffield, and zero to 80 mm for Timimoun. Thus, although Sheffield receives more rain, the variability is less than for Timimoun.

Again this type of data may be very useful. It may be known, for example, that a particular crop may only grow satisfactorily if the annual rainfall exceeds 600 mm. From information on annual rainfall frequencies it is possible to determine the likelihood of receiving this amount of precipitation. If the probability is, say, 90 percent the farmer may

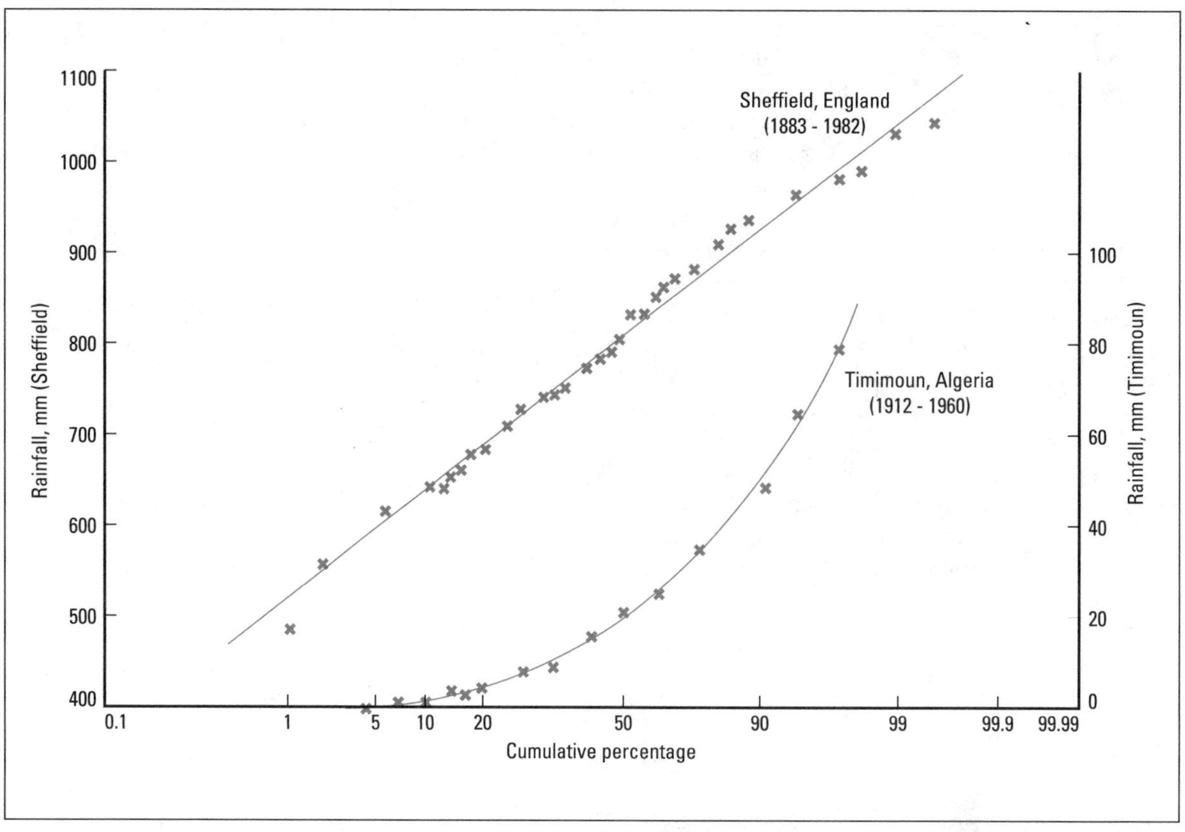


FIGURE 11.19 ▼ Cumulative percentage frequency graphs of annual precipitation at Sheffield, England (1883–1982), and Timimoun, Algeria (1912–60)

well think it worthwhile to grow the crop; if it is only 20 percent, then it is unlikely to be worth the risk. Similarly, it is possible to determine in the same way how often, on average, it will be necessary to irrigate crops.

Rainfall variability may also be expressed statistically by the **coefficient of variation**. This is calculated from the formula

$$CV = \frac{s}{\bar{x}} \times 100 \text{ percent}$$

where $\bar{x}$ is the average rainfall and $s$ is the standard deviation.* This defines the variability relative to the mean. With a standard deviation of 200 mm and a mean annual precipitation of 1600 mm, the coefficient of variation would be 12.5 percent, but with the same standard deviation and a mean of only 400 mm, the coefficient of variation would rise to 50 percent. This is a useful measure since it gives an indication of the importance of the variability. The pattern of variability, measured in this way, is shown for the Canadian Prairies in Figure 11.20.

# Spatial variations in precipitation

## ATMOSPHERIC CAUSES OF VARIATION

We all know that annual rainfall totals vary from one part of the world to another, even when altitude is allowed for. Locally, however, it seems likely that annual totals will be fairly consistent. It is also clear that, in the short term, quite marked differences in rainfall may occur within short distances, depending upon the route taken by a particular storm or cyclone; indeed, it is sometimes possible to see it raining on one side of the street and dry on the other.

* The standard deviation is a measure of dispersion of the data. It is calculated from the formula $s = \sqrt{\frac{\Sigma (x - \bar{x})^2}{n}}$ where $n$ is the number of observations in the data set and $\Sigma$ means 'sum of.'

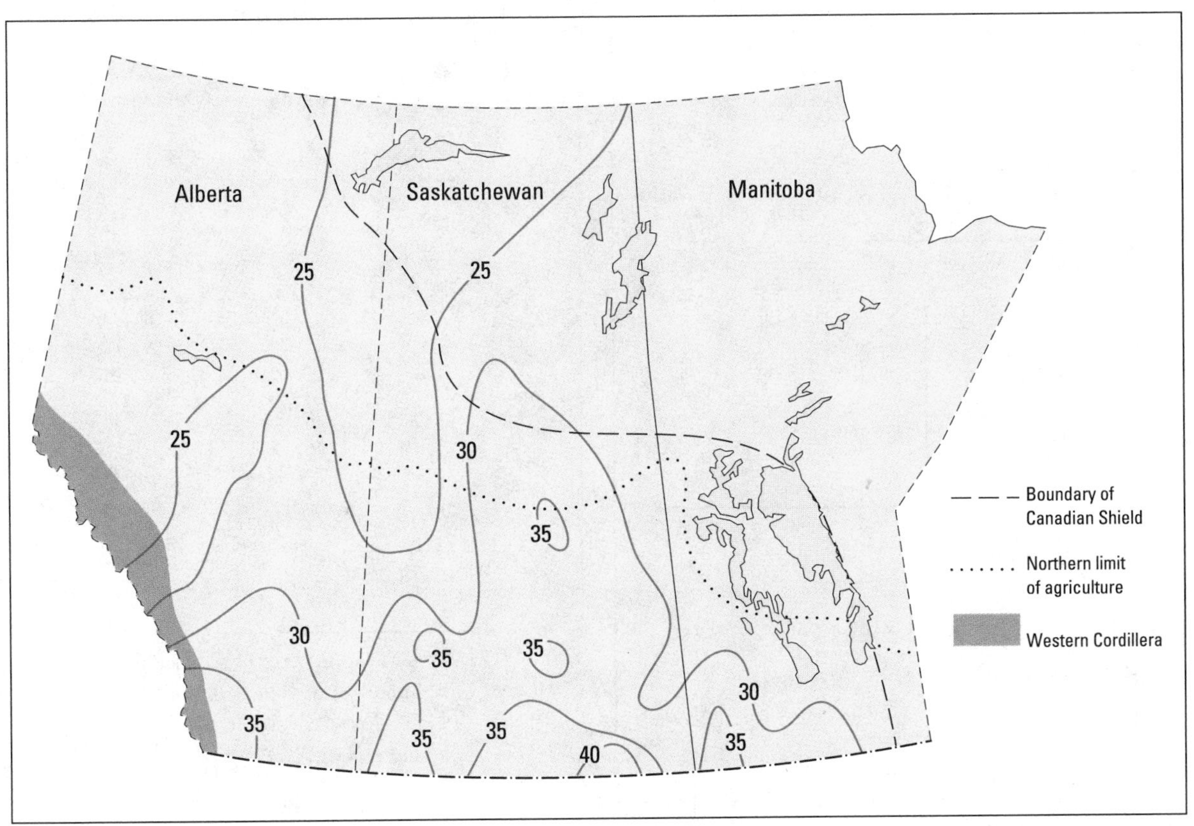


FIGURE 11.20 ▼ Coefficient of variation of summer precipitation for the Prairie region of Canada 1957–87 (after Purves, unpublished thesis, 1988)

In order to study spatial variations at a local scale, we need a dense network of recording stations, for otherwise individual storms may be missed as they pass between the rain gauges. One such investigation was carried out in Illinois, where fifty recording rain gauges were set up in an area of 1400 km$^2$ of flat rural land. The experiment was maintained for five years measuring individual storms, and for thirteen years for monthly and seasonal analyses. Comparisons were made by correlating rainfall at a gauge at the centre of the area with all other gauges. Correlation is a statistical measure that provides an index of the strength of the relationship between two variables; a value of +1.0 indicates a perfect positive linear relationship, a value of –1.0 shows a perfect negative linear relationship, and a value of 0.0 shows no relationship (Figure 11.21). The terms positive and negative refer to the direction of the slope of the trend line drawn through all the points on the scatter diagram.

For the shortest time period studied (one minute) the degree of correlation fell rapidly with distance from the central gauge (Figure 11.22). Thus, at a distance of only 8 km from the central gauge, the rainfall pattern is different; in many cases it would have been raining at the central gauge but not 8 km away. This is what we would expect if rainfall was produced by localized summer convection storms, each affecting an area of only a few square kilometres.

At a longer time scale, the degree of correlation is better. Taking rainfall totals for whole storms (Figure 11.23A), it is apparent that the gauges close to the central station are quite strongly correlated. Nevertheless, at a distance of about 16 km the degree of correlation is low. Again, this is probably due to the effect of local variability caused by the passage of small summer convection storms. If, instead, we look at frontal storms or storms associated with low pressure systems, we get a different picture (Figure 11.23B). Now, most of the area shows a close correlation with the central gauge; indeed, almost a perfect correlation—indicating that these more general storms affected the whole area equally.

These results indicate some of the atmospheric factors controlling rainfall variability. Convectional

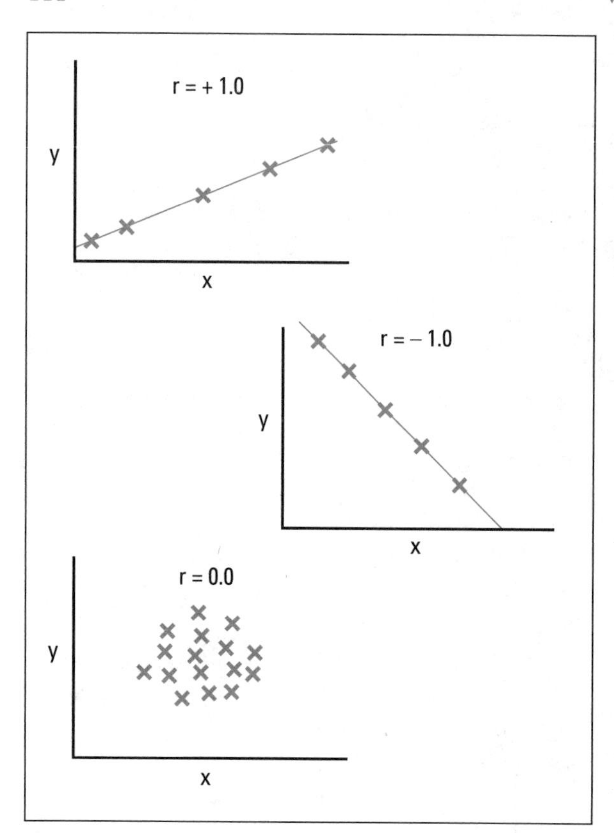


FIGURE 11.21 ▼ Scatter diagrams demonstrating different correlation coefficients

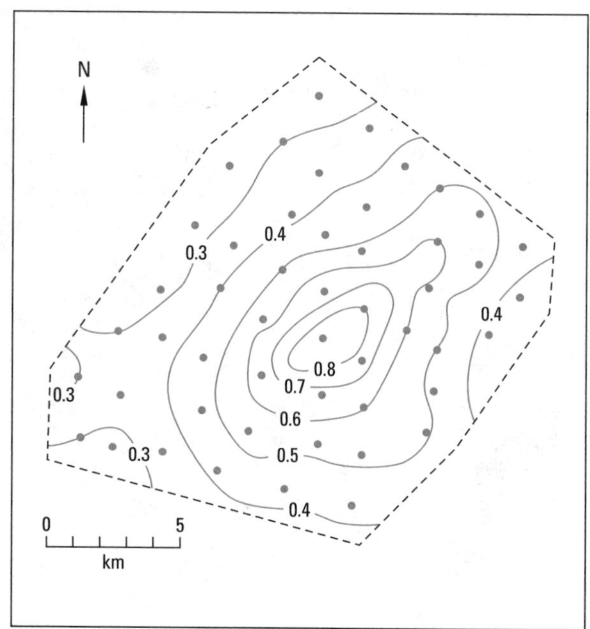


FIGURE 11.22 ▼ Correlation patterns associated with one-minute rainfall rates in warm season storms in Goose Creek, central Illinois (after Huff and Shipp, 1969)

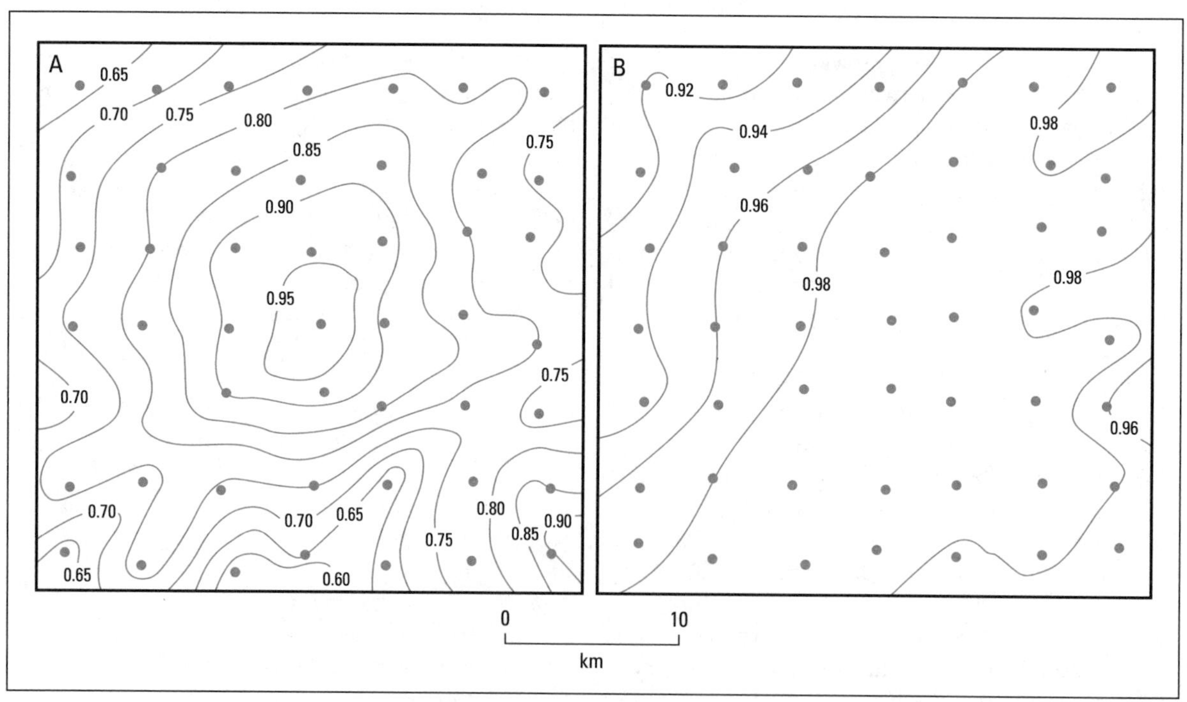


FIGURE 11.23 ▼ Correlation patterns associated with (A) air mass storms and (B) low pressure centres, during May–September in Illinois (after Huff and Shipp, 1969)

storms give high levels of spatial variation, while cyclonic rainfall is spatially much more uniform. In the tropics, where a great proportion of the rainfall comes from convectional storms, the spatial variation is particularly marked. Table 11.5 illustrates this point. It gives rainfall totals at rain gauges only 3.2 km apart in a level area of Dar-es-Salaam in Tanzania. It is clear that the totals are very different. The reason is that most of the rainfall is derived from individual cumulonimbus clouds that produce intense precipitation over an area of about 2 to 60 $km^2$. The storms often build up without any significant movement, so areas just beyond the limits of the cloud may receive no rainfall at all. Sometimes the storms develop over a wider area, perhaps 500 $km^2$, but even so they do not give rain everywhere. Using the correlation method, we find that the relationship between rain gauge totals falls to zero within 100 km and is negative beyond. In other words, if rainfall were high for a particular period in one area, it would tend to be low beyond 100 km.

## Surface modifications of precipitation

So far we have considered rainfall variability over essentially flat terrain. Few areas of the world are extensively flat, however, and surface irregularities interfere with atmospheric processes to give even more complex spatial patterns of variation in rainfall. Even relatively small hills can have a marked effect. The Chiltern Hills in southern Britain, for example, only rise some 90 m above the surrounding land but they receive appreciably more rainfall.

Within any climatic region, the relationship between rainfall and altitude is generally quite consistent. In most cases, precipitation increases with

**Table 11.5**

*Daily rainfall (mm) at two stations 3.2 km apart near Dar-es-Salaam, Tanzania, April, 1967*

| | *April 6th* | *7–8th* | *10th* | *12th* | *13th* | *24th* | *25th* | *Month* |
|---|---|---|---|---|---|---|---|---|
| Station A | 54.9 | 33.3 | 2.3 | 13.5 | 13.5 | 14.2 | 21.1 | 315.2 |
| Station B | 100.3 | 3.3 | 31.8 | 64.8 | 5.1 | 8.4 | 0.8 | 437.6 |

After I. J. Jackson, *Climate, Agriculture and Water in the Tropics* (1977), Longman.

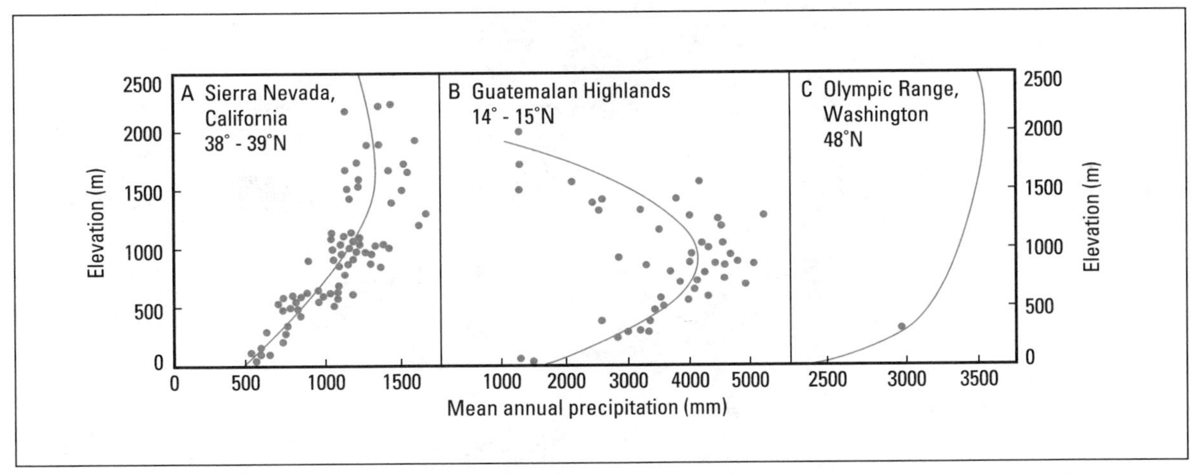


FIGURE 11.24 Generalized curves showing the relationship between elevation and mean annual precipitation for west-facing mountain slopes in North and Central America (after Barry and Chorley, 1982)

increasing altitude, even in relatively arid areas. At Grand Canyon, for example, average annual precipitation increases from less than 250 mm on the canyon floor at 760 m, to 400 mm on the southern rim of the canyon at 2100 m. On the forested northern rim, 2600 m above sea level, rainfall totals over 600 mm.

Nevertheless, the progressive increase in rainfall with altitude does not always extend to the summits of the mountains. The Sierra Nevada in California are no wetter on the summit than they are 1200 m lower (Figure 11.24A). In Hawaii, the peaks of Mauna Loa and Mauna Kea receive far less than the windward slopes, even close to sea level. It is also apparent that the relationship between altitude and precipitation varies from one part of the world to another. In the tropics, much of the precipitation is produced by warm clouds whose upper limit is only 3000 m above the ground, thus the mountain effect is reduced. The height of maximum precipitation is reduced as shown in Figure 11.24B. In contrast, in temperate areas, a large proportion of the rainfall comes from deep stratiform clouds that extend through a considerable part of the troposphere. Here

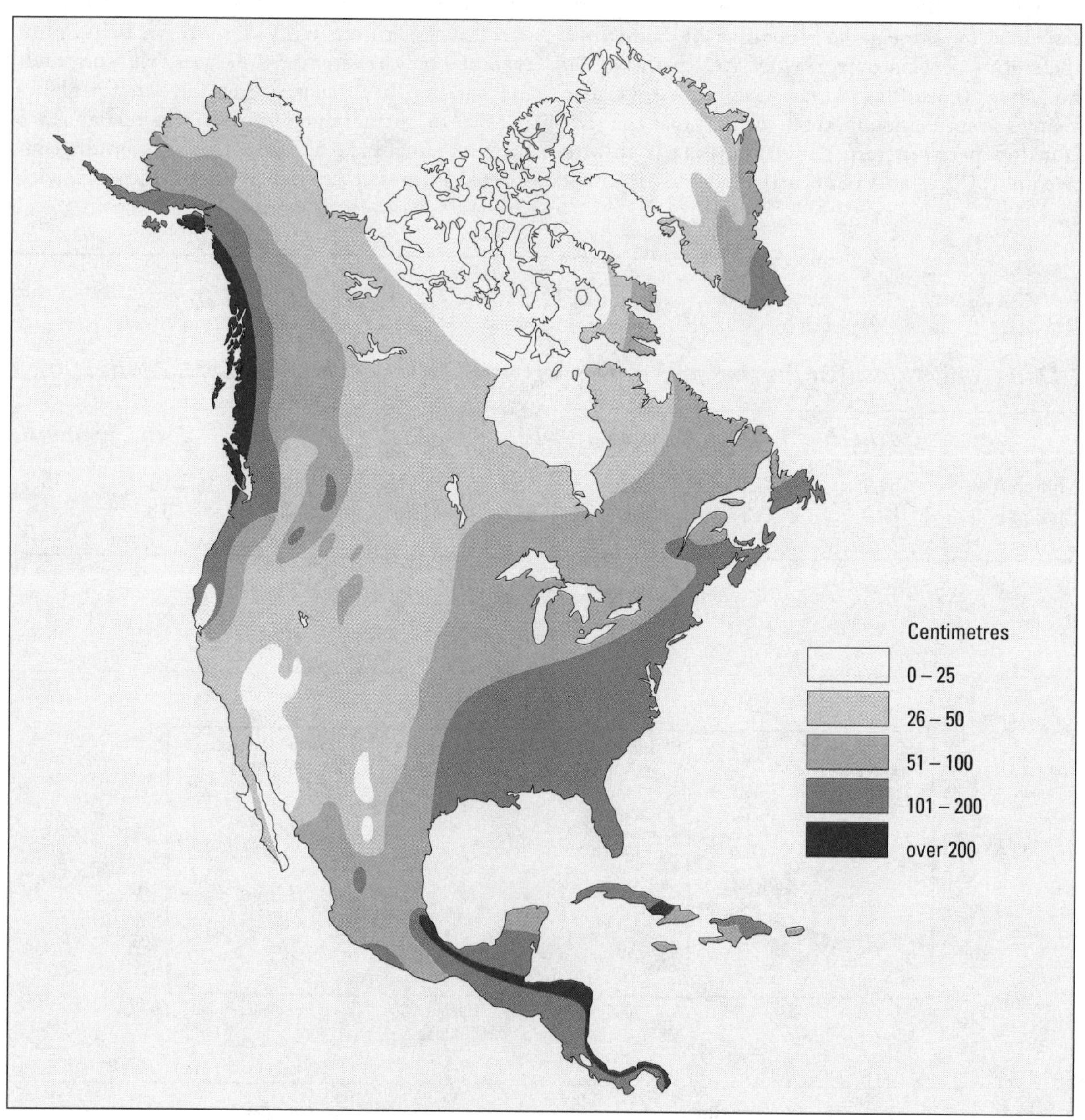


FIGURE 11.25 ▼ Mean annual precipitation in North America

the effect of altitude on rainfall is more marked (Figure 11.24C). Comparisons are difficult, however, because much of the precipitation on the mountains in temperate areas falls as snow and, as we have seen, this is impossible to measure accurately.

The importance of surface topography on precipitation is indicated at a general scale for North America by Figure 11.25. We can see the pattern of rainfall is appreciably modified by the Western Cordillera, especially in the northwestern United States and the province of British Columbia. This gives rise to higher totals on the westerly slopes, and a marked rain shadow effect on the east, as shown in Figure 11.11. Note how rapidly the amount of precipitation decreases on the eastern slopes of the mountains out onto the Great Plains: the prevailing winds have lost much of their moisture by the time this region is reached.

# Conclusion

Precipitation input to the terrestrial portion of the hydrological cycle is probably one of the most important regulators of the system, for it determines the intensity and distribution of many of the processes operating within the system. As we will see in the next chapter, it is closely related to the rate of evapotranspiration, and as we will find in later chapters, it also influences the pathways of runoff and underground flow and the magnitude of streamflow. Through these processes, and through the direct effects of the impact of rainfall on the ground, it also takes part in many geomorphological processes; it causes rainsplash and soil erosion, and plays a vital role in weathering and rock breakdown. The distribution of rainfall across the globe therefore controls to a large degree the operation of the landscape system. Precipitation is similarly a vital input to the ecosystem, and the distribution of vegetation, fauna, and human settlement owes much to the pattern of rainfall.

For these reasons, and because of their ultimate importance for us, a great deal of attention is given to measuring, mapping, and predicting precipitation. As we have seen, scarcity of data, particularly in less accessible parts of the world, limits our ability to gain a complete picture of precipitation inputs. On the whole, however, rainfall is one of the easiest components of the hydrological cycle to measure, and for that reason it provides an ideal opportunity for coming to grips with real-world hydrological processes.

# Evapotranspiration

## Evapotranspiration processes

Evaporation and transpiration form the major flows of moisture away from the earth's surface. Because we can rarely see the processes taking place, it is easy to neglect this component of the hydrological cycle, but it is an extremely important one. It returns moisture to the air, replenishing that lost by precipitation, and it also takes part in the global transfer of energy. In this chapter, we will examine the processes involved in evaporation and transpiration, and show their significance at a global and local scale.

### EVAPORATION

Evaporation can be defined as the process by which a liquid is converted into a gaseous state. It involves the movement of individual water molecules from the surface of the earth into the atmosphere, a process occurring whenever there is a vapour pressure gradient from the surface to the air. Thus, evaporation requires that the humidity of the atmosphere be less than that of the ground. The process also requires energy: $2.48 \times 10^6$ J to evaporate each kilogram of water at 10°C. This energy is normally derived from the sun, although sensible heat from the atmosphere or from the ground may also be significant. However, when the air reaches saturation (100 percent relative humidity) evaporation cannot take place. At this point the number of molecules entering and leaving the surface is balanced, a condition known as **saturation equilibrium**.

### TRANSPIRATION

Transpiration is a related process involving water loss from plants. It occurs mainly by day when small pores on the leaves of plants, called stomata, open under the influence of sunlight. These expose the moisture in the leaves to the atmosphere and, if the vapour pressure of the air is less than that in the leaf cells, water is transpired. As a result of this transpiration, the leaf becomes relatively dry and a moisture gradient is set up between the leaf and the base of the plant. This draws moisture up through the plant and from the soil into the roots (Figure 12.1).

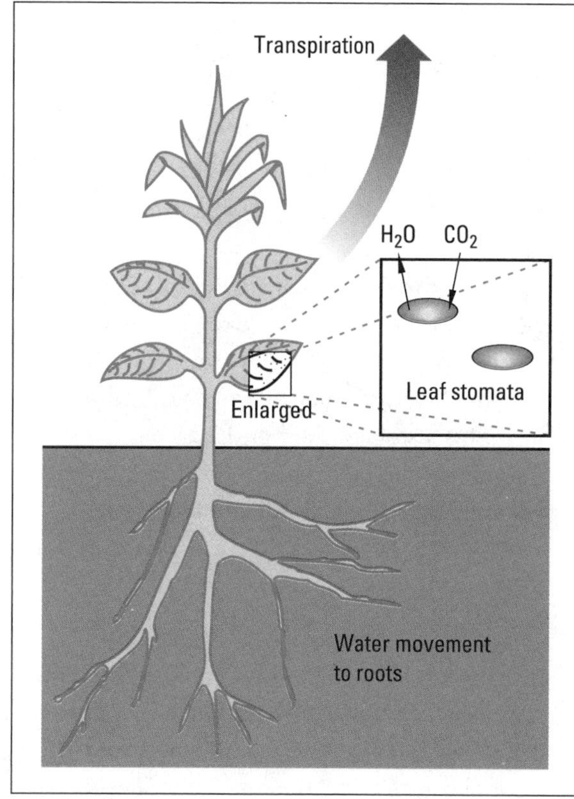


FIGURE 12.1 ▼ Exchanges of water and gases by transpiration in plants

As far as the plant is concerned, this is a passive process; it is controlled largely by atmospheric and soil conditions and the plant has little influence over it. Consequently, transpiration results in far more water passing through the plant than is needed for growth. Only 1 percent or so is used directly in the growth process. Nevertheless, the excessive movement of moisture through the plant is of great importance, for the water acts a solvent, transporting vital nutrients from the soil into the roots and carrying them through the cells of the plant. Without this process, plants could not live.

### EVAPOTRANSPIRATION

In reality, it is often difficult to distinguish between evaporation and transpiration. Wherever vegetation is present, both processes tend to be operating together, so the two are normally combined to give the composite term evapotranspiration.

Evapotranspiration is governed mainly by atmospheric conditions. Energy is needed to power the process, and wind is necessary to mix the water molecules with the air and transport them away from the surface. In addition, the state of the surface plays an important part, for evaporation can only continue so long as there is a vapour pressure gradient between the ground and the air. Thus, as the soil dries, the rate of evapotranspiration declines. Lack of moisture at the surface often acts as a limiting factor on the process.

We can therefore distinguish between two aspects of evapotranspiration. **Potential evapotranspiration** (PE) is a measure of the ability of the atmosphere to remove water from the surface assuming no limitation of water supply. **Actual evapotranspiration** (AE) is the amount of water that is actually removed. Except where the surface is continually moist, actual evapotranspiration is significantly lower than PE.

## Potential evapotranspiration

### ENERGY INPUTS

The main variable determining potential evapotranspiration is the input of energy from the sun, and it has been estimated that this accounts for about 80 percent of the variation in PE.

The amount of radiant energy available for evapotranspiration depends upon a number of factors, including latitude (and hence the angle of the sun's rays), day-length, cloudiness, and the amount of atmospheric pollution. Thus PE is at a maximum under the clear skies and long hot days of tropical areas, and at a minimum in the cold, cloudy polar regions. In the short term, however, rates of PE may vary considerably at any single place. Daily variations in radiation inputs cause marked fluctuations in PE, so that very little evapotranspiration occurs at night. Even subjectively we can get some idea of this by noting how long the ground stays wet after a shower of rain during the night, yet how quickly it dries during the day. Similar patterns occur seasonally. PE reaches a peak during the summer months and declines markedly during the winter (Figure 12.2). The diagram is the same as those used to illustrate water balance in Chapter 7. Wilmington, Delaware, represents a wet, relatively warm midlatitude station on the east coast of North America. Moisture supply through precipitation is relatively constant through the year. The major cause of variability in the moisture budget is the temperature. Note that there is a surplus from October to June, and the shortfall due to higher summer temperatures is generally supplied by soil moisture.

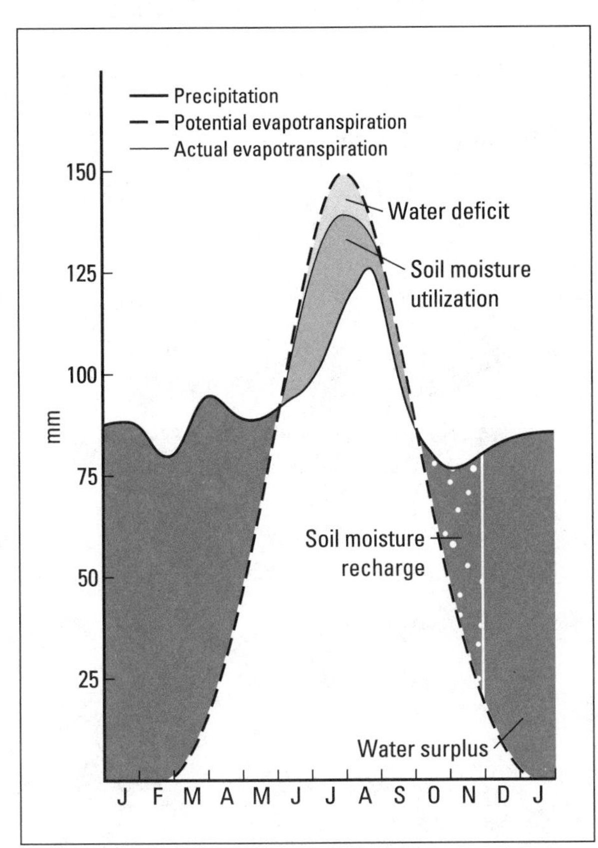


FIGURE 12.2 ▼ Average moisture budget for Wilmington, Delaware. The difference between potential and actual evapotranspiration is small because the distribution of precipitation is relatively uniform throughout the year (after Mather, 1974)

## WIND

The second important factor is the wind. Wind enables water molecules to be removed from the ground surface by a process known as **eddy diffusion**. This maintains the vapour pressure gradient above the surface. Wind speed is obviously one of the variables determining the efficiency of the wind in removing water vapour, but it is not the only one. The rate of mixing is also important, and this depends upon the turbulence of the air and the rate of change of wind speed with height (Figure 12.3).

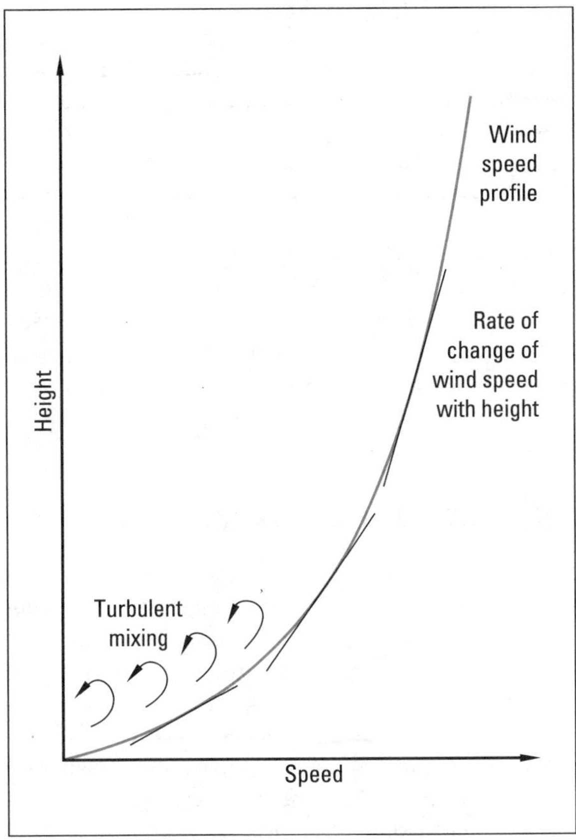


FIGURE 12.3 ▼ Wind speed profile and its relationship with the rate of change of wind speed with height

In the diagram, the rate of change of wind speed with height is shown as straight gradient lines (broken lines). The wind profile is shown as a continuous, smooth curve indicating the gradual increase of wind with height. Compare this with the Ekman spiral concept shown in Figure 5.9. Turbulence in the lower atmosphere occurs up to about 3000 m above the surface and is indicated by the looping arrows.

## VAPOUR PRESSURE GRADIENT

Third, evapotranspiration is related to the gradient of vapour pressure between the surface and the air. Unfortunately, the vapour pressure gradient has proved very difficult to measure precisely in the layer immediately above the surface, so wherever possible, methods of calculating PE use measurements of vapour pressure at one level only.

## ACTUAL EVAPOTRANSPIRATION

Actual evapotranspiration only equals PE if there is a constant and adequate supply of water to meet the atmospheric demand. This situation exists over moist, vegetated surfaces and it is also approximated over water surfaces such as the open sea or large lakes, but most land surfaces experience significant periods when water supply is limited. As a result, AE falls below PE. We can get some idea of the importance of surface conditions by considering evapotranspiration in a variety of situations. Let us start by examining evapotranspiration from an open water surface.

## EVAPOTRANSPIRATION FROM WATER SURFACES

Because there is an unlimited supply of water to maintain evaporation, and because there is no vegetation to complicate the process, the surfaces of oceans or large lakes provide the simplest situation in which to study evapotranspiration. In this situation, of course, transpiration does not occur and water loss is entirely by evaporation. The main factors determining water loss are therefore the atmospheric conditions, and there is generally a close relationship between AE and PE.

Nevertheless, the relationship is not perfect, and the reason is that the water is able to absorb a large amount of energy which is not used in evaporation. The energy is expended in heating the water, and much of it is recirculated through the water body by advection and convection.

Evaporation is greatest when the sea is warm in comparison to the air. In general this is the case, as air temperatures are slightly below those of the sea over much of the globe for much of the time. Where upwelling of cold water from the sea bottom occurs, however, surface temperatures are greatly reduced and the difference between sea and air temperatures becomes small; in some cases the sea may even be cooler than the atmosphere. An example of this phenomenon occurs off the coast of Peru where the cold Humboldt Current brings bottom waters to the surface. As a result, the air is warm relative to the sea, it retains more moisture and so the humidity gradient above the surface is low. This greatly reduces the rate of evaporation, and, as Figure 12.4 shows, the effect continues some way out into the Pacific.

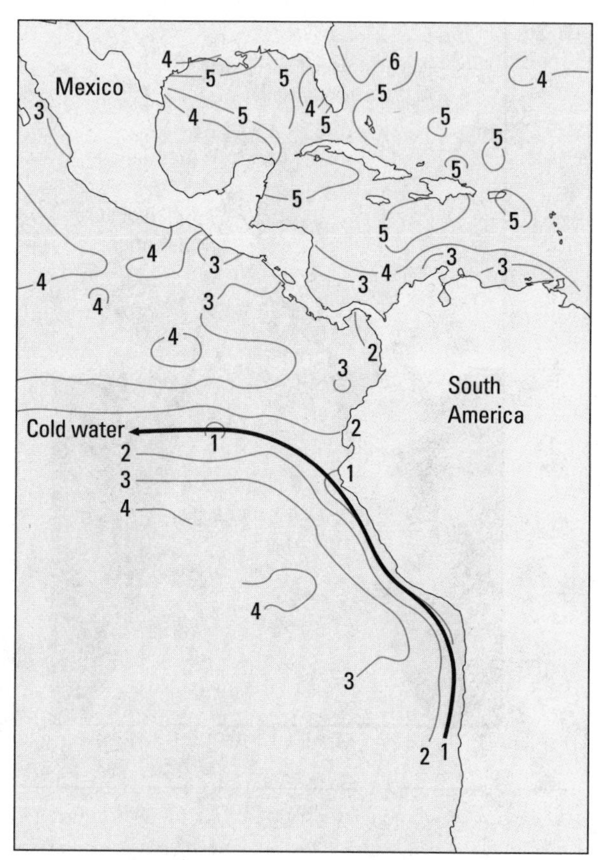


FIGURE 12.4 ▼ Mean evaporation in millimetres per day from the tropical Atlantic and eastern Pacific oceans (after Hastenrath and Lamb, 1978)

At the oceanic scale, there have been few studies of evaporation. Smaller, freshwater bodies, such as Lake Ontario, have been investigated more fully, however, and Table 12.1 lists the water and energy budget components for the lake on a monthly basis. During the spring, as air temperatures are rising, the water surface remains cold and evaporation becomes negative (i.e., there is net condensation of water on the lake). All the available radiation and heat from the air goes into warming up the deeper water of the lake. During the cooling phase of the cycle, from late August to about March, the situation is reversed and heat is released from the depths of the lake. Evaporation can continue, maintained by the subsurface heat.

This discussion of the energy budget may sound unimportant to all but the fish of the lake. However, it has vital implications for the climate of the surrounding areas. The very high rates of evaporation in the autumn and early winter mean that water is being added to the atmosphere in large quantities. With low temperatures, much of the moisture is returned to the ground as snow on the southeastern shores of the lakes (Figure 12.5). The snowbelt is a term for regions downwind of the open water that receive heavier snowfall. It is a phenomenon found at other lakes but often has different names. Winds blowing from the northwest pick up moisture from lakes Winnipeg and Manitoba and create a snowbelt extending to the southeast known as 'streamers.' The size and orientation of the Great Lakes relative to the prevailing winds make the effect especially noticeable. In spring, the low water temperatures of the lakes keep temperatures low along the lake shores and delay blossoming of the orchards. This reduces the chance of severe frosts damaging the fruiting of the trees, one of the rea-

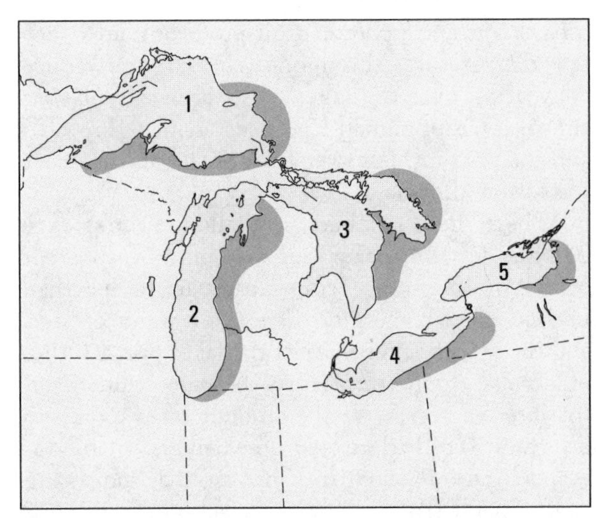


FIGURE 12.5 ▼ Snowbelts of the Great Lakes (after Eichenlaub, 1970)

## Table 12.1

*Water and energy budget of Lake Ontario*

| | *J* | *F* | *M* | *A* | *M* | *J* | *J* | *A* | *S* | *O* | *N* | *D* |
|---|---|---|---|---|---|---|---|---|---|---|---|---|
| Net income of energy (W m$^{-2}$) | +2 | +35 | +85 | +140 | +185 | +200 | +200 | +160 | +110 | +50 | +10 | –5 |
| Exchange of heat between lake surface and deep water (W m$^{-2}$) | +250 | +160 | –15 | –165 | –245 | –190 | –130 | –70 | 0 | +65 | +145 | +200 |
| Energy available at lake surface (W m$^{-2}$) | +246 | +192 | +67 | –25 | –59 | +10 | +70 | +89 | +109 | +115 | +153 | +190 |
| Surface temperature (°C) | +3 | 2 | 2 | 2 | 4 | 10 | 19 | 20 | 17 | 11 | 7 | 5 |
| Evaporation (mm day$^{-1}$) | –3.2 | –2.6 | –1.4 | +0.6 | –0.3 | –0.1 | –1.6 | –2.6 | –3.6 | –3.0 | –3.0 | –2.9 |
| Latent heat flux (W m$^{-2}$) | –90 | –75 | –40 | +15 | –10 | –5 | –50 | –75 | –105 | –85 | –85 | –85 |

*Note* Plus and minus signs indicate a net gain or loss for each category.

After D.H. Miller, *International Geophysics*, no. 21 (1977), Academic Press.

sons this is an important fruit-producing area. The low temperatures throughout the summer reduce convection over the lakes to such an extent that annual totals of rainfall have been estimated to be 6 percent (80 mm) less over Lake Michigan than over the surrounding land.

Where the water body is shallow, heat storage becomes less important and may only be apparent on the diurnal scale. Temperatures are higher than in deep water bodies, for there is less water to heat, and the thermal properties of the lake approach that of a moist ground surface. As the proportion of soil to water increases, so the amount of evaporation decreases, very little stored heat being available for evaporation. Measurements in Australia, comparing evaporation from a water surface and wet soil, showed that the rate for the soil was only 86 percent of that from the water surface.

*Evaporation and sublimation from snow surfaces*

Snow covers large areas of the land surfaces of the northern hemisphere during the winter months, and some areas throughout the year. Recent satellite studies show that the extent of snow cover varies considerably from year to year. Snow surfaces have different energy and water balance patterns which have implications for global climate.

Here we are concerned with the movement of moisture between the atmosphere and the snow surface. It is difficult to determine an accurate water balance for snow because it is difficult to determine the energy balance. Each varies as the amount of water movement within the snow varies and the water changes phase. For example, surface melting would result in percolation of meltwater into the snow, and resulting heat transfer. Phase changes such as melting, freezing, evaporation, or condensation require energy absorption or release within the snow.

A snow surface in below-freezing air with no energy input from the sun will experience inconsequential evaporation, although some **sublimation** could occur. The amount of energy required for sublimation, the change of phase of an ice crystal directly to water vapour, is large, and is usually not available, especially in the winter months.

If the air is warm or the snow is melting, different energy conditions exist, therefore different evaporation and sublimation conditions exist. Energy is available for evaporation and sublimation. If the air is dry, that is, below saturation, water will evaporate from the melting snow surface. However, if the air is saturated, its vapour pressure will usually exceed that of the melting snow and moisture will condense onto the surface. If the surface is at or below freezing, this condensation will form frost. When it occurs on a snow or ice surface it will not be very visible, but on a darker surface it is seen as **rime** or **hoar frost**. People often say that the frost is coming out of the ground, but as we can see this is not the case, it just looks that way. Rime and hoar frost can create very slippery and hazardous conditions.

Rain can be an important input of heat energy to an ice or snow surface. The usual sources of energy for melting are radiation and convection. Rain during this process can thus accelerate the rate of melting considerably. This was probably a major factor in the final melting stages of the continental glaciers of the Pleistocene; we believe these were more rapid than standard energy balance models would allow.

The rapid melt by rain is also critical for flooding during snowmelt in the spring. Every flood forecaster worries about rain during spring thaw because it adds water, but it also increases the rate of snowmelt.

## EVAPOTRANSPIRATION FROM LAND SURFACES

Because of the importance of the energy and water balance to growing crops, there has been a large number of studies of evapotranspiration from vegetated surfaces. The presence of a vegetated surface complicates the energy exchanges, however, for the plants intercept radiation and rainfall inputs, they affect the temperature and wind profiles near the ground, and they also modify humidity. The degree of these effects varies with the character of the vegetation, so evapotranspiration from a vegetated surface often differs markedly from PE (Figure 12.6).

Within a mature crop we can identify three layers: the upper layer or canopy, the main stem zone, and the ground surface. During the day most incoming radiation is absorbed by the canopy. The air space between the canopy and the ground acts as an insulator so that it is the top of the vegetation rather than the soil that is the active surface. Consequently, transpiration rather than evaporation takes place.

As long as moisture is available in the soil, plants are able to transpire at or very close to the potential rate. Thus, in a moist soil evapotranspiration pro-

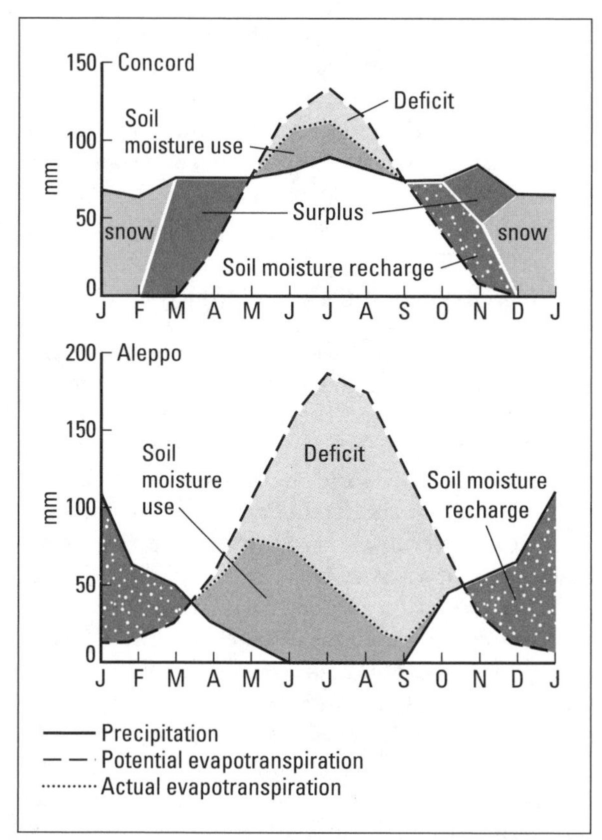


FIGURE 12.6 ▼ Moisture budget diagrams for Concord, New Hampshire, and Aleppo, Syria. The method assumes that 50 percent of soil water surplus runs off in the first month, 50 percent of the remainder in the next and so on, unless additional surplus forms (after Barry, 1969)

ceeds unhindered, water being drawn up the plant from the soil to replace that lost from the leaves. As the soil dries, however, the plants experience increasing difficulty in extracting moisture, and the rate of transpiration cannot be maintained. Several changes take place. The plants start to suffer from moisture stress and nutrient deficiencies, and in some cases the stomata in the leaves may close, reducing transpiration further. But the drain upon the soil moisture store continues, so the moisture stress gets worse. Progressively, the rate of actual evapotranspiration falls below PE (Figure 12.7).

It now seems clear that the effect of declining moisture availability depends upon a variety of conditions, including vegetation type, rooting depth and density, and soil type. In a heavy clay soil, for example, it seems that evapotranspiration rates fall

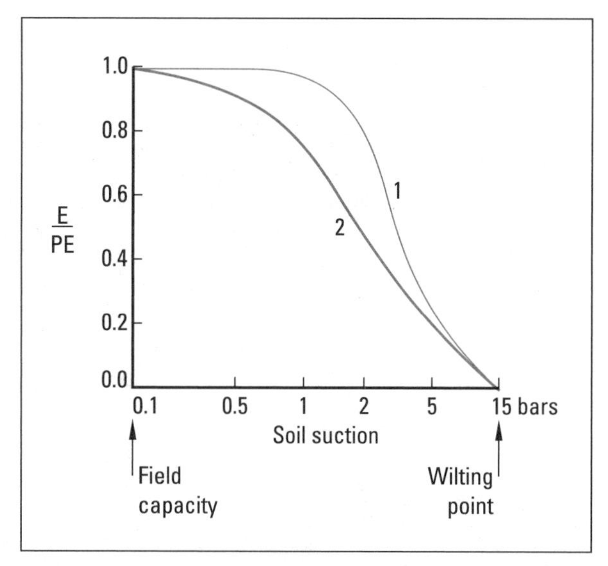


FIGURE 12.7 ▼ The relationship between the ratio of actual to potential evapotranspiration E/PE and soil moisture. Curve 1 is a vegetation-covered clay loam under low evaporation stress; curve 2 is a vegetation-covered sandy soil under high evaporation stress (after Barry, 1969)

only slightly as the soil dries, until the point is reached where no more water is available to plants. Evapotranspiration then falls rapidly. Conversely, in a sandy soil, the decline in actual evapotranspiration rates is much more regular (Figure 12.7) as the sandy soil's capacity to retain moisture is less than that of a clay.

## PLANT RESPONSES TO MOISTURE STRESS

The reduction in evapotranspiration as the soil dries has a number of implications. Eventually, of course, the plants experience severe nutrient deficiencies and yields are reduced. Thus we often see a close relationship between the degree of moisture stress and crop yields (Figure 12.8). In addition, moisture in the plant helps to control its temperature; the energy used in transpiration cannot heat the plant. As transpiration declines, more energy is available for heating, and the leaves get warmer. Initially, this may encourage growth, but ultimately, if high temperatures are reached, it may damage the plant. The soil, too, is heated more effectively so surface temperatures rise. In one study in Wisconsin, with an air temperature of 28°C, the surface of a dry sandy

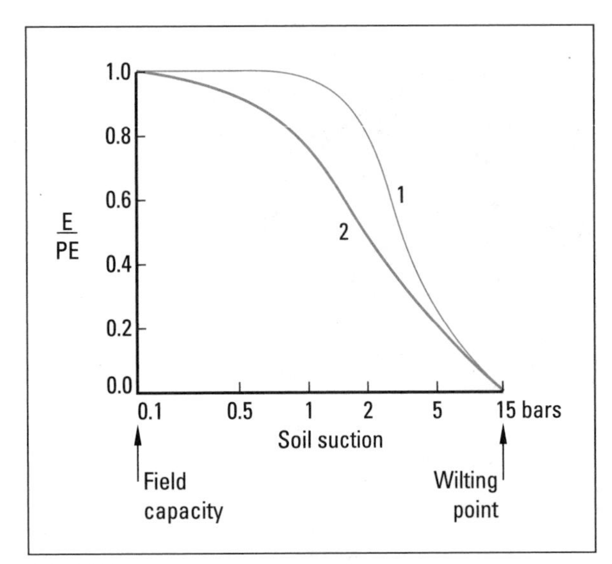


FIGURE 12.8 ▼ Relationship between yields of corn and the number of days of critical soil moisture tension (soil moisture below 12 percent on a loess soil) (after Rubin and Bielorai, 1957)

soil was 44°C, while the same soil kept moist reached only 32°C; more energy was used in evaporation from the wet soil.

The ultimate effect of continued drying of the soil is that plants can no longer obtain any water, and transpiration ceases. Since the water in the plant cells is important in keeping the plant rigid (turgid), as the plants dry out they become limp and wilt. At this stage, therefore, the soil is said to be at **wilting point** (see Chapter 13).

Of course, evapotranspiration does not always continue until the wilting point is reached. Instead, renewed rainfall generally wets the soil and rejuvenates water uptake by the plants, and increases transpiration. Thus the soil acts as an important store of moisture, as we will see in the next chapter, supplying water for transpiration between each storm.

Where the intervals between each period of precipitation are long, the ability of the soil to supply water may be stretched to the limit and moisture stress may be a common occurrence. In these circumstances the vegetation often adapts to the hydrological conditions, by developing deeper roots, or by regulating water use in a variety of ways. For example, the plants may have a dormant period during the dry season, completing their growth cycle during the brief wet period. The sight of deserts blooming after a storm is a most remarkable one. Other plants adapt by controlling their stomata, closing them during periods of dryness in order to reduce water loss. Others are able to alter the orientation of their leaves so that the stomata are more sheltered from the hot sun. Yet others, like the mesquite of southwest USA, have thick waxy leaves that protect them from the radiation and slow down water losses.

## EVAPOTRANSPIRATION AND IRRIGATION

One way of combating the effects of excessive evapotranspiration is through the use of irrigation. People may try to make good the water deficit during dry months by importing water from elsewhere. In many parts of the world, agriculture is dependent upon irrigation; the fertile farmlands of southern Australia, for example, would be little more than desert without a supply of water to crops in this way.

Many irrigation methods have been developed over the centuries, some supplied by groundwater (see Chapter 13) and some using surface waters. Underground waters often need to be raised to the surface, and in primitive systems buckets, winches, and levers may be used. In other cases, people have developed ingenious methods of exploiting the topography to feed water to the fields by gravity. The **qanats** of Chile and Iran (see Chapter 13) and the **galeria** found in the Tehuacan Valley of Mexico (Figure 12.9) are examples. More sophisticated systems use pumps and hydraulic rams. Once at the surface, the water has to be carried to the fields, either in canals or pipes. It is then distributed by a variety of methods to the plants. Common distribution systems are illustrated in Figure 12.10.

Irrigation of crops is a highly skilled task if it is to be successful. The farmer must not only be able to predict when crops will need water to counteract the effects of excessive evapotranspiration, but also apply the right amount of water, and control its quality. Failure to do these things may lead to serious problems.

Applying water when it is not needed, or in too great a quantity, is a waste of what, in arid areas, is a valuable and scarce commodity. It can also damage the soil and the crop. It may compact the surface and lead to saturation of the soil. This encourages shallow rooting of the plants, which are able to obtain all the water they need from the surface. When irrigation ceases, the water level in the soil falls and the plants start to suffer from moisture stress. In addition, surplus water may leach salts from the soil. To some extent this is beneficial, for

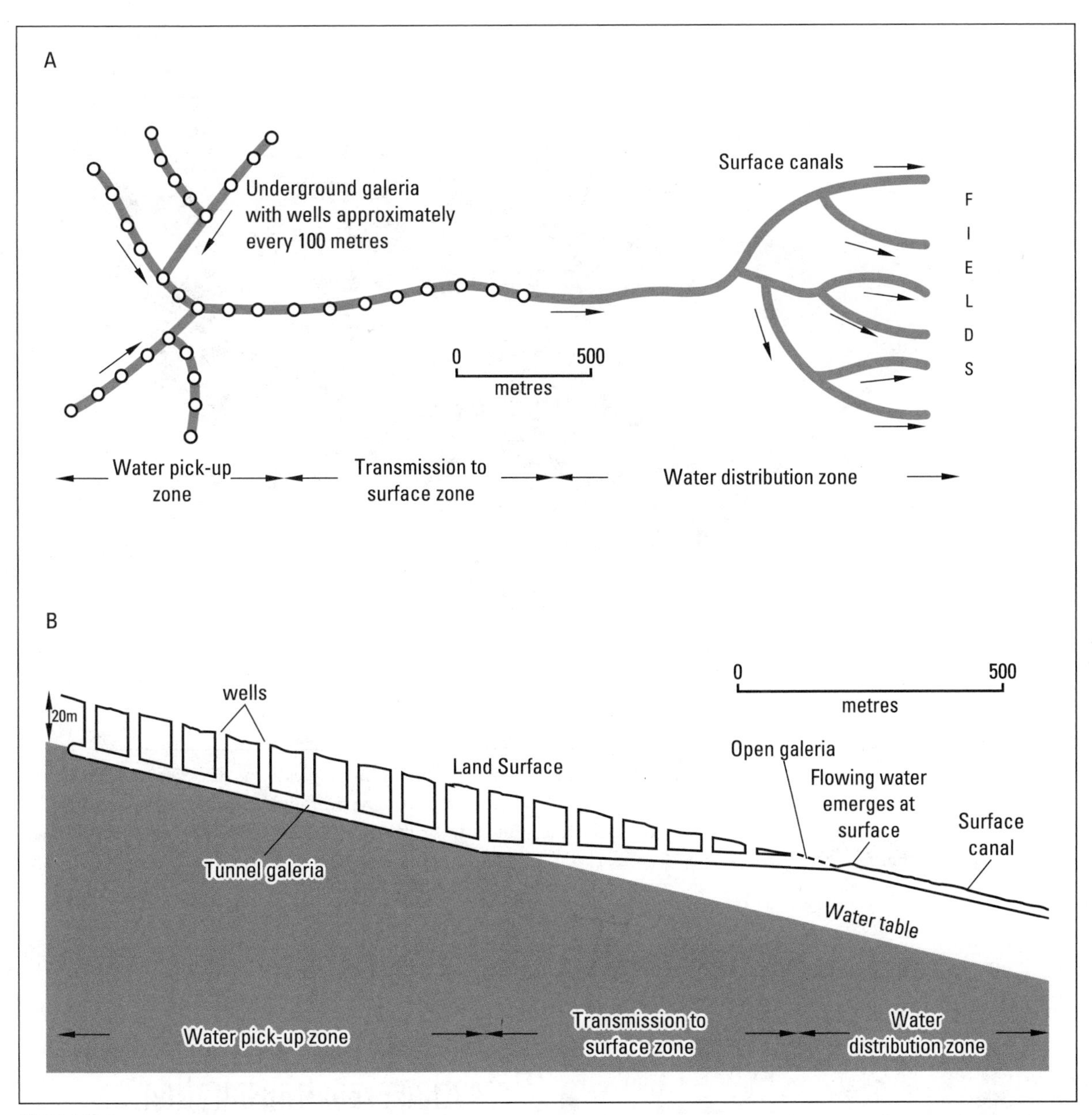


FIGURE 12.9 ▼ Typical galeria irrigation system from the Tehuacan Valley, Mexico; (A) plan view and (B) cross section (after Kirkby, 1969)

there are often excessive quantities of sodium and calcium in the surface layers; irrigation with large amounts of water washes them into the lower layers, away from the roots. But it is not only the harmful salts that are removed; vital plant nutrients may also be washed away, reducing soil fertility.

The application of too little water may also cause problems. Evaporation quickly dries out the surface soil, drawing water to the surface and causing accumulation of salts in the root zone. The plants suffer from high salinity (salt toxicity), and in severe conditions they wilt as the sodium draws the sap from the roots (a process known as plasmolysis). In severe conditions, a salt pan may even form at the surface (Figure 12.11).

Perhaps the most serious problem of irrigation is the supply of water of suitable quality. In arid areas, where excess evapotranspiration is most common, water resources are scarce and may suffer from high concentrations of soluble salts. During conveyance of water to the fields, considerable evaporation takes place, and this too increases the salt concentration.

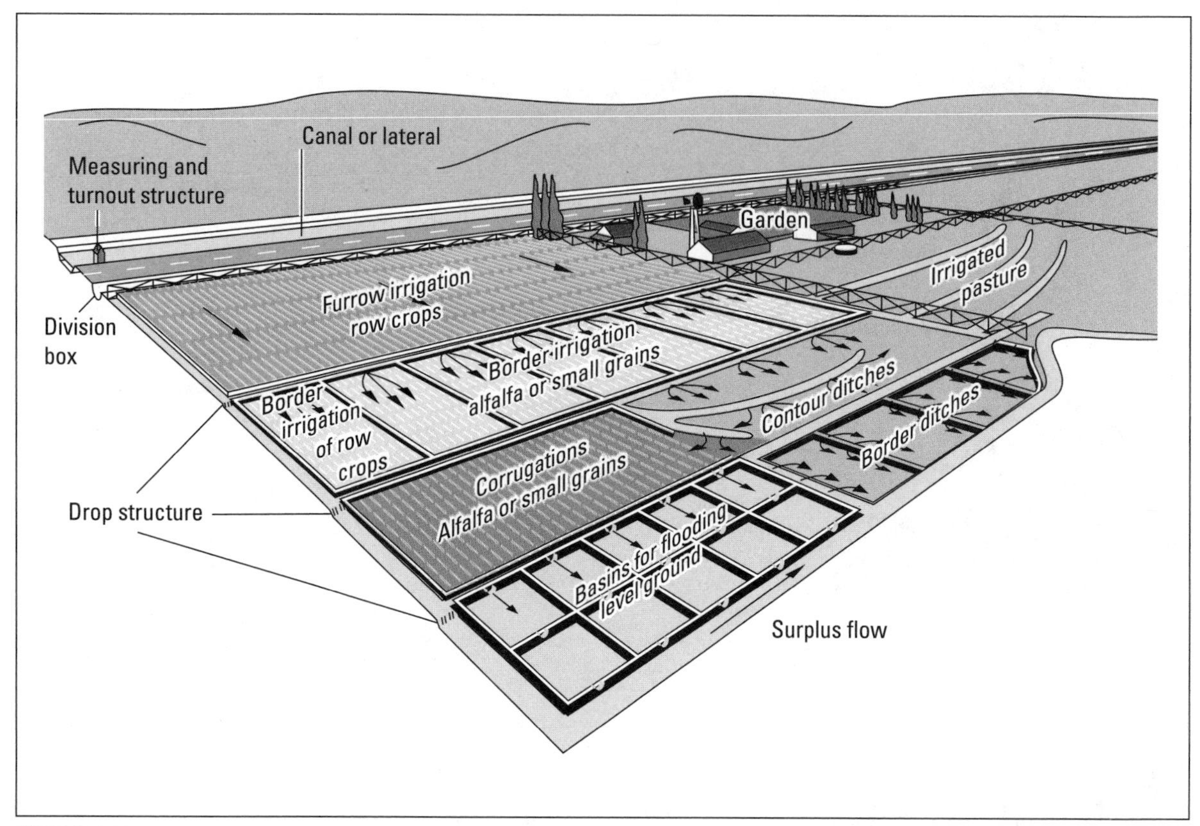


FIGURE 12.10 ▼ Irrigation techniques for field crops using the channel system. Arrows indicate the direction of water flow. Water is released to the entire area from the canal at the top. It moves through channels to the drainage ditch at the bottom, then out through the waste ditch. It is difficult to achieve even distribution with this technique (after Israelson and Hansen, 1962)

FIGURE 12.11 ▼ Saline groundwater has been drawn to the surface by summer evaporation near Mankota, Saskatchewan, 1985. The salt deposit visible here has killed the vegetation (photo courtesy of Saskatchewan Ministry of Agriculture and Food)

Application of saline water to the crop rapidly leads to a variety of problems.

# Measuring evapotranspiration

One of the main needs of the farmer or irrigation engineer is to be able to predict when plants will suffer from moisture stress and how much water must be applied. This involves being able to measure or calculate the rate of evapotranspiration. Knowledge of evapotranspiration losses is also required to plan water management policies; the hydrologist needs to know what proportion of the precipitation will be available to replenish groundwater or runoff into streams. Therefore, the measurement of evapotranspiration is important. Unfortunately it is also difficult. Several systems of measurement have been developed:

1. Direct measurement (e.g., with evaporation pans and lysimeters)
2. Meteorological formulas
3. Moisture budget methods

## DIRECT MEASUREMENT

Possibly the most widely used method of direct measurement is the **evaporation pan**. This consists of a shallow pan filled with water. The rate at which the water is lost through evaporation is measured with a gauge. This procedure measures only potential evaporation, for it does not allow for limitations of moisture supply, nor does it directly determine transpiration losses. In addition, the results seem to vary according to the size, depth, colour, composition, and position of the pan, so it is not always easy to compare results from different sites.

An alternative system of direct measurement is the **lysimeter** (Figure 12.12). This may be employed to measure either potential or actual evapotranspiration. To measure PE, the column of soil is kept constantly moist so that water deficiencies do not occur. To measure AE, the column is allowed to respond naturally to atmospheric conditions. Regular weighing allows the moisture content to be determined. If the amount of precipitation is known, the moisture loss through evapotranspiration can be calculated (Table 12.2). Commercial lysimeters may weigh up to 60 tonnes, and for really accurate results large columns and precise yet robust weighing instruments are required. But it is quite easy to construct a simple

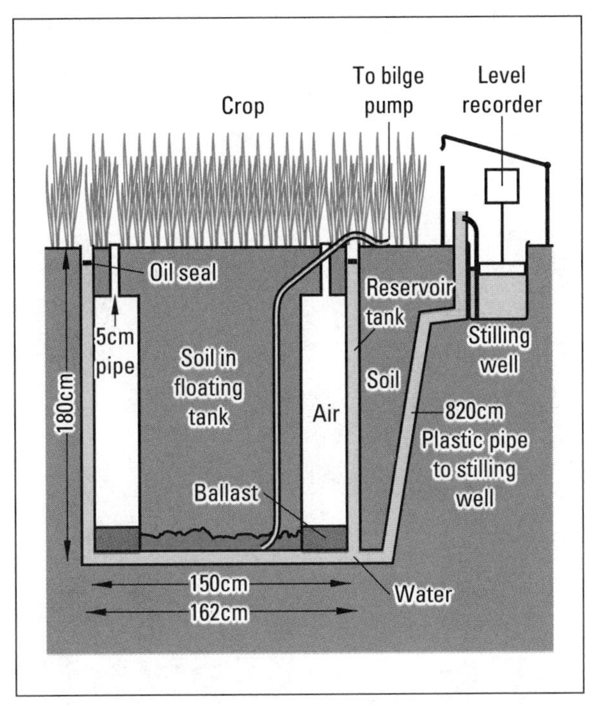


FIGURE 12.12 ▼ Lysimeter installation at Hancock, Wisconsin. The soil block floats in a tank of water. Changes of water level are recorded instead of weighing the block (after Barry, 1969)

## Table 12.2

*Calculation of evapotranspiration through lysimeter moisture measurements*

| Date | 1<br>Precipitation (cm) | 2<br>Weight of precipitation (g) | 3<br>Weight of lysimeter (g) | 4<br>Previous weight of lysimeter (g) | 5<br>Change in weight (g) (3 – 4) | 6<br>Weight transpired and evaporated (g) (2 – 5) | 7<br>Water transpired and evaporated (cm) (6 ÷ surface area) |
|---|---|---|---|---|---|---|---|
| 1.8.81 | 0.24 | 75.36 | 9110.35 | 9062.75 | +47.60 | 27.80 | 0.09 |
| 2.8.81 | — | — | 9097.21 | 9110.35 | –13.14 | 13.14 | 0.04 |
| 3.8.81 | — | — | 9042.94 | 9097.21 | –54.27 | 54.27 | 0.17 |
| 4.8.81 | — | — | 8986.32 | 9042.94 | –56.62 | 56.62 | 0.18 |
| 5.8.81 | 0.51 | 160.14 | 9124.67 | 8986.32 | +138.35 | 21.79 | 0.07 |

*Note* Surface area of lysimeter is 314 cm$^2$.

and effective lysimeter from a tin can and a piece of nylon mesh (Figure 12.13).

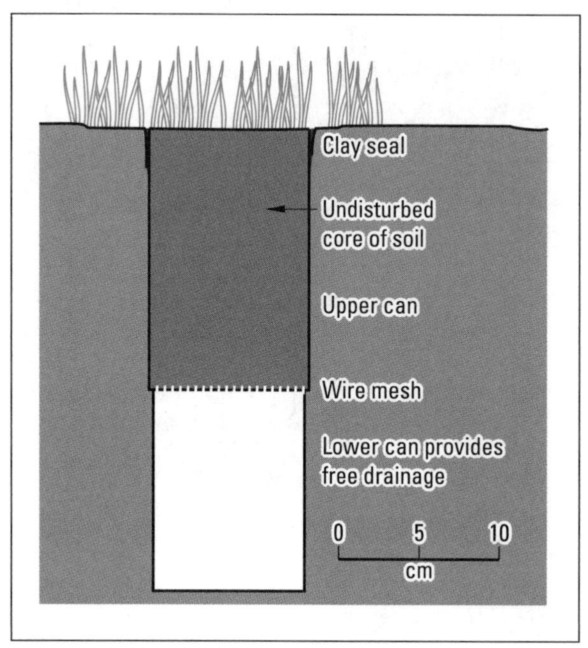


FIGURE 12.13 ▼ Field weighing lysimeter. The can is removed and weighed in the field (after Atkinson, 1971)

## METEOROLOGICAL FORMULAS

Because evapotranspiration is greatly dependent upon atmospheric conditions, it is possible to derive good estimates of PE from data on meteorological conditions. Formulas have been developed to do this, some of them so complex that it is almost impossible to use them under normal circumstances; the necessary data just do not exist.

This problem is illustrated by what at first seems to be a simple approach. As we saw in Chapter 3, the energy budget at the surface can be expressed as follows:

$$Q^* = Q_H + Q_E + Q_G$$

where $Q^*$ is the net radiation, $Q_H$ is the sensible heat flow, $Q_E$ is the heat usage through evaporation, and $Q_G$ is the heat flow into the ground. If we could determine all the other components of the equation we could find $Q_E$ by difference, and conversion to water depth would tell us how much evaporation was occurring. Unfortunately, we rarely know the value of the other components; $Q_H$, in particular, is difficult to determine. Consequently, to solve this equation, we would have to use further indirect methods to allow us to estimate $Q_H$.

Because of this problem of obtaining data, a number of simpler, more empirical formulas have been produced. These are much easier to use; they are, however, based not on physical principles but on the observed relationships between evapotranspiration and one or more climatological variables. Lack of data is a problem for most parts of the world and in most time periods. Often the only available technique is the empirical. A farmer's knowledge of the local climate is empirical; it is understanding built up over years of observation and experience without formal training in the mechanisms involved. The work of Thornthwaite represents one of the best known examples of an empirical approach to assess water balance. He developed a formula to determine PE for regions south of 50°N latitude. He assumed that once the mean monthly temperature was below freezing, nothing moved in the water cycle. This underlines a major problem with the empirical approach: often the relationships that have been obtained under one particular climatic regime may not be applicable elsewhere.

We encountered this when discussing climatic classification in Chapter 7. In simple terms, PE is calculated from the formula:

$$E = 1.6\,(10T/I)^a$$

where $E$ is the unadjusted value of PE, $T$ is the mean monthly temperature in °C, $I$ is an annual heat index† at the measuring station, and $a$ is a constant that varies in relation to $I$. The value of $E$ is adjusted to allow for day-length to give PE.

In this formula, Thornthwaite is using temperature as a substitute for radiation, and it therefore works reasonably well where the two are closely correlated. In the tropics, however, the equation underestimates PE because temperatures lag behind radiation inputs. In addition, the method takes no account of wind, although this may be locally important. Nevertheless, the relative simplicity of the method makes it popular, and despite its shortcomings and its inevitable inaccuracies, it is one of the more widely used methods of assessing PE.

The amount of moisture lost from the surface is the actual evapotranspiration (AE). It is limited by the amount of precipitation and moisture stored in

† $I = \sum (T/5)^{1.514}$, where $\sum$ means 'sum of' for all months.

the soil. If PE exceeds AE then moisture must be taken out of the soil to offset the deficit. In this case, if the deficit continues, a point will be reached where the soil moisture that is available to plants is depleted. As we learned, this is the wilting point.

In Canada, The **Penman formula** is the main method for calculating PE. It is less empirically based than the Thornthwaite method. Studies of drought conditions in western Canada use the more sensitive **Palmer Drought Index**. One advantage of Penman's method, compared with some others, is that it requires meteorological measurements to be made at one level only. Unfortunately, net radiation is rarely measured directly and has to be derived by the method outlined above. Penman's method appears to be the most appropriate approach for conditions in Canada. The pattern of PE across Canada calculated by this method is shown in Figure 12.14.

## MOISTURE BUDGET METHODS

An alternative method of obtaining AE is by the moisture balance equation. The components of the movement of moisture over time on a land surface are given in the water budget equation:

$$P = ET + \Delta ST + S$$

where $P$ is the amount of precipitation, $ET$ is loss by evapotranspiration, $\Delta ST$ is the gain or loss of moisture from the soil, and $S$ is the amount of surplus. A surplus is achieved if soil moisture is at capacity and there is more than sufficient precipitation to satisfy evapotranspiration.

If the soil moisture is at less than capacity, then it will be recharged from the surplus. If there is still surplus it will go to runoff or percolate down to the water table. This will be discussed at length in Chapter 13.

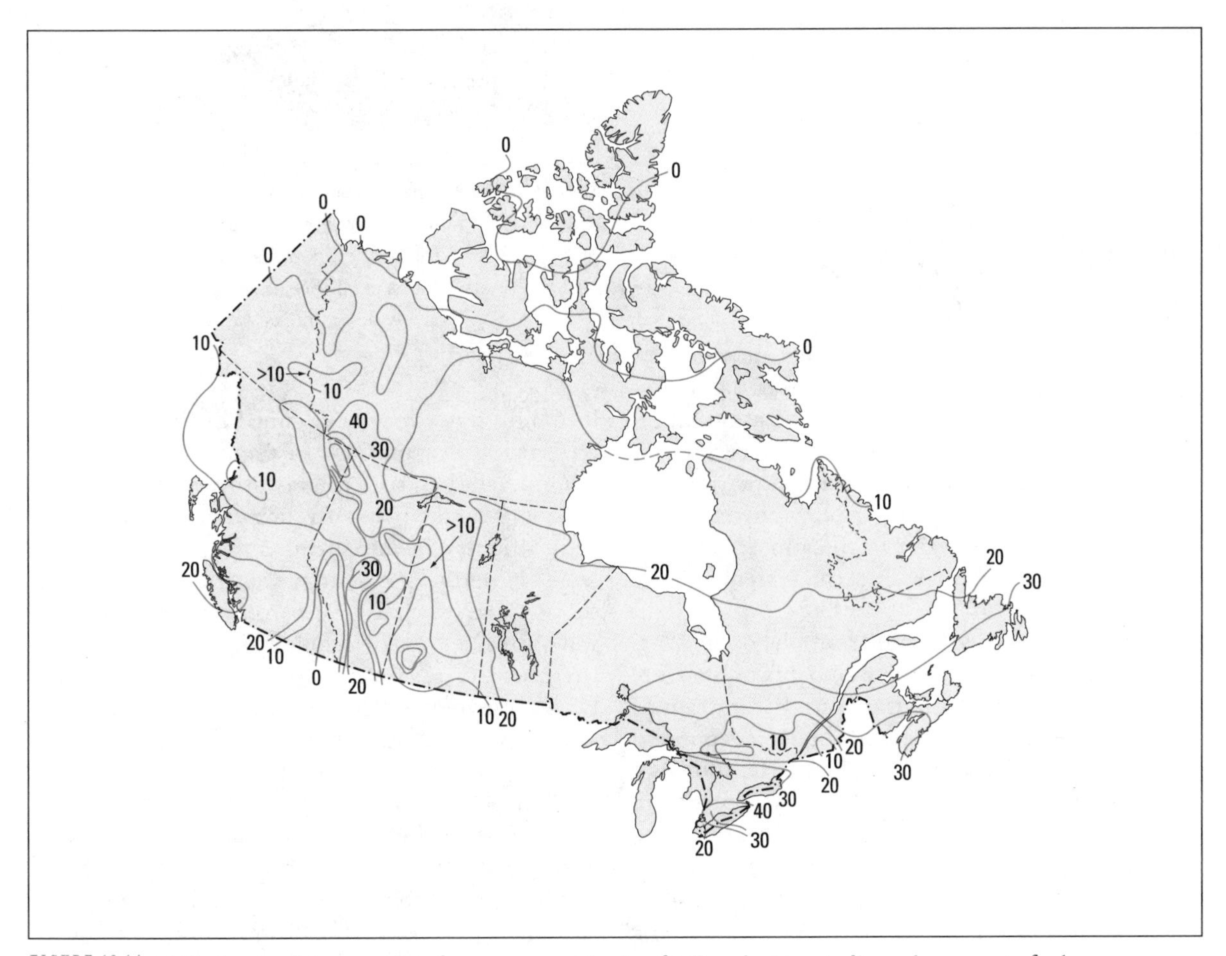


FIGURE 12.14 ▼ Estimate of the annual surface evapotranspiration for Canada. Lines indicate the amount of solar energy ($Q_E$) used in the process, measured in kilolangleys per year (after Hare and Thomas, 1979)

At a site, the moisture balance can be expressed as:

$$P + I = E + R + D + \Delta S$$

where $P$ is precipitation, $I$ is irrigation, $E$ is evapotranspiration, $R$ is runoff, $D$ is drainage to bedrock, and $\Delta S$ is the change in soil moisture content. As before, if we knew all the other elements in this equation, we could calculate $E$ by difference. Several of the components present little problem, for precipitation and irrigation inputs can easily be measured, as can runoff. But drainage to bedrock ($D$) and changes in soil moisture content ($\Delta S$) are rarely known. As a result, the method can only be used on a large-scale, longterm (e.g., annual) basis where it can be assumed that drainage to bedrock is balanced by release from spring seepage, and where changes in soil moisture content are negligible.

It is clear from what we have said that evapotranspiration remains one of the most difficult aspects of the hydrological system to measure. For this reason, our knowledge of the processes involved remains uncertain. For the same reason, it is difficult to give precise figures to the global pattern of evapotranspiration. Nonetheless, it is useful to consider the role of evapotranspiration within the global hydrological cycle.

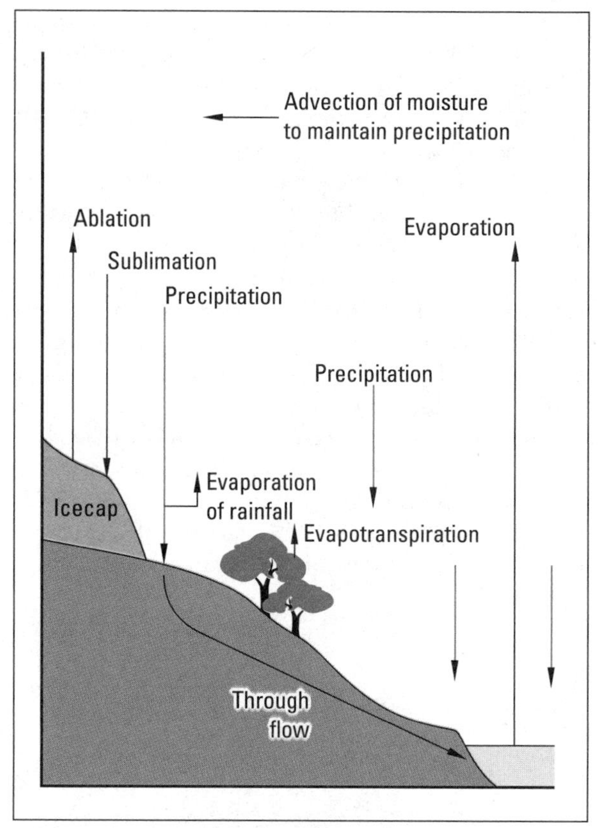


FIGURE 12.15 ▼ The hydrological cycle, showing the main losses of water to the atmosphere. Ablation includes evaporation and sublimation losses from the surface. Sublimation with arrow down indicates frost added to the ice surface

## EVAPOTRANSPIRATION IN THE HYDROLOGICAL CYCLE

As we have noted, evapotranspiration provides the main output of moisture from the terrestrial hydrological system, returning water to the atmosphere. So far we have discussed some of the processes, but it is important to appreciate that evapotranspiration occurs in many different stages of the hydrological cycle. Thus, losses of water to the atmosphere may take place at any point within the system (Figure 12.15).

One of the major losses, for example, occurs during precipitation. Considerable amounts of moisture may be evaporated during rainfall, and small droplets in particular may be totally evaporated before they reach the ground. Similarly, moisture intercepted by vegetation is also susceptible to direct return to the atmosphere by evaporation. The amount of moisture retained on vegetation during a storm varies according to the character of the storm, the species of plant, leaf density (and therefore the time of year), and, of course, the vegetation density. In the case of woodlands, as much as 50 percent of incoming water may be retained in the canopy and returned as evaporation. In the case of more low-lying vegetation, such as grass, the amount of interception is not known with such certainty, because of the difficulty of measuring interception in such crops. Nevertheless, again it seems likely that interception may reach 10–30 percent; we only have to walk across a grass field following rain to appreciate the quantity of water trapped in the vegetation. Once more, most of this is lost through evaporation.

A proportion of the water that reaches the soil is also returned by evaporation, for in heavy storms rainfall often collects in surface depressions and these are gradually dried by the sun. Similarly, some of the rainfall flows across the surface as runoff, and further evaporation losses may occur at this stage. Rates are generally low, however, for turbulence mixes the water and disperses the heat from solar radiation through the water body. Thus much greater inputs of energy are needed to heat the water and less is available to carry out evaporation.

Nevertheless, losses at this stage are often important. Open canals and reservoirs may lose considerable quantities of water through evaporation (Table 12.3), and, in arid areas especially, this may represent an irretrievable loss of an important resource.

In vegetated areas, the major process of moisture return is by transpiration. Rates of transpiration vary according to the character of the vegetation and therefore change over both space and time (Figure 12.16). They are at a maximum when the vegetation is in full leaf and the soil is moist; they decline as the plants lose their leaves or the soil

## Table 12.3

*Mean daily evaporation from an open water surface (mm day$^{-1}$), Molato, Italy*

| *J* | *F* | *M* | *A* | *M* | *J* | *J* | *A* | *S* | *O* | *N* | *D* |
|---|---|---|---|---|---|---|---|---|---|---|---|
| 0 | 0 | 1.8 | 2.7 | 3.0 | 6.2 | 7.9 | 5.7 | 4.3 | 4.1 | 2.3 | 1.3 |

Source H.L. Penman, 'Natural evaporation from open water, bare soil and grass,' *Proceedings of the Royal Society*, 193.

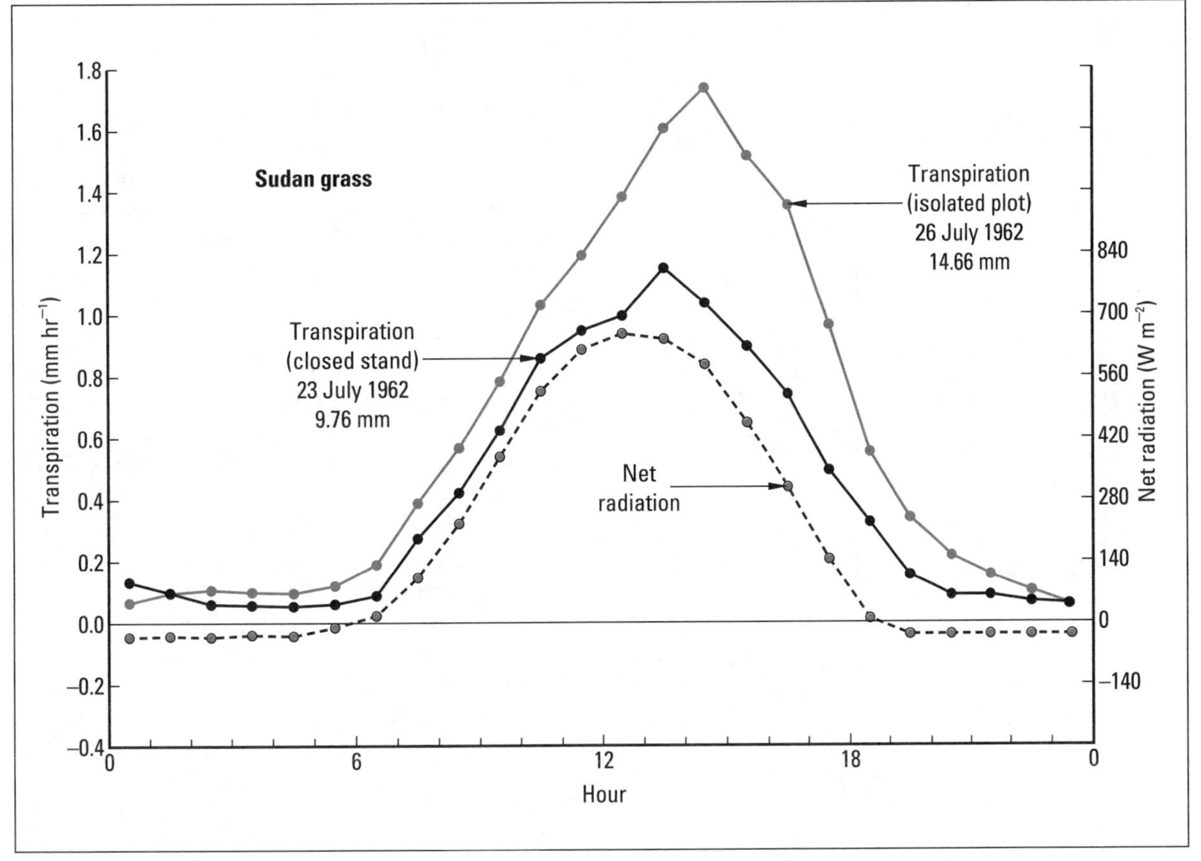


FIGURE 12.16 ▼ Transpiration from a closed stand of Sudan grass at Phoenix, Arizona, on 23 July, 1962, and from an isolated 1 m$^2$ plot of the same grass three days later. Average net radiation for the two days differed by only 5 percent (after Van Bavel *et al*, 1963)

dries out. During the course of a single year, therefore, transpiration losses may show complex fluctuations in response to prevailing conditions.

Without doubt, the major evaporative losses occur from the sea—possibly 85 percent of the global return to the atmosphere is from the oceans. The reason is not only the great extent of the oceans—some 70 percent of the world's surface—but also that evaporation can continue at or near the potential rate. Unlike evapotranspiration from the land, the process is unhindered by water shortages. Even so, seasonal and regional differences in evaporation can be seen due to the effect of changing meteorological conditions (Figure 12.17).

Finally, evaporation may occur from the other main storage component of the hydrological cycle—the cryosphere. In general, losses are small, for it requires large amounts of energy to convert ice to water vapour—the process known as sublimation; some $2.83 \times 10^6$ J are needed to evaporate 1 kg of ice at 0°C. Sublimation does occur in the marginal areas of glaciers and icecaps, however, where seasonal inputs of solar radiation may be high, and perhaps 2 percent of the moisture is returned to the atmosphere each year in this way.

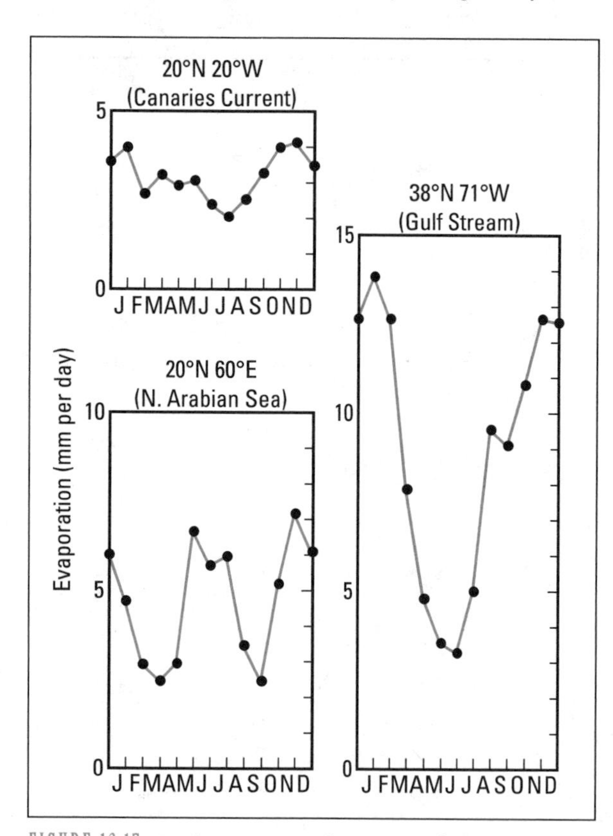


FIGURE 12.17 ▼ Average annual variation of evaporation at selected oceanic locations (after Sellers, 1965)

# Conclusion

The importance of evapotranspiration to the hydrological cycle is apparent. It provides a vital link between the surface waters and the atmosphere, a link operating at almost every stage of the cycle (Figure 12.15). In the process of returning water to the atmosphere, evapotranspiration involves a major exchange of energy; thus it is also an important part of the energy cycle of the atmosphere. In addition, evapotranspiration has an overriding influence upon vegetation growth. Transpiration provides the means by which nutrients are absorbed by plants, and by which plant temperatures and turgidity are maintained. When evapotranspiration is prevented, for example when the plant is kept in a totally saturated atmosphere, or when there is no available moisture in the soil, plant growth ceases.

Through the control evapotranspiration exerts on the hydrological cycle, and through its effects on plants, it also has a fundamental influence upon human life. In many parts of the world, high rates of evapotranspiration constrain agriculture, and farmers have been forced to adapt to the conditions by using irrigation or by growing drought-resistant crops. Excessive evapotranspiration, like lack of rainfall, is also responsible for the hardship and famine so apparent in many arid areas. As we have seen, evapotranspiration remains one of the most problematic parts of the hydrological cycle to study. Much more research is needed before we fully understand the processes and their implications, or can do much to plan for their consequences.

# Runoff and storage

## The hydrological crossroads

Alfred the Great, an early king of England, was not just a great ruler, he was also something of a philosopher and scientist. He had what would today be regarded as a curious notion about the working of the hydrological cycle. In a book called *Boethius*, he claimed that all rivers flowed to the sea, but that from there the waters seeped into the rocks and made their way back through unknown passages to the same stream from which they emerged! We know now that this is wrong, that streams obtain their water from rainfall. But what happens between the rainfall and the stream? How does the water reach the channel, and what happens to it on the way? These are some of the questions we will be answering in this chapter.

They are questions of considerable importance, for what happens to the rainfall after it has reached the ground influences the way streams behave, the supply of water for human needs, and many processes of erosion in the landscape.

Where, then, does the water go? There are four possible routes (Figure 13.1). It may go straight downward at the surface, infiltrating the soil and

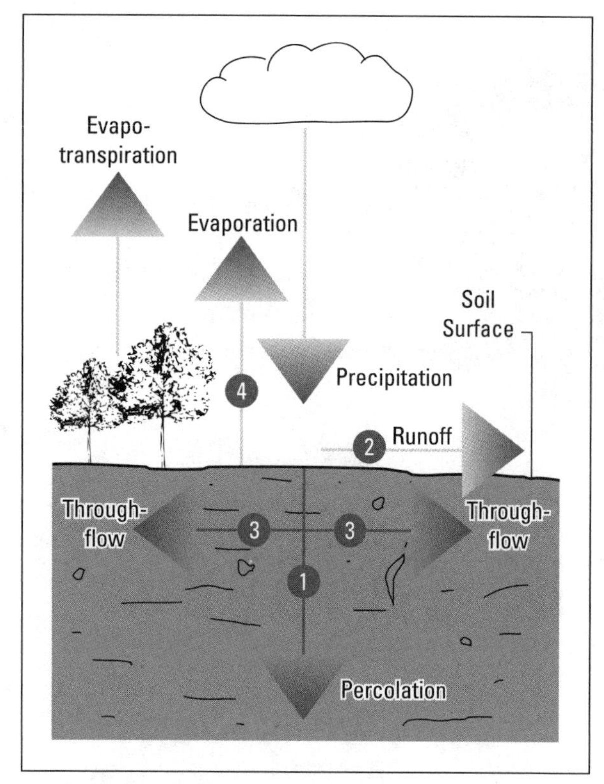


FIGURE 13.1 ▼ The hydrological crossroads

percolating down to the bedrock. It may be deflected laterally, flowing either in the soil or on top of it downslope toward stream channels and then to the sea. Or it may be trapped and then returned whence it came, by evaporation (or transpiration) to the atmosphere.

Thus the ground surface acts like a hydrological crossroads, diverting the rainfall four ways. The relative proportions going in each direction are extremely significant, for people obtain much of their water from groundwaters; these need to be replenished by percolation (route 1 in Figure 13.1). Similarly, streamflow is maintained in part by the water that moves laterally in or on the soil (routes 2 and 3); if this supply of water ceases then the streams may dry up. Plants obtain their moisture from the soil, and if sufficient water is not retained close to the surface then vegetation cannot continue to transpire and plants may wilt and die.

It is not only the amount of water that is significant; the rate of movement is also critical. In particular, the speed at which the water flows laterally into rivers and streams affects the rate at which these respond to rainfall, and their liability to flooding. In many arid parts of the world the extremely rapid runoff of water into streams during rainfall may result in a torrent of water developing in the channels. People have been drowned as a result; campers on what were once dry river beds have been swept away by the flash flood before they could escape.

# Overland flow

## THE GENERATION OF OVERLAND FLOW

The streets of our local town are soon awash during a heavy storm, as water collects in the gutters and pours into the drains. Off the streets, in the parks and gardens, the effect of the storm is less dramatic; very much more of the rainfall disappears. The difference is that in the street the rainfall cannot soak into the ground. The road surface is almost impermeable and much of the water gathers on the surface, then runs off as **overland flow**. Soil, however, is far more permeable; much of the rainfall infiltrates and relatively little remains on the surface. In many cases it is only during the heaviest storms, or the most prolonged periods of rain, that overland flow occurs on the soil.

These observations provide us with the key to understanding the process by which overland flow is generated. Two sets of factors govern the process: the permeability or, more strictly, the **infiltration capacity** of the surface, and the intensity and duration of the storm. Overland flow occurs when the amount of rain reaching the surface is greater than the ability of that surface to absorb the water.

The reasons for this imbalance between rainfall and infiltration vary. Two theories have been developed to explain it, and there has often been conflict between geographers who support the two views. We will see, however, that both are probably valid under different circumstances; indeed, there is less difference between them than is often implied.

## HORTONIAN OVERLAND FLOW

The traditional concept of overland flow generation is attributed to R.E. Horton, and runoff that seems to fit his theory is often referred to as **Hortonian overland flow**. Horton worked mainly in the south and west of the United States, and as we will see this influenced his ideas of what happened during a storm. He argued that overland flow occurred when the intensity of the storm was so great that the soil could not absorb the water quickly enough. In other words, when rainfall intensity exceeded infiltration capacity, overland flow was produced.

Because the infiltration capacity of many soils is high—that is to say, the soil can rapidly absorb large quantities of water—Hortonian overland flow occurs mainly in areas that experience very intense and sudden storms, such as the semi-arid areas of southwest USA.

### SATURATED OVERLAND FLOW

Elsewhere, Hortonian overland flow is less common. Instead, much of the water infiltrates the soil, then passes through the soil layers as throughflow, or percolates to the groundwater (Figure 13.2). Nevertheless, even in these areas, surface runoff does occur, but it is explained by rather different processes. Overland flow in these instances is known as **saturated overland flow** and it occurs as follows.

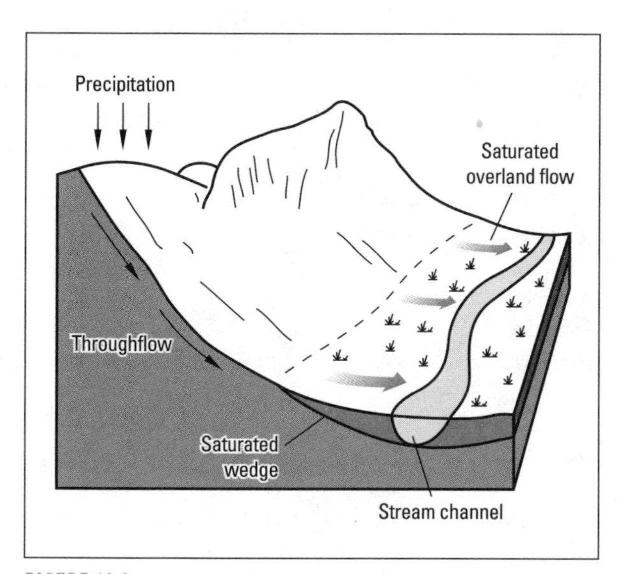


FIGURE 13.2 ▼ The generation of saturated overland flow

During the early part of the storm, rainwater infiltrates the soil and drains slowly downward. The rate of percolation is often slow, however, especially at depth where the soil is more compact and where channels created by roots and soil animals (e.g., earthworms or termites) are absent. At the same time, inflow of water may come from adjacent areas so that water enters the soil more quickly than it can escape. Ultimately, the surface layers of the soil become saturated. No more water can infiltrate and overland flow occurs.

## Rainfall intensity and duration

In both cases, one of the critical factors influencing the generation of overland flow is the amount of rainfall. Hortonian overland flow is produced mainly during very intense storms, which may be relatively rare and short-lived. As we saw in Chapter 11 they occur mainly in association with unstable air in which convective clouds can form and lead to very localized, but often heavy rainfalls. Saturated overland flow tends to occur following more prolonged rainfall. It develops mainly in slope foot or bottom land sites where slope angles are low, the soil is naturally wet, and high inputs of water from upslope take place (Figure 13.2). Although the intensity of precipitation may be low, therefore, gradual saturation of the soil occurs. Ultimately the soil becomes waterlogged and any additional input of water, either from upslope or direct rainfall, spills across the surface.

Both forms of overland flow are often localized, for rainfall intensity varies from one place to another. Often this variation is topographically controlled, and exposed slopes catch much of the rain, while the sheltered areas are left in a slight rain shadow.

### THE EFFECT OF VEGETATION

Vegetation modifies rainfall so that the intensity and distribution of rain at the ground surface are not the same as above the vegetation canopy. We can see the effects of **interception** most clearly in woodlands. The tree canopy may intercept as much as 30 percent of the rainfall in heavy storms, and much more in light showers (Figure 13.3). Much of this may be trapped on the leaves until returned to the atmosphere by evaporation. Much of the remainder is diverted down the tree trunk, where it is known as **stemflow** (Figure 13.4). Often 10 percent or more of the precipitation may reach the ground as stemflow, and it is clear that this water is concentrated into a very small area around the base of the trunk. There is also a tendency for part of the rainfall to be shed at the edge of the canopy (canopy

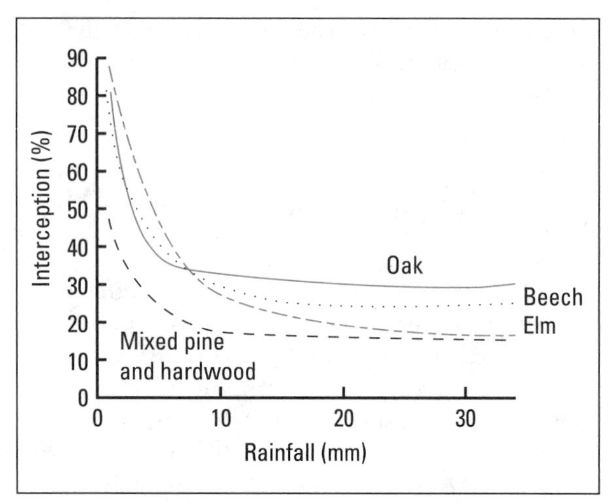


FIGURE 13.3 ▼ Relationships between rainfall amount and interception for different tree species (from Kittredge, 1973)

drip), giving a zone of high intensity around the fringe of the tree (Figure 13.4). As anyone who has sheltered from a storm knows, relatively little rainfall passes through the canopy by **throughfall**. At least during the early parts of the storm the canopy acts as an efficient umbrella. As the leaves become saturated, however, more and more of the rainfall makes its way through the canopy as **leaf drip** and this umbrella effect is reduced.

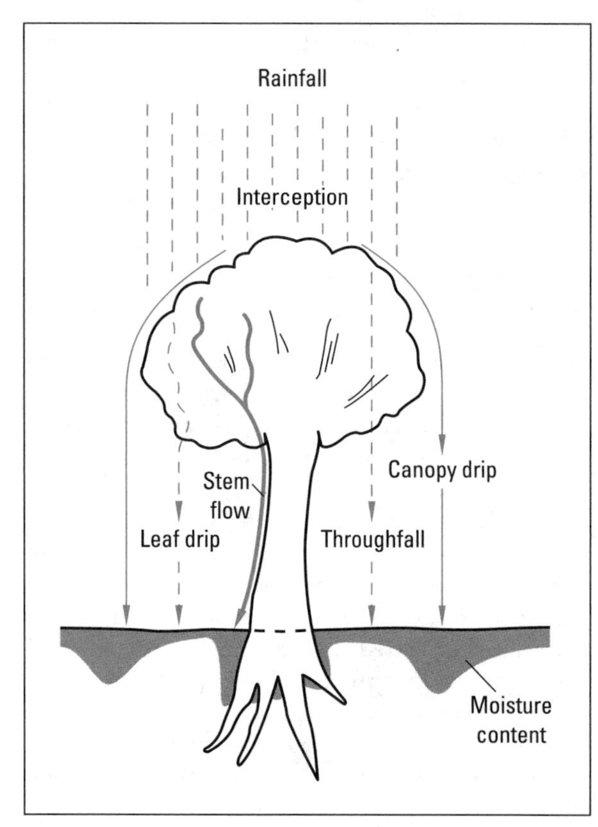


FIGURE 13.4 ▼ The routing of rainfall by trees

The ability of trees to intercept rainfall in this way varies according to species and time of year. Clearly, far less is intercepted when the trees are leafless. Other types of vegetation also intercept rainfall in a similar way, and thus during a single storm inputs of rainfall beneath a vegetation cover are almost always less than in the open land (Table 13.1). In addition, the distribution and intensity of the rainfall beneath the canopy is markedly irregular, with the result that overland flow may be produced locally (e.g., around the tree trunk) even in relatively gentle storms. This, in fact, is one reason why the soil is often bare and eroded next to the trunk.

**Table 13.1**

▼

*Rainfall interception by different types of vegetation*

| *Vegetation cover* | *Condition* | *Interception (percent)* |
|---|---|---|
| Douglas fir | 25 years old | 43 |
| Spruce fir | Mature | 37 |
| Lodgepole pine | Mature | 32 |
| Hemlock | Mature | 30 |
| Heather | Different ages | 35–66 |
| Mixed grass sward | Full cover | 26 |
| Clover | Full cover | 40 |
| Corn | During growing season | 16 |
| Corn | Full cover | 40–50 |
| Soybean | Full cover | 35 |
| Oats | During growing season | 7 |
| Oats | Full cover | 23 |

# Infiltration capacity

The second factor influencing the generation of overland flow is the infiltration capacity of the soil—the ability of the soil surface to absorb water.

This depends upon a variety of conditions, some of them, such as the texture and structure of the soil, relatively permanent, and others, such as the existing moisture content, relatively transient.

## THE EFFECT OF SOIL TEXTURE AND STRUCTURE

Water enters the soil by flowing into the small holes and cracks (the pores and fissures) in the surface. The pores occur mainly between the individual particles of soil because the particles do not fit tightly together. In addition, large pores or microchannels are created by organisms and plant roots that burrow or push their way into the soil. The soil particles, however, are grouped together into aggregates or peds separated by thin fissures and planes of weakness. These provide extensive, large voids within the soil into which water can flow. Clearly, the extent, number, and size of pore spaces (which together define the porosity of the soil) vary considerably. Many factors affect these properties. The size of the particles making up the soil (soil texture) are important. In general, fine-grained soils are more compact and have smaller voids than coarse-grained soils. As a result, water infiltrates less rapidly. Soil management is also important. Ploughing breaks up and overturns the surface soil, increasing the porosity and encouraging infiltration. Compaction, by machinery (e.g., in the ruts caused by tractors) or by animals, reduces porosity and inhibits infiltration. Vegetation has a further effect. A dense network of roots increases the infiltration capacity by providing channels in the soil, while the debris from the plants may provide a loose, porous surface layer into which the water can readily infiltrate.

## ANTECEDENT MOISTURE

The other main factor of importance is the existing moisture content of the soil (the **antecedent moisture**). As long as pores and fissures in the soil are free of water they can absorb the rainfall and infiltration can continue. But what happens when the pores become filled with water? The result is that infiltration can no longer occur and, as we have seen, saturated overland flow takes place. Thus, as the soil becomes saturated and the water infiltrating the soil builds up in the upper layers, the infiltration capacity falls until it equals the rate of downward percolation through the soil. At the point where the infiltration capacity falls below the rainfall intensity, overland flow is produced (Figure 13.5).

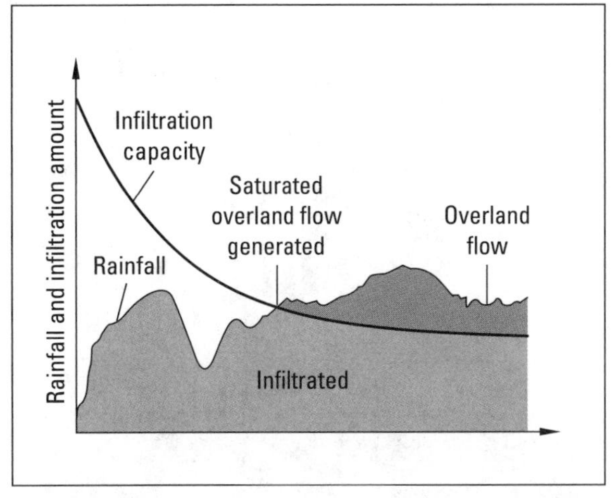


FIGURE 13.5 ▼ Infiltration capacity and rainfall during a storm and the generation of saturated overland flow

One factor complicates this picture. During rainfall, as the soil becomes wetter, changes in soil structure tend to occur that act to reduce infiltration capacity even more. The impact of raindrops on the surface may detach particles that lodge in the pores and seal the soil, producing a surface crust. In some cases this crusting is very severe and it leads to rapid reduction in infiltration. Clayey soils may also swell as they become wet; water becomes bound up within some of the complex clay particles in the soil, causing them to expand. As a result, the pores are closed up and infiltration capacity falls. When the soil dries again the clay shrinks and cracks may develop. Clearly, in this state, the soil has a very high infiltration capacity.

## MEASUREMENT

It is apparent that the infiltration capacity of the soil varies markedly over time and space. Variations in infiltration capacity also account for localized development of overland flow. For this reason, infiltration capacity is an important regulator within the hydrological system, and a vital property to measure when we are trying to monitor the system. In fact, measurement is fairly easy and the simple equipment shown in Figure 13.6 gives reasonably good estimates of infiltration capacity.

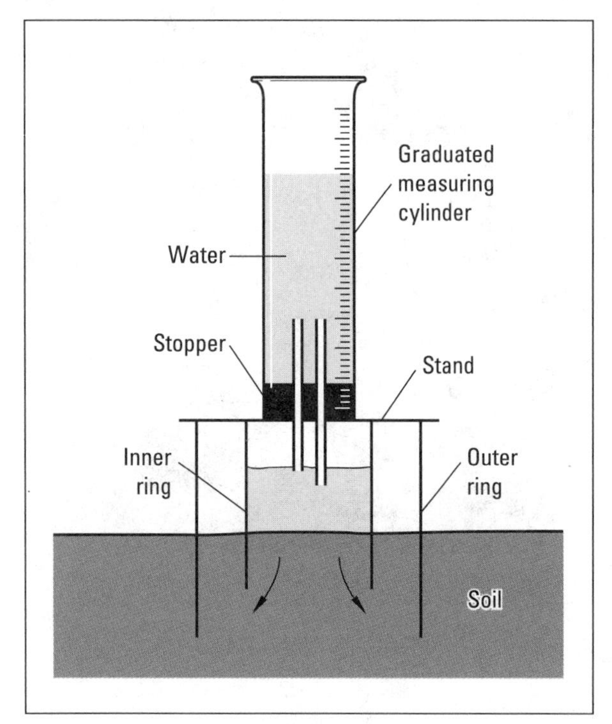


FIGURE 13.6 ▼ The ring infiltrometer for measurement of infiltration capacity. Measurement is normally made in mm $hr^{-1}$

## RILL AND CHANNEL FLOW

When the infiltration capacity is exceeded and water gathers on the soil surface, the first effect is normally for very localized flow to occur as water moves into small surface depressions and gathers there. It may be stored here for some time as **depression storage**, and if rainfall ceases very little overland flow may occur. Afterward, the water slowly infiltrates into the soil as the underlying layers drain, or it may be evaporated back to the atmosphere. If rainfall continues, however, the depressions become full and the water overflows to make its way downslope. It is at this stage that extensive overland flow becomes visible.

Initially, flow is often widespread and the water forms a thin sheet over the surface. This is known as **sheetflow**, and is most commonly seen in areas where the soil surface is relatively smooth and on the rounded crests of slopes. Sheetflow, however, rarely persists for any distance and it soon gives way to **rill** or **gully flow**. Rill development may start around stones or the low points of depressions where the flow is concentrated, or where surface unevenness or the effect of raindrops cause the sheetflow to become turbulent. In both cases localized erosion occurs and micro-channels develop that further concentrate the water. Eventually, the rills may become quite large and persistent and at this stage they are known as gullies. We will examine the processes involved in gullying in Chapter 21.

Both sheetflow and gully flow are relatively rapid, and rates of flow of as much as a metre per second are common on steep slopes. Flow is most rapid in channels, however, because the water then is deeper so the effect of friction with the bed is reduced. Even so, it is clear that flow rates vary markedly according to slope and surface conditions. Velocities tend to increase with slope angle so that flow is most rapid on steep hillsides. Friction increases with surface roughness, so flow is reduced on ploughed soils or where there is a dense cover of vegetation.

# Throughflow

The water that enters the soil does not all drain downward to the bedrock. Much of it flows laterally, eventually escaping back to the surface as seepage and contributing to streamflow. This is known as **throughflow** and it takes place mainly when the soil is fully saturated.

Compared to overland flow, movement in the soil is slow and often irregular. The reason is that much of the flow occurs in very small pores and fissures, and here, close to the soil particles, the water is affected by forces that hinder movement. We will examine these forces later.

Rates of movement reach a maximum in soils on steep slopes and in large channels or pore spaces. Much of the flow occurs through the coarser pore spaces in the soil (those greater than about 1 mm in diameter), but flow is most rapid in the tunnels created by rodents and other soil animals, and the moving water is often able to erode these tunnels to produce extensive systems of pipes within the soil. Rates of flow similar to those on land may be achieved in these pipes. In contrast, rates of movement through the fine pores in a heavy clay soil may be almost imperceptible, perhaps no more than 1 mm per day. Thus, the rates of throughflow are highly dependent upon soil conditions and vegetation, as well as topography.

Water moving through the soil tends to gather at the slope foot so that the soil there becomes totally saturated (Figure 13.2). As it does so, the water starts to emerge along seepage lines and produce overland flow. In addition, seepage may occur directly into the stream through the bed and banks. In this way, most of the throughflow reappears to join with the water moving by sheetflow and in rills and gullies and enter streams. It contributes to the next stage in the runoff process: streamflow.

# Streamflow

Streams represent permanent and relatively large features of the landscape, as we will see in Chapter 21. They also play a vital role in the hydrological system, for they conduct water across the land to the sea.

Streamflow varies markedly over time in response to the inflow of water from the surrounding land. Thus, if we monitored streamflow we would find that the volume of water passing through the channel—the stream discharge—fluctuated from day to day and even from hour to hour (Figure 13.7).

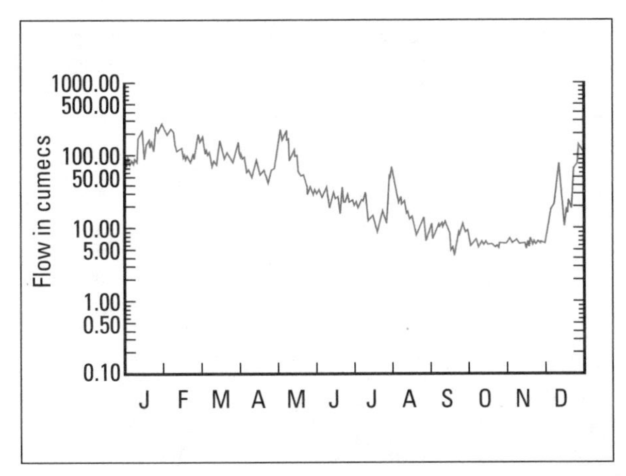


FIGURE 13.7 ▼ A stream hydrograph: River Thames, England, at Teddington, 1978 (from DoE Water Data Unit, 1983)

## MEASUREMENT OF STREAMFLOW

**Stream discharge** represents the volume of water passing through the channel in a given period of time. It is the product of the width and depth of the channel and the velocity of the water:

$$Q = x \times d \times v$$

where $Q$ is the discharge, $w$ is width, $d$ is depth, and $v$ is velocity. Normally discharge is measured in cubic metres per second (sometimes called cumecs).

Because width, depth and velocity all tend to vary together as discharge increases, measurement of discharge is complex. One-off measurements may be made by computing the depth and width of the flow at a cross section, and then measuring the velocity either with a current meter or by assessing the time taken for floats to pass over a measured distance (Figure 13.8). When we need to obtain a more continuous record of streamflow, however, this approach is clearly too time consuming, so permanent, automated measuring sites are set up.

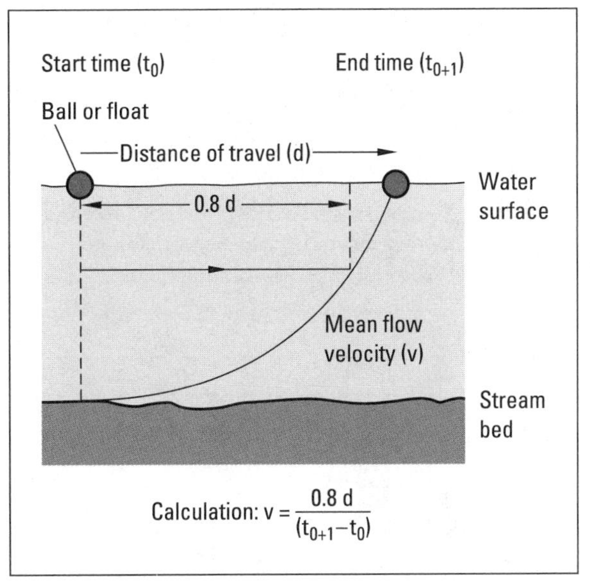


FIGURE 13.8 ▼ Measurement of stream velocity by the use of floats. The time taken for the float to pass over a measured distance is determined and the surface velocity computed. This is then converted to an estimate of mean flow velocity by multiplying by 0.8 (the ratio of mean flow to surface flow velocity)

In general, the relationship between width, depth, and velocity of flow at these sites is calculated at different discharges. Then only the depth of flow needs to be measured; the other variables can be calculated from these relationships. Commonly, a flow recorder is installed, consisting of a float attached to a pen. As the depth of the water varies the float moves the pen up or down and the pen traces the changes on a rotating chart (Figure 13.9). In this way a continuous record of streamflow is

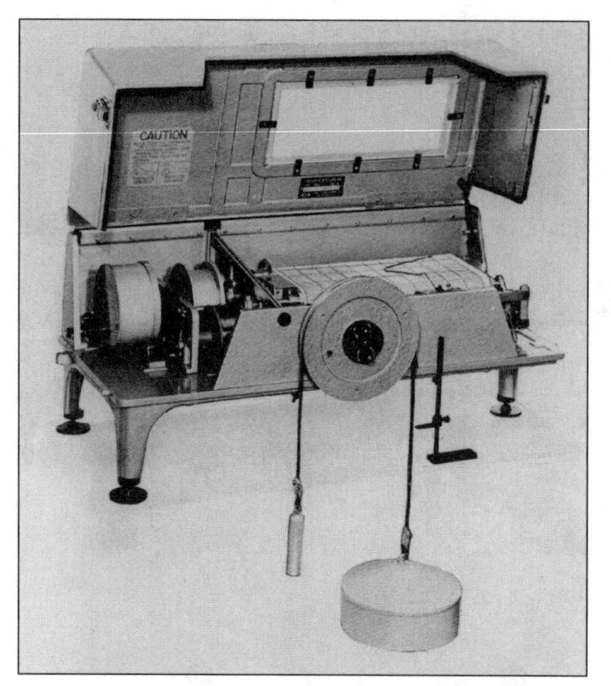


FIGURE 13.9 ▼ A Stevens type A-35 water level recorder used for tracing a hydrograph (photo: Ecosystem Sciences and Evaluation Directorate, Environment Canada)

gained. When this record is expressed as stream discharge it allows us to see the change in streamflow during an individual storm. The graph of flow is then called the **storm hydrograph**.

## THE STORM HYDROGRAPH

Typically, the storm hydrograph consists of a curve something like that in Figure 13.10. Soon after the onset of rainfall, the hydrograph rises rapidly, showing an increase in discharge. The time between the start of rain and the rise in the hydrograph (or the rainfall and streamflow peaks) is known as the **lag**; it represents the time it takes for the water to move overland and through the soil into the stream channel. Following this steep rising limb of the hydrograph the curve peaks, then starts to decline as discharge falls. This falling limb of the hydrograph tends to be less steep than the rising limb, showing that the decline in discharge occurs more gradually. Ultimately, the hydrograph settles at a low level and it characteristically continues at this level until the next storm. This is known as the base flow stage.

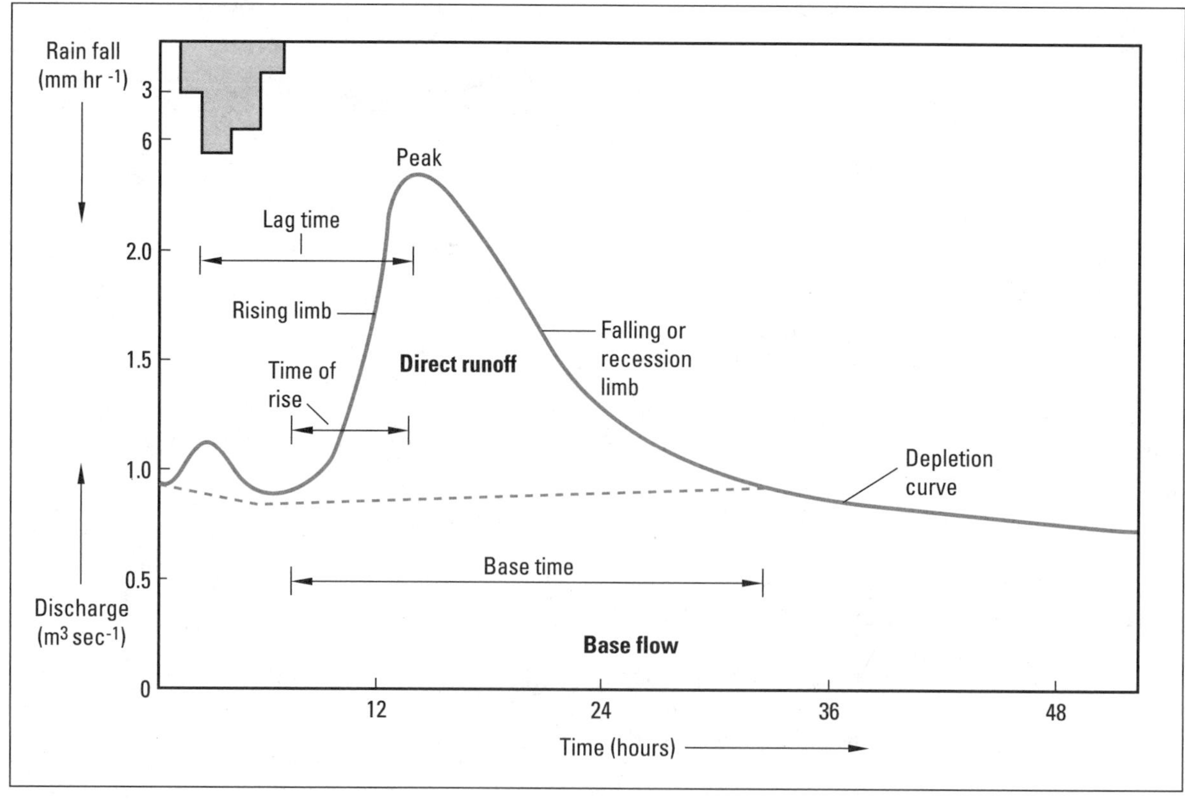


FIGURE 13.10 ▼ The storm hydrograph

## CONTRIBUTIONS TO THE STORM HYDROGRAPH

The shape of the storm hydrograph reflects two important factors: the character of the surrounding land (the catchment area of the stream) and the nature of the storm. If we compare hydrographs for different storms or from different streams, therefore, we can detect subtle variations in the shape of the hydrograph. Let us look at the processes involved in creating the storm hydrograph by looking at events during a single storm. We will assume that there has been no rain for several days so the stream is at its base flow stage. We will also assume that the storm is sudden and very intense.

The first evidence we see of the storm within the hydrograph is a small 'blip' (Figure 13.10). What causes it? The answer is rain falling directly onto the stream—**channel precipitation**, as it is known. Clearly this involves only a very small quantity of water, but in an intense storm it leaves its mark on the hydrograph.

As we have seen, during the early part of the storm much of the rain seeps into the ground, so very little overland flow occurs. That which does often ends up trapped in depressions on the surface and does not reach the channel. For some time, therefore, the stream does not receive any water from the surrounding land. This is what creates the lag. Only after it has been raining for a while is overland flow generated, and there is then a short delay while this water moves to the stream channel. Flow is rapid on the steeper slopes of the valley side, but on the flat land beside the channel flow is slow, so often it is an hour or so before this water reaches the stream. As water starts to enter the channel, the storm hydrograph starts to rise. It goes on rising as the pulse of water from around the stream flows into the channel. When the rate of inflow is at a maximum, the peak of the hydrograph is reached.

By this stage some of the water that had soaked into the soil may also have reached the channel. In particular, water that flows more rapidly through pipes and large pores in the soil may emerge in the stream banks as the soil around the channel becomes saturated. This contributes to the rising limb of the hydrograph.

The inflow of water flowing overland continues for a short time after rainfall ceases, but very soon all this water has washed into the channel and the input of surface water declines. If this were the only input of water then the hydrograph would fall as quickly as it had risen. Indeed, this tends to happen where Hortonian overland flow occurs. Where throughflow is occurring, however, the fall in the hydrograph is slowed down, for water continues to seep into the channel as the flow within the soil makes its way downslope to the stream. It may take many hours or even days for this flow to cease. When it does, the hydrograph returns to its base flow stage.

As we have seen, the interplay of these processes depends very much on the characteristics of the storm and the catchment area of the stream. For this reason the pattern we have identified gives only a general picture of the contributions to the hydrograph. Figure 13.11 indicates the types of variation that may occur in different conditions.

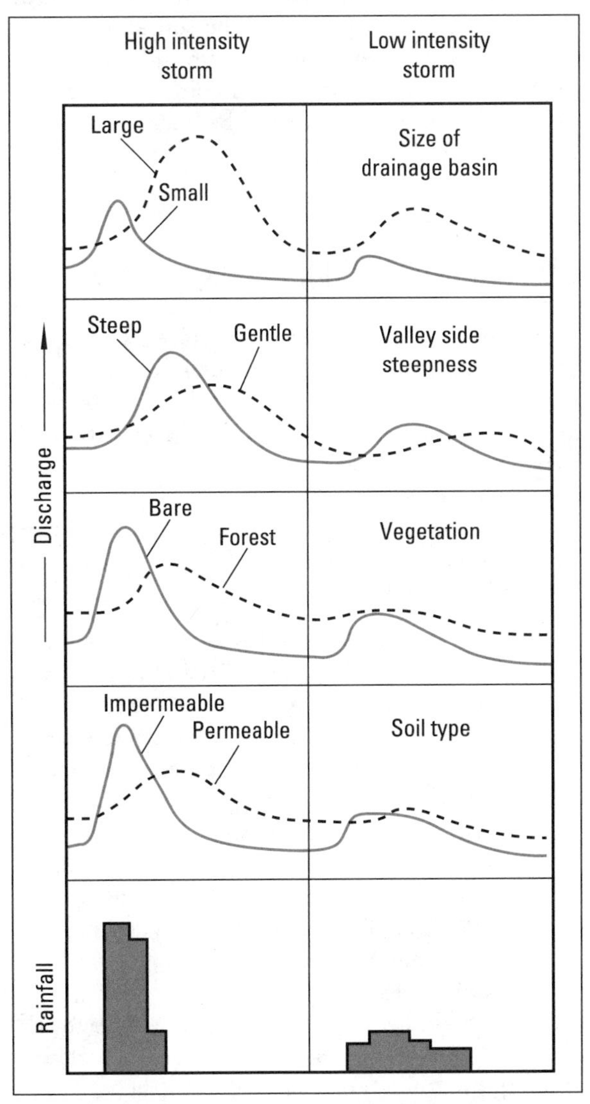


FIGURE 13.11 ▼ Hydrographs for different catchments under two different storm conditions

## BASE FLOW

So far we have looked at the shape of the storm hydrograph, but what about the flow that occurs between storms—the **base flow**? Where is this water coming from and what controls it?

The intriguing thing about base flow is that it may continue for many months after rainfall; even during periods of prolonged drought it persists with little variation. Thus it is more or less independent of short-term rainfall events, although there may be a slight seasonal fluctuation. As this implies, base flow represents water that is being released from relatively longterm storage, rather than water being cycled through the landscape following rainfall. There are two main contributions to base flow. The first is the water held within the soil and released by excessively slow drainage through the small pores in the soil. The second, and often most important, contribution is from the bedrocks—from the groundwater.

# Overland flow and humans

Water that runs off over the surface, whether as sheetflow, rill and gully flow, or streamflow, has many implications for us. It may erode soils, wash away fertilizers and seeds, flood the countryside, towns, and cities. Let us look at some of the effects of this hazard.

The terror induced by a flood can only be imagined. Consider the words of an eyewitness in Trinidad, Colorado:

> Thursday (17 June, 1965) the water was running about normal. . . . I had pulled the blinds down and then I got up to look out the front door, and there it was coming up as fast as everything. By then it was already around the house. It started coming in under the door and I knew it was going to be bad, but by then it was too late to get out.
>
> I tried to call the Sheriff. . . . When I placed the phone back on the stand, I was in water almost up to my waist. . . . Right then with the water pouring in and the noise and everything, I really thought I was a goner.
>
> I put my dog on the bed and I got up there too. I held tight on the head of the bed and it started jumping around, floating here and there. The water just kept coming in and filling things up. I was afraid it would get so high I would drown.
>
> Just about then . . . a part of the wall right in the bedroom broke, and then water was running out as fast as it ran in. I could hear windows breaking and doors were crashing and furniture was swimming around. . . . I began to get awful cold and was shivering wet. It smelled awful, that mud and dirt in the water.

Not everyone survives to tell the tale. In August, 1969, Hurricane Camille swept through Virginia, causing floods that killed 152 people. In fact, between 1903 and 1972 an average of 100 people were killed each year by floods in the USA. These figures are insignificant compared to the great floods of history, such as those in Hunan Province, China, in 1887, which killed an estimated one million people.

The effect on people is not only direct. Floodwater may bury or ruin crops, destroy roads and buildings, sweep away vehicles, and cause untold ecological damage. A flood in Rapid City, South Dakota, on 9 June, 1972, for example, not only killed 237 people and injured a further 3057, but destroyed 5000 cars, 1335 homes and did a total of $160 million-worth of damage. Often in the wake of the floods come famine and disease.

In the face of floods of this nature, people have learned to take precautionary measures. We may avoid the most flood-prone areas, and try to build flood-proof houses and factories. We may try to control the stream itself. In these ways we can avoid loss of life and greatly reduce flood damage. In the Lismore area of New South Wales in Australia, for example, it is estimated that flood-proofing has reduced damage to less than 20 percent of its original level (Table 13.2). In the Fraser River valley in British Columbia the construction of dykes, dams, and reservoirs has been proposed to regulate flow and to cope with discharges of as much as 17 000 $m^3 s^{-1}$, a flow that is expected to occur, on average, only once in 150 years. As the Tennessee Valley scheme has shown (Figure 13.12) not only can flooding be controlled by such measures, but the reservoirs may provide hydroelectric power, recreational facilities, and a better water supply.

Not all attempts to control flooding are successful, however, and there have been several examples of human intervention exacerbating the flood problem. The flood at Rapid City occurred because heavy rainfall resulted in the failure of a reservoir dam. In 1864 the Dale Dyke dam near Sheffield, England, collapsed, causing floods that killed almost 250 people. The consequences of stream control also extend well beyond the effect on flooding. Changing the stream regime may alter the rate of sediment transport with serious effects on deposition and erosion downstream. Ecological conditions, too, may be affected. When the Hoover Dam was built in the Colorado River, for example, changes in the river bed were seen as far as 560 km down-

## Table 13.2

*Effects of flood-proofing on flood damage, Lismore, NSW, Australia*

| *Flood height (m)* | *Potential damage (without flood-proofing) A$* | *Actual damage (with flood-proofing) A$* | *Actual damage as percent of potential* |
|---|---|---|---|
| 13.0 | 25 630 000 | 5 156 000 | 20.1 |
| 12.0 | 12 735 000 | 2 350 000 | 18.5 |
| 11.0 | 3 614 000 | 688 000 | 19.1 |
| 10.0 | 291 000 | 53 200 | 18.1 |

Data D.I. Smith, 'Actual and potential flood damage: a case study in urban Lismore, NSW, Australia,' *Applied Geography* 1 (1981), pp. 31–9.

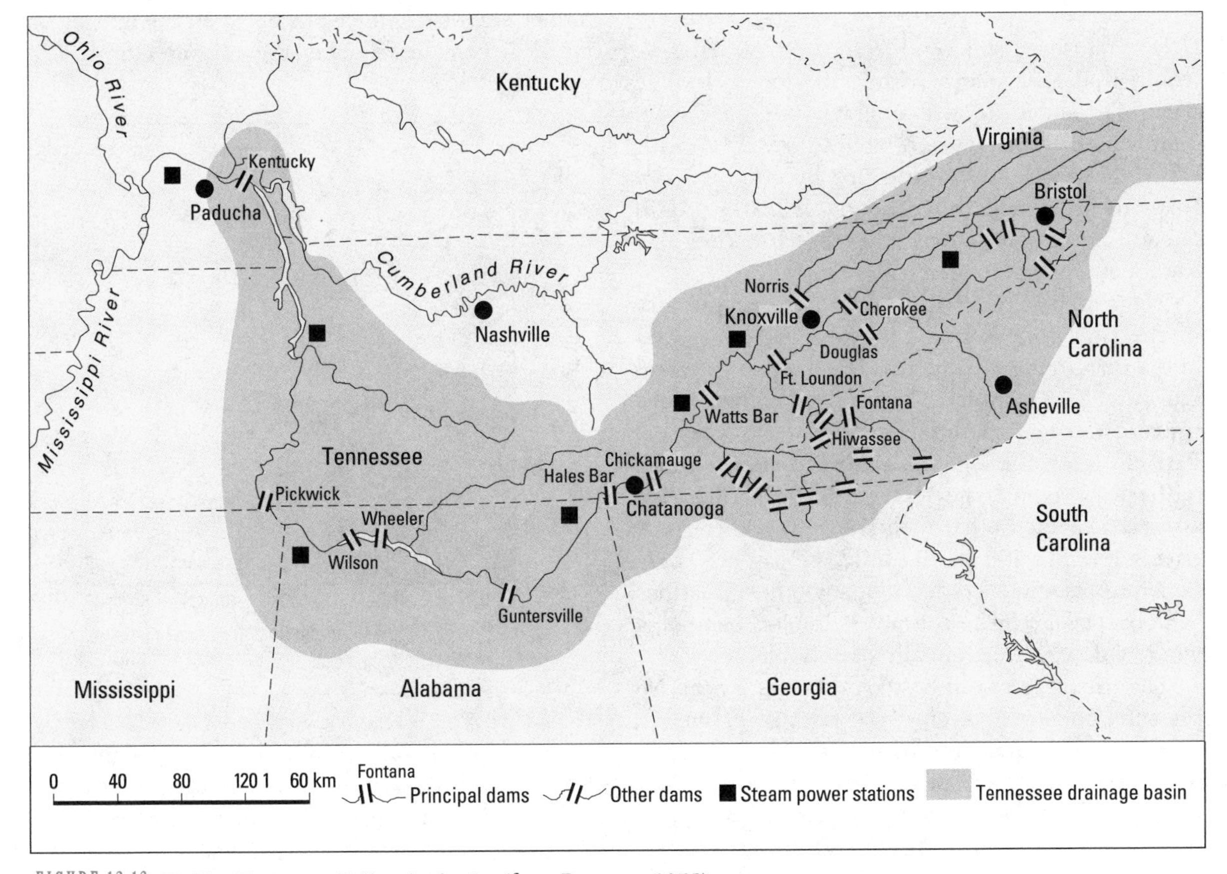


FIGURE 13.12 ▼ The Tennessee Valley Authority (from Paterson, 1975)

stream. Construction of the Aswan Dam and Lake Nasser in Egypt had even more far-reaching effects. Sediment, which once fertilized the soils in the lower reaches of the Nile, was trapped in the lake, causing siltation behind the dam and reducing soil fertility downstream. To counteract this, farmers have had to buy fertilizers, the cost of which outweighs the value of the electricity produced by the dam. In addition, reduction of the amount of sediment reaching the Nile Delta has led to erosion of the sand bar, which once protected the brackish estuarine water, and the consequent collapse of Egypt's sardine industry. As if this were not enough, the change in the regime of the Nile has resulted in the creation of stagnant water in dykes and pools in which water snails breed. These snails are vital in the life cycle of the Bilharzia fluke which ultimately attacks internal organs of the human body, leading to general debilitation and death. Clearly, we should attempt to manage the hydrological cycle with caution.

# Water storage in the soil

If we monitored soil conditions after rainfall, we would find that water in the soil slowly drained downward under the effect of gravity for a day or so, after which drainage more or less ceases. At this stage, however, soil would not be entirely dry. Instead, a considerable amount of water would remain in the soil. Why is this so? How does this water defy the pull of gravity?

The reason is that within the matrix of the soil, in the tiny pore spaces, the water is affected by forces that counteract gravity and operate to retain moisture. The main force involved is known as the **matric force** and it is due to two processes. Particles attract water molecules to their surface (adhesion), while the water molecules are also attracted to each other (cohesion). Together these effects retain water in the soil. We can see these same processes in operation in many other situations as well; they are the reason water droplets hang on a window pane during rainfall, for example.

The matric force can be thought of as a negative pressure, one acting against the force of gravity to draw or suck water upward into the soil. For this reason we often speak of it as **soil suction** or **soil tension**. Traditionally this has been measured in bars (just as we used to measure atmospheric pressure), but now we use Pascal units (1 Pa = $10^{-1}$ bars). Close to the surface of the soil particle the matric force is extremely strong, equivalent to about $1 \times 10^9$ Pa (almost 10 000 times atmospheric pressure), but away from the particle surface it declines rapidly (Figure 13.13). Within a distance of only 0.06 mm it has shrunk to about $5 \times 10^7$ Pa and is almost negligible. Thus, it is only the water held very close to the surface of the particles that is affected to any significant extent by the matric force. Water beyond about 0.06 mm from the particle is free to drain under the influence of gravity.

Because of the matric force, the soil remains wet for a long period after rainfall, thin films of water forming around the particles and persisting in the finest pores in the soil. This water is almost immobile, but not quite. It cannot move under the effect of gravity, but it can move through the effect of the matric force itself. Where there are differences in the matric force, the water is drawn from areas where the force is low to areas where it is high. These differences arise because of localized drying of the soil. As the soil dries, either by evaporation or because plants withdraw moisture from the pores, the water films become thinner. As they become thinner, the matric force operating on the outside of the film increases. Thus water migrates from wet

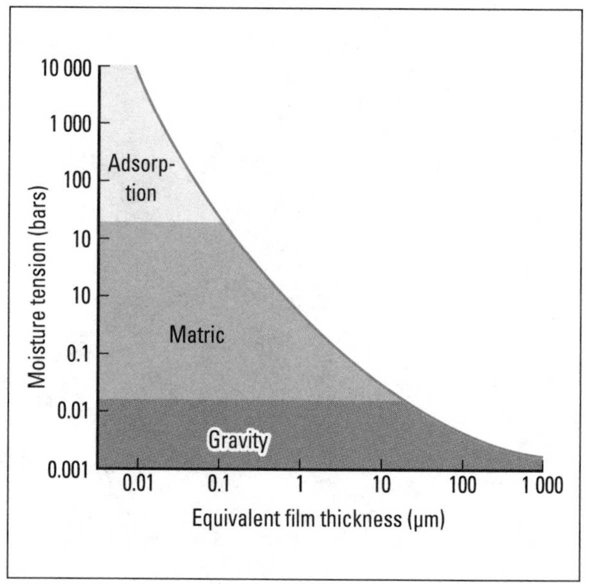


FIGURE 13.13 ▼ The relationship between soil moisture tension and thickness of the water film. The moisture tensions shown represent the force operating to hold water to the soil particles at the outer limit of the water film; 1 bar = $10^4$ Pa (from Briggs, 1977)

areas of the soil where the water films are thick to the drier areas, where the films are thin. This movement can be in any direction, upward, downward or sideways, and it is entirely independent of gravity.

Movement in this way is known as **capillary movement** and we can see it in operation when we dip a thin glass tube into water (Figure 13.14). The water is drawn up the tube. It goes on rising until the weight of the water in the tube is just balanced by the force drawing it upward. It is an extremely important process, for although it operates relatively slowly in the soil it is one of the main ways that water moves upward toward the roots of plants. Not all the water retained in soil is able to move in this way. Water that is very close to the particles, closer than about 0.0002 mm, is almost wholly static. It is only removed when the soil is heated and evaporation takes place.

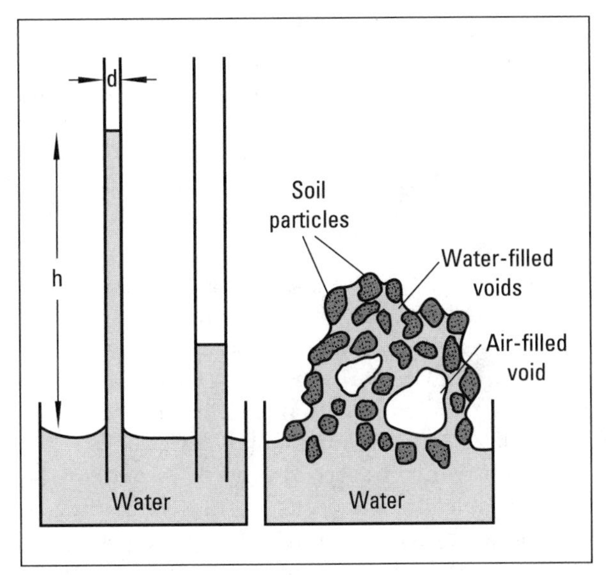


FIGURE 13.14 ▼ Capillarity; the height of capillary rise (h) is related to the diameter of the tube (d) as follows: h = 0.3/d

It follows from what we have said that we can define three main types of water in the soil: that which moves freely under the effect of gravity; that which moves more slowly through capillary processes; and that which is totally immobile. These are known as the **gravitational**, **capillary**, and **hygroscopic water**, respectively. Following rainfall, we find that the gravitational water drains away within about two days. The soil is then said to be at **field capacity**. Afterward, if there is no more rainfall, the capillary water is slowly redistributed and drawn toward the surface where it is removed by plants and by evaporation. Eventually, almost all the capillary water is lost and the soil is said to be at wilting point. Plants cannot obtain the remaining moisture and they wilt and die. As we can see from Figure 13.15, the tension in the soil at this point is about $15 \times 10^4$ Pa. Most of the water that is left is referred to as hygroscopic water.

## THE SOIL WATER TABLE

Let us now look at what happens to the gravitational water, the water that was free to drain downward through the soil. Where does it go?

We can see the answer in many cases if we dig a hole in the soil. Ultimately, we reach a point where

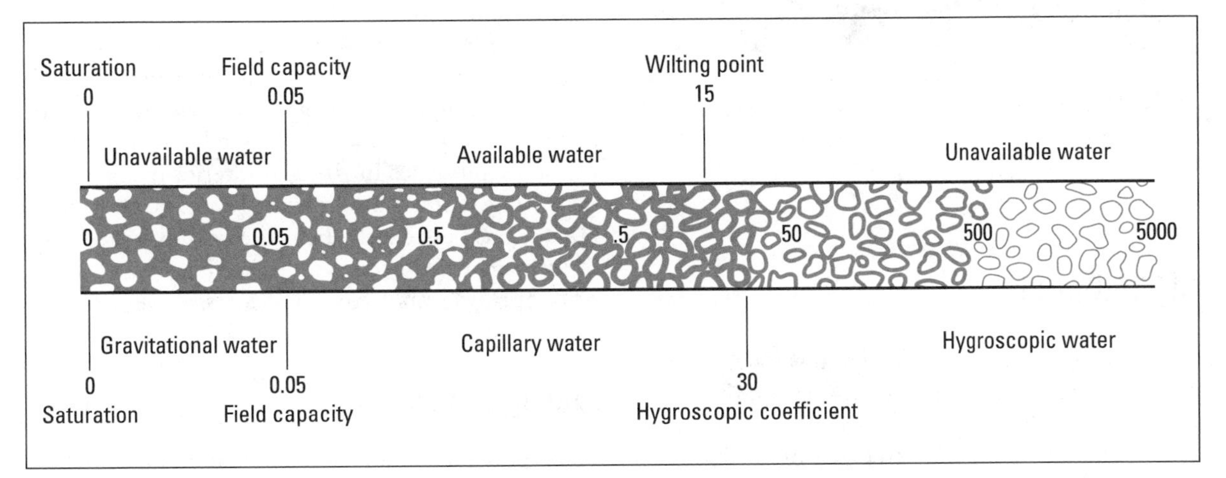


FIGURE 13.15 ▼ Classification of soil moisture according to biological (upper) and physical (lower) criteria. The figures represent soil moisture tensions in $10^4$ Pa (from Briggs, 1977)

the hole starts to fill with water. We have reached the soil **water table**. Beneath this level the soil is totally saturated, and it is to this level that the gravitational water drains.

The fact that this saturated zone develops deep in the soil indicates that something must prevent the water from draining downward, and in many cases the reason is the presence of a dense, impermeable layer in the soil or the underlying bedrock. As we have already seen, the water can only move through the relatively large pores and fissures in the soil, and if a layer develops that has no large pores, then the water is trapped. Drainage ceases and the soil becomes perpetually waterlogged. A layer like this which prevents water movement is said to be impermeable. It may occur in the soil because of compaction caused, for example, by machinery.

The water below the water table is not entirely immobile, and over time it tends to flow laterally into lower areas, eventually emerging in the valley bottoms to contribute to streamflow. It is this water that is partly responsible for the base flow of streams. This movement is slow, however, and if we mapped the height of the water table in an area we would generally find that it roughly followed the surface topography (Figure 13.16). Thus, the water is trapped here for a considerable time, and soil water storage represents one of the main storage components of the hydrological system.

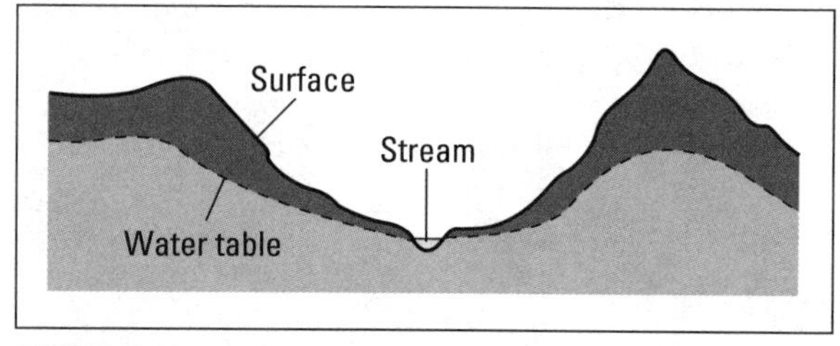


FIGURE 13.16 ▼ Relationship between the regional water table and surface topography

# Groundwater

As we saw in Chapter 10, groundwater comprises some 25 percent of the total fresh water in the terrestrial system. Although much of this is not available for development, groundwater is a major source of supply for many parts of the world. It is also considered a vast untapped reserve by many investigators.

The term groundwater applies to all water, in all its forms, beneath the surface. For example, in Canada much of the groundwater is frozen and referred to as permafrost. The chemistry and quality of water varies from region to region depending upon the minerals in the local rocks. Similarly, the rate of movement of water varies depending upon the type of subsurface material. In some sand and gravel glacial deposits the movement can be quite rapid.

As we have noted, the zone of saturation below the water table develops because downward movement of water is impeded by an impermeable layer. The same situation exists in bedrocks. Water percolates downward through pores and fissures in the rocks until it reaches an impermeable layer. It then becomes trapped and a water table forms in the rocks. The water beneath is known as **groundwater**.

Rocks that store water in this way are termed **aquifers**. Chalk, sandstone, and some limestones provide exceptionally good aquifers, and it is from these that people derive much of their groundwater. On the other hand, rocks such as clay, shale, and many igneous rocks are impermeable and they are known as **aquicludes**. It is these rocks within the strata that halt the downward percolation of water and create the aquifer.

Aquifers develop in a variety of situations (Figure 13.17). In some cases alternating beds of permeable and impermeable rocks occur, giving rise to one or more **perched aquifers**. Water from these escapes as springs along a clearly defined spring line (Figure 13.18). In other cases, the aquifer may be warped into a synclinal basin and sealed by an overlying impermeable layer. Water then collects in the aquifer under considerable pressure, and where the water escapes back to the surface (in a well or along a natural fault) it emerges with considerable force. **Artesian basins** of this type are found adjacent to mountain ranges, and they may be very important in supplying water in the arid interior of the continent. The name comes from the city of Artois, France, but perhaps the best known and largest area of artesian wells occurs in Australia. The town of Alice Springs stands around an artesian well.

## GROUNDWATER AND HUMANS

Much of our incessant need for water is met by the natural underground reserves of the aquifers. Over the years our ability to extract these waters has

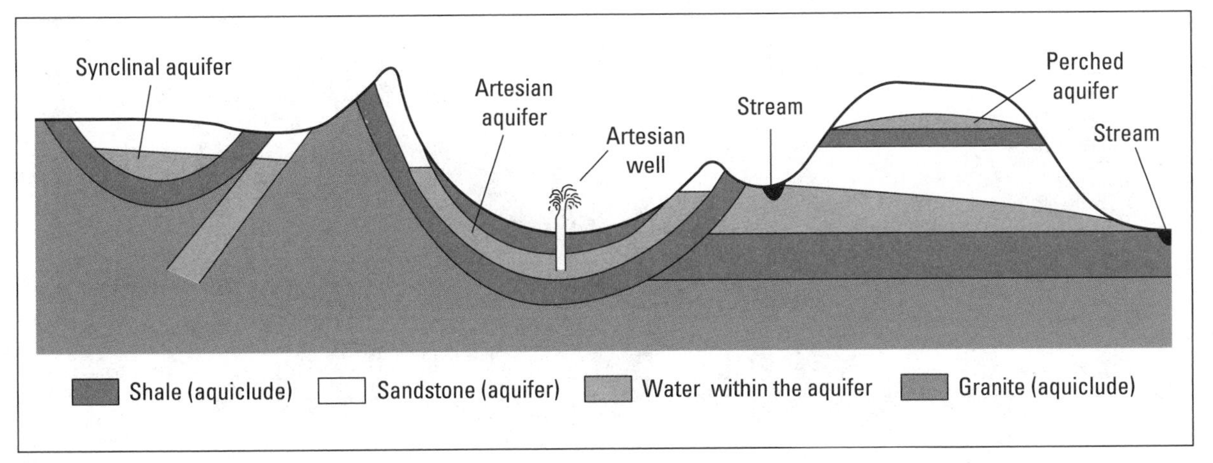


FIGURE 13 17 ▼ Types of aquifer

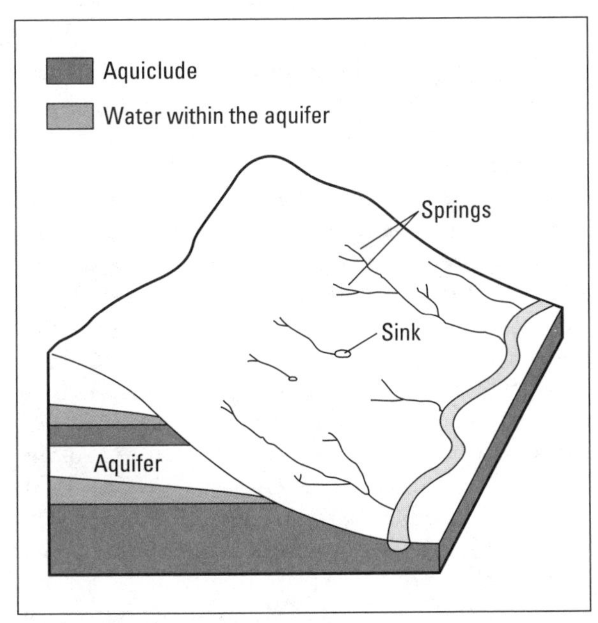


FIGURE 13.18 ▼ Spring lines developed along a perched aquifer. Note that where a stream passes off the impermeable aquiclude back onto the aquifer it may disappear underground at a sink

improved. Simple stone-lined, hand-pumped wells and **qanats** (Figure 13.19) have given way to deep tube-wells from which the water is raised by pumps and hydraulic rams. As a result, the rate at which we are abstracting the groundwaters has increased; along with this, the problems of groundwater exhaustion and falling water tables have increased.

When we draw water from a well, the water table around the well falls, through a process known as **drawdown**. This creates a **cone of depression** around the well (Figure 13.20). Water moves through the aquifer to replace that removed by flow through the rocks. Where the rocks are porous and flow is rapid, replenishment of the well is efficient and the cone of depression is small. Where the aquifer is relatively impermeable and flow is slow, replenishment is much less efficient and the cone of depression is deep and increases in area. In either case, if we draw more water than can be replaced by lateral flow, then the cone of depression becomes deeper and larger and eventually the well runs dry. Its owners then have two options: dig it deeper or drill another well elsewhere. If the well is deepened, then so is the cone of depression, and eventually—if the depth of the well does not make it too costly to raise the water in any case—the well reaches the bottom of the aquifer (Figure 13.20). It then has to be abandoned. If a new well is dug elsewhere, the process repeats itself.

Over time, of course, it is not only the lateral flow of water through the aquifer that is important. This may supply water to the well, but it is only redistributing water within the aquifer. If the water table is not to fall overall, then the aquifer itself must be recharged. As we have seen this takes place mainly through deep percolation of water from the surface. It is the relative rates of this recharge and our abstraction of the water that determine whether or not, in the long run, the aquifer is being depleted.

There are many examples of this happening. The apparently vast and limitless bodies of groundwater are often the result of thousands of years of slow percolation with no real extraction. Then people extract

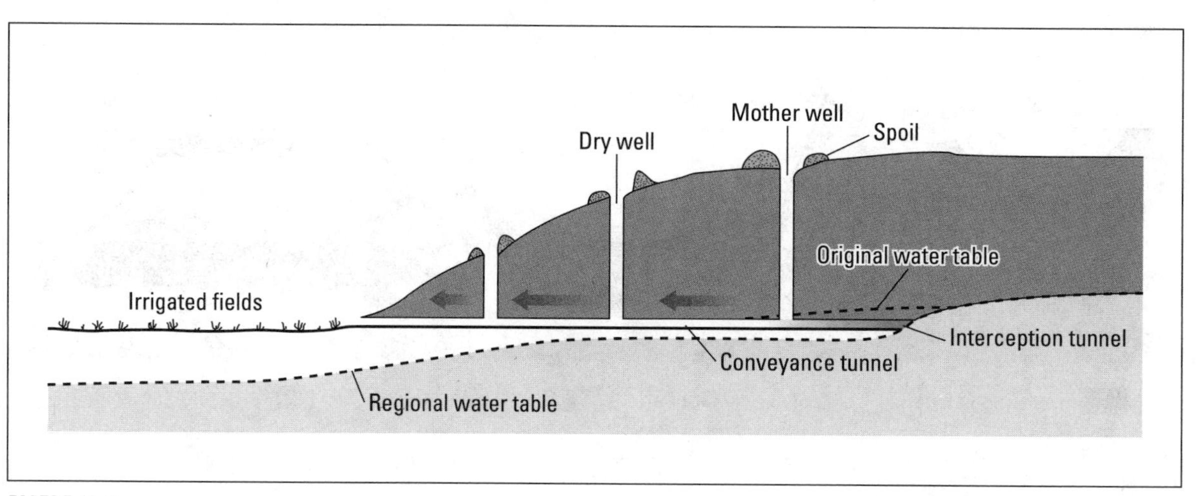


FIGURE 13.19 ▼ A cross section through a typical qanat system, used throughout the Middle East

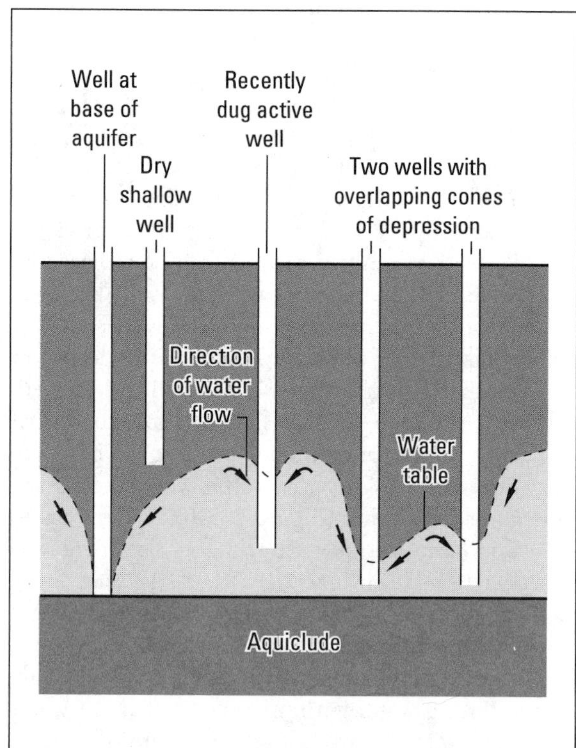


FIGURE 13.20 ▼ Cones of depression developed around wells. Water extraction in London, England, has resulted in serious drawdown in this way over the last century. More recently a decline in the rate of water use has caused the reverse problem—a rise in the water table and a flood risk to subsurface structures such as the London Underground

the water much more rapidly than it is currently being replenished and the water table falls. In London, England, this has been taking place for several centuries and it is estimated that the water table has fallen by up to 45 m in 33 years. The same is happening in the Ogallala formation in the High Plains of Texas. The groundwaters here are ancient waters that probably built up over a period of several million years. In 1938, before large-scale pumping started, they were believed to have a volume of about 600 000 km$^3$. Since then these reserves have dropped by about 110 000 km$^3$. People have been extracting the water far more rapidly than the scanty precipitation can replenish it.

The effect of a falling water table can be disastrous. Where the water table lies close to the surface quite a small fall may lead to drying up of streams and lakes and problems of drought for farmers as water contents in the soil decline. Near the coast, water may be drawn into the aquifer from the sea; saline water will contaminate the groundwaters and make them useless for consumption and even irrigation (Figure 13.21). This is particularly acute in Malta, which obtains almost all of its water from aquifers connected to the sea, and under extreme conditions can turn large areas of good agricultural land into virtual desert.

It is not only through abstraction that people affect the groundwaters; they are influenced as well by population. Agricultural fertilizers in particular may seep into the ground and accumulate in the aquifers, and in time toxic levels of nitrate and phosphate have been known to build up. In parts of

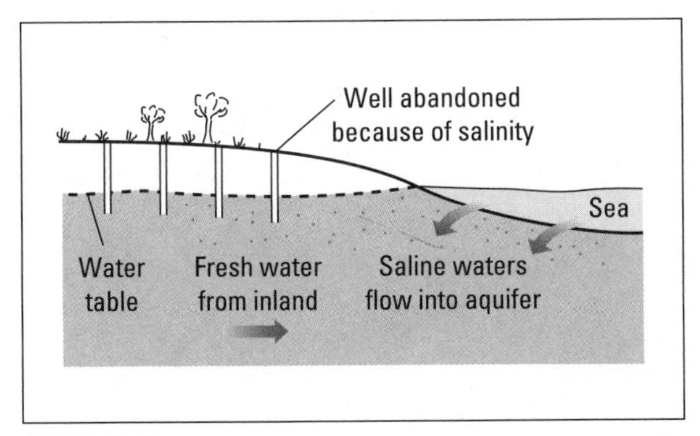


FIGURE 13.21 ▼ Salinization of groundwaters due to falling water table

Canada it is possible to detect a gradual increase in pollutant levels in recent years as fertilizers, which have been applied in increasing quantities since the 1950s, make their way into the aquifers. Unfortunately, the rate of percolation of this water into the aquifers is so slow that the damage may have already been done by the time the first signs are detected.

# Conclusion

Runoff, infiltration, and groundwater storage are vital parts of the hydrological system, governing the way rainfall is routed through the landscape and controlling the time it takes for the water to reach the sea.

Waters that are diverted over the surface as overland flow move quickly and may enter the sea within days or even hours of reaching the ground. Water that soaks into the ground and reaches the water table may take months, years, or even centuries to complete the cycle.

In addition to the control on the hydrological cycle, runoff and storage also affect many other aspects of the global system. Water that runs across the land surface is a major agent of erosion and transport and is responsible for cutting many features of the landscape, as we will see in Chapter 21. The water that soaks into the soil takes part in weathering reactions and it is vital to plants; without it vegetation would die and the earth would be barren.

# Oceans and their circulation

## The character of the oceans

In recent years we have discovered a great deal about the oceans, but they still represent the 'great unknown' compared with other parts of the global or solar systems. Scientists have set foot on the moon, sent spacecraft to Mars and Venus, and photographed Jupiter and Saturn, but the pressure of the water has restricted exploration of the oceans to the uppermost layers. To delve beyond 100 m we must depend upon instrumental technology. Some refer to the deep oceans as 'inner space,' although the parallels with outer space are tenuous. For example, the lack of atmosphere is a problem, but humans can cope better with a vacuum than they can with the pressures and environment of water. Far more people have been to the top of Mount Everest (8839 m) than have been to the bottom of the Marianas Trench (11 022 m). Slowly, progress is being made to discover the nature of the deepest ocean, its animal life living in permanent darkness, its low temperatures and peculiar currents. Even on the surface we know less about the controls of water movement than we do about the atmosphere, and yet the oceans are the greatest stores of water on earth, contain vast quantities of minerals and nutrients that affect our food supply, and play a fundamental role in moisture and energy cycling.

If so little is known about the oceans, why should they be included in a book about physical geography? Some of the reasons would become clear if we tried to contemplate the world without oceans and imagine what the effect might be. With no oceans, evaporation would become minute on a global scale and the hydrological cycle merely a small exchange between the freshwater store of the atmosphere and evapotranspiration from the ground. It is doubtful if there would be enough water to produce runoff. The earth's climates would all show continental qualities, with hot summers and cold winters. In fact the changes would be so great that we cannot really imagine life as we know it without the oceans. Let us look, then, at the characteristics of the oceans before considering the actual processes that affect them and the consequences they have on the oceanic circulation.

## THE SHAPE OF THE OCEANS

Geological evidence from sediments suggests that the oceans have existed for at least the past 3000 million years. But we now know from sea floor spreading (Chapter 17) that the ocean basins we have today are relatively young features of the earth—less than 250 million years old—formed by movement of the crust away from the mid-oceanic ridges. As a result of this, the positions of the continents and therefore the sizes of the oceanic basins have also changed through time. See Figure 9.9, which shows the varying positions of the continents in geologic time. Currently the ocean basins are interconnected like spokes from the hub of the circumpolar Southern Ocean, northward to the Pacific, the Atlantic, and the Indian oceans, in total covering some 70 percent of the earth's surface. During the last Ice Age, sea level was as much as 200 m lower than at present. Water was removed from the oceans to build the vast ice sheets on the land and consequently sea level was reduced. With lower oceans more land was exposed so that the edge of the oceans was along the edge of the continental shelves. The proportions of land and water, currently 30/70 percent, were then about 35/65 percent. The extent of water over the surface of the planet, its importance to life, and the fact that it is the only planet in the solar system with water suggest that a name change is in order. The planet should be renamed "Water."

The area of the oceans varies greatly, from the 165 million km$^2$ of the Pacific to the mere 14.3 million km$^2$ of the frozen Arctic Ocean (Table 14.1). In terms of depth, the oceans are peculiar. It might be expected that they would gradually deepen toward the farthest point from land, but this is far from true. Some of the deepest parts of the ocean, such as the Marianas Trench, are quite close to land (Figure 14.1). These **oceanic trenches** are most common in the Pacific, running parallel to the island arcs. However, almost all the oceans show relatively shallow seas near the continental coasts, then a sudden deepening down the continental slope before reaching the abyssal plains.

If we plot the frequency with which certain depths would be found in the ocean we obtain the graph shown in Figure 14.2. Nearly a third of the ocean floors are between 4000 and 5000 m below sea level, and 72 percent are between 3000 and

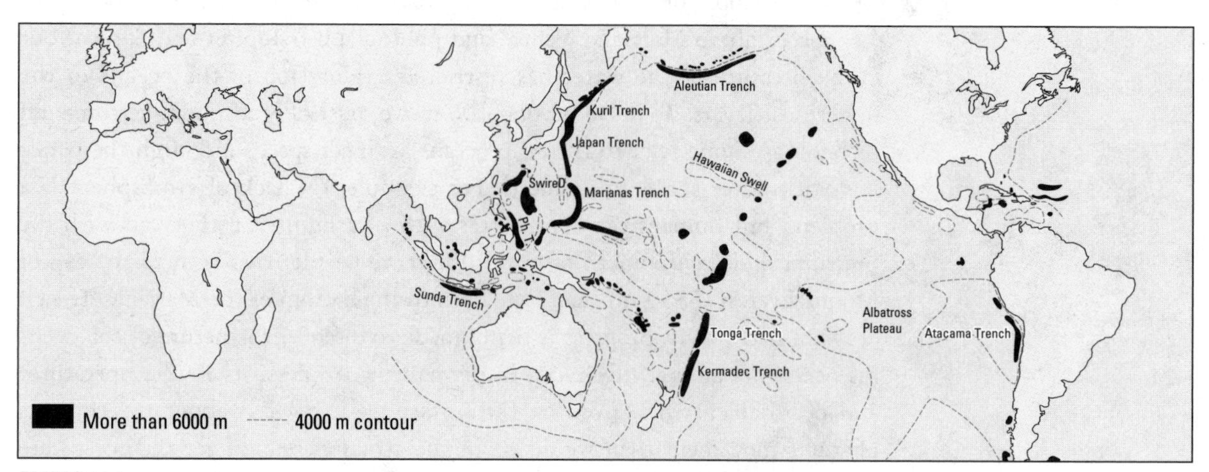


FIGURE 14.1 ▼ The Pacific sea floor and the location of the main trenches

6000 m. The shallow continental shelves, less than 200 m deep, make up about 8 percent of the oceanic area.

Table 14.1

*Properties of ocean basins*

| | *Area ($km^2 \times 10^6$)* | *Mean depth (m)* |
|---|---|---|
| Total oceanic area | 361 | 3650 |
| Pacific | 165 | 4270 |
| Atlantic | 81 | 4270 |
| Indian | 75 | 3960 |
| Arctic | 14 | 1330 |
| Minor seas | 26 | — |

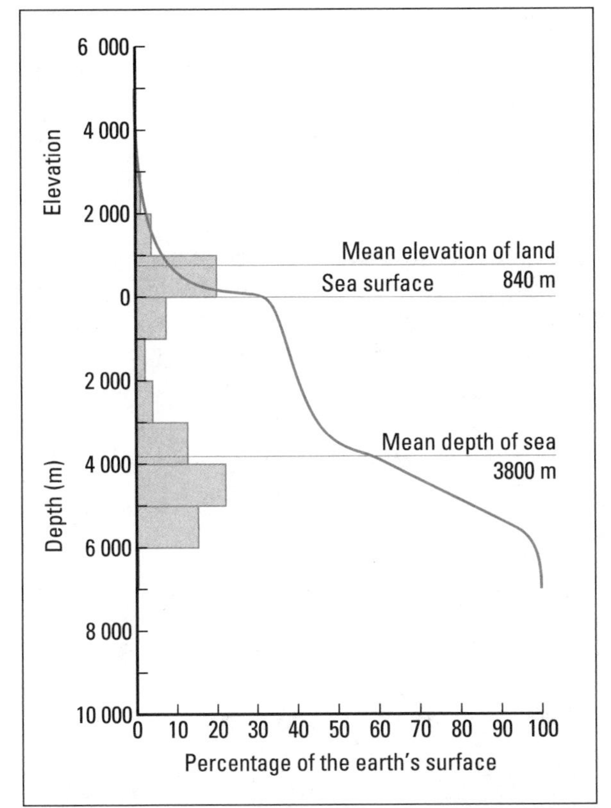


FIGURE 14.2 ▼ Hypsographic curve showing the area of the earth's solid surface above any given level of elevation or depth. At the left is the frequency distribution of elevations and depth for 1000 m intervals (after Lagrula, 1966)

## TEMPERATURE

Ocean temperatures vary considerably both with depth and laterally. We lack detailed data on ocean temperatures so it is difficult to get a complete picture of the global pattern. There are few parts of the world where surface temperatures are taken regularly, and even fewer where subsurface temperatures are recorded. Until the development of satellite photography we were dependent upon ships' logs, which gave a dense network of readings along the main shipping routes and few elsewhere. Even now, areas with a high frequency of cloud cannot be monitored by satellites, so sea surface temperatures are not well documented. Figure 14.3 shows sea surface temperatures for the tropical Atlantic and eastern Pacific based mainly on ship observations over a 60-year period.

At a global scale we might expect that ocean temperatures would show a general decline away from the equator in response to reduced radiation inputs. To some extent this is true (Table 14.2). The

Table 14.2

*Mean surface temperatures of the oceans by latitude (°C)*

| *Latitude* | *Atlantic Ocean* | *Indian Ocean* | *Pacific Ocean* |
|---|---|---|---|
| (°N) | | | |
| 70–60 | 5.6 | — | — |
| 60–50 | 8.7 | — | 5.7 |
| 50–40 | 13.2 | — | 10.0 |
| 40–30 | 20.4 | — | 18.6 |
| 30–20 | 24.2 | 26.1 | 23.4 |
| 20–10 | 25.8 | 27.2 | 26.4 |
| 10–0 | 26.7 | 27.9 | 27.2 |
| 0–10 | 25.2 | 27.4 | 26.0 |
| 10–20 | 23.2 | 25.9 | 25.1 |
| 20–30 | 21.2 | 22.5 | 21.5 |
| 30–40 | 16.9 | 17.0 | 17.0 |
| 40–50 | 8.7 | 8.7 | 11.2 |
| 50–60 | 1.8 | 1.6 | 5.0 |
| 60–70 | −1.3 | −1.5 | −1.3 |
| (°S) | | | |

Data H.U. Sverdrup, *et al*, *The Oceans, their Physics, Chemistry and General Biology* (1970), Prentice-Hall.

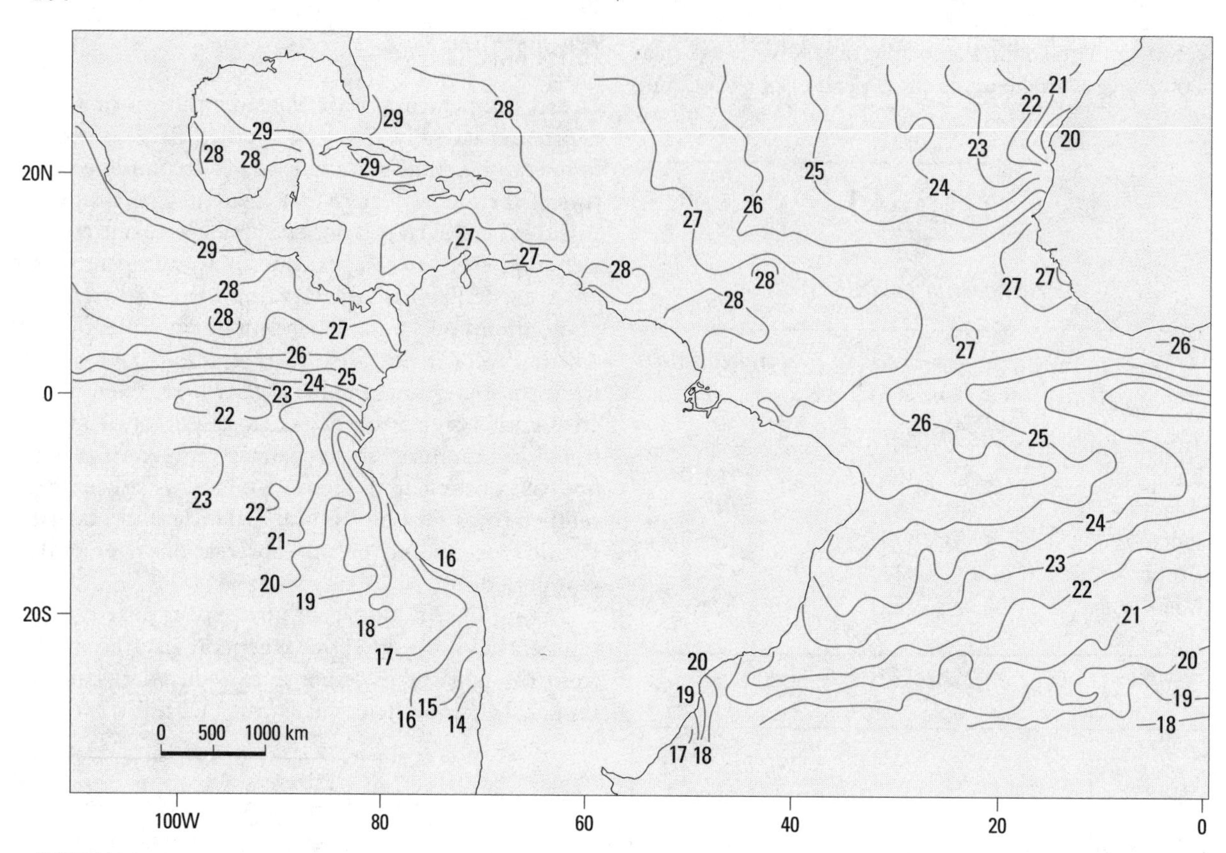


FIGURE 14.3 ▼ Sea surface temperatures (°C) in July over the tropical Atlantic and eastern Pacific oceans. Note water temperatures on the west coast of South America, the region where El Niño events occur

Arctic Ocean is frozen because it does not extend into parts of the world receiving high inputs of radiation, and it is also more or less cut off from the Atlantic and Pacific. The Red Sea, conversely, is almost entirely enclosed by hot desert lands with high energy receipts; it would be surprising if the waters were not warm. But several other factors affect sea temperatures, so many of the other oceans show a much more complex pattern.

One factor that modifies the effect of radiation is advection. This occurs mainly at depth and it involves the movement of water from one place to another without any mixing with the atmosphere. At the surface, several other processes influence temperatures. Loss of heat by evaporation is the most important process, though sensible heat convection occurs between the ocean and atmosphere, transferring heat from the air to the water when the sea is colder than the air, or from the oceans to the atmosphere when the air is colder. Condensation of water vapour liberates latent heat into the oceans. In addition, surface winds have an important effect. They can produce a mixing of the upper layers to give a marked discordance in the temperature profile (Figure 14.4); winds blowing offshore move the waters seaward, allowing cool water from the ocean bottom to rise to the surface. This reduces surface temperatures and weakens the temperature gradi-

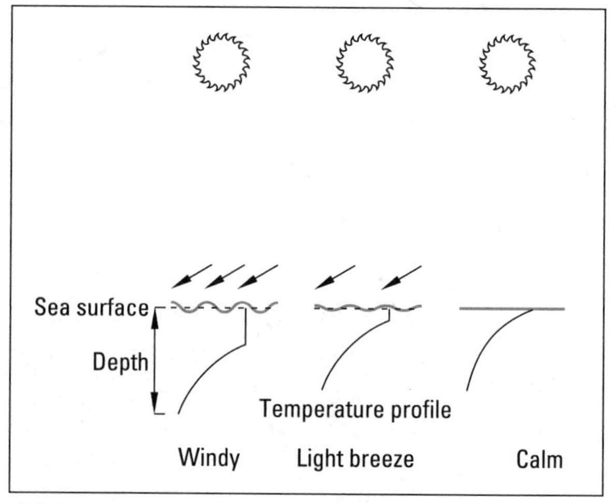


FIGURE 14.4 ▼ The effects of wind speed on surface water temperature profiles on sunny days

ent. It may also have important effects on the regional climate, as we saw in Chapter 7. A thermometer is lowered from the surface to record temperature variations with depth. These are plotted to produce a **bathythermograph**, which shows the temperature profile in the same way as for the atmosphere.

The effects of some of these processes can be seen in Figure 14.3. The influence of upwelling cold water is seen off Chile and Peru, the Canary Islands, and southern Africa, for example, and the anomalously cold surface temperatures are found on the east side of both the Atlantic and Pacific ocean basins. Apart from modifying the regional climate in these areas they also bring nutrient-rich waters to the surface, stimulating organic activity. Many of the world's major fishing zones are associated with these upwelling waters.

Looking at the vertical pattern, we find a general decline in sea temperatures with depth (Figure 14.5). This occurs for two main reasons. Radiation is absorbed almost entirely in the upper few metres of the ocean, so little heating occurs at depth. In addition, cold water tends to subside, gathering in the lower part of the ocean and reducing temperatures there. Thus, at about 1000 m, sea temperatures are only about 5°C and they continue to fall below this depth to only 1°C or 2°C. This temperature structure makes the ocean very stable. Unlike the atmosphere, it experiences very little vertical movement, and much of the mixing takes place by small eddy currents and molecular diffusion. The nature of these deep movements of water is discussed later in this chapter. It is important to note that these deep layers are very slow-moving. It is estimated that some take as much as 10 000 years to complete a cycle. This has implications for global climate. Heat energy within one of these circulations could be stored in the oceans for a long time, to be released and create climate change at a much later date.

Nevertheless, there are variations to the temperature profile, as we can see if we look at a north–south cross section of the Atlantic Ocean. Temperatures at the surface are higher in the tropics and the warm water extends to greater depths between 20°N and 30°N of the equator. This is largely due to the intense solar warming in this area.

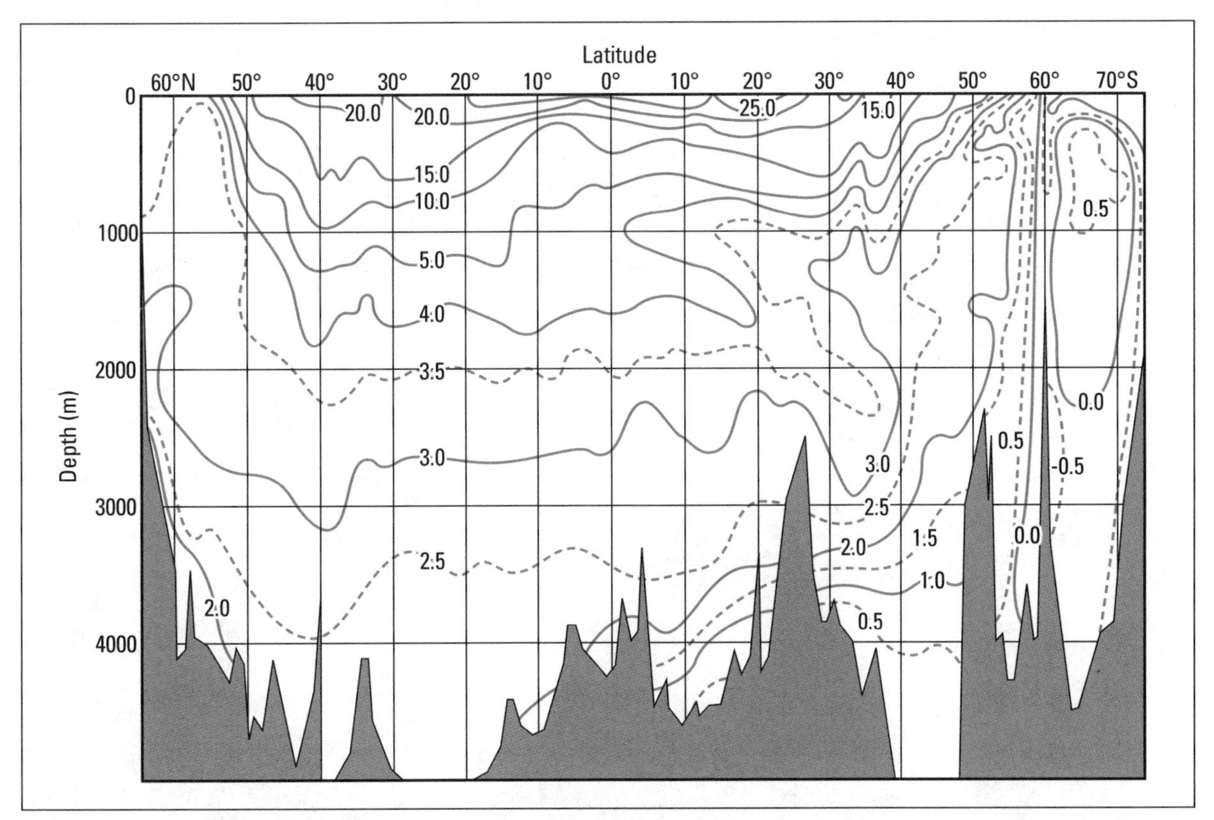


FIGURE 14.5 ▼ Longitudinal cross section of the Atlantic Ocean showing the temperature structure in relation to depth (°C) (after Ingmanson and Wallace, 1973)

## SALINITY

As anyone who has swum in the sea knows, the sea water is salty. It contains a variety of dissolved minerals (Table 14.3), but by far the most abundant is common salt (sodium chloride or NaCl).

On average, sea water contains about 35 g of sodium chloride in each litre, but this concentration varies considerably from one part of the oceans to another. Much of the salt has been derived from the land, for over the eons of geological time streams have carried sodium and other solutes into the sea, where they are trapped. It has been calculated that if all the minerals dissolved in the sea were precipitated out and returned to the land, they would produce a layer over 150 m thick. Locally, however, this slow buildup of salt is being counteracted by inputs of fresh water, by precipitation, by ice melt or by runoff, which dilute the concentration. Thus salinity tends to be relatively low near the coast and higher in the mid-ocean areas. It also tends to increase with depth, since salty water is generally more dense and sinks.

At a global scale, the highest concentrations of salt are found about 20°N and 20°S of the equator. This narrow belt of high salinity occurs because of

### Table 14.3

*Concentrations of the major components of sea water*

| | *Parts per thousand (⁰/₀₀)* | *Percent of total minerals* |
|---|---|---|
| Chloride | 18.980 | 55.04 |
| Sodium | 10.556 | 30.61 |
| Sulphate | 2.649 | 7.68 |
| Magnesium | 1.272 | 3.69 |
| Calcium | 0.400 | 1.16 |
| Potassium | 0.380 | 1.10 |
| Bicarbonate | 0.140 | 0.41 |
| Bromide | 0.065 | 0.19 |
| Boric acid | 0.026 | 0.07 |
| Strontium | 0.013 | 0.04 |
| Fluoride | 0.001 | 0.00 |
| Total | 34.482 | 99.99 |

Data D.E. Ingmanson and W.J. Wallace, *Oceanology: An Introduction* (1973), Wadsworth.

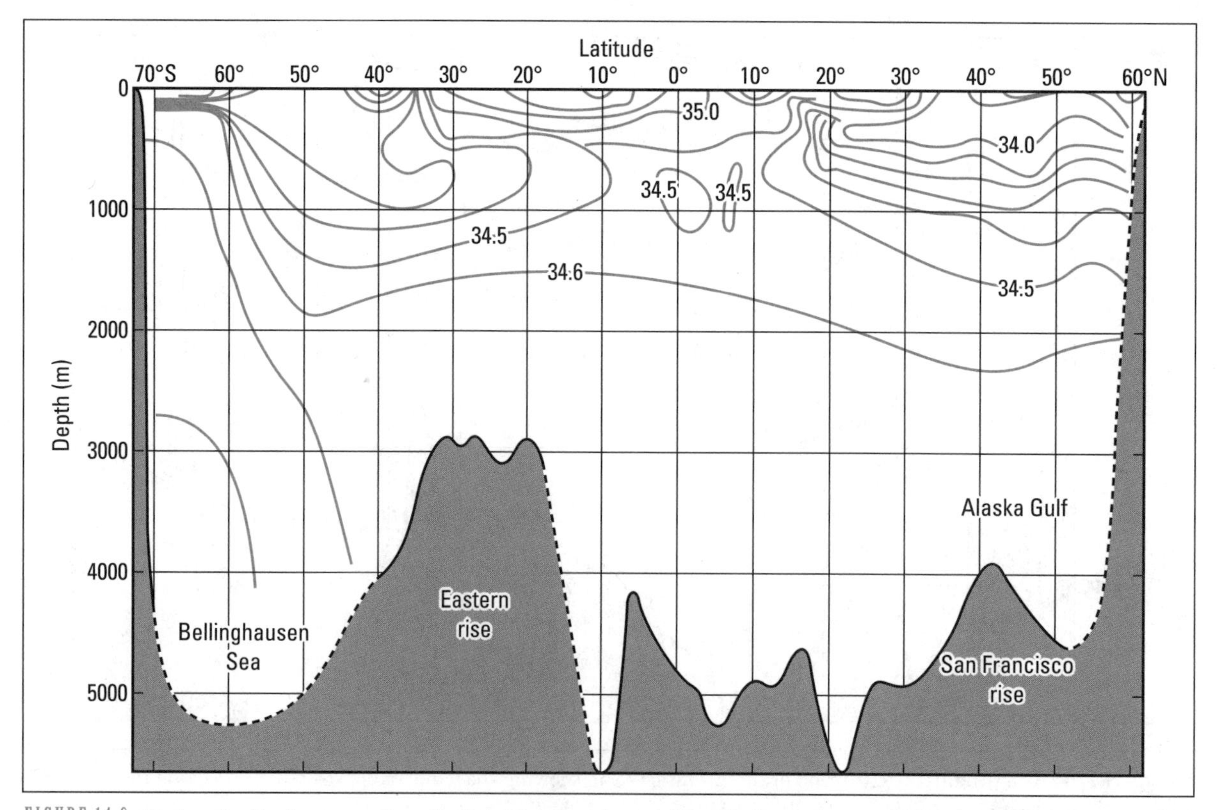


FIGURE 14.6 ▼ Longitudinal cross section of salinity for central part of Pacific Ocean (parts per thousand) (after Gerard, 1966)

the imbalance between the input of fresh water from precipitation and runoff and the output by evaporation. In these areas of the subtropical anticyclones, precipitation is low and evaporation is high. There is a net loss of fresh water, so salt concentrations increase. In high latitudes, the melting of ice causes a more complicated pattern and a marked seasonal variation in salinity. Where seas are enclosed we find extreme levels. In the Red Sea and Persian Gulf, both precipitation and runoff are low and evaporation high; no mixing occurs, so salinities reach 40‰ (parts per thousand). Indeed, the Dead Sea is so saline and the waters so dense that it is possible to float upon it without sinking. The Arctic is the least salty of all the oceans. It is a large ocean, a fact that is lost in the distorted projections used for most maps and atlases. It is also deep, with an average depth of about 3000 m; however, it is essentially an enclosed basin with very little inflow or outflow of water. The deepest channel connecting it to the Atlantic is a narrow valley that lies on the east side of Greenland and which is approximately 800 m deep. If we imagine the Arctic Basin with the water removed we would see that there is a ridge across the Bering Strait, and a similar ridge from Greenland to Iceland and across to the continental shelf surrounding the British Isles.

Large rivers such as the Ob, Yenisei, and Lena in Russia and the Mackenzie in Canada contribute large volumes of fresh water onto the surface of this large ocean. Low temperatures and ice cover mean that there is very little evaporation, thus the average salinity is 30‰. The low salinity levels are important for the formation of the ice pack. An increase in salinity which might follow from reduced flow of rivers into the ocean could hinder freezing and reduce the size of the ice pack. This would have an effect upon the temperature difference between the equator and the pole and bring about a change in the dynamics of atmospheric flow.

In the Baltic Sea, in contrast, precipitation and runoff are appreciable and evaporation is relatively low so we find salinities of as little as 5‰ (e.g., in the Gulf of Bothnia).

As with temperature, the vertical variations of salinity are influenced by the origin of the waters (Figure 14.6). The main feature is a zone of minimum salinity between 700 and 800 m. The source of this less saline water is the cold, dense, and relatively fresh surface waters of the Antarctic south of 45°S, which slowly sink and extend northward. This tongue of cool, low salinity water can be traced in the Atlantic as far as 20°N of the equator, and to the equator in the Indian and Pacific oceans.

Below this layer, salinity increases to a maximum between 1500 and 4000 m. Deeper again, we find water of slightly lower salinity: cold water which originated in the Antarctic area. Superimposed on this general pattern, however, are smaller-scale variations. The Mediterranean Sea has above-average salinity levels. Across the Strait of Gibraltar (Figure 14.7) there is a surface flow from the Atlantic bringing in cool, relatively fresh water, while at depth warm saline waters flow in the opposite direction. This lower flow involves water volumes over 100 times those of the Mississippi, and it is not surprising that its effect can be detected well out into the Atlantic, at depths of 1000–2500 m.

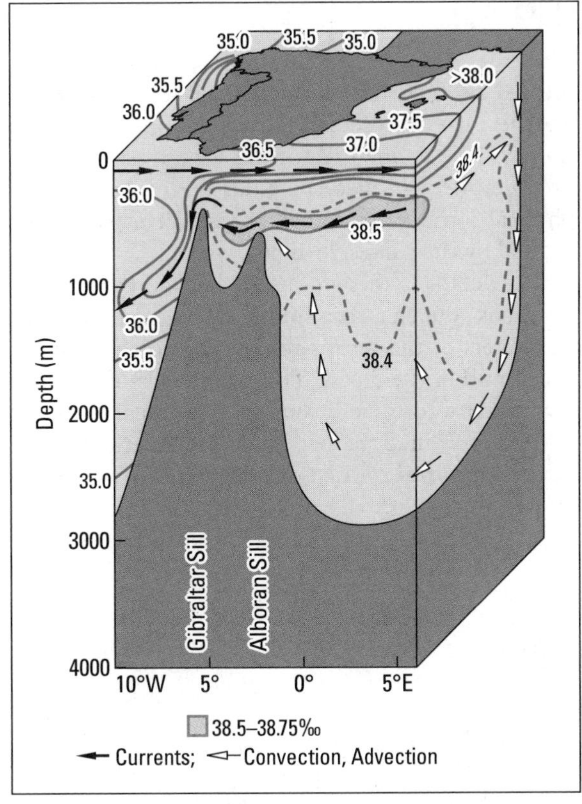


FIGURE 14.7 ▼ Vertical circulation and distribution of salinity in the Strait of Gibraltar during winter (after Gerard, 1966)

## Ocean tides and currents

For any visitor to the seashore it is the waves and the tides that are most apparent. Compared with other water bodies, the sea appears to be much more dynamic and mobile, and the seashore is a zone of great activity. Waves are continually breaking against the cliffs and, for many seas, the point at which the waves break on the shore varies during the course of the day, depending upon the state of the tide. However, in many seas the tidal range is small. It is less than 1 m in parts of the Mediterranean. Elsewhere the range can be enormous; it is 12.2 m at Avonmouth on the Severn Estuary, England, and up to 15.4 m in the Bay of Fundy, Nova Scotia. This tidal activity is clearly a widespread and powerful process, so powerful that it has been suggested that much of the world's energy could be obtained from it. But how does it happen?

## TIDES

The tide refers to the rhythmic diurnal or semi-diurnal rise and fall of the sea surface. Unlike the weather, we can predict the tides—and how high they will be at any given time. The reason for this predictability is that the tides are caused by the physical forces of our solar system, primarily those exerted by the sun-earth-moon system.

The details of these forces are beyond the scope of this book, but in a very simple way we can visualize the waters of the oceans responding to the gravitational pull of the moon. This counteracts the earth's own gravity to some extent and, in essence, creates two points where the net gravitational effect of the earth is diminished, one directly beneath the moon and the other on the opposite side of the earth (Figure 14.8). The differences are small but they are sufficient to cause the water to move down the 'gravitational gradient' and it is this movement that we see in the tides.

If the earth rotated more slowly, and if the moon remained above the same point on the earth, the oceans would adjust to this gravitational pull and they would have a permanent slight bulge. There would be a mound about 358 mm high in the middle of the gravitational basins, and a trough 179 mm deep along the intervening zone. This pattern is known as the equilibrium tide. But the speed of the rotation of the earth and the movement of the moon mean that the focus of attraction sweeps around the globe at 3000 km per hour. The equilibrium tide never gets a chance to establish itself, and the oceans constantly pour one way and then the other following the moon's effect. It takes 24 hours and 50 minutes for the moon to return to the same point above the earth, and in this time two tidal rises occur: one when the moon is overhead and the second when the moon is on the opposite side of the earth. It is this that gives the almost twice-daily tides.

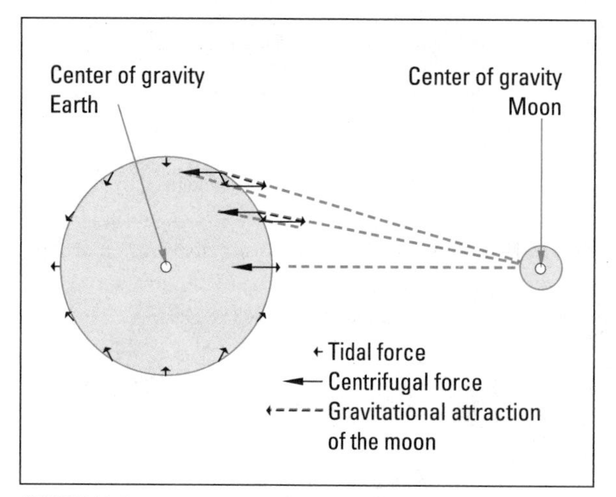


FIGURE 14.8 ▼ The tide-generating forces between the earth and moon (after Weyl, 1970)

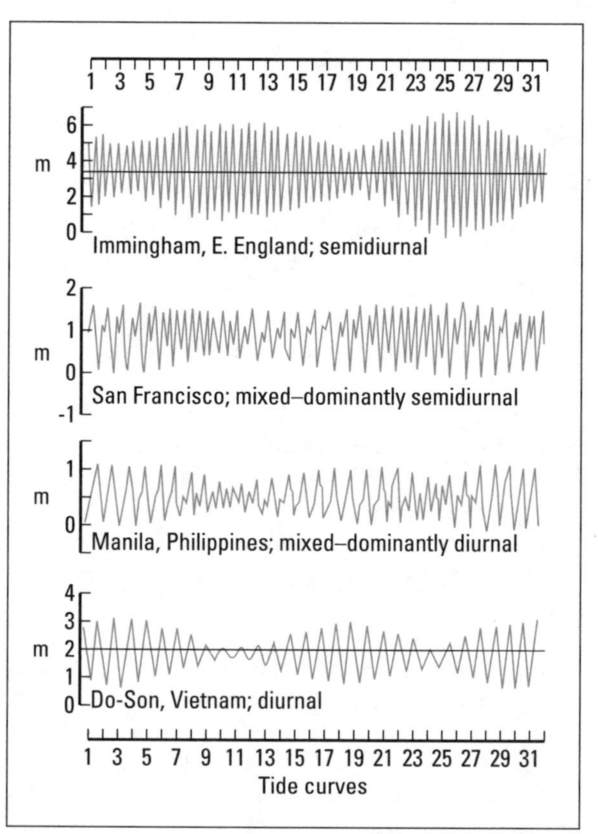


FIGURE 14.9 ▼ Representative tidal curves (after King, 1962)

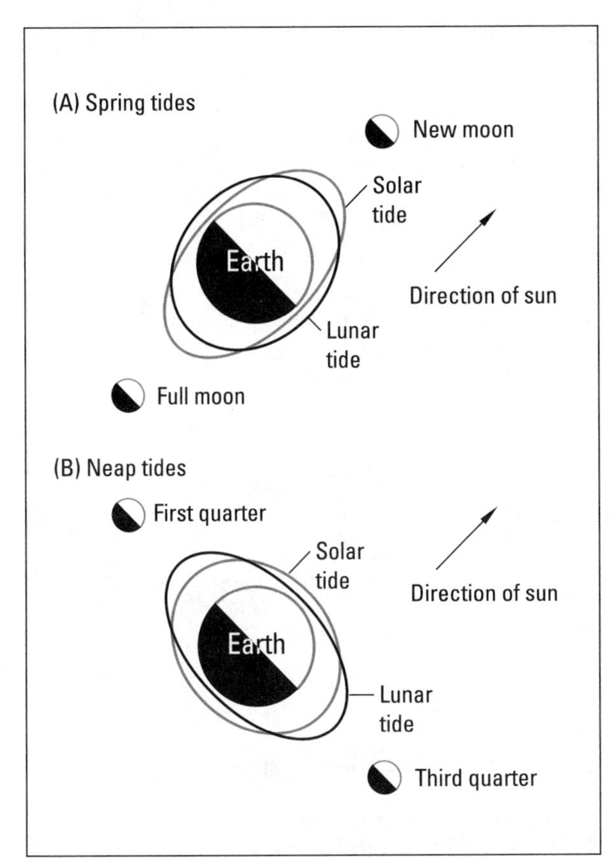


FIGURE 14.10 ▼ The effects of the gravitational pull of the sun and the moon on tides. When both bodies are pulling in the same direction, spring tides result (A); when at right angles, neap tides are experienced (B) (after Harvey, 1976)

If we plot tide levels for several weeks, we would notice that the tidal range is not constant (Figure 14.9). Over a period of about seven days the range builds to a maximum value, called the **spring tide**. It then declines over the next seven days to give a low value, called the **neap tide**. This modification of tidal pattern is produced by the interaction between the lunar and solar gravitational forces. When the sun and moon are pulling together (Figure 14.10) we get a greater attractive force operating on the oceans and a larger tidal range. When they are at right angles to each other (at first and third quarters of the moon) the effects are counterbalanced to a certain extent and so the tidal range is less. In even more detail, the tidal patterns over the globe vary slightly depending upon the interactions between the solar and lunar forces and the coastal and seabed configuration.

## SURFACE CURRENTS

As well as the flux of tides around the earth, we find that the oceans are also affected by net flows of water in one direction—ocean currents. In the surface layers, the currents are developed mainly by wind stresses, the frictional interaction between air movement and the sea surface. These surface movements may also disturb the waters at depth, but more generally deep currents develop due to differences in temperature and salinity.

When a wind blows across the ocean surface it causes a movement of the upper layers of the water, to a depth of about 100 m. Because of the deflection caused by the earth's rotation, this moves at right angles to the direction of the wind, to the right in the northern hemisphere and to the left in the southern hemisphere. The general pattern of oceanic circulation in the north, therefore, produces a piling up of water at about 30°N and a depression at about 60°N (Figure 14.11). A similar pattern develops in the southern hemisphere and, between the two, at the equator, a more complex depression occurs.

Water movement around these **gyres**, as they are known, is similar to air movement around high or low pressure cells. Due to the effect of Coriolis force, the water flows at right angles to the surface gradient, giving a more or less circular (geostrophic) flow around the gyre (Figure 14.12). In the northern hemisphere this is clockwise around the high at 30°N and counterclockwise around the depression. In the southern hemisphere the flows are reversed, while at the equator the two gyres merge to give a flow shaped rather like a figure-eight.

Because the earth's surface is not covered entirely with sea, and because of the irregular configuration of the continents, we do not find quite such a simple pattern in reality, but the basic elements are nevertheless visible in the global ocean currents (Figure 14.13).

*The Pacific Ocean* The Pacific Ocean, the largest and most regular of the world's oceans, displays the pattern most clearly. In the north a clear subtropical gyre can be seen, consisting of the westward flowing North Equatorial Current and the eastward flowing Kuroshio and North Pacific currents. The flow is completed by the California current. This gyre therefore carries warm water northward along the Japanese coast and then generally eastward to California. Cooler water flows southward down the California coast.

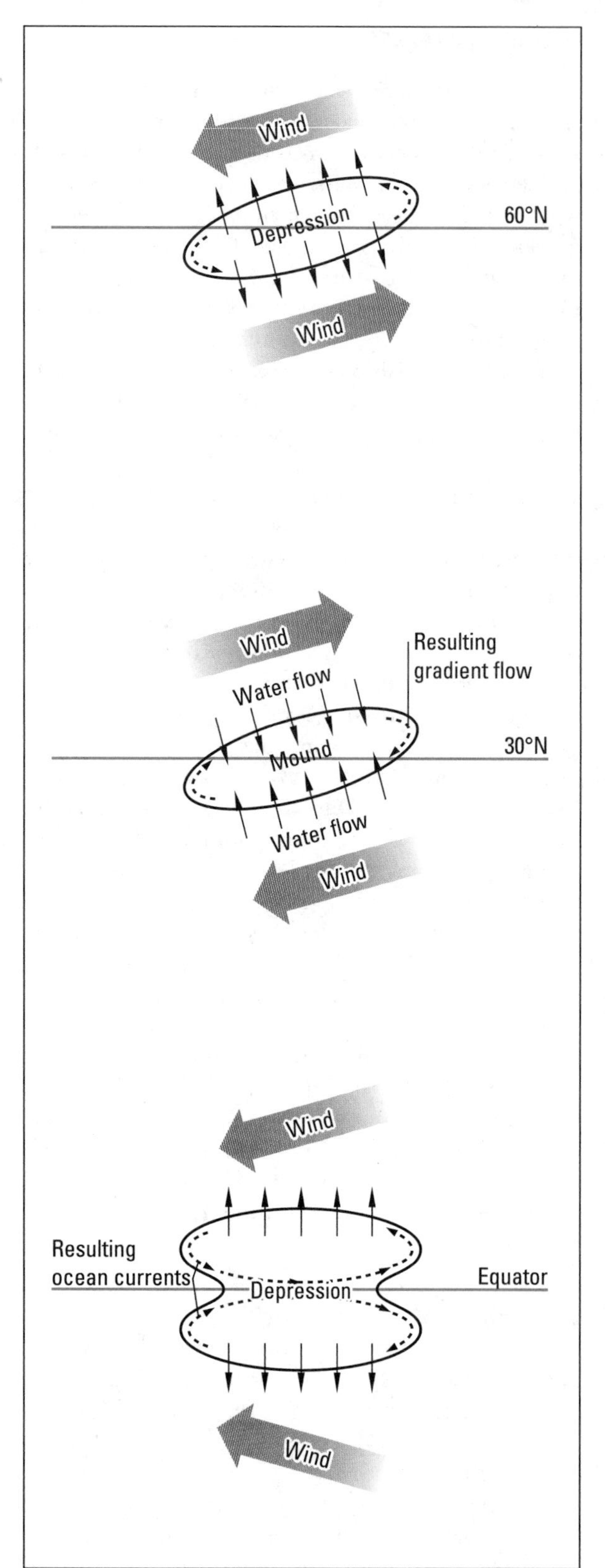


FIGURE 14.11 ▼ Creation of oceanic low and high pressures by the prevailing winds (after Pirie, 1973)

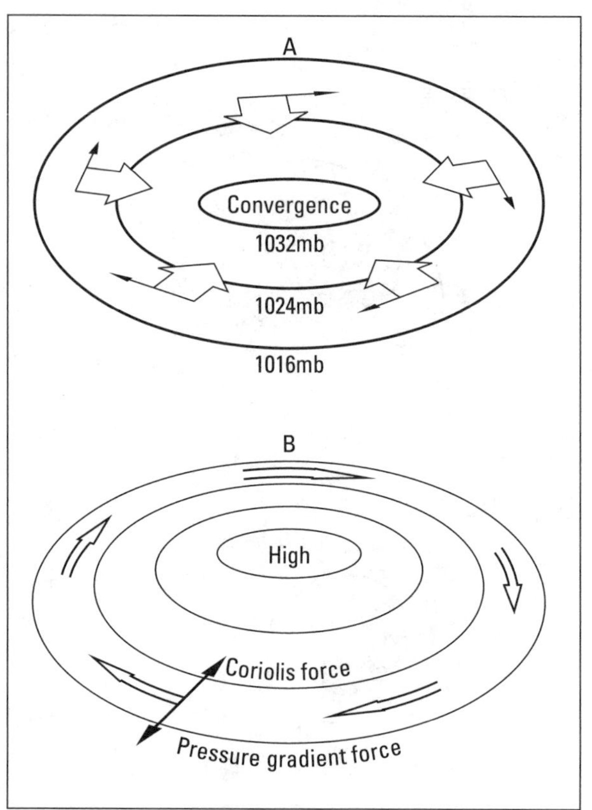


FIGURE 14.12 ▼ Water movements associated with anticyclonic winds in the northern hemisphere: (A) atmospheric pressure and winds and the associated frictional transports to the right; (B) resultant topography of the sea surface and associated gradient currents below the depth of frictional influences (after Harvey, 1976)

A similar subtropical gyre is also seen in the South Pacific. This comprises the warm South Equatorial and East Australian currents and the cool Peru Current which flows northward along the Peruvian coast.

The equatorial gyre is also visible, although it tends to be displaced slightly north of the geographic equator due to the similar displacement of the atmospheric circulation here. The Equatorial Counter Current provides the central line of this double cell, which is completed by the North and South Equatorial currents.

The subpolar gyres are less clearly developed in the Pacific. In the north the Bering Strait restricts the circulation, and no warm water passes through the strait to complete the gyre. All that can be detected is the southward flowing Oyashio Current

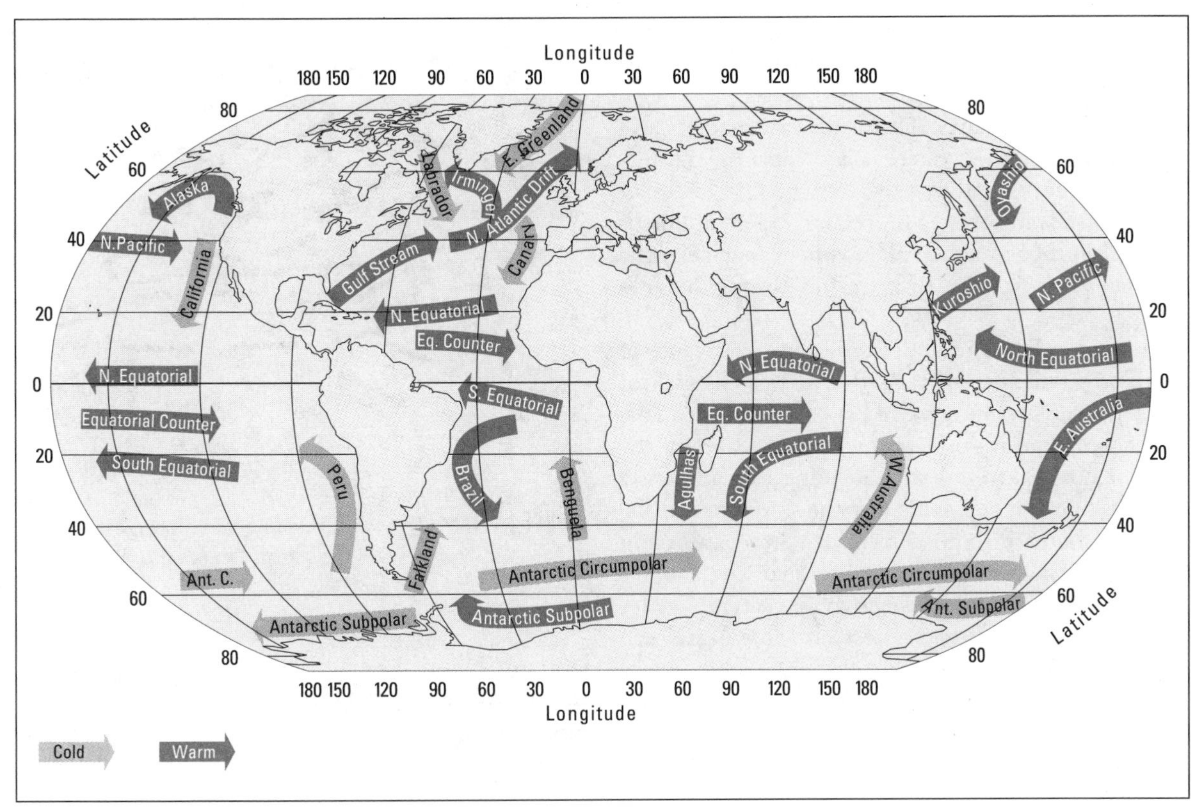


FIGURE 14.13 ▼ Main oceanic currents during the northern hemisphere winter

and the almost separate cell produced by the Alaska Current. In the South Pacific the Antarctic Circumpolar and Antarctic Subpolar currents give a more complete gyre at 60°S.

*The Atlantic Ocean* In general, the same pattern of circulation can be seen in the Atlantic. The northern subtropical gyre is possibly one of the best known, to people who live around the North Atlantic. Driven by the tropical easterlies, the waters of the North Equatorial Current are heated under the tropical sun as they move westward parallel to the equator. The Gulf Stream is created where the waters of the North Equatorial Current have piled into the Gulf of Mexico, and are then whirled out toward the northeast. They have been heated more in the shallower waters of the Gulf. In this way the Gulf Stream becomes a major transporter of the surplus heat energy of the tropics to the energy deficit polar regions.

Coriolis force is now pushing the current to the right, and off the coast of the eastern United States and Canada it is propelled eastward by the prevailing westerlies and changes its name to the North Atlantic Drift. Here it is also met by the cold surface flows out of the Arctic Basin formed by the subpolar gyre. The major currents here are the Labrador and East Greenland currents, with the latter flowing between Iceland and Greenland in the deepest channel (approximately 700 m) of the Arctic Basin. They meet the warm North Atlantic waters in the region of the Grand Banks. The prevailing airflow here is from the south so warm air blows across the cold currents. Cooling is rapid, condensation occurs, and so the dense fogs are produced for which the Grand Banks fishing area is notorious. It has the highest number of days of fog per year of any region currently monitored.

Icebergs calving from the icecaps of Greenland, Baffin Island, and Ellesmere Island are carried south in the cold water of the Labrador and East Greenland currents. These huge ice islands are a threat to shipping, as the *Titanic* discovered in 1912. They also pose a danger to the growing number of oil drilling platforms on the Grand Banks. Where the cold, dark, green-black polar waters meet the warm, light, blue-green tropical waters of the Gulf Stream a distinct line in the ocean is visible, especially from the air, and there is a significant change in the air temperature over the two bodies of water.

The North Atlantic Drift continues northeastward driven by the westerlies and carrying the surplus heat energy from the tropics. The effect of these warm waters is felt in Iceland and all of northwestern Europe as far as 69°N, where they keep the Russian port of Murmansk ice-free throughout the year.

Benjamin Franklin understood the significance of the North Atlantic Drift. As Postmaster General for the United States he advised his postal ships to take water temperatures to determine if they were in the drift or not and to sail with it going east and away from it going west. In this way the ships crossed the Atlantic faster in both directions, thus expediting the mail. In Figure 14.13 note that the Canary Islands Current completes the gyre, carrying cooler waters south to rejoin the westward flowing North Equatorial Current off the coast of Africa. These currents carried Columbus to the New World—but they also carried the slave ships from Africa to the Americas.

*The Indian Ocean* The Indian Ocean produces an interesting anomaly to this general pattern of circulation. This occurs because of the marked reversal of atmospheric circulation associated with the seasonal progress of the trade winds (the monsoon). This reversal of wind directions causes a related change in the oceanic circulation (Figure 14.14). In winter the northeast monsoon blowing across the Indian Ocean generates a current moving in the same direction toward East Africa, termed the Northeast Monsoon Current. This eventually merges into the Equatorial Counter Current just south of the equator. In summer, a clockwise pattern of circulation develops. A weaker reverse flow exists off Sumatra, complicated by the configuration of the land. The flow as a whole, however, is generally counterclockwise.

In summer, the pattern reverses and a clockwise circulation develops. Off Africa we find the strong Somali Current appearing in April and reaching its peak in July and August (heavy arrows, Figure 14.14). This turns eastward into the Southwest Monsoon Current in the Arabian Sea (remember that currents, like winds, are named according to where they have come from, not where they are going to) and then merges into the South Equatorial

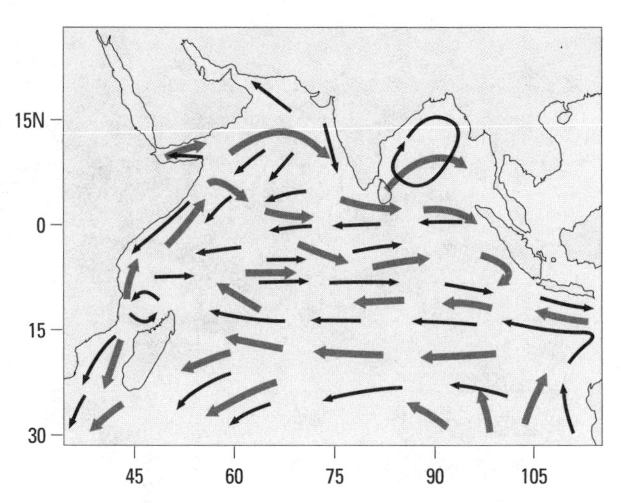


FIGURE 14.14 ▼ The seasonal variation of the surface currents of the Indian Ocean (after Weyl, 1970)

Current (Figure 14.14). By November, this circulation is weakening and the flows start to reverse once more.

## DEEP OCEAN CURRENTS

Although the surface flows are the most obvious aspects of the oceanic circulation and, certainly in the days of sailing ships, were perhaps the most important, beneath the surface there are other flows. These are usually much slower and we know much less about them. They include both horizontal flows in the ocean depths and also vertical movements.

*Subsidence and upwelling* We have already mentioned that the temperature structure of the oceans means that they are much more stable than the atmosphere. Convection currents of the form we find in the air do not occur. Nevertheless, considerable vertical motion must take place if the circulation of the oceans is to be maintained. There are several situations in which these occur. Subsidence takes place because of differences in density of the waters, produced either by differences in salinity or temperature (or, indeed, a combination of the two). Ascent of water (**upwelling**) occurs where surface winds blow coastal waters seaward.

The main areas of subsidence occur in the high latitudes, where cold polar waters are created by the input of ice from the icecaps and by the radiational cooling of the surface. The cool waters sink and, as we have noted, gather in the lower levels of the oceans where they help maintain low temperatures. The effect of these general zones of subsidence can be seen particularly in the Atlantic (Figure 14.15), but they also occur, less markedly, in the Pacific.

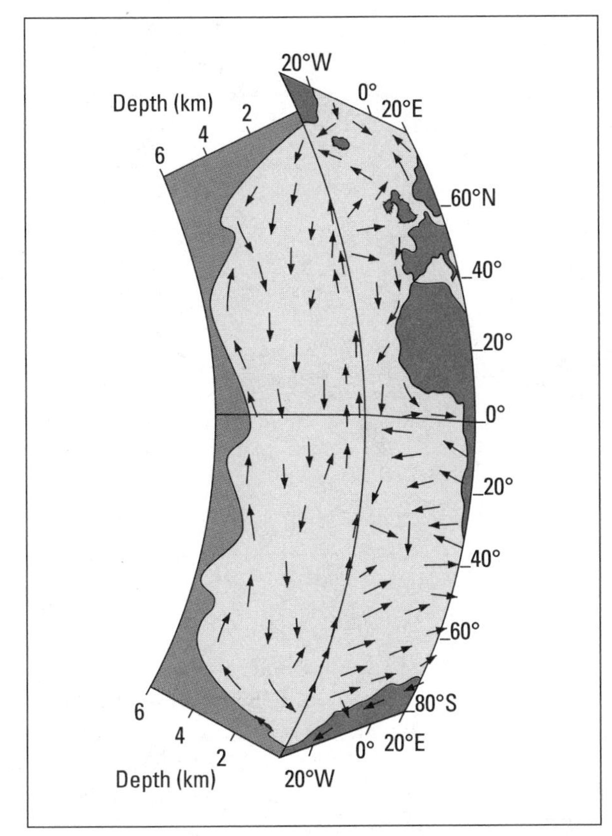


FIGURE 14.15 ▼ Surface and subsurface currents in the Atlantic Ocean (after Harvey, 1976)

More localized subsidence occurs in a number of areas. Cool sediment-laden water from the continents often subsides to some extent as it enters the sea, while the saline waters from the Mediterranean subside as they enter the Atlantic. In addition, in zones of convergence, where surface currents meet, subsidence tends to occur. The cool Labrador and Greenland currents, for example, sink beneath the warm North Atlantic Current in the Newfoundland area.

Upwelling occurs mainly along the deep coastal areas where offshore winds blow water seaward. Water removed in this way is replaced by cool bottom waters that rise from the ocean depths. As we have seen (Chapter 7), these greatly modify the local climate and they also influence the ecosystem.

*Horizontal flows at depth* We know very little about flows deep in the ocean, although we can, to some extent, deduce what happens from the patterns we find at the surface. Clearly, to complete the circulation, these deep flows must feed the patterns of surface movement. Because the oceans are a closed system, whenever water moves from one region to another it is replaced. The evidence is that cold water is the dominant force dictating movements in the deep ocean. Cold water is either moving into an area displacing warm water, or it is moving out from an area and allowing warm water to move in. (The same is true in the atmosphere, where the denser cold air determines the pattern of movement.) This is shown by the fact that although the polar waters are less saline, the cold Arctic and Antarctic waters sink to the ocean floor and generally move toward the equator, as illustrated by the dark heavy arrows in Figure 14.16. In addition, by studying temperature and salinity profiles of the oceans we can get a broad picture of subsurface movements.

Without doubt the most important flow is that formed from the Antarctic and Arctic zones of subsidence. In the Atlantic, the cold Antarctic waters seem to flow at the very bottom of the ocean, north to 20°–40°N. The Arctic waters flow southward at a slightly higher level (3000–4000 m) to almost 60°S of the equator. In the Pacific the flows are slower and less apparent, in part because the Bering Strait prevents deep, cool bottom water spreading from the Arctic Ocean. The circulation, as far as we know it, is illustrated in Figure 14.16.

Nearer the surface, in what is often called the central zone, more complex patterns of flow occur, with water from a variety of sources mixing and converging. In the Atlantic, water from the Mediterranean cuts across the general southerly drift of the subsurface circulation and extends as much as 2500 km into the ocean. In the Pacific, water from the North Atlantic and Indian oceans merges with that from the Antarctic to give sluggish, ill-defined flows.

*El Niño and the Southern Oscillation* An example of a major feedback problem in the atmosphere is a feature known as the Southern Oscillation. As well as the Hadley cell around the intertropical convergence zone, there is also a circulation in the Pacific between the Indonesian equatorial low and the southeast Pacific subtropical high. In general

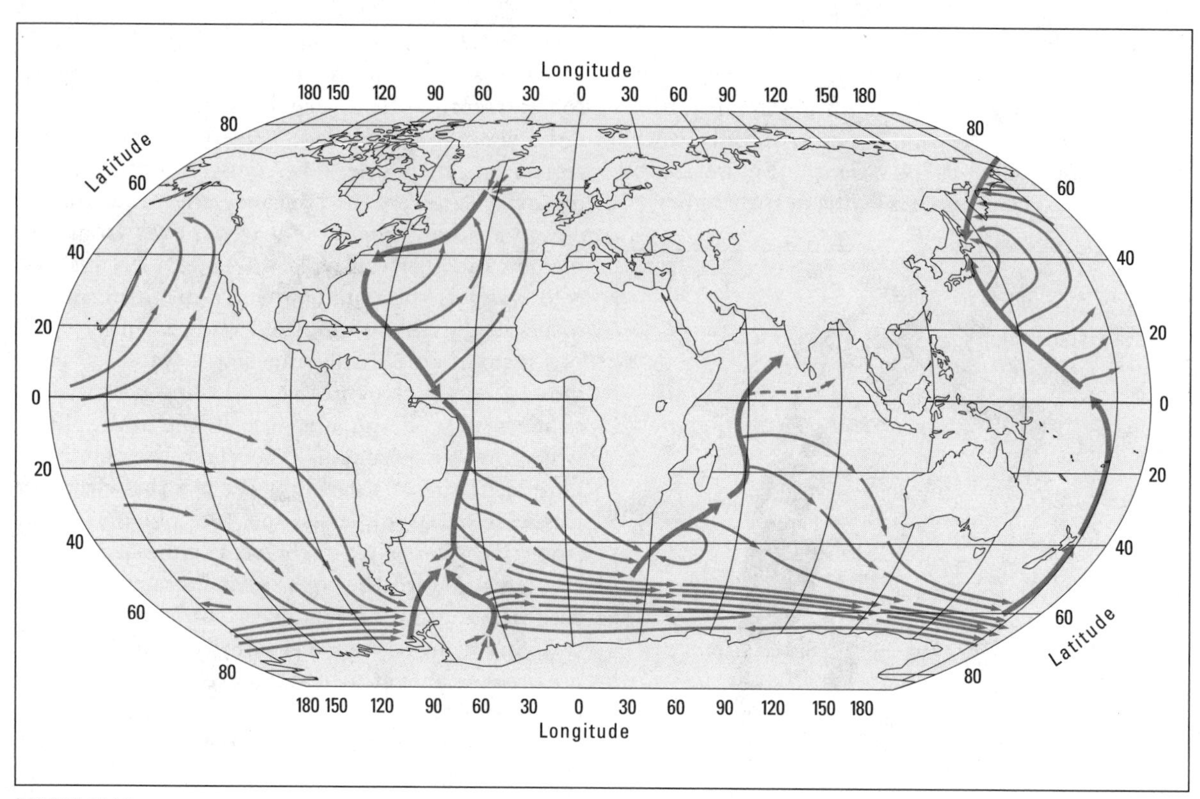


FIGURE 14.16 ▼ Deep oceanic circulation below 2000 m (after Ingmanson and Wallace, 1973)

terms, when pressure is high in the Pacific Ocean, it tends to be low in the Indian Ocean from Africa to Australia. Where pressure is low, rainfall is high and vice versa. The pressure gradient tends to maintain strong equatorial easterlies, which in turn maintain upwelling of cold water, so the process is self-sustaining—a positive feedback. Periodically the situation reverses, to give high rainfall in the eastern Pacific and low rainfall in Indonesia. The effect was pronounced in 1983: coastal Ecuador received 300 percent of average precipitation between November, 1982, and May, 1983, amounting to 3258 mm, whereas Indonesia had received only 30 percent of its average.

The cause of the persistence is reasonably well established and is a good example of the feedback

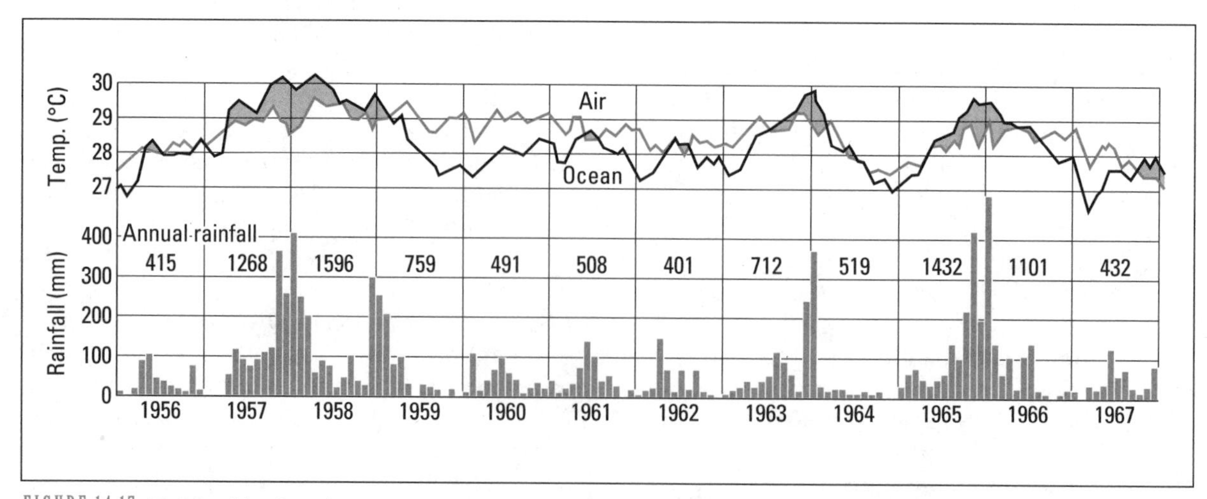


FIGURE 14.17 ▼ Monthly air and sea temperatures with rainfall data for Canton Island (2°48'S, 171°43'W)

mechanism. The problem is to understand what causes a breakdown of the system to give a reversal, with its characteristic period of between two and five years (Figure 14.17). J. Berknes suggested in 1969 that when the equatorial easterlies weakened, the upwelling cold water would be reduced and warmer water from the western Pacific would extend over the surface waters off the coast of Peru and Ecuador. People in those regions call this warm current El Niño, or the Christ Child, because it usually arrives close to Christmas. Crews of Spanish vessels sailing into these waters in the sixteenth century learned of these reversals of currents, and that the effect becomes much stronger about every ten years.

It is important to note that El Niño is initially an effect of climate, but it also then affects the climate, particularly in those regions bordering the Pacific in equatorial latitudes. Heavy rain in Peru, Ecuador, and the southern part of North America as well as droughts in Australia, southeast Asia, and South Africa are usually attributed to the warmer waters. The interconnection between ocean and atmosphere is clearly demonstrated in these two phenomena; but we must remember that something must trigger the change in atmospheric pressure that in turn causes the oscillation and warmer water.

# Conclusion

Considering their size, we know remarkably little about the oceans, especially the magnitude of their exchanges. However, we have been able to outline the general aspects of their circulation. This pattern is a vital part of the hydrological system, transferring waters across the globe, supplying waters to areas of deficit, removing them from areas of excess. The oceanic circulation also involves a transfer of energy that is important to the global energy system.

But possibly the most significant aspect of the oceans is their size. They contain 97.6 percent of the world's water; they are the major store within the hydrological cycle. Compared with the oceans, the atmosphere, the cryosphere, and the groundwaters contain negligible quantities of stored water. As the foremost store of the world's water, the oceans exert a dominating control over the hydrological cycle. They influence the amount of water available for circulation, they provide the main surface for evaporation, they influence climate and thereby affect rates of precipitation. They may be the last unknown, but they are a part of the world we need to know much more about.

# Landscape form and process

## Introduction

The landscape is not static. Over the years and eons it changes and develops. New landscapes are formed; old ones are destroyed or modified. Sometimes these changes are dramatic. New islands, such as Surtsey, may rise from the ocean; volcanic activity, earthquakes, and land tremors may rip apart and recreate whole regions almost overnight. Sometimes the consequences of these events are immediate and terrible for humans. More commonly, the changes are slow and almost imperceptible. It is the processes of gradual wear and movement, operating over thousands or even millions of years, that carve and mould the landscape.

The landscape that we see, therefore, is merely a fragment in time, rather like a still from a film. It represents just one moment in an endless sequence of development. To understand the landscape we need to consider the processes that have been, and still are, acting upon it. It is not enough merely to describe what we see, to classify the landforms; we must try to grasp the 'logic' behind the landscape. The questions we should ask, therefore, are not 'what is it?', but 'why and how?' Why is the landscape that shape; how did it form; how is it changing; what will it be like in the future?

These questions raise a myriad of more detailed questions. We may try to explain a landform as the work of ice or frost, of the action of rain or rivers, of wind or gravity, but such explanations are too general. In reality, the landform does not develop through the operation of such general forces; it is a result of countless minute events, many working in opposition to each other. It is the result of the formation of individual ice crystals, of the impact of individual drops of rain, of individual gusts of wind. If we are to understand our landscape fully, it is these details that we must ultimately comprehend.

To tackle the details, however, it is useful to have a more general framework within which to operate. We need a model of landscape development to guide us. In this context, we might start by noting that the landscape is carved out of, and built from, rock materials, and that all landscape-building processes involve in some way the reorganization, alteration, or movement of these materials across the earth's surface. As a basis for our model, therefore, we can define what is referred to as the cycle of rock materials; it is this cycle that underlies the development of the landscape.

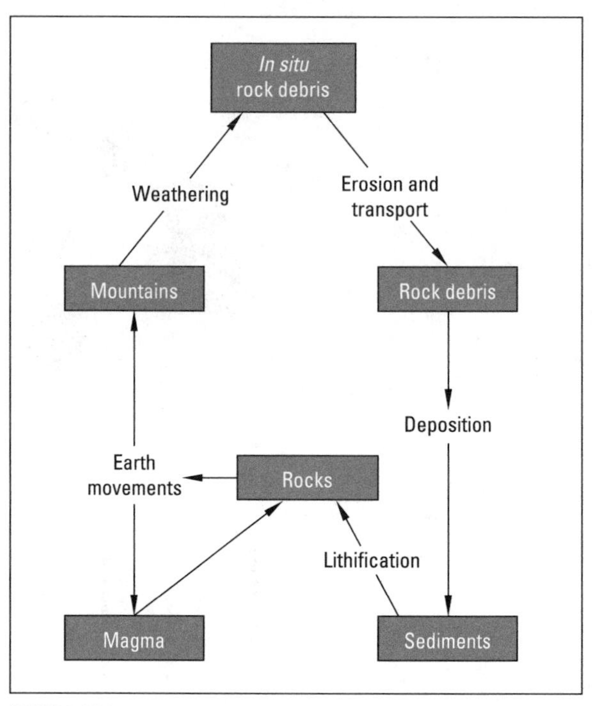


FIGURE 15.1 ▼ A general model of the cycle of rock materials

## The rock material cycle

The cycle of rock materials is illustrated in general terms in Figure 15.1. Magma, molten rock from deep in the earth's interior, is intruded into the earth's crust or escapes onto the surface, where it cools and **lithifies** (i.e., forms rock). The rocks formed in this way are often subjected to powerful earth forces that lift and fold them, while, once exposed at the earth's surface, they are also subjected to attack by wind, water, and ice. Thus the rocks are gradually weathered and the debris is carried through the landscape by streams and glaciers, by wind and gravity. Ultimately, the material is laid to rest and then, through the effects of burial and compression, by heat and chemical reactions, it is transformed into new rock types. Then the whole cycle can start anew.

The model we have outlined in Figure 15.1 is, of course, extremely simple. It is important to appreciate that the rock materials do not necessarily follow very precisely the idealized sequence; they may miss some steps and repeat others. They may, for exam-

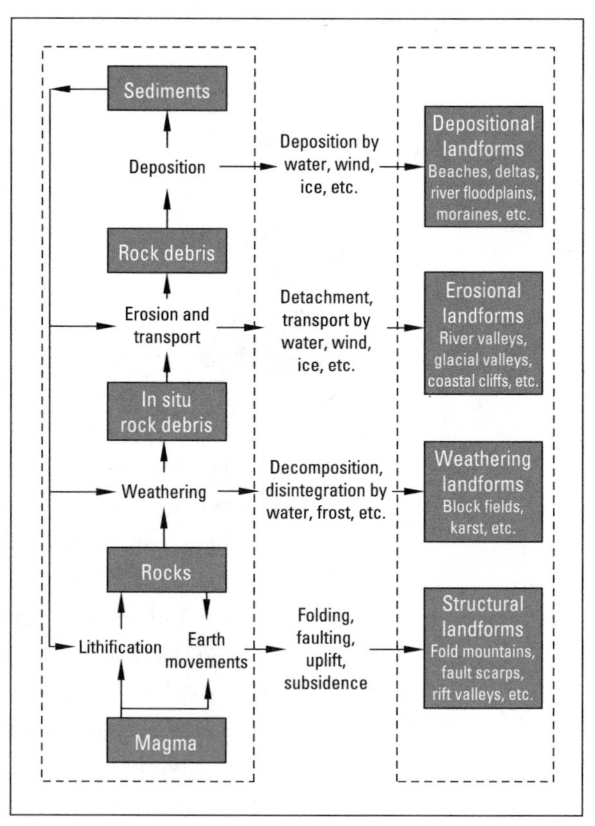


FIGURE 15.2 ▼ The cycle of rock materials (left) and landforms

ple, be subjected to many phases of earth movement before being exposed at the surface and weathered, or they may be repeatedly eroded and redeposited before coming to rest for long enough to be lithified again into rock. Thus the progress of rock materials through this cycle is not continuous, nor is it rapid. It may take many millions of years for the cycle to be completed. In the process, however, the forces acting upon the materials shape them into the landforms that make up our landscape: the continents and oceans, the mountains and valleys, the hillslopes and plains (Figure 15.2).

Earth movements play a fundamental role. They are responsible not only for the creation of relatively local or regional features, such as fold mountains like the North American Rocky Mountains or the South American Andes Mountains, or faults like the San Andreas Fault in California or the Cabot Fault in Atlantic Canada, but also for the broad configuration of the oceans and continents, and for the worldwide distribution of mountains, marine trenches, and volcanic activity. All these features are related to the slow but massive displacement of the earth's surface known as plate tectonics. The vast plate-like areas that are the basis of the continents are gradually migrating across the surface, powered by the deep circulation of heat and molten rock in the earth's interior. Where these plates are separating, new ocean floor is being created as magma wells upward and escapes through fissures in the crust. Where the plates are converging, the continental rocks are being subjected to huge compressional forces which fold and lift them into mountain chains, and release volcanic activity from deep in the mantle, occasionally with great violence, as in the explosion of Mount St. Helens (see Chapter 17). Also in these zones of convergence, the crust is being forced downward, to be consumed within the earth's interior. Thus, these massive earth movements not only produce **structural landforms**, but they also create and consume crustal materials, and thereby play a vital role in the natural cycling of rock materials.

By comparison, the effects of weathering and erosion might seem mild. Nevertheless, together they provide the main means by which the earth's surface is stripped and lowered—a process known as **landscape denudation**—and by which the materials are redistributed across the earth's surface. In general, weathering provides the preparation for this movement by weakening the rocks and providing debris for transport. Water plays a major part in this process, carrying substances to and from the weathering rocks, acting as an important solvent, and also setting up significant physical stresses within the materials. The effectiveness of different weathering processes varies with environmental conditions. Rates of denudation are determined by a combination of weathering and erosion processes; for example, the rate of denudation in high mountains is much greater than that of lowland plains. As rocks decompose and disintegrate, landforms of weathering are produced, such as the spectacular karstic landscapes of limestone areas in Yugoslavia and western Ireland.

The weathered materials, however, are subjected to other forces. Water, wind, ice, and gravity detach and transport the rock fragments. Material from cliffs and hillsides tumbles down the slope; rain splash moves small particles on the ground surface; running water picks up the debris and carries it farther; winds transport the fine material. As it is transported the material acts as an agent of erosion in its own right, abrading the surface and causing new debris to be detached and removed. Thus, **erosional landforms** are developed. Streams and glaciers cut valleys; the wind strips the surface and carves the resistant rock masses; on the coast, storm waves etch the cliffs.

Ultimately, the debris transported by these processes comes to rest, and new landforms are built. Fans of debris accumulate at the foot of the slope; beaches are formed at the margins of the sea; glaciers retreat and leave behind masses of material in an irregular and chaotic landscape. Sometimes these **depositional landforms** are short-lived. The next flood may remove the material, or a new ice advance may override its former deposits. Occasionally, however, the deposits remain for long periods. They may be buried; the earth's surface may rise or the sea level may fall, and the marine sediments may be left high and dry. Buried deposits are compressed, altered by pressure, heat, and chemical reactions, cemented by compounds washed through them by percolating water. They are turned into sedimentary rocks; as we have seen, the whole cycle may start again.

## The debris cascade

We can illustrate the cycle of rock materials more anecdotally by following an individual quartz pebble on its journey through the landscape. The pebble started, perhaps, as a piece of vein quartz in the volcanic rocks of the Canadian Shield. Long ago, as the climate began to change drastically during the buildup to one of the Quarternary Ice Ages, it was pried from its cliff by frost. Released, it bounced down the slope and came to rest at the slope foot. There it lay for many centuries, attacked by the sands carried in the fierce winds and by the lichen that grew on the exposed ground. An ice sheet began to develop and incorporated the pebble with the debris at its base. So, chipped and scratched by constant scraping against the bedrocks, the pebble was carried to the ice sheet terminus where it was deposited in a moraine.

Even there the journey was not over, for meltwaters from the now decaying ice eroded the moraine and washed the pebble downstream. Over the years, hopping from one gravel bar to the next, becoming rounded by wear as it did so, the pebble made its way into the estuary of the St. Lawrence River. There, it was caught by the tides, for a period became entangled with seaweed, and was washed far along the coast. Eventually, the waves shuffled it onto the beach and then, during a severe spring gale, it was flung high onto the storm ridge, where it now lies, smooth and spherical.

The journey is, of course, a hypothetical one. In reality we cannot know the detailed history of an individual pebble in this way. But it is probably a representative example of the history that many pebbles in our landscape have had. And it illustrates what we refer to as the **debris cascade**—the movement of rock materials through the landscape following uplift. It is an apt term, for it indicates the generally downward course of this movement, from source to final resting place, which is often in the sea.

The journey we have described also illustrates several important aspects of this movement. As the pebble—indeed as any material—passed through this cascade, it was affected by many different processes: by the action of wind and water and ice; weathering, transport, and deposition. These processes modified the pebble. It was fractured and chipped by the ice, abraded by the water; it ended smaller and rounder than it started. In addition, the pebble itself took part in the formation of many different landforms. Now it is part of a talus slope, now a glacial moraine. Next it becomes part of a series of gravel bars, and ends as a component of a beach. Even when it was mobile, it was affecting the landscape, wearing the bedrock beneath the glacier, battering against other pebbles as it moved down the stream. Thus, the movement of this material through the debris cascade contributes to and affects the landscape through which it passes. The processes operating in the debris cascade act to shape the landscape.

## The landscape as a process-response system

This relationship between the processes operating within the landscape, the materials on which they are acting, and the shape or character of the landforms that develop is central to our understanding of geomorphology, and it is a theme we will follow in the rest of this section. In the language of systems theory, we are considering the landscape as a **process-response system**: one in which a **morphological system** (the landforms) and a **cascading system** (the debris cascade) interact. More simply, the morphology of the landscape is a function of the processes acting upon it.

It is important, however, to consider some of the implications of this concept—implications that are not always remembered. First, this is not a one-way relationship. Not only do the processes affect the landscape, but so does the landscape influence the processes that operate on it. Thus, glaciers and streams mould the valleys, but the shape of the valleys also influences the way the glaciers and streams behave. Similarly, tides and waves act to build beaches, but the shape of these beaches modifies the flow of the currents and the pattern of the waves. In other words, feedback processes occur. Landscape form and process are intimately and reciprocally related.

Second, though, we must remember that few landforms are simple. Many are both complex and composite. That is to say, landforms are often produced not by a single process acting in isolation, but by a whole range of processes acting together. Waves and wind act in unison to mould a sandy shore. Water, ice, and many hillslope processes work together to carve the valleys. Thus, many landforms are **polygenetic**: they owe their origin to several different processes. Moreover, these processes often change over time, so that individual landforms go through several cycles of development. We saw this in the case of the pebble we discussed earlier, and the same is true of larger landforms and whole landscapes. Periods of weathering are followed by phases of glacial action; these in turn are succeeded by fluvial action or the effects of wind. Landforms, therefore, are often **polycyclic**. What we see now is simply one stage in their history, possible not even an important stage. The landscape as a whole is, in part, a legacy of the past. If we are to understand the landscape we need to understand not only the processes operating at present, but also the factors that have operated in the past to make the landforms the way they are.

It is here that geomorphology and geology come together. We use our geomorphological knowledge of present-day processes as a basis for interpreting more ancient landscapes. We use modern landforms as analogues of the past. But we also need an understanding of past environmental conditions and events as a framework within which to place our interpretations. Thus, as background to our study of the landscape, we need a broad understanding of geological history.

# The geological column

The general history of the earth is portrayed by what is known as the **geological column**. A simplified version is shown in Table 15.1. The column is divided into four major eras: the Precambrian, Paleozoic, Mesozoic, and Cenozoic. These eras are subdivided into periods—phases when the earth is believed to have experienced relatively stable conditions, or changed in a consistent manner. These are further divided into epochs, each of which represents a specific chapter of earth history.

This history began some 4600 million years ago, but the early events are obscure, for no rocks from that time have been found. The oldest known rocks, in fact, are about 3800 million years old, from Minnesota in the USA. Life did not appear until considerably later: the earliest evidence comes from rocks in Australia, some 3000 million years old.

The old rocks of Quebec are part of a large geologic feature known as a **continental shield**. Shields form the major component of the continents, and are composed of a complex mixture of igneous and metamorphic rocks. Today, the shield areas tend to be low relief, after millions of years of erosion, and large parts are covered and surrounded by sedimentary rocks that are the result of that erosion. The Canadian Shield occupies the northeast corner of North America; it makes up most of the surface from the Atlantic coast to a line forming an arc from Great Bear Lake through Lake Winnipeg and the Great Lakes and along the St. Lawrence River, and includes northern New York State and the New England states (Figure 15.3). The sedimentary rocks that make up the rest of the continent overlap the edges of the Shield and stretch across to the Rocky Mountains. Here the sedimentary rocks have been folded into a series of mountain chains.

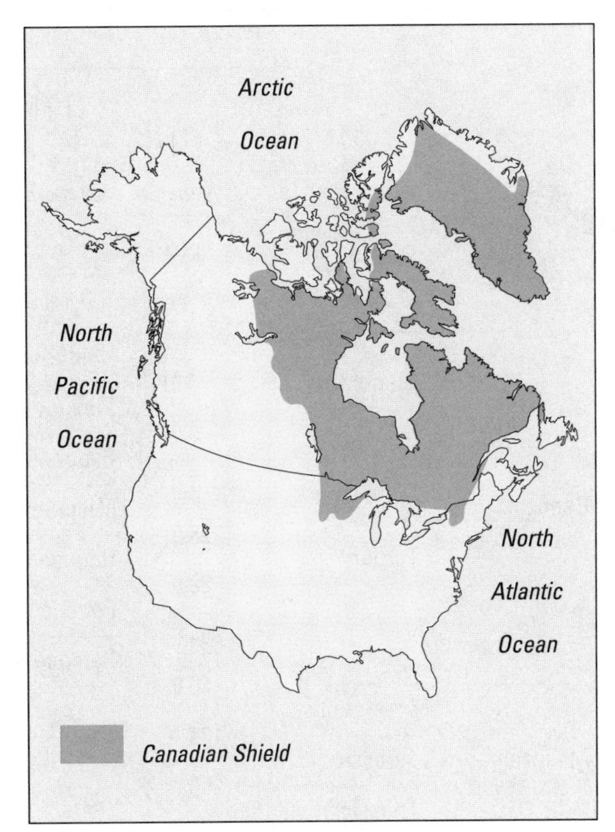


FIGURE 15.3 ▼ The Canadian Shield

The Canadian Shield is an example of an exposed shield, a part of the earliest chapter of the geologic evolution of the continent, the Precambrian Era. We can read the story as it continues in the Grand Canyon and other deeply eroded trenches of the Colorado Plateau in Arizona, Utah, Colorado, and New Mexico. At the lowest levels in Grand Canyon, Precambrian rocks are exposed. Unlike the Canadian Shield, these are overlain by the sedimentary layers of a shallow **geosynclinal sea**. The sea disappeared several million years ago. The deposits were subjected to **epeirogenic movement**, or rising and falling of continents over large areas. Combined with environmental changes, these movements resulted in the deposition of many different sedimentary layers. For example, the vast potash deposits of Saskatchewan were formed when a desert environment existed in what are now the Prairies. These major changes and movements occurred in the Cambrian, Ordovician, and Silurian periods.

Throughout geologic time, there have been periods of intense crustal deformation known as **orogenies**, which were characterized by extensive folding and faulting (Table 15.1). In North America, the Appalachian orogeny, at the end of the Paleozoic Era during the Permian Period, formed the Appalachian Mountains. More recently, the Nevadian and Laramian orogenies in the Mesozoic Era, and the Cascadian in the Cenozoic Era, saw the formation of the major portion of the Rocky Mountains. On a global scale, the mountains formed through the middle of the Tertiary Period are classified as the alpine system.

The Cretaceous Period, which marks the end of the Mesozoic some 65 million years ago, was a time

## Table 15.1

*A general outline of the geological column*

| *Era* | *Period* | *Age (m yrs)* | *Epoch* | *North American Orogenies (mountain-building periods)* |
|---|---|---|---|---|
| Cenozoic | Quaternary | 0.01 | Holocene | |
| | | 3.0 | Pleistocene | |
| | Tertiary | 7.0 | Pliocene | Cascadian |
| | | 26.0 | Miocene | |
| | | 38.0 | Oligocene | |
| | | 54.0 | Eocene | |
| | | 65.0 | Paleocene | |
| Mesozoic | Cretaceous | 135 | | Laramian |
| | Jurassic | 200 | | Nevadian |
| | Triassic | 240 | | |
| Paleozoic | Permian | 280 | | Appalachian |
| | Carboniferous | 370 | | Acadian |
| | Devonian | 415 | | Caledonian |
| | Silurian | 445 | | Taconian |
| | Ordovician | 515 | | |
| | Cambrian | 590 | | |
| Precambrian | | | | Grenville |
| | | | | Hudsonian |
| | | | | Kenoran |

of dramatic change; there is strong evidence that this was caused by an asteroid striking the earth and throwing up sufficient dust to cause catastrophic cooling. There is agreement that something happened to bring about one of the most significant and rapid changes in the faunal sequence, the demise of the dinosaurs. The extinction of species since the **Pleistocene** epoch is less drastic. Unlike the Cretaceous transition from the Mesozoic to the Cenozoic, the Pleistocene saw widespread changes in the surface, with vast ice sheets covering large areas of the higher latitudes in the northern hemisphere. The modern desert regions were humid and covered with vegetation, as illustrated in the rock paintings of hunters in the Sahara region of about 20 000 years ago.

During the subsequent **Holocene** epoch, our own time, geomorphic processes have perhaps been less active—the climate is temporarily more benign—but even so old landforms have been further modified and new ones created.

The lesson is clear. Looking at the present landscape, we see a system that is dynamic, changing, but which is also in large part ancient and inherited. To understand it, we need to comprehend the events and processes of the past. We need to appreciate the intimate relationship between process and form in landscape: the way processes determine form and, in turn, the way landform influences process. We will examine this relationship in subsequent chapters. Initially (Chapters 19–21), we will consider the main stages in the cycle of rock materials. Then we will look specifically at the landscape as a process-response system, and discuss how the processes of the debris cascade (weathering, erosion, and deposition) affect different morphological systems (hillslopes, drainage basins and streams, glaciers, aeolian systems, lakes, and seas).

# The formation of rocks

## The role of rock formation

The wide variety of landforms and landscapes we see at the earth's surface are due to many different factors. One of the most important is the variation in rock type that occurs, often over quite short distances. This affects the resistance of the surface to weathering and erosion, and thus influences the rates at which these processes operate. In a general fashion, therefore, the landscape reflects the character of the bedrock in which it is carved. Rocks that are resistant to weathering and erosion, for example, tend to survive as upstanding areas of countryside, while less resistant materials are eroded to form vales or low-lying plains. The residual materials produced by weathering and erosion—the debris that is carried by wind, water, and ice across the earth's surface and that is ultimately deposited and converted to rock—owe their characteristics to the rocks from which they were derived. The nature of these materials, in turn, affects the changes that take place during transport, and thus the character of the depositional landforms. Hence, rock type exerts a fundamental control over both the geomorphic processes operating at the earth's surface, and the landscapes that are formed.

# The properties of rocks

The term rock means many things to many people. In geomorphology it is given a relatively wide definition, including the hard rocks, such as granite, sandstone, or schists, and many relatively soft materials, such as consolidated clays and sands. A distinction is made, however, between rock and sediment.

**Sediment**, often the raw material of rocks, is the loose, uncompacted material laid down by agents of deposition such as water, wind, or ice. Rock is different in that its materials have undergone alteration by pressure, heat, or chemical processes, and it has therefore adopted a different structure.

Three main types of rock are commonly recognized:

1. Igneous: formed by solidification of molten magma.
2. Sedimentary: formed by consolidation of old rock debris or sediments at the earth's surface.
3. Metamorphic: formed by alteration of existing rocks by intense heat or pressure.

We will consider the processes by which these are formed later. For the study of rocks in the landscape, it is useful to examine three main properties: composition, texture, and structure.

## COMPOSITION OF ROCKS

Most rocks are composed of **minerals**. These are amalgamations of inorganic compounds arranged according to a fairly specific and definite pattern. Over 2000 minerals have been recognized in the earth's rocks, but almost all the commonly occurring ones are related to ten major mineral groups (Table 16.1).

The minerals that make up the rocks are composed of **elements**. A large number of elements occur naturally in the earth, but only eight are abundant. These eight—oxygen, silicon, aluminum, iron, calcium, sodium, potassium, and magnesium—account for about 99 percent of the total weight of the earth's rocks.

Thus, eight major elements, arranged to give ten main mineral groups, constitute the majority of the earth's crust. Nevertheless, there are a large number of rocks, for it is clear that the relative proportions of these and the less abundant minerals may vary considerably, and it is these variations that greatly influence the geomorphological significance of the

### Table 16.1

*Major rock-forming minerals*

| *Minerals* | *Average abundance in selected rocks (percent)* | | | |
|---|---|---|---|---|
| | *Granite* | *Basalt* | *Sandstone* | *Limestone* |
| Quartz | 31.3 | 0.0 | 69.8 | 3.7 |
| Feldspars | 52.3 | 46.2 | 8.4 | 2.2 |
| Micas | 11.5 | 0.0 | 1.2 | 0.0 |
| Pyroxenes | trace | 36.9 | 0.0 | 0.0 |
| Chlorites | 0.0 | 0.0 | 1.1 | 0.0 |
| Amphiboles | 2.4 | 0.0 | 0.0 | 0.0 |
| Olivine | 0.0 | 7.6 | 0.0 | 0.0 |
| Carbonates | 0.0 | 0.0 | 10.6 | 92.8 |
| Clay minerals | 0.0 | 0.0 | 6.9 | 1.0 |
| Iron ores | 0.5 | 2.8 | 0.3 | 0.3 |

*Data* A. Holmes, *Principles of Physical Geography* (1965), Nelson.

rocks. Minerals such as quartz are highly resistant to weathering, and tend to abound in the residual materials of weathered and eroded rocks. Others, such as the olivines and pyroxenes, are far less resistant to attack and are more readily destroyed during weathering (see Table 18.1).

### TEXTURE

Not only does the composition of the rock-forming minerals vary; so does the size of the crystals that they form. Rocks may be composed of crystals, or particles composed of more than one crystal, which vary in size from microscopic, less than 0.0001 mm in diameter, to many centimetres in diameter. Sedimentary particles are classified according to their size. Grain size classifications and descriptions do have some overlapping groups, but the most commonly used classifications are given in Table 16.2. Grain size, or **texture**, is therefore an important feature in classifying rocks. It is important because it affects characteristics such as the porosity, susceptibility to weathering, and engineering properties of rocks.

### ROCK STRUCTURE

The physical arrangement of the grains is also important. In many cases rocks are not composed of a random agglomeration of grains, but are structured; the grains are arranged in layers or groups, between which occur lines of weakness. These lines of weakness may be **bedding planes**, formed during deposition of the sedimentary materials of which some rock is composed, or **joints** and **cleavage** lines formed during the drying, cooling, or compaction of the rock. Along these planes of weakness, rocks split or weather more readily.

## Formation of igneous rocks

We have already seen that rocks can be divided into three main groups, and certainly the most abundant within the earth's crust is the igneous group. **Igneous rocks** are produced by cooling of magma, the molten rock material that originates deep in the earth and rises upward into the crust. The nature of igneous rocks varies, however, depending upon both the initial composition of the magma and the conditions during cooling. In this latter context, an important distinction can be made between **extrusive igneous rocks**, which are formed from magma escaping onto the earth's surface, and **intrusive igneous rocks**, which are formed by magma cooling in the crust (Figure 16.1). As we will see, these vary considerably in their character and composition.

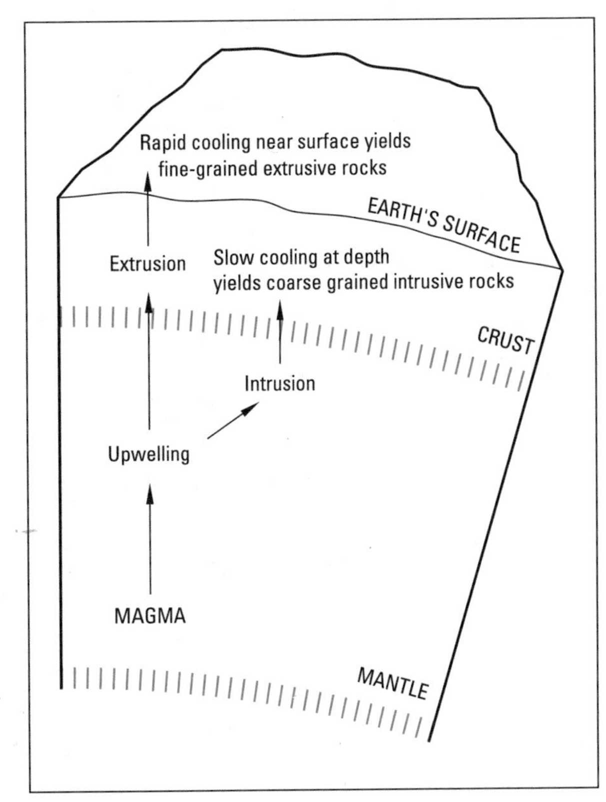


FIGURE 16.1 ▼ The formation of igneous rocks

### THE ORIGINS OF MAGMA

Most of the igneous rocks are derived from **silicate magma**. This forms at depths of about 15 to 25 km in the earth, where temperatures are in the range of 500°C to 1200°C, while pressures may be more than 10 000 times those experienced at the earth's surface. Under these conditions, the peridotite that comprises the mantle undergoes partial melting. In the process, a magma is formed of rather different composition from the initial materials: it is richer in silica but depleted in iron and magnesium. As this magma rises along lines of weakness in the crust, it melts and dissolves some of the rocks it encounters, so that its composition changes further.

# Table 16.2

*Grain size scales for sediments**

| *U.S. standard sieve mesh number* | | *mm* | *Microns* | *Phi (ϕ)* | *Wentworth size class* | |
|---|---|---|---|---|---|---|
| | | 4096 | | –12 | | |
| | | 1024 | | –10 | Boulder (–8 to –12ϕ) | |
| Use | | 256 | | –8 | | |
| wire | | 64 | | –6 | Cobble (–6 to –8ϕ) | |
| squares | | 16 | | –4 | Pebble (–2 to –6ϕ) | GRAVEL |
| 5 | | 4 | | –2 | | |
| | | 3.36 | | –1.75 | | |
| | | 2.83 | | –1.5 | Granule | |
| | | 2.38 | | –1.25 | | |
| 10 | | 2.00 | | –1.0 | | |
| | | 1.68 | | –0.75 | | |
| | | 1.41 | | –0.5 | Very coarse sand | |
| | | 1.19 | | –0.25 | | |
| 18 | | 1.00 | | 0.0 | | |
| | | 0.84 | | 0.25 | | |
| | | 0.71 | | 0.5 | Coarse sand | |
| | | 0.59 | | 0.75 | | |
| 35 | 1/2 | 0.50 | 500 | 1.0 | | |
| | | 0.42 | 420 | 1.25 | | |
| | | 0.35 | 350 | 1.5 | Medium sand | SAND |
| | | 0.30 | 300 | 1.75 | | |
| 60 | 1/4 | 0.25 | 250 | 2.0 | | |
| | | 0.210 | 210 | 2.25 | | |
| | | 0.177 | 177 | 2.5 | Fine sand | |
| | | 0.149 | 149 | 2.75 | | |
| 120 | 1/8 | 0.125 | 125 | 3.0 | | |
| | | 0.105 | 105 | 3.25 | | |
| | | 0.088 | 88 | 3.5 | Very fine sand | |
| | | 0.074 | 74 | 3.75 | | |
| 230 | 1/16 | 0.0625 | 62.5 | 4.0 | | |
| | | 0.053 | 53 | 4.25 | | |
| | | 0.044 | 44 | 4.5 | Coarse silt | |
| | | 0.037 | 37 | 4.75 | | |
| | 1/32 | 0.031 | 31 | 5.0 | | |
| | 1/64 | 0.0156 | 15.6 | 6.0 | Medium silt | |
| Analyzed | 1/128 | 0.0078 | 7.8 | 7.0 | Fine silt | |
| by | 1/256 | 0.0039 | 3.9 | 8.0 | Very fine silt | |
| Pipette | | 0.0020 | 2.0 | 9.0 | | MUD |
| or | | 0.00098 | 0.98 | 10.0 | Clay | |
| Hydrometer | | 0.00049 | 0.49 | 11.0 | | |
| | | 0.00024 | 0.24 | 12.0 | | |
| | | 0.00012 | 0.12 | 13.0 | | |
| | | 0.00006 | 0.06 | 14.0 | | |

* Sediments are analyzed for grain size according to the dimensions of the sieve mesh through which they will pass; fine sediments are classified according to their settling time from suspension. The Wentworth size classes, based on millimetres of mesh, was most commonly used in the past. Currently the phi (ϕ) scale, a logarithmic transformation of the Wentworth scale, is frequently used.

From R.L. Folk, *Petrology of Sedimentary Rocks* (1968), Hemphills Publishing.

## COOLING PROCESSES

The magma that enters the crust, or escapes onto the surface, therefore varies in composition depending upon the conditions it experienced as it rose from the mantle. Further changes also take place during subsequent cooling. Some of the constituents then escape in gaseous form, while those that are left solidify to form minerals. What happens at this stage, and the character of the rocks that develop, depends particularly upon the rate of cooling, and this in turn is dependent upon the environment in which cooling takes place. Typically, extrusive magmas cool rapidly, for the temperatures at the surface are lower, and dispersal of heat through the surrounding rocks and into the atmosphere or oceans is relatively efficient. The presence of water will considerably aid cooling, and magmas that are released into the ocean, for example, solidify very rapidly, producing highly characteristic rocks. Conversely, magmas that are trapped within the crust tend to cool more slowly; the temperatures and pressure in the surrounding rock are higher and the opportunity for dispersal of heat is much less.

## REACTION SERIES AND MINERAL FORMATION

The differences in cooling environment and rate have fundamental effects on the nature of the rocks that are formed. These effects occur because of the process of **magma fractionation**. Imagine a large mass of magma trapped beneath the surface but cooling gradually. The magma will tend to cool more rapidly at the outer edges and also in that area closer to the surface of the earth. Because the constituents of the magma tend to form crystals at different temperatures as the magma cools, a sequence of different minerals is formed. The range of minerals will reflect the pattern and rate of cooling, but will also depend upon the initial temperature of the magma. Not all these minerals are stable, and some may melt again or be altered before the rock finally solidifies.

The sequence in which minerals form during cooling of the magma was studied by N.L. Bowen in the 1920s. He found that there existed two separate sequences, which have been termed the **Bowen Reaction Series**. One sequence tends to occur in discrete steps; as the temperature alters there are abrupt changes in the nature of the minerals being formed, one mineral disappearing and another appearing. This sequence is called the **discontinuous series**. The other sequence is more transitional; a fall in temperature is accompanied by a progressive change in mineral composition, with one mineral grading into another. This is referred to as the **continuous series**.

Both the continuous and discontinuous reaction series may operate simultaneously, and they tend to converge, at relatively low temperatures, with the formation of a mineral known as orthoclase, one of the feldspar-type minerals which is rich in sodium. At lower temperatures still, muscovite (a form of mica) and quartz are produced.

At higher temperatures, the two series produce very different minerals. The discontinuous series commences with the formation of olivine, which is replaced during cooling by minerals of the pyroxene group. This in turn is replaced by amphiboles and then biotite (another form of mica). All these minerals are dark in colour and contain relatively high proportions of magnesium and iron. They are referred to as **mafic minerals**.

The continuous series, in contrast, produces light-coloured minerals dominated by feldspar. Initially, feldspars rich in calcium are formed (calcic feldspar), but these grade into sodium-rich forms (plagioclase). These minerals, together with the low-temperature products (orthoclase, muscovite, and quartz), are known as **felsic minerals**, since they are dominated by feldspar and silica.

One question is raised: how do the minerals formed early in the continuous reaction series survive during cooling? After all, we would expect from what we have just said that all the olivine, for example, would have been altered before the magma had solidified. This is a difficult question, for we do not know exactly what happens inside the magma. It appears that during cooling some of the minerals become segregated due to their different densities. Heavy minerals such as olivine and pyroxene may sink, forming a layer at the base of the magma that protects the crystals from alteration; lighter minerals such as the feldspars may float and similarly avoid alteration. In the case of magmas that cool slowly at depth, a further process may operate. Slow cooling allows the gradual growth of very large crystals, or the development of clusters of crystals, and the centres of these may again be protected from alteration. Thus, factors such as the rate of cooling and the relative significance of the two reaction series determine the character of the rocks that form.

In addition to these variations in composition, the rate of cooling affects the texture of igneous rocks. Those formed by slow cooling tend to be dominated by coarse-grained minerals, for ample time exists for growth of the crystals. Rapid cooling, on the other hand, generally results in finer-grained and glassy rocks such as obsidian.

This is a major simplification of a very complex process. As a generalization, however, it is a useful basis for understanding igneous rocks.

## THE NATURE OF IGNEOUS ROCKS

The almost infinite range of rock types produced by igneous activity makes any rigorous classification highly complex. It is possible to devise a very crude classification (Table 16.3) that identifies the main forms of igneous rocks.

At one extreme in this classification we may see the granite rocks. These are coarse-grained, acidic rocks containing relatively large quantities of quartz and orthoclase. Granite is one of the igneous rocks formed at some depth in the earth, where slow cooling allows the growth of large crystals and a predominance of low-temperature minerals.

A gradation can be seen from granite through to more basic coarse-grained igneous rocks. The proportions of quartz and orthoclase decline and the quantities of other minerals, such as plagioclase and hornblende (one of the amphibole minerals), increase as we progress along this series. Thus granite grades to granodiorite, and thence to diorite and ultimately gabbro. This, unlike granite, is a relatively dark-coloured rock, for it contains little or no quartz and orthoclase (both felsic minerals), but is dominated instead by calcic feldspar and the darker, mafic minerals. Some samples of igneous rocks are shown in Figure 16.2.

The other main difference in igneous rocks relates to grain size, and the end members of the series we have just outlined grade to fine-grained equivalents. These represent the rapid cooling at the surface of the same magma types. Granite, for example, grades to rhyolite, an extrusive igneous rock of similar mineral composition, but much finer texture. Gabbro grades to basalt, a very dark, fine-grained material produced by lava flows from volcanic activity (see Chapter 19).

## ECONOMIC SIGNIFICANCE OF IGNEOUS ROCKS

It is interesting to note that igneous activity produces many of the world's most important minerals. These are generally associated, not with the main body of the igneous rock, but with the more volatile substances related to them. These **volatiles**, moving

### Table 16.3

*Classification of igneous rocks*

| Mineral composition (Acid ↑ Quartz increasing; Ferro-magnesian minerals increasing ↓ Basic) | | *Grain size* | | |
|---|---|---|---|---|
| | | *Coarse* | *Medium* | *Fine* |
| Acid | Potassium feldspars | Granite | Micro-granite | Rhyolite |
| | Potassium feldspars | Syenite | Porphyry | Trachyte |
| | Sodic feldspars | Granodiorite | Micro-granodiorite | Dacite |
| | Sodic feldspars | Diorite | Micro-diorite | Andesite |
| | Calcic feldspars | Gabbro | Dolerite | Basalt |
| Basic | Calcic feldspars | Olivine gabbro | Olivine dolerite | Spilite |

FIGURE 16.2 ▼ Igneous rocks. (A) Basalt, an extrusive igneous rock, is formed in lava flows and volcano cinder cones; note holes formed by gas bubbles; (B) diorite, an intrusive igneous rock, consists of feldspar and pyroxene; (C) large white crystals of feldspar are embedded in porphyry, a phase of rhyolite (photos: Richard K. Herd, courtesy Geological Survey of Canada); (D) lithified volcanic ash from Bennett Lake, BC, is pyroclastic rock (photo: M.B. Lambert, courtesy Geological Survey of Canada)

through the surrounding rocks as gases or liquids, may become trapped in cracks and fissures within the rock where they solidify to give veins. Lead, copper, zinc, and many other metalliferous minerals are produced in this way. Other rocks of economic value are produced by the segregation of crystals during cooling; heavier minerals may sink to the bottom of the cooling magma to create a deposit of relatively pure material: iron- and nickel-rich ores may be formed in this way.

One of the main features of our exploitation of this planet over the years has been the attempt to locate and mine some of these products of igneous activity. It is clear that a knowledge of igneous processes plays a part in finding suitable locations.

## Formation of sedimentary rocks

Sedimentary rocks are produced in many different ways in many different circumstances. Unlike igneous rocks, they are all formed at the earth's surface, and they are all composed of old rock debris that has been eroded, transported, and redeposited. A very wide range of rocks is produced in this way.

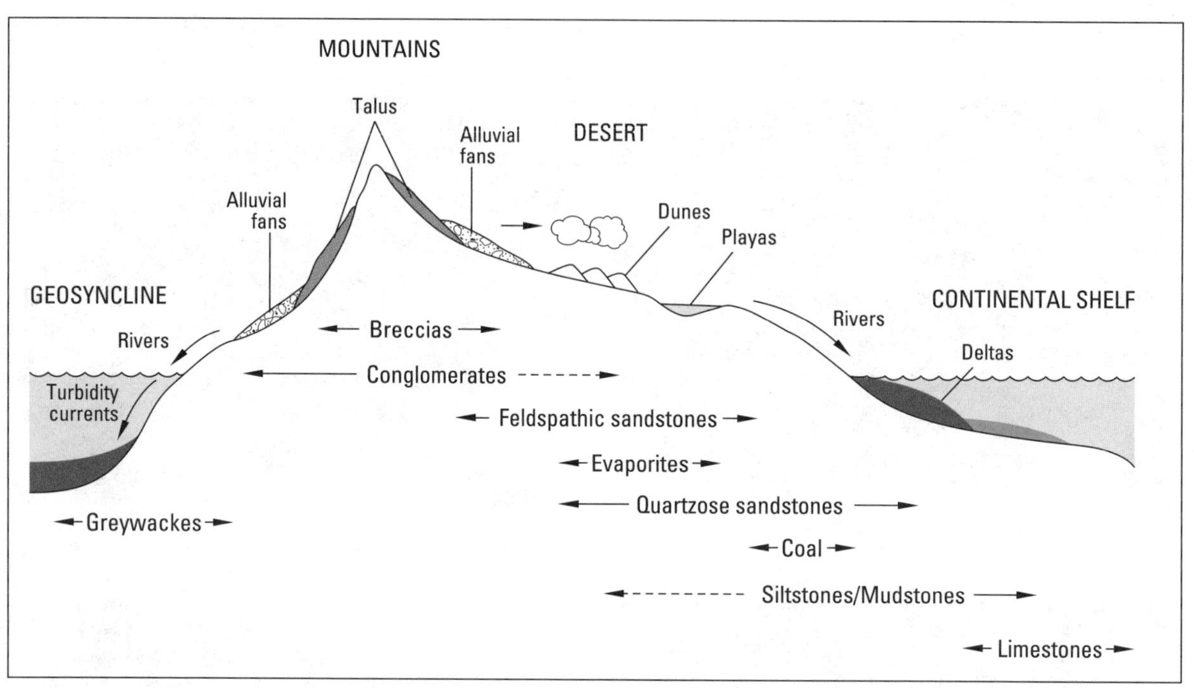


FIGURE 16.3 ▼ A general model of the environments of sedimentary rock formation

We can get some idea of the factors controlling their character from the simple model in Figure 16.3. As this indicates, the nature of sedimentary rocks is a function mainly of the duration and distance of transport, and their environment of deposition.

## COMPOSITION OF SEDIMENTARY ROCKS

A large proportion of sedimentary rocks are produced by the compaction and alteration of **detrital material**. This is the residual debris laid down at the end of the erosion process.

Quartz is one of the most resistant detrital minerals, and it is certainly the most common. It is composed of silica and oxygen ($SiO_2$), and is commonly in the form of an almost amorphous material, lacking any lines of cleavage and any major lines of weakness. The grains may survive many cycles of erosion, sometimes being converted to sedimentary rocks, then eroded and redeposited several times. In addition to quartz, mica (especially muscovite), calcite, and feldspar are relatively common detrital minerals, although none are so resistant to weathering. Small quantities of **ancillary minerals** may also be contained in the detrital material. Some of these are of high density and are referred to as **heavy minerals** (e.g., garnet, tourmaline, and rutile), and several of these are quite resistant to weathering. The occurrence of these ancillary minerals is often revealing, since they may tell us something about the provenance, or origin, of the sediments.

Almost all the constituents of sedimentary rocks are detrital. Only small quantities of minerals are formed in sedimentary environments. One of these is calcium carbonate ($CaCO_3$). This is commonly produced by living creatures, which assimilate calcium from the sea into their bodies to produce shells and skeletal material that sink and collect on the sea floor when the creatures die. Calcium carbonate and other minerals may also be produced by precipitation from water containing dissolved substances. This produces a hydrogenic sediment. Salt (sodium chloride, NaCl) is another common example, formed by evaporation of saline waters.

Two other types of sedimentary material may be mentioned. One is comprised of the decomposition products of weathering, the minerals that are formed by alteration of the less stable original minerals. Clays are an example of this process, for many **clay minerals** are derived from the breakdown of detrital minerals. Many of the silicate minerals are altered in this way during weathering to give a range of clay minerals. As we will see when we consider weathering processes in more detail, the nature of the clay depends to a great extent upon the type and intensity of weathering. The other type of material is biological remains, such as vegetation transformed into coal.

## LITHIFICATION

In their original form, detrital materials are the unconsolidated sediments that must be consolidated and altered to form rock. This process of conversion from raw sediment to rock is known as **lithification**, and it comes about in a variety of ways. One set of processes is by drying or **dessication** of the material. Another process is **compaction**, usually caused by the accumulation of overlying materials. In the latter, water is expelled, forcing the grains closer together. This produces a more permanent bonding between the particles.

Lithification is produced by chemical as well as mechanical processes. Changes in the surrounding environment may result in chemical reactions, to produce compounds that bind the particles

FIGURE 16.4 ▼ Jointing in sedimentary rocks. (A) Rectangular jointing in shales at Kettle Point on Lake Huron, Ontario. (B) Structure of a 'kettle' concretion and deformation of shales at Kettle Point. (C) Sediments with a high angle of dip exposed on the north shore of the St. Lawrence River, Quebec, showing the bedding planes. (D) Laminated silts in a section of modern ice-dammed lake deposits, Dusty Glacier, St. Elias Mountains, Yukon (photos: P.G. Johnson)

together. **Oxidation** of the material, due to loss of water from the sediment and exposure to air, may lead to the creation of iron and other oxides that coat the grains and form a bond between them. **Cementation** also occurs because of the precipitation of other substances from water percolating through the material, or by evaporation of the water. These substances may be washed in from outside, or from overlying sediments, or they may be dissolved from the particles themselves and redeposited around them. In addition to iron oxides, calcium, silica, and aluminum compounds may all act as cements; two of the most common are calcium carbonate and silica.

Lithification of sedimentary rocks tends to alter the original composition of the materials to some extent. It also imposes on the materials a new structure; for example, joints may be produced through shrinkage of the sediments during drying (Figure 16.4). Not all the structures found in sedimentary rocks are produced during lithification; many are the results of the original depositional processes. Most sedimentary rocks, however, retain evidence of their depositional history in the form of bedding planes, laminations, and various sedimentary structures (e.g., ripple marks, cross-bedding, mud cracks, and the impressions left by organisms that once lived in or on the accumulating sediment).

## THE NATURE OF SEDIMENTARY ROCKS

To get an idea of the main types of sedimentary rocks we find in the earth, let us return to Figure 16.3. As this shows, we can start by assuming a relatively simple situation in which a large mountain range is being attacked by mechanical and chemical weathering. The debris produced by these processes is transported by the action of streams, wind, and ice, and ultimately much of it is carried into the sea. As we will see in Chapter 19, during transport the debris is worn and sorted according to size, so that the material becomes finer the further it is transported. Deposition of the debris en route, and its eventual compression and compaction, produces a range of sedimentary rocks.

Close to the mountains, on the valley sides and floors, the coarsest debris accumulates. It consists of angular rock fragments, transported only short distances and hardly worn at all. We can see similar material today accumulating as talus on steep valley slopes (see Chapter 20). When lithified, angular rock fragments form a rock known as breccia. Material that is transported slightly farther and worn more intensively accumulates in the form of rounded boulders and pebbles. As finer material filters into the spaces between these large particles, and as compounds such as iron oxides, silica, or calcium carbonate are precipitated around the particles, the debris is cemented into a rock known as conglomerate (Figure 16.5).

FIGURE 16.5 ▼ Sedimentary rock: a meta-conglomerate (photo: Richard K. Herd, courtesy Geological Survey of Canada)

Farther from the mountains, the material is more worn and finer-grained, for it has been transported greater distances. Nevertheless, many of the relatively fragile minerals may have survived, so in addition to resistant grains of quartz there may remain considerable quantities of minerals such as feldspar. Sand size material (0.6–2.0 mm in diameter) that is deposited close to the source areas produces a special form of sandstone known as feldspathic sandstone (or arkose). Rocks of this sort are often found fringing the old mountain ranges (Figure 16.3). Beyond them, or in areas where the sediments have been more intensively worn before or during transport, occur sandstones composed almost wholly of more resistant materials, particularly quartz. As Figure 16.3 indicates, the sandstones are often found in the old coastal areas, for it is here, in the estuaries and in the shallow sea beyond, that the sands carried by the rivers are deposited and then moved around by the tides. But in continental areas, vast areas of sand may accumulate in deserts, and these too may ultimately be lithified into rock.

The finest material—the silt (0.002–0.06 mm diameter) and the clay (<0.002 mm diameter) produced by weathering or by prolonged attrition during transport—is generally carried the greatest distance. Much of it reaches the sea and is washed into the deeper waters where it very slowly accumulates on the ocean floor. The ocean floor deposition zones are called **geosynclines**. The silt forms siltstone when compressed and dried, the clay is changed first to shale (when it is partially compacted) and then to mudstone.

Where deep ocean basins occur close to mountain areas, a relatively wide range of materials may be transported down the slope and accumulate in the basin (see Chapter 24). The poorly sorted sandstones that are formed are known as greywacke (or, more generally, **turbidites**) and are common sedimentary rocks from Lower Paleozoic strata (e.g., in western Newfoundland, Wales, and the Lake District of England).

All these rocks are known as **clastic rocks** (from the Greek word meaning broken), for they are produced from fragments of old rocks. However, as we have mentioned, some of the material eroded from existing rocks may be dissolved and carried away as solutes. This material is ultimately precipitated by chemical or biological processes to produce **non-clastic rocks**.

One of the most common examples of chemical precipitation is the formation of **evaporites**. Salts dissolved in lake waters are precipitated as the water is evaporated, and they accumulate on the lake bed (Figure 16.6). They may be dominated by a variety of minerals such as sodium (in which case halite is formed), or calcium (which produces gypsum or anhydrite). In addition, similar (though more complex) processes may result in the creation of surface crusts on land. The silcretes (formed of silica), calcretes (calcium carbonate) and ferricretes (iron compounds) that occur in many subtropical areas are examples of these.

Biological precipitation of calcium carbonate is also a widespread process, and this gives rise to limestone rocks. These vary considerably, according to their exact nature of formation. **Chalk**, like that which comprises the famous White Cliffs of Dover in England, is produced by the accumulation of microscopic skeletal remains of marine organisms, particularly foraminifera. **Coral** or **reef limestones**, such as the Carboniferous limestone of the Pennines, is built by the accumulation of carbonates secreted by corals and algae growing in clear, shallow, littoral waters (see Chapter 24). Other limestones, however, may be formed by the deposition of materials eroded from older rocks; such detrital limestones are really clastic rocks.

FIGURE 16.6 ▼ Evaporite deposits on the underside of a rock from below the thin soil and rock glacier, Maxwell Creek, St. Elias Mountains, Yukon; the rock in the photo is about 20 cm across (photo: P.G. Johnson)

Within limestones, it is common to find bands or nodules of chert and flint. These are composed of silica, and are produced by precipitation both biologically and chemically. Water percolating through the chalk derives silicon from fossil organisms. This is precipitated in highly concentrated form as **flint** nodules. However, not all flint is formed after deposition of the chalk; some, it is argued, may form during deposition by the precipitation of silica around the remains of marine organisms. **Chert**, too, is produced in this way; deposits of siliceous radiolarian remains (minute simple organisms rich in silica) are cemented by precipitated silica to produce a compact, splintery rock.

Biological accumulation also produces rocks such as coal. Most coal deposits originated as deltaic swamps in which the vegetation died and accumulated. Because of the anaerobic (oxygen deficient) conditions within the swamp, the organic material does not break down very rapidly, but builds up to form peat. A rise in the sea level (or subsidence of the land) ultimately floods the area, killing the vegetation and allowing muds and silts to be deposited above the peat. Continued burial results in compaction and desiccation of the peat, which is gradually changed into lignite. This is brownish in colour and woody in appearance and contains 30–40 percent water by volume. Further compaction and heating, however, convert the lignite to bituminous coal, which in turn is altered to anthracite, or hard coal.

### ECONOMIC SIGNIFICANCE OF SEDIMENTARY ROCKS

Apart from the obvious use of many unconsolidated sediments and sedimentary rocks for building (clays and shales for bricks, sandstones and limestones as stone), many of the rarer products of sedimentary processes have economic value. Flint was used by prehistoric peoples to make tools (e.g., arrowheads and axes). Many mineral deposits (e.g., phosphates, gypsum, potassium) are used as fertilizers in agriculture; rock salt (halite) is used both for human consumption and in industry; sulphur and sodium are employed extensively in the chemical industries. In addition, several of the sedimentary products are used as a source of energy. Coal is the most widespread example, but oil and gas, both associated with the accumulation of organic materials in deep-sea areas, are clearly vital to modern societies. Oil and gas, produced from biological material in the sediments, accumulate in reservoirs within the sedimentary rock (Figure 16.7).

## Formation of metamorphic rocks

Igneous activity and lithification create new rocks. Metamorphism, on the other hand, is a process of alteration of existing rocks within the crust rather than a creative process. It is convenient to discuss it here since it shares many characteristics of these processes. As before, the character of the rock that develops depends to a great extent upon the nature of the original materials (the pre-existing rock) and the conditions in which metamorphism occurs.

### PROCESSES OF METAMORPHISM

Metamorphic alteration of rocks is brought about by excessive heat and pressure, and by chemical changes resulting from the action of hot gases or liquids passing through the rock. As such, metamorphism tends to be associated with igneous activity, since the intrusion of magma into the crust clearly results in considerable changes in the surrounding environment. The magma exerts a pressure upon the adjacent rock, it heats it, and volatile substances escaping from the magma permeate the surrounding material.

It is apparent that in many instances all three processes may operate together, and they may affect rocks over a large area. Such **regional metamorphism** tends to show a distinct pattern, however, for the effects are most intense close to the source of the heat and pressure, and die out farther away. Close to the source, in fact, the effect of the hot gases and liquids is at its strongest, and local, **contact metamorphism** occurs. This may produce what is known as an **aureole** around the intrusive magma. With increasing distance, this effect becomes less important, and heat and pressure cause the major changes. Heating of the original **country rock** may lead to localized melting and recrystallization of the minerals, while pressure alters the texture and structure of the rocks. Even if temperatures are not sufficient to cause melting, changes in the composition of the rocks may occur due to chemical reactions brought about by the heat.

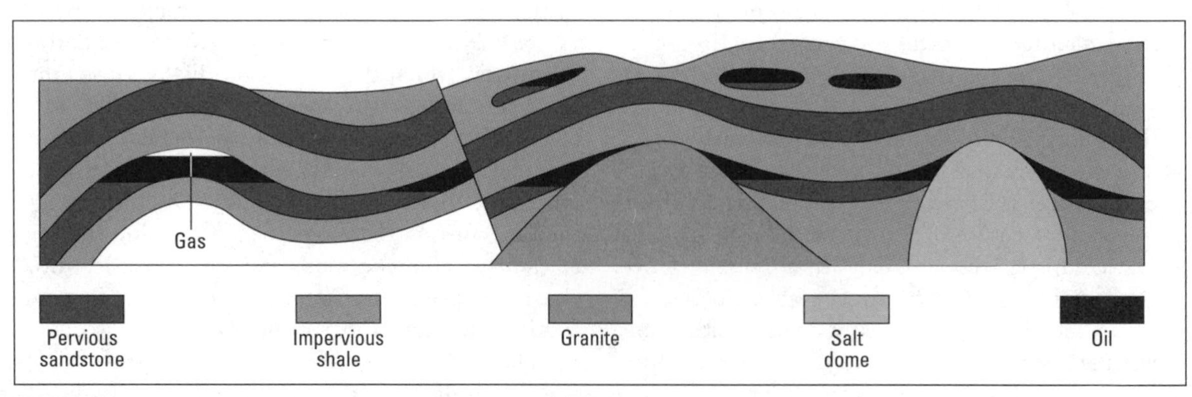


FIGURE 16.7 ▼ Geological situations in which oil and gas are formed

In some cases, of course, the three processes may operate independently. Localized heating of the rocks may result only in **thermal metamorphism**. Pressures, exerted perhaps by intense mountain-building processes (see Chapter 17), may lead to structural change with little or no alteration in composition. This is known as **dynamic metamorphism**. Alternatively, active gases and liquids permeating through the rocks may result in chemical changes such as the replacement of elements within the minerals through the process of **metasomatic metamorphism**.

## THE NATURE OF METAMORPHIC ROCKS

These processes of metamorphism can be considered as a sequence, grading from relatively weak processes which leave intact many of the original features of the country rock, to intense processes which obliterate almost all the initial characteristics. It is here that the tripartite classification of rock formation into igneous, sedimentary, and metamorphic runs into difficulties, for it is clear that metamorphism grades imperceptibly into the other two processes. Very weak metamorphism is akin to the process of lithification; it may involve no more than pressure upon the surrounding sedimentary rocks. Intense metamorphism, with remelting and recrystallization of the original materials, is similar to certain stages in the formation of igneous rocks.

In the case of more extreme metamorphic change, the nature of the original rock is relatively unimportant. Impregnation of these rocks by igneous materials and associated metasomatism may result in rocks totally unlike the original forms. Granitization of country rocks produces a rock similar to granite (which is an igneous rock) from a wide range of original materials. Clays, sands, and even basalts may all be altered to a similar end product.

More commonly, metamorphism is less intense and some characteristics of the original rocks are

A

B

C

FIGURE 16.8 ▼ Metamorphic rocks. (A) Marble, produced by the recrystallization of limestone or dolomite under pressure; the dark veins here are jade-green. (B) Gneiss, a metamorphic product of igneous rocks. (C) Crystalline limestone, a non-clastic rock (photos: Richard K. Herd, courtesy Geological Survey of Canada)

preserved. Commonly, the structure is more or less altered, with the imposition of cleavage and various flow structures such as **foliation** and **banding** (Figure 16.8). Foliation refers to a layering of the rock (like the sheets of paper in a book) brought about by reorientation of the particles by pressure and shearing. Banding represents a layering of the rock due to the segregation of minerals into discrete zones. The difference is of some importance, since foliation tends to produce weaknesses in the rock that may be exploited by weathering; banding may leave a quite strong, massive rock.

In addition to these structural changes, there may, of course, be changes in composition. Less stable minerals may be altered and new minerals formed. Such alterations, therefore, do not necessarily affect the whole rock, and many of the original constituents may be preserved.

Metamorphic rocks vary in character, then, in response to differences in the original materials and in the intensity and process of metamorphism. A broad classification is possible by relating the rocks to these factors (Figure 16.9).

One of the most commonly found series of metamorphic rocks is that resulting from the alteration of clays and shales. Under conditions of relatively weak metamorphism these change to slate, a compact, fine-grained rock with a distinct cleavage. This grades into phyllite as metamorphism becomes more intense, and thence into schist. This latter is frequently a beautiful rock, with plate-like grains of mica set in thin foliations; within these may occur scattered larger crystals of minerals such as garnet.

The metamorphic sequence exhibited by sandstone produces quartzite at all but the most intense levels of alteration. Quartzite is a hard, compact rock produced by the recrystallization of silica. The Cambrian Eriboll quartzites of northwest Scotland are fine examples.

Limestone, subjected to relatively low levels of metamorphism, is altered to marble, a rock of immense variety. Many so-called marbles are, in fact, merely very fine-grained, highly polished limestones. However, when subjected to heat, the calcium carbonate in the original limestone recrystallizes to give crystals of roughly equal size. Pressure may distort these to give a beautiful range of internal patterns to the rock (Figure 16.8).

Another common metamorphic rock is gneiss. This, again, is a term that is used rather loosely. Many gneisses are intermediate between granite and other rocks, and, as this implies, it is a product of intense metamorphism, often of pre-existing igneous rocks. It is formed of alternating bands of minerals such as quartz and feldspar with micas and amphiboles. It often grades into schist.

## ECONOMIC SIGNIFICANCE OF METAMORPHIC ROCKS

As we have seen, several of the metamorphic rocks are considered to be beautiful, and thus one of their main commercial uses is for ornaments and craft work. Metamorphism also produces a variety of ore deposits such as copper, zinc, and lead. These are often developed in the zone of contact metamorphism, where alteration of the country rock by gases and liquids derived from the magma are most intense. Many precious stones are similarly produced

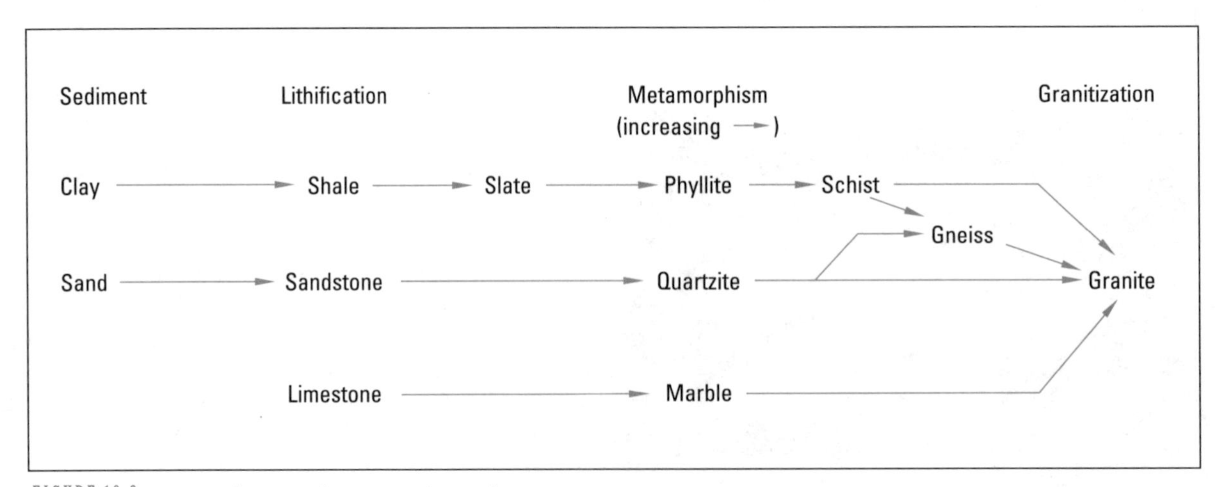


FIGURE 16.9 ▼ Classification of metamorphic rocks

by contact metamorphism; diamonds, sapphires, rubies, and garnets are all formed in this way. In addition, slate is an important building material, and the finely laminated yet strong slates of Wales have long been quarried as roofing materials.

## ROCK TYPE AND LANDSCAPE FORM

As we have noted, the nature of the rocks exerts a major influence upon the character of the landscape, and even a cursory analysis shows a general relationship between geology and topography in many areas (Figure 16.10). In some cases this relationship is so faithful that specific types of landscape are seen to be associated with specific rocks. In humid temperate regions, at least, clay lands are characteristically gentle and undulating, while areas of Carboniferous limestone, such as the Green River and Mammoth Cave regions of Kentucky or the Rocky Mountains of Alberta (Figure 16.10A), are often typified by karstic scenery. Granite areas, such as New England or the Laurentian region of Quebec, generally comprise rounded hills capped by craggy tops. In older land-

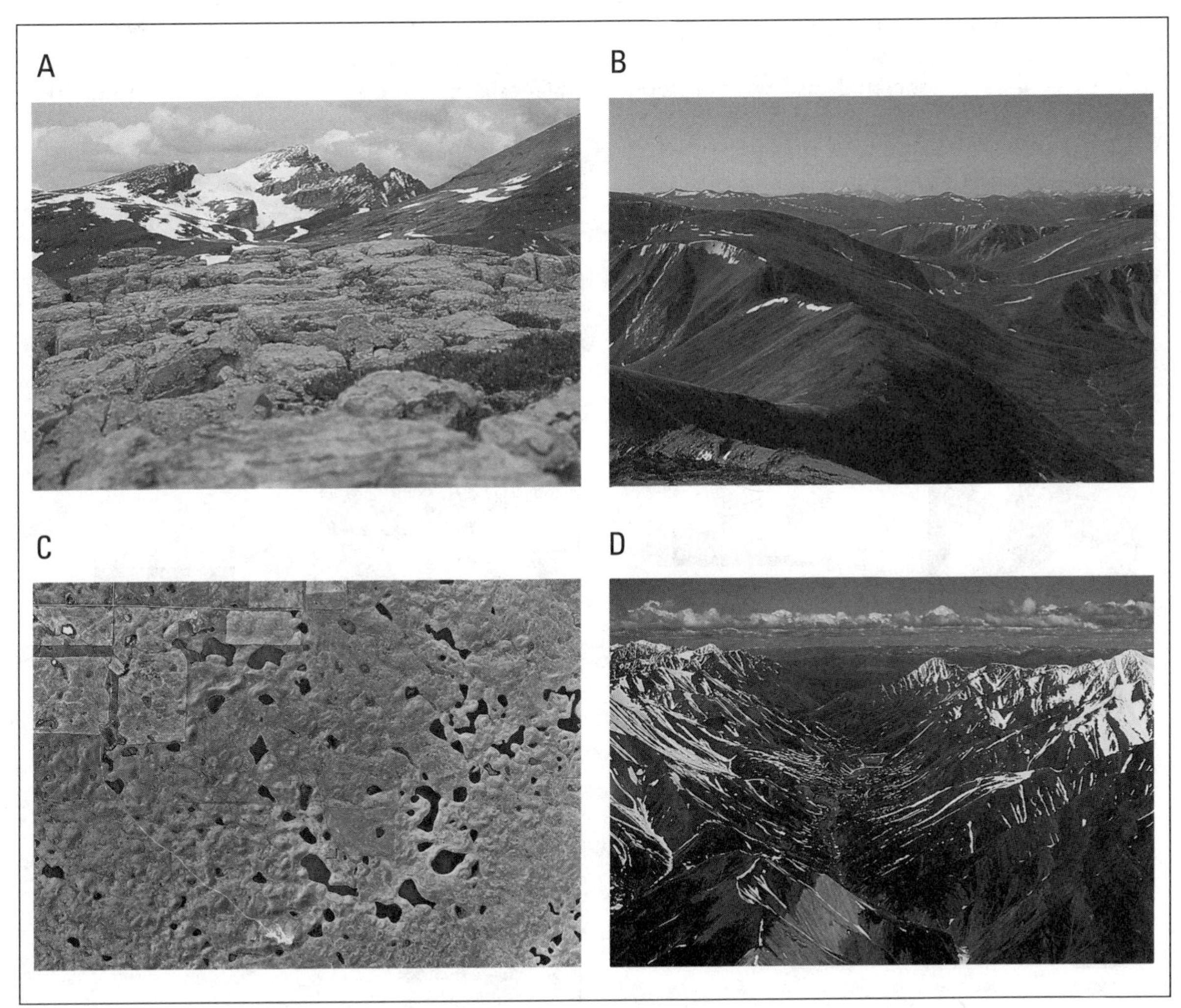


FIGURE 16.10 ▼ The relationship between geology and landscape. (A) A limestone karst landscape: Icefield Parkway, Alberta (photo: P.G. Johnson); (B) the granodiorite landscape of the Ruby Range, southwest Yukon, glaciated in the early Pleistocene, then subjected to intense subaerial erosion, forming tors (photo: P.G. Johnson); (C) near Gleichen, Alberta, these lakes formed in the irregular topography of the moraine that was left behind when the area was deglaciated about 13 000 years B.P.; scale 1:30 000; (D) glacially eroded metamorphosed sediments of the Dalton Range, St. Elias Mountains, southwest Yukon (photo: P.G. Johnson)

scapes that have not been glaciated or were glaciated early in the Pleistocene, such as the Ruby Range, southwest Yukon, these craggy caps are more pronounced and are called tors. Unconsolidated glacial deposits produce very irregular topography (Figure 16.10C). In contrast, the glacially eroded metamorphosed rocks of the St. Elias Mountains, Yukon (Figure 16.10D), produce very rugged topography.

On a smaller scale, variations in rock type control the details of the landscape. Local differences in slope form, for example, are often related to the outcrop of rocks of different lithology, while valley side slope-angles frequently vary from one rock type to another (Figure 16.11). Similarly, rates of denudation differ. Varying rates of denudation and different rock types are associated also with different processes of erosion. Limestone and chalk, known as nonclastic rocks, are susceptible to chemical weathering, and much of this involves loss of material in solution. Limestone is particularly vulnerable through the weathering process of hydrolysis, or as it is more commonly called, the process of carbonation. Clastic sedimentary rocks are those made up of particles broken from a parent-material rock. Sandstone and mudstone are examples. They tend to be attacked by mechanical weathering and the solid debris is removed by streams. Thus, rock type is one of the major exogenous controls on the processes operating within the debris cascade.

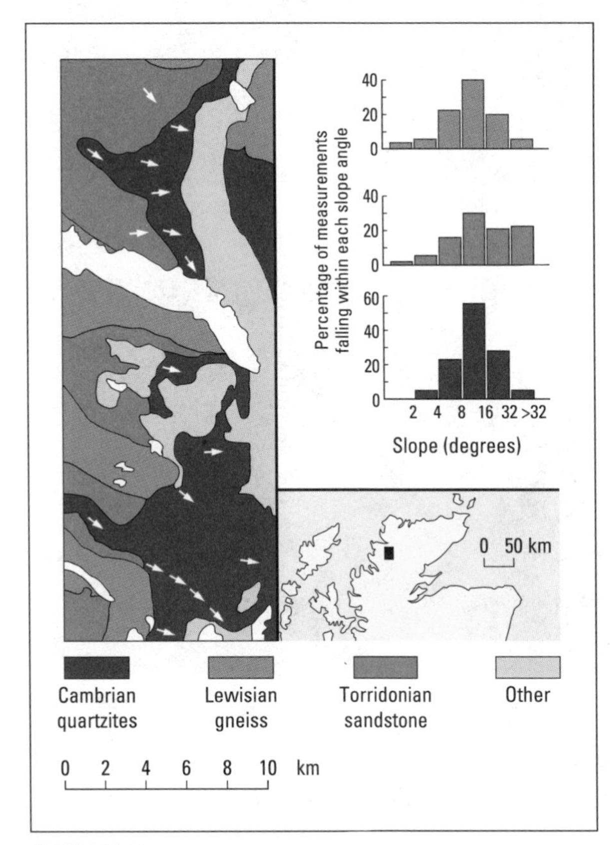


FIGURE 16.11 ▼ The relationship of geology to slope angle: the Loch Assynt area of Scotland (after Gardiner, 1983)

# Conclusion

Although the processes of rock formation are complex, and a detailed examination lies beyond the scope of this book, a general picture can be gained by considering the processes and relationships that occur within three broad zones of the earth: the interior (mainly the mantle), the crust, and the surface.

Formation of igneous rocks takes place in all three zones. Magma from the mantle moves upward into denser crustal rocks and may cool there to form intrusive igneous materials. Further upwelling may bring the magma to the surface, where rapid cooling results in the formation of extrusive igneous

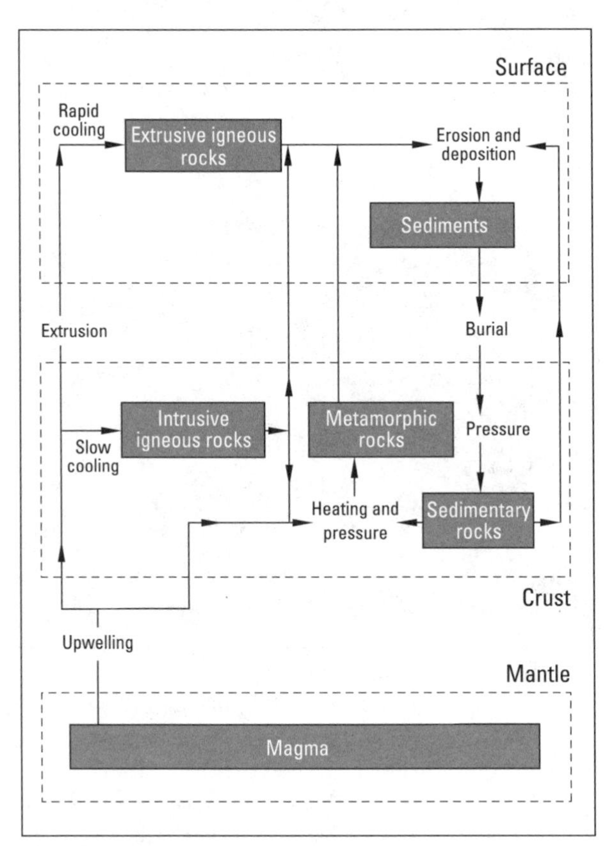


FIGURE 16.12 ▼ A general model of rock formation

rocks. Factors such as the amount of water and the depth of cooling play an important part in both cases.

Sedimentary rock formation results from a much more varied set of processes. The initial materials collect at the earth's surface, either as clastic sediments, as biogenic accumulations, or as evaporites. Burial, compaction, and desiccation of these materials result in the imposition of new structures and some mineral alteration. In general, these processes of lithification take place in the crust.

Metamorphic rocks generally result from the interaction of igneous activity with existing crustal rocks. The gases, heat, and pressure produced in the crust by upwelling magma alter the surrounding rocks and produce both chemical and structural changes.

Looking at the processes of rock formation as a whole, we can see how these various activities interact (Figure 16.12). It is also apparent that the exhumation of rocks within the crust rejuvenates the processes, breaking down the fresh rocks and supplying new materials as sediments for ultimate lithification. In all cases, the rocks that are formed influence the nature of the landscapes that develop at the surface.

# Earth-building

## The structure of the earth

Before we can discuss the question of how the processes of landscape formation occur, it is important to discuss the more fundamental issue of the structure of the earth. Many of the earth-building processes originate deep within the earth, and owe their effects to the inner character of our planet.

The earth is an oblate spheroid: that is to say, it is a sphere, slightly flattened at the poles. It is composed of a number of different layers. The outer ones have been observed either directly or by deep borings (e.g., the Moho Project), or can be deduced from the study of materials reaching the surface. The characteristics of the inner layers are deduced from seismic evidence, that is, the behaviour of shock waves, and from astronomical measurements.

A highly simplified diagram of the earth's structure is given in Figure 17.1. This shows that the innermost layer of the earth—the **inner core**—is believed to be solid, and to be separated from the liquid outer core by a transitional zone, referred to as the **F layer**. The density of the inner core is assumed to be in the order of 14 g $cm^{-3}$; the **outer core** is believed to have a density of about 10 g $cm^{-3}$. These densities imply that the core is composed of iron and small portions of nickel.

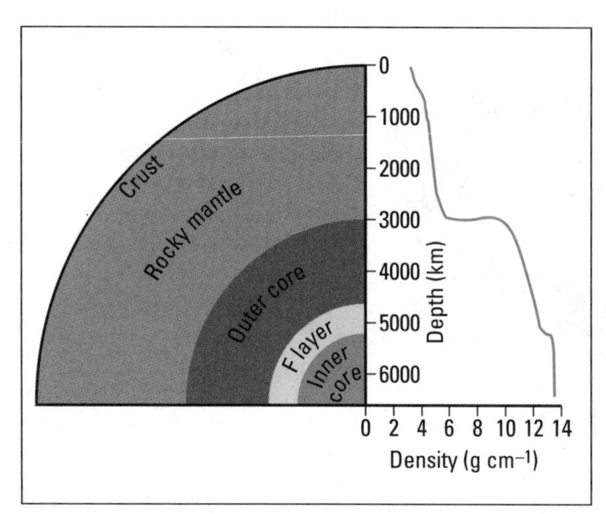


FIGURE 17.1 ▼ The internal structure of the earth, showing changes in density with depth

Around the core is the **mantle**. There appears to be an abrupt change at the boundary of these two layers, and the mantle is assumed to consist of solid rock dominated by the mineral olivine. The density of this is about 5 g $cm^{-3}$, decreasing to about 3–4 g $cm^{-3}$ in the upper mantle. This change in density seems to be related to a change in composition, and peridotite, a rock made up of the minerals olivine and pyroxene, probably dominates in the upper layer.

At the top of the mantle occurs another marked discontinuity. This is referred to as the **Mohorovicic boundary** (or Moho) after the Yugoslavian seismologist who discovered it early this century. This separates the mantle from the crust, although the zone of separation is irregular. Beneath the continents, the crust is relatively thick and the 'roots' of the land masses sink into the underlying mantle. Beneath the oceans, the crust is much thinner. Until recently it was thought to be only 5–6 km thick, but now it is believed to average about 8–9 km. The continental crust is relatively light, with a density of about 2.85 g $cm^{-3}$, while the oceanic crust has a density of 3 g $cm^{-3}$ on average. Thus the continents can be envisioned as floating upon the denser mantle, rather like ice cubes in water, with the bulk of their volume below the surface.

On the basis of recent evidence, however, it is believed that the upper layers of the earth are rather more complex than this simple model suggests. The crust and part of the mantle beneath are thought to form large, plate-like areas of rigid **lithosphere** which float upon the soft and plastic **asthenosphere** below. Thus, the crust and uppermost part of the mantle act as a single unit. As we will see, these lithospheric plates play a fundamental role in earth-building processes.

# Plate tectonics

## THE DISCOVERY OF PLATE TECTONICS

It was realized as long ago as the sixteenth century, before the world was even fully mapped, that the coastlines of the continents showed a remarkable conformity, and Francis Bacon was probably the first to indicate that they could be fitted together to form a single super-continent. It was during the twentieth century, however, that these ideas developed into a clear concept of what became known as **continental drift**. Two geologists, F. B. Taylor in the USA and Alfred Wegener in Germany, came independently to the same general idea: that the continents were not fixed, but were adrift.

It is to Wegener that the concept of continental drift is normally attributed. Like Bacon, he realized that the continents could be fitted together in the manner of a jigsaw puzzle (Figure 17.2). He conjectured that from the Carboniferous Age, some 250 million years ago, until sometime in the Pleistocene, about 1 million years ago, a single continent, which he called Pangaea, slowly broke up and drifted apart. Thus were born the continents of the world.

A wide range of evidence seemed to support this interpretation. Rare, identical fossils, such as the Cambrian coral-like organism **Archaeocyatha**, are found in rocks on different continents, now separated by thousands of kilometres of ocean. Areas affected by glaciation in the Carboniferous have broken apart and drifted thousands of kilometres. Similarly rocks, mountain ranges, and geological structures can be matched across oceans.

For a time, the concept of continental drift was widely accepted, but gradually doubts reemerged. The main problem was that no mechanism was known that could have caused Pangaea to break up and that could have moved the continents such vast distances, in different directions. Over time geologists abandoned the notion; or rather, because such is the way of scientific progress, the new generation of geologists simply refused to accept the ideas. No alternative hypothesis, however, was advanced to replace these ideas.

An answer began to emerge during the 1950s and 1960s, derived from studies of the ocean floors.

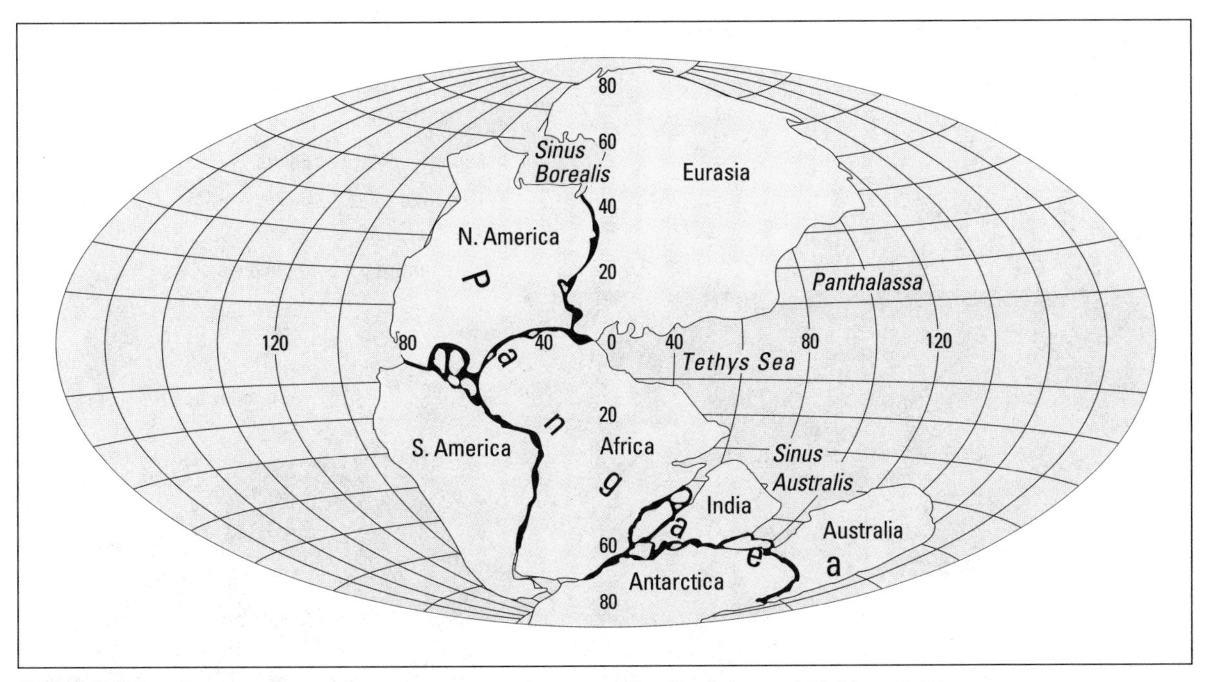


FIGURE 17.2 ▼ A reconstruction of Pangaea, 200 million years ago (after Dietz and Holden, 1970)

As volcanic rocks cool, they adopt a magnetic orientation which reflects that of the earth's magnetic field. At times, however, the magnetic field of the earth reverses (the positive and negative poles change places) so that rocks formed at different times show opposite magnetic orientations. Taking a transect across the floor of the ocean, geologists found that not only did the paleomagnetism of the rocks vary, as if the ocean floor had been formed at different times, but the patterns either side of the **mid-oceanic ridge** were almost identical (Figure 17.3). Moreover, it became clear that rocks relating to each magnetic phase could be traced linearly along the ocean floor like a band parallel to the mid-oceanic ridge. In addition, the width of each band was found to be proportional to the duration of the paleomagnetic phases established from studies of rocks of known age on land. Comparisons of the sequences on land and on the ocean floor thus enabled the rocks either side of the mid-oceanic ridges to be dated. And, startlingly, it was found that they became progressively older farther away from the mid-oceanic ridge. The implication was clear: as the crust at the mid-oceanic ridges split, and the ocean floor moved apart, new rocks were being created in the gap.

So, the idea of **sea floor spreading** developed. Further evidence supported the existence of the process. Drilling in the ocean floor, for example, showed that the sediments above the volcanic rocks increased in thickness away from the mid-oceanic ridges, indicating that longer periods of time had been available for deposition. But what was the mechanism of spreading? One suggestion was that the earth as a whole was expanding, and that, as it did so, the rigid crust fractured and was pulled apart. Other geologists argued that the size of the earth was relatively constant, and that the crust was in fact being carried across the ocean floor. In this case, somewhere, the crust must be destroyed at about the same rate at which it was being formed. One possibility was that this destruction took place at the deep marine trenches, such as those that fringe the western Pacific. But, on the other hand, the fact that the continents, many of them formed of rocks millions of years old, had clearly retained their general shape (as indicated by the way they could be fitted together) showed that the continents themselves were not being destroyed.

Tentatively, explanations began to emerge. What if the crust consists of huge, dense plates on which the lighter continents are perched? Could it be these plates that are moving across the surface, carrying the continents with them, then sinking into the mantle along the line of the ocean trenches, to leave the land masses bobbing at the surface? It

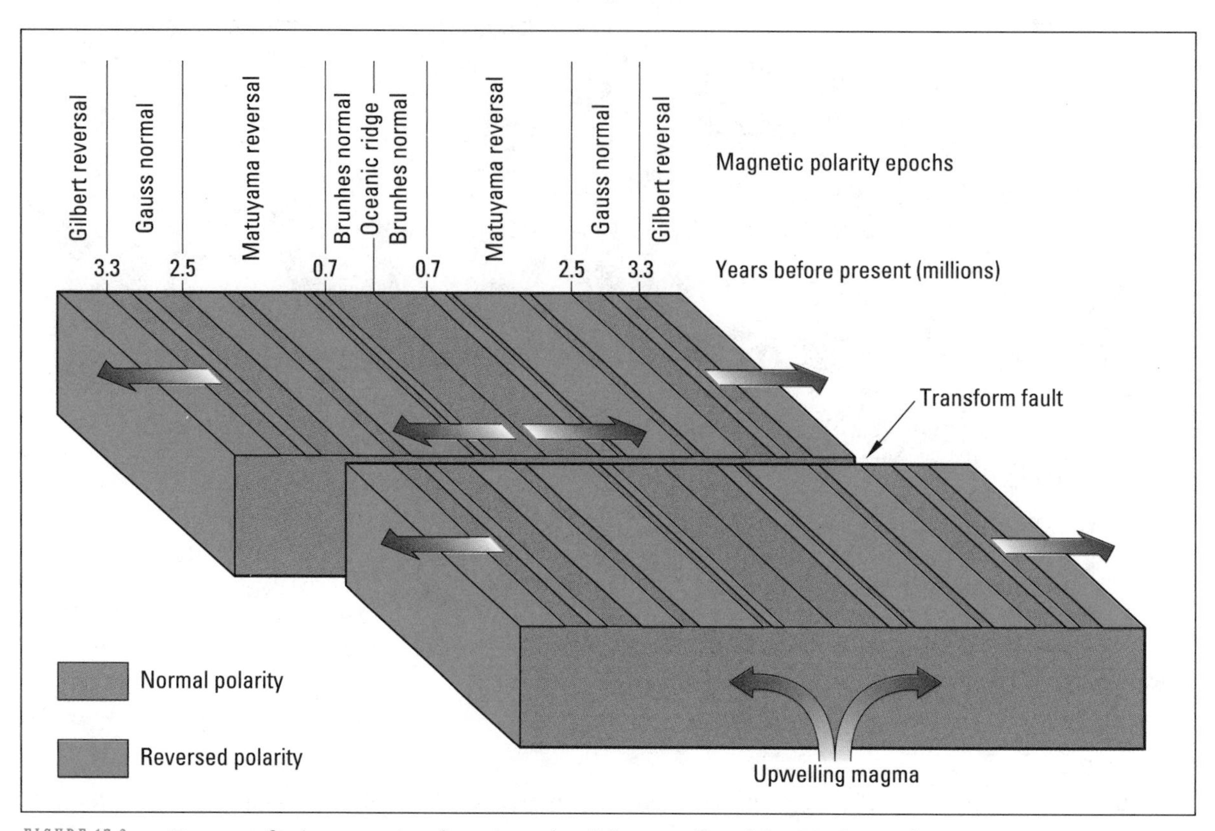


FIGURE 17.3 ▼ Patterns of paleomagnetism from the rocks of the ocean floor (after Hurley, 1968)

might seem a wild idea, but it was also an attractive one, for it seemed to offer explanations of many phenomena. It explained the regular pattern of paleomagnetic bands on the ocean floor, and the distribution of sediments. Simply, as the plates moved apart, new magma welled up along the mid-oceanic ridges and cooled, adopting the magnetic orientation of the earth at that time and creating a new band of ocean floor. On either side, the crust moved away, gathering sediment as it did so. It also seemed to explain other major features of the earth's surface structure. Where two plates converged and collided, immense compressive forces must be generated that would buckle the continental rocks and produce fold mountains. Here, too, we might expect volcanic activity. Such indeed is the case, for the marine trenches of the Pacific are associated with some of the world's most active volcanoes.

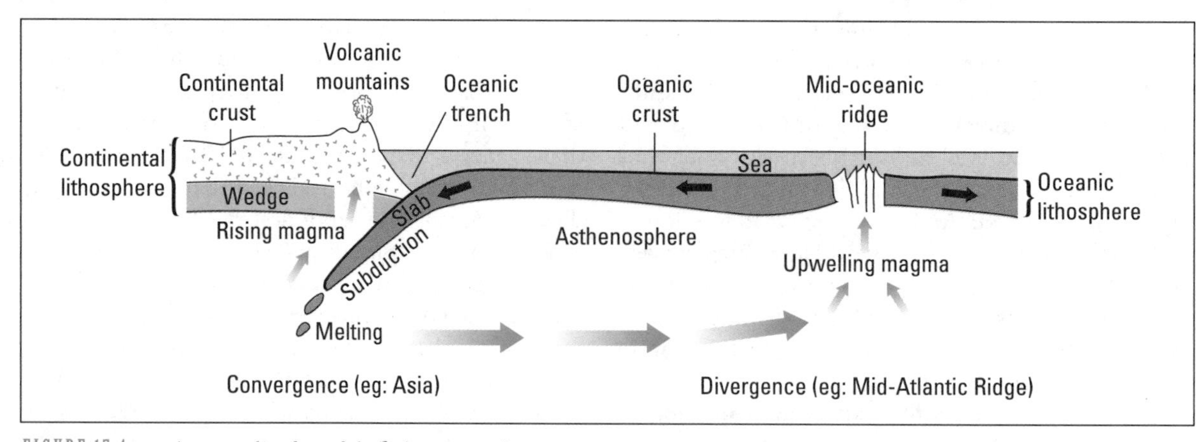


FIGURE 17.4 ▼ A generalized model of plate tectonics

*An outline of plate tectonics* Thus was born a new concept: **plate tectonics**. A highly simplified picture of this is given in Figure 17.4, showing the way the lithospheric plates ride above the semi-plastic asthenosphere. At the mid-oceanic ridges the earth is splitting, allowing magma to rise through the asthenosphere. On either side, the lithospheric plates are moving apart, across the asthenosphere. Where two plates converge, one (referred to as the slab) sinks—a process known as plate **subduction**. In the process, the slab experiences massive increases in pressure and temperature until, at a depth of 400–600 km, it melts and is absorbed in the mantle. Melting and friction release pockets of basaltic and andesitic magma, which rise through the overriding plate (the **wedge**) and create volcanoes. The stresses caused by subduction also trigger earthquakes: shallow ones where the crust is stretched as it dips beneath the surface, and deeper ones where friction and compression start to destroy the slab.

Where two plates collide in this manner, it may be expected that the denser plate will subside and the lighter, more buoyant plate will remain at the surface. Where two plates carrying only thin rocks of the ocean floor meet, therefore, either plate may be subducted. On the other hand, where an oceanic plate meets a plate carrying continental crust, the tendency will be for the oceanic plate to subduct. As we have seen, this is because the continental crust is deeper and lighter than the oceanic crust, and is thus more buoyant. It is largely for this reason that the continents have survived subduction.

In some cases, however, two continents must collide. What happens then? The answer is that neither subsides. In fact, subduction tends to cease and the continents become fixed together. If this happens, though, rates of subduction across the earth as a whole no longer match rates of formation of new crust. Consequently, adjustments in subduction rates must occur elsewhere to maintain a balance within the system.

One other situation must also be mentioned, for in some circumstances adjacent plates slide past each other. Here, interactions between the two plates are less severe, but the tearing action at the plate boundaries causes the development of transform faults, at right angles to the mid-oceanic ridges but parallel to the direction of movement (Figure 17.5).

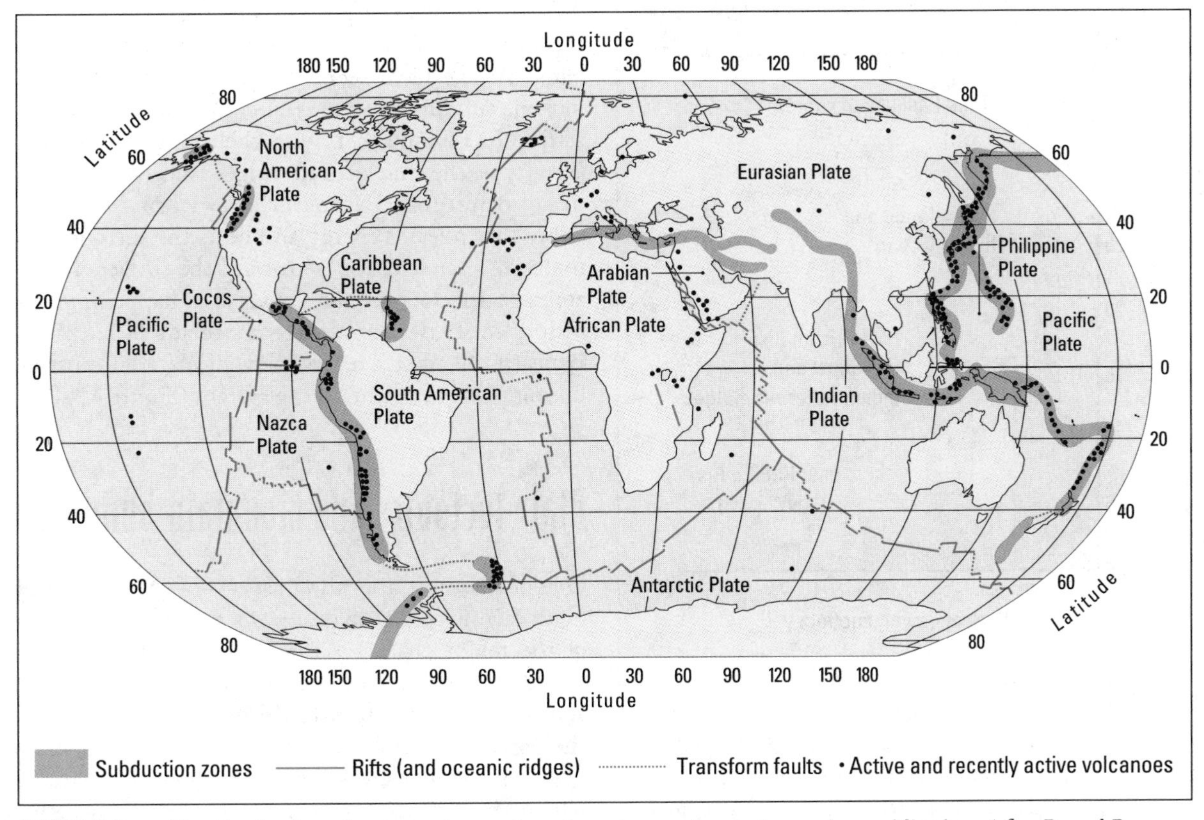


FIGURE 17.5 ▼ The distribution of active and recently active volcanoes in relation to the world's plates (after R. and B. Decker, 1982)

## PROCESSES AND MECHANISMS

If this general picture of plate tectonics is accurate, it should be possible to identify the earth's surface features and associated seismic activity. There is some disagreement about this evidence, but the broad layout indicated in Figure 17.5 is widely accepted. This, of course, is not a static picture: it represents the current state of play in plate tectonic activity. Over the millennia, the plates have been moving and, in the forseeable future, they will continue to do so. Rates of plate movement can be calculated from the age–distance relationship of the rocks either side of the mid-oceanic ridges. These indicate rates of movement up to 8–9 cm $y^{-1}$, highest in the Pacific and least in the Atlantic and Indian oceans (Figure 17.6).

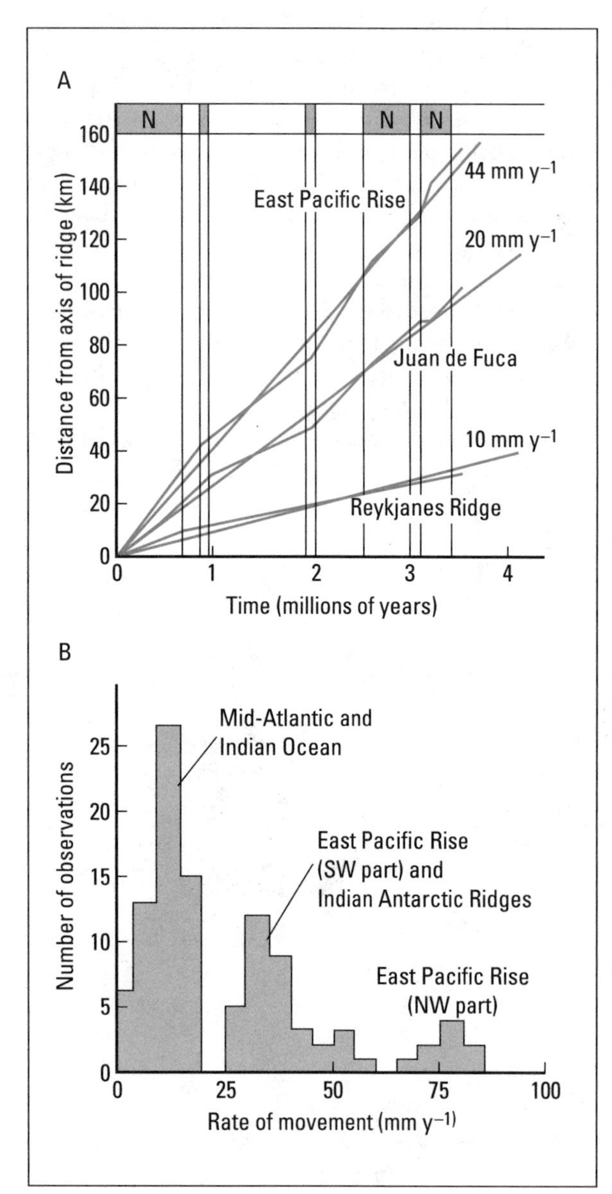


FIGURE 17.6 ▼ Rates of sea floor spreading as shown by age–distance relationships for (A) transects across three mid-oceanic ridges; (B) measurements at each 5° length of ridge from the Atlantic, Pacific, Indian, and Antarctic oceans (after Oxburgh, 1974)

This movement has certainly been going on for many millions of years, for the rocks of the ocean floor, although nowhere near as ancient as those of the continents, date from at least Jurassic times. As Wegener postulated, therefore, Pangaea probably started to break up some 200–250 million years ago; since then the continents have been transported to their present positions (Figure 17.7). Questions nevertheless remain: What is the mechanism for plate movement? What caused the breakup of Pangaea? And why did Pangaea not break up before?

The answers appear to lie in the existence of convection currents within the mantle. If they do exist, these currents are certainly not regular or permanent. They are affected, for example, by surface conditions, so that the positions of the plates modify the distribution of the currents—a process of feedback. As we have seen, when two plates become locked, subduction ceases and the convection pattern must be modified. Nonetheless, it appears that activity at the mid-oceanic ridges is associated with decay of radioactive materials within the earth, decay that releases energy and heats the surrounding materials. These then rise toward the surface. In the zones of subduction the converse is happening, for older, cool, dense surface material is sinking. Beneath the crust a reverse flow of material must be taking place to maintain the system (Figure 17.4).

# Plate tectonics and mountain-building

One of the most appealing aspects of the concept of plate tectonics is that it appears to explain so many of the major characteristics of the earth. We have already noted that the distribution of both volcanic activity and mountain-building can be related to the pattern of plate movement. More detailed studies of individual areas, however, show the ways in which these processes operate. The Himalayas, for example, appear to have been developed initially by the collision of the Australian and Indian plates

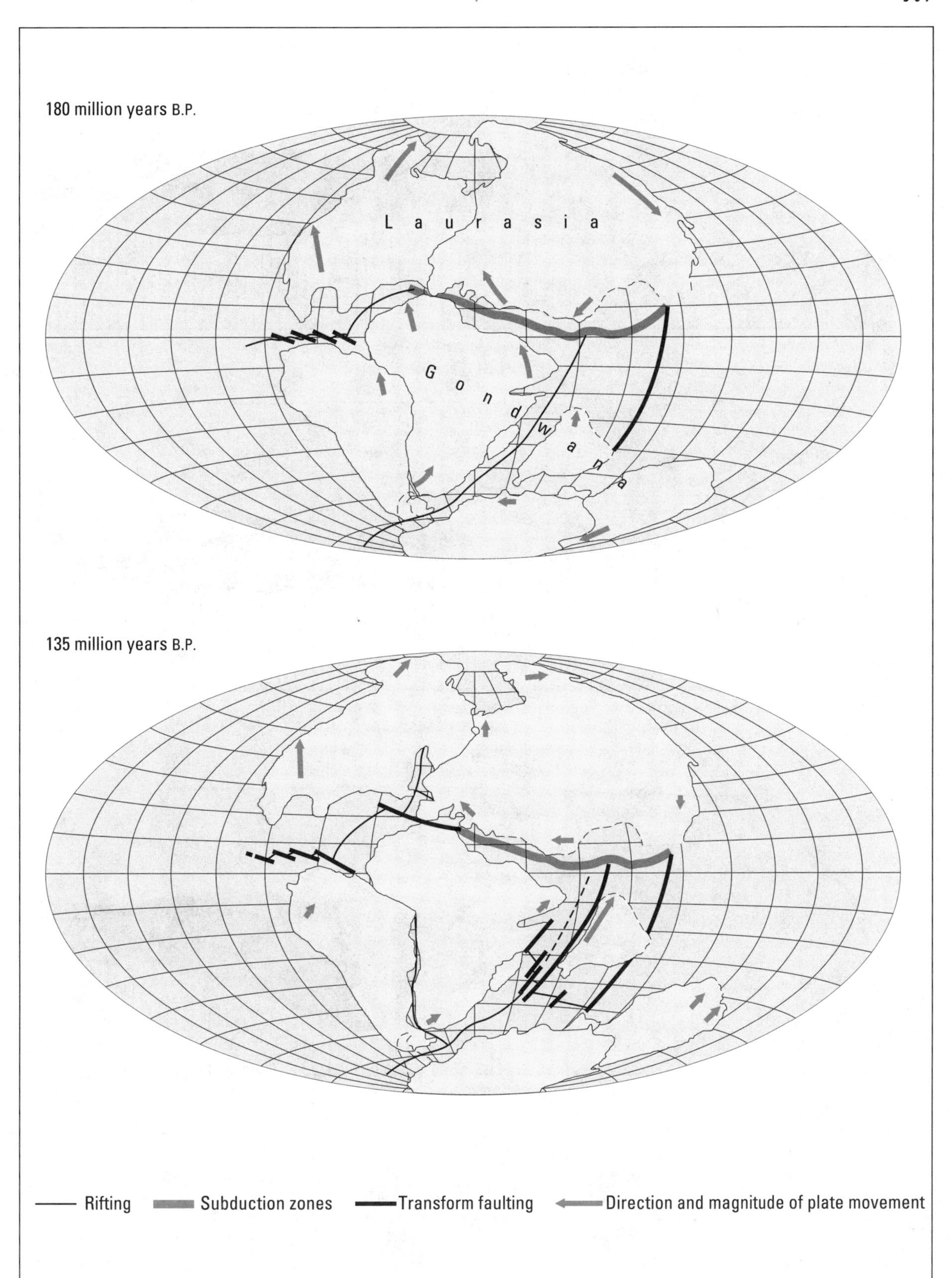


FIGURE 17.7 ▼ The breakup of Pangaea (after Dietz and Holden, 1970)

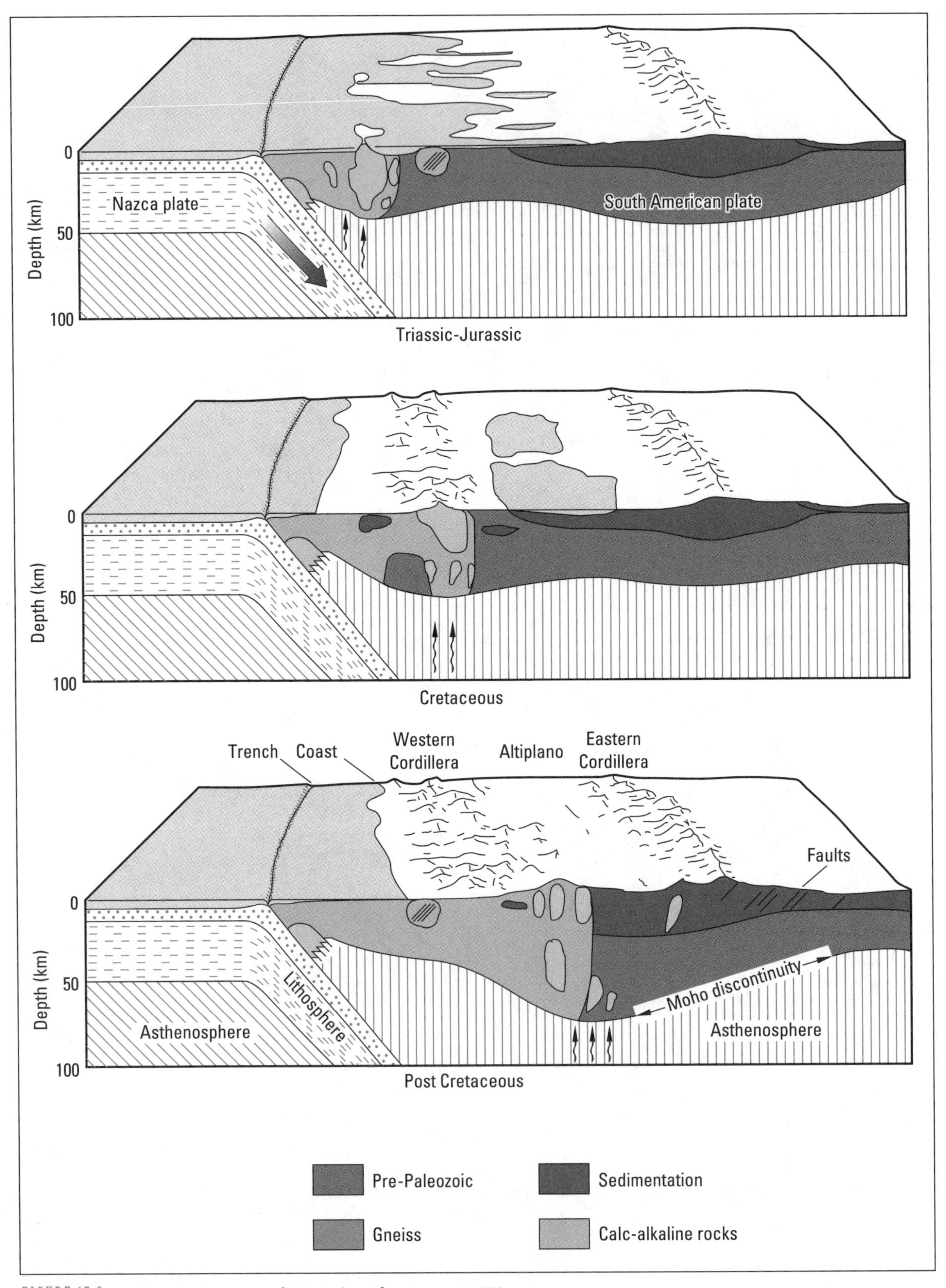


FIGURE 17.8 ▼ The development of the Andes (after James, 1973)

which thus became locked and now, as a single plate, are being forced against the Eurasian Plate to the north. Similarly, the eastern Alps resulted from the collision of the Eurasian and African continents. In both cases, the continental boundaries are not wholly conformable, so that while the land masses are in contact and undergoing folding and uplift in some places, gaps exist between them elsewhere. The Mediterranean and Black Sea are possible gaps of this type.

In the case of the Andes and Rockies, rather different processes must have been at work, for in neither case do the converging plates both carry continents. The Andes, for example, lie at the junction of the oceanic Nazca Plate and the continental edge of the South American Plate. Their history seems to have been complex. It seems that initially subduction of the Nazca Plate during Triassic and Jurassic times led to the creation of an arc of volcanoes off the South American coast. As subduction continued, during the Cretaceous period, volcanic activity became more intense and huge bodies of andesite were intruded into the sedimentary rocks, raising the surface into what is now the Western Cordillera. At the same time, the forces associated with this activity rippled eastward, folding the rocks and thrusting up the fold mountains of the Eastern Cordillera. Since then, erosion of the two ridges has led to deep infilling of the intervening area to produce the altiplano (Figure 17.8).

Plate tectonic activity can thus be related to the formation of most of the younger mountain ranges of the world—those which have formed since the breakup of Pangaea. But what of the older periods of mountain-building? Were they, perhaps, formed during earlier phases when the continents were adrift before their accretion into the Pangaean land mass? Have there been several phases of continental accretion and breakup? The processes of plate tectonics are not confined solely to the last 250 million years. What we have is possibly a process that has been operative for much of the earth's history. It has been suggested, for example, that the Appalachians were formed by subduction and continental collision during the Paleozoic era, possibly as Pangaea came together.

One thing is certain: plate movement has been taking place for a long time and, at one place or another, continental collision and mountain-building have been occurring at intervals throughout this

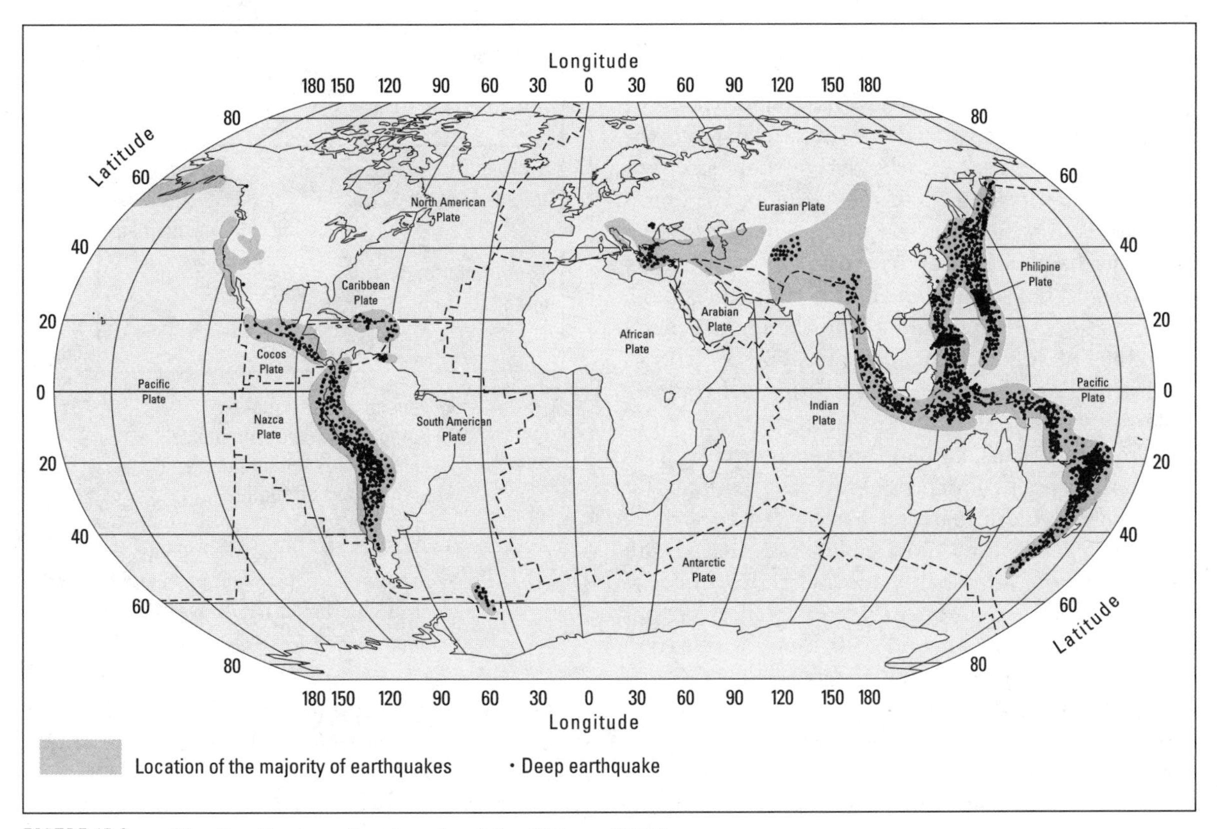


FIGURE 17.9 ▼ The distribution of earthquakes (after Toksoz, 1975)

period. Such phases of mountain-building are nevertheless often brief, a mere 50–100 million years, and they do not occur everywhere at the same time. Thus, the traditional concept of phases of mountain-building (**orogenies**) as synchronous, worldwide events is no longer tenable.

# Plate tectonics and earthquakes

As an understanding of plate tectonics has developed, it has become clear that many of the world's main earthquake zones are associated with plate boundaries. Shallow earthquakes (<70 km depth) occur along the marine trenches where plate subduction is taking place, along the mid-oceanic ridges, and in the slip-fault zones where two plates are sliding past each other. Intermediate (70–300 km) and deep (>300 km) earthquakes, on the other hand, are related almost wholly to areas of subduction (Figure 17.9). Indeed, analyses of the focal points of earthquakes in the subduction zones have enabled us to identify the actual shape of the subsiding slab of oceanic plate, and have shown the various ways that subduction occurs (Figure 17.10).

An earthquake is a sudden and massive fracturing and movement of the rock along a **failure plane**. This failure plane tends to develop where the stresses within the rock are greatest relative to the strength of the rock. Rocks derive their strength from the interlocking and cementation of the grains or crystals of which they are built. The stresses, on the other hand, are set up by tension or compression due to differential movements of the lithosphere. Where plates are descending into the mantle along the lines of subduction, for example, the stresses within the slab build up with depth until they are able to overcome the inherent strength of the rock. The rock then fails, and an earthquake occurs. This generally happens at depths of less than 700 km, for by then the plate material has heated to around 2000°C and has been more or less assimilated into the mantle: it is probably too plastic to undergo brittle failure. Consequently, earthquakes do not seem to originate beyond 700 km. Most earthquakes occur at a depth of 5–15 km.

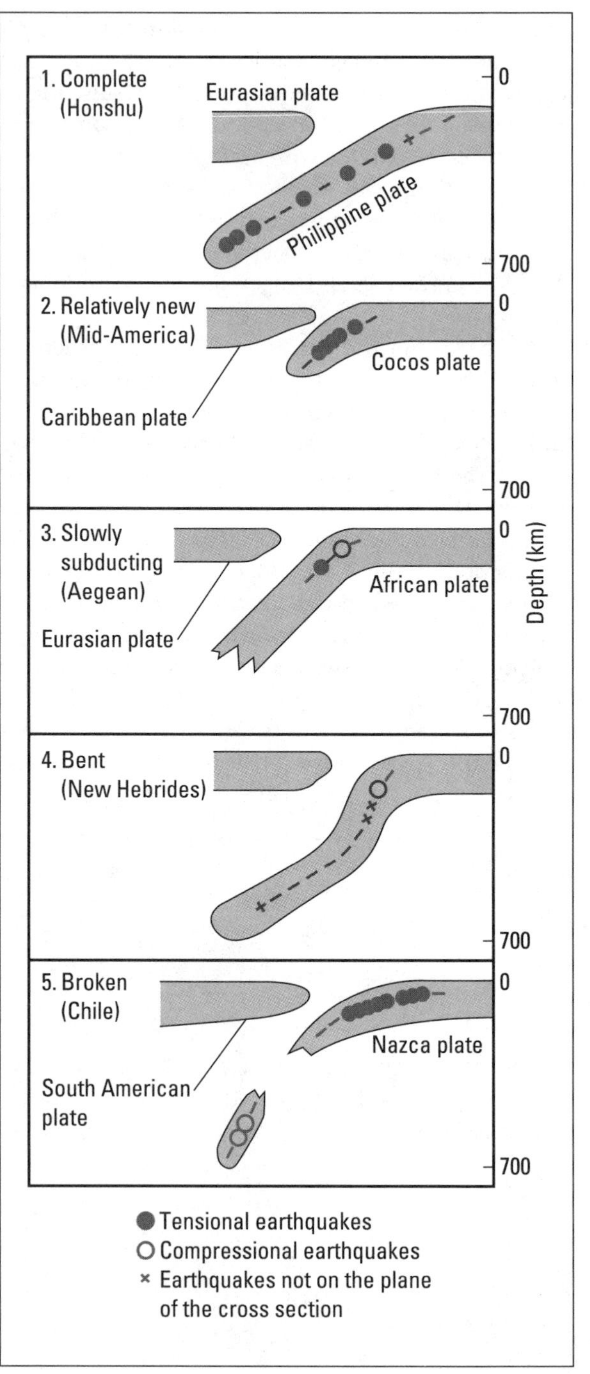


FIGURE 17.10 ▼ Earthquake depths and postulated patterns of subduction at five convergent plate boundaries (after Toksoz, 1975)

Earthquakes are surprisingly common. Fortunately, only a few make news through the destruction they cause, and most occur in remote areas where they do not affect humans. The energy released by earthquakes may be vast, however, and the largest events may produce as much as $10^{19}$ Joules. The magnitude of earthquakes is normally measured, in practice, on the Richter scale (named for the seismologist who devised it). This rates earthquakes on an open-ended scale beginning at 0 and is proportional to the logarithm of the amplitude of the seismic wave. An earthquake of magnitude 6 is, therefore, 10 times as great a disturbance as one of magnitude 5. The largest earthquake recorded, in Chile in 1960, reached a magnitude of 8.9, and three events of 8.6 have been measured since 1900. Most major earthquakes, however, have magnitudes of about 6.5; structural damage to buildings occurs at magnitude 5 or greater.

The energy released by earthquakes is expended in the form of shock waves. These travel through the rocks, from the centre of the earthquake (the **focus**), through the earth's interior and along its surface. **Foreshocks** due to the shattering of obstructions or bonds along the failure plane precede the main shock. The **principal shock** that follows is the most severe, but it may last for no more than a few seconds or, at the most, minutes. **Aftershocks**, occurring for hours or days following the primary shock, are minor movements triggered by the main displacement.

The shock waves produced by the earthquake travel through the earth in three main ways. Deep waves travel rapidly through the inner earth in the form of primary P waves and secondary S waves. Longer wavelength L waves travel at the surface of the earth. The P waves involve the vibration of particles in the direction of wave propagation. The S waves involve transverse motion of the particles. Whereas P waves can pass through solids, liquid, and air, the S waves can only pass through solids. S waves travel at about 60 percent of the speed of P waves. L waves, which have the longest wavelengths, are the slowest. Both P and S waves are reflected by the discontinuities within the earth, and as a result there develops a shadow zone that does not feel the effect of these waves following an earthquake. L waves, however, may travel right around the earth (Figure 17.11).

The focus of the earthquake can be as deep as 700 km (Figure 17.10). The point at the surface vertically above the focus is know as the **epicentre**. Often, severe earthquakes occur so deep in the earth that no surface displacement is seen. Nevertheless, as the shock waves erupt at the surface they cause immense disturbance of the land; huge ripples may

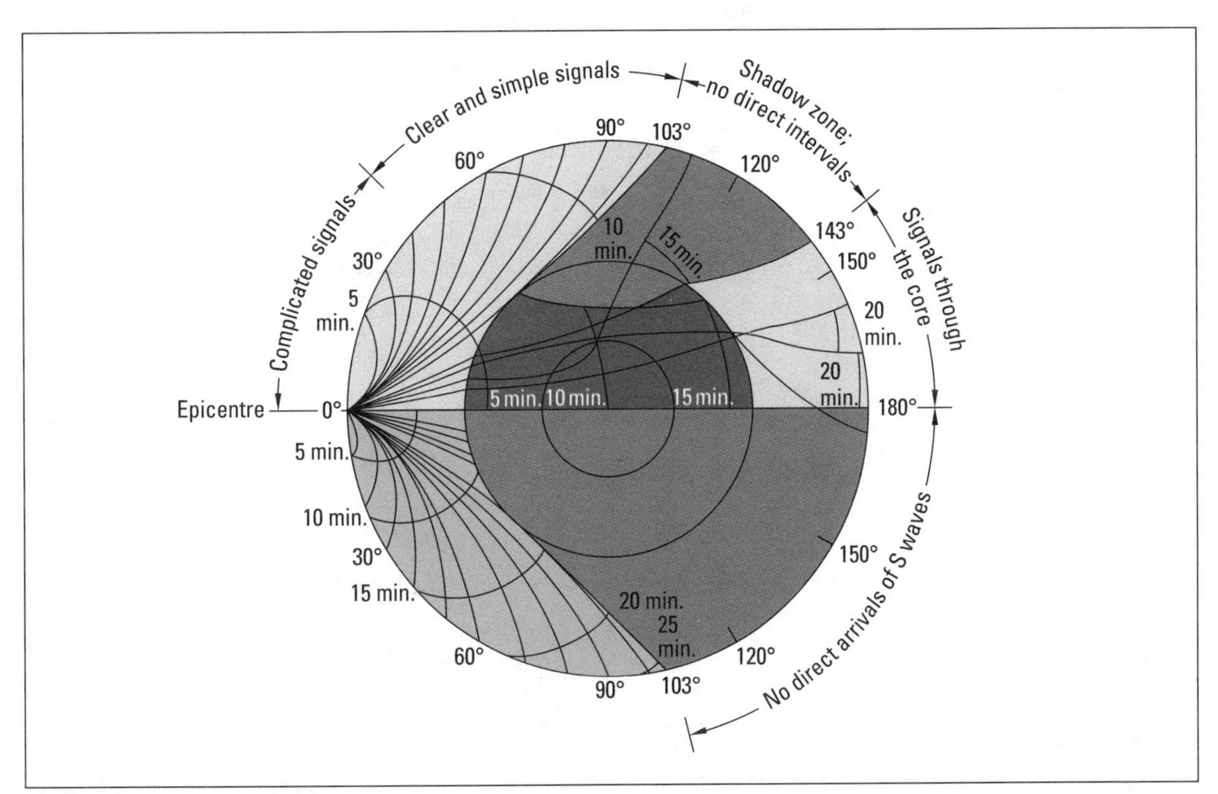


FIGURE 17.11 ▼ Seismic wave velocity as a function of depth. The circular graph plots locations on the surface where seismic waves from a common source arrive simultaneously. This diagram shows how the interior structure of the earth affects travel paths and times, and was deduced from the consistent arrival times of earthquake waves (from *Geology Today*, CRM Books, 1973)

spread out as the waves run across the surface; the land may bulge and crack, then close up again. It is these land waves, occurring close to the epicentre, that occasionally may trigger rockfalls, landslides, and avalanches. The earthquake or related geomorphic processes may collapse buildings and bridges, and rupture dams and utilities. Subsequent fires and floods may add to the devastation. The San Francisco earthquake of 1906 was one of the best-known examples, but in our century there have been a number of devastating earthquakes (Table 17.1).

Another consequence of earthquakes may be the generation of huge sea waves or **tsunami**. These are produced by displacements in the ocean floor, either as a direct effect of the displacement caused by the earthquake, or indirectly, by submarine slumping of sediments around the coast, and especially on the edges of the submarine trenches. In 1958 a tsunami some 30 m in height approached the Alaskan shore at a speed of over 200 km per hour and devastated large areas. Whole series of waves may be set up by these events, and they may have wavelengths of several hundred kilometres. Their effect is most intense where they enter shallow water, and thus they are most damaging where the coastline slopes gently, allowing the energy that is being transmitted to create a steep-fronted wave.

## EXTRUSIVE AND INTRUSIVE IGNEOUS ACTIVITY

As we have seen, the earth's subduction zones and mid-oceanic ridges are areas of intense volcanic activity. Between them, they account for about 95 percent of all the active volcanoes, the remaining 5 percent occurring in areas associated with what are called **mantle plumes** in the interior of the plates. The types of volcanic activity occurring in these three areas vary markedly. In the case of the mantle plumes, it seems that basaltic magma rises to the surface from permanent hot spots beneath the crust, possibly associated with stable convection cells in the mantle.

With mantle plumes in oceanic areas, where the crust is thin, the magma breaks through to the sea floor where it is rapidly cooled and produces **pillow** or **columnar lavas** (Figure 17.12). Typically, the lavas build up to form a gently sloping **shield volcano** and, in some instances, this may rise high enough to reach above sea level. A volcanic island is

### Table 17.1

*Major earthquakes 1900–1980*

| *Date* | *Location* | *Richter magnitude* | *Death toll (where known)* |
|---|---|---|---|
| 1906 | San Francisco | 8.3 | 315 (+ 352 unaccounted for) |
| 1906 | Andes of Colombia, Equador | 8.6 | |
| 1906 | Valparaiso, Chile | 8.4 | |
| 1911 | Tien Shan, Sinkiang, China | 8.4 | |
| 1920 | Kansu, China | 8.5 | |
| 1923 | Tokyo, Japan | 8.3 | 140 000 |
| 1933 | Japanese Trench | 8.5 | |
| 1949 | Garm, Tadzhikistan, central Asia | 7.9 | 12 000 |
| 1950 | North Assam, India | 8.6 | |
| 1960 | Chile | 8.3–8.9 | |
| 1964 | Alaska | 8.6 | 115 |
| 1975 | Haicheng, China | 7.3 | |
| 1976 | Antigua, Guatemala | 7.5 | |
| 1976 | Tangshen, China | 8.5 | 750 000 |
| 1978 | Oaxaca, Mexico | 7.8 | 0 |
| 1980 | Southern Italy | 6.8 | 3 000 |

FIGURE 17.12 ▼ (A) Columnar, jointed, basaltic lava flows, Rusty Lake, NWT; (B) pillow lava formed in a submarine eruption, Iceland; (C) Ash eruption from a cinder cone volcano, Eldfell, Iceland (photos: M.B. Lambert, courtesy Geological Survey of Canada)

thus created. Over time, as the plate drifts across the hot spot, a succession of such volcanoes is formed, producing an island chain. The best-known is the Hawaiian chain, the oldest member of which is Midway (about 18 million years old) in the northwest; the youngest member of the group, Hawaii, in the southeast, is still active.

Volcanic activity is less common where mantle plumes lie beneath continents, for the magmas there are more likely to be trapped in the crust. In these cases, intrusive igneous features are formed, as we will discuss later. Nonetheless, where the magmas do break through to the surface, vast areas of basalt may be produced, forming extensive plateaus. Possibly the most spectacular example is the Deccan plateau of India, which covers 700 000 km$^2$, while similar **flood basalts** form an area about 300 000 km$^2$ in the Columbian plateau of the northwest USA and the interior plateau of British Columbia.

Volcanoes and extensive lava flows are also associated with the mid-oceanic ridges. Here, magma rises into the long fissures created by separation of the plates that make up the ocean floor. Commonly, the site of eruption is only temporary, and over time it migrates along the fissure so that a series of volcanoes is formed. Nowhere are the effects better seen than in Iceland. This country lies astride the Mid-Atlantic Ridge and is build almost entirely of basaltic lavas of Tertiary and Quarternary age. It is still highly active and, in 1783, for example, a series of eruptions occurred along a 32 km fissure at Laki, creating a line of over a hundred small **cinder cones**, built up from **tephra** ejected during the eruption. Tephra are particles of solidified magma. In 1964, eruptions occurred in the Vestmannaeyjar off the south coast. Ash and lava piled up until they rose above the surface of the sea and created an entirely new island—Surtsey. A few years later an eruption occurred on the nearby island of Heimaey. Vast amounts of ash and lava were ejected, burying large parts of the island's only town and almost blocking the entrance to the harbour (Figure 17.12C).

If we examine a map of the world's volcanoes, we find that about 80 percent lie along the zones of subduction—notably in a great arc of islands around the western side of the Pacific and along the line of the Andes and western cordillera of North America (Figure 17.5). Volcanoes along the island arc of the Pacific seem to coincide with areas where the subducted plate is 100–200 km deep. Whether activity is a result of melting of the plate during subduction, or whether magma from the mantle escapes along fissures created as the plate subsides into the mantle is not clear. It is apparent, however, that island arcs like this are common where two oceanic plates collide. In these cases, the volcanoes that are formed are typically **strato-volcanoes**, consisting of alternating layers of ash and lava. They are composed of a range of rock types, including basalt, andesite, and rhyolite.

On occasion, these volcanoes are explosive. In 1883, for example, Krakatoa, a volcanic island in Indonesia, exploded, ejecting 75 km$^3$ of material into the air and leaving a huge **caldera**, some 7 km by 6 km in diameter. The event is vividly described by A. Holmes in *Principles of Physical Geology*, 1964, p. 336:

> The climax was reached during the last week of August. On the 26th formidable detonations were heard every ten minutes. Dense volcanic clouds reached a height of 17 miles, and ashes, converted into stifling mud by incessant rain, fell over Batavia (now Djakarta) which was plunged into thick darkness, relieved only by vivid flashes of lightning. On the morning of the 27th came four stupendous explosions, the greatest of which was heard 3000 miles away in Australia, and a vast cloud of incandescent pumice and ashes rose 50 miles into the air.

Giant tsunami were generated by the explosion and 36 000 people died in Java and Sumatra.

Several other calderas of this nature are found close to the zones of plate subduction, all of them testifying to massive volcanic explosions. Crater Lake in Oregon is a spectacular example, some 15 km in diameter. Even larger is Lake Toba in Sumatra (about 1000 km$^2$). Mount St. Helens is a good example of a relatively small-scale caldera eruption, and is discussed in detail at the end of this chapter.

As all these examples show, volcanoes vary considerably both in the nature of their eruptions and in the character of the ejecta they produce (Table 17.2). Depending upon these two factors, the morphology of the volcano also varies.

## INTRUSIVE IGNEOUS ACTIVITY

We noted earlier that magma rising from the mantle does not always reach the surface of the earth but may be confined within the crustal rocks. Within

## Table 17.2

*Types of Volcano outputs*

| *State* | *Name* | *Description* |
|---|---|---|
| Gas | Fume | |
| Liquid | Aa | Rough blocky lava |
| | Pahoehoe | Smooth ropy lava |
| Solid | Dust | <0.06 mm |
| | Ash | 0.06–2.0 mm |
| | Cinders | 2.0–64 mm |
| | Blocks | >64 mm, solid |
| | Bombs | >64 mm, plastic |
| Flows | Pyroclastic | Hot, liquid flows |
| | Mudflows | Flows liquified by rainfall, ice melt, snow, or water released from crater lakes (these are not made up of tephra) |

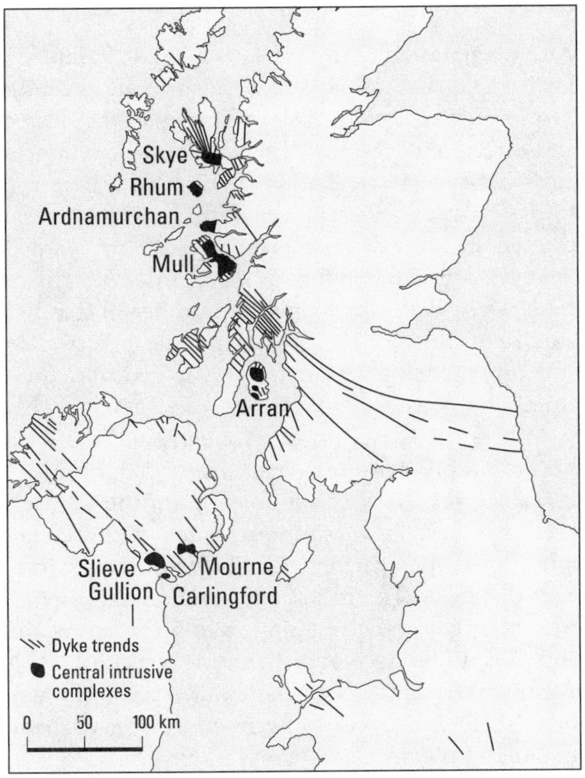

FIGURE 17.13 ▼ Dyke swarms emanating from Tertiary igneous intrusions in northern Britain (from Sparks, 1971, and Richey, 1964)

the crust, the magma cools relatively slowly to form intrusive igneous rocks (see Chapter 16).

The intrusive magmas may produce many different features. Thin vertical veins within the crust may be exploited by the magma, forced apart by the pressures it generates, and ultimately filled with the solidified rock. These **dykes**, as they are known, often spread out from the centres of intrusion (Figure 17.13). Because the dykes have a large surface area relative to their volume, the magma in them cools fairly quickly, and fine-grained rocks predominate. Sometimes the dykes feed into surface vents, such as volcanoes, and they thus act as conduits by which magma reaches the surface.

Where magma seeping along dykes is able to exploit a horizontal plane of weakness within the rock, a feature known as a **sill** may develop. Occasionally, these sills remain concordant with the strata for considerable distances, but, where vertical discontinuities or weaknesses occur, they may break through the strata and exploit different levels in the rock (Figure 17.14).

More viscous magma may spread less rapidly through the rocks and may, instead, create dome-shaped masses of more limited extent. **Laccoliths** of

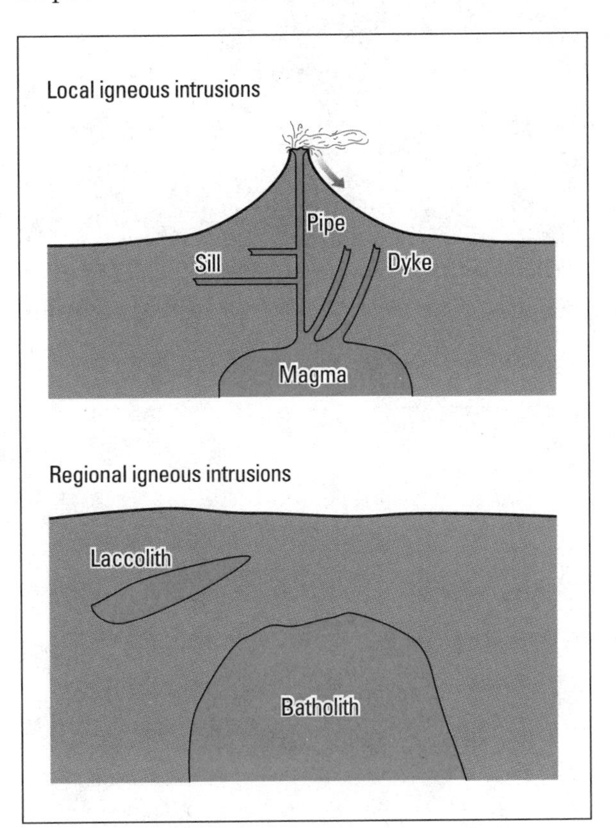


FIGURE 17.14 ▼ Types of igneous intrusion

this form occur in the Henry Mountains of southern Utah. Larger-scale masses formed by the upwelling of magma are known as **batholiths** (Figure 17.14). These are typically formed of granitic rocks and they incorporate and displace large quantities of the surrounding country rocks. The Ruby Range, east of Kluane Lake in Yukon, is an example of this type of feature exhumed by erosion.

All these subsurface igneous forms may influence the surface topography when they are exposed by erosion. Dykes may be either more or less resistant than the surrounding rocks, and may thus form thin walls or trenches across the landscape (Figure 17.15). Sills, such as the Fairhead Sill, in Northern Ireland, similarly form dominant structural landforms (Figure 17.16). Exhumed laccoliths and batholiths may also impose their character upon the landscape.

FIGURE 17.15 ▼ An intrusive dyke exposed by differential erosion on the ocean shoreline; the North Star Dyke near Ballycastle, Northern Ireland (photo: P.G. Johnson)

FIGURE 17.16 ▼ Intrusive sills form the cliffs at Fairhead, near Ballycastle, Northern Ireland (photo: P.G. Johnson)

# Folding and faulting

The upheavals created by plate movement, subduction, earthquakes, and volcanic activity set up huge stresses in the earth's crust. These often deform the rocks, either by **folding** where the rocks are able to deform plastically, or by **faulting** where brittle failure occurs. Within each rock type there are natural layers as in sedimentary rocks, and planes of weakness as in granite and metamorphic rocks. Sedimentary rocks, usually level when formed, are often warped into a different position.

It is necessary to determine the trend of these layers under the ground for accurate geologic maps or to follow the line of an important mineral deposit. The angle formed between the horizontal plane and the actual plane of the rock layer is called the **dip angle**. The line drawn parallel to the earth's surface perpendicular to the dip of the layer is called the **strike**. The orientation of the line of strike can then be determined relative to magnetic and true north.

## FOLDS

At its simplest, a fold represents a single 'crumple,' rather like that which can be formed by pushing gently on a tablecloth. Depending upon the degree and direction of pressure, this fold may be symmetrical or asymmetrical (Figure 17.17). Where the fold produces an arch-like form, the feature is termed an **anticline**; where the rocks are warped downward it is referred to as a **syncline**. A series of folds often develops to give a number of parallel anticlines and synclines. Minor folding may also be superimposed upon larger features to give anticlinoriums and synclinoriums (Figure 17.17).

Where the lateral pressures are greater, the folds become more complex. Anticlines may be overturned as one limb of the fold passes the vertical; they may become **recumbent**, when the rocks are forced over upon themselves. This folding may fracture the rocks. The rocks may shear along a line of weakness, generally running through the axis of the anticline, to produce **overthrusting** (Figure 17.17). All these features vary greatly in size, and folds may be as small as a few centimetres or as large as many kilometres in width and height.

In landscapes that have not yet been drastically modified by erosion, these processes may be reflected by the topography; anticlines form hills, and syn-

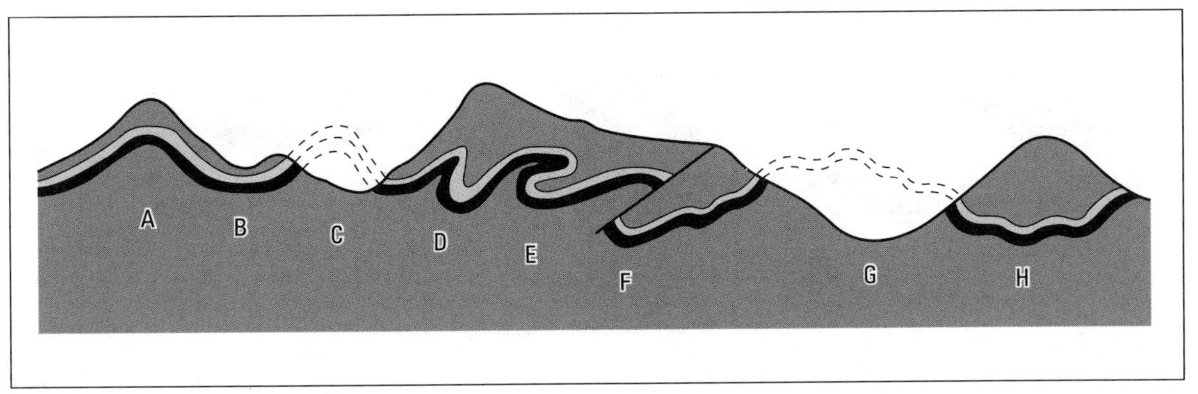


FIGURE 17.17 ▼ Types of fold: (A) symmetrical anticline; (B) symmetrical syncline; (C) asymmetrical anticline; (D) inclined anticline; (E) recumbent fold; (F) overthrust fold; (G) anticlinorium; (H) synclinorium

clines, valleys. In time, however, erosion modifies the original surface expression of the folds by preferentially removing weaker rocks. Indeed, in some cases, inversion of the landscape may occur as the anticlines are worn away and the more compressed and resistant strata in the syncline are preserved as hills (Figure 17.17). In areas of complex folding, such as the European Alps and the fold mountains of South Africa, it may be very difficult to identify the original pattern of folding. In addition, because folding may entirely reverse the sequence of rocks exposed at the surface, it can lead to problems in reconstructing the chronology of the area.

Although folding is often closely associated with major orogenic activity, there are many examples of folding produced by smaller-scale events. The intrusion of magma into the crust may cause localized folding, while at an even smaller scale, salt domes, created by the rise of relatively low density salt into denser overlying rocks, may produce similar features. Localized folding also arises from the subsidence of dense material into less dense substrata (load structures), and from the growth of ice wedges and lenses beneath the ground surface.

## FAULTS

When the relatively rigid rocks of the crust are subjected to stress, they tend to fail along distinct planes, which are called **faults**.

Various types of faults occur, depending upon the nature of the stresses that caused them. The simplest form, the **normal fault**, is created by tensional forces operating in opposite directions (Figure 17.18A) Characteristically, the rocks break along a steeply inclined fault plane and one mass of rock slips vertically relative to the other. Such faults give rise to simple **fault scarps**. These may show a considerable range in both height and length. **Reverse faults** may be produced by compressional stress (Figure 17.18B). Movement occurs along a steep plane. Here, the tendency is for one side of the fault to be upthrust relative to the other. As with normal faults, these may occur together, and blocks of land may be upthrust to give a feature known as a **horst**. On a large scale, the Black Forest area of Germany represents a horst (Figure 17.18C).

More complex faults arise where failure occurs not along a single plane, but along a number of parallel planes (Figure 17.18D); this produces a steplike scarp. In addition, tensional stresses may result in the subsidence of a block of land, as the rocks on either side are pulled apart, creating a steep-sided trench in the landscape, known as a **graben**. On a large scale, these features form **rift valleys** (Figure 17.18E).

Where the stresses are exerted parallel to each other, **transform faults** may develop (Figure 17.18F). A well-known example is the San Andreas Fault in California; another is the Aspy Fault of Nova Scotia (Figure 17.19). As we have seen, transform faults are also common in the mid-oceanic ridges, where differential horizontal movement of the oceanic crust is occurring.

Intense compressional stress may result in low angle thrust faults. The overthrusting of rocks involved in intense folding is an expression of this process, but overthrusting need not always be associated with folding. Characteristically, **overthrust faults** have a gently dipping fault plane (Figure 17.18G).

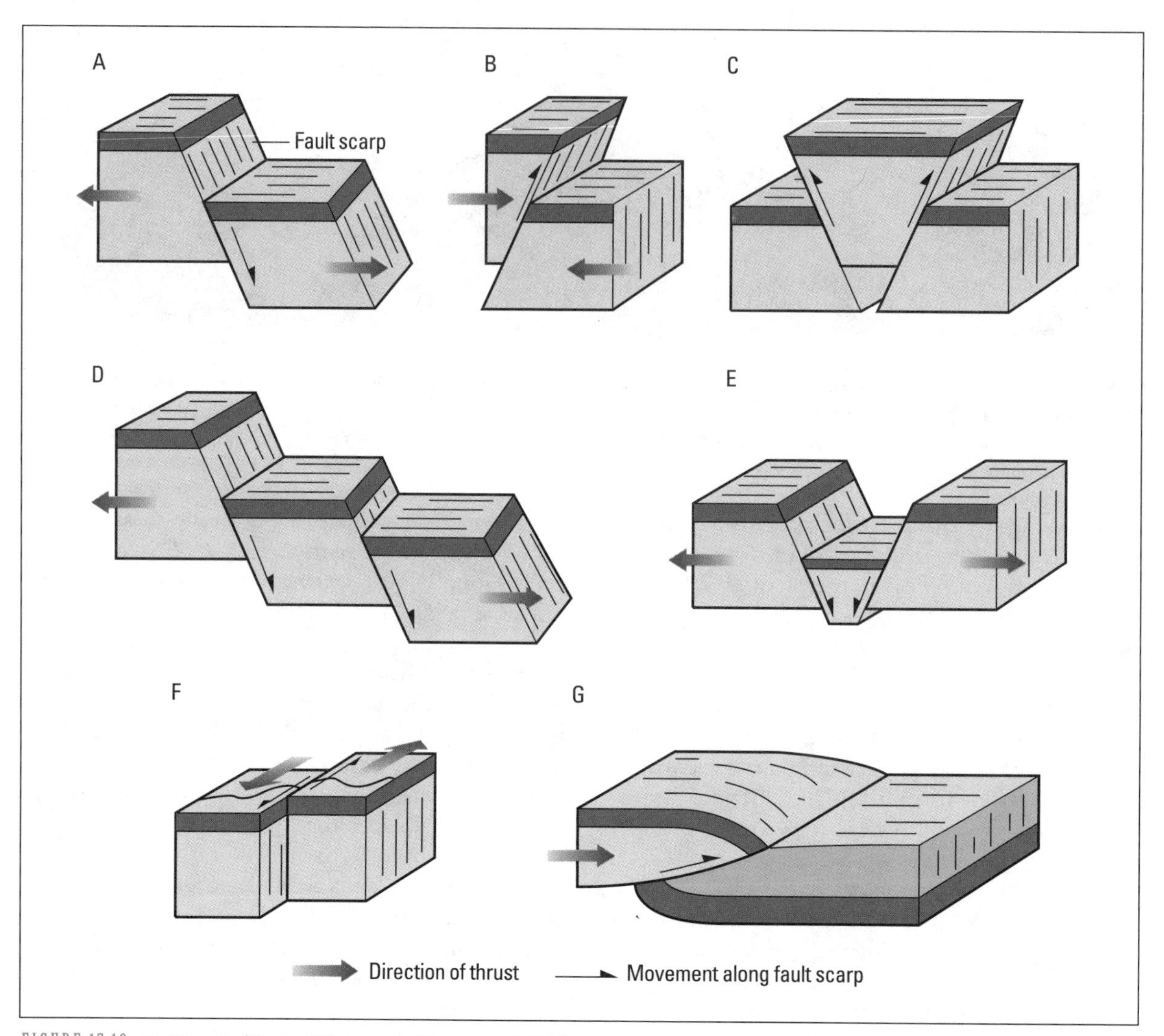


FIGURE 17.18 ▼ Types of fault: (A) normal; (B) reversed; (C) horst or upthrust; (D) stepped; (E) rift; (F) transform; (G) overthrust

FIGURE 17.19 ▼ The Aspy Fault, Nova Scotia (photo: D.R. Grant, courtesy Geological Survey of Canada)

# Earth-building processes and humans

## PLATE TECTONICS AND MINERAL RESOURCES

The processes of earth-building have considerable significance for us. In particular, it is now apparent that processes of plate tectonics account for the global distribution of many mineral resources. Many of the world's major deposits of metallic minerals, for example, are associated with former zones of subduction or geosynclines along the margins of the plates. The reactions that occur as the subducted plate descends into the mantle release mineral-rich solutions that rise into the crust and precipitate in the form of mineral deposits. Metallic sulphides (i.e., metals combined with sulphur) are especially significant in these areas, and a large proportion of the world's iron, copper, and other base metals are located in zones of plate convergence. Gold is often associated with these reserves. The deep marine trenches created by subduction and the geosynclines along the margins of the continents favour the accumulation of organic debris which is subsequently buried by sediments and converted to oil (Figure 17.20).

Plate convergence, as we have seen, often occurs either at continental margins or in association with island arcs. In both cases, the mineral resources formed in these areas tend to be relatively accessible, and have been known and exploited for many centuries. The concept of plate tectonics, however, has drawn attention to a second area of rich mineral resources: the mid-oceanic ridges. Here, too, volcanic activity results in the release of hydrothermal solutions which have deposited large concentrations of minerals, including iron, copper, lead, uranium, and mercury. Because of the problems of exploration on the sea floor, the full worth of these resources is not yet known, but an indication of their potential is given by the mineral resources found in the Troodos Massif of Cyprus. This is believed to be an old mid-oceanic ridge that has subsequently been uplifted. It contains enormously rich deposits of copper (from which the island gets its name) as well as iron, zinc, gold, and silver. It is not certain that these formed while the area was part of the mid-oceanic ridge, but if so they indicate the mineral wealth that may be awaiting discovery beneath the sea floor.

## EARTH-BUILDING AS HAZARD

Not all the effects of earth-building are beneficial. As we have seen, earth-building processes involve immense upheavals of the earth, and in some cases they may be serious hazards. In 1980, the inhabitants of the area around Mount St. Helens in Washington State saw some of these dangers most forcibly. Few are likely to forget the experience of the mountain erupting in a deluge of rock and cinders and sulphurous smoke. The mountain, which had last erupted in 1857, began to stir on Thursday, 20 March. The first sign was an earthquake on the north side of the mountain, but soon earthquakes became frequent. On 27 March, a small eruption occurred, producing a new crater 65 m wide on the summit. Small eruptions followed, and five days later a second fresh crater appeared beside the first. Slowly, the north side of the mountain began to bulge; by mid-April it had risen 100 m. On 12 June, and again on 22 July, vast explosions occurred as the mountain, split by a magnitude 5 earthquake, was suddenly relieved of the confining pressures. Over 400 m of the summit was removed and the cloud of dust and debris flattened forests over 32 km away, scattering ash as far afield as Portland, and laying waste in total to 560 km$^2$ of the surrounding countryside (Figure 17.21).

Afterward, the full impact of the event became clear. The mountain had changed shape. An estimated 2.7 km$^3$ of material had been blown off; 600 000 tonnes of ash had fallen on Yakima, to the northwest; one town had been abandoned and in all

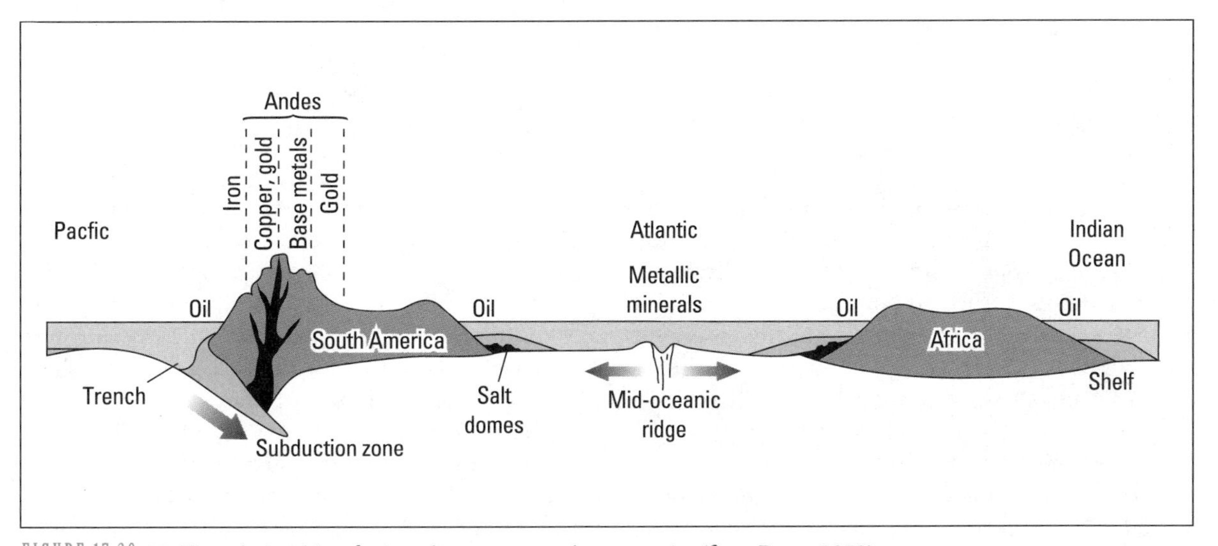


FIGURE 17.20 ▼ The relationship of mineral resources to plate tectonics (from Rona, 1973)

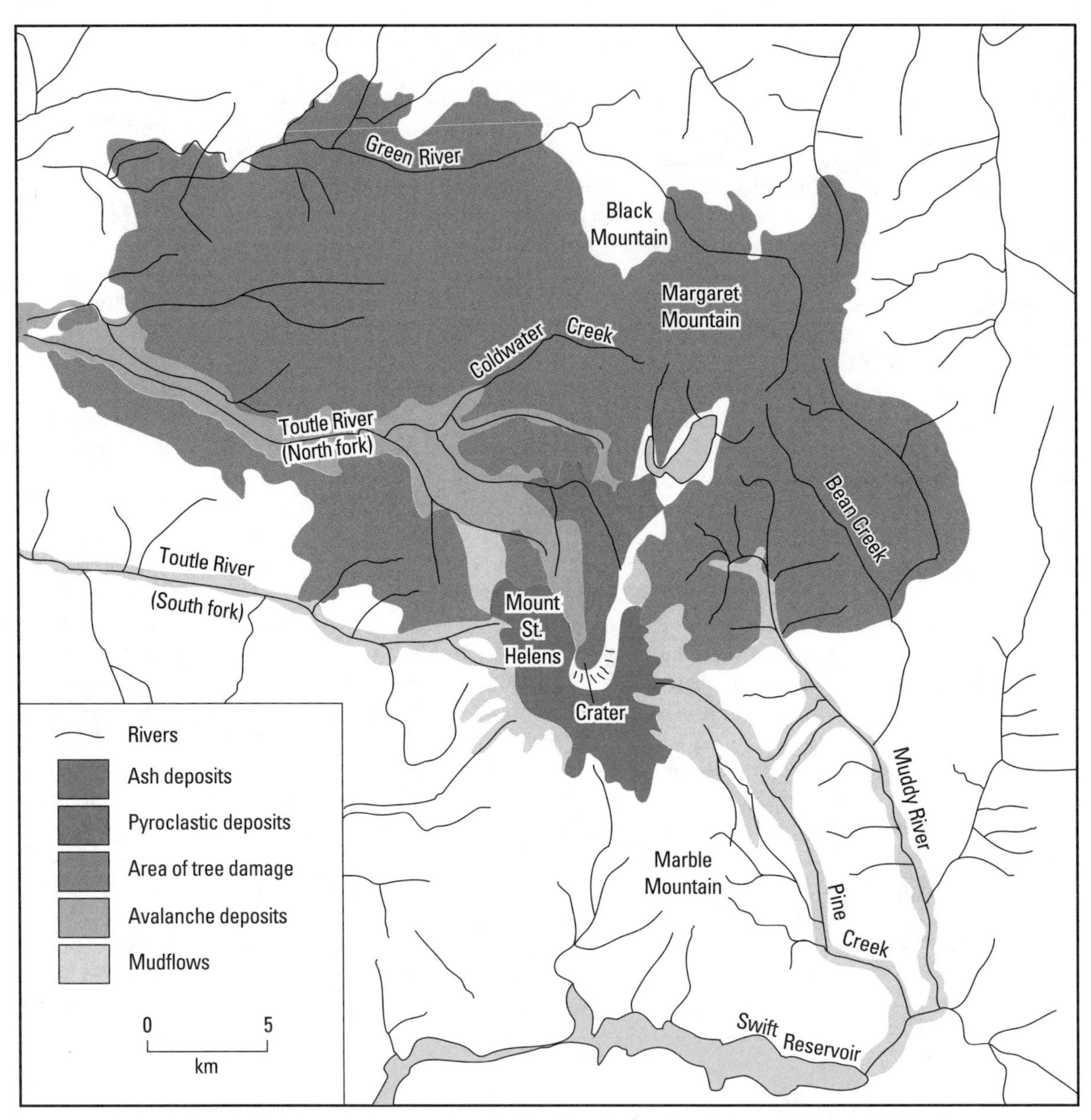


FIGURE 17.21 ▼ The area affected by the Mount St. Helens eruption (based on Decker and Decker, 1982)

sixty-three people were dead. All around, dams had burst, rivers had changed course, roads had been blocked, and crops smothered. Both the direct and indirect effects of such an eruption are clearly catastrophic, yet this, by geological standards, was a relatively minor event!

The same is true of earthquakes. As all who live in the San Francisco area know, this great city is built upon an active fault line. The fault moved in 1906, and San Francisco was all but destroyed. In an attempt to minimize the effects, water is being pumped into the fault on the hypothesis that this will lubricate the slip line and encourage it to move in a series of small, gradual slips rather than a single, catastrophic one. An earthquake, magnitude 6.9 on the Richter scale, occurred in 1989 during a telecast of a World Series baseball game, and there was serious damage in urban areas.

Given the threat they impose, it would clearly be useful to predict events such as earthquakes and volcanic eruptions. To some extent, we can. In many cases they take place with a rough sort of regularity. Earthquakes, for example, require the buildup of critical stresses in the rock, and these can be computed from a knowledge of the rock structure and the forces operating in the crust. That is why we know that another is likely in California. Moreover, we also know now that the spatial distribution of earthquake and volcanic activity is related to the pattern of plate tectonics, so we can define areas where such hazards are most likely to occur. Nevertheless, we are still a long way from predicting such events with any precision. The processes involved are not understood clearly enough, and the data needed for prediction are not normally available.

There are, however, signs that prediction is possible. The Chinese, for example, managed to predict the earthquake that wiped out Haicheng in 1975. They did so not only by monitoring the buildup of pressures in the rocks, but by analyzing changes in the magnetic and gravitational properties of the area, variations in the water levels in wells, and even unusual behaviour in animals. As a result, 100 000 people were moved out of the city before the earthquake, and disaster was averted. Unfortunately, they had a serious failure of predictive techniques just months later when an earthquake causing great loss of life (some estimates put the figure over 100 000) occurred in the same region.

Human ability to trigger earthquakes, although inadvertent, holds both a threat and a promise. Studies in Colorado have clearly demonstrated that by changing the fluid levels in the rocks, small earthquakes can be 'turned on' and 'turned off.' The discovery of the connection was fortuitous. The need to dispose of a nerve gas created for military use led to the drilling of deep wells on an army base in the Denver area. It was noted that when the fluid was injected, the number of small earthquakes increased. When injection stopped, the earthquakes were fewer. This connection was confirmed in controlled experiments carried out in the Rangely, Colorado, oilfield. It is believed that the fluid reduces the friction between contact surfaces of a fault, and allows the rocks to move more easily.

Other evidence shows that the construction of dams leads to an increase of earth tremors in the immediate area. Conversely, in some areas the removal of fluids reduces the slippage between rock surfaces, thus preventing earthquakes. The potential to induce or reduce earthquakes is exciting, but fraught with danger. There are no guarantees that the earthquake created will be small. Similarly, the prevention of earthquakes might only allow tectonic pressures to build toward even more powerful and devastating events. Earthquakes occur in relatively concentrated regions of the globe (Figure 17.9) where plate movement is continuing. It is argued that a series of smaller induced tremors is better than allowing locked plates to build toward potentially devastating releases of energy. Setting natural forces in motion is increasingly within our capability. Controlling and predicting the outcome of such interference, not to mention the legal consequences, is another matter. If such activities are to be pursued, a great deal of social responsibility is required.

Earthquakes and volcanoes vary considerably in character, and methods that work in one case do not necessarily work everywhere. But the necessity to understand these tremendous processes of the earth is clear.

# Weathering

## The basis of weathering

### THE ROLE OF WEATHERING IN THE ENVIRONMENT

Weathering is the term used for all the processes by which the physical and chemical characteristics of rocks at or close to the earth's surface adjust to their environment. In general, **weathering** processes break down rock material.

The implications of weathering are manifold. The products of weathering become available for erosion; they are thence transported through the landscape, deposited, and ultimately lithified to form new rocks. Weathered residues may lie on the surface of the ground and contribute to the soil. Substances extracted from the rocks by weathering may be dissolved in the waters that percolate into the ground, and then carried to streams or taken up by plants. Weathering processes, therefore, are fundamental to many other aspects of the environment.

### THE CAUSES OF WEATHERING

The disequilibrium that initiates weathering processes arises for a number of reasons. One of the most common we have discussed in the previous chapter. Earth movements, such as folding and faulting generally associated with the collision of crustal plates, force the rocks into mountain chains. Rocks that may have been buried deep within the crust or sediments laid down deep beneath the sea may be exposed at the surface; the change in conditions—the exposure of these rocks to the water, gases, heat, and dissolved substances at the surface—results in changes within the rocks. Constituents that are unstable in these conditions are altered to more stable forms, or are destroyed.

Another common cause of disequilibrium is volcanic activity. The magmas that well up from the mantle are similarly exposed, and become subject to weathering. Erosion, too, may uncover rocks that once lay deep in the crust, and again a state of disequilibrium may be induced, while a fall in sea level may expose marine sediments to subaerial conditions.

It is not only change in the relative position of the rocks that stimulates weathering, however; the surrounding environment may also change. Fluctuations in climate, such as increased rainfall or reduced temperature, may lead to alterations in the nature and rate of weathering. Even changes in the vegetation may have an effect, for many of the substances that take part in weathering are derived, directly or indirectly, from the vegetation. Human activity also changes the weathering environment; pollutants entering the atmosphere may be particularly effective as agents of weathering.

As a rule, this disequilibrium is most acute at the start of weathering, and diminishes thereafter. As a result, the motive power behind the process tends to decline over time, and as the rocks approach their equilibrium forms, the rate of change declines (Figure 18.1). This is only part of the story, for changes in the initial character of the rocks brought about by weathering may provide more suitable conditions for the operation of other processes.

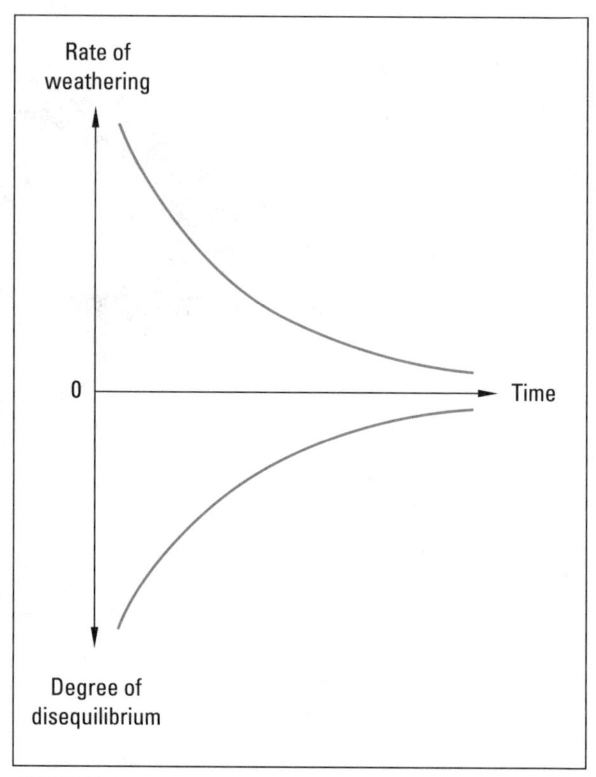


FIGURE 18.1 ▼ The general relationship between rate of weathering and disequilibrium

# The location of weathering

Weathering, in general, is confined to a relatively narrow zone of the earth's crust. The forces involved in weathering are largely related to the atmosphere, and thus it is mainly rocks that are in close contact with the atmosphere that are subject to weathering. The **weathering zone** extends from the soil surface to the maximum depth of penetration of mobile water, gases, and energy derived from the atmosphere. There are exceptions, as we will see, but in most cases this zone is between a few centimetres and several metres in thickness. Within it, weathering tends to be most intense in the upper layers and may operate irregularly throughout the rest of the zone, depending upon local conditions.

This weathering zone comprises several different environments (Figure 18.2). It includes the soil, which is largely a product of the weathering process acting upon the original materials, a zone, in some cases, of unsaturated material; an upper zone of seasonal or occasional saturation which fills during precipitation or snowmelt; and a lower zone of temporary saturation from upward movement of groundwater or downward movement of rainfall or snowmelt. The base of the lowest level of the fluctuating zone is generally marked by the water table, beneath which the ground remains permanently saturated. Beneath the water table weathering may operate, but it tends to do so at a much reduced rate, largely because of the slow turnover of the groundwater (Chapter 13) and the consequent stability of conditions. If weathered material is not removed, the ground waters and the rocks almost reach a state of equilibrium. A **weathering front** can be identified, below which there is no change in the physical or chemical characteristics of the rock.

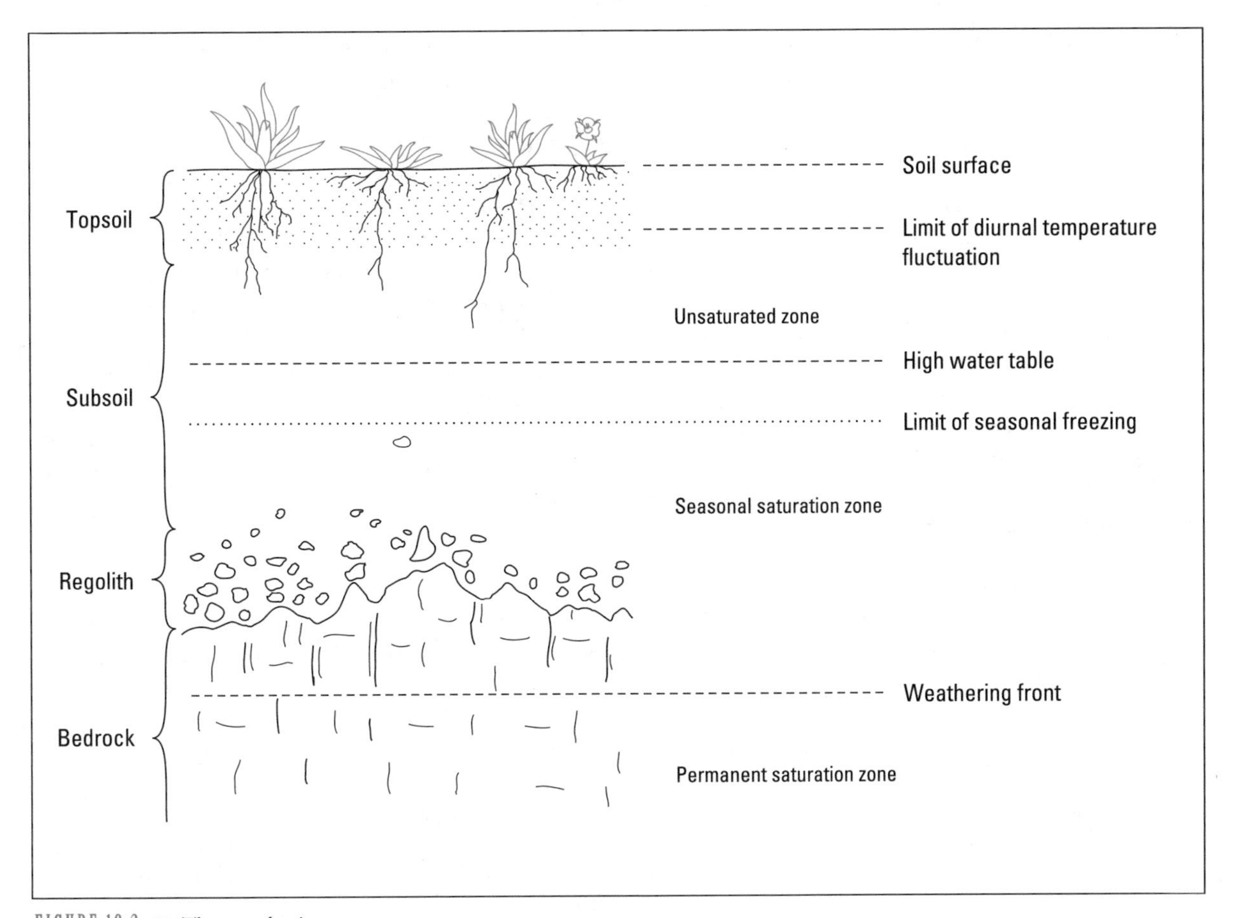


FIGURE 18.2 ▼ The weathering zone

# Weathering agents

The rocks and soils within the weathering zone are acted upon by a number of agents. These may be considered the inputs to the weathering system.

The main weathering agents are water, heat, and dissolved substances (reactants). Most of these are initially derived from the atmosphere, but there is clearly a tendency for all three to work their way downward through the weathering zone (Figure 18.2). In the process, changes occur, and materials washed from one layer are transported to lower layers. This may have considerable importance, particularly because substances dissolved in percolating waters often act as significant agents of weathering. It is a situation that can be seen, somewhat artificially, on some buildings. Water draining off a copper roof and dripping onto the stone of the building increases the weathering of the stone by the actions of substances dissolved from the roof.

## HEAT

One of the main agents of weathering is heat. It operates in two ways: directly, through fluctuations in temperature which exert stresses upon the rocks; and indirectly, through its influence on chemical reactions, most of which are accelerated by an increase in temperature.

The action of heat is most intense close to the surface, for few soils or rocks are good conductors of heat, and extreme temperatures are rarely felt below a few centimetres depth. In fact, it is this lack of heat conductance that encourages weathering processes due to temperature fluctuations. The inability of rocks to transmit heat downward means that the rock surfaces heat up considerably during hot periods, and cool rapidly during cold periods. Thus, the skin of exposed boulders and rocks in desert areas may be subjected to considerable stresses induced by these temperature changes. Even a shallow covering of soil or weathered material tends to reduce the effect by insulating the rocks from the heat of the sun.

In polar and subpolar environments, heat plays a rather different role, for extremely low temperatures may allow the development of a layer of permanently frozen soil (permafrost) near the surface of the ground. In this zone, stresses may be set up by the freezing of water, which expands by about nine percent of its volume as it changes to ice. In confined areas, this expansion may set up considerable stresses. Above the zone of permafrost there is the active layer, in which seasonal or shorter-term melting may occur. Here, the fluctuations in temperature around freezing point, and the consequent alterations of ice formation and melting, may cause even more intense weathering.

In alpine areas, freezing and thawing of water in rock on valley sides can be particularly effective in the breakdown of the rock.

It is also important to note that there is a wide region of transition from temperate to polar climates where fluctuations of temperature above and below freezing can play a role in weathering. This can again be illustrated by reference to the urban environment, where paved road surfaces disintegrate due to freeze–thaw cycles.

## WATER AND DISSOLVED SUBSTANCES

As this implies, it is often the combination of heat and water that is responsible for weathering. Most water entering the weathering zone of the earth is derived from rainfall, and it is the chemical as well as physical character of this water that is important.

Pure water is composed of hydrogen and oxygen in the ratio of two hydrogen ions to one oxygen ion ($H_2O$). An ion is an atom or molecule bearing a net electrical charge. This may be positive, e.g., $H^+$ (hydrogen) or $NH_4^+$ (ammonium), or negative, e.g., $HCO_3^-$ (hydrogen carbonate) or $O^-$ (oxygen). Under natural conditions, the combination of these elements is not always stable, and the water molecules tend to separate (dissociate) to produce individual hydrogen ($H^+$) and hydroxyl ($OH^-$) ions. The concentration of these dissociated or free hydrogen ions determines whether the water is acid or alkaline. Water that contains an excess of free hydrogen ions is acid, and it is generally able to weather rock minerals more readily. Acidity is measured on the **pH scale**. This ranges from 0 to 14; liquids with a pH below 7 are acid, those with a pH above 7 are alkaline. We will encounter this scale again when we discuss soils in Chapter 26.

Water is rendered acid either by the introduction of extra hydrogen ions, or by the dissociation of the water, which increases the concentration of free hydrogen ions. These processes are brought about in a variety of ways. Hydrogen ions, for example, may be released by vegetation; consequently, rainwater dripping through vegetation and percolating through the soil tends to be made slightly acid. Similarly, carbon dioxide acquired from the atmosphere tends to react to produce free hydrogen ions in rainwater:

$$\underset{\text{water}}{H_2O} + \underset{\text{carbon dioxide}}{CO_2} \rightarrow \underset{\text{free hydrogen ion}}{H^+} + \underset{\text{hydrogen carbonate ion}}{HCO_3^-}$$

This converts the rainwater to a weak solution of carbonic acid, which is able to attack many rock minerals.

Many other substances may also become dissolved in rainwater, increasing its weathering ability. Pollutants such as sulphur dioxide may convert the water to a weak sulphuric acid, while organic compounds washed from vegetation produce weak organic acids. As the water percolates through the soil and surface rocks, it picks up various other organic and inorganic substances which may similarly make it more active as an agent of weathering.

Against the inclusion of these reactants within the water must be set the tendency for the water to absorb, and eventually become saturated with, the products of weathering. Water percolating through limestone, for example, tends to dissolve the calcium carbonate. Ultimately, it will contain so much calcium that it will be unable to absorb any more. It thus loses its ability to weather the limestone with which it subsequently comes into contact. Indeed, the water may end up precipitating the dissolved substances, a phenomenon that contributes to the formation of the beautiful features such as **stalagmites** and **stalactites** that adorn many cave systems (Figure 18.3). The details, both of limestone weathering and of deposition, are more complex than this simple picture indicates, but it does demonstrate the principle that the percolating waters may ultimately reach chemical equilibrium with the rocks they are encountering.

The ability of water to carry out weathering is therefore dependent to a great extent upon its chemistry. As we have seen, however, the chemistry

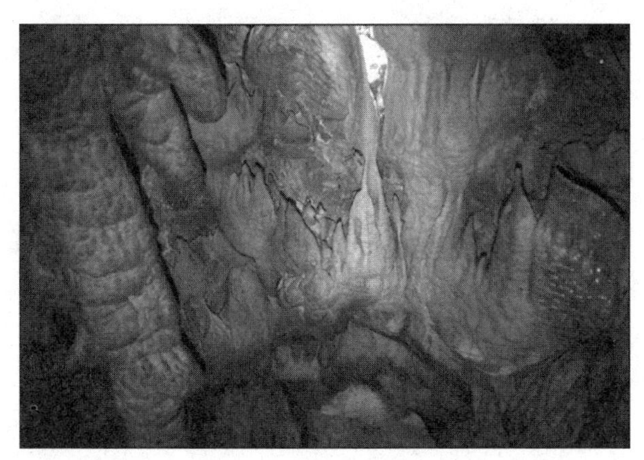

FIGURE 18.3 ▼ Stalactites and columns in Mammoth Cave, Kentucky (photo: P.G. Johnson)

of the water changes as it carries out weathering, and this introduces an important principle of weathering. This is that active weathering requires the constant regeneration of the weathering agent. Thus, water must have reasonably free entry to, and egress from, the materials that are being weathered. It is essential that the weathering agent does not remain around so long that it comes into equilibrium with the rocks it is attacking; if it does so, the rate of weathering declines.

Another important factor relating to the role of water in weathering is clearly the quantity of water passing through the weathering zone. This is controlled partly by climatic conditions—particularly the balance between rainfall and evapotranspiration—and partly by the nature of the surface rocks and soils. On the whole, rocks and soils that allow ready and rapid percolation of water are more susceptible. Nevertheless, the quantities of water percolating through the weathering zone tend to fluctuate over time, and this undoubtedly alters the rate of weathering. The relationship is not simple, however, for as we shall see later it is often the alteration of wet and dry conditions that encourages change in the rock and soil materials.

## Earth materials and weathering

The processes of weathering are influenced by the nature of the materials being weathered, as well as the agents of weathering. There are three main properties of the materials that may be mentioned, although these often interact closely, and other factors may be locally important.

As our discussion of the percolation of water implies, one of the constraints upon the rate and nature of weathering is the **structure** of the rocks, and a second is the **permeability** of the material.

### STRUCTURE

Joints, bedding planes, and faults in rocks often control the pathways taken by percolating water, and the quantities of water that can move through the rock. Similarly, as we noted in Chapter 13, the movement of water through soil is largely controlled by structures within the soil.

The importance of structure is twofold, however. It influences the rate or quantities of water passing through the material, and also determines the surface area of that material that is available to be weathered. The surface area exposed to weathering in a well-jointed rock is far greater than that exposed in a massive rock. The joints and fissures allow removal of the weathering products, and prevent the water coming into equilibrium with the rock.

### TEXTURE

We must remember that water may also move through the rock, and the relative importance of structure and permeability varies with rock type. Particle size and the arrangements of the particles determine the permeability and **porosity** of the rock. Where there is free movement of water, small particles present a larger surface area in relation to their volume to the agents of weathering than do large particles. All else being equal, therefore, finer-grained materials weather more rapidly than coarser-grained ones. In reality, all else is rarely equal, and rock or soil texture is often related to both structure and composition. Indeed, fine-grained materials are frequently composed of relatively stable materials which resist weathering, and are often tightly bound together so that there are few pore spaces between the particles through which water can flow. The effect of texture, therefore, cannot always be distinguished from related effects of other factors.

### ROCK COMPOSITION

To be precise, it is not the rocks or soils that undergo weathering, but the mineral grains of which they are composed. The chemical composition of these minerals, therefore, and the proportions of the different minerals in the material, influence the overall nature of weathering processes.

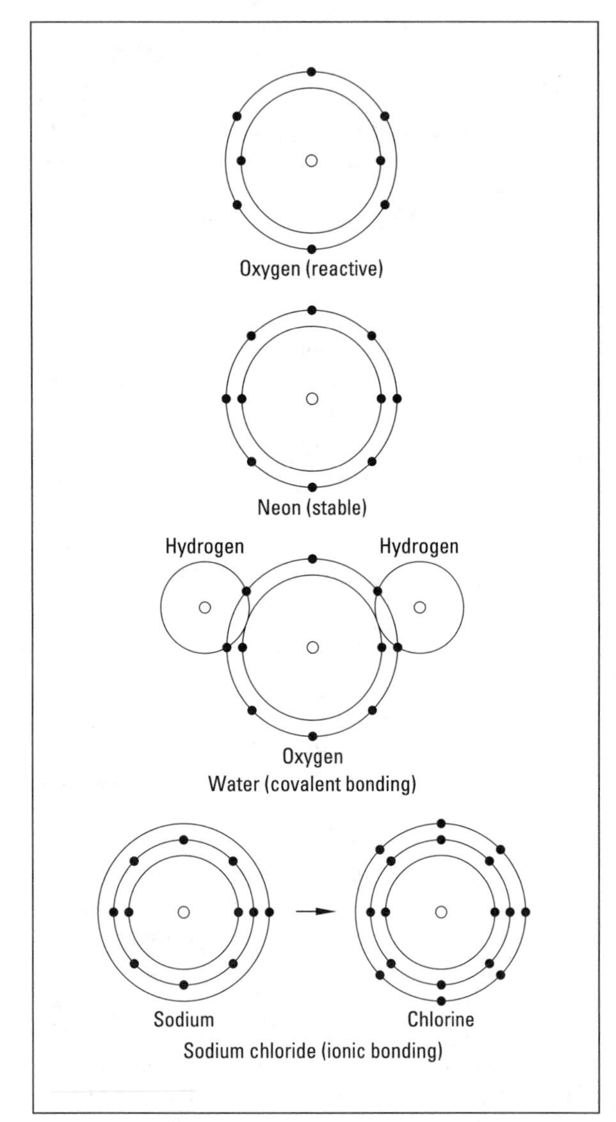


FIGURE 18.4 ▼ Structure of the atom. Atoms consist of a nucleus surrounded by a series of shells containing electrons. The first shell contains up to two electrons; outer shells contain up to eight electrons. The atom is most stable when the outermost shell is full. Thus, the oxygen atom, with only six electrons in the outer shell, is reactive, while neon is stable. Bonding between atoms occurs either by sharing electrons (covalent bonding, as in water) or by transferring electrons (ionic bonding, as in sodium chloride)

To some extent, the principles of weathering of minerals are similar to those we have mentioned previously. The presence of structural weaknesses within the minerals, for example, is likely to encourage weathering either by allowing breakage to occur more easily, or by allowing the penetration of water and reactants into the mineral grain. Beyond this, however, it is the detailed chemistry of the mineral, and the arrangement of the molecules and the structure of the atoms of which it is composed, that are critical to weathering.

Considered very simply, the constituent elements of a mineral can be seen to occur as atoms. Each atom is composed of a nucleus, consisting of protons and neutrons. Around the nucleus there circulates one or more electrons, arranged in a number of 'shells' (Figure 18.4). It is the character of these outer shells that determines how active the element is, for it is through the interaction of the electron shells that elements react chemically. Thus, minerals composed of the more reactive elements are relatively unstable, but superimposed upon this is the effect of the structural arrangement of the atoms within the mineral grain. In general, these two factors determine the susceptibility of different minerals to weathering.

Several attempts have been made to define sequences of mineral stability. In detail, specialized series have to be defined because there are many different forms of weathering, and each mineral reacts rather differently according to the weathering environment. Nevertheless, a general picture of mineral stability can be attained (Table 18.1) and, as this implies, variations in the composition of the material play a major role in weathering processes. It is notable that this sequence is approximately the reverse of the sequence of crystallization seen in igneous rocks (Chapter 16).

## INTERACTIONS BETWEEN WEATHERING AGENTS AND MATERIALS

The processes of weathering depend upon the interaction between the agents of weathering and the materials with which they come into contact. Different processes of weathering therefore occur, depending upon these two factors. It is useful, however, to define two main types of weathering: mechanical and chemical. As we will see, these often interact closely, but they provide a basis for discussing the more detailed processes involved in rock mass, and, particularly where previous chemical weathering has weakened them, the mineral grains themselves may be shattered.

## Table 18.1

*Mineral weathering series: the relative resistance of common minerals to weathering*

| | *Heavy minerals* | *Light minerals* |
|---|---|---|
| Most resistant | Rutile | |
| ↑ | Zircon | Quartz |
| | Tourmaline | ↑ |
| | Garnet | |
| | Staurolite | Micas |
| | Kyanite | ↑ |
| | Amphiboles | |
| | Sphene | |
| | Pyroxenes | Feldspars |
| Least resistant | Olivine | |

weathering. Biological weathering, which has both mechanical and chemical effects, is often put forward as a third group.

# Mechanical weathering

The result of all forms of **mechanical weathering** (also known as physical weathering) is the fracturing of the rock. As we noted in the previous chapter when discussing faulting, rocks are able to withstand stresses to a certain extent, due to the bonds within and between their constituent particles. When the stresses exceed the strength of the rock, however, failure occurs.

Unlike the situation in faulting, failure in weathering reactions tends to occur somewhat irregularly within the rock mass, rather than along discrete and continuous failure planes. This is because most stresses exerted by weathering processes are more randomly and locally distributed within the material. In addition, these stresses are relatively small, and it is only at particularly weak points within the rock that failure occurs.

As we can imagine, the main weaknesses in most rocks are associated with joints and bedding planes, and it is along these that failure most often occurs. Nevertheless, stresses caused by mechanical weathering may also pry individual minerals from the

## FREEZE–THAW

The stresses applied to rocks during mechanical weathering derive from a number of sources. As we have seen, water is often a main agent of weathering, especially when it freezes. Freeze–thaw activity is thus a major process of mechanical weathering. Freezing is a desiccation process; it dries the rock locally, creating a moisture tension gradient from the frozen zone (high moisture tension) to the unfrozen zone (low tension). Water moves along this gradient toward the area of high tension, and is thus apparently attracted toward the growing body of ice. In this way, ice crystal growth is encouraged and, under suitable conditions of ample moisture supply, large crystals or lenses of ice may develop. The increase in volume of the water as it freezes may set up stresses of as much as $7 \times 10^7$ Pa on the surrounding materials.

In practice, these pressures are rarely reached, for freezing does not normally occur in a wholly confined space. The access the water uses to enter the crevice in the rock, for example, provides an escape for the growing ice body, and relieves some of the pressure that would otherwise develop. Nevertheless, ice growth is often at least partly confined, for freezing generally takes place from the surface of the crevice inward. Thus, the surface ice seals the crevice, and subsequent freezing of the remaining water may build up considerable pressures on the surrounding materials.

If freezing is a one-time event, the amount of disintegration is likely to be small, for most rocks are able to yield to such pressures to some extent without rupturing. Repeated freezing and thawing, however, increase the degree of rock disintegration. Each time the ice thaws, more water is able to enter the crevice, refilling it, and ensuring that during the next freezing cycle stresses will be greater. Because of this, it is generally the number of freeze–thaw cycles, rather than the intensity of freezing, that controls the degree of weathering. In addition, of course, disintegration is encouraged by an abundant supply of water, so that freeze–thaw processes tend to be most active in moist sites, such as those fed by lateral movements of groundwater, and in porous rocks which are able to absorb large quantities of water.

### THERMAL WEATHERING

Desert areas are often characterized by an abundance of shattered blocks, and by rounded boulders from which skins of material appear to be flaking (Figure 18.5). It has been hypothesized that stresses may be set up by the expansion of rocks upon heating. This process is believed to operate mainly in desert environments where great extremes in air temperature occur, and where there is little or no vegetation to protect the surface. Under these conditions, diurnal surface temperature ranges of 100C° are not uncommon. During the daytime, the outer skin of exposed rock may be heated intensely, while the interior of the rock remains cool. The differential expansion that results may cause flakes of rock to crack away from the surface, producing the effect of onion-weathering or **exfoliation** (Figure 18.5). Repeated heating and cooling intensifies this process. Differential heating and expansion may also occur on a smaller scale, for differences in the thermal properties of dark and light materials mean that mafic minerals may be heated preferentially with the result that stresses are set up between adjacent grains. In this case, individual grains may be pried from the rock.

The effect of these processes has not been confirmed. In possibly the most thorough experimental investigation of the phenomena, Griggs (1936) subjected granite to 89 400 heating and cooling cycles (equivalent to about 245 years of thermal weathering)—with no detectable effect! What is apparent, however, is that the process is more effective in the presence of small amounts of water. The formation of dew on the rock surfaces during overnight cooling is responsible for the efficacy of the process.

## Biological weathering

Plant roots have also been held responsible for mechanical weathering. Roots develop down joints and along bedding planes, and, as the lifting of paving stones and asphalt demonstrates, quite large stresses are exerted on the surrounding rock. If the confining pressure of the rock is great, then this force is limited, but if the root can grow and continue to exert its pressure, the effect is very dramatic. If rock is close to the surface and joints are available, roots will penetrate and be a very dynamic force in the mechanical weathering process. Once the root has opened up passages and lifted the rocks, it allows weathering agents access to areas beneath. In large areas of North America, particularly in the Shield areas of the northeast with very little glacial drift covering the bedrock, biological activity is a major part of weathering (Figure 18.6). It must be remembered that much of eastern Canada and the United States was once forested, and therefore this process was probably more active previously.

Many other biological agents are important in enhancing weathering. For example, in marine areas, molluscs bore into rocks, weakening the rock and also allowing weathering agents, particularly water, to penetrate. Burrowing animals also penetrate directly into the rock or loosen the soil to allow water to penetrate more easily.

A

B

FIGURE 18.5 ▼ Exfoliation in sandstones at Cassidy Arch, Utah. (A) General view of the landscape; (B) detail of exfoliated block (photos: P.G. Johnson)

FIGURE 18.6 ▼ Tree roots penetrating joints in limestone at the Parliament Buildings in Ottawa promote the physical weathering of the rock (photo: P.G. Johnson)

## SALT WEATHERING

As we mentioned earlier, the distinction between mechanical and chemical weathering is not always clear-cut. This is seen in particular in the case of salt weathering (also known as salt crystallization), a process of considerable importance in many desert areas. In part, this is a purely mechanical process in that, where saline waters are evaporating, crystals of salt may grow in crevices within the rock. As with the development of ice crystals, these tend to grow from the surface downward so that crystallization occurs in partly confined conditions. Again, stresses exerted on the surrounding materials are maximized by repeated saturation and drying of the rock, for this allows continual increments of salt to be added, permitting the growth of large crystals which fill the available space. The effectiveness of the process is also dependent upon the degree of supersaturation of the water, for this controls the ability of the crystals to continue growing against the confining pressure of the surrounding materials.

Experimental studies have shown that the rate of rock disintegration by salt weathering varies markedly between rocks. As with freeze–thaw effects, salt weathering is most effective in porous rocks such as chalk and open-fabric limestones or sandstones. More massive rocks such as diorite and granite are almost wholly resistant to the process (Figure 18.7). It has also been found, however, that rates of disintegration vary when rock specimens are treated with different types of salt solution: sodium sulphate, for example, is far more effective than sodium chloride. In part, at least, this indicates that salt weathering is not such a simple process as is often assumed, and involves chemical reactions as well as mechanical stresses. Sulphate solutions seem to accelerate weathering because the sulphur initiates complex chemical reactions (e.g., hydration, described later) which weaken the rock and make it more susceptible to the effects of crystal growth. Similarly, failure of buildings in desert areas is often the result of chemical reactions due to the inclusion of salt in the mortar.

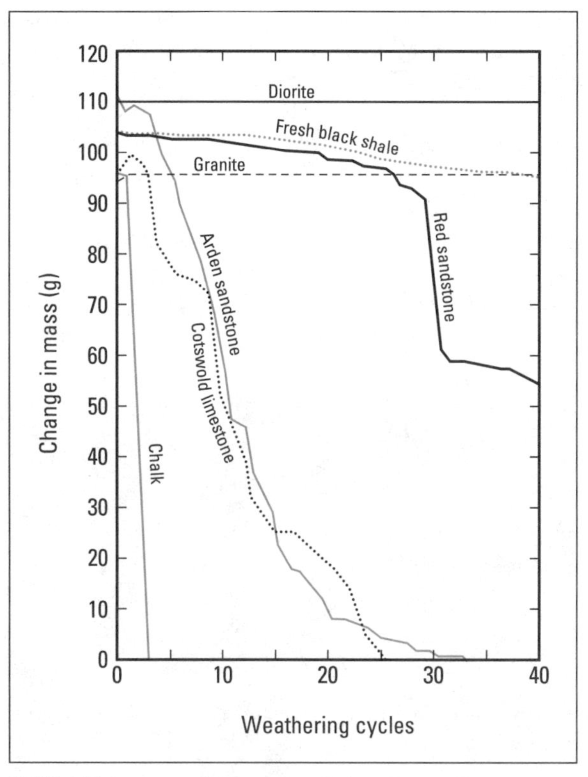


FIGURE 18.7 ▼ Breakdown of rocks by salt weathering. Each of the rock samples was subjected to up to forty salt-weathering cycles in a solution of sodium sulphate (from Goudie *et al*, 1970)

### WETTING AND DRYING

Similar complexity occurs in the processes of wetting and drying. This involves a number of different effects; one is the slaking action of water. Simply, as the soil or a soft, porous rock is wetted, the water traps air bubbles within the pore spaces. When wetting is rapid, the air cannot diffuse outward but ultimately explodes, causing minute ruptures in the material. Chemical reactions (such as hydration and hydrolysis, described later) also occur, however, so that the process is not simply one of mechanical weathering.

### PRESSURE-RELEASE

Finally, we might mention the process of pressure-release (or **dilatation** as it is sometimes called). In many ways, this has been an underrated process, though it tends to occur on a larger scale than the other processes we have discussed. It results from the unloading of rocks during exhumation or during deglaciation. As the overlying materials are removed, the consolidating pressures are released, and the rock tends to rebound. The stresses created open up joints and bedding planes. Only rarely does the process result in direct disintegration, but it seems to be a widespread means by which rocks are weakened and made more susceptible to other weathering processes. Many areas of the Canadian Shield have areas of pressure release weathering (Figure 18.8). As we will see later, for example, it is believed to be a process involved in the formation of tors.

FIGURE 18.8 ▼ Pressure release fractures on the Canadian Shield, Lake Temiscaming, Quebec (photo: P.G. Johnson)

## Chemical weathering

Chemical reactions within the soil or rock normally involve water. Three general types of reaction can be identified:

- those in which ions from the rock mineral are removed and enter the water
- those in which ions from the water enter and combine with the mineral
- those in which an exchange of ions takes place

*Solution* The first process is often referred to rather loosely as **solution**. In reality, it involves a number of slightly different reactions. Some substances are soluble in pure water (salt is an example) and dissolve by the migration of ions from the mineral into a dispersed form in the water (simple dissolution). Other substances enter into solution largely through the action of reactants dissolved within the water. In one way or another they are combined with ions or compounds in the water, and essentially can be considered to leave the mineral to take part in this combination (Figure 18.9). One example of this is the process known as **chelation**. In this, ions leave the mineral to combine with organic compounds within the water.

*Hydrolysis and hydration* **Hydrolysis** is possibly one of the most important chemical reactions involved in weathering. It occurs when free hydrogen or hydroxyl ions in the water enter into the mineral structure and create a new compound. The process operates to some extent in pure water, for this dissociates to yield $H^+$ and $OH^-$ ions, but the presence of carbon dioxide greatly accelerates the reaction by forming a weak carbonic acid:

$$\underset{\text{carbon dioxide}}{CO_2} + \underset{\text{water}}{H_2O} \rightarrow \underset{\text{carbonic acid}}{H_2CO_3}$$

This immediately dissociates to form hydrogen carbonate (once referred to as bicarbonate) and hydrogen ions:

$$\underset{\text{carbonic acid}}{H_2CO_3} \rightarrow \underset{\text{hydrogen carbonate}}{HCO_3^-} + \underset{\text{hydrogen ion}}{H^+}$$

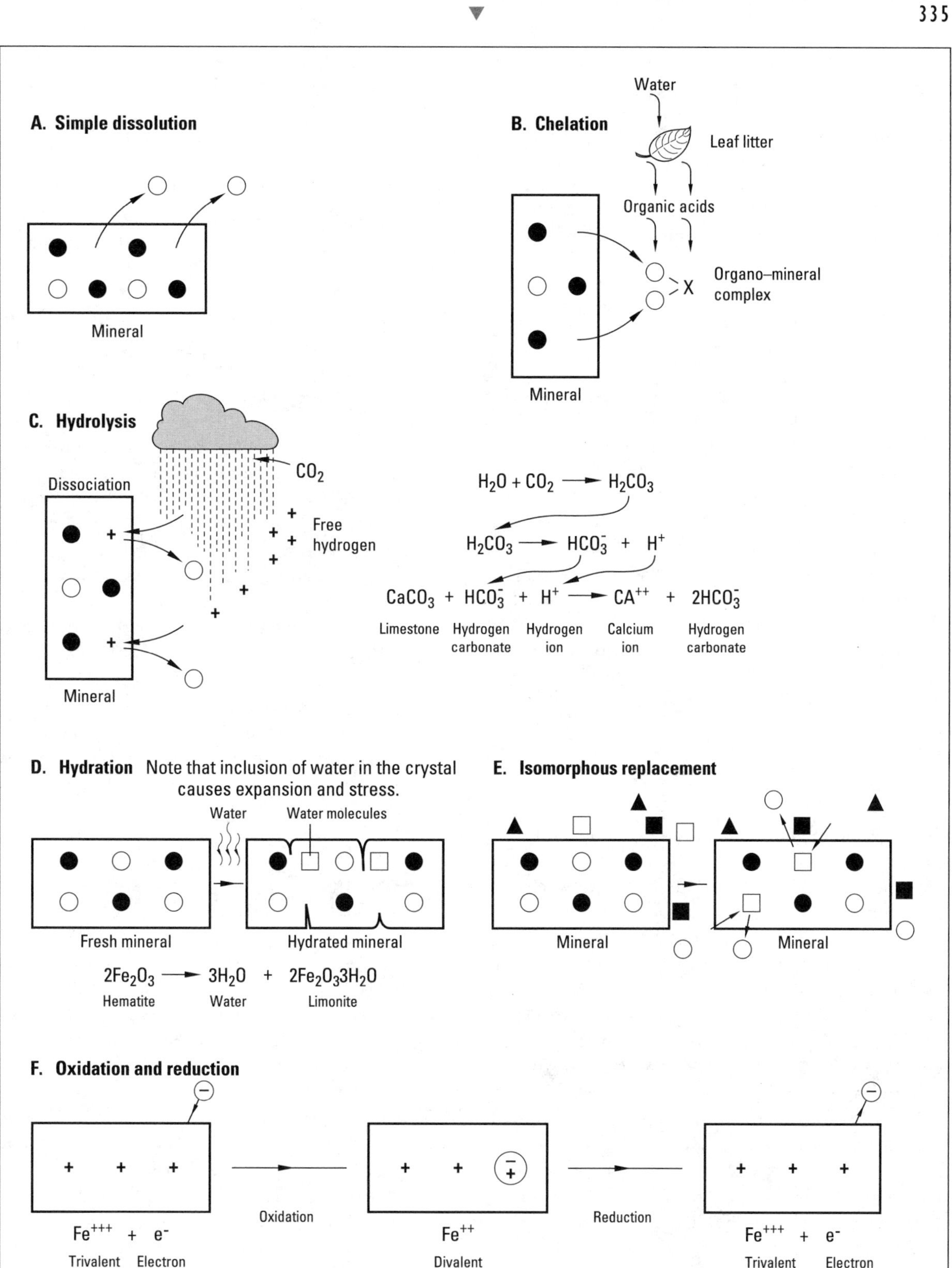


FIGURE 18.9 ▼ Processes of chemical weathering

These ions, in turn, take part in mineral weathering. The weathering of limestone by this process (sometimes referred to as carbonation) is illustrated in Figure 18.9. A similar process also occurs in the weathering of orthoclase feldspar to kaolinite:

$$2KAlSi_3O_8 + 2H_2O + CO_2 \rightarrow Al_2Si_2O_5(OH)_4 + K_2CO_3 + 4SiO_2$$

orthoclase feldspar + water + carbon dioxide → kaolinite + potassium carbonate + silica

The potassium carbonate produced in the process is removed in drainage waters.

The carbon dioxide necessary to start this process comes either from the free atmosphere or, in soils, from respiration by plants and soil organisms. As a result, hydrolysis is favoured by biological processes, and is most active in or just beneath the root zone, where carbon dioxide concentrations are high.

**Hydration** is a related process in which not just the hydrogen or hydroxyl ion, but the whole water molecule combines with the mineral. It is exemplified by the conversion of hematite to limonite (Figure 18.9), and, as this shows, involves the development of considerable physical stresses due to expansion of the mineral.

*Isomorphous replacement* Ion exchange processes, in which ions are lost from the surface of the mineral and are replaced by others from the water, are common in mainly chemical reactions, but they are readily reversible and cannot be considered true weathering reactions. On the other hand, **isomorphous replacement** of ions within minerals may occur, and this process is less readily reversed and fundamentally alters the composition of the mineral. It is a basic process in clay formation. In general terms, it operates through the exchange of ions held within the structure of silicate minerals with similar-size ions from the water. The control upon the process is the size relationship of the ions involved. [Ionic size is normally measured in nanometres (1 nm = 0.000001 mm).]

Silica, for example, with a diameter of 0.39 nm, may be replaced by aluminum, which is slightly larger, 0.57 nm. This, in turn, may be substituted by magnesium (0.78 nm). In the process, the chemical composition of the mineral changes, but, in addition, the inclusion of ions of somewhat different size may result in stresses that rupture the mineral. Physical breakdown as well as chemical alteration therefore occurs.

*Oxidation and reduction* These are two processes of widespread significance. **Oxidation** is a process by which compounds lose an electron (negatively charged particles). Under conditions of high oxygen availability, these electrons become attached to oxygen within the air, and there is a tendency for a progressive release of electrons from compounds within the minerals. Iron, for example, may release electrons in this way; in the process it is said to be oxidized.

The reverse process is known as **reduction**. Here, electrons, derived in many cases from the respiration of plants, become attached to compounds such as iron in the soil. The compound is thus reduced (Figure 18.9).

We will examine the full implications of oxidation and reduction, together with many other weathering processes, in Chapter 27; the point to note here is that oxidation and reduction are closely related (they often occur simultaneously within the weathering material), and they represent processes by which the composition of minerals is altered. On the whole, the compounds produced by oxidation are relatively stable; those produced by reduction are less stable and may be further altered by, for example, solutional processes.

## Products of weathering

The processes of weathering result in the loss of certain compounds from the material, and the alteration of others to new forms; certain more stable compounds remain in their original state. The residues of weathering consist of the altered and unaltered materials left by the processes.

These residues take a variety of forms. Among the most common unaltered residues are the more resistant minerals, particularly quartz. It is partly because of the stability of quartz in the face of many weathering processes that it dominates many detrital sediments. It derives this stability from its atomic and mineral structure.

The other more stable minerals may also escape weathering under all but the most intense conditions. They tend, however, to be less abundant within the environment anyway, and although enrichment of weathered residues with minerals such as garnet, tourmaline, rutile, and zircon may occur, these are rarely more than accessory minerals within the residues.

Many of the residues of weathering are alteration products of the processes that were active. Even quartz, under extreme weathering conditions, may be dissolved and reprecipitated in the form of non-crystalline opal or chalcedony. The less resistant silicate minerals tend to break down more readily, especially under the effect of chemical weathering, to produce **clay minerals** (Figure 18.10). The nature of the clay mineral that forms depends upon the intensity and duration of weathering, as well as the nature of the original mineral.

The nature of the residual products also depends upon the nature of the weathering process, and for this reason broad generalizations are difficult. Embleton and Thornes (1979) proposed a very detailed diagram of weathering processes and products (Figure 18.11). Some of the terms Embleton and Thornes used have not been introduced here, but we can follow the simple concepts we have used through the flow diagram. In general, it is clear that mechanical weathering tends to produce chemically unaltered residues, but results in changes in the size and shape of the materials; chemical weathering influences the composition of the materials. In reality, of course, a combination of both chemical and mechanical weathering occurs, and the residues are characteristically a mixture of shattered and rotted material.

The products of weathering that are dissolved within water, or released as gases to the atmosphere, are lost more or less irretrievably from the materials. They thus represent outputs from the weathering system. They consist, in most cases, of the more soluble and least stable components of the original material. Most of these products are carried away as dissolved materials within the water, and thereby

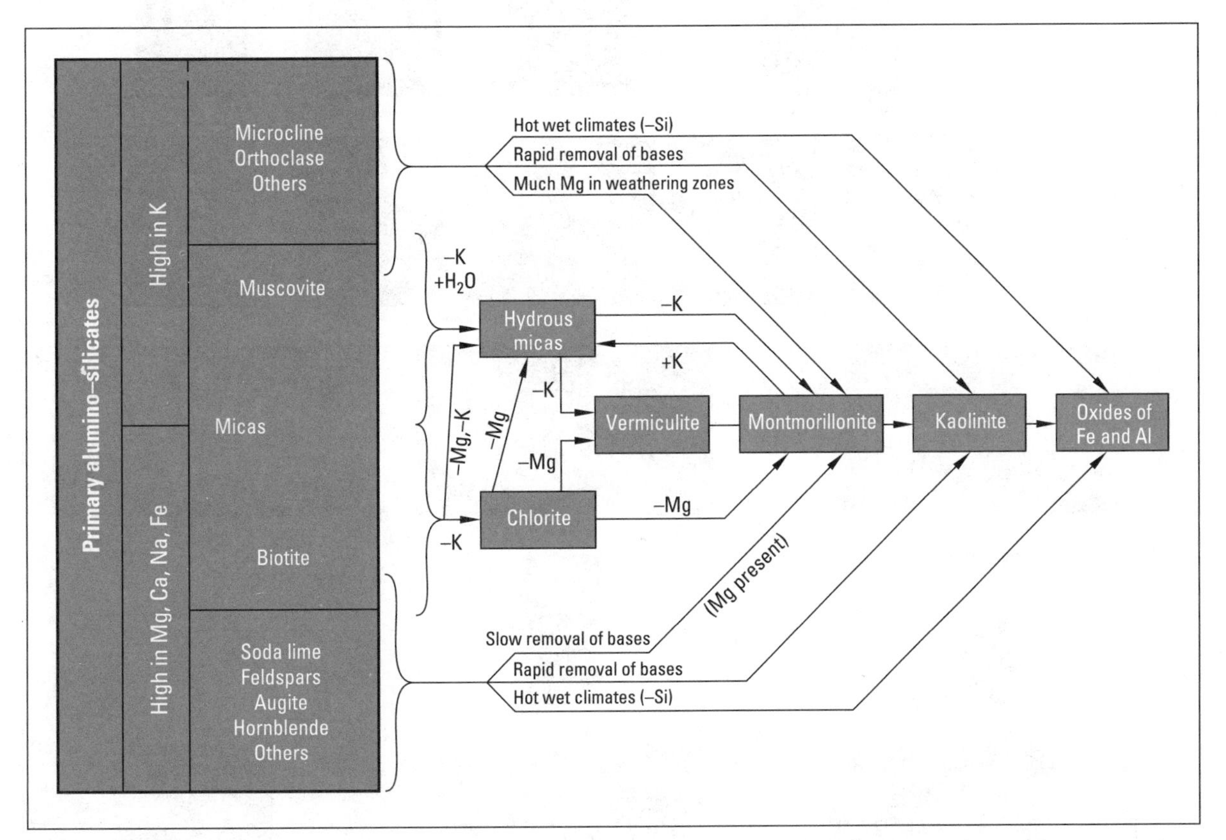


FIGURE 18.10 ▼ Processes of clay mineral formation. Under relatively mild or brief weathering conditions, the primary alumino-silicate minerals tend to break down to hydrous micas or chlorite. Under more intense weathering, vermiculite and montmorillonite tend to form, while kaolinite and oxides of iron and aluminum are generally produced by intense or prolonged weathering. In each case, clay formation involves the loss of soluble elements such as potassium, sodium, calcium, and magnesium (from Brady, 1973)

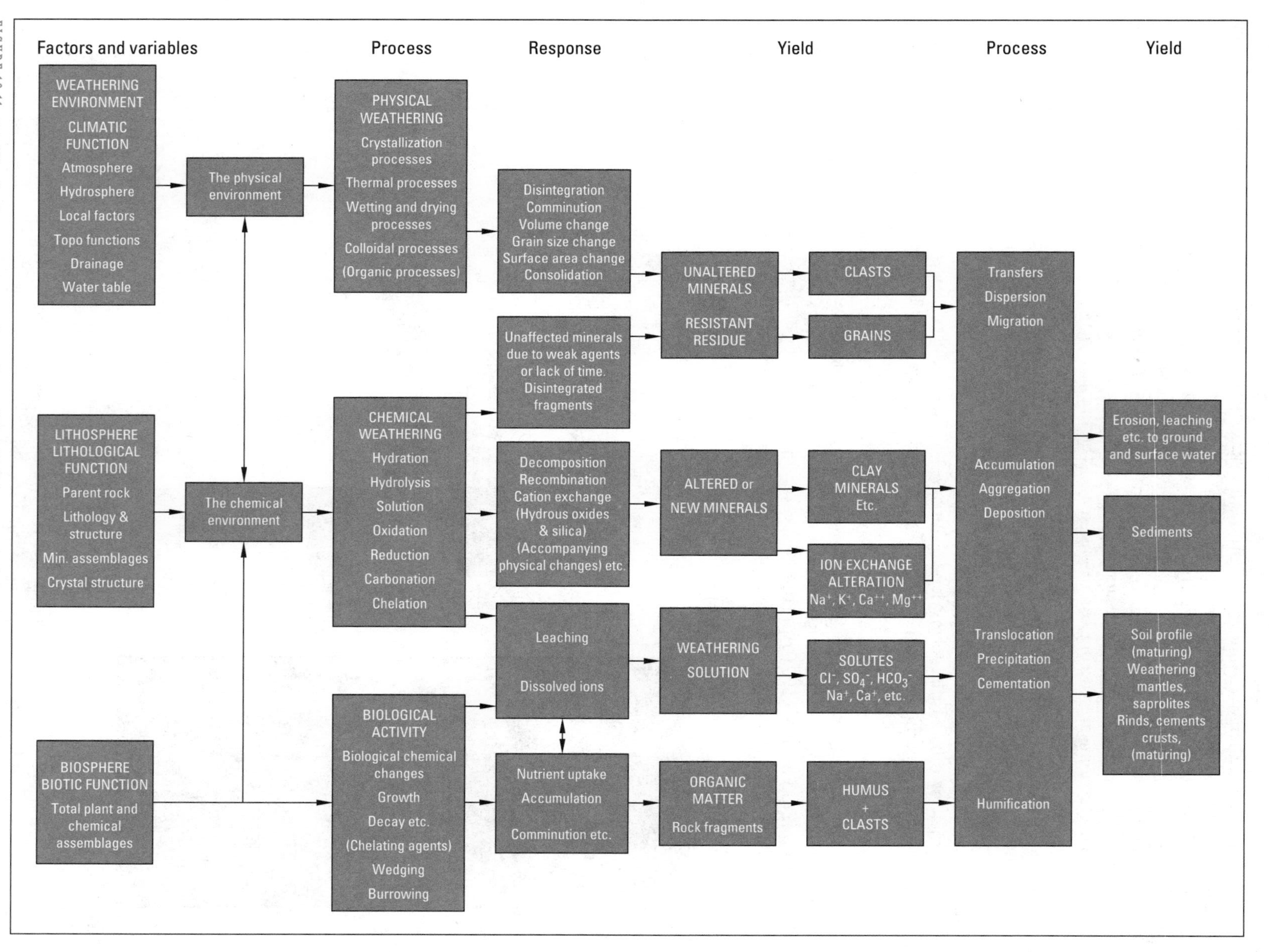


FIGURE 18.11 ▼ Weathering processes and products (from Embleton and Thornes, 1979)

enter streams, lakes, and groundwaters. Here they may act as a pollutant, as far as humans are concerned; high levels of lead and cadmium, for example, may be washed into streams from the weathering of ore deposits.

More importantly, in most cases, these weathering products become inputs to other parts of the environment. They may be taken up by plants as nutrients; they may become concentrated and eventually accumulate as precipitates in lakes or in the sea, or in cave systems beneath the ground.

# Landforms of weathering

Weathering also alters the morphology, or shape, of the zone within which it operates, and thereby produces weathering landforms. In truth, many of these landforms are only expressed at the surface through the subsequent effects of erosion. A major factor is the rate of removal of the material created by weathering. Erosion and transport depend on climatic conditions and topography. For example, in areas of high relief and high precipitation erosion and transport are very effective; in areas of freeze–thaw activity on moderate slopes, transport occurs slowly; in areas of low precipitation, the effectiveness of wind action must be considered. The various processes will be examined in subsequent chapters.

Many areas of the world present a variety of forms because of changing climate and geologic activity. For example, the icecaps that moved across large areas of northern North America during the Pleistocene removed previously weathered material and left new unweathered surfaces. The upland regions of New England, or the Laurentians, represent such areas. Because world systems are dynamic, it is very unusual to find a region that has experienced unchanging conditions of geologic activity and climate for extended periods of time. The result is that the landforms of a region are more likely to be polygenetic, that is, having several causes of origin. Similarly, other regions can show the production of similar forms by different processes; this is known as **equifinality**. Most landforms are a result of two dominant inputs, the geology from below and the climate from above. At one time, geomorphologists placed great emphasis on the study of climogenetic regions (where the emphasis was on the climate producing distinctive morphological assemblages) or morphogenetic regions (where the similarities of form are the dominant elements of classification).

Some examples of landforms and landscapes that are dominated by weathering processes will illustrate the discussion. The story of a professor showing slides of desert landforms in Australia while discussing African landforms demonstrates the general similarities, but closer examination reveals many differences. Similar morphology does not necessarily mean similar processes.

## TORS

**Tors** are residual blocks of bedrock that stand above the general surface of the surrounding terrain. They usually occur on the summits of hills. Most theories of origin of tors involve weathering processes plus erosion processes, but there are two main ideas; one proposes that the rock is differentially weathered by tropical climatic conditions and the weathered material is removed by erosion processes, the second proposes that the weathering and erosion occurs under periglacial climatic conditions. These can be illustrated by the tors of the Ruby Range in Yukon (Figure 18.12) and the tors of Dartmoor in southern England (Figure 18.13). It was argued that the deep weathering of the Dartmoor area occurred when the continental plate was closer to the equator, and continental drift has subsequently moved the plate to its present location. With the periglacial forms, the rounding of the rock form occurs with weathering after exposure of the rock mass. With the tropical forms, the rounded blocks are produced by deep weathering processes.

FIGURE 18.12 ▼ Tor produced under periglacial climate conditions in the Ruby Range, southwest Yukon (photo: P.G. Johnson)

FIGURE 18.13 ▼ Hay Tor on Dartmoor, England, an area not glaciated during the Pleistocene. Tors here were formed by deep weathering of the granite (photo: P.G. Johnson)

This is, therefore, an excellent example of the role of weathering processes, and an illustration of the concept of equifinality. It also indicates, however, that the combination of weathering processes with erosion and transport of the detritus produced by weathering are inextricably linked in the production of landforms and landscapes.

## LIMESTONE LANDFORMS

Landforms produced on and in limestone are excellent illustrations of the role of weathering. Chemical weathering is the principal process in limestone bedrock regions, and it results in distinctive landform assemblages called **karst**. The name derives from the Karst area of Yugoslavia where these features are particularly well developed. In North America, the Green River of Kentucky cuts down through the limestone bedrock in a region that has excellent examples of karst landforms, including the famous Mammoth Caves. Other important regions of karst include the Yucatan Peninsula and Puerto Rico, where the warm weather and heavier rainfalls, as on the Adriatic coast of Yugoslavia, increase the effectiveness of chemical weathering to create classic karst features. Karst is also produced in cold environments, but the processes of limestone dissolution are much slower in northern Canada. The caves of the northern Yukon are excellent examples of cold region karst, and small karst landforms occur in all limestone regions of Canada, for example, in the Ottawa Valley region and in the Interlake region of Manitoba. The Manitoba karst depressions are referred to as snake pits by local people, because they provide a constant-temperature environment for snakes, which emerge in thousands in the spring.

Rainwater is a weak carbonic acid, created by carbon dioxide absorbed into water droplets in the atmosphere. The water may also acquire humic acids from vegetation and soil systems. The result is an active solution that reacts with carbonate rocks such as limestone and marble, dissolving the calcium carbonate. The penetration of joints and bedding planes produces a variety of features such as **clints** and **grikes** (Figure 18.14) that characterize exposed limestone surfaces called **karst pavements** (Figure 18.15). On the surface between the joints, various **solution basins** and **runnels** also develop. The major action is beneath the surface, where the solution of limestone results in extensive cave networks. The rainwater infiltrates down to the water table, with the result that surface streams are unusual in karst regions. Through solution of rock in the larger joints, caverns develop, especially in the saturated zone at the top of the water table, where flow tends to be more rapid. If the water table is lowered, either by drier climate or by tectonic uplift of the whole region, then caves formed at that level will become dry. These become the

FIGURE 18.14 ▼ Weathering of a limestone surface at Peyto Glacier, Alberta. Solution activity is concentrated along joints and bedding planes in the rock, producing clints and grikes (photo: P.G. Johnson)

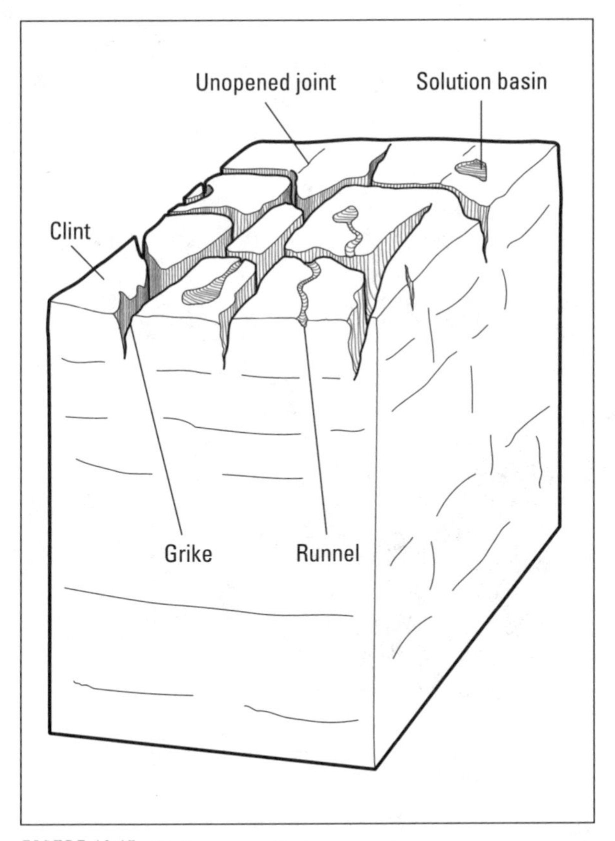


FIGURE 18.15 ▼ Features of a karst pavement

caves that are so intriguing to visitors, such as the Carlsbad Caverns of New Mexico, the Mammoth Caves of Kentucky, or the caverns of the Black Hills of South Dakota. Cave systems can become perched above the water table because of regional uplift or a drop in the water table due to erosion of river systems. The extent of this change can be illustrated by the gorge of the Nahanni River in northwest Canada. Here, caves are found in the sides of the gorge, hundreds of metres above the river.

# Weathering and climate

Between the polar extremes and the tropics, the deserts and humid oceanic regions of the world, there are great variations in climatic conditions. Since water and heat play such vital roles in weathering, we might expect that these variations are reflected in the pattern of weathering processes and associated landforms across the globe.

To some extent this is the case. At a broad scale, we can indeed define gradients in the rates of effectiveness of individual weathering processes, related to changes in factors such as the amount of annual rainfall or temperature conditions. In Yukon, for example, the probability of freeze–thaw cycles shows major differences from north to south in different months of the year (Figure 18.16). This probability is based on mean monthly maximum and minimum temperatures, and some cycles may be expected in most months. Similarly, it has been argued that the significance of chemical weathering varies geographically in relation to rainfall and mean annual temperature (Figure 18.17). On the basis of such relationships, attempts have been made to identify weathering regions. The most noted is the scheme devised by Peltier, illustrated in Figure 18.18 (page 344). As might be anticipated, this shows that mechanical weathering is most effective in cold, moist regions, while chemical weathering is at its most intense in humid tropical regions. All forms of weathering tend to decline with increasing aridity.

Peltier's classification of weathering regions has often been quoted, and it is possibly valid in broad terms. However, it is at best a highly intuitive and generalized description of the distribution of weathering processes. At a smaller scale, it has limited significance, for other factors—including topography, geology, vegetation cover, soil conditions, and microclimate—all affect rates of weathering. A transect across a limestone coast, for example, will reveal marked variations in the intensity and character of weathering within distances of a few metres, biological weathering being active in the lower intertidal zone, wetting and drying and salt weathering being dominant near the tidal limit. Similarly, considerable differences in weathering processes may exist between slopes of different aspect. The western cordillera of North America shows considerable variation in frost action with exposure, and in chemical weathering with the precipitation gradients across the ranges. Finally, we also have to remember that the climate has itself changed fundamentally during the Quaternary, and many of the effects of weathering that we see in the landscape around us are legacies from previous episodes, when weathering processes may have been different. Thus, relationships between present-day climatic conditions and landforms of weathering are often relatively weak.

January

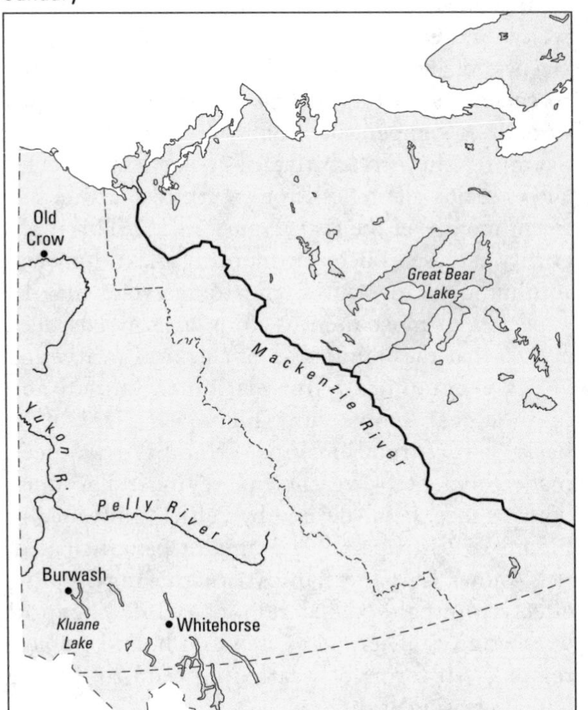


April

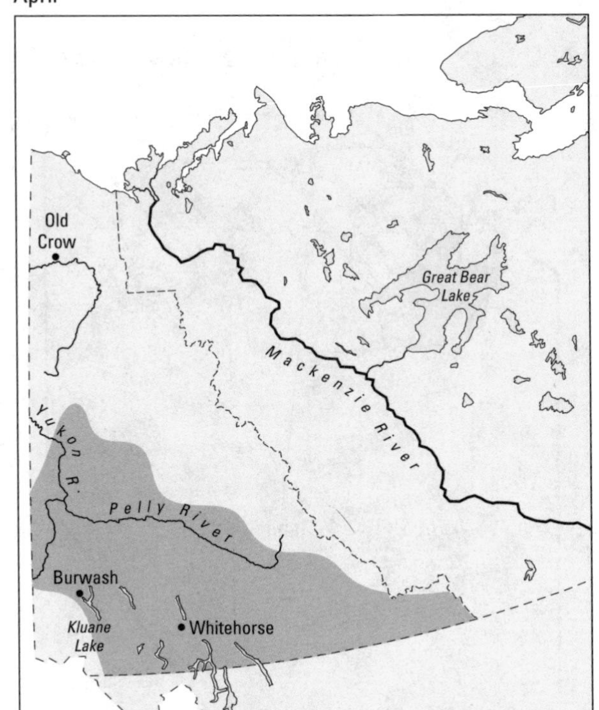


July

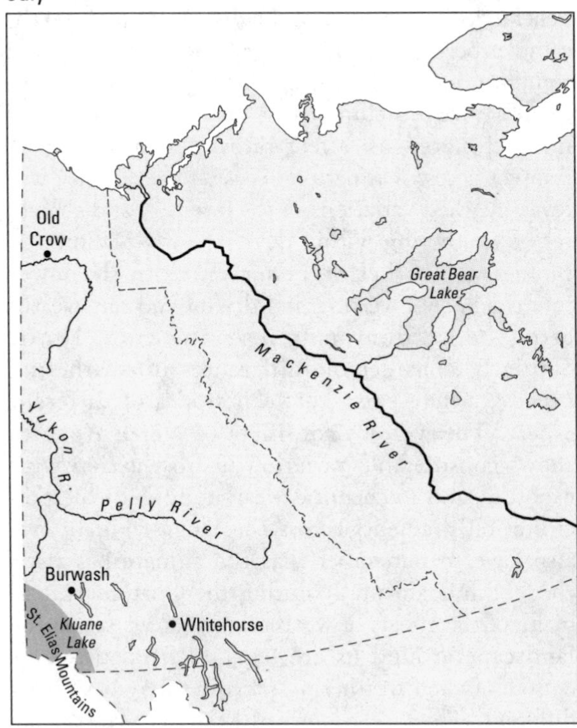


October

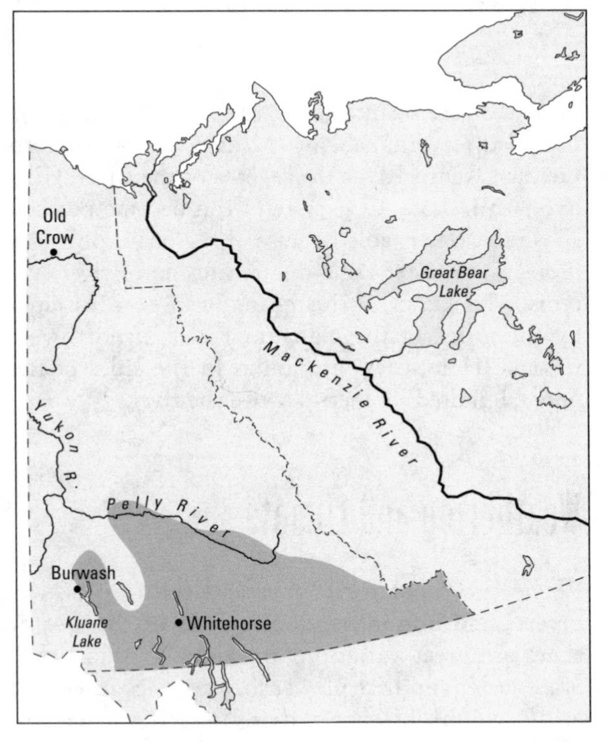


Area of high probability of freeze-thaw cycles

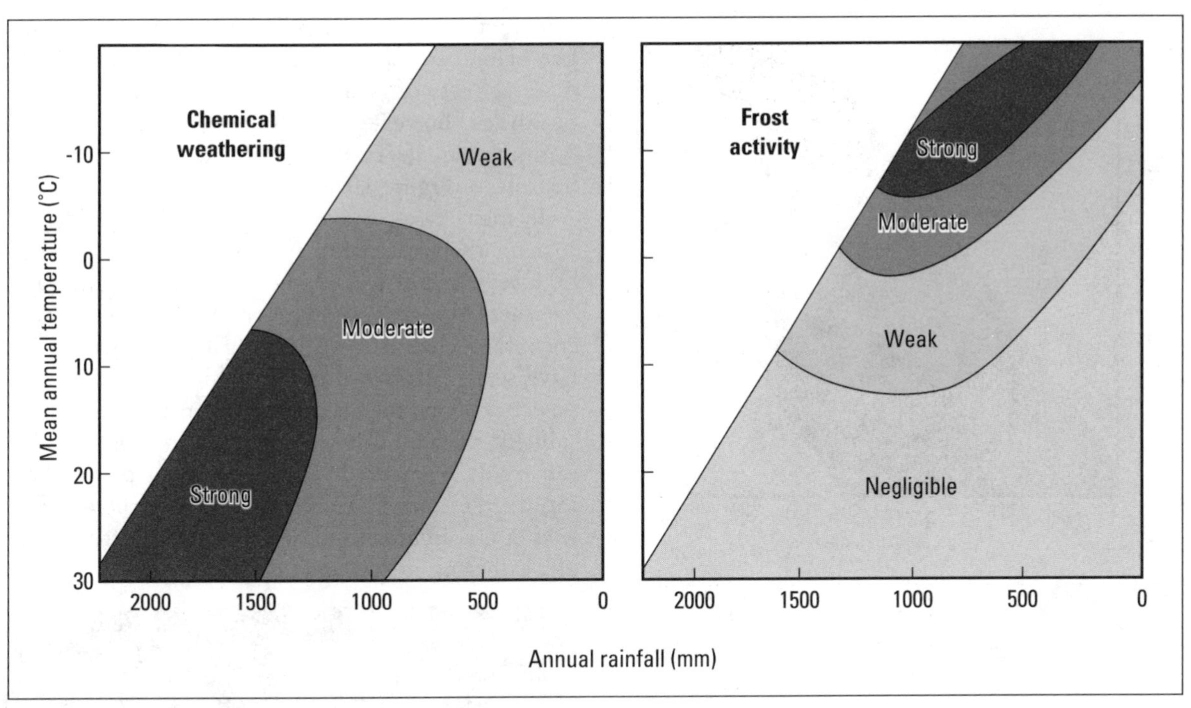


FIGURE 18.17 ▼ The intensity of weathering processes in relation to annual rainfall and mean annual temperature (after Peltier, 1950)

# Weathering and humans

The products and processes of weathering have considerable implications for us. Several of the materials produced by weathering have economic importance, among them bauxite. This is a product of hydrolysis of plagioclase feldspars, and is the main source of the world's aluminum. Kaolinite, produced by the rotting of granite, apparently by hot, briny waters rising through the rock (hydrothermal weathering), is used in the paper and ceramics industries (it is sometimes referred to as china clay).

Weathering is also a fundamental process in soil formation, and thus much of our use of land for food production is influenced by the products of weathering. In addition, the weathered mantle of the earth is important in relation to engineering, for it is necessary to construct buildings on a firm foundation. The effect of weathering is often to weaken the surface materials, and thus, particularly where the residues are deep, the creation of stable foundations may present a problem.

FIGURE 18.16 ▼ (Opposite) Probability of freeze–thaw cycles in Yukon in January, April, July, and October. This is based on the assumption that for the cycles to be effective in weathering, temperatures must rise above 2 or 3°C and fall below –1°C. The maps are based on monthly mean daily maximum and minimum temperatures. In January, the probability of a freeze–thaw cycle is very low throughout the Territory, because mean daily maximum temperatures are well below 0°C. In April, there is a high probability across most of the south. In July, there is only a high probability in the St. Elias Mountains, as mean daily minimum temperatures are above 0°C in the rest of the Territory. By October, the south again has a high probability. Note that the data are based on mean monthly temperatures, thus the map identifies only areas of high probability. A freeze–thaw cycle is possible at any time in the year. For example, temperatures above 0°C have been recorded in Whitehorse in January; the latest spring frost has been July 4 and the earliest fall frost was July 30 (Wahl *et al*, 1987). Burwash airport has recorded its last spring frost on July 15, and its earliest fall frost on July 16. Old Crow, in the north, has corresponding dates of July 1 and July 31

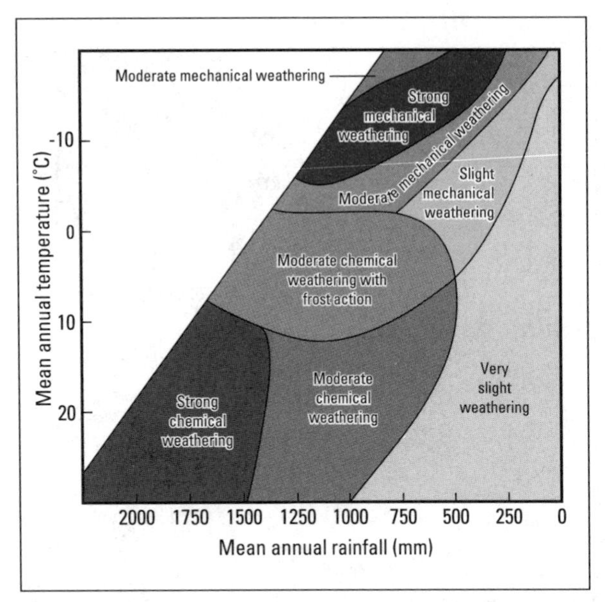


FIGURE 18.18 ▼ Weathering regions of the world in relation to mean annual temperature and rainfall (after Peltier, 1950)

Road cuts and embankments must also be planned with careful regard to the effects of weathering upon the stability of the materials. There are a number of occasions when failure of slopes has occurred because slopes cut through weathered rocks were too steep. A similar problem may arise if weathering continues after construction; this may, in extreme cases, undermine foundations sufficiently to cause collapse, or destabilize slopes.

Fortunately, weathering processes are generally so slow that it is rare for human constructions to suffer from their subsequent effects. It has been shown, for example, that weathering of clays in roadside embankments is unlikely to reduce the strength of the material enough to cause its collapse within the normal life of the feature. To put it another way, our constructions generally fall down or become obsolete before they are destroyed by weathering! More direct effects of weathering can be seen in many buildings, however, on a smaller scale; many old buildings are worn by the effects of solution and hydrolysis (Figure 18.19).

In many cases, too, we are responsible for accentuating the weathering process. Salt used to keep highways free of ice splashes against buildings and weathers the stonework. Atmospheric pollutants encourage chemical weathering by rainfall. As we have seen, sulphur dioxide, which is emitted by power stations for example, may produce a dilute solution of sulphuric acid in the atmosphere. This can result in markedly increased rates of weathering. In these ways, we are encouraging the destruction of our architectural heritage and affecting rates of weathering in the landscape as a whole.

FIGURE 18.19 ▼ Weathering of the sandstone on the Parliament Buildings, Ottawa. The rock has been discoloured by chemical alteration of the constituent minerals. Stone used at the corners and around windows is more resistant to mechanical weathering than the sandstone

# Erosion and deposition

## Principles of erosion

### THE NATURE OF EROSION

The evidence of erosion is all around us. The hills and valleys, the polished or carved surface of exposed rocks, the shape of the coastline—all are products of erosion by wind, water, ice, and debris acting under gravitational stresses. Not only do the landforms testify to the effects of processes caused by these agents in the past, but we can observe these processes at work today. If we look into a stream after a period of heavy rain, we see the murkiness of fine particles of soil and sediment being washed down-valley by the swollen waters. On a windy day we may be able to feel the dust and grit in the air. We can hear the crunch and rattle of pebbles on the beach as the ebb and flow of the sea sorts and moves the shingle. All around us, the weathered residues of the earth's surface are in motion.

The processes of erosion and transport caused by these agents are an important part in the sequence of landscape development. They redistribute the products of weathering; they wear down the features of earth-building; they reestablish an equilibrium between the shape of the landscape and the forces acting upon it. They also have immense significance for human life. One of the main impacts of erosion today is borne by mature soils as a result of agricultural systems and silviculture techniques. The removal and destruction of this soil, from which we derive most of our food, may be catastrophic; it can all but destroy societies, as we will discuss at the end of this chapter; it can cause untold suffering (see, for example, John Steinbeck's description of conditions in the 'Dust Bowl' in his novel *The Grapes of Wrath*).

## THE ENERGY OF EROSION

The disequilibrium between the earth's landscape and the forces acting at the surface arises largely from the potential energy imparted to the materials of the surface during mountain-building. Uplift of the rocks provides them with the potential for downward movement under the influence of gravity. Water and ice on these uplifted surfaces also has a potential energy; it tends to flow downhill, and so its potential energy is converted to kinetic energy. It is this kinetic energy—the energy of movement—that provides the water and ice with the ability to transport rock particles, and to erode.

In addition to potential energy derived from the position of the surface materials, there is another source of kinetic energy that acts upon these materials. This is derived from atmospheric processes. Rainwater has a kinetic energy; due to the lifting of water into the atmosphere through the processes of evaporation and convection, raindrops have potential energy that is expended in motion through the atmosphere. The impact of raindrops upon the ground is a vital process in erosion, as we saw in Chapter 11; the energy of impact is derived from the kinetic energy of rainfall. Similarly, the air itself has kinetic energy, which is expressed by the action of winds. These winds are able to pick up and transport rock particles; they act as agents of erosion.

## THE SEQUENCE OF EROSION

The energy available for erosion operates in a variety of ways. Much of the energy is used in carrying rock particles through the landscape in the process of transport. This, however, is only one step in the overall cycle of erosion. Before transport can begin, the particles have to be picked up; they have to be entrained. Before they can be entrained, in many cases, they have to be detached from the rock or soil masses. Thus erosion can be seen as three processes: **detachment**, **entrainment**, and **transport** (Figure 19.1). As we will discover, these three processes are often closely related, and a single particle moving through the landscape may undergo repeated detachment, entrainment, and transport. For these reasons, perhaps, it is not entirely valid to visualize them as a true sequence of events. For the sake of simplicity, it is useful to consider each step in the process separately and in sequence. We must remember that the material may be deposited and re-entrained a number of times in the transport process before final deposition in sedimentary basins.

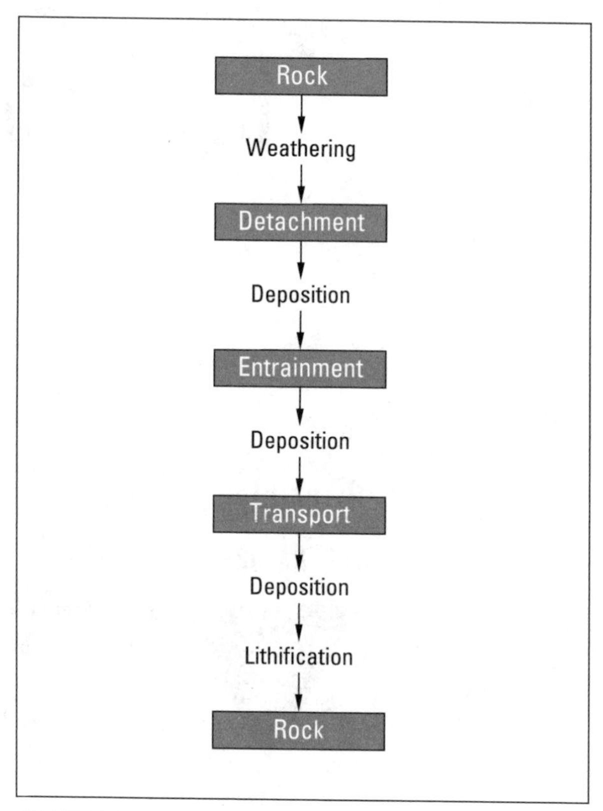


FIGURE 19.1 ▼ General model of the erosion cascade

# Detachment

## THE FORCES OF ATTACHMENT

The ability of wind, water, or ice to pick up and transport rock particles is controlled to a great extent by the size of those particles relative to the energy available. In general, smaller particles are eroded more readily than larger ones (although we will qualify this statement later), and it is only when the rocks are broken down into relatively small fragments that erosion proceeds. Ice does, however, have the capacity to erode much larger fragments. Thus, the detachment of materials is a vital step in the process of erosion.

Detachment involves the breaking of bonds that hold the particles together. The exact nature of these bonds depends upon the type of material we are dealing with. In unconsolidated sediments, the bonds are weak and may be a function of water content or packing of the particles. In many sedimentary rocks the bonds arise from the cementing effect of compounds such as iron oxides, silica, or calcium carbonate, which coat the grains and fill the interstices between them. In many igneous rocks, these bonds are derived from the intricate intergrowth of crystals, and from the fusion of the crystals during cooling. In both cases, the bonds are relatively strong. By contrast, materials that have been weathered (such as the soil) are weak. Chemical reactions and the mechanical effects of frost action, wetting and drying, and heating and cooling have broken or weakened the bonds between particles. Often all that holds the particles together is cohesion, provided by thin films of water, the weak cementing effects of organic compounds, calcium carbonate, and iron oxides, and the minute electrochemical forces that occur between clay particles. Many of these forces are impermanent, and are considerably reduced by wetting of the material, which dissolves some of the cements and ultimately starts to force the particles apart. Consequently, weathered materials are generally more susceptible than fresh rock to detachment, and it is partly for this reason that the soil is prone to erosion.

## THE FORCES OF DETACHMENT

From what we have said, it is clear that detachment of particles from rocks and soil masses is often accomplished by weathering. In addition, however, the agents of erosion exert their own forces of detachment upon surface rocks and soil. The exact nature of these forces depends upon the character of the erosive agent, but two main processes can be defined:

1. **Quarrying:** the detachment of particles by the action of the erosive agent itself.
2. **Abrasion:** the removal of particles by material carried by the agent of erosion (i.e., by the sediment load).

*Quarrying* Although it might logically be considered the major cause of detachment, quarrying is often relatively ineffectual. Even in glaciers, the ability of the ice itself to tear particles from the rock surface is limited, although it does operate in the process known as **plucking**. This occurs when ice freezes onto the surface, particularly in cracks and crevices, and plucks weakened fragments from the surface of the rock.

Flowing water is also able to carry out quarrying to a limited extent. Where the flow is confined, and flow velocities are high, a process known as **cavitation** may occur. In turbulent conditions in water, minute bubbles, which are vacuum, are formed; when these implode they exert considerable force upon the adjacent surface. Evidence for the operation of these processes is limited.

*Abrasion* Probably the most effective force for detachment is abrasion. This results from the impact and friction exerted by the material being transported by the agent of erosion. Abrasion occurs at the base and edge of glaciers; for example, rock fragments held in the ice scrape the surrounding rock. Similarly, boulders or sand grains bouncing along the stream detach fragments from the bed and banks. Wind also causes abrasion, for sand transported by the wind tends to detach particles as it bounces on the soil surface.

The strength of this force relates to the velocity of movement of the particles, their size (mass), and their concentration at the rock surface. The force seems to be particularly active in glaciers, where the particles may be very firmly trapped by the ice. However, its effect may be seen in a variety of other environments; witness, for example, the shattering caused by pebbles thrown by the sea against a cliff or beach—there is a record of a 60 kg boulder being thrown 30 m onto a beach in Oregon—or consider the force that larger boulders being dragged along the bed of a stream must impose.

*Raindrop impact* Possibly one of the most widespread and important processes of detachment is that caused by **raindrop impact**. This process has dramatically increased in importance due to the clearance of natural vegetation from large areas for agricultural purposes, removing the material that acts as a shock absorber between the raindrop and unconsolidated soil deposits. The force of a raindrop falling onto a soil or weathered rock surface is often sufficient to break the bonds that link the particles. The importance of this process is derived from two factors: first, the surprising strength of the force and, second, its frequency and extent of operation. The strength of the force (i.e., the kinetic energy) is related to the velocity of the raindrop when it reaches the ground (its terminal velocity) and the mass of the raindrop. This relationship is expressed by the equation

$$E = \frac{1}{2}V^2 \times M$$

where $E$ is the kinetic energy
$V$ is the terminal velocity
$M$ is the mass

The terminal velocity of large drops often approaches 8 or 9 metres per second, and in heavy storms droplets of as much as 5 mm in diameter with a mass of about 0.05 g may be generated.

The frequency and extent of raindrop impact is related to the character of the storm. Intense storms often affect only small areas, but the predominance of large droplets and the large number of droplets makes them particularly effective. Gentler rain that may affect much wider areas has little ability to detach particles since the droplets tend to be small and to have a lower terminal velocity ($2 - 3$ m $s^{-1}$).

# Entrainment

When a particle is picked up by the agent of erosion, the process is called entrainment. In many cases, entrainment and detachment are so closely related that they become one. The force needed to detach a particle from a rock or soil mass is considerably more than that necessary to cause entrainment; thus if conditions are sufficient to allow detachment, entrainment automatically follows. Not all erosion, however, involves the removal of small fragments or particles from larger rocks. Often it is unconsolidated sediments or soils that are being eroded. In these cases, entrainment may act independently; loose particles are lifted from the surface by the agents of erosion.

*Forces resisting entrainment* Several forces provide non-cohesive particles with resistance to entrainment. The most important in many cases is frictional force, which arises from the interaction between the particle and its surroundings. Frictional resistance only operates when the particle is being placed under stress, and the magnitude of the force increases as the applied stress increases. Thus, a particle at rest on a horizontal surface has no frictional strength. If stress is applied (for example, by a force trying to push the particle), frictional resistance will prevent the particle from moving, but as the applied stress increases, there will come a point at which frictional resistance reaches a maximum value. Additional forces applied to the particle will cause movement (Figure 19.2).

The maximum frictional strength of particles can be expressed as an angle, since it is clear that by tilting the surface on which the particle rests a point will be found at which spontaneous sliding occurs. This angle is referred to as the **angle of sliding friction**. It depends upon two factors: the weight of the particle—and thus the downward force it exerts on the surface—and the roughness of

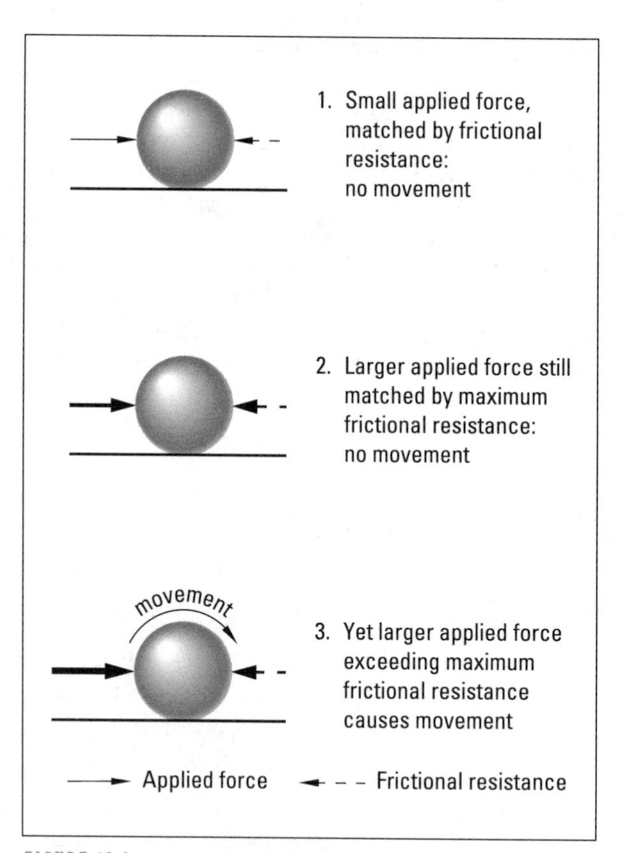


FIGURE 19.2 ▼ Initiation of particle movement

the contact between the particle and the surface. Thus, heavier particles have a greater frictional resistance to movement, while rougher particle–surface contacts also increase the frictional resistance. This latter factor becomes particularly important in many erosional processes, because particles do not normally lie on a perfectly flat surface, but tend to be lodged within depressions. In this case, the roughness of the particle–surface contact is expressed as the ratio of the radius of the depression in which the particle lies to the particle radius (Figure 19.3). We will see the significance of this, in particular, when we consider erosion in streams.

In addition to the effects of friction, particles derive a degree of resistance to entrainment from cohesion. This arises from the electrochemical forces that operate between the particle and surrounding particles or the surface on which it lies. Water may play a part in this cohesive strength, because thin veneers of water around and between particles may help to hold them together by surface tension effects (Figure 19.4).

FIGURE 19.3 ▼ The effect of surface roughness on frictional resistance. For particle A, the relative roughness is 0.7 and frictional resistance is high; for particle B, the relative roughness is 0.3 and frictional resistance is low

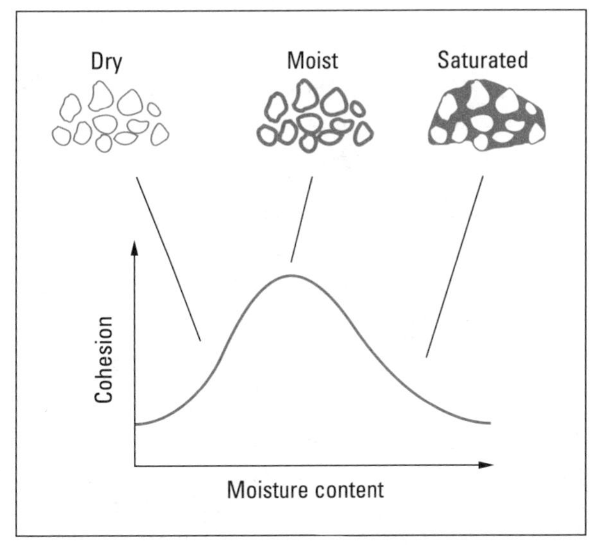


FIGURE 19.4 ▼ The effect of moisture content on cohesive strength

## FORCES OF ENTRAINMENT

Forces causing movement vary according to the nature of the eroding agent. In general, two main forces can be seen to operate: the gravitational force, which is a function of slope angle and particle weight, and the drag exerted by a medium (such as water, air, or ice) moving past the particle.

*The effect of gravity* A body on the earth is subject to a gravitational 'pull' which acts to hold it on the surface. The magnitude of this gravitational force is related to the weight of the body.

When a particle rests on a horizontal surface, this gravitational force acts downward into the earth; it can be seen, in fact, that it is involved in giving the particle a frictional resistance to movement, as we have just explained. If the surface is tilted, however, part of the gravitational force operates downslope. It can be demonstrated that the magnitude of the force acting in a downslope direction is proportional to the sine of the slope angle:

$$F = w \sin \varnothing$$

where $F$ is the downslope force
$w$ is the weight of the particle
$\varnothing$ is the slope angle

There comes a point, therefore, where the downslope force derived from this gravitational effect is sufficient to overcome the frictional resistance of the particle: at this point, the particle will slide. This angle is referred to as the **critical angle of sliding** ($\emptyset_{crit}$). It is clear that the critical angle of sliding equals the angle of sliding friction ($\mu_{crit}$):

$$\emptyset_{crit} = \mu_{crit}$$

In other words, particles under the effect of gravitational force will move when the downslope component of that force is just sufficient to overcome frictional resistance.

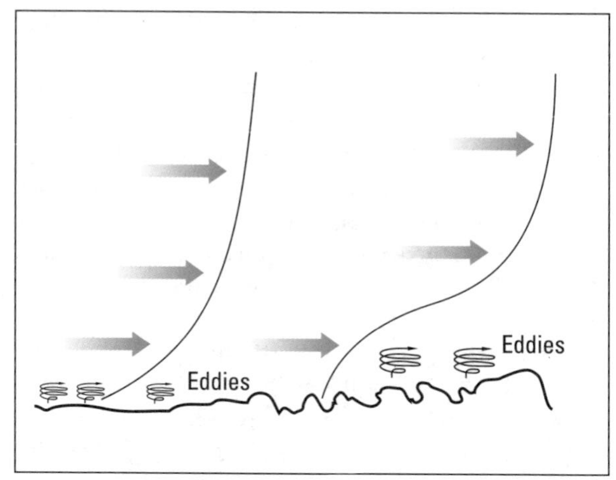


FIGURE 19.5 ▼ Velocity profiles and eddying above a smooth and rough bed. Although the rougher surface reduces velocity immediately above the bed, it may encourage entrainment because of turbulence that lifts particles into the flow, and because of the steep velocity gradient above the eddy layer

*Fluid forces* However, particles do not only move due to the effect of gravity. Flowing water or air play a major part in entrainment. These exert both a horizontal drag on the particle and a vertical lift. The horizontal force is derived from the 'push' of the agent against the particle. If this push is sufficient to overcome the frictional and cohesive resistance, then the particle rolls or slides forward. The strength of this horizontal force is related to flow velocity, and thus faster flowing air or water is able to move larger (or more resistant) particles. In addition, the strength of the fluid forces depends upon the density of the medium itself. Water, for example, is some 9000 times as dense as air and can therefore exert forces 9000 times as great.

The vertical lift is derived from turbulence and from the buoyancy given to the particles. Turbulence tends to occur due to irregularities in the surface which cause eddies within the flow (Figure 19.5). These eddies tend to carry the grain upward. Buoyancy occurs because the density of the fluid is, to some extent, able to reduce the weight of the grain. Thus buoyancy acts to diminish the resistance of the particle by apparently giving it a degree of lift. If it were possible slowly to increase the density of the fluid, there would, of course, come a point at which the particle would float. Once the particle is lifted from the surface, its frictional and cohesive resistance derived from contact with the bed declines to zero; only the gravitational resistance remains.

In the case of transport by fluids, there exists a critical condition, therefore, at which entrainment starts. In reality, this depends upon many factors, but a general and dominant relationship exists between entrainment and flow velocity. This allows us to define a critical entrainment velocity; that is, the velocity at which entrainment occurs.

The **critical entrainment velocity** varies according to the resistance of the particles. Thus, critical entrainment velocity can be related to particle size. For entrainment by flowing water, the relationship is similar to that shown in Figure 19.6. This assumes that the particles have a constant density of 2.65 g $cm^{-3}$ and that the density of water is 1 g $cm^{-3}$.

Of course this relationship is highly generalized. In practice it does not hold true exactly. The critical entrainment velocity is affected by a variety of other factors, such as grain shape and slope angle (both influence frictional resistance). In addition, most sediments consist of a mixture of materials of different size; this too influences the point at which entrainment occurs. The diagram does, however, illustrate one point of particular interest. We can see that, although there is a tendency for smaller particles to require lower flow velocities to cause entrainment (as we might predict from what we have said), this relationship is not true for very small particles. Clearly other factors must be important here.

The explanation relates to the nature of resistances to movement in fine-grained material. Fine silt and clay particles tend to have a strong cohesive resistance, which overshadows the effect of frictional resistance. The cohesion makes them more resistant to entrainment. In addition, very small particles do

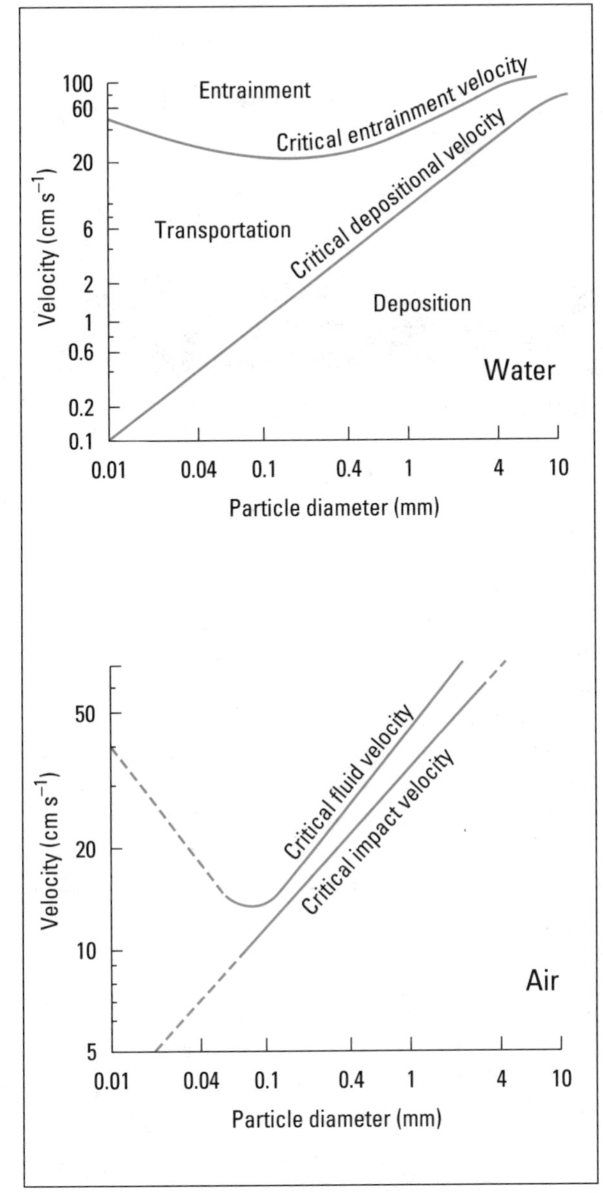


FIGURE 19.6 ▼ The relationships between particle diameter and critical entrainment velocities for water and wind. Note that these relationships assume spherical particles and uniform sediment size (adapted from Hjulstrom, 1935 and Bagnold, 1954)

not protrude so far into the flow, and thus protect each other from the forces of entrainment; they present an essentially smooth surface to the flow.

These relationships between particle size and flow velocity are valid, in broad terms, for both water and wind (Figure 19.6). Several factors complicate the picture in reality, however. As we have noted, turbulence is important in controlling entrainment and in very turbulent conditions entrainment may occur at much lower velocities, simply because the turbulence tends to lift the particle upward; it reduces, in effect, the weight of the grain and thus its resistance to entrainment. Moreover, the impact of particles already in motion may encourage entrainment. In the case of wind transport, this may be very important (Figure 19.6). On a stream bed, boulders bouncing along the surface tend to push others into movement. Raindrops falling through a thin wash of water may have the same effect. In practice, this means that once particles are brought into motion, there is a tendency for further entrainment to occur relatively easily.

Once a particle is in motion it requires a lower force (i.e., a lower velocity) to keep it in motion than it does to initially entrain it. This phenomenon arises because a body in motion has a lower resistance to motion than a static one.

# Transport

## PROCESSES OF TRANSPORT IN A FLUID

It follows from what we have just said that once a particle is entrained, it tends to be transported, at least for some distance. Transport can occur in four ways:

1. in suspension
2. by saltation
3. by traction
4. in solution

The suspended particles are carried along by the water and do not touch the stream bed.

Movement by **saltation** is analogous to bouncing, in which particles are lifted off the surface, but tend to fall back again (Figure 19.7). Immediately or soon afterward they may be picked up again and carried forward. It is clear that this process may be important in causing entrainment, for we have seen that the impact of bouncing particles tends to throw other material into motion.

Movement by **traction** refers to the rolling, sliding, and shuffling action of particles almost permanently in contact with the stream bed. Solutional transport occurs in aqueous environments, where substances dissolved in the water are carried along as individual ions.

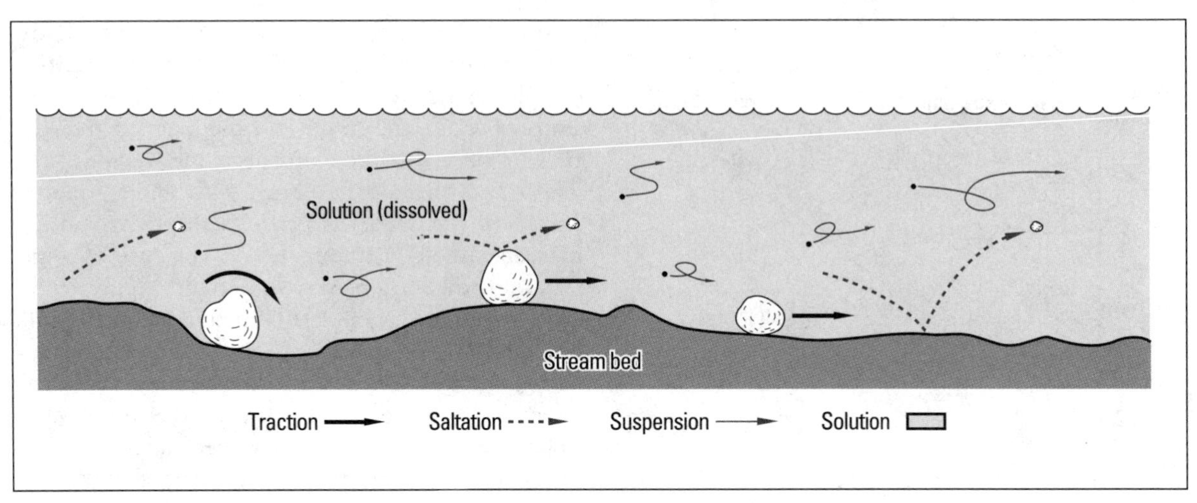


FIGURE 19.7 ▼ Processes of transport. The diagram shows characteristic paths of particles moving by traction, saltation, suspension, and solution

## FACTORS AFFECTING TRANSPORT

The weight, size, and shape of particles play an important part in determining which of these processes operates. Larger particles tend to move by traction, and only the finest particles are normally light enough to be carried in suspension. This has some significance, for it is clear that material moving in suspension or by solution tends to move faster than either saltation or, slowest of all, traction material. The nature of transport thus determines how fast, and how far, the particles are carried. The different processes lead to a sorting of the material by size; soluble, fine, or light particles

### Table 19.1

*Sediment loads, discharge, and area of major rivers*

| *River* | *Climate* | *Discharge ($m^3\ s^{-1}$)* | *Area ($km^2 \times 10^3$)* | *Total load ($t\ y^{-1} \times 10^6$)* | *Dissolved (percent)* | *Suspended (percent)* |
|---|---|---|---|---|---|---|
| Amazon | Tropical rainy | 175 000 | 6 300 | 498 | 37 | 63 |
| Congo | Tropical rainy | 39 200 | 770 | 79 | 30 | 70 |
| Colorado | Cool dry desert | 640 | 635 | 55 | 3 | 97 |
| Nile | Hot dry desert | 2 830 | 3 000 | 111 | 13 | 87 |
| Brahmaputra | Mountain—seasonal tropical | 19 300 | 580 | 795 | 9 | 91 |
| Danube | Moist cool temperate | 6 430 | 805 | 68 | 47 | 53 |
| Mississippi | Moist warm temperate | 2 830 | 3 000 | 307 | 30 | 70 |
| St. Lawrence | Moist cool boreal | 10 700 | 1 025 | 5 | 91 | 9 |
| Volga | Moist cold boreal | 8 400 | 1 350 | 26 | 75 | 25 |
| Yukon | Moist cold boreal | 6 200 | 770 | 79 | 30 | 70 |

*Data* A.D. Knighton, *Fluvial Forms and Processes* (1984), Edward Arnold.

are generally carried farther and faster than large or heavy particles.

In addition, however, the nature of the agent of erosion and the regional climate are significant. On the whole, transport by solution is most important in temperate and tropical areas, while solids are dominant in colder regions. In the case of transport by winds, there is a tendency for relatively larger quantities of material to be moved by saltation. In transport by water, rather less moves by saltation, and suspension movement is more important, largely because turbulence may be greater, and because the water itself gives the particles a degree of buoyancy. Much movement also occurs in solution (Table 19.1). Transport by ice, in glaciers, is less easy to define, and much more difficult to monitor. Probably most of the material transported by glaciers is dragged along the bed of the ice sheet as traction load. In the case of movement under the effect of gravity, most transport occurs in contact with the surface. On steep slopes, however, material may bounce in a form of saltation.

Material moving by traction, saltation, and suspension influences the processes of transport by exerting a frictional drag upon the transporting agent. Particles carried by the wind, for example, cause friction with the air. This has the effect of damping down turbulence, and reducing to some extent the flow velocity. Thus, as the concentration of sediment being carried increases, so the ability of the agent to transport material declines. Clearly the two tend to approach a condition of dynamic equilibrium in which, as additional material is brought into motion, equivalent quantities of sediment are dropped from the load. The maximum quantity of material that can be transported is referred to as the competence of the transporting agent. In the case of transport by solution, competence relates to the amount of the material that can be dissolved before saturation of the water occurs.

# Deposition

We only have to watch sand being blown across a dune, or washed across the beach by a shallow stream to appreciate that transport of material is rarely continuous. Instead, individual particles are entrained, carried a short distance, and deposited. They rest there a short while and may then be swept on again. Why does the material acts in this way?

## THE BALANCE OF FORCES

The main reason for this behaviour is that transport depends upon an appropriate balance of forces within the transporting medium. A reduction in the competence of the stream or wind, or an increase in the resistance of the particles, may upset this balance and cause deposition.

Reductions in competence come about in a variety of ways. Velocity or turbulence may be diminished locally by the sheltering effect of large boulders, walls, hedges, or other obstructions. Additionally, competence changes quite markedly over time.

Wind velocity, for example, varies considerably over a period of seconds, and as a result its ability to transport material is constantly changing. In the longer term, the wind may die completely as the storm passes, or stream discharge may fall as the flood wanes, and again competence declines. Wherever the ability of the transporting medium to carry material falls in this way, deposition occurs (Figure 19.6).

Increases in the resistance of material to transport may also occur. Particles may become lodged between larger particles on the surface, for example, or trapped by vegetation. By chance, they may also adopt a more stable orientation, governed by their shape. Rod-shaped particles tend to be transported mainly by rolling about their long axis; if they fall into a position with their long axis parallel to the flow they are less easily moved and become stabilized.

### *Longterm changes in transporting conditions*

The processes of deposition we have mentioned so far are short term in their influence. The next gust of wind or the next storm may again dislodge the particle and transport will start once more. In the longer term, deposition may occur because of much more general changes in the erosional environment. If the climate becomes drier, rivers may lose their power to carry material (although this is not always the case since this change may be compensated by other changes in conditions). More impressively, a warming of the climate may lead to widespread melting of glacial ice, with mass deposition of the debris it carries. This occurred at the end of the Quaternary Period, and it is responsible for the widespread glacial deposits found in the northern hemisphere.

*Precipitation and flocculation* We must also mention precipitation and flocculation as processes of deposition. Material carried in solution is deposited mainly through precipitation, a process brought about in most cases by marginal changes in the character of the water. Thus, changes in the chemistry, temperature, or gas content of the water may lead to precipitation of dissolved substances. These processes are responsible for many of the beautiful depositional features seen in caves. It is not, as is often supposed, evaporation of water that accounts for the growth of stalagmites and stalactites; rather it is a process of degassing, as carbon dioxide escapes from the water and alters its chemical equilibrium (Figure 18.3).

**Flocculation** is also a chemical process. Many of the fine muds and clays deposited in estuarine or near-shore areas are produced by this process. Salt within the water leads to the agglomeration of minute clay particles into fluffy masses which are heavy enough to sink and collect on the bed. Were it not for flocculation, much of this material would be so small that it would remain in suspension almost indefinitely.

### SEDIMENT SORTING DURING EROSION AND DEPOSITION

The processes of erosion and deposition tend to lead to a marked change in the character of the material being carried through the landscape. Part of this change is due to the wear and breakage of the particles during transport. The continual collision of particles against each other and against the rock surface leads to their slow attrition. During transport, therefore, the particles tend to become smaller and more rounded as sharp corners and protrusions are knocked off.

In addition—and often more important—there is a change in character due to preferential transport and deposition of certain particles. Both shape and size may be important in this sorting process. Smaller particles, as we have seen, tend to be transported faster and farther; they therefore accumulate in more distant areas. Larger particles travel slowly and are deposited more readily; consequently they accumulate close to their source.

Certain shapes are also more susceptible to transport than others. Spherical or near-spherical particles may roll more readily over the surface, while flat, disc-shaped particles tend to be more buoyant. These may be carried farther than very irregular, angular particles.

Changes in composition may similarly occur as a result of transport. Due to the constant physical wear of the particles, only the most resistant survive long distance transport, and softer rocks and minerals are often destroyed close to their source.

All these processes operate together, of course, and it is often difficult to separate their effects. By studying changes in the nature of the materials as they are transported, it is possible to distinguish some of the effects of transportation agents upon the sedimentary materials.

## Landforms of erosion and deposition

### EROSIONAL LANDFORMS

We will look in detail at many erosional landforms in later chapters, but a few generalizations are in order here.

First, we might note that erosional landforms are typically composed of relatively fresh rock. The weathered residues are removed, and only the more resistant materials remain. Thus, the glacial landscapes of northern Canada and the fluvially dissected topography of the Appalachian granites have a common feature; the upstanding blocks represent the resistant cores from which more weathered material has been stripped.

By the same token, many erosional landforms are guided by the structure of the original rocks. The wind-carved landforms of arid and semi-arid areas often show this effect clearly, with the ancient bed-

FIGURE 19.8 ▼ Percé Rock, Percé, Quebec, shows the influence of jointing in rock on the formation of sea stacks (extreme right) and caves (right of centre) (photo: P.G. Johnson)

ding of the rocks picked out by the abrasive effect of the sand. Coastal erosion tends to attack the joints and fissures in the rocks preferentially, cutting deep inlets along these lines of weakness and sometimes carving a magnificent scenery of stacks and caves and arches (Figure 19.8). Similarly, the karstic scenery of limestone country often shows the effect of structural controls on erosion.

## DEPOSITIONAL LANDFORMS

*Dynamic features* Many of the landforms associated with sediment deposition are relatively transient in character, but often very active. They represent equilibrium forms in which material is temporarily stored as part of a natural cycle of erosion and deposition. Thus, during the transport of materials down a stream channel, there is a tendency for temporary storage of sediment in a variety of features. With an increase in stream discharge, due to heavy rainfall or melting of snow for example, the materials may be reworked more generally and carried farther down the stream. Similarly, material is laid down within the floodplain of the river, subsequently to be reactivated as the channel shifts laterally and erodes the alluvium. In a stream, therefore, the progress of material down-valley tends to occur not as a continual movement from one end of the valley to the other, but in the form of discrete steps. This is particularly true of material carried by traction or saltation.

The same principle is true of most other environments. Material blown by the wind is alternately deposited and transported, even within a single gale. Between gales, the sediments may lie at rest for some time. Erosion and deposition by ice, by the sea, and by gravity operate in a similar cyclical fashion. The landforms produced by these cyclical events may persist for some time, since they are often being constantly renewed and regenerated, but they are active. Sometimes, as in the case of dunes, ripples, and bars in streams, the features migrate while retaining their form. They are, in the context of the system, an expression of the dynamic equilibrium in the landscape.

*Relict landforms* These active landforms may be contrasted with relatively inactive forms which were created, normally under different conditions, during the past. Our landscape is cluttered with such features: river terraces, glacial landforms, the extensive wind-blown loessial plains of China and much of the midwest USA, the vegetated sand dunes of the Carberry Desert east of Brandon, Manitoba, the fossil raised beaches that surround much of the coastline of Canada (Figure 19.9). All these are relict depositional landforms that have been preserved since they were deposited.

FIGURE 19.9 ▼ A raised beach platform on the north side of Bottle Cove, west coast of Newfoundland, evidence of a previous higher relative sea level after deglaciation, before isostatic recovery raised the level of the landmass (photo: P.G. Johnson)

The fact that such features have been preserved indicates, in most cases, that environmental conditions must have changed since they were formed. Otherwise, the fluctuations in the competence of the transporting agent would have led to reworking of the deposits. Thus, one of the main prerequisites for the longterm survival of depositional landscapes is that they form under conditions of waning erosive power. This may be due to climatic changes, such as those that caused the retreat of the ice sheets that left behind the extensive glacial landforms; it may be due to morphological changes, such as a diversion in the course of the river due to erosion elsewhere in the valley; or it may even be due to human activity. In all cases, however, it is the change in the depositional and erosional environment that allows these features to be preserved.

# Erosion and humans

Needless to say, we are greatly affected by erosion. **Soil erosion**, in particular, represents a major threat, and it is a process in which we ourselves often play a major part. Through clearance of vegetation for agriculture, and through repeated tillage, soil has been exposed to the forces of wind and water; in many cases, we have reduced the resistance of the soil to these forces. The implications are manifold. Erosion may take away people's livelihoods; the eroded particles may damage crops already in the soil; sediment accumulates in roads, ditches, and streams and causes a problem of pollution. Once initiated, erosion often spreads, feeding on itself, for sand grains carried by the wind encourage the entrainment of more material in neighbouring areas, while deposition of the sands may bury the vegetation.

In desert margins and semi-arid regions, such as the Great Plains region and the Canadian Prairies, erosion and increasing alkalinity represent serious threats to the potential of the soil. Many believe that soil erosion is the most threatening longterm problem facing humankind (Figure 19.10). Even in relatively humid environments the problem can be severe. The problems are not new. What is new is the scale and extent of the problem. Throughout history, civilizations dependent upon agriculture have run into the problems of erosion and alkalinity. The first major centre of agriculture, in the Tigris and Euphrates valleys (modern day Iraq), experienced the problems some 9000 years ago. Soil erosion, accelerated by irrigation, led to continual degradation. We have not learned any lessons yet, because exactly the same problems exist in many parts of North America. Water is a major transporter of the soil, but it is the wind that creates the greatest problem as it removes the fine material essential to the fertility of the soil. The 'Dust Bowl' of the 1930s was a particularly extreme example of the damage caused by wind erosion. In most examples of the destruction of soil by a civilization, it has been caused by overuse without regard to environmental limits.

FIGURE 19.10 ▼ A gully in glacial till on the north shore of Lake Erie, Port Stanley, Ontario (photo: P.G. Johnson)

It is difficult to imagine the severity of the situation in the 1930s; a four day storm in May, 1934, carried approximately 300 million tons of soil 2500 kilometres across the United States and darkened the sky in New York and other cities. In recent years there have been more dust storms in many places in western North America. These are partly

FIGURE 19.11 ▼ Beach erosion: retreat of the sand dune coastline has exposed the remains of a concrete bunker built during World War II on the beach at Ainsdale, Lancashire, England (photo: P.G. Johnson)

FIGURE 19.12 ▼ A blowout through seacoast dunes has exposed a forest buried during the formation of the dunes; Shallow Bay on the west coast of Newfoundland (photo: P.G. Johnson)

due to climatic changes, but it is also clear that we have forgotten or ignored many of the lessons of the 1930s. The continued breaking of marginal land, the removal of rows of trees deliberately planted as windbreaks, overgrazing in semi-arid areas, all have aggravated and accelerated the problems of alkalinity and erosion.

However, erosion is not only a hazard in relation to agriculture. Coastal areas, for example, are prone to rapid erosion (Figure 19.11). Often, from our attempts to control such processes in one area, the problem is made worse elsewhere, for the materials being washed down the coast are part of a delicate, interrelated system. Coastal dunes are also subject to severe wind erosion, often induced by use of these areas for recreation (Figure 19.12). Trampling of vegetation may leave loose sand open to winds that blow in from the sea, and rapid deflation may occur.

In arctic regions, the soil and vegetation are even more vulnerable to the effects of human behaviour. The 'tundra buggy,' a vehicle designed to travel with a footprint equal to a person's weight, leaves distinctive tracks. Vegetation is killed and the soil altered by the wheels passing over the thin soil overlying the permafrost. As we expand our activities into these regions, the extent of the erosion and disturbance are bound to continue, especially if we pay no heed to the knowledge we already have.

# Hillslopes

## Introduction

Hillslopes are a vital part of our landscape. The whole land surface, in fact, can be thought of as a mosaic of slope units, ranging from vertical mountain and marine cliffs to gentle or almost horizontal plains. Each of these units acts as an important system, through which energy and matter are moved (Figure 20.1), often under the direct influence of gravity. The outputs from these hillslope systems in turn become inputs to other systems—to streams, glaciers, or the sea—and thus processes acting on slopes exert fundamental controls on other parts of the landscape. Moreover, the form of the hillslope is itself dependent on the processes acting upon it. Hillslopes can therefore be thought of as process-response systems, the morphology of which responds to, and also affects, the processes operating within the system.

Hillslopes are also important in practical terms. They exert a direct effect on human activity, governing the type of agriculture that can be carried out and the suitability of land for construction. The use of tractors and combine harvesters, for example, is limited by slope angle, so cultivation of land steeper than about 11° for cereal crops is normally not possible. In some countries cultivation of other crops is carried out on much steeper slopes: up to 25° in Spain

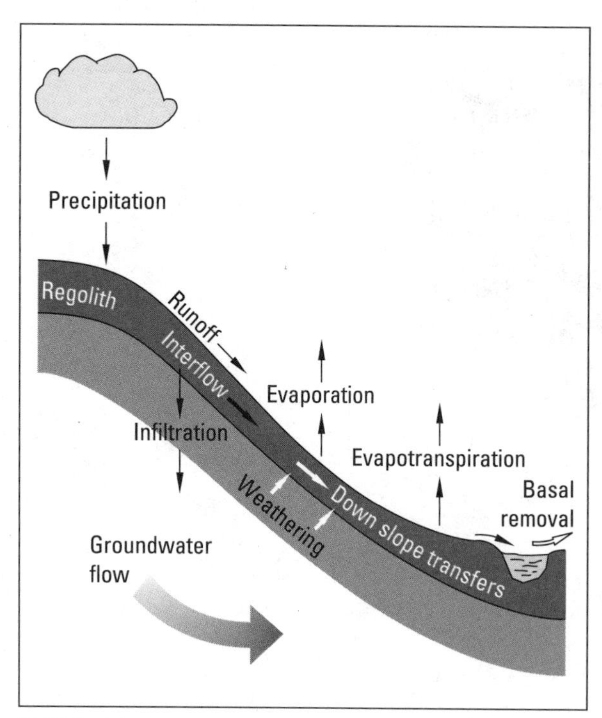


FIGURE 20.1 ▼ The hillslope system

and 18.5° in Malaysia. These differences reflect combinations of the stability of the soils in the different areas, the differences in farming methods and population pressure, and climatic conditions.

As this implies, slope conditions also affect land use indirectly, for hazards such as soil erosion and hillslope failure are important constraints. In general, as slope angles become steeper the amount of runoff and the intensity of erosion increase at an increasing rate (Figure 20.2). This, too, exerts a major control on the suitability of land for cultivation, for where excessively steep slopes are ploughed, gullying often occurs. Similarly, the likelihood of slope failures such as landslides and rockfalls depends on slope conditions. This is significant in relation not only to our use of natural slopes, but also to the ways we construct artificial slopes such as dumps of mine tailings, road cuts, and embankments. If the slope is too steep, failure may occur, with disastrous results.

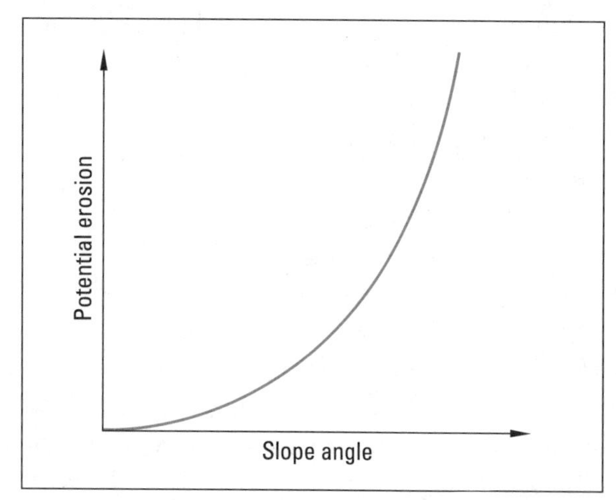


FIGURE 20.2 ▼ The relationship between slope angle and potential rate of erosion

# The hillslope system

## INPUTS AND OUTPUTS

The hillslope system illustrated in Figure 20.1 receives inputs of energy and materials from a variety of sources. Inputs include solar radiation, precipitation, and dissolved substances and solids washed from the atmosphere, as well as debris derived from weathering of the parent materials. Outputs occur by evapotranspiration, by percolation of water and dissolved substances into the bedrock, and most important, by removal from the slope foot in streams, by glaciers, or by the sea. In high latitude areas seasonal snow cover, or semipermanent snow banks in polar regions, can promote slope erosion by **nivation** (see Chapter 22).

The magnitude of these inputs and outputs depends upon a number of factors, including the geology, climate, and relationship of the slope to the wider landscape (Figure 20.3). Inputs of rock materials, for example, are controlled to a great extent by rates of weathering and these, as we have seen, depend in turn upon the character of the bedrock and the local climate. Weathering on resistant, acid crystalline rocks such as quartzites is slow and as a consequence inputs of debris to the slope system are limited; the quantity of debris moving down the slope and being released at the slope foot is therefore limited also. Conversely, in many unconsolidated or soft rocks, weathering is rapid, inputs are high, and the slopes are much more dynamic. Slopes in arid areas may also be relatively less active than those in polar or humid temperate regions due to the lower rates of weathering and the restricted inputs of debris to the system.

Outputs of debris from hillslope systems are controlled primarily by slope foot conditions. The presence of an active stream, for example, encourages removal of debris from the base of the slope and

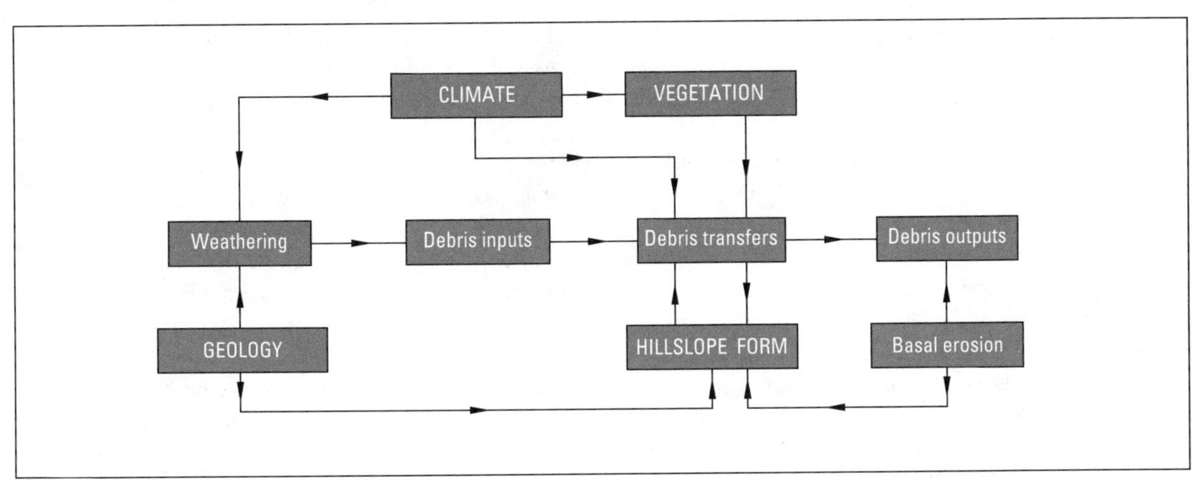


FIGURE 20.3 ▼ The hillslope as a process-response system

helps to keep the hillslope active. Where removal of debris from the slope foot is limited, perhaps because the stream is too small or too distant to carry away the material, debris accumulates at the base of the slope. This clearly alters the slope form, as we will see later in the chapter, but it also affects the overall sediment budget of the slope system, and may ultimately affect the processes of weathering and debris transport operating on the slope. As the material builds up, for example, it blankets the bedrock and protects it from weathering. Thus, negative feedback occurs and inputs of debris to the slope are reduced.

As this implies, the balance between inputs to and outputs from the hillslope system exerts a major control over slope form and process. In time, a balance between inputs and outputs tends to develop, determined by the more limiting of the two processes. Where inputs are the controlling factor—as on hard rocks which weather very slowly—the slope is said to be weathering limited. Conversely, where the potential for weathering is high but outputs from the hillslope are restricted (e.g., due to the lack of an active slope-foot stream) the system is said to be transport limited.

## TRANSFER PROCESSES AND HILLSLOPE STABILITY

The processes by which materials are moved through the hillslope system are many and varied. Running water plays a significant role, and we will discuss this more fully in Chapter 21 when we examine stream systems. More important in many cases, however, are the group of processes known as **mass movement**. These involve the transport of debris for the most part not as individual particles but en masse, under the direct influence of gravity.

The operation of mass movement processes depends upon the development of instability in the hillslope. Under these conditions, failure of the slope materials occurs, either as a sudden, spontaneous event or as a slow, imperceptible, and more or less continuous process. The type of failure that occurs, and the speed at which it operates, is a function among other things of the degree of instability that is generated. Rapid, large-scale failures tend to occur when the stresses exerted on the slope materials exceed their strength—a condition often produced by the operation of short-term trigger mechanisms. Movement normally occurs when the stresses only just exceed the strength but are maintained over a long time period.

What, then, are the sources of the stresses and strength acting within hillslope materials? A major source of stress is gravitational force. We saw in the last chapter that the magnitude of this force is related to the slope angle and the weight of the material:

$$F \infty w \cdot \sin \phi$$

where $F$ is the downslope gravitational force, $w$ is the weight of the material acting at any point on the slope, and $\phi$ is the slope angle.

A little thought will show that the magnitude of this force is not constant over the whole slope surface, but increases downslope and down into the soil. This is because the weight of the overlying material increases with its thickness, making the value of $w$ in the equation progressively larger. It is also clear that gravitational force can be increased by loading of the hillslope—for example, by the construction of a building or by the passage of a vehicle.

The strength of the hillslope materials, on the other hand, varies according to the character of the rock or soil. In the case of unconsolidated materials such as gravels or sand, the main source of strength is derived from frictional resistance. This depends upon the size, shape, and packing of the particles. Soils, fine-grained sediments such as clays, and solid rocks, however, obtain their strength largely from cohesion. As we learned in the previous chapter, this arises in part from surface tension effects associated with the presence of thin water films between the particles. More important, though, are the effects of interparticle fusion, chemical cements, and electrochemical bonds.

In fresh, massive rocks, the cohesive strength may be extremely high and if this were the only factor involved such rocks would be able to support vertical cliffs several thousand metres high. In practice, however, most rocks contain joints, bedding planes, and faults across which the cohesive strength is relatively low. These therefore act as lines of weakness along which failure tends to occur, and which limit the stresses the hillslope can stand. In addition, the strength of the material is often diminished by weathering, while saturation of the rock leads to the development of positive pore water pressures as the water films grow and push the particles apart. Water may also dissolve some of the cements binding the minerals, so that after prolonged rain their strength may fall markedly.

The stability of the hillslope depends upon the relationship between the stresses and the strength of the material. These vary with depth, due to differences in the weight of the overlying material and the degree of compression (Figure 20.4). They also vary according to the nature of the material and the slope form. More important, however, both stress and strength tend to fluctuate over time, and it is these fluctuations that often trigger slope failures.

Many factors may act as trigger mechanisms. One of the most common, as we have indicated, is prolonged heavy rainfall, for this results in saturation of the hillslope materials and a consequent loss in strength. Short-duration heavy rainfall will result in high runoff rates but low infiltration, and thus it is not as effective as a trigger mechanism. In the longer term, weathering also reduces the strength of the materials. In northern environments snowmelt can raise pore water pressures in the spring sufficient to cause failure. Changes in the magnitude of hillslope stresses similarly occur. Earthquakes or volcanic activity, for example, may cause tremors that give rise to short-lived but large increases in stress. A nearby landslide or rockfall may have the same effect, as may explosions such as quarry blasting. Even the passage of a heavy vehicle may generate minor shock waves that trigger slope failure. In addition, as we have noted, loading of the slope by buildings may increase the stress, while oversteepening of the hillslope—by natural erosion or by excavation (e.g., for a road cut)—increases the gravitational force operating on the material and reduces slope stability (Figure 20.5).

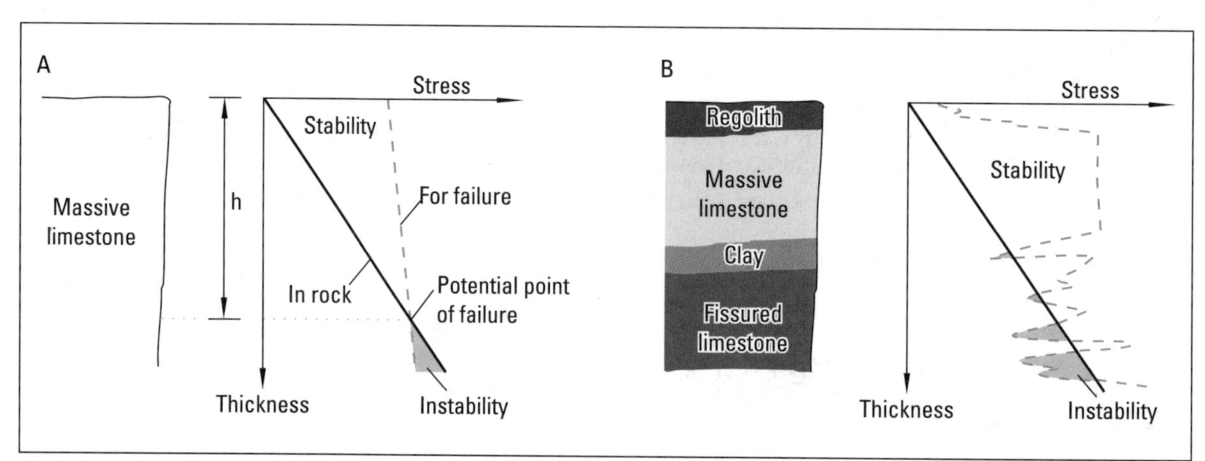


FIGURE 20.4 ▼ Generalized stress–strength relationships in vertical slopes. (A) The simple situation of a massive limestone slope. If the rock had no weak points the stable height of the cliff could be up to 8000 m. (B) A variety of materials in a composite sequence alters the vertical stability of the cliff. Note that the massive limestone is still stable, but other layers are weaker and much more likely to break down, thus they lower the stability

FIGURE 20.5 ▼ Landslide in Leda clay, Gatineau, Quebec, caused by a construction excavation into the terrace. Leda clay was deposited in the proglacial Champlain Sea about 10 000 years B.P. It is thixotropic: normally a stiff clay, it is liable to flow rapidly if it is disturbed by shock, sudden change of moisture content, or construction activity (photo: P.G. Johnson)

In reality, these various effects do not act wholly independently, and slope failures are often a result not of a single trigger mechanism but of the cumulative influence of several different processes. Thus, the development of instability may be associated with a specific combination of events: for example, a major earthquake following a period of prolonged rainfall (A in Figure 20.6); or oversteepening of slope materials whose strength has already been reduced by intense weathering (B in Figure 20.6).

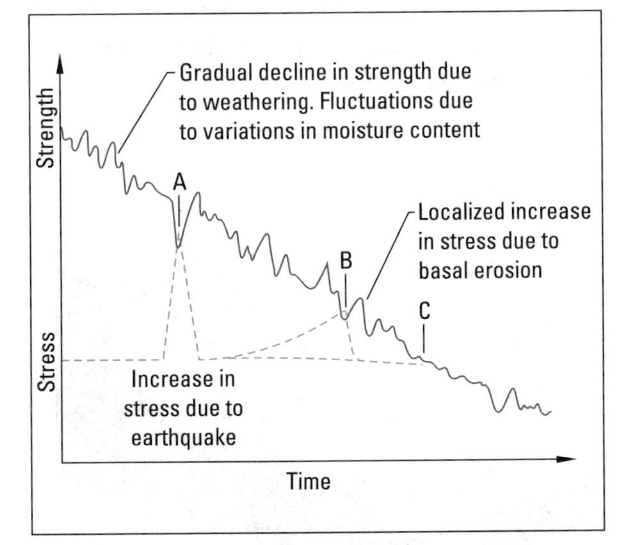


FIGURE 20.6 ▼ Changes in hillslope strength and shear stress over time, and the initiation of landslides. A, B, and C = potential landslide episodes

# Hillslope processes

As we have noted, numerous processes are involved in the transfer of debris through hillslope systems. These processes vary considerably in relation to the character of the slope materials, and it is therefore useful to consider separately slopes formed in non-cohesive materials (e.g., gravels, sands), semi-cohesive materials (e.g., soils and clay), and hard rocks. In the first, strength is derived largely from frictional resistance and is therefore low; as a result, failure often occurs at relatively low slope angles. In the latter two, cohesion provides an additional strength and thus even steep slopes may be relatively stable.

## SLOPE PROCESSES IN NON-COHESIVE MATERIALS

Slopes formed from non-cohesive materials are characteristic of many landforms constructed of coarse-grained sediments, such as alluvial fans, talus slopes, sand dunes, and glacial outwash features. Many artificial slopes are similarly composed of non-cohesive materials: a mine tailings dump, for example, is often in this form.

On slopes of this nature movement occurs largely through the sliding or rolling of individual particles, as localized instabilities arise, or through a process of shallow sliding. In the latter, planes of weakness develop near the surface; these reduce the **shear strength** of the material and create instability. Failure of the material along this plane may be triggered by minor events such as rainfall or vibration, and the surface material rapidly slides en masse downslope. The **shear plane**, along which the material moves, tends to be relatively planar and shallow (Figure 20.7).

Materials tend to adopt a relatively constant slope angle after failure. This angle is dependent upon the nature of the material and is referred to as the **angle of repose**, or the **angle of residual shear**. As the slope angle is steepened the material remains stable for a period. Then individual grains start to move as local instabilities are created (e.g., due to the random arrangement of the particles). Finally, again at a constant angle, mass sliding occurs. This starts near the base of the slope, where the gravitational force is at the critical level. The angle at which movement is initiated is referred to as the **angle of initial yield**.

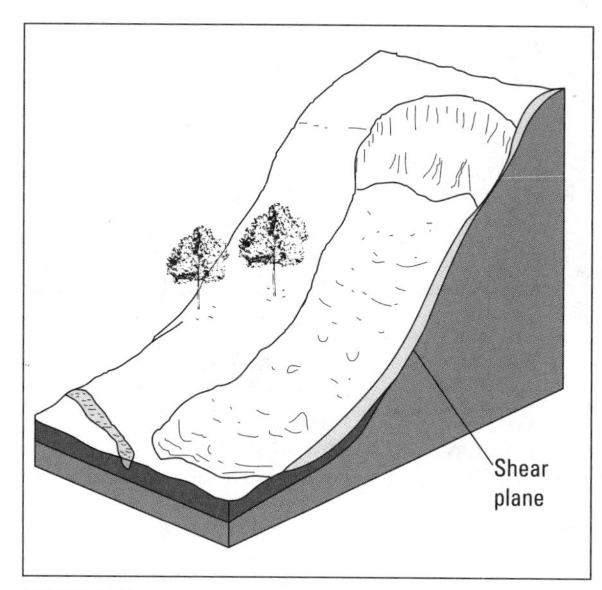


FIGURE 20.7 ▼ A debris slide

The reason for these different angles is related to the character of the strength of the material. When the material is static (as it is before sliding), its frictional resistance is relatively high. When the particles are moving, however, their frictional resistance is slightly reduced. This results in a lower angle of stability; that is to say, the material only restabilizes when the stresses, which are a function of slope angle, are relatively low.

## SLOPE PROCESSES IN SEMI-COHESIVE MATERIALS

Slopes formed in clays, or covered with a mantle of soil, display rather different processes of mass movement. In these cases, the materials have a degree of cohesion that makes them potentially more stable than non-cohesive material (though markedly less so than many solid rocks). This cohesion is derived mainly from electrochemical bonds between the fine particles and the surface tension effects of water films contained in the pore spaces.

Both these sources of cohesion are dependent upon moisture content, and, consequently, the strength of these materials tends to be controlled to a large degree by moisture conditions. This has considerable significance, for it provides a means by which stability may be reduced in the short term, without any change in slope form (e.g., basal undercutting). Repeated failure may therefore be possible as moisture contents increase and the strength of the material falls. This gives rise to a range of slope processes, some of which only operate slowly, while others occur rapidly.

In regions where the soils or clays are seasonally frozen they are most unstable during the thaw. Ice formation in the clays breaks the electrical bond between particles, so that in the spring the strength of the materials is at its lowest and, frequently, the water content is at its highest due to the spring snowmelt.

*Rotational slips and mudflows* Among the forms of rapid movement are rotational slips and mudflows. **Rotational slips** occur along clearly defined planes of weakness which develop concave to the surface (Figure 20.8). The generation of these failure planes is due to the distribution of stresses within the material. Rotational slips are common forms of failure, especially in clays and shales (Figure 20.9), although they are rarely perfect in form. Clearly, structural weaknesses within the material result in the deflection of the failure along these planes (Figure 20.8).

In some cases failure may not occur until the material is so saturated that it acts almost like a fluid. In this condition a variety of processes, generally referred to as **mudflows**, may occur. The saturated material flows like a thick slurry downslope until, as the water is lost through seepage, the flow comes to a halt (Figure 20.10). Mudflows may operate on very low slope angles, for the high moisture content of the material reduces the frictional resistance and the cohesion almost to zero. Thus, only small gravitational stresses are necessary to initiate

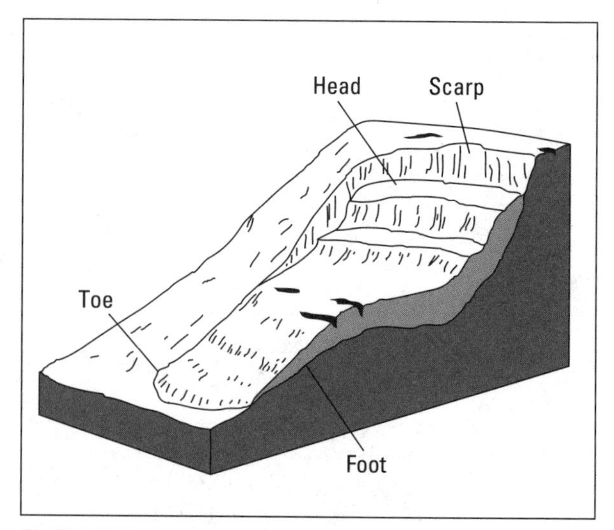


FIGURE 20.8 ▼ A rotational slip

FIGURE 20.9 ▼ A rotational slip in Leda clay on the South Nation River, Ontario. The slip developed as a retrogressive series of failures, visible as lines on the surface, and it temporarily blocked the river. These slips have been common events in Leda clay in historic and recent time. The clay (named for a species of small clam found in the deposits) was laid down in the Champlain Sea along the margin of the retreating ice sheet 10 000 years B.P.; it is 30 m thick in places

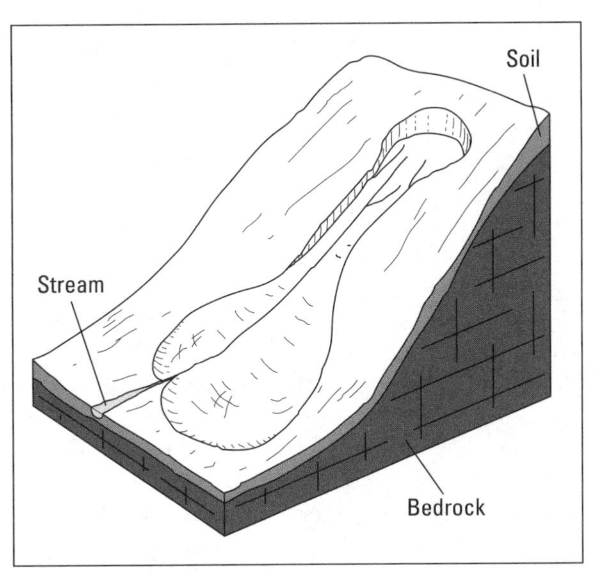


FIGURE 20.10 ▼ A mudflow

movement. In addition, rates of movement may vary considerably, depending upon slope angle and moisture content. In some cases almost instantaneous failure may occur as the material is rapidly saturated; in other cases slower, longterm movement may take place, often during the wetter periods of the year. In Chile, in 1949, over 1.5 million m$^3$ of material was swept down the southern slopes of Carro Oadillal into the sea within a day. On the other hand, many coastal mudflows operate much more slowly, with flow rates of no more than 10–20 m per year (Table 20.1).

## Table 20.1

*Rates of mass movement*

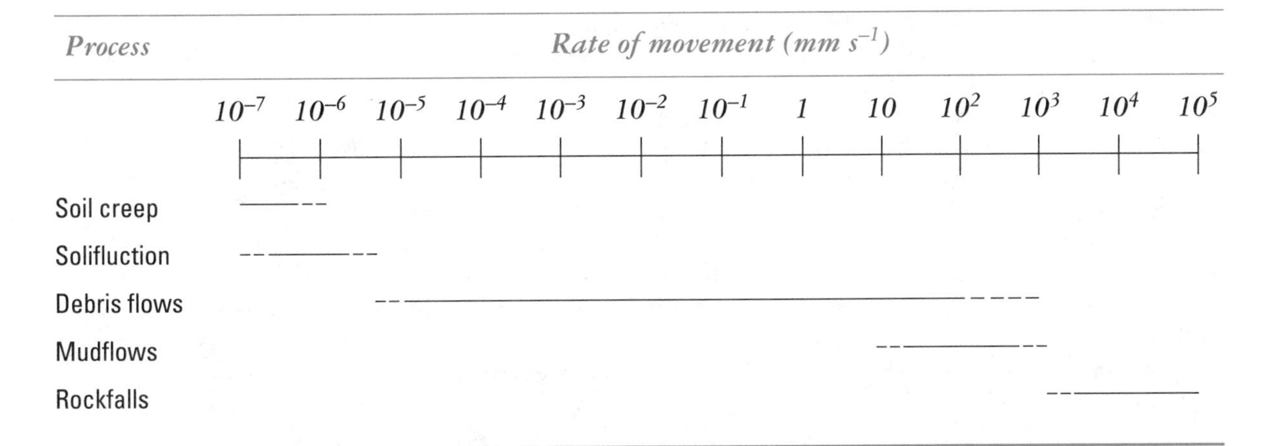

Cohesive materials, like unconsolidated sediments, may be subject to shallow sliding when surface materials are weakened by weathering or saturation. Indeed, it can be seen that shallow slides represent an intermediate form between the deeper rotational slip and the superficial mudslide.

*Debris flows* In many mountain regions saturation of unconsolidated sediments of variable composition can result in rapid flow of the sediments. These occur along existing stream or debris-flow eroded channels. The saturation of sediments during the spring snowmelt is the prime cause of these events. They have occurred at Howe Sound, British Columbia, on a regular basis for as long as settlement records exist. The location of settlements on the debris flow fans, as the only building land along the Sound, exposed them to considerable hazard. Massive engineering structures have been put in place to divert or channel these flows. Measured velocities of debris flows have a very wide range (Table 20.1) as a result of the great variations in sediment characteristics, water content, and slope angles.

*Flows in permafrost areas* Thermokarst processes, which involve mudflows and debris flows due to the melt of massive ground ice, are considered in Chapter 22 on periglacial processes.

*Soil creep and solifluction* Many of the slope processes acting upon cohesive materials operate more slowly. One of the most widespread and intriguing processes is **soil creep**. Material in the weathered surface layers moves gradually downslope in a series of jerky steps (Figure 20.11). The process may be powered by a variety of mechanisms. The effect of gravity, temperature fluctuations, and variations in moisture content within the soil may all act to cause displacement of particles. We can distinguish between various types of soil creep. Often the process acts discontinuously and seasonally and is therefore known as discontinuous creep. The two most important mechanisms in this case are the heave and settlement of particles due to freeze–thaw activity, and the shrinkage and expansion of material due to wetting and drying (Figure 20.12).

Continuous soil creep is largely independent of moisture or temperature fluctuations, and is due, instead, to gravitational force. It occurs when the stress due to gravitation effects is between the yield strength and residual strength of the material; in this case a form of viscous failure occurs. The process operates, therefore, where stress and strength are almost in balance. Rates of movement are in the order of a few centimetres per year, and, as with dis-

FIGURE 20.11 ▼ Slow soil creep on the slope of this valley in glacial deposits near Gloucester, Ontario, is indicated by the small terraces (photo: P.G. Johnson)

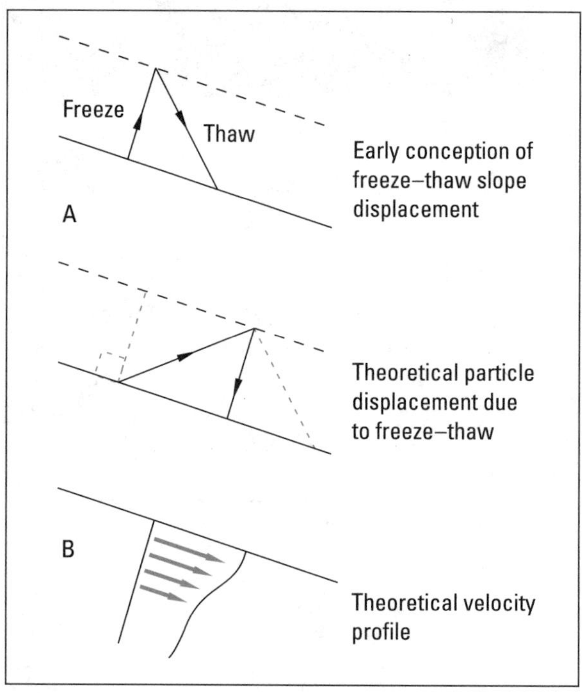


FIGURE 20.12 ▼ Discontinuous creep. (A) The theoretical pathway of a particle during a single expansion-contraction cycle. (B) Theoretical deformation profiles with depth

continuous creep, there is a tendency for movement to be most rapid at the surface, where the strength of the material is generally lowest (Figure 20.13).

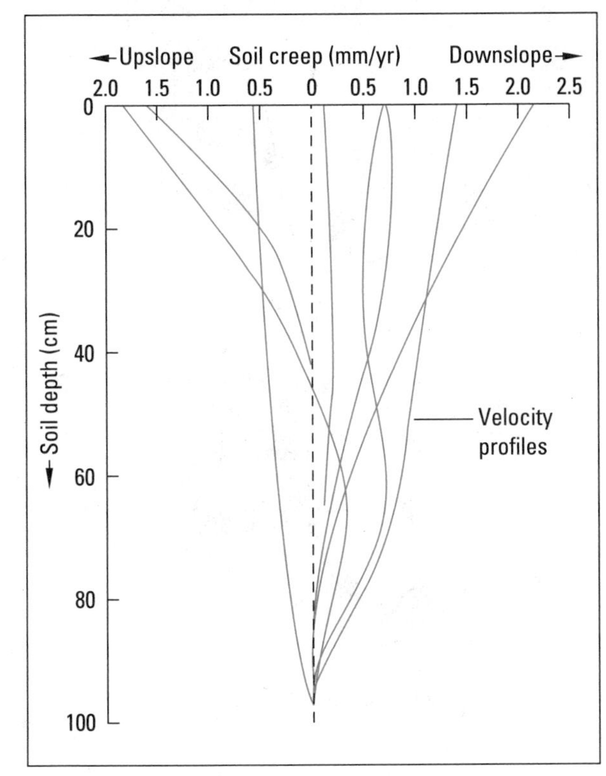


FIGURE 20.13 ▼ Velocity profiles of soil creep in the Mendip Hills, Somerset, England (from Finlayson and Statham, 1980)

Slow movement caused by freeze–thaw processes is known as **solifluction**, and it is a widespread and important phenomenon in polar and subpolar areas (see Chapter 22). Solifluction tends to occur in conjunction with a variety of related processes. Measurements of solifluction show it to be a slow process. In Alaska, rates of 2–6 cm $y^{-1}$ have been determined; elsewhere rates range from less than 1 cm to over 30 cm $y^{-1}$ (Table 20.1). It should also be noted that solifluction may occur on relatively low slope angles.

## SLOPE PROCESSES IN HARD-ROCK SLOPES

Failure of hard-rock slopes is often spectacular and rapid. The reason is in part that such rocks are able to form relatively stable slopes of considerable steepness and height. The spectacular amphitheatre in the Drakensberg Mountains of South Africa, for example, is formed of basalt cliffs over 500 m high. When instabilities in these materials do develop, therefore, the potential for failure is considerable.

As we have seen, hard rocks derive their strength almost entirely from the strong inter-granular bonds, but weaknesses occur along bedding planes and joints. Rock structure consequently exerts a fundamental control over processes of mass movement in these rocks. In relatively massive rocks, small fragments may be pried loose by gravitational stresses, aided perhaps by freeze–thaw processes, to give **rockfalls**. The material then builds up a talus slope at the foot of the slope.

Larger-scale failure may occur where movement occurs along well-defined joints or bedding planes. Toppling failure takes place where joints are vertical, slab failure and wedge failure where the weaknesses are inclined (Figure 20.14). In addition to the original joints and bedding planes in the rock, tension cracks may develop due to the pressure release during erosion. In the case of relatively massive rocks, failure often involves the extension of irregular, disconnected weaknesses, as in Figure 20.14D. Massive **rockslides** may then occur (Figure 20.15).

Not all movements in solid rocks are rapid, and a variety of slower, more continuous processes may operate. Among these are **cambering** and associated processes. These occur most commonly where solid (competent) rocks overlie clays or other incompetent materials. Under the weight of the overlying materials the clays may be deformed and squeezed out into the valley floor. At the same time, oversteepening of the slope allows slabs of rock to slide slowly downslope. The joints in the rock are gradually opened up to create gulls, and the rocks appear to dip into the valley (Figures 20.16 and 20.17).

In extensively glaciated areas, the valley sides have been stressed by successive glacier advances. The removal of stress during deglaciation induces unloading fractures along the valley summits. Failure along these fractures has produced massive rockslides in mountain regions and continues to pose a hazard today.

# The role of water on hillslopes

As we noted earlier, water plays a significant role in the development of many slopes, especially in humid temperate regions. Rainsplash is a particularly wide-

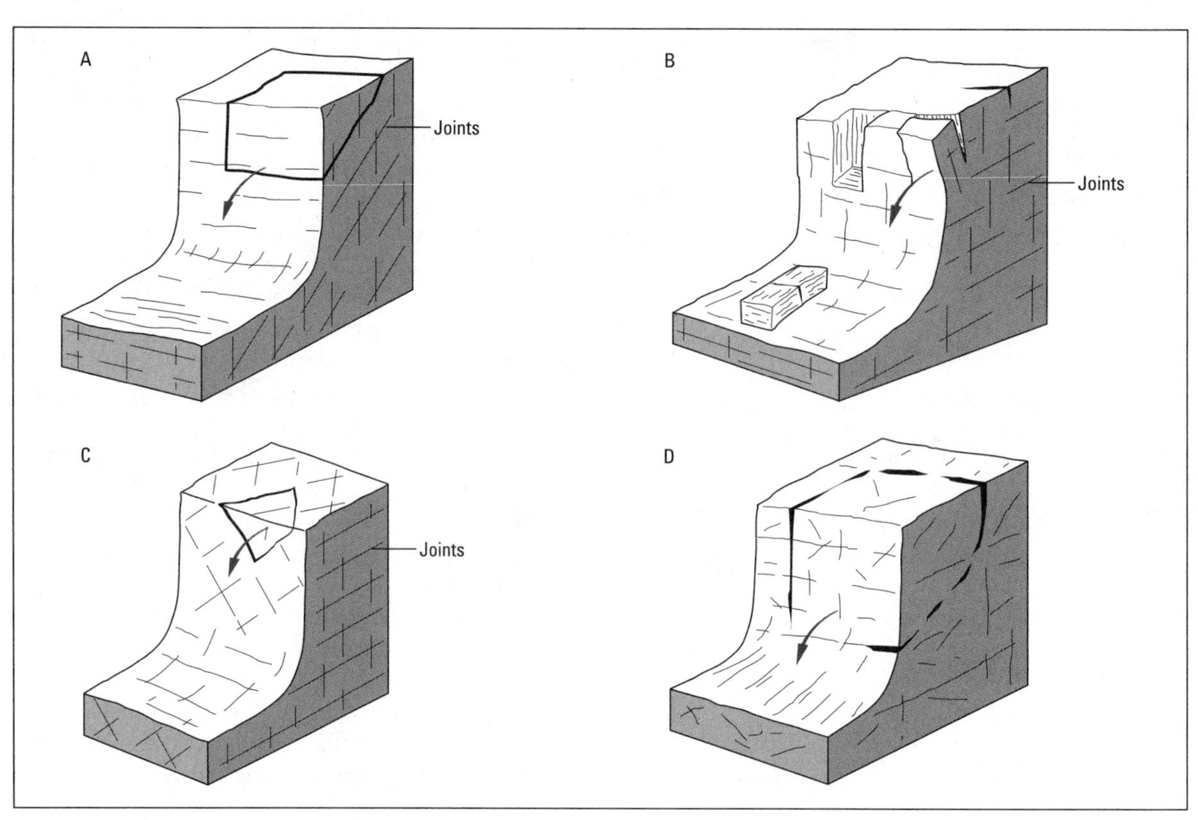


FIGURE 20.14 ▼ Types of large-scale failure in hard rocks: (A) slab failure; (B) toppling failure; (C) wedge failure; (D) rock-fall controlled by jointing

FIGURE 20.15 ▼ The Turtle Mountain landslide, Frank, Alberta, 1903 (photo courtesy of Frank Slide Interpretive Centre)

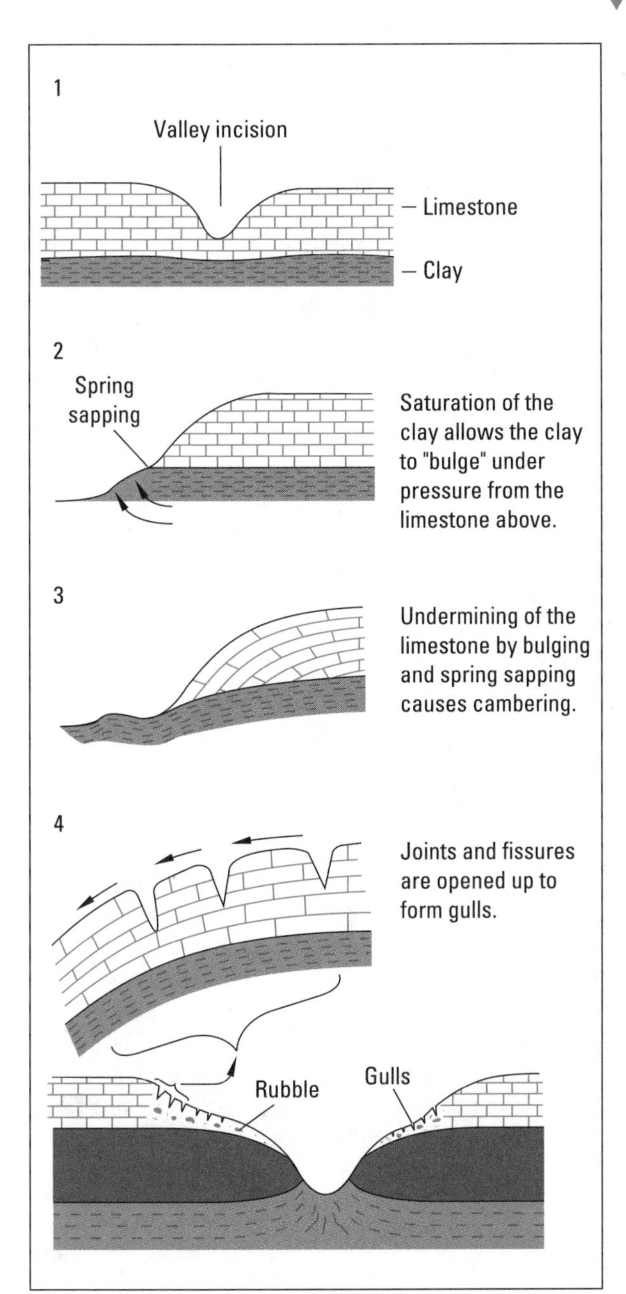


FIGURE 20.16 ▼ Gulling and cambering

FIGURE 20.17 ▼ Tension fissures have formed at the top of a cliff undergoing sagging or cambering in Newfoundland; a network of tension fissures is called a rock labyrinth (photo: D.R. Grant, courtesy Geological Survey of Canada)

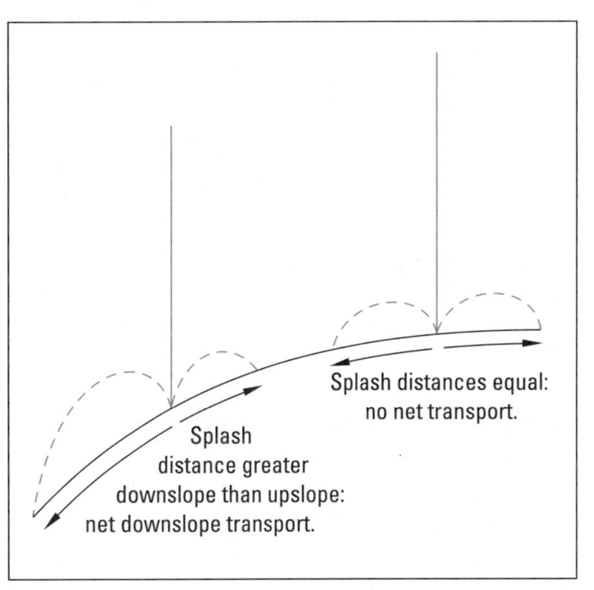


FIGURE 20.18 ▼ Particle movement by rainsplash

spread and effective process. The impact of raindrops on the soil detaches individual particles and throws them into the air. On a horizontal surface, the effect is to redistribute the material without any net transport, but on a slope there is a tendency for more material to be splashed down-slope than upslope (Figure 20.18). As a result, rainsplash leads to a net downslope movement of the material. Indeed, on slopes of about 25° or more, almost all the splash occurs in a downslope direction.

In addition, considerable transport of material takes place by surface runoff. On relatively smooth surfaces, runoff occurs in the form of **sheetwash**. The erosive power of this is normally limited because the sheet of water is shallow and non-turbulent and cannot readily entrain material. Rain falling through the flow, however, may detach soil particles and encourage some transport.

Sheetwash is rarely stable over long distances or for long periods of time because the turbulence caused by rainfall, or by local topographic irregularities, stimulates scour and leads to the development of small depressions. These rapidly extend into **rills** which, by a process of positive feedback, further concentrate flow and grow to form **gullies**. Over time, coalescence of gullies may result in large, permanent channels along which large quantities of sediment may be transported. Much of this material is disgorged at the slope foot, where the gullies open out onto the valley floor, and here the debris may build up to form an **alluvial fan**.

# Slope development

If we look at the landscape around us, we can detect many different types of slope. Their angles vary; so does their curvature. Some are straight (or rectilinear), some are concave, some convex. Many are composite in form.

At first, there may seem to be little logic to the distribution of slope forms. Yet, if hillslopes truly represent process-response systems, there should exist a close relationship between the processes acting on them and the character of the slope form that develops. Different processes should give rise to different slope forms. Over time, slopes should develop toward predictable, equilibrium forms. The question therefore arises: how do slopes evolve, and what governs their evolution? The search for an answer has occupied geomorphologists for many years. Three general models or concepts of evolution have been postulated: slope decline, parallel retreat, and slope replacement.

## SLOPE DECLINE

The concept of **slope decline** was proposed by W.M. Davis at the beginning of this century as part of a general model of landscape development. He argued that landscapes undergo a cyclical pattern of evolution, commencing with tectonic uplift and progressing through a phase of fluvial incision into a final phase of a low-altitude, low-relief plain called a **peneplain** (Figure 20.19). During fluvial incision, relatively steep slopes are formed, but as the river reaches a uniform concave profile (the **graded profile**) adjusted to sea level (the **base level**), downcutting ceases and hillslope processes lead to a gradual waning of the valley sides. Davis called the three stages of this development youth, maturity, and old age. He was never explicit about the exact nature of these processes, but he showed that the consequence is the development of concavo-convex slopes of progressively lower angle (Figure 20.19).

Although Davis's model of landscape development is now generally dismissed, such slope forms are, in fact, common in many humid temperate regions. It is also apparent that different hillslope processes tend to operate on different parts of the slope, and together these may result in slope decline (Figure 20.20). Thus, soil creep is active on the convex crest area. Here there is no input of material from upslope, while rates of transport increase as the slope gets steeper. Thus, soil creep is able to transport the debris as it weathers, little material builds up and, over time, this zone undergoes gradual lowering. By contrast, the steeper, rectilinear mid-slope receives inputs of material from upslope as well as

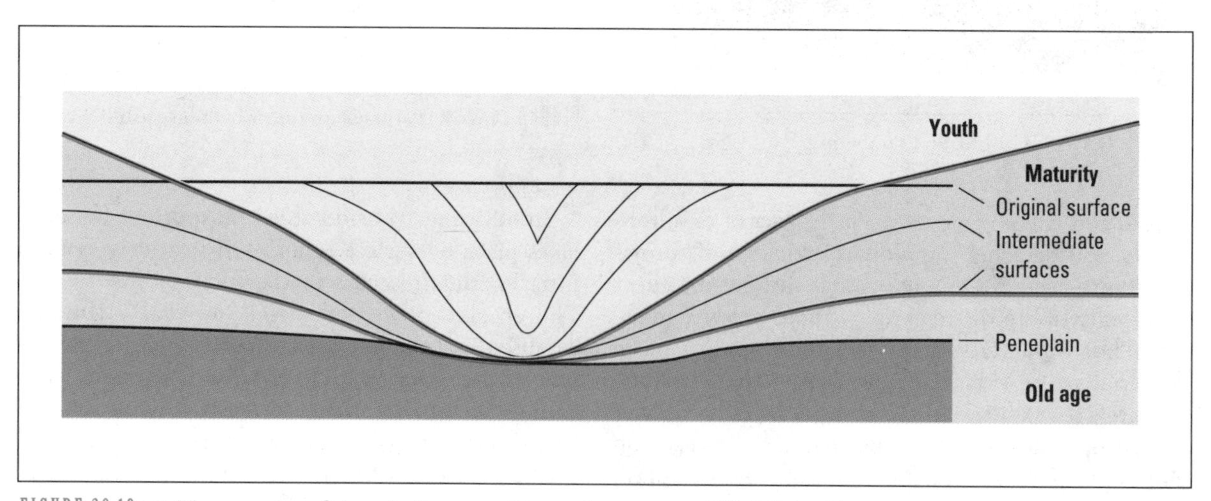


FIGURE 20.19 ▼ The sequence of slope decline according to the concept of W. M. Davis

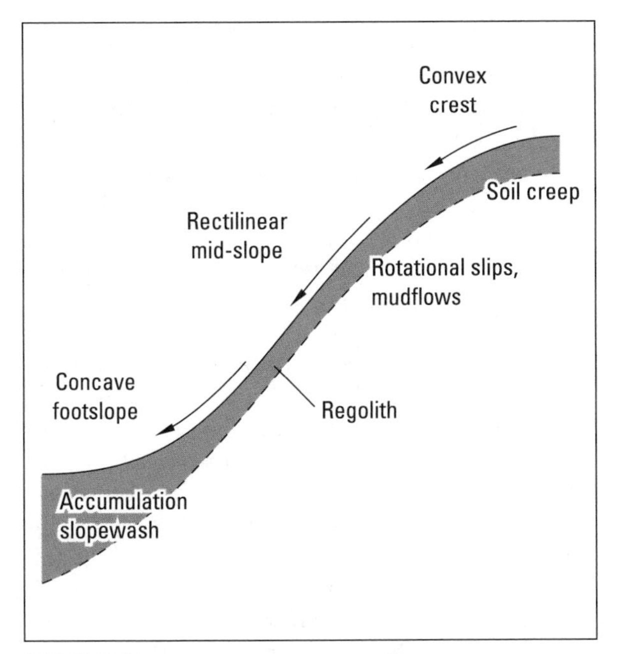


FIGURE 20.20 ▼ Processes operating on concavo-convex slopes

from weathering of the underlying bedrock. Larger quantities of material must be transported through this zone, therefore, and rapid mass movements often occur due to periodic oversteepening of the slope and the development of thick, unstable mantles of debris. Over time, this segment is also lowered while maintaining its essentially straight form.

At the slope foot, a basal concavity evolves due to the slower rate of decline. This segment receives all the debris removed from the whole of the slope length, which accumulates on the footslope and protects the underlying bedrock from weathering. Transport of material through this zone is primarily by sheetwash and by removal of debris at the base of the slope, usually by stream action (**basal erosion**). Where basal erosion is active, the concavity does not develop, but where erosion is limited inputs to the footslope exceed outputs and the concavity is marked (Figure 20.20).

## PARALLEL RETREAT

Concavo-convex slopes are, of course, by no means ubiquitous. In many semi-arid areas, for example, more complex profiles are common, comprising an upper convexity, a steep cliff or **free-face**, a rectilinear **debris slope**, and a gently concave **pediment** (Figure 20.21).

The evolution of these slopes has been considered to take place by a process of parallel retreat. This idea was most strongly advocated by the South African geomorphologist L.C. King. According to King, each of these slope segments is characterized by different processes. Thus, rainsplash and soil creep are the main processes on the crest, while rockfalls and landsliding are active on the free-face. On the debris slope, sliding and avalanching of the unconsolidated materials occur, so that the slope angle is determined by the angle of residual shear of the debris. The pediment is maintained largely debris-free by the process of sheetwash.

Slope development in this situation, King argued, is controlled by loss of material from the free-face. As this takes place, the debris slope also moves back (though by a process that is not always clear), with the result that the whole slope under-

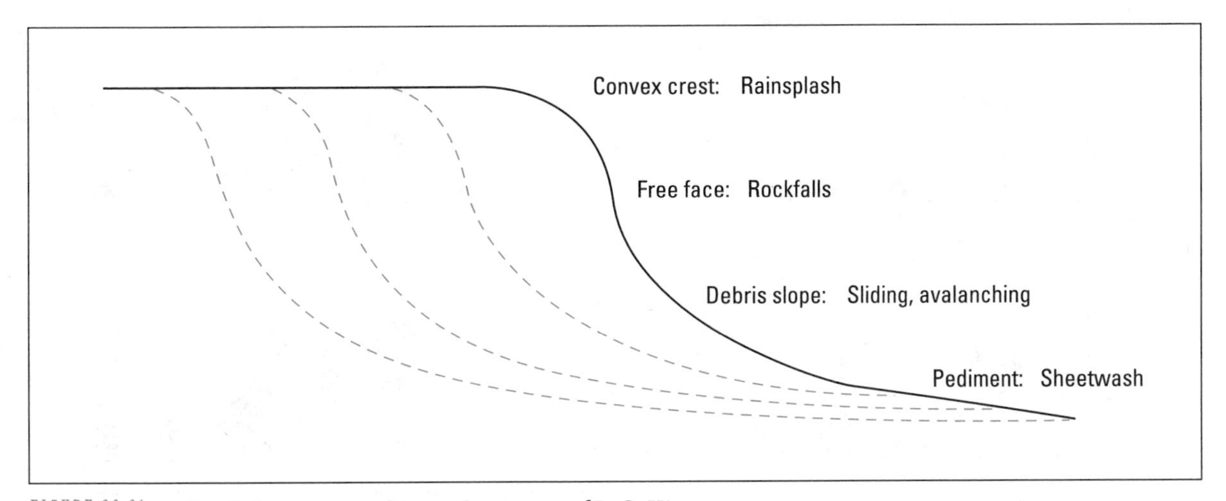


FIGURE 20.21 ▼ Parallel retreat according to the concept of L. C. King

goes parallel retreat (Figure 20.21). As a consequence, the pediment is gradually extended while the hilltops diminish. Late in the cycle of erosion the hills are left as isolated, steep-sided relicts, called **monadnocks** in eastern North America (Figure 20.22), **buttes** or **mesas** in the west, and **inselbergs** in Africa.

FIGURE 20.22 ▼ A monadnock: Main Topsail in Newfoundland is an erosion remant of deep weathering during the Tertiary warm period; the mountain has been further smoothed by recent glacier activity (photo: D.R. Grant, courtesy Geological Survey of Canada)

King's model of slope evolution appears to be broadly valid in the semi-arid and geologically stable environment of southern Africa. Like Davis's model its relevance has been challenged, although King himself has claimed that parallel retreat and 'pedimentation' is an almost universal mechanism of landscape development. Whether or not this is so, parallel retreat can certainly be envisaged in other situations, notably where cliffs or rectilinear hillslopes are being affected by lateral basal erosion. This might take place, for example, where a stream is undercutting a hillside (Figure 20.23). This process is most likely to occur in sections of the valley as the stream migrates across the valley floor. It may also take place where the sea is attacking a coastal cliff. As long as the rate of undercutting is balanced by the rate of debris transport on the slope, and greatly exceeds the rate of weathering and slope lowering, a form of parallel retreat may occur.

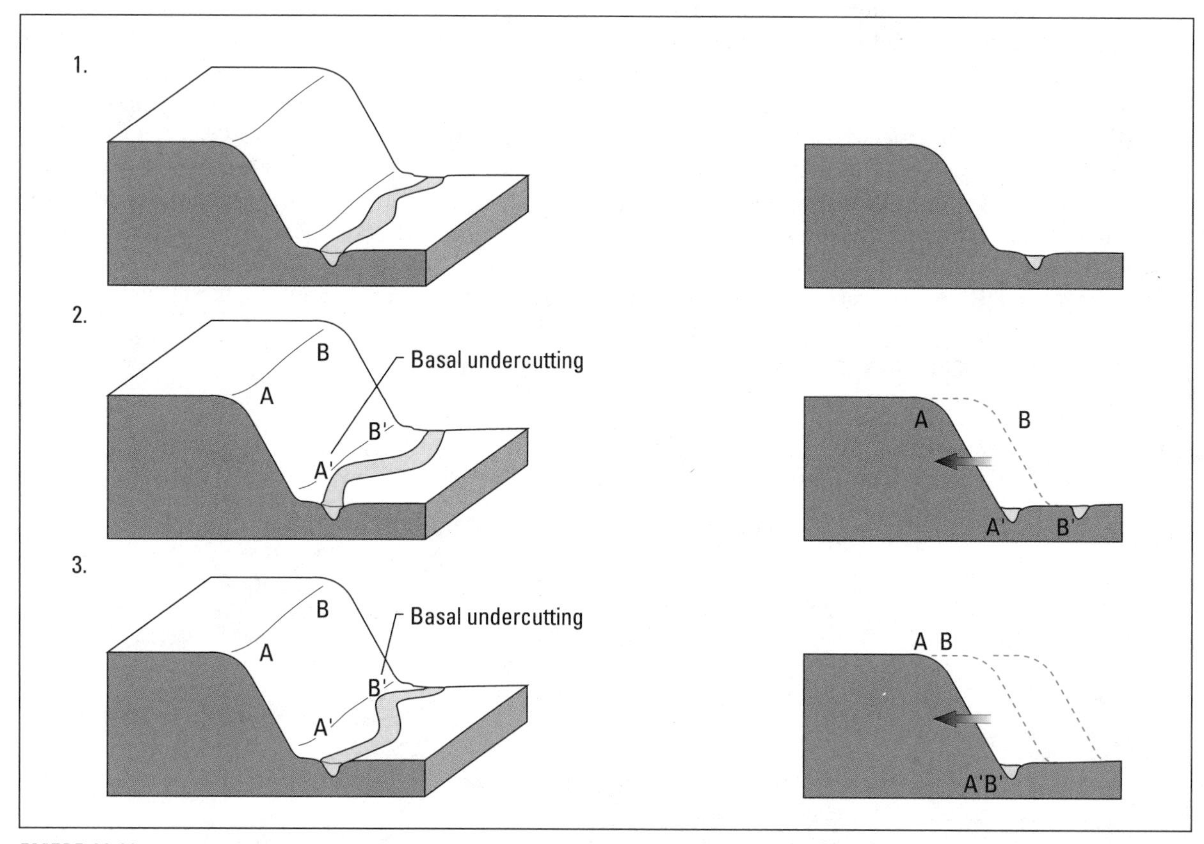


FIGURE 20.23 ▼ Parallel retreat due to basal undercutting of a valley side-slope by a stream

## SLOPE REPLACEMENT

One of the weaknesses in King's model of slope development is the process by which the debris slope retreats in pace with the free-face. In the absence of active basal erosion it is difficult to see why this should happen, and modern studies of cliff-talus slopes indicate that in many cases it does not. Instead, a process of slope replacement occurs. Simply, as the cliff retreats, the talus (equivalent to King's debris slope) extends until it totally covers the surface.

The mechanism by which this happens is shown in Figure 20.24. Initially, material falling from the cliff face accumulates as an ill-sorted pile of debris at the base. Usually, this material forms a cone shape along the base of the cliff, called a **talus cone**. As these cones grow they tend to overlap each other until a general slope is formed known as a **talus slope** or scree slope (Figure 20.25). This is comparable to King's debris slope. As this pile grows, debris falling onto it is able to bounce and roll away from the cliff due to the high impact velocity of the particles. This results in the development of a concave, low-angle talus slope. In time, as the cliff continues to weather and retreat, the fall height—and thus the impact velocity—of the material declines. More debris therefore gets trapped on the talus slope, steepening the angle and eliminating the concavity. Ultimately, as the cliff height declines to zero, debris is released onto the talus with a very low impact velocity; movement then takes place by sliding and avalanching, and the talus tends toward the angle of residual shear.

A critical element in the transfer processes of debris on the slope in most mountain regions is the occurrence of snow avalanching and slush flows as the snow melts. There is a transition in slope angle of debris accumulation from high-angle talus to lower-angle debris flow fans to low-angle alluvial fans, depending on which process dominates the debris transfer. In many paraglacial and periglacial areas the presence of ice in the deposits results in flow of the debris. The forms produced by this flow are called rock glaciers (Figure 20.26).

These rock glaciers are an excellent illustration of why the Davis and King type models have limited

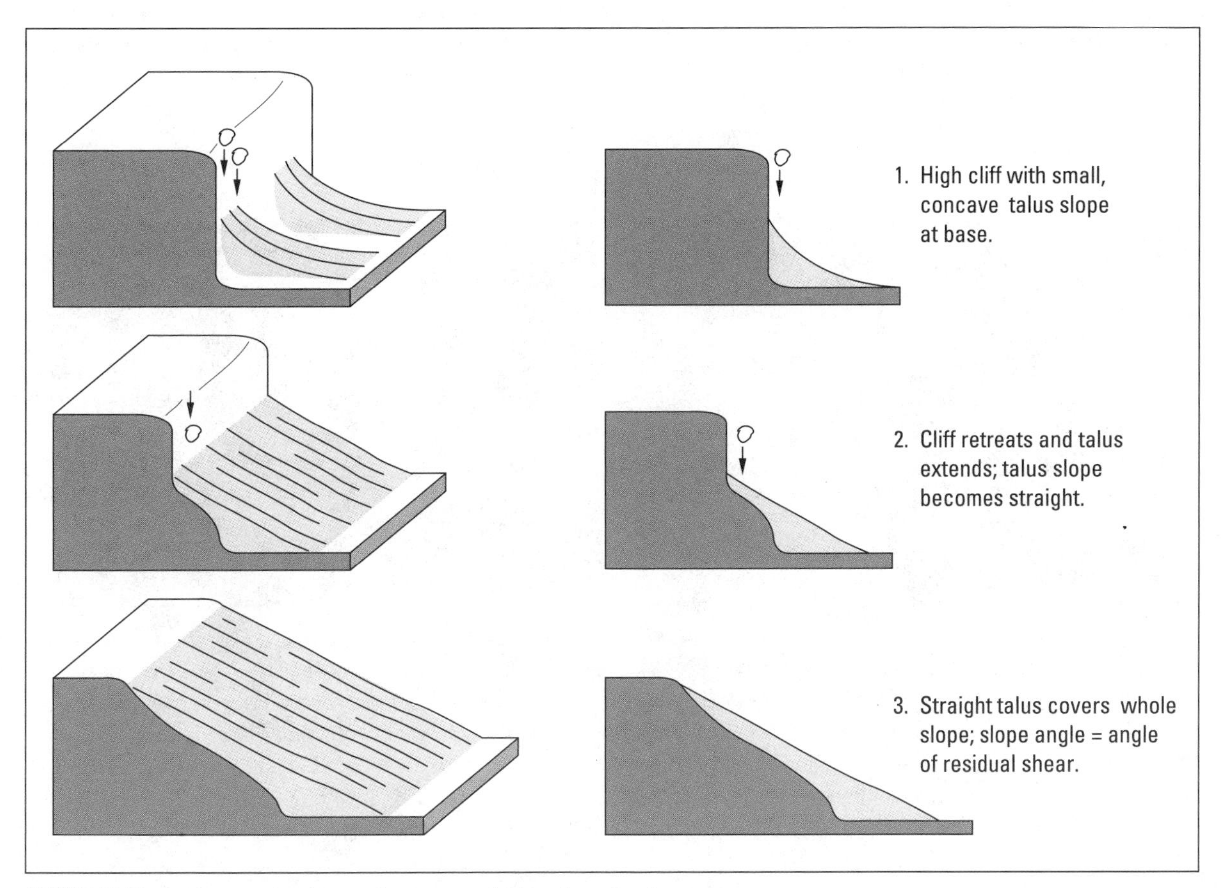


FIGURE 20.24 ▼ Slope replacement: the process of talus slope development

FIGURE 20.25 ▼ Talus slope development in granodiorite, Rockslide Creek, Ruby Range, southwest Yukon. Two phases of talus accumulation are visible here: the darker slope is older, weathered, high-angle talus; the lighter talus cones are more recent deposits at the base of chutes from below the remaining rock exposure at the hill top (photo: P.G. Johnson)

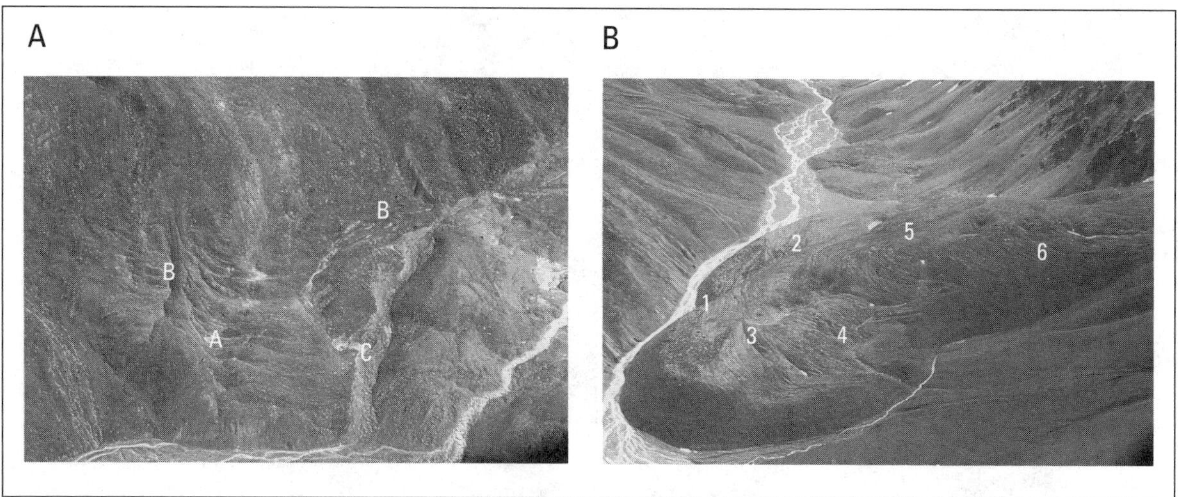


FIGURE 20.26 ▼ Rock glaciers, Grizzly Creek, St. Elias Mountains, Yukon. *A.* A rock glacier developed from the base of a talus slope; flow is caused by formation of ice in the rock under periglacial conditions. Note the closely spaced arcuate flow ridges (A), lateral levees from different periods of flow (B), and the area where the neoglacial advance of a glacier eroded part of the rock glacier (C); the last indicates that the rock glacier has not been active for hundreds of years. *B.* A rock glacier has developed in the terminus area of a glacier, where a moraine with a glacier-ice content continues to flow after the retreat of the glacier. These forms typically have large lobes with some minor flow ridge detail. Numbers 1 to 6 indicate progressively younger flow lobes (photos: P.G. Johnson)

applicability to northern environments, where major adjustments to recent (Pleistocene and Holocene) extensive glaciations are still taking place. Neither Davis nor King considered any ice-related processes in their models.

### ENVIRONMENTAL INFLUENCES ON HILLSLOPE DEVELOPMENT

As mentioned earlier, much attention has been given to finding a universally applicable model of slope development. None of the models so far proposed, however, seems to have general validity, and different slopes appear to evolve along various pathways to different equilibrium forms. Why is this so?

One reason is that processes of hillslope development vary with climate. In humid temperate regions, for example, chemical weathering tends to produce a deep soil mantle which supports a dense soil and vegetation cover. Moreover, rainfall intensities are low. As a consequence, infiltration capacities are high relative to rainfall inputs, little overland flow is generated, and the surface is protected from rainsplash. On the other hand, retention of water in the soil favours processes such as soil creep and shallow sliding. In contrast, in arid regions there is little vegetation cover, soils are often thin, and processes such as rainsplash and sheetwash predominate. In periglacial environments, seasonal saturation of the slope materials encourages processes such as mudflows and solifluction.

In different climatic regions, therefore, different processes tend to dominate, and these give rise to different slope forms. Nevertheless, climate is not the only control on slope development. Geology also exerts an important influence. As we noted earlier, this governs the susceptibility of the material to weathering and thus influences the rate of debris input. Rock structure also controls the strength of the material and the range of angles over which it will form stable slopes.

It should also be remembered that slopes take a long time to develop, and that environmental conditions themselves do not always remain stable. The effects of climatic variation, tectonic activity, or changes in sea level may all be reflected in the present form of hillslopes. Indeed, because of repeated environmental changes some slopes may never achieve complete equilibrium with the prevailing conditions.

In general, therefore, we need to consider slope development not in terms of universal processes, acting toward a single form, but as the adjustments of hillslope systems to spatially and temporally varying inputs and outputs. In other words, as in all process-response systems, the form that hillslopes progress toward is a function of the processes acting upon them. These, in turn, are dependent upon the interaction between the hillslope materials and the surrounding environment.

## Slopes and humans

It is apparent that slope failure may have terrible consequences for humans. On a small scale, collapse of embankments and localized landslips may cause inconvenience and endanger life (Figure 20.27); but on a larger scale the effects may be catastrophic. In 1970, for example, in Peru a spate of small rockslides was triggered by an earthquake and these caused local damage. More tragically, a major failure was initiated on the near-vertical face of Mount Huascaran. The material fell a distance of 650 m and crashed into loose glacial debris at an altitude of about 6000 m above sea level. Much of this debris was picked up as the material then tumbled down the mountainside, taking only two minutes to travel 14.5 km—a speed of 400 km $hr^{-1}$. The slide destroyed all in its path, including the lives of 21 000 people. Even after it had slowed to a mere 25 km $hr^{-1}$, it continued for a further 50 km (Figure 20.28).

FIGURE 20.27 ▼ Landslide at Besham, on the Karakoram Highway, north Pakistan (photo: P.G. Johnson)

FIGURE 20.28 ▼ Mount Huascaran, Peru, site of the devastating 1970 debris flow. The slide travelled down a steep valley (A) onto an ancient debris-flow fan (B); the ridge (C) is the edge of the 1970 flow (photo: R.J. Wesche)

A similarly horrific story can be told of the Langarone landslide in northern Italy. The rocks of the Langarone valley dip inward, toward the valley, with the result that they have very low shear strengths. (The shear strength is the same as stress strength as defined in Chapter 16. It refers to the ability of a slope to remain stable at its angle of repose.) Despite this, a huge reservoir—Lake Vaiont—had been constructed. Under the effect of the positive pore water pressures of this lake, and lubricated by prolonged and intensive rainfall, part of the valley side began to slide in the spring of 1963. Although attempts were made to monitor the slide to assess the problem, the scale of the event was initially underestimated, for the slide was so big that it was taking with it the supposedly fixed markers against which its progress was being measured! The danger was realized too late and attempts were made to avert disaster by draining the reservoir. On 9 October, 1963, an area of the valley side covering 1.8 by 1.6 km collapsed into Lake Vaiont, creating a flood wave that overtopped the dam and swept down the valley. At least 2600 people were drowned.

As this example indicates, hillslopes can be hazards, and people are often responsible for increasing the hazard by mismanagement of the land. This is also illustrated by the landslide at Turtle Mountain in the Canadian Rockies, in 1903 (Figure 20.15). The mountain is composed of Devonian and Carboniferous limestones which have been thrust-folded so that they dip at 50°–65°. The main joint system is inclined at 25°–40° into the valley, and the valley sides have been oversteepened by glacial erosion. The whole area is thus potentially unstable; but despite this, a coal mine was opened in the hillside in 1901. By October of the following year, repeated small collapses were occurring and gangways and chambers had to be constantly repropped or even abandoned. Nevertheless mining continued and chambers up to 120 m in height were excavated. Then, on 29 April, 1903, the roofs began to cave in. The miners escaped to the surface just in time, while behind them 9 million tonnes of rock fell almost 800 m down the mountainside. The rock hit the ground and rebounded over a sandstone ridge into the Old Man River. There it ploughed on, climbing almost 120 m up the opposite valley side and covering an area of 2.5 km$^2$. Seventy people were killed.

The lesson is clear: we need to understand how hillslopes function if we are to cope with the hazards they present.

# Streams

## The hydrological cycle

To understand the development of landscape, particularly in high latitude and midlatitude continental regions, we must develop the discussion of runoff and storage from Chapter 13. The hydrological regimes in areas dominated by snow and ice, and the geomorphic process associated with them, are fundamentally different from those in temperate midlatitude regions and low latitude regions. We will compare the hydrological regimes dominated by seasonal snowmelt, snowmelt and glacier melt, and direct runoff, and consider the role of surface ice formation in the winter on rivers, lakes, and oceans.

### THE UNIFYING CONCEPT OF GEOMORPHOLOGY

The transfer of sediment on the earth's surface, introduced in the discussion of landscape form and process, is the basis of geomorphology. Sediment is generated from bedrock by weathering and erosion, then transported and deposited by water, ice, and air to create a wide variety of landforms; as part of the process, sediment is returned to the geosynclinal basins, where new rock may be formed. This transfer of sediment is controlled by the presence or absence of water, by water runoff, by the formation and movement of glaciers, and by the promotion of aeolian activity. We have already seen how presence or absence of water, in conjunction with other climate variables, controls physical, mechanical, and biological weathering processes.

A landform is the physical expression of the current stage of movement of material between initial weathering and erosion through a number of transport and storage situations. A landform may be the residual bedrock, with its own distinctive form, or residues from the weathering and erosion of the bedrock. Landforms may be the result of the action of different processes over time (for example, glacial erosion and transport modified by fluvial or coastal processes); or the result of the same process a number of times (for example, cycles of fluvial erosion, transport, and deposition). The hydrological cycle is, therefore, the concept that provides unity to geomorphology. All the subdisciplines of geomorphology, such as glacial, fluvial, coastal, periglacial, and karst geomorphology (which are the subjects of subsequent chapters), can be related to each other by reference to the hydrological cycle.

The hydrological cycle portrays the movement of water from the oceans, through atmospheric systems, through terrestrial and subterranean systems and back to the ocean reservoirs. As we have seen, the concept can also incorporate the lack or absence of water in desert environments. The standard representation of the hydrological cycle is the temperate region cycle; perhaps this reflects the predominance of human activity in these regions. Diagrams of the temperate cycle have been the accepted format in texts adopted in high latitude countries as well as in low latitude countries. This leaves out important elements of the high latitude cycle, in particular the fact that ice is a very important element of all components of the high latitude cycle. Ice occurs in various seasonal and semi-permanent forms in high latitudes, and these regulate the movement of water through the physical environment.

Over 95 percent of Canada is snow covered for part of the winter. Thus water is stored in varying quantities and for varying lengths of time, depending on geographic location, until the spring melt when it is released into the river systems. This seasonal storage characteristic is a fundamental component of northern hydrology. The storage and release regimes are also critical to the operation of geomorphic processes. Due to the low temperatures experienced in northern latitudes and at very high altitudes, there is also widespread development of ice cover on lakes, rivers, and the ocean; this regulates physical processes through the winter, and particularly in the spring during ice breakup. There are also many areas of semi-permanent storage of water on the earth's surface in glaciers and icefields, and beneath the surface in periglacial ice.

We must look, therefore, at both the temperate and northern hydrological cycles, and an introduction to hydrology as the basis of geomorphology.

## THE TEMPERATE HYDROLOGICAL CYCLE

The temperate hydrological cycle is illustrated in Figure 21.1 This shows a range from regions of abundant precipitation to regions of low precipitation. You will remember, from the discussion of climate, that the distribution of precipitation through the year ranges from well-distributed to highly seasonal regimes, so that geomorphological processes depend both on amount and distribution. Water is routed back to the ocean by surface runoff into rivers, or by infiltration and groundwater flow. The runoff response depends on a number of factors:

- the topography of the basin
- the slope of the basin
- the immediate precipitation history
- the characteristics of the precipitation event, that is, the areal extent, duration, intensity, and direction of the storm in comparison with the basin orientation

Water infiltrating into ground materials may be routed close to the surface (interflow) or deeper in geological structures (groundwater flow). Within geological structures, it may be stored for hundreds or thousands of years in porous rock structures called aquifers. Groundwater may reemerge at the surface as springs, or by more general release called groundwater recharge, into lakes or rivers.

Within the cycle, there are many feedback loops to the atmosphere, by evaporation from water bodies and the ground, and by evapotranspiration from plants. Feedback loops may be local or regional in scale.

Figure 21.1 shows desert areas as low precipitation environments where watercourses are intermittent, evaporation from groundwater increases salt

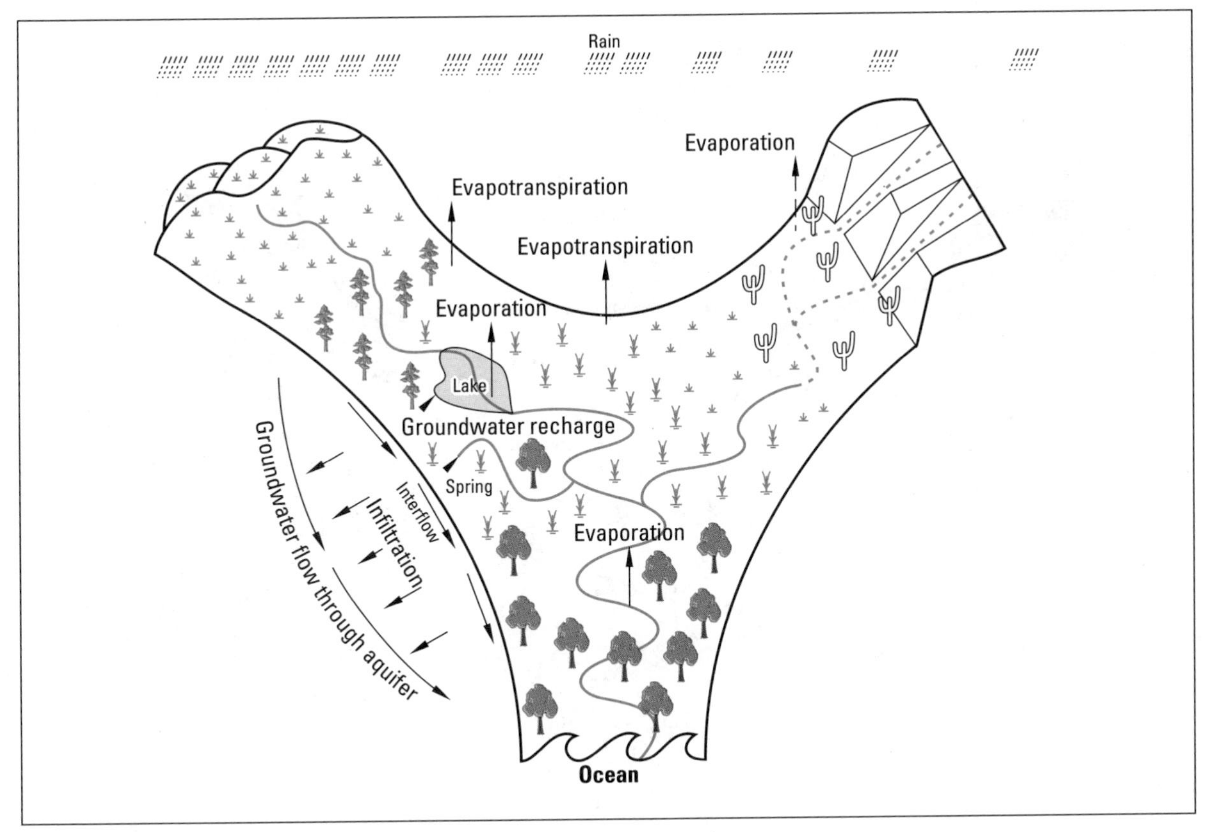


FIGURE 21.1 ▼ The temperate region hydrological cycle

concentrations in the soil, and geomorphic processes are dominated by aeolian conditions. The occasional flash flood, due to high magnitude rainstorms with a frequency of 10 to 20 years, has major geomorphological effects on the transfer of sediment and migration of stream channels.

## THE NORTHERN HYDROLOGICAL CYCLE

The snow and ice components of northern regions create significant differences in the hydrological cycle. The northern cycle is shown in Figure 21.2. The northern environment can be divided into three overlapping zones.

1. A zone of semi-permanent storage of water in the form of glacier ice, periglacial ice, or multi-year sea ice.
2. A zone of seasonal storage with snow, river ice, lake ice, and sea ice that melt in the spring.
3. A zone where there is no solid storage.

The presence of ice in the ground precludes, or interferes with, the movement of water into, and in, the subsurface system. It also holds summer melt at the surface and in the active layer, increasing evaporation rates. Glaciers and ice fields store large volumes of water over hundreds and thousands of years. Some release in the summer depends on the seasonal climatic conditions. Snow accumulates through the winter, and is then rapidly released during the spring melt. In addition to evaporation and evapotranspiration return to the atmosphere, sublimation transfers directly from the solid form.

## HYDROLOGICAL REGIMES

We can see the effects of the northern hydrological cycle by looking closely at three examples of river regimes in Canada.

1. One that is primarily pluvial in character, with immediate response to rainfall through the year: the Sarita River.
2. One that is nival in character with a major spring snowmelt peak: the Pelly River.
3. One that has a glacier component to the basin, so the regime is a combination of nival and

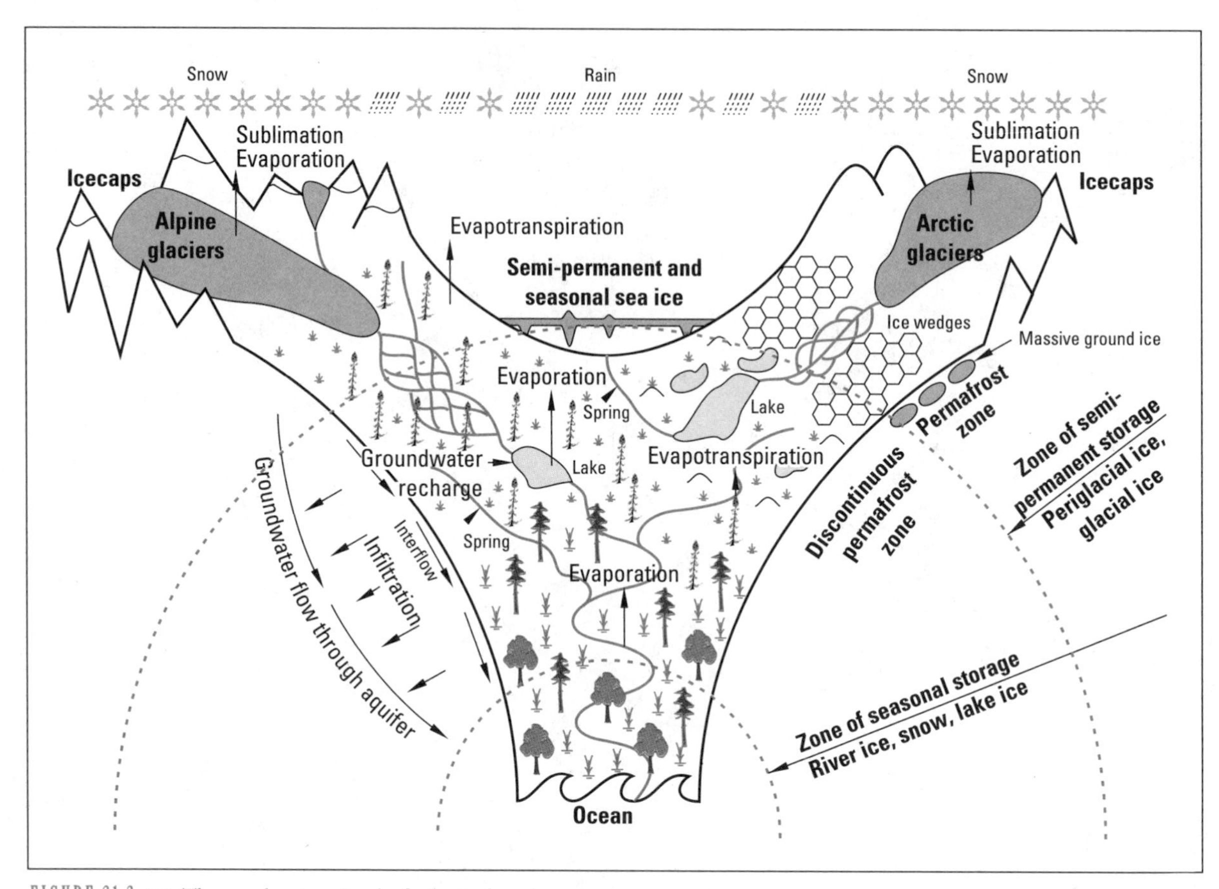


FIGURE 21.2 ▼ The northern region hydrological cycle

glacial: the Lillooet River, which has snowmelt from the non-glacierized part of the basin, snowmelt from the glacier, and glacier ice melt.

The Sarita River, a small river near Bamfield on the west coast of Vancouver Island, illustrates the pluvial basin. The West Coast islands receive most of their precipitation in winter, and the peak flow of the river occurs during the period October to March. The monthly mean flows (Figure 21.3) show the basic annual regime, and the monthly maximum and minimum indicate that the flow is very variable, reflecting the large fluctuations in precipitation regimes on the West Coast. The monthly means mask some important characteristics that are critical to physical processes. The river discharge varies dramatically from day to day (Figure 21.4) in response to precipitation events. This is only a small river basin, 162 km$^2$ in area, so the response to precipitation is very rapid, but the 1986–87 daily means illustrate the variability. What are the implications of this? The main importance perhaps is

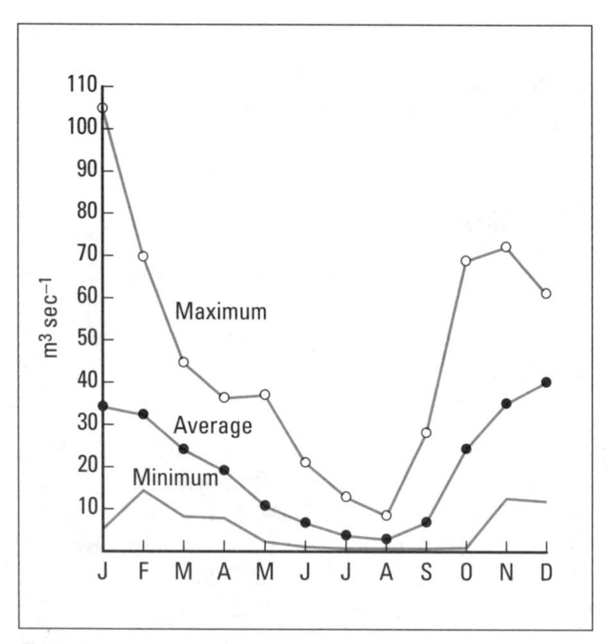


FIGURE 21.3 ▼ Monthly mean flows of the Sarita River, Vancouver Island

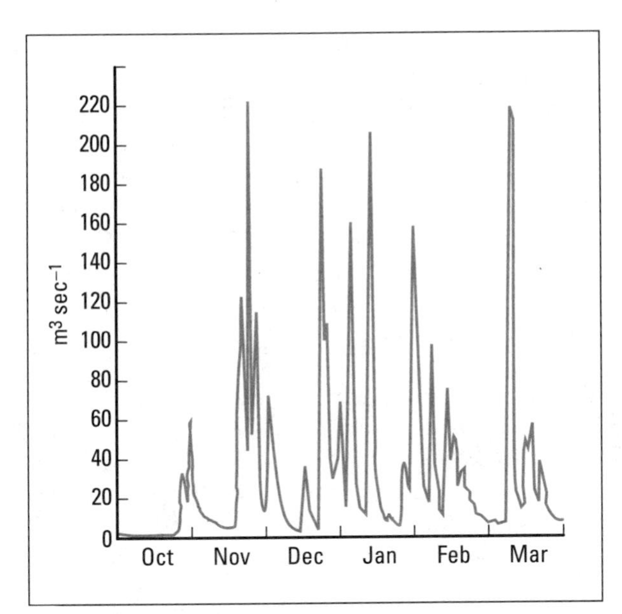


FIGURE 21.4 ▼ Daily mean flows of the Sarita River, 1986–87

that coarse sediment movement occurs as a series of mobile and stationary events, it is not a continuous transport from source to the final ocean sink.

The Pelly River, Yukon, illustrates the nival, snowmelt-dominated regime. Again, monthly mean figures (Figure 21.5) show the general annual pattern, with very low discharge through the late winter, the period of snow accumulation. The snowcover melts rapidly in May and June, producing a very sharp regime peak which then declines rapidly through the rest of the summer with the end of the snowmelt. The flow in July to October declines to the winter level except for some major rainstorm inputs. The summer records of 1988 and 1989 show the daily changes. The snowmelt peak finishes in a few days, so that the average flow in June (700 $m^3 sec^{-1}$) may have peaks of one or two days of over 1000 $m^3 sec^{-1}$ and the maximum June average of 985 $m^3 sec^{-1}$ had a peak daily flow of 1200 $m^3 sec^{-1}$. In both 1988 and 1989, the snowmelt occurred in two peak events (Figure 21.6) in early May and in early June. In 1989, there was then a steady decline in flow for the rest of the summer. In 1988, the summer decline was interrupted by a very wet period in July, a regional weather system that covered most of the northwest.

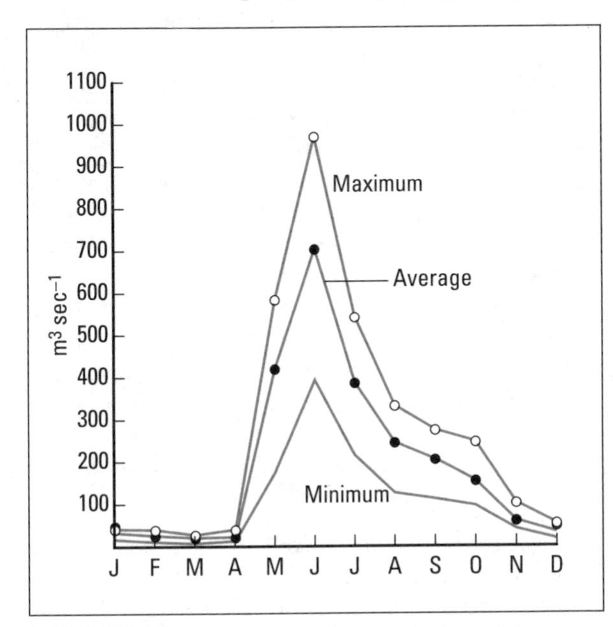


FIGURE 21.5 ▼ Monthly mean flows of the Pelly River, Yukon

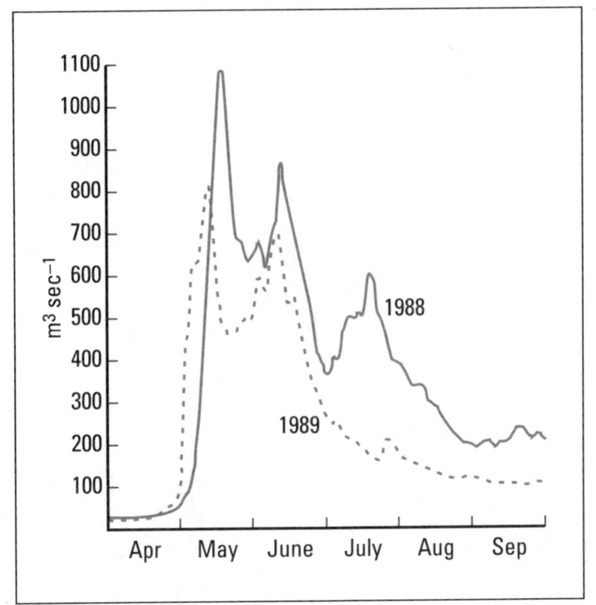


FIGURE 21.6 ▼ Daily mean flows of the Pelly River, 1988–89

The Lillooet River of southern British Columbia is the combined glacierized/nival basin. The mean monthly figures (Figure 21.7) show the essential characteristics of a low winter flow and a spring snowmelt rise in May and June, but then a continuation of water supply from snowmelt on the glacier ice and melt of the glacier ice itself. This extends the summer peak flow into July and August. The daily fluctuations (Figure 21.8) show the strong variations of snow and icemelt, with high runoff peaks correlated either with periods of strong insolation or with rainstorms on the glacier snow and

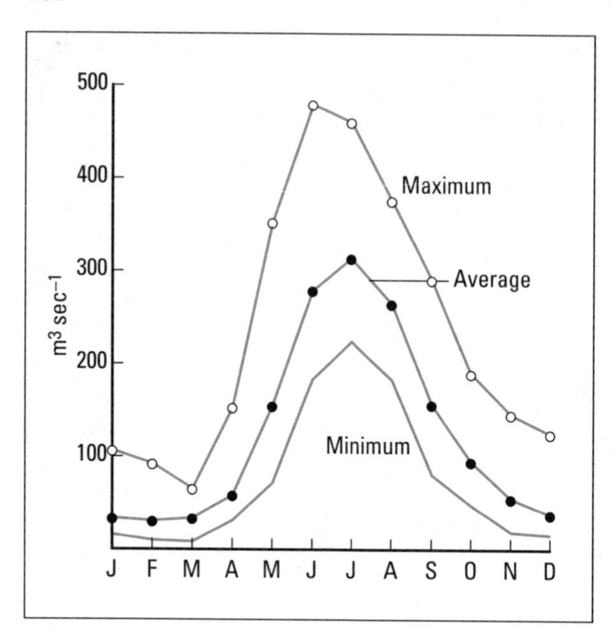


FIGURE 21.7 ▼ Monthly mean flows of the Lillooet River, British Columbia

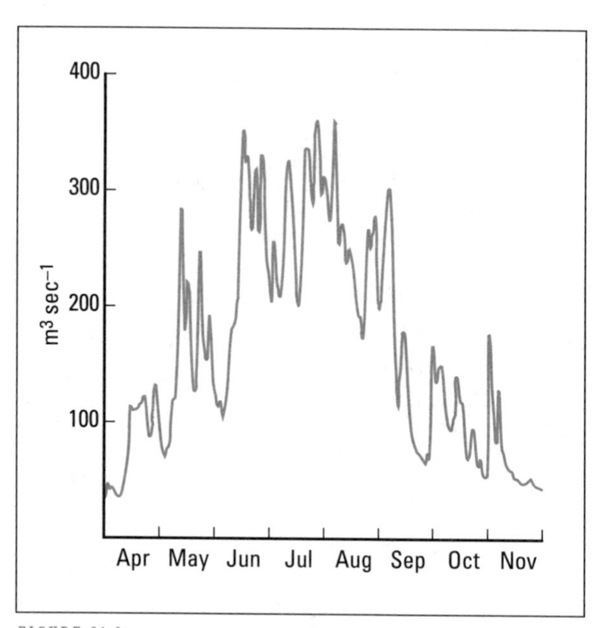


FIGURE 21.8 ▼ Daily mean flows of the Lillooet River, 1988

ice which promote rapid response. In addition, these basins will also have a high degree of fluctuation of discharge over 24 hours.

These are three very different regimes. As we progress north into the Arctic, the snowmelt regime can become even more concentrated, into a few days, and much of the runoff is over snow, so there is little potential geomorphic impact. As we progress into these northern latitudes, the snowmelt peak comes later in the year and the period of glacier melt is very short. Consequently, in the Arctic Islands, 99 percent of stream discharge occurs during a three-month period. Similar trends are apparent in higher altitudes with shorter periods of snow and glacier ice melt.

### SEDIMENT TRANSPORT

The sediment transport regimes are also concentrated into short periods, corresponding to the main spring flood event, for nival basins. In most cases, more than 75 percent, and frequently more than 90 percent of sediment transport occurs in less than 30 days. In glacierized basins, the sediment transport is concentrated in the ice melt period, and particularly during the period when the glacier drainage system is subglacial (This will be discussed in Chapter 23).

## Streams and the landscape

### THE RIO GRANDE

> In all its career the Rio Grande knows several typical kinds of landscape. . . . It springs from tremendous mountains. . . . It often lies hidden and inaccessible in canyons. . . . From such forbidding obscurities it emerges again and again into pastoral valleys of bounty and grace. . . . In such fertile passages all is green, and the shade of cottonwoods and willows is blue and cool, and there is reward for life in water and field. But . . . the desert closes against the river, and the gritty wastelands crumble into its very banks, and nothing lives but creatures of the dry and hot; and nothing grows but desert plants of thirsty pod, or wooden stem, or spiny defense. But at last the river comes to the coastal plain. . . . After turbulence in mountains, bafflement in canyons, and exhaustion in deserts, the river finds peaceful delivery into the sea, winding its last miles slowly through marshy bends, having come nearly one thousand nine hundred miles from mountains nearly three miles high.

That, in the words of a traveller, is the Rio Grande between Texas and Mexico. And as these words demonstrate so vividly, this mighty river—the second largest in the United States—is an active and intimate part of the landscape. It acts as a major

FIGURE 21.9 ▼ A laboratory flume and a compound V-notch weir construction used to study stream processes (photo courtesy Canada Centre for Inland Waters)

FIGURE 21.10 ▼ Extreme pothole development in the Maligne River canyon, Alberta (photo: P.G. Johnson)

artery for water moving from land to sea, and as a vital component of the debris cascade. It is not difficult to accept, when we see the Rio Grande, that rivers are important agents in carving and moulding the landscape.

Yet we do not need to go to the great rivers of the world to see these processes in action, for, within broad limits, the action of running water in the landscape is similar whatever its scale. A trickle of water across a pile of sand, a gully on a hillside, a mountain stream, or a huge river such as the Mississippi or Rhine are all governed by the same principles, and they all interact with the landscape in a similar fashion. Indeed, much of our knowledge of fluvial process has been derived from small streams, and even from artificial streams created in laboratory flumes (Figure 21.9). The natural small stream is much easier to study. The flume allows for simulation of stream process, but it is important to remember that these are very small artificial streams, and may tell us little about big streams.

## THE WORK OF STREAMS

Moving water and sediment alter the landscape. The stream erodes its bed and banks, undercutting cliffs, carving deep channels, gorges, and canyons. It picks out weaknesses in the rocks and creates potholes, rapids, and waterfalls (Figures 21.10 and 21.11). It deposits vast spreads of alluvial silt and clay at its margins to produce extensive, flat, often boggy floodplains (Figure 21.12). It constructs smaller features within its channel—gravel bars, dunes, ripples (Figure 21.13). In northern regions, rivers are ice-covered in the winter and discharge is low so that there is little geomorphic activity. With the spring melt and ice breakup, the physical activities of iceflows on the shore can generate sediment. Ice jams can cause flooding and major physical effects on the river banks. Over time, the action of the streams, ever changing, with alternate periods of erosion and deposition, results in the gradual movement of the channel across the surface. In areas of high relief this may be primarily incision of the channel. In areas of low relief, depending on sediment load and river regime, this may result in the development of braided systems, the development of meanders, the diversion of streams to leave oxbow lakes and abandoned meanders. Over even longer periods of time, net aggradation may be superseded by net erosion of the stream bed, and the old flood-

FIGURE 21.11 ▼ Takakkaw Falls, Yoho National Park (photo: P.G. Johnson)

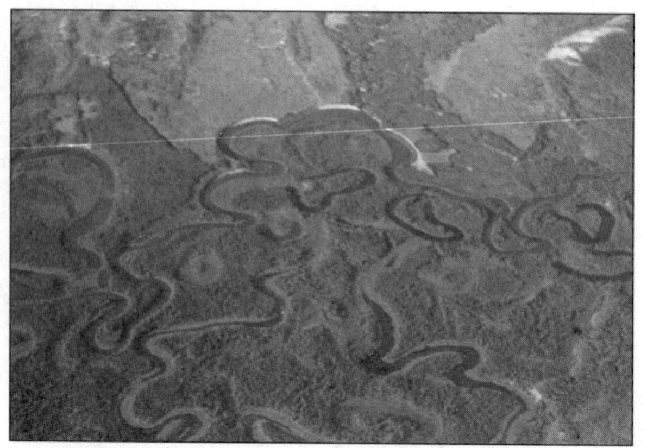

FIGURE 21.12 ▼ The floodplain of the Takhini River, Yukon; the river migrates across the floodplain over time, producing the intricate old and active channel network (photo: P.G. Johnson)

plains may be converted into extensive river terraces. These and many other features are characteristic of fluvial landscapes—we will examine them in detail later in this chapter. Many of the landforms we see around us, even in the most arid regions of the world, owe their origins, at least in part, to the action of running water.

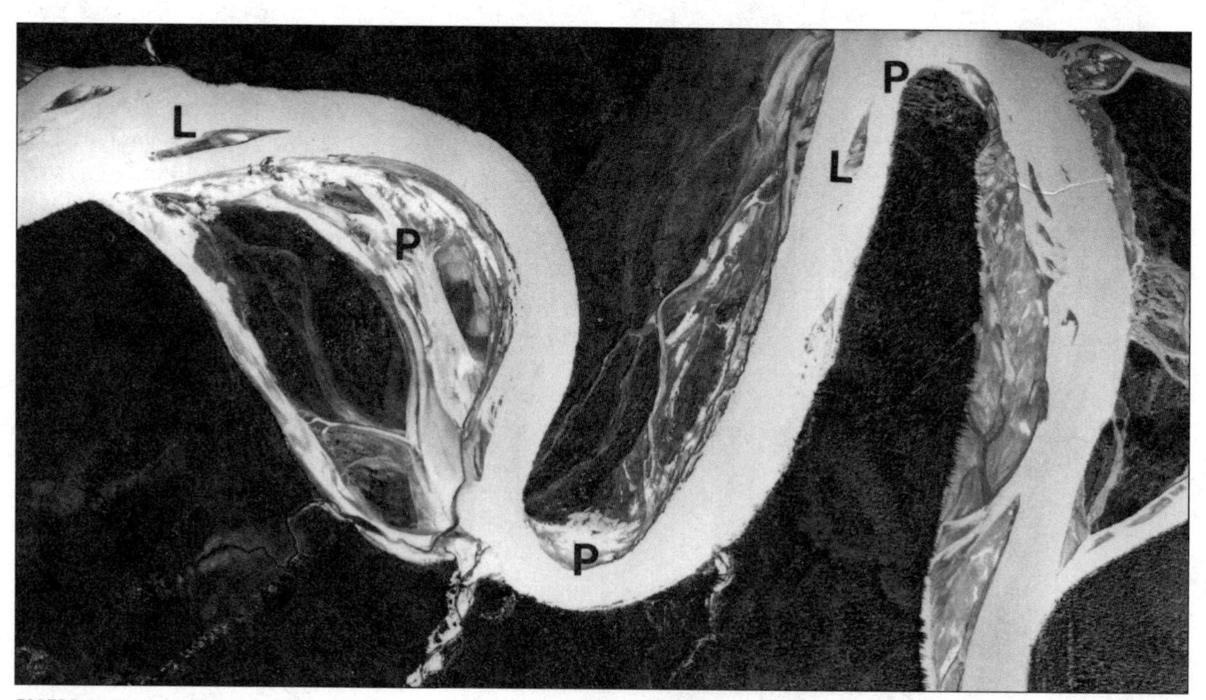


FIGURE 21.13 ▼ Point bars (P) and longitudinal bars (L) on the Donjek River, close to the Alaska Highway, Yukon (photo: P.G. Johnson)

## STREAM CHANNEL PROCESS

*The stream channel system* If we follow a stream to its mouth, we can see the subtle changes in the form and size of the channel. It becomes larger, deeper, and wider. Typically, its slope declines, and in many cases the channel becomes increasingly sinuous as braids (Figure 21.14) or huge meanders develop (Figure 21.15). All these changes are intricate adjustments to the processes operating within and upon the channel.

Stream channels are systems. They receive, transport, modify, and lose material. Water enters the channel by direct precipitation, by inflow from tributaries, and by seepage from the banks. It brings with it dissolved and solid materials. Additional debris is eroded from the stream banks and bed. These sediments are carried down the channel toward the sea; on the way, they are occasionally deposited and stored. Losses of water occur throughout the length of the channel, by seepage into the bedrock and by evaporation to the atmosphere (Figure 21.16). The channel system, like almost all environmental systems, is open to both inputs and outputs.

FIGURE 21.14 ▼ The braided stream system of the Donjek River, Yukon (photo: P.G. Johnson)

## STREAMFLOW

One of the main processes controlling the nature of the channel system is the flow of water through the channel. This is defined in general by the stream discharge. As we saw in Chapter 12, stream discharge is measured in cubic metres per second (cumecs), and corresponds to the velocity of flow multiplied by the cross-sectional area of the channel:

$$Q = v\, w\, d$$

where $Q$ is the discharge
$v$ is the mean velocity
$w$ is the average width
$d$ is the average depth of the flow

FIGURE 21.15 ▼ Meanders in a Prairie river system; note the oxbow lake in the centre (photo © 1956, Her Majesty the Queen in Right of Canada, from the National Air Photo Library, Energy, Mines and Resources Canada)

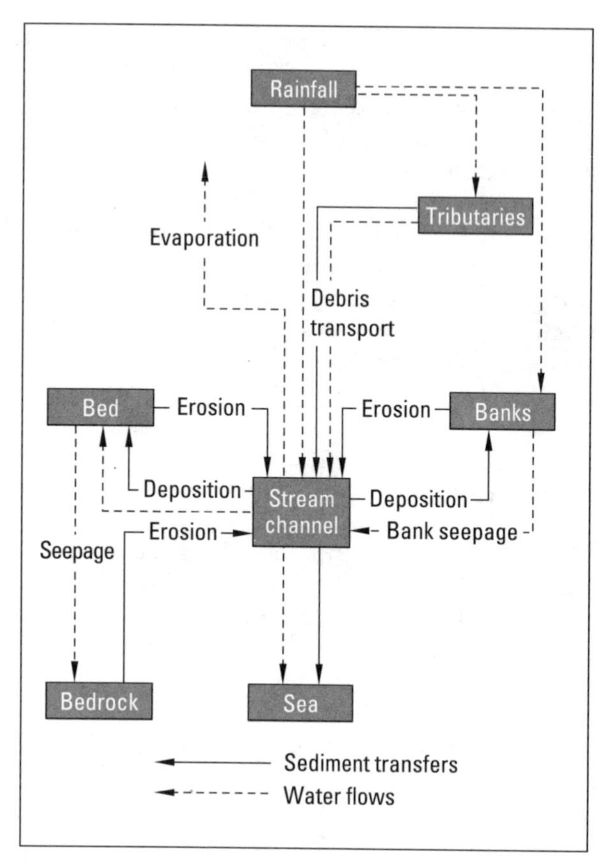


FIGURE 21.16 ▼ The stream system

On larger streams, the velocity is measured in a series of vertical sections across the stream channel and water level is continuously recorded (Figure 21.17).

The discharge of the stream varies over both time and space. Generally, it increases downstream as more water enters the channel from tributaries and from bank seepage. This increase in discharge results in corresponding adjustments in velocity, depth, and width. It is commonly believed that stream velocity declines downstream, the waters becoming slower and more sluggish, but this is rarely true. It is a misconception that arises because streams tend to be less turbulent and more muddy in their lower reaches, disguising the swiftness of flow. In most cases, depth and width of channels increase downstream (Figure 21.18 A and B). There is no strong trend for velocity to increase down-

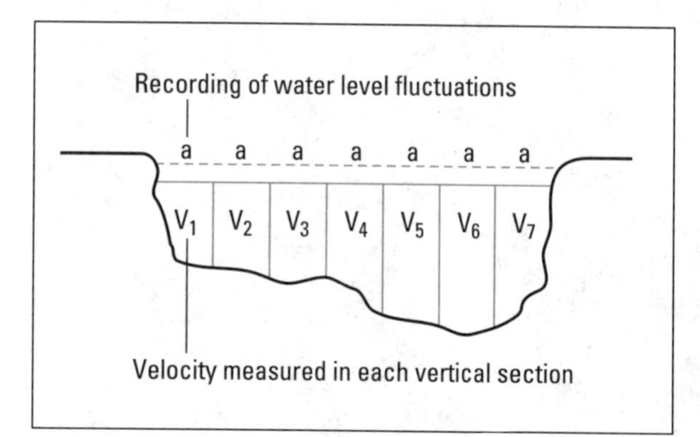


FIGURE 21.17 ▼ Measurement of velocity in a number of sections across a stream profile

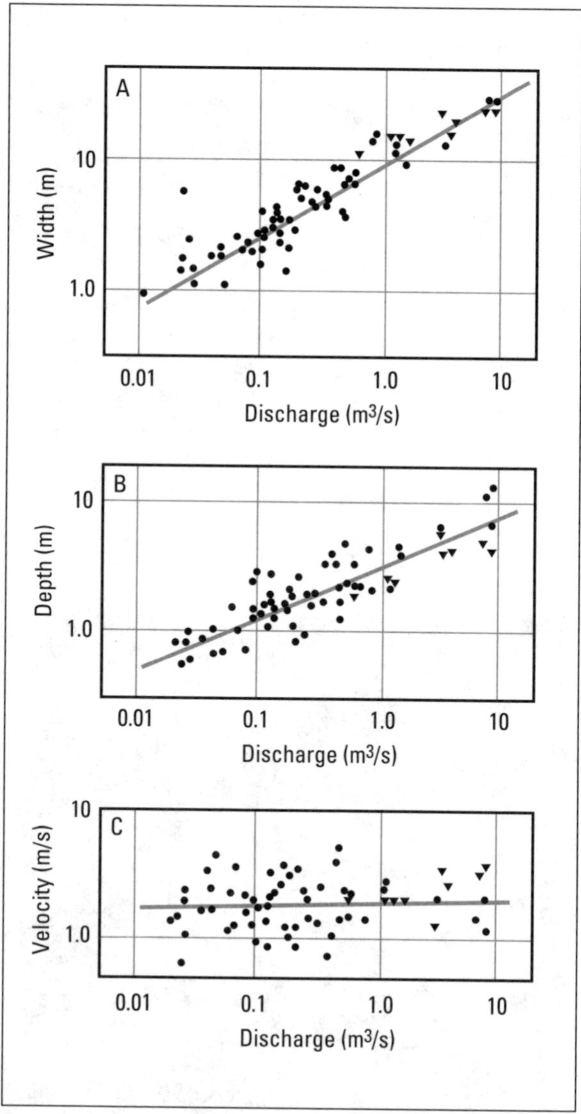


FIGURE 21.18 ▼ Relationship between bankfull discharge and channel width (A), depth (B), and velocity (C) as discharge increases downstream (from Wolman, 1955). ▼ = main sampling points; ● = single measurement sample

stream due to controls of stream gradient (Figure 21.18C).

Variations in discharge also occur at any single point along the stream over time.

These variations can be seasonal, due to changes in inputs from rainfall, spring snowmelt, or summer glacier melt, or due to artificial controls such as reservoirs. Variations can also be intra-seasonal or daily. Examples of the former are due to extreme storms in the river basins, such as the Peace River Flood of 1972 (Figure 21.19). We have only to watch a stream during a rainstorm to see the way this increase in discharge occurs. Velocity and depth both increase dramatically. Width changes only slightly, for most stream channels are steep-sided, so even as they fill up they do not become much wider. However, when the banks are overtopped and the water spills out onto the floodplain, it is clear that the width expands suddenly and often catastrophically (see Figure 21.20).

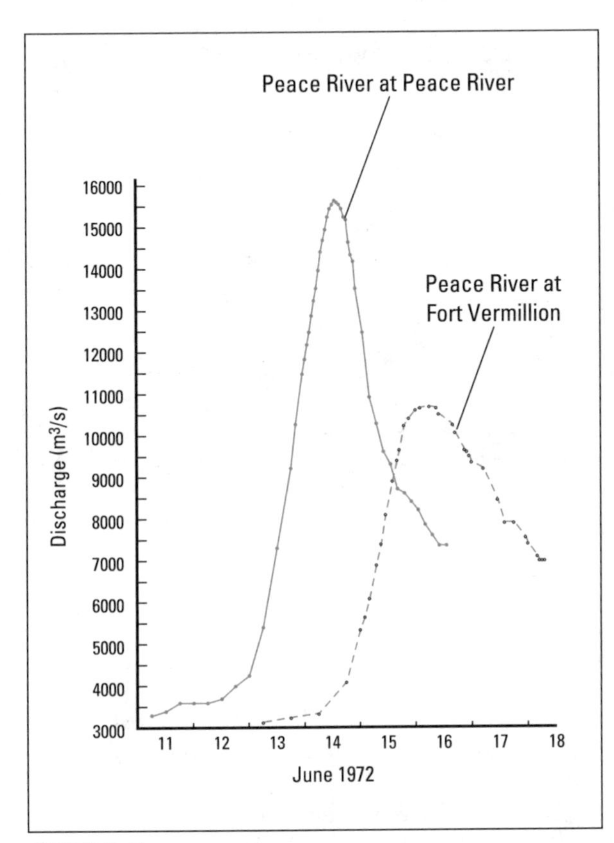


FIGURE 21.19 ▼ The hydrograph of the flood, Peace River, Alberta, 1972

FIGURE 21.20 ▼ Floods on the South Nation River, eastern Ontario (photo: P.G. Johnson)

## VELOCITY AND TURBULENCE

Looking at a stream in cross section, we can see a number of other characteristics of the flow. If we measured stream velocity at different depths, we would find that velocity was at a maximum close to the surface and at a minimum near the bed; a similar pattern would occur laterally (Figure 21.21). The reason for this is that the material in the channel bed and banks, together with the sediment being dragged or washed along the channel, exerts a frictional drag on the water, which reduces its velocity. The coarser the material, the rougher the channel, or the greater the concentration of sediment being transported by the stream, the greater is the reduction in velocity.

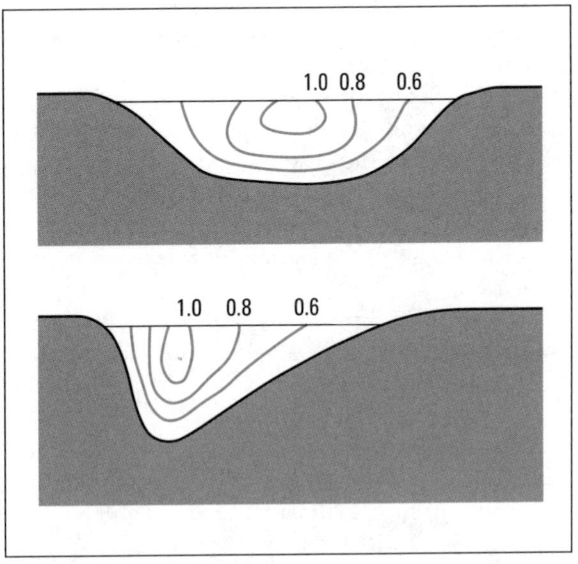


FIGURE 21.21 ▼ Velocity distribution in stream channels. The velocity isolines, or isovels, are shown in m $s^{-1}$

Flow in channels can be described as laminar flow or turbulent flow. With laminar flow there is no mixing in the vertical plane, that is, all motion is in the axial direction; in turbulent flow, there is rapid vertical movement of water in eddies or vortices (Figure 21.22). The criterion for laminar flow is that the Reynolds number is less than 500; for turbulent flow, the Reynolds number should be greater than 2000. The Reynolds number ($N_R$) can be calculated as follows:

$$N_R = \frac{VR}{\mu}$$

where $V$ is the average flow velocity
$R$ is the hydraulic radius of the channel
$\mu$ is the density

In a similar way, the Froude number is calculated as follows:

$$F = \frac{V}{\sqrt{gd}}$$

where $V$ is the average velocity
$g$ is the gravitational constant
$d$ is the depth

The Froude number can be used to indicate the change from tranquil flow $F<1$ to rapid flow $F>1$. The relationship between the Reynolds number and the Froude number can be shown with respect to velocity and channel depth for flow in a broad open channel (Figure 21.23).

Laminar flow is an efficient transporter of sediment. Turbulent flow causes the dissipation of the stream's energy. However, because the molecules are moving in many directions they are much more effective at eroding and lifting material from the walls and bed of the channel. (Note that gases also exhibit turbulent and laminar flow, thus the atmosphere has a frictional layer where it is in contact with the ground and a laminar flow layer away from the surface.)

Seen in three dimensions, the line of maximum velocity (the thalweg) weaves from side to side, wriggling snake-like down the channel as it is deflected from one bank to another (Figure 21.24).

Over smooth beds, such as a clay bed, a thin laminar flow layer will develop—this explains the low erodibility of such sediments. Over rough beds, the turbulent layer will extend to the bed.

At the stream bed, stresses on particles or rocks extending into the flow will promote entrainment of the material and the development of turbulence.

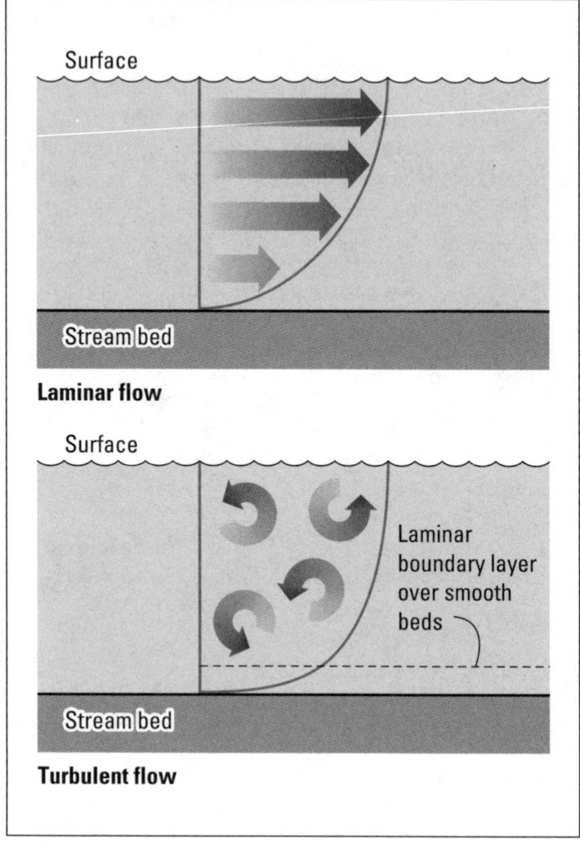


FIGURE 21.22 ▼ Laminar and turbulent flow velocity profiles

Over very rough beds, the flow can occur over a surface of separation (Figure 21.25A), inducing eddies at the bed. The importance of these eddies can best be illustrated by the formation of a pothole (Figure 21.25B), where flow across the top of the hole enhances eddies that rotate any rock particles in the hole, promoting continued erosion of the hole.

## OVERBANK FLOW

Flow is not always contained within the channel, and during periods of excessive stream discharge **overbank flow** may occur. Water then inundates the floodplain that borders the stream, and complex patterns of flow may develop around obstacles and features on the surface.

Vegetation on the floodplain helps to reduce flow velocities, but even so the force of these flood waters may be surprisingly strong, and walls and buildings may be demolished, while the floodplain sediments are eroded and carried downstream.

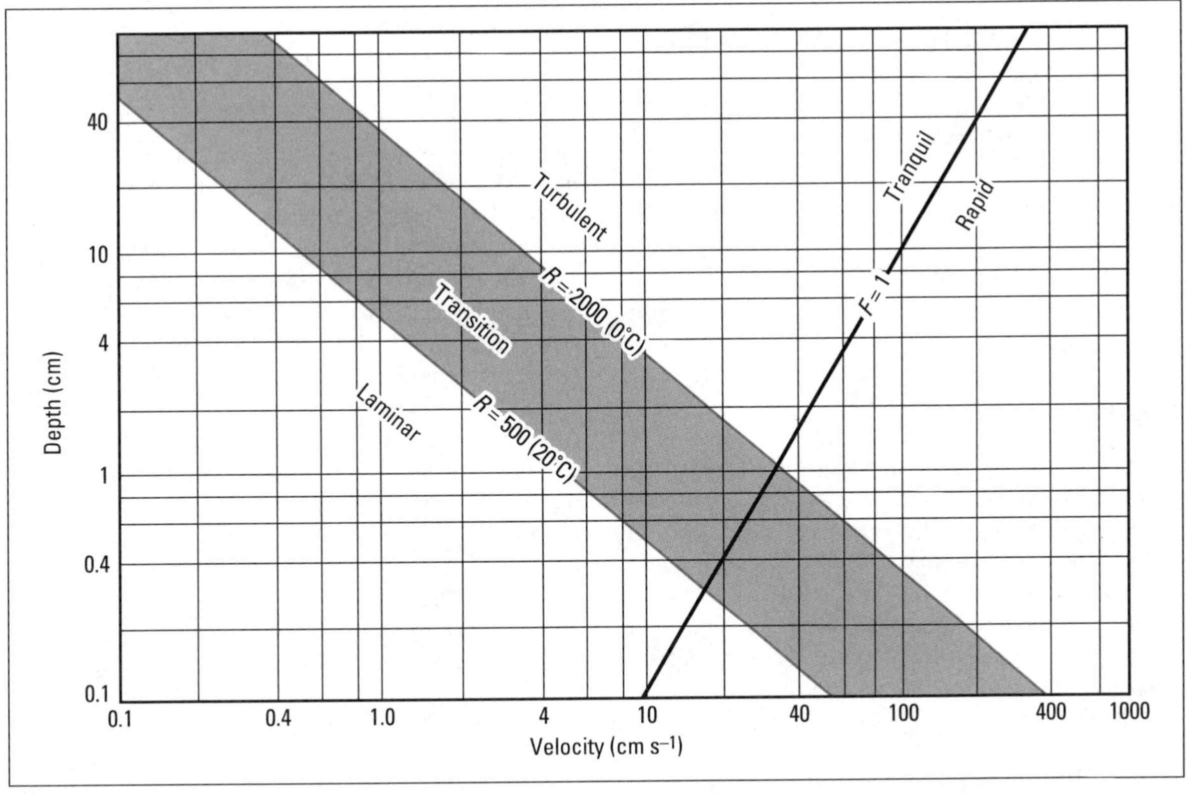


FIGURE 21.23 ▼ Regimes of flow in a broad open channel (after Sundborg, 1956)

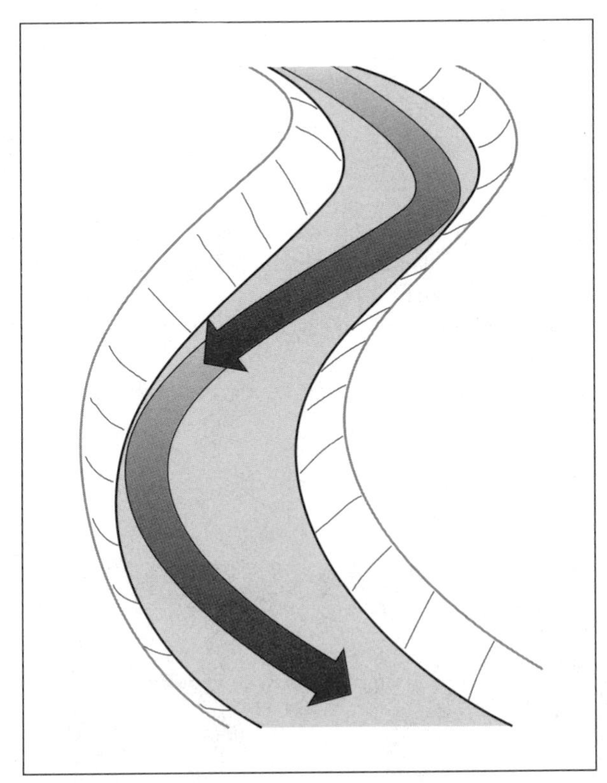

FIGURE 21.24 ▼ The thalweg of a stream channel

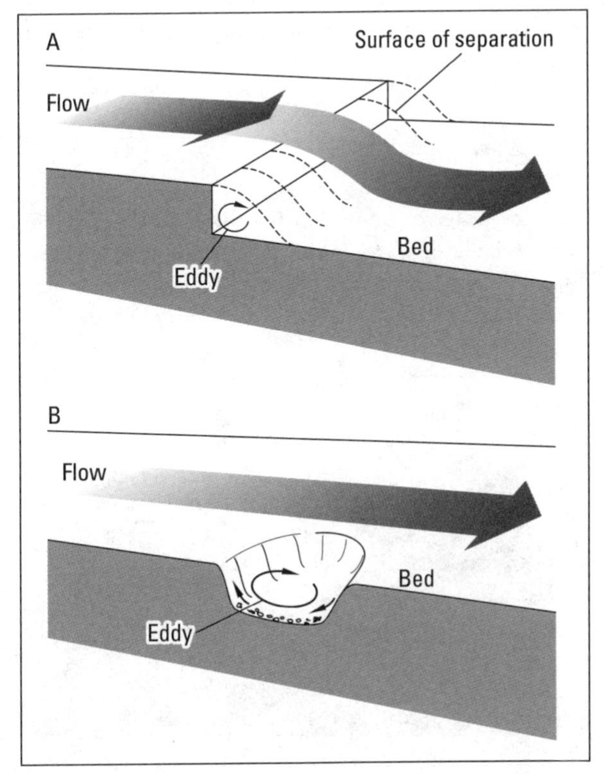


FIGURE 21.25 ▼ Formation of eddies along a stream bed

# River ice formation

Ice is a critical element of northern rivers, both in the formation of ice in the fall, and in the breakup of ice in the spring. River ice forms in three ways:

- in turbulent water by the production of frazil ice
- in calm or slow laminar flow conditions by growth at the supercooled surface
- from slush produced by snow

## FRAZIL ICE

In turbulent flow the supercooling of the surface layer of water is mixed through the stream and small ice crystals, called **frazil ice**, form through the water column. These crystals a) float to the surface, where they accumulate into pancakes and eventually contribute to the surface ice cover, b) adhere to the stream bed, where the accumulation produces **anchor ice**, or c) accumulates beneath already formed surface ice cover (Figure 21.26). The process is cyclic in that the formation of ice crystals in the water raises the temperature, and the supercooling ceases. The anchor ice and the accumulation beneath the plate ice restrict the channel flow, and during spring discharge can cause ice jam formation and flooding.

## PLATE ICE

In calm water or under laminar flow conditions, the surface layer of water becomes supercooled. Ice crystals will appear, initially along the bank or around rocks that break the surface of the water. These crystals will grow at random orientations away from the bank or rock and down into the water, depending on the temperature, and under prolonged cold weather and non-turbulent water conditions will form a continuous ice cover over the river. This cover will grow in thickness, to maintain the water temperature at close to 0°C, during prolonged cold spells. The growth of ice will be inhibited by snow accumulation on the surface, which provides an insulating blanket for the river. The structure of the ice has two main layers, a lower layer produced by freezing of the water, and an upper layer of snow that will be flooded by water injected through cracks in the ice. These layers are referred to as black ice and white ice, respectively. The progression of climate in winter will determine the characteristics of the ice. Cold conditions without snowfall will promote the development of thick plate ice. Large snowfalls during ice formation will produce thin ice because of the insulating effect of the snow. In both cases, from the point of view of the snowmobiler and the skier, the surface will be uniform, but the ice strengths will be very different. There are numerous cases each year of recreational snowmobilers breaking through weak snow-covered ice. In Ottawa, the Rideau Canal is cleared for skating each winter. To promote rapid formation of ice that is strong enough for trucks and tractors, early snowfalls are cleared by snowblowers to remove the insulating blanket.

## SLUSH ICE

Snow falling into open water at 0°C will accumulate as a structureless slush layer. If the air temperature is sufficiently low, this may freeze and develop into a solid ice cover on the river.

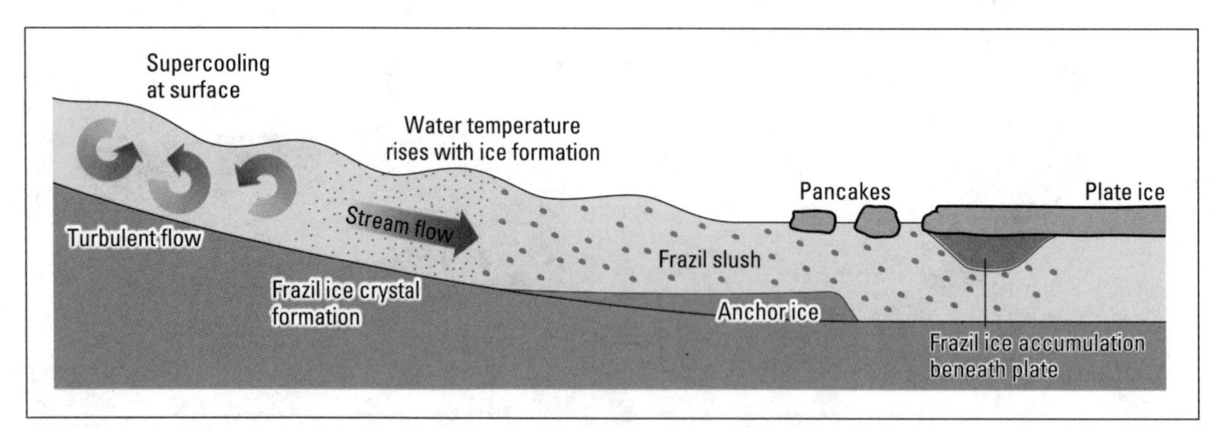


FIGURE 21.26 ▼ Frazil ice formation (modified from Michel, 1978)

## WATER LEVEL VARIATIONS

Variations in water level of the stream during freeze-up produce a number of complexities in the structure of the ice. Two basic patterns may be considered—formation with a water level drop, or a water level rise (Figure 21.27), although combinations of rising and falling water levels have been recorded.

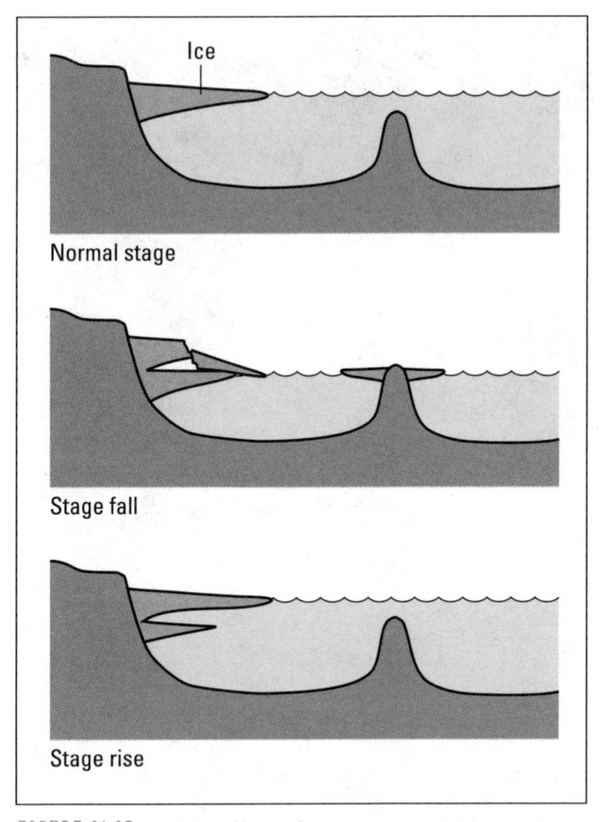


FIGURE 21.27 ▼ The effects of river stage variations on ice formation

## SPRING BREAKUP

As the climate warms in the spring, the ice will start to melt and river discharge will rise as snow in the basin melts and runs off. The combination will cause the ice cover to break up and be carried downriver by the increasing discharge. Under normal circumstances, the ice floes are broken up and cleared out of the river without major hydrological problems. As well, there is the geomorphic effect of sediment generation with the movement of material frozen into the base of the ice, and the impact of ice on the banks of the river. On occasion, however, the ice becomes jammed, forming an obstruction across the river. This may be caused by a number of factors:

- a very rapid rise in river levels before the ice has weakened sufficiently to break into small pieces
- restriction in the river channel by anchor ice on which the ice floes get jammed
- natural restrictions in channel width or depth
- artificial barriers such as bridge piers

Once ice gets jammed, more floes moving downriver will ride up against the ice barrier or be swept under the barrier with the risk of closing off the channel completely (Figures 21.28, 21.29). The water rises rapidly behind the dam and may cause serious flooding and movement of ice blocks onto the surrounding terrain. The jam will break when there is sufficient hydrostatic pressure behind it to move the ice, or when the ice weakens. It is often necessary to remove the blockage with explosives to relieve flooded areas. The city of Ottawa has a unique program of removing ice from the Rideau River before normal breakup to eliminate the risk of

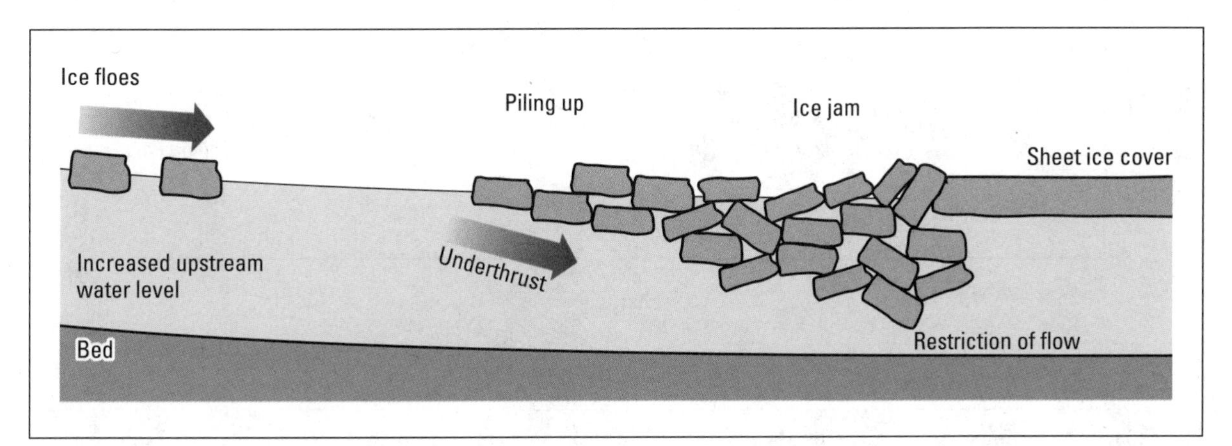


FIGURE 21.28 ▼ Formation of an ice-jam blocks the river

FIGURE 21.29 ▼ Ice jam on Greens Creek, Ontario, caused by a rapid increase in discharge in the early spring (photo: P.G. Johnson)

FIGURE 21.30 ▼ Ice stranded above river level after flood waters receded from the Rideau River, Ottawa, Ontario (photo: P.G. Johnson)

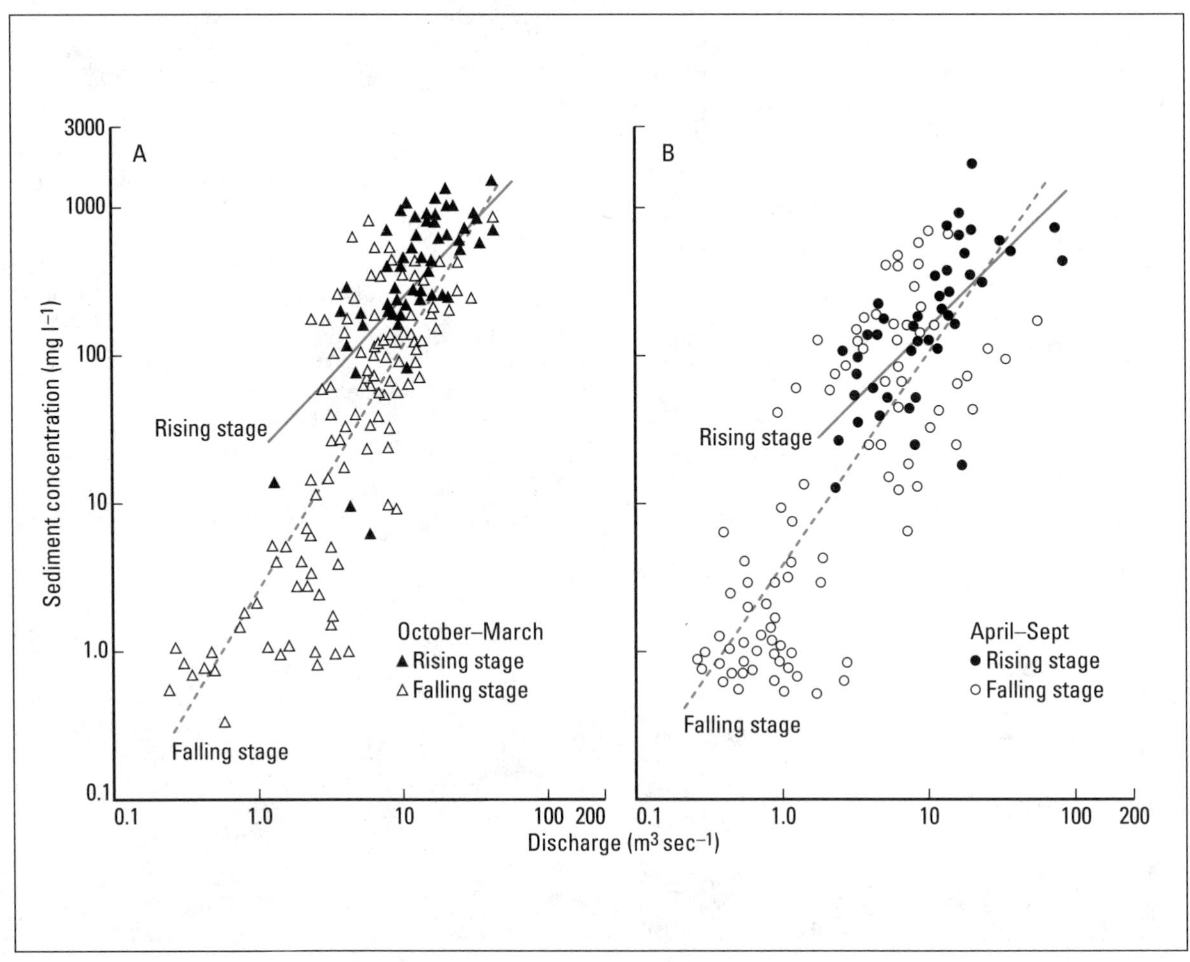


FIGURE 21.31 ▼ Sediment rating curves for the River Creedy, Devon, England, in winter (A) and summer (B). Note that the relationship varies both with season and with stage (from Walling, 1978)

flooding. The ice is blasted from the river progressively upstream from the confluence with the Ottawa River. At bridges and utility crossings sections of ice are separated from the banks by a large rotary saw and the whole section blasted out and then broken up. An ice-jam flood on the Rideau River is shown in Figure 21.30.

### ICINGS

Another ice formation process that causes hydrological and geomorphological changes is ice accumulation on stream beds where there are springs or groundwater seepage. These can become a few metres thick and are responsible for stream diversion during the spring runoff. They are particularly important in braided streams, where they promote migration of the braided system.

## Sediment in stream channels

At one time or another, most streams carry sediments that have been washed into the channel from the surrounding land, eroded from the channel bed and banks by the stream, or contributed by glacier drainage systems. The quantity of this material varies considerably from time to time due to changes in discharge. Normally, as discharge and velocity increase, the amount of sediment being transported rises correspondingly. A relationship between the discharge and sediment load can be calculated for any single station and a sediment **rating curve** can be produced (Figure 21.31). A rating curve can be used to predict sediment concentration at any discharge. Significantly, however, the changes in concentration are not simple, but vary markedly from one storm to another (Figure 21.32). The sediment rating curve shows only the overall relationship.

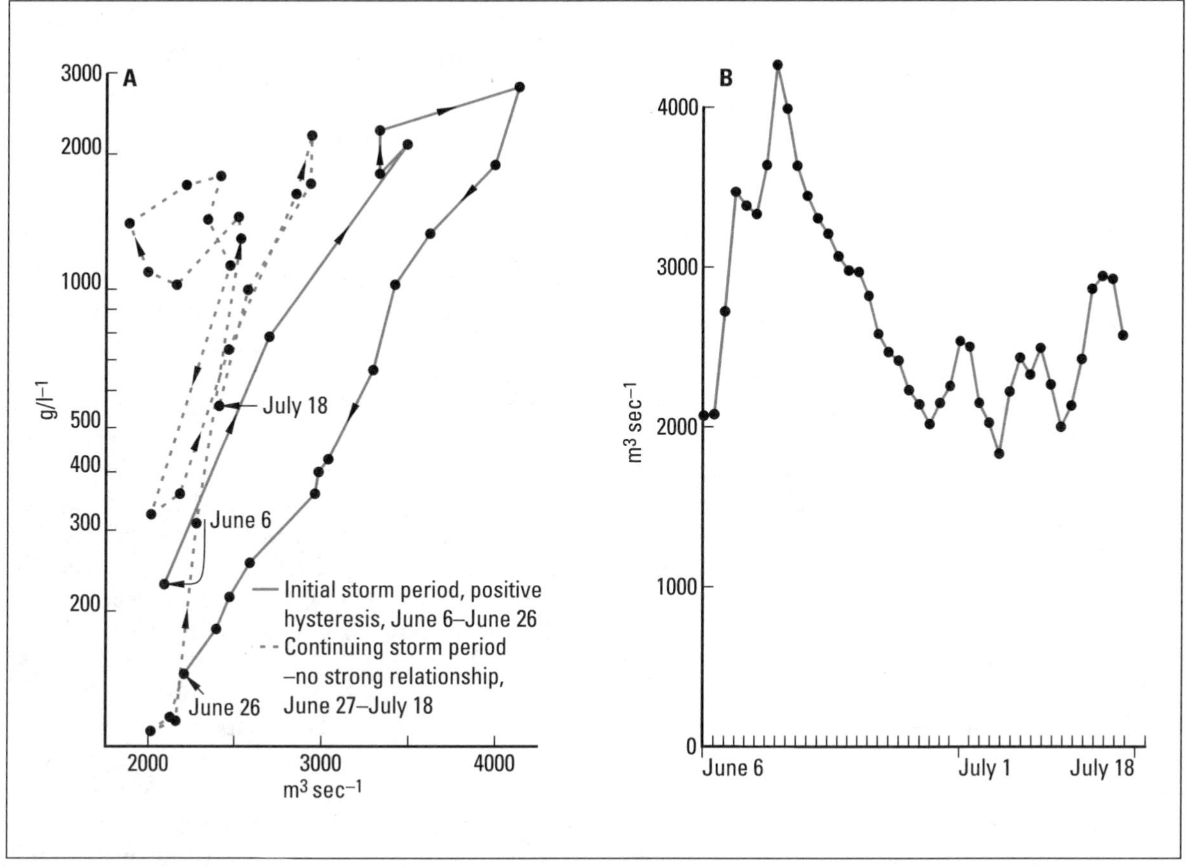


FIGURE 21.32 ▼ (A) Sediment concentration discharge relationships for a series of storm runoff events on the Peace River, Alberta, 1988, measured at Peace River. (B) Hydrograph for floods, June and July, 1988 (Source Inland Waters Directorate, Environment Canada)

Streams fed by glacier discharge show little relationship between discharge and suspended sediment because of glaciological controls on the availability of sediment (Figure 21.33). In nival basins, the suspended sediment discharge relationships are different in the pre-snowmelt, snowmelt, and post-snowmelt periods (Figure 21.34).

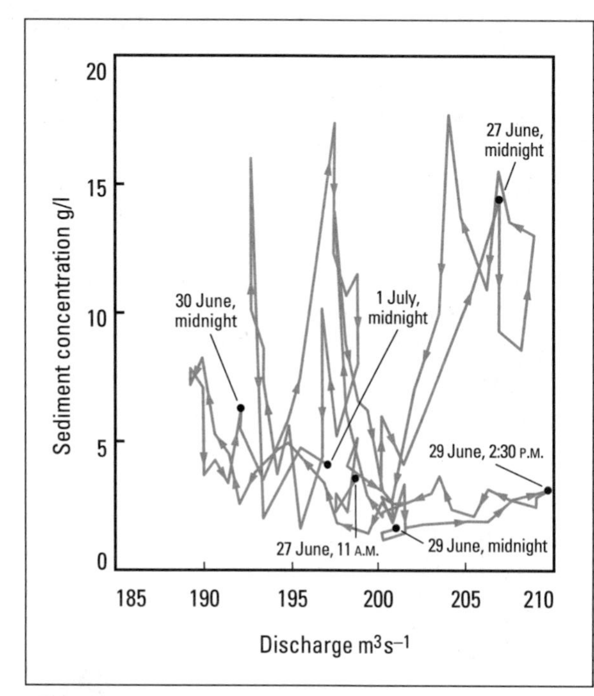


FIGURE 21.33 ▼ Glacierized basin suspended sediment discharge relationships, Slims River valley, Yukon

During periods of low flow, relatively little sediment movement takes place; as a consequence, river channels tend to be stable in these phases. Conversely, as the flow increases, more and more sediment is entrained from the floor and stream banks. Loose sediment on the bed is picked up largely by the fluid drag exerted by the flowing water. Material from the stream bank is eroded mainly by the action of bank-caving (Figure 21.35). The waters in the channel may undermine the bank and cause collapse; but more important in many instances are the processes of freeze–thaw and wetting and drying that pry material from the bank during periods of low flow. Later, when the water level rises, it sweeps this material into motion.

Much of the material that is carried during periods of high discharge is therefore old material reworked from the bed or banks. Most of it moves intermittently, with often long periods of storage in the channel or in the floodplain interrupted by short periods of transport.

New material is also moved. Undercutting of bedrock slopes may create cliffs from which boulders are eroded; rocks may tumble and bounce down talus slopes into the channel; slower processes of creep and solifluction carry material down the hillside into the stream. During storms, overland flow may also develop on the valley sides, and sediment may be transported by sheetwash, and in rills and gullies, into the channel.

## SEDIMENTS AND SORTING

The stream carries pebbles and sand along the bed (bedload or traction material), silts and clays in suspension (suspended load), and material in solution (dissolved load). Both the absolute quantities and the relative proportions of these components vary from one stream to another (Table 19.1) and, perhaps more significantly, from one time to another. The coarser particles, for example, move only when the discharge is very high, and many of the larger boulders entering the channel may be too big to be moved by all but the greatest floods. They may accumulate as a **lag deposit**, the material remaining after all the finer, more easily transported sediments have been removed. On the other hand, the finer silts and clays may be in almost perpetual motion; certainly dissolved materials, such as fertilizers, animal wastes, and soluble compounds washed or weathered from the soil are constantly being carried downstream.

There are two important implications of this phenomenon. The first is that the material is actively sorted by size during transport; the second is that different components of the stream sediment are in equilibrium with different flow conditions. For this reason, there are great difficulties in defining a single equilibrium condition for the channel as a whole.

The downstream sorting of sediment can be observed along the course of almost any stream. From the headwaters to the mouth there is a decrease in the size of sediments. Because the finer particles travel more frequently and more rapidly, they are washed farther downstream, while the slow, infrequently moving coarser materials are left in the upper reaches. It should be noted again that the commonly held idea that material is progressively deposited—coarser particles first and finer particles later, because the velocity declines downstream—is

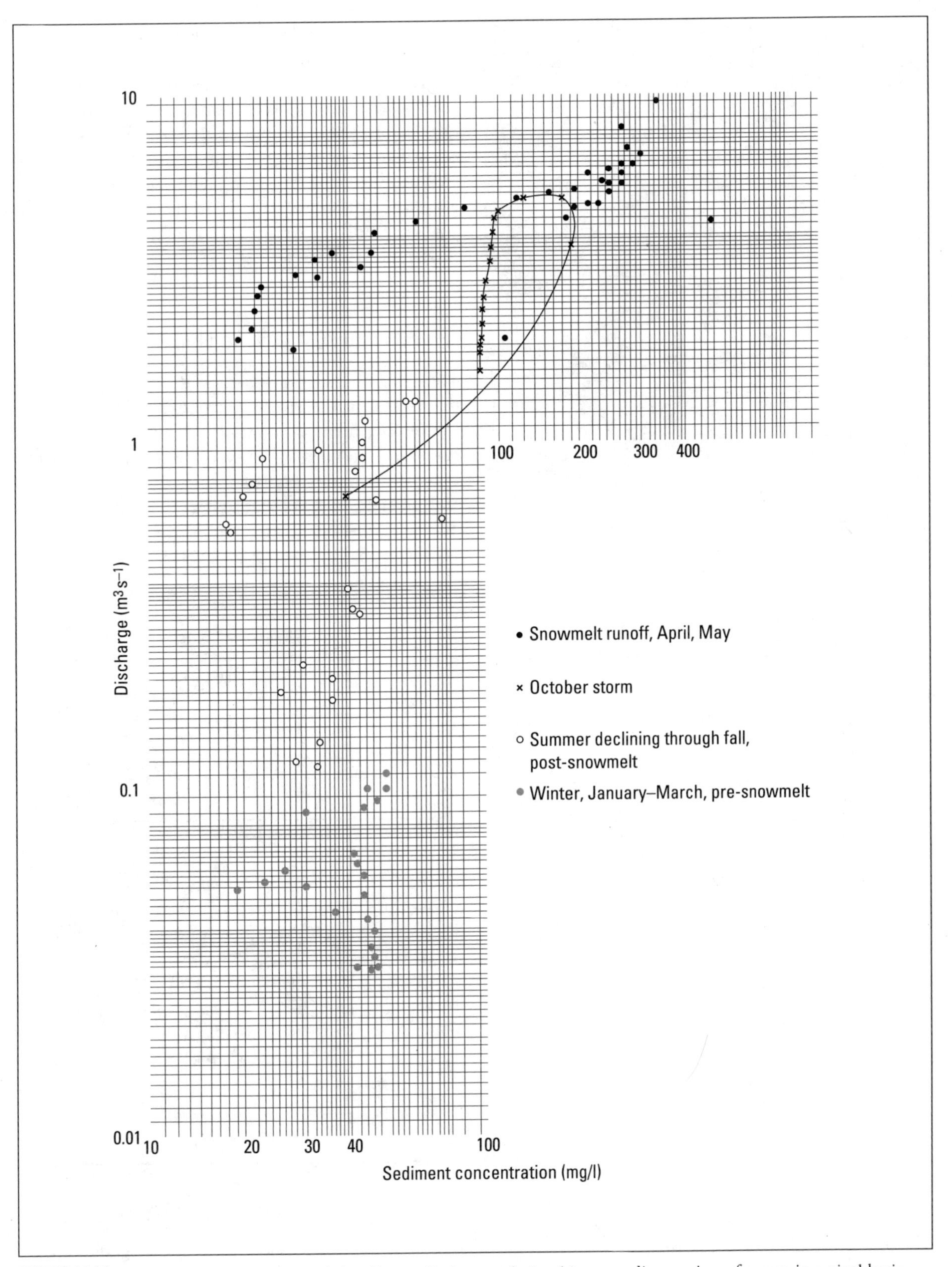


FIGURE 21.34 ▼ Variations in the suspended sediment discharge relationships according to time of season in a nival basin, Halfway Creek, Ontario, 1983 (Source Inland Waters Directorate, Environment Canada)

FIGURE 21.35 ▼ Bank caving in the Brudenel River, Prince Edward Island (photo: P.G. Johnson)

generally false, because the velocity varies considerably along the length of the stream.

As we have noted, movement of sediment is not continuous, but involves intermittent transport and deposition. Even during a single flood, erosion and deposition may be occurring simultaneously as the sediment-laden waters drop some of their load and pick up new material. This results in complex processes of **scour and fill** during peak flows (Figure 21.36).

## SEDIMENT DEPOSITION IN CHANNELS

The intermittent movement and deposition of sediments within the stream channel is associated with the development of a variety of sedimentary structures and bedforms. Many of these are dependent upon the complex interplay between stream velocity and particle size, so that as the discharge increases during a storm or spring runoff, the character of the bedforms may change (Figure 21.37); the bedforms,

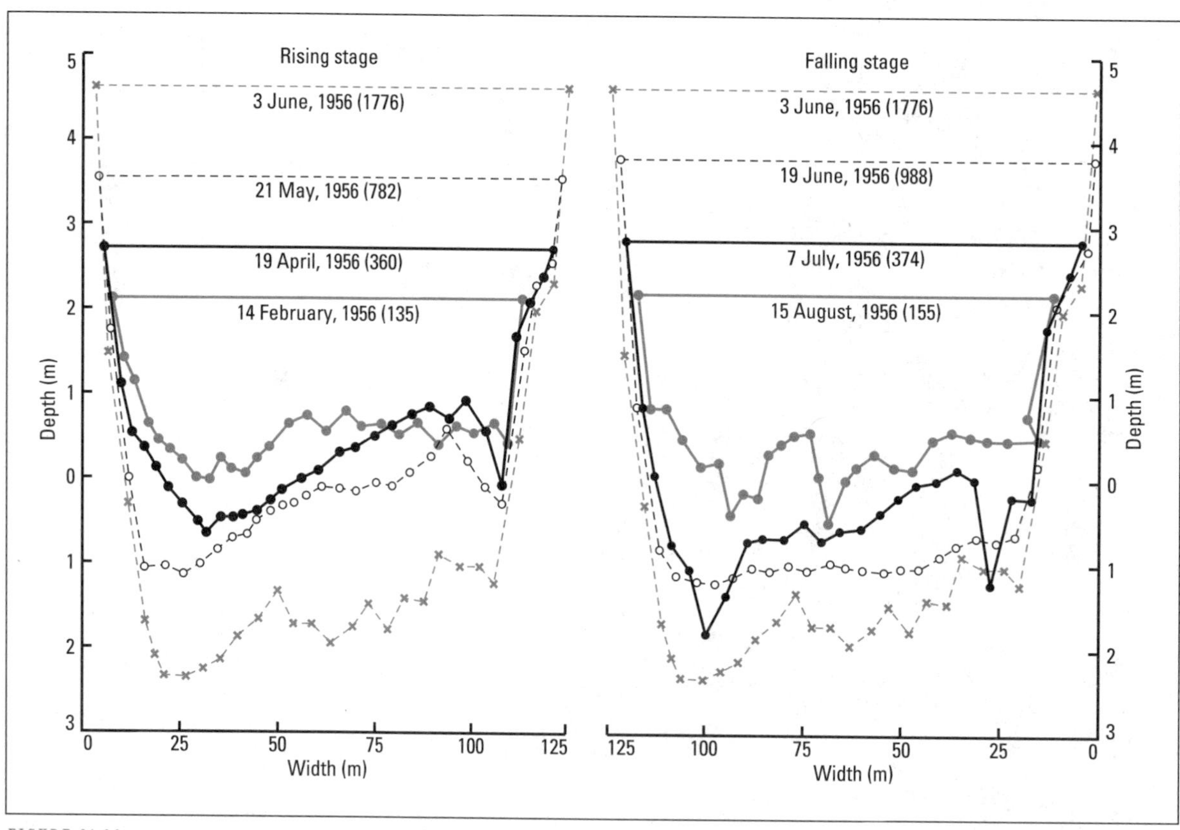


FIGURE 21.36 ▼ Changes in the channel profile of the Colorado River in Arizona, 1956, show the effect of scour during the rising hydrograph and filling during the declining hydrograph. Figures in brackets represent discharge in $m^3 s^{-1}$ (from Leopold *et al*, 1964)

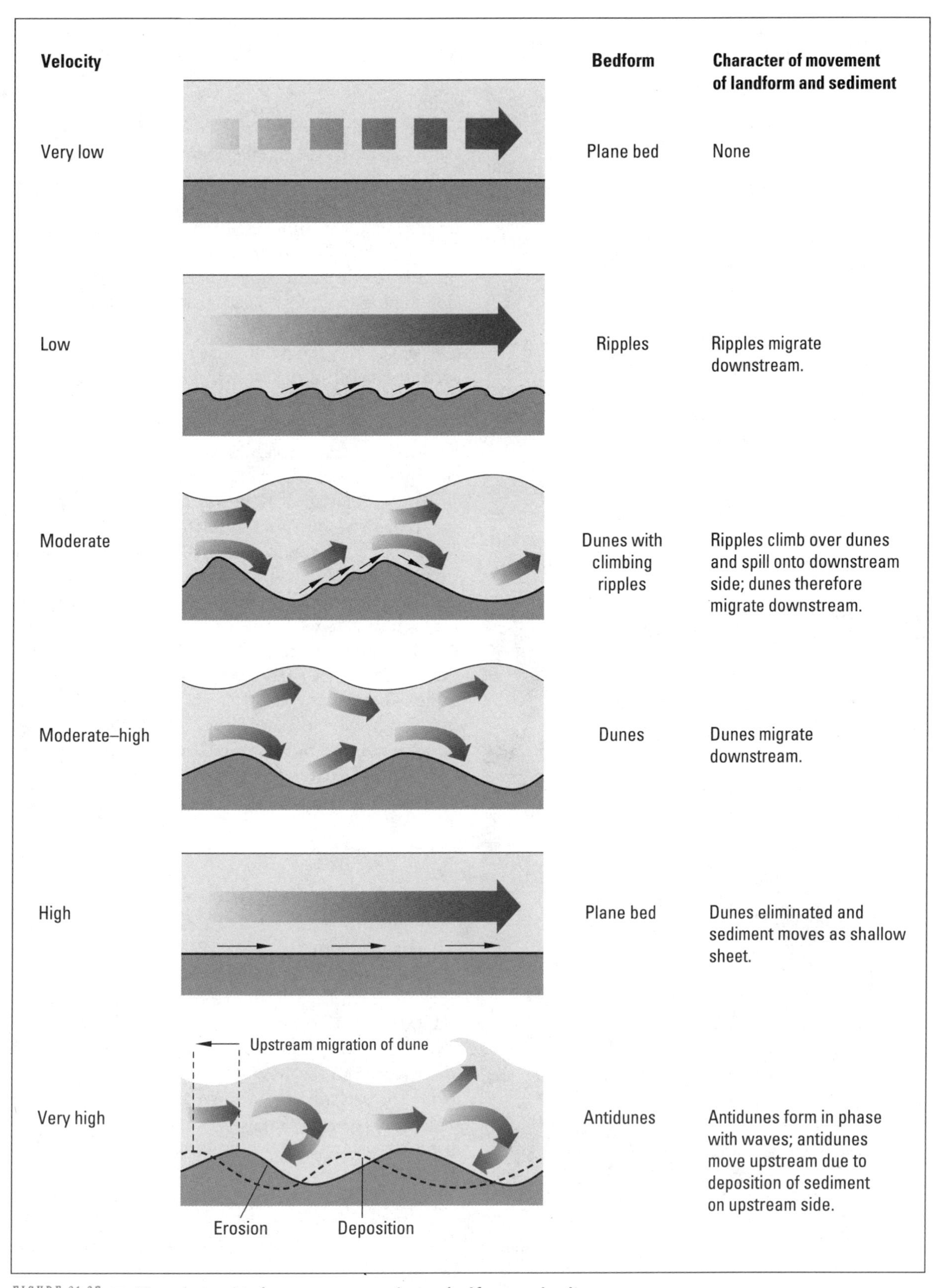


FIGURE 21.37 ▼ The relationship between stream velocity, bedform, and sediment movement

like other features of the channel, adjust to equilibrium with prevailing flow conditions.

*Dunes and ripples* In channels composed mainly of finer materials—sand and silt—the main bed features are **dunes** and **ripples**. These are similar in formation, but whereas the dunes are normally in the order of 10 or more centimetres in height (and some may be several metres) and often a metre or so apart, ripples are much smaller (often no more than a few centimetres in height and spacing).

The development of dunes and ripples is complex, for they involve the interaction of the sediment being carried by the stream, the feature itself, and the character and velocity of the flow (Figure 21.38). As the feature develops, it tends to disturb the flow and create conditions that favour the growth of further dunes or ripples. It is clear that, because of this close relationship, the size and character of the dune tends to be closely associated with flow conditions. Many dunes, for example, are related to water depth. One implication of this relationship is that it is possible to estimate the depth of water from fossil dunes preserved in fluvial sediments.

FIGURE 21.38 ▼ The sand dune pattern is observed after a test in a sediment transport flume (photo courtesy of Canada Centre for Inland Waters)

Dunes and ripples are dynamic features. They tend constantly to receive material from upstream and lose material downstream. The exact nature of this gain and loss is interesting, and it results in a migration of the dune or ripple. In general, material on the gently sloping **stoss-side** of the dune (the upstream side facing into the flow) creeps up the slope under the effect of water flow, until it reaches the crest. The particles then avalanche down the steeper **lee-side** and collect at the foot. A degree of sorting according to size occurs here, for the coarser particles tend to roll farther. Finer material may also be swept off the dune and hop downstream; as it lands it sets other particles into motion. Over time, the dune appears to travel downstream.

As the velocity of flow continues to increase, all the sediment becomes mobile and the bedform returns to a planar form. This planar form differs from the initial planar form because the sediment is mobile. At very high flow velocities, dune forms redevelop, but with the turbulence of flow, sediment is eroded from the face of the dune and deposited on the upstream side of the downflow dune. This has the effect of an upstream migration of the position of the dune, while continuing a downstream migration of sediment. These dunes are called **antidunes**. **Standing waves** are often formed above the antidunes in extreme flows.

*Gravel bars* In streams carrying mainly coarse materials—gravels and coarse sands—bars develop. These take a range of forms. **Point bars** can be seen on the inside of meander bends in many streams (Figure 21.13). They develop where the flow is inhibited by the increased frictional resistance and reduced depth of the water, and by the tendency of the thalweg to be thrown toward the outside of the bend. As a result, the ability of the water to carry sediment (its competence) is reduced, and coarser particles accumulate on the inside of the bend. The feature that is formed is characteristically lobe-shaped and shows an interesting distribution of particle sizes; coarser material is concentrated on the outer margins of the feature where the flow is most active, while finer particles accumulate on the inside of the bar.

*Pools and riffles* In straight streams, bars tend to develop in a rather different fashion. Material builds up in regularly spaced **riffles** along the channel bed. These typically slope alternately toward one bank and then the other, and the thalweg winds between them. Between the riffles occur deeper **pools**; the whole sequence is known as a pool-and-riffle (Figure 21.39). The spacing of these features is associated with the size of the channel; the distance between the top of one riffle and the next is, on average, in the order of five to seven times the channel width.

The reason for the development of pool-and-riffle sequences is not entirely clear. It is apparent that the material on the riffle tends to be coarser than that in the pool. It is also clear that the form of the features tends to remain constant, although material moves into and out of each riffle. This has been demonstrated by placing painted pebbles on the surface of riffles; during periods of high flow they are seen to hop downstream and are commonly found resting on the next riffle. In other words, the features are in a condition of steady-state equilibrium with stream flow.

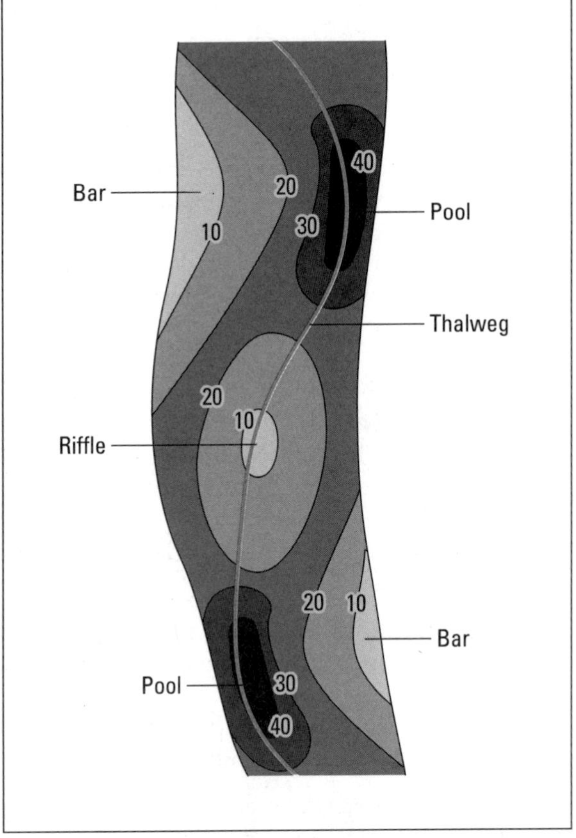


FIGURE 21.39 ▼ Overhead view of a pool-and-riffle sequence showing depth of bed below water surface (contours labelled in centimetres)

# Floodplain development

When rivers flow across areas of low relief, periods of high discharge, caused by spring snowmelt, excessive glacier-melt in source areas, or as a response to storms, frequently overtop the channel banks. As a result of the extra channel flow, sediments are deposited across the inundated low-lying regions, building up areas called floodplains.

As an example, in northern California a storm on 21 December to 23 December had a rainfall of 280 mm (Figure 21.40). One small stream, Coffee Creek, swollen by the runoff from this storm, had a peak runoff of about 600 $m^3$ $sec^{-1}$. This discharge peak has a recurrence interval of 100 years, that is, on the basis of the statistical record for this stream,

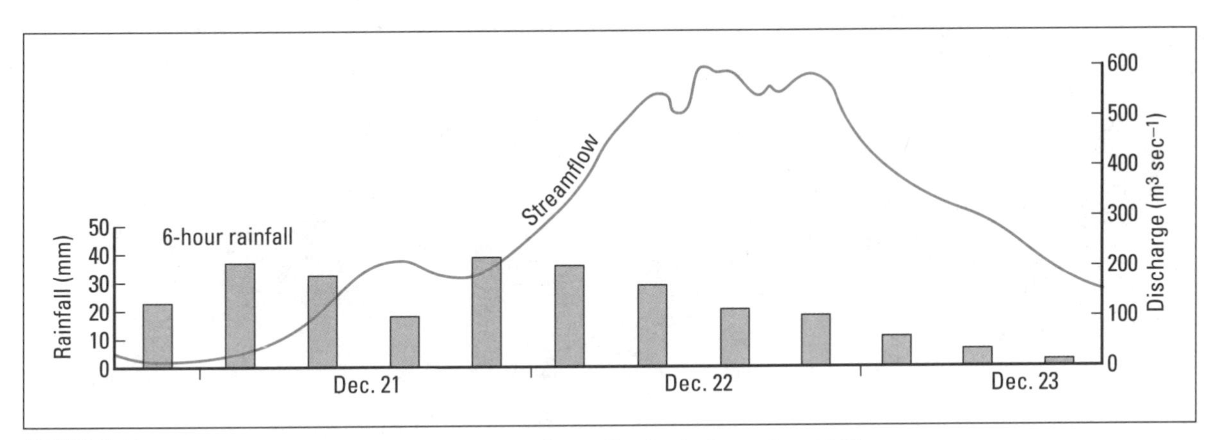


FIGURE 21.40 ▼ Rainfall (6-hour totals) and discharge for Coffee Creek, December, 1964

this runoff would only be expected to occur once in every 100 years. The stream overtopped its banks and flooded the surrounding land (Figure 21.41). The waters scoured out new channels, filled old ones, and caused extensive erosion of the valley floor. As the flood subsided, the eroded sediments were deposited across the floodplain, covering 70 percent of the surface to an average depth of 0.4 m. The power of the stream to erode and transport debris, and to build its floodplain, was amply demonstrated.

This occurrence was an exception; more commonly, floods are less dramatic, less catastrophic; normally it is only finer material that sweeps across the floodplain in this way. Nevertheless, the principle is clear: when streams overtop their banks, they act to shape their floodplains.

Four stages in floodplain inundation can normally be recognized:

1. Initial spilling of waters onto the floodplain.
2. Development of continuous flow across the floodplain.
3. Waning of the flood and retreat of the water.
4. Drying of the floodplain surface.

During the first stage, erosion is particularly active. As the waters rise and spill over their banks they come into contact with the **levées** that border the channel. These are sandy or gravelly ridges, usually one half to four times the channel width in diameter, which run parallel to the stream. They are formed when the river floods and the water flowing outside the channel along the riverbank deposits sediment. This builds a barrier that slopes away from the river, often preventing the flood waters from flowing back. When the floodplain is wide and flat, the levées also tend to prevent tributaries flowing directly into the river. These streams flow down the major river valley running parallel to the main river; they are known as yazoo streams. The levées are also intersected at intervals by narrow gaps called **crevasses**, and it is through these that the waters pour onto the floodplain. Erosion in these crevasses is rapid, and large parts of the levée may be destroyed. Sediments deposited beyond the crevasses are called crevasse **splays**. (Students should note the dual use of the word 'crevasse,' in this sense and in reference to glacier ice structure, discussed in Chapter 23.)

Gradually the floodplain becomes drowned as the water flows through the crevasses across the levées. Depressions and abandoned channels become filled and then these are connected as the water level rises and spills over the intervening ridges and mounds. To some extent, vegetation protects the floodplain itself from erosion, but where flow is rapid and turbulent, plants may be ripped from the soil and the sediment beneath scoured out and washed away. Erosion also occurs where obstacles such as large trees, walls, or buildings cause the flow to eddy or restrict its passage downstream. On the whole, however, the second phase of flooding is marked by widespread deposition on the floodplain surface. Erosion continues at the margins of the channel, for there the water remains turbulent, but elsewhere the flow is generally less turbulent, shallower and constrained by vegetation. Fine materials, carried in suspension across the levée, start to fall to the bottom of the water; clay, silt, and sand accumulate on the floodplain. This material is known as **alluvium**.

In time, as the flood wanes, the waters subside and the higher parts of the floodplain reemerge. In the isolated pools and ponds that are left in the depressions, the finest materials continue to settle

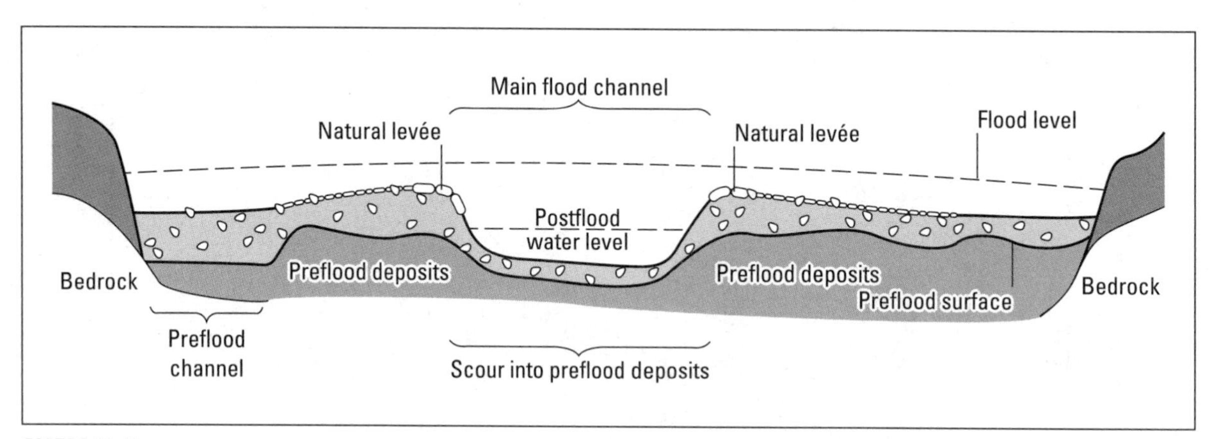


FIGURE 21.41 ▼ Cross section of Coffee Creek and changes in the channel as a result of the flood in December, 1964

out. Winds disturb the surface of these waters and the material on the floor is moulded by the turbulence into ripples, while at the margins of the pools small beaches are formed. Finally, evaporation and drainage empty the pools and the floodplain surface dries out. Vegetation takes root in the new material and soil development commences. A new floodplain surface is created.

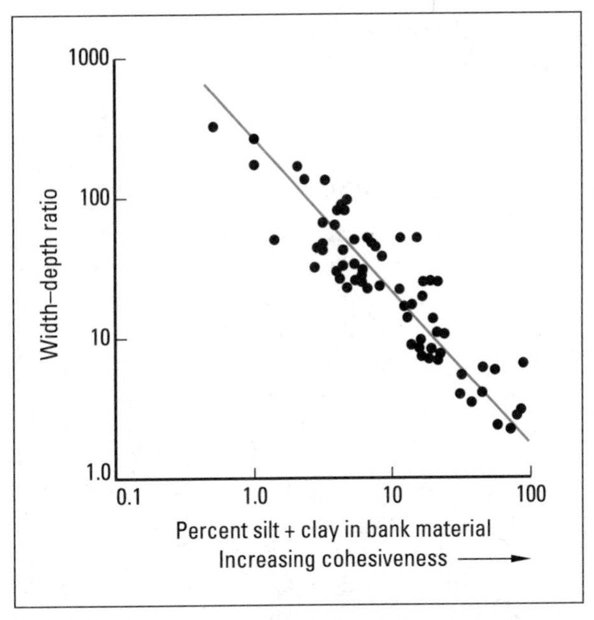


FIGURE 21.42 ▼ The relationship between channel shape and the clay–silt content of the bank materials (after Schumm, 1970)

# Stream channel form

Overbank floods are only occasional events, even though they may result in large-scale modification to the river form and landscape. During regular periods of fluctuating discharge, due to lower intensity rainstorm runoff or gradual snowmelt runoff, the channel form may change slowly.

## CHANNEL SHAPE

Water and sediment moving down the stream channel interact with the bed and bank materials to modify the shape of the channel. Thus, the form of the system adjusts in response to the internal processes. In hydrodynamic terms, a perfect channel is one that is parabolic in shape. Few channels, however, are perfectly parabolic; instead they range from deep, narrow forms to wide, shallow ones.

In general, wide, shallow channels develop where the materials that make up the bank are loose and non-cohesive. Sands and gravels, for example, are unstable at steep angles. They collapse if the stream cuts deeply into them. Where the banks are composed of more cohesive materials, such as silt and clay, the channel is often deeper and narrower. The ratio of width to depth of stream channels is therefore inversely related to the silt and clay content of the bank material (Figure 21.42). For the same reason, channels cut through bedrock tend to be deep and narrow; these are slot gorges, such as those cut by tributaries of the Colorado River gorge (Figure 21.43).

The shape of the channel is not a fixed feature of the stream, but varies according to flow conditions. As we have seen, scour and deposition may occur during a single flood, altering the shape of the channel in a complicated manner. In most cases, the channel returns to a stable form during low flow conditions, when discharges are too small to under-

FIGURE 21.43 ▼ Rainbow Bridge Gorge on the Colorado River, Utah (photo: P.G. Johnson)

mine the banks, and too little sediment is being transported to significantly modify the shape of the channel.

## MEANDERING CHANNELS

Stream channels usually have some degree of sinuosity; only rarely are they straight for any distance. A measure of the sinuosity can be calculated as the ratio of the actual channel length to the straight-line distance between two points (Figure 21.44). Channels in which sinuosity, a relatively regular pattern of side-to-side movement, is occurring are referred to as **meandering**. The word derives from the River Meander in Turkey. The Takhini River in the southern Yukon (Figure 21.45) has extensively developed meanders in glaciolacustrine sediments.

A number of explanations for the development of meanders have been proposed over the years, including the occurrence of random obstacles (such as resistant materials in the floodplain or large boulders), the need for the stream to lose energy as its sediment size declines but its discharge increases, and even the effect of the earth's rotation.

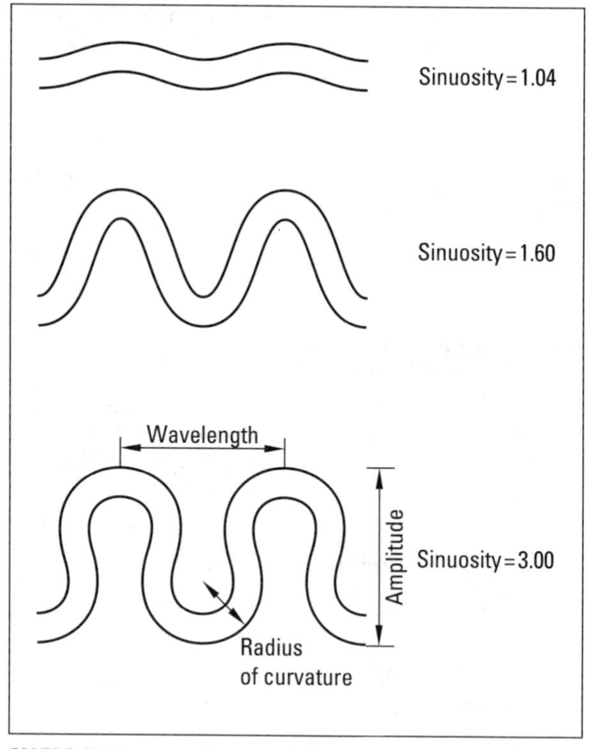


FIGURE 21.44 ▼ Sinuosity of channels

FIGURE 21.45 ▼ Meanders in glaciolacustrine silts on the Takhini River, Yukon; we can see point bars (A), an incipient cut-off (B) which could potentially form an oxbow lake (C), and erosion of the outer bank of the meander (D) (photo: P.G. Johnson)

Typically, a meandering stream consists of a series of curves of approximately similar form. The character of the meanders can be described in terms of their average amplitude, wavelength, and radius of curvature (Figure 21.44). Numerous studies have shown that these variables are closely related to the size and character of the stream. For example, meander wavelength increases as the radius of curvature or width of the channel increases (Figure 21.46). These relationships seem to reflect a more fundamental association between the size of the meanders and the stream discharge.

The consistency of the pattern shown by meanders indicates that they cannot be regarded as products solely of random conditions, such as changes in the resistance of the bed or bank material. Instead, it seems to be some internal property of the channel system that controls their development. One of the most likely explanations is that they develop in response to an excess of free energy in the stream. Downstream, the discharge tends to increase, while the sediment becomes finer. Thus the stream has more energy, but relatively less work to do. This results in more free energy within the system; in a sense the stream is underworked. Meanders are a means of carrying out work and expending this free energy uniformly throughout the stream, a way of maintaining a balance between the capacity to do work and the amount of work done. Interestingly, the shape of curve that best achieves this is what is known as a sine-generated curve. Many meanders approximate to curves of this nature (Figure 21.47).

One reason for favouring this explanation is that it helps to account for the occurrence of meanders

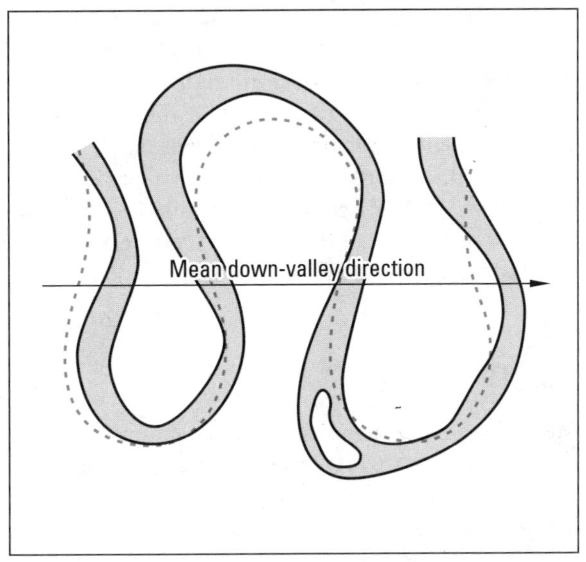


FIGURE 21.47 ▼ Relationship between meander pattern and a sine curve: the Mississippi River near Grenville, Mississippi (from Leopold and Langbein, 1979)

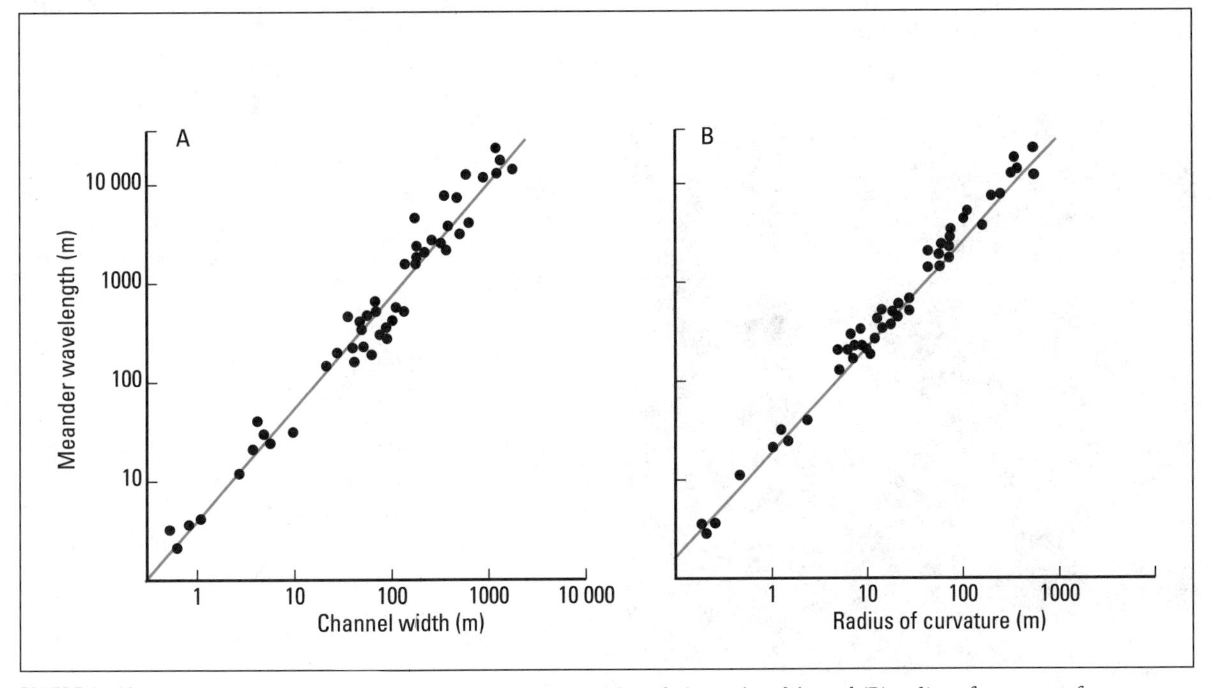


FIGURE 21.46 ▼ Relationships between (A) meander wavelength and channel width, and (B) radius of curvature for a range of natural and flume streams (from Leopold *et al*, 1964)

under a wide variety of circumstances. Since meanders are found not only in alluvial streams but also in streams developed on glaciers and in the Gulf Stream, it certainly appears that meandering is a fundamental response of flowing water, independent of sediment or bank conditions.

The constant adjustment of the channel to changes in discharge and suspended sediment load produces a migration of the meanders. This is evident in the erosion of the outside of the bend of the stream and the deposition of sediment on the inside of the bend, creating what is known as a point bar (see Figure 21.13). Extreme development of the sinuosity results in cut-off bends in the river, producing **oxbow lakes** (see Figure 21.15).

## BRAIDED CHANNELS

Many streams do not have a single sinuous channel but are characterized by the separation of the channel into a number of small interlocking channels. These are called **braided channels** (Figure 21.48). The streams draining ice sheets and glaciers frequently develop this form because of the variability of the discharge and the larger calibre of the sediment in the channel. The braided system is highly active and unstable. The river regime, with a rapid increase in discharge in the spring (see Figures 21.5 and 21.7), mobilizes sediment in the channel. As the discharge falls, the coarsest sediment is deposited on the channel bed (Figure 21.49). This produces scour near the banks, and in the deeper channels the centre channel deposits will be above water level. Differential scour between the two channels will produce an asymmetrical bar (Figure 21.49). The channel bifurcations and bar formations will be repeated constantly downstream until the sediment size load decreases to fine sediments. Ice accumulation in the channels, or icings, can also promote the migration of the channels when the spring flow occurs before the winter ice accumulation has melted.

FIGURE 21.48 ▼ The braided stream system of the Donjek River, Yukon (photo: P.G. Johnson)

There is, moreover, a marked contrast between the generally coarse sediments accumulating in braided stream systems and the finer sediments associated with meandering channels.

Conditions favouring development of braided channels include easily erodible banks (low silt and clay content), a high discharge relative to slope (Figure 21.50), a high sediment load, and, in some cases, a variable discharge. These are all conditions represented by streams draining from glaciers and icecaps, for they commonly cross wide, flat expanses of gravel and sand, and are subject to marked changes in discharge as the ice undergoes seasonal or longer periods of melting and advance. It is not surprising, then, that braided streams are common in these environments, and there is reason to believe that they were more widespread during the Quaternary Period, when many rivers in what are now cool, temperate regions were affected by meltwaters.

Transitions between braiding and meandering streams occur as flow regimes and sediments are modified downstream. This kind of transition can be seen on the Donjek River, Yukon (Figure 21.13).

## THE LONG PROFILE OF STREAMS

We are all acquainted with the change in slope displayed by most streams. In their headwaters, the channels are steep; as they get closer to the sea they become more gently sloping until, as far as the eye can tell, they flow across a flat surface. In other words, the generalized **long profile** of most streams is concave-upward (Figure 21.51).

The gradient of the stream is a reflection of the energy available for transport of sediments and the size of the sediments. In general, in the upper reaches of the stream the discharge is low and the

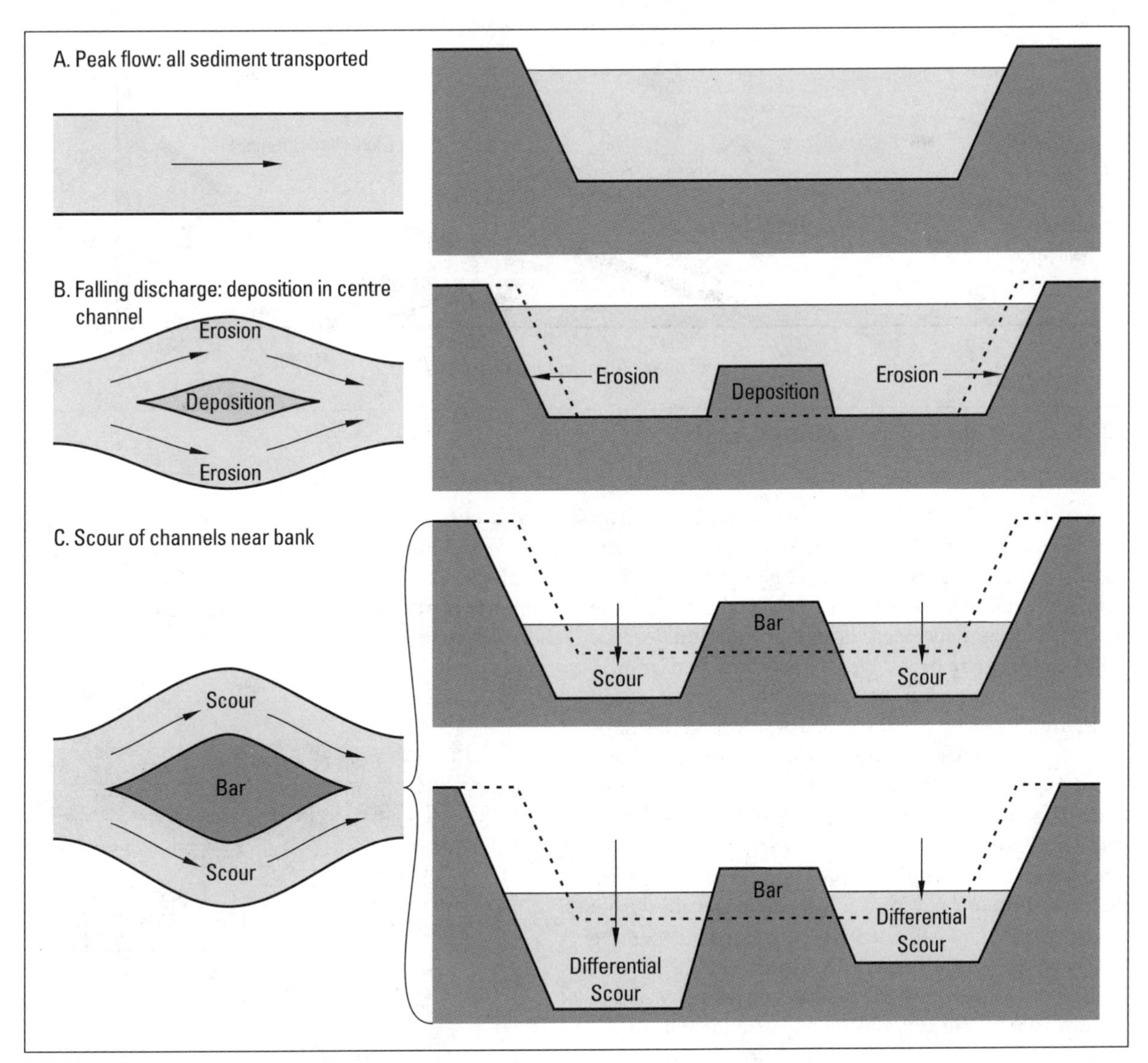


FIGURE 21.49 ▼ The formation of a braided channel system

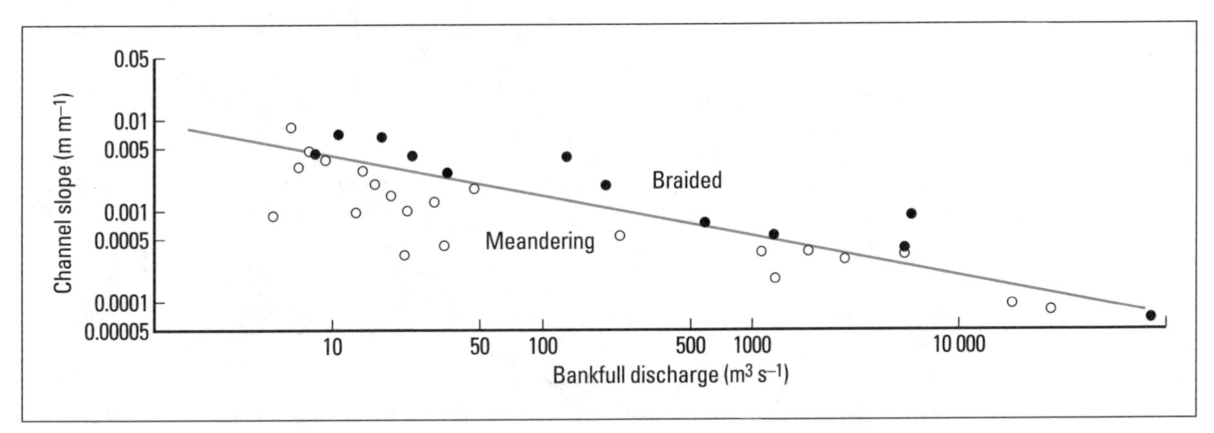


FIGURE 21.50 ▼ The relationship between bankfull discharge and channel slope for meandering and braided streams (after Leopold and Wolman, 1957)

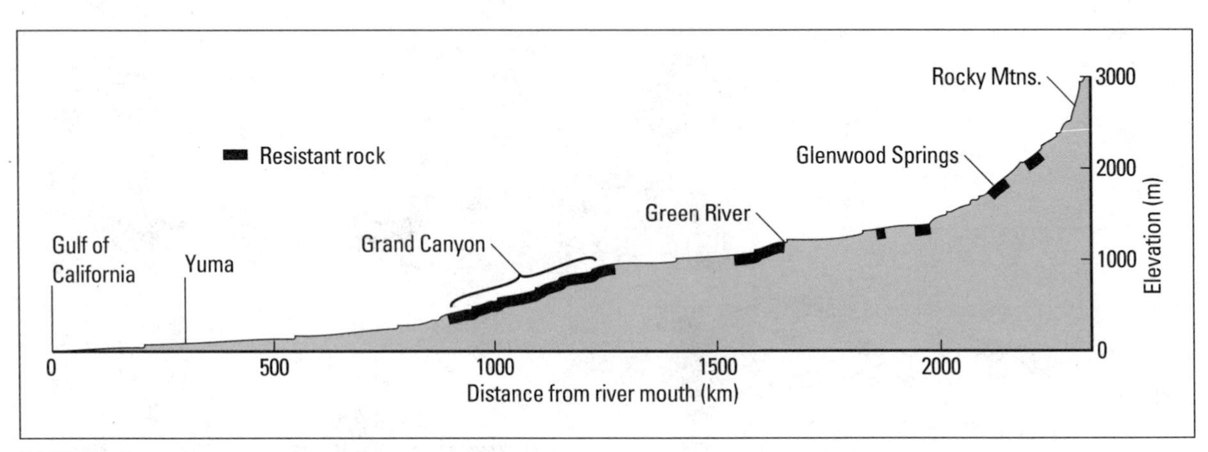


FIGURE 21.51 ▼ Profile of the Colorado River, USA

size of the sediment is large so that high velocities are necessary for transport of material, predominantly as bed load. As the discharge increases downstream, with the runoff from larger areas, and the sediment size decreases, larger proportions of the load are transported as suspended sediment. Although suspended sediment can be transported with much lower stream velocities, a large amount of the stream energy is used in the transport. Lower slope angles are required for the river. One can visualize the effects on stream gradient by considering the confluence of two streams with different flow conditions. If a high slope-angle stream with a heavy sediment load joins a low-angle stream with a heavy sediment load, then the combined flow at the lower slope angle cannot transport the combined sediment, and deposition will occur. If a low-angle stream with a heavy sediment load joins a high-angle stream, then the combined flow is capable of transporting more sediment than it has available, and the excess energy will be used to degrade the bed. The stream maintains a balance between discharge, sediment load, and slope.

In fact, there is no fundamental reason why streams should adopt a concave long profile, and in some conditions, where the relationship between stream energy and sediment load is suitable, straight or even convex profiles may develop. Where discharge declines downstream—due, for example, to excessive evaporation, seepage into the bedrock, or extraction by humans—then the degree of concavity diminishes and convex profiles can occur. It has also been shown that if sediment size increases markedly downstream—as happens when coarse material is introduced in the lower reaches from a cliff or talus slope, or from an active tributary—the long profile tends toward a convex form. On the Colorado River (Figure 21.51), the convex elements of the profile are due to the occurrence of more resistant bedrock.

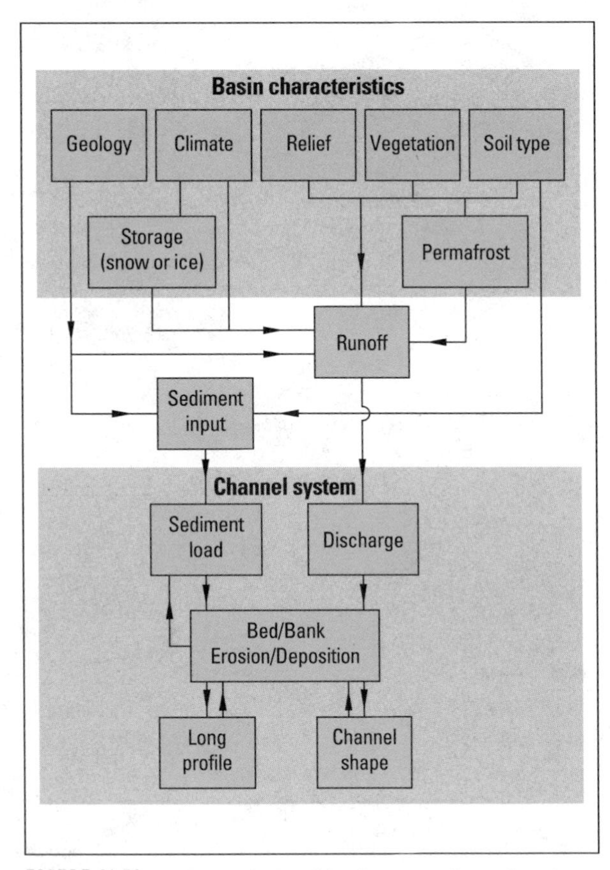


FIGURE 21.52 ▼ Interrelationships between channel variables and basin characteristics

As these examples indicated, the long profile of the stream owes much to the relationship between discharge and sediment size, or, more generally, available energy and work rate. Indeed, the whole stream system can be conceived as a set of interdependent variables that respond mutually to external conditions such as climate, geology, and regional relief (Figure 21.52). The fact that there are so many variables, all interacting, means that it is difficult to predict precisely the ways the stream system will respond, and as Leopold and Langbein (1979) have stressed, introduces a degree of indeterminacy to stream systems. Stated simply, the large number of variables, the almost infinite number of streams and environmental conditions, and the wide range of possible adjustments that each may make to a particular event mean that it is impossible to forecast in any single situation what the outcome will be.

## Streams and drainage basins

So far we have looked at what goes on within the stream channel and its immediate floodplain. Streams, however, can be viewed in relation to more general systems. These are most commonly defined as all areas that contribute runoff into a stream above any point defined on the stream, and are known as drainage basins. The topographic height that marks the division between one system and another is called the watershed.

The drainage basin and the streams within it are intimately related, for the basin represents the gathering ground (or **catchment area**) for the water that flows through the stream. Variations in the character of the drainage basin—in its geology, climate, or topography, for example—affect the character of the stream. In this context, one of the most important aspects of the drainage basin is the drainage density, that is, the average length of stream channel per unit area of drainage basin:

$$D_d = \Sigma L / A$$

where $D_d$ is the drainage density ($km/km^2$)
$L$ is stream length (km)
$A$ is the area of the drainage basin ($km^2$)

Drainage density influences the time it takes for water to concentrate in the stream channel, and affects stream discharge and channel characteristics.

One of the most important controls on drainage density is exerted by the infiltration capacity of the soil. As we saw in Chapter 13, this influences the amount of water available for surface runoff, and thus has an effect on stream channel development. Where infiltration capacity is high, little overland flow occurs, and channel development is inhibited; drainage density is low. Where infiltration capacity is low, most of the water runs off as overland flow; rills, gullies, and major streams develop and a dense network of channels emerges. Drainage density is therefore high.

Many different factors influence infiltration capacity, so these processes vary considerably from one area to another. One of the main influences is that of geology. Rocks such as limestone, gravel, sand, or well-jointed igneous rocks tend to have high infiltration capacities, so they are associated with low drainage densities. Impermeable rocks, such as shale, clay, or massive igneous rocks, give rise to rapid runoff and higher drainage densities.

Topography is also important. Runoff tends to be encouraged on steep slopes; it is reduced where the land is flat. Vegetation has an additional effect. Dense vegetation helps to intercept rainfall, to increase the infiltration capacity of the soil and to prevent overland flow. Consequently, it inhibits channel development. Where vegetation is sparse, drainage densities are much higher. The effect is often seen on agricultural land, and, indeed, reflects a major problem in intensively farmed areas. Where vegetation is removed, dense networks of eroded gullies may be formed.

In northern latitudes, these relationships can be further complicated by the presence of permafrost and ground ice. These conditions inhibit infiltration of water.

Geology, relief, and vegetation affect what happens to the water when it reaches the ground, but the input of water itself clearly depends upon rainfall. Climatic factors, also, influence drainage basin character. We might guess that drainage density is highest in areas of high rainfall, and to some extent this is true. But is is not invariably so, for three reasons. The first is that it is not simply rainfall, but the balance between rainfall and evapotranspiration that is important. Where evapotranspiration is high, less rainfall is available for erosion and channel formation. The second reason is that rainfall and vegetation tend to be closely related, so that areas of high rainfall are often characterized by a dense vegetation cover; as we have seen, vegetation encourages infiltration and reduces runoff. Finally, it is not so much the total amount of rainfall that is critical, as the intensity of rainfall. Short, intense storms lead to more overland flow than prolonged, light rain, as we discovered in Chapter 11. Thus, the relationships between climate and drainage density are often complex.

The storage of water, either seasonally as snow, or semi-permanently as glacier ice, are major influences in high alpine or high latitude environments. As mentioned in the discussion of river regimes the runoff is then concentrated into much shorter time periods. Frozen ground beneath snow also controls the characteristics of runoff during the spring in many of these areas.

### DRAINAGE PATTERN

It is not only the drainage density that varies in response to basin characteristics. So, too, does the shape of the drainage basin. Drainage basins in areas of high relief, for example, tend to be relatively long and thin; broad, short basins are typical of newly exposed coastal areas. In addition, the **drainage pattern** within the basin varies markedly in relation to factors such as geology, climate, and relief. For many years the description of fluvial systems was based on the arrangement of the stream tributaries (Figure 21.53). **Trellised** drainage patterns tend to develop where there is strong structural control upon streams, the main channels being aligned parallel to the 'grain' of the country with minor tributaries feeding in at right angles. **Rectangular** patterns evolve where faults or joints guide stream channels, producing a characteristic 'gridiron' pattern with right-angle bends and a distinct and regular arrangement. **Parallel** patterns are often associated with areas of steep relief or non-cohesive materials (e.g., gullies on mine dumps often adopt this form), while **dendritic** patterns are typical of better adjusted systems on uniformly dipping strata. Finally, **deranged** drainage patterns are found in areas where the original drainage network has been disturbed, for example, by glacial activity. It is apparent that relatively rarely do real-world drainage patterns fit exactly the idealized forms! This type of approach has been replaced by a more process-oriented approach since the 1950s.

Lakes must be considered an integral part of the river system. Their role as a storage component on the river regime extends periods of higher discharge but usually damps extremes. They are also major sediment sinks in the system, and change the hydraulics of the system as sediment is lost from the flow. In northern environments, the winter lake ice cover, which develops in a similar way to plate ice on rivers, promotes deposition of finer suspended sediment by reducing water turbulence. The lake ice itself can promote erosion by the physical effect of grinding on the shoreline, by thermal expansion of the ice, particularly in small bays, and can exert stress on artificial structures along the shore sufficient to pull piles out of the lake bed.

## Fluvial landscapes

### LANDSCAPES OF FLUVIAL EROSION

In 1858, Lieutenant Joseph Christmas Ives entered the Grand Canyon. In his diary he wrote,

> The famous 'Big Cañon' was before us; and for a long time we paused in wondering delight, surveying that stupendous formation through which the Colorado and its tributaries break...the corresponding depth and gloom of the gaping chasms into which we were plunging imparted an unearthly character to a way that might have resembled the portals of the infernal regions. Harsh screams issuing from aerial recesses in the cañon sides and apparitions of goblin-like figures perched in the rifts and hollows of the impending cliffs gave an odd reality to this impression.

Without doubt, the landscapes produced by fluvial erosion include some of the most spectacular and awe-inspiring scenery in the world. Gorges and canyons such as the Grand Canyon of the Colorado or the Iron Gates on the Danube represent remarkable examples of the power of streams as agents of erosion. But how do these landscapes develop?

One reason is mountain-building and uplift. As the land surface is raised, the potential energy of streams is increased and they erode their beds in the attempt to regain equilibrium with their surroundings. Where uplift involves little tilting or warping of the rocks, a more or less flat plateau is formed. Waters flowing across this surface tumble over the margins to create waterfalls and rapids. At the base of the waterfall they erode a deep, smooth **plunge pool**. One of the most famous waterfalls in North

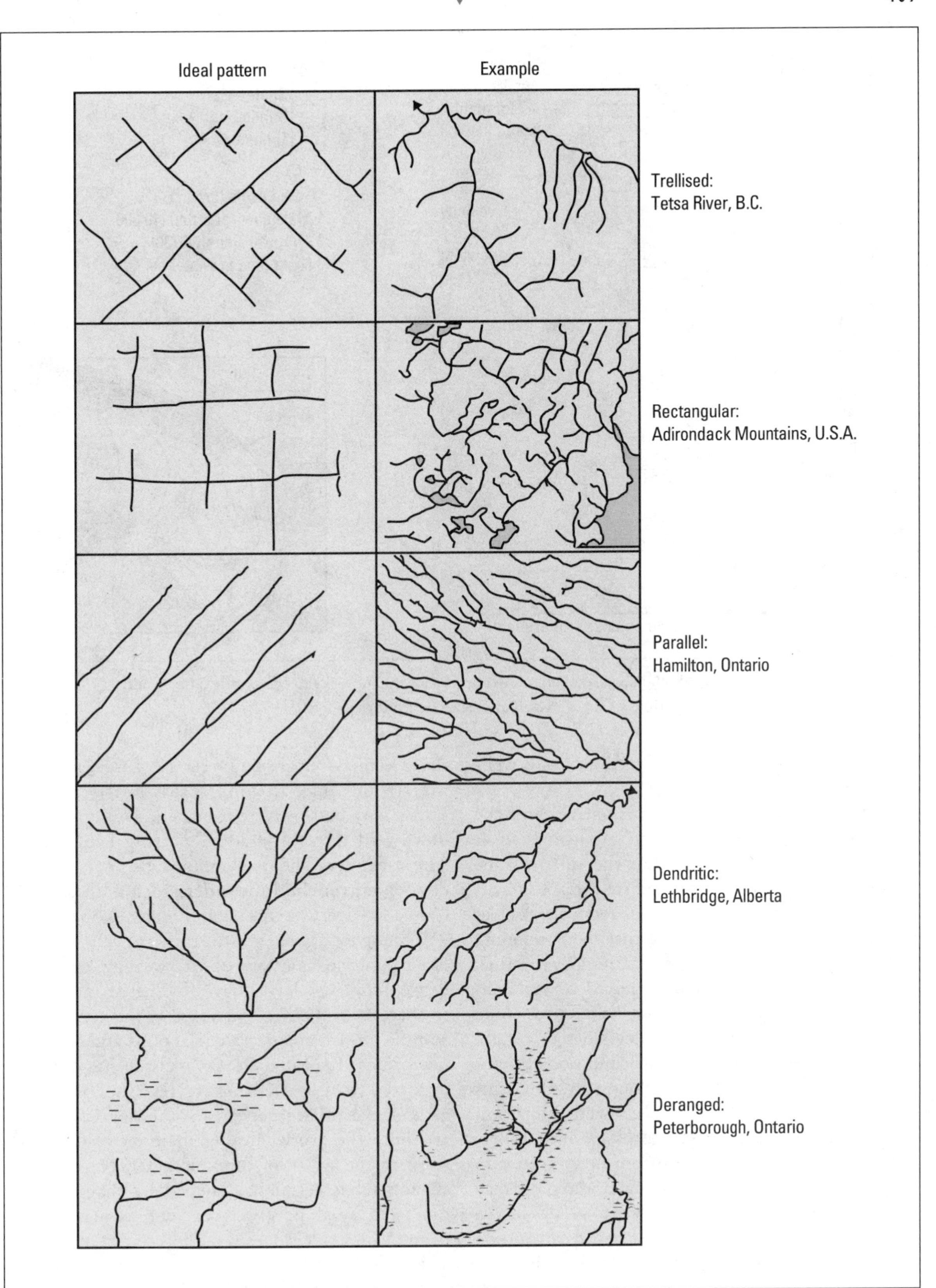


FIGURE 21.53 ▼ Types of drainage pattern: theoretical and actual

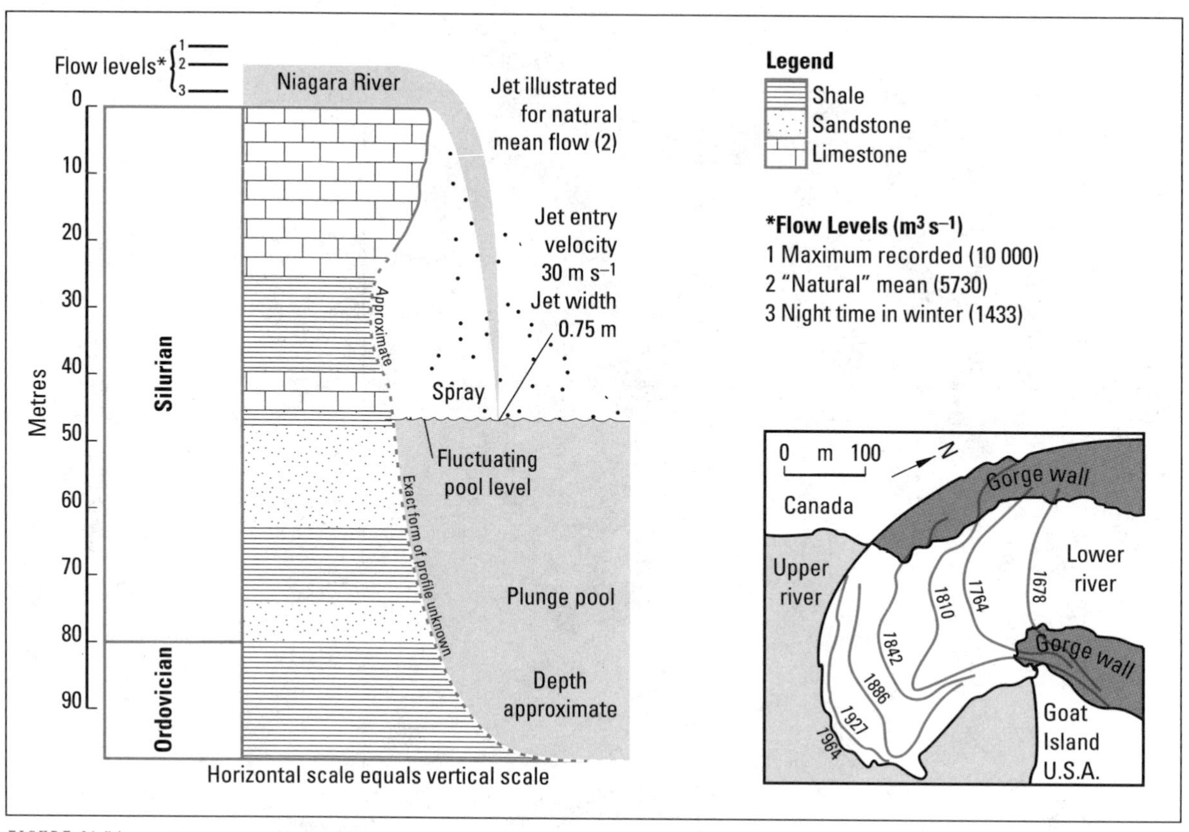


FIGURE 21.54 ▼ Cross section through Niagara Falls showing relationship of rocks to the profile (after Tinkler, 1986). Approximate positions of the waterfall are based on Philbrick, 1970

America, Niagara Falls (Figure 21.54), has a well-documented retreat history.

Erosion at the plateau edge is active, and the waterfall eats back into the scarp. As it does so, it exposes a range of rocks within the valley; some may be soft, others resistant, and the stream attacks them at different rates. The harder rocks themselves produce waterfalls and rapids so that the developing gorge has an irregular long profile. Given time, these irregularities might be removed by the constant action of the water; given time, also, slope processes may act upon the valley sides to reduce them to more gentle forms. But conditions do not always remain constant long enough to allow this smooth profile to develop. Renewed mountain uplift may stimulate the processes once more; changes in the course of the stream may cause it to abandon its old route, and rapids and waterfalls may be left literally high and dry. What we see in the Grand Canyon is a landscape that began 65 million years ago and has been gradually rising ever since.

Mountain uplift is not always rapid, and slow earth movements may allow the initial character of the stream system to be preserved. Large, meandering streams once flowing across low-lying landscapes may be slowly rejuvenated, so that they cut downward into the gradually rising land surface. The meanders themselves may be incised; thus develop **entrenched meanders** such as those of the Colorado River system, Utah (Figure 21.55).

Throughout geological history, the earth's surface has experienced constant uplift, warping, and subsidence, while sea levels have fluctuated in response to changes in the size and shape of the ocean basins and the growth and decay of the polar and continental icecaps. Repeated fluvial incision has therefore occurred. These events have left their mark on almost every landscape; to some degree, we can attribute the overwhelming majority of our landscape to the action of running water. Nevertheless, streams rarely act alone. The valleys they carve are modified by slope processes such as mass movement. Valley side slopes wane and retreat as a result. The ultimate form of the fluvial landscape therefore owes as much to the work of slope action as it does to the work of streams.

FIGURE 21.55 ▼ Entrenched meanders on the Colorado River system, Utah (photo: P.G. Johnson)

## DEPOSITIONAL LANDFORMS

We have already discussed many of the depositional features that occur within the stream channel and the development of the floodplain. In addition, extensive deposition often occurs where rivers enter large bodies of water such as lakes or the ocean (Figure 21.56). The stream's capacity to carry sediment is a function of its velocity. The river is slowed when it enters the lake or ocean and it thus deposits the heavier sediments. These deposits of sand and silt build up as a **delta**. The finer silts and clays are carried farther out into the deeper water where they gradually settle out to form the very fine bottom ooze. The delta shape is formed because the stream moves from side to side depositing sediment in ridges stretching out into the water. Once a ridge has formed, the stream moves to the side, to deeper water, following the line of least resistance. Another ridge builds and the stream moves again. The stream will continue to build ridges and move sideways until it comes up against the shoreline; it then moves back to the other side. The result is a feature in the form of the Greek letter Δ. The Nile delta in Egypt is a good example of this classic form. Other forms are created by the nature of the shoreline, the types of currents in the larger body of water, and a variety of other factors. The Mississippi River delta represents another classic form known as a bird's foot. This is produced by rapid deposition of sedi-

A

B

FIGURE 21.56 ▼ Examples of deltas. (A) Delta forming into Peyto Lake, Alberta. The stream is glacier-fed and contains a high suspended-sediment load, clearly visible in the plumes into the lake (photo: P.G. Johnson). (B) Part of the Mackenzie Delta, NWT. The main channel is in the background of the photo, and interchannel lakes can be seen in the foreground (photo: John De Visser/Masterfile)

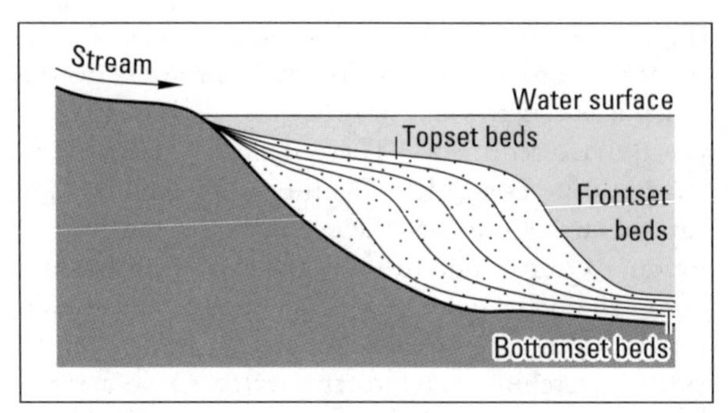


FIGURE 21.57 ▼ The structure of a simple delta

FIGURE 21.58 ▼ Alluvial fans in the Donjek Valley, Yukon (photo: P.G. Johnson)

ments along the margins of the fresh water flow due to the formation of large agglomerations of silts and clays. This process is called **flocculation** and is caused by electrical bonding between clay particles due to the salts in sea water.

When studied in cross section, the deltas all have a similar structure. Sediment swept onto the delta by the stream slumps and avalanches down the delta-front to produce steeply-dipping beds of material that has settled to the bottom. Over time, the flat upper surface of the foreset beds is overlain by topset beds (Figure 21.57).

The same processes occur on land where streams emerging from the mountain cascade to the foot of

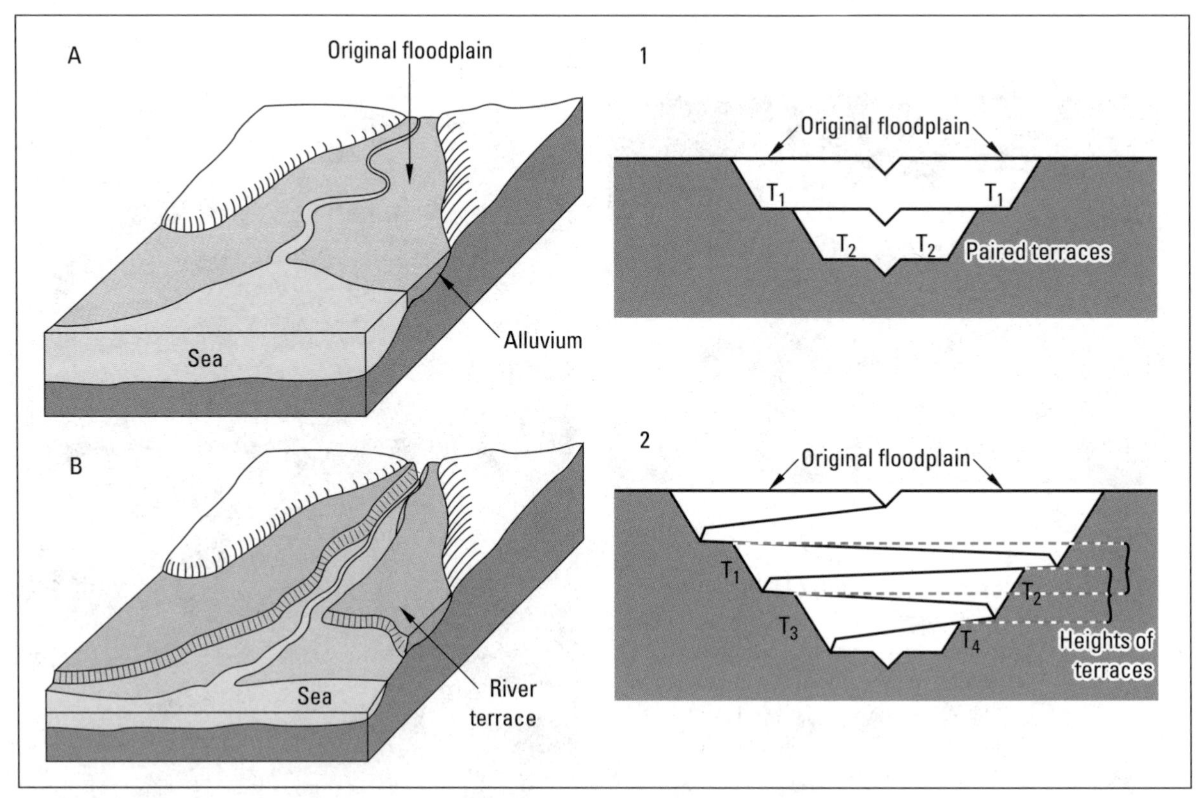


FIGURE 21.59 ▼ Development of thalassostatic river terraces: (A) stream builds up an alluvial floodplain adjusted to sea level; (B) sea level falls and the stream cuts down through the alluvium to a new level, leaving the old floodplain as a river terrace. If the stream cuts vertically into the sediments (1), it creates paired terraces; if the stream migrates across the valley (2), non-paired terraces are produced

the slope. The sudden slowing causes deposition of the coarser material, resulting in features that could be called 'land deltas,' but are correctly called **alluvial fans** (Figure 21.58). The fan continues to accumulate until the sediment supply is cut off or the stream is abandoned, at which stage the fan becomes fossilized. The main period of formation of the Donjek Valley alluvial fans was during the deglaciation of much of the area in the late Pleistocene, when large volumes of sediment were released from the ice.

Many valleys are flanked by sequences of **river terraces**, which provide flat, fertile land for agriculture and ideal conditions for settlement. As a result, they have often been sites for the development of towns; Terrace, British Columbia, is a fine example.

Terraces develop in a variety of ways. Some are produced by changes in sea level. A rise in the sea level causes the stream to deposit its load (aggradation) as it adjusts to the lower slope angle. When the sea level falls, the stream cuts through the sediment, leaving a bench-like feature (Figure 21.59A). This is known as a **thalassostatic terrace** (from the Greek word *thalassa*, meaning sea). If the incision of the stream is vertically into the sediments, as in Figure 21.59B, it leaves paired terraces, that is, terraces at the same level on each side of the stream. If the incision is by migration of the stream across the valley, the changes in the direction of migration may produce non-paired terraces (Figure 21.59B).

River terraces are also formed by changes in the supply of sediment and water to the stream. During periods of abundant sediment supply, the streams may become choked with debris, so that, over time, they aggrade the bed. This happened widely in the glacial periods when frost activity encouraged weathering of rocks and solifluction transported the material down the valley sides. For much of the time the streams were unable to cope with this input of sediment and aggradation occurred. When the climate changed and temperatures increased, however, and water locked up in the snow and ice cover was released into the streams, discharge increased, and so did the competence of the streams. At the same time, the establishment of vegetation upon the slopes stabilized the valley sides and sediment inputs diminished. The streams eroded their beds and cut through the sediments to produce what are known as **climatic terraces** (Figure 21.60).

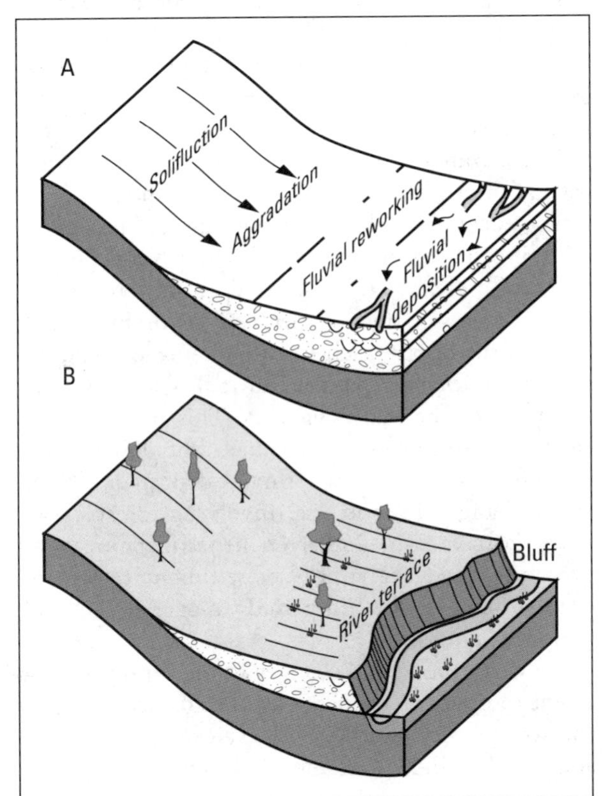


FIGURE 21.60 ▼ Development of climatic river terraces: (A) a braided stream, fed by solifluction and mass movement of hillslope materials in a periglacial environment, builds up a floodplain of coarse debris; (B) warming of the climate results in the development of a vegetation cover, stabilization of the hillslopes and concentration of the stream into a single meandering channel. During the relatively short period of climatic change, the stream is able to cut downward, leaving the old floodplain as a river terrace

# Streams and humans

The relationship between streams and humans is intimate. Over the centuries, streams have exerted a fundamental influence upon our economies and politics. They have helped to fashion the distribution of our settlements; they have acted as barriers to cultural diffusion and as routeways for trade and movement. They have been both a focus of civilization and a threat to it. In return, we have modified the streams, sometimes deliberately, often accidentally.

We have already discussed the hydrological aspects of this intimate relationship between the streams and ourselves (Chapter 13). As we saw then, we derive water from streams and also use them as a method of removing and diluting waste products. But the landscapes of fluvial action are also important, as are the geomorphological processes operating within stream channels.

It was to the extensive flat and fertile floodplains of major rivers that prehistoric peoples were attracted, and it was here that the great civilizations developed: the Babylonians and Sumerians on the Tigris and Euphrates, the Egyptians on the Nile, the Chinese dynasties on the Yellow River. It was not just the presence of water that drew these peoples and permitted them to establish permanent, sophisticated settlements. It was also the agricultural potential of the land, a potential maintained by the silts the rivers deposited on the floodplain each time they flooded. Even today, a large proportion of the world's population is directly dependent upon this process for their livelihood. Where natural flooding is too irregular or too limited, artificial means of inducing floods are used. Irrigation channels and ditches are cut across the floodplain, and the silt-laden waters are diverted into these, then dammed and allowed to flood the land. In this way, the soil is revitalized by the sediment washed from lands upstream.

Elsewhere, as we have seen, it has been to the drier, higher river terraces that people have been attracted. These, again, provided fertile agricultural land, especially in temperate regions where the floodplains were often boggy and inaccessible. They also provided firm foundations of gravel and sand upon which to build, and today these same deposits are excavated for the construction industry around many major towns.

Today, also, fluvial landscapes attract us for tourism and recreation. Features of fluvial erosion, such as the gorges of the Grand Canyon and the Rhône, waterfalls like Niagara Falls and Victoria Falls, or the rapids of Cascade Canyon are spectacular and beautiful.

But streams are also hazards. The danger of floods we have already noted in Chapter 13. Stream erosion and sudden channel changes may be similarly catastrophic. In 1938, for example, three days of continuous rain resulted in extensive floods in California. The rate of erosion is estimated to have been as much as 287 500 cubic metres of sediment per square kilometre, most of it fertile soil from agricultural land. Seven hundred new gullies were formed within an area of about 1600 km$^2$. The loss to agriculture was devastating. In November and December of 1990, a series of heavy rainstorms hit the state of Washington and the province of British Columbia, causing extensive damage and erosion. There have also been numerous spring runoff floods on the Peace River, Alberta, as in the spring of 1992.

During events of this type, the stream may abandon its channel and carve itself a new route. In China, the Huang or Yellow River—'China's Sorrow'—has done so on numerous occasions with terrible cost to human life. It has been estimated that the river has changed course some twenty times during the last 4250 years, and many of these changes have involved substantial alterations (Figure 21.61). In 1855, a quarter of a million people died as the river abandoned one course and found a new one. These **avulsions**, as they are known, may occur with surprising rapidity.

The hazards of streams are not always entirely natural in origin. Often they derive from the effects we have had upon channel conditions and processes. Gullying, for example, induced by overgrazing or vegetation clearance in the steep headwater areas, results in erosion and extension of the drainage network in the upper reaches of the drainage basin; we saw this in the last chapter. But it also delivers extra sediment to the main channels. This may initially encourage erosion of the channel, but downstream the effect is to accelerate fluvial deposition. Over time, changes in land use, involving forest clearance, cultivation, and even urbanization, have greatly altered the supply of sediment to stream channels and resulted in periods of aggradation and periods of erosion.

As we can imagine, changes in the supply of sediment to channels influence all the characteristics of the stream—its width, depth, velocity, turbulence, and profile. Under extreme conditions, the input of new sediment may convert a meandering channel to a braided form. But human effects upon the landscape do not only cause changes in sediment inputs; they also influence the rate and quantity of water supply to the channel. Urbanization, for example, may significantly increase the amount of surface runoff and lead to a marked rise in peak flows. This too, will cause adjustments in the stream channel.

Changes in channel form produced by human influences have been called **stream metamorphosis**. Changes in the shape of the channel of Watts Branch in Maryland occurred during building operations in the area. The channel became smaller as sediment

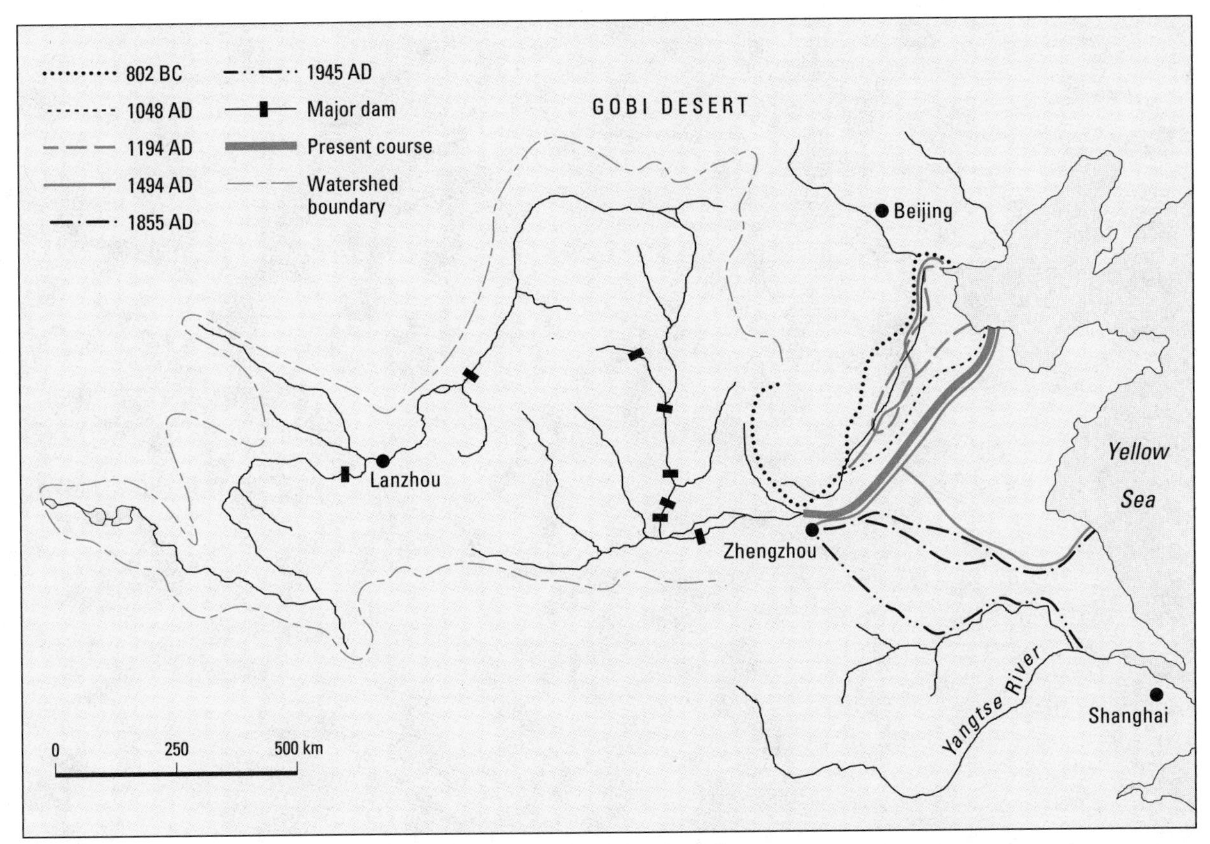


FIGURE 21.61 ▼ Changes in the course of the Huang River, China, since 800 B.C. (from Tsung-Lien Chan, 1976)

disturbed during construction was washed into the stream. Subsequently, the capacity of the channel increased as the sediment supply declined, but the rate of surface runoff rose in response to the more extensive area of impermeable urban land. During severe storms, the flood of water from urban land may greatly increase stream erosion, and the channel may adjust accordingly. Reduction of sediment downstream of dams can also increase the energy available for erosion. It has been shown that 9.87 million $m^3$ of bottom sediment were scoured from the channel of the Colorado River downstream from the Glen Canyon Dam between 1963 and 1965, before the channel stabilized.

Not all the effects of human interference with stream channels and drainage basins are as dramatic as these examples indicate. Often, the consequences are subtle. Nevertheless, our use of the drainage basin almost invariably has some impact on the stream system, and frequently these impacts feed back to affect us in many ways. The need, therefore, is to understand the ways that stream systems operate and to manage them not in a piecemeal fashion but as integrated parts of whole drainage basins.

# Periglacial systems

## Periglacial environments

The term **periglacial** describes the environmental conditions in all regions dominated by intense **frost action**, regardless of their present or past proximity to glaciers. This is a rather vague definition since it is quite difficult to assess the importance of freeze–thaw processes in the field. A simpler indication that a site is periglacial is the absence of glaciers and the presence of **permafrost**. Permafrost is a thermal condition in which the temperature of earth materials remains at or below 0°C for two or more years. This condition is relatively easy to measure and map. A complication, however, is that while all non-glacierized permafrost areas are periglacial, not all periglacial areas possess permafrost. Cold environments with strong maritime influences, such as northern Newfoundland, are outside the margins of the permafrost zone but retain a periglacial character. Equally, some high-altitude areas in temperate latitudes are regarded as periglacial regions but do not possess permafrost (e.g., the southern French Alps).

It is estimated that about 20 percent of the world's continental area is characterized by periglacial conditions. Most of these vast tracts of land have low population densities or are uninhabited. Because periglacial areas are so inaccessible, misconceptions about them abound. For example, some believe that the Arctic is covered by snow year-round. In fact, winter snow accumulation over most of the Canadian High Arctic is limited by aridity, nearly all the snow melts in a period of two to three weeks starting in mid-June, and even at a latitude of 80°N, maximum air temperatures in July can exceed 10°C.

Another common misconception about periglacial areas is that they are all the same. There is actually considerable variability in climate within the periglacial domain, and this is reflected in the vegetation. In the High Arctic, some parts are virtually barren, while periglacial areas farther south exhibit a complete vegetation cover that ranges from tundra to northern boreal forest. Similarly, periglacial terrain within Canada varies from the flat, lake-dominated landscape of the Mackenzie Delta to the steep, high-altitude slopes of the Rocky Mountains.

Periglacial systems include many processes other than frost action. While some of these occur only in periglacial environments (e.g., the formation of certain types of ground ice) most also operate in nonperiglacial areas. This chapter concentrates on the unique aspects of the periglacial environment, especially permafrost and ground ice, but includes descriptions of more widely distributed processes when they are influenced by periglacial conditions.

# Permafrost

Permafrost is defined by temperature, not by the state of any included soil moisture (i.e., whether it is ice or liquid water). In some materials, temperatures may fall below 0°C while much of the pore water remains liquid. When describing permafrost, the words frozen and unfrozen refer only to the state of the water. The thermal condition is expressed in two other terms: **cryotic**, meaning at a temperature equal to or less than 0°C, and **noncryotic**, meaning at a temperature greater than 0°C. Thus a section of core from the bottom of the Beaufort Sea might be perennially cryotic at a temperature of –2°C, but due to saline pore water, could also be unfrozen.

Permafrost underlies about 25 percent of the earth's total land area, although some of this is beneath cold-based glaciers and so cannot be regarded as periglacial. In the northern hemisphere, extensive permafrost is found in Canada, Alaska, Russia, the People's Republic of China, Greenland, and Scandinavia (Figure 22.1). Permafrost occupies

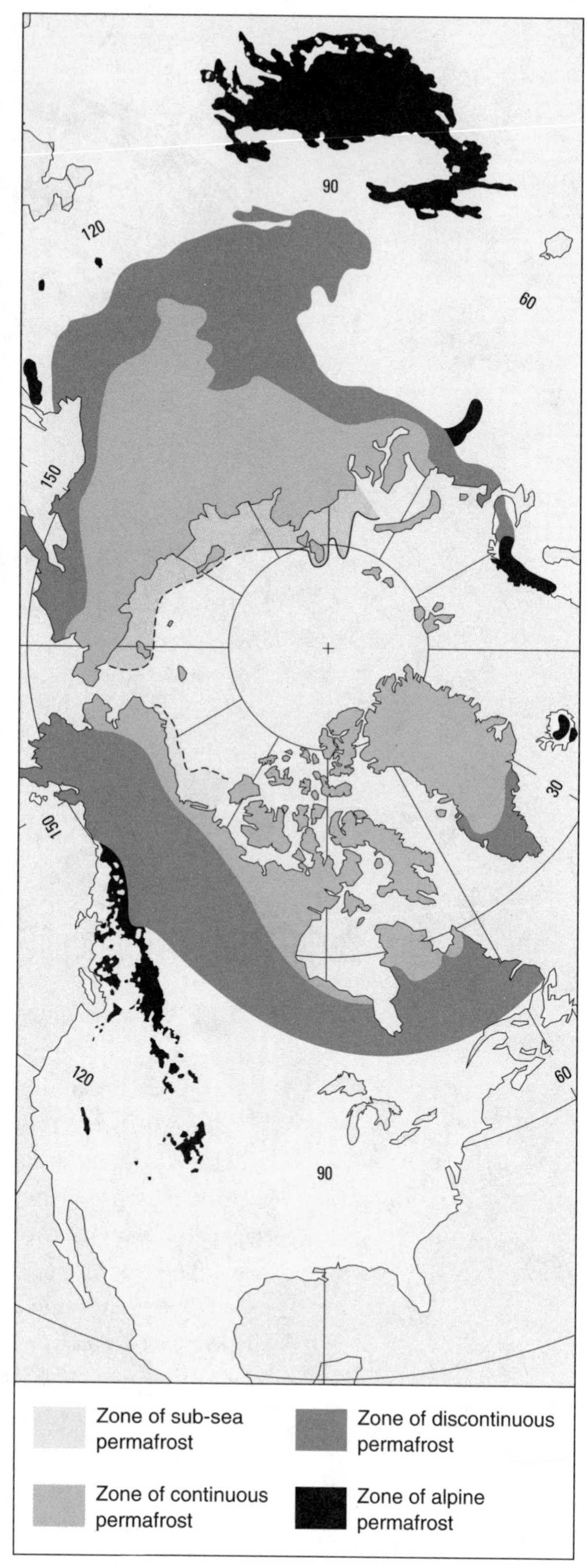


FIGURE 22.1 ▼ Distribution of permafrost in the northern hemisphere (modified from Péwé, 1983)

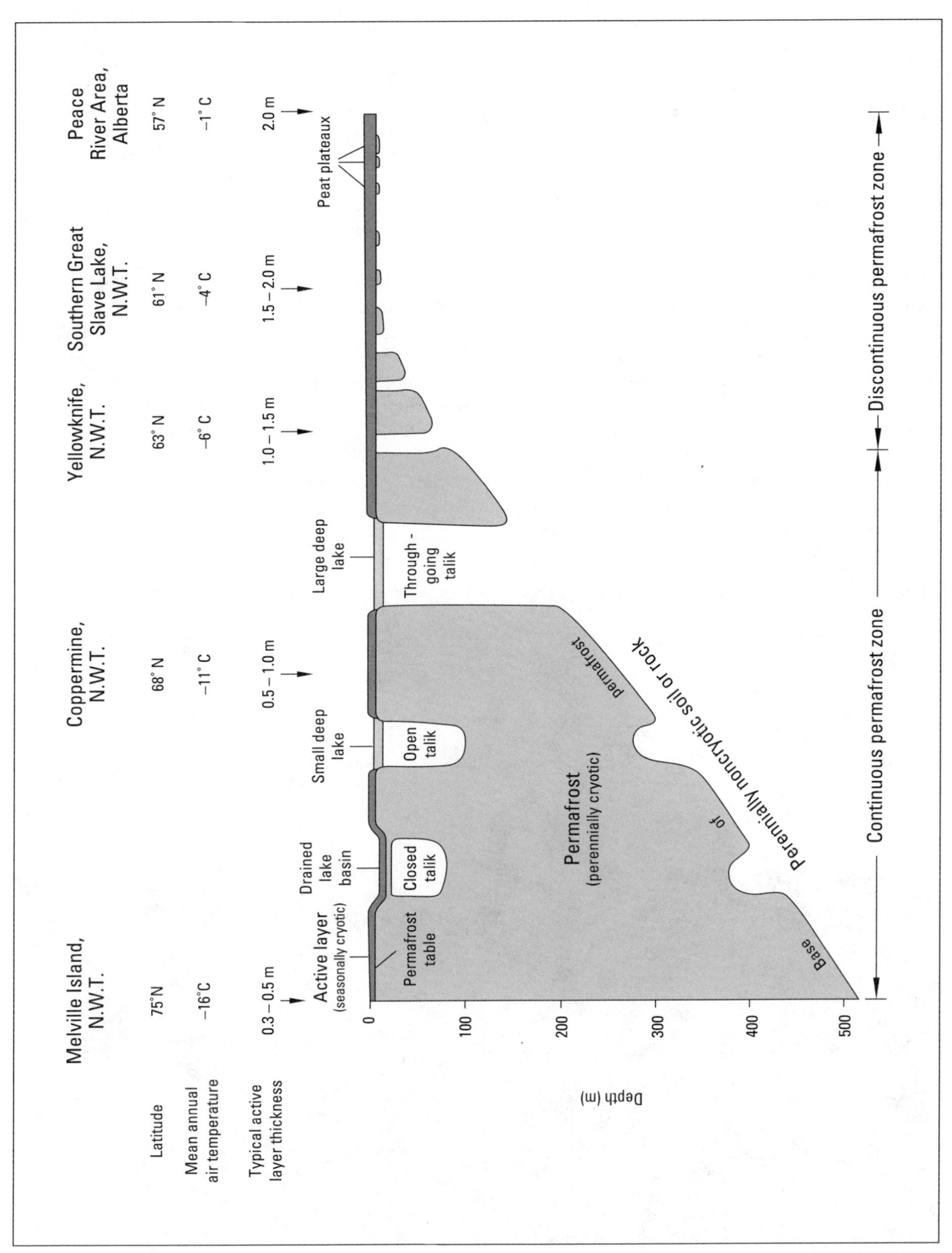


FIGURE 22.2 ▼ Transect of permafrost from north to south in Canada along longitude 115°W showing thicknesses in the continuous and discontinuous zones. Taliks are produced by the thermal effects of lakes that do not freeze through in winter. Note: active layer not to scale

about 50 percent of Canada and occurs in four zones: continuous, discontinuous, alpine, and sub-sea.

The continuous permafrost zone in Canada extends from the Arctic Islands as far as southern Hudson Bay (Figure 22.1). Permafrost is present everywhere except in newly deposited sediments or beneath deep lakes or rivers. Some of it has been in existence continuously for more then 40 000 years. Thicknesses range from more than 500 m in the north to about 100 m in the south (Figure 22.2).

At the northern margin of the discontinuous zone, permafrost is absent in particularly warm soils, such as on south-facing slopes, at locations with deep snow accumulation, and at well-drained sites. Permafrost diminishes southward in area and thickness, until at the southern limit it is present only sporadically, usually in areas of peat accumulation. At these sites permafrost may be only a few metres thick (Figure 22.3).

The uppermost layer of ground in a permafrost area is the **active layer** and it is bounded at its base by the **permafrost table**. The active layer is seasonally cryotic, and thaws in summer to depths ranging from about 0.3 m in parts of the High Arctic to several metres at the southern margin of the discontinuous zone. Perennially noncryotic layers or bodies termed **taliks** may be present within permafrost as a response to groundwater flow, or the thermal influence of a water body that is too deep to freeze completely in winter (Figure 22.2).

Alpine permafrost occurs because of the reduction in air temperatures with altitude. The zone of discontinuous alpine permafrost typically occupies an elevation range of about 1500 m within the Rocky Mountains and at higher elevations permafrost is continuous (Figure 22.4). The minimum elevation of permafrost rises in a southerly direction, from about 600 m near Banff, Alberta, to over 2000 m in New Mexico at a latitude of 35°N.

Sub-sea permafrost is known to occur in the Beaufort Sea off the northern coasts of Canada and Alaska, and in the Kara, Laptev, and East Siberian

FIGURE 22.3 ▼ Permafrost mound east of Hveravellir, central Iceland. Permafrost in this area is discontinuous and is present in the mound but not beneath the stream or surrounding terrain (photo: A.G. Lewkowicz)

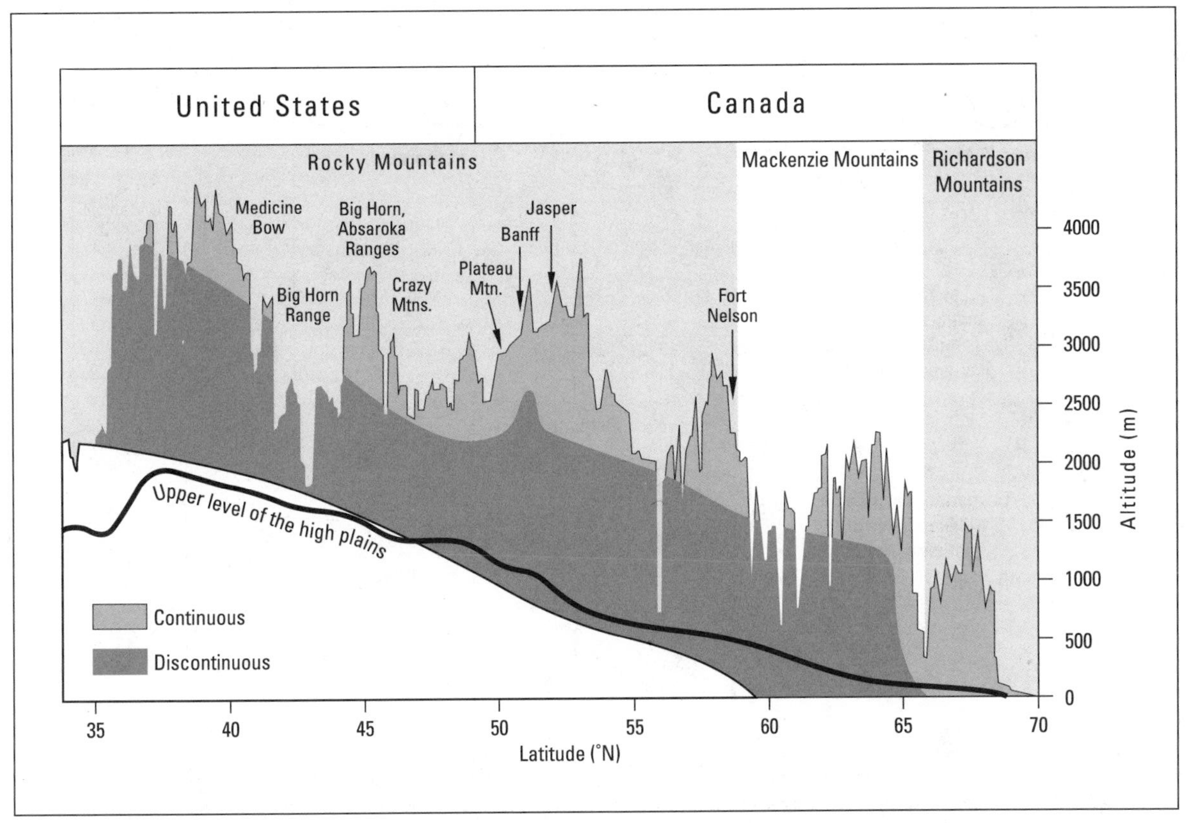


FIGURE 22.4 ▼ Alpine permafrost zonation along the eastern ranges of the Rocky Mountains (modified from Harris, 1986)

seas adjoining Russia (Figure 22.1). This type of permafrost is either a response to bottom sea water temperatures below 0°C, or a relic from earlier times when sea levels were lower and the present seabed was exposed to cold air temperatures.

During the Pleistocene Epoch, permafrost and periglacial zones were shifted far south of their present-day boundaries. The casts of ice wedges, for example, provide evidence that permafrost existed in southwestern Ontario immediately following the retreat of the Late Wisconsinan glacial ice. The former presence of frozen ground may be of more than academic consequence. Landslides within certain clays in southern England where permafrost was also extensive can only be understood if it is recognized that previous freezing reduced their normal shear strengths.

Permafrost is the result of the annual inputs and losses of energy to the ground surface balancing at a temperature below 0°C. There is a fairly good correlation between air temperature and ground temperature, but the latter is up to three to four Celsius degrees warmer because snow cover reduces heat losses from the ground in winter and there is no similar insulator to keep out summer heat inputs. The extreme southern boundary of the discontinous permafrost zone in Canada broadly correlates with the –1°C mean annual air isotherm, and between –1°C and –4°C permafrost is quite restricted in extent. The continuous permafrost zone starts between the –6°C and –8°C mean annual air isotherms (Figure 22.2); this boundary often coincides with the tree line.

Permafrost can be in equilibrium with the present surface energy balance, or can be aggrading (growing) or degrading (declining). If climatic conditions remain constant for a long period, perhaps 1000 years or more, permafrost may stabilize, with a fixed temperature profile and thickness (Figure 22.5). Since the climate is rarely so constant, permafrost is usually in some state of disequilibrium. It can take a very long time for deep permafrost to thaw, and it is believed that some of the frozen ground in Alaska and Siberia is a relic from the last glaciation.

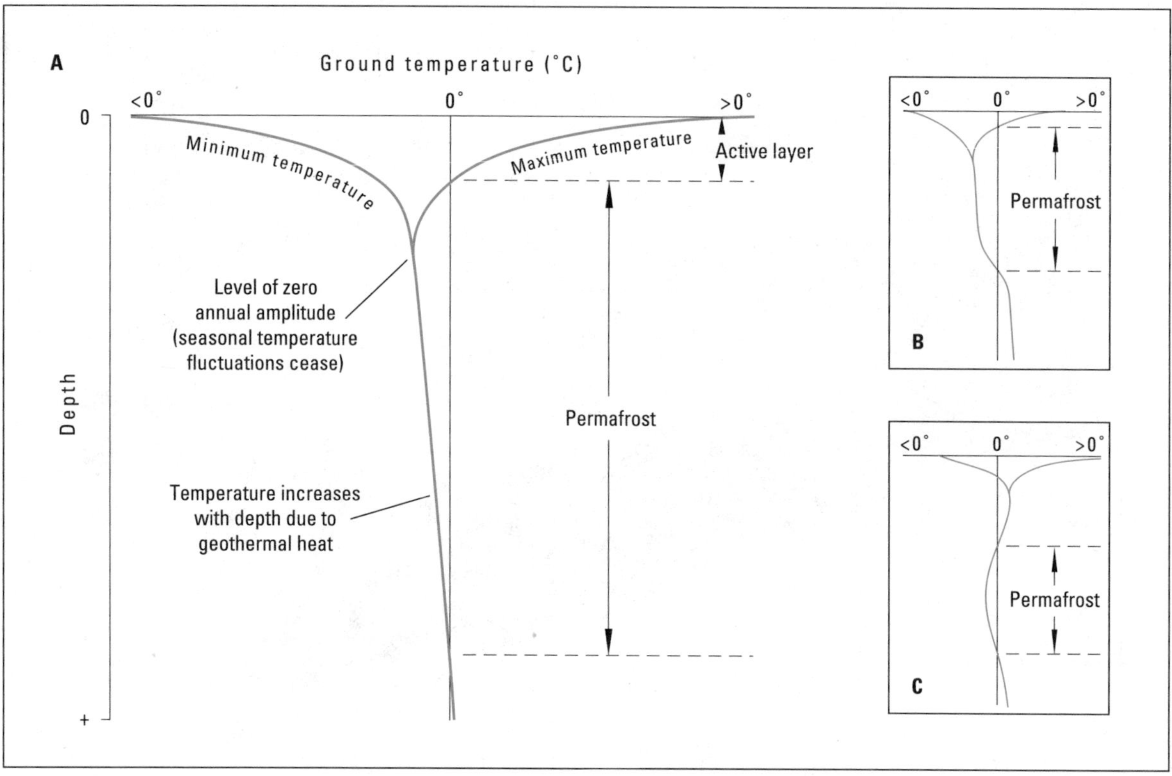


FIGURE 22.5 ▼ Permafrost temperature profiles: (A) stable permafrost, as indicated by the uniform geothermal gradient; (B) aggrading permafrost, shown by the sharp inflection in the geothermal gradient that in time will even out at an increased permafrost thickness; (C) degrading relic permafrost, indicated by the perennially noncryotic zone between the base of the active layer and the top of the permafrost

# Ground ice

## DISTRIBUTION AND TYPES

Frozen moisture that exists beneath the surface of the earth is called **ground ice**. It is extremely variable spatially and can range from almost 100 percent by volume at locations where massive ice is present (Figure 22.6) to virtually zero in dry permafrost.

Ground ice is usually concentrated in the uppermost layers of permafrost and becomes less important with depth. In one part of the Mackenzie Delta, for example, average ice volumes are 60–70 percent at depths of 0.5–1.5 m below the surface and decline to 40–50 percent at depths exceeding 5 m. Most ground ice develops in one of the following five ways:

1. **Pore ice** consists of subsurface water frozen in place and is the most widespread ice type. It is present in the pore spaces of the vast majority of cryotic sediments.
2. **Segregated ice** develops as a result of suction forces that attract water from unfrozen parts of a soil to the **freezing front**. The forces result from capillarity and adsorption and become stronger the smaller the pore size, so that segregated ice is generally limited to soils containing silt and clay. The resultant ground ice varies in form from lenses and layers to massive ice (Figure 22.6). It is ice segregation that causes much of the **frost heave** in both permafrost and non-permafrost areas.
3. **Intrusive ice** is formed as a layer or mass by the bulk freezing of a body of water injected under pressure into sediments. It constitutes some of the ice within **pingos** (see below).
4. **Wedge ice** develops when surface water enters thermal contraction cracks and freezes. The process revolves around an annual cycle that starts in the autumn with ground freezing. By

FIGURE 22.6 ▼ Massive ground ice about 8 m thick exposed near Tuktoyaktuk on the Beaufort Sea coast, Northwest Territories. Note the stratified sediment layers separating bands of almost pure segregated ice and the person for scale (photo: A.G. Lewkowicz)

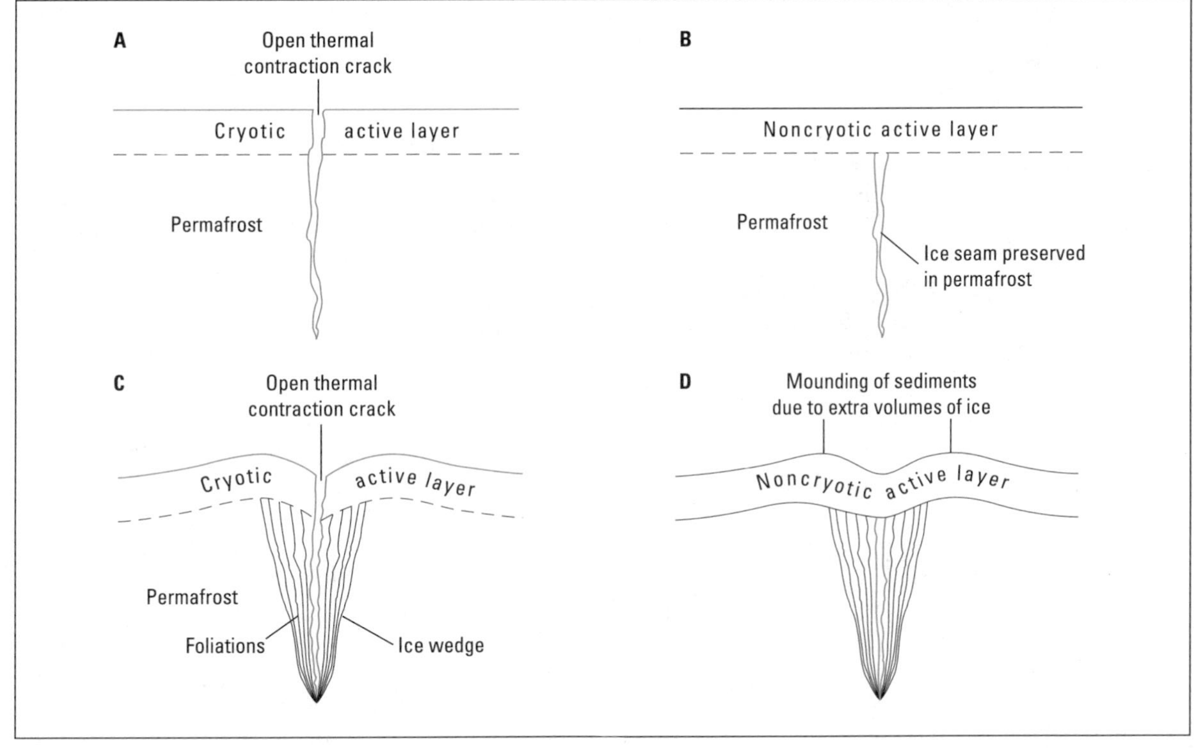


FIGURE 22.7 ▼ Development of an epigenetic ice wedge: (A) first winter; (B) first autumn; (C) 500th winter; (D) 500th autumn (modified from Lachenbruch, 1962)

midwinter, the frozen soil has cooled and contracted sufficiently that tensional stresses exceed its frozen strength and linear cracks develop that are about 1 cm wide and penetrate up to 8 m into the permafrost (Figure 22.7A). Winter snow and spring meltwater enter the thermal contraction crack and freeze as a single vein of ice, preventing the complete closure of the crack as the ground warms and expands in the spring. Ice within the active layer thaws during the summer but any within the permafrost is preserved (Figure 22.7B). The following winter, the ice vein acts as a plane of weakness so that the crack tends to develop at the same place in the soil and the cycle is repeated. Typical annual vein growth is less than 1 mm, but after many years of cracking an ice wedge will have developed (Figure 22.7C and D; Figure 22.8). The additional volume of ice within the soil results in upturning of sediments on either side of the wedge. Within the wedge itself, narrow bands of sediment-laden ice termed **foliations** represent increments of growth.

Two types of wedges are recognized. **Epigenetic** wedges develop after the enclosing sediments have been deposited and their foliations are continuous. In North America, wedges are typically epigenetic and attain dimensions of 3–4 m in width and 5–10 m in depth. **Syngenetic** wedges grow as the enclosing sediments are being laid down, for example in an aggrading floodplain. This type of wedge is common in Siberia where widths of up to 10 m and depths exceeding 50 m have been reported (Figure 22.9). In syngenetic forms, foliations near the top of the wedge terminate before reaching the bottom.

5. **Buried ice** includes glacier, lake, river, and sea ice, as well as snowbanks that developed on the surface but have been covered by sediment and preserved in the permafrost. In most cases, the processes responsible for burial (rapid mass movements, for example) are localized and thus occurrences of buried ice tend to be quite small in extent. The exception to this rule is glacier ice buried under a layer of ablation till or outwash deposits. Some extensive massive ice in the western Canadian Arctic, formerly believed to be segregated, is now thought to be buried glacier ice of Wisconsinan age.

FIGURE 22.8 ▼ Large epigenetic ice wedge exposed near Tuktoyaktuk, Northwest Territories. Because the wedge is almost pure ice, it thaws more slowly than the surrounding icy sediments, leaving it raised relative to the face on either side (photo: A.G. Lewkowicz)

## LANDFORMS ASSOCIATED WITH GROUND ICE

The most common landforms that can be attributed to ground ice are ice-wedge polygons. Linear thermal contraction cracks form networks of quadrilaterals, pentagons, and hexagons that cover extensive areas of lowland tundra. Polygons in poorly drained localities are frequently low-centred and surrounded by a raised rim on either side of the ice-wedge trough (Figure 22.10). Those in better-drained sites or on sloping terrain are usually high-centred and the position of the ice wedges is indicated by shallow linear depressions.

A **pingo** is a second landform resulting from ground ice. It is a conical or elongate hill that can vary in height from a few metres to over 60 m, and can be up to 300 m in diameter. Its core includes pore, segregated, and intrusive ice and may also contain sediment. Pingos have been the subject of much research although they occupy a very small percentage of the periglacial landscape. In North

FIGURE 22.9 ▼ The upper portion of a syngenetic ice wedge, Yamal Peninsula, Russia. The decrease in width toward the top of the wedge indicates that this part developed recently as overlying sediments were deposited: it has not cracked as many times as the rest of the wedge beneath (photo: A.G. Lewkowicz)

FIGURE 22.10 ▼ Ice wedge polygons, 10–20 m in diameter, Fosheim Peninsula, Ellesmere Island, Canada. Those on the right of the photo are better drained and high-centred. Those nearer the valley are wetter and low-centred (photo: A.G. Lewkowicz)

America, they are concentrated in the Mackenzie Delta area where there are about 1500 of the closed-system type (Figure 22.11), and in the unglaciated areas of central Alaska and the Yukon where more than 700 open-system pingos have been identified.

**Closed-system pingos** form as taliks close up. Where a lake or river channel is sufficiently deep (usually more than 2–3 m), it will not freeze to the bottom in winter. Hence a layer of water will be in contact with the underlying sediment throughout the year and its thermal effect maintains a talik within the permafrost (Figure 22.12A). If the lake fills in or drains due to erosion, or if the river channel migrates, the talik begins to freeze from the surface downward and inward from the surrounding permafrost (Figure 22.12B). The pore water in the talik undergoes expansion by 9 percent as it changes to ice and since the saturated sediments cannot accommodate this volumetric increase, water is expelled in advance of the freezing front into the remaining unfrozen zone. Pressures resulting from pore-water expulsion are relieved by doming of sediments in that part of the former lake or river bottom which has the least strength, generally the site of a shallow residual pond (Figure 22.12C). The pingo ice core is formed of segregated and intrusive ice. Pingo growth can continue for hundreds of years until all water in the former talik is frozen. Alternatively, growth can result in the sides of the pingo becoming unstable as the sediments overlying the core are stretched. If the ice core is exposed, it melts and the pingo collapses to form a distinctive doughnut-shaped landform with ice-cored ramparts.

**Open-system pingos** occur in thin continuous permafrost or in the discontinuous zone, in valley bottoms, or on the lower portions of valley sides. Unlike closed-system pingos which are found mainly in areas of low relief, the open-system forms require elevation differences because these supply the hydraulic head that causes much of the doming of sediments. Sub-permafrost or intra-permafrost groundwater under artesian pressure moves toward the surface where it freezes (Figure 22.13). The

FIGURE 22.11 ▼ Ibyuk, one of the few closed-system pingos with a name. It is probably the most-photographed pingo in the world because of its large size (48 m high), classic shape, and proximity to Tuktoyaktuk. Note the slumping along the steep sides that may ultimately lead to the ice core being exposed and to collapse of the pingo (photo: A.G. Lewkowicz)

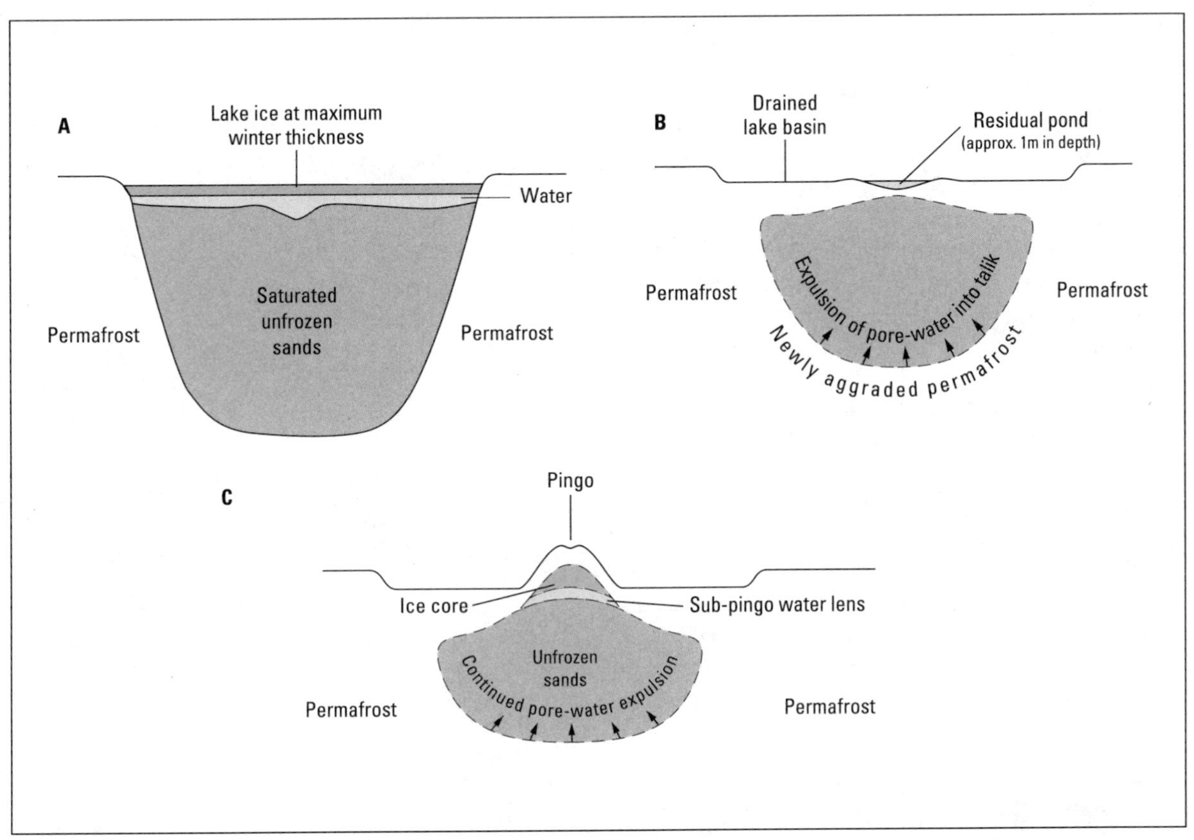


FIGURE 22.12 ▼ Formation of a closed-system pingo. (A) talik beneath a deep lake; (B) drainage of the lake leaves a residual pond and contraction of the talik results in pore-water expulsion into the unfrozen zone; (C) pingo forms at the site of the pond where frozen ground is thinnest and weakest. A sub-pingo water lens originates from pore water explusion. Note: active layer omitted for clarity (after Mackay, 1979)

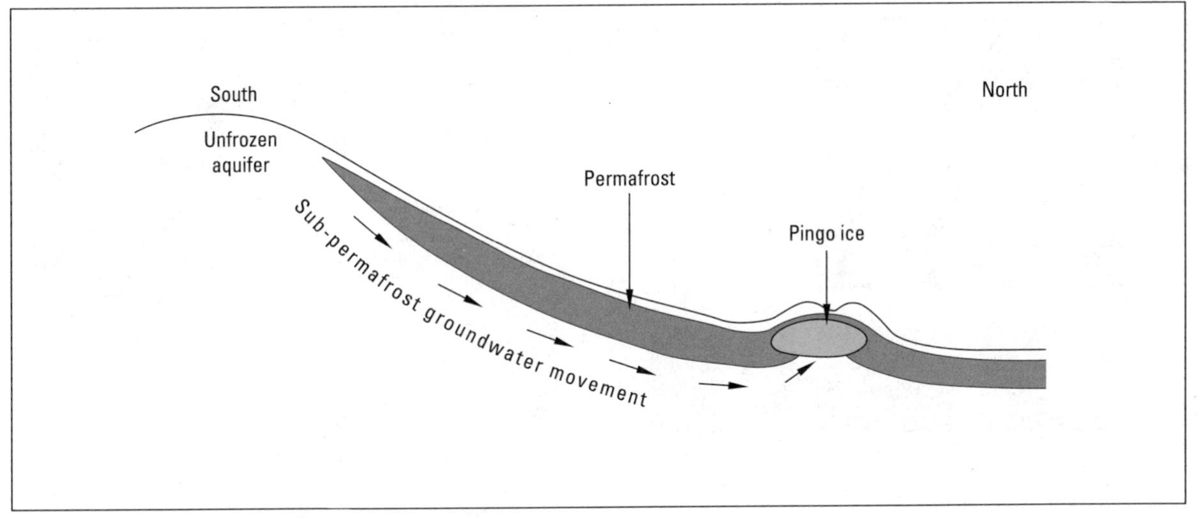


FIGURE 22.13 ▼ Formation of an open-system pingo in the discontinuous permafrost zone. Sub-permafrost groundwater under artesian pressure freezes close to the surface, leading to doming of the overlying sediments (after Muller, 1968)

forces that cause mounding of the sediments are derived partly from these hydraulic pressures and partly from the ice segregation that takes place during freezing. Since this process does not necessarily relieve the artesian pressures, open-system pingos may rupture and give rise to a spring.

**Palsas** are ice-cored mounds that develop in areas where peat is present. Most palsas are found in the discontinuous permafrost zone within bogs or areas of poor drainage. Their appearance varies from round to complex amoeba-like mounds and they range in size from 0.5 to 7 m high and 5 to 50 m in diameter. Both pore and segregated ice are present in palsas, the latter in the form of thin bands of ice interbedded with the mineral soil that is usually present beneath the peat cover.

Palsas in the discontinuous zone often occur as isolated islands surrounded by unfrozen soils. They form as a result of positive feedback following a random reduction in snow accumulation at a site within a peat bog (Figure 22.14). Frost penetration and heaving due to ice segregation are enhanced at the site, producing a small mound. The greater depth of frost makes it unlikely that the ground will thaw completely during the following summer. Summer thaw is also slowed due to the reduced thermal conductivity of the peat caused by evaporation drying the elevated surface. The mound that remains at the end of summer develops a thinner snow cover the following winter and growth accelerates. Measurements show that up to 35 cm of vertical growth is possible in a single year. The positive feedback loop operates until breaks appear in the peat cover as it is stretched over the enlarged core, or the peat is deflated by wind action. The ice within the palsa then thaws and the mound collapses.

# Thermokarst

**Thermokarst** processes are those involving the melt of ground ice, but the term thermokarst can also be used to describe the irregular, hummocky terrain that results. The cause of thermokarst is dis-

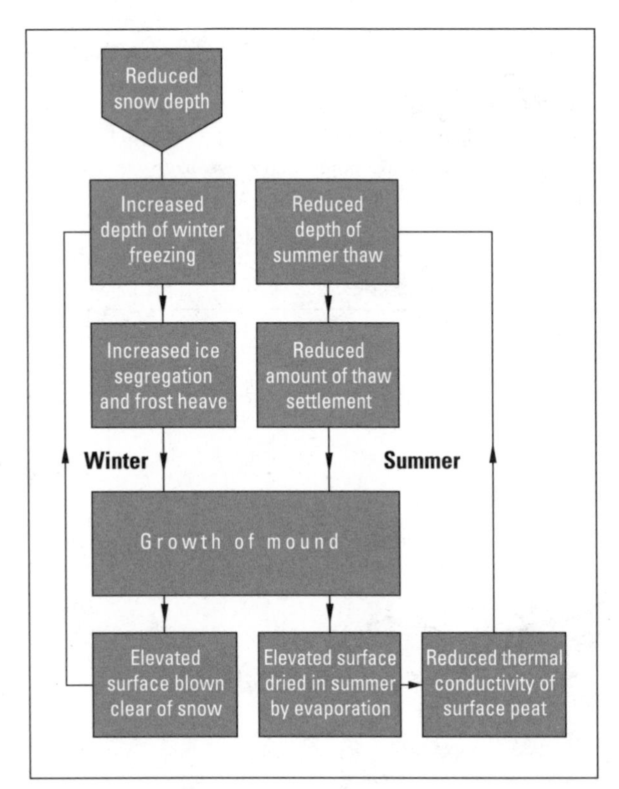


FIGURE 22.14 ▼ Growth of a palsa after an initial random reduction in snow depth. Positive feedback loops operate in both winter and summer to cause growth of the mound

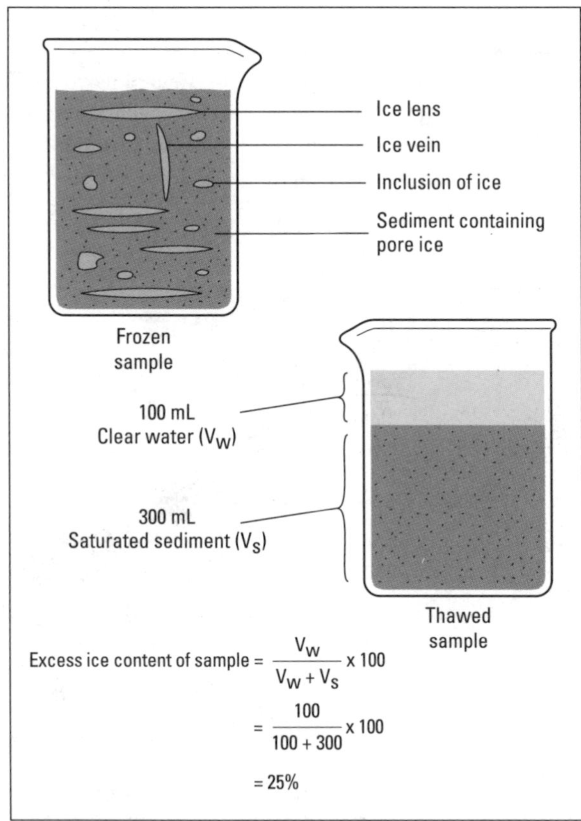


FIGURE 22.15 ▼ Calculation of excess ice content by thawing a frozen soil sample

ruption of the surface energy balance of ice-rich permafrost terrain. Most disturbances result in an increased active layer depth and thaw of the top of the permafrost. A key factor in assessing the potential effects of disturbance is the **excess ice content** of the permafrost. This is the volume of ground ice that exceeds the pore space that would be present if the soil were unfrozen. It is assessed by allowing a sample to thaw in a measuring cylinder: the excess ice content is equal to the volume of water that develops on top of the saturated thawed sediment, divided by the total volume of the sample (Figure 22.15). In the field, water derived from melting excess ice ponds, flows away, or evaporates, so that the ground surface is lowered irreversibly.

The upper part of permafrost often has a high excess ice content and a relatively small initial disturbance can have major consequences (Figure 22.16). Disturbances that cause localized thermokarst include a change in geomorphic conditions (e.g., undercutting of a slope by a river resulting in exposure of ice) or in vegetation cover (e.g., destruction of insulating mosses in a forest fire). Climatic warming may also initiate thermokarst at a regional scale. Such a change must have taken place in Western Europe following retreat of the Pleistocene ice sheets and could occur in permafrost areas in the future as a result of greenhouse-gas induced climatic warming.

Thermokarst processes can be subdivided into those that accomplish thermal subsidence (or downwearing) and thermal erosion (or backwearing). Thermal subsidence operates on flat terrain by thawing from above and gives rise to a variety of ground ice degradation forms. These range from depressions along ice wedge troughs, to extensive flat-bottomed valleys and hollows. These landforms are common in certain parts of Yakutia in Russia but are not found to the same extent in North America.

The most common thermokarst forms in the lowland permafrost areas of Canada and Alaska are **thermokarst lakes**. These relatively shallow lakes are often rounded and result from a combination of subsidence and bank erosion associated with melting of ground ice. Once started, a lake grows at the

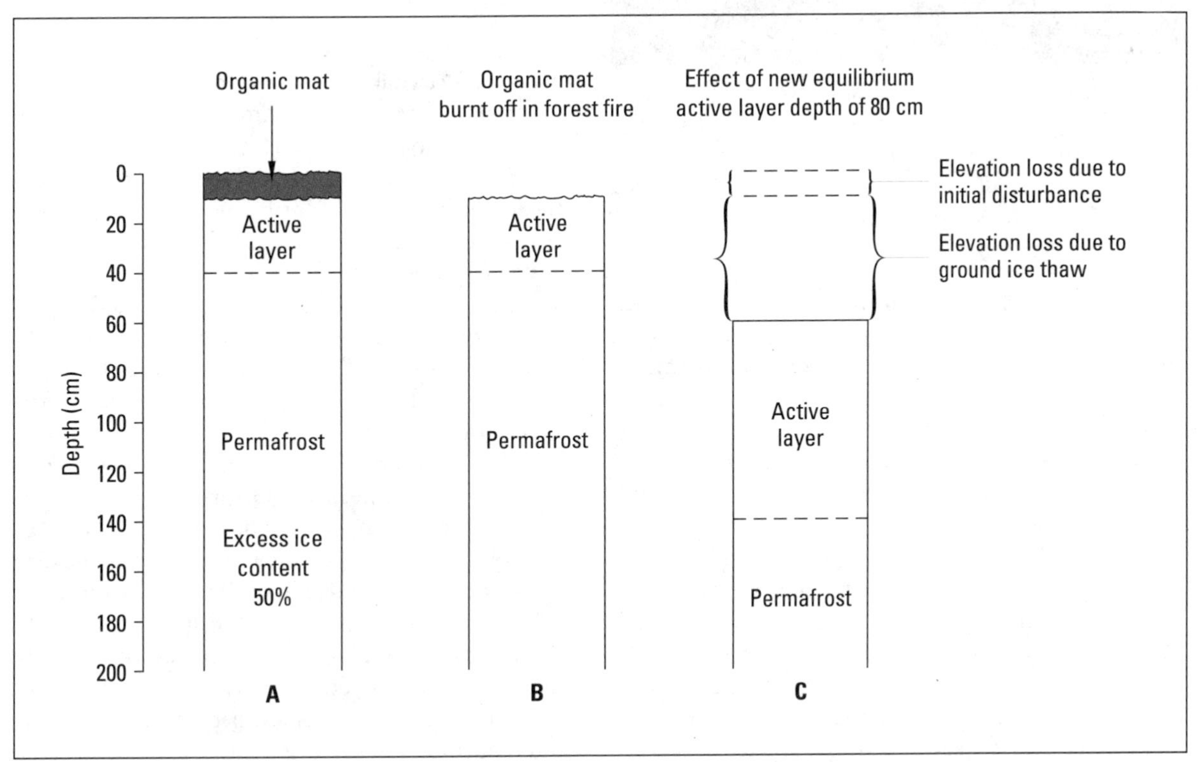


FIGURE 22.16 ▼ Thermokarst caused by a forest fire. (A) before disturbance, 10 centimetre-thick organic mat insulates ground so that active layer extends to a depth of 40 cm; (B) forest fire burns off organic mat; (C) new equilibrium: active layer is 80 cm thick, so 1 m of permafrost thaws to provide 50 cm of additional unfrozen soil (after Mackay, 1970)

surface through wave action and bank undercutting, and its talik enlarges and deepens, thereby thawing any surrounding ice-rich sediments. Frequently, thermokarst lakes capture one another as they enlarge, a process that is evident in their outlines (Figure 22.17).

FIGURE 22.17 ▼ Thermokarst lakes, western Banks Island. Note the way larger lakes tend to capture smaller ones by expanding through wave erosion and thermokarst subsidence (photo: A.G. Lewkowicz)

Thermal erosion is confined to ice-rich slopes, and the forms produced include **retrogressive thaw slumps** and **thermo-erosional niches**. Viewed from the air, a thaw slump is a horseshoe-shaped landform with a melting ice face at its headwall (Figure 22.18). They are started by erosional processes (e.g., fluvial or coastal) which expose the icy permafrost to summer energy inputs. They can also be initiated by people building highways or railway cuts. As the ice ablates, it undermines the active layer material which collapses onto the face (Figure 22.19). The collapsed soil mixes with the meltwater and moves off the ice downslope as a mudflow. Once initiated, a thaw slump is self-perpetuating and enlarges upslope until either the ground-ice body ends, or the slope angle becomes too low to evacuate the mudflow so that the ice face becomes covered by debris. In one part of southern Banks Island where retrogressive thaw slumps are

FIGURE 22.18 ▼ Retrogressive thaw slump initiated by fluvial erosion, Fosheim Peninsula, Ellesmere Island. The ice face is about 5 m high. One year after this photo was taken, the headwall had retreated 15 m upslope and the headwall had increased in height to 8 m (photo: A.G. Lewkowicz)

initiated by coastal erosion, maximum rates of headwall retreat are about 14 m yearly and typical slumps are active for 10 to 15 years.

Thermo-erosional niches develop along coasts and rivers by a combination of thaw of icy permafrost and the removal of sediment by normal mechanical processes. The result is a notch at water level that can extend back under the bank up to 10 m. The frozen sediments overlying the niche eventually break along a line of weakness (often an ice wedge) and the block collapses into the water where it disintegrates.

The importance of thermokarst in permafrost areas is very variable. In landscapes of consolidated or resistant rocks which have low ice contents (e.g., much of the eastern Canadian Arctic), thermokarst processes may be quite insignificant. On the other hand, in ice-rich lowlands (e.g., the western Canadian Arctic and Northern Siberia), thermokarst processes can be extremely active agents of erosion. Understanding the operation of these processes under natural conditions is important, because many of the disturbances associated with the economic development of permafrost regions lead to increased active layer depths and the potential for thermokarst.

FIGURE 22.19 ▼ Headwall of a retrogressive thaw slump, Fosheim Peninsula, Ellesmere Island. Ablation of the ice face (left) leads to undercutting of the dry active layer which collapses by falling outward or in a semicircular fashion. Collapsed material mixes with water from the melting ice to form mud which flows downslope. Equipment in the photo is monitoring energy exchanged with the melting ice surface (photo A.G. Lewkowicz)

# Frost action and patterned ground

One of the processes in the periglacial environment that has been extensively studied is freeze–thaw or frost action. This includes the effects of the volumetric expansion of water that occurs on freezing, as well as processes associated with migration of water to a freezing plane to form segregated ice.

When freezing and expansion of water takes place within the confines of a rock, pressures build up and these may be sufficient to break the rock into angular pieces. In some periglacial areas that were not glaciated during the Pleistocene (e.g., the central Yukon), large parts of the terrain are covered by blockfields, the result of frost-shattering over many thousands of years. At a smaller scale, talus slopes of frost-shattered debris and individually shattered boulders (Figure 22.20) are features typical of many periglacial areas.

FIGURE 22.20 ▼ Frost-shattered boulder, central Banks Island. The expansion of water on freezing has broken the boulder along its bedding planes (photo: A.G. Lewkowicz)

The susceptibility of a rock to frost-shattering is related to its moisture content, permeability, porosity, and tensile strength. The extent to which it is broken down also depends on the number and intensity of freeze–thaw cycles to which it is subjected (see Chapter 18). Numerous laboratory experiments have been carried out under controlled conditions of temperature and moisture supply to isolate the effects of these factors. Massive igneous rocks have been found to be the most resistant to frost action because of their low water content and high strength, while rocks that have the opposite characteristics (e.g., shale) are easily broken down. For some rock types it is now possible to predict without destructive testing both how susceptible a rock will be to particular freeze–thaw cycles and the size of the resulting fragments. It has not been easy to transfer these laboratory results back to the field, however, in part because the environmental conditions under which shattering occurs on rock surfaces have not been widely investigated.

Ice segregation processes are an important part of frost action. Segregation occurs when moisture moves through fine-grained soils toward a **freezing**

FIGURE 22.21 ▼ Frost-heaved stone, southern Banks Island. A combination of frost-pull and frost-push processes has tilted the rock into a near-vertical position and thrust it out of the soil (photo: A.G. Lewkowicz)

FIGURE 22.22 ▼ Sorted circles about 3 m in diameter at Kvadehuksletta, Spitsbergen, Norway. Circular patterned ground forms of this type are attributed to frost heave and thaw settlement within both the fine centres and the coarse rims (photo: A.G. Lewkowicz)

**front**. In non-permafrost areas and in the discontinuous permafrost zone, this water movement is upward in the autumn toward the freezing front which advances from the surface. Ice lenses and layers form roughly parallel to the slope, with the greatest amount of ice (and hence frost heave) immediately below the surface. In soils of the continuous permafrost zone, however, water also moves downward in the active layer toward the top of the cold permafrost, where a second freezing front advances upward in the autumn. This produces ice lensing and frost heave at the base of the active layer as well as beneath the soil surface. Ice segregation processes result in the progressive upward movement of stones by a combination of pushing and pulling in both permafrost and non-permafrost soils. By the time the stones emerge at the surface they have often been tilted to vertical or near-vertical positions (Figure 22.21).

The processes of ice segregation and frost heave are responsible for one of the most unusual phenomena found in periglacial environments—**patterned ground**. Patterned ground is the term used to describe various regular geometric forms such as circles (Figure 22.22), polygons (Figure 22.23), nets, steps, and stripes (Figure 22.24) that commonly develop on the ground surface in periglacial regions. All of these may be non-sorted or sorted. The former are outlined by vegetation, while the latter involve the concentration of particles of different grain sizes to reveal their shapes. Patterned ground forms range in size from a few centimetres to more than 20 m in diameter. While frost action processes, particularly cracking and sorting, are believed to be critical to the development of larger polygonal forms, processes such as desiccation cracking and slopewash may be partly responsible for the development of other types of patterned ground (e.g., small polygons and stripes).

The overall importance of patterned ground in the periglacial environment is a matter of debate. The fascination of the early explorers was trans-

FIGURE 22.23 ▼ Non-sorted polygons of two different sizes, southern Melville Island, Canada.The larger mesh, outlined by vegetation, is about 2 m in diameter, while the smaller is less than 15 cm across (notebook for scale). The larger forms are probably due to frost action, while the smaller ones result from desiccation cracking (photo: A.G. Lewkowicz)

FIGURE 22.24 ▼ Sorted stripes at an elevation of 3100 m in a non-permafrost area, La Mortice, southern French Alps. At this site, small sorted polygons exist on flat terrain and degrade into stripes on slopes greater than 5° (photo: A.G. Lewkowicz)

formed into considerable research effort by modern scientists. Nevertheless, it has proven difficult to isolate the most significant processes in patterned ground formation, probably because the forms are polygenetic. While patterned ground is widespread, it can be argued that it represents a decoration of the ground surface and that more significant elements of the periglacial landscape have been neglected and deserve greater attention.

## Mass movement processes

Slow and rapid mass movement processes have been described earlier (see Chapter 20), but some details specific to the periglacial environment warrant additional discussion.

One slow mass movement process that has been recognized recently by engineers and geomorphologists is **permafrost creep**. This is the slow, continuous deformation of permafrost on slopes and is comparable to continuous soil creep in non-permafrost areas. Movements are concentrated in ice-rich horizons, and at temperatures close to 0°C, the mixture of soil, ice, and unfrozen moisture deforms plastically. Rates are slow (less then 5 mm yearly) but because the motion can be deep-seated, permafrost creep may be an important longterm denudational process. It could also be significant to the stability of any structure built on steep slopes in permafrost.

A second and better-known slow mass movement is **solifluction**, which incorporates frost creep and gelifluction. Frost creep results from soil particles being heaved during freezing in a direction at right angles to the surface, and settling back during thawing in a more vertical manner (Figure 22.25A). Gelifluction is the flow downslope in summer of thawed, saturated soil overlying seasonally or perennially frozen ground. The two processes operate even on very low-angle slopes and give rise to a number of forms including solifluction sheets, lobes (Figure 22.26), and terraces.

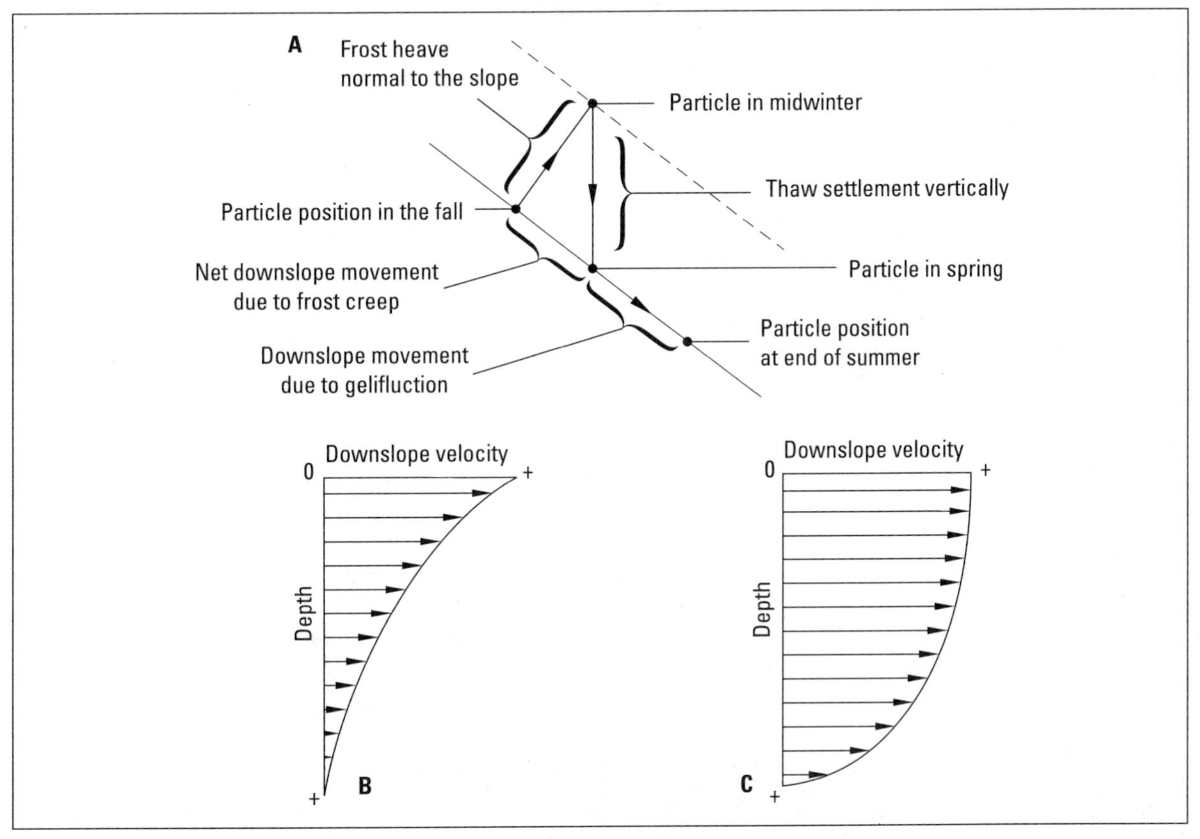


FIGURE 22.25 ▼ Frost heave and solifluction. (A) Movement of a particle on the soil surface due to frost creep and gelifluction; (B) concave downslope velocity profile in an area of one-sided freezing; (C) convex downslope velocity profile in an area of two-sided freezing

FIGURE 22.26 ▼ Solifluction lobe in seasonally cryotic soils at an elevation of 2300 m, Lanserlia area, southern French Alps. Movement of the lobe is being monitored by measuring the angles of tubes inserted into the ground (photo: A.G. Lewkowicz)

The overall importance of solifluction in geomorphic terms depends on the velocity of surface movement (typically 0.5 to 5.0 cm yearly) and the way that velocity decreases with depth. As described above, in non-permafrost areas and the discontinuous permafrost zone, frost heave is concentrated near the surface. Frost creep movements and gelifluction decline rapidly with depth, producing a velocity profile that is concave downslope (Figure 22.25B). In the continuous permafrost zone, however, the growth of ice lenses at the base of the active layer promotes deep-seated heave during freezing and shearing immediately above the permafrost during thaw. The resultant velocity profile is convex downslope (Figure 22.25C). For the same surface velocity, therefore, total denudation will usually be greater in an area of two-sided freezing than in an area of one-sided freezing.

Most of the rapid mass movements described in Chapter 20 can occur in periglacial areas. One type of shallow landslide, termed an **active-layer failure**, is confined to permafrost slopes. The surface thawed layer detaches from the underlying frozen material and moves downslope over a distinct failure plane that roughly parallels the pre-existing surface topography. Depending on the degree of saturation, the soil type, and the velocity of movement, the detached mass may move by sliding with little internal deformation, by flowing, or by a combination of both. The resultant form is typically long and narrow, with a crescent-shaped scar, a track bordered by lateral berms, and a spreading toe zone exhibiting transverse compressional ridges (Figure 22.27). Active-layer failures vary in size from a few square metres to 700 m long and 200 m wide.

The underlying cause of active-layer failures is a reduction in the shear strength of the slope and/or an increase in shear stress. Extra shear stress can be provided by the weight of water in the active layer from a particularly rapid snowmelt or summer rainstorm. An important mechanism that can reduce shear strength is thaw consolidation. If soil containing ice thaws faster than water is able to flow away, positive pore-water pressures may develop, with the result that the soil particles are pushed apart by intervening water films. This is especially likely in areas of two-sided freezing late in the summer, when ice lenses at the base of the active layer thaw. Active-layer failures are also common following forest fires or other disturbances when ice in the top of the permafrost melts.

Another rapid mass movement process in periglacial areas is **rockfall**. It is primarily a vertical transport process and is important locally where resistant rocks form outcrops with vertical or near-

FIGURE 22.27 ▼ Active-layer failure about 500 m long on a 10° slope, Fosheim Peninsula, Ellesmere Island. The unfrozen material slid downslope over ice thawing within the basal part of the active layer. Note the bare scar area upslope and transverse compressional ridges in the toe of the slide in the foreground (photo: A.G. Lewkowicz)

vertical free faces. Although other processes (e.g., pressure release) may contribute to rockfall in periglacial areas, studies have shown that a seasonal cycle of rock release exists. Rockfall frequency in summer peaks earlier on south-facing slopes than on north-facing ones, indicating that the thaw of ice that has wedged apart rocks on the face is a significant part of the process.

Only one longterm study has attempted to compare the relative importance of slow and rapid mass movements in periglacial areas. The results from the Kärkevagge, a valley in northern Sweden, showed the latter to be more important than the former and this is probably a reasonable conclusion for periglacial areas where slopes are steep. In many periglacial regions, however, slopes rarely exceed 10° and in these areas it seems likely that sporadic and isolated rapid mass movements are of secondary importance compared to continuous, widespread, slow mass movement.

## Fluvial processes

The processes by which streams entrain and transport sediment in periglacial and non-periglacial parts of the world are identical. However, certain stream discharge regimes and stream forms are typical of periglacial areas and warrant discussion here.

Most rivers in the periglacial environment exhibit a particular runoff regime termed **nival**. Instead of the river showing a number of periods of high flow throughout the year due to rainfall events, the bulk of runoff occurs over two or three weeks in response to spring snowmelt (Figure 22.28). During this period, small rivers exhibit a daily cycle of discharge that corresponds to the melt of snow in their drainage basins. Despite the small total annual precipitation in many periglacial areas, therefore, high discharges do occur in rivers because much of the year's precipitation runs off in a few

FIGURE 22.28 ▼ The confluence of the Sachs River and a small tributary during peak flow, southern Banks Island, 31 May, 1984. Snowmelt produces the largest discharge of the year as part of the river's nival regime (photo: A.G. Lewkowicz)

days. Moreover, in the permafrost zone, infiltrating water is prevented from recharging deep groundwater by the presence of impermeable frozen ground. Following the melt period, discharges decline (Figure 22.29), and many small rivers in the Arctic dry up completely later in the summer.

A second group of rivers in periglacial areas drain basins containing glaciers or ice sheets. Their regime is termed proglacial and is similar to the nival regime during snowmelt, but is sustained later in the season by glacier melt.

The variation in discharges, both seasonal and diurnal, is one of the factors that contributes to the channel form of small periglacial rivers. Most are braided (Figure 22.30), a condition that is attributed to the effects of (a) their nival or proglacial regimes, (b) the non-cohesive nature of stream banks (especially where vegetation is absent and thermal erosion is active), (c) an abundance of sediment to transport (especially in proglacial or recently glaciated areas), and (d) relatively steep channel slopes (resulting from isostatic rebound following deglaciation), which give rise to high-energy conditions in the stream. Large rivers (e.g., the Mackenzie) often meander because they flow above taliks year-round and do not have the same degree of hydrologic variation.

FIGURE 22.30 ▼ A typical braided river, northeast Axel Heiberg Island, Canada. Note the numerous branching channels, many of which are unoccupied late in the summer (photo: A.G. Lewkowicz)

FIGURE 22.29 ▼ The Sachs River in late summer, 25 July, 1984; same location as Figure 22.28. The river is confined to a single channel on the left and the small discharge is derived from slow drainage of the active layer, late-lying snowbanks, lakes, and occasional precipitation events. The tributary stream has completely dried up (photo: A.G. Lewkowicz)

FIGURE 22.31 ▼ Beaded drainage in a small stream, Yamal Peninsula, Russia. The pools represent locations where the stream crosses ice wedges (photo A.G. Lewkowicz)

A specific channel form that exists in small streams in permafrost areas is beaded drainage. This occurs when a stream crosses a network of ice-wedge polygons. Thermal erosion by the stream and the changed surface energy conditions result in thawing of the ice wedges, which become deep pools interlinked by much shallower drainage channels (Figure 22.31). Beaded drainage is usually visible on aerial photographs and can be a useful indicator of ice-rich terrain.

FIGURE 22.32 ▼ Terrain disturbance resulting from the passage of a single bulldozer, Melville Island. Photo taken in 1976, about five years after the initial disturbance (photo: A.G. Lewkowicz)

# Interaction of humans and periglacial systems

## TERRAIN DISTURBANCE PROBLEMS

Terrain disturbance occurs when human activities in permafrost areas lead to a change in the ground surface energy balance and thaw of ice-rich permafrost. In most cases, the scientific knowledge and technology now exist to avoid such disturbances, but the financial costs of doing so can be considerable. The greatest problems are in the discontinuous permafrost zone, where slight alterations of the ground thermal regime can lead to complete degradation of permafrost. In the continuous zone, the same degree of disturbance would result only in a thicker active layer.

Two examples of human-induced thermokarst are shown in Figures 22.32 and 22.33. The first occurred in the early 1970s on the Sabine Peninsula of Melville Island in relation to hydrocarbon exploration activity. A gas well blowout took place during the summer and all available vehicles were driven to the drill site to assist. The passage of a single bulldozer across sensitive terrain resulted in compression of the soil, increased heat transmitted into the ground, thaw of the underlying ice, and concentration of drainage and erosion along the line of travel. Five years later, a large gully had developed that ran across the landscape for many kilometres (Figure 22.32).

Figure 22.33 shows the effect of a greater initial disturbance, the removal of surface materials. When the Sachs Harbour airstrip on Banks Island was being built at the beginning of the 1960s, sand and gravel were stripped from adjoining areas and bulldozed onto the runway. Ice-wedge polygons underlying the borrow areas subsequently thawed, resulting in an irregular pattern of hummocks and depressions containing standing water. After about 15 years, the form of the thermokarst area had stabilized, but it will remain visible for many decades.

## SOLUTIONS TO TERRAIN DISTURBANCE

To avoid terrain disturbance careful construction techniques are required. The planned town of Inuvik on the Mackenzie River was built on ice-rich terrain more than 25 years ago and very few problems resulted. The key techniques were the construction of gravel work pads by rear-end dumping so that heavy machinery did not cross unprotected terrain, and the use of wooden piles to raise buildings above the ground surface and permit the circulation of cold air in the winter (Figure 22.34). The distribution of water and sewage services in

FIGURE 22.33 ▼ Terrain disturbance in borrow pits adjacent to the Sachs Harbour airstrip, Banks Island. Much of the irregular hummocky terrain formed by thawing of ground ice is covered in ponds. Photo taken in 1985, about 25 years after the initial disturbance (photo: A.G. Lewkowicz)

FIGURE 22.34 ▼ Housing at Inuvik, Northwest Territories, Canada. Row houses are supported by wooden piles inserted into the permafrost. The air gap permits heat loss from the ground in winter so that there is no progressive thaw beneath the building. Water and sewage services are contained in heavily insulated utilidors, which are also supported on piles (photo: A.G. Lewkowicz)

Inuvik is handled in insulated closed tunnels called **utilidors**, themselves elevated above the surface on piles. They are expensive to build, and suffer occasional freezing in winter, but avoid the problem of permafrost degradation.

In the oil and gas industry, terrain disturbance was generally prevented during construction and operation of the Trans-Alaska oil pipeline on ice-rich permafrost. This pipeline carries oil a distance of 1300 km at temperatures of 65°C from Prudhoe Bay on the north coast of Alaska to Valdez on the south coast. Where the pipeline route encounters ice-rich terrain it proved impossible to bury the pipe: even with thick insulation, the heat from the oil would have thawed the permafrost, leading to settlement of the pipe and ultimately rupture. Instead, the pipe was mounted above ground on a horizontal support beam that extended between two vertical support members (VSM). To withstand the stresses developed in the pipe during seasonal expansion, contraction, and earthquakes, the VSMs had to be made of steel. This posed problems because the heat flux down the steel piles was enough to thaw the permafrost. The solution was to use heat pipes inside the VSMs (Figure 22.35). Heat pipes are sealed tubes containing anhydrous ammonia with radiator fins at their upper ends. When the ground is warmer than the air, as it is in winter, the ammonia evaporates from the lower end of the pipe and condenses at the upper end. This removes heat from the ground and transfers it to the upper part of the pipe, where it is lost to the atmosphere. During the summer when the air is warmer than the ground, the heat pipes become inactive. The action of the heat pipes is sufficient to maintain permafrost around the VSMs and thereby ensure the integrity of the pipeline.

The complicated solution used on the Trans-Alaska oil pipeline had a price, particularly since the initial estimates of how much of the pipeline route would be in ice-rich permafrost were optimistic. In the end, almost 50 percent of the pipeline

FIGURE 22.35 ▼ The Trans-Alaska oil pipeline near Fairbanks, Alaska, showing the vertical support members, heat pipe radiator fins, and the pipe mounted on a teflon-coated shoe. The pipe zigzags to allow it to move laterally in response to temperature changes and earthquakes. The pipeline is buried beneath the hillslope in the distance because the ground does not contain significant amounts of ground ice (photo: A.G. Lewkowicz)

was elevated and costs for the project escalated from an estimate of $900 million in 1968 to more than $7 billion by the time oil began flowing in 1977.

## FROST HEAVE PROBLEMS

The effects of terrain disturbance in the form of linear scars representing seismic exploration lines across much of the Mackenzie Delta and the North Slope of Alaska are still visible many years after their formation. A greater problem impeding the development of natural gas resources in these areas, however, is frost heave. Gas would be transported at subzero temperatures in any pipeline that is built. In the discontinuous permafrost zone, this could result in freezing of previously non-permafrost soils with water migrating toward the freezing front (the pipe). Since frost heave would not be equal along the length of the pipe, the potential for rupture would be very high. It is possible to avoid cumulative frost heave by elevating the pipe above the ground on a gravel berm, but the costs are high.

Zero-frost-heave designs do not solve all the problems for gas pipelines. In the discontinuous zone, permafrost, where present, is very thin. If terrain disturbance results in thaw, the pipe is subject to settling. Contrarily, if the pipe cools the soil, additional freezing will occur at the base of the permafrost and cause frost heave. Moreover, heaving can continue to occur within cryotic soils (even those that are not saline) because measurable amounts of unfrozen pore water can be present at subzero temperatures. While the permeabilities of cryotic soils are very much lower than soils above 0°C, the pressures associated with ice formation increase as the temperature declines and become very great if new ice is formed at –5°C or –10°C. The construction of a cold gas pipeline from northern sources through the discontinuous permafrost zone to southern markets clearly will represent a delicate balancing act between the problems of thaw settling and frost heave.

## Conclusion

Most of the periglacial areas of the world are located far from present-day glaciers. They are subject to cold climatic conditions, and in many cases are underlain by permafrost. Deep seasonal frost or perennially cryotic conditions result in the dominance of frost action as a geomorphic process. One of the results of this process is the widespread occurrence of patterned ground phenomena including polygons, circles, and stripes. As in the case of many other processes, frost action also operates in nonperiglacial areas, but at lesser frequencies and magnitudes compared to the periglacial landscape.

The only geomorphic processes and landforms unique to the periglacial landscape are those that occur as a consequence of the formation or melt of ground ice in permafrost areas. Ground ice aggradation gives rise to features such as pingos and ice wedges, while degradation results in thermokarst lakes and retrogressive thaw slumps. The presence of permafrost also enhances the operation of several mass movement processes, including solifluction, which moves significant volumes of material even on very low-angle slopes. Active-layer detachments, shallow landslides, also occur on slopes that would be stable if permafrost were not present.

Most technological problems of construction on permafrost terrain can be overcome, as shown by the relatively few problems encountered during the development of the town of Inuvik. However, the costs of construction are much greater wherever ice-rich permafrost is present. The results of past mistakes in the form of terrain disturbance will remain as scars on the landscape for many decades to come.

# Alpine and continental glaciation

In the study of glaciation and its effects a distinction is made between alpine and continental glaciation. Although much of the ice physics is the same the scale is vastly different and the added contrasts between ice confined in valleys and ice existing over extensive areas of lower relief imply that different landscape forming processes are at work. Alpine glaciation was the first to be scientifically examined because it was accessible in the mountains of Europe; the very name alpine reflects this. Many of the terms we use in glaciology and glacial geology are European in origin.

Glantz had written a geography of Greenland in the eighteenth century, but his observations were restricted to what was visible from the ocean and the narrow strips of land in the south. Antarctica was unknown until the nineteenth century, when sailing vessels penetrated the dangerous waters south of 50° latitude. It was to be some time before the idea that huge ice sheets had once covered very large areas of the earth's surface was brought forward in the scientific community. Louis Agassiz, the famous Swiss glaciologist, was converted to glacial theory during his career; he visited North America at the end of the nineteenth century and drew attention to the evidence of past glaciation in many areas. For example, he noted that there had been a vast lake formed behind the retreating ice that had occupied a great triangular area across parts of South and North Dakota, Minnesota, Manitoba, Saskatchewan, and northwestern Ontario (see Figure 23.40).

Today it is the practice to distinguish between a range of glacier forms as well as between alpine and continental glaciation. Embleton and King (1975) proposed a classification of ten types (Table 23.1) and other terms for some of the types have been suggested. This classification is based on a combination of size, topographic location, and location of part of the glacier. These classifications do not include some very important elements of glacier systems. For example in the Himalaya/Karakoram there is a regional differentiation between glaciers that are supplied by snow accumulation on the glacier in the higher altitudes and glaciers that are supplied by snow avalanching off high peaks into the valleys. In this chapter we will discuss glacial systems, that is, the properties of and the processes caused by ice in glaciers. **Glaciology** is the study of all types of ice; here we will be restrictive in the type of ice studied. Distinction will be made between landforms created by erosion as the ice forms and moves, and those created by deposition as the ice melts. Terminology is a constant problem because each region has evolved its own names. For example, the lake that occupies the hollow left at the origin of a melted valley glacier is called a tarn, from the Scottish name, or a **cirque lake**. As our

## Table 23.1

*Classifications of glaciation proposed by Embleton and King, 1975*

| | *Other possible terms* |
|---|---|
| 1. Niche, wall-sided, or cliff glacier | Glacieret |
| 2. Cirque glacier (Figure 23.1) | |
| 3. Alpine-type valley glacier | |
| 4. Outlet-type valley glacier | |
| 5. Transection glacier (Figure 23.2) | Icefield |
| 6. Piedmont glacier | |
| 7. Floating glacier tongues and ice shelves | Tidewater glaciers |
| 8. Mountain icecap | |
| 9. Glacier cap or icecap | |
| 10. Continental ice sheet | |

FIGURE 23.1 ▼ A cirque glacier on the south side of the Kaskawulsh glacier valley, St. Elias Mountains, southwest Yukon. The deeply eroded basin surrounded by steep slopes is typical of these forms (photo: P.G. Johnson)

FIGURE 23.2 ▼ The St. Elias Icefields, looking southeast to the Hubbard Glacier, St. Elias Mountains, Yukon. The central icefields are more than 1500 m deep (photo: P.G. Johnson)

understanding of processes has developed it has become apparent that forms such as drumlins that were long thought to be depositional landforms can now be explained by erosion or deposition, and even by the action of vast subglacial floods.

## Distribution of glaciers

Ice covers 10 percent of the world's land surface. In Greenland and the Antarctic, for example, the landscape is dominated by almost uninterrupted expanses of ice in the form of huge ice sheets. Smaller icecaps occur in Iceland and the Canadian Arctic islands, while in the European Alps, the North American Cordillera, the Andes, and the Himalaya/Karakoram system icefields feed spectacular valley glaciers. Many smaller areas of glaciation remain in Scandinavia, the Tien Shan Mountains of China, and even within the tropics on Mt. Kenya.

In the past, there have been many periods when the icecaps extended far beyond their present limits, and when periglacial conditions reached across a large area of the middle latitudes (Figure 23.3). We find evidence of past glacial periods in the Permian and Carboniferous tillites (ancient glacial deposits) of South Africa and Brazil, for example. More recently, during the Pleistocene epoch, there were

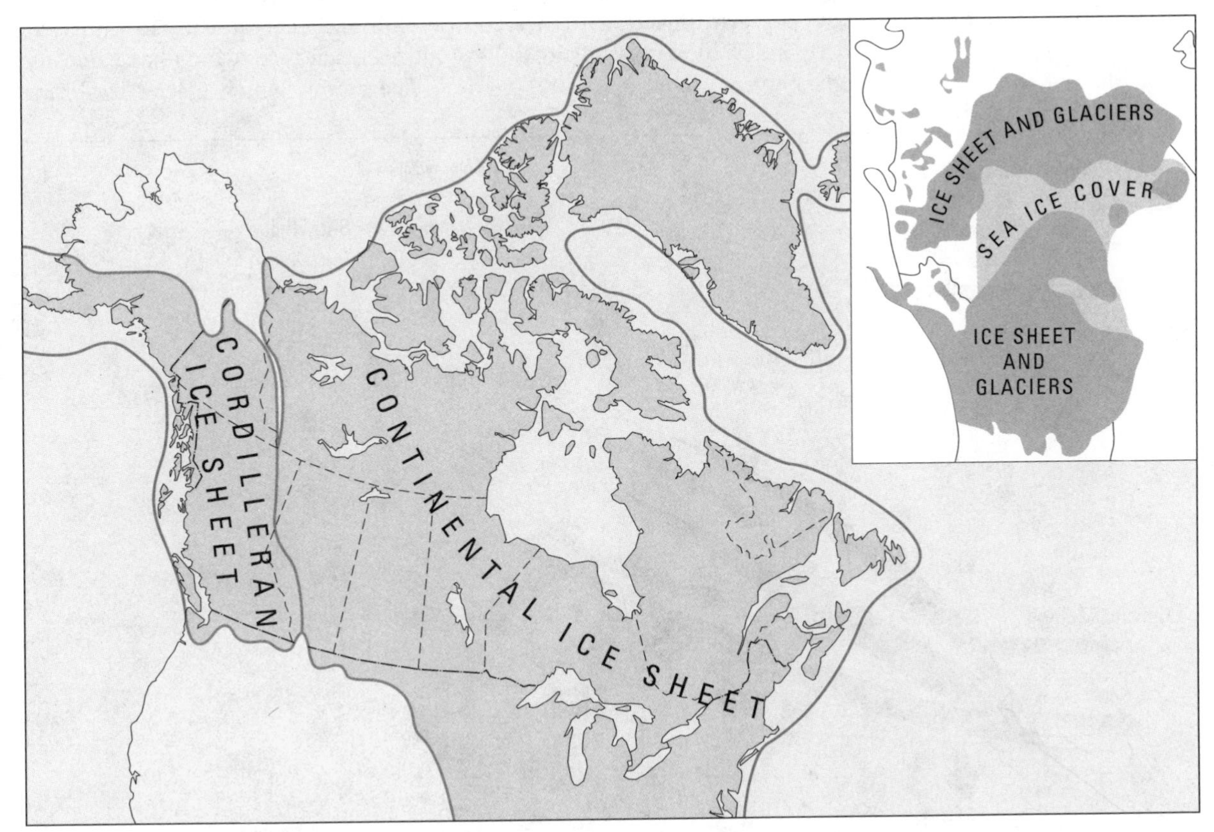


FIGURE 23.3 ▼ Quaternary ice limits in Canada and adjoining areas. The inset map shows the extent of ice in the northern hemisphere in the last phase of the Ice Age. Note that there is little difference between the two limits over Canada (inset map after A. Dawson, *Ice Age Earth*, Routledge Publishing)

long periods when much of Canada, the northern USA, Britain, Scandinavia, and northern Europe lay beneath vast ice sheets. These Pleistocene glacial periods have left a clear imprint on the present landscape; to understand this legacy of glaciation we need to examine the processes and landforms of modern-day glacial regions.

# The glacial system

## ACCUMULATION

Ice sheets and valley glaciers differ considerably in scale, shape and, as we will see, the detailed processes of movement, erosion, and deposition. As a consequence, they are associated with rather different landforms. But at a simple level, both can be thought of as open systems that receive inputs in the form of snow and rock debris and lose outputs through melting and sediment deposition (Figure 23.4).

The main input is normally by direct snowfall onto the ice surface, though small quantities may also be added by freezing of runoff (**superimposed ice**) or rainwater, by rime, and by snow blowing from adjacent areas. Avalanching snow from mountain peaks and valley walls can also be an important input. In the Himalaya/Karakoram Mountains, avalanching is the primary source of accumulation on some glaciers. Winter snowfall usually occurs over the whole glacier surface. Typically, however, summer melting removes the snow cover from the lower areas, so that net accumulation is restricted to higher parts of the glacier or icecap (Figure 23.5). Apart from some high-altitude and high-latitude glaciers some seasonal melting occurs. Those icecaps and ice sheets where no melting occurs are very important in the study of climate change because each year's snowfall retains gases and particles from the atmosphere. These can be measured to build up a picture of changes in atmospheric conditions.

At the end of the summer where snowmelt occurs the residue of the snow cover has changed to a granular mass of crystals called **firn**. Because of the effects of compaction and partial melting and refreezing, the firn is denser than the original snow, with a density of about 0.4–0.6 g $cm^{-3}$ compared to 0.1 g $cm^{-3}$ for newly fallen snow.

Over time, the firn accumulates and undergoes further compaction and recrystallization until it ultimately produces glacier ice, which has a density of 0.9 g $cm^{-3}$. The rate at which glacier ice forms

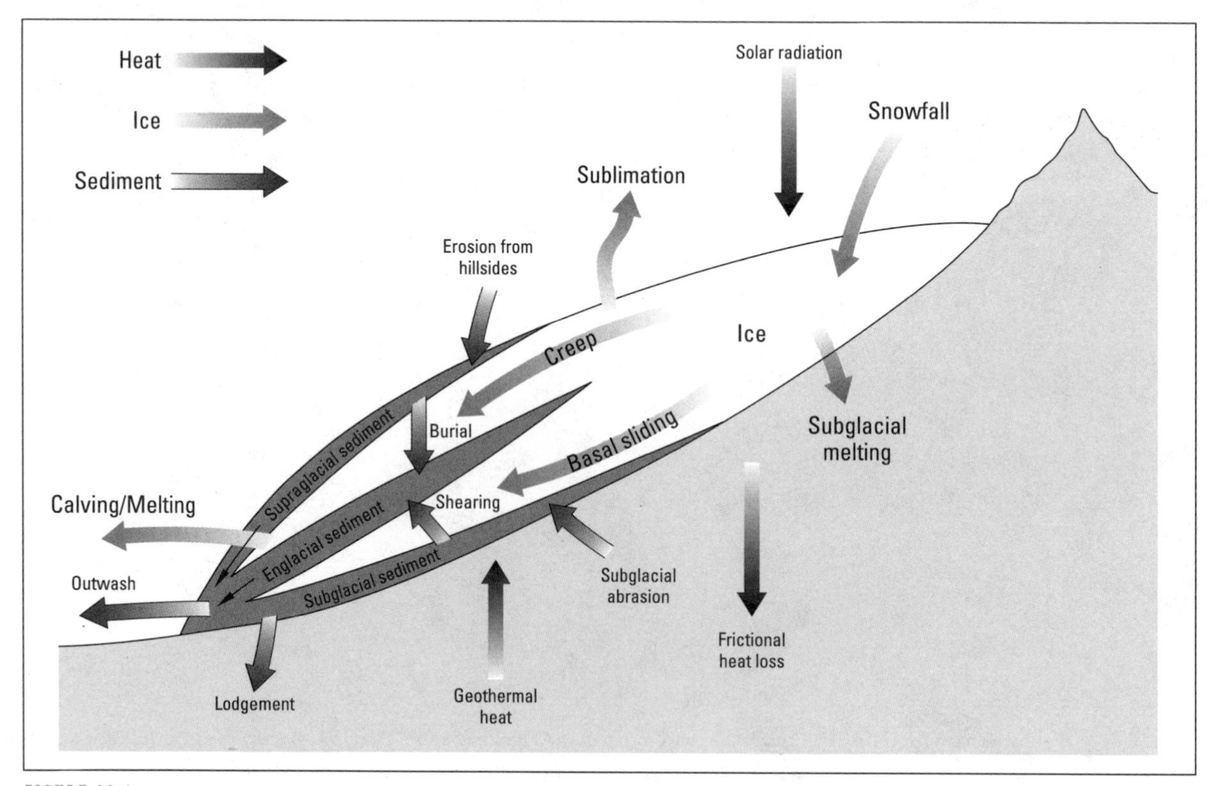


FIGURE 23.4 ▼ The glacial system

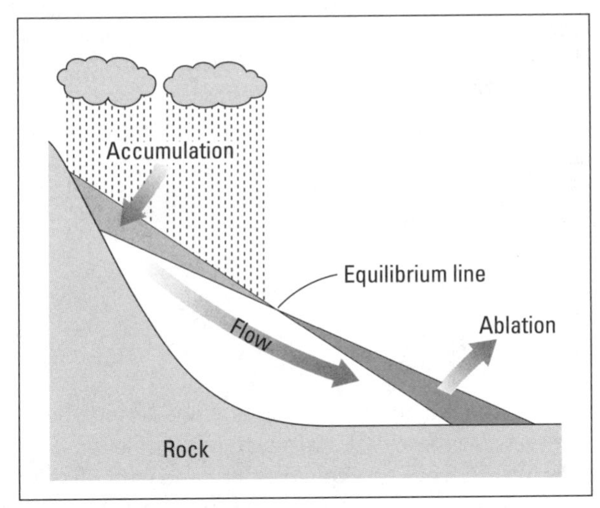


FIGURE 23.5 ▼ The ice budget of a glacial system. In the upper part of the glacier net accumulation takes place; in the lower zone net ablation occurs. The equilibrium line falls where accumulation and ablation are in balance. Glacial flow maintains the balance either side of the equilibrium line and maintains an equilibrium surface profile

varies from one area to another depending on the rate of snowfall and melting and the temperature of the snow. On Seward Glacier in Alaska, for example, the firn–ice boundary lies at only 13 m depth, and ice is formed in three to five years. In contrast, at the much colder and drier Plateau Station in Antarctica, ice formation takes about 3500 years and the boundary is at 160 m.

## ABLATION

During the summer, much of the snow that falls onto the glacier, and some of the existing glacier ice, is lost by **ablation**. This involves two processes: sublimation and melting. **Sublimation** is the direct conversion of ice to water vapour. In most cases it is almost negligible, for it requires much more energy to evaporate ice than to melt it.

Melting occurs through a variety of effects. Surface melting takes place largely because of energy from the sun and atmosphere, though water flowing over the ice surface may also be warm enough to cause some melting. The snow conditions produced by ablation (Figure 23.6) change from no-melt (in the dry snow zone) to melt (in the percola-

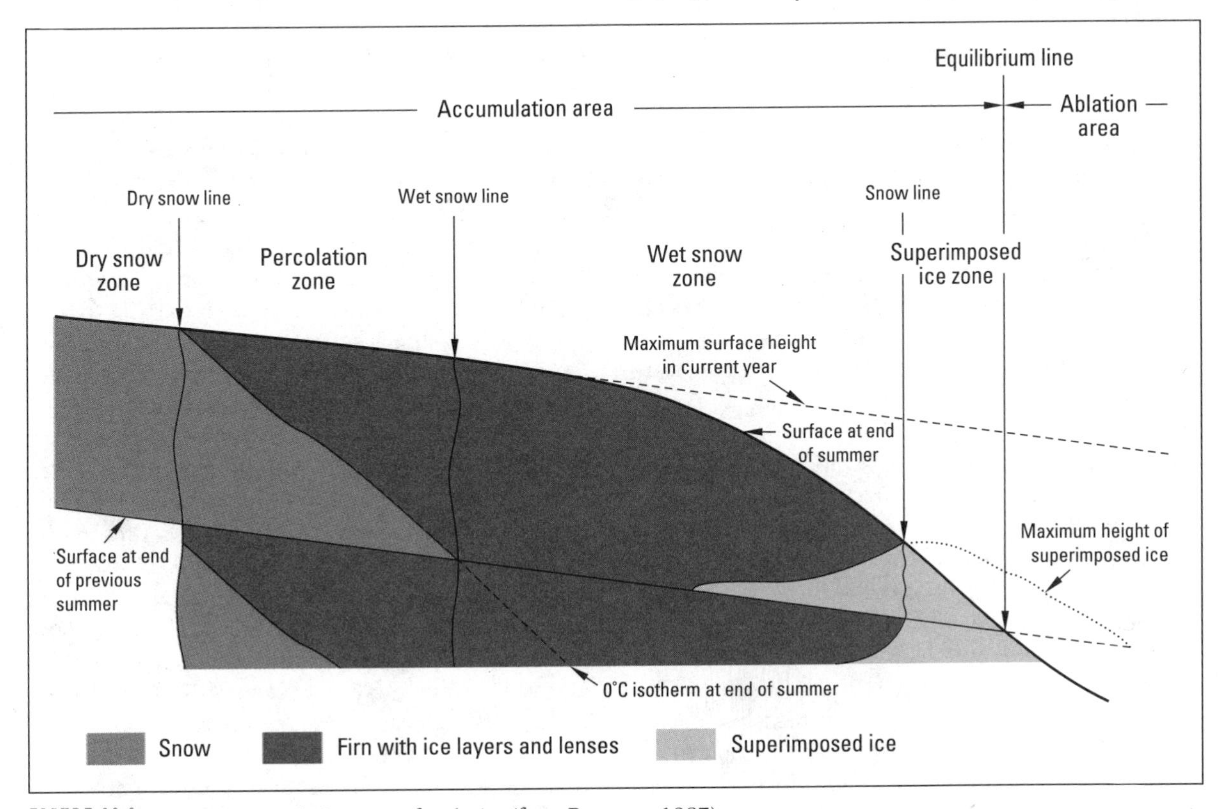


FIGURE 23.6 ▼ The accumulation area of a glacier (from Paterson, 1987)

tion zone) until all the previous winter's snow is saturated, creating the wet snow zone. In the lower wet snow zone some of the meltwater may freeze at the base of the snow and eventually be exposed in the superimposed ice zone. There are two important limits: the **snow line** (also called the firn line), above which there are some remnants of the previous winter's snowfall; and the **equilibrium line**, which falls between zones of net gain in mass and net loss in mass of the glacier in one year. The snow line is more visible on the glacier.

Subsurface melting is also caused by flowing water that percolates into the ice and flows along shear planes, crevasses, and tunnels. As well, **pressure melting** occurs at the base of glaciers, when the weight of overlying ice reduces the melting point of the ice. Thus in places the ice temperature can be below 0°C but may be at the **pressure melting point**, with the result that melting occurs. The ice temperature, in turn, is a function of the atmospheric temperature regime at the surface. If the air temperature is below 0°C for most of the year the cold can be transmitted through the ice, and the temperature drops below the pressure melting point. If melting occurs in the snow and water percolates down through the snow and refreezes, the ice temperature can be raised to 0°C. At Byrd Station in Antarctica, for example, pressure melting occurs at a temperature of –1.6°C at the base of the ice, which is 2164 m deep.

Ice temperature and the occurrence of basal melting are also influenced by inputs of heat from geothermal sources and by friction. On average, geothermal heat flows into the base of the ice at a rate of about 60 mW $m^{-2}$, sufficient to melt about 6 mm of ice at pressure melting point each year. Locally, however, geothermal inputs may be much greater. Many Icelandic icecaps, for example, are developed above active volcanoes and, at times, these release vast quantities of heat into the ice, causing massive and often catastrophic melting. The term **jokulhlaup** (glacier burst) is used to describe the flood of water these subglacial volcanic events produce. (Just such an event occurred in Colombia once, when the Nevado del Ruiz volcano melted only a small portion of its ice field. The resulting meltwater mixed with volcanic mud and rushed down the Lagunilla River valley, killing 23 000 people.) By comparison, release of heat by friction between the bed and ice base is small, though ice flowing at a rate of 20 m $y^{-1}$ produces as much energy as the average geothermal input.

*Calving* Where glaciers or icecaps terminate in the sea or in deep lakes, losses also occur through the process of **calving**. Large masses of ice break away as **icebergs**, which ultimately melt or sublimate. In the case of icecaps such as the Antarctic, which occupy permanently cold regions and are bounded on all sides by the sea, calving is by far the major process of ice loss.

## THE MASS BALANCE

Over a period of time, the glacier or ice sheet receives and loses ice through accumulation and ablation. Its volume changes; this process of change is known as the **mass balance** of the glacier. A layer of ice 1 metre thick is equivalent to a layer of water 90 cm thick. A layer of firn (density 0.5 g $cm^{-3}$) 1 metre thick is equivalent to 50 cm of water. (Measures of ice thickness are converted to water equivalent because of the variable density of ice.) At any point on the glacier, the balance may be negative or positive. Where accumulation exceeds ablation, a positive net balance is said to exist; where losses are greater than gains the net balance is negative. Over the icecap or glacier as a whole, a negative balance results in shrinkage or retreat, while a positive balance causes an expansion or advance of the ice. Within the glacier or icecap, however, a consistent pattern tends to exist, with a positive balance in the higher, **accumulation zone** and a negative balance in the lower, **ablation zone**. These two zones are divided by the equilibrium line, where the net balance is zero. Glacier flow maintains the system by transferring excess ice from the accumulation to the ablation zone (Figure 23.6).

The areas of the accumulation and ablation zones are not fixed, but vary from year to year in response to fluctuations in snowfall and temperature. Thus the position of the equilibrium line also varies around a mean position. Similarly, the mass balance of the glacier may vary from year to year as a result of annual climatic variations. Figure 23.7, for example, shows the mass balance of some Norwegian glaciers and, as we can see, there have been some years with a positive balance although the overall tendency is for a negative mass balance. The cumulative mass balance curves for North American glaciers have a similar negative trend. The anomalous curve for Sentinel Glacier is in question (Figure 23.8).

In the longer term, progressive changes in ice volume may occur due to a bias toward a negative

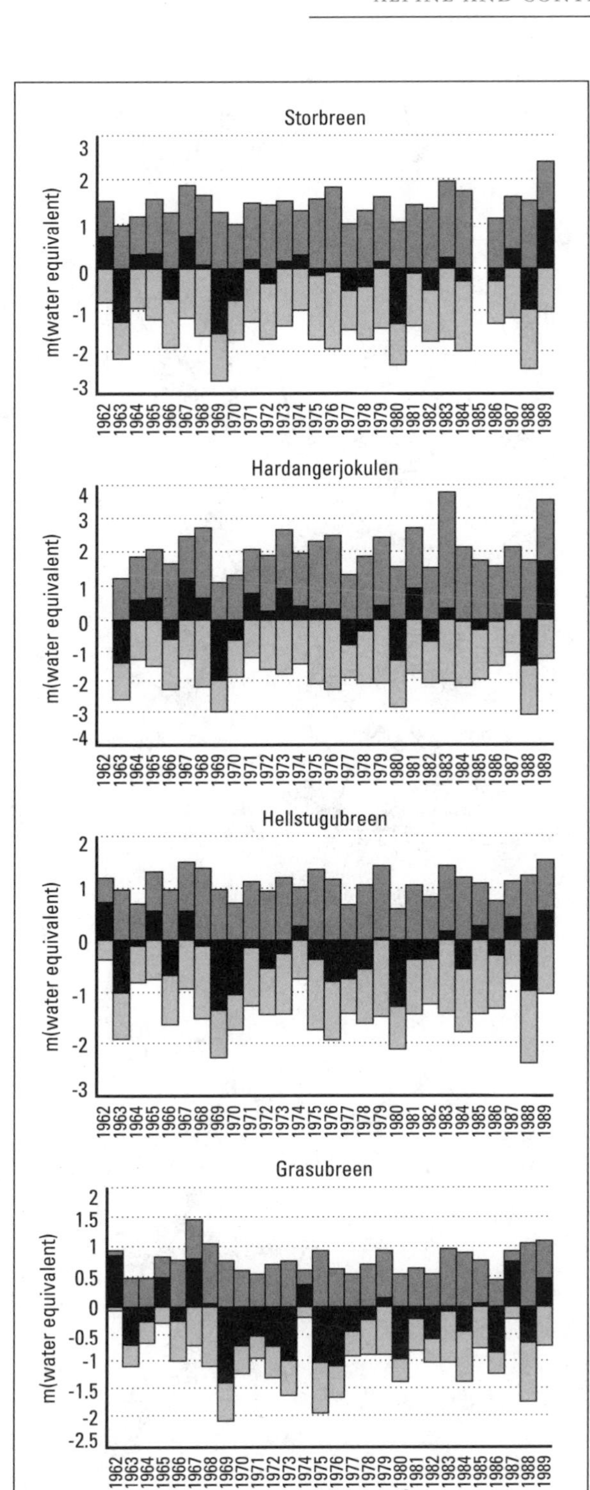


FIGURE 23.7 ▼ The mass balance of four Norwegian glaciers; note the variability from year to year and the general negative trend in glacier mass (after Ostrem *et al*, 1991)

or positive balance. This was seen on a grand scale in the Quaternary, when there were periods of several thousands of years during which positive net balances predominated and icecaps extended far into what are now temperate regions. Subsequently, the mass balance became negative for long periods and, during the interglacial phases, the icecaps retreated. The reasons for these large-scale variations were discussed in Chapter 9. It is worth pointing out, however, that expansion and retreat of ice sheets tend to occur at different rates depending on the source of imbalance in the glacial system. Where imbalances occur due to a change in snowfall amount (that is, in accumulation rate), a considerable time may elapse before the effect is felt at the ice margin, for the change has to be transmitted through the length of the ice sheet. Conversely, when the imbalance is due to a change in the ablation rate, the response is felt immediately at the ice margin and there is very little time lag. In other words, the position of the ice margin is much more responsive to fluctuations in ablation rate than it is to variations in accumulation rate. The position of the ice margin moves back and forth as the balance changes. A positive balance causes an advance, a negative balance causes a retreat. But overall the ice continues to flow toward the ice margins. The rate of response to the changes in mass of glaciers is partly a function of glacier size. The larger glaciers respond more slowly to change.

## ICE MOVEMENT

The mechanisms of ice movement are complex. They are also very difficult to measure or observe. Getting inside the glacier is expensive and the rate of movement is generally slow. Glacier flow involves three main processes: creep, basal sliding, and fracture.

Creep takes place due to the pressure of the overlying ice and gravitational forces related to the surface gradient of the glacier. Ice acts rather like a plastic, and its deformation under stress is complex. With increasing stress the rate of deformation first increases and then decreases. With a constant stress the rate of deformation decreases with time. Deformation involves the movement of individual crystals, internal changes in the crystal, and migration of the crystal boundaries so that over time the crystals become larger. The processes are aided by the production of water along the crystal boundaries and along planes within the crystal. It is obvious,

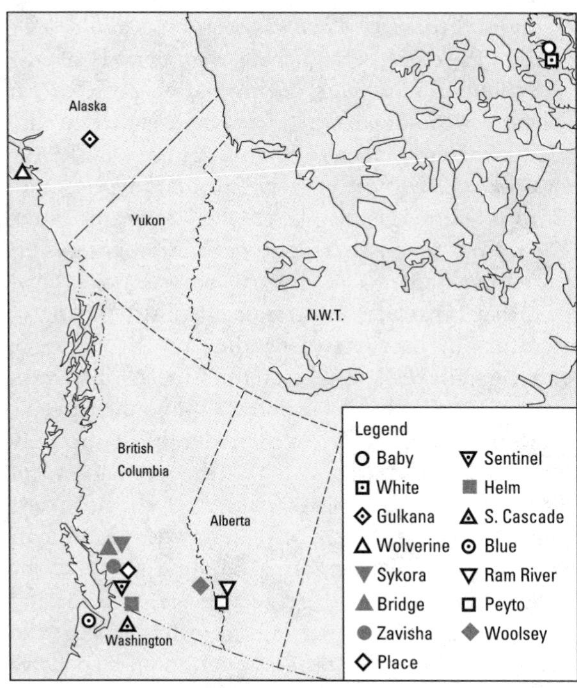


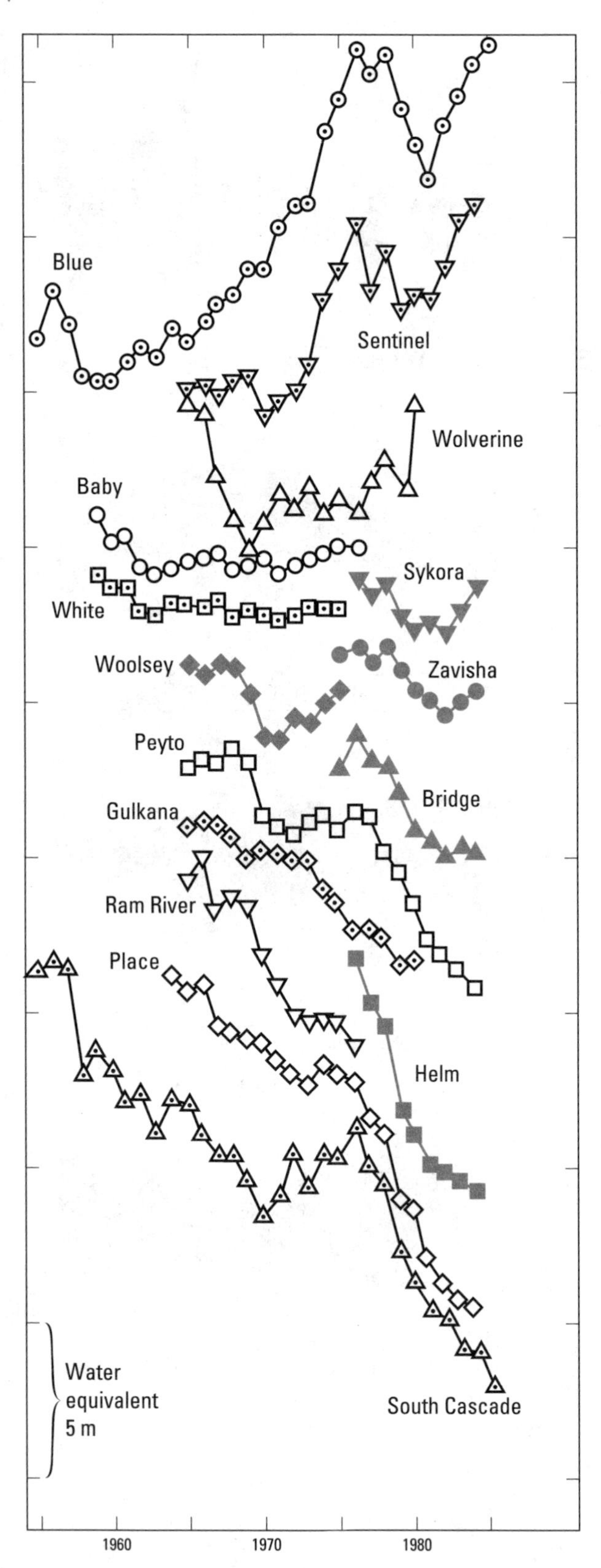


FIGURE 23.8 ▼ Cumulative mass balances of North American glaciers (right). The positive trends for Blue and Sentinel are known to be *in error*; the terminus position of Sentinel Glacier in particular continues to retreat. The negative trend of the glaciers from 1960 to the present is very strong. Note the vertical axis, "metres of water equivalent," is not continuous; it appears here for scale (from Letréguilly and Reynaud, 1989)

then, that the warmer the ice, that is the closer its temperature is to pressure melting point, the greater will be the deformation. Cold ice, with temperatures below pressure melting point, will deform much more slowly. Rates of creep vary through the ice mass, but are normally greatest at the base, where pressure melting provides water to lubricate the crystals and where the stresses from the overlying ice are greatest.

Water also plays a major part in **basal sliding**. This process is partly caused by the slippage of ice over a thin water layer at the ice-bedrock interface. The water reduces friction between the rock surface and the ice, and allows the glacier to slide under the influence of gravitational and overburden pressures. Locally accelerated rates of basal sliding tend to occur wherever basal melting of the ice takes place. One factor that encourages this is the presence of irregularities on the bedrock surface. Increased pressure on the upslope side of small obstacles (less than about 1 m in height) causes melting. The water lubricates the base of the ice and facilitates flow across the obstacle. Larger obstacles inhibit basal sliding, but result in enhanced basal creep. The increased pressure on the upslope side of irregularities also causes increased rates of deformation; this is called **enhanced plastic flow**.

Because of variation in bedrock topography and discontinuities in the rate of ice flow, considerable internal stress may develop within the glacier which cannot always be accommodated by gradual deformation. Under these conditions, the ice may fracture along clearly defined **shear planes** (Figure 23.9). These are typically concave upward and result in the transport of basal ice toward the surface of the glacier. They are most common where the ice is relatively thin and where creep is limited. The colder the glacier ice becomes, the less likely it is to deform by creep, and fracture becomes more likely.

FIGURE 23.9 ▼ The Donjek Glacier; shear planes at the glacier terminus produced during a period of surge. Glacier movement is from right to left (photo: P.G. Johnson)

## RATES OF ICE MOVEMENT

Together, the processes of creep, basal sliding, and fracture account for the movement of ice from the accumulation zone to the ice margin. As we have seen, however, rates of movement vary within the ice mass due to differences in the stresses to which the ice is subject and to the availability of water. The role of pressure melting in providing water is fundamental and results in marked differences in flow rate between glaciers that are warm enough at the base to experience enhanced creep and sliding, and those that are frozen throughout their depth. The former are called warm-based glaciers; the latter are cold-based glaciers. These terms are slightly misleading; as noted earlier, pressure melting occurs largely as a function of ice thickness. Thus variations in ice thickness within a single glacier result in transitions from warm-based to cold-based ice (Figure 23.10).

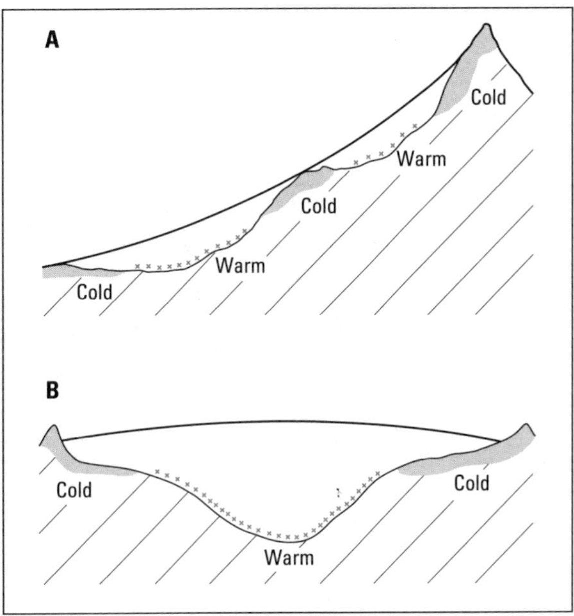


FIGURE 23.10 ▼ Distribution of warm- and cold-based ice in a glacier in response to variations in ice thickness: (A) longitudinal section down a glacier; (B) cross section across a glacier or icecap covering an existing valley. Crosses show zones of pressure melting; shaded areas show where the glacier is frozen to the bed

The implication is that creep is most active near the base of the ice, where the sliding occurs, where pressure melting and ice stresses are at a maximum. It may seem paradoxical, therefore, that rates of ice movement are greatest at the surface of the glacier (Figure 23.11). The reason is quite simple; although the amount of creep declines toward the ice surface, the total rate of flow increases because the ice is being carried along by the movement of the underlying layers: the amount of internal deformation is additional to the basal flow. In most glaciers the proportions of the flow due to sliding and internal deformation can vary dramatically over very short distances. With the Athabasca Glacier the proportion of flow due to basal sliding can vary from 10 percent (as shown in Figure 23.11) to 90 percent of the total. Movement is also restricted at the sides of valley glaciers by friction with the rock wall (Figure 23.12).

FIGURE 23.11 ▼ (Right) The distribution of flow velocity with depth in the Athabasca Glacier, Alberta (after Sugden and John, 1977, data from Savage and Paterson, 1963)

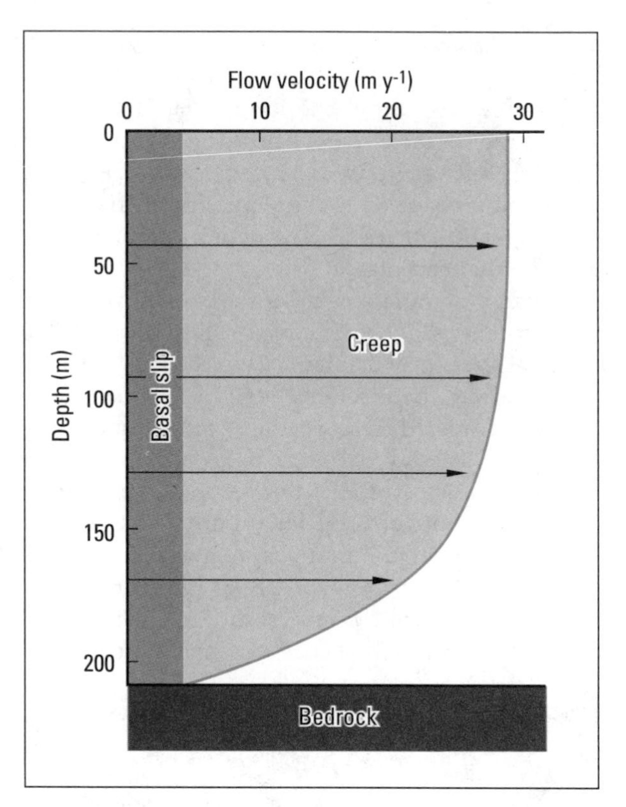


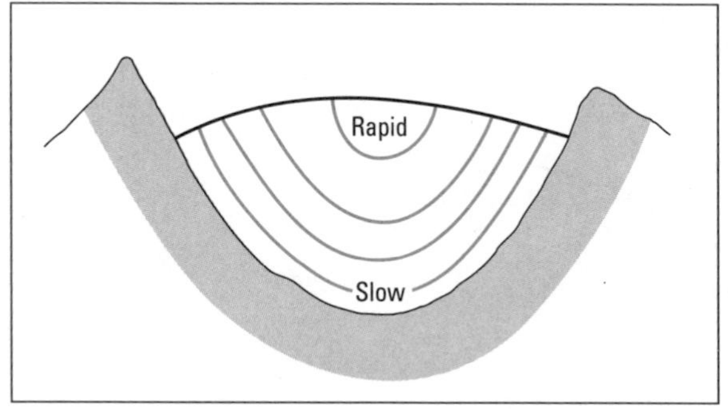


FIGURE 23.12 ▼ Cross-sectional distribution of flow velocity in a glacier; lines show bands of equal flow velocity

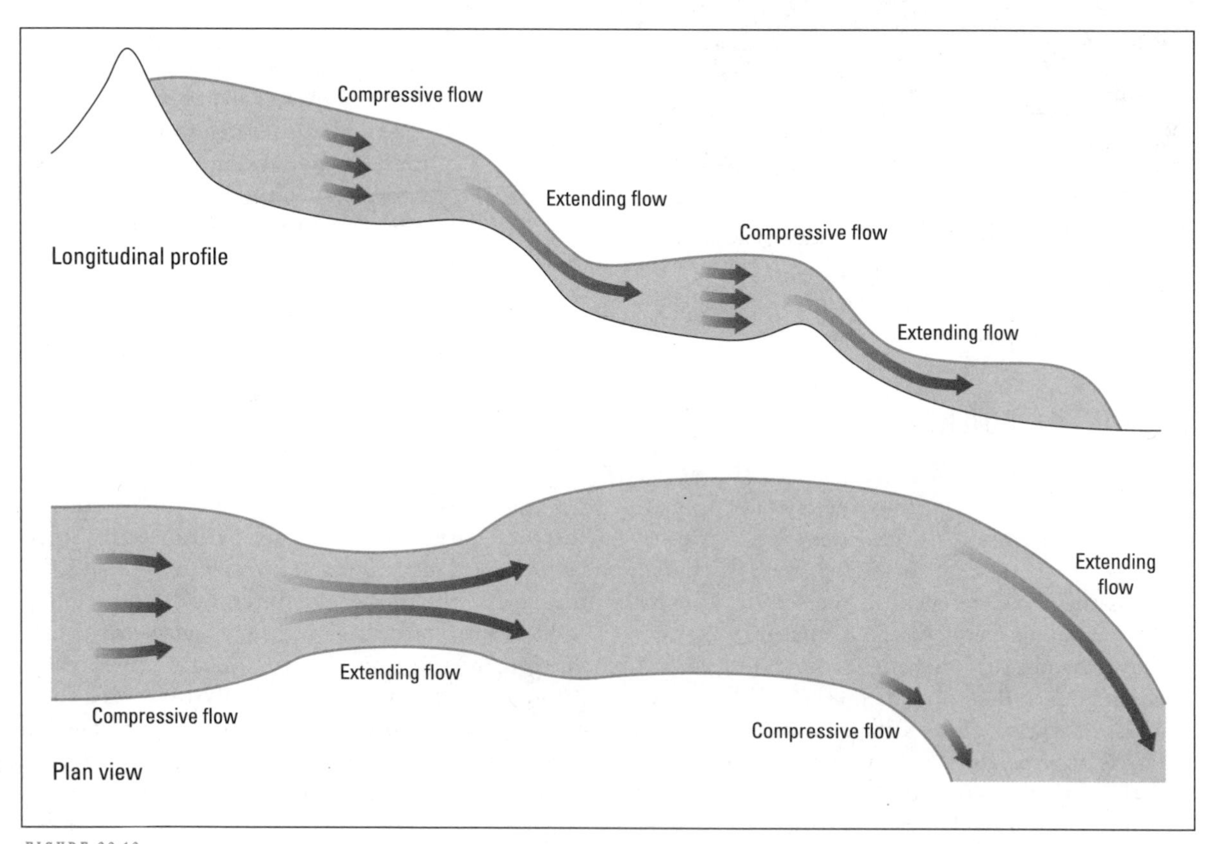


FIGURE 23.13 ▼ The patterns of extending and compressive flow in glaciers due to topographic variations

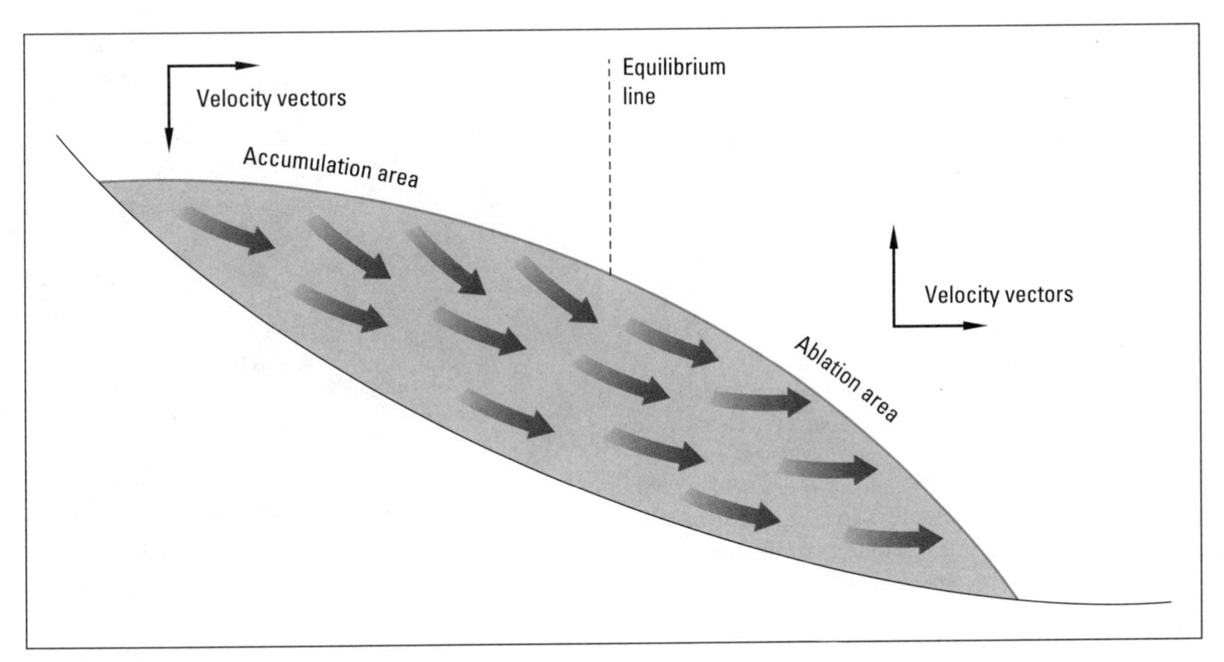


FIGURE 23.14 ▼ Idealized flow pattern in a glacier

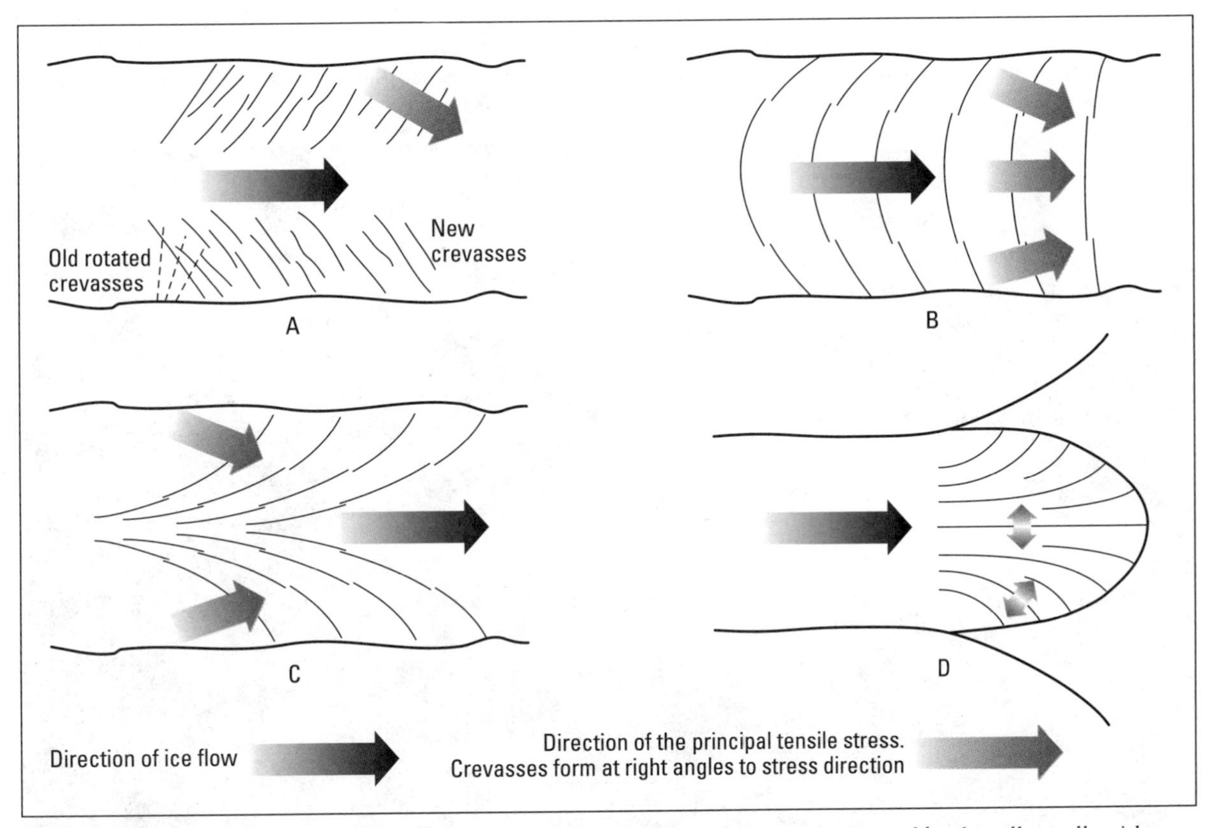


FIGURE 23.15 ▼ Principal types of crevasses in valley glaciers. (A) Effect of shear stresses caused by the valley walls with no change in the velocity of glacier flow along the centre line; (B) effect of shear stresses caused by extending flow, an acceleration of the glacier; (C) effect of shear stresses caused by compressive flow, a deceleration of the glacier; (D) radial splaying in the glacier at the terminus

The mass of the ice and variations in the topography of the surface over which the ice is flowing can also produce zones of increased velocity (**extending flow**) and reduced velocity (**compressive flow**) (Figure 23.13). In general we tend to see extending flow in the accumulation area, as stresses increase down-glacier, and compressive flow in the ablation area, as the stresses decrease with ablation. The vertical flow vectors are downward in the accumulation area and upward in the ablation area (Figure 23.14). These variations encourage the development of **crevasses** in the glacier ice. **Transverse crevasses** are produced across the centre of the glacier in extending flow when the stresses exceed the strength of the ice. At the margins of the glacier the lower flow velocities caused by friction with the valley wall rotate the angles of **principal stress** within the glacier ice so the crevasses lie at an angle to the valley side (Figures 23.15 and 23.16).

FIGURE 23.16 ▼ Transverse and marginal crevasses on the Grizzly Creek glacier, St. Elias Mountains, Yukon (photo: P.G. Johnson)

During the 1960s it was realized that marked changes in flow rate occur over time. The Steele Glacier in Yukon was observed to be moving much faster than normal and the terminus of the glacier moved 5 km down-valley. This rapid movement occurred for about two years and then the glacier returned to its normal slow rates of flow. These rapid movement events were called **glacier surges**, and in the last 30 years it has been observed that they are very common in all glacierized regions. They are caused by imbalances between the amount of ice in the accumulation zone and the amount in the ablation zone. As a result, more ice tends to be transferred down-glacier. The periods of rapid flow

FIGURE 23.17 ▼ The folded medial moraines of the Kluane Glacier, Yukon, produced by periods of glacier-surge (photo: P.G. Johnson)

FIGURE 23.18 ▼ The terminus of the Donjek Glacier during a surge. Note the wedge of the old stagnant terminus in front of the reactivated ice (photo: P.G. Johnson)

FIGURE 23.19 ▼ The debris-covered terminus of the Kaskawulsh Glacier, Yukon (photo: P.G. Johnson)

occur in a regular periodic cycle for each individual glacier, but they range from eight years to more than 30 years. The periods of rapid flow produce folding of the medial moraines on the glacier (Figure 23.17) and a steep active terminus (Figure 23.18). Rates of flow may be increased 10–100 times during the surge, and the glacier may advance at a phenomenal rate: in 1963–64, the Bruarjökull in Iceland advanced up to 8 km at speeds of as much as 5 m $h^{-1}$.

Because of these variations, it is difficult to generalize about glacier flow rates. Well-lubricated, warm-based glaciers on the west coast of Greenland, however, have average flows of 4–5 m $d^{-1}$, whereas glaciers in the Alps commonly flow at rates of about 0.1 m $d^{-1}$. Modern ice sheets have even lower flow rates, generally in the order of only 0.01 m $d^{-1}$. The same is not necessarily true of icecaps during the Pleistocene, however, for the vast thicknesses and the different climatic conditions probably led to much greater pressure melting and much greater flow rates.

# Glacial erosion

## PROCESSES OF EROSION

The sediment load of glaciers varies considerably. Sediment occurs in the glacier system at the surface (**supraglacial sediment**), within the ice (**englacial**), and at the base of the glacier (**subglacial**). Supraglacial load can cover all of the glacier surface (Figure 23.19) or be almost absent (Figure 23.20). The subglacial load can also vary from almost nonexistent to a few metres in thickness. The amount and distribution of the englacial load depends on processes of transfer of the sediment from supraglacial or subglacial locations. Formation of shear planes, water freezing onto the base of the glacier, and sediment falling or being washed into crevasses are three of the processes by which this occurs. The size of this material varies from small particles to large blocks. But how was it incorporated into the glacier?

*Detachment processes* We can think of glacial erosion as a three-stage process involving detachment, entrainment, and transport of debris. Detachment is a result of preglacial or interglacial processes and of abrasion during glaciation. Preglacial processes such as chemical weathering, dilatation (pressure release), and frost shattering loosen the bedrock and provide debris that can relatively easily be picked up by the glacier. Abrasion occurs when debris held in the base of the glacier grinds against the bed, prying away small fragments of rock that add to the basal load. The efficiency of this process is determined primarily by the concentration of debris in the base of the glacier, the velocity of ice movement, and the thickness of the ice. Clearly the amount of abrasion tends to increase as basal sediment concentration and ice flow increase, for a greater area of the bedrock surface will be affected in any period. However, over time the abrasive capacity of the ice declines unless new material is being added to the base of the glacier, for

FIGURE 23.20 ▼ Peyto Glacier in the Rocky Mountains of Alberta has a surface that is almost free of sediment (photo: P.G. Johnson)

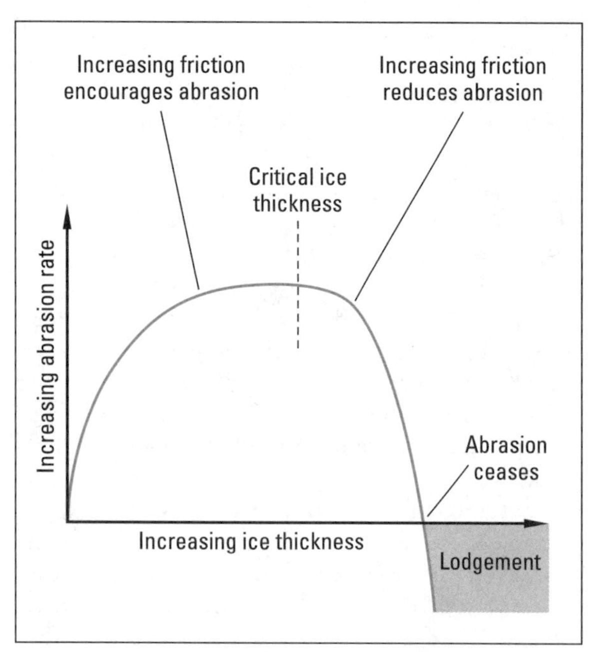


FIGURE 23.21 ▼ The effect of ice thickness on rates of abrasion and lodgement of till (from Sugden and John, 1977)

the particles protruding from the ice are gradually worn by attrition. Renewal of the basal load occurs both by entrainment of new debris and by the down-melting of material through the ice, toward the bed. Abrasion also tends to increase with mounting ice thickness due to the increase in applied pressure. Beyond a critical thickness, however, the relationship is reversed: first, because pressure melting may produce a sufficiently deep layer of basal water to buoy up the glacier and reduce the extent of contact with the bed; and second, because very high pressures at the base of the ice may cause particles to lodge in the bed, increasing friction, reducing ice velocity, and ultimately causing deposition of the basal load (Figure 23.21).

Detachment may also occur at the glacier bed by processes of dilatation and freeze–thaw. The idea of pressure release operating beneath a huge glacier may seem unlikely, but it is theoretically possible. As the ice cuts into the bedrock it ultimately reaches a position where the weight of material removed is greater than the weight of the overlying ice. From then on, the rocks are subjected to progressive unloading as erosion continues, with the result that dilatation occurs, opening up joints and bedding planes and encouraging detachment. Freeze–thaw operates only beneath warm-based glaciers, where pressure melting and percolation of surface meltwaters provide water that is able to flow into joints in the bedrock and refreeze. Because pressure melting only occurs where the ice is relatively thick, freeze–thaw is probably most active beneath the main body of the glacier and declines in significance toward the ice margin.

*Entrainment* Entrainment involves a large number of processes. Blocks falling onto the glacier surface from the valley walls or carried onto the ice by streams or wind are automatically entrained by the moving ice. Basal entrainment also occurs as a result of traction and regelation. Ice sliding across a bedrock surface exerts a tractive force on material on that surface and, if this is sufficient to overcome the frictional resistance of the debris, entrainment takes place. This process seems to account for most of the coarse debris in the base of the ice.

**Regelation** is probably most effective in warm-based ice, where water produced by pressure melting refreezes to create a **regelation layer**. Fine debris caught up within this layer may be plucked from the surface and carried along by the ice.

At the ice front entrainment also takes place through the ploughing action of the snout during glacier advance, and through squeezing of material from beneath the ice. Both processes are most effective when the ice front is advancing over relatively soft or saturated materials.

*Transport* Once entrained, glacial debris is transported supraglacially, englacially, and subglacially. Debris falling onto the glacier from the valley walls commonly forms a long ridge of material at the glacier edge, known as **lateral moraine**. Where two or more ice streams merge, these moraines may coalesce in the centre of the glacier to form a **medial moraine**. A very small amount of sediment will promote ice melt because of the absorption of solar energy by the sediment. The transmission of heat through sediment is very poor, however, so that thicknesses greater than a few millimetres will reduce ice melt. Therefore, medial moraines are usually ridges down the glacier which are far larger than the amount of sediment would produce (Figure 23.22). Similar features develop where ice diverges and then rejoins around a rock outcrop.

FIGURE 23.22 ▼ The prominent medial moraines of the Kaskawulsh Glacier are produced by the confluence of a number of valley glaciers. Compare the undeformed pattern of these moraines on a glacier that does not surge to the folded moraines of Kluane Glacier, Figure 23.17 (photo: P.G. Johnson)

## LANDFORMS OF GLACIAL EROSION

The processes of glacial erosion commonly leave a clear imprint on the landscape. At the detailed scale, particles caught in the glacier base cut gorges in the bedrock, referred to as **striations** (Figure 23.23). **Chatter marks** may occur where rocks protruding from the ice have pried thin flakes or chips from the bedrock surface. In addition, the continual wear of the rock produces a smooth, polished surface. Wear is often accentuated along existing joints, and concentration of the abrasive action of the debris in this way may create **glacial grooves** (Figure 23.24). In contrast, detachment of blocks by regelation and freeze–thaw processes often creates angular, plucked surfaces. These are particularly common in the lee of obstacles protruding into the base of warm-based ice.

The role of subglacial water flow, either in distinct channels or as sheetflow, is lately believed to be extremely important. Grooves and potholes have been compared to forms produced in bedrock by fluvial action. Massive sheet floods during deglaciation of ice sheets are thought to be responsible for the formation of many oriented landforms such as drumlins and flutes (to be discussed later).

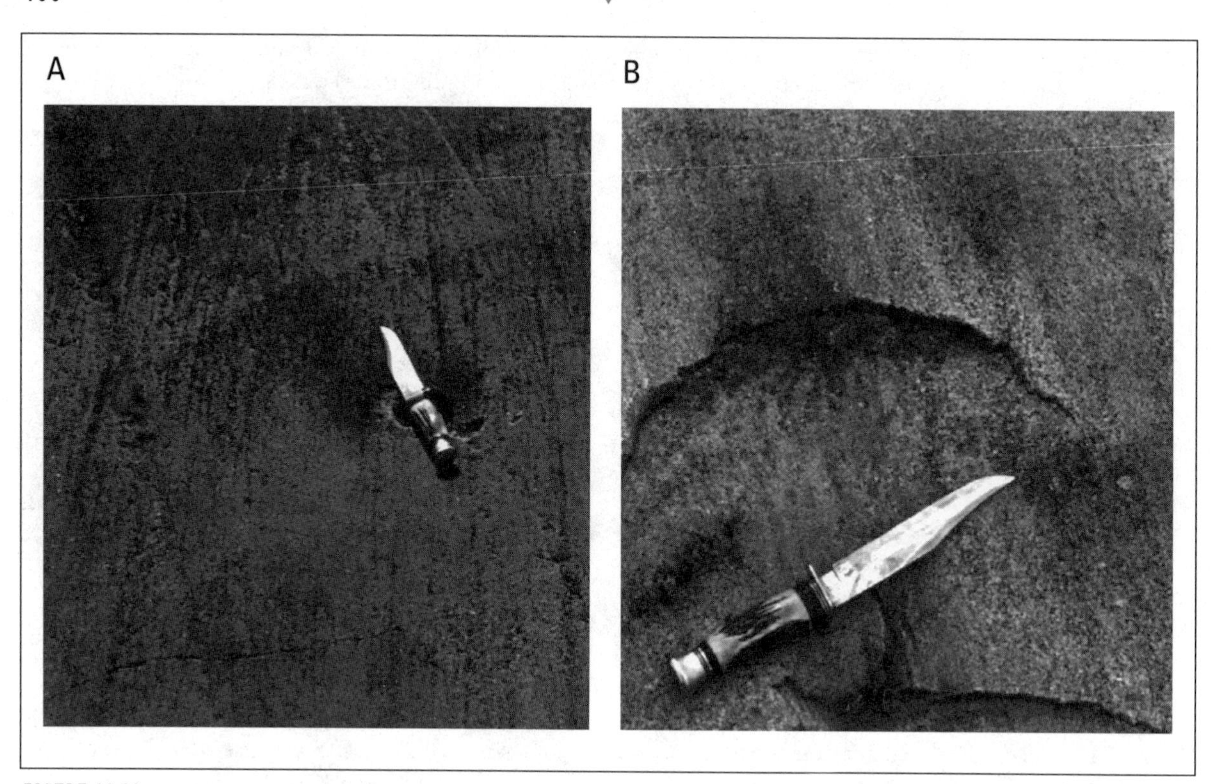


FIGURE 23.23 ▼ (A) Striations on the Canadian Shield; Ile du College, Lake Temiscaming, Quebec; (B) chatter marks on Shield rock, Lake Temiscaming (photos: P.G. Johnson)

FIGURE 23.24 ▼ Glacier grooves in limestone bedrock at the terminus of the Peyto Glacier, Alberta (photo: P.G. Johnson)

*Landforms of alpine glaciation* The most prominent glacier forms of alpine glaciation are icefields, valley glaciers, icecaps, and cirques. This landscape is dominated by highly eroded mountain forms, with deep glacier troughs as the major valleys. The deep glacier troughs have a broadly U-shaped cross section and are referred to as U-shaped valleys (Figure 23.25). These tend to be relatively straight, compared to river valleys, because the glacier erodes any ridges that extend into the valley, producing **truncated spurs**. The glacier erodes at all points of the ice-bedrock contact, thus widening and deepening the valleys. The valley sides are steep, with the watersheds frequently sharp, peaked ridges called arêtes (Figure 23.26). Differences in the rates of erosion along a valley, due to variations in the hardness of rock or to glaciological changes, can lead to some sections of the valley being formed into basins, which become lake sites upon deglaciation. The over-deepening of the troughs together with changes in sea level during and after glaciation result in flooding of the troughs near the coastline. These **fjords** are characteristic features of the coastlines of British Columbia, the state of Alaska, Norway, and New Zealand (Figure 23.27). Fjords characteristically have a bedrock lip near the ocean shoreline due to the reduction in erosion as the glacier ice floated as it reached the sea (Figure 23.28). Initially, icebergs would calve into the open ocean. As deglaciation progressed the fjord would flood with water and float the glacier tongue. While the glacier tongue was grounded on the lip, icebergs

FIGURE 23.25 ▼ The broad, U-shaped trough of a glaciated valley, the Mistaya Valley of the Rocky Mountains. View to the south from the Peyto Lake lookout on the Icefield Parkway (photo: P.G. Johnson)

FIGURE 23.26 ▼ The sharp arête ridges formed by glacial erosion are visible above the level of the icefields in the Centennial Range, St. Elias Mountains, southwest Yukon (photo: P.G. Johnson)

would still calve into the ocean but as deglaciation continued and the glacier terminus retreated into the fjord the larger icebergs would be trapped in the fjord. Eventually the glacier would retreat above water level. Throughout deglaciation, sediment from the glaciers would be deposited in the marine and fjord environments.

FIGURE 23.27 ▼ (Left) The fjord coastline of British Columbia; Howe Sound north of Vancouver (photo: P.G. Johnson)

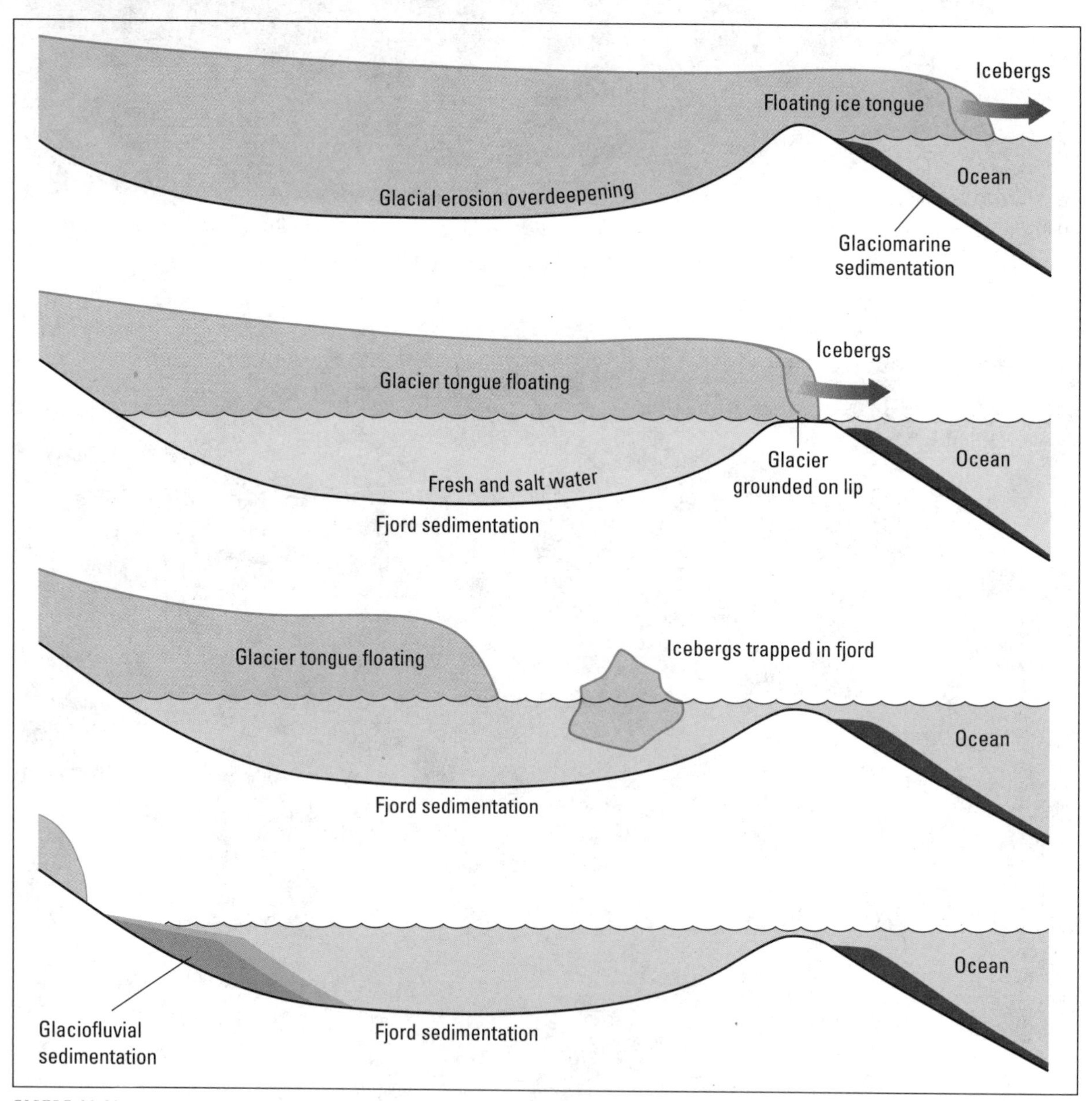


FIGURE 23.28 ▼ The formation and deglaciation of fjords

The amount of valley erosion is partly a function of ice thickness, so the large valley glaciers erode much more deeply than smaller tributary glaciers. After deglaciation the tributary valley floor is much higher than the floor of the main valley. This leaves a **hanging valley** perched on the hillside (Figure 23.25).

In the less intensively glaciated areas, for example toward the margins of the icefields, small glaciers form on mountain slopes from small semi-permanent snow patches. As these glaciers increase in size they erode basins, called **cirques** (Figures 23.29 and 23.1), out of the mountainside. The flow of ice concentrates erosion at the thickest part of the ice, producing a basin (Figure 23.30); after deglaciation the basin can become the site of a small lake, commonly called a **tarn** in England. One permanent feature of cirque glaciers is the Bergschrund crevasse, which forms close to the headwall. This is the main zone of detachment of the glacier from the rock, but frequently it does not intersect the bedrock except at the bottom of the crevasse. The importance of the Bergschrund to the erosion of the cirque is therefore limited; freeze–thaw activity in the crevasse is restricted because the temperatures just below the glacier surface tend to remain constant. The crevasse functions as a route for rockfall from the headwall to reach the base of the glacier.

FIGURE 23.29 ▼ Horn peak and arêtes, Mount Lucania, northern St. Elias Mountains, Yukon (photo: P.G. Johnson)

Cirques tend to develop just above the snow line, with the result that in any area distinct levels of cirque formation can be seen. These cirque levels are not necessarily horizontal over any great distance, however, for the height of the snow line varies in relation to aspect and distance from the sea (continentality). In the Alps, for example, cirque levels tend to be lower on north-facing slopes than on

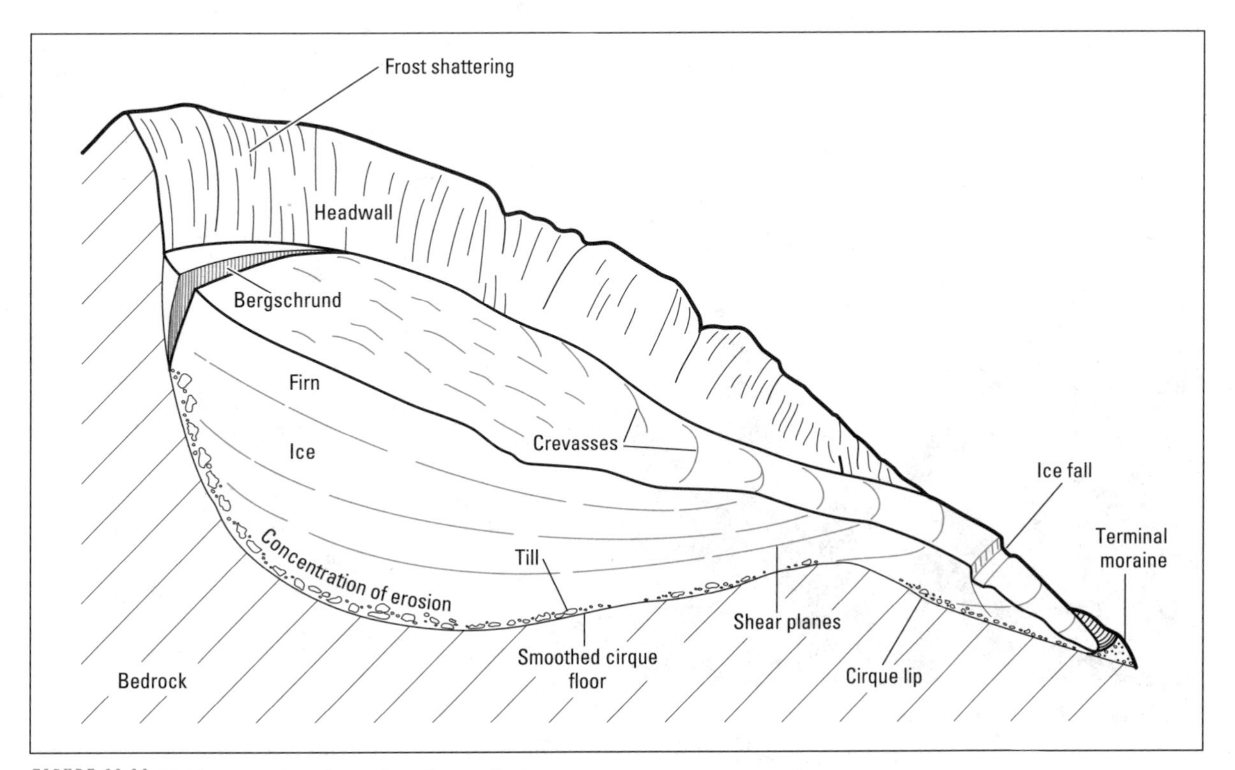


FIGURE 23.30 ▼ Cross section through a cirque glacier

south-facing slopes. Changes in the level of the snow line occur over time due to variation in climate, so that several different cirque levels may exist in a single area, each representing development at a different period. Two levels of cirque formation can be detected on the Dingle Peninsula in southwestern Ireland, for example, reflecting a lowering of the snow line as the climate cooled during the last glacial period. As the climate warmed at the end of the glaciation, however, the snow line rose again, the lower cirques were abandoned, and the higher ones were briefly reactivated.

The combined action of valley glaciers and cirques produces the striking landscapes of **horn** peaks and arêtes as a mountain is eroded on all sides (Figure 23.29). The European Alps are famous for horn peaks such as the Matterhorn and Mont Blanc.

*Landforms associated with ice sheets* The rock surfaces abraded by ice sheets also have striations and chatter marks, and some of the small-scale landforms, such as roches-moutonnées and whalebacks, are also formed (Figure 23.31). Even glacial troughs may be produced where erosion is concentrated along existing valleys or areas of weaker strata. Nevertheless, the broader character of the landscape formed beneath ice sheets is highly distinctive. Areal scour operates where the ice is thick enough to cause pressure melting, producing extensive, smooth, abraded surfaces (Figure 23.32). Where the ice is thinner, erosion is limited and is affected to a much greater extent by existing relief. Pressure melting occurs only in valley bottoms, where the ice is deep, and these are over-deepened to form troughs. Elsewhere, basal ice temperatures are low, movement is slight, and erosion may be negligible.

FIGURE 23.31 ▼ A roche-moutonnée in glaciated cols between valleys, Kvitvandet, central Norway (photo: P.G. Johnson)

FIGURE 23.32 ▼ The areally scoured landscape of the Canadian Shield, a result of Pleistocene glaciation by the Laurentide Ice Sheet (photo: P.G. Johnson)

**Rock flutes** are parallel, elongated ridges and furrows with a regular amplitude and wavelength. They are formed by rotation of rock particles within the basal layers of the ice, producing uniform differential erosion. Meltwater erosion at the base of ice sheets is believed to be an important erosion agent. The water is either concentrated in channels or may extend as a regional flood. Under the pressure of the ice sheet, these flows are capable of rapid erosion of grooves.

# Depositional processes and landforms

The stripped and scoured relief of glaciated uplands testifies to the massive erosive power of glacial ice. This is only half the story of glacial impact on the landscape, for all the debris that has been scraped, chipped, and plucked off the surface is deposited elsewhere. This material can be deposited in three ways: directly from the ice, by meltwater flow (these are called glaciofluvial deposits), and in lakes (glaciolacustrine deposits). In each case the sediments and the landforms produced are distinct.

## DEPOSITION DIRECTLY FROM THE ICE

Material that is directly deposited by the ice is called **till**. It is composed of the widest range of sediment sizes, from clays to boulders, and geological provenances in the basin. Three types of primary till formation can be recognized, **ablation till, basal till**, and **lodgement till**. Ablation till is composed of the sediment that was transported supraglacially and englacially, and is deposited as the ice melts. The till is not consolidated and has a high water content during deposition. With the uneven melt out of the ice this material has a tendency to flow, producing **flow tills**. Basal till is the sediment that was transported subglacially, and is deposited when the ice becomes inactive. The characteristics of basal till depend on the concentration of material in the basal layers of the ice and the stresses it has undergone during transport. Lodgement till is also sediment transported subglacially, but it is progressively deposited under stress at the glacier bed. This lodgement till is very highly consolidated material. The larger particles in the till, called **clasts**, if they are elongate in form, will be arranged with a preferred orientation due to the stress of transport and deposition at the base of the glacier. This preferred orientation is called the **till fabric**, and it can be used to deduce the direction of ice movement (Figure 23.33).

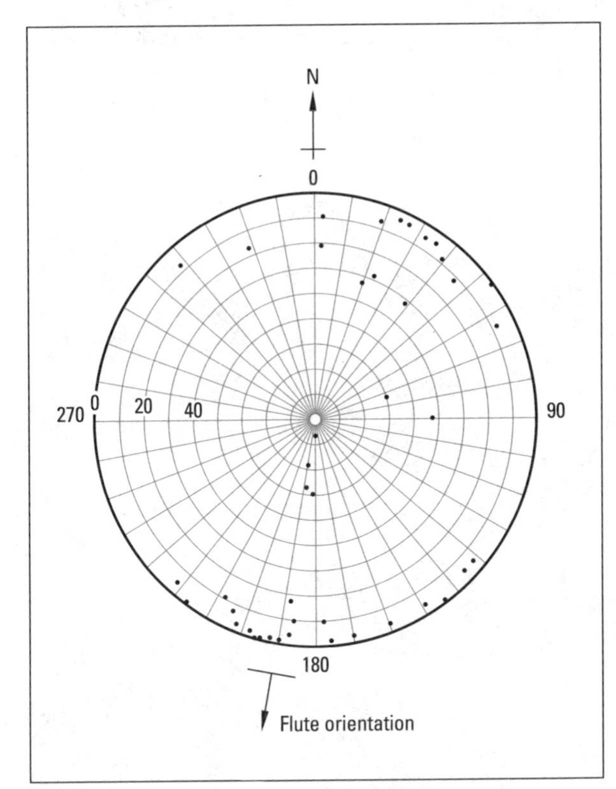


FIGURE 23.33 ▼ Till fabric from a fluted moraine, Norway. Each dot shows the angle of dip (on the radius) and long axis orientation (on the circumference) of a rock clast. The preferred orientation of the clasts is clearly almost parallel to the orientation of the flute (from Briggs, 1976)

## GLACIOFLUVIAL DEPOSITION

The meltwater of snow and glacier ice flows along supraglacial, englacial, and subglacial channels to the glacier terminus. The water picks up sediment along the way; when it emerges from the glacier it is carrying a large amount of suspended sediment and moving a large bedload. Suspended sediment concentrations of 17 g l, and averages of 10 g l $d^{-1}$, have been measured in the Slims River discharging from the Kaskawulsh Glacier, Yukon. The bedload that is transported ranges in size up to coarse gravel and boulders. Thus there is a large volume of sediment that can be deposited in the proglacial area. In addition, the glacier discharge may also erode and transport existing deposits in the proglacial zone. All these sediments laid down by glacier meltwater are glaciofluvial deposits.

## GLACIOLACUSTRINE DEPOSITION

Lakes are very common in the proglacial environment and the ice marginal environment. Hollows in the glacial deposits, dams produced by glacial landforms, and concavities in the bedrock profiles of valleys are common sites of these lakes (Figure 23.25). The lakes are sediment sinks for the large amounts of sediment carried by glacial streams. Large deltas are formed at the input and sediment is transmitted through the rest of the lake by **overflows** (if the input water and sediment is less dense than the lake water), **interflows** (if there is a density contrast in the lake), and **underflows** (if the input water and sediment is denser than the lake water) (Figure 23.34). Sediment may be deposited directly on the lake bottom or settle from the overflows and interflows. There is a difference in the settling rates and types of sediment deposition between winter and summer. In summer the lakes are ice-free and tur-

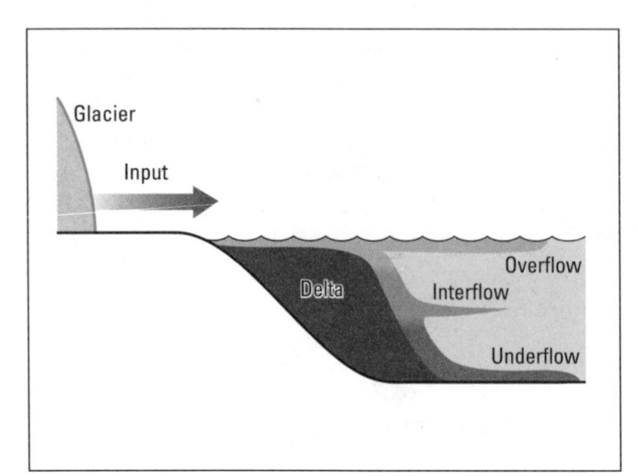


FIGURE 23.34 ▼ Sediment input to proglacial lakes

bulent due to wind action and high inputs, so only the coarser sediments will be deposited. In winter the lakes are ice-covered, which reduces turbulence, and the input is reduced. This allows the finer particles to settle. Thus we find rhythmic deposits of coarser (summer) and finer (winter) layers; these are called **varves** (Figure 23.35).

FIGURE 23.35 ▼ Varves in glaciolacustrine deposits near Guigues, western Quebec (photo: P.G. Johnson)

FIGURE 23.36 ▼ A spindle-shaped drumlin at Bish Lake, northern Saskatchewan; glacier flow was from the northeast, that is, from left to right in this photo. There are large subglacial meltwater channels (tunnel channels) associated with drumlins in this area, and meltwater that eroded the tunnel channels has cut into some drumlins. The tombolo connecting the drumlin to the mainland and the cuspate spit formed by wave action (lower right) are modern features (photo: J. Shaw)

# Landforms produced by deposition

## SUBGLACIAL LANDFORMS

The deposits laid down at the base of glaciers are frequently shaped into distinctive landforms, ranging from small flutes to large drumlins. Small ridges of till (less than 1 metre high) are called **flutes**; note the problem with terminology, that *flute* is used as a description for small depositional forms and large erosional forms. Larger, streamlined hills are known as **drumlins** (Figure 23.36) or **drumlinoid landforms**, depending on their elongation (width-to-length ratio). Characteristically, drumlins are up to 50 m high and up to 1000 m long.

Drumlins are among the most interesting glacial landforms. Early in this century it was accepted that they were depositional landforms produced by differential deposition of the subglacial load of the glacier. Places where deposition occurred were then streamlined into drumlins. Over the last thirty years it has been proposed that drumlins can also be produced by differential erosion of sediments, by erosion of bedrock, and most recently, produced by subglacial meltwater floods that carved hollows in the base of the ice, which were subsequently filled by sediments. It is certain that these shaped hills consist of almost all types of glacial deposits: till, glaciofluvial sediment, and glaciolacustrine sediment.

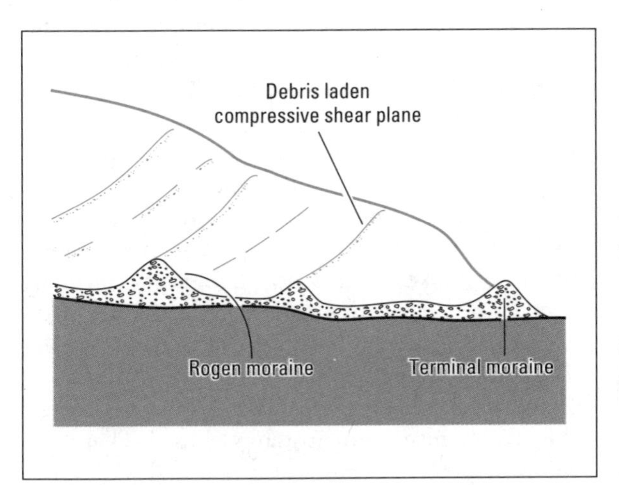


FIGURE 23.37 ▼ The formation of rogen moraines. Compressive shear planes in the ice (e.g., behind an area of relatively 'dead' ice at the snout) allow particles to migrate upward, creating a transverse ridge of debris on the glacier bed

Among the other common subglacially produced landforms are:

- **crevasse fillings**, sediment that has been squeezed into the base of the crevasse or worked in from above
- **rogen moraines**, small ridges aligned perpendicular to the flow direction of the ice, apparently formed by changes in compressive flow at the glacier terminus (Figure 23.37)
- **eskers**, long, sinuous ridges produced by glaciofluvial deposition in subglacial channels, which may extend for hundreds of kilometres

# Landforms of the glacier margin

Moraines are the landforms that accumulate at the glacier margin. They have a variety of forms, for example **lateral moraines** (Figure 23.38) and **terminal moraines**, depending on location. They are also classified according to the processes by which they are produced, for example **push moraines**, which are built up by advancing glaciers bulldozing material in front of them; or by their structure, such as **ice-cored moraines** (Figure 23.39), produced when debris accumulates over glacier ice at the terminus. Small moraines produced each year during gradual retreat of the terminus position of the glacier are known as **washboard** or **deGeer moraines**.

FIGURE 23.38 ▼ Little Ice Age lateral moraine along the north margin of the Kaskawulsh Glacier, Yukon (photo: P.G. Johnson)

FIGURE 23.39 ▼ Ice-cored moraine of the Donjek Glacier, Yukon (photo: P.G. Johnson)

It is obvious that moraines are complex landforms, found in a number of different environments. Lakes form between closely spaced ridges or in **kettle holes** as the ice core melts. The higher remnants between the kettles are known as **kames**. The complete melting of the ice core results in a very subdued moraine ridge compared to the original landform. The Donjek Moraine (Figure 23.39) stands up to 80 m high above the surrounding landscape and only 10 percent of it appears to be sediments.

## EXTENSIVE STAGNATION OF ICE

As the glacier or ice sheet gets thinner during deglaciation, due to increased melt and/or decreased ice input, it may reach a point where it is too thin to continue moving. This produces extensive areas of stagnant ice. As the ice continues to melt the sediments on the surface are continuously moved around. At the final stage of ice melt this leaves extensive areas of very irregular terrain with numerous kettle lakes and kames. This type of landscape is very common across the prairies of North America where it is often known as **hummocky moraine** or **dead-ice topography**.

## THE GLACIOFLUVIAL LANDSCAPE

The glaciofluvial landscape is composed typically of extensive braided stream deposits (see Chapter 22). Extensive deposits at the margins of ice sheets are called **sandar** (singular **sandur**). The coarse sediment load and the highly variable discharge from the glacier are the perfect conditions for the braiding channel form. In alpine glacial valleys, where braiding is restricted by the valley walls, the formations are called valley sandar. Terraces produced by meltwaters flowing along ice margins, confined between the ice and the valley side, are known as **kame terraces**; these are left as perched forms on the valley sides after deglaciation.

## ICE-DAMMED LAKES

One characteristic of glacierized regions that has important consequences for landscape development and as a natural hazard is the formation and potential catastrophic drainage of ice-dammed lakes. Lakes have always formed around ice margins where drainage has been impeded by the mass of the ice. In alpine regions these may be small marginal lakes or large lakes dammed in deglaciated tributary valleys. In regions, of continental glaciation, massive lakes formed, for example along the southern margin of the Laurentide Ice Sheet in North America, where glacial Lake Agassiz was one of a series of ice-dammed lakes (Figure 23.40, see also Figure 9.19). Glacial Lake Missoula, in the northwest United States, was a relatively small glacial lake, but it drained catastrophically, with discharges estimated at up to $13.7 \times 10^6$ $m^3$ $sec^{-1}$. To put this into perspective, this would be one thousand times the maximum instantaneous discharge measured on the lower Fraser River at Mission, British Columbia.

Ice-dammed lakes in alpine regions may drain either by surface channels or subglacial channels. Surface channels across bedrock drain slowly. But surface channels across the ice are quickly incised by rushing water, and can cause partial or complete drainage of a lake in a few days. Subglacial channels, caused by a combination of floating of the ice margin and the expansion of subglacial channels by the effect of waterflow, can cause drainage in less than 48 hours. Lake George in Alaska, which drained in 1958, had a maximum instantaneous discharge of 10 000 $m^3$ $sec^{-1}$; Summit Lake in British Columbia drained in 1965 and 1967 with a maximum instantaneous discharge of 3260 $m^3$ $sec^{-1}$. These floods, called glacier lake outburst floods or jökulhlaups (glacier bursts), can transport large volumes of sediment, and pose serious threats to economic activity within the basin.

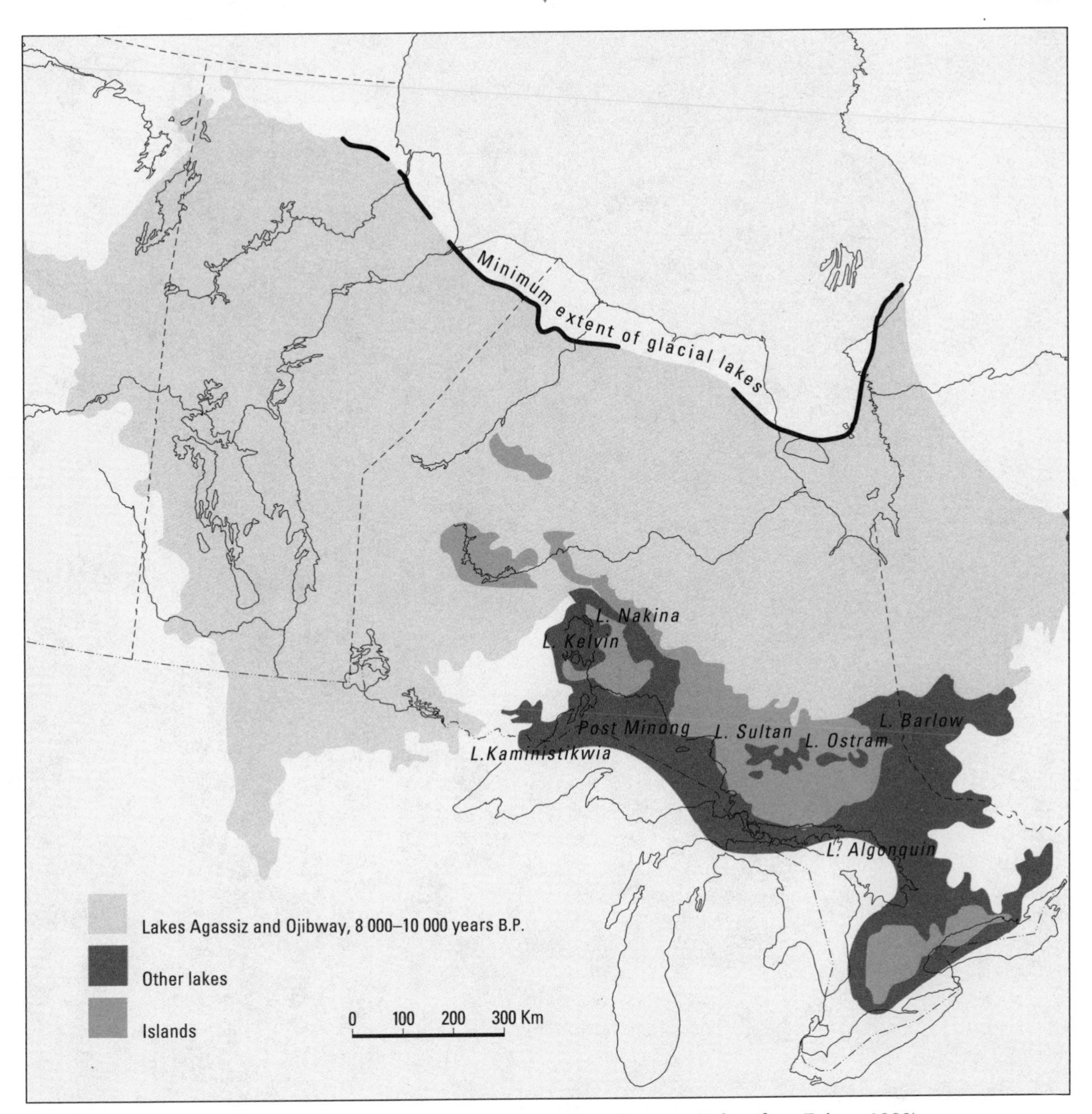


FIGURE 23.40 ▼ Areas of the southwestern Canadian Shield inundated by glacial lakes (from Fulton, 1989)

# Aeolian systems

## Distribution of aeolian activity

Almost 10 percent of the world's land surface may be described as hot desert (Figure 24.1). Within this area, the action of the wind is all-pervading. It plays a vital part in landscape development, carrying abrasive fragments of rock across the surface, and redistributing the materials to leave vast expanses of land bare and rocky and cover others with thick layers of sand. It directly influences human activities; it constrains our use of the land; it poses barriers to movement and communication; it irritates lungs and eyes and creates a constant health risk.

A further 10 percent of the continents consists of polar deserts (Figure 24.1). Again, wind activity is important in shaping the land. As we shall see, fierce polar winds strip sediment from the river plains and carry it many kilometres across the surface. In the past, when the polar ice sheets extended far into lower latitudes, these polar deserts occurred in what are now temperate areas. They have left behind the legacy of widespread wind-blown deposits known as loess.

Wind activity is not confined to the desert areas, however. In many semi-arid and even humid temperate areas, the effects of wind erosion are of considerable import. Coastal areas are often subject to the action of winds from the

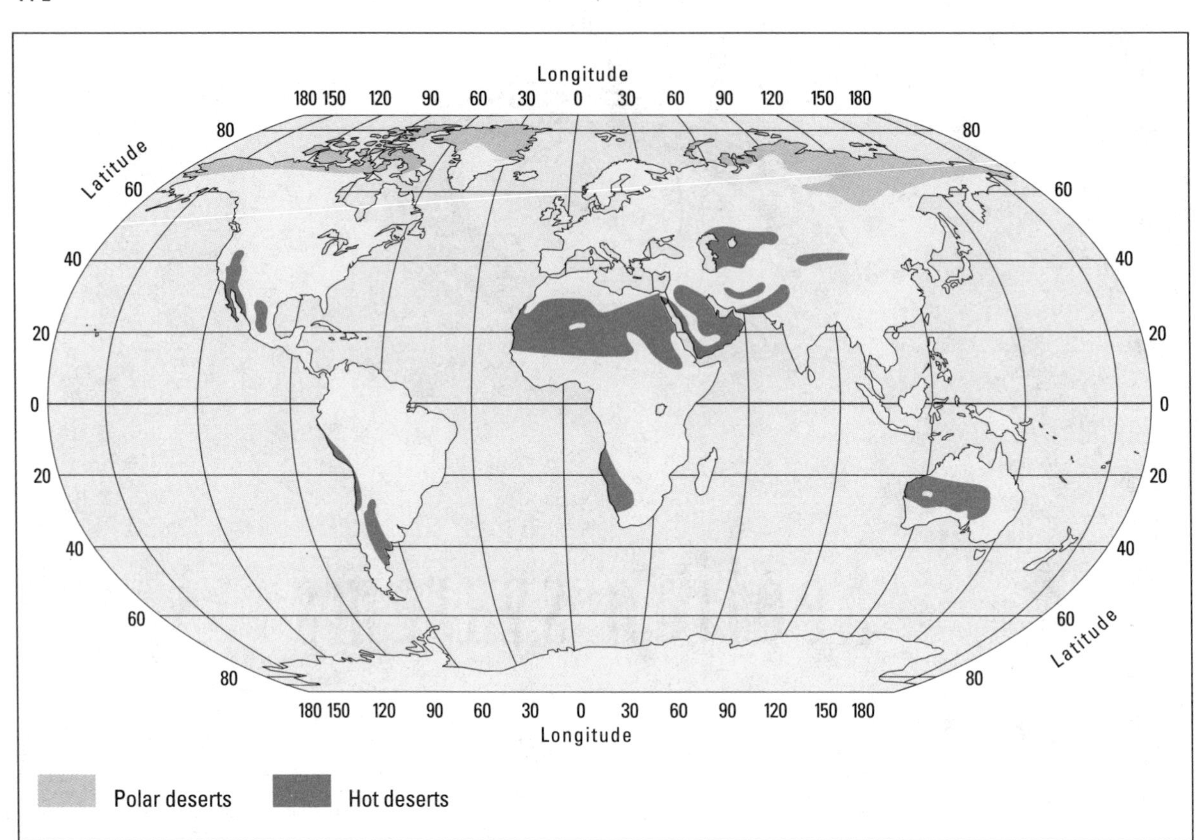


FIGURE 24.1 ▼ The distribution of hot and polar deserts

sea, and aeolian landforms lie inland of many sandy beaches. Many agricultural areas are also prone to wind activity. The North American 'Dust Bowl' is the most often-cited example, but wind erosion occurs in most intensively cultivated countries. Human attempts to clear the land for cropping, and our repeated tillage of the soil, have often produced unprotected, powdery surfaces, susceptible to the erosive power of the wind.

On the other hand, once the air has picked up sediment, its erosive potential increases enormously, for the sand grains bouncing on the surface have a kinetic energy sufficient to move particles as much as six times their own diameter. Wind erosion is consequently a process that is subject to threshold constraints: once threshold conditions controlling the initial entrainment of material by fluid drag are reached, the rate of erosion increases swiftly.

# Aeolian processes

## ENTRAINMENT

The ability of the wind directly to attack and erode the land surface is severely limited. Air has a very low fluid density; only about $1.22 \times 10^{-3}$ g cm$^{-3}$ at a temperature of about 20°C, roughly one nine-hundredth that of water. As a result it is able to exert only a relatively small force against objects in its path. Entrainment of material by fluid drag (i.e., the direct force of the wind) is therefore restricted.

*Fluid drag* What are the factors controlling the action of erosion by fluid drag? A major determinant is clearly wind speed. In general, as wind speed rises, the entrainment potential of the wind increases. It is not, however, the mean wind speed that is most important, for over short periods—often fractions of a second—great variations in velocity may occur. It is the extreme gusts that are instrumental in entraining sediment, so the erosivity of the wind is related more closely to the duration and magnitude of these brief spells of high wind speed.

Wind conditions at the ground surface, where particle entrainment occurs, are not the same as those in the open atmosphere. The surface exerts a frictional drag against the air which reduces wind velocity and often creates a thin layer in which wind speed is zero (Figure 24.2). The depth of this dead air layer depends upon the surface roughness, but is generally about one-thirtieth of the average height of the obstacles. Over a bare sand surface, the particles create a degree of roughness that acts to shelter grains lying in the depressions between the larger particles. Where vegetation is present, however, a much deeper dead air layer may be created, which protects the whole surface from erosion. For this reason the presence of vegetation is a major constraint on erosion, and maintaining a good vegetation cover provides one of the most effective means of erosion control.

Surface roughness also helps to generate turbulence in the wind. As we saw in Chapter 19, this is important, because turbulent eddies give a vertical lift that helps to raise particles off the stream bed, and in the same way they help to raise particles off the ground surface and keep them buoyant in the air. In addition, bedforms such as ripples and dunes, produced on the surface by the wind itself, modify the wind flow at ground level (Figure 24.3). Turbulent eddies may be created in the lee of such features. These eddies help to keep the area between ripples or dunes free of sediment and play an important part in the dynamics of sediment transport and deposition.

*Fluid impact* As we have noted, the entrainment of particles by fluid drag represents only the first stage in the initiation of transport. Most material carried by the wind is entrained by the impact of particles on the surface. These impacts cause individual particles to bounce upward, off neighbouring grains. Often, several grains may be set in motion at once by the impact of a single particle on a loose sand bed.

*Frictional resistance* The frictional resistance of the surface materials to entrainment is largely a function of their weight. Larger or heavier grains

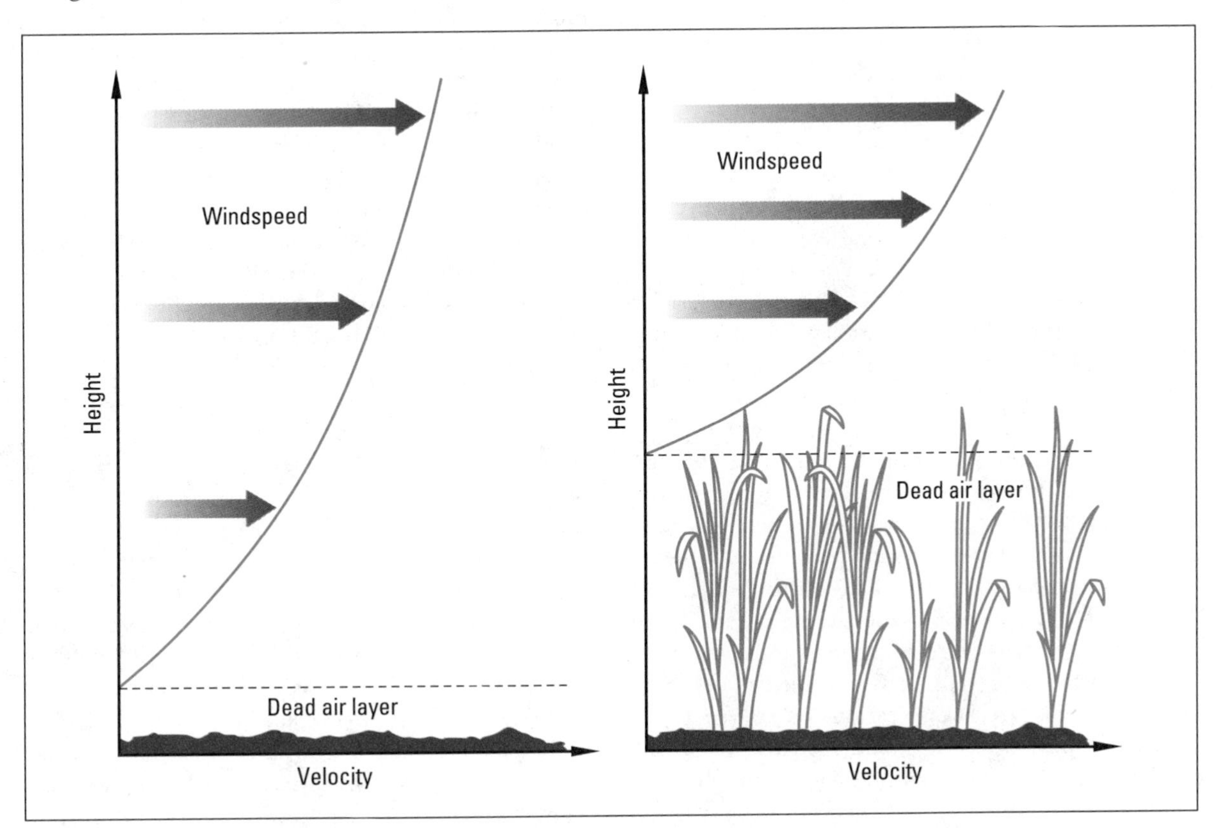


FIGURE 24.2 ▼ The effects of surface roughness on the distribution of wind speed with height: (A) flow above a smooth sand bed; (B) flow above a densely vegetated surface (after Embleton and Thornes, 1979)

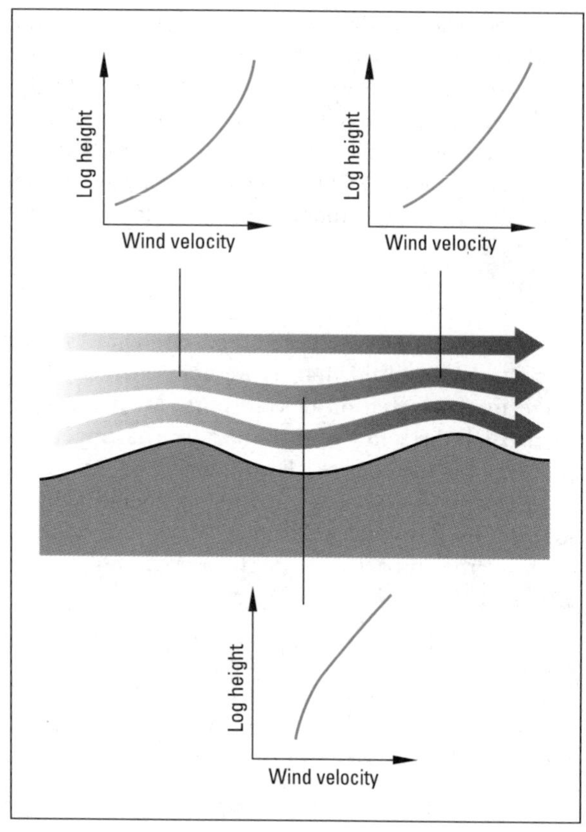


FIGURE 24.3 ▼ Streamlines and velocity height profiles of wind flow over sand ripples

require higher fluid drag to initiate movement than smaller or lighter particles. However, very fine material in the silt and clay size fractions has greater resistance to entrainment because of its cohesiveness; as well, the surface produced has such a low surface roughness that the particles do not protrude into the active wind flow. Thus the critical entrainment velocity for fluid drag increases below a particle diameter of about 0.1 mm (see Figure 19.6). In materials that are predominantly silt and clay it is frequently the coarser particles that control the entrainment of the finer particles. There has to be sufficient fluid drag to set sand particles in motion.

Moisture is also important in controlling entrainment, for water films provide a source of intergranular cohesion that increases the frictional resistance of the surface materials. It is not until thawing has occurred and evapotranspiration has started to dry out the soil at the start of the growing season that erosion becomes a serious problem in the northern half of North America. More generally, it is the lack of moisture that is primarily responsible for the high rates of wind erosion in arid and semi-arid areas. The effect operates both directly and indirectly: it makes the soil non-cohesive and susceptible to erosion, and it inhibits vegetation development so that the surface is frequently bare and unprotected.

The salt content of the ground water also controls entrainment. During evaporation salts are deposited in the surface sediments, eventually cementing them together and preventing entrainment. In this situation entrainment is only possible for a short period during the drying of the sediments after rain or inundation by floodwater.

## TRANSPORT

When grains are lifted off the surface their frictional resistance declines, for they are no longer in contact with adjacent particles, and they come under the influence of stronger horizontal wind flows. The horizontal forces drive the grains forward, while vertical eddies help to keep them airborne. Together, these forces counteract the gravitational pull that draws the particles back to the ground. In most cases, however, the vertical currents are insufficient to keep the particles aloft for long and they fall slowly to the earth along a characteristic curved path (Figure 24.4). The length of this path and the height the particles reach are governed by the relationship between wind speed, turbulence, and particle size, but under reasonably uniform conditions of sediment supply and wind velocity the particles tend to follow a fairly consistent trajectory. As we will see, this has considerable significance for the development of aeolian bedforms.

The wind is a very effective agent of sediment sorting. Large particles are not normally moved: the largest material that can be entrained by fluid drag is about 3 mm in diameter; particles of this size bouncing on the surface can thus dislodge grains up to about 18 mm in diameter. Larger particles can be moved—in fact there is evidence of huge boulders sliding across desert surfaces under the effect of wind—but these are exceptional and the extent of transport is extremely limited. More commonly, coarse debris is left as a lag on the surface, where it protects underlying material from erosion.

The material that is entrained is sorted according to both size and shape. Larger particles (above about 1 or 2 mm in diameter) move almost entirely by creep and are thus subject to considerable frictional retardation, with the result that they move rela-

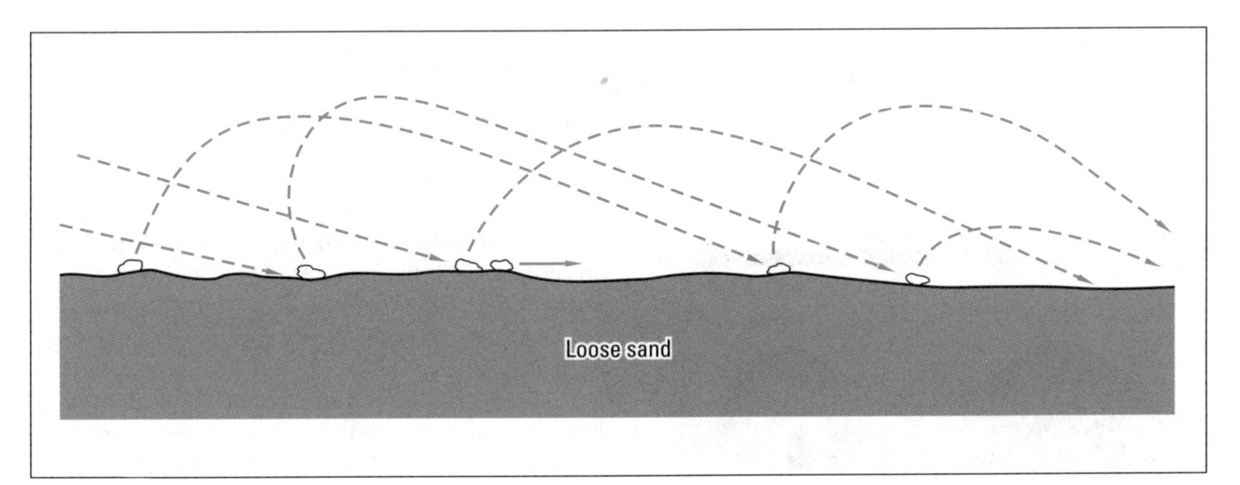


FIGURE 24.4 ▼ Trajectories of saltating aeolian sand grains

tively slowly. Material between about 0.1 mm and 1.0 mm in diameter moves largely by saltation, for it is too heavy to be retained in the airflow for long. Finer material, however, may be carried to heights of several hundred metres and, under convective atmospheric conditions, may be transported thousands of kilometres before deposition. Silts transported this way form vast spreads of **loess** in areas such as Northern China (Figure 24.5); during cold, arid phases in the Quaternary, similar silts were laid down over much of Kansas and the midwestern United States.

The effect of shape is more complex. Spherical particles tend to have relatively high settling velocities, and consequently do not remain airborne as long as flatter, discoidal grains. On the other hand, spherical grains tend to be thrown steeply into the air when entrained by particle impact, whereas discoidal grains rise more obliquely. As a result, spherical particles adopt somewhat steep, short trajectories when moved by saltation, or are transported preferentially by creep. Conversely, discoidal particles move mainly by saltation along shallow, flat trajectories. Because of these effects, flatter particles often move farther than spherical grains, especially at low wind velocities. As wind speed increases, however, this pattern is reversed due to the tendency of more spherical grains to bounce back into the airflow after impact.

Together, these processes lead to marked patterns of sorting within aeolian deposits. At a large scale, materials become finer with increased distance from their source. At a small scale, distinct size grading occurs on individual bedforms such as ripples and dunes. Coarser particles tend to accumulate on the

FIGURE 24.5 ▼ Erosion in a loess landscape: the Fen River valley in northern China. More than 125 m of windblown silt has accumulated during recent glacial periods; the darker layers are soils that developed during interglacial periods (photo: R.J. Fulton, courtesy Geological Survey of Canada)

crest of ripples, for example, while finer material collects in the more sheltered troughs between the ripples (Figure 24.6).

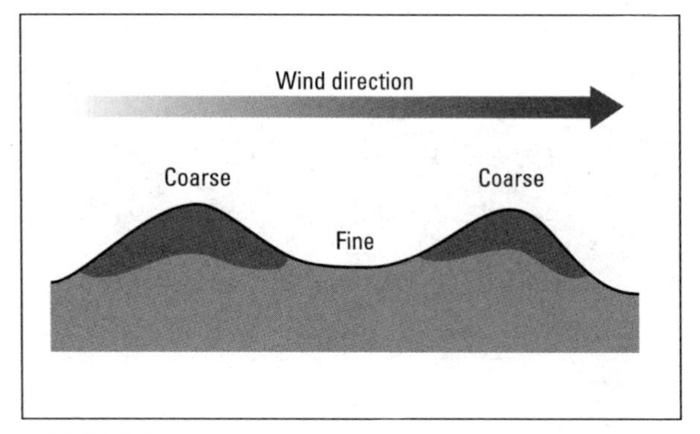


FIGURE 24.6 ▼ Particle size variation across sand ripples

### DEPOSITION

The sand and silt carried by the wind are ultimately deposited. Deposition occurs for a variety of reasons and in a variety of circumstances. Mass deposition of the sediment load occurs when wind speed drops at the end of a storm. During a storm, material may be trapped against an obstacle or held by a moist patch of ground; local deposition also takes place in sheltered areas provided by small surface irregularities. Deposition against permanent obstacles produces sand drifts, which are static features. Many features of aeolian deposition, however, are dynamic, steady-state forms, dependent upon a balance between rates of entrainment, transport, and deposition (Figure 24.7).

## Aeolian landforms

### LANDFORMS OF HOT DESERTS

To many people, the term 'desert' conjures up visions of limitless expanses of sand. This impression is generally far from true. Sand deserts (**ergs**) are not ubiquitous in arid regions, and it has been estimated that about 85 percent of aeolian sands are confined to a relatively small number of large ergs, each covering at least 32 000 km$^2$. Even within these areas, the sand cover is not continuous, and between the sand areas there are expanses of bare bedrock. Elsewhere, the hot deserts consist of extensive bedrock plains, spreads of gravel (**regs**), or coarse boulder pavements (**hamadas**). All these landscapes reflect the combined effects of the removal (**deflation**) and deposition of the sand by the wind. In addition, however, it is important to remember that water plays a significant part in sculpting the desert landscape. Rainfall may be rare but when it occurs it is often intense and flash floods develop. The water cuts deep gullies and trenches, known as **wadis**, and transports vast quantities of sediment. Much of this material ends up in inland **playa lakes** where the water evaporates to leave spreads of saline sediment.

*Sources of aeolian sand* Sand deserts may not be ubiquitous, but they contain vast quantities of sand. The Rub al Khali erg in Arabia covers 560 000 km$^2$, with sand deposits up to 300 m in depth. Where does all this sand come from?

FIGURE 24.7 ▼ The front of an advancing parabolic dune in the Great Sand Hills region southwest of Lancer, Saskatchewan (photo: D.A. St-Onge)

FIGURE 24.8 ▼ Sand ripples on an active dune surface, Great Sand Hills, Saskatchewan (photo: D.A. St-Onge)

This is a pertinent question, for we have already noted that the wind itself is not very efficient at eroding bedrock surfaces. Most of the sand must be derived from other sedimentary deposits. Much, in fact, comes from alluvial fans where streams draining the surrounding mountain areas dump their debris at the desert edge. Dry playa lakes and wadi floors also supply sand, while some is derived from in situ weathering of the bedrock surface. Considerable controversy still exists about the importance of this last source, for it has been argued that weathering processes in desert areas are relatively ineffective. Nonetheless, salt weathering, heating and cooling, and thermal contraction may all encourage rock disintegration and provide at least small quantities of material.

*Depositional features of hot deserts* Starting with the small-scale features of deposition, **sand ripples** are common features of almost all sand surfaces (Figure 24.8). Typically they are a few centimetres in height and are spaced at intervals of several centimetres to metres. Their long axes run at right angles to the wind direction and they tend to migrate downwind. They may be symmetrical or

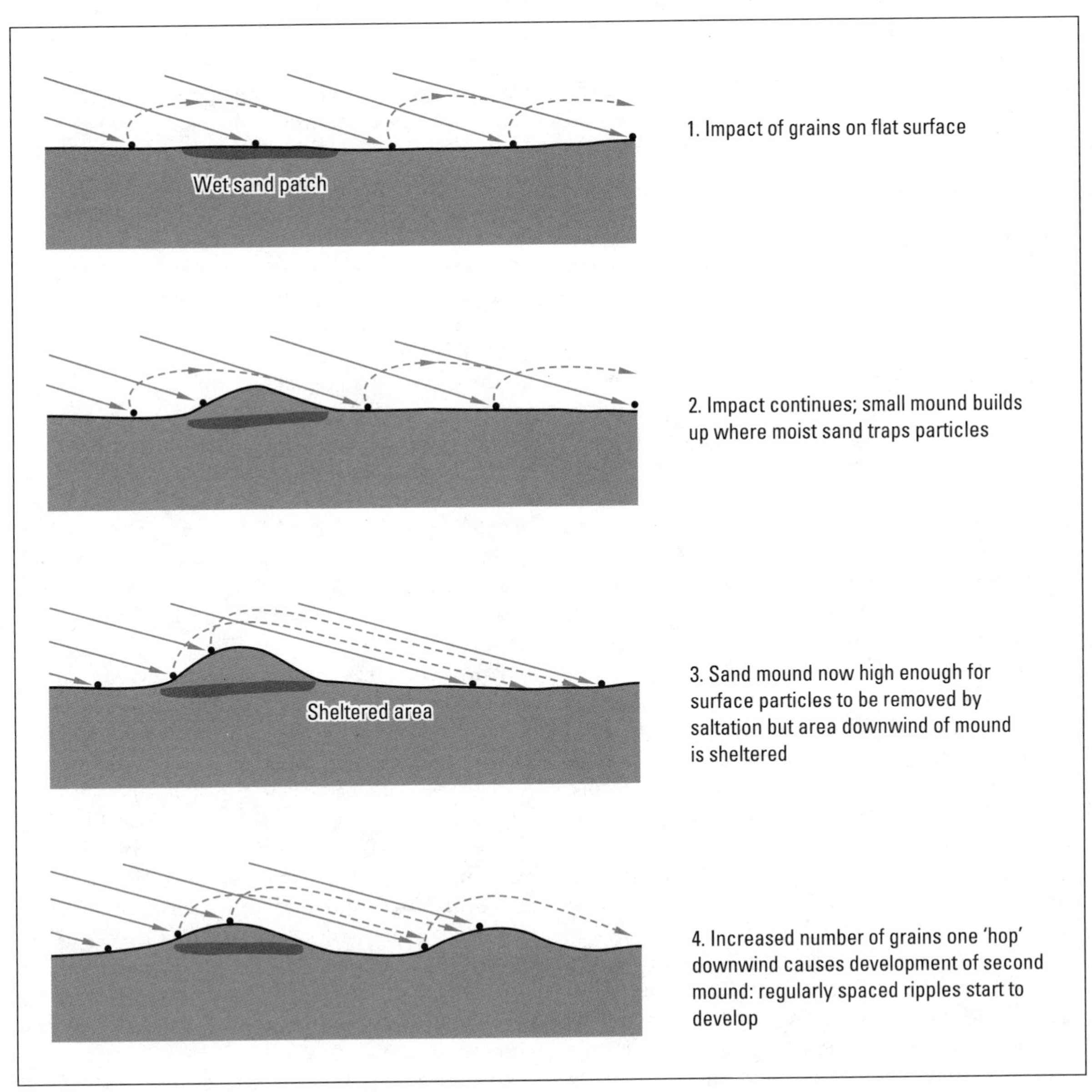


FIGURE 24.9 ▼ The development of sand ripples

asymmetrical in cross section. Their form, however, depends to a great extent on wind conditions. Over a wide range of conditions their wavelength-to-height ratio (known as the **ripple index**) is between 15 and 20, but as wind velocity increases the ripples become flatter and more widely spaced, and the index rises to as much as 50 or 60.

The reasons for ripple formation are related to the dynamic interaction between wind flow and sediment movement. Initiation is often due to a random factor, such as local variations in sediment size or wind speed, patches of moist sand, or even the presence of a small obstacle on the surface, which results in the accumulation of a small mound of sand. This mound creates, on its lee side, an area that is protected to some extent from the saltating sand grains, whereas the windward side tends to be subject to higher rates of particle impact (Figure 24.9). The result is that material starts to collect on the upwind side of the mound.

In time, however, the mound becomes large enough that the grains on the surface are no longer held by whatever initially trapped the sand. Thereafter, as grains land on the upwind side of the mound, the impact throws new material into motion. The dislodged particles tend to move a similar distance due to their uniform saltation trajectories, and they thus create a second zone of intensified impact one 'hop' downwind. These impacts in turn set further particles moving, excavating a small depression in the process and triggering yet further saltation downwind. A single ridge creates a wave in the airflow, which as it damps out downwind initiates more ridges, eventually producing a succession of ripples. In this way, ripples develop that are spaced at intervals equivalent to the average path length of the saltating particles. Ripples formed in this way are referred to as impact or **ballistic ripples**. Their height depends on grain size, larger ripples being associated with coarser or more poorly sorted sediments. As a ripple forms, sediment moving by creep is transported on the windward side of the ripple, where fluid drag is sufficient to set particles in motion, and deposited on the leeward side. In this case the ripples will migrate in the direction of airflow.

Under conditions of extremely poor sorting or a relatively coarse sediment relative to the wind speed, **sand ridges** form. The wind is only able to transport the finer fraction by saltation, with the result that the coarser particles are moved by creep or accumulate as a lag deposit. In time, the largest particles become concentrated at the surface, protecting the underlying sediment, and trapping particles that are being transported along the ground by creep. Gradually a ridge builds up of sediment that is too coarse to be removed by saltation (Figure 24.10). Sand ridges of this sort may reach considerable heights and appear to be essentially immobile; ridges up to 60 cm are seen in the Libyan Desert, for example.

Large-scale features such as the huge dunes that characterize the sand seas of the Libyan and Arabian deserts are some of the most spectacular aeolian landforms. They may be up to 200 m high and 5 km or more in wavelength. They vary greatly in genesis and size, however, and in recent years it has become common to divide them on the basis of scale into **dunes** (with wavelengths up to 0.5 km) and **draas** (wavelengths of 0.5 to 5.0 km).

In practice, dune forms show immense variation and tend to merge into each other, both in terms of morphology and genesis, but four main types are extensively found in desert areas: seif, barchan, transverse, and aklé dunes. Each seems to represent a different relationship between sand supply, wind speed, and wind direction.

**Seif dunes** are named from the Arabic term meaning 'sword edge' and as this implies they are long, steep-sided, sharp-crested features (Figure 24.11). They are also called in some areas **longitudinal dunes** for they are oriented parallel to the wind. They are found in many sand deserts, including the Arabian, Libyan, Namibian, Australian, and Arizona deserts. In Iran, they are up to 200 m high and they are commonly 60 to 100 km in length. Often, seif dunes join to create so-called tuning-fork features (Figure 24.11).

How seif dunes form is not entirely certain. It was once thought that they were produced by crosswinds, blowing at 90° to each other. Recent observation, however, suggests that they are due instead to the development of helical airflows between the dune ridges. Probably because of excessive heating of the sand surface, convection cells develop that result in two parallel vortices (rollers) circulating in opposing directions between each pair of dunes (Figure 24.12). These sweep sand off the desert floor and move it obliquely up the dune face. Evidence for this type of movement is sometimes seen in the strings of debris and small-scale sand ridges running up the flanks of seif dunes.

In contrast, **barchans** often form as isolated features. They are crescentic mounds of sand, oriented

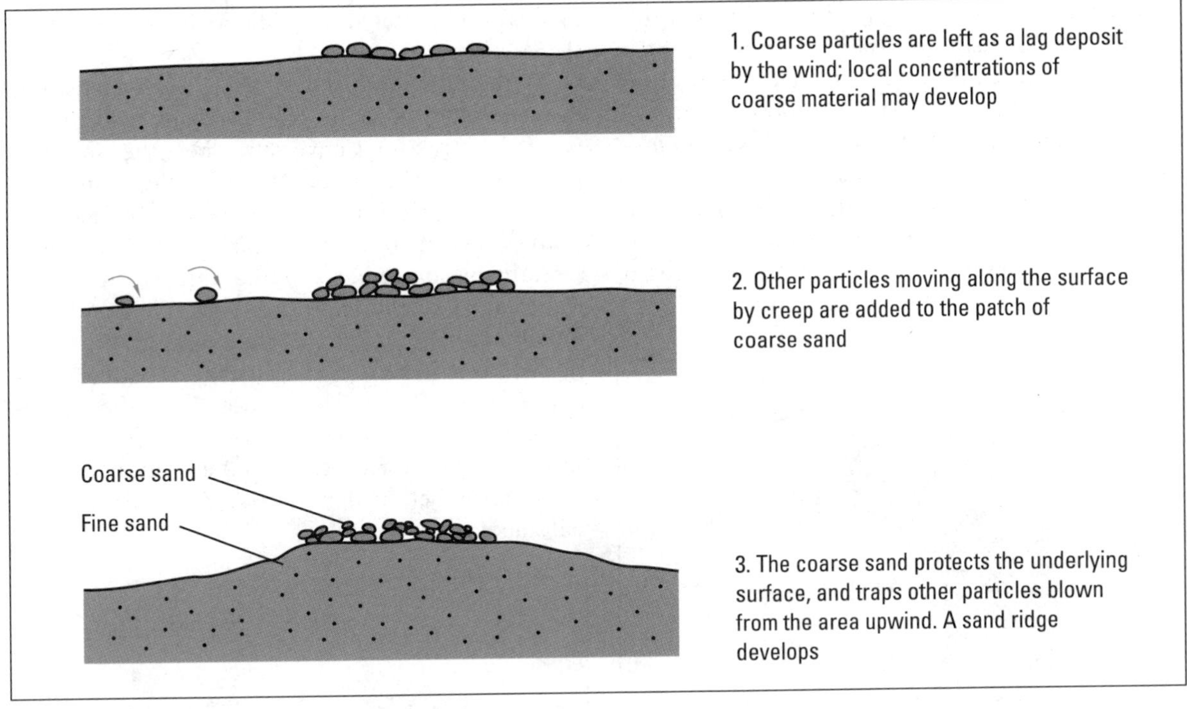


FIGURE 24.10 ▼ The development of sand ridges

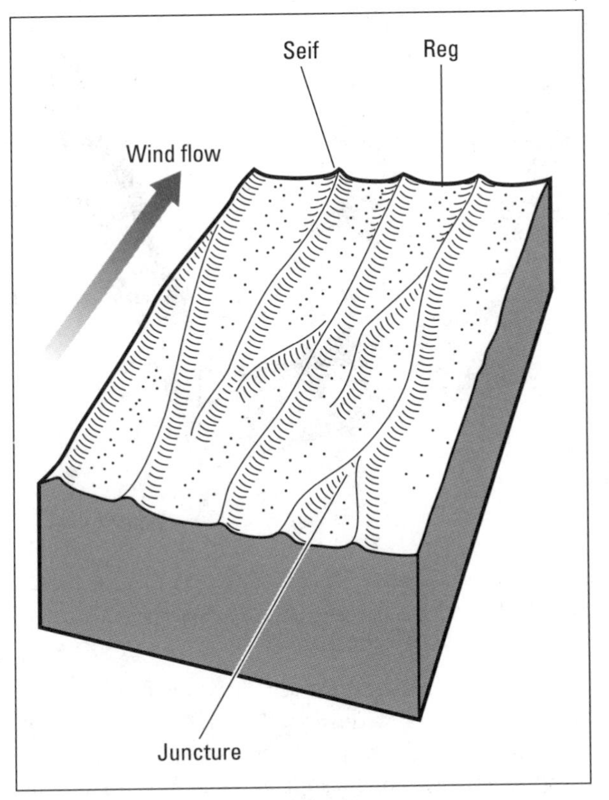


FIGURE 24.11 ▼ Typical system of seif dunes showing tuning-fork forms

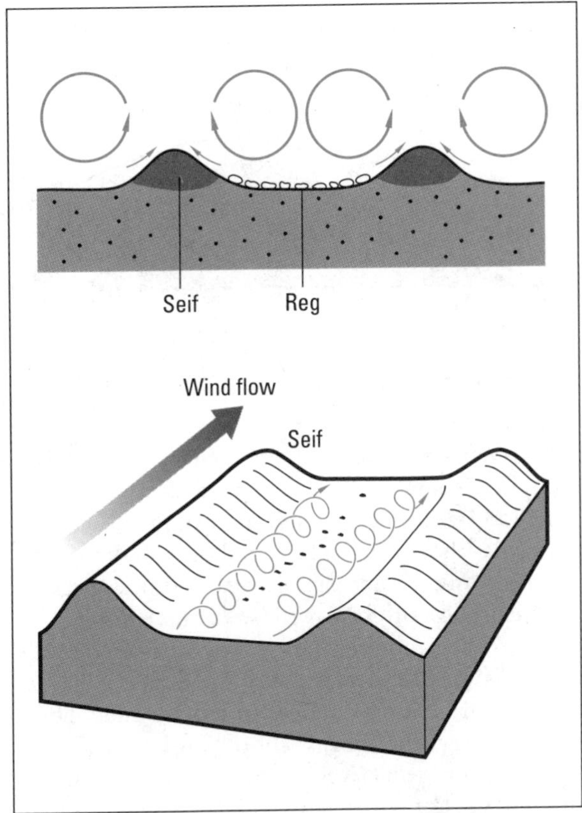


FIGURE 24.12 ▼ The development of seif dunes

with their 'horns' pointing downwind (Figure 24.13). The steep slip face that points downwind is typically at an angle of about 34°, the angle of repose of dry sand. Barchans vary greatly in size, reaching diameters of 400 m and heights of 30 m—though some may be as high as 150 m. The most perfect examples tend to be found as isolated dunes developed on bedrock surfaces under conditions of limited sand supply, but they commonly occur in groups as rather less well-developed forms.

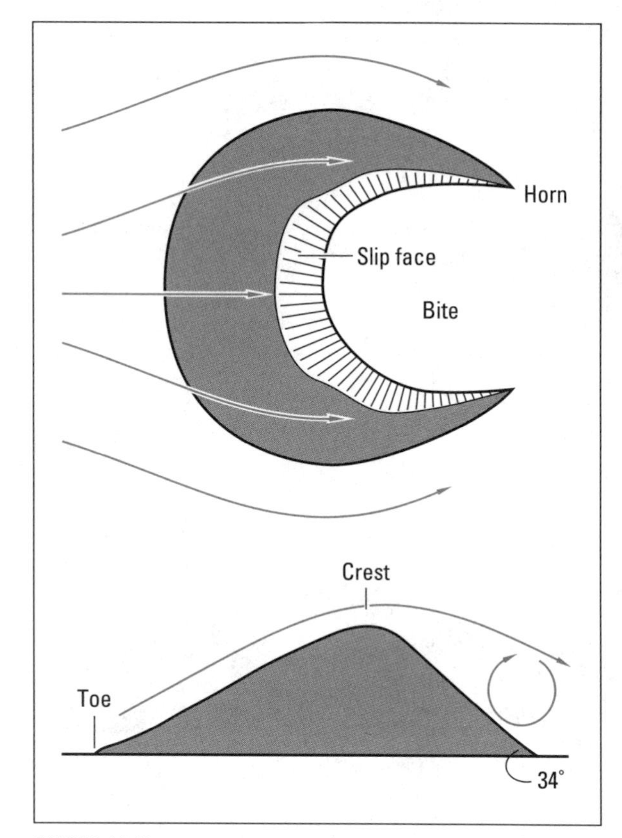


FIGURE 24.13 ▼ A barchan dune

Barchans are common in many sand deserts, including those of Peru, Egypt, and the Imperial Valley in California, but significantly none occur in Australia: they seem to be found only under conditions of limited sand supply, flat bedrock surfaces, and relatively uniform winds. In these situations, the barchans represent an aerodynamically stable form. As the wind passes over the dune, it divides slightly and eddies develop on the lee side. These tend to keep the 'horns' short, while within the 'bite' the air is relatively calm. As a result, sand is moved up the windward side of the dune and avalanches down the slip face. In this way, the dune gradually migrates downwind without altering its shape.

As the sand supply diminishes with increasing distance from its source, the barchans tend to disperse and break up. Conversely, where the sand supply is greater the dunes tend to coalesce and lose their crescentic form, ultimately merging into **transverse dunes**. These are long, relatively straight or somewhat serrated ridges oriented at right angles to the wind direction. They are commonly arranged at regular intervals and separated by sediment-free areas or coarse gravel-strewn surfaces (Figure 24.14). They appear to be the characteristic form in areas of high sand supply in which surface heating (or other effects) are insufficient to produce the helical flow necessary for formation of seif dunes.

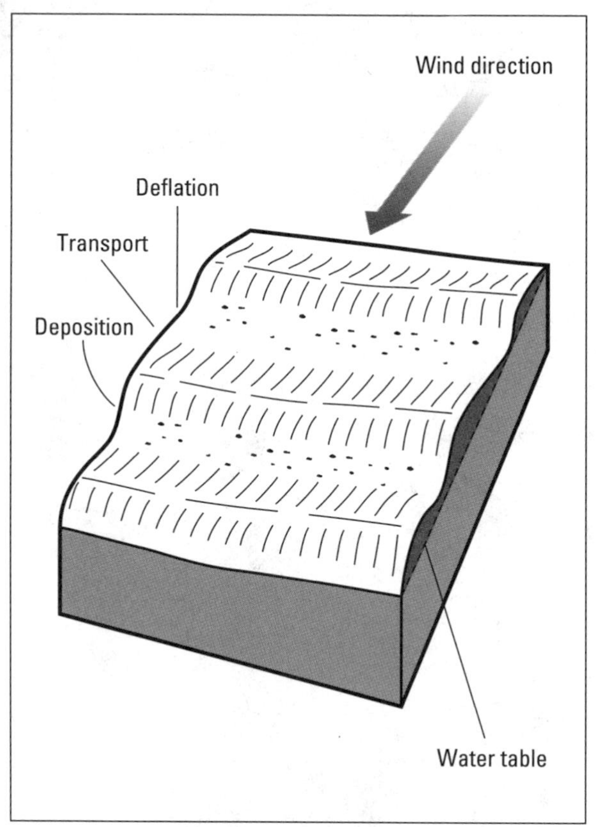


FIGURE 24.14 ▼ The development of transverse dunes

In practice, none of these relatively simple forms of dune are very widespread, and the majority of sand deserts are probably characterized by more complex forms known as **aklé**. These consist of sinuous dune ridges, aligned transverse to the wind direction, and composed of crescentic sections, pointing alternately upwind and downwind. The way they form is far from certain, but they appear to result from the development of complex patterns of secondary wind flow due to interactions between the bedform and the wind.

*Erosional features of hot deserts* The sediment-laden winds of a desert sandstorm operate like a sandblasting machine, wearing away at anything in their path. In the process they produce a range of erosional landforms. Pebbles exposed on the surface, for example, are polished and faceted by the incessant abrasion, and where the winds are from a consistent direction they are shaped into **dreikanter** (German: 'three edges'). Larger boulders and hills are smoothed and etched, weak beds being picked out to create complex shapes (Figure 24.15). At a larger scale, the wind may ultimately open up huge **deflation hollows** in the bedrock surface. Erosion often starts along a line of weakness (such as an area of intense jointing) or along a wadi that cuts through the cemented surface rocks, and continues until it reaches the water table. One spectacular example is Al Fugaha in Libya, an almost circular depression 2 to 3 km in diameter and over 60 m deep.

As we have already seen, deflation also leaves eroded desert surfaces: rock pavements and gravel plains. Indeed, it is landscapes such as these that make up the majority of many desert areas. In the Sahara, for example, 70 percent of the surface area consists of sandless rock plains, regs, and hamadas. Often the pebbles and boulders in these areas are coated with so-called **desert varnish**, a smooth, dark mineral layer containing iron, manganese, and various trace elements. How this varnish forms is not known for certain. One theory suggests that it is due to the development of a thin film of dew on the pebble surfaces which encourages ions to diffuse from the interior of the rock. In recent years it has proved possible to obtain carbon-14 dates on rock varnish and derive information on Paleo-environments, such as age and rates of formation of the varnish. In addition, external materials, particularly windblown particles, become incorporated into the varnish.

FIGURE 24.15 ▼ Wind-eroded forms known as the Three Sisters, Goblin Valley, Utah. Differential etching of the sandstone beds by aeolian activity produces these dramatic forms (photo: P.G. Johnson)

## LANDFORMS OF COASTAL REGIONS

Aeolian processes can operate wherever there is an adequate supply of sand, a dry and unvegetated surface, and high wind speeds; such conditions characterize many sandy coasts. Here, the source of sand is the beach: prevailing winds or local onshore breezes are often strong, and, because of the exposure, the salinity, the coarseness of the sediments, and the impact of human trampling, vegetation growth is often limited. As a result, the winds winnow the beach material and transport it inland to form transverse dunes, running parallel to the shoreline (Figure 24.16).

The formation of coastal dunes often follows a distinct cycle in which soil formation and vegetation development play significant parts. Initially, sand is transported from the beach and accumulates a short distance inland, often where it is stabilized by vegetation, in the form of transverse **foredunes** (Figure 24.17). The dune shape is governed by aerodynamic factors, and typically has steeper slopes on the landward side. During transport, marked sorting of the sediment by size and shape occurs, and the dunes are often composed of material that is finer and more rounded than that on the beach (Figure 24.18).

Over time, the foredune may migrate inland as sand creeps up the windward slope and avalanches down the leeward side. As this happens, the area of available sand supply increases and ultimately new foredunes may form. In either case, the effect is to reduce the rate of sediment supply to the older dunes, allowing them to become stabilized. Gradually, the sand weathers, soil starts to form,

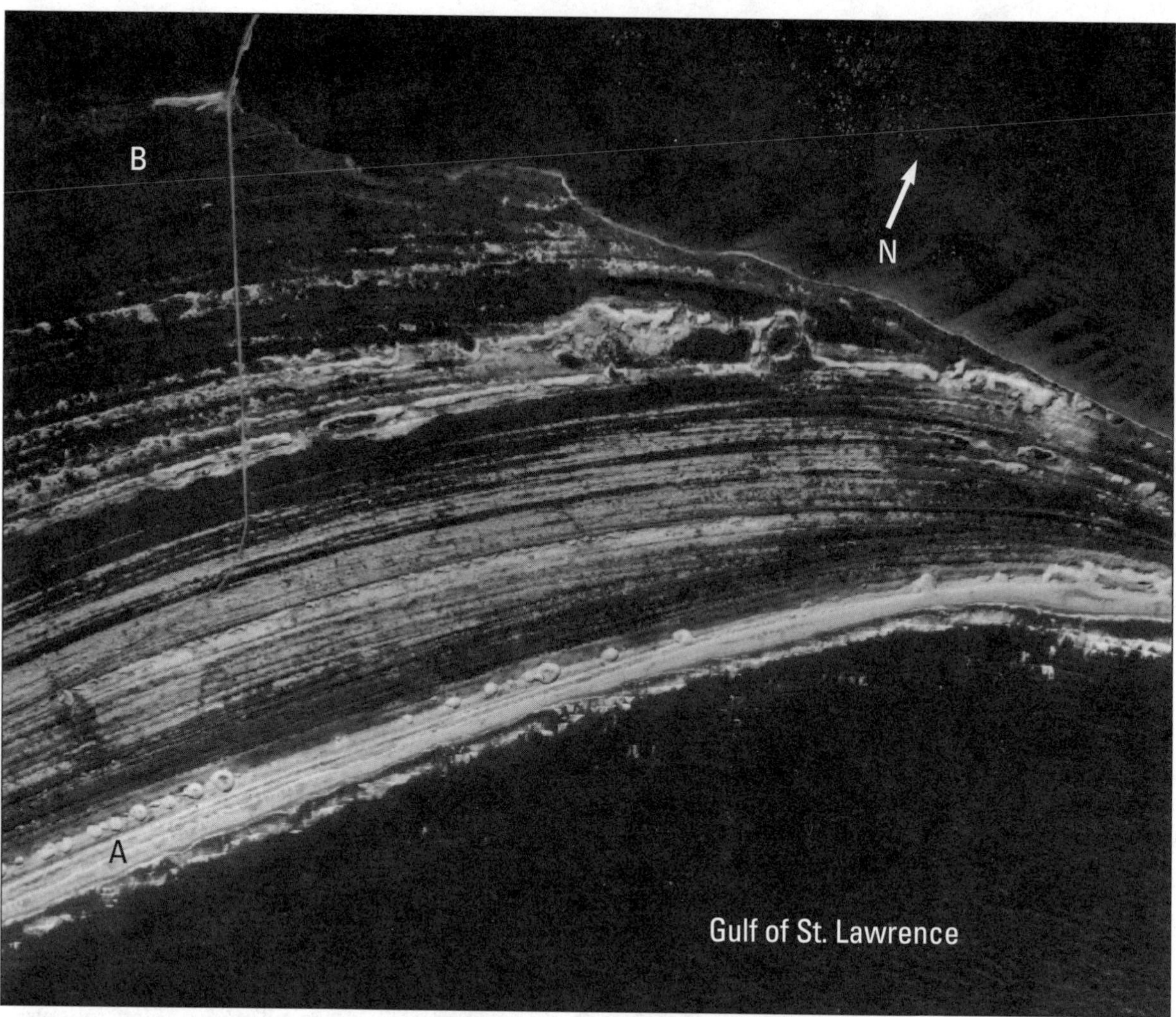


FIGURE 24.16 ▼ Aerial view of transverse dune ridges in the Magdalen Islands, Quebec, from an altitude of roughly 2500 m; scale 1:16 000. Note blowouts at *A*, old dunes under vegetation cover at *B* (photo copyright 1956 National Air Photo Library, Energy, Mines and Resources Canada)

FIGURE 24.17 ▼ Foredune formation at the top of the beach slope; west Newfoundland (photo: P.G. Johnson)

and vegetation starts to invade. Ultimately, a whole series of parallel dunes of increasing age may form inland, separated by low-lying areas called swales (Figure 24.19).

During the evolution of the stable dune, a number of additional changes may occur. Marram grass is one of the main forms of vegetation to become established because it tolerates—indeed it requires—continual burial by sand (Figure 24.20). The grass provides protection to the dune, helps to trap sediment, and reduces erosion. When the dune becomes isolated from the beach and sediment supply falls, however, the grass may start to die. Bare patches develop on the old dune surface and erosion may be reactivated. This, in turn, increases the supply of sediment to dunes farther inland.

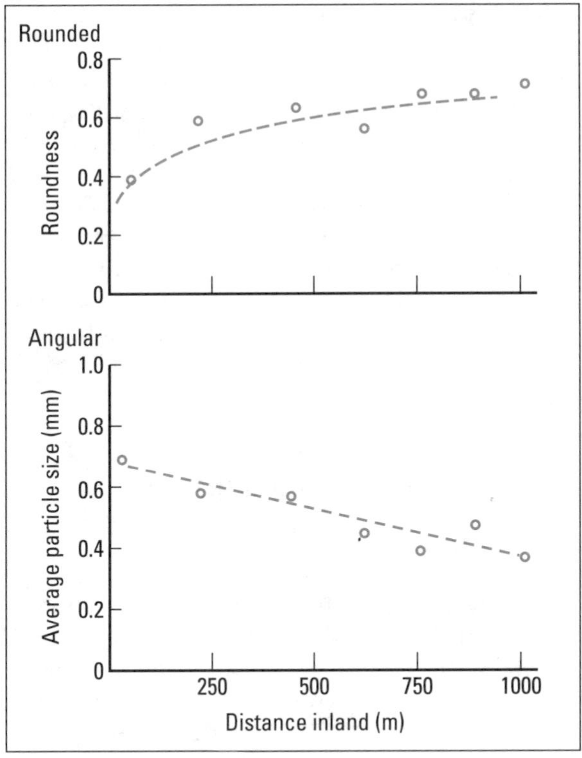


FIGURE 24.18 ▼ Typical variations in particle roundness and average size across a coastal dune sequence; samples are from the crests of dune ridges

FIGURE 24.20 ▼ Marram grass exposed in a blowout in dunes at Southport, England (photo: P.G. Johnson)

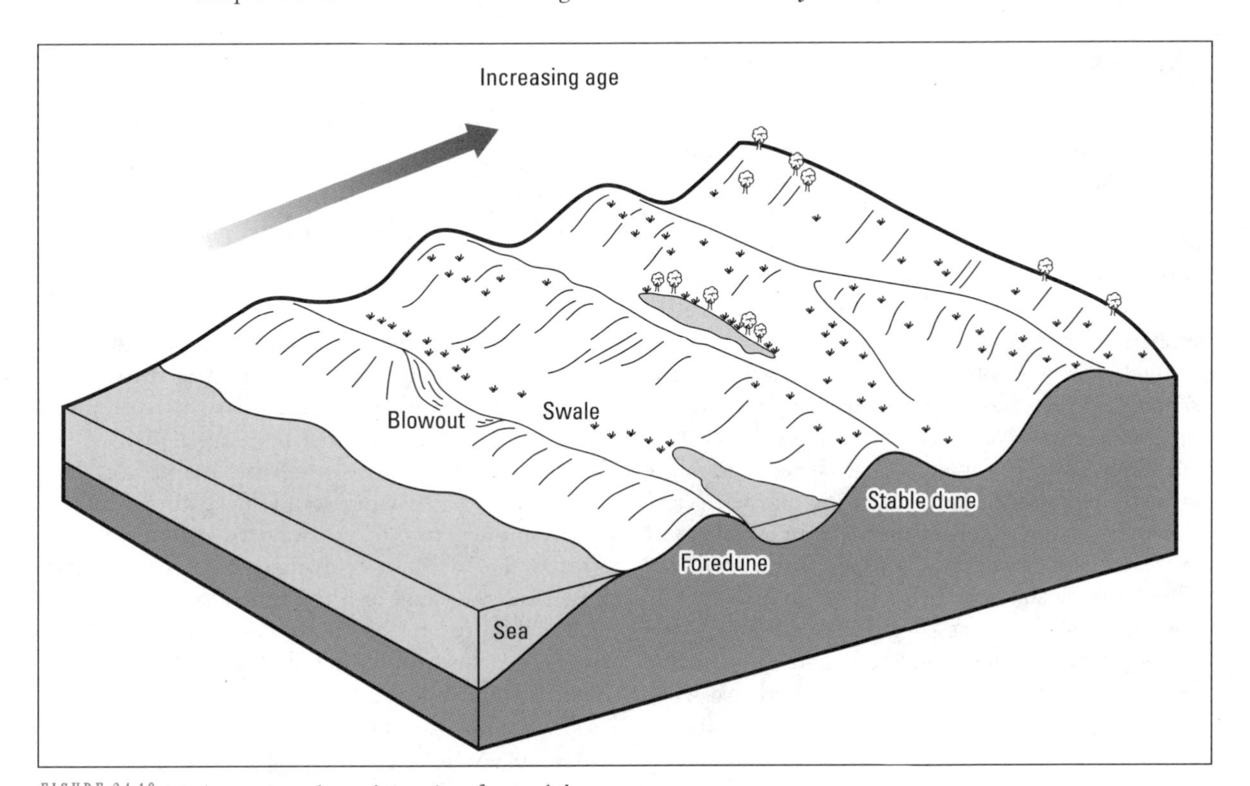


FIGURE 24.19 ▼ A sequence through a series of coastal dunes

More drastic and localized erosion may occur where the vegetation is destroyed. These **blowouts** (Figure 24.21) often result from overgrazing by rabbits or sheep, or trampling by humans. The material is carried inland and deposited as a longitudinal or barchan-type dune or on existing transverse dunes.

FIGURE 24.21 ▼ A blowout in sand dunes near Southport, England. The destruction of the dunes started with public use of the area, which destroyed the deeply-rooted marram grass (photo: P.G. Johnson)

## LANDFORMS OF POLAR DESERTS

Polar deserts are characterized by extensive exposed surfaces with little or no vegetation. Combined with these conditions, however, there exists a climate dominated by low temperatures and prolonged freezing. Physical weathering of the rocks by freeze–thaw processes produces sands that may be transported by winds, but in addition the debris washed from icecaps and glaciers contains fine-grained sand and silt ground up beneath the ice. When this material is deposited with the coarser material as extensive sandar the surfaces are left bare and exposed to the wind. Wind speeds are high, with strong katabatic winds blowing off the icecaps. Moreover, at low temperatures the air is relatively dense so that it exerts a greater fluid drag; the velocity necessary to move a sand grain 2 mm in diameter is about 5 m $s^{-1}$ less in the Antarctic than in a subtropical desert. Thus, sand dunes often develop at the margins of the outwash plains, while the finer material, mainly silt, is transported farther afield.

These wind-blown silts are often deposited in the extensive sheets of loess previously discussed. During the Quaternary, when ice extended into what are now the cool temperate zones, widespread loess deposits were formed. These include silts up to 30 m deep in some parts of North America and northern Europe. In addition, thinner loess deposits were laid down over even wider areas, and these have become incorporated into the soils to give an additional silty component. In some cases, the direction of the winds bearing the loess can be deciphered, for the silts become finer downwind.

# Aeolian processes and humans

## SOIL EROSION BY WIND

> Now the wind grew strong and hard and it worked at the rain crust in the cornfields. Little by little the sky was darkened by the mixing dust, and the wind felt over the earth, loosened the dust, and carried it away. . . . The wind grew stronger, whisked under stones, carried up straws and old leaves, and even little clods, marking its course as it sailed across the fields. The air and the sky darkened and through them the sun shone redly. . . .

So John Steinbeck, in *The Grapes of Wrath*, describes the beginning of a dust storm in Oklahoma. After the storm, the dust settles:

> All day the dust sifted down from the sky, and the next day it sifted down. An even blanket covered the earth. It settled on the corn, piled up on the tops of the fence posts, piled up on the wires; it settled on roofs, blanketed the weeds and trees.

Such were conditions time and time again across much of the North American prairie during the 1930s. So extensive was the erosion, so widespread the damage, that the area became known as the Dust Bowl. It provided a terrible example of how the activity of the wind could affect human life. It is a reminder, too, of the way we encourage wind erosion. For while the ultimate cause of the Dust Bowl was the wind and the inherent susceptibility of the soils to erosion, it was human activity that triggered the problem. For decades, people had been ploughing the fragile land, removing the crops, leaving the soil bare and unprotected to the wind. In the 1930s, following several years of dry weather, the parched soil started to erode. Crops were ripped out, seeds and fertilizer blown away, the fine fertile clays and silts lost. As it moved across the fields, the dust lacerated the seedlings; as it landed, it filled ditches, blocked railroad tracks, and buried other crops.

The devastation of the Dust Bowl caused much suffering, but it was a lesson well-learned. Since then, soil scientists and agriculturalists have tried to understand more clearly what controls soil erosion by wind. Much of this work has been carried out by W. Chepil. He and his colleagues have analyzed results from hundreds of experiments, both in the field and in wind tunnels. Based on this research, they have devised the so-called wind erosion equation showing the relationship between wind erosion and environmental and management factors. The equation states:

$$E = f(c, i, l, k, v, p)$$

That is, the amount of erosion $E$ is a function of climate $c$, soil erodibility $i$, the exposed length of the field $l$, the roughness of the surface $k$, the vegetation cover $v$, and the management practice $p$.

Each of these factors can be defined and analyzed. The climate factor, for example, is dependent on wind speed and soil moisture. Soil erodibility is dependent upon the proportion of the soil made up of stable aggregates greater than 0.84 mm in diameter. The equation is used to assess the amount of erosion and to help define ways of controlling soil loss. For example, the agriculturalist can calculate the effect of changing the vegetation cover (e.g., by growing a different crop), or reducing the length of field exposed to the wind (by planting hedges and windbreaks), or altering the surface roughness by tillage. Based on this information the farmer can learn the best way of avoiding wind erosion. Even with our knowledge of the causes of this erosion, the continuing periods of drought that affect parts of the North American prairie result in some soil erosion each year.

## DESERTIFICATION

The experience of the Dust Bowl has focused attention upon the sensitivity of semi-arid areas and raised concern about longterm changes in these areas. In recent years a more general question has been asked: Are the deserts advancing? Some observers certainly think so; one reported in 1975 that the desert in the Sudan had advanced 90 to 100 km south during the preceding 17 years. But the rate and extent of such advances are difficult to measure. There are numerous reasons for this. The first is that desert advance is often a slow and subtle process. It rarely occurs in the form of a clearly advancing dune, but as a gradual and irregular deterioration in vegetation. The second is that from year to year marked changes take place in response to variation in rainfall, disguising the general trend. Third, desertification does not always take place at the desert margins. Often it affects sensitive areas hundreds of kilometres from the present deserts.

What causes desertification? In part, perhaps the process results from climatic change. More prolonged, more frequent, or more intense periods of drought may prevent the vegetation reestablishing itself, so that the land becomes bare and susceptible to erosion. In addition, human activity undoubtedly plays a vital part, as it did in the Dust Bowl. Overgrazing, especially during dry years, reduces

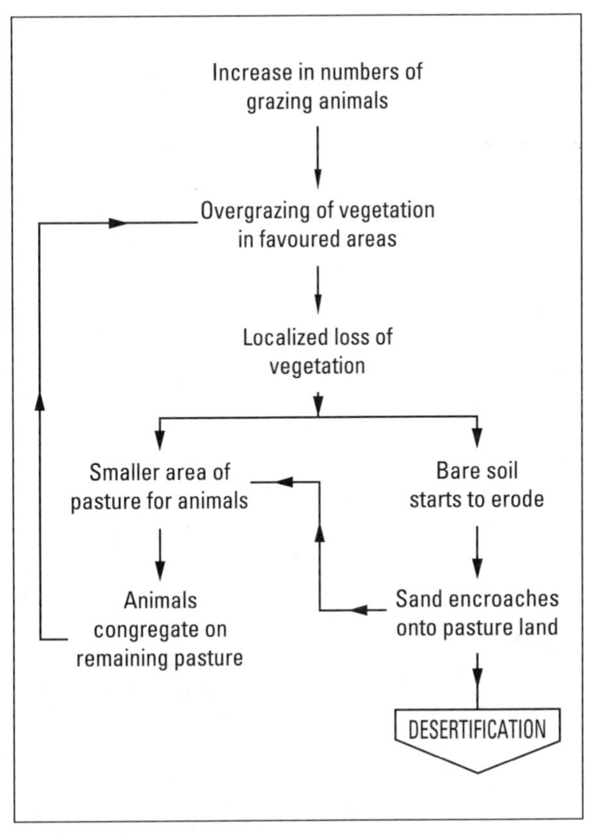


FIGURE 24.22 ▼ The effects of increasing grazing intensity on desertification

the vegetation cover. Thus a series of positive feedback effects occur that lead to desertification. As the vegetation disappears the animals become concentrated on the surviving pasture; this too is overgrazed. As the soil becomes exposed, it is subject to wind erosion; consequently the vegetation cannot regenerate easily. The process speeds up and, in the end, land that was once pasture becomes desert (Figure 24.22). Over time the degraded patches link up to carry the process over extended areas. The present problems in the Sahel area of Africa demonstrate this point vividly.

# Lacustrine, coastal, and marine systems

## Water at the earth's surface

Water covers 70 percent of the earth's surface. The overwhelming proportion of this area is ocean; only a fraction—less than 0.02 percent by volume—is made up of freshwater lakes and saline inland seas.

All of the Arctic Ocean, large areas of the southern oceans around Antarctica, and most high-latitude lakes are also affected by semi-permanent or seasonal ice cover.

As we saw in Chapter 14, movement of water through the ocean is one of the main processes by which energy is transferred from one part of the globe to another. In addition, the oceans are important components of the hydrological cycle, acting as huge stores for water running off the land and as the major sources for evaporation and the return of water to the atmosphere. They are also vital parts of the landscape system, for much of the debris washed, blown, or scraped from the continents is carried into the oceans. At the margins of the land, waves and ocean currents and ice attack the rocks, eroding and depositing debris along the coastline. Ocean currents transport the material over long distances, along the coast and out to sea. In the still, dark depths of the oceans, the debris slowly accumulates. It is trapped there, with no ready escape. It lies there until plate movements and continental collisions force the ocean floor into mountain chains, or subsume the material into the mantle.

# Lacustrine systems

## LAKES AS MODELS OF OCEAN SYSTEMS

The oceans are clearly major features of the earth's surface, but we know remarkably little about them. The reason for our ignorance is simple; we can see the struggle between land and sea at the shoreline, measure the rates of cliff erosion or beach formation, and trace the movement of sediment along the coast, but in the mid-ocean areas geographic study is almost impossible. The waters are too impenetrably deep to be observed directly by divers; all we can do is to take observations with instruments lowered into the ocean, and extract samples of the sediment on the ocean floor with boring equipment mounted on ships. Otherwise we are dependent upon the clues we can pick up from geological studies of ancient sea floor sediments now preserved in the rocks of the continents.

It would be useful if we could gain at least a general impression of ocean systems by studying a simpler, more manageable environment—a natural

FIGURE 25.1 ▼ Examples of lakes. (A) A proglacial lake at the terminus of a plateau glacier; the lake formed during retreat of the glacier from its Neoglacial maximum position. Airdrop Lake in the St. Elias Mountains, Yukon. (B) The Gladstone Lakes in the Ruby Range, Yukon. The lakes are in a deep, glacially eroded valley, and are separated by large alluvial fans coalescing across the valley floor. (C) Ice-covered Lac Ste-Marie in the Canadian Shield north of Hull, Quebec. (D) Lakes in the glacially scoured Canadian Shield north of Lake Superior, Ontario (photos: P.G. Johnson)

model, as it were, of the oceans. And indeed we can, for many of the processes operating within the vast oceans take place in smaller scale in lakes. The analogy is by no means perfect; their mode of formation is different; the magnitude of the forces acting within lakes is much less than in oceans, yet the sediments may be of a similar size. Nevertheless, they form a useful introduction to ocean systems, and they provide a model that can be studied almost everywhere.

## THE FORMATION OF LAKES

The word 'lake' conjures up different images to different people: the vast water bodies of the Great Lakes or Lake Victoria, the mountain cirques and pools found in glaciated uplands, the spectacular caldera-rimmed lakes such as the Lac d'Issarlès in France. As this indicates, lakes are formed by a variety of processes: by subsidence of the land, volcanic action, warping and folding of the surface, differential erosion, or by damming of river valleys with sediment. Lakes formed by different processes are shown in Figures 25.1 and 25.2.

## INPUTS AND OUTPUTS OF LACUSTRINE SYSTEMS

Rivers draining the surrounding land provide the main inputs to lakes; they supply water, sediment, and dissolved substances. Rainfall, too, introduces material, not only water but also dust, gases, and solutes washed from the atmosphere. In addition small quantities of water may seep into the lake from the surrounding rocks, or bubble up through springs in the lake floor.

Losses occur mainly through drainage and evaporation. Not all lakes have outlet streams, but all experience loss of water through evaporation to the atmosphere. Small quantities of water and dissolved substances may also escape by seepage into the bedrocks (Figure 25.3).

It is clear that, in general, a rough balance must exist between inputs of water to the lake and outputs from it. However, this balance is not always maintained, and for quite long periods of time an imbalance may exist. This leads to changes in the storage within the lake; the amount of water held in the lake changes and the lake level rises or falls. When inputs exceed outputs, the lake level rises; when outputs exceed inputs the lake level falls.

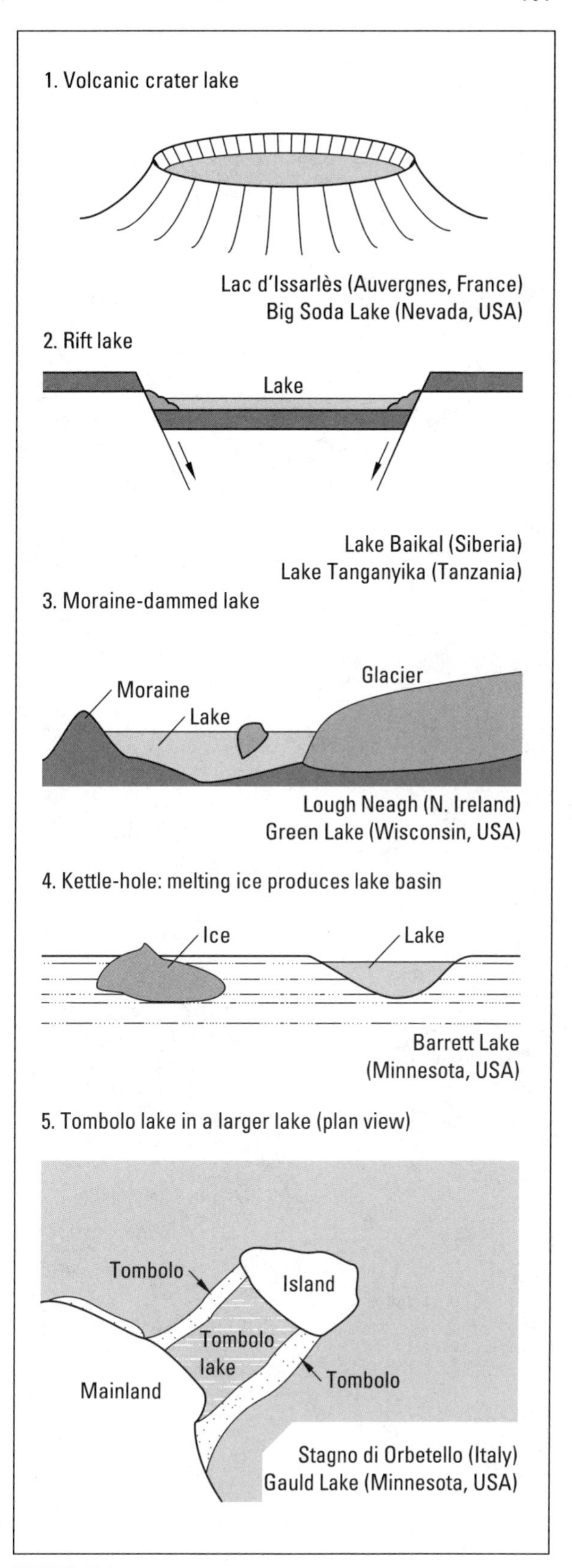


FIGURE 25.2 ▼ Common types of lakes

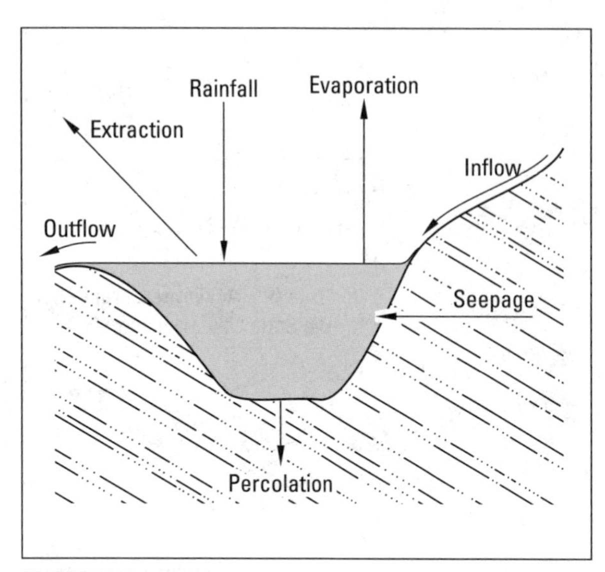


FIGURE 25.3 ▼ The water budget of a lake

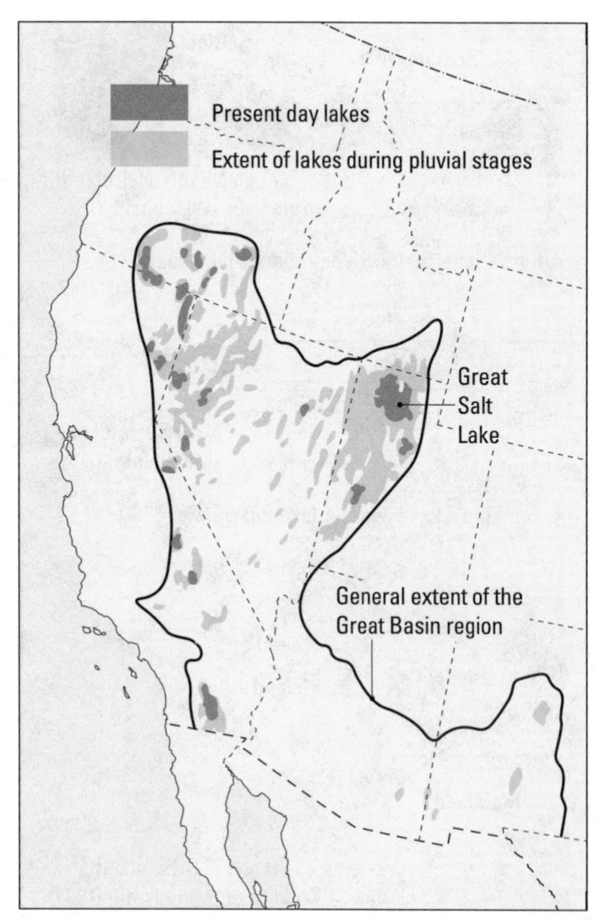


FIGURE 25.4 ▼ Changes in the extent of the lakes of the Great Basin region, USA (from Thornbury, 1954)

This has happened on a large scale. In many parts of the southern USA and Africa, for example, the Quaternary Period saw dramatic fluctuations in lake levels due to variations in climatic conditions. At times, areas that are now desert or semi-desert were much more humid, for disruption of atmospheric circulation by changes in the extent of the polar ice sheets brought rain-bearing winds to these regions. Rainfall increased, and the small desert playas were transformed into vast water-bodies (Figure 25.4). These periods were known as **pluvials** (from the Latin *pluvius*: rain). In between times, in the inter-pluvials, arid conditions returned; losses by evaporation may have increased while inputs from rainfall fell. The lakes shrank or disappeared.

Disparities between inputs and outputs also occur in the much shorter term, due to yearly or seasonal variations in rainfall and evaporation. In northern latitudes, dominated by cold winters with snowfall, the lakes may be low throughout the winter with no input from the precipitation remaining stored on the ground. In spring there is a great influx of water from snowmelt and subsequent

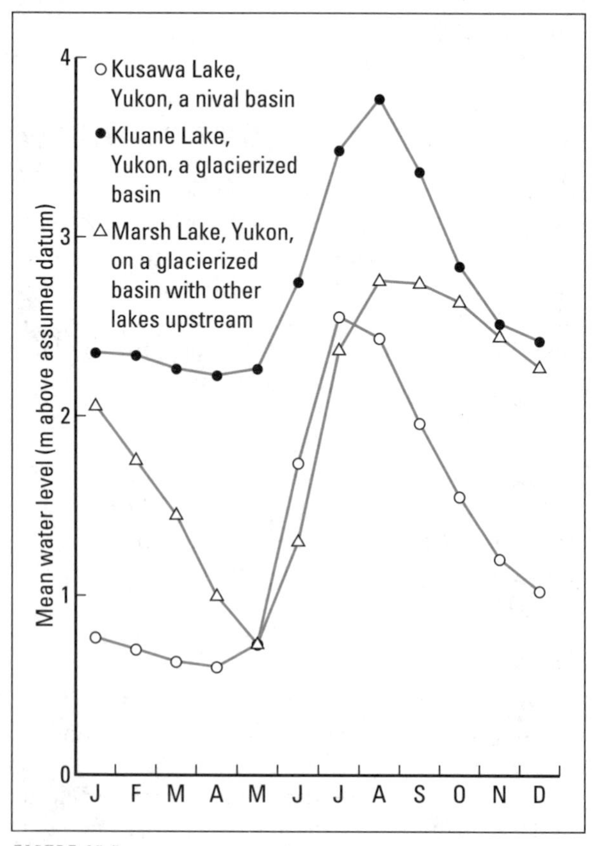


FIGURE 25.5 ▼ Regimes of northern lakes; Yukon Territory (Source Inland Waters Directorate)

decrease in water levels through the rest of the summer. Glaciers in the basin will maintain high input throughout the summer, with lake level drop in the fall (Figure 25.5). In many temperate regions, there is a tendency for lake levels to fall during the summer months, when rainfall declines and evaporation increases. Some lakes even vanish completely, for the water table, too, may fall, so that the lake drains away into the underlying, permeable bedrock.

Short-term climatic oscillations also produce variation in lake levels. The level of Lake Victoria in East Africa, for example, fluctuates in close harmony with rainfall inputs. For a long time these variations coincided with sunspot activity, and it was postulated that this controlled rainfall in the area, and therefore lake level. But more recent studies have shown that the fluctuations are in fact much more complex, although still related to rainfall (Figure 25.6). The Great Salt Lake in Utah (the Ice Age parent of this lake is called Lake Bonneville, Figure 25.4), and Lake Eyre in Australia, show similar patterns of variation (Figure 25.7).

The regulation of lakes by dams, together with natural variation, can produce a number of trends. The Great Lakes system of North America between 1952 and 1973 went through a period of regulated lowering of the lake levels, which was only reversed because of adverse effects on shorelines (Figure 25.8). Lake Superior is not regulated to any extent and shows only minor longterm trends.

*Sedimentation in lakes* The waters flowing into lakes often carry vast amounts of debris swept from the surrounding land. As long as the water is flowing rapidly, this material is carried in suspension, but as the waters enter the lake their velocity declines, the sediments start to settle out, and material accumulates on the lake floor.

Sedimentation produces a variety of features. At the mouth of the inlet stream, **deltas** develop as coarser debris is deposited. As we saw when considering fluvial processes, the material builds up as dipping layers of sediment, each new flood tending to erode debris from the top of the delta and to carry it for deposit at the delta front. The rate of accumulation may be rapid. The Rhine is estimated to add 2 790 000 $m^3$ of sediment each year to the Lake Constance delta.

In glacierized basins where glacier discharge carries very high suspended sediment loads, delta formation can also be very rapid. The Slims River that drains from the Kaskawulsh Glacier deposits in excess of $210 \times 10^6$ kg of silt per day in the peak discharge season.

Farther away from the stream mouth, the finer materials collect: clays and silts which sink slowly to the bottom of the lake and produce layer upon layer of new sediment. Often, where seasonal variations in the input of material occur, these form laminated clays or **rythmites**. As we saw in Chapter 23 the seasonal rythmites that form in proglacial lakes are called **varves**. Each spring, as the snow and ice melt, the suspended sediment load consists of clays, silts, and fine sands. Due to turbulence in the lake caused by the input and by wind action across the surface, only the coarser fractions may settle to the lake bottom in the summer. In winter, with very low input of water and sediment and the reduction of turbulence due to ice formation on the lake, the finer sediment settles out of suspension.

With high suspended sediment concentrations the density of the input is frequently sufficiently high to cause an underflow where all of the sediment

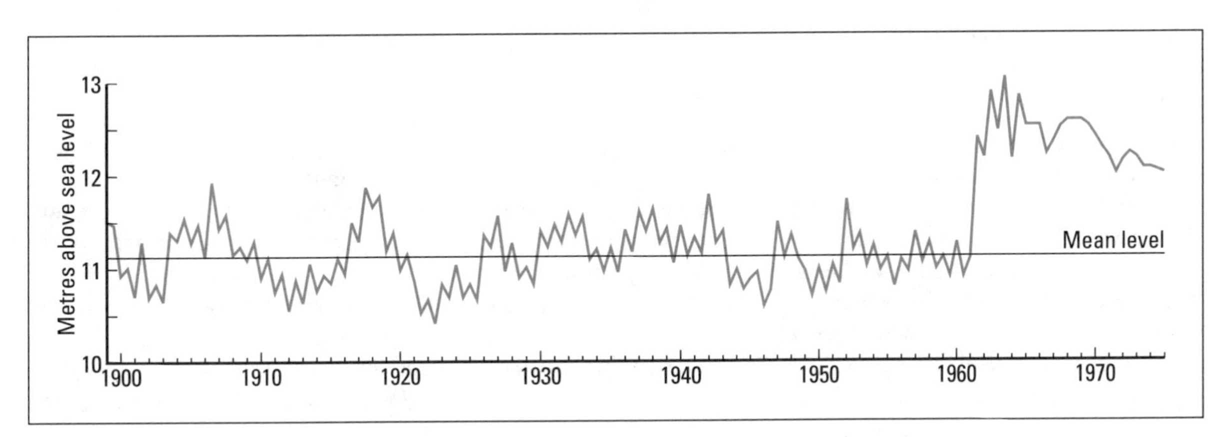


FIGURE 25.6 ▼ Variations in the level of Lake Victoria, Africa, 1899–1975 (from Lamb, 1977)

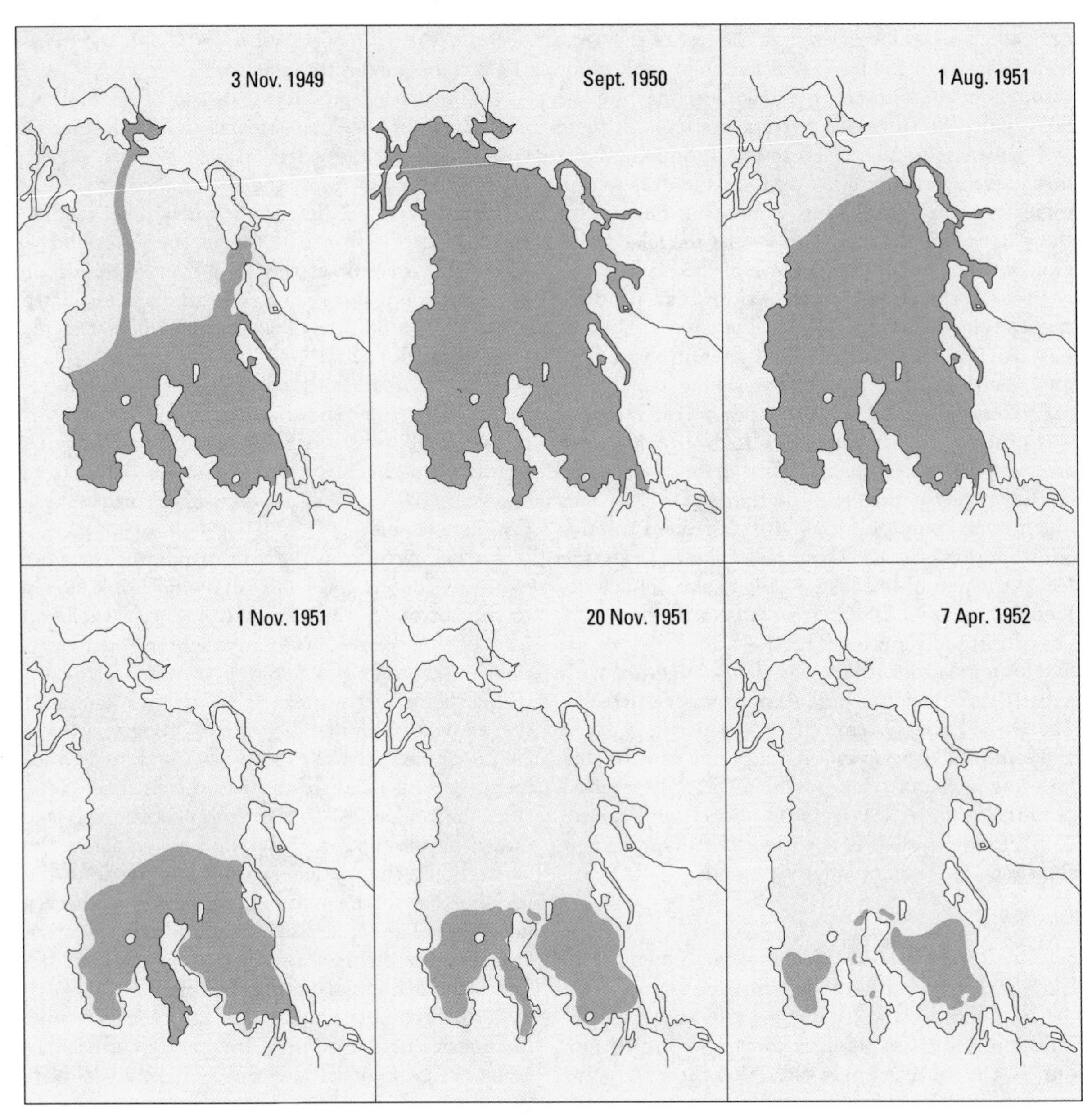


FIGURE 25.7 ▼ Variations in the extent of Lake Eyre, Australia, 1949–52 (from Bonython and Mason, 1953)

is transported along the lake bottom. Each year, therefore, a pair of laminae are deposited, one coarse and one fine. Over the years a rhythmic sequence of these varves is formed. So regular are these sequences that varved clays and silts can be used to date the age of the lake, simply by counting the number of varves in the sediments. Moreover, changes in the thickness and particle size of the varves can give clues to the climate at the time of deposition. Rather like tree rings, these sediments preserve a record of annual changes in the environment.

Coarser material deposited at the lake shore and material eroded along the lake shore is frequently reworked and redistributed by waves and by ice. It accumulates at the lake margin as **beaches**, and it is carried down the shoreline by currents and deposited as **spits** and **forelands** (Figure 25.9). At the same time, the waves and currents attack the

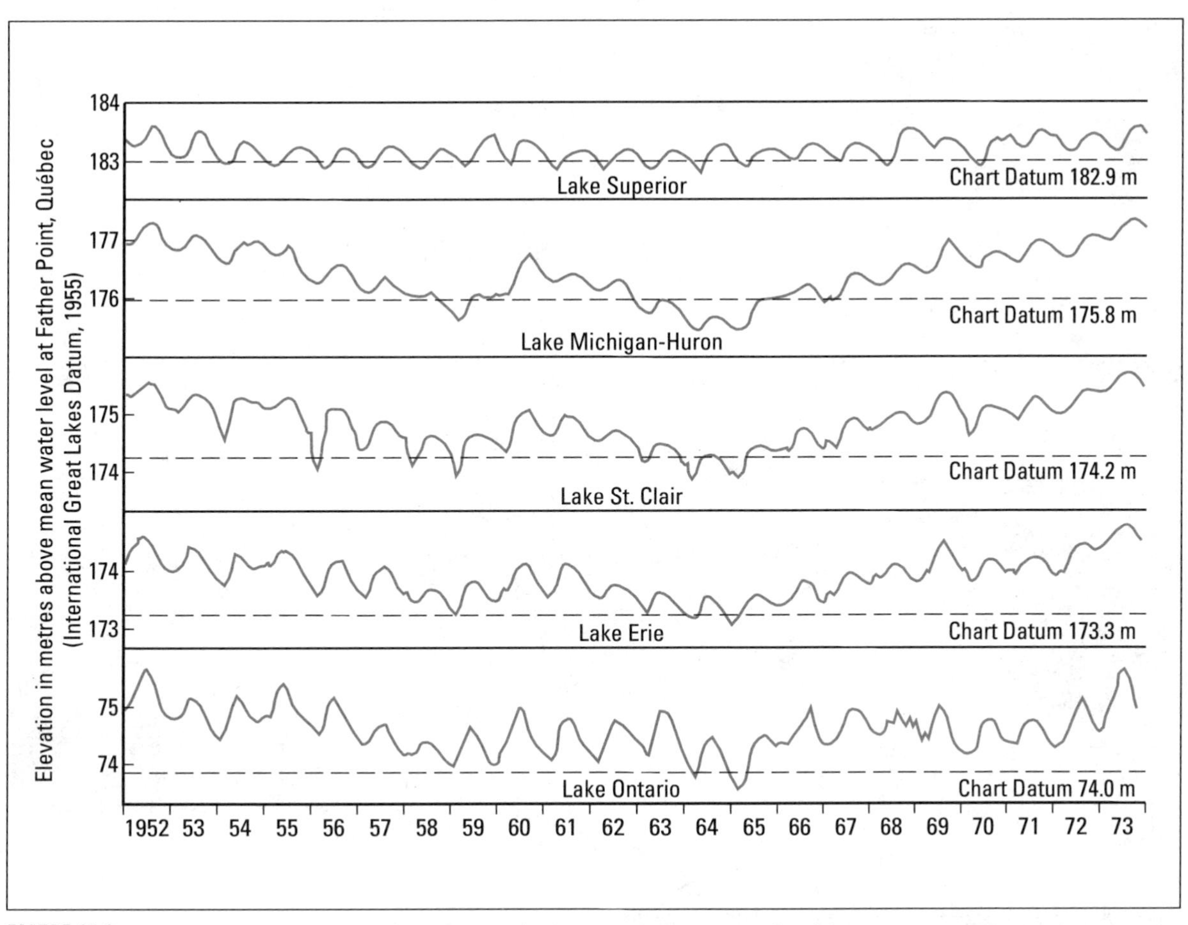


FIGURE 25.8 ▼ Regulated water levels of the Great Lakes system, showing annual and longer-term trends

shoreline, dislodging sediment and cutting **cliffs** and **wave-cut notches** in the more resistant rocks.

The seasonal ice cover that forms over lakes in high-latitude winters has an important role in erosion and sedimentation processes. The development of a total ice cover prevents wave activity on the lake, eliminating the wave erosion and transport processes for much of the year. We have already seen that the very low turbulence under ice cover allows the finer suspended sediment particles to settle. In spring, during breakup, the ice can be rafted onto the shoreline, producing large amounts of impact erosion. On smaller lakes the change in volume of ice with temperature causes erosion. The ice forms a complete cover of the lake in early winter and as the temperatures decline in midwinter the ice contracts. Cracks in the ice are filled from below and freeze. In the spring, with warmer temperatures, the ice expands and pushes up the shoreline, causing erosion. Rocks frozen into the ice may also scour the lake bed as the ice expands (Figure 25.10).

We mentioned earlier that a rough balance tends to exist between inputs and outputs of water in the lake system. But what about the sediment? What happens to this material?

The answer is that most of it remains trapped within the lake. Unlike the water it has no escape. Over time it accumulates in the lake; the storage of sediment continually increases. We can see the process happening in our local pond or pool; we know that it happens in reservoirs. Eventually, the sediment builds up, extending from the margins, reducing the capacity of the lake until it is nothing but a boggy, muddy depression. As it does so, as land is produced where once there was water, vegetation becomes established, and the old lake becomes no more than a memory. In the once glaciated lowlands of Canada and the United States

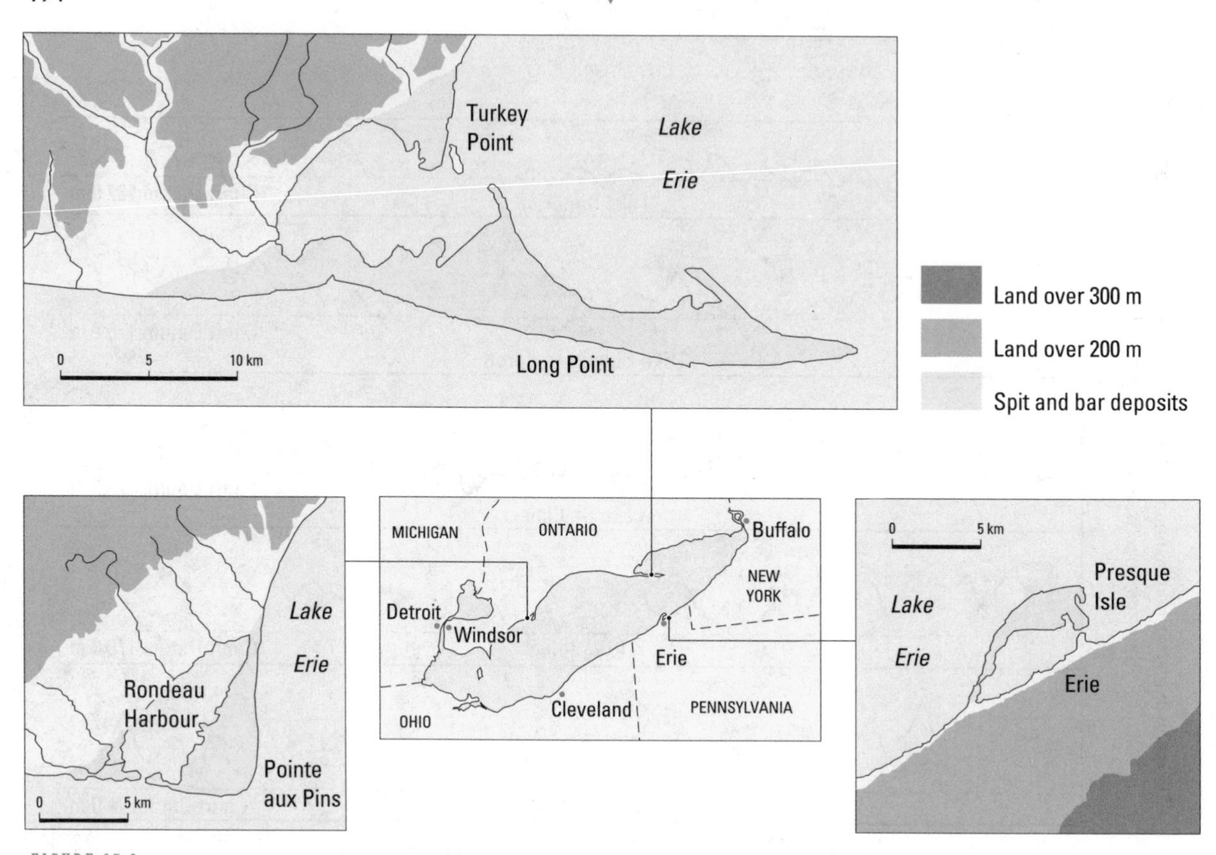


FIGURE 25.9 ▼ Lacustrine spits and bars, Lake Erie, Canada and USA

FIGURE 25.10 ▼ Boulders, frozen into lake ice during freeze-up, scour the shoreline when ice expands in the spring. Bruce Peninsula, Ontario (photo: P.G. Johnson)

the evidence of these processes is common. Old kettle holes, produced by melting of stagnant ice during the glacial period, were once lakes. In the centuries since they were formed they have been slowly filled by sediment and peat growth, and invaded by vegetation.

# The world's oceans

We have a picture, now, of the processes operating within lakes. The broad structure of the lacustrine system can be represented as in Figure 25.11. To what extent does that picture provide a model of marine systems? Can we now apply it to that 70 percent of the earth's surface made up of ocean?

Clearly the scale is rather different. Lake Superior is a mere 83 300 km$^2$ in area; the huge inland sea of the Caspian 436 400 km$^2$. The Pacific, on the other hand, is an estimated 165 384 000 km$^2$, and the Atlantic about half that size (Table 25.1). The depths of the oceans, too, are orders of magnitude greater. The deepest known lake, Lake Baikal in

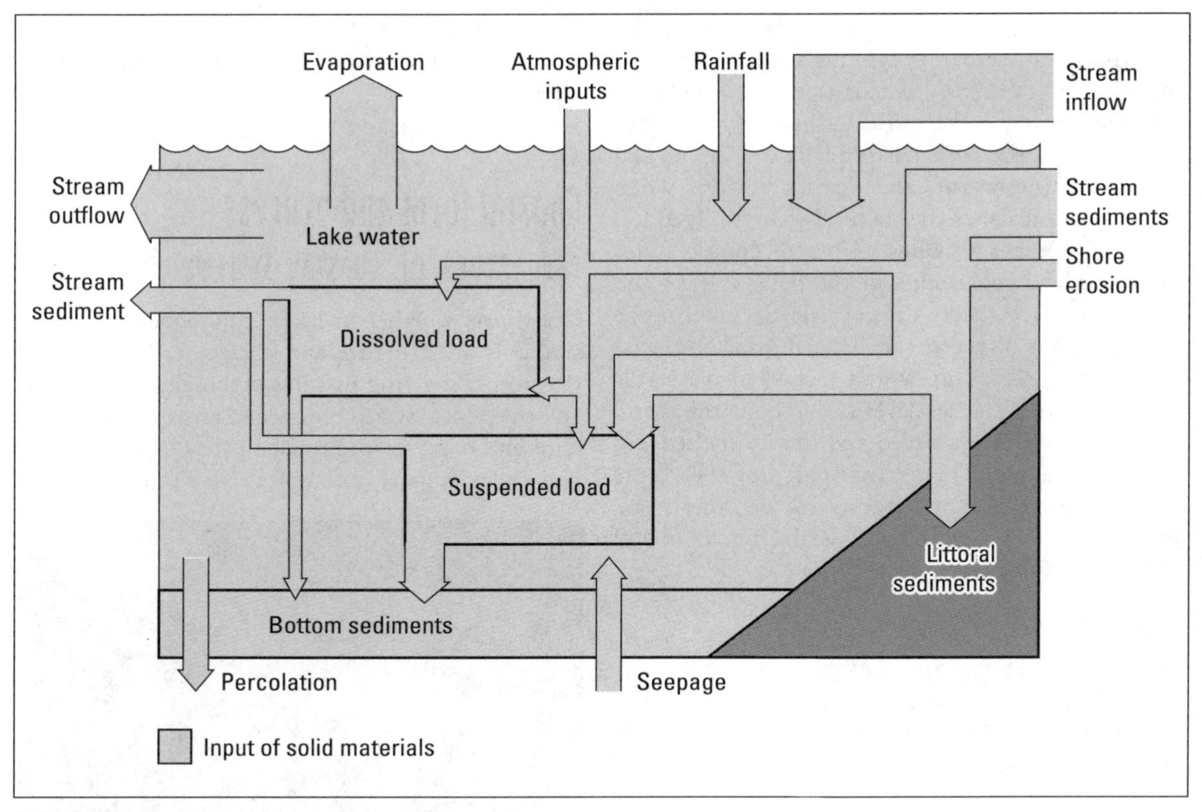


FIGURE 25.11 ▼ A model of the lacustrine system

Russia, reaches a depth of about 1740 m; the Pacific approaches depths some six times that value (Table 25.1). Different processes also operate within the world's oceans. As we have seen, crustal formation is occurring in the mid-ocean ridges, as magma wells to the surface and as the ocean floor spreads as a result of plate movement. Simultaneously, in the deep marine trenches, the crust is subsiding and the lithospheric plates are being consumed. The interconnections between the main oceans mean that major transfers of water occur between them. It is also inconceivable, of course, that oceans should suffer infilling, as do the smaller lakes. There are limits to the validity of our model.

## Table 25.1

*The size of the world's lakes and oceans*

| *Name* | *Origin* | *Area ($km^2$)* | *Maximum depth (m)* |
|---|---|---|---|
| Pacific Ocean | – | 165 384 000 | 11 730 |
| Atlantic Ocean | – | 81 484 800 | 9 225 |
| Indian Ocean | – | 75 110 400 | 7 450 |
| Caspian Sea | tectonic | 436 400 | 946 |
| Lake Superior | glacial erosion | 83 300 | 307 |
| Lake Victoria | tectonic | 68 800 | 79 |
| Lake Baikal | tectonic | 31 500 | 1 740 |
| Crater Lake, Oregon | volcanic | 55 | 608 |

Nevertheless, the general principles remain valid. Precipitation, streamflow, and seepage provide inputs in the same way; evaporation and percolation into the crustal rocks cause losses of water. Over time, as well, the same balance between inputs and outputs is important, and during periods when there is an imbalance the sea level changes. It did so in the Quaternary, when climatic changes and expansion and contraction of the polar icecaps and continental ice sheets caused fluctuations in the water budget of the oceans. During cold, glacial periods the icecaps grew and locked up vast volumes of water. The sea level throughout the world fell. As the climate warmed and the ice melted, sea levels rose again. These variations, however, were superimposed upon a longer-term decline in sea level related to a gradual increase in the volume of the ocean basins.

### THE STRUCTURE OF THE OCEANS

The oceans can be subdivided into three main zones (see Figure 14.2). In the open ocean areas, we find the ocean basins. These are many kilometres deep—much of them between 3 and 6 km—and they are underlain by crustal material that is both thinner and younger than that beneath the continents. In these areas the processes of sediment accumulation are slow and poorly understood. Over the last 2 to 3 million years, depths of no more than a few metres of sediment have built up in many places. The main exceptions are at the margins of these zones and the adjacent continental shelves, where vast accumulations have occurred. These zones act as the final resting place for much of the sediment swept off the continents.

The **continental shelves** are in most areas only a few kilometres wide; the mean width is about 75 km. They are also relatively shallow. On average they are about 130 m deep at their outer margins. They occupy about 8 percent of the ocean area. Most of the sediment carried into the ocean collects initially on the continental shelves; some remains there almost indefinitely, until continental collisions and crumpling force it up into mountain chains; but a proportion is stored there before being carried over the **continental slope** and into the ocean basins.

The continental shelves fringe the continents, and it is on the inner margins of the shelf that most activity probably occurs. Here are the main dumping grounds for the streams that enter the sea; here the oceans and the land interact. And it is in the coastal zone that we can see landform processes at work.

## Coastal form and process

### THE NATURE OF COASTAL SYSTEMS

Coastlines, whether in lakes or in oceans, vary considerably in structure and character. They include the steeply shelving margins of mountainous regions and the gentle sweep of lowland areas. They include cliffs and bays cut into resistant rocks (Figure 25.12) and the extensive salt marsh and mud flats that

FIGURE 25.12 ▼ A hard-rock coastline, Bottle Cove, west coast of Newfoundland (photo: P.G. Johnson)

FIGURE 25.13 ▼ Cole Harbour salt marsh, Nova Scotia (photo: P.G. Johnson)

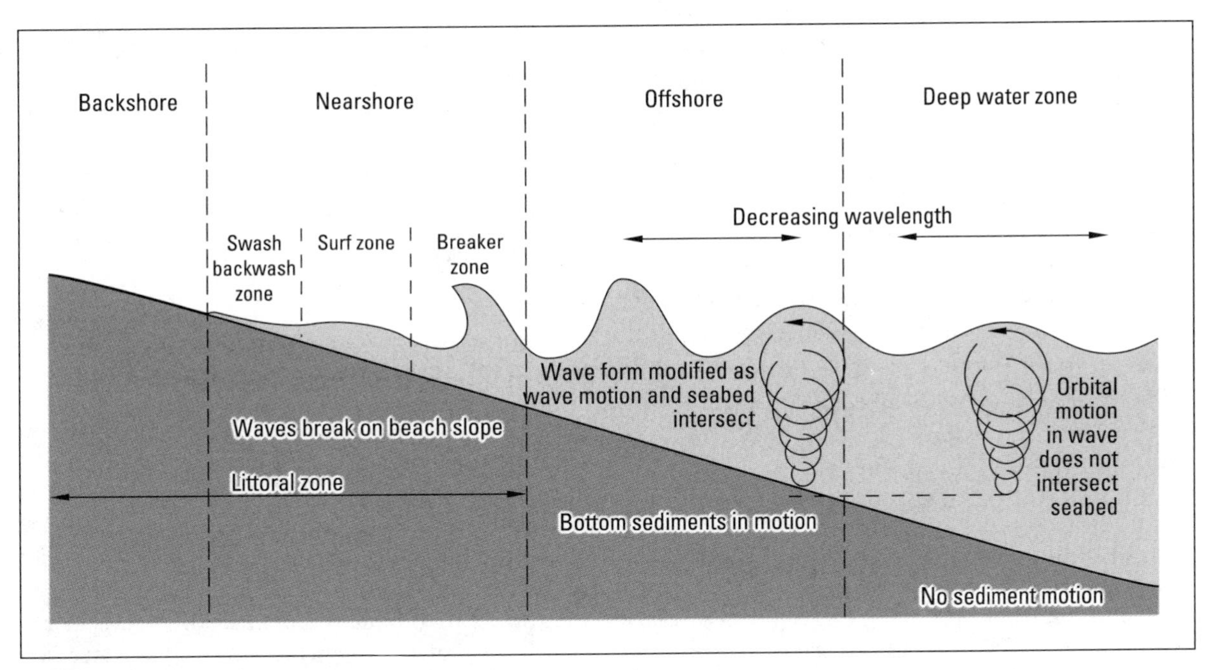


FIGURE 25.14 ▼ Main zones of coastal areas

characterize coastlines such as the Atlantic coast of Nova Scotia (Figure 25.13). Generalizations about the structure of coastal systems must therefore be made with caution.

In many coastal areas, however, it is possible to identify four main zones (Figure 25.14). At some distance from the coastline there occurs a **deep-water zone**, where the action of the waves does not reach to the sea floor. This gives way landward to what is often called the **offshore zone**—the region in which the waves start to interact with the bottom. Within this zone a mutual reaction occurs; the waves affect the sea floor, which in turn modifies the waves. Eventually, the waves start to break and enter the **nearshore zone**. Above the level of normal wave activity is the **backshore zone**.

## PROCESSES IN THE DEEP-WATER ZONE

The coastal areas of the world are dominated by the progress of energy landward from the open sea. It is this energy that transports material to and along the coastline, and it is this energy that attacks the land and erodes the coastal rocks. It is within the deep-water zone that this energy gathers.

The energy is derived mainly from the waves, which are generated by wind. The friction and pressure of the moving atmosphere set up an oscillation within the water, by which individual water particles move in a circular manner. It is important to appreciate that the water particles themselves do not travel forward with the wave. The apparent motion of the wave is produced by transmission of energy through the water, from one particle to another (Figure 25.15). The orbital movement of

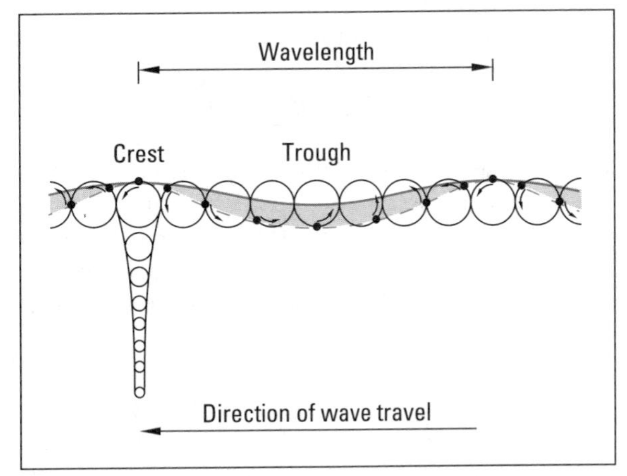


FIGURE 25.15 ▼ Water movement in an ocean wave. Each circle represents the orbit of individual water particles. Note that the diameter of the orbit declines with depth, such that at a depth equivalent to half the wavelength the diameter of orbit is only 4 percent of that at the surface

the water declines with depth, the diameter of the orbit being halved for every increase in depth of about 0.11 L (where L is the wavelength).

The character of the waves can be described by their wavelength, height, and rate of progression. The **wavelength** is largely a function of the wind speed and the **fetch** (uninterrupted travel distance) of the waves. **Wave height** is also controlled by wind speed and fetch, as well as the gradient of the water surface. This develops because of the piling up of water in the deep-water zone by tidal currents. In general, wave height gives a good indication of the wave energy, the two being related as follows:

$$e \text{ varies as } h^2$$

where $e$ is the energy and $h$ the mean height of the waves.

The oscillation of a water particle takes place within a single wave. It rises to the wave **crest** and sinks within the **trough** (Figure 25.15). The speed of the wave thus depends upon the rate at which this orbit occurs. In general, wave speed is closely and positively related to wavelength, although there is a tendency for higher waves to travel faster than low waves.

It is this wave motion that is responsible for the main energy transmission through the deep-water zone and into the offshore zone. There are, however, other processes operating within the deep-water zone that may supplement this energy. **Tidal currents** may be important, although their force is generally rather limited. In many cases they travel at little more than 0.2 m $s^{-1}$, and they affect only the surface of the ocean. Locally, however, more active tidal currents develop, particularly where they are constricted in some way. Off the New South Wales coast of Australia, a southerly current reaches 2.0 m $s^{-1}$, while tidal currents sweep through Hell's Gate in New York's East River at over 2.5 m $s^{-1}$, and currents of 8.3 m $s^{-1}$ have been recorded along the British Columbia coast.

Tidal currents may therefore act as a means of energy transmission into the offshore zone, and, where they are deep, they may disturb and entrain sediments on the floor of the deep-water zone. Certainly within the English Channel there is evidence of marked and consistent movements of sediment on the sea floor, well beyond the reach of surface waves.

The tides have an additional effect, however, for by piling up water in the coastal zone they extend the influence of coastal processes over a considerable height range. The **tidal range** is generally no more than a few metres, but where the tides are channelled into a narrow, exposed inlet, much greater tidal ranges are found. At the head of the Bay of Fundy, Nova Scotia, the tidal range reaches 15.4 m; in the Bristol Channel in England the tidal range is in the order of 12 m.

## PROCESSES IN THE OFFSHORE ZONE

Within the offshore zone, the interaction between sea floor and wave action takes effect. Significant disturbance of the floor only occurs when the depth of water is less than half the wavelength. At this point, the horizontal motion of the water particles at the bottom starts to be retarded by friction with the bed. This produces a marked difference in the velocity of water movement between the water surface and the sea floor that results in the wave breaking. The first signs of the wave breaking occur when the water is about twice the depth of the wave height; at this point the waves become asymmetrical and peaked. At a depth of about 1.3 times wave height, they break. At the same time, as the wave becomes increasingly influenced by the sea floor, the wavelength decreases, and increased disturbance of the bottom sediment occurs (Figure 25.14).

It is within this zone, therefore, that waves are able to entrain the bottom sediment and carry it landward. Only finer material is moved in this zone, because the energy available to carry the sediment is relatively low at the bottom of the water, and because only little turbulence occurs. Removal of the finer particles from the bottom deposits leaves a coarse-grained lag material.

## THE NEARSHORE ZONE

Breaking of the incoming wave represents the boundary between the offshore and nearshore zones. In the nearshore zone the oscillatory motion of the particles gives way to a forward movement, creating a **wave of translation**. This runs landward, expending its energy by friction with the rising slope of the bottom and by collision with the backwash from earlier waves.

The nearshore zone is often divided into a number of more specific areas related to the character of this water movement and its effects upon beach morphology. At its outer margins, where the waves break, is the **breaker zone** (Figure 25.14). This grades landward into the **surf zone**. Here, the action of the incoming waves is often accompanied by longshore currents, and by seaward movement of the water. Particularly important are **rip currents**. These are localized streams resulting from the buildup of water on the beach which flows back in the form of fast-flowing narrow currents. They may cut through the breaker zone and attain speeds of over 1 m $s^{-1}$. They represent a form of output from the landward zones of the coastal system by which water and sediment are returned to the open sea (Figure 25.16).

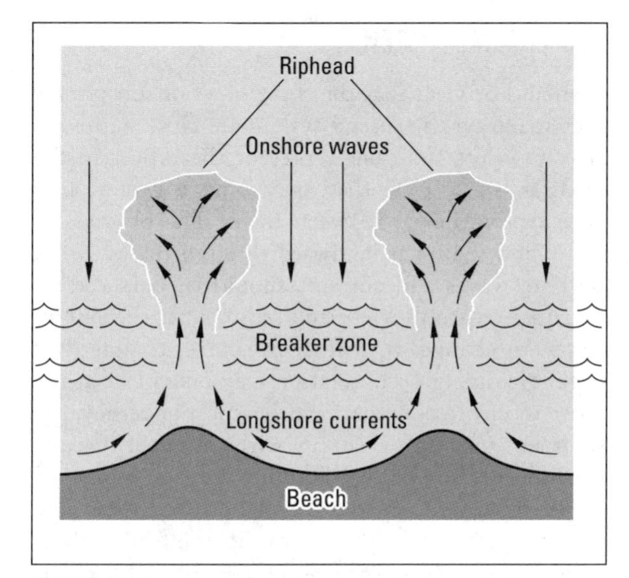


FIGURE 25.16 ▼ The development of rip currents (plan view)

Landward of the surf zone is the **swash zone**, within which the water runs up the beach as an increasingly shallow film of turbulent water. We only have to watch this process on the beach to see that, again, there is a reverse process in operation—the **backwash**. As the incoming wave loses energy in the swash zone the water starts to run back down the slope. Much of it percolates into the beach, but some flows back into the surf zone where it collides with the next incoming wave. This zone of collision is often known as the **transition zone**.

As ever, processes within the nearshore zone must be in some form of equilibrium, such that the rate of input of water to the beach is balanced by the return of water to the sea. As we have seen, backwash and the action of rip currents perform this function to some extent. But not all the water returns directly. Often there is a local buildup of water at the coast, which results in a lateral flow down-coast. In this way **longshore currents** are created. In addition, tidal currents may be channelled along the coast, particularly in the narrow zone landward of the breaker zone; here they may reach velocities of several metres per second. Longshore currents transport sediment in the offshore and nearshore zones that is called **longshore drift**.

One of the most common causes of longshore movement, however, is the oblique angle of approach of waves to the beach. Incoming waves are slowed by friction with the sea floor, and as a result they become progressively slower as the water becomes shallower. Where incoming waves approach the shore obliquely, this leads to bending of the waves parallel to the coast, called **wave refraction** (Figure 25.17). Even so, the waves still

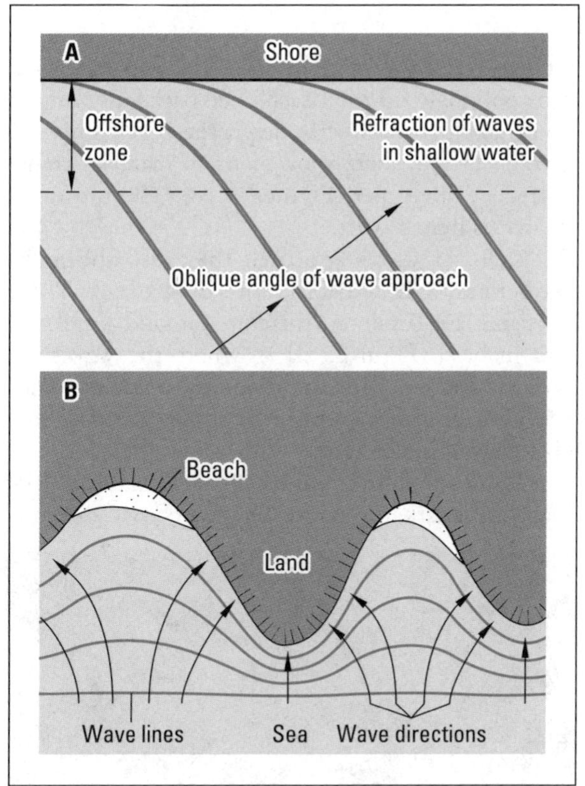


FIGURE 25.17 ▼ Wave refraction along a straight coastline (A). Wave refraction along a headland and bay coastline. The effect of refraction concentrates wave energy on the headlands and results in erosion of the cliff; in the bay, wave energy is less and deposition tends to occur (B)

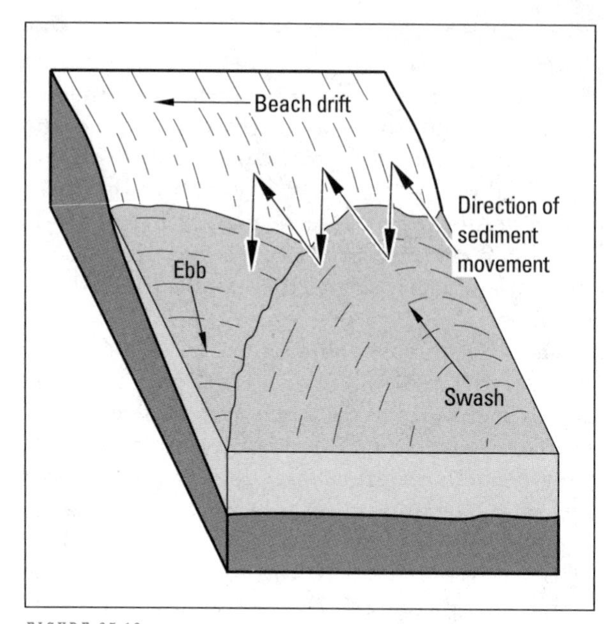


FIGURE 25.18 ▼ Sediment movement by beach drift

tend to reach the shore at an angle, and the swash runs obliquely up the beach. The backwash, on the other hand, runs directly down the beach slope, to give a net longshore movement to material transported by the water (Figure 25.18). The process is known as **beach drift**.

Wherever waves approach the coast obliquely, wave refraction and longshore drift occur. These processes are common, therefore, around headlands and in bays (Figure 25.19), where the waves are bent by friction with the changing coastline. They are also common where onshore winds blow obliquely onto the coast.

The movement of material by longshore drift may be remarkably rapid. On coasts in California, the effects of wave refraction give rise to longshore movements of material at rates of around 30 cm $s^{-1}$, and rates of 1.2 m $s^{-1}$ have been recorded. The pathway followed by these particles tends to be irregular, however, so that not all the movement represents longshore progress. Even so, considerable volumes of material may be moved annually by longshore drift: at Santa Barbara on the California coast, for example, over 200 000 $m^3$ of sediment may be transported in a single year; 40 million $m^3$ of sediment are estimated to move northward along the coast of the Netherlands each year.

FIGURE 25.19 ▼ Wave refraction patterns in a marine embayment near North Rustico, Prince Edward Island (photo: P.G. Johnson)

## THE COASTAL SYSTEM

It should be clear that the three zones or subsystems that make up the coastal system are closely interactive. Outputs from one represent the inputs to the next. It is also clear that, over time, a general balance tends to exist between them. The progress of water and sediment landward through the action of onshore waves and currents should be balanced by an equal return of water and sediment if the system is not to change. In reality, of course, change does occur. Sediment accumulation may take place at the coast to create aggradational beaches; or removal of sediment may exceed supply with the effect that the beach is eroded and the shoreline retreats. Moreover, as we have seen, movement of both sediment and water occurs not only in an onshore–offshore direction, but also along the shoreline. Longshore currents result in the circulation of water and sediment down-coast. It is the interaction of all these processes that determines the shape and the landforms of the coastline.

## ICE IN THE COASTAL ZONE

In most high-latitude regions, such as the Canadian Arctic, the presence of seasonal or semi-permanent sea ice plays a major role in the formation of the coastal zone. At the average sea water salinity of 35 ‰ (parts per thousand), ice will start to form at a temperature of –1.8°C. During the initial formation of the ice, salt water is trapped within the crystal structure. The presence of the small pockets of salt water makes the **first-year ice** relatively weak. During one winter a continuous layer of ice may develop over the Arctic Ocean. The movement

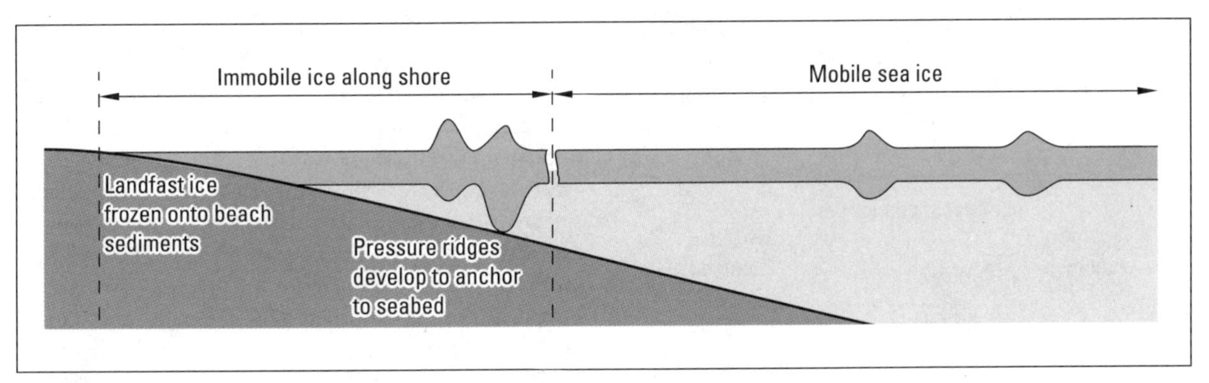


FIGURE 25.20 ▼ Development of ice in the coastal zone

caused by currents and wind breaks up the ice and causes collisions between parts of the ice sheet, called **ice floes**, and **pressure ridges** develop along the zones of collision. Along the shore, ice will form along the beaches. This **landfast ice** will only extend a short distance from the beach and may have a zone of large pressure ridges formed where the more mobile open water ice collides with it (Figure 25.20). The first-year ice will develop to about 2 m to 2.5 m thick. During the summer the ice will melt but some may survive, due to severe climatic conditions or movement into colder regions; the following winter it will continue to develop. In this period most of the salt water is lost from the ice and it becomes much stronger. Ice that lasts through one summer and continues to form is called **second-year ice**, and if it continues beyond two years it is called multi-year ice and may reach thicknesses of 4 m to 5 m.

In addition to the sea ice formed by the freezing of sea water, the high-latitude oceans are also affected by icebergs that calve off the large tidewater glaciers, particularly in Greenland and Antarctica. These icebergs range in size from the small **bergy bits** just a few metres across, to the massive **tabular icebergs** which may be many kilometres across and which are broken off the major ice shelves of Antarctica. In the Arctic, tabular icebergs from ice shelves such as the Ward Hunt Ice Shelf on the northern coast of Ellesmere Island are often caught in the currents of the Arctic Ocean. These are called **ice islands** and are significant in two ways, as a threat to oil exploration in the Beaufort Sea, and as the sites of major research camps that have collected climatological, meteorological, and oceanographic data while drifting around the Arctic.

The importance of ice in the coastal zone is in its role of protecting the shoreline from wave erosion during the winter, but also in promoting erosion due to the physical effects of the ice during breakup.

## ISOSTATIC AND EUSTATIC CHANGES

Of particular importance to the development of the coasts of northern high-latitude regions have been the isostatic and eustatic changes associated with the Pleistocene glaciations. Over Canada the weight of the ice during the Pleistocene glaciations was sufficient to depress the level of the continental landmass. During and after deglaciation, the landmass rebounded. These **isostatic changes** followed a specific pattern (Figure 25.21). As the ice sheet began

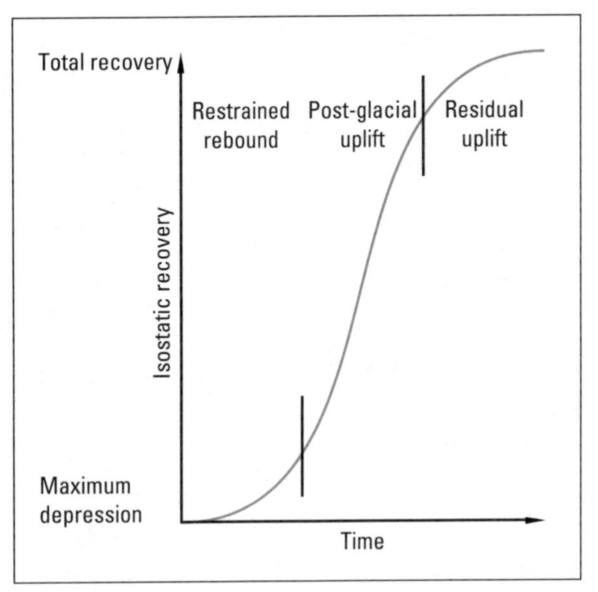


FIGURE 25.21 ▼ The pattern of isostatic recovery from the maximum depression due to loading during glaciation

to lose mass, there was an immediate elastic response, a period called **restrained rebound**, with a slower period of non-elastic response, **post-glacial uplift**, which decelerated with time. Finally, there was a period of final adjustment, **residual uplift**, which was very slow.

Superimposed on these changes are **eustatic changes**, the changes in sea level as a result of the uptake and release of water by the glaciers. If we create a model of the combination of eustatic and isostatic changes during a period of glaciation we can illustrate their importance to coastline development (Figure 25.22). The sea level will be at its lowest at the maximum of glaciation. At the start of deglaciation the water level response is more rapid than the landmass response, so that the water floods the landmass. The highest shoreline produced at this time is called the **marine limit**. With no further changes in the sea level after deglaciation and with the continuation of post-glacial uplift, the relative sea level drops and a series of raised shoreline features are produced (Figure 25.23).

**Start of glaciation**

| **Eustatic changes** | **Isostatic changes** |
|---|---|
| Immediate sea level lowering | Gradual isostatic depression |

Sea level at lowest — **Maximum of glaciation** — Isostatic depression continuing

**Start of deglaciation**

| Eustatic | Isostatic |
|---|---|
| Immediate sea level response | Restrained rebound |
| **Maximum deglaciation** | |
| Highest level of water on landmass "marine limit" | |
| | Post-glacial uplift |
| Falling "relative sea level" | |
| Production of sequences of shoreline features | Residual uplift |
| Stable sea level | Full recovery |

FIGURE 25.22 ▼ Isostatic and eustatic changes during glaciation and deglaciation

FIGURE 25.23 ▼ An oblique aerial view of the shoreline of Melville Peninsula, Canadian Arctic. Lines across the lower part of the photograph are raised beaches produced by isostatic uplift of the land after deglaciation; upper part of the photo shows terrain that lay above the marine limit, the highest water level after the melting of the ice sheet and before isostatic uplift (photo © 1948 Her Majesty the Queen in Right of Canada; courtesy National Air Photo Library, Energy, Mines and Resources Canada)

# Coastal landforms

## LANDFORMS OF DEPOSITION

Most of us are familiar with sandy beaches. Beaches of this type form in relatively low-energy environments where there is an ample supply of fine, transportable material. Where the material is coarser, and the wave energy higher, pebble beaches tend to form. Both sand and pebble beaches consist in general of a sloping area which is covered by water at high tide and one or more almost parallel ridges or **berms**.

Both the shape and the composition of beaches are products of the action of waves and currents operating within the offshore and nearshore zones, and both tend to change over time as the balance between deposition and erosion alters. If we were to monitor these changes, by repeatedly surveying the beach profile or analyzing the character of the sediments, we would discover that beaches are highly dynamic features (Figure 25.24). The maximum and minimum heights of all points along a profile measured over a number of years is called the **sweep zone**. Figure 25.24A shows the sweep zone for Scripps Pier between 13 October, 1937, and 10 August, 1945. Figure 25.24B shows the seasonal changes with the summer of 1941 and the winter of 1941–42. This illustrates the general accretionary trend on the upper beach slope in summer, with lower energy conditions, and the accretion on the lower beach slope in winter, with higher energy conditions. Figure 25.24C shows the changes that can take place over a period of a few days as a result of one storm. Detailed studies would also show that there is a close interaction between the beach profile, the size and shape of the sediments, and the processes operating at any point on the beach.

At the outer margin of the surf zone, for example, where the waves are breaking, there is a relatively

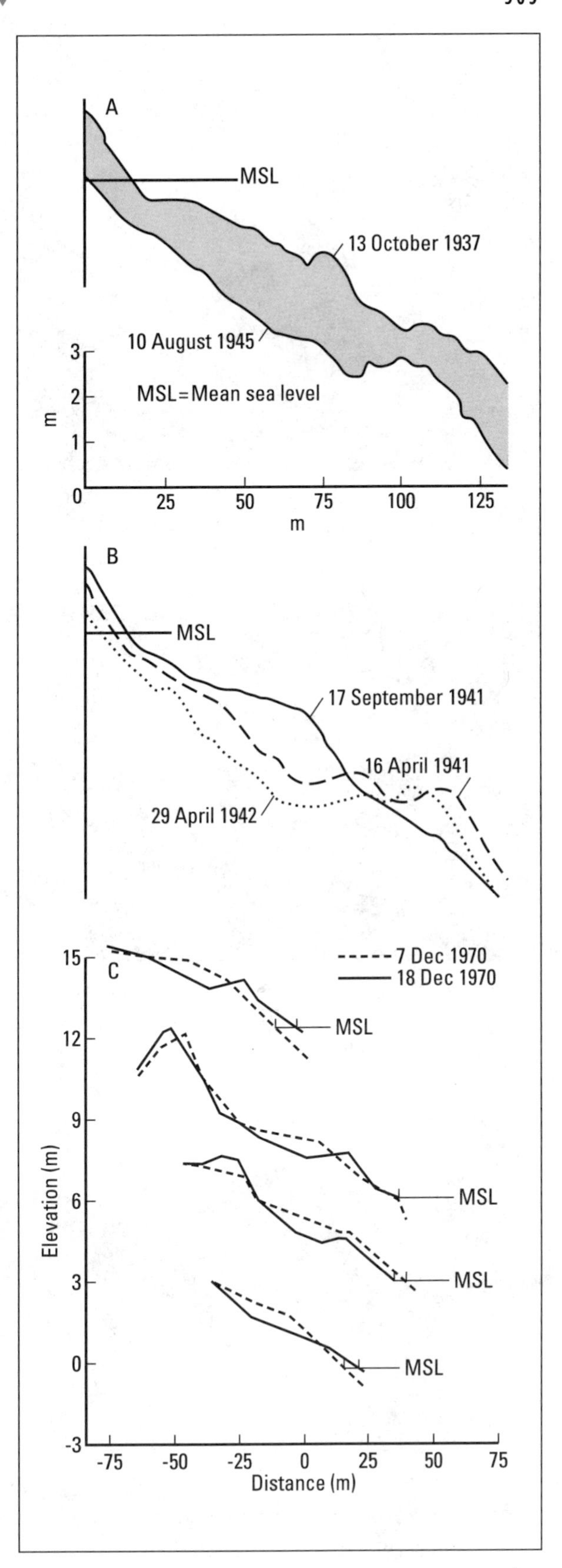


FIGURE 25.24 ▼ (A) The sweep zone for the Scripps Pier beach, New Jersey, between 13 October 1937 and 10 August 1945. (B) Seasonal changes at Scripps Pier, summer 1941 and winter 1941–42. (C) Beach profiles at Long Beach Island, New Jersey, before and after a storm in December 1970. This shows the magnitude of beach changes that can occur during one storm (from DeWall *et al*, 1977)

FIGURE 25.25 ▼ Storm berm on a beach near Lockeport, Nova Scotia (photo: P.G. Johnson)

Beach cusps on the Pacific Ocean shoreline, northwest of Victoria, Vancouver Island (photo: P.G. Johnson)

high-energy area in which there is considerable turbulence. This allows the water to pick up material from the bottom, and all but the coarsest particles are carried landward. The result is the development of coarse lag material in the breaker zone. As the water moves inland, the friction with the bed and the gradual shallowing of the water result in a reduction in wave competence, and the progressive deposition of the coarser material. The sediments therefore become finer.

Near the landward margin of the surf zone the collision of swash and backwash again creates a high-energy, turbulent zone, and coarser materials accumulate. Beyond, in the swash zone, the waters run relatively gently onto the beach, losing energy rapidly as the beach becomes steeper, and transporting only the finer sediment.

The berms are created mainly from material thrown onto the beach by storm waves (Figure 25.25). They are, therefore, composed of coarse material which has been transported at times when wave energy is high. Often two berms occur, a lower summer berm and a higher winter berm, reflecting the more intense storms of winter. In addition, on sandy coasts, wind action may redistribute the berm materials and create a series of sand dunes inland (Chapter 24).

All these processes interact to produce the characteristic profile of the beach and the sequence of sediments shown in Figure 25.14. Superimposed upon this general pattern, however, are more local features. **Beach cusps** form because of the fan-like intrusion of waves onto the beach (Figure 25.26); accumulations of shell material may occur at the upper tidal limit.

FIGURE 25.27 ▼ Formation of coastal spits and bars: (1) cuspate bar; (2) spit; (3) bayhead bar; (4) tombolo; (5) recurved spit; (6) mid-bay bar and spit

A

B

FIGURE 25.28 ▼ Examples of coastal landforms. (A) A spit with fishing shacks, Three Fathom Harbour, Nova Scotia; (B) a bar across Petit Etang, west coast of Cape Breton, Nova Scotia; (C) a tombolo at Percé Rock, Quebec (photos: P.G. Johnson)

Material moved by longshore drift tends to accumulate in areas where the current is impeded. This may occur for a variety of reasons: due to the occurrence of shallower water, or to the intrusion of a headland into the flow, or to the meeting of two opposing currents. The sand and silt deposited as a result often form a spit or sand bar. **Spits** develop through the accumulation of sediment outward from the shore. They tend to be self-producing to some extent, because once deposition has started, the effect of the spit is to reduce current velocity further and encourage further deposition (Figures 25.27 and 25.28).

**Sand bars** take a variety of forms. They often develop across bays or at the mouths of rivers, and a specific type—the **tombolo**—may be formed between the mainland and an island (Figure 25.27). Coastal bars are not only formed of sand; gravel may accumulate in the same way. Such features do not always protrude above the water, and submerged sand bars are a traditional hazard to shipping.

Elsewhere, the accumulation of fine sediments in very sheltered inlets and estuaries, or behind spits, results in the formation of **mudflats**. In time, as these build up, vegetation may become established to produce **salt marshes**. The plants that invade these areas need to be highly tolerant of salt and of occasional flooding, but once established they trap the sediment and accelerate deposition. Eventually, the salt marsh builds up until it stands above all but the highest tides (Figure 25.13).

## LANDFORMS OF EROSION

Rocky coastlines such as those of Newfoundland and Vancouver Island present an exciting contrast to the calm of many sandy beaches. As the roar of storm waves on the rocks indicates, these coastlines develop in relatively high-energy environments—those exposed to waves travelling over long distances, and facing strong onshore winds.

One of the dramatic features of these coasts is the **marine cliff** (Figure 25.29). This is formed by the constant attack of the sea, particularly against resistant, massive rocks. It is not only the power of the breaking waves that erodes the cliff, but also the sediment carried by the water. Boulders and large pebbles may grind incessantly against the foot of the cliff, while, during periods of storm, coarse material may be thrown against the rock face and chip material from the surface. Large boulders at the base of the cliff help to dissipate wave energy, thus reducing the rate of coastal erosion. This has been copied in engineering by the use of material known as rip-rap—everything from large boulders to shaped concrete forms used to protect cliffs from erosion. These processes are supplemented by more subtle effects. Wetting and drying of the rock face, growth of salt crystals in minute cracks and joints in the rock, and even the boring of marine molluscs may help to erode the cliff. As the cliff is undermined, and as the debris falls into the sea, the waters wear the material down and carry away the fragments, which join the attack upon the cliff.

The vigor of this attack is often greatest just below the water surface, and thus erosion of the cliff is at a maximum here. Over time, a **wave-cut notch** may form at the base of the cliff. In time, as well, the cliff tends to retreat as erosion at the base undermines the rock face and causes slumping and rockfalls. Cliff retreat leaves behind a level or gently sloping shore platform (Figure 25.30). This platform can reduce the power of the incoming waves in the way a beach does, so it eventually protects the cliff from further erosion. For this reason, there is a limit to the width most shore platforms reach. This is a case of negative feedback; features develop that damp down the initial changes.

FIGURE 25.29 ▼ Marine cliff in sandstone, Cape Blomidon, Nova Scotia (photo: P.G. Johnson)

FIGURE 25.30 ▼ Erosion platform and wave-cut notch at Cavendish, Prince Edward Island (photo: P.G. Johnson)

The shore platform may itself be attacked. When the tide is out, erosion of its outer margins may occur, while the exposed inner parts are subject to wetting and drying and salt weathering. Marine molluscs are also particularly effective in the intertidal zone, and it is possible to identify a sequence of erosional features along many rocky coasts that reflect the interaction of biological and physical weathering and erosion.

Variations in the intensity of wave attack, or in the resistance of the rocks along a shoreline, result in differential erosion. A headland-and-bay coastline therefore develops. The softer rocks are cut back, leaving the harder rocks as headlands. Again it is possible to detect in the evolution of these coastlines a form of negative feedback, for the embayments are eventually sheltered from the incoming waves by the headlands, and longshore drift tends to carry material into the bays, slowly silting them up. At the same time, the exposure of the headlands to active wave attack results in their slow destruction. Thus, a form of dynamic equilibrium is established.

## EROSION BY SEA ICE

Sea ice causes erosion along the shoreline and in deeper water. In deeper water (depths of 50 m and more), pressure ridges, icebergs, or ice islands may ground on the sediment. This grounding erodes

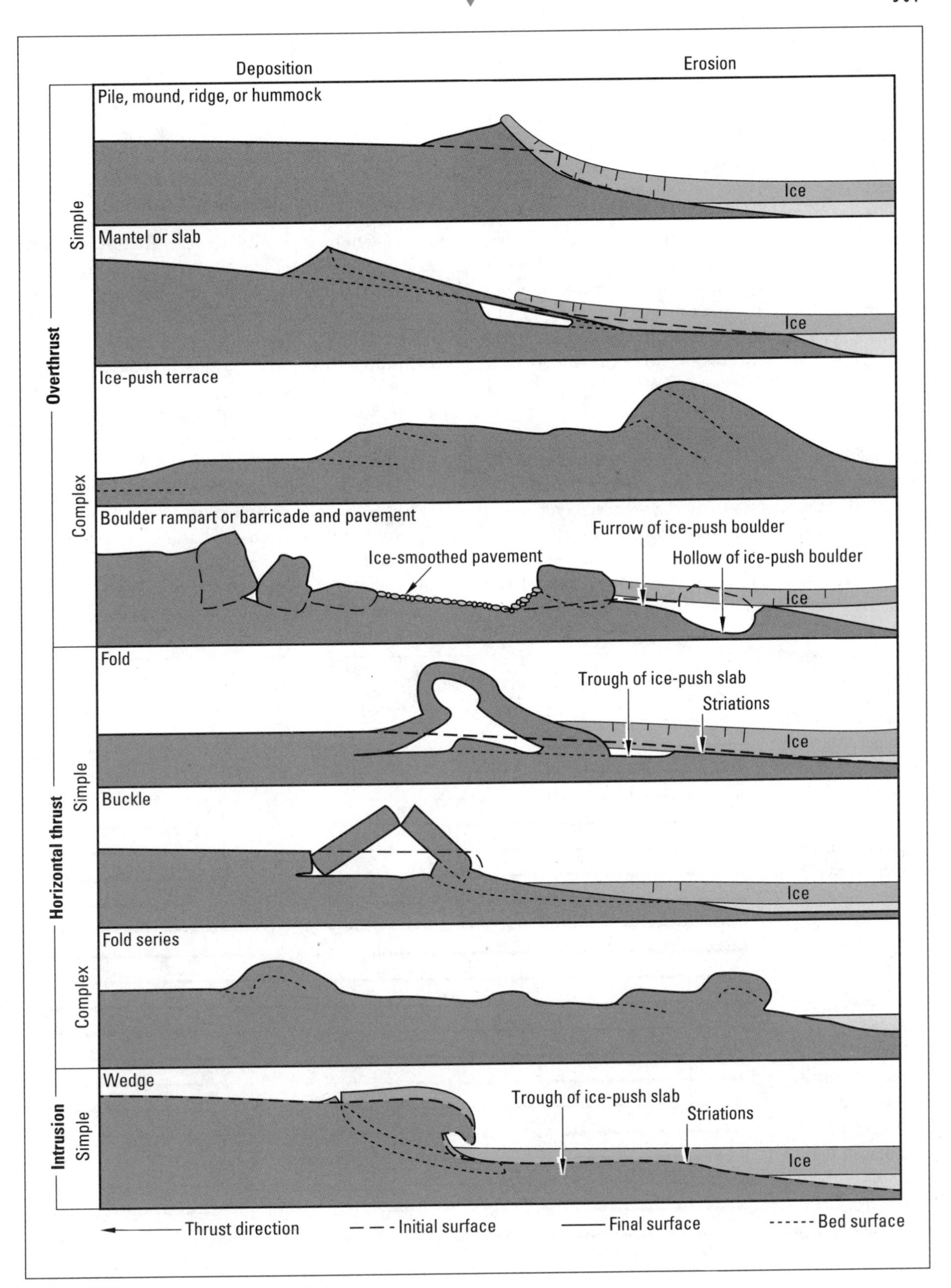


FIGURE 25.31 ▼ Ice-thrust shore morphology

gouges into the soft bottom sediments. On the Beaufort Sea coast, the seabed is criss-crossed with gouges produced by ice grounding as a result of wind action or current action. This ice grounding process is of critical concern to the oil industry in its considerations for seabed oilwells and for the construction of pipelines in shallow water.

The ice in the offshore and nearshore zones can cause erosion and deposition in a number of ways. The effects of the ice grounding on the beach slope depend on whether there is overthrusting, horizontal thrusting, or intrusion of the ice (Figure 25.31). This activity by the ice during breakup can contribute to increased erosion rates during ice-free conditions by disturbing the sediments in the coastal zone.

Any large boulders frozen into the ice can be gouged across sand flats or erosion surfaces by the movement of ice pushed by winds or currents. Along the shores of the St. Lawrence River many boulders 2 to 3 m long are moved annually across the mud flats.

It must be emphasized that large areas of the Canadian coast are affected by ice for part of the year. This ranges from eleven months in the High Arctic, to three and a half months at Summerside, Prince Edward Island, to virtually no ice at all on the West Coast (Figure 25.32).

## CORAL COASTLINES AND BARRIER REEFS

In areas where wave energy conditions are low, where coastal waters are clear, unpolluted, warm, and rich in nutrients, **coral reefs** may develop. The association of corals and algae results in the secretion of calcium carbonate in the form of intricate growths attached to the sea floor. Colonies of corals produce extensive deposits, which build outward into the sea in the form of a **fringing reef**. The coral is attacked by the waves and broken down into coralline sand which accumulates as beach material, and is often cemented into **beach rock**. The waves carry clean water with abundant food materials that feed new coral growth. A balance between the growth and destruction of the reef is attained.

The growth of coral reefs is slow, and at intervals changes in sea level may occur that affect the devel-

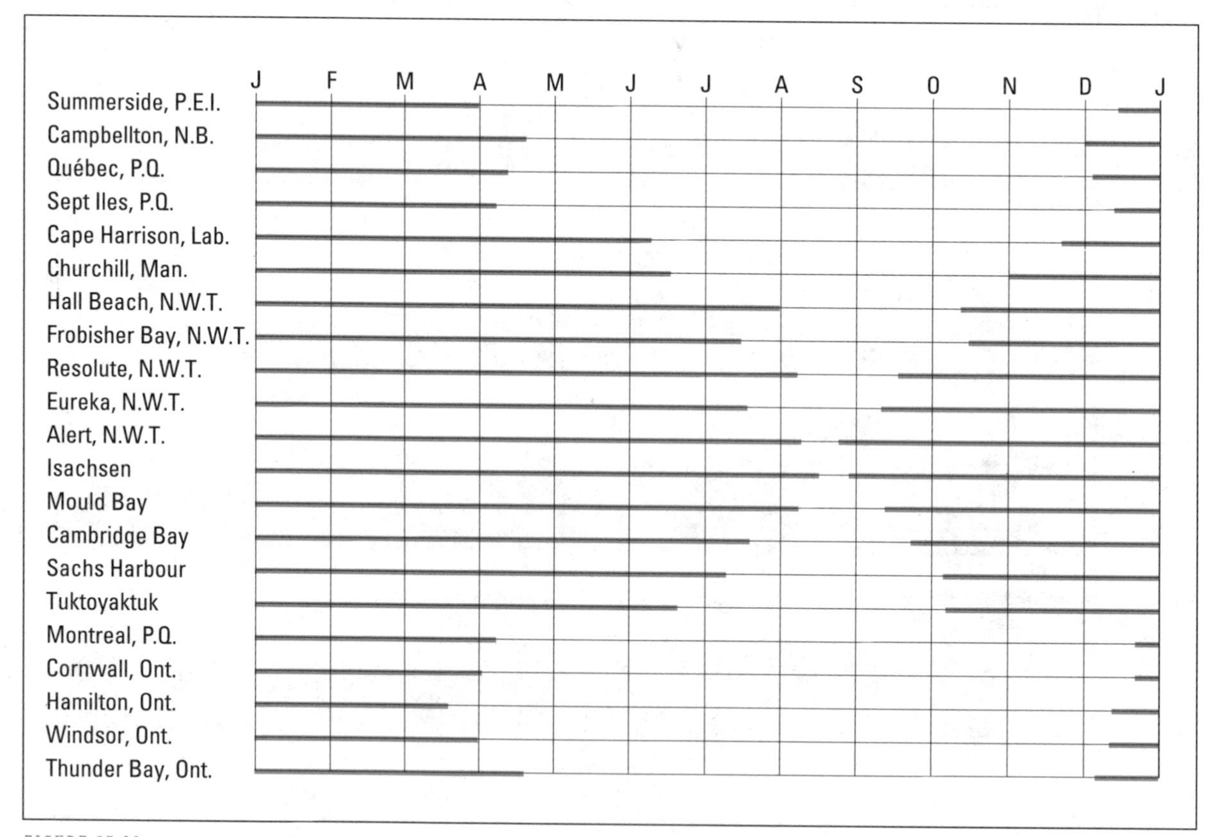


FIGURE 25.32 ▼ Duration of ice on beaches or in the nearshore zone at selected Canadian sites (from Bowen, 1981)

opment of the reef. A slow rise in sea level results in the drowning of the old, dead inner reef and the formation of a **lagoon** behind the actively growing fringe. In this way a **barrier reef** is formed. The lagoon is linked to the open sea by channels cutting through the barrier reef. Possibly the most famous examples are found along the eastern coast of Australia—the Great Barrier Reef.

Small barrier reefs have formed around isolated volcanic islands in some parts of the world, particularly the Pacific. These islands are interesting features, and their existence has long puzzled geomorphologists. Today, it is believed that they are a result of volcanic activity along the mid-ocean ridges. Fringing reefs have developed at their shorelines and, over time, the sea level has risen, possibly due to melting of the polar icecaps. The coral has grown to keep pace with the rising waters, but ultimately the island itself has been drowned. An **atoll** is therefore created: an almost circular reef enclosing an empty lagoon.

As we have mentioned, coral development is closely dependent upon coastal conditions. Temperature is particularly critical, and coral growth is only active where water temperatures are above 20°C. In the geological past, these conditions have been widespread, and rocks produced by coral development—coralline limestones—are extensive. Today, however, the distribution of active growth is much more restricted, and in many places it is being further inhibited by human activities. Coastal pollution, and the attack on coral by marine organisms accidentally introduced or encouraged by humans, pose a threat to several reef areas.

# Coasts and humans

For centuries, people have watched as the sea has eroded some coastlines and created others. In many areas, human activity has caused these changes; until quite recently we have been powerless to do much about it. The eastern seaboard of the USA, with its low-lying coastal barrier, is particularly prone to changes from natural and human activities. Many schemes and constructions have been developed to remedy the situation. Some of these have been successful; others, such as the dune stabilization schemes in North Carolina, have created additional problems. On the west coast, installations such as the breakwaters in Santa Barbara and Santa Monica, California, have been very successful in altering the depositional pattern and maintaining usable harbours. Attempts to reduce coastal erosion have not been as intensive in North America as in Europe. European populations are greater and there has been more of a threat to settlements.

Over the last 200 years, we have attempted to arrest the action of the sea and protect the coastline from erosion. In many areas, **groynes** have been constructed to catch the material swept away by longshore drift (Figure 25.33). The groynes are normally arranged in series, often of different lengths to give a tapering sequence. Unfortunately, there is considerable disagreement about which way the taper should run—whether the groynes should become shorter up-current or down-current, so they are not always as effective as intended. Moreover, trapping the sediment in one place may encourage erosion elsewhere, for the areas down-coast of the groynes are deprived of material that once protected them. In other words, the coastline acts as an integrated unit; we cannot change one part of it without also affecting other parts.

It is not only erosion that threatens the coast; sedimentation is also a problem. In the past, ports of considerable commercial importance have been left stranded by the incessant encroachment of the land into the sea. Often these processes have been exacerbated by human activity, sometimes far removed from the coastline. In the Persian Gulf, for example, the delta of the Tigris River has advanced over 100 km during the last 2500 years, as silts washed down the rivers have accumulated along the coast. Vegetation clearance, canal-building, and

FIGURE 25.33 ▼ Old groynes on Cavendish Beach, Prince Edward Island. Note that they have not been effective in controlling erosion because of the continuing relative rise in sea level (photo: P.G. Johnson)

drainage in the Tigris and Euphrates valleys were largely responsible for this.

Similar effects are produced today by river management. The building of dams, diversion of streams, and control of flooding may all disrupt the sediment supply to the coast and upset the balance between erosion and deposition. In the Nile Delta, major changes in coastal conditions occurred because of the Aswan High Dam. The reduction in the amount of sediment entering the Mediterranean resulted in the destruction of offshore bars that had been important in keeping the coastal waters brackish. When they were destroyed, salt water invaded the coastal lagoons and, among other things, killed off the fish that were the basis of Egypt's sardine industry.

In North America there has been discussion of diverting the headwaters of the Mackenzie River south to the USA, and damming the Liard River for hydroelectric power. Either project would have a major effect on the Mackenzie Delta, where the river enters the Beaufort Sea.

Siltation of coastal waters is often controlled by dredging. This is necessary to keep channels open and prevent flooding in the river valleys. The sediment is normally taken out to sea and dumped, but the procedure is not always successful. Tidal currents may return much of the material within a short time. In the Humber Estuary in eastern England, the movement of sediment was monitored using radioactive tracers. It was discovered that the material was being washed back into the estuary within a matter of days.

The story of our intervention in coastal systems is not always one of defeat and error, of course. In many cases erosion has been successfully controlled, and in some instances some of the land that has been lost has been regained by coastal reclamation. One of the most remarkable examples is in the Netherlands, where reclamation has been going on since the Middle Ages. Despite the natural tendency for the area to be drowned due to the slow rise in sea level and the settling and subsidence of the land, the Dutch have managed to reclaim half a million hectares. The work has not been without its problems; in 1953 violent coastal storms led to the loss of almost two-thirds of the reclaimed area. But almost all this land has again been salvaged, and, protected behind the extensive Zuyder Zee dam, the sediment of the IJssel and Vecht rivers is being trapped in huge lakes. As these materials build up, and as the lakes are pumped dry, huge areas of land called **polders** are created. Once drained and treated with fresh water to remove the salt, the land provides invaluable fertile soil (Figure 25.34).

In North America, large areas of coastal marsh have been reclaimed for industry, settlement, and particularly airports. These have had major effects on wildlife, because the marshes are one of the most productive ecological zones in the world, and the developments have also altered current and sedimentation patterns.

In some areas, the coastline has been extended out to sea or into a lake. The Lake Ontario shoreline of Toronto was originally at the site of the railway tracks at Union Station. This is now about 750 m from the waterfront.

We are increasingly aware of the need to manage coastal areas as a whole, taking account of interactions within the system. One of the most intriguing examples of this new philosophy is the California Coastal Plan. This is aimed not only at controlling erosion and sedimentation along the whole

FIGURE 25.34 ▼ The Dutch polders; the IJsselmeer is fresh water

California coast, but also at managing wildlife resources, agriculture, recreation, and industry in a coordinated fashion. Plans of this type are exciting, but it is important to remember they will only succeed if our understanding of the system we are trying to manage is reasonably accurate. And at that scale, failure could be highly expensive.

# The continental shelves and ocean basins

## THE CONTINENTAL SHELVES

The continents may extend out beyond the shoreline for distances of over 200 km, as the continental shelf, before there is a rapid drop to the deep ocean basins. The width of the continental shelf is very variable; in North America it is narrowest off the west coast and widest off the Maritimes and New England coast, where it forms the very rich fishing grounds of the Grand Banks. The continental shelf is an important zone, for it collects much of the sediment washed into the sea, and processes operating upon the shelf control what happens to this material in the long term.

Sediment is carried onto the continental shelves by the action of waves and tidal currents. Most of the material is fine, for the coarser debris tends to be deposited close to the coast. However, in the past, glacial deposits have also accumulated on the shelf areas, for the Pleistocene ice sheets extended across these surfaces as they were exposed by the falling sea level. It is estimated that sea levels may have fallen as much as 150 m, sufficient to expose most of the earth's shelf areas. During these periods, weathering and erosion must have modified the shelves, and they must have been partly buried by sediments swept over them by the action of wind and water. Deltas and stream alluvium must have accumulated as the rivers extended toward the retreating sea, while in some areas the ice sheets themselves plastered glacial deposits across their surface.

During periods of rising sea level, transgression of the sea across these areas introduced coastal deposits. We may visualize the slow encroachment of the rising sea over the continental shelf; in the process all the activities we associate with modern coastlines must have acted upon the surface. Gravel and sandy beaches, cliffs—with the debris eroded from them lying at the base—and numerous other coastal features must have formed. The materials on the continental shelves are consequently varied.

### *Processes acting on the continental shelves*

Today, three main processes affect the continental shelves. Deep tidal currents carry sediment across their surface. These are particularly active around the mouths of rivers, where quite rapid currents may extend from the estuaries some way out to sea. They carry with them silts and clays washed from the land. Slower-moving currents associated with larger aspects of oceanic circulation may also be important (Chapter 14). Cool polar water moving into lower latitudes at depth may transport fine shelf sediments, while currents such as the Gulf Stream may be active on the continental shelf. As shown in Figure 25.35, a distinctive pattern of sediment movement seems to occur around Britain due

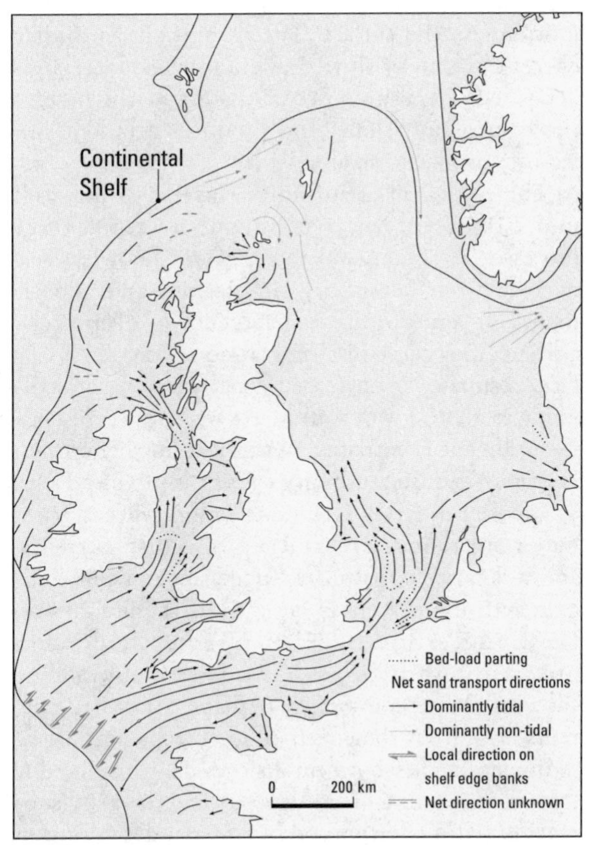


FIGURE 25.35 ▼ Sediment movement around the British Isles. Arrows show directions of new sediment transfer (from Stride, 1982)

to the circulation of the North Atlantic Drift through the Irish Sea and southward down the North Sea.

The outer margins of the continental shelves merge into the continental slopes, which dip gently (mean slope 4°) for a kilometre or so, before plunging more steeply to the **abyssal plain** (Figure 14.2). Within this area there occur many **submarine canyons**, some of immense size. The processes responsible for the formation of these canyons have been the source of much argument, but it is now believed that they were formed by the movement of the sediment-laden sea water down the continental slope. Here, therefore, a third, and most important, process is in operation: the action of turbidity currents.

**Turbidity currents** are composed of dense masses of silt and mud moving over the ocean floor under the effect of gravity. As the materials roll and sludge down the continental slope they gather speed. The particles erode the surface of the slope as they bounce along, pulling more material into motion. As the surface flattens out, the turbidity current starts to slow down and the material is deposited. Deposition occurs mainly at the base of the continental slope, and produces a fan-shaped feature, similar to an alluvial fan.

The initiation of turbidity currents is not well understood. Clearly it is difficult to study these processes in action, and much of our knowledge is derived from laboratory simulations and experiments. It appears that convergence of deep ocean currents may cause localized erosion of the shelf surface, resulting in increased concentrations of sediment in the bottom waters. As we have seen, once the sediment is entrained in this way, further erosion is encouraged; the turbidity current is produced.

Cold and often sediment-laden waters from major rivers may also initiate turbidity currents, and it is notable that many submarine canyons coincide with the mouths of large rivers (Figure 25.36); the Ganges, Indus, Nile, Niger, Rhône, and Mississippi are all associated with submarine canyons. More commonly, perhaps, turbidity currents arise from the effects of localized slumping of sediment on the continental slope. This may be due to over-steepening of the deposits laid down by submarine sedimentation, or by the trigger action of earth tremors or severe storms. The effect of turbidity currents is illustrated by just such an example. In 1929, an earthquake is known to have triggered a major slump in the Grand Banks area south of Newfoundland. This became a turbidity current that swept down the continental slope and severed the trans-Atlantic telephone cables that lay in its path. Over a period of 12.25 hours, the current travelled almost 1000 km and cut five cables (Figure 25.37).

## THE OCEAN BASINS

The deeper parts of the ocean basins—the abyssal plains and the ocean trenches—are probably the least known parts of the earth. The processes that go on there are largely a matter of speculation. Only at their margins (where they receive inputs of sediment from the continental slopes) and in the vicinity of the mid-ocean ridges (where they are subject to volcanic activity) are there likely to be any great depths of sediment; elsewhere they must be dark, silent, and still. All that happens is the slow, perpetual accumulation of fine detritus that settles from the water above: microscopic remains of marine organisms (radiolaria) and volcanic debris and dust carried by winds from the distant land. These materials comprise what is known as the marine ooze.

Close to the mid-oceanic ridges, where the ocean floor is still young, the ooze is thin. Farther away on the older rocks, it has been gathering for hundreds of thousands of years, more or less undisturbed. Here it reaches thicknesses of several metres, and contains a record of recent earth history, of the periods of volcanic activity when the rain of dust was more intense, and of the fluctuations in climate during the Quaternary Period, when the marine organisms changed in response to variations in temperature and salinity. As we saw in Chapter 9, by analyzing the foraminifera preserved in these sediments, we can reconstruct changes in global temperatures over the last two and three million years.

These are not the only materials on the ocean floor, however. In addition, there are rich deposits of ores that have precipitated from the ocean waters. Manganese is particularly abundant, mainly in the form of oxide nodules, crusts, and pavements at depths of 3500 to 4500 metres. Some estimates suggest that there may be as much as $4000 \times 10^9$ tonnes of manganese on the sea floors in this form. Many of the nodules also contain copper, nickel, and cobalt, while copper and zinc are also locally abundant in areas where hot, salt-rich brine has leached minerals from the volcanic rocks beneath.

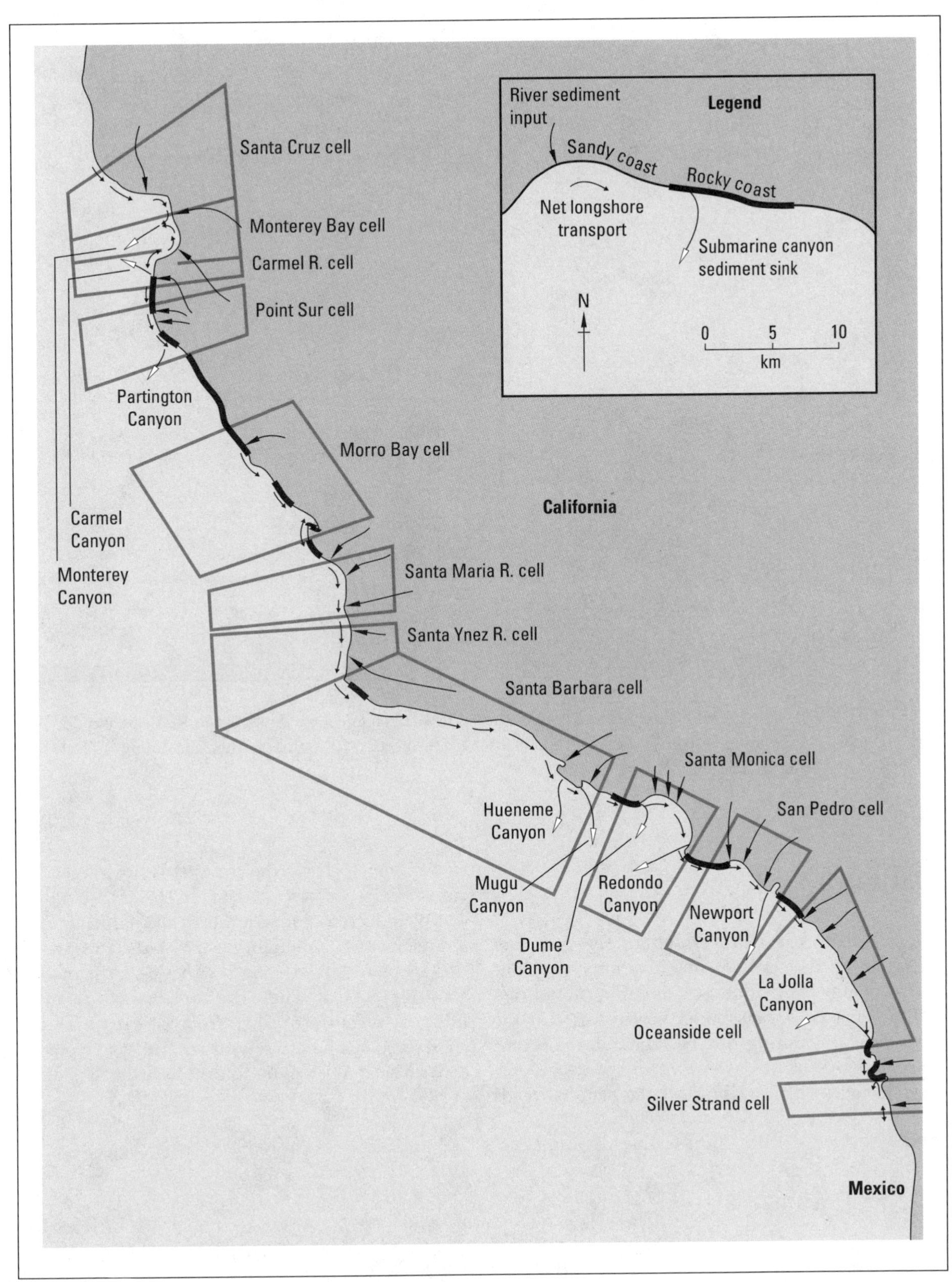


FIGURE 25.36 ▼ Littoral cells, and submarine canyons and rivers in southern California (adapted by J. D. Hansom from Department of Navigation and Ocean Developments, 1977)

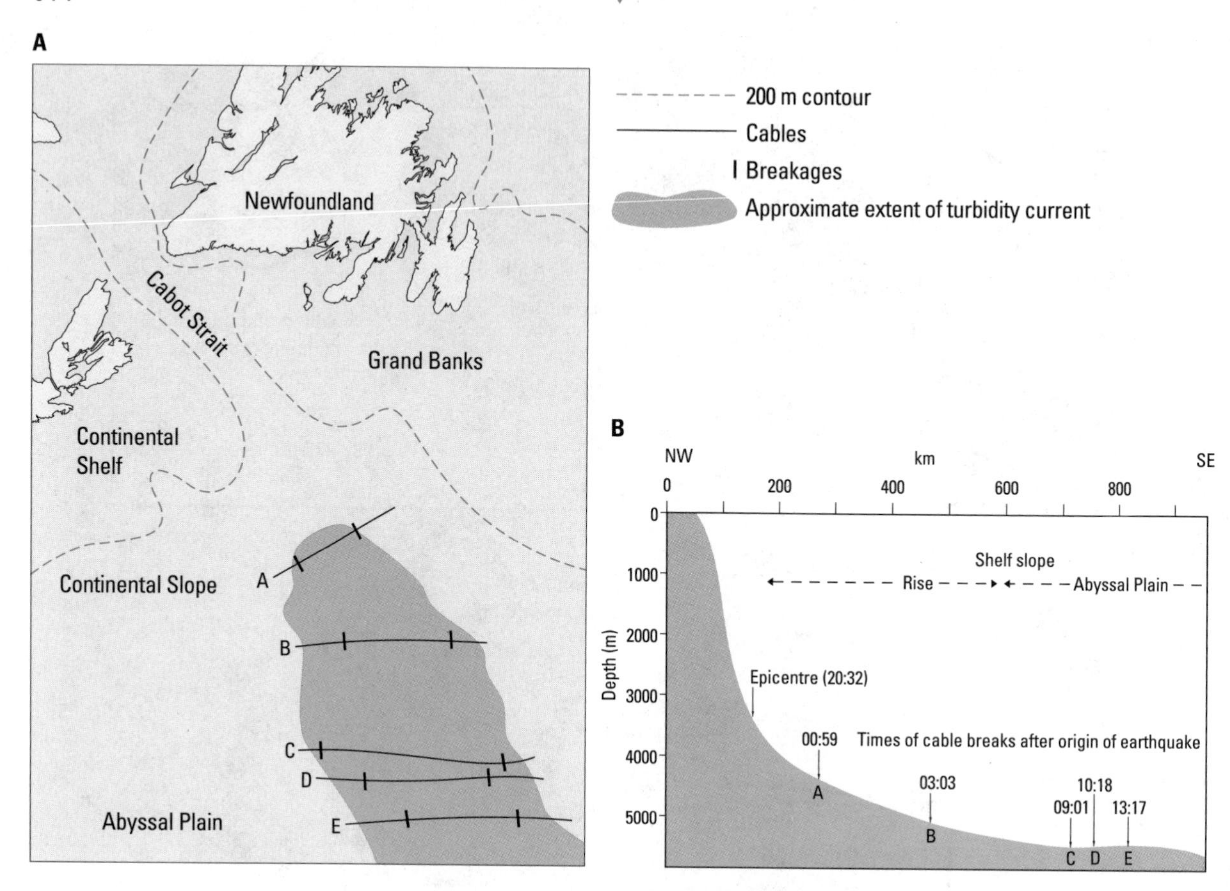


FIGURE 25.37 ▼ The effects of the Grand Banks turbidity current in 1929 on telegraph cables in the North Atlantic: (A) positions of some of the cables and locations of the breaks; (B) section showing the timing of breaks (after Holmes, 1965)

# Conclusion

We still have much to learn about the processes operating on the sea floor, and about the distribution of sediments by tides, ocean currents, and turbidity currents. What is clear, however, is that the movement of sediment into and through the oceans represents one of the last stages in the debris cascade, for once on the ocean floor the material tends to be trapped. Nevertheless, this is not the end of the story. As we saw in Chapter 16, earth-building and plate tectonics may result in the uplift of ocean sediments into mountain chains or their consumption by plate subduction. In either case, the materials are recycled: when the mountains are again lifted, or when the magmas from the earth's interior are spewed again onto the surface, the whole process of weathering, erosion, and deposition is started once more.

# Biogeography: study of the biosphere and ecosystems

## What is biogeography?

In 1957, in his book *Biogeography: An Ecological Perspective*, Canadian Pierre Dansereau defined **biogeography** as the study of the origin, distribution, adaptation, and association of plants and animals within a temporal and spatial context. **Ecology**, the study of the relationships between organisms and their environment, is the core of biogeography. Biogeographical studies place the organism under study within its environmental context, and therefore include many aspects of its **habitat** (the environment in which it is found—its 'address') and its **niche** (its function in the ecosystem or its ecological role—its 'job').

Traditionally, biogeography has been concerned with the themes of both ecology and evolution (origins) of organisms, but now it also includes the influence of humans on their environment. After all, humans, the species *Homo sapiens*, are just another of earth's organisms gaining sustenance from and depending on the earth's biosphere for their existence.

### THE BIOSPHERE

The **biosphere** is the layer on planet earth, only a few kilometres thick, in which life can be supported. It extends into the crust wherever there are living organisms, and includes the air, soil, and water that support life. Earth's atmosphere extends to 1600 km, but life is restricted to this thin surface zone, where organisms are major components in **ecosystems**.

## General structure of an ecosystem

### HUMAN LIFE AND THE ECOSYSTEM

We derive our food, much of our pharmaceuticals, and many building supplies from the plants and animals that occupy the earth. We also try to manage many of them, to control them to increase or improve their quantity and quality for human use. In all these cases, we operate within the biosphere, which can be further subdivided into earth's **biomes**, which in turn are composed of ecosystems. Biomes include major plant and animal groupings, like the taiga or boreal forest, where environmental conditions such as soils and climate are similar over large regions. Biomes are composed of many ecosystems that are defined by the complex interactions among living organisms (**biotic** components) and their nonliving environment (**abiotic** components).

Our role in these ecosystems is a unique one. More than any other biotic component, we tend to control them. We act as a regulator. Because of this influence on the biosphere, it is vital that we understand it, although the lesson of history is that we do not always do so. Through our ignorance, we often damage ecosystems, disrupting them at a cost to our own well-being and to the integrity of the ecosystems themselves. The consequences, such as species' extinctions, famine, and disease, have been seen in all parts of the world on innumerable occasions.

What, then, is the nature of the biosphere on which we so clearly depend? What are the principles that govern ecosystem function?

### COMPONENTS OF AN ECOSYSTEM

We can best answer these questions by considering a simple example. Let us examine the relationships and processes involved in a small, isolated, treed bog. The bog includes many different components: the plants, including the trees; the understorey shrubs and the plants growing at the ground surface; and the animals (mostly invertebrates) that live in the deep moss and soil layers. These are all closely associated (Figure 26.1). Many of the animals derive their food from the plants; others are predators that prey on fellow animals. Some of the plants are parasites that live off other plants.

The relationships extend further. The plants themselves obtain their energy from the soil, the sun, and the atmosphere. Water and nutrients are taken up from the soil; vital elements are extracted from the atmosphere; the whole process is powered by energy from the sun. The network of relationships, therefore, includes the processes operating within the soil and the atmosphere of the bog. The pathways of energy flow determine the boundaries of the ecosystem, and the amount being transferred within an ecosystem is normally much greater than

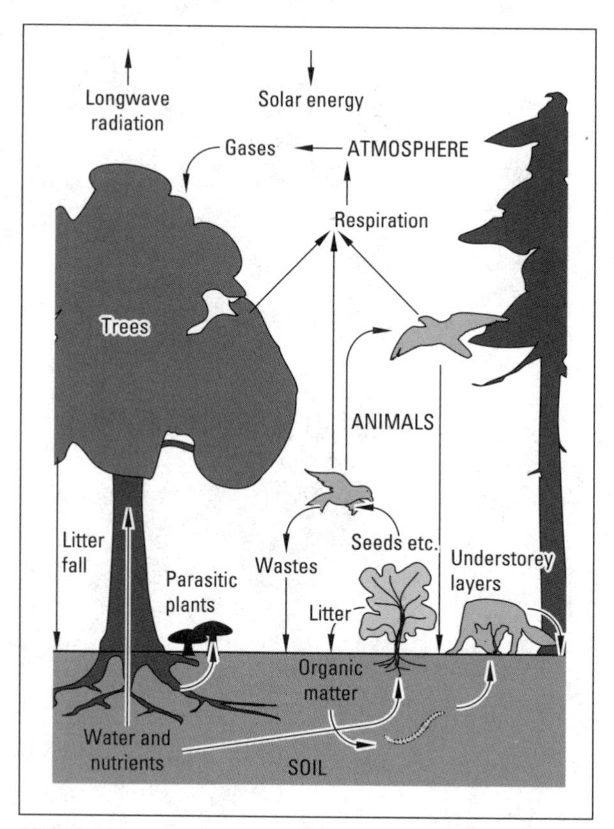


FIGURE 26.1 ▼ Generalized structure of an ecosystem

that being exchanged between ecosystems. In the soil there are countless organisms that take part in the conversion and transfer of material to the plants; they also break down the debris that accumulates on the ground as the plants die or lose their leaves. In the atmosphere, similar cycling processes remove products from the vegetation and return vital materials to the plants. Overall, therefore, we must consider the woodland as an ecosystem including not only the plants and the surface animals, but also the soil and all the organisms it contains, as well as the immediate envelope of the atmosphere with which the bog interacts (Figure 26.1).

Ecosystems can be defined at any scale, as long as one is dealing with a self-perpetuating unit in which biotic and abiotic components interact through the cycling of energy. Plants obtain their foodstuffs from the soil and atmosphere, and may be eaten by animals that may be hunted by other animals (perhaps humans). The remains or waste products of these animals and of the plants return to the soil. They are attacked and decomposed and mixed into the soil. Ultimately, the residual materials are converted to a form that can again be taken up by plants.

## Inputs and outputs of ecosystems

Of course, the processes operating within the ecosystem do not function in isolation. We have already mentioned that they are powered by energy from the sun, and thus solar radiation represents one of the basic inputs into the ecosystem. There are, however, many other inputs. Precipitation provides a supply of water and minerals; much of the incoming water is stored in the soil and is eventually used by plants and animals. Mineral portions of the soil can be broken down by weathering to supply inputs to the soil. The organisms also may, in part, enter the ecosystem as inputs from outside. Migrating animals may move into the area; winds and streams may carry seeds which germinate into plants. Humans may introduce various inputs. We may bring in seeds or seedlings to replenish the vegetation; we may also introduce animals, often accidentally. In addition, we apply fertilizers to encourage growth, or pesticides to control creatures considered pests. All these represent inputs from outside the ecosystem (Figure 26.2).

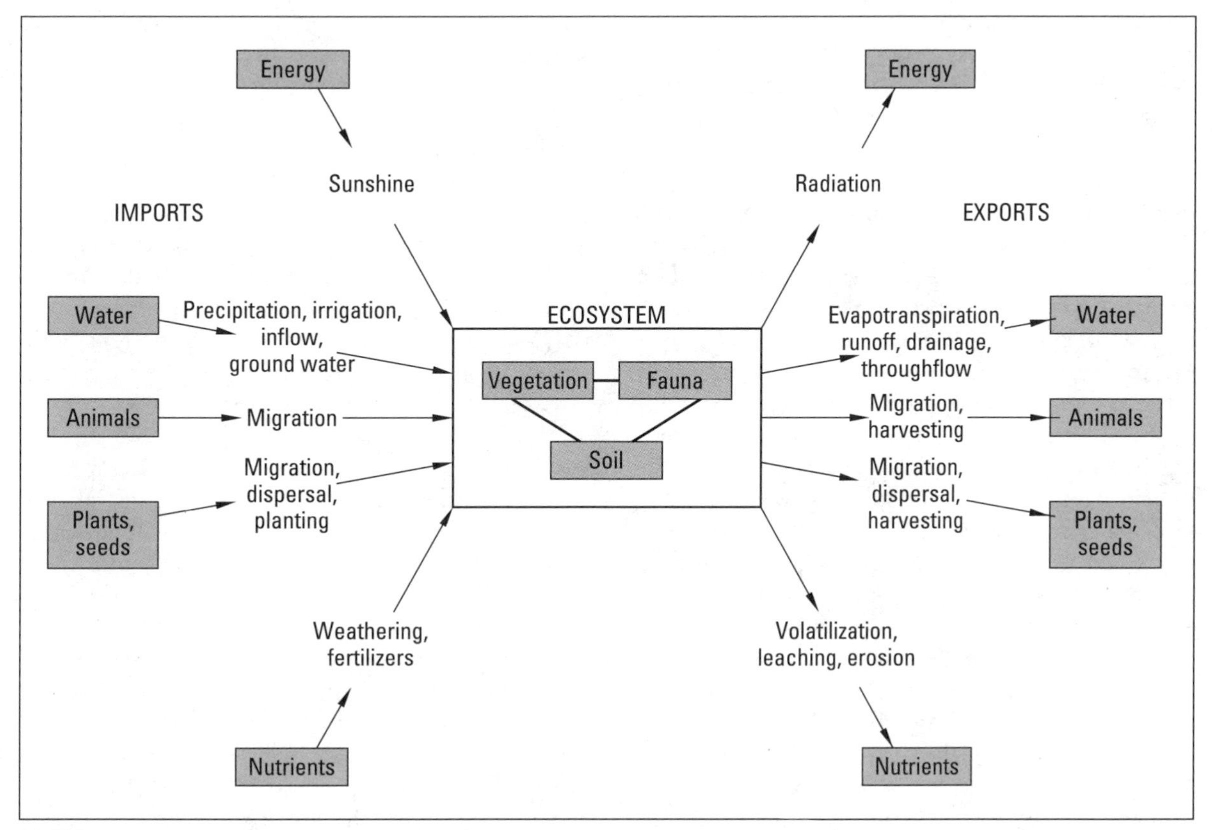


FIGURE 26.2 ▼ Inputs and outputs of an ecosystem

By the same token, the ecosystem loses outputs (Figure 26.2). Water drains through the soil and flows out of the area or may be lost through evaporation; animals move away; predators consume prey and then leave; the wind picks up and carries the dry leaves; humans remove timber and agricultural products; gases and heat escape back to the atmosphere; soil nutrients may be removed by leaching or by soil erosion. The ecosystem is consequently an open system; it receives and loses both energy and matter; however, it is the cycling of energy within the ecosystem that dominates energy flow. It is the balance between the inputs and outputs that governs the state of the system. Again, we can see this by considering our bog example. If humans extract groundwater from the bog's source aquifer so that the local water table is depressed, or if a new road dams surface discharge from the bog, then fundamental changes can occur to the bog ecosystem. Since all components of the ecosystem are interrelated, a change in one component has the potential to affect the entire system. Water table drop may remove water from the rooting zone of certain plants, and when they decline in production the herbivores dependent on them will be affected. Lower plant production will affect the rate of **litter** accumulation, and therefore the soil will be altered over time. As ever, the integration of the system makes it respond as a whole to changes in external conditions.

It is also apparent that the inputs to and outputs from any ecosystem represent a link with other parts of the environment. The ecosystems themselves are associated in this way, for the outputs from one ecosystem become inputs into an adjacent ecosystem. For example, tens of thousands of barren-ground caribou (*Rangifer tarandus groenlandicus*) migrate between taiga and tundra ecosystems in central Canada each year. In addition, ecosystems are linked to other environmental systems. The debris from the forest floor may be carried by streams into fluvial systems, while the gaseous losses and evaporation from the vegetation become inputs to atmospheric systems. In the same way, ecosystems often act as controls upon other processes in the environment. The vegetation, as we have seen (Chapter 13), intercepts rainfall and controls inputs of water into the soil and streams; it acts as a control within the hydrological cycle. Vegetation also protects the land surface from erosion and thereby acts as a control on many geomorphological processes. On a global scale, plants produce the oxygen we breathe and are a major $CO_2$ sink, reducing the impact of this greenhouse gas as it affects global change. Indeed, it is partly because of these interactions that ecosystems are such an important feature of our environment—and this is why we need to manage them with care and understanding.

# Cycling processes in ecosystems

## ECOSYSTEM STRUCTURE AND CYCLING

Both energy and matter are cycled through ecosystems. Most of the energy moves in the form of chemical energy stored in the tissue of **biota** as carbohydrates, proteins, etc.; whereas the matter that is cycled includes water and a wide range of mineral substances. These latter are referred to as nutrients, and their transfers from one component of the ecosystem to another make up the nutrient or biogeochemical cycle.

These transfers of energy and nutrients do not occur haphazardly; they tend to follow distinct and consistent pathways. This is because the ecosystem, despite its awesome complexity, is structured and organized. Thus, the relationships within it follow certain clear patterns.

One of the fundamental patterns we can see in any ecosystem is the **food chain**. This illustrates the links between the different organisms as shown by the energy transfers. Each organism tends to fulfill a specific role within the ecosystem—its **niche**—and to maintain specific relationships with its neighbours. Thus, energy passing through the ecosystem does so along well-defined paths.

We will look at details of these routeways later; first, it is useful to sketch the broad picture of food chains. All organisms are classified into one of two groups: **autotrophs** or **heterotrophs**. The autotrophs are those organisms that manufacture their own food, using energy from the sun, carbon dioxide from the air, water, and minerals from the soil. By far, the most important members of this group are the green plants, although many micro-organisms, including algae, are capable of the same function. Autotrophs are often referred to as producers.

The heterotrophs are those organisms that derive their energy from the autotrophs or other heterotrophs. That is, they either eat the plants (**herbivores**) or each other (**carnivores**), or they feed on the dead tissues of other organisms (**scavengers**), or decompose the materials (**decomposers**). Thus, heterotrophs include primary consumers, secondary consumers, and decomposers.

As we can see, this classification of organisms provides a framework for understanding the flow of energy through the ecosystem. Producers (autotrophs) convert solar energy to chemical energy and provide food for the primary consumers. These are, in turn, devoured by the secondary consumers, while all three groups are consumed by the decomposers. In this way, energy flows through the system (Figure 26.3).

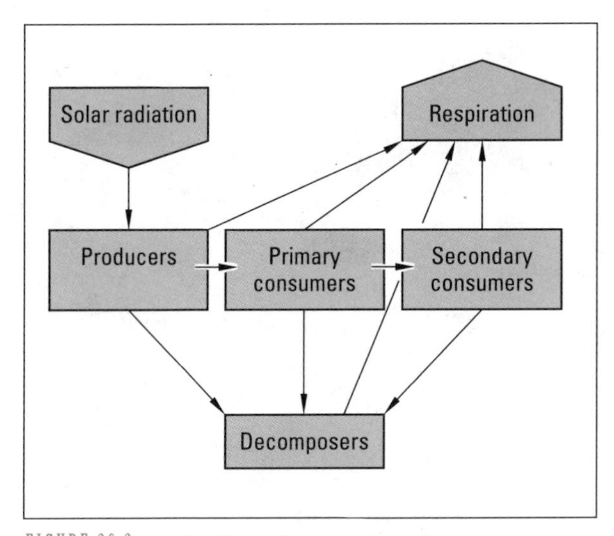


FIGURE 26.3 ▼ The flow of energy through an ecosystem

If we consider specific ecosystems, we find that particular organisms fill the roles we have identified in this diagram. The links between them represent a food chain; put simply, food flows from one link to the next in the chain. In fact, the term 'chain' is a little misleading. It is better to think of it as a food web, for the links are not unidirectional; many interactions occur between the various members (Figure 26.4)

## ENERGY FLOWS IN THE ECOSYSTEM

The food web provides a simple picture of energy flow within an ecosystem. The inputs are derived from solar energy, but most of the transfers among levels occur in the form of chemical energy. The energy is transferred from one **trophic level** to another, either to a higher level or back to a lower trophic level when, for example, an organism dies. Thus energy is constantly cycling throughout the ecosystem.

The transfer of energy in this way is, however, far from perfectly efficient. Considerable losses in energy occur during each transfer. Much of this energy is expended by the living animal in the form of body heat and respiration; much is also lost because of the incomplete breakdown of the compounds that are eaten by each organism. Thus, leakage of energy from the system occurs. Some of this energy becomes outputs from the ecosystem, in the form of heat lost to the atmosphere or as kinetic energy and chemical energy when organisms leave the ecosystem. In ecosystems where seasonal migrants are present for only a portion of the year, there may be significant energy imports and exports associated with animal movements. In addition, some energy becomes concentrated in forms that are unavailable to organisms, as when organic matter accumulates as peat. It is temporarily withdrawn from the energy cycle.

Two processes are of particular importance to energy flow and cycling within ecosystems. These are photosynthesis and respiration. **Photosynthesis** is the process by which the autotrophs convert solar energy, in the form of sunlight, to chemical energy. It is a complex process, but in general terms can be seen as a reaction involving carbon dioxide, water, and light energy. The carbon and water combine to form carbohydrate, one of the basic components of living organisms, and oxygen is given off as byproduct. The energy involved in the reaction is stored within the carbohydrate:

$$\underset{\text{(water)}}{H_2O} + \underset{\text{(carbon dioxide)}}{CO_2} + \text{sunlight} \rightarrow \underset{\text{(carbohydrate)}}{-CHOH-} + \underset{\text{(oxygen gas)}}{O_2}$$

In this way, plants convert solar energy into a form that can be used later.

The energy is released through the process of **respiration**, during which the carbohydrate is converted to water and carbon dioxide. This requires the combination of the carbohydrate with oxygen, and thus involves the intake of oxygen by the plant. As we will see, most of the oxygen is absorbed from the soils by plant roots (oxygen is also absorbed

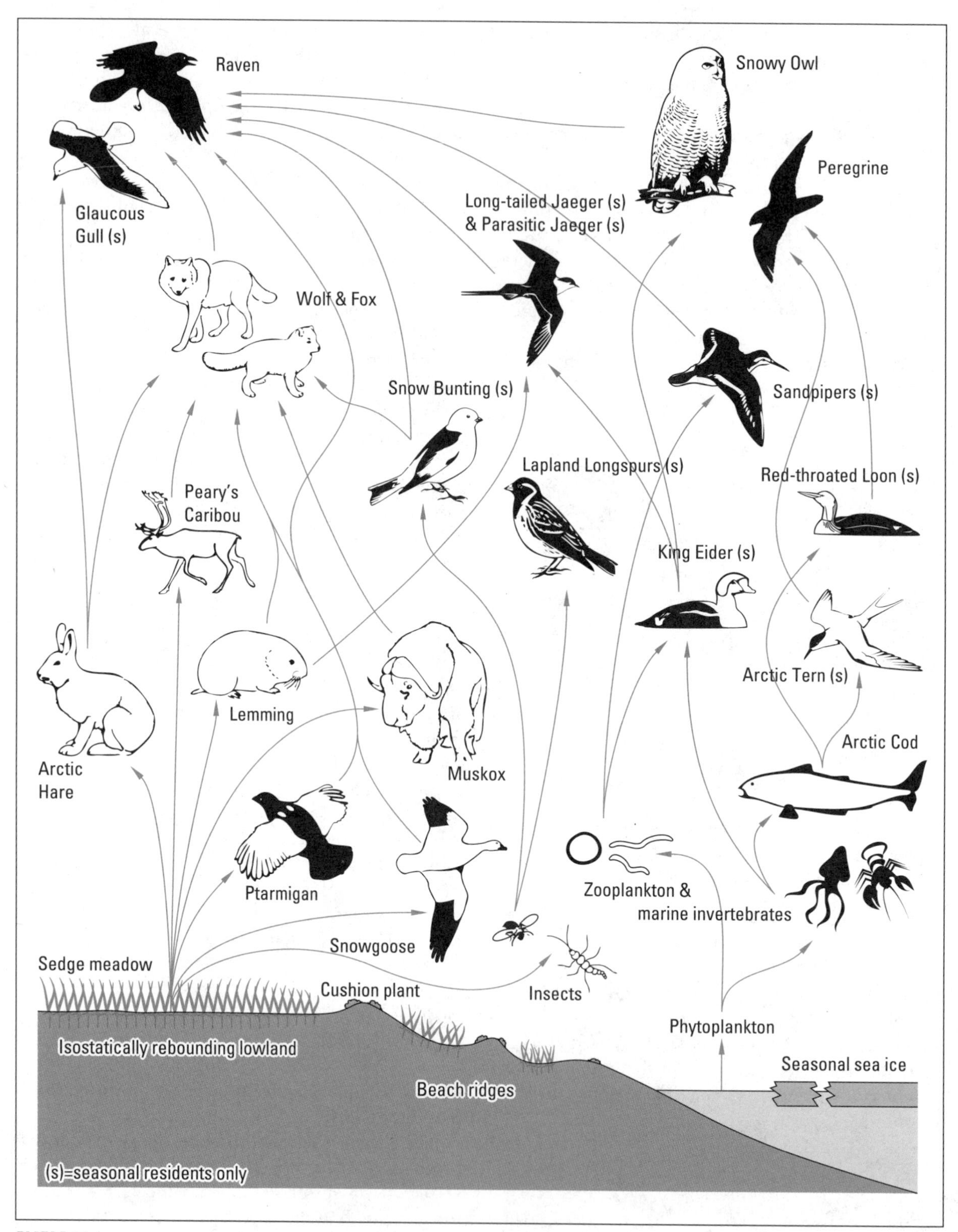


FIGURE 26.4 ▼ Part of the food web in a High Arctic oasis: the Truelove Lowland, Devon Island. Sedge meadows are restricted to less than 2 percent of the area of the Canadian Arctic Archipelago, but because they are the most productive ecosystems in the region, they represent biological oases within the polar desert and semi-desert (from Bliss, 1977)

through the stoma on leaf surfaces). Respiration is an oxidation process, in which the carbohydrate breaks down to simpler molecules, releasing energy in the process:

$$\text{–CHOH–} + O_2 \rightarrow H_2O + CO_2 + \text{chemical energy}$$

This chemical energy is stored within specific energy-carrying molecules in the plant. It is subsequently used to produce the other compounds required by the plant, such as amino acids. Both photosynthesis and respiration are necessary to plant survival. They require the supply of mineral nutrients (e.g., nitrogen), and can therefore be considered within the context of nutrient cycling (Chapter 29).

By the time animals eat the plants, a large part of the initial energy input has been lost. Typically, little more than 10 percent of the energy is passed on to the next trophic level, although in special circumstances as much as 50 percent may be transferred. Thus, the efficiency of energy transfers can be estimated by comparing the proportion of organic matter in each level of the ecosystem (Figure 26.5). Not all light energy is available for plants, since foliage reflects an average of 20 percent of incoming radiation (Table 3.2). Not all energy within a trophic level is available for use by the next level, since it may be stored in locations or forms that cannot normally and readily be accessed. Furthermore, not all weight (**biomass**) is absorbed. The indigestible components are not assimilated, but passed through and defecated. These energy sources become the basis of the detrital food chain. As we see in Figure 26.5, the proportion of organic matter in each level of the ecosystem is normally low; as well, it is rare for more than four trophic levels to exist within an ecosystem, since the loss of

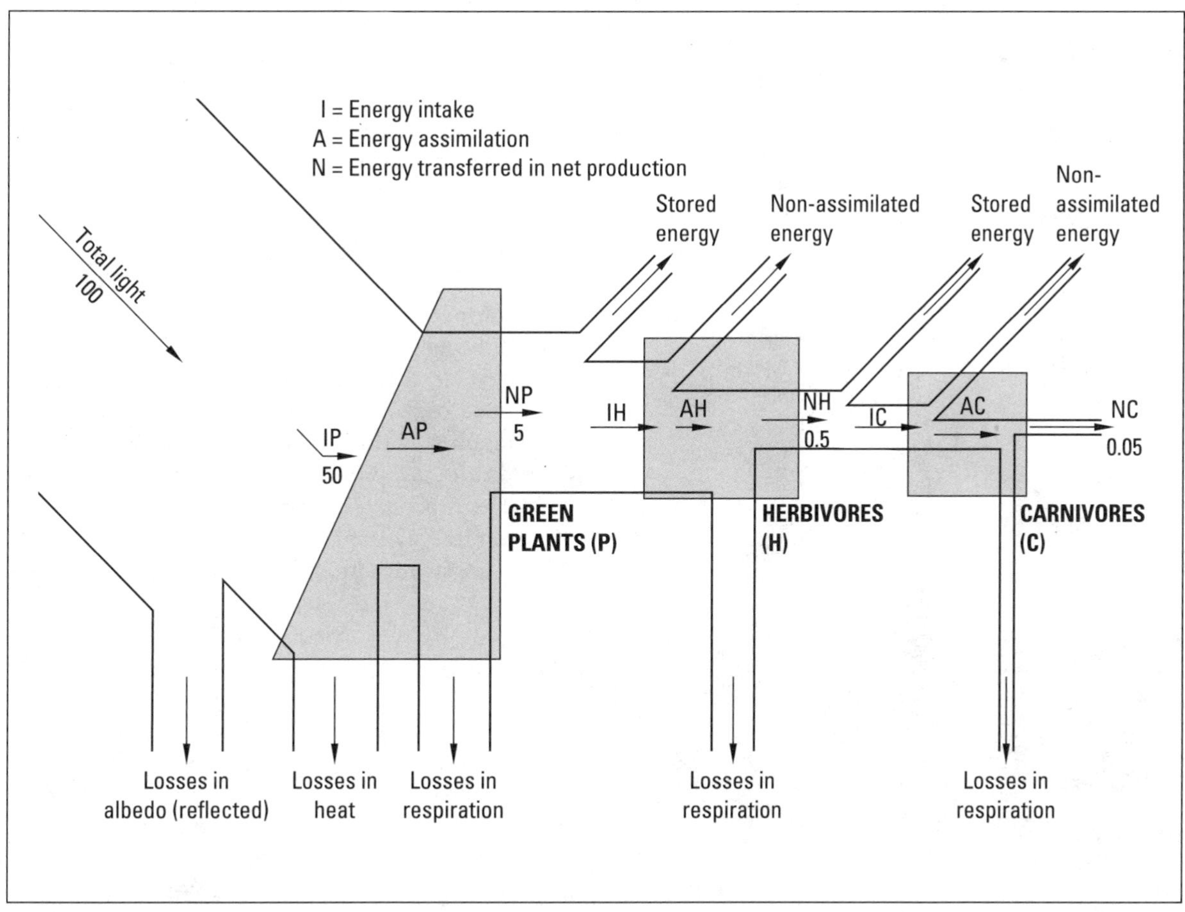


FIGURE 26.5 ▼ Energy transfers through a typical woodland ecosystem. Thickness of the bars is generally proportional to the amount of energy flowing along the specified route; size of boxes relates to the amount of energy stored in the specified component. Figures indicate the percentage of incoming energy at each trophic level compared with the amount available as total light

energy during each transfer is such that more levels cannot be sustained.

The inefficiency of these transfers has a further implication, for it means that each successive level within the ecosystem is by necessity much smaller. It requires a very large population of producers to support the total pyramid. As a consequence, if we plot **trophic pyramids** for different ecosystems, we find a very consistent pattern (Figure 26.6); the total biomass of the organisms in each trophic level declines rapidly upward.

As we will see later, this rapid decline has been used as evidence for the inefficiency of many of our agricultural systems. It is argued that meat production involves an unnecessary loss of energy; far more food energy could be made available if the food chain was shortened and we ate the plants, rather than the primary consumers.

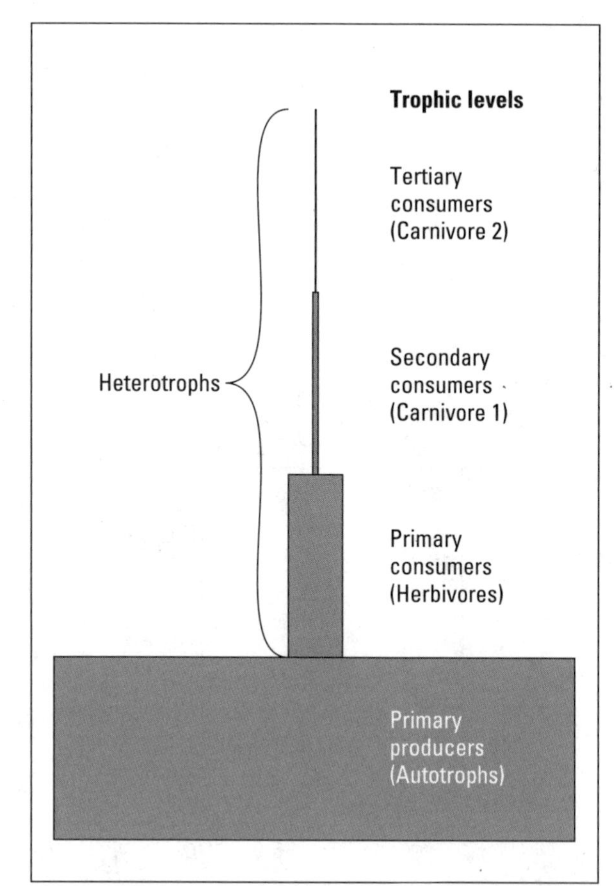


FIGURE 26.6 ▼ A typical pyramid of biomass or energy within an ecosystem. Scavengers would utilize biomass from any of the upper trophic levels, and decomposers would utilize detrital materials generated at any level

## BIOGEOCHEMICAL CYCLING

In addition to a cycling of energy, the movement of materials through the ecosystem involves a transfer of mineral nutrients. Organisms require these nutrients for their healthy growth. Minerals are essential for the formation of new cell material and the production of compounds such as proteins and amino acids. Many nutrients are involved in these processes; the most important are compounds of carbon, oxygen, nitrogen, potassium, and phosphorus.

The inputs of nutrients to the ecosystem occur in a number of ways. Weathering of the mineral component of soils releases inorganic compounds into the soil; rainfall introduces material from the atmosphere; soil organisms and plants absorb substances from the air; people apply fertilizers and pesticides. Within the ecosystem, these compounds are altered and transferred, passing through food chains, and breaking down through chemical reactions in the soil.

Within the food chains, it is common to recognize two main cycles. The **grazing food chain** or **web** involves the plants, the herbivores that feed on them, and the carnivores that feed on the primary consumers. This operates mainly in the vegetation layer (Figure 26.7). It forms a basis, as we will see later, for examining biogeochemical cycling in the vegetation and the animal communities. The other main component of the food web is the **detritus food chain** or **web**. This is dominated by the heterotrophic decomposers, the organisms that break down dead organic material and release the nutrients contained within it. Decomposers often convert the organic compounds back to inorganic form through a process of mineralization. It is a process that operates mainly within the soil, and it will form the basis for our examination of biogeochemical cycling (Figure 26.7).

The cycling of nutrients is not only a biological phenomenon. As the term 'biogeochemical' implies, physical and chemical processes also take part. Organic debris may be decomposed purely by physical and chemical activity; nutrients within the soil may be stored by physio-chemical forces operating on the surfaces of the minute colloidal particles (clay and microscopic organic fragments). Water percolating through the soil and moving both laterally and vertically under the effect of capillary forces also transports nutrients through the soil. Overall, however, these cycles are dominated by biological and chemical processes.

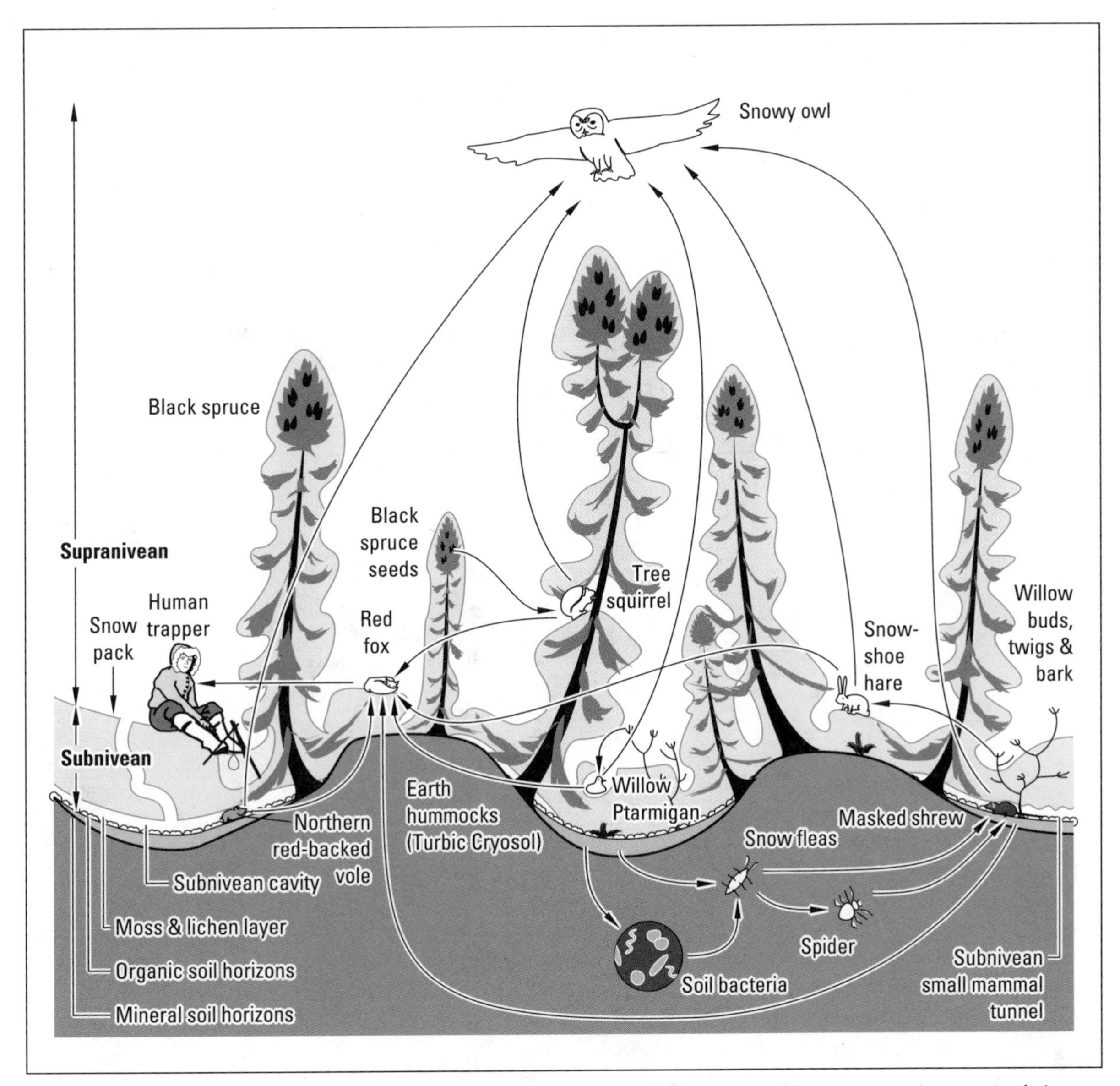


FIGURE 26.7 ▼ The grazing and detritus food chain of the subarctic needle-leaf forest. In winter, animals are active below and within the snow (the subnivean environment), as well as above the snowpack (the supranivean environment). Plants are mostly dormant during winter; however, in the subnivean environment, temperatures remain close to 0°C and some insects can be active. In the snow-free season, the complexity of these food chains increases with the arrival of migratory animals and birds

# World ecosystems

## FACTORS AFFECTING ECOSYSTEMS

It is apparent that the character of the world's ecosystems varies considerably. The range of organisms involved, and their relationship with their environment, differ from one area to another. These variations arise mainly from the differences in the external factors that control ecosystem structure and behaviour.

Three main factors will be discussed here. Climate is clearly of major importance, since this directly controls many of the inputs to the ecosystem (in particular, solar energy and rainfall) and indirectly influences many other ecosystem processes. The rate of weathering, for example, is related to climatic conditions, and thus climate influences the inputs of nutrients. The operation of many chemical and biological reactions is also climatically dependent; temperature and water availability, for example, influence the rate of primary production (i.e., organic material formation) in the ecosystem (Figure 26.8). These effects are also evident in the global distribution of biological productivity (Figure 26.9).

Geology and geomorphic history also are important, since they tend to influence the inputs of nutrients. In general, ecosystems are more productive on parent materials derived from rocks that weather rapidly and release large quantities of nutrients than on resistant, nutrient-deficient rocks.

Finally, humans also play a vital role, since we control the structure of many ecosystems through our management of the environment. In some cases, we create artificial ecosystems as part of our agricultural exploitation of the environment. Urbanized areas are the ultimate in ecosystem alteration, since many components in more natural systems are insignificant in terms of energy flow (e.g., soils, producers). We can also accidentally and indirectly alter natural ecosystems through our production of pollutants and our manipulation of other ecosystems.

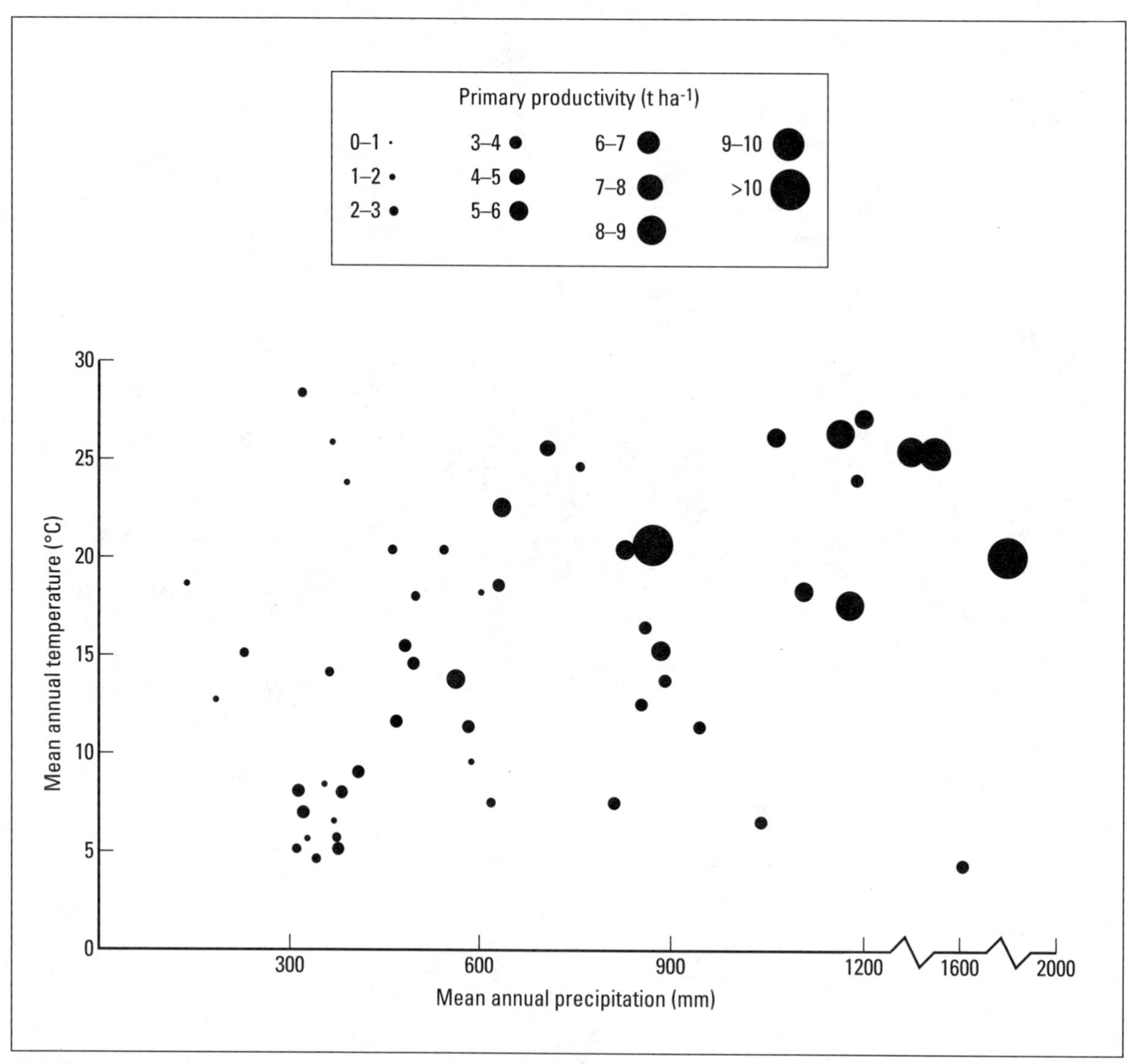


FIGURE 26.8 ▼ The relationship between primary productivity of natural grasslands and climatic conditions (after Lavenroth, 1979)

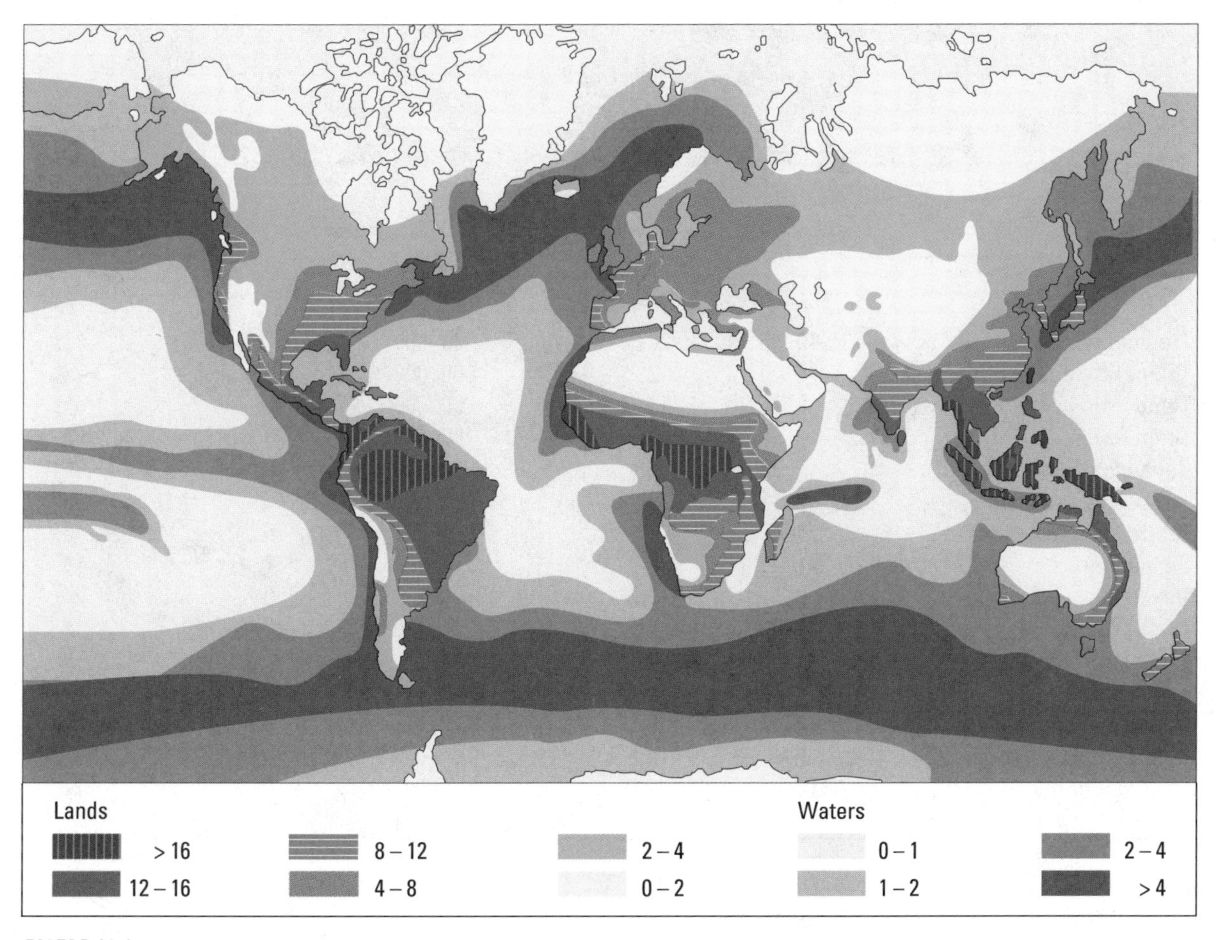


FIGURE 26.9 ▼ The primary productivity of terrestrial and oceanic systems. Units are $t\ ha^{-1}\ y^{-1}$ (after Lieth, 1965)

## TERRESTRIAL ECOSYSTEMS

We can compare ecosystems on the basis of their **net annual primary productivity** (the quantity of new organic matter produced each year). Three main terrestrial ecosystems will be discussed here. We will discuss others in more detail later (Table 26.1).

Woodland (or forest) ecosystems represent the natural state of much of the earth's land surface. That such ecosystems only occupy about one-quarter of that surface today is to a large degree a result of human activity, mainly in the pursuit of agricultural production. Woodland systems differ markedly in their detailed structure, but on the whole they represent the most complex terrestrial ecosystems. They contain many different structural levels (Figure 26.10), and tend to have a very high rate of primary production. In tropical areas, for example, the rain forests have a net annual production of about 20 $t\ ha^{-1}$. Even in the midlatitudes, a mixed oak forest is capable of producing as much as 15 $t\ ha^{-1}$ of organic matter each year.

In addition to their high annual rate of primary production, forest ecosystems are distinctive in that they store a disproportionate quantity of nutrients and organic matter in the vegetation. As we will see, this is a feature that causes some difficulty when people come to exploit these areas, for in clearing the vegetation they remove most of the accumulated nutrients within the system; the system that is left is therefore relatively unproductive.

In contrast, natural grassland ecosystems have a relatively low annual productivity, with average rates of about 5 $t\ ha^{-1}$. This is somewhat less than the average achieved by many agricultural systems. Cereal farming, for example, can attain productivities of several thousand tonnes per hectare each year, while the average productivity in midlatitudes is

## Table 26.1

*Net primary productivity and total biomass of selected ecosystems*

| Ecosystem | Net primary productivity ($t\ ha^{-1}\ y^{-1}$) Range | Mean | Biomass ($t\ ha^{-1}$) Range | Mean | Area $10^6\ km^2$ |
|---|---|---|---|---|---|
| Tropical rain forest | 10–35 | 22 | 60–800 | 450 | 17 |
| Temperate deciduous forest | 6–25 | 12 | 60–600 | 300 | 7 |
| Temperate evergreen forest | 6–25 | 13 | 60–2000 | 350 | 5 |
| Taiga forest | 4–20 | 8 | 60–400 | 200 | 12 |
| Savanna grassland | 2–20 | 9 | 2–150 | 40 | 15 |
| Temperate grassland | 2–15 | 6 | 2–50 | 16 | 9 |
| Arctic and alpine tundra | 0.1–4 | 1.4 | 1–30 | 6 | 8 |
| Semidesert | 0.1–2.5 | 0.9 | 1–40 | 7 | 18 |
| Desert | 0–0.1 | 0.03 | 0–2 | 0.2 | 24 |
| Arable land | 1–40 | 6.5 | 4–120 | 10 | 14 |
| Lakes and streams | 1–15 | 4 | 0–1 | 0.2 | 2 |
| Estuaries | 2–40 | 15 | 0.1–40 | 10 | 1.4 |
| Upwelling ocean zones | 4–10 | 5 | 0.05–1 | 0.2 | 0.4 |
| Continental shelves | 2–6 | 3.6 | 0.01–0.4 | 0.01 | 26.6 |
| Open ocean | 0.02–4 | 1.25 | 0–0.05 | 0.03 | 332 |

*Data* R.H. Whittaker and G.E. Likens, 'The biosphere and man,' in H. Leith and R.H. Whittaker (eds.), *Primary Productivity of the Biosphere* (1975), Springer Verlag, pp. 305–28.

between 5 and 10 t ha$^{-1}$ yr$^{-1}$. In agricultural ecosystems, of course, human influence is of fundamental importance. We control and maintain the system, often by using relatively high levels of inputs. Thus, primary net production figures may be misleading; they may be achieved at the cost of considerable inputs from other systems. In addition, agricultural ecosystems are often highly simplified, consisting of extensive monocultures of grains or other food crops. The food chains are deliberately shortened, and unwanted components such as predators and competitors are eliminated. Thus, they tend to be dominated by a very small number of species, sustained by the application of fertilizers and pesticides, and exploited through the use of machinery requiring fossil fuels. Moreover, as we will see later, the outputs from the system are much greater than those of natural ecosystems, and it is not always clear that the system is being maintained in a condition of longterm stability; exhaustion of nutrient reserves, and damage to the soil, may threaten many agricultural ecosystems.

## AQUATIC ECOSYSTEMS

Aquatic ecosystems cover more than two-thirds of the earth's surface, and although they have a relatively low annual productivity, they are clearly important due to their large areal extent (Table 26.1).

Within the oceans, the character of the ecosystem tends to vary according to water depth and the proximity to land. In general, shallow, littoral waters are more productive than deep, mid-ocean waters, largely because of the greater food supply available from the land. Estuarine areas, for example, may be as productive as the most productive terrestrial ecosystems, for they receive vast inputs of nutrients from the rivers. The continental shelves are less productive, but are nevertheless important for humans as the source of most fishery products. Open ocean areas, which cover about 60 percent of the globe, are generally the least productive, save where convergence of the deep currents results in upwelling cool waters in otherwise warm areas. This produces active plankton growth, which gives rise

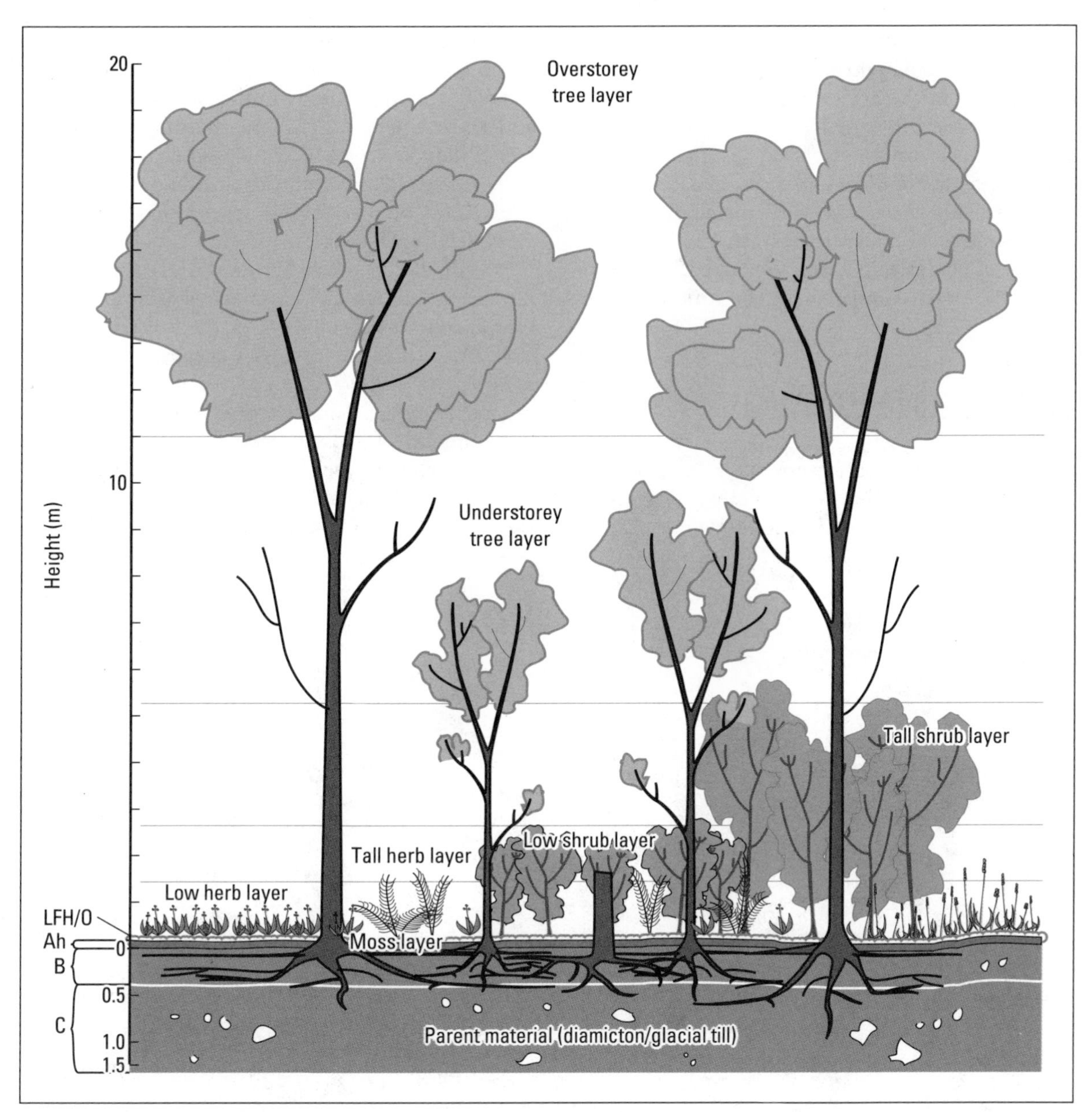


FIGURE 26.10 ▼ Vertical structure of a typical aspen (*Populus tremuloides*) woods. Some overlap occurs among structural levels (or synusia); for example, the tall shrubs may extend into the understorey tree layer

in turn to increased productivity of other marine animals. Many important fishing areas, such as Peru and southwest Africa, are based upon this phenomenon. High productivity in marine ecosystems also results in the formation of large terrestrial deposits of **guano**, the waste excretions of the sea birds that feed on the fish. Guano is rich in phosphate and is exploited as a fertilizer.

Lacustrine ecosystems may also be productive, again mainly due to high inputs of nutrients from the surrounding land. These inputs may in fact become a problem, for excessive quantities of nitrogen and phosphorus compounds from farmland and human waste often accumulate in inland waters, leading to **eutrophication**. The nutrients stimulate intense algal growth that, upon dying, is attacked by organisms in the decomposer food chain. These decomposers consume oxygen, and levels may drop below what some animal species require, so that they migrate or die. Beneath the ice in winter, major fish kills may occur when oxygen being depleted by decomposers cannot be replenished by diffusion from the atmosphere. Indeed, one of the problems of lacustrine ecosystems is their isolation; they may act almost as closed systems, in which inputs may accumulate and force major ecosystem changes. This is the case with the acidification of lakes throughout eastern North America as a consequence of acid precipitation generated by automobile and industrial emissions to the atmosphere.

## An approach to ecosystems

One way of studying ecosystems is to concentrate upon the biogeochemical cycles. This is the approach we will adopt in the following chapters. As we have seen, two main components to these cycles can be identified: those operating within the soil and related primarily to the detritus food chain; and those operating within the plant layer and related largely to the grazing food chain. Later, we will trace the flow of nutrients through these main subsystems before looking at specific examples of ecosystems in the world.

We should be aware, however, that the cascade of nutrients through an ecosystem is associated with morphological development of the soil and vegetation. Over time, the processes operating at the surface of the earth act to create specific soil and vegetation types. Thus, we need also to analyze the development of ecosystem form and structure. In the next chapter, we will consider one aspect of this: the nature of soil formation. We can use this as a starting point for our study of nutrient cycles.

# Soil formation

## Introduction

When rocks are exposed to the atmosphere, they undergo weathering. In Canada, much of the landscape has been subjected to glaciation, and as a consequence, the bedrock is often deeply mantled by unconsolidated deposits dating from the Quaternary Period. Weathered bedrock and unconsolidated surficial deposits collectively form a veneer of material known as **regolith**, which is chemically similar to its source or **parent material**. In many cases, the regolith is able to support plants and animal life, and in time, organic remains accumulate and become mixed with the mineral matter. In this way, soil is formed. **Soil** is the mixture of parent material and organic matter that develops at the earth's surface and is capable of supporting plants.

In many areas of the world, soils have been developing for perhaps millions of years. In Canada, much of the landscape has only recently been exposed to weathering processes (less than 15 000 years). Soils over most of northern North America are therefore very young, compared to other areas of the globe. For example, northern Russia has similar climate and vegetation, but soil-forming processes (**pedogenesis**) have affected the soil over a much longer time period.

Soil is not a disorganized mixture of material; it has an internal organization and, in the words of the early Russian soil scientist V.V. Dokuchaev, it is 'an organized natural body.' We can see the truth of this if we dig a pit and look at the soil (Figure 27.1). Normally, we find that it consists of **horizons** or layers of material of different colours, composition, and structure. These horizons are not normally depositional features—they have not formed by the sequential accumulation of sediments—but are the result of pedogenic processes operating within the soil itself. Together, they form the **soil profile**, and they reflect the combined action of many different and often conflicting pedogenic processes. In this chapter we will consider the nature of these processes and the way they influence the development of the soil profile.

It is important to stress the role of soils within ecosystems. They are a fundamental component of terrestrial ecosystems. They provide a substrate for plants, a medium within which their roots can take anchorage, and a source of water and nutrients for the vegetation. Soils provide a habitat for many of the organisms involved in cycling nutrients through the ecosystem. In addition, outputs from the soil provide inputs to aquatic ecosystems: lakes, rivers and coastal waters derive many of their nutrients from the soils of neighbouring areas. The character of the soil therefore exerts a major control upon ecosystem processes and, more particularly, upon the productivity of the ecosystem.

## COMPOSITION OF THE SOIL

In addition to the solid material—the organic and inorganic matter derived from weathering and plant decay—the soil consists of water and air. It is

FIGURE 27.1 ▼ A soil profile. This Turbic Cryosol lies under tundra vegetation in the Richardson Mountains of the NWT, at kilometre 53 of the Dempster Highway. The permafrost table forms the base of the soil pit here; cryoturbation has resulted in the burial of surface organic matter; gleying is common in the B horizon (photo: G.P. Kershaw)

referred to as a three-phase medium, implying not only that it consists of three main components, but also that significant transfers of matter and energy take place between them. The relative proportions of the three phases vary considerably, although in many soils the solid phase makes up about 50 percent of the total volume, and the air and water make up the remainder (Figure 27.2). Air and water compete for a place in the spaces between the solid particles, and their relative quantities change over time in response to precipitation inputs and losses by drainage, transpiration, and evaporation.

The composition of the three components also varies. Air within the soil is similar to that of the open atmosphere, but is enriched with carbon dioxide and deficient in oxygen (Table 27.1), due to plant respiration. The composition of air in the soil changes over time, depending on the rate of organic activity and the ease with which the gases can diffuse through the soil to the open atmosphere. The soil water also differs somewhat from the water we find in lakes or rivers. It tends to be much richer in dissolved substances washed from the soil and vegetation and, because it is attracted to soil particles, is much less free to move.

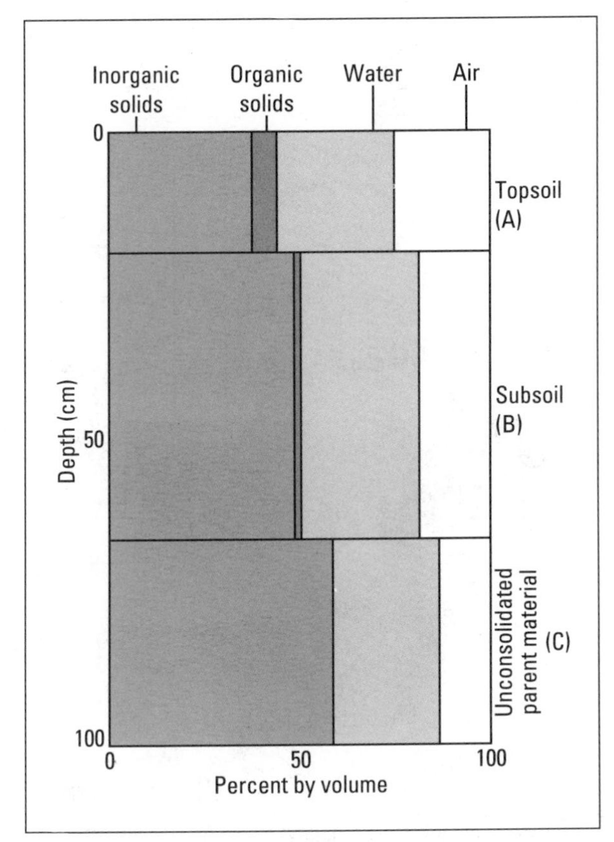


FIGURE 27.2 ▼ Composition of the soil: typical proportions of inorganic and organic solids, water, and air in the soil profile

Table 27.1

*Composition of the soil air relative to the open atmosphere*

| | *Percent by volume* | |
|---|---|---|
| | *Soil air* | *Open atmosphere* |
| Nitrogen | 79.0 | 79.01 |
| Oxygen | 18.0–20.8 | 20.96 |
| Carbon dioxide | 0.15–0.65 | 0.03 |

Both the organic and inorganic matter that make up the solid material of soil vary considerably. The inorganic fraction consists, in the main, of partly weathered rock fragments and minerals, including in particular the more resistant materials such as quartz, feldspar, clay minerals, and compounds of iron and aluminum. The composition of the inorganic material depends upon the nature of the parent material, the climate, and the time available for weathering, so many other minerals may also be present. These materials vary considerably in form, ranging from large stones or boulders to minute clay particles. The finer particles—the sand, silt, and clay fractions of the soil—rarely occur as individual grains, but are bound together by cohesion and various cementing agents into aggregates.

The organic fraction of the soil is similarly variable. It consists of the living and dead cells of animals and plants and the organic acids (e.g., fulvic and humic acid) formed by decomposition of these materials. In some cases, organic compounds occur as discrete horizons or masses of organic matter; in peat, for example, essentially all of the soil is composed of plant debris. But, in most cases, the organic matter is intimately mixed with the mineral material. Organic material plays an important part in the soil, helping to cement the particles together and supplying a vital store for nutrients.

Although much of the organic matter in the soil is relatively transient, decomposing to form organic acids or being liberated into the atmosphere as gases, some of it—the portion called **humus**—is persistent. Humus and the inorganic solid fraction are the main constituents of the soil profile, and differences in the character, arrangement, and proportions of these components give rise to different soil profiles.

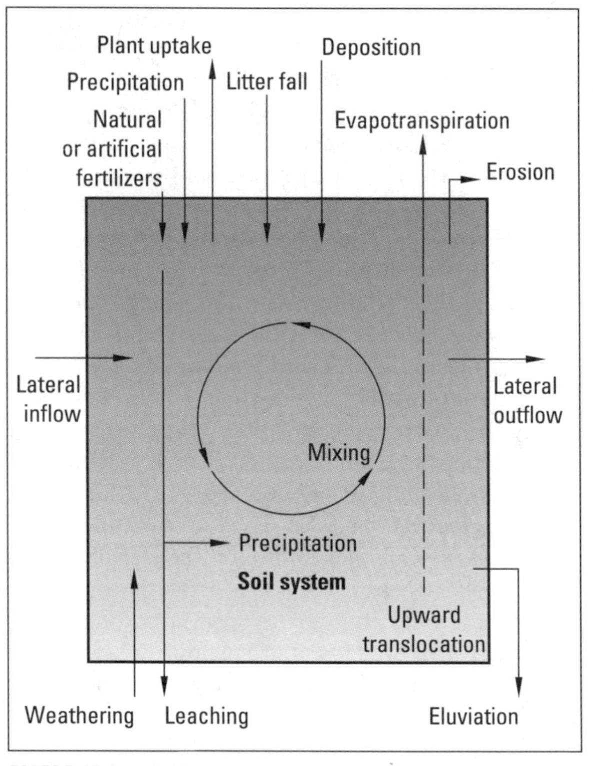


FIGURE 27.3 ▼ The soil system

# The soil-forming system

## INPUTS TO THE SOIL SYSTEM

If we think of the soil profile as a system, we can readily appreciate that, like other systems, its form depends upon three groups of processes: the inputs of raw materials to the soil, the loss of materials from the profile, and the internal transfers and reorganization of these materials within the profile (Figure 27.3).

The main inputs of matter come from the parent materials of the soil. These are released by weathering at the base of the soil profile, and contribute to the lower horizons. We can often see evidence of this input at the bottom of the profile: joints are opened

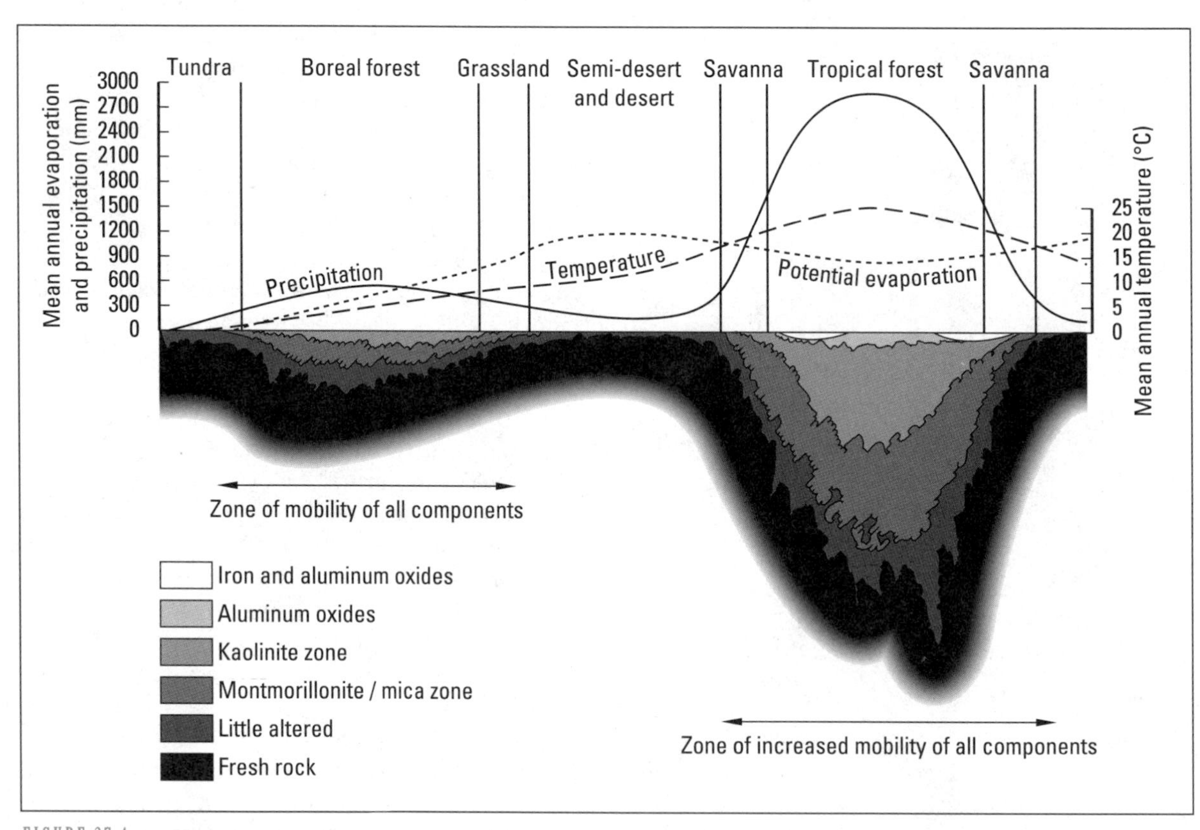


FIGURE 27.4 ▼ Variations in soil composition and depth in relation to climate (after Strakhov, 1967)

up and rock fragments occur in a state of partial detachment. The rate at which these materials are released depends upon the rate of weathering, which is a function of parent material composition and climate. On massive, resistant rocks, for example, the rate of weathering is often slow, with the result that soils are thin. On softer, less resistant lithological types, much more rapid weathering may occur and larger quantities of material are released, resulting in deeper soils. Similarly, where climatic conditions favour active weathering, deeper soils develop. As we saw in Chapter 18, rates of chemical weathering tend to be greatest under conditions of high temperature and humidity, and, not surprisingly, we find that soils in tropical areas are often several metres deep, while those in polar regions are shallow and poorly developed (Figure 27.4).

In addition to the inputs from parent material, significant inputs of material come from surface accumulations. If amounts of this material are large enough, then buried soils or **paleosols** form. One of the most important situations in which this occurs is in alluvial valleys, where flood waters deposit silt on the floodplain. Indeed, many early civilizations became established in alluvial valleys such as the Indus, Nile, and Ganges, for the annual floods helped to replenish the soil and renew the nutrients removed by agriculture. In natural systems, alluvium accumulation is no less important (Figures 27.5 and 27.6). Less welcome accumulations also occur on soil in the desert fringes, for, as the deserts expand, winds carry sand onto the surrounding

FIGURE 27.5 ▼ Loess derived from outwash and deglaciated surfaces has buried a soil at this location near Carmacks, central Yukon. After the deposition of the loess, soil developed on the new surface (photo: G.P. Kershaw)

FIGURE 27.6 ▼ Sedimentation associated with flooding along one of the channels of the Mackenzie River Delta, NWT, produces unstable surfaces for plants. Flooding at this site is an annual event, and soil development is limited by rapid rates of silt accumulation. Plants that can keep pace with the sediment accumulation are rhizomatous species such as horsetails (*Equisetum* spp.) and sedges (*Carex* spp.) (photo: G.P. Kershaw)

agricultural lands, resulting in gradual encroachment of the deserts. During the Quaternary Period, widespread surface accumulations also took place in the northern hemisphere as winds blew fine-grained glacial debris across the frozen ground, forming extensive loess sheets that in some areas are tens of metres thick (see Figure 24.5). At times, when wind activity declined and the surface became more stable, these deposits were weathered and reorganized to form soils, so that many loess deposits now consist of alternating sequences of loess and soil. On a local scale, surface accumulation occurs in many footslope areas, when mass movement and surface runoff carry material downslope.

As we have seen, organic inputs to the soil are also important. These are derived mainly from the vegetation, which provides an annual input of debris in the form of leaves, plant litter and dead roots, which accumulate at the surface. The character of this material is strongly influenced by the nature and type of plant cover. In coniferous forests, the organic debris is typically acidic and resistant to decomposition, with the result that it accumulates as a distinct layer of humus. Debris from deciduous trees is generally broken down and mixed with the soil more rapidly.

The rate of organic inputs also depends upon the density of the vegetation. In the tropics, the annual input of plant debris may total as much as 0.02 t $ha^{-1}$, while in the sparsely vegetated desert margins, as little as 0.001 t $ha^{-1}$ of organic carbon may accumulate each year.

Another source of inputs is through **lateral inflow** of solute-rich groundwater or percolating soil water. Where deeply rooted plants can tap these sources of water and nutrients or where discharge occurs at the surface, these inputs may be important to the soil.

## OUTPUTS FROM THE SOIL SYSTEM

Losses from the soil occur through erosion and leaching. Erosion is often the more dramatic process, since it acts at the surface and has direct effects upon soil fertility. Moreover, it is often encouraged by agricultural practices that leave the soil bare and unprotected from the wind and rain. Thus, rates of soil erosion on cultivated land may be many times greater than those from grassland or woodland ecosystems (Table 27.2). Where winter snowpack develops, spring runoff can be extremely effective in eroding soil, since all the precipitation collected over winter is released during the short melt season.

The susceptibility of the soil to erosion is a function of many factors, including climate, soil conditions, vegetation cover, topography, and disturbance. We saw the way these factors interact to influence wind erosion in Chapter 24, and similar interactions affect erosion by rainfall (Figure 27.7). In both cases it is clear, however, that agriculture plays a significant role in causing and controlling losses by erosion.

### Table 27.2

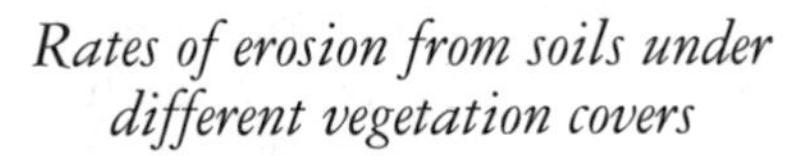
*Rates of erosion from soils under different vegetation covers*

| *Vegetation cover* | *Rate of soil loss (kg $ha^{-1}$)* | *Runoff (percent of rainfall)* |
|---|---|---|
| Woodland | 0.82 | 0.09 |
| Grassland | 4.2 | 0.29 |
| Rotation crops | 4 440.8 | 8.8 |
| Cotton | 10 320.9 | 10.5 |
| Bare soil | 26 604.1 | 29.1 |

*Data* H.H. Bennett, *Soil Conservation* (1939), McGraw-Hill.

**Leaching** is a more insidious process, for it is less readily visible and often less catastrophic in its effect. It involves the loss of soil materials in solution, and is most active under conditions of high precipitation, rapid drainage, and a plant cover that produces acidic litter (e.g., spruce trees and Labrador tea). These conditions favour the development of abundant hydrogen ions in the soil water, which displace other nutrient from the soil through hydrolysis. The percolating waters carry these substances downward through the soil profile, depositing some in the lower layers, but removing the most soluble elements entirely through lateral outflow.

In montane, temperate, and arctic environments where the climate is cool or where the soil is frozen for much of the year, decomposition is slow. This is due, in part, to the short period during which soil

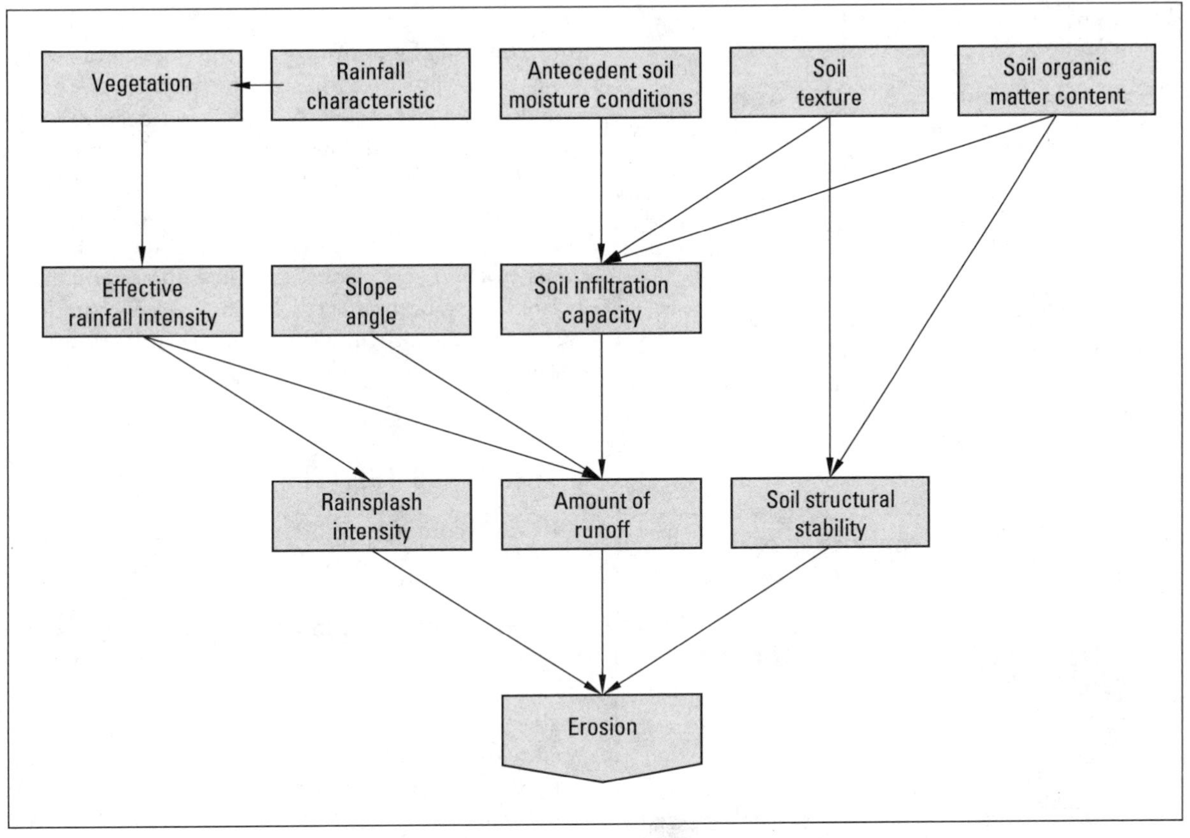


FIGURE 27.7 ▼ Factors affecting soil erosion by rainfall

biota are active. In addition, physio-chemical processes are less effective at lower temperatures. Fire oxidizes organic material, and with severe ground fires these stored nutrients can be lost. However, when the fire is less severe, burning can release nutrients to the soil if the ash remains on the surface. Furthermore, heat penetrating the thinner organic layers can enhance decomposition, and release nutrients, increasing the rate of nutrient cycling. Thus, depending on the nature and type of wildfire, the soil can experience a range of effects from losses to gains of nutrients.

Many other processes also remove material from the soil profile. Plants use many substances, and in some cases, people have a direct effect. For example, in the Republic of Ireland, almost one-third of the country's energy is derived from peat. Significant losses may occur in the form of gases. Organic matter, in particular, is lost in gaseous form, as we will see in Chapter 29, for it decomposes in the soil to form carbon dioxide, hydrogen, and oxygen which escape to the atmosphere. This process is encouraged by cultivation, for during tillage the soil is loosened and aerated so that organic matter decomposition is speeded up and the rate of gaseous diffusion is increased. As a result, cultivated soils often have less organic material than the grassland or woodland soils that would have existed before agricultural development (Table 27.3).

## Table 27.3

*Organic matter content of soils under different vegetation covers*

| *Vegetation cover* | *Organic matter content (percent)* |
|---|---|
| Woodland | 5–15 |
| Grassland | 8–10 |
| Grass–arable rotation | 2–5 |
| Continuous arable | 1–3 |

# Soil profile development

The relative magnitudes of inputs to and outputs from the soil determine whether or not the soil profile continues to develop. During the early stages of soil development, inputs characteristically exceed outputs and the soil depth increases. Under conditions of excessive erosion, for example when steep slopes are cultivated, outputs may exceed inputs and the soil profile becomes truncated (Figure 27.8). The character of the soil profile, however, depends upon the processes of reorganization within the soil. These involve movement of soil materials (either upward or downward) by water, soil organisms, and cultivation. Some of these processes segregate the material and favour the development of horizons, while others cause mixing of the soil and destruction of horizons.

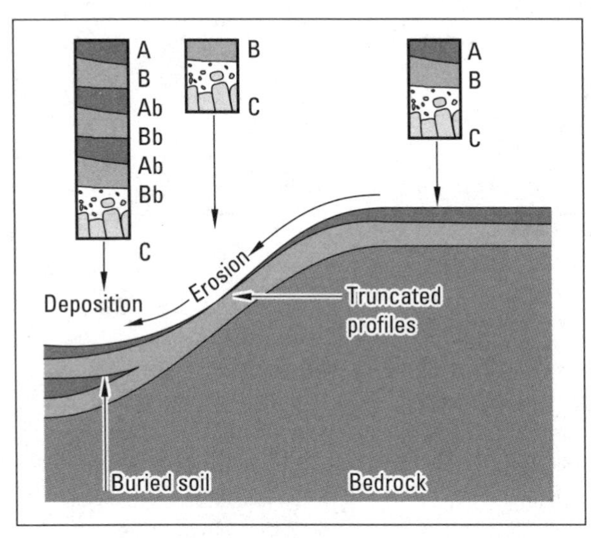


FIGURE 27.8 ▼ Profile truncation and the formation of buried soils on steep slopes

## CLAY TRANSLOCATION

Water is one of the main agents of soil profile development, and downward percolation of water through the soil transports material in both suspension and solution. Movement of material in suspension is common under many conditions. Clay movement, or **lessivage**, involves the detachment and mechanical movement of individual clay particles in the upper layers of the soil and their transport downward through the profile. The process is most active in porous, slightly acid soils containing moderate quantities of kaolinite clay. Clay translocation occurs more readily in soils containing kaolinite because the kaolinite does not expand when wet. Less kaolinite means the permeability is reduced and hence less translocation occurs. Clay particles are washed down root channels, worm tunnels, fissures, and pore spaces and are deposited in the subsoil. Here, it tends to form thin veneers or skins (cutans) around larger particles and along the walls of the pores. Many soils in humid temperate regions have been affected by this process, particularly under conditions of high precipitation and acidic vegetation.

## LEACHING AND CHELUVIATION

Chemical compounds are also washed downward through the soil profile in solution. The process of leaching is of major significance in soils throughout the temperate and humid tropical regions. Rainwater or snowmelt water picks up substances from the atmosphere and vegetation, which makes it slightly acid; in many cases, the precipitation is a weak carbonic acid, due to the presence of carbon dioxide dissolved from the atmosphere (Chapter 18). When this water enters the soil, it dissolves substances such as calcium carbonate, and carries the materials into the subsoil. It also removes nutrients from the surfaces of clay and organic particles by a process of cation exchange (see Chapter 29) and washes these materials downward.

**Cheluviation**, or **chelation** (from the Greek word meaning 'to claw'), is a similar process. Rainwater or snowmelt water, picking up organic compounds from the vegetation or the surface layers of the soil, is converted to a weak organic acid. These organic compounds are then able to react with mineral elements in the soil to form organo-mineral complexes. These complexes of organic and mineral materials are highly mobile and are readily washed down through the soil to accumulate at depth. As with leaching, cheluviation is a process encouraged by plants that produce acidic litter.

Both leaching and cheluviation remove the more soluble elements from the surface layers of the soil. Under extreme conditions, this can leave the topsoil acidic and relatively infertile. The most mobile compounds may also be removed from the profile, while less soluble materials are precipitated in lower horizons. The depth at which accumulation occurs depends upon the relative solubility of the substance, the amount of water passing through the soil

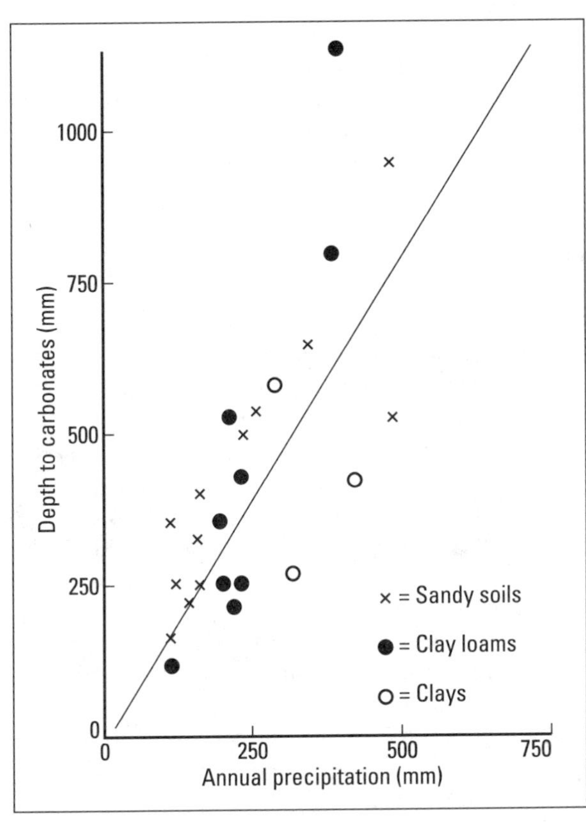


FIGURE 27.9 ▼ The relationship between depth to calcium carbonate and annual precipitation (after Arkley, 1963)

(and hence the climate), and specific chemical conditions within the soil profile. For example, calcium carbonate precipitation depth varies with precipitation (Figure 27.9), while aluminum and iron carried downward by cheluviation seem to accumulate largely as a response to changes in the degree of aeration and the availability of organic substances. Under certain conditions, iron may be precipitated in a distinct layer (an ironpan) within the soil.

Leaching and related processes operate to some extent in most soils. They remove many nutrients, particularly nitrogen, calcium, and phosphorus, from agricultural soils. This contributes to the process of **eutrophication**, when nitrogen and phosphorus are washed into rivers and become concentrated in lakes.

## UPWARD TRANSLOCATION

Net movement of compounds upward through the soil by water occurs mainly in arid and semi-arid areas, where evapotranspiration exceeds precipitation. This results in a tendency for water to be drawn to the surface of the soil along the capillary pores (Chapter 12). At the surface, the water evaporates, and the salts are precipitated. Sodium compounds are particularly affected by this process, and they give rise to alkalization; salt crusts develop in the surface layers (producing solonetzic soils), which can cause considerable problems for farming (Figures 27.10, 27.11; see Figure 12.11). Calcium,

FIGURE 27.10 ▼ A Brown Solodized Solonetz soil from Saskatchewan, in which the solonetzic B horizon is angular blocky; the salt layer appears as a white crust below the knife (photo courtesy Land Resource Research Centre, Agriculture Canada)

FIGURE 27.11 ▼ Solonetzic soils develop where high rates of evapotranspiration draw mineral-rich water to the surface of the soil. Salts of calcium and magnesium are left behind, and in extreme cases a surface crust develops, preventing plant growth. This area is in southwestern Alberta, where moisture deficits can exceed 300 mm (photo: G.P. Kershaw)

iron, and even silica may also be transported in this way. Under tropical conditions, where extreme seasonal variations in rainfall occur, the process is particularly active. Tropical vegetation, high temperatures, and heavy rainfall during the wet season encourage the leaching of iron and silica within the soil; then, during the drier months, water is drawn back to the surface and the iron is precipitated in the upper layers. This process is important in the formation of laterite soils.

## BIOLOGICAL PROCESSES

Various biological processes redistribute soil materials within the profile. Plants, in particular, cause upward translocation of compounds, for their roots (which can reach several metres in depth) extract water from the lower levels in the soil, and draw the dissolved substances to the surface. Since most plant debris is eventually returned to the soil surface, this leads to a net upward movement of nutrients.

Many animals are also active in disturbing and rearranging the soil. The burrowing of rodents, the tunnelling of earthworms, ants, termites, and the smaller scale activity of micro-organisms all aerate and mix the soil.

## PHYSICAL MIXING

Mixing of the soil is also helped by the development and growth of ice crystals. Due to seasonal frost and shallow freeze–thaw activity, soil water freezes, and the resulting ice segregation and expansion mixes the soil. In periglacial environments, where permafrost is present, mixing or frost churning at depth is referred to as **cryoturbation**. Since permafrost underlies as much as half of Canada, cryoturbation is an important pedogenic process (Chapter 22). The formation and growth of ice crystals can also assist in the breakdown of soil aggregates. As the ice melts, fine material is washed into the spaces, so that the soil is reorganized. Under extreme conditions, as in tundra regions, the effect is to sort the soil into patterned ground features (see Chapter 22); where it is less intense, the main effect is to mix the soil and destroy the horizonation boundaries.

People, too, cause physical mixing of the soil. Ploughing, to create a seedbed for crops and to remove weeds and bury stubble, disturbs the soil to a depth of 20 cm or more, eradicating the original horizons and creating an evenly mixed, ploughed layer. Disturbances caused by root growth and uprooting of trees (**tree throw**) (e.g., by wind or animals) also mix the soil.

## GLEYING

**Gleying** results from the effects of oxidation and reduction, but involves little real movement of the soil constituents. During periods of waterlogging, the soil usually experiences anaerobic conditions, for the pore spaces become filled with water and air cannot enter the soil. It is important to remember that even under fully saturated conditions, soil water can be highly oxygenated and oxidizing conditions can prevail. Just because a soil is saturated with water does not mean that reducing conditions prevail. The oxygen that normally acts as a sink for electrons released by plants and organisms in the soil is lacking, and other compounds, such as iron and aluminum oxides, accept the electrons. As we saw in Chapter 18, the result is chemical reduction of these compounds.

Ferric iron is converted to ferrous iron:

$$Fe^{+++} + \text{electron} \rightarrow Fe^{++}$$

and then to a ferrous oxide:

$$2Fe_2O_3 \rightarrow 4FeO + O_2$$

Ferrous oxide is characteristically grey or green in colour. Aluminum, sulphides, sulphates, nitrates, and many other compounds may be reduced in the same way. The soil, as a result, becomes duller in colour, and many of the reduced compounds, which are typically more soluble, may be washed from the soil.

During drier periods, air enters the soil and a reversal of this process occurs. Oxygen accepts the electrons and combines with the reduced substances to form higher valency compounds. The ferrous iron is converted back to the ferric state, which is typically red or yellow. The soil thus becomes brighter in colour.

Alternate phases of reduction and oxidation are common in soils with fluctuations in the water table, and this results in the creation of varied soil colours, known as **mottles**. Oxidation is often greatest along the larger pores, such as root and earthworm channels, due to the improved aeration.

# Soil formation and soil properties

During the course of soil formation, the original raw materials—the parent material and organic matter—are redistributed and altered. In the process, the physical, chemical and biological properties of the soil develop, resulting in the creation of distinct horizons. As shown in Appendix III, it is on the basis of these features that we recognize, describe and classify soils and can deduce both the nature of their formation and their potential use. The number of properties we can consider is almost infinite, and each can tell us something different about the soil's history or management. Let us look, therefore, at some of the main properties of the soil profile, and examine the ways in which they develop.

## SOIL TEXTURE

If we examine a soil sample closely—or better still if we moisten it and rub it between our fingers—we can discern one of the fundamental properties of the soil: its **texture**. Soil texture refers to the relative proportions of sand, silt, and clay. It can also be affected by factors including the size and shape of the individual particles, the moisture content, the amount of organic matter, and the soil structure. **Soil consistency** refers to the way a soil feels, that is, its handling properties. **Bulk density** is a measure of the mass of the dry material per unit volume. As such, it indicates the potential water-holding capacity of the soil and is important in determining the rate of heat gain or loss in the soil. Bulk density will also influence the penetrability of the soil by roots and burrowing animals.

By far, the most important aspect of texture is **particle size**. The material in a soil ranges from large boulders to minute clay particles too small to be discerned by the eye. For the purpose of description, the soil particles are often divided into size classes. Numerous systems defining particle size have been developed. The method used in the United States defines six size grades: boulders, cobbles, pebbles, sand, silt, and clay, on the basis of particle diameter. In Canada, the classes are: stones (over 250 mm in diameter), cobbles (75 to 250 mm), gravel (2.0 to 75 mm), sand (0.05 to 2.0 mm), silt (0.05 to 0.002 mm), and clay (less than 0.002 mm). The most important of these in many ways are the sand, silt, and clay fractions, for these are chemically most active and are able to hold both nutrients and water in the soil. Together, they are referred to as the **fine earth fraction**, and they contribute vitally to soil fertility and water-holding capabilities. Consequently, it is with the fine earth that the soil scientist is mainly concerned when analyzing the soil, and soil texture is classified on the basis of the relative proportions of sand, silt, and clay (Figure 27.12). Each textural class represents a combination of particle sizes that not only feels different, but also reacts differently to environmental conditions.

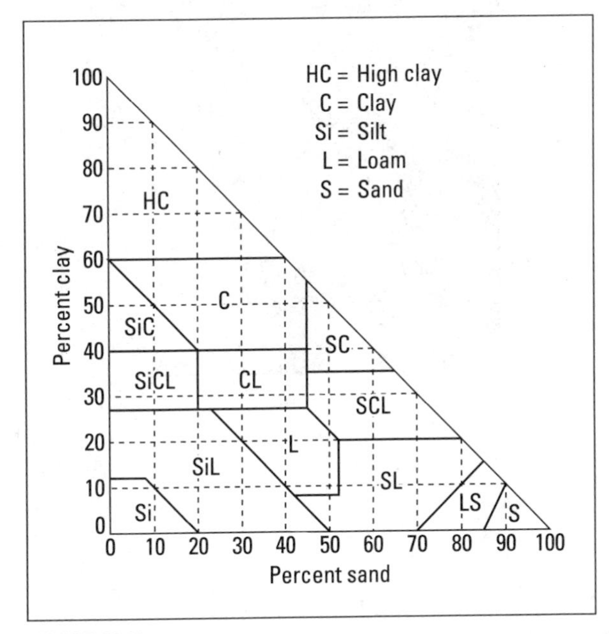


FIGURE 27.12 ▼ Soil texture classes

The texture of soil depends to a great extent upon the character of the original parent material. It is also influenced by weathering. More intense or more prolonged weathering generally results in finer-grained soils. In addition, texture is affected by the processes of soil horizon development. Lessivage, for example, may reduce clay content in topsoil and add clay to the subsoil (Figure 27.13). By examining the vertical pattern of texture in a soil profile, therefore, we get an indication of the history of that soil.

## CLAY MINERALOGY

Of all the constituents of the inorganic fraction of the soil, probably the most important is the clay component. Clay has a fundamental influence on

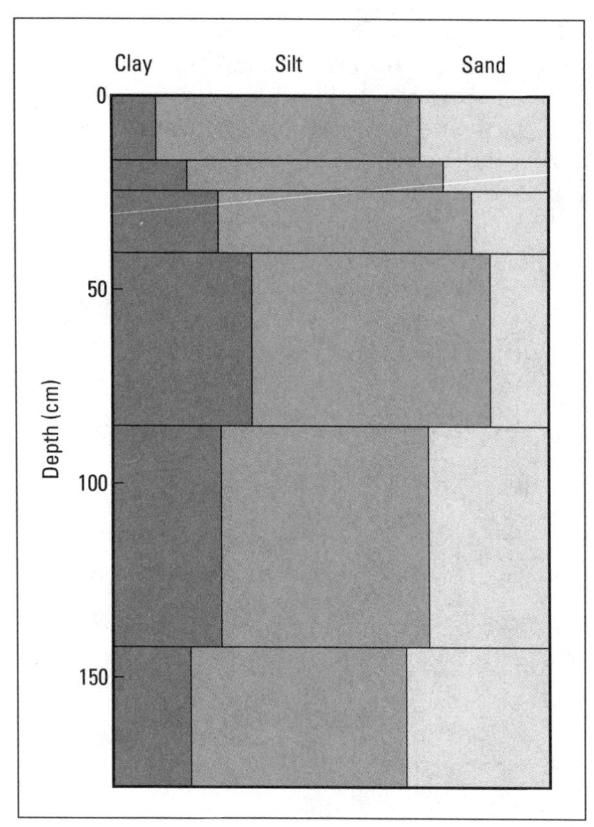


FIGURE 27.13 ▼ Textural profile of a Luvisolic soil, showing the increased clay content in the subsoil as a result of clay translocation (from Jarvis, 1968)

both the physical and chemical character of the soil. It helps to give the soil structure, it holds water in the soil, and it provides the capacity to store and exchange plant nutrients.

These vital attributes arise from the physical and chemical structure of clay particles. Clay comprises particles less than 0.002 mm in diameter. The small size of the particles means that they have a very large surface area relative to their volume, and this makes them highly reactive. Some clay particles (those less than about 0.001 mm) can behave as **colloidal materials**, with the ability to disperse spontaneously in a solvent such as water. This property, in part, accounts for the tendency for clay particles to be translocated by acidic, percolating waters.

In addition, many clay particles are composed of aluminum and silica, arranged in a lattice-like structure which is, to some extent, flexible and plastic. This allows the clay particles to shrink and swell, and it also permits them to absorb water and other substances into their structure. Moreover, clay particles have a net negative electrical charge, giving them the capacity to attract and hold (adsorb) positively charged ions, and to repel negatively charged surfaces.

*The structure of clays* Most clay minerals are silicates. In the past, they were believed to be amorphous, with no regular structure, but since it has been possible to examine them with Xrays and electron microscopes, we see that they do, in fact, have a definite crystalline structure.

The basic building blocks of this structure are atoms of silica, aluminum, oxygen, and hydrogen. Silica and oxygen are arranged in the form of a single $Si^{++++}$ atom surrounded by four $O^{--}$ atoms, giving a four-sided molecular structure. This is known as the silica tetrahedron (Figure 27.14). Adjacent tetrahedra are linked by shared oxygen ions to produce a sheetlike layer. At the same time, each aluminum atom is surrounded by six oxygen or hydroxyl ($OH^-$) ions to give an eight-sided molecular structure, called the aluminum octahedron (Figure 27.14). Again, these are linked by shared oxygen atoms, producing an octahedral sheet.

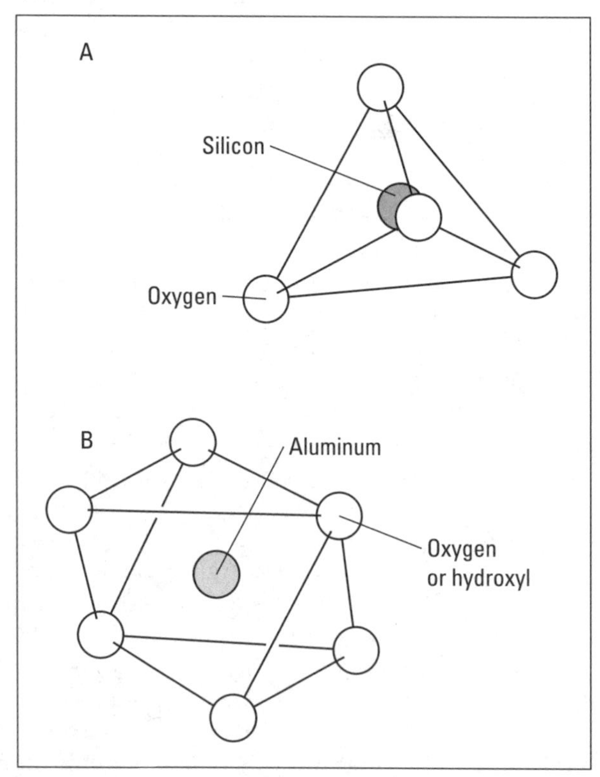


FIGURE 27.14 ▼ The structure of the silica tetrahedron (A) and aluminum octahedron (B)

The silicate clays are built from various combinations of three sheets: the silicon tetrahedra, aluminum octahedra, and magnesium octahedra. Alternating sheets of silicon tetrahedra and aluminum octahedra produce what are called 1:1 clays, the most widespread example of which is kaolinite. Each pair of sheets is held together by hydrogen ions, making a relatively rigid and stable structure. In the 2:1 clays, sheets of aluminum octahedra are sandwiched between two silicon tetrahedra sheets. In minerals such as montmorillonite, each sandwich is, in turn, linked by weak oxygen bonds. As a result, these 2:1 clays are relatively flexible and are able to expand. Moreover, their internal surfaces are open to chemical reactions, making them highly reactive. Not all 2:1 clays are of the same structure, however, and in illite, for example, the crystal units (the 'sandwiches') are held much more rigidly by potassium ions. This produces a more stable, nonexpanding clay.

*Clay formation* As we saw in Chapter 18, clays are produced by the weathering of alumino-silicate minerals. Two dominant processes occur: neoformation and alteration. **Neoformation** involves the precipitation of new, secondary minerals from solutions containing the dissolved products of the weathering of primary alumino-silicates. A wide range of clay minerals may form, depending upon the chemical composition of the solution and the weathering environment.

**Alteration** involves the selective removal and substitution of certain ions from the primary minerals and a gradual loosening of the crystal structure. **Isomorphous replacement** plays a major role in this process (Chapter 18). Ions of silicon and aluminum are replaced by other ions of similar size but lower electrical charge. It is this process that gives the clay its negative charge. For example, the silicon ion has four unbalanced protons, and thus has a valency (a charge) of four. If silicon is replaced by a divalent ion, such as magnesium ($Mg^{++}$), then the clay effectively loses two positive charges. This must leave two unsatisfied negative charges in the clay. Similarly, if the magnesium is replaced by potassium ($K^{+}$), which is monovalent, another unbalanced negative charge is produced. Because of this, the magnitude of the electrical charge on the clay depends on how many and which ions are available to take part in substitution. In alkaline environments, where there are abundant potassium, magnesium, and other bases, clays with a high charge form. In more acid environments, clays with a lower charge tend to develop. On the whole, the 2:1 clays, such as montmorillonite and vermiculite, have a higher negative charge than the 1:1 clays, such as kaolinite. It should also be noted that some poorly formed crystalline (i.e., amorphous) clays do occur. How they form remains uncertain, but they tend to have a lower electrical charge, and to be less plastic and less cohesive than the crystalline clays.

## SOIL STRUCTURE

We only have to look at the surface of a ploughed field to appreciate that the soil rarely consists of a collection of loose particles. Instead, the particles are arranged into aggregates called **peds**. A number of forces form peds. The initial impetus for aggregation comes from the cohesive forces in the soil associated with clay formation and organic matter decomposition. As we have seen, these lead to the development of small colloidal particles with minute negative electrical charges at their surfaces.

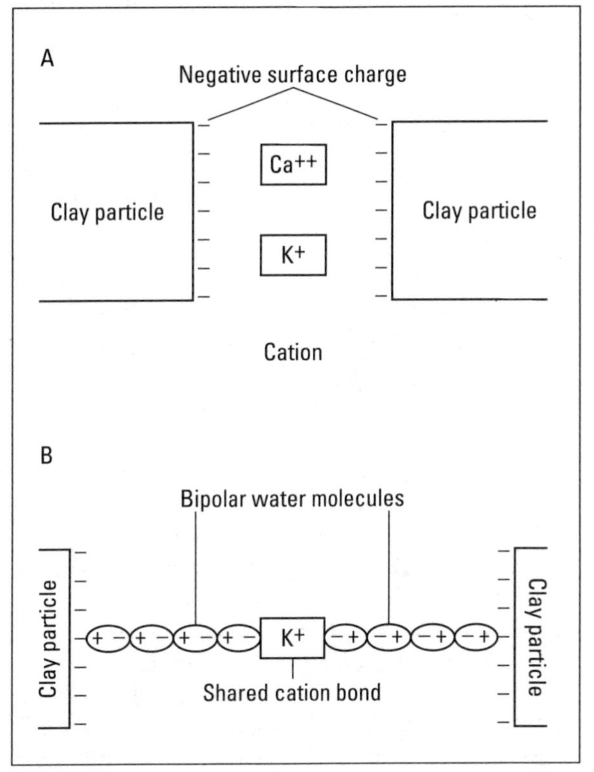


FIGURE 27.15 ▼ Cohesion due to (A) cation bonding, and (B) development of chains of bipoplar water molecules

When these particles are drawn very close together, these electrical charges form a mutual attraction with positive charges inside the clay crystalline structure. In addition, water molecules and positively charged ions, such as calcium ($Ca^{++}$) or potassium ($K^{+}$), may form bridges between the colloids and create small aggregates (Figure 27.15). The colloidal particles may also form 'domains,' which envelop and bind together larger sand and silt grains (Figure 27.16).

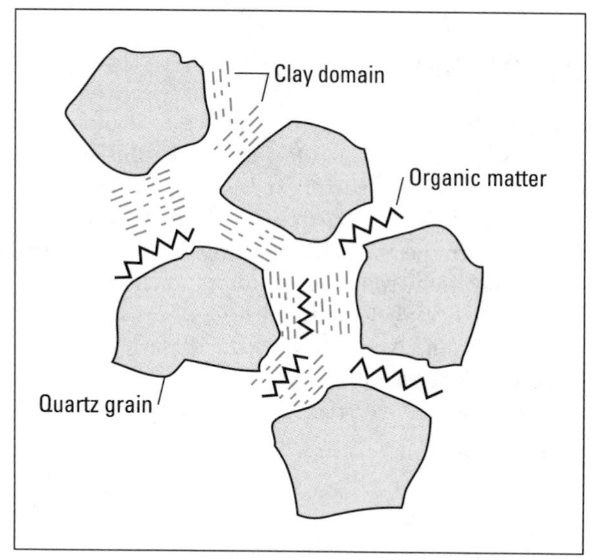


FIGURE 27.16 ▼ The binding of quartz sand grains by clay domains and organic matter. Each 'domain' is composed of a group of tightly linked clay particles

Cohesion plays an important role in generating soil structure, but it is not permanent; when the soil is wetted, for instance, water enters the pore spaces between the particles and may push them apart. The cohesive forces are then reduced. Stability is given to the aggregates by the action of cements in the soil. These include both chemical compounds, such as calcium carbonate and iron oxides, and organic compounds formed during the decomposition of plant materials. Complex organic polymers called **polysaccharides** are particularly important in this respect since they are water-repellent. They therefore act like a waterproof coating on the aggregates, and make the soil resistant to structural disintegration by wetting. In addition, fungi and plant roots may enmesh the particles and bind them together.

It is clear that the soil structure depends to a great extent upon both soil texture and organic activity in the soil. In this context, earthworms play a major part, for they help to decompose the plant debris and mix it with the other soil components. Consequently, structure is often best developed in soils with an active earthworm population. Different types of soil structure are shown in Appendix III.

## SOIL pH

The ability of the soil to support plant life depends upon a wide range of chemical properties. One of the most critical is the **soil reaction** or **pH**. This relates to the concentration of free hydrogen ions in the soil. Hydrogen ions are provided by the dissociation of water, by root activity, and by many weathering processes. As the concentration of hydrogen ions increases, the soil becomes more acid.

Acidity, the concentration of active hydrogen ions, is measured on the pH scale, thus:

$$pH = \log_{10}\left(\frac{1}{H}\right)$$

That is to say: the pH is the logarithm of the reciprocal of the molar concentration of the hydrogen ions in the soil solution. As Table 27.4 shows, the soil is considered acidic when the pH is below 7.0;

### Table 27.4

*The pH scale*

| *pH* | *Hydrogen ion concentration (moles/litre)* | *Hydroxyl ion concentration (moles/litre)* | *Description* |
|---|---|---|---|
| 3.0 | 0.001 | 0.00000000001 | Excessively acidic |
| 4.0 | 0.0001 | 0.0000000001 | Strongly acidic |
| 5.0 | 0.00001 | 0.000000001 | Moderately acidic |
| 6.0 | 0.000001 | 0.00000001 | Slightly acidic |
| 7.0 | 0.0000001 | 0.0000001 | Neutral |
| 8.0 | 0.00000001 | 0.000001 | Alkaline |
| 9.0 | 0.000000001 | 0.00001 | Strongly alkaline |
| 10.0 | 0.0000000001 | 0.0001 | Excessively alkaline |

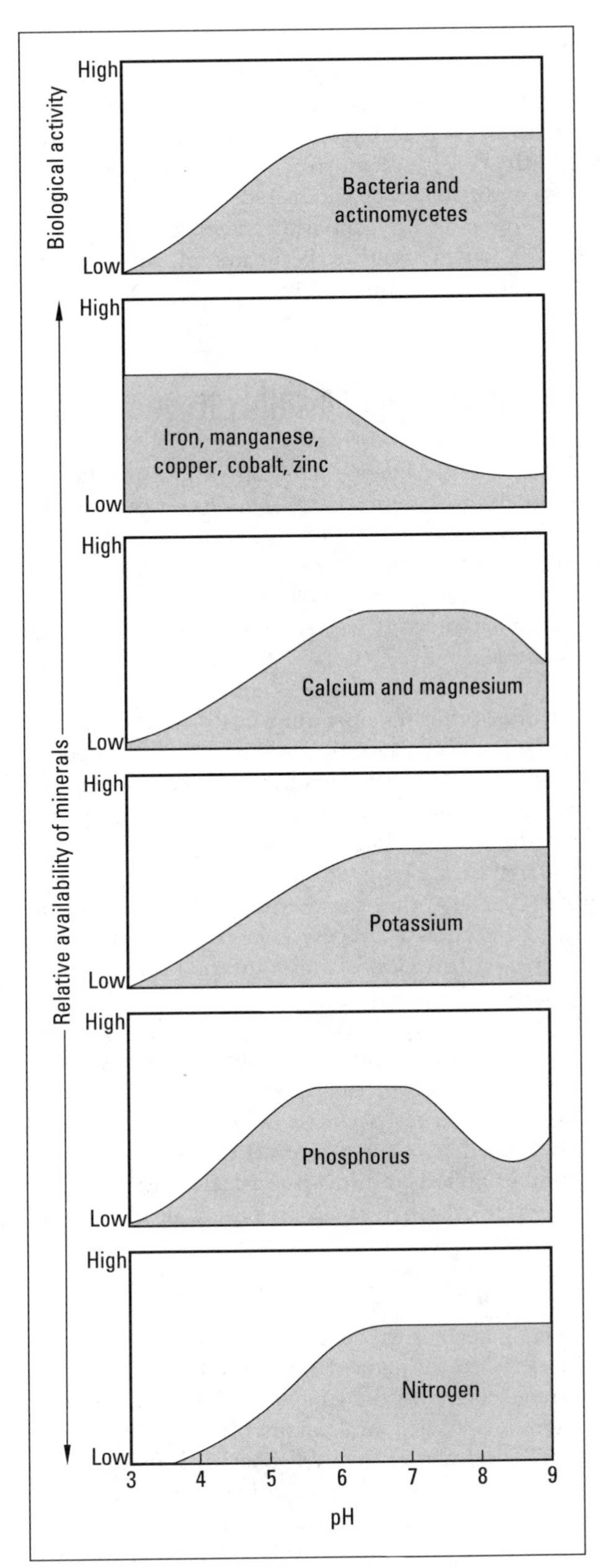


FIGURE 27.17 ▼ The effects of pH on nutrient availability and biological activity

it is alkaline when the pH is above 7.0. Note two things: that acidity increases as pH decreases; and that the scale is logarithmic, so a change of one unit in pH represents a ten-fold change in hydrogen ion concentration.

The importance of pH to soil processes and fertility is mainly that it influences the solubility of many nutrients. At low pH, there is a tendency for most nutrients to be very soluble, so much so that they are readily washed from the soil and become unavailable to plants. At high pH, the nutrients are often insoluble, and plants cannot extract them from the soil. As this indicates, the optimum conditions for plant growth lie between a pH of about 6.0 to 7.0. Soil pH also affects the activity of some soil organisms; in this way it influences the biological cycling of nutrients (Figure 27.17).

Soil pH is very sensitive to soil-forming processes. In calcareous soils, pH is high because the calcium carbonate dissociates to form free carbonate ions:

| $CaCO_3$ | → | $Ca^{++}$ | + | $CO_{3-}$ |
|---|---|---|---|---|
| (calcium carbonate) | | (calcium ion) | | (carbonate ion) |

The carbonate ions then combine with free hydrogen ions to form hydrogen carbonate which is highly soluble and is readily removed from the soil in drainage waters:

| $CO_{3-}$ | + | $H^{+}$ | → | $HCO_{3-}$ |
|---|---|---|---|---|
| (carbonate ion) | | (hydrogen ion) | | hydrogen (carbonate ion) |

Leaching reduces soil pH (makes the soil more acidic) because it removes calcium and leaves the hydrogen ions free in the soil.

## CALCIUM CARBONATE CONTENT

The calcium carbonate content is an important chemical property of the soil, affecting pH and many other soil properties. In particular, it influences soil structure, for calcium ions act as bridges between the colloids, and calcium carbonate is a strong cementing agent. As with pH, the content of calcium carbonate is influenced by leaching; as the intensity of leaching increases, the depth to calcium carbonate in the soil also increases.

## ORGANIC MATTER CONTENT

One of the most important and diagnostic properties of the soil is its organic matter content. As we have seen, organic debris enters the soil from plants and animals, and is then broken down to form humus. In the absence of soil organisms—for example, in very wet or acidic conditions, in cold Arctic and Subarctic environments—the organic debris breaks down slowly and accumulates on the surface. In acidic environments, such as peaty soils, it leads to the development of a **mor** humus layer, consisting of a clearly defined sequence of fresh plant litter, partly decomposed organic matter (the **fermentation layer**), and fully decomposed humus. Where soil organisms are active, however, the organic matter is mixed in with the soil and becomes intimately bonded with the mineral particles. In this state, it is referred to as **mull** humus, and it acts as a cementing agent that stabilizes the soil structure.

The organic matter content of the soil is naturally a reflection of the quantity and character of organic inputs. It is also closely associated with the activity of soil organisms. On one hand, the organisms are the major agents of decomposition; on the other hand, the organic material provides their basic foodstuff. It is also notable that the organic matter is affected by climatic conditions, for rates of decomposition are greatest under high temperatures and low moisture content, whereas accumulation of organic matter tends to occur in cool, wet, or acidic environments (Figure 27.18).

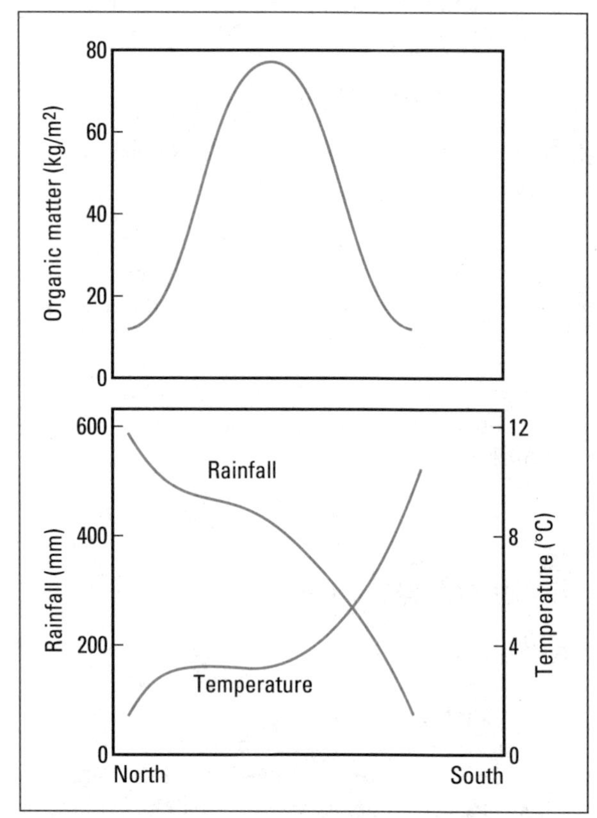


FIGURE 27.18 ▼ The relationship between organic matter content and climate in a transect across central Asia (based on Kononova, 1966)

# Soil formation and soil type

As any farmer knows, soils can vary markedly over short distances. At a broader scale, we can identify more fundamental variations in soil type—from the deep, red lateritic soils of tropical areas, for example, to the ashen grey, podzolized soils of cool temperate latitudes (Chapter 30). What causes these variations?

The immediate answer is differences in the soil-forming processes operating in different places: in the inputs of material by weathering and deposition; in the redistribution of the material within the soil by leaching, cheluviation, lessivation, upward translocation, gleying and mixing; in the losses of material by leaching, erosion, and plant uptake. The soil, like any other component of the environment, can be seen as a process-response system, whose form is a function of both internal and external processes (Figure 27.19).

This, however, merely raises another question: what controls the intensity of these processes? One attempt to answer this question was provided by the American soil scientist H. Jenny in the 1940s. He argued that the soil profile is a product of a number of soil-forming factors, all operating over time. He expressed this relationship in the form of an equation, as follows:

$$S = f(cl, o, r, p, t, \ldots .)$$

where $S$ denotes the soil, $cl$ the climatic factor, $o$ the biotic factor (organisms), $r$ the relief factor, $p$ the parent material, $t$ the time factor, and the dots represent unspecified local factors.

To some extent, Jenny's equation is useful, for it indicates the main factors influencing processes of soil formation, but it is far too generalized to be of much value in trying to explain the detailed distribution of soil types. There is no way we can solve this equation as it stands. We need to consider more closely the relationships between the individual processes of soil formation and the external environmental conditions, so that we can see how the different processes interact under any given set of environmental conditions.

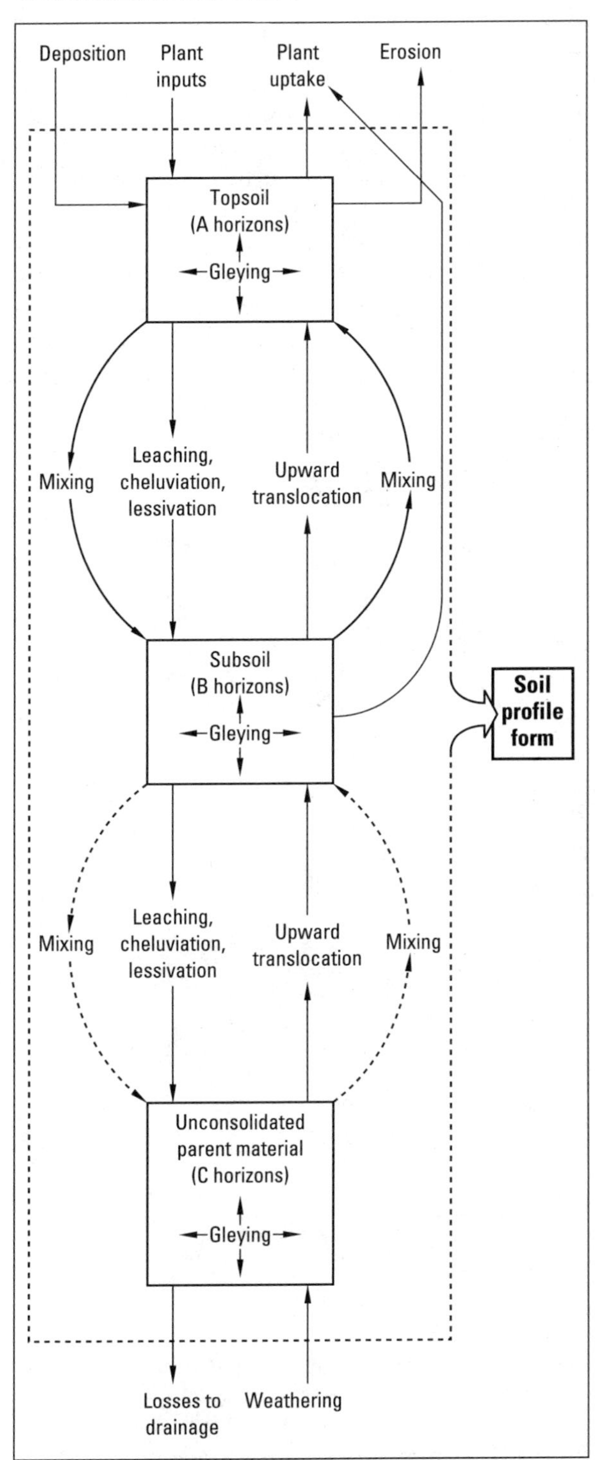


FIGURE 27.19 ▼ The soil as a process-response system

Some indication of these relationships is given in Figure 27.20. As this shows, a major control on soil formation is the availability of water, for water is involved in processes such as leaching, cheluviation, lessivation, and erosion. Water availability, in turn, depends upon the balance between precipitation and evapotranspiration. Similarly, the rate at which water moves through the soil is important, and this is influenced by factors such as bedrock permeability, slope angle, and slope position. Leaching and cheluviation, for example, are inhibited where water flow is restricted, but gleying and organic matter accumulation are enhanced. Water quality is also important, and rates of chemical weathering, as well as many processes of redistribution, are enhanced by increased acidity. This depends, among other things, on the character of the vegetation.

All these relationships are reflected in the character of the soil. It is therefore possible to develop a

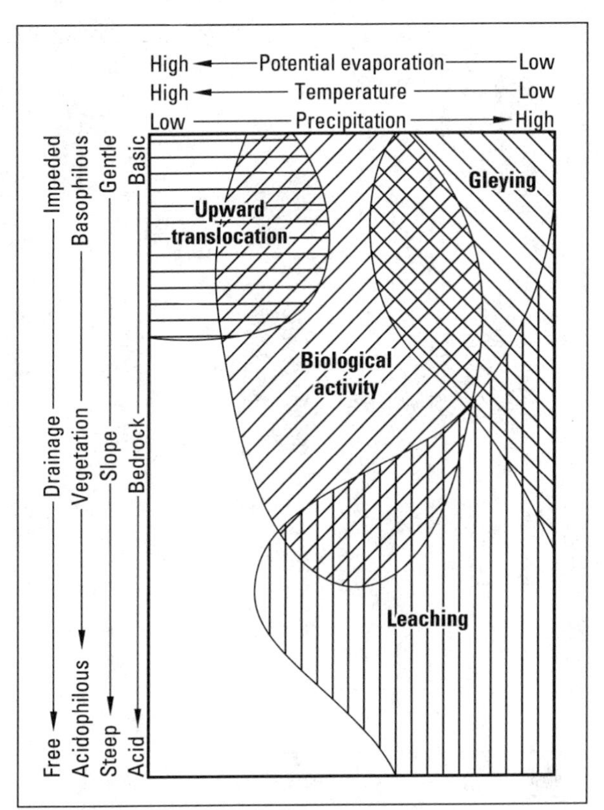


FIGURE 27.20 ▼ The environmental domains of four main sets of soil-forming processes. The diagram shows the zones in which each set of processes is at an optimum in relation to climatic (horizontal axis) and vegetational, geological, and topographical conditions (vertical axis)

general model of soil formation as shown in Figure 27.21. We would expect to find, for example, intensively leached, acid soils associated with areas characterized by some combination of high precipitation (P) relative to evapotranspiration (ET), acidic vegetation cover, acid bedrock, low temperatures, and high rates of water flow. Alkaline and saline soils, on the other hand, tend to occur where the P : ET ratio is low, where temperatures are high, and/or where the vegetation and bedrock are base-rich. Organic soils are associated with wet or acid conditions; shallow, stony soils with dry and cold environments and hard, resistant parent materials.

We will see the implications of these relationships at a global scale in Chapter 30, when we consider the distribution and character of the world's major biomes. An indication of the general validity (and limitations) of the model can be seen if we compare the map of Canadian soils regions (Figure 27.22) and the environmental zones in Canada (Figure 27.23). (See also the map of soils of the world, Figure 30.4.) Soil type is based on our classification using a limited number of criteria. However, there is a close relationship between pedogenic processes and the resulting soil type (Figure 27.24). For example, sod formation dominates in the region of Chernozemic soils on the Prairies, but several other processes occur as well; it is just that they are less important in the development of the soil. At an even more detailed level, we can see sim-

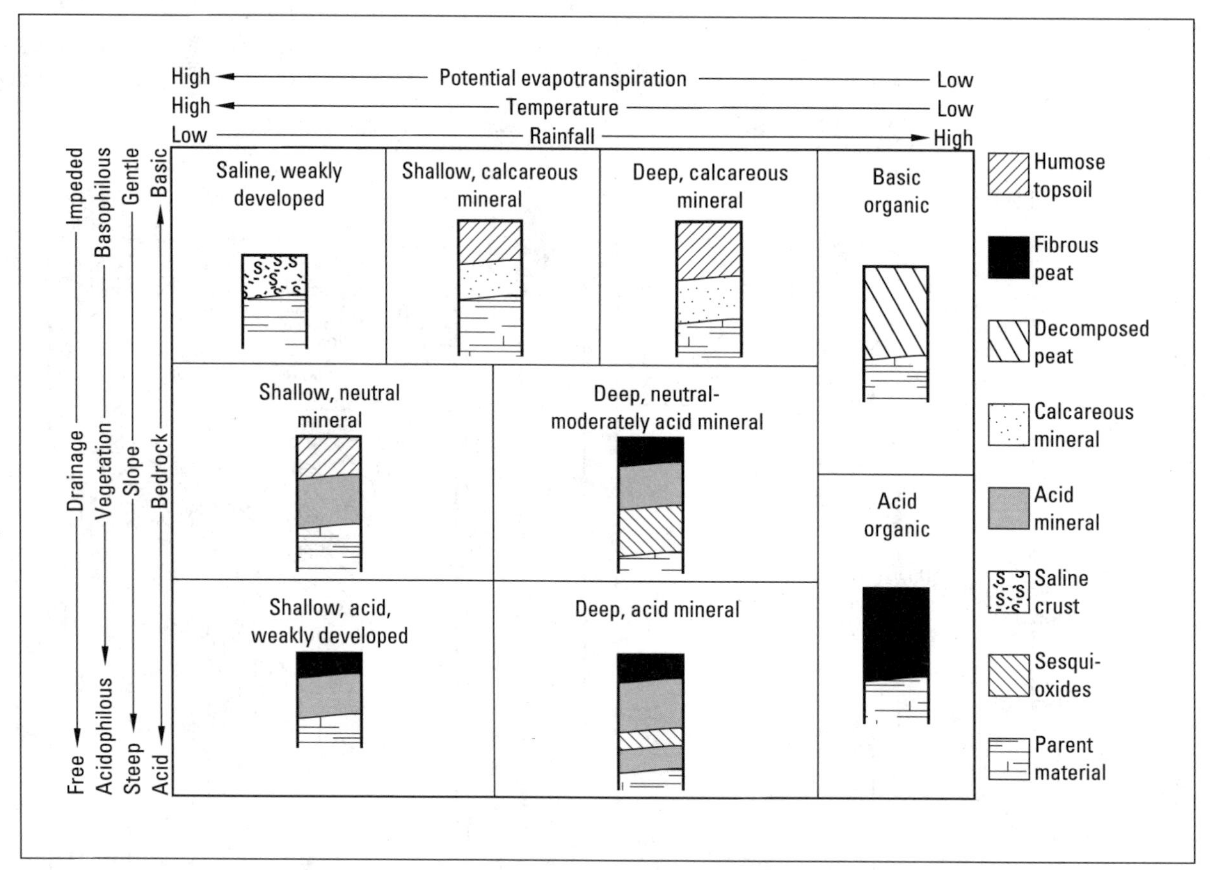


FIGURE 27.21 ▼ Relationships between environmental conditions and general soil types

FIGURE 27.22 ▼ (Right) Soils of Canada. There is a close association between soil type and ecozone in Canada, e.g., Chernozemic soils are found in the Prairie ecoregion, and Cryosolic soils in the Taiga, Boreal, and the Arctic ecoregions. In general, the structural complexity of vegetation is reduced poleward; the Pacific Maritime and Mixed Wood Plains ecoregions have the greatest species richness (map courtesy of J.A. McKeague, Agriculture Canada)

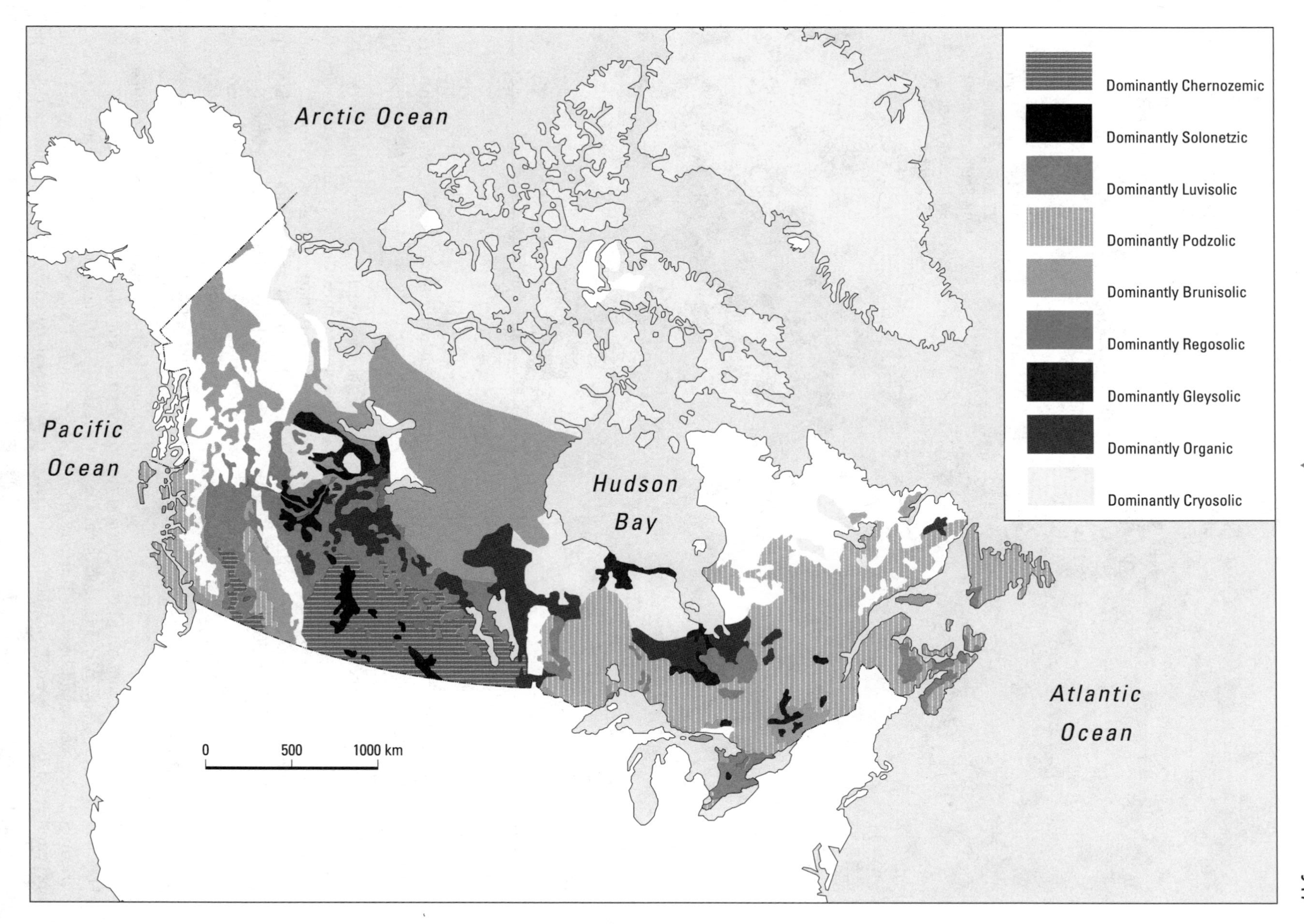
Dominantly Chernozemic
Dominantly Solonetzic
Dominantly Luvisolic
Dominantly Podzolic
Dominantly Brunisolic
Dominantly Regosolic
Dominantly Gleysolic
Dominantly Organic
Dominantly Cryosolic
Arctic Ocean
Pacific Ocean
Hudson Bay
Atlantic Ocean
0
500
1000 km

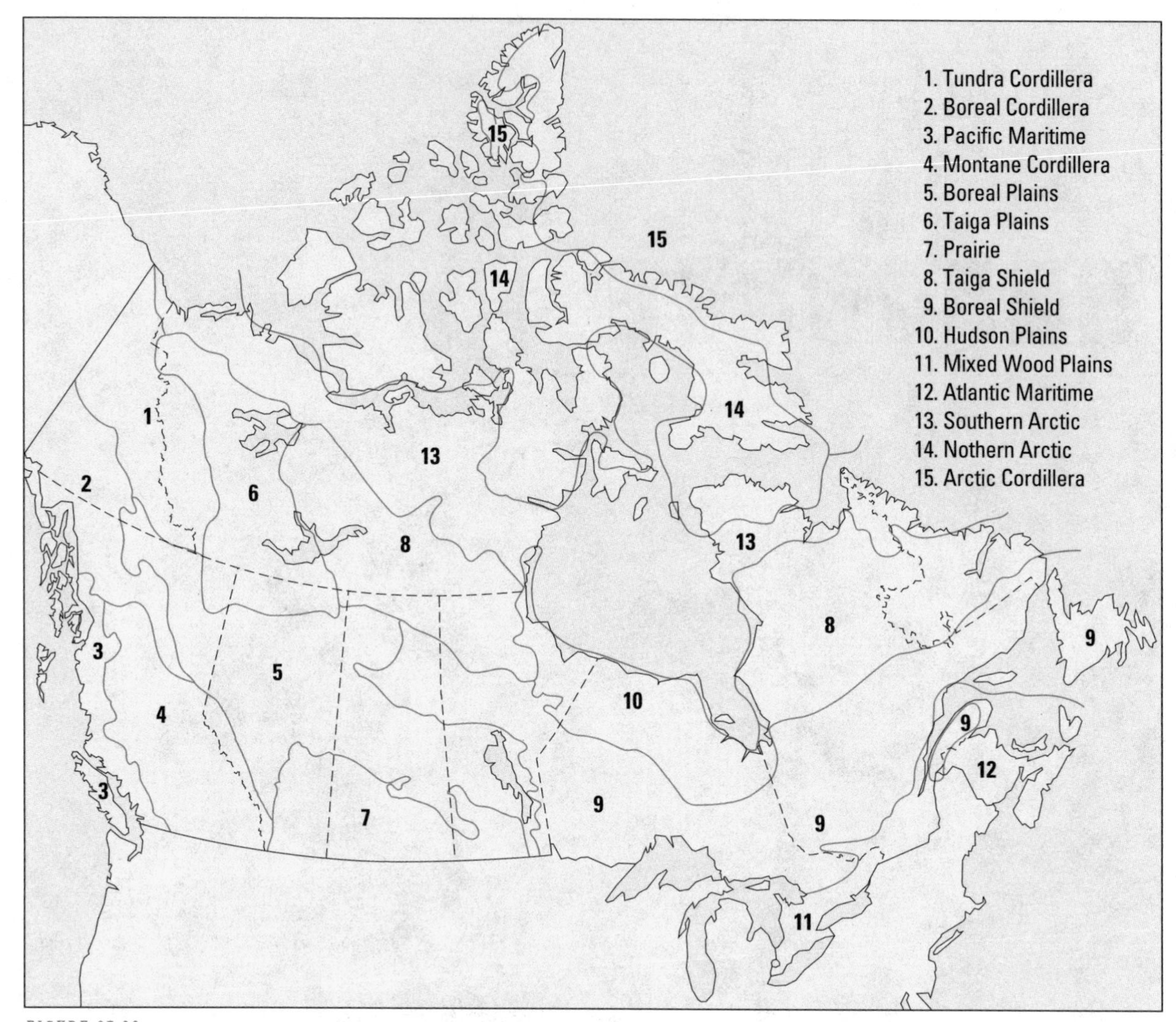


FIGURE 27.23 ▼ Ecozones of Canada

ilar relationships affecting the pattern of soils within a single valley or slope or **catena**: soil conditions vary with differences in slope angle, geology, aspect (and therefore microclimate), and vegetation (Figure 27.24). Within one ecosystem of British Columbia, the mountain hemlock zone, several plant assemblages occur in association with specific soil types (defined at the series level) (Figure 27.25).

# Soil description

Any field description of soils will be customized to the objective of the study. A description of soils for agricultural purposes would be quite different from one conducted by a forester or plant ecologist. However, there are a number of common characteristics used in the field description of a soil profile (Appendix III).

Horizons are usually initially defined on the basis of colour and texture. Colour is determined with the aid of a colour chart such as the Munsell colour guide, whereby samples from each horizon are matched to coloured tiles and the colour, hue, and chroma can be identified and the appropriate code recorded. Each horizon is measured to determine thickness and horizontal variability. The presence of structures such as mottles within a layer would necessitate separate colour descriptions for these features within any horizon.

Texture is initially described by hand texturing, where the feel of the soil is used by experienced researchers to determine the dominant particle type

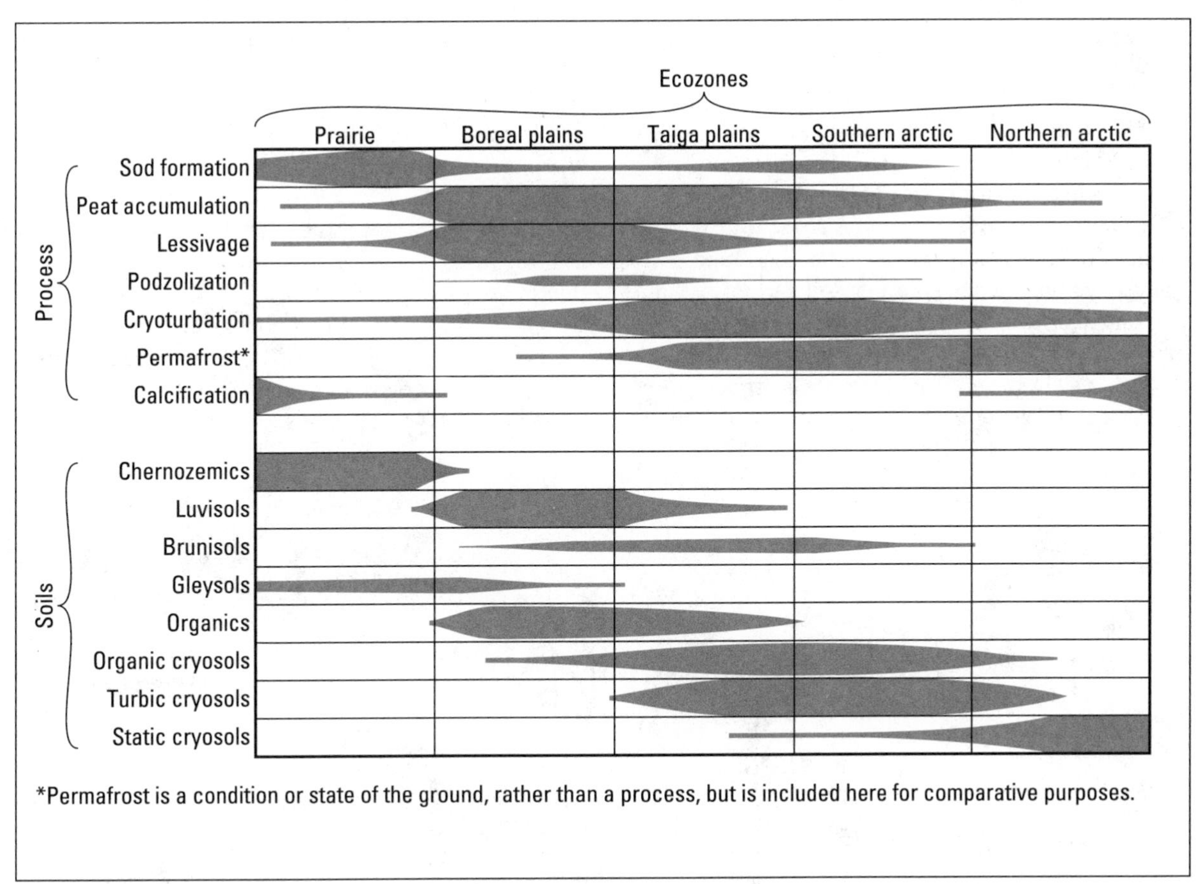


FIGURE 27.24 ▼ Pedological processes and soils associated with ecological regions in northwestern Canada (see Figure 27.23 for ecozones) (from R. Olson, R. Hastings, F. Geddes, eds., *Northern Ecology and Resource Management*, University of Alberta Press)

(sand, silt, or clay). Final textural descriptions are based on laboratory analysis of the actual sand, silt, and clay content.

A provisional classification of each horizon is made (Appendix III). The Canadian soil classification system employs a capital letter code to designate the horizon, and lower-case suffixes to denote specific processes that have affected the horizon. Generally, the first soil layer is a litter layer composed primarily of leaves, twigs, or mosses. L designates a relatively unaltered litter layer, in which plant parts are readily identified. H denotes an organic layer in which the original structure of the litter is unrecognizable, while F falls between L and H. In bogs or wetlands, the organic horizon (O) usually consists of more than 17 percent organic carbon by weight. The A horizon is usually the uppermost mineral soil horizon, with organic carbon comprising less than 17 percent of its weight. A darker-coloured A horizon, reflecting organic matter or humus enrichment, is an Ah horizon, while one that has been lightened by leaching or eluviation is an Ae horizon. The next horizon down is usually a mineral soil horizon enriched by organic matter (Bh), sesquioxides or clay, or with distinct structures or colour changes suggesting hydrolysis, reduction (Bg), or oxidation (Bm). Below this, the C horizon is little altered from the original parent material, and is relatively unaffected by the pedogenic processes affecting the A or B horizons.

Most soils have horizontally developed horizons. However, soils affected by cryoturbation may have a cyclic soil body, with patterns repeated horizontally and layers mixed vertically. Many other characteristics of the horizons can be determined (Appendix III) to give a more comprehensive description of the soil, but analysis would depend on the purpose of the investigation (e.g., pH, effervescence, conductivity, and so on).

# Conclusion

As we have seen, the soil is a complex system. Because it lies at the interface between the lithosphere, the atmosphere, and the biosphere, it is influenced by a wide range of factors, and it is the product of continual conflict and adjustment at this boundary. We cannot consider soil formation independently of geology, topography, climate, and vegetation.

The soil, however, is not simply a product of the processes operating at this interface. It also affects the processes and influences the other components of the environment. It influences the rate of weathering and the processes of sediment movement and deposition, for example, and thereby affects geomorphological processes. It influences the absorption and reflection of energy, and the release of water into the atmosphere, and thus has an effect on microclimate. It affects the movement of water as it flows into streams and seeps into the bedrock, and thus it is an important component of the hydrological cycle. Above all, the soil exerts a fundamental control on processes of plant growth, and thereby affects the development of vegetation.

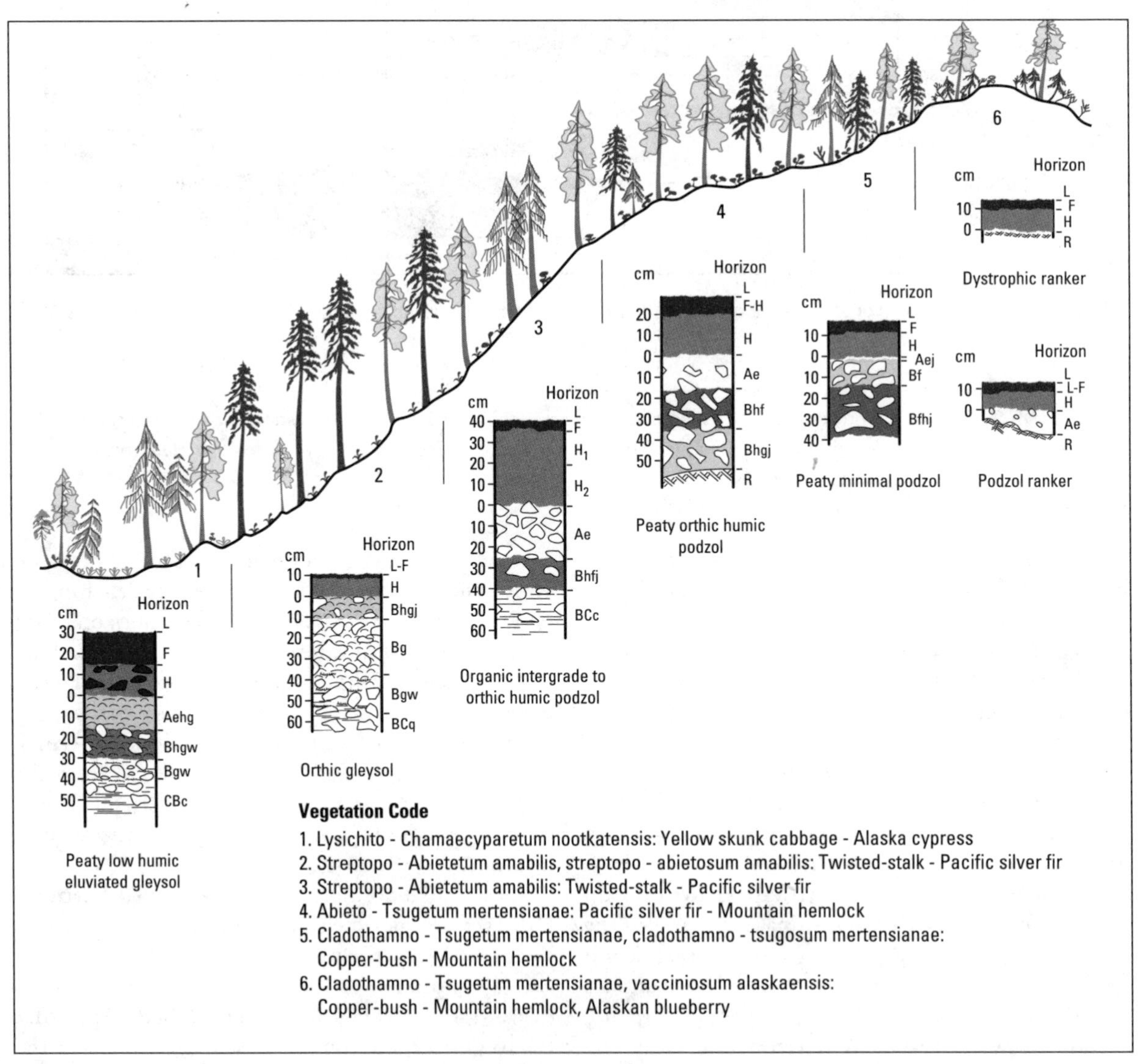


FIGURE 27.25A ▼ The relationship between vegetation and soil conditions. The Forest Subzone, Subalpine Mountain Hemlock Zone, showing a representative sequence and the relationship of ecosystem units (after Krajina and Brooke, 1969)

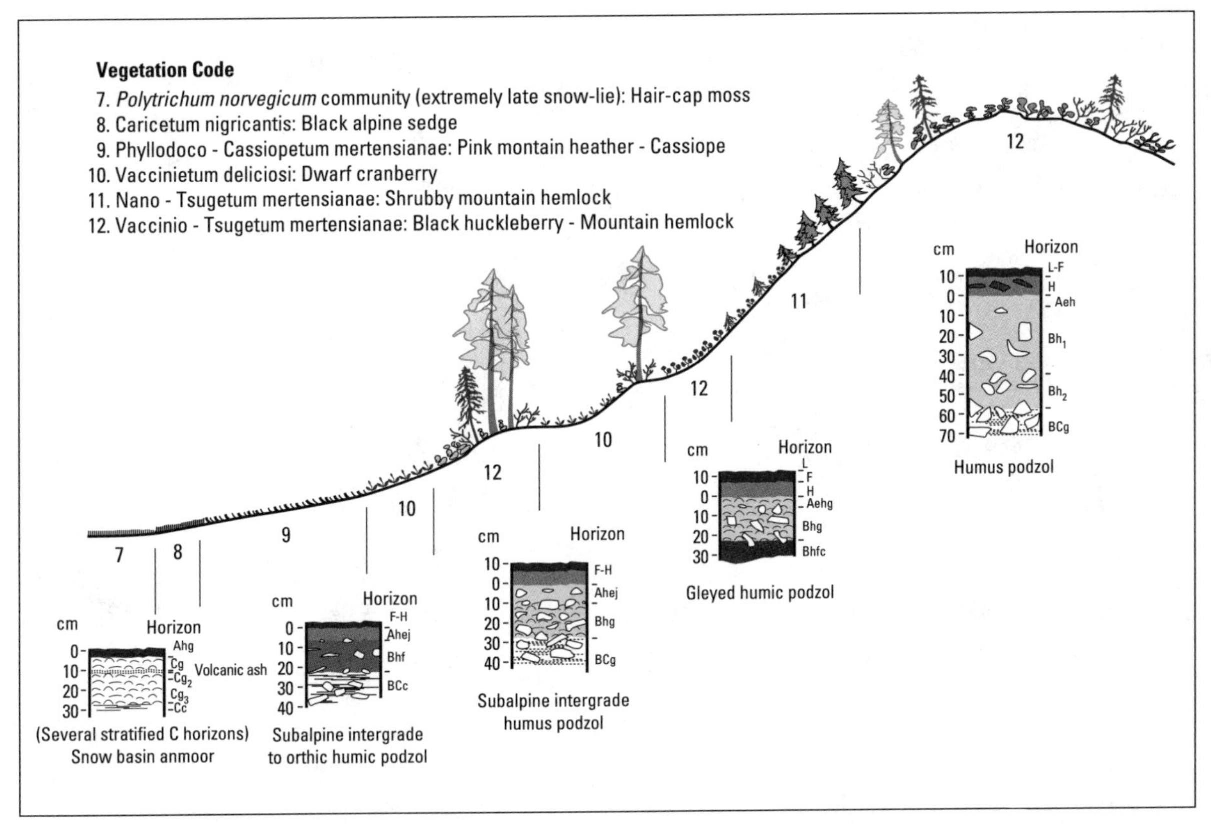


FIGURE 27.25B ▼ The Parkland Subzone, Subalpine Mountain Hemlock Zone, extends upslope almost to the treeline (after Krajina and Brooke, 1969)

# Development of vegetation

## The structure of vegetation

With the exception of recently created land surfaces (e.g., new lava flows, mobile sand dunes or recent landslides), plants grow on all types of terrestrial surfaces. Even on glaciers, there are specialized algae that gain nutrients from the dust blown onto the icefields. Individual plants of the same **species**, living together in the same region, are part of the same **population**. Populations of plants that grow together in recurring patterns form **plant communities**, which together comprise a region's **vegetation**. Vegetation varies considerably from site to site (and over time). For example, plants may cover less than 1 percent of the ground and only achieve heights of a few millimetres in High Arctic polar desert tundra, or may be extremely vigorous and lush, reaching heights of tens of metres, even providing a substrate for other plants, as in tropical rain forests.

Within the vegetation, there typically exist several layers or **synusia**. The roots represent the lowest level, penetrating into the soil, and providing the main link between the soil and vegetation systems. Immediately above the ground surface are the low-lying plants, often only a few millimetres in height, occupying the ground cover layer or lowermost strata (synusia). They are commonly overshadowed by taller herbs and grasses, which, in turn, are overtopped by intermediate-height shrubs. Above these layers are the tree synusia; the young saplings are lowermost in this zone, followed by the secondary or understorey trees, and finally the dominant or overstorey trees (see Figure 26.10). The whole structure represents a vertical stratification; it has significance, not only in relation to the visual character of vegetation, but also to processes operating within vegetation. We have seen earlier (Chapter 8) that specific micro-climatic conditions develop within the vegetation layers (**bioclimate**); in addition, competition among different plant species (**interspecific competition**), or among individuals of the same species (**intraspecific competition**), is associated with this structure. For example, the amount and quality of light and water reaching plants in lower synusia is affected by the plants above.

Of course, not all types of vegetation consist of all these layers. As we have mentioned, in some extreme environments only the lowest strata are present. Where the growth of plants is favoured, however, as in many moist tropical areas, even more synusia may be found (Figure 28.1).

It is clear that this layering of vegetation represents a distinct structure of different plant types; each layer tends to be dominated by specific plant species. Within any one layer, however, there may be considerable diversity. Thus, within the overstorey layer it is common to find three or four main tree species in temperate areas, while in tropical forests this synusia may contain a dozen or more.

Vegetation is an assemblage of plants, each with a particular niche within the overall structure. One of the fascinating aspects of vegetation is the way these different plants, each within its own specific niche, interact and are interdependent.

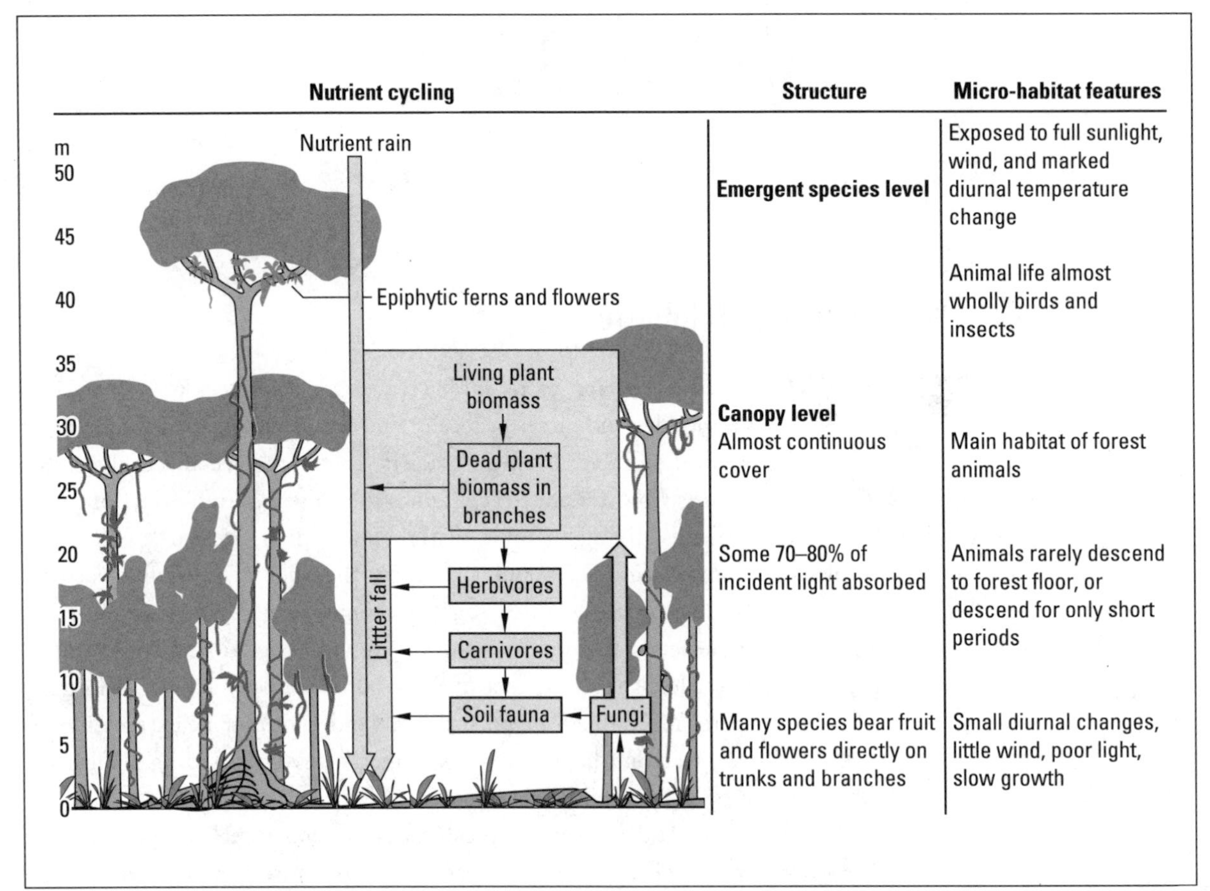


FIGURE 28.1 ▼ The relationships among vegetation structure, nutrient cycling, and micro-habitat in the tropical rain forest. Nutrients are concentrated in the plant biomass, while the rapid nutrient cycling results in relatively low soil fertility (from Collinson, 1977)

## THE ANIMALS

Within the vegetation, we often find a varied and active animal **community** that includes invertebrates and vertebrates (although our bias is to concentrate on the larger animals and group them with such terms as 'wildlife'). Animal species tend to form groups associated with the strata of the vegetation. Some live in the canopy of the trees (**arboreal** species). Some are found mainly in the understorey and within the stem zone, some live at ground level, and others live in the soil (**fossorial** species). These animals show a distinct spatial zonation, inhabiting specific territories and/or niches. However, like the plants, they exhibit a network of interactions in procuring food, and through the relationships of predators to prey and prey to scavengers.

# Changes in the biosphere

## THE EVIDENCE FOR CHANGE

The **biosphere** is a dynamic system. Both vegetation and animals are constantly changing. We can readily see some of the short-term changes if we wander through a woodland. Trees die, fall, and decay, animals migrate into or out of the area, the forest itself may be expanding onto neighbouring abandoned farmland. In the long term, the whole structure of the vegetation may change as a result of changes in external and/or internal conditions. We know, for example, that the climatic variations during the Quaternary Period led to marked adjustments in vegetation; the evidence is preserved in the pollen records of peat bogs and lakes (Figure 28.2). Irrespective of climate, the invasion of vast areas by glaciers, and the consequent removal or burial of the original soils, led to major adjustments in the vegetation and animal assemblages. In North America, some plants and animals are still responding to the last glaciation; for example, migration is still occurring, with species like lodgepole pine (*Pinus contorta*), found in the western mountains and plains, and jack pine (*Pinus banksiana*), found in central Canada. These species' ranges have only recently overlapped, as they slowly spread to colonize areas deglaciated in the last 10 000 years. They are now hybridizing in areas of Alberta and Saskatchewan.

Humans have modified vegetation by clearing forests, cultivating land, and producing pollution.

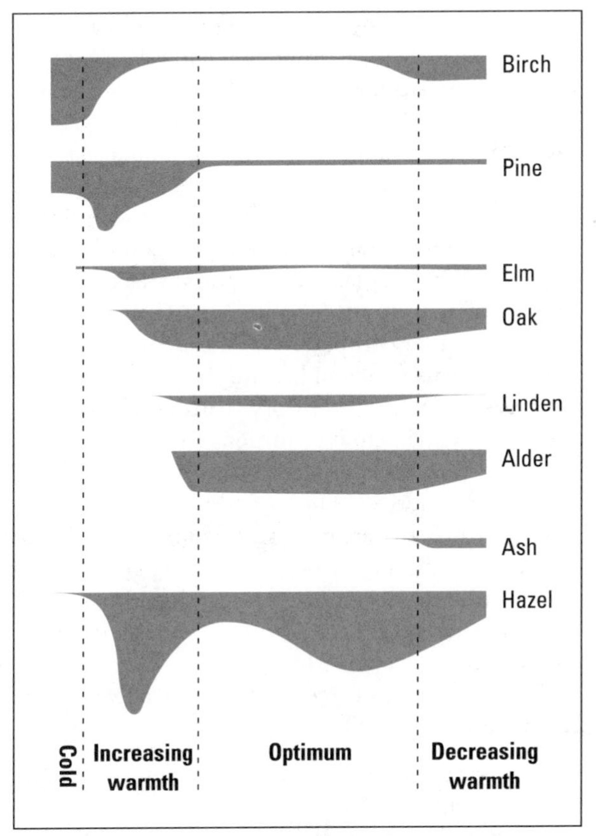


FIGURE 28.2 ▼ Generalized pollen diagram showing changes in vegetation during the transition from a cold (glacial) to a warm (interglacial) phase. Changes like this occurred following the last glaciation in eastern North America. The initial abrupt change from the cold (glacial) period to the next phase of increasing warmth produced marked changes in the composition and dominance of plant cover, as reflected in the pollen record (from Godwin, 1975)

In North America, some animal species, like the coyote (*Canis latrans*) and white-tailed deer (*Odocoileus virginianus*), have responded to these environmental changes by expanding their ranges, but others, like the grizzly bear (*Ursus arctos*) and woodland caribou (*Rangifer tarandus caribou*), have been severely reduced in numbers and distribution.

## PROCESSES OF CHANGE

Plants and animals are constantly responding to fluctuations in their environment. Their adjustments are yet another example of an equilibrium response, as organisms attempt to maintain an equilibrium with their environment. However, internal relationships are not the only factors affecting the processes of change. Change in any ecosystem invariably means adjustments in the inputs and outputs of the ecosystem. New animals and plants appear, and thus the interactions with these individuals become important. Competition, both interspecific and intraspecific, is an important factor in the process of ecosystem development. Clearly, if one species is more competitive than another, it will tend to oust its rival and become established. This is the basis of Darwinian theory: natural selection for the new environmental conditions.

In addition, other aspects of the plants or animals must be considered. For an individual to become established in any area, it is not sufficient for it to be just a good competitor; it must first get there by some means of dispersal or migration. Plants and animals that reach a suitable area first are most likely to become established.

We can identify three main factors affecting the development of the biota, therefore:

1. Interactions between organisms and the abiotic environment.
2. Interactions between the plants and animals.
3. Means of dispersal and migration.

We must consider these factors before examining the broader question of ecosystem development.

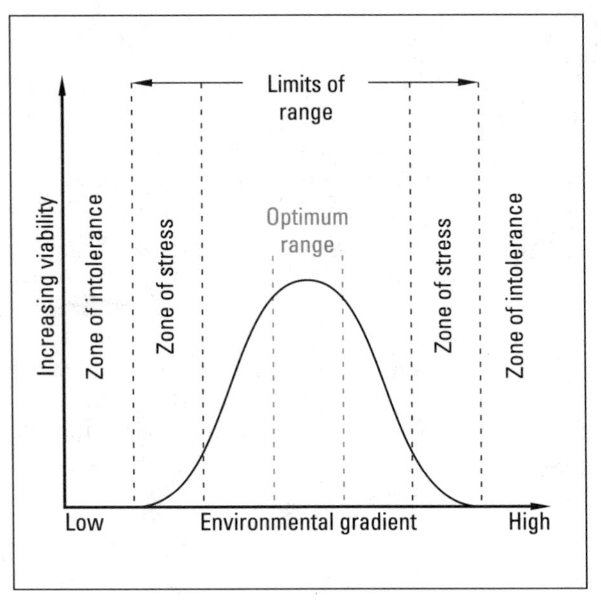


FIGURE 28.3 ▼ The viability of an individual organism in relation to changing conditions along an environmental gradient. The gradient shown may represent, for example, increasing pH (see Figure 28.8)

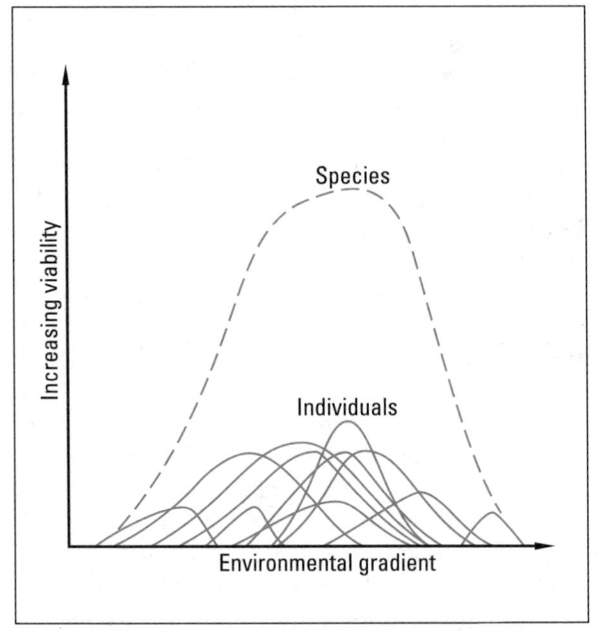


FIGURE 28.4 ▼ The relationship between the viability of a species and the individual members of that species along an environmental gradient. Each individual has a slightly different range of tolerance but most individuals have optimum viability near the centre of the species range

# Environmental relationships

## ENVIRONMENTAL FACTORS

Plants and animals respond to two main abiotic factors: the climate and the soil. The prevailing state of these two factors determines to a great extent whether an individual will survive. Most organisms can tolerate a range of conditions. At the extremes of their range, they become stressed and survive less well; but near the middle of the range, they find optimum conditions (Figure 28.3). The tolerance levels of individuals within a species vary (Figure 28.4), and if one individual is not able to survive, another member of the species might.

Plants and animals respond not to the general nature of the climate or the soil, but to specific aspects of the environment. Important climatic factors include precipitation, temperature, the amount and nature of the radiation, wind, and humidity. Collectively, these factors control rates of plant growth, transpiration (and hence nutrient uptake), respiration, and photosynthesis. One of the main factors is the balance between heat loss and energy assimilation. Similarly, in the soil, it is not the general character of the soil that is critical, but rather the specific conditions, such as soil depth, nutrient reserves, pH, water retention and drainage, and soil structure.

## THE LAW OF THE MINIMUM

Consequently, inadequacy in a single climatic or soil property can limit the growth of plants and the survival of animals. It matters little, for example, that precipitation is adequate and the duration of sunlight ideal if the temperature is too low to allow plant growth. In this instance, temperature acts as a limiting factor. This principle is embodied in the **Law of the Minimum**, developed in the nineteenth century by the German agricultural chemist, Justus von Liebig, and illustrated by Figure 28.5. In simple terms, the Law of the Minimum states that it is the most limiting factor that controls the response of the individual. In Figure 28.5, nitrogen requirements limit plant growth.

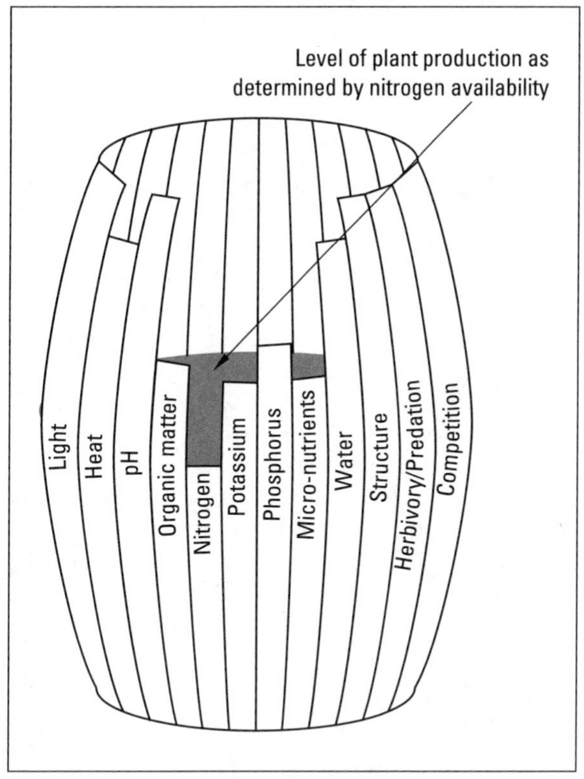


FIGURE 28.5 ▼ The Law of the Minimum (after Brady, 1973). In this example, nitrogen limits plant production. If more nitrogen becomes available, then potassium may become limiting

This idea seems simple and logical, but it has interesting implications. One implication is that changes in any but the limiting factors have no real effect. Thus, if we see that a plant is dying, it will have no substantial effect to give it fertilizer, sunray lamps, and encouraging words, while its real need is for more water. Another implication is that if we remove the effects of one limiting factor, another will take its place.

Nevertheless, there is reason to doubt the validity of this principle when extreme conditions do not exist, for to some extent, the factors governing growth are substitutable. Within limits, it is possible to compensate for inadequacies in one factor by improvements in another. Thus, an increase in the supply of certain nutrients may enable the plant to compensate for low temperatures; increasing the food supply of an animal may enable it to withstand harsh climatic conditions.

Many of the factors controlling plant and animal responses are closely related. This, in part, explains the substitutability of different factors. It also means that changes in one factor can have complex reciprocal or exacerbating effects on other factors. For example, if the climate becomes warmer, not only do temperatures rise, but evaporation and water loss also increase. At the same time, soil conditions may change as increased temperatures stimulate more vigorous biologic activity and faster nutrient cycling. Many of these responses may counteract one another. As we have seen in many cases, it is often the balance between opposing feedback reactions that determines the response of the system. Whatever the outcome, an individual factor can trigger many other responses through the system.

## ENVIRONMENTAL GRADIENTS

We need to take limiting factors into account when considering the relationships between organisms and their environment. The response of an individual to a single environmental factor approximates a bell-curve (Figure 28.3). The optimum response typically occurs somewhere near the centre of the range. In the case of a species, the overall range encompasses the cumulative responses of many individuals (Figure 28.4). Under extreme conditions, the survival of a species can depend on the ability of just one of the individuals that makes up the species to tolerate the environmental conditions. Plant and animal species respond to environmental gradients. Most individuals grow best under optimum conditions, but a few may survive or do well in extreme environments. If we imagine an area in which a critical environmental property varies from one extreme to another, we should find that the number of surviving individuals is small at the extremes and large in the central optimum area. This, in a simplified form, is the concept of environmental gradients. Each species responds to a variety of environmental conditions, and as these factors vary from one area to another, different assemblages of plants and animals develop.

# Competition

## PLANT AND ANIMAL ASSEMBLAGES

As we have noted, plants and animals do not act entirely as individuals, nor do they live in isolation. Rather, they are forced to compete with one another for a position in the world. This involves both intraspecific competition (competition among individuals of the same species) and interspecific competition (competition among different species).

In natural ecosystems, many different plants and animals live together in assemblages. Specific sets of environmental conditions tend to be associated with a specific plant and animal assemblage, so that it is possible to talk about a particular type of ecosystem occurring under certain environmental conditions. We talk, for example, about tropical rain forests and savanna grassland. In neither case will the vegetation be the same everywhere, but in general terms the same species will be represented in similar relative quantities.

Vegetation patterns arise partly from the competitive advantage of certain individuals and species under certain conditions. If we consider a simple environmental gradient and the responses of individual members of different species (Figure 28.6), at some point one species will have a competitive advantage, enabling it to capture a larger amount of the resources available (and to respond favourably to these conditions). In other situations, another species may have the advantage. When two or more species compete for the same niche, the partitioning of resources ultimately leads to one species benefitting at the expense of the others. This illustrates the **competitive exclusion principle**.

Competition among individuals and among species relates to their degree of adaptation to specific environmental conditions. This is not an absolute parameter, and as conditions change, so do individual success rates.

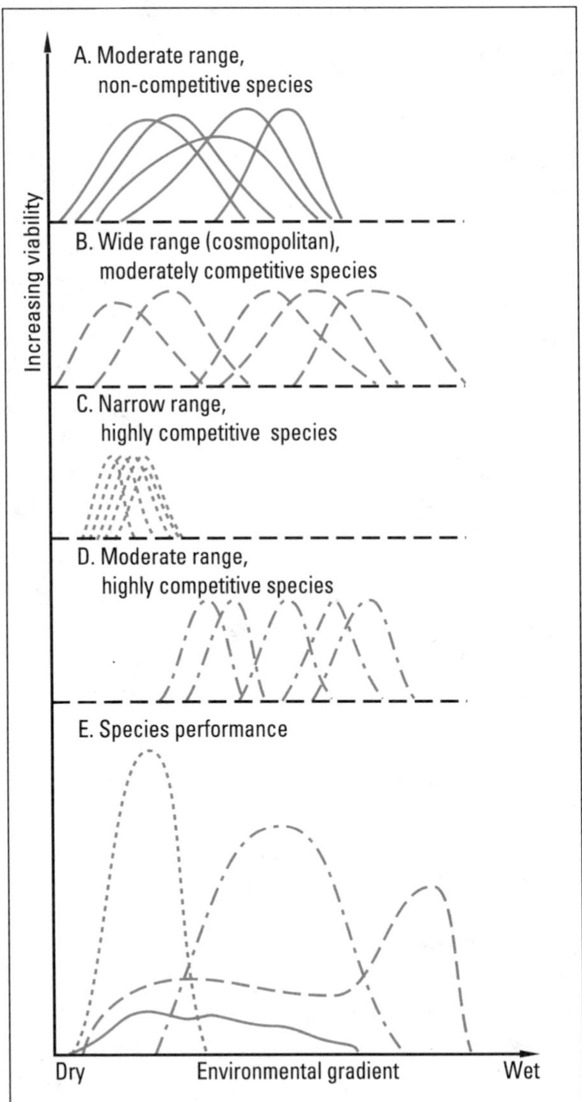


FIGURE 28.6 ▼ The viability of individual members of four species of differing competitive ability (A – D) in relation to soil moisture, and the effects on species distribution (E). (Viability is any measure of performance, e.g., biomass, production, or growth rate)

## REPRODUCTIVE/LIFE STRATEGIES AND COMPETITION

MacArthur and Wilson (1967) proposed that species can be placed along a continuum according to their reproductive strategies. Species with a high potential for population growth but with poor competitive ability were ***r*-selected**, whereas species that were slower to reproduce, longer-lived, and more competitive were ***K*-selected**. In the wild, deer mice (*Peromyscus maniculatus*) are capable of producing 1 to 9 young in each of 4 to 8 litters a year, can live up to 2 years (most are preyed upon in their first year), have a pronounced annual population fluctuation (reflecting climate and forage changes), and are common across North America. This is typical for an *r*-selected species. On the other hand, polar bears (*Ursus maritimus*) are capable of producing 1 litter of 1–4 young (average of 2) every 2 years, probably live 25 to 35 years, and are uncommon and widely dispersed in their pack-ice habitat. The bear is a *K*-selected species.

## THE NATURE OF COMPETITION

Competition is expressed in different ways. Plants often compete for light, with the more competitive species or individuals growing rapidly and shading their rivals. The shaded plants may find their growth restricted and competition for water and nutrients could further limit their growth. Eventually, a process of negative feedback may develop, resulting in the elimination of less competitive rivals (Figure 28.7). One of the critical factors in competition for light is the rate of upward growth, and there is a tendency for fast-growing, tall plants to dominate their smaller competitors. However, changes in the environment brought about by competition for light may create conditions highly suitable for some plants. For example, a substrata of shade-tolerant plants may develop beneath a dense, light-restricting canopy.

Competition for water and nutrients is also important. In competition for water, root density, root depth, and transpiration rate are clearly impor-

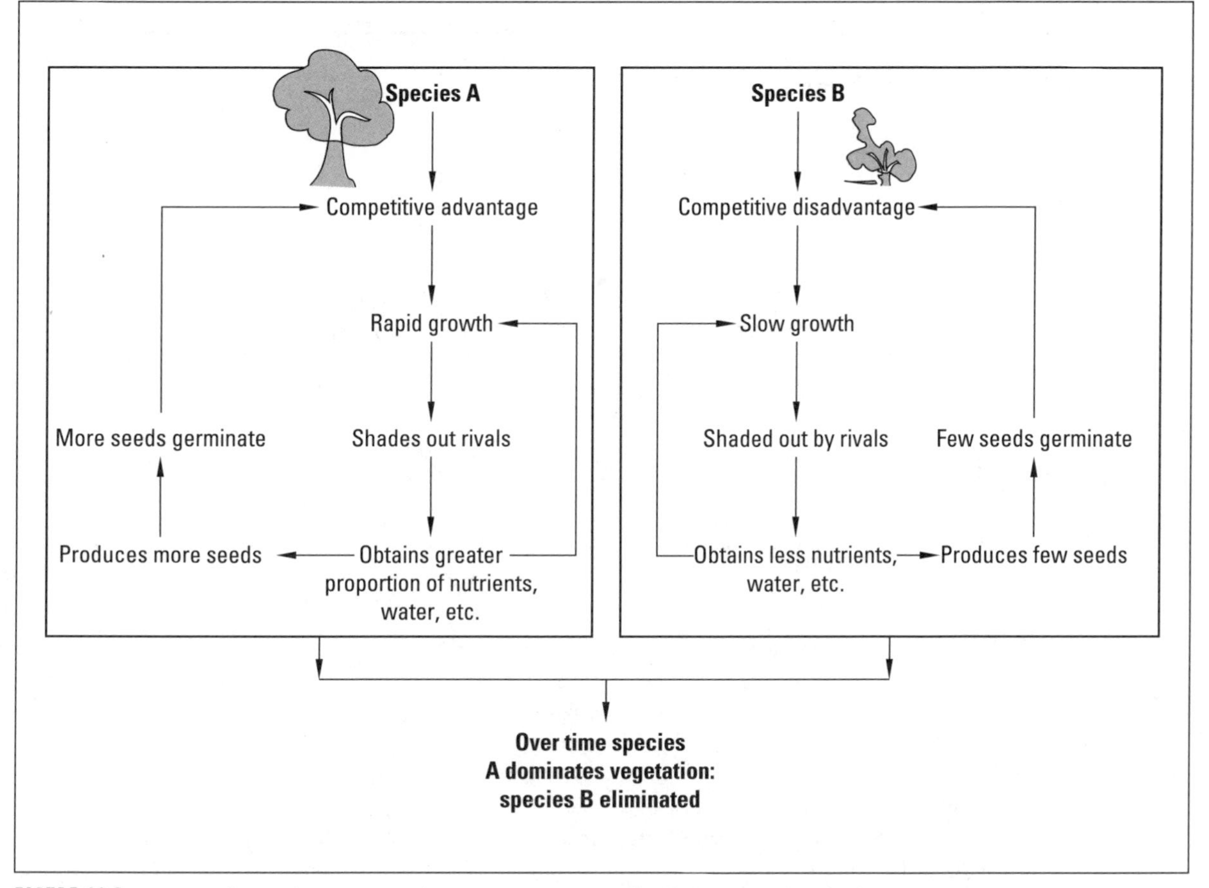


FIGURE 28.7 ▼ The effects of competition between two species of differing competitive ability

tant. Plants that draw their water from similar levels may face severe competition, and those with the greatest root density may be most successful. In areas of water shortage, plants that have longer roots may be more competitive than those drawing water from near the surface.

Competition for nutrients also relates to these factors, but microbial relationships may also be important. **Mycorrhizal** associations with fungi and symbiotic relationships with bacteria may give one plant an advantage over another. For example, in soils with low phosphorus content, plants with symbiotic bacterial associations tend to be more competitive. Also, some plants exude substances or encourage microbial activity that specifically inhibits other plants. In the subarctic and high boreal regions of

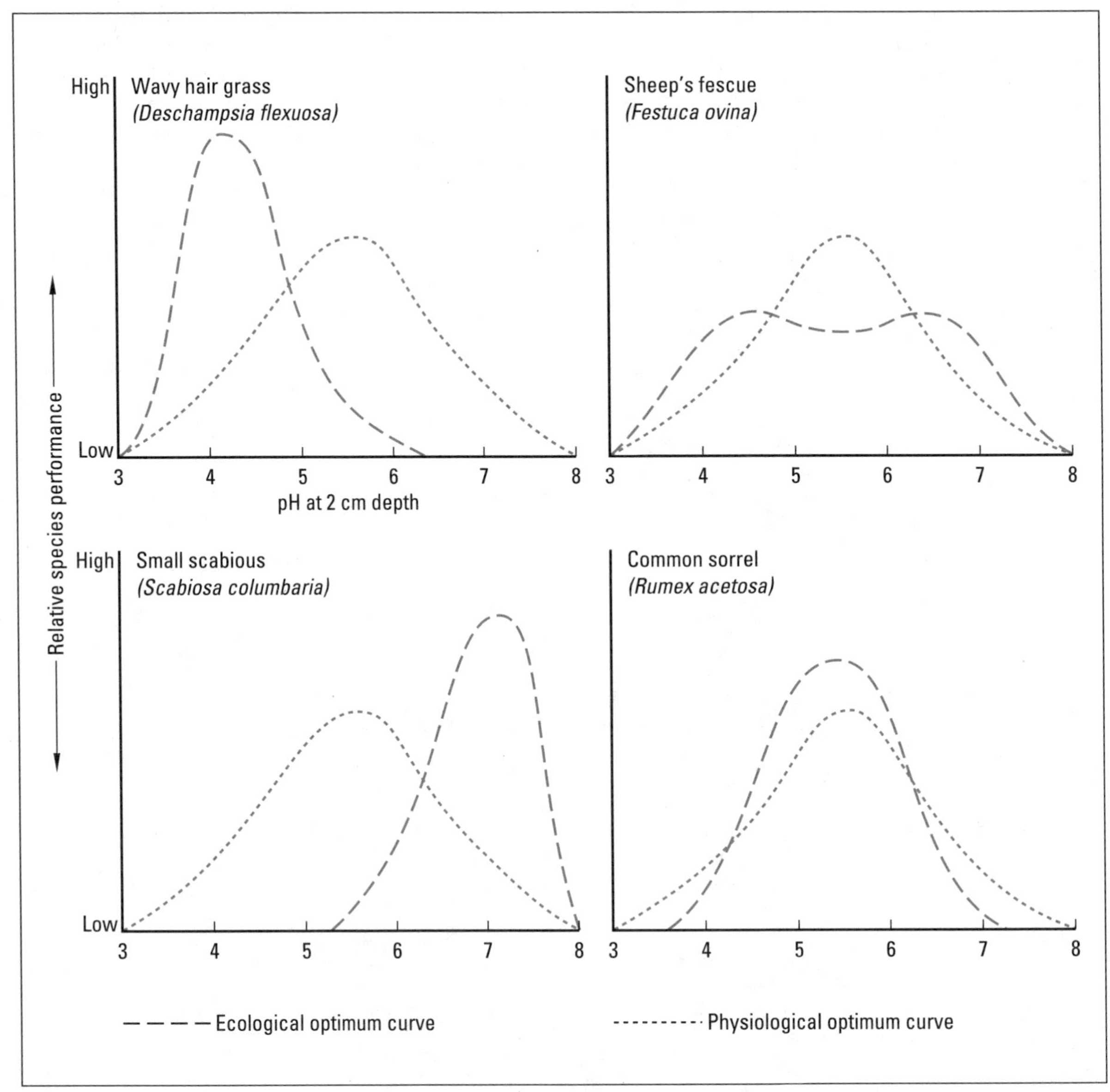


FIGURE 28.8 ▼ The response of different species to soil pH in pastures near Sheffield, England. The physiological optimum curve shows their relative performance under noncompeting conditions; the ecological optimum curve shows their frequency under conditions of competition in the field, as determined from studies of 340 quadrats (1 $m^2$) (from Rorison, 1969). Note that wavy hair grass is confined to more acid sites than it would otherwise favour; small scabious inhabits more alkaline sites; and only common sorrel (the most competitive species) occupies its physiological optimum range

Canada, lichens often form continuous carpets in the Lichen Woodland. These lichens, of the genera *Cladina* and *Cladonia*, produce **allelopaths** to deter the mycorrhizal fungi that are essential for some plants to establish in these areas. Plants lacking these mycorrhiza either cannot germinate seeds, or grow poorly, and therefore are easily outcompeted.

Competition among individuals tends to be keenest among members of the same species, as they have almost identical requirements. In terms of the assemblage as a whole, however, it does not matter much which individual of the same species wins the competition since they are the same species. Ultimately, the composition of the assemblage will be dominated by the most successful members of that species. Interspecific competition is most intense when species have similar niche requirements (Figure 28.8). Less competitive species may die or be forced to leave optimal areas, and occupy sub-optimal niches where competition is less severe, but in which they nevertheless have an advantage over rivals. On a larger scale, these differences in the ability of plants to compete successfully under different environmental conditions account for the variations in ecosystems we see across the world. Some examples are illustrated in Figure 28.9.

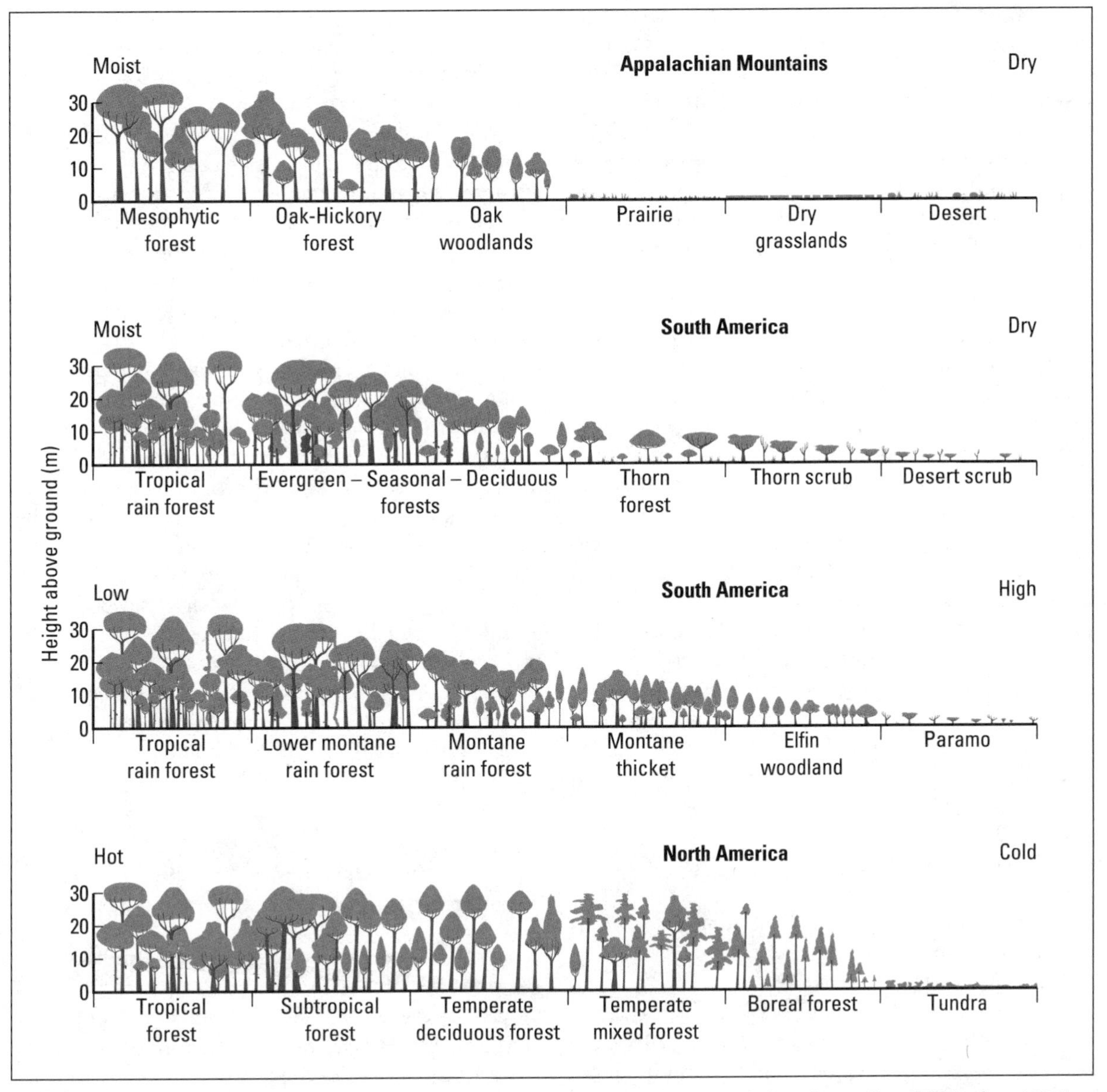


FIGURE 28.9 ▼ Changes in the structure of the vegetation along selected environmental gradients (from Whittaker, 1975)

# Dispersal and migration

## PROCESSES OF MIGRATION

Animals move from one area to another through physical migration. An individual wanders into a new area, forsaking its old home and finding a new one.

The reasons for migration vary. In many cases, minor changes in environmental conditions may be important. For example, the development of a slight imbalance between predator and prey in an area may cause some predators to migrate to new areas in search of food. In other cases, migration is a result of changes in the supply of plant foods. The loss of food due to overgrazing of grasslands may cause herbivores to move to new pastures. In both cases, more subtle and more general changes may be involved. Overgrazing or the movement of the prey may be a response to gradual climatic changes. Vast migrations must have occurred during the Quaternary in response to fluctuating environmental conditions. For any individual, the change may have been abrupt, but for populations as a whole migration was almost certainly more gradual.

Migration has numerous implications. If one part of the food chain moves, other parts are affected, creating imbalances, and in time the whole ecosystem may change. Alternatively, rival species may move in to fill a vacant niche. In recent times humans have played an important part in animal migration, deliberately or accidentally introducing new species to parts of the world that once lay beyond the reach of the animals concerned. In several cases, the migrants found a ready niche, but often their arrival created grave imbalances in existing ecosystems and occasionally led to catastrophic adjustments. For example, the introduction of rabbits to Australia had serious repercussions: they caused dramatic change to the nature of the vegetation and outcompeted native animals occupying the same niches.

## PLANT DISPERSAL

The ability of plants to adjust to changes in environmental conditions is restricted by their immobility. Some plants may be able to adapt their life forms to new conditions, but in general plants cannot respond as directly as animals. Instead, they live, die and regenerate as individuals, only the individual's vigour with which they perform these functions changes as their environment is altered. Migration reflects survival patterns under different environmental conditions. Improvement in one part of their habitat does not lead to a physical movement of plants toward it, but survival of seeds or young plants may be better in that area. Over time, the differences in survival produce shifts in distribution.

One of the critical factors in the development of vegetation is the dispersal of plants. Plants that are able to spread their seeds over a wide area can respond to environmental changes more readily than plants that distribute their seeds locally or reproduce by vegetative propagation. For example, many of the common weeds which infest large areas of the world are dispersed by wind and are *r*-selected species. The European dandelion (*Taraxacum officinale*) is an excellent example. Dandelion seeds are characteristically light, are produced frequently and in great numbers, and often with a high degree of genetic variability. This last factor gives them the ability to tolerate a wide range of conditions.

The means and efficiency of dispersal of seed-bearing plants vary considerably. Some plants produce heavy, robust seeds that tend to travel only short distances, but survive well due to their internal food supply. The timing of seed dispersal is often important to seedling survival. The jack pine (*Pinus banksiana*) has **serotinous cones** that can stay on the tree for decades and remain closed so that the seeds within are viable. When conditions are right, usually after a wildfire, the resinous seal on the cones is volatized, and the cones open. The small, winged seeds can then be broadcast from the burned snags that remain standing after the fire. Many plants produce seeds that are transported by animals, especially birds. Humans are possibly the most effective agent of dispersal the world has known. The tomato provides an interesting example, for its seeds are very robust, and have been widely dispersed in sewage and domestic refuse.

# Plant community succession

## THE BASIS OF PLANT COMMUNITY SUCCESSION

As new land becomes available to plants and animals, the processes of dispersal and migration bring individuals to the area. Then, environmental conditions and competition determine whether individuals with specific tolerances and competitive characteristics are able to survive. The vegetation that develops, and the animals that it supports, depend on the integration of these three factors. As environmental conditions change, the reduced vigour of one species may allow others, with slightly different requirements, to become established. As a result, the vegetation changes.

Typically, changes in the vegetation are slow and progressive. Moreover, as the vegetation changes, it causes related adjustments in the abiotic environment (in the soil and microclimate) so that complex feedback mechanisms start to operate. Understandably, plant community succession is a complex and controversial topic.

## DEVELOPMENTS DUE TO ENVIRONMENTAL CHANGES

Two environmental factors account for many of the changes in vegetation patterns over the last 2 million years. One is the fluctuation in climate associated with the growth and decay of the continental and alpine icecaps and the other is the effect of humans, especially through the use of land for agriculture.

Climatic changes during and following the glacial periods of the Pleistocene are a fascinating topic in themselves (Chapter 9). Their effects on vegetation varied throughout the world. Broad patterns of change suggest that the glacial phases were associated with a much more restricted pattern of vegetation zones, and the virtual disappearance of some (Figure 28.10). It seems, for example, that the tropical rain forests were all but eliminated in much of the equatorial region. Only isolated remnants remained, and it was from these that the rain forest vegetation dispersed as climatic conditions ameliorated. Interestingly, this may be one reason for the relative diversity of the flora and fauna of these areas today. The periods of isolation allowed the development of different species and subspecies in different areas—organisms that subsequently have intermingled in the expanding tropical forests.

During the last 5000 years, however, the effects of climatic change have been overshadowed by the activities of humans. We have cleared large areas of forest, burned the natural vegetation, introduced new plant species, eliminated competition in some systems, and restricted the range of many species. We still continue to effect changes today.

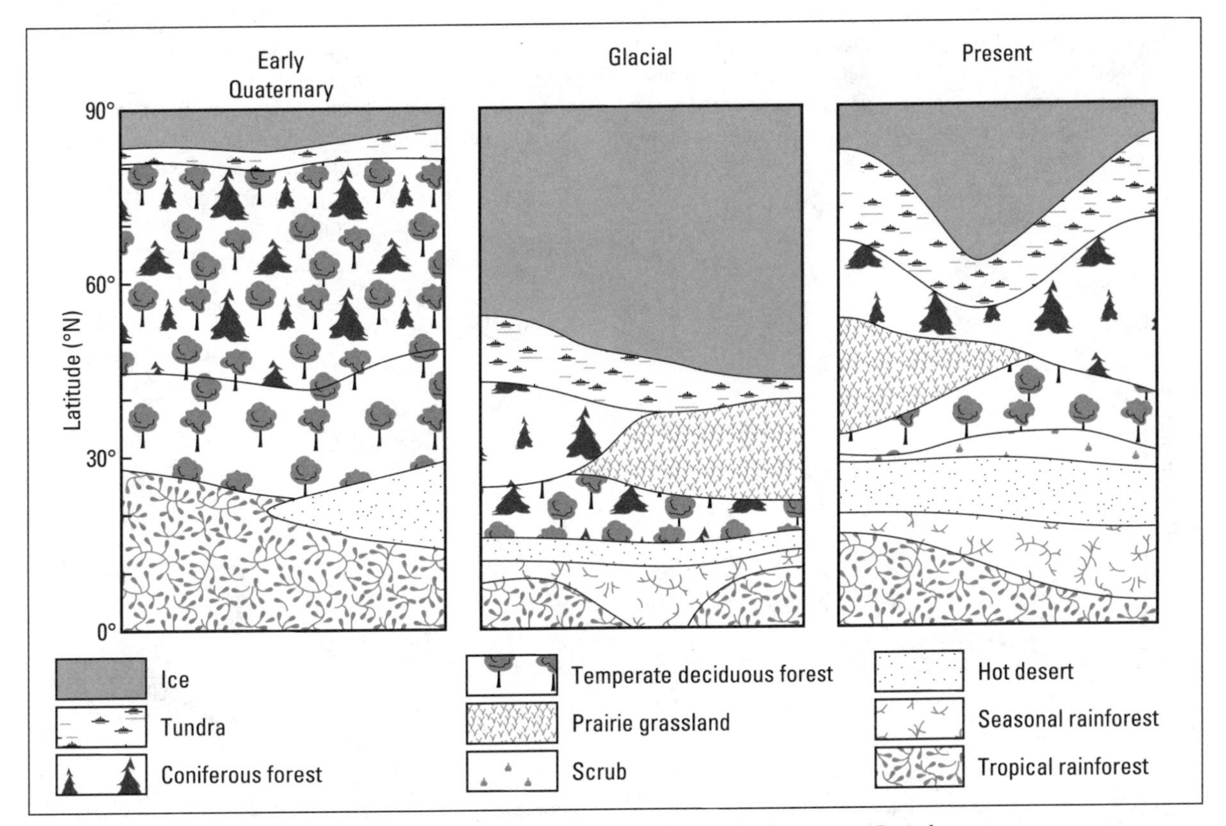


FIGURE 28.10 ▼ Variations in the distribution of vegetation zones during the Quaternary Period

## PLANT COMMUNITY SUCCESSION AND THE CLIMAX CONCEPT

The changes that have accompanied climatic fluctuations have themselves been accompanied by developments in soil and microclimatic conditions. Similar relationships occur whenever the vegetation changes, for given sufficient time, plants will influence their environment. We can see this most clearly in the events following the invasion of virgin land, irrespective of climatic change. This is known as a **seral succession**.

During the early stages of plant invasion, as we have seen, the most readily dispersed plants are usually first to arrive—the *r*-selected species. These will include locally derived species and long-distance, but efficiently dispersed plants. The survival of these pioneers depends on their degree of adaptation to the prevailing conditions. Often, conditions in new areas are relatively harsh, with exposed, infertile soils, and little available water. Consequently, a small number of more hardy pioneers becomes established.

In time, these pioneers help to weather the soil and release nutrients, and they also alter site conditions and provide shelter, enabling the establishment of other plants. Indeed, by altering their environment, they create the conditions for their own elimination. Subsequent arrivals usually include more competitive species, capable of exploiting these new conditions (Figure 28.11).

It has been argued that this process moves toward an ultimate and characteristic form of vegetation, known as the **climatic climax vegetation**. The composition of this type of vegetation is said to depend mainly on gross climatic conditions. Soils and other environmental factors become less important as they are increasingly modified by the vegetation. Thus, irrespective of the original parent material, a deep, well-weathered soil ultimately develops. In most cases, the climax plant community consists of woodland: tropical rain forest in

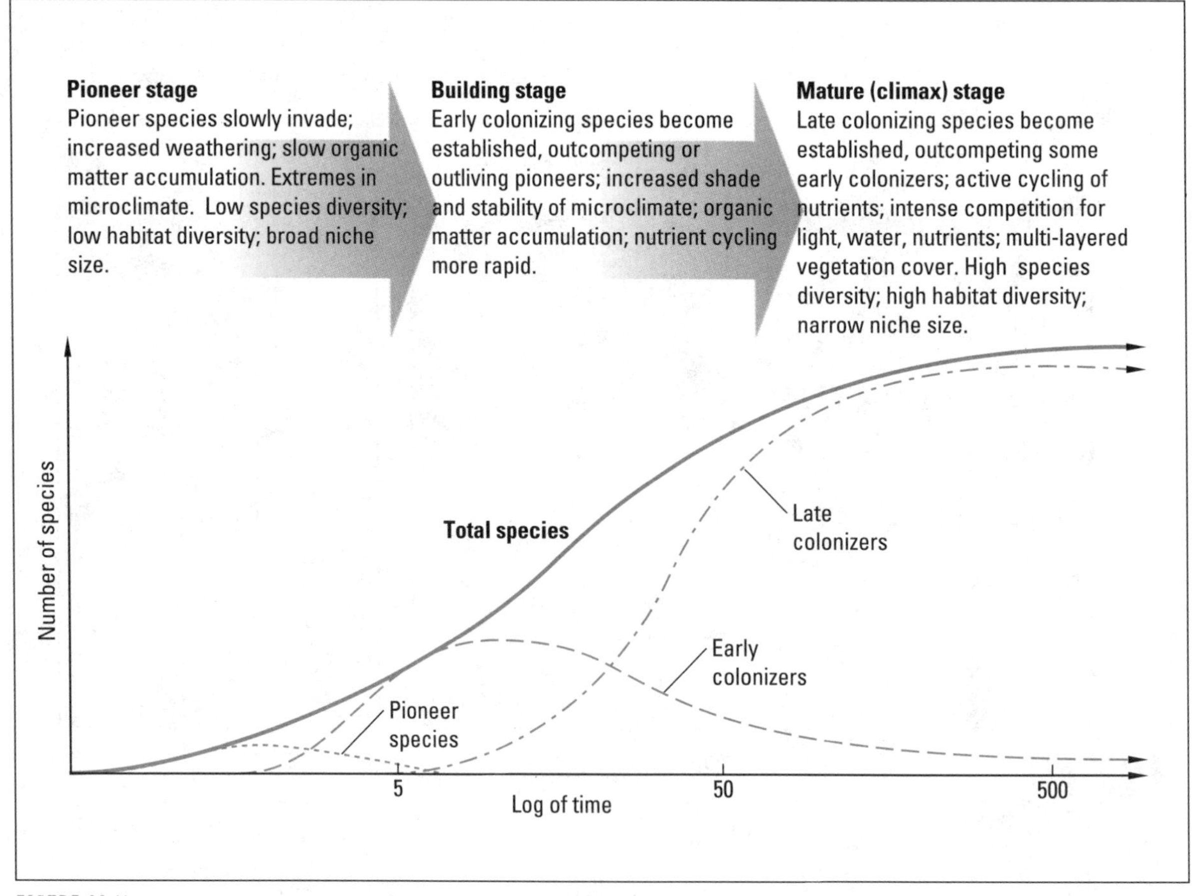


FIGURE 28.11 ▼ A general outline of changes involved in a seral succession on a newly available substrate such as a previously unstable sand dune, a recent lava flow, or an emerging seabed

equatorial areas, mixed oak woodland in many warm temperate areas, northern softwood (coniferous) forest in the cooler temperate zones.

The validity of this concept is questionable, and there are many reasons for doubting its practical significance. For example, climatic climax vegetation develops so slowly that few areas remain stable for the length of time required. Throughout the Quaternary, climatic fluctuations have been too rapid for the climax vegetation to develop over large parts of the globe. Moreover, it is no longer accepted that climate plays such a dominating role in the development of vegetation.

Nevertheless, the principle of vegetation succession remains valid. Given the creation of a new substrate for a plant community—whether it be a sand dune, a bare rock surface, a mine tailings dump, or an area reclaimed from the sea—a sequence of soil and vegetation changes is bound to occur, during which increasingly complex plant communities become established. The presence of animals is determined, in large part, by the plants of a region, and thus plant community succession drives the evolution of the ecosystem.

# Ecosystem dynamics

## THE EXAMPLE OF THE LITHOSERE

In order to summarize the processes of ecosystem change, it is useful to consider two or three examples of seral succession, examining the environmental factors and relationships involved in each. Let us begin with succession on a bare rock surface in the Niagara Escarpment on the Bruce Peninsula of Central Ontario (Figure 28.12). The limestone is first colonized by **saxicolous crustose lichens**, which are short-lived, as the soluble bedrock provides an unstable substrate. Drought-tolerant, **poikilohydric** mosses, such as hair-cap moss (*Polytrichum juniperinum*), appear in cracks and fissures where water may collect. The moss expands outward from these pockets and provides a rooting zone for fragile fern (*Cystopteris fragilis*) and juniper (*Juniperus communis*). At this point, eastern white cedar (*Thuja occidentalis*) becomes established, along with tall shrubs such as choke cherry (*Prunus virginiana*). Later, forests dominated by balsam fir (*Abies balsamea*), sugar maple (*Acer saccharum*), and

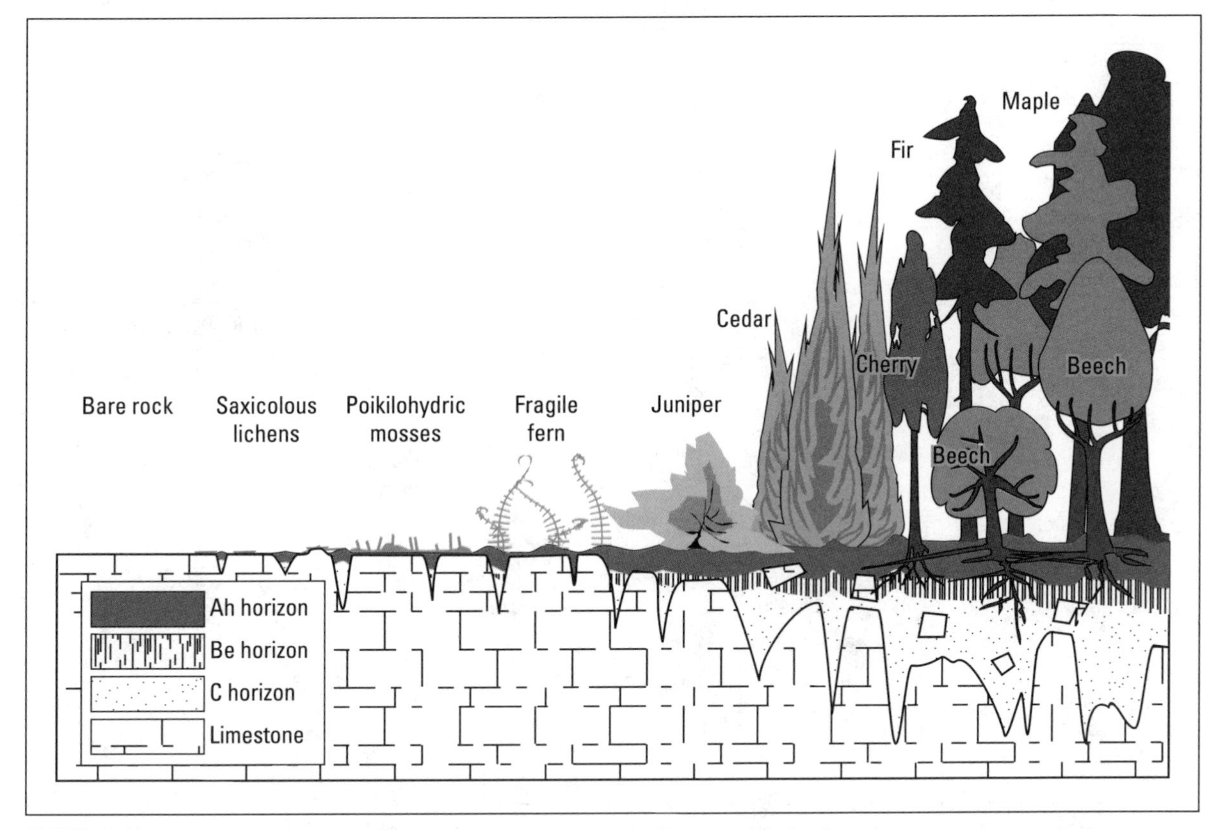


FIGURE 28.12 ▼ A lithosere sequence on the Niagara Escarpment, Bruce Peninsula, Ontario

beech (*Fagus grandifolia*) occupy the area. This type of sequence, beginning with a bare rock surface, is known as a **lithosere**.

The lack of nutrients and the exposure of the surface create conditions where few plants can survive. Consequently, only **epiphytes** (plants that obtain their water and nutrients from the atmosphere) and particularly hardy species are able to extract nutrients from the unweathered rock and become established. These pioneers (mainly lichens and mosses) provide a substrate on which other plants can root, and subsequently, perennial herbs (e.g., bracken fern) may become established. As they die and decay, they add to the litter and the humus and help to weather the underlying rocks. Animals and micro-organisms gradually appear, including herbivores that graze on the plants and decomposers that break down the plant and animal residues. In time, dwarf shrubs such as poison ivy (*Rhus radicans*) and common juniper (*Juniperus communis*) appear, and these, in turn, provide a basis for taller shrubs, such as staghorn sumach (*Rhus typhina*). Ultimately, sumach gives way to white cedar (*Thuja occidentalis*), then to a mixed sugar maple (*Acer saccharum*) and beech (*Fagus grandifolia*) forest.

## THE EXAMPLE OF THE PSAMMOSERE

A similar sequence is followed in the development of vegetation on sand dunes (a **psammosere**). As we saw in Chapter 24, the young dune surface is unweathered and subject to frequent burial by new sand. Both water and nutrients are lacking, and combined with the instability of the substrate, few plants can become established (Figure 28.13). Dune crests shelter hollows from the wind, and these substrates thereby become stable enough for sand heather (*Hudsonia tomentosa*) and kinnikinnick (*Empetrum nigrum*) to become established. These plants further anchor the shifting sand, and because of their drought tolerance, they are able to persist. As the surface becomes more stable, two **non-vascular plants** colonize the dune—hair-cap moss (*Polytrichum piliferum*) and woolly lichen (*Stereocaulon tomentosum*). Soon bluestem wheatgrass (*Agropyron smithii*) and huron tansy (*Tanacetum huronense*) become established. Later, jack pine (*Pinus banksiana*) seedlings take hold, and eventually the stand becomes an open jack pine woodland with yellow reindeer lichen (*Cladina mitis*) forming extensive mats in open areas. This vegetation type is

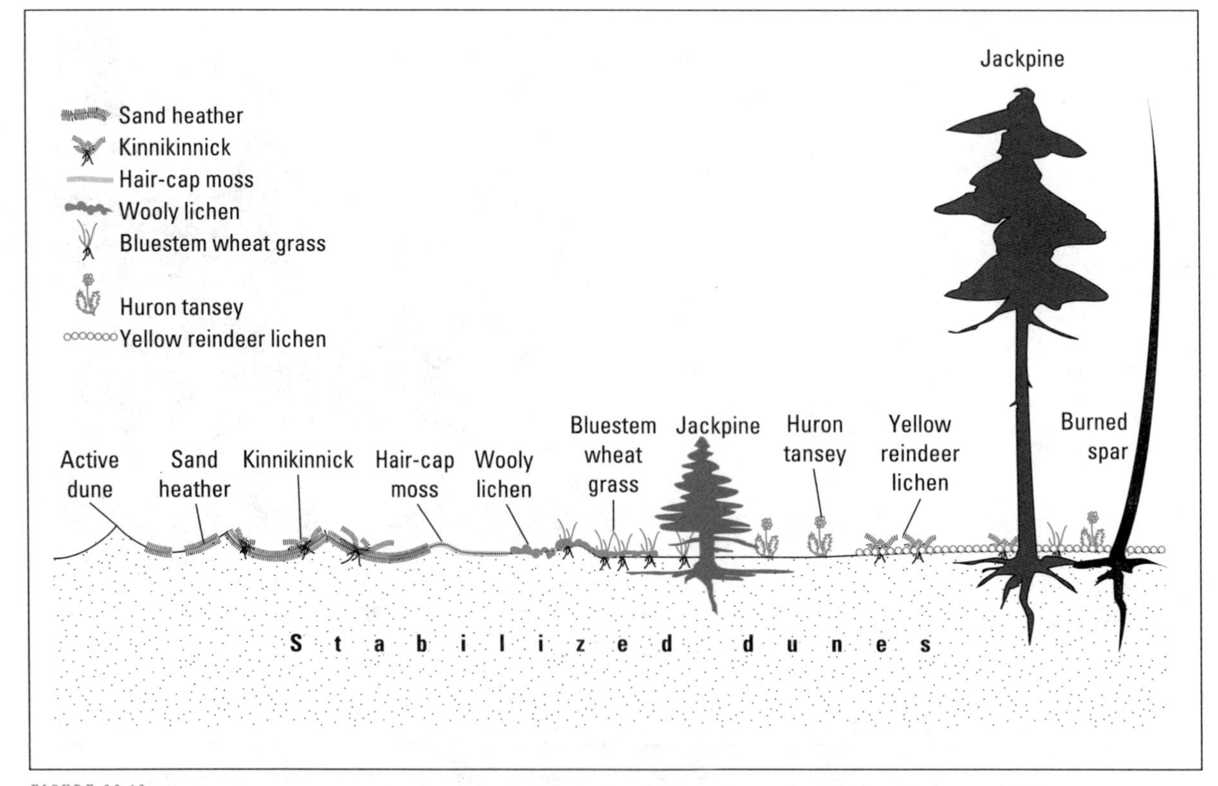


FIGURE 28.13 ▼ A psammosere sequence from the Athabasca Sand Dunes in northern Saskatchewan and Alberta

highly susceptible to fires, which sweep through these areas every 60 years on average. The stabilization of the sand, the accumulation of humus, and the opportunity for weathering result in the slow formation of poorly developed soils with little organic matter.

## THE EXAMPLE OF THE HYDROSERE

Both the lithosere and the psammosere are examples of dryland successional gradients, known as **xeroseres**. In many cases, however, the development of vegetation occurs on wet or flooded sur-

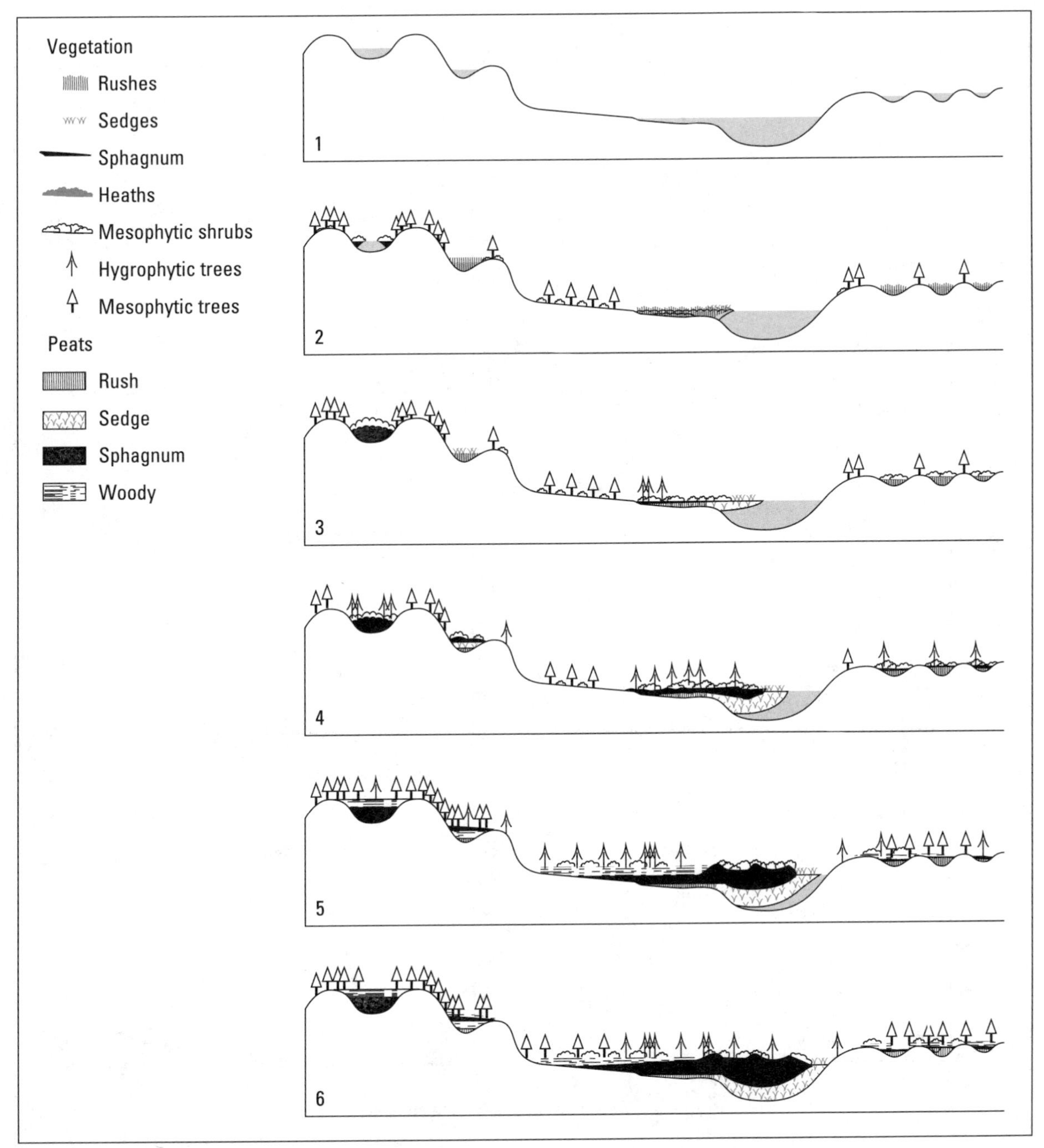


FIGURE 28.14 ▼ Bog succession on the Laurentian Shield: six stages show the nature of the peat deposit and the vegetation on different types of topography. This is typical autogenic succession (from Dansereau and Segadas-Vianna, 1952)

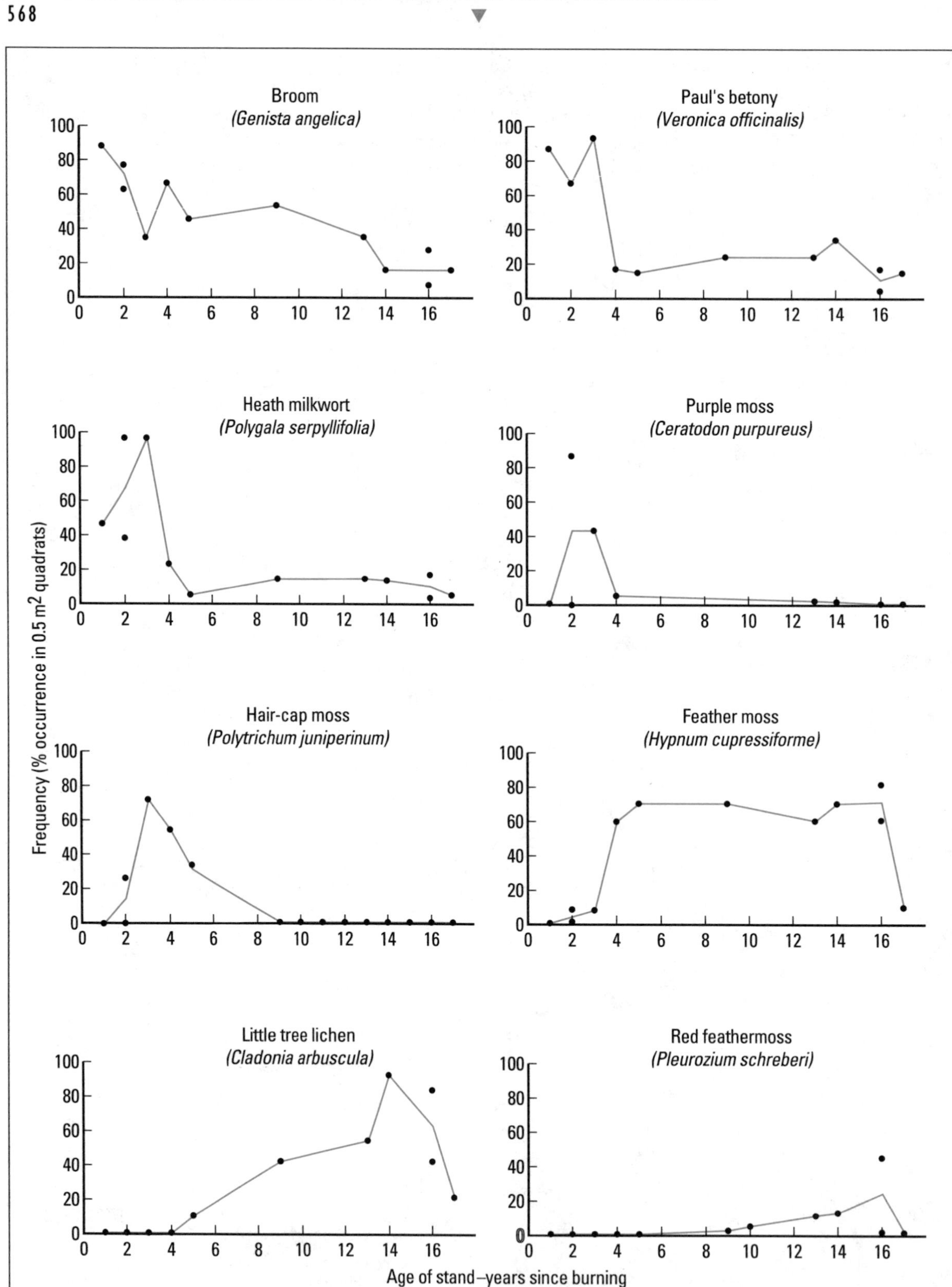


FIGURE 28.15 ▼ Changes in the vegetation of a heathland after a deliberate burn (from Gimmingham, 1972)

faces, such as lakes that are gradually filled with organic accumulations. This type of successional sequence is known as a **hydrosere**.

On the Canadian Shield, there are many closed depressions containing wetlands (Figure 28.14). Mesophytes, including trees such as aspen (*Populus tremuloides*) and shrubs such as bunchberry (*Cornus canadensis*), are most common on the higher, well-drained ground where moisture is not excessive. Hygrophytes are more tolerant of high soil moisture content and include trees such as black spruce (*Picea mariana*) and shrubs such as bog-rosemary (*Andromeda glaucophylla*). Sedges that form floating mats, extending out from the edges of ponds, encroach on the water. The accumulation of organic material (largely the products of generations of sedges) forms peat. This poorly decomposed material provides stable rooting for rushes and eventually for ericaceous or heath shrubs. These plants also add their annual litter, raising the substrate level slightly above the water table. Sphagnum mosses (*Sphagnum* spp.) grow rapidly and may form cushions that provide slightly elevated micro-sites where woody plants can root. As the surface rises with the peat accumulation, black spruce (*Picea mariana*) trees establish and contribute their litter to an ever-increasing woody peat. If the site continues to infill and rise above the water table, mesophytic species such as aspen (*Populus tremuloides*), paper birch (*Betula papyrifera*), and white spruce (*Picea glauca*) may be able to outcompete the hygrophytic plants that previously dominated the site. At any point, fires could pass through the area and cause enough damage to the plants and the organic soil to force the system back to an earlier plant community. If the fire burns deep enough and peat is removed, the site may be dominated once again by **herbaceous** hygrophytes.

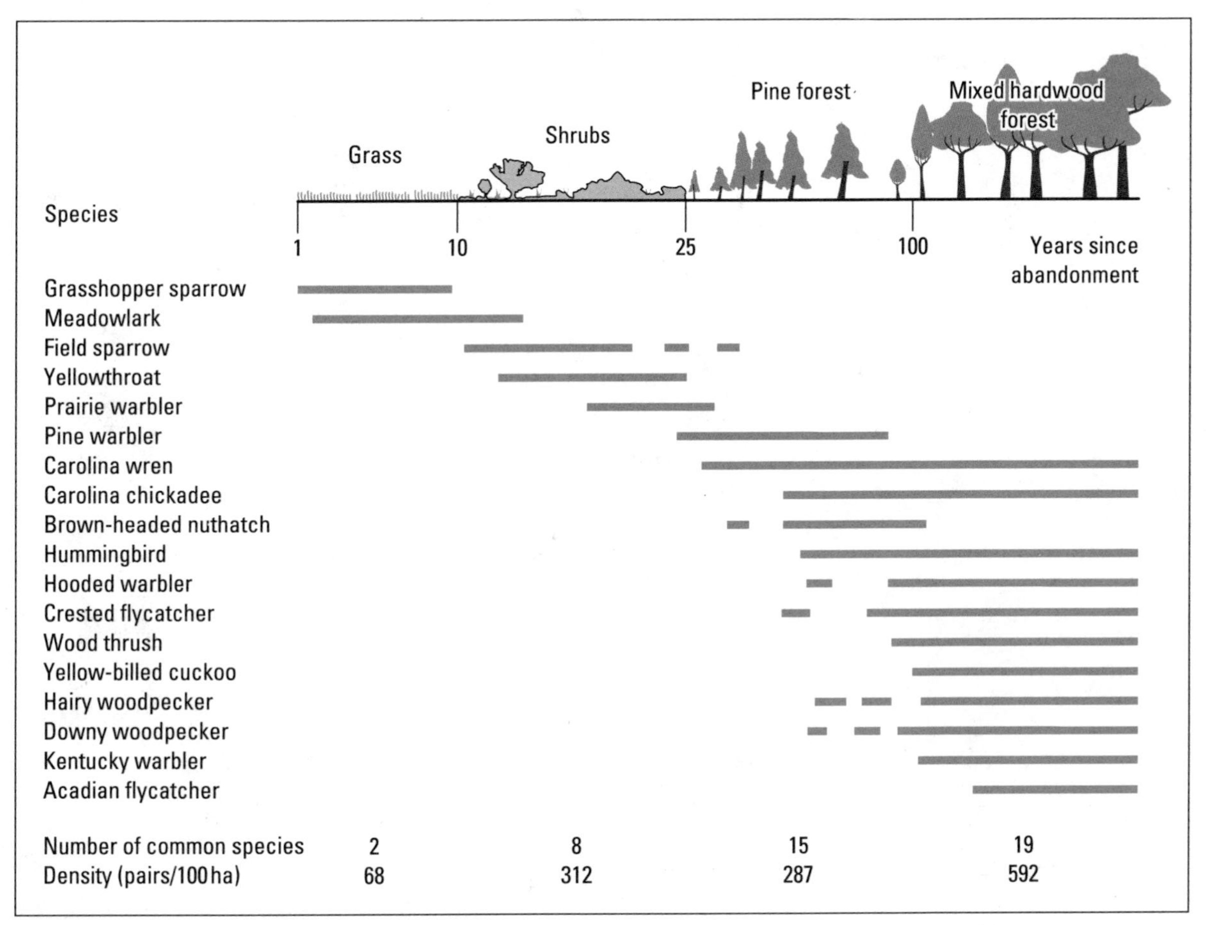


FIGURE 28.16 ▼ Changes in bird populations during a succession on abandoned farmland in southeast USA. Not all species are listed (from Johnston and Odum, 1956)

# Conclusion

The changes that take place during the development of vegetation represent a natural progression toward equilibrium: an adjustment toward a state of balance with environmental conditions. As in all such developments, the rate of change tends to be most rapid at the beginning and slows with time, as equilibrium is approached. One of the main problems in trying to manage vegetation is predicting the rate and direction of responses to change following interference. For example, after a forest fire or logging operation, we need to know how quickly the vegetation will recover and what its composition will be. Similarly, if we remove animals from an area of pasture, we need to assess the speed and character of the changes in the vegetation that will follow. Sometimes a system may revert to its pre-disturbance state, but often when we interfere with conditions, we initiate a whole new sequence of development (Figure 28.15).

We must also realize that it is not only the vegetation that changes. Vegetation and animal communities are intimately related, so that as one changes the other does also. As the vegetation changes from pioneer stage to climax forest, the animal assemblages vary accordingly (Figure 28.16). Given sufficient time, the soil, too, will be altered. The vegetation, and the accumulation of organic debris, for example, may encourage weathering and leaching, leading to deeper soils. Leaching, in turn, may make the soils more acid, so that calcicole species are unable to survive. In time, as a forest cover develops, the increased interception of rainfall may reduce leaching, and changes in the soil may be reversed. Soils, vegetation, and animals are intimately related in the ecosystem, therefore, each affecting the other.

# Biogeochemical cycling

## Nutrients in the biosphere

### THE DEVELOPMENT OF THE CONCEPT OF NUTRIENT CYCLES

During the seventeenth and eighteenth centuries, a great controversy raged among agricultural chemists: what was the food of plants? Many ideas were put forward. The chemist and diarist, Evelyn, suggested that there was some 'nitrous spirit' in the air from which they fed, though Jethro Tull (a famous agriculturalist) complained that since this 'nitrous spirit' could seemingly rust iron bars, he could not conceive that it could be the food of delicate plants. At about the same time, von Helmont carried out an experiment, in which he grew a willow tree in a sealed tub of soil, to which nothing but water was applied. By weighing both soil and tree before and after the experiment, he showed that while the tree gained weight, from only 5 pounds to 165 pounds, the soil lost no more than a few ounces of weight. Tull, himself, claimed that plants devoured the soil particles and, soundly attacked a Dr. Woodward who disagreed with him, saying that Woodward was like the old woman who could see a needle on a barn but could not see the barn. It was all furious and bitter stuff!

Only in the mid-nineteenth century did the truth begin to emerge, and thanks to the work of chemists like Sir Humphrey Davy and Justus von Liebig, and agriculturalists like William Lawes, the secrets of plant nutrients were discovered. Following years of careful experiments, these and other scientists showed that plants required a range of nutrients, including nitrogen, phosphorus, potassium, and calcium. Many of these the plants obtained from the soil, others from the atmosphere.

These discoveries had widespread implications. From them arose the fertilizer industry and the modern approach to agriculture. From them, too, developed the concept of biogeochemical cycles. The nutrients required by plants, it was realized, flowed through the ecosystem along far-reaching and complex pathways; from soil to plants, from plants to animals, from animals back to soil (Figure 29.1). It was probably the German scientist Ebermayer who first established the concept of the nutrient cycle in 1876, but it seems to have been a Russian ecologist, Vernadskii, who coined the term 'biogeochemical cycles'. Since then, the concept has become a fundamental part of ecological studies.

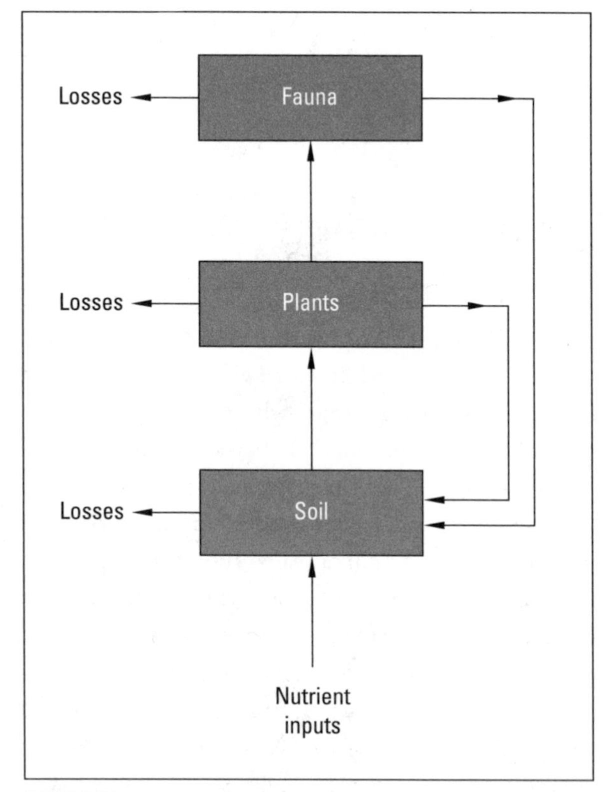


FIGURE 29.1 ▼ The general structure of the biogeochemical cycle of an ecosystem

### THE NUTRIENTS

It is time, perhaps, to ask a fundamental question: what exactly are nutrients? Almost any chemical element may, in fact, act as a nutrient; the term is simply applied to elements or compounds which are required by plants to fulfill their growth functions. Eighteen essential nutrients are commonly recognized, and these are split into two main groups: macro-nutrients which are required in large quantities, and micro-nutrients which are used in much smaller proportions. The main macro-nutrients are carbon, oxygen, nitrogen, calcium, potassium, phosphorus, sulphur, and magnesium. The main micro-nutrients are manganese, iron, silica, sodium, and chlorine, together with several trace elements such as boron, copper, zinc, and molybdenum (Table 29.1).

All these nutrients are derived, ultimately, from rocks or from the atmosphere. But they enter ecosystems by a variety of processes: by weathering, in precipitation, by biological processes (including human activity), and by deposition. They may also be lost from ecosystems. Erosion may remove nutrients from the soil, people remove them when harvesting crops, waters percolating through the soil wash away essential elements. As in all systems, therefore, the flow of nutrients depends upon the relative magnitudes of inputs and outputs.

## Nutrient inputs to ecosystems

### WEATHERING

On a global basis, one of the main, longterm inputs of nutrients comes from rock weathering. In Canada, large areas of the country are dominated by glacially related landforms, and these surficial deposits are the parent materials that, upon weathering, provide nutrient inputs to ecosystems. The process is slow, and the quantities of nutrients released into the soil in this way are often small, but it is nevertheless vital to the maintenance of the biogeochemical cycle. Numerous processes are involved. Mechanical weathering breaks up the rocks, and chemical processes such as hydration and carbonation bring the substances into solution (Chapter 18). Possibly the most important processes, however, are hydrolysis and oxidation. Together, these alter primary rock minerals to secondary, clay minerals. As we saw in Chapter 27, isomorphous replacement then results in the release of certain atoms from the internal

## Table 29.1

*Plant nutrients*

| *Nutrient* | *Amount needed (mg/litre of soil)* | *Function in plants* | *Sources* |
|---|---|---|---|
| Oxygen | – | Respiratory processes | Atmosphere |
| Carbon | – | Carbon used in photosynthesis | Atmosphere |
| Nitrogen | 15 | Protein synthesis, formation of nucleic acids, amino acids, vitamins | Soil, as ammonium, nitrite or nitrate; often fixed by bacteria in plant roots |
| Sulphur | 1 | Protein and vitamin synthesis | Soil, as sulphates, gypsum, or pyrites; as hydrogen sulphide in waterlogged soils |
| Calcium | 3 | Metabolic processes; vital constituent of cell membranes | Soil, from limestone, feldspars, augite, gypsum |
| Potassium | 5 | Protein synthesis and transphosphorylation (conversion of sugar to phosphate) | Soil, from feldspars, micas, and clay minerals |
| Magnesium | 1 | Constituent of chlorophyl; enzymic reactions | Soil, from dolomitic limestone, montmorillonite clays, biotite, augite, hornblende |
| Phosphorus | 2 | Component of many organic molecules; major source of energy through conversion of ATP to ADP | Iron, aluminum, and calcium phosphates in soil; dissolved phosphates in soil solution |
| Iron | 0.1 | Oxidation–reduction processes in respiration | Iron oxides, sulphates, and silicates; often chelated with organic acids |
| Manganese | 0.01 | Small quantities used in enzymic reactions | Iron–magnesian minerals |
| Copper | 0.0003 | Respiratory metabolism | Igneous vein minerals |
| Zinc | 0.001 | Enzymic reactions | Igneous vein minerals |
| Boron | 0.05 | Cell division during growth | Soluble borates (mainly from marine sources) |
| Molybdenum | 0.0001 | Nitrogen fixation and assimilation | Igneous vein minerals |
| Cobalt | 0.00001 | Nitrogen fixation in root nodules | Igneous vein minerals |
| Sodium | 0.05 | Unknown | Sodium chlorides; sea spray |
| Silica | 0.0001 | Unknown | Silicate minerals |
| Chloride | 0.05 | Regulates osmotic pressure; balances cation concentration in cells and sap | Dissolved in rainwater entering soil |

structure of the clays. Silica, aluminum, manganese, potassium, and sodium are all made available as nutrients for plants in this way.

## ATMOSPHERIC INPUTS

Significant quantities of nutrients are derived from the atmosphere, either in precipitation or by biological processes. The animals, for example, obtain their oxygen for respiration from the open atmosphere, while plants absorb both carbon dioxide and nitrogen through their leaves. In addition, large quantities of atmospheric nitrogen may be fixed in the soil by bacteria and algae. We will examine this process more fully later in this chapter, but its importance is illustrated by the fact that these atmospheric inputs account for about one-half of the nitrogen circulating in agricultural ecosystems.

Precipitation, too, acts as a major source of nutrients. Nitrogen, calcium, phosphorus, and sodium are all washed into the soil in this way (Table 29.2). The amounts involved are often small, but in low-productivity ecosystems, such as bogs, they may be vital to plant growth. Epiphytes (plants that grow on others but derive their moisture and nutrients from the atmosphere) like bromeliads/air ferns (*Bromelia* spp.) are wholly dependent upon atmospheric sources for their nutrients.

## FERTILIZER INPUTS

In our modern world, we govern many ecosystems and significantly interfere with others. A major way we do so is by the application of fertilizers. Most intensive farming ecosystems are today based upon high inputs of fertilizers—in particular, of nitrogen, phosphorus, and potassium. In addition, large quantities of calcium are applied as a liming agent to control soil pH. Total world consumption of fertilizers was estimated as 112 million tonnes in 1980 (Table 29.3), and in some ecosystems (e.g., in arable farming ecosystems) this may account for 80 percent or more of the total inputs. Even where fertilizers are not used directly, they may act as a major source of nutrients, for a significant proportion of the nutrients washed into the soil from the atmosphere is derived from agricultural fertilizers—either as dust blown away during fertilizer application or gases released into the air during chemical decomposition of the substances.

## BIOLOGICAL INPUTS

We introduce nutrients to ecosystems in many other ways. Young livestock, seeds and seedlings, for example, act as important inputs of nutrients, while imported feedstuffs—such as hay or corn bran for

### Table 29.2

*Chemical composition of rainfall reflects air quality, with stations in North America generally having much lower values than in the UK and Czechoslovakia*

| | Concentration ($mg\ l^{-1}$) | | | | |
|---|---|---|---|---|---|
| *Nutrient* | *Rothamsted, UK* | *Saxmundham, UK* | *Hubbard Brook, New England, USA* | *Czechoslovakia* | *Nova Scotia, Canada* |
| Sodium | 1.9 | 4.4 | 0.06 | 0.18 | 0.08 |
| Potassium | 0.7 | 0.6 | 0.07 | 0.08 | 0.06 |
| Calcium | 1.8 | 1.6 | 0.16 | 0.62 | 0.13 |
| Magnesium | 0.4 | 0.7 | 0.03 | 0.07 | 0.10 |
| $NH_4$–Nitrogen | 1.7 | 1.8 | 0.21 | 0.85 | 0.01 |
| $NO_3$–Nitrogen | 1.1 | 1.2 | 0.22 | 0.68 | 0.18 |
| Phosphorus | 0.13 | 0.02 | 0.003 | 0.006 | n.d. |
| Chloride | 5.9 | 8.8 | 0.45 | 0.41 | 1.44 |
| Sulphur | 3.3 | 3.3 | 0.9 | 1.72 | 0.47 |

## Table 29.3

*World consumption of fertilizers (1939–80)*

| Year | Consumption ($t \times 10^6$) Nitrogen | Phosphate ($P_2O_5$) | Potassium ($K_2O$) |
|---|---|---|---|
| 1939 | 2.6 | 3.6 | 2.8 |
| 1960 | 9.7 | 9.7 | 8.6 |
| 1970 | 28.7 | 18.8 | 15.5 |
| 1980 | 57.2 | 31.1 | 23.5 |

*Data* G.W. Cooke, *Fertilizing for Maximum Yields* (1983), Granada

cattle—can be a major source of nutrients. In intensive grazing ecosystems in the USA, it is common for 25 percent of nitrogen and 10 percent of potassium to be derived from these sources; in the Netherlands, where dairy farming is even more intensive, the values may reach 60 percent and 40 percent respectively.

Natural processes of migration may also provide inputs. Animals wandering into a new area carry with them nutrients which they release either in their feces or when they die and their bodies decay. We see evidence of this on a small scale on the farm, for cattle and sheep tend to graze in certain parts of the range or field and deposit their feces elsewhere. This results in a gradual transfer of nutrients from one part of the system to another (Figure 29.2). On a larger scale, the mass migration of animals such as barren ground caribou (*Rangifer tarandus groenlandicus*), either seasonally or in response to longer-term climatic change, may lead to quite marked inputs of nutrients in some areas.

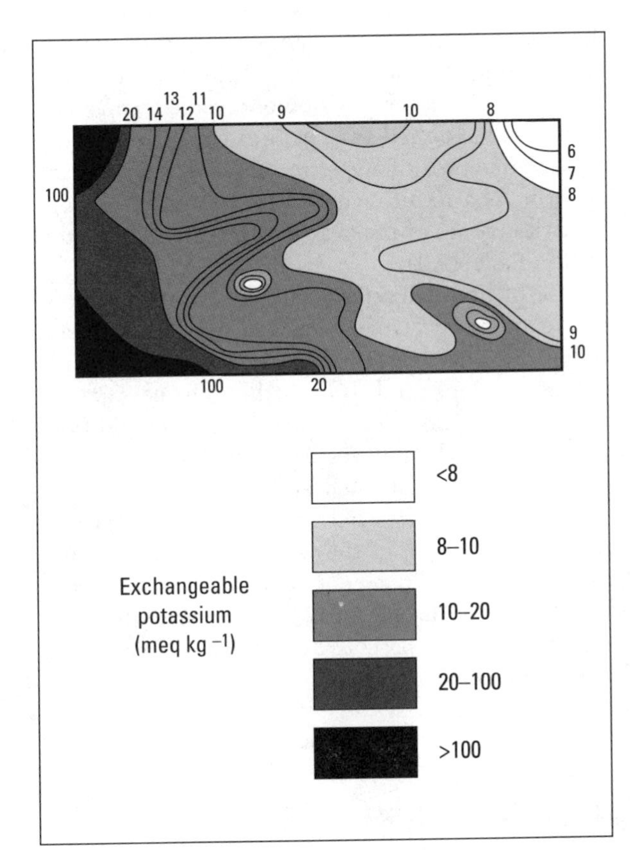


FIGURE 29.2 ▼ The distribution of potassium across a field after grazing by sheep. Higher values appear in areas where animals defecate more frequently—sheltered areas where they move after grazing (based on Hilder, 1964)

# Nutrient outputs from ecosystems

If the ecosystem is in equilibrium, the rate of nutrient inputs is balanced by the outputs, so that the total quantity of nutrients in circulation remains the same. This is not always the case, of course, and in many agricultural ecosystems outputs may exceed inputs so that nutrient reserves fall and the system degenerates. In other cases, the reverse may be true; lakes, for example, may act as nutrient sinks—the nutrients that enter them may be unable to escape, with the result that the substances build up in the waters, eventually choking them. This is the process of eutrophication. In most situations, however, nutrients are lost through a number of processes.

## EROSION

One of the most important means by which nutrients are lost is soil erosion. The process is particularly active in agricultural ecosystems, where cultivation leaves the soil bare and unprotected, so that wind and water can carry away the finer materials. It is these finer fractions—the clays and organic compounds especially—that store many of the nutrients in the soil, so when the particles are lost the nutrients go with them. In this way, erosion results in the selective removal of soil nutrients, in particular, phosphorus, potassium, and nitrogen.

## LEACHING

Leaching is another important process of nutrient loss. As we saw in Chapter 27, water percolating through the soil carries with it nutrients in solution. Many of these are ultimately washed out of the soil profile and carried into streams and rivers, where they contribute to water pollution; so the outputs from one ecosystem become the inputs to another (Figure 29.3). It is a process that is encouraged by high precipitation and acidic plant litter (e.g., pine), but leaching losses are often at their greatest in arable farming ecosystems, for arable soils contain large quantities of readily soluble fertilizer nutrients.

There is, however, an interesting dichotomy with leaching. If the nutrients in the soil are to be available to plants, they must be soluble, for plants obtain most of their essential elements as solutes in the water they take up. On the other hand, if the nutrients are soluble, they are also highly susceptible to leaching. Thus, increasing the availability of nutrients to plants may increase leaching losses.

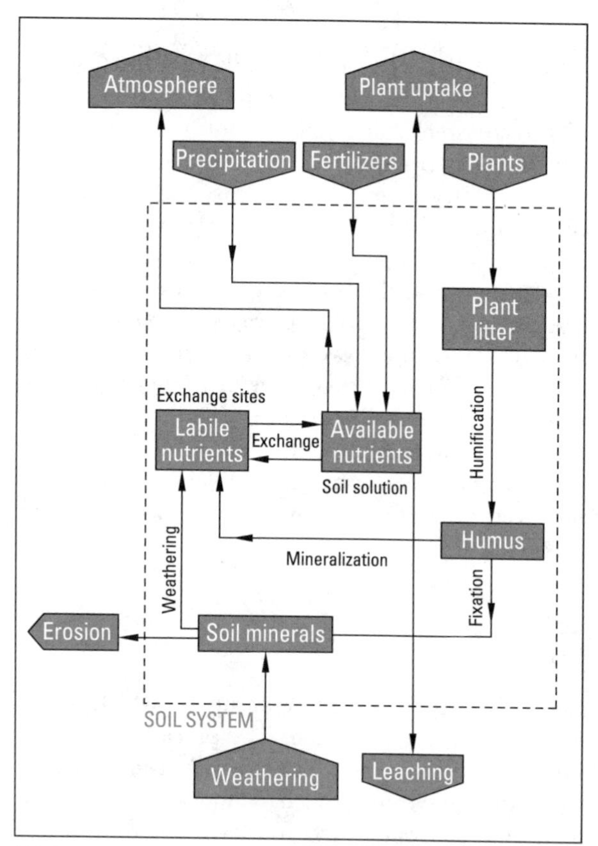


FIGURE 29.3 ▼ The nutrient cycle in the soil, showing the main inputs, outputs, and pathways for nutrient transfers

## GASEOUS LOSSES

Significant losses of nutrients often occur in gaseous form. Especially when the soil is wet and air is lacking (anaerobic), many compounds are reduced to gas, which can then diffuse through the water and through the soil pores into the open atmosphere. Nitrogen, in particular, is lost in this way, and it has been estimated that in the Netherlands as much as 80 percent of the nitrogen applied to the soil as fertilizer may be lost by this process of denitrification. This represents a significant waste of fertilizer, of course, but other effects may also be felt. Some of the gases produced by reduction of carbohydrates, for example, are highly toxic. Ethylene, which is formed by the decomposition of organic compounds under waterlogged (anaerobic) conditions, may damage roots with long-term consequences for plant growth. As a result, even a few days' waterlogging at critical times of the year may greatly reduce crop yields (Table 29.4).

The plants themselves contribute to the gaseous loss of nutrients. During respiration, plants release carbon dioxide through their roots. Carbon dioxide is highly soluble and diffuses through the water in the soil and escapes to the atmosphere. In addition, oxygen is lost through the leaves of growing plants.

Fire oxidizes plant matter, and if severe enough, may combust organic-rich soil horizons. In much of the forested region of the world, fires are a natural and recurring phenomenon. Gaseous losses induced by burning from one ecosystem will eventually be returned, but not necessarily to the same ecosystem. Frequent burning may limit some ecosystems because of these losses.

## EMIGRATION AND HARVESTING

Just as material may be introduced to ecosystems by migration, so too may it be lost. The emigration of animals, and the removal of vegetation by humans, are both processes by which outputs occur from the ecosystem.

## Table 29.4

*The effects of seven days' waterlogging on crop yields in Hungary*

| Crop | J | F | M | A | M | J | J | A | S | O | N | D |
|---|---|---|---|---|---|---|---|---|---|---|---|---|
| | *Percentage reducation in yield due to waterlogging in month* | | | | | | | | | | | |
| Grass | – | – | – | 10 | 15 | 20 | 20 | 10 | – | – | – | – |
| Sugar beet | – | – | 50 | 50 | 50 | 40 | 40 | 40 | 40 | 10 | – | – |
| Potatoes | – | – | 80 | 80 | 90 | 100 | 100 | 100 | 40 | – | – | – |
| Winter wheat | 5 | 5 | 15 | 25 | 40 | 50 | – | – | – | – | – | – |

*Data* B.D. Trafford, 'Field drainage,' *Journal of the Royal Agricultural Society of England*, 131 (1970), pp. 129–52.

People play a particularly important part in these processes, since we use much of the land for agriculture or forestry, and consequently remove large quantities of the material during harvesting. Crops are cut, animals are taken for slaughter; milk is taken from cows, wool from sheep. The nutrients in these are thus removed (Table 29.5)

## Table 29.5

*Nutrients retained in crops and livestock that could become outputs when harvested*

| Crop | Yield ($t\ ha^{-1}$) | Nutrient content ($kg\ ha^{-1}$) N | P | K | Ca | Mg | S |
|---|---|---|---|---|---|---|---|
| Wheat | 6 | 120 | 25 | 80 | 20 | 15 | 25 |
| Barley | 5 | 100 | 18 | 60 | 15 | 8 | 20 |
| Potatoes | 50 | 180 | 25 | 200 | 10 | 20 | 25 |
| Grass | 10 | 250 | 30 | 250 | 70 | 20 | 20 |
| Rice | 2.2 | 26 | 8 | 8 | 2 | 4 | – |
| Tea | 1.3 | 60 | 5 | 30 | 6 | 3 | – |
| Cattle | 2.5/ha | 50 | 30 | 15 | 25 | 10 | – |

*Data* G.W. Cooke, *Fertilizing for Maximum Yields* (1983), Granada.

In addition, many of our agricultural activities cause significant losses. In many agricultural ecosystems, we burn the vegetation, either to clear the land for tilling or to remove crop residues and weeds. Considerable quantities of plant nutrients are lost in the smoke, and the remainder in the ash. Not all these nutrients are retained in the ecosystem as a whole. The losses to the atmosphere may be blown out of the area; the losses to the soil may be removed rapidly by leaching. Thus, repeated burning can lead to a decline in the fertility of the ecosystem.

# Biogeochemical cycling in the soil

The biogeochemical cycle in the soil is complex. The main inputs to the soil come from weathering, precipitation, fertilizers, atmospheric sources, and plants. Under natural conditions, inputs from plants are the most important, including not only nutrients released by organic matter decay, but also substances washed in from the above-ground parts of plants (see Figure 28.1). Losses are by leaching, erosion, gaseous loss, and plant uptake. Within the soil, nutrients are stored on the particles or in chemical compounds, and circulated both in the water and in the organisms.

### ORGANIC MATTER DECOMPOSITION

The input of nutrients through the decomposition of organic matter is dependent upon the action of soil organisms. It is, therefore, a process which is closely associated with the detritus food chain. The initial attack on plant debris often occurs before the material reaches the soil. The leaves of trees, for example, may be partly decomposed by fungi while still on the tree. This process releases some of the more unstable compounds such as the sugars, and rainfall washes them into the soil. Once the leaf falls to the ground, larger soil organisms attack it. Earthworms, arthropods (such as ants, beetles and termites), and gastropods (slugs and snails) macerate (chew up) the leaf material and make it more susceptible to attack by the smaller, micro-organisms. General purpose organisms—mainly fungi and heterotrophic bacteria—tend to wage the next attack on the debris, and during this stage more resistant carbohydrate compounds (e.g., polysaccharides and starches) are decomposed. Finally, actinomycetes and certain more specific fungi and bacteria attack the remaining substances: mainly cellulose and lignin (Figure 29.4). At the end of the process, the resistant fulvic and humic acids remain. It is these which comprise the humus of the soil.

The whole procedure may take only a few months; the leaves of many deciduous trees, for example, rot within nine months. In tropical areas, the breakdown of organic matter may be completed within a matter of weeks, because of the abundance of organisms such as termites and the thriving populations of micro-organisms. On the other hand, pine and spruce needles, in cool temperate areas, may take a decade or more to decompose fully.

The remaining humus decomposes even more slowly since it breaks down mainly by chemical oxidation. Only in tropical conditions, where soil temperatures are sufficiently high to speed up the chemical processes, does the humus disappear at all rapidly; in temperate areas, dating of the material by radiocarbon methods has shown that some of the humus in the soil may be as much as 1000 years old, and in the Subarctic and Arctic, it is often possible to find organic matter in soil horizons created by the first plants to invade the site following deglaciation (8000–9000 years B.P.).

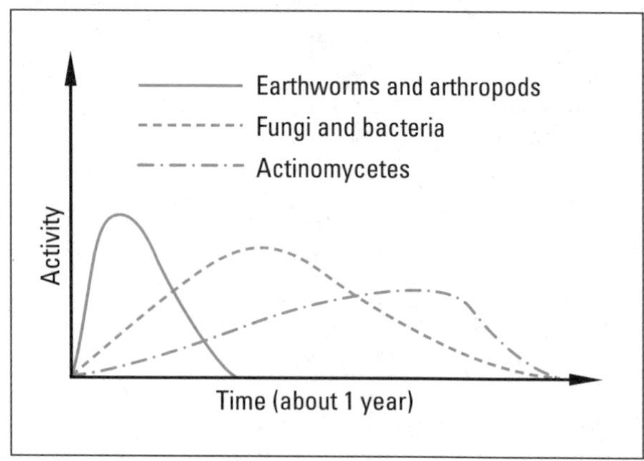


FIGURE 29.4 ▼ The changing activity of different groups of soil organisms during decomposition of organic matter (from Briggs, 1977)

Organic matter decomposition releases nutrients into the soil and is a process of considerable significance to the nutrient cycle. The nutrients are not always released immediately in an available form. Many are bound up within the bodies of the consumers and decomposers. They are made available only when these organisms themselves decay and the organic compounds are mineralized into inorganic forms. This effect of organic assimilation of the nutrients may be important, for during active decay, the populations of the organisms may increase markedly, and in the process they may take up nutrients that would otherwise have been available to plants. For a short period, therefore, nutrient deficiencies may occur, and plant growth may be inhibited (Figure 29.5).

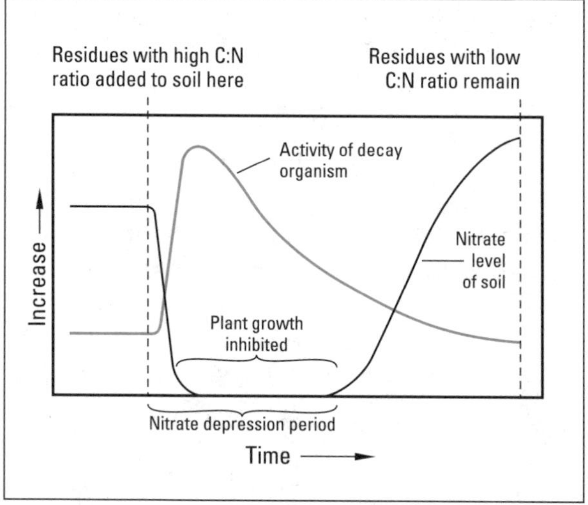


FIGURE 29.5 ▼ The effects of raw organic residues on soil biological activity, nitrate levels, and plant growth. The peak in activity of decay organisms temporarily depletes nutrients, but when the organisms die, the usable forms of the nutrients are available for plant growth (after Brady, 1973)

The compounds (potential nutrients) that are released in an available form are not all stored; some may be lost from the soil before they can become attached to the soil particles. Carbon, oxygen, and hydrogen may all be lost in gaseous form; calcium and magnesium may be dissolved in the soil water and lost by leaching. Water itself is another common decomposition product. Thus, the input of nutrients to the soil nutrient store is not totally efficient; considerable losses may occur in the process.

## THE NATURE OF SOIL NUTRIENTS

Nutrients are stored within the soil in a variety of forms. Most occur as ions; that is, as atoms or groups of atoms bearing an electrical charge. Positively charged ions are known as **cations** (e.g., calcium, $Ca^{++}$, or sodium, $Na^{+}$); negatively charged ions are **anions** (e.g., nitrate, $NO_{3-}$, or phosphate, $PO_{4-}$). Some are bound up within the structure of the soil particles—for example, within the clay minerals. Others, such as phosphorus, are held in chemical compounds (e.g., calcium phosphate). These nutrients are thus unavailable to plants unless the minerals are weathered and the nutrients are released. They are called **labile nutrients** (Figure 29.3). Many of the cations are also held in an exchangeable form; they are bound loosely to the surface of colloidal clay or organic particles. Although they are not immediately available to plants, they may be released relatively easily by the process of cation exchange. For this reason, they are referred to as **exchangeable nutrients**. Finally, there are the **available nutrients**—those nutrients that are dissolved in the soil water (or soil solution, as it is commonly known)—which may be removed directly by plants.

The exchangeable nutrient reserves in the soil have particular importance. If growth is to be maintained, the nutrients removed from the soil solution by plants must be replenished by release of labile or exchangeable ions. Release from labile forms is slow, and for this reason nutrients such as phosphate, which are held mainly in this form, cannot be replaced rapidly. On the other hand, most of the cations in the soil are held in exchangeable form, and these may be released quickly, ensuring that deficiencies of these nutrients rarely occur. Let us look at the processes involved.

## CATION ADSORPTION

The ability of soils to store nutrients in an exchangeable form arises from the colloidal properties of the clay and organic materials. As we have seen, isomorphous substitution during clay formation results in the replacement of certain atoms in the clay structure by others of similar size. In most cases, this process involves atoms of decreasing valency. For example, silicon, which has a valency of four ($Si^{++++}$), is replaced by aluminum, with a valency of three ($Al^{+++}$). This, in turn, may be replaced by divalent magnesium ($Mg^{++}$). In this way, the clays progressively lose positive charges, and this leaves unsatisfied negative charges within the mineral. Similar unsatisfied negative charges may also be created by breakage of the clay minerals during weathering. In both cases, the result is to give the particles a net negative charge.

Generally, similar processes occur during the decomposition of organic matter. During the oxidation of organic matter, the hydrogen and carbon compounds tend to be released, leaving unsatisfied negative charges associated with the oxygen atoms exposed at the surface of the material. This negative charge is similar to that produced in clay minerals.

As we know from our own experiments with two magnets, opposite poles attract each other. Thus, the negative charges on the colloidal particles attract positively charged cations. In this way, the colloids collect around themselves a swarm of cations (Figure 29.6). These cations are said to be adsorbed to the colloids. The strength of adsorption depends upon a number of factors, but is greatest close to the surface of the particle and declines with distance; within the equivalent of a few molecular layers, it is negligible.

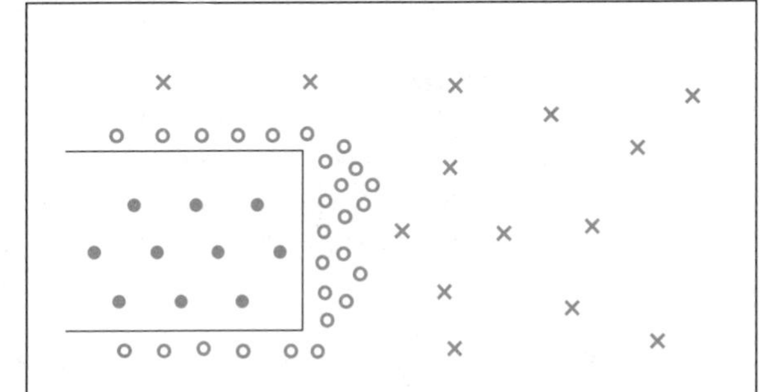

• Ions held within colloidal particles (unavailable)
○ Ions adsorbed on colloidal surfaces (exchangeable)
× Ions in solution (available)

FIGURE 29.6 ▼ The distributions of cations in relation to colloidal particles (from Briggs, 1977)

## CATION EXCHANGE

Release of the cations from the particles occurs mainly because of a disequilibrium between the concentration of ions in the soil water which surrounds the colloid and the concentration on the colloidal surface. The equilibrium conditions are to some extent a function of the valency of the particular nutrient cation under consideration (Figure 29.7). However, in general, if the concentration of ions in the soil water is increased, a net migration of those ions onto the colloidal particles occurs, with the consequent release of others; if the concentration of ions in the soil water is reduced, a net movement of those ions away from the colloids occurs (Figure 29.8).

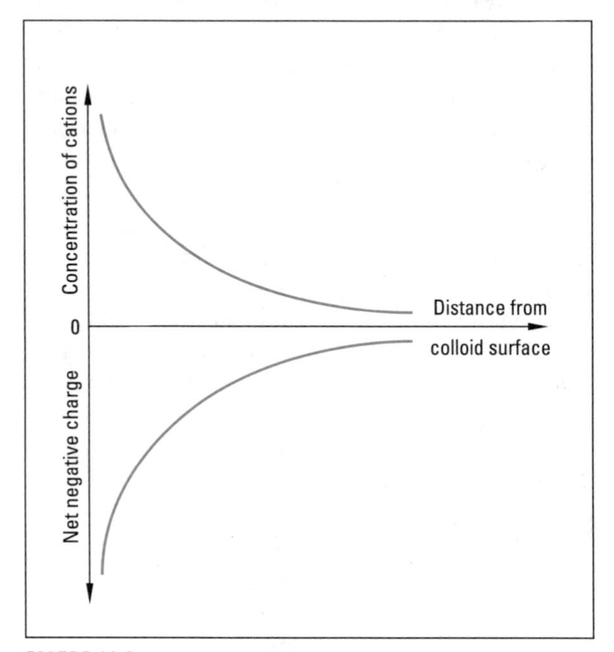


FIGURE 29.7 ▼ The distribution of adsorbed cations and strength of negative charge with distance from a colloidal surface. The graph relates to a distance of only a few molecular layers

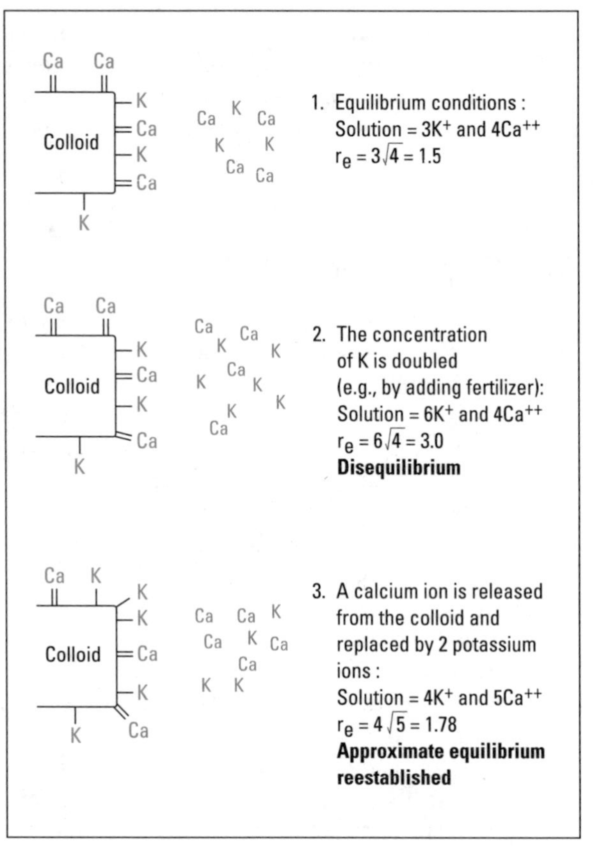


FIGURE 29.8 ▼ Cation exchange. In conformance with the ratio law, a change in the absolute concentrations of two cations of different valency upsets the chemical equilibrium of the soil solution and causes cations to be released from the colloids

It is through this process that plants obtain many of their nutrients. Roots of plants tend to release hydrogen ions, which become dissolved in the soil water. Over time, therefore, there is a tendency for the concentration of hydrogen ions in the water to increase; this is counteracted by the adsorption of hydrogen onto the colloids and the release of other ions into solution. The ions which are released may be taken up by the plants. This is the process of **cation exchange**.

## CATION EXCHANGE CAPACITY

Colloidal particles vary considerably in the density or degree of the negative charge which develops on their surfaces. In general, the densest charges occur in organic compounds, although these are not always stable, for in time the organic matter decomposes further and is lost. In the case of the clay minerals, the charges are greatest on the 2:1 clays, such as montmorillonite (Chapter 27). This is because these minerals contain relatively weak bonds within their lattice structure, which are able to expand to some extent and allow ions to enter within the lattice. Thus, these clays have both a surface and an internal charge that is available to attract the nutrient ions. In other words, in relation to their volume, they have a very large active surface area. The lowest charges are found on the 1:1 clays of the kaolinite group. These are much more rigid, they have a smaller surface area, and they are consequently less active.

The charge developed by the colloids is measured in terms of milli-equivalents per 100 g. This, simply, is the weight of hydrogen ions that could become attached to the surface of 100 g of soil if all the negative charges were occupied by hydrogen. The range of the charges developed on different colloids is indicated in Figure 29.9. These values are often referred to as the **cation exchange capacity**.

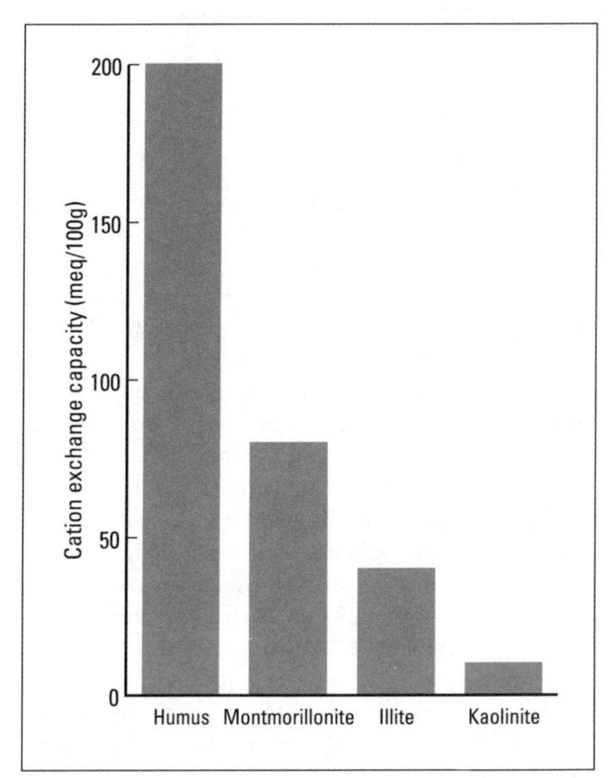


FIGURE 29.9 ▼ The cation exchange capacity of different soil colloids

# Biogeochemical cycling in vegetation

## PLANT UPTAKE

Plants require nutrients for a variety of functions (Table 29.1). Nitrogen is a vital component in amino acids, proteins and chlorophyll. It thus acts as a major control upon photosynthesis and energy production. Phosphorus is an important component of all living cells; potassium plays a role, which is not fully understood, in cell production and growth, and thereby influences the uptake of other elements and the processes of respiration and transpiration; calcium is used in the cell walls and is an important control upon the growth of the meristem—the cells in the root that allow it to extend into the soil.

Most nutrients, as we have seen, are derived from the soil, and thus a good place to start a discussion of the nutrient cycle within the vegetation is the uptake from the soil. This involves a variety of processes, not all of them perfectly understood. Certainly, mass flow represents one of the main means by which the plants obtain their nutrients; the plants transpire moisture from their leaves, creating a moisture gradient within the plant. Moisture then moves up the plant, generating a tension within the roots, and water migrates into the roots from the soil. In the process, the nutrients in solution are transported into the roots.

This process is supplemented by diffusion. Nutrients tend to move through the soil solution in response to concentration gradients; that is, they move from areas of high concentration to areas of low concentration. As plants remove nutrients from the soil water around their roots, new nutrient ions move in to take their place. In this way, a slow flow of ions occurs through the soil water.

Both mass flow and diffusion operate over relatively small distances, and plants seem to be dependent upon root extension for locating both the water and the nutrients. They also obtain nutrients through direct contact with the colloidal particles; ions—in particular, hydrogen—are then released from the roots and replaced by others from the colloid.

The exact process by which nutrients pass through the cell wall of the roots is not clear. It has been suggested that it is a process aided by enzymes which act as carriers for the nutrient ions (Figure 29.10). In addition, soil organisms, in particular bacteria and fungi, may transport the nutrients into the plants. As we will see later, these rhizosphere associations are particularly important in relation to nitrogen and phosphorus.

Within the plants, the nutrients enter the xylem—the cells within the plant along which the solutions from the soil move. The nutrients thus rise through the plant by capillary action. This is essentially a process of mass flow. It leads to a concentration of nutrients in the younger parts of the growing plant. This arises largely because, as the solutions evaporate from the leaf surface, the dissolved substances are left within and on the leaf.

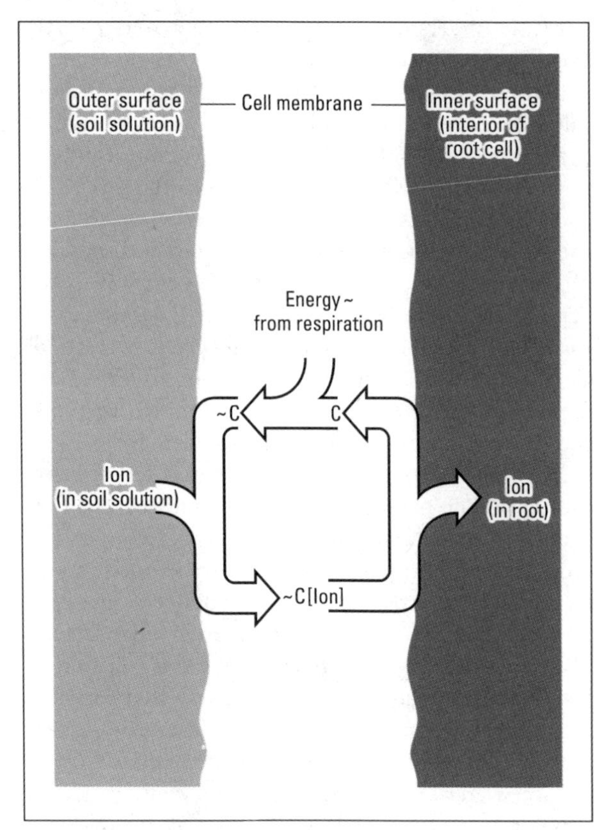


FIGURE 29.10 ▼ The role of enzymes in the uptake of ions by plant roots. The enzymes act as carriers (C) specific to particular groups of ions. Energy is derived from respiration, allowing the ion to be taken up by the enzyme, which then transports it across the cell membrane. Inside the root cell, the ion is released, the enzyme obtains new energy from respiration, and the process is repeated (after Brady, 1973)

Rain or snowmelt is then able to wash the nutrients from the leaves, and when they die, the leaves return the nutrients to the soil surface. Only a small proportion of the nutrient content of the vegetation is therefore held for any length of time in the plants. This storage takes place mainly in the boles and branches of trees and shrubs which may persist throughout the life of the plants.

The rate at which nutrients move through the plant has been measured by 'labelling' nutrients with radioactive substances. These are monitored as they emerge from the plant leaves in leaf drip and stemflow, as they are released by organic matter decomposition of plant residues, and as they are taken up and further released by other plants in the ecosystem. From studies of this sort, it can be shown that nutrients such as phosphorus, injected into the roots of trees, may emerge from the leaves within 28 hours and be taken up and released by other plants in about 3 days. Clearly, not all the nutrients are moving at this rate; some are retained and released by the much slower decomposition of the leaves during the autumn. Nevertheless, it is clear that the nutrient cycle within the plant may be very rapid.

The return of nutrients to the soil, therefore, occurs partly through rainwash and snowmelt and partly through organic matter decomposition.

## CYCLING WITHIN THE ANIMAL COMPONENT OF ECOSYSTEMS

The plants, as we have indicated, are the primary producers in terrestrial ecosystems. They convert solar energy to chemical forms and extract water and nutrients from the soil to form plant tissue. They also act as the foodstuff for the consumers and decomposers; they are the foundation of the food chain. From the plants, nutrients are transferred through the animals, mainly via the grazing food chain.

Animals typically browse or graze off the younger parts of plants and thus use the more nutritious components. In fact, in some agricultural systems, specific attempts are made to encourage the growth of young plant material so that animal nutrition is improved. Thus, in grasslands, the animals are allowed to eat the young grass; as the plants get older and start to flower, their nutritional value declines, for they have higher carbohydrate contents (i.e., more green plant) but lower concentrations of nutrients and proteins. Similarly, controlled burns of forested areas encourage regeneration; this creates a young and more nutritious vegetation cover.

The nutrients taken in by the primary consumers are, to some extent, stored within their bodies. A large proportion, however, is lost in the excretion products. Solid and liquid components are attacked by the decomposers—including various flies, beetles and micro-organisms—and the nutrients either released into the soil or atmosphere, or stored within the bodies of the decomposers. At the same time, the primary consumers may be hunted and eaten by secondary consumers, which in turn may be devoured by tertiary consumers (often humans) (see Figures 26.4 and 26.7). Detritus from the prey at each stage in this process is eaten by further decomposers, and thus enters the detritus food chain.

If it were possible to monitor the movement of nutrients through a natural ecosystem in this way, it would be found that the food web created by these organisms is typically complex. Even within very simple situations, a wide range of animals may be involved (Figure 26.4). It is also apparent, however, that each step in the process involves considerable losses of energy and nutrients from the system (Figure 26.5), so that it is rare for more than four trophic levels to exist. Energy losses occur in the form of respiration; nutrient losses occur mainly through excretion.

# Nutrient cycles

## THE NITROGEN CYCLE

The nitrogen cycle represents one of the most interesting nutrient cycles (Figure 29.11). Nitrogen (N) is possibly the most important, and often most limiting, nutrient for plant growth. Moreover, it follows a cycle in which soil organisms play a fundamental role. It is N that is the first number in commercially available fertilizers (N-P-K).

Most nitrogen is derived from the atmosphere. Small proportions enter the soil in precipitation or through the effects of lightning. The majority, however, is fixed within the soil by micro-organisms. These are of two general sorts. Some live freely (non-symbiotically) in the soil and build up nitrogen from the atmosphere in their bodies, releasing it as they die and decay. A wide range of fungi, bacteria, and algae take part in this non-symbiotic fixation. The other group consists of the symbiotic nitrogen-fixers. These are mainly bacteria which live in close and mutually beneficial association with plant roots. They irritate the roots and produce small nodules in which they live. They absorb nitrogen from the air and pass it into the plant. These symbiotic relationships are restricted to specific plants (in particular, legumes [the pea family *Leguminosae*]), but they represent a major source of nitrogen in the soil.

In the case of the symbiotically fixed nitrogen, the nutrient enters the soil store mainly following death and decay of the plant, or as a return to the

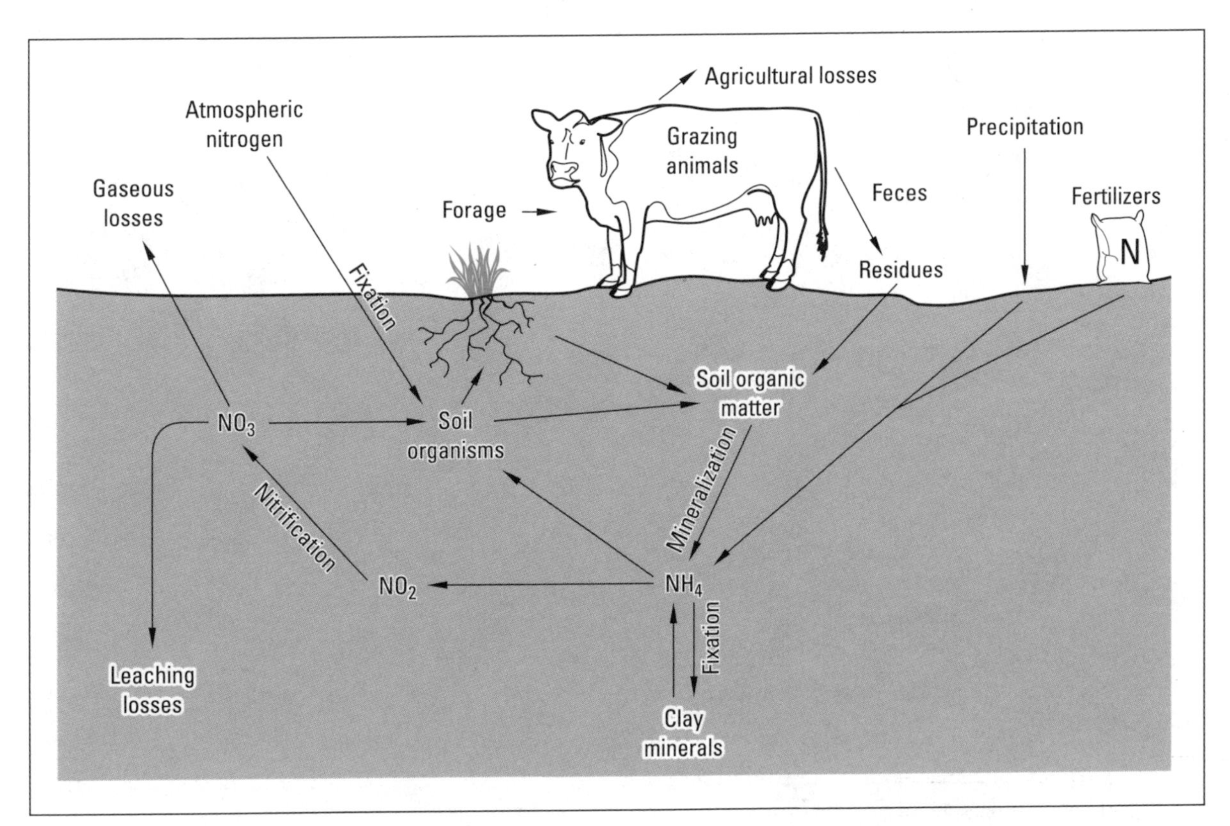


FIGURE 29.11 ▼ The nitrogen cycle

soil in animal wastes. The non-symbiotic fixation of nitrogen requires the death and decomposition of the soil organisms before the nitrogen is released. In both cases, decomposers within the soil attack the organic compounds and the nitrogen is converted from organic ammonia ($NH_3^+$) to inorganic ammonium salts ($NH_4^+$). This process is known as mineralization, and it is carried out by a variety of bacteria, actinomycetes, and fungi.

In the form of ammonium, the nitrogen may be adsorbed onto the clays, and thus enters the exchangeable nutrient store. When it is released through cation exchange, it may be altered by a small range of autotrophic bacteria into nitrite. *Nitrosomonas* is one of the main bacteria genera to take part in this process:

$$\underset{\text{(ammonium)}}{2NH_4^+} + \underset{\text{(oxygen)}}{3O_2} \rightarrow \underset{\text{(nitrite)}}{2NO_2^-} + \underset{\text{(hydrogen ions)}}{4H^+} + \text{energy}$$

This process is essentially one of oxidation.

Further alteration by a restricted range of bacteria, including members of the genus *Nitrobacter*, oxidize the nitrite to nitrate, a process known as nitrification:

$$\underset{\text{(nitrite)}}{2NO_2^-} + \underset{\text{(oxygen)}}{O_2} \rightarrow \underset{\text{(nitrate)}}{2NO_3^-} + \text{energy}$$

In the form of nitrate, the nitrogen is available to plants, and it may be captured through root absorption. However, it is also very soluble and may be removed by leaching (Figure 29.3). In addition, it may be reduced to gaseous nitrogen and then lost to the atmosphere, or again taken up by soil organisms and recycled. It is mainly in the form of nitrate, therefore, that nitrogen losses from the system occur. Nitrogen taken up by the plants is returned in plant debris and released into the soil during organic matter decomposition.

## THE PHOSPHORUS CYCLE

Phosphorus (P), like nitrogen, occurs mainly as an anion (e.g., $PO_3^-$, $PO_4^-$). It is derived not from the atmosphere, however, but from the weathering of

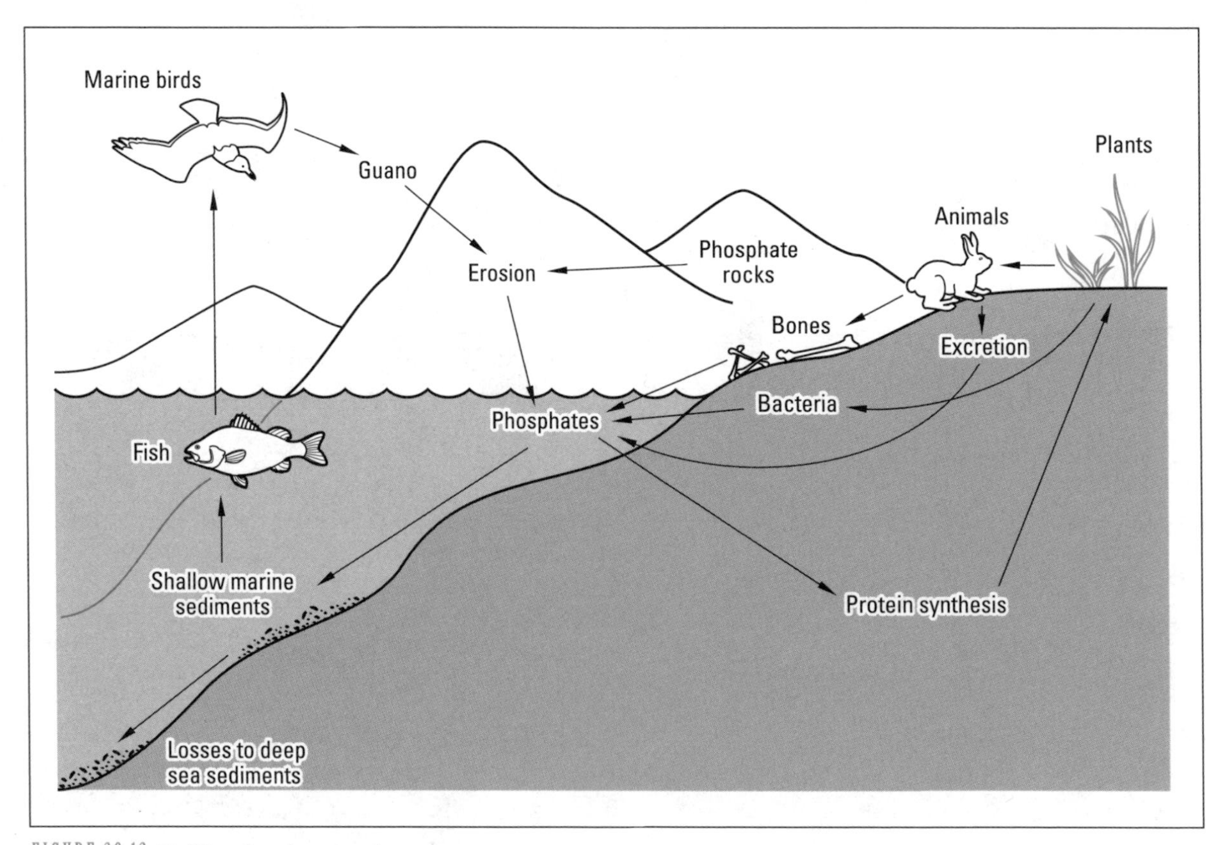


FIGURE 29.12 ▼ The phosphorus cycle

phosphatic minerals, such as apatite and many clay minerals. It occurs within the soil in three main forms: as inorganic phosphate in the original minerals, or as compounds coating mineral particles; as organic phosphate bound up within the plant debris and soil fauna; and as soluble phosphate within the soil solution. On the whole, phosphorus is highly insoluble, so the last of these is quantitatively the least important. Nevertheless, it is mainly from the soluble component that plants obtain their phosphates. Thus, it is essential that there is a constant turnover of phosphorus between the other two forms and the soluble phosphate (Figure 29.12).

The importance of this can be appreciated when it is seen that the total phosphorus content of the soil solution would be exhausted within a few hours if constant replenishment by weathering did not occur.

Phosphorus is taken up by plants in the form of phosphate. The process is often aided by soil organisms, including fungi and bacteria. From the plants, the phosphorus may return directly to the soil, mainly through leaf fall and organic matter decomposition, or it may be passed along food webs. In the latter case, much of the phosphorus is returned in animal wastes, particularly dung. Considerable proportions of phosphate may be fixed within the animal bones, however, and this is only returned to the soil when the animal dies, and the decomposition of bones is complete. Indeed, in the past bones were a major source of phosphate 'manure'. On the prairies, bison jumps used by native people for thousands of years were mined at the turn of the century for the bison bones, and as a consequence few of these archeological sites remain untouched.

As this implies, phosphates are also returned to the soil in considerable quantities as fertilizers. In addition to bone manure, 'artificial' fertilizers are used, often derived from phosphate-rich rocks. **Guano** is another source, and this illustrates a further aspect of the phosphorus cycle (Figure 29.12), for the nutrients which are washed from the land enter streams, are carried to the sea, absorbed by fish, eaten by seabirds, and eventually deposited as guano. People then collect this for fertilizer.

The phosphorus that is returned to the soil in organic form is slowly released during organic matter decomposition. Bacteria play a vital part in this process, mineralizing the phosphorus and converting it to inorganic forms. It may then be taken up again by plants, or if combined with other elements, such as iron and calcium, forms the insoluble phosphate compounds in which it is stored. Small amounts are also adsorbed onto colloids.

## THE POTASSIUM CYCLE

Potassium (K) is the third essential plant macronutrient and, like phosphorus and nitrogen, is commonly applied as a fertilizer in order to ensure adequate plant nutrition in agricultural systems. Unlike these other two nutrients, however, potassium is a cation and thus is subject to rather different processes of cycling (Figure 29.13).

The main natural sources of potassium are potassic minerals such as K feldspars and micas. Together, these account for between 90 and 98 percent of the total potassium in the ecosystem. These primary minerals slowly weather, producing potassium-containing clay minerals such as illite and vermiculite, and releasing potassium into the soil. Muscovite, for example, weathers to illite as follows:

$$\underset{\text{(muscovite)}}{K_2Al_4(Al_2Si_6)O_{20}(OH)_4} + Si^{++++} \rightarrow$$

$$\underset{\text{(illite)}}{K_{0.2}(K_{0.8})Al_4(AlSi_7)O_{20}(OH)_4} + K^+ + Al^{+++}$$

Most of the potassium held in clay minerals (shown as $K_{0.8}$ in the above equation) is not readily available to plants, but is held within the clay lattice. In the long term, it is released by gradual breakdown of the clay. A small proportion (shown as $K_{0.2}$ above), however, is held in an exchangeable form on the surface of the clay minerals. This is released by exchange with other cations (e.g., $H^+$) in the soil solution, and is thus available for plant uptake. These exchangeable reserves are themselves replenished by potassium from within the clay lattice. Thus, a three-way transfer of potassium takes place:

$$\text{non-exchangeable K} \rightleftharpoons \text{exchangeable K} \rightleftharpoons \text{soil solution K}$$

The processes of transfer are interdependent, so that these three reserves of potassium are kept in more or less constant ratios.

At any time, about 10 percent of the readily available potassium in the soil is dissolved in the soil solution and is directly accessible to plants. This potassium is also available for leaching, however, and a small proportion is consequently carried away in drainage waters, much of it ultimately reaching the sea. Losses by leaching are greatly dependent upon pH, and almost four times more potassium may be lost from an acid soil (pH *c.* 4.5–5.0) than from a neutral soil (pH 7.0). In addition, soil erosion results in the loss of mineral and exchangeable potassium held on the soil minerals.

The potassium taken up by plants is used in photosynthesis, starch formation, and translocation of sugars. Some is subsequently removed in harvested material and stored in human bodies before it passes into the sewage system. Much, however, is returned to the soil in plant residues, animal wastes, and manures. Together with inputs from inorganic fertilizers, these provide the main source of soil potassium. Within the soil, these inputs are released by biological decomposition and chemical reactions, then recycled, leached, adsorbed or fixed within the clays by incorporation into the mineral structure (Figure 29.13).

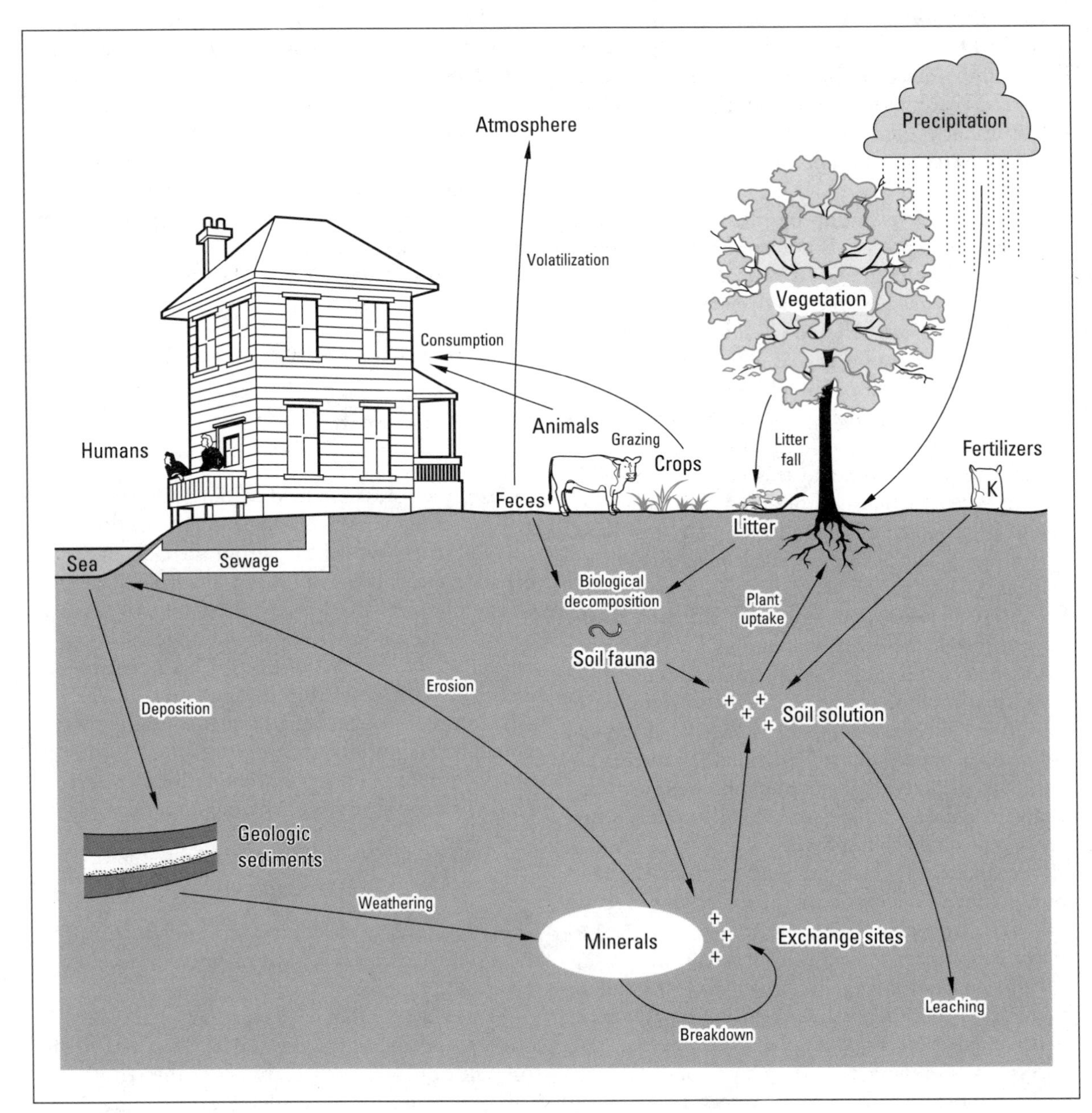


FIGURE 29.13 ▼ The potassium cycle

# Conclusion

The importance of biogeochemical cycles within ecosystems is clear. It is through this circulation of nutrients that life-giving substances are transferred. It is a process that ties the components of the ecosystem intimately together. The significance of the process extends far beyond the ecosystem, however, for the movement of nutrients also involves the hydrological and landscape systems. Water is one of the main agents of nutrient cycling; it carries the nutrients to the plants, it washes them from the leaves and animal wastes, it takes part in the decomposition of organic matter. Water is also a fundamental agent in the initial weathering and release of nutrients from rocks, and it carries nutrients out of the system by leaching and erosion. For this reason, many of the outputs from the ecosystem become inputs to the surface hydrological system, as confirmed by pollution of streams and lakes by fertilizer residues.

In the same way, there is a close relationship between biogeochemical cycling and landscape processes. The majority of the nutrients are released by weathering of rock materials; the nutrient inputs to the soil and vegetation are thus the outputs from landscape processes. Similarly, losses of nutrients by wind erosion and rainfall erosion are a result of geomorphological processes. As ever, nature recognizes no clear boundaries between the systems we try to identify.

# World biomes

## Soil and vegetation patterns

### THE CLASSIFICATION OF SOILS AND VEGETATION

As we journey across the world—even over short distances—we can observe changes in the environment. Most of these changes are seen as we move through different ecosystems. We can pass from forest to grassland, from agricultural land to seminatural or natural areas. Often we see new animals associated with changes in vegetation, as different animals (heterotrophs) rely on different plants (producers) for their niche and habitat requirements. We also notice variations in the soils, including changes in colour and differences in surface expression. Some create large, cloddy fields, while others give fine, smooth surfaces. Some show evidence of erosion, while others look peaty and fibrous. We can see these changes, and to some extent categorize what we see, but if we are to be more specific and analytical about the different types of ecosystems, we need to classify these features much more definitively.

In fact, a great deal of effort has gone into developing methods to classify the world's soils and vegetation. Most soil classification systems have been developed to aid agriculture. Information about the type of soil in an area makes it possible to develop better plans for agricultural management. In developing new areas for agriculture, a survey of the soil is essential. Soil surveys also provide a means of collecting information on one of the world's most important resources and of developing an inventory of this resource. Soil surveys and classification systems also have scientific value, as they facilitate the transfer of information. Using brief descriptions of the soils, it is possible to convey a great deal of information from one soil scientist to another.

The same general principles are true of vegetation surveys and classification systems. The aim, again, is to collect information to improve our understanding and management of vegetation, particularly with regard to the conservation and protection of ecosystems. Again, this is achieved by field surveys, carried out by trained field workers.

Although it is possible to map the location and identity of each plant, this would take an inordinate length of time. Soil does not consist of series of individuals, but rather is a continuum. We can dig holes and look at the soil profile, but each hole is merely a cross-section of a laterally continuous and variable medium. In both cases, we are faced with major problems of classification. If we are to produce maps of the soil and vegetation, we must be able to classify these components as general types and to draw lines separating one type from another.

This presents two major problems. First, we are trying to classify and map a continuum. On our map, we must draw lines separating one soil or vegetation type from another, but in reality, these lines delineating abrupt boundaries do not exist. Similar problems confront any mapping project, be it a surficial geomorphology or climatic map. The second problem is that in trying to identify different soil or vegetation types, we are dealing with multivariate phenomena: there are countless different properties that we could use to classify them. Colour, texture, structure, depth, mineralogy, chemistry, organic properties, and many other properties could be used to describe soil. The occurrence and relative frequencies of plant species, plant form, plant size, and many other parameters could be used to describe vegetation. Clearly, no system of classification is perfect.

Too often, we assume that the classification method we are familiar with is best in all cases. This is not so. Appendix III outlines two methods used to classify soils, but these are only examples from a long list of classification systems.

# World soils

On a world scale, it was once common to relate soils to climatic conditions. The Russians—in particular V. V. Dokuchaev and N. Sibertzev—were among the first to devise a systematic classification of soil based on climate late in the last century. Then, in the 1930s, C. F. Marbut extended their approach, and subsequently it became widely used in many other parts of the world, including Europe. This classification system was based on the principle that the soil is a product of the prevailing environment, within which climate and vegetation are dominant influences.

This approach recognized zonal, azonal, and intrazonal soils. **Zonal soils** had well-developed characteristics that reflected the influence of the active pedogenic factors dictated by climate and biota (particularly vegetation). **Azonal soils** were poorly developed because slope, parent material, or insufficient time had limited their formation. **Intrazonal soils** were associated with poor drainage, salinity, or

| | | | COOL CLIMATES | WARM CLIMATES | |
|---|---|---|---|---|---|
| | | | Tundra soils | | |
| MOIST CLIMATES | Pedalfers | Podzolic | Podzols<br>Brown forest soils<br>Prairie soils | Podzols<br>Red & yellow Podzolic soils | Increased leaching ↑ |
| MOIST CLIMATES | Pedalfers | Lateritic | — | Tropical red soils<br>Soils with Laterite horizons | |
| DRY CLIMATES | Pedocals | | Chernozems<br>Chestnut-brown soils<br>Brown soils<br>Sierozems | Various tropical Pedocals | Increased leaching ↑ |

FIGURE 30.1 ▼ The zonal classification of world soils according to Marbut (from Eyre, 1968)

calcareous parent materials in localized sites. These soils had well-developed characteristics, but did not reflect the regional climate or vegetation.

Each of these broad classes of soil was subdivided into a hierarchy of great soil groups, subgroups, and so on (Figure 30.1). However, several problems were encountered with this approach. In particular, recognition of a soil type depended on correct interpretation of the soil formation processes, which is not always possible. Also, as new information on soil types became available, it was clear that many of the world's soils did not fit readily into this system.

New approaches were tried, and in the United States a series of provisional classification systems was introduced. These led to a working method known as the **Seventh Approximation** in 1960. It was given this name because it was the seventh attempt to produce a workable system. Since then, further work has taken place, and the method has been refined into the **Comprehensive Soil Classification System**. Many other countries, including Canada in 1978, have produced classification systems that are similar in structure. The **Canadian System of Soil Classification** (CSSC) (Appendix III) recognizes soil types on the basis of measured properties of the profile, and uses a hierarchical approach, with multiple levels of classification, moving from the most general to the more specific (Figure 30.2). This allows a large number of locally significant soil series to be grouped into a few, more general groups, great groups, or orders for use at a broader scale.

Most classification schemes have inherent problems in application. For example, soil classification is based on strict quantitative definitions of many properties, and slight differences in key criteria can lead to separation of soils that are similar in many other respects. Also, soil classification is hierarchical, yet soils are not really arranged in that way and groups of soils do not belong to some 'family' of soils. Each soil profile develops through the interaction of specific sets of environmental processes. We cannot think of soils as being related, as are animals and plants that have evolved from a common ancestry.

Nevertheless, the CSSC has much to commend it, and it introduces a number of significant principles. It confines itself to measurable properties of the soil and avoids classification on the basis of assumptions about genesis or environmental relationships. It also stresses the hydrological properties of the soils, a

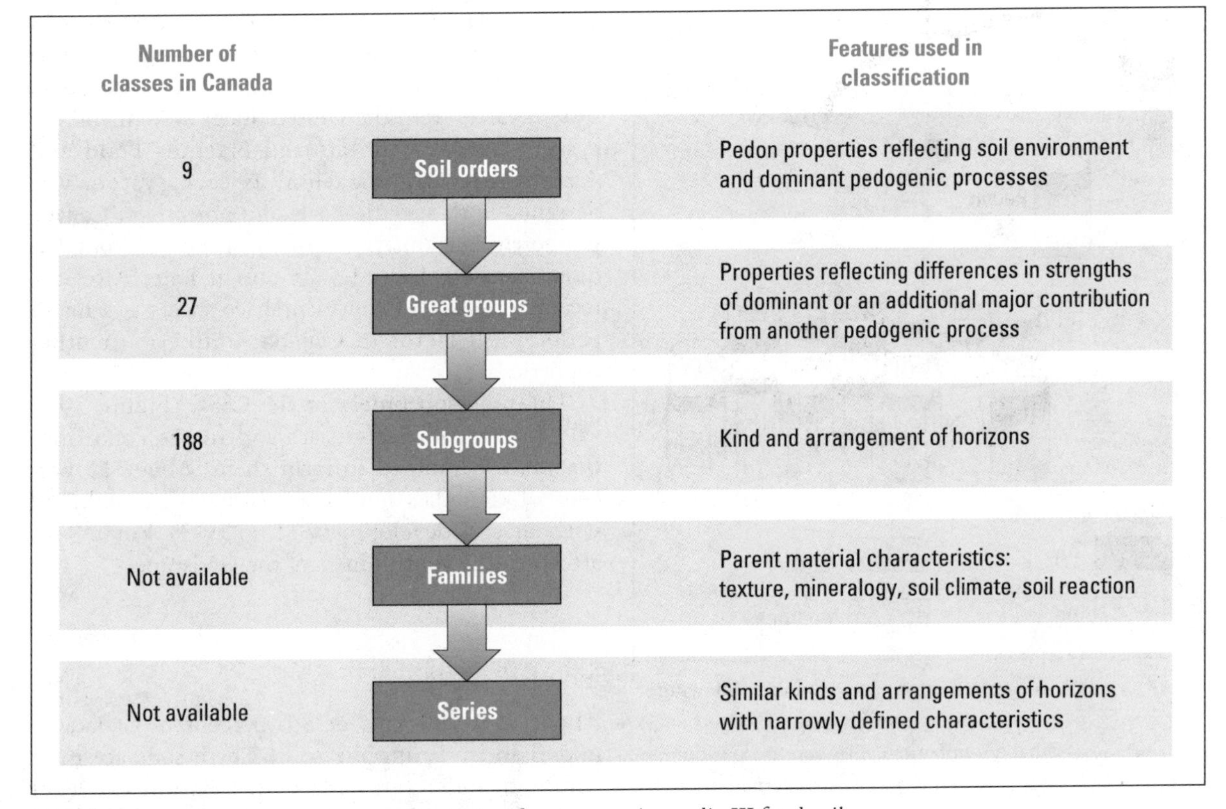


FIGURE 30.2 ▼ The Canadian System of Soil Classification; see Appendix III for details

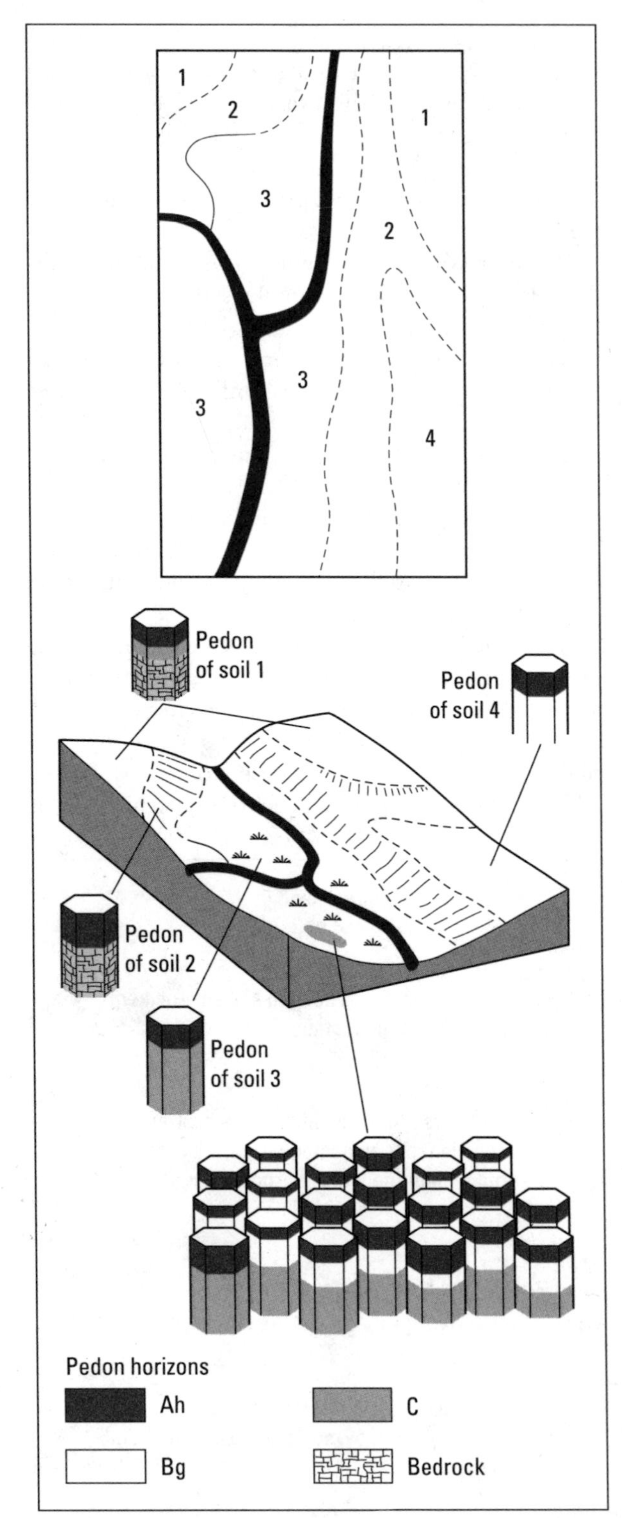


FIGURE 30.3 ▼ Soil distributions on a landscape, showing representative profiles (pedons), and the way that a number of contiguous pedons can be grouped to form a polypedon (note the variability between pedons)

factor of great importance to agriculture. The CSSC is based on the concept of the **pedon**, a three-dimensional soil column (often considered to be hexagonal) extending to the base of the soil. This is the smallest unit of soil recognized, and it is considered to be uniform in character. Over space, soils can be arranged into **polypedons**, contiguous groups of similar pedons. Inevitably, such pedons vary slightly, and 'rogue' pedons may be included in the polypedon. Polypedons are rarely uniform. Nevertheless, the polypedon represents an area of soil that can be mapped as a unit (Figure 30.3).

On the basis of this classification, general maps of the soils of the world can be produced (Figure 30.4). At this scale, only suborders can be shown, but the importance of climatic factors, including temperature and moisture regime, becomes apparent. Nonetheless, classification at this scale has only the most general value.

# Canadian soils

Commencing in 1914, Canadian scientists began surveying soils in order to classify and map them. The first classification system was in use by the end of World War II, and the most recent revision was released in 1978 (see Figure 27.21). The Canadian system varies from the United States system and the system used by the United Nations Food and Agriculture Organization, as each system was designed with specific goals and objectives. Canada is a northern country, and many of the soils found at more southern latitudes are absent here. Also, the recent influence of continental ice sheets is a major pedological factor in Canada, unlike most other countries.

The nine soil orders in the CSSC (Figure 30.2) vary in their characteristics and in the conditions that have combined to form them. About 15 percent of Canada's land mass consists of rockland without soil development (1 375 031 $km^2$). This attests to the youthfulness of the landscape.

## BRUNISOLIC ORDER

About 789 780 $km^2$ or 8.6 percent of Canada is underlain by Brunisolic soils. These soils are most common under forested conditions but do occur through a broad range of ecosystem types. They

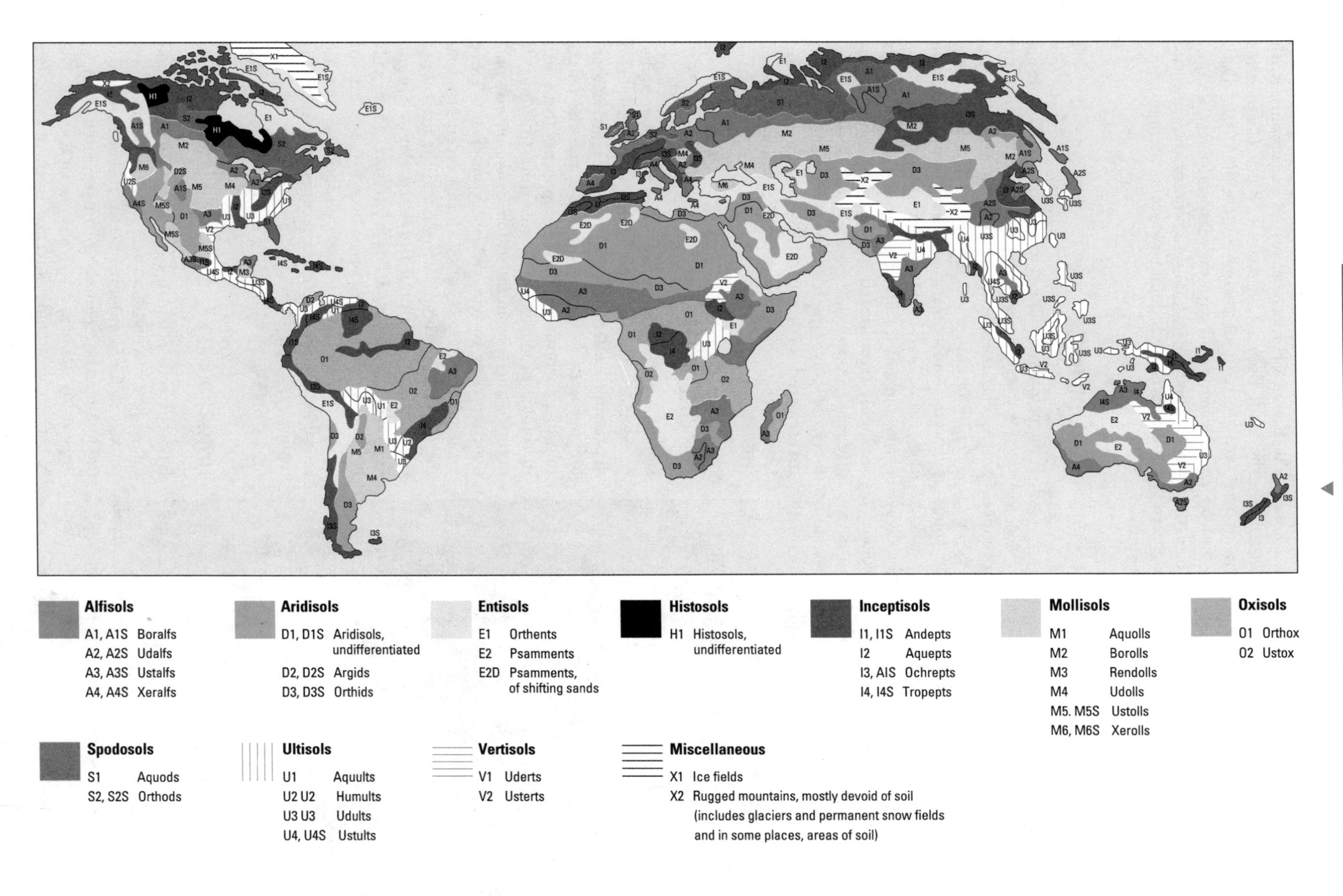


FIGURE 30.4 ▼ Soils of the world: the Comprehensive Soil Classification System (after Brady, 1973)

have a B horizon (Bm) that is brownish in colour, with insufficient clay accumulation to be classed as Luvisolic and insufficient amorphous material to be classified as Podzolic. Thickness of the Ah horizon and soil reaction (pH) determine in which of the four great groups these Brunisolic soils are placed: Melanic Brunisols, Eutric Brunisols, Sombric Brunisols or Dystric Brunisols.

## CHERNOZEMIC ORDER

Prairie or grassland ecosystems usually produce Chernozemic soils. These cover about 468 190 km$^2$, or 5.1 percent of Canada, and are some of the country's most productive agricultural soils. The dark A horizon is rich in organic matter, produced by decay of a dense mat of fibrous grass roots. The colour of the A horizon is the main criterion used to determine in which of the four great groups these soils are classified. Shortgrass prairie usually occurs with the Brown Chernozems, mixed grass prairie with the Dark Brown Chernozems, Black Chernozems with the tallgrass prairie (Figure 30.5), and Dark Gray Chernozems with the grassland-forest ecozone, where woody plants are a significant component of the otherwise herb-dominated plant communities.

FIGURE 30.5 ▼ A Black Chernozem profile from the tallgrass prairie (photo courtesy Land Resource Research Centre, Agriculture Canada)

## CRYOSOLIC ORDER

If the permafrost table is within 1 m of the soil surface (or 2 m if soils are strongly cryoturbated), the soil is classified as a Cryosol. These soils cover 3 672 080 km$^2$ or 40 percent of Canada. Tundra, subarctic, and boreal forest ecosystems dominate these areas. Static Cryosols form where the soil is permeable and ice is not a factor. Turbic Cryosols are highly cryoturbated (see Figure 27.1), and usually result in patterned ground, such as earth hummocks or polygons. Horizontal horizons are often interrupted with a cyclic soil body in Turbic Cryosols. Organic Cryosols are predominantly peat.

## GLEYSOLIC ORDER

Soils that are frequently flooded or waterlogged due to poor drainage or topographic position (closed depressions) often develop into Gleysols. Waterlogging leads to oxygen depletion, and the reduction of previously oxidized components (brownish in colour) produces a grey colouring (gleying). If drying occurs later in the season, oxidation may produce mottling (bright yellow-brown spots). Three great groups are recognized in this order, based on the thickness of the Ah horizon and the presence of a B horizon containing illuviated clays—Humic Gleysols, Gleysols, and Luvic Gleysols. These soils have been estimated to occur over 117 143 km$^2$ or 1.3 percent of Canada. However, because they are azonal and locally restricted, areal estimates are difficult.

## LUVISOLIC ORDER

Luvisolic soils occur in coniferous, deciduous, and mixed forest ecosystems over 809 046 km$^2$ or 8.8 percent of Canada (Figure 30.6). Parent material is predominantly medium- to fine-textured, well to imperfectly drained, calcareous glacial deposits. The A horizon is strongly eluviated in these soils. The

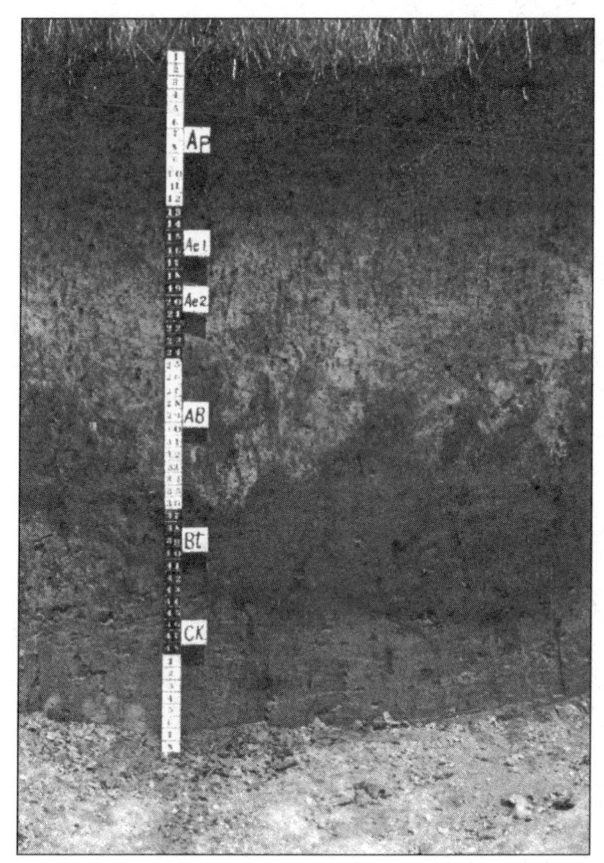


FIGURE 30.6 ▼ A Brown Brunisol-Grey Luvisol found in the spruce and aspen forests of Alberta and Saskatchewan (photo courtesy Land Resource Research Centre, Agriculture Canada)

removal of clay (lessivage) produces greyish Ae horizons, and the silicate clay is translocated to the Bt horizon, which becomes brownish. Soils in the Gray Brown Luvisol great group have organic and mineral soil mixed in the upper horizons, usually as a result of earthworm activity. Soils of the other great group in this order, the Gray Luvisols, occur in the cooler climates of western Canada.

## ORGANIC ORDER

Organic soils form mainly in wetland, heath, or bog ecosystems, and are composed predominantly of organic material in various states of decomposition. Organic soils must be at least 40 cm thick (or 10 cm if overlying rock) and be composed of at least 30 percent organic matter by weight. There are four great groups in this order. Fibrisols have at least 40 percent of their volume consisting of readily recognizable fibrous plant fragments. Mesisols are 10 to 40 percent fibrous material by volume, and Humisols have up to 10 percent fibrous content. Folisols consist of mats of forest litter overlying bedrock. Organic soils cover 373 804 km$^2$, or 4.1 percent of Canada.

## PODZOLIC ORDER

Podzolic soils are found in heathlands and coniferous forest ecosystems and cover 1 429 111 km$^2$, or 15.6 percent of Canada (Figure 30.7). The surface layers of these soils are usually poorly decomposed litter, followed by an eluviated A horizon and a B horizon in which combinations of organic matter, aluminum, and iron are illuviated. Calcium is leached from the A horizon by water that has become acidic as it passed through the litter layer

FIGURE 30.7 ▼ A Placic Podzol common in the evergreen forest of eastern Canada (photo courtesy Land Resource Research Centre, Agriculture Canada)

and consequently the horizon becomes light grey in colour. There are three great groups in this soil order—Humo-Ferric Podzol, Ferro-Humic Podzol, and Humic Podzol.

### REGOSOLIC ORDER

As the name implies, these soils are little altered from their parent materials, perhaps due to the young age of these surfaces or to recent geomorphic disturbance. Regosols may have a dark A horizon, but otherwise they exhibit no horizon development. They are associated with many different ecosystems and cover 73 442 km$^2$, or 0.8 percent of Canada. The two great groups in this order are the Humic Regosols, which have an Ah horizon of at least 10 cm, and the Regosols, which have an Ah layer less than 10 cm thick.

### SOLONETZIC ORDER

Solonetzic soils develop on saline parent materials, mostly in grassland or parkland ecosystems. The B horizon contains sodium ions that, upon wetting, cause dispersion and swelling of clay particles. Water percolation and throughflow is therefore restricted, as pores are closed off by the clay. The consequent concentration of salts may inhibit plant growth, especially when there is insufficient flushing of the soil profile (see Figure 27.10). Solonetzic soils are common in southern Alberta, where saline parent materials, low precipitation, and high evapotranspiration leave little opportunity for flushing of the soil by water. Soils of this order occupy 72 575 km$^2$, or 0.7 percent of Canada. The four great groups within the Solonetzic order reflect the degree of salt and sodium ion translocation to depth. Saline parent material first develops into a Solonetz, then into a Solodized Solonetz, and finally into a Solod, which may eventually evolve into one of the other soil types given sufficient leaching.

### VEGETATION OF THE WORLD

The classification of vegetation presents rather different problems. Again, a natural continuum must be divided into artificial classes, but now the continuum is composed of an almost infinite number of discrete individuals. Thus, vegetation classification depends on the spatial arrangement of individual plants in recognizable and repetitive associations or communities.

Many vegetation classification systems assume that plant assemblages are distinctive due to the response of the constituent plants to the environmental conditions. French biogeographers, for example, have attempted to classify vegetation on the basis of characteristic or faithful species, arguing that certain plants provide a clear indication of ecological conditions, based on their environmental preferences and tolerances. In some cases, classification is more quantitative, based on the overall similarity of plants within the assemblage—**floristic classification**—which requires the use of complex statistical procedures and computers to analyze a vast array of data dealing with the abundance or distribution of constituent species.

These approaches are useful when examining vegetation at a local level, but they are not really suitable for presenting the broader picture. At this scale, it is common to classify vegetation in a much more intuitive way, based on the appearance of the plant communities—**physiognomic classification**. One of the most successful examples of this approach was developed by Küchler in the United States and has been tested in many parts of the world and found to be reliable. The physiognomic approach recognizes **formations**—vegetation types with distinctive and uniform structure, appearance, and composition. These are typically grouped into larger classes called formation-types or biomes (Figure 30.8).

## Ecological relationships and the biome

### THE NATURE OF SOIL–VEGETATION RELATIONSHIPS

If we compare the maps of world soil and vegetation types, the broad similarity between them becomes apparent. This reflects the influence of climate and major physiographic features such as high mountain chains. However, the comparison is to some extent misleading, as each of the main soil and vegetation types recognized at this scale represents a variable and complex unit. Even on a local scale, however, we can see relationships between the distribution of soils and vegetation (see Figure 27.25).

This is not surprising, as these two components are closely associated within the ecosystem. The biogeochemical cycles by which plants gain and lose

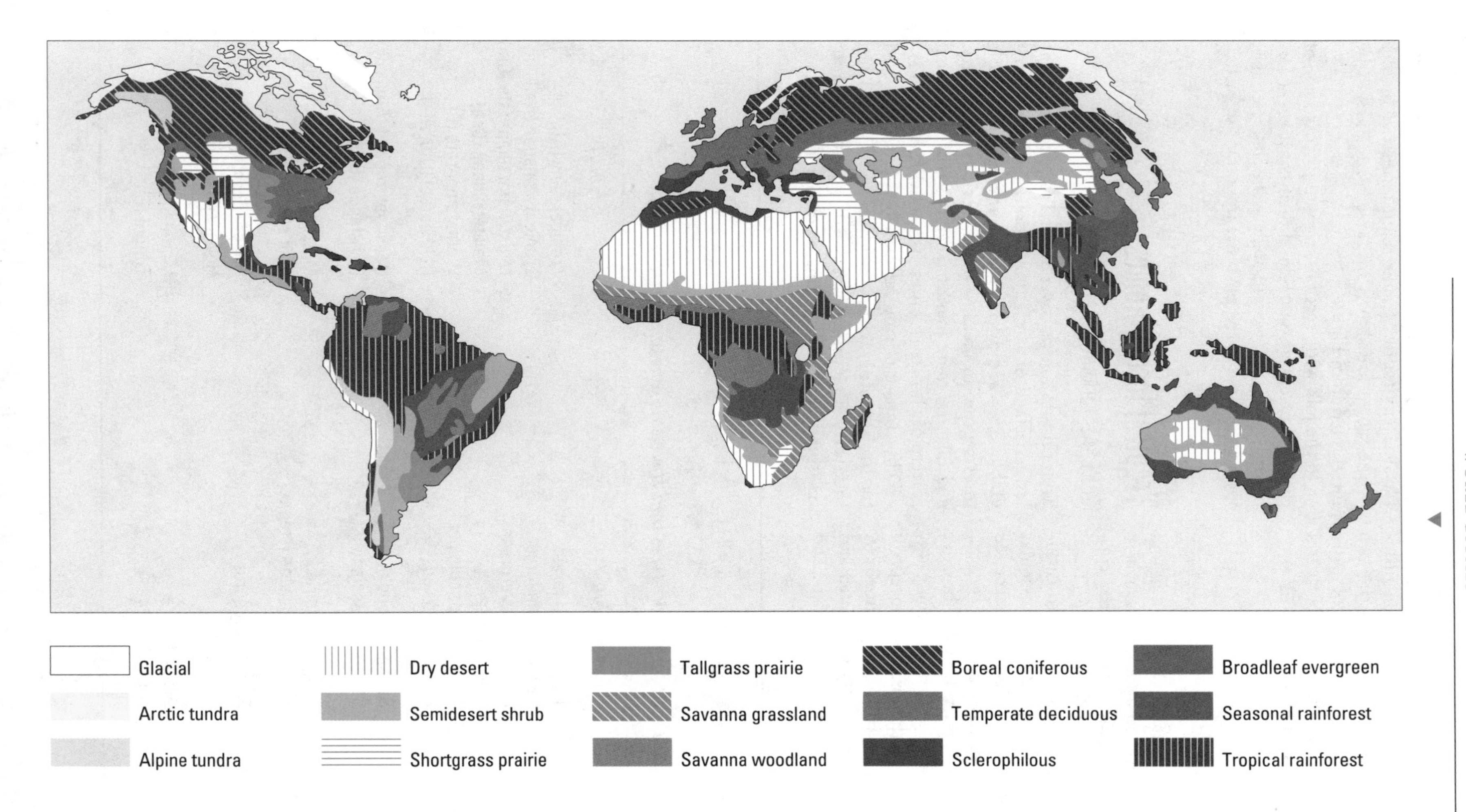


FIGURE 30.8 ▼ Formation-types of the world

nutrients represent a major link. Vegetation influences soil development and vice versa, so the two evolve together. Consequently, some ecologically integrated units can be found, in which both soil and vegetation are reasonably uniform.

### RELATIONSHIPS WITH ANIMALS

Animals are intimately associated with the soil and vegetation. This is apparent on a local scale, where the structure of the food chain dictates that consumers be tied closely to their food source. Thus, strong links join vegetation to herbivore to carnivore. Many decomposers are specific to particular types of organic debris, and therefore are associated with specific plant and animal assemblages.

These relationships are less valid on a broader scale, as generalization leads to numerous complications and exceptions. Nevertheless, the concept of the **biome** has developed within this context. A biome is an area that is ecologically integrated and relatively uniform in its soil, plants, animals, and physical environment. The main terrestrial biomes of the world are shown in Table 30.1 and can be interpreted from Figure 30.8.

In the following sections, we will discuss some of the more extensive biomes and their constituent formation-types, and we will analyze the character and, more importantly, the processes important to each.

## Tropical forest biomes

### GENERAL CHARACTER

Tropical rain forests cover about 13.2 percent of the land surface of the earth, about $17 \times 10^6$ km$^2$ (Tables 26.1 and 30.1). These complex and variable forests grow in areas with no distinct seasonality, and in some more marginal regions with a seasonal pattern of climate (Figures 30.9, 30.10, and 30.11).

The climate of tropical rain forests is dominated by high temperatures, high rainfall, and high humidity. Precipitation exceeds potential evapotranspiration, and luxuriant vegetation develops.

Table 30.1

*Major biomes and formation-types and their environmental relationships*

| *Biome* | *Formation-type* | *Climate** |
|---|---|---|
| Forest | Tropical rain forest | Tropical rainy (Af) |
| | Seasonal rain forest | Tropical monsoon (Am) |
| | Broad-leaf evergreen | Moist warm temperate (Cfa) |
| | Sclerophilous | Mediterranean (Csa) |
| | Temperate deciduous | Warm temperate (C) |
| | Boreal coniferous | Cold boreal (Dfb, Dfc) |
| Savanna | Savanna woodland | Seasonal tropical (Aw) |
| | Savanna grassland | Dry steppe (BSh) |
| Grassland | Tallgrass prairie | Moist warm temperate (Cfa) |
| | Shortgrass prairie (steppe) | Cold steppe (Bsk) |
| Desert | Semidesert scrub | Hot dry desert (BWh) |
| | Dry desert | Hot dry desert (BWh) |
| Tundra | Alpine tundra | Mountain (H) |
| | Arctic tundra | Tundra (E) |

*See Appendix II

FIGURE 30.9 ▼ Tropical rain forest in Central America: Corcovado National Park, Costa Rica. The forest is multi-layered with many plant species at all levels (photo: H. Jungius/World Wildlife Fund Canada)

The soils associated with these forests are characterized by intense and perhaps prolonged weathering with active leaching. Decomposition is so rapid that, despite high inputs of plant debris, the soils rarely develop a distinct organic surface layer.

FIGURE 30.10 ▼ A remnant of rain forest on North Island, New Zealand. Kauri trees in the emergent layer extend tens of metres above the canopy layer (photo: G.P. Kershaw)

Moreover, due to intense weathering and leaching, more soluble constituents are totally removed. Iron and even silica may be mobilized. Also, these soils have a low cation exchange capacity and a limited supply of bases such as calcium and potassium (i.e., a low base status). They show marked oxidation, which leads to redder soil colours. Using the USDA classification described in Appendix III, these soils are mainly **oxisols** and **ultisols**. They also include the ferricretes and laterites, which are frequently defined as tropical forest soils.

Tropical rain forest vegetation is typically diverse both in species composition and structure. Whereas temperate forests may contain only three or four tree species per hectare, tropical forests often include as many as 100 different tree species per hectare. A comparison of Canada's native tree diversity to that of one national park in Australia's rain forest illustrates how diverse these ecosystems can be. In Canada, with an area of $9.970 \times 10^6$ km$^2$, there are approximately 116 native tree species (of which only 23 can be considered widely distributed). In Australia's Palmerston National Park,

FIGURE 30.11 ▼ This temperate rain forest is found in a narrow coastal area near Tlell, British Columbia, in the Queen Charlotte Islands (photo: Elaine Freedman)

with an area of 25.56 $km^2$ ($0.25 \times 10^{-5}$ percent the size of Canada), there are over 500 tree species (19.6 species $km^{-2}$ vs. $1.16 \times 10^{-5}$ species $km^{-2}$ in Canada).

Most tropical tree species are evergreen, and those that are leafless for any period shed their leaves at irregular intervals. There is no autumn in the sense that we know it in more temperate areas. Moreover, intense competition and the diversity of plants leads to complex structuring of the forest, with five or more strata recognizable and a significant **epiphytic** component (Figure 28.1). Species diversity or richness is high, but species' distributions are often restricted to small areas.

Many different types of animals are present, taking advantage of the diverse niches provided by the vegetation. A large majority of the animals are **arboreal** (living in trees). Those at the canopy level rarely descend to the forest floor, but those at lower levels in the trees—the middle-zone fauna—come to the ground more frequently. Both large and small animals live on the ground, but there are few subterranean animals, compared to other environments. In general, the high habitat and niche diversity results in a high species richness. Often, many animals and plants are ultimately dependent on a few plant species for their existence. Because these diverse and complex ecosystems can change frequently over short distances, relatively minor disturbances, such as small logging operations, can cause species extinctions.

## ENVIRONMENTAL RELATIONSHIPS AND VARIATIONS

The history of the rain forests is more complex than was once thought. These forests are no longer considered ancient, almost unchanged biomes that have survived since the Tertiary. Instead, it seems that they experienced dramatic changes during the Quaternary Period, and probably owe much of their present diversity to the periods of isolation they

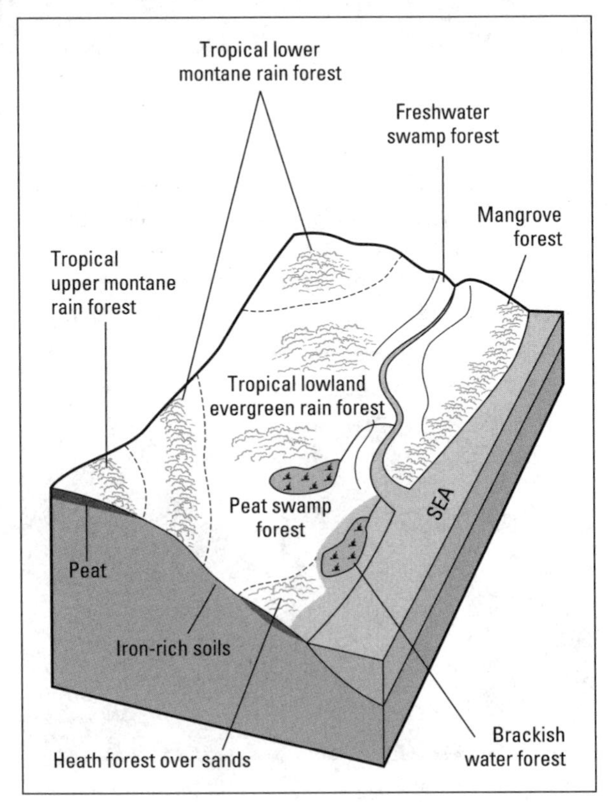


FIGURE 30.12 ▼ Relationships between tropical rain forest biomes and environmental conditions (based on Collinson, 1977)

experienced at that time. This isolation led to the evolution of many **endemic species**, each found exclusively in the area in which it speciated.

It is also clear that distinct variations related mainly to climatic and geological factors occur within these biomes. An idealized picture of these patterns is shown in Figure 30.12.

### BIOGEOCHEMICAL CYCLING

One of the main features of tropical forests is the huge concentration of nutrients stored within the vegetation (Figure 30.13). The soil has a low nutrient storage capacity, and nutrients are retained mainly in the dense, lush vegetation. This has considerable significance, for when the forests are cleared, by artificial or natural processes, most of the nutrient store is lost. The remaining soil is very infertile and therefore regeneration is slower than might be imagined. If crops are planted, yields decline rapidly without massive injections of fertilizer.

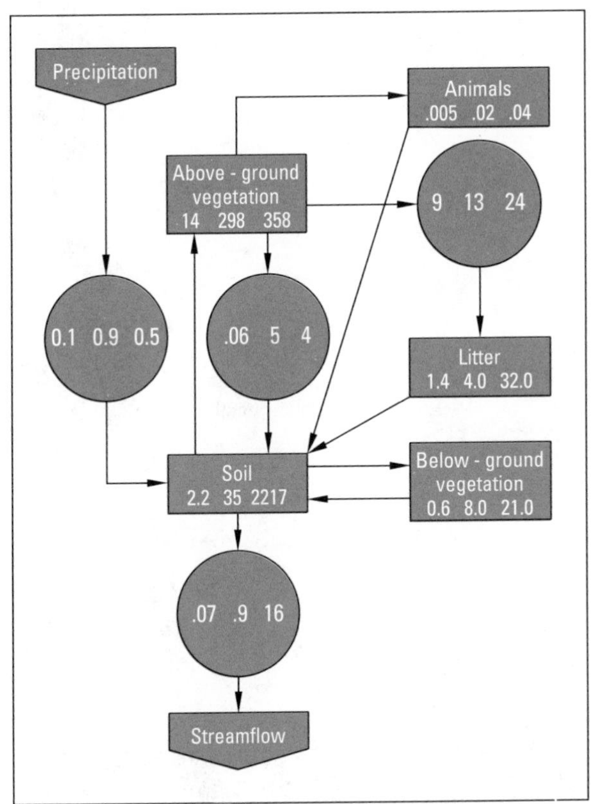


FIGURE 30.13 ▼ The nutrient flow of a tropical rain forest, Panama. Figures in boxes represent total stores of phosphorus, potassium, and calcium in g $m^{-2}$; figures in circles represent annual flows of P, K, and Ca in g $m^{-2}$ $y^{-1}$ (from Golley, 1978)

Under undisturbed conditions, large quantities of nutrients and energy cycle rapidly through the tropical forest, despite the fact that food chains are long and extremely complex. It has been estimated that annual biomass production (not net primary productivity) in a central Amazon rain forest near Manaos is in the order of 1100 t $ha^{-1}$. A high proportion of this annual production is associated with the plants—leaves, fruits, and flowers of the trees, ground flora, and epiphytes—which typically account for 4 to 9 percent of the total biomass (compared with 1 to 2 percent in temperate forests). In contrast, animals account for only 0.02 percent of the biomass at this Amazon site. The annual litter fall is also high, but with rapid decomposition and leaching there is little surface accumulation. Most decomposition is carried out by fungi, and soil animals, instead of feeding on the organic matter itself, tend to feed on the fungi. Earthworms are confined to the upper rooting zone in the soil, and there is little mixing of the soil by animals. Consequently, the soil horizons are distinct.

In the seasonal forests, biomass production is considerably slower, averaging just over 200 t $ha^{-1}$. Most of the processes we have just described operate more slowly, so leaf litter accumulates at the surface, and more specialized relationships between litter and soil organisms are found. Fire is also an important factor in these areas. Fire releases nutrients to the soil and reduces the storage of nutrients in the vegetation. However, leaching is less marked and more nutrients are retained in the soil.

The nutrient budgets of tropical rain forests are interesting in that annual turnover is much higher than that of almost any other biome. The rate of cycling is perhaps three to four times that of temperate forests, and large quantities of nutrients are returned each year to the soil (Figure 30.13). However, leaching is active, and the loss of these nutrients from the soil to surface water bodies can be intense.

# Savanna biomes

### GENERAL CHARACTER

**Savanna** biomes cover approximately 11.6 percent of the land surface of the earth (Table 26.1), including many of the subtropical regions fringing the rain forests of Africa and South America. Although it is often referred to as a grassland biome, the savanna is an open woodland in many cases, with widely spaced and rather scrubby trees.

The formation of savanna is of considerable interest and dispute. Climatic factors alone cannot account for the character of these areas. Although they experience a distinct dry season during which many plants **aestivate** (become dormant), and although precipitation is variable, in many cases it appears that the climate could support a much more luxuriant and diverse flora. One possible reason for this disparity is that the savanna represents a form of **plagioclimax**, one which has been severely curtailed by human activities. Human-induced wildfires, in particular, have played a major part in the development of savanna, and many of the trees are fire-resistant. With the action of fire, and the voracious appetite of termites, seeds rarely survive. In response, trees produce enormous numbers of seeds

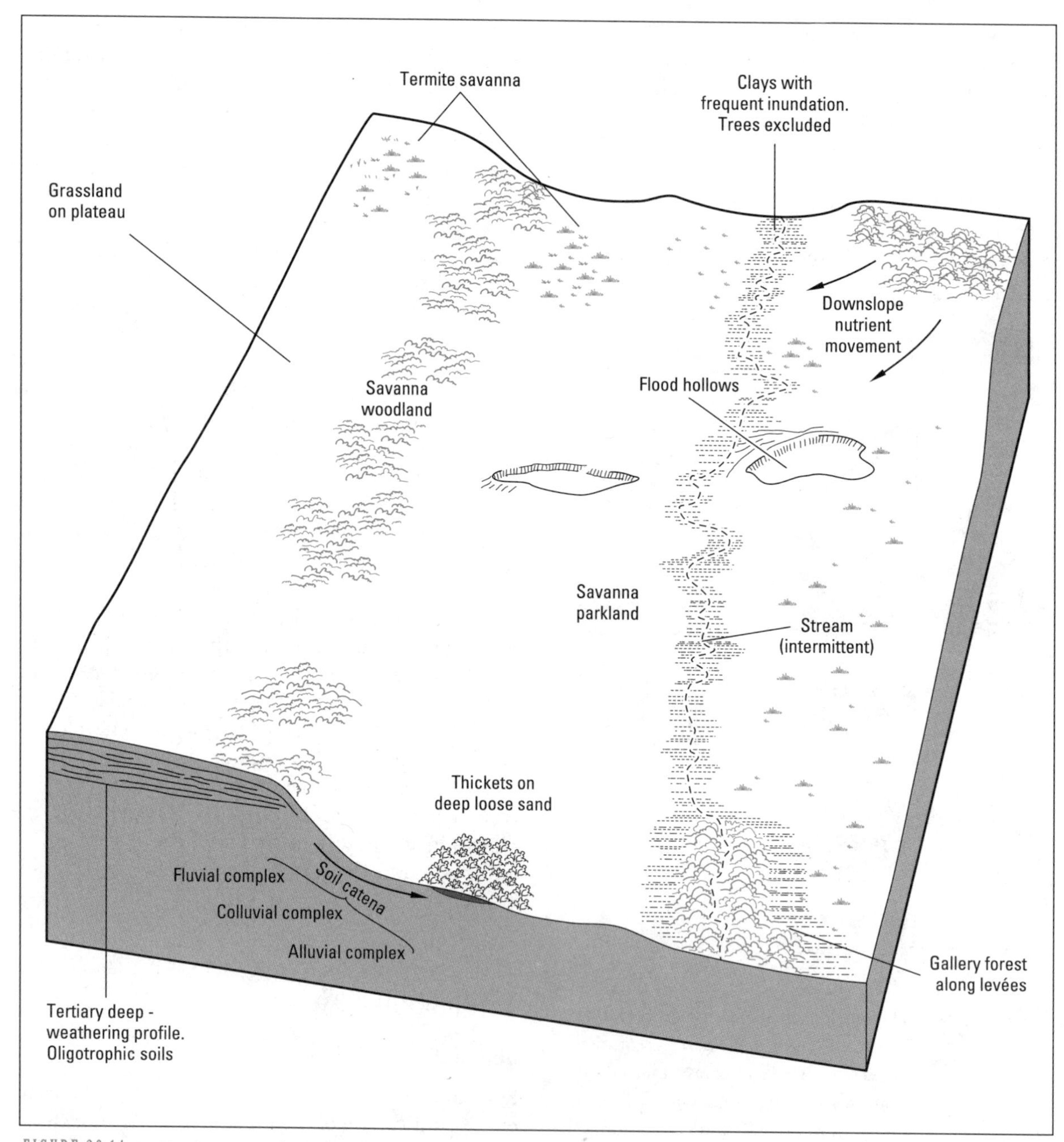


FIGURE 30.14 ▼ Environmental relationships in a savanna area (after Collinson, 1977)

each year. *Acacia karoo*, for example, releases as many as 20,000 seeds, of which about 90 percent are typically fertile. Few survive to grow into trees, however, as the scattered nature of the arboreal vegetation shows.

Many savanna trees are xerophytes. Their morphological and physiological resistance to water loss, and their ability to maximize the uptake of water, allow them to survive dry periods. They also have deep roots and flattened crowns. Some shed their leaves during the dry season in order to reduce transpiration. Savanna trees are often stunted, and may be overtopped by the tall grasses of this biome. Browsing by the savanna animals is a major constraint on tree growth and survival, and overgrazing is one of the main causes of savanna degradation. Some plants have developed thorns as protection from browsing.

Herbaceous savanna plants are dominated by a few species. African elephant grass is sometimes abundant, and may reach heights of several metres. The density of trees relative to grass is, to some extent, climatically controlled, and trees become scarcer in the drier margins of the savanna. However, there are often subtle local variations in vegetation related to topography and drainage (Figure 30.14), particularly in the drier areas, where vegetation is attuned to the short growing season, and herb-layer plants grow rapidly once the rains come. Although the soils are relatively dry at the end of the arid season, there is no need for the rain to replenish soil moisture before plants can extract the water efficiently. Instead, the plants transpire at their full rate immediately, as much of the rainfall seems to be absorbed by the plants before it can be moved into the finer pore spaces in the soil.

Savanna soils are variable, including oxisols, ultisols, vertisols, and ustalfs (Appendix III). Soil distribution is related to climatic, geological and geomorphological conditions. Slope processes are active, as plant cover is often insufficient to prevent erosion and down-slope washing of nutrients. Consequently, marked catena sequences develop on the hillslopes, grading from shallow stony soils to deeper, less well-drained, base-rich alluvial soils (Figures 30.14 and 30.15).

Red, non-swelling acidic kaolinitic soils
Lateritic crust
Deep, moderately nutrient-rich soils
Black, nutrient-rich, swelling, montmorillonitic soils

FIGURE 30.15 ▼ Catena sequence in a savanna area

FIGURE 30.16 ▼ Termite mounds in the semi-desert interior of Australia show the importance of insects in this biome. The mounds are only the minor surface expression of the vast system of tunnels and galleries that lie beneath the ground (photo: G.P. Kershaw)

Compared to tropical forests, animal species diversity in savannas is low, but large populations of these animals are found. Surprisingly, interspecific competition seems to be limited, and the food chains are relatively short, with few secondary consumers. Most carnivores prey directly on herbivores. For example, lions attack mainly zebras, wildebeests, antelopes, and giraffes. However, many scavengers and decomposers, including mammals and insects, also feed on the lions' kills. Termites are very abundant, and their mounds are a major feature of the savanna landscape (Figure 30.16). These insects attack and macerate plant debris, making it more readily available for decomposition by other organisms. They also eat growing plants, especially during periods of drought.

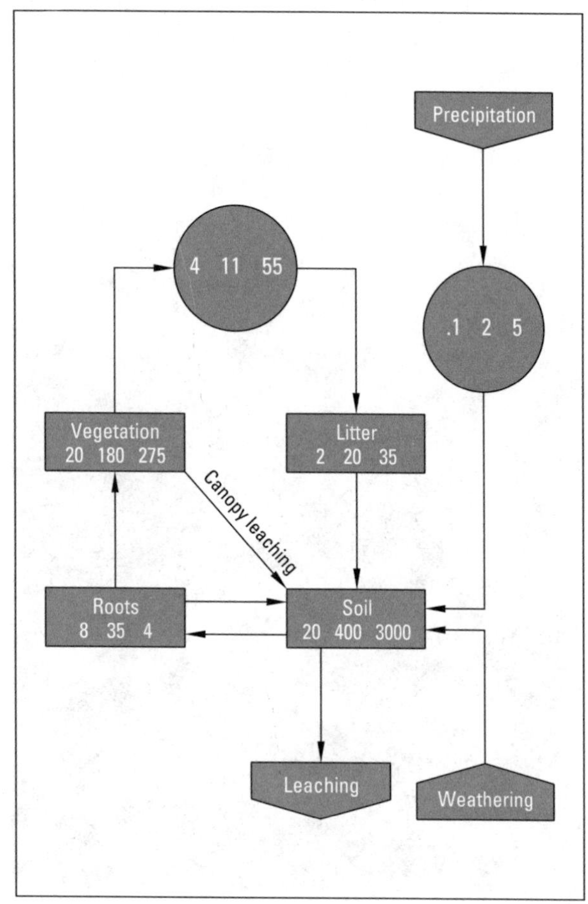


FIGURE 30.17 ▼ The main elements of the nutrient cycle in a savanna area. Figures in boxes represent amounts in phosphorus, potassium, and calcium in kg $ha^{-1}$ $y^{-1}$ (data derived from Nye and Greenland, 1960, and Rodin and Bazilevich, 1965)

## BIOGEOCHEMICAL CYCLES IN SAVANNA AREAS

Relatively little is known about the nutrient cycles of savanna areas. It is clear that net primary productivity is much less than that of the tropical rain forests. An annual total of 2 to 20 t $ha^{-1}$ is typical, but quantities vary considerably with climate and tree cover.

Nutrient cycling is rapid, due to the speedy breakdown of organic materials by soil organisms. High temperatures encourage chemical activity, and silica appears to be particularly soluble in these areas, often making up a major part of the total nutrient budget. Indeed, plant uptake of silica is so great that clots of amorphous silica may form in some leaves.

It is also apparent that the nutrients which are returned to the soil are removed less readily by leaching (Figure 30.17). This occurs for two related reasons. Lower rainfall leads to less intense weathering and leaching and the soils with clays that have a higher cation exchange capacity form stronger bonds with nutrients. Most nutrient loss in many of these areas results from soil erosion, a major problem facing agriculture in the savanna, where overgrazing and tillage leave the soil unprotected. The sudden, intense storms characteristic in the savanna can result in intense rainsplash and erosion of unprotected surfaces.

# Desert and semidesert biomes

## GENERAL CHARACTER

The common picture of desert areas as vast expanses of barren, shifting sand is false for all but a small part of this biome. Most deserts and semideserts support widespread, relatively sparse vegetation with a distinctive array of wildlife.

Arid and semi-arid land covers almost one-third of the land surface of the globe (Table 26.1). Almost 60 percent of this is true desert. The remainder varies from steppe grassland to thorny scrub. In all cases, however, potential evapotranspiration greatly exceeds rainfall (Figure 12.6). This condition arises for a variety of reasons. In coastal areas bordering cold, upwelling waters, cool onshore winds are warmed and cause evaporation as they blow onto the land. This occurs in Peru and in the Baja California desert. Rainshadow deserts, such as the

FIGURE 30.18 ▼ Hot desert, Owens Valley, California, the Sierra Nevada Range in the distance. Before the construction of the Los Angeles Aqueduct, there were productive fruit orchards in this area (photo: Elaine Freedman)

Mojave and Patagonian desert, occur in the lee of mountain barriers. The subtropical deserts of the Sinai and Sahara are a result of stable anticyclonic atmospheric circulation. The interior regions of the continents support deserts because winds blowing inland from the sea deposit most of their moisture en route. This is typical of the deserts of central Asia and Arizona.

Desert vegetation consists mainly of short perennial grasses and thorny scrub (Figure 30.18). Only in extreme cases, such as rocky hamadas and regs, and the shifting sand dunes and sand seas of the Sahara, is vegetation absent. Even in these areas, locally developed lines of vegetation occur along wadis, with lusher growth around oases. In all cases, plants must be able to survive periods of drought, and thus xerophytic plants predominate. The adaptation of plants to desert conditions varies. For example, the saguaro cactus develops a widely spreading root system; the mesquite has roots that may reach depths of over 50 m; and many cacti and agaves store water in their roots, stems, and leaves. Some plants reduce water loss through evaporation by controlling their stomates, while others have long dormant periods, growing and flowering briefly and irregularly when moisture is available.

The soils associated with desert conditions are typically little weathered, and lacking in humus. In the most extreme cases, no true soil exists, but even where sufficient plant growth does occur to provide a surface accumulation of plant debris and a food base for soil fauna, the lack of leaching and chemical weathering leaves soils relatively infertile. Salinity is also a major problem, as constant evaporation from the surface draws water from the lower layers of the soil and leads to the accumulation of salts in the upper horizons (the Canadian Solonetzic Order) (see Figure 27.11). These salts are derived in part from the groundwater, and salty sea water seeps into aquifers in many coastal areas. Winds blowing from the sea also may introduce salt in rainfall. Finally, if the parent material is rich in salts (marine origin lithologies), Solonetzic soils may develop. Practically no leaching occurs, so even though the salt is soluble, it accumulates in the soil.

The animal life of desert areas is more diverse than might be imagined. Nevertheless, relatively few large mammals can survive under these conditions, and insects and arachnids predominate. Flies, scorpions, crickets, locusts, and grasshoppers are among the most common inhabitants. Several vertebrates also are found, of which the ostrich, the sandgrouse, the red kangaroo, the emu, and various rodents are most abundant.

### BIOGEOCHEMICAL CYCLES

Little information is available on which to base a discussion of general nutrient cycles in deserts. Total biomass is small and it is apparent that cycling is slow and involves very small quantities of nutrients. Net annual primary productivity is closely related to rainfall, and probably ranges from about 0.05 to 0.2 t $ha^{-1}$. About 80 percent of the organic material in a desert is underground, and cycling mainly occurs through the decay of root material. Leaching losses are negligible, but gaseous losses associated with the chemical oxidation of plant materials can be significant. Soil erosion may also represent a significant loss. Due to the low rainfall and the limited weathering, the more insoluble compounds are present only in small quantities. Sodium and calcium tend to dominate in desert nutrient budgets (Figure 30.19).

## Temperate grassland biomes

### GENERAL CHARACTER

Steppe and prairie grasslands are essentially the same, but geographically and historically, **steppe** has referred to Asian grasslands while **prairie** has referred

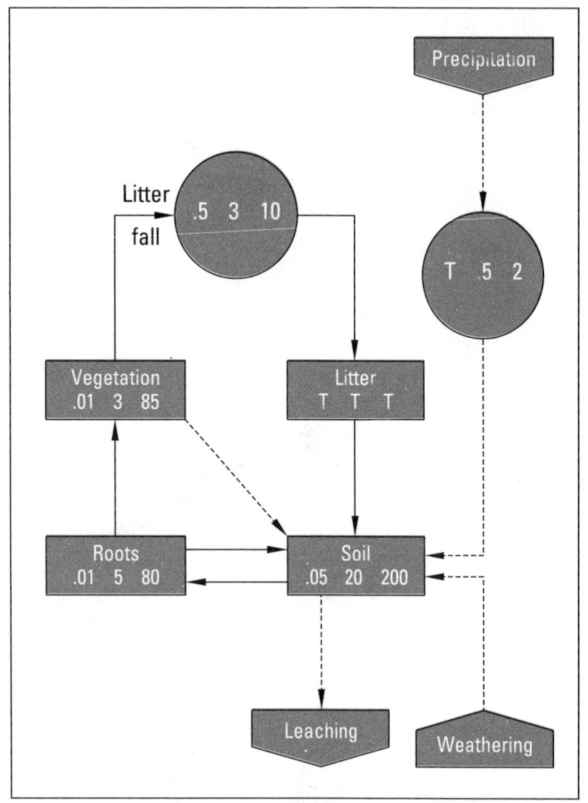


FIGURE 30.19 ▼ The nutrient cycle of a hot desert. Data on such biomes are rare, and only very general estimates can be given. Figures in boxes represent the amounts of stored phosphorus, potassium, and calcium in kg $ha^{-1}$; figures in circles represent annual flows of P, K, and Ca in kg $ha^{-1}$ $y^{-1}$ (T = trace). Broken lines show minor nutrient flows

to North American grassland. Both are grass- and herb-dominated vegetation types. Temperate grasslands occupy only 7 percent of the earth's land mass, but represent one of the most important agricultural zones in the world. Many seminatural grasslands have been taken over for intensive farming. Only marginal areas are not used in agriculture, and these merge into the desert biomes.

Characteristically, grasslands experience a significant soil moisture deficit due to a long period of drought in the late summer and autumn. The stress associated with this probably prevents trees from growing in these areas. Compared to the savanna areas, climatic conditions are often less favourable for plant growth. Canadian shortgrass prairie is dominated by spear grass (*Stipa comata*) and grama grass (*Bouteloua gracilis*). Mixed grass prairie is dominated by wheat grass (*Agropyron* spp.) and fescue grasslands are dominated by rough fescue (*Festuca scabrella*) and parry oatgrass (*Danthonia parryi*). Tallgrass prairie occurs in parts of North America with higher precipitation (e.g., in the more eastern states of the US midwest such as Iowa and Missouri). Here, big bluestem (*Andropogon gerardii*) and little bluestem (*A. scoparius*) grasses predominate.

Chernozemic soils are typical of the steppe and prairie areas (Figure 30.5). Organic activity is intense, and these deep unleached soils have high organic matter content and a high nutrient supply. The pH ranges from neutral in wetter areas to slightly alkaline (7–8) in drier regions. When these soils are tilled, they are prone to erosion.

Temperate grasslands represent a transition zone for many animals. Species that inhabit the margins of forests and deserts come together here, between the two more extreme biomes. In historical times, the North American prairies were dominated by vast herds of bison (*Bison bison*), wapiti (*Cervis elaphus*), and the pronghorn (*Antilipcapra americana*), but overhunting and habitat loss with the incursion of the early settlers, and the spread of cattle and sheep, resulted in the virtual extinction of these animals over much of their range.

## BIOGEOCHEMICAL CYCLES

The productivity of the prairies is relatively high, even under seminatural conditions. The total biomass of these grasslands approaches 50 t $ha^{-1}$, with net annual production averaging 6 t $ha^{-1}$, and ranging from 2 to 15 t $ha^{-1}$ (Table 26.1). In the drier, shortgrass steppe, production is lower. In some of the better steppe land of Ukraine, total plant biomass is estimated to be 20 to 50 t $ha^{-1}$ with the annual production of new material averaging about 5 t $ha^{-1}$.

The cycling of nutrients in this ecosystem is dominated by the close interaction between the grasses and the soil. Uptake of nutrients by the grass is balanced by high rates of return, both from the roots and from the above-ground parts of the vegetation. Litter accumulates at the surface, as the rate of decomposition is rather slow. This is due, in part, to the long cold winters, which leave only a short active season for the soil fauna. As much as 40 percent of the carbon content of grassland systems may be stored in the litter layer. This has significant effects on soil processes. It provides a protection from erosion, absorbs heat, and holds moisture, often preventing deep percolation of water. This restricts leaching, but can also create problems during drought, for the water cannot be stored at depth, away from the effects of intense solar radiation and drying winds.

## Temperate deciduous, broad-leaf evergreen, and sclerophilous forest biomes

### GENERAL CHARACTER

The temperate deciduous, broad-leaf evergreen, and sclerophilous forests collectively occupy about 9.3 percent of earth's land surface (Table 26.1 and Figure 30.8).

The **deciduous** forests (Figure 30.20) occur mainly in warmer, temperate areas. They are typically dominated by oak (*Quercus* spp.), beech (*Fagus* spp.), and hickory (*Carya* spp.), with smaller quantities of understorey trees such as birch (*Betula* spp.), hazel (*Corylus* spp.), sycamore (*Platanus* spp.), and maple (*Acer* spp.). The broad-leaf evergreen forests are concentrated in more humid zones. These forests are composed mainly of oaks (*Quercus* spp.), magnolias (*Magnolia* spp.), hollies (*Ilex* spp.), and sabal palms (*Sabal* spp.). **Sclerophilous** forests are found in areas with a Mediterranean climate. They are characterized by drought-tolerant species such as olives (*Olea* spp.), sessile oak (*Quercus petraea*), and Aleppo pine (*Pinus halepensis*). However, where Mediterranean forests are disturbed by overgrazing and clearing, they tend to degrade into scrublands called **maquis**, with thorny shrubs, or **garigue** with waxy-leaved shrubs such as juniper (*Juniperus* spp.) and broom (*Cytisus*, *Genista*, or *Spartium* spp.).

Like the tropical forests, these forests have a distinct vertical structure. Usually, three main layers

FIGURE 30.20 ▼ The temperate deciduous forest of eastern North America is dominated by broad-leaf trees that have a winter dormant period (photo courtesy of Forestry Canada)

can be recognized, but in some cases (e.g., in coniferous forests) understorey trees can be almost absent, as the canopy intercepts as much as 99 percent of the incoming sunlight. The competition for light is a major factor determining success in these areas. Solar radiation is limited, particularly during the winter months. Consequently, there is a marked seasonality related to temperature and sunlight, and one of the major factors controlling the forest growth (especially in the more northern regions) is the length of the growing season. Deciduous trees lose their leaves during the winter months, and thereby reduce their rate of transpiration. This limits their energy requirements in the winter and reduces susceptibility to frost damage. Plants in the understorey and herb layer have adapted to take advantage of the increased light available in the spring and fall when overstorey trees have dropped their leaves.

Temperate forest soils range from Luvisols to Podzols in needle-leaf evergreen forests (Figure 30.21), to deeply weathered and leached oxisols and ultisols in broad-leaf evergreen forests. Soil type is related to climatic pattern, in part, but it is also affected by the vegetation itself. The litter produced by coniferous trees is relatively acid and resistant to decomposition. It also provides organic acids which encourage leaching and cheluviation. Consequently, there is a tendency for increased nutrient loss and active translocation of iron, clay, and organic matter in the soils of coniferous forests. In contrast, the moist, humid conditions of the broad-leaved evergreen forests are more akin to those of the tropics. Here, soils have low base status and organic matter decomposes rapidly, due to the high temperatures and intensified chemical weathering processes.

The animal life of these forests is far more restricted in terms of species diversity than that of the tropical forests. Numerous tree-dwelling animals such as tree squirrels (*Sciuridae*) and porcupines (*Erethizon dorsatum*) are found, while deer (e.g., *Odocoileus virginianus*, *Cervus elephus*, and *Capreolus* spp.), foxes (*Vulpes* spp.), pumas (*Felis concolor*), lynxes (*Lynx lynx*), and various mice and voles live on the ground. Soil organisms are more abundant in deciduous forests than in the evergreen forests, where plant materials are less nutritious and more acid conditions prevail in the litter.

FIGURE 30.21 ▼ Ferro-humic Podzol soils are common in temperate forest regions. They are strongly influenced by acidic litter at the soil surface; the dark upper layer is a rusty colour (photo courtesy Land Resource Research Centre, Agriculture Canada)

## BIOGEOCHEMICAL CYCLES

Many studies have focussed on nutrient and energy cycling in forests of the temperate latitudes, and there is no shortage of relevant data. It is known, for example, that the annual biomass of these forests varies considerably over time and space. The biomass of evergreen forests can reach about 2000 t $ha^{-1}$ in some cases, but the average for temperate deciduous forests is only 300 t $ha^{-1}$ (Table 26.1). These values change over time, and ecosystem biomass seems to reach a maximum at about 200 years (Figure 30.22). Annual rates of assimilation of organic matter also vary.

Nutrient cycle monitoring in temperate forests has shown that large quantities of nutrients may be stored in the vegetation. Most of this is retained in the leaves and young shoots (Figure 30.23), and therefore the annual turnover of nutrients is large. Considerable quantities of nutrients are also retained in the soil, except where intense leaching

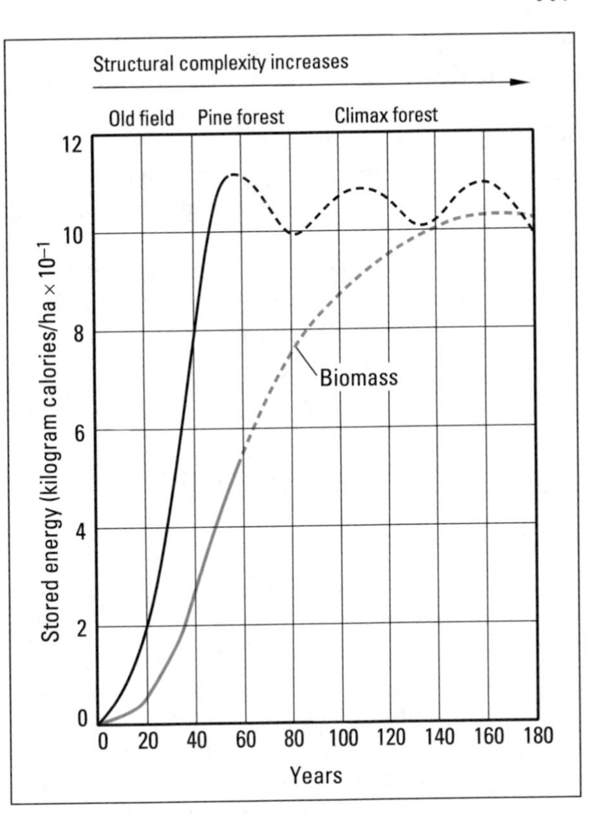


FIGURE 30.22 ▼ (Right) Changes in total energy storage of a forest over time (from Woodwell, 1963)

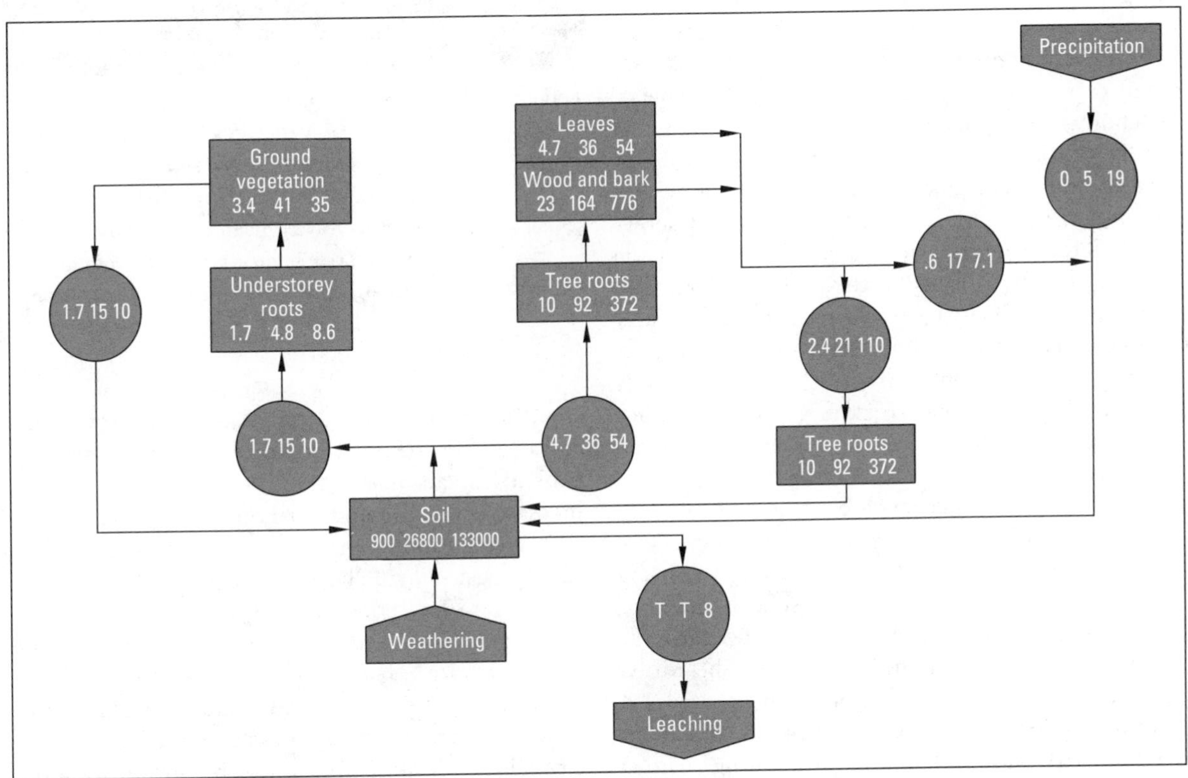


FIGURE 30.23 ▼ The nutrient cycle of a mixed oak forest. Figures in boxes represent the amounts of phosphorus, potassium, and calcium stored in the mature vegetation and soil in kg $ha^{-1}$; figures in circles show the annual flows of P, K, and Ca in kg $ha^{-1}$ $y^{-1}$ (T = trace) (from Duvigneaud and Denaeyer-De Smet, 1970)

had led to rapid removal. One of the controls on this nutrient cycle is the release of mineralized material from organic compounds. Decomposition is slow in many of the cooler region forests, and consequently uptake by trees is limited. Nitrogen and phosphorus, in particular, may be limited in such situations.

The distribution of nutrients in these forests can be very important, for humans often exploit these systems, with drastic consequences for the ecosystem (see Chapter 31).

# Taiga or boreal forest biomes

## GENERAL CHARACTER

Boreal forest or taiga ecosystems occupy approximately 9.3 percent of the world's land surface. This biome, unlike most we have discussed, is largely in a natural state today (Figure 30.24). Some of the coldest temperatures on earth have been recorded in the taiga, and temperatures can dip into the –45 to –50°C range for prolonged periods of time. During the growing season, maximum temperatures commonly range from 25 to 30°C. Wildfires can become very prevalent under warm conditions, and large tracts of forest can be burned each year, depending on the amount of snowmelt and rain.

FIGURE 30.24 ▼ The boreal forest of the taiga biome: the Mackenzie Valley in winter. The trees are black spruce (*Picea mariana*) and the much less abundant larch (*Larix laricina*), which loses its leaves in winter. Snowfall accumulates on tree branches when wind cannot dislodge it; up to 40 percent of winter precipitation may be held in the trees (photo: G.P. Kershaw)

One of the most significant ecological factors for animals and plants alike is the length of the winter season and the type, amount, and timing of snowfall. Winter can persist for 50 to 70 percent of the year. Most snow collects on the ground surface and in spring, much of the resulting snowmelt water flows overland as the frozen soil is unable to absorb it. The runoff from snowmelt has significant impacts on aquatic and riparian ecosystems, since approximately half of the annual precipitation can be released in a 1.5 to 2 week period.

The snowpack is not just a bank of snow waiting for spring. Animals require it for overwinter survival, especially those that cannot survive the harsh conditions that prevail in the atmosphere above the snow. These **subnivean** animals stay beneath the snow to benefit from the insulation, the moist atmosphere, and the protection from predators it provides (Figure 30.25). Changes in the light regime, with snowpack thinning in spring, trigger breeding. The snowpack is the key to overwinter survival of subnivean animals such as the voles (*Clethrionomys* spp., *Phenacomys* spp., and *Microtus* spp.), lemmings (*Lemmus* spp.), shrews (*Sorex* spp.), and deer mice (*Peromyscus* spp.). **Supranivean** animals, living above the snow, include waders, such as moose (*Alces alces*) and caribou (*Rangifer tarandus*), that have legs long enough to walk through all but the deepest snow. There are also floaters, animals that have a low enough **weight-load-on-track** to stay on the surface of the snow. These include the snowshoe hare (*Lepus americanus*) and the wolverine (*Gulo gulo*). Most animals in this biome are migratory, and move south or to lower elevations in winter and then back in spring for the breeding season. These include many waterfowl species.

Overwintering, supranivean animals are very well adapted to this environment. They switch food sources with supply, develop protective coating and colouration, and have numerous physiological and behavioural adaptations to cope with this environment. Even in extreme cold, these animals are little bothered, provided they are in good condition.

Plants also have to cope with winter. In the more extreme conditions of the taiga, where **subarctic** forests merge with tundra areas, the deciduous softwood, larch (*Larix* spp.), tends to dominate the tree component in Asia, while the evergreens, black spruce (*Picea mariana*) and white spruce (*P. glauca*), dominate in North America. The larch combines the advantage of both needle-leaf and deciduous trees. Its needles give it a low rate of transpiration and the ability to reduce this further during periods of low energy. However, the advantages of the deciduous strategy must be traded off against the time needed to develop leaves in a relatively short growing season. Evergreen trees can break dormancy and commence photosynthesis earlier each growing season, and also have an advantage if the summer is unusually short.

Throughout the taiga, soils are cold. Often, Cryosolic soils have permafrost beneath a shallow active layer that is frozen during part of the winter, causing most biologically related soil processes to cease. Generally, nutrient cycling is slow because of the slow rates of decomposition under cold tempera-

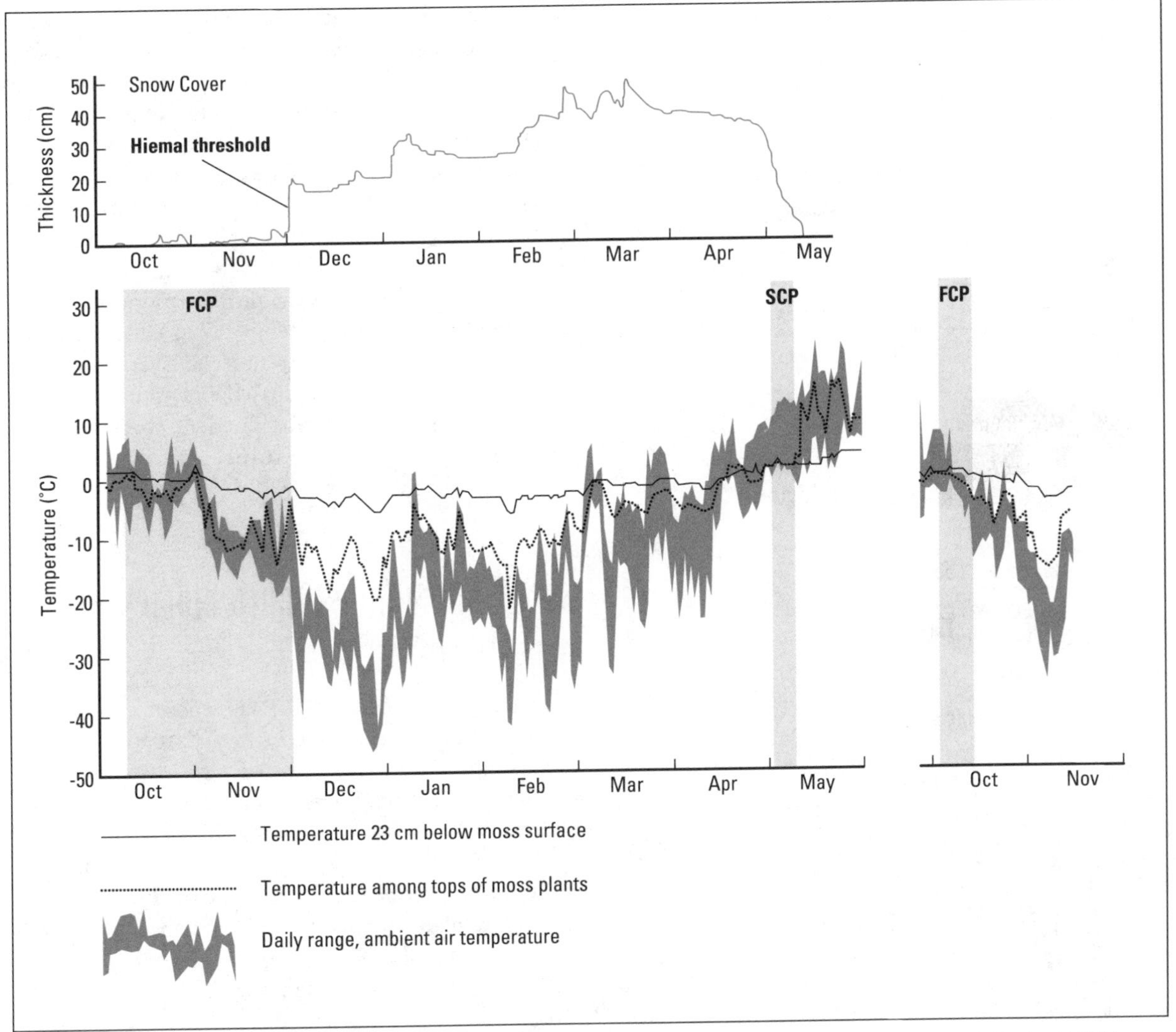


FIGURE 30.25 ▼ Some environmental characteristics of subarctic white spruce (*Picea glauca*) taiga, central Alaska (65°N). Note the first fall critical period (FCP), left: because of temperature fluctuations around the freezing point, and the length of the FCP, the forage quality for small mammals is poor; the shorter FCP the next year suggests more forage was available. SCP = spring critical period (from Pruitt, 1957)

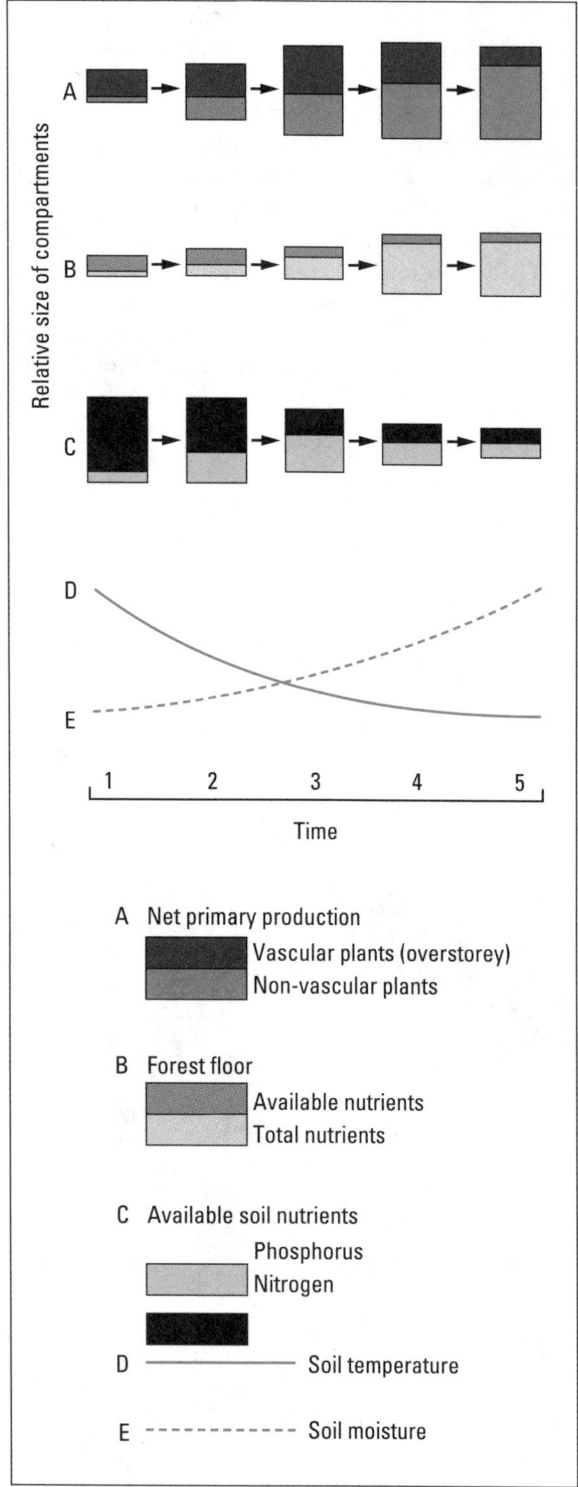


FIGURE 30.26 ▼ Successional changes in a stand of burned black spruce being revegetated by more black spruce, from early successional stages through mature 300-year-old stands (from Zasada *et al*, 1977)

tures year-round. Peaty surface layers of relatively undecomposed litter represent a significant nutrient store. This is tapped only after light surface fires, with subsequent elevated near-surface soil temperatures which may persist for years until the moss layer is reestablished (Figure 30.26). Organic soils are common, dominating the large tracts of muskeg where anaerobic conditions, combined with low pH and low temperatures, result in little decomposition.

### BIOGEOCHEMICAL CYCLES

The productivity of the taiga is relatively high, but at 8 t $ha^{-1}$ $y^{-1}$, it is the lowest of the forested biomes. The standing biomass is relatively high, with large amounts of material tied up in the overstorey trees (Figure 30.27) which can survive for well over 200 years. The cold soils, combined with low pH, moist to waterlogged soils (permafrost prevents percolation), a long period of seasonal frost and cool air temperatures, all slow nutrient cycling. Much of the nutrient pool remains tied up in the peaty soils. Weathering is slow for many of the same reasons that decomposition is slow. Furthermore, the presence of permafrost in some areas can effectively isolate the soil from parent materials at depths as shallow as 50 cm. As discussed previously, wildfire can effectively mobilize the nutrient stores in these peaty soils, and it is omnipresent in the taiga biome.

## Arctic and alpine tundra biomes

### GENERAL CHARACTER

**Tundra** biomes occupy 6.2 percent of earth's land surface (Table 26.1). Our earlier discussion of tundra regions, from the point of view of climate and topography, showed that tundra is characterized by intensely cold conditions and the development of permafrost in the soil. Only for a brief period in the summer do temperatures rise above freezing point and allow plant growth to occur. Then, with incredible rapidity, the plants complete their life cycle before the next cold season curtails activity (Figure 30.28).

The term tundra means treeless, and this biome is defined by the **timberline** in the mountains and the **treeline** at high latitudes. Tundra vegetation is composed mainly of a few hardy species of low-

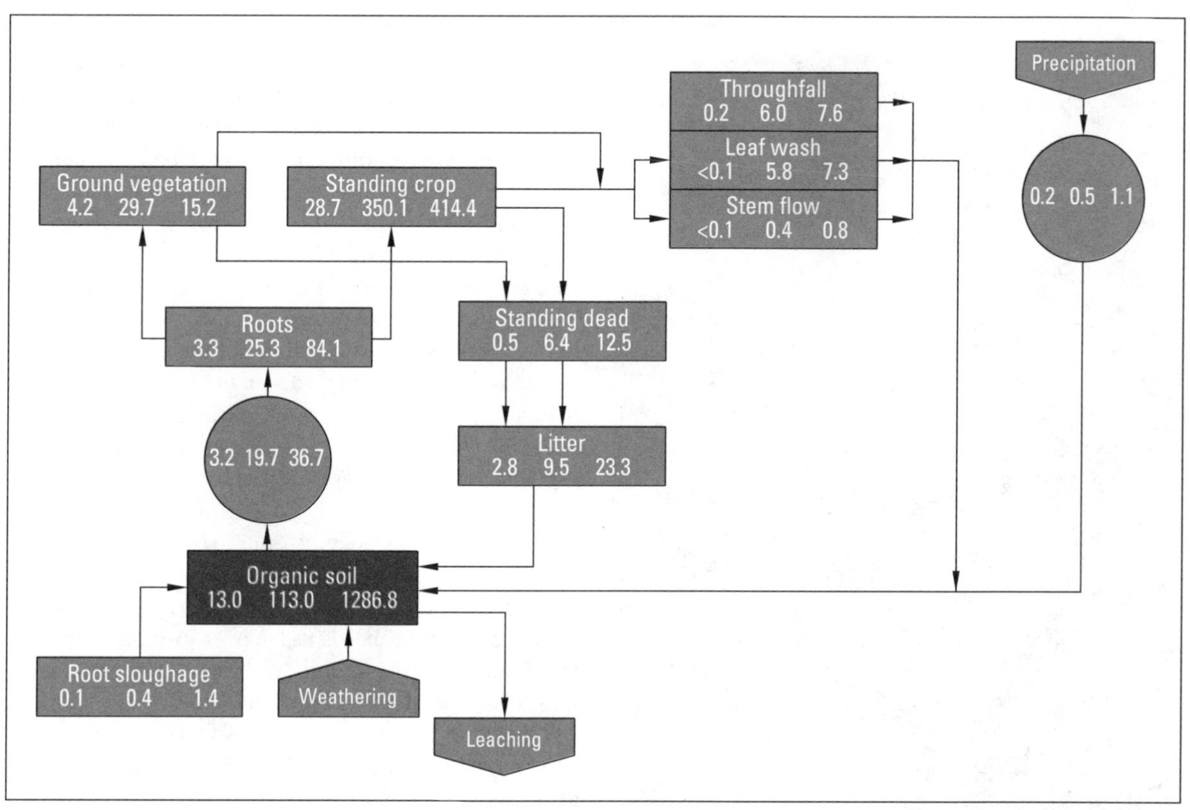


FIGURE 30.27 ▼ The nutrient cycle of a taiga black spruce forest on organic soils (peat). Figures in boxes represent the amounts of phosphorus, potassium, and calcium stored in the various producer components and the soil in kg ha$^{-1}$; figures in circles show the annual flows of P, K, and Ca in kg ha$^{-1}$ y$^{-1}$ (from Gordon, 1983)

FIGURE 30.28 ▼ Woodland caribou (*Rangifer tarandus caribou*) cows and calves graze in a tundra region within the Mackenzie Mountains, NWT. This tundra area extends along the mountains and north to the North Slope of Yukon, where it merges with low Arctic tundra. The floristic composition and low-growing form of the plant cover (<50 cm) are the main characteristics defining tundra (photo: G.P. Kershaw)

growing sedges, grasses, dwarf shrubs, lichens, and mosses. Woody plants such as the dwarf birch (*Betula nana*) and dwarf willow (*Salix herbacea*) in Europe, and the ground birch (*B. glandulosa*) and arctic willow (*S. arctica*) in North America, are important components of shrub-dominated tundra. Saxifrages (*Saxifraga* spp.) are also common, but non-vascular plants such as mosses and lichens can be very important components of the plant cover.

Snowcover is extremely important in controlling the characteristics of the plant cover. Snow mantles the surface and protects plants from the effects of ice crystal abrasion when the wind blows. Snow insulates the rooting zone, and when it melts provides moisture for growth. In the Canadian High Arctic, where much of the landscape receives so little precipitation that it is classified as a polar desert, the winter snowpack represents a store of water accumulated over 8 to 10 months of subzero temperatures. Late-lying snow drifts melt, and areas downslope receive this moisture well into the growing season.

FIGURE 30.29 ▼ This Brunisolic soil profile was in shrub tundra near Macmillan Pass, NWT. The severity of the thaw season limits pedogenic processes in tundra ecosystems. The white layer 5 cm thick is White River volcanic ash dating from 1220 B.P. Its depth of only 5 cm or less indicates that the rate of A horizon development is very slow, about 0.04 mm $yr^{-1}$ (photo: G.P. Kershaw)

The soils of tundra areas are poorly developed, mainly because of low mean annual temperatures. Physical weathering dominates with little chemical breakdown. Also, many of these landscapes are very young (Figure 30.29), and the permafrost layer (Cryosols) prevents downward movement of moisture, so waterlogging and gleyed soils result. On higher gravel ridges, better-drained, stony soils develop (Regosols), and in depressions peat accumulates.

Animal species diversity, like that of the flora, is limited. Animals have adapted to the cold temperatures in a variety of ways. Many are large, with a small surface-area-to-volume ratio, which reduces heat loss during cold periods. Some animals, such as the Arctic fox, change their colour to white in winter and become dark-coated in the summer. This is important for camouflage. The fur and feathers of different animals have evolved to combat the cold. The polar bear has white fur for camouflage in hunting, but that fur (composed of hollow hairs) is an extremely efficient insulator, even allowing the bear to swim for hours at a time in sea water that is below freezing. Muskoxen do not change colour, but rather shed their heavy winter fur in the brief summer season. Ptarmigan turn white in the winter and can fluff their feathers to create an insulative covering that resists temperatures well below –40°C.

Most of the animals in the tundra biome are summer residents that migrate, some of them over considerable distances, to less severe areas in the winter. Not all migrations are made to avoid the cold. The caribou herds move over great distances in the Canadian Arctic and subarctic, but rarely have relief from harsh climates. Their migration is probably more related to stress associated with heat, insect pests, and food availability. The polar bear also migrates, not to avoid the cold, but because its preferred home on the seasonal pack ice disappears in the summer. Anyone who has seen polar bears along the southwestern shores of Hudson Bay in the late summer knows how uncomfortable these animals are in the heat.

Alpine tundra is structurally similar to Arctic or Antarctic tundra, and many plant and animal species are common to both types of tundra. Many other environmental characteristics are also similar. The main differences result from the elevation and evolutionary history. Mountains, with their steep slopes and great elevation differences, have a high-energy environment where geomorphic processes are very effective in moulding the landscape. Higher elevations result in thinner air and greater exposure to ultraviolet radiation, conditions that require responses from biota that live there. In the mountains, changes can occur on a daily basis. For example, there may be frost every night, not just at certain times of the year. Mountains can be viewed as islands, isolated within non-tundra biomes. This isolation has led to speciation along new lines made possible by genetic isolation. Although there are structural similarities between altitudinal and latitudinal tundra, there are also significant differences in a host of ecosystem processes and components.

### BIOGEOCHEMICAL CYCLES

Few attempts have been made to monitor nutrient cycles in tundra areas. It is clear, however, that the total productivity of tundra ecosystems is small, and that nutrient cycling is slow. The slow rate of organic decomposition and slow weathering of minerals severely limits the quantity of available nutrients. Nitrogen dominates in the nutrient cycles and is assimilated mainly through the activity of the lichens and shrubs, most of which have nitrogen-fixing capabilities.

Measurements of the total biomass of tundra biomes have revealed relatively low biological activity. Studies in the Arctic indicate a total biomass of 14.3 t $ha^{-1}$ at a latitude of 63°N, and of only 3.4 t $ha^{-1}$ at 75°N. Most of this organic material lies below the ground, and annual productivity probably averages about 1.4 t $ha^{-1}$, with annual values typically ranging from 0.2 to 1.8 t $ha^{-1}$.

## Conclusion

We are still far from fully understanding the processes operating in the world's ecosystems. As we noted in earlier chapters, the problems associated with obtaining data are formidable, particularly under the inhospitable conditions of the tropical rain forest, desert, or tundra. Nonetheless, our ignorance cannot be condoned. Many attempts to exploit and manage these ecological resources have encountered difficulties, because of our lack of understanding of the systems we are manipulating. Today, humans are exploiting large areas of tropical rain forest and extending their influence into remote desert and tundra regions. Chapter 31 will reveal how far-reaching these effects can be. It is uncertain whether our species will be able to deal with the consequences of our interference with so many natural ecosystems.

# Humans and the biosphere

## History of human activities

### AGRICULTURE AND ECOSYSTEMS

Throughout most temperate regions and in much of the tropical and subtropical world, it is difficult to find landscapes that do not bear the imprint of agricultural activities. We have noted already that the savanna grasslands owe their existence, at least in part, to human activities. Even the tropical rain forests have been widely affected by clearance and cultivation. We have also seen that the present extent of the arid scrubland and desert owes much to the activities of farmers. Only remote environments—tundra regions, central deserts (such as the Australian Desert), the interior of large, impenetrable tropical rain forests, and relatively inaccessible mountain areas—have escaped the impact of direct human activities. However, virtually all areas of the globe have been indirectly affected through the global spread of atmospheric and aquatic pollution. For example, the Chernobyl nuclear reactor explosion and Arctic haze have measurable effects on many remote Arctic and Subarctic ecosystems in the northern hemisphere. Even in these remote areas of the globe, there are people living. However, their lifestyles (limited agriculture, hunters and gatherers) place them within the ecosystem structure, as part of the consumer trophic levels, rather than as all pervasive controllers of the food chain.

## THE HISTORY OF FOREST CLEARANCE

Human intervention has often been operating for many centuries, but the character of human impact in different environments varies. We can recognize three early stages in history when humans began to make and use tools, and from this, to shape and control the ecosystem to suit their purposes. In the earliest stage, during the Paleolithic (Old Stone Age), the impacts were probably limited, as technology was very simple and human numbers were small. As technology advanced, however, and populations increased, human influence became more far-reaching. Thus, during the Mesolithic (Middle Stone Age), people began to settle in more permanent camps, and to clear and manage the surrounding area for grazing and fuel. In the last phase of the Stone Age, the Neolithic, human influence grew even greater.

Throughout the early periods of development, people affected the ecosystem partly as hunters and gatherers, and partly as settlers and farmers. In the former role, they may have hastened the extinction of certain animals, in some areas. In the latter role, they had a more fundamental impact, as they cleared large tracts of woodland, and set in motion marked changes in the ecosystem. There is evidence, for example, that Mesolithic forest clearance in upland Britain led to the decline of certain tree species, particularly the linden (*Tilia*), and the opening of the forest canopy resulted in major changes in soil development. Many of the upland peats developed as a result of these human influences (Figure 31.1).

Later, forest clearance extended and intensified these effects. During the Neolithic, and even more so during the Iron Age and Bronze Age, the development of a relatively sophisticated and widespread

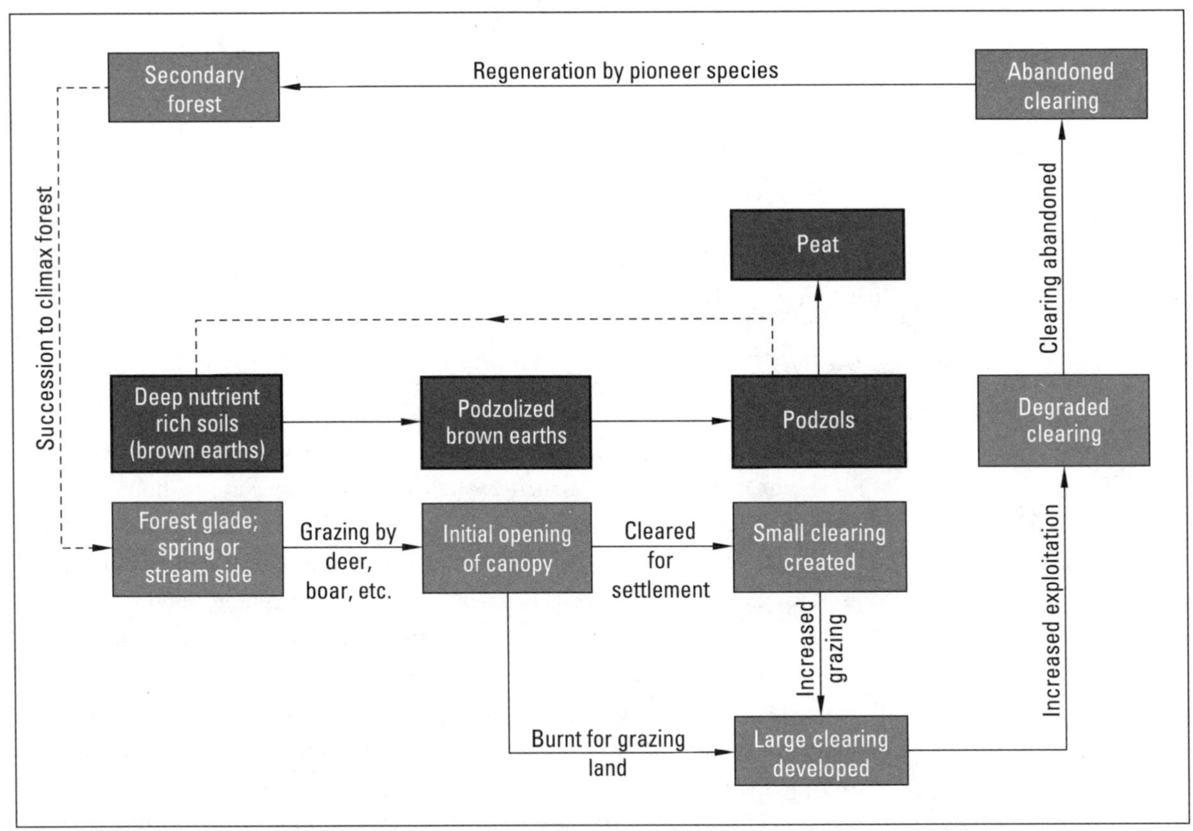


FIGURE 31.1 ▼ The effects of prehistoric peoples on temperate forest ecosystems (based on Simmons, 1964). Changes in vegetation affected pedogenic processes, and if maintained long enough, affected soil type. Broken line depicts a reversal of the process following abandonment of clearings and forest regeneration

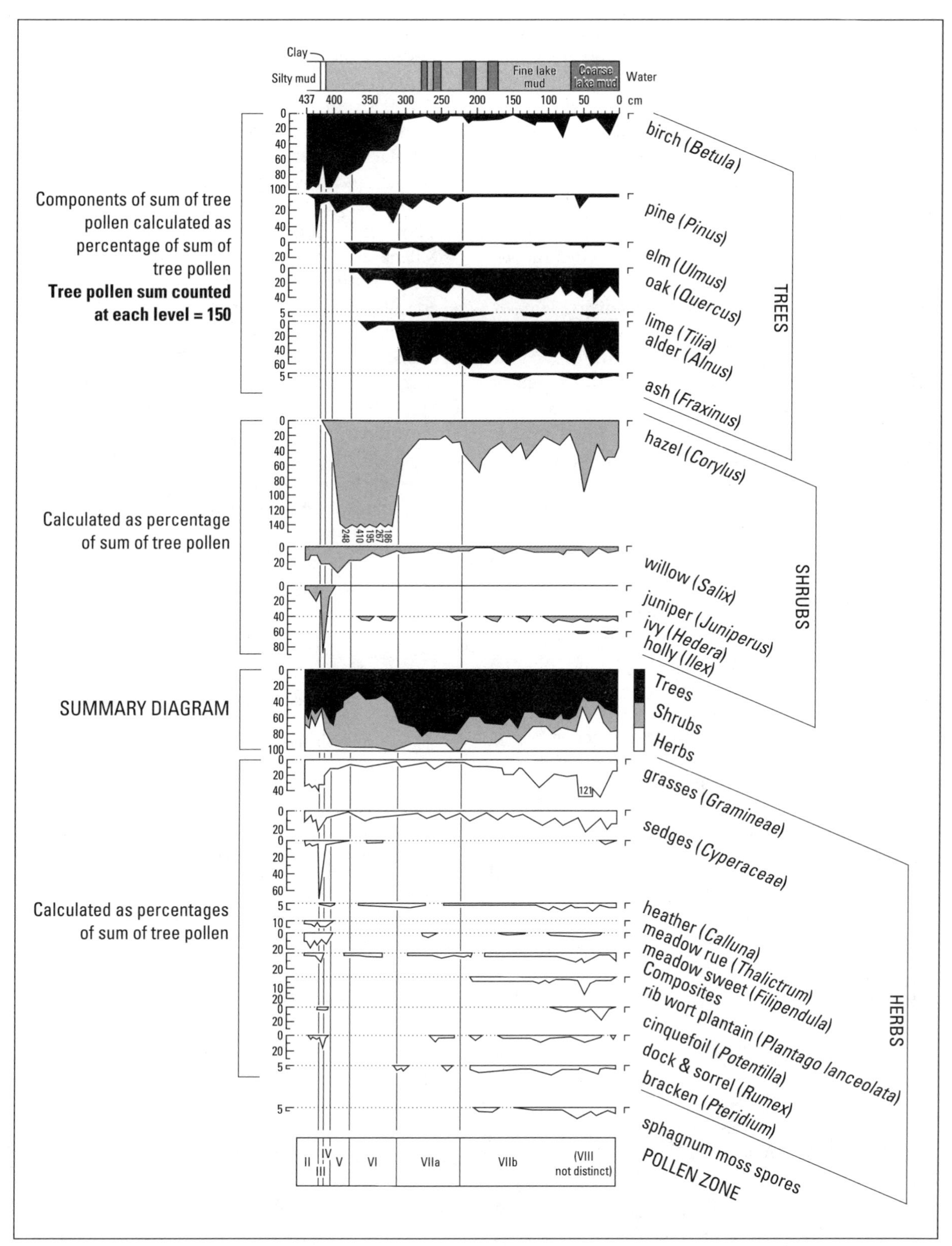


FIGURE 31.2 ▼ A pollen diagram from Blelham Tarn, English Lake District. Not all pollen types are represented. Note the decline in elm and pine, and the increase in herb pollen at the zone VIIa-VIIb boundary (from Evans, 1970). After this time, herbs become more and more important as the rate of forest clearing increases

form of settled agriculture led to the removal of vast areas of woodland in Europe. About 4000 years ago, during the Bronze Age, there was a marked decline in elm (*Ulmus*), and at about the same time, fossil pollen from peat bogs shows an increase in weed and cereal plants (Figure 31.2). As well as removing the natural vegetation, early humans helped new species to become established and introduced several for their own use.

Episodes of human activity were not synchronous everywhere. We know, for example, that human cultures progressed at different rates in different parts of the world, and even today some civilizations have similarities with the stone-age communities of Africa 10 000 years ago. The spread of cultural innovations was slow, and often it was many centuries before new techniques were adopted in distant areas. In the Middle East, settled agriculture, with tillage and systematic cropping, was established about 10 000 years ago, probably in Iran. In subsequent millennia, agriculture spread through diffusion of the ideas, but it was almost certainly reinvented, quite independently, by different peoples. In Britain and in the Americas it emerged independently about 5000 years B.P. There is evidence that agriculture was practiced about 7000 years B.P., and possibly earlier. However, scanty clues from tropical Asia hint at an even earlier history, possibly before 42 000 years B.P.

Whenever it started, the implications of farming are clear. Cultivation required land free from natural vegetation. With population growth and improved agricultural techniques, the area of cleared land expanded and so did the intensity of the changes.

Many early agricultural systems were based on the principle of slash-and-burn. Areas were cleared by burning, then cultivated for a few years until yields declined and the natural vegetation started to reinvade. Then the plot was abandoned in favour of a new site. As long as population densities were low, this

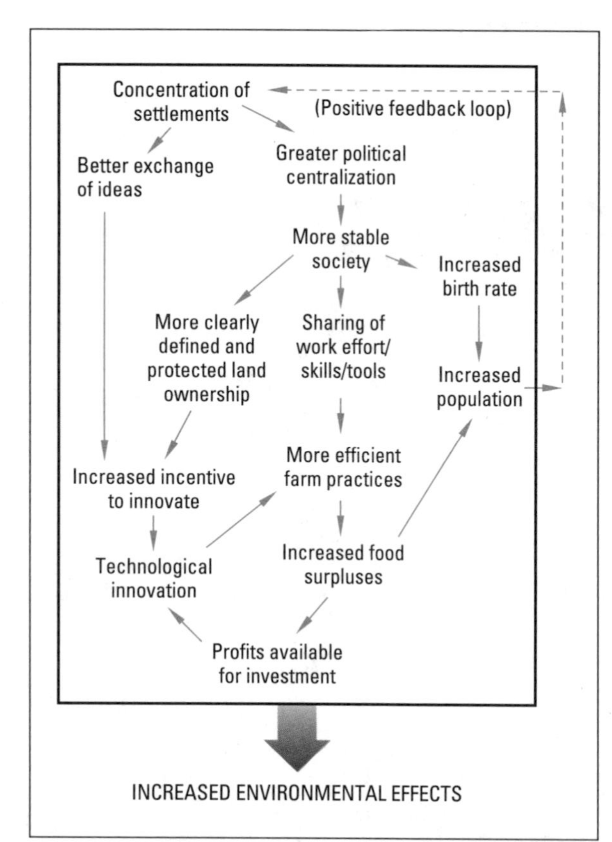


FIGURE 31.3 ▼ The social and technological effects of concentration of population into permanent farming communities. The broken line indicates a positive feedback loop, intensifying the relationships

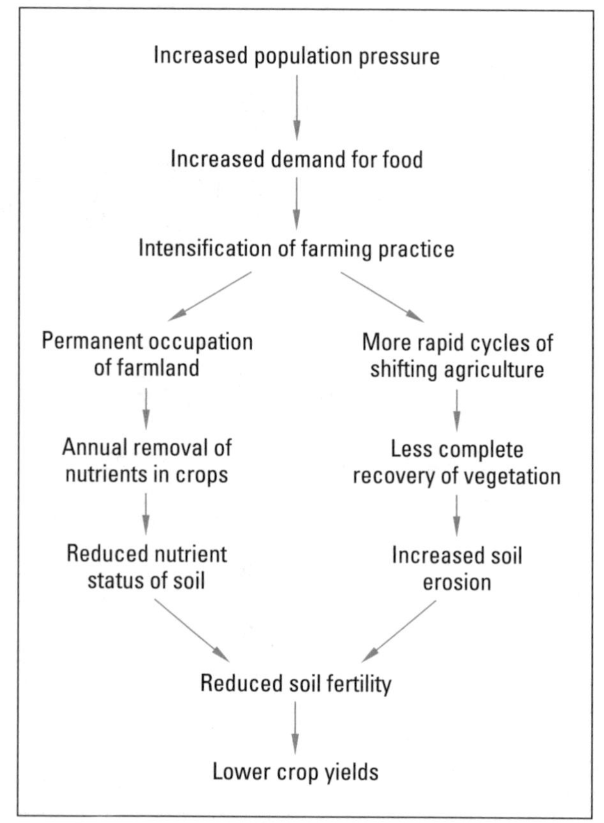


FIGURE 31.4 ▼ The longterm effects of increased population levels on soil fertility and crop yield. The result could be local famine, farm abandonment and migration, unless technological innovations modify the cycle

system could be maintained, and natural vegetation could regenerate. However, as the frequency of clearing increased, the ability of the natural vegetation to recover decreased, and the ecosystem declined.

The concentration of the population into more or less permanent communities initiated a cycle of technological and cultural advancement that increased our ability to manage the environment (Figure 31.3). At the same time, the tendency for more settled farming, or for more rapid cycles of slash-and-burn, led to a progressive decline in the fertility of soils and the energy status of the ecosystem (Figure 31.4).

In the past four or five centuries, agricultural progress has speeded up considerably. The development of sophisticated ploughs and seed drills by people such as Jethro Tull; the generation of new strains of plants and animals; the discovery of the principles of plant nutrition and the consequent evolution of the fertilizer industry; the increased use of machinery in agriculture; and the recent upsurge in the use of pesticides—all have fundamentally altered the agricultural ecosystem.

### INDUSTRIALIZATION AND ECOSYSTEMS

Of course, people have progressed not only through the development of agricultural technology. Humans have also created a complex and sometimes frightening array of other technologies. Industry and war, in particular, have been mirrors of this progress. Both have had a major impact on the world's ecosystems.

The development of industry is commonly regarded as a recent phenomenon, starting with the Industrial Revolution in western Europe during the nineteenth century. Since then, industry based on the use of fossil fuels, particularly coal and, more recently, oil, has become one of the main consumers of the earth's resources. The extraction and exploitation of these resources has disrupted many natural ecosystems. Some effects are direct. The creation of large holes in the ground and the dumping of waste material on surrounding land is clearly disruptive at a local level. But there are also many more indirect and more pervasive effects. In particular, pollutants released by the exploitation of these resources have spread through most of the world. Airborne dust and gases, waterborne substances, and accumulations of waste materials in the soil, all have affected natural plant and animal life.

Radioactive fallout from nuclear weapons testing during the 1960s led to widespread dispersal of substances such as strontium-90 and cesium-137. The latter has been found in Arctic food chains, where it is absorbed by lichens, and thus passed to the caribou, and then to human consumers. Other predators, such as wolves and foxes, have abnormally high levels of cesium-137 in these areas.

Clearly, human impact on the biosphere is significant. Sometimes this impact is deliberate, and the consequences are calculated, but often it is not. The history of human activities indicates that we need to understand the ecosystems we are dealing with to avoid causing damage to them and ultimately, to ourselves. Our survival is dependent on the health of the earth's biosphere.

## Agricultural ecosystems: cereal cropping

### THE GENERAL NATURE OF CEREAL CROPPING

One of the most intensive and extensive forms of agriculture is cereal cropping. Cereal cultivation began early in farming history, when wild wheat crossed with a natural goat grass to form a hybrid wheat plant. The result was a grass known as **emmer** which had plumper, more nutritious grains. Later, emmer was crossed with another goat grass, and bread wheat developed. This plant was the foundation of many cereal-growing cultures in the world. Domestication of corn probably occurred at the same time in the New World (perhaps in Mexico), and became the focus for cereal-growing in other areas. Since then, deliberate crossbreeding has led to the development of many different types of corn and wheat, but the basic principles of their cultivation are similar, and the effects on the ecosystem are comparable.

In most cases, cereal cropping involves the creation of a seedbed on which the crop is sown by artificial means. The seedbed is prepared by various tillage practices, such as ploughing, harrowing and rotovating, to break up the soil aggregates, remove weeds, and bury or chop up the residues of previous crops. In all but the more primitive cereal-growing practices, sowing involves the use of tractors and seed drills. Following sowing, the land may be rolled and harrowed to bury the seed and create a more compact soil environment for germination. As the plants grow, fertilizers and pesticides may be applied, and when the crop is mature, harvesting is often carried out by combines.

All these practices have considerable effects on the soil. With repeated passage of vehicles over the land, as much as 90 percent of the surface may be directly affected by vehicle wheels (Figure 31.5). If the soil is wet, wheel slippage and compaction can damage the soil structure. Tillage can also alter internal soil processes, killing soil organisms (particularly earthworms), encouraging chemical oxidation of the humus by improving aeration, enhancing wind and water erosion as root binding of soil particles is lost, and ultimately reducing nutrient retention and leading to instability in the soil structure.

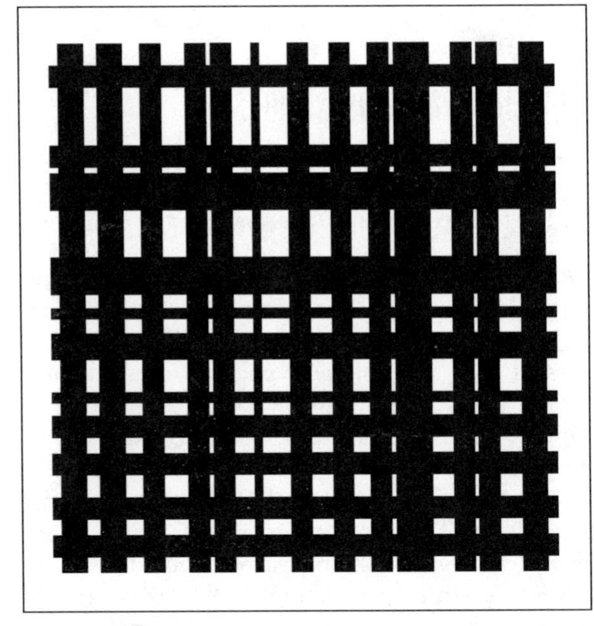

FIGURE 31.5 ▼ The extent of soil surface coverage by tractor wheels during a typical annual sequence of cultivation

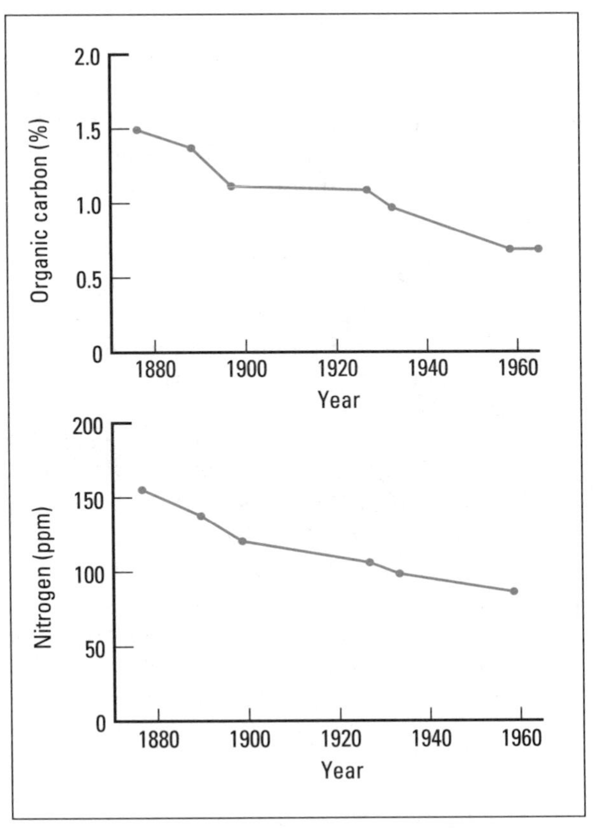


FIGURE 31.6 ▼ Changes in soil conditions under longterm intensive farming (after Mattingly *et al*, 1975). Plots had received NPK fertilizer and had been under wheat and barley continuously since 1876

As agriculture becomes more intensive, in an attempt to feed the growing world population and to provide a higher standard of living, cereal-growing tends to move toward a system of monoculture, where the same crops are grown repeatedly. Even where different crops are rotated, they may have similar requirements. Over time, therefore, marked changes in the soil can occur (Figure 31.6).

Many of these changes relate to the effects of tillage, and in recent years attempts have been made to introduce systems such as **zero-tillage** and minimal cultivation that involve less frequent tillage. In these, ploughing and seedbed preparation are greatly reduced or totally eliminated. The seed may be sown directly onto undisturbed soil. Weeds and crop residues are controlled by herbicides, and insect pests—which may build up in undisturbed soil—are killed with insecticides. In an attempt to reduce soil erosion and moisture loss in drought-prone regions of the Prairies, the practice of leaving stubble is employed to trap winter snow and anchor the soil with the roots of last year's crop.

The accumulation of pesticides in the soil and in runoff water can be worrysome. As much as 50 percent of compounds sprayed onto the vegetation may be lost by leaching and runoff. Within the soil, numerous detrimental side effects can occur. Many insecticides kill both pests and beneficial soil organisms. More active organisms are particularly vulnerable to the toxic effects of pesticides, as they cover greater distances and come into contact with larger quantities of the chemical. Since predatory organisms usually are more active, insecticide may preferentially kill the very creatures that naturally control parasitic organisms and agricultural pests.

The effects of disturbing the delicate balance of the soil fauna may be the opposite to what is desired (Figure 31.7). In time, the fauna may recover, but repeated applications of pesticides can permanently suppress certain organisms, many of them beneficial. This can have a number of consequences. It may lead to increased numbers of pests; it may disrupt the detrital food chain so that the structure of the soil ecosystem is altered; and it may inhibit decomposition and nutrient cycling.

Two factors critical to the effect of pesticides in the soil are persistence and toxicity. Persistence is measured on the basis of the half-life of the substance: the time it takes for its concentration to be halved. The half-lives of some common pesticides are shown in Table 31.1. On the whole, insecticides have longer half-lives, and are therefore more persistent, than herbicides. Toxicity is measured by the $LD_{50}$ parameter. This is the dose necessary to kill 50 percent of the target population. Examples are given in Table 31.1.

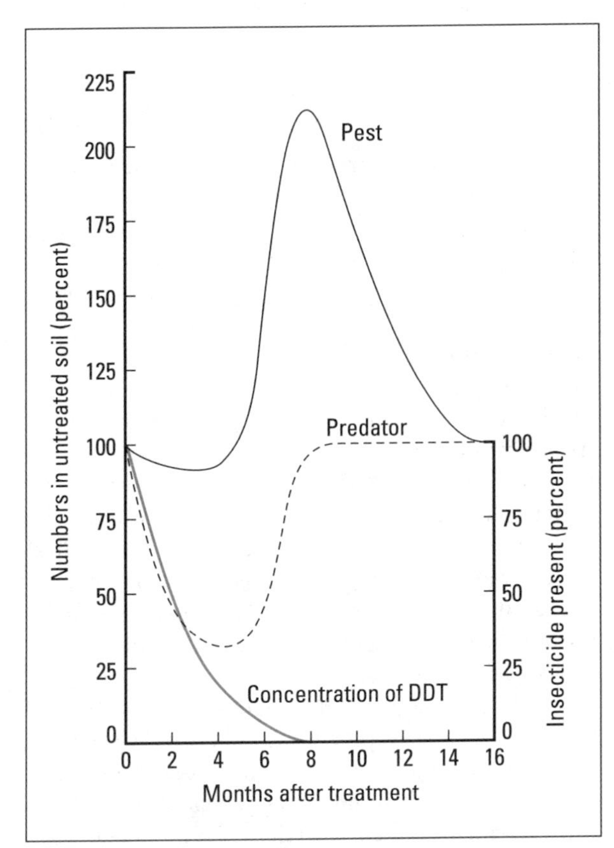


FIGURE 31.7 ▼ Effects of DDT on predator and prey organisms in the soil (after Edwards, 1969). One hundred percent would be the 'natural,' pretreatment level (left) and the amount of DDT remaining after treatment (right)

## Table 31.1

*The persistence and toxicity to animals of common pesticides and herbicides*

| *Pesticides* | *Half-life (months)* | *Lethal dose (mg kg$^{-1}$)* |
|---|---|---|
| DDT | 27.5 | 113 |
| Lindane | 21.2 | 125 |
| Dieldrin | 22.0 | 90 |
| Aldrin | 8.3 | 60 |
| Parathion | 1.5 | 3.5 |
| Phorate | 0.3 | 3.7 |
| Simazine* | 1.8 | 5000 |
| Atrazine* | 2.7 | 3080 |
| 2,4,5-T* | 2.5 | 300–500 |
| MCPA* | 0.7 | 700 |

* Herbicides; all others = insecticides

In the past, very persistent pesticides of the organochlorine type, such as DDT, dieldrin, and lindane, were widely used. These broke down slowly in the soil, became concentrated within the food chain, and led to unacceptable pollution, not only in the soil but also in animals at higher trophic levels. DDT, for example, has been responsible for the near extinction of the peregrine falcon during the last few decades. The accumulation of DDT in these birds resulted in thinner eggshells, and consequently relatively few chicks survived to hatch (Figure 31.8). More recently, many organochlorine compounds have been banned or strictly controlled, and organophosphate pesticides have taken their place in North America. These chemicals are more volatile and therefore decompose much more rapidly, with half-lives measured in terms of weeks or months rather than years (Table 31.1). Organophosphates also are more toxic, so smaller quantities need to be applied. Unfortunately, raptors such as the peregrine falcon are migratory, and it is possible that they are still exposed to the damaging chemi-

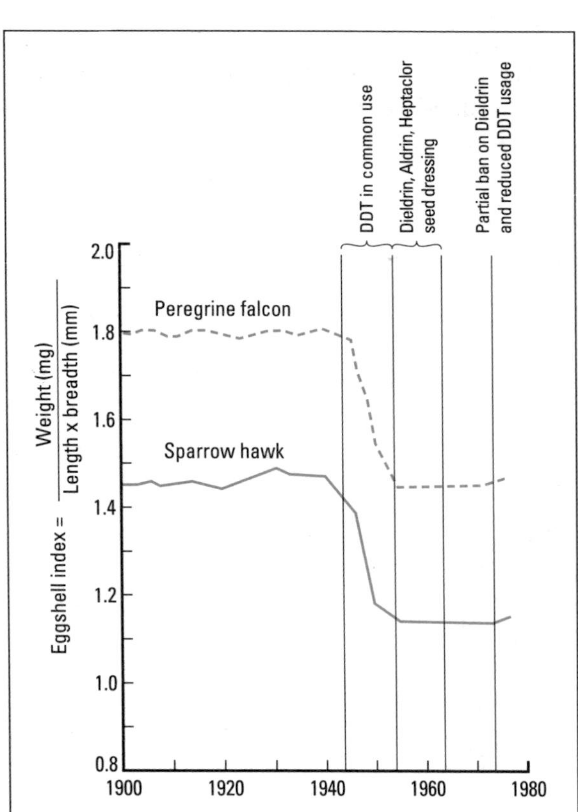


FIGURE 31.8 ▼ Effects of organochlorine insecticides on eggshell thickness of peregrine falcons and sparrow hawks in Britain (from Brown, 1976). Seed dressings were used to deter predation on stored and sown seeds

cals in their winter range, since many countries have not banned the use of these compounds.

Four main processes break down pesticides. Volatilization reduces the compound to gaseous form, and it escapes to the atmosphere. Biological decomposition, through the action of certain soil organisms, converts the pesticides to different forms. In most cases, this conversion produces a less harmful compound, but in some instances it can produce a more lethal substance. DDT, for example, may break down to produce DDD, which is highly toxic. Chemical decomposition may also take place. Photodecomposition is the breakdown of compounds under the effects of sunlight. This occurs either at the soil surface or on the leaves of the plants.

Pesticides may also be adsorbed onto colloidal particles in the soil and stored there, in much the same way that plant nutrients are held. Erosion of the soil can remove these substances. Also, leaching may carry soluble pesticide compounds out of the soil and into streams or groundwater. This removes the pesticide from the soil, but pollutes aquatic ecosystems.

There is certainly cause for concern about the effects of pesticides on the environment. In Australia, for example, there is evidence that organisms may become immune to pesticides. When DDT was first used commonly, after World War II, it was applied at rates of 1 kg $ha^{-1}$, but by the late 1960s, doses of 100 kg $ha^{-1}$ were necessary.

Pesticides do provide us with many benefits. In some parts of the world, they have been responsible for huge increases in agricultural output, and in a hungry world this is important. Pesticides have also helped to rid vast areas of insect-carried disease. Malaria has been eradicated from many thousands of square kilometres through the use of DDT. However, it is critical to weigh the advantages and disadvantages before using pesticides, and this requires an understanding of the ecosystems involved.

## NUTRIENT CYCLING IN CEREAL GROWING

One of the most important aspects of cereal-growing systems is the harvest of the crop. The loss of this material represents a major loss of biomass and nutrient reserves in the system. Farmers need to replenish soil fertility by applying fertilizers, as continued cropping leads to lower levels of organic matter in the soil (Figure 31.6) and eventually to soil exhaustion.

The removal of the crop and the input of fertilizers are two vital components of cereal-growing systems that distinguish them from most other natural or agricultural ecosystems. The amount of nutrients introduced by fertilizers far exceeds the natural inputs from either the atmosphere or from weathering, while losses in removal of the crop are normally greater than those associated with leaching or erosion (Figure 31.9).

Within the agricultural ecosystem, cycling of nutrients is often rapid, for although there tend to be fewer soil organisms than in natural ecosystems (Table 31.2), tillage encourages oxidation and the mineralization of the nutrients. The cereal crop ecosystem involves very few plant species, often only one, and competitors are ruthlessly removed by herbicides. There are almost no animals outside the soil, and therefore, the internal food chain is very simple. Apart from a few pests that may attack the

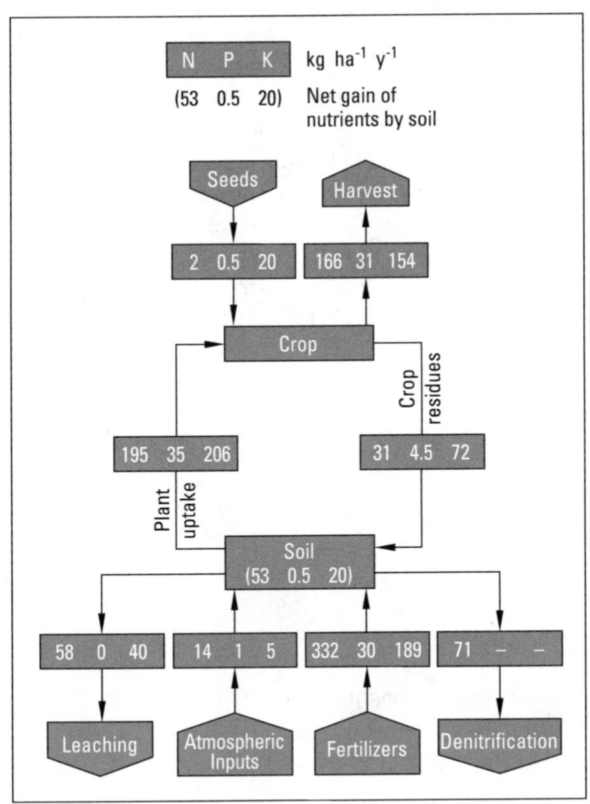


FIGURE 31.9 ▼ The nutrient cycle of an intensive farm in which crop residues are removed from the land: potato-wheat-sugar beet rotation on clay soil in the Netherlands (from Henkens, in Frissel, 1978). Figures show flows of nitrogen, phosphorus, and potassium in kg $ha^{-1}$ $y^{-1}$

## Table 31.2

*Earthworm numbers in arable and grassland soils*

| *Land use* | *Earthworm nos. ('000 $ha^{-1}$)* | *Earthworm weight (kg $ha^{-1}$)* |
|---|---|---|
| Permanent pasture | 700–1100 | 500–700 |
| Ley grass | 550–800 | 475–700 |
| Young arable | 575–825 | 400–825 |
| Old arable | 175–300 | 70–150 |

*Data* E.W. Russell, *Soil Conditions and Plant Growth* (1974), Longman.

crop, only decomposers are at work. Humans and their livestock are the only significant consumers. Indeed, the attempt to shorten food chains and thereby avoid waste is one of the main characteristics of agricultural ecosystems. However, these systems still transfer relatively small amounts of energy to the consumer. For example, the energy absorbed by humans from potatoes probably averages no more than 0.2 percent of the initial solar inputs to the agricultural ecosystem (Figure 31.10).

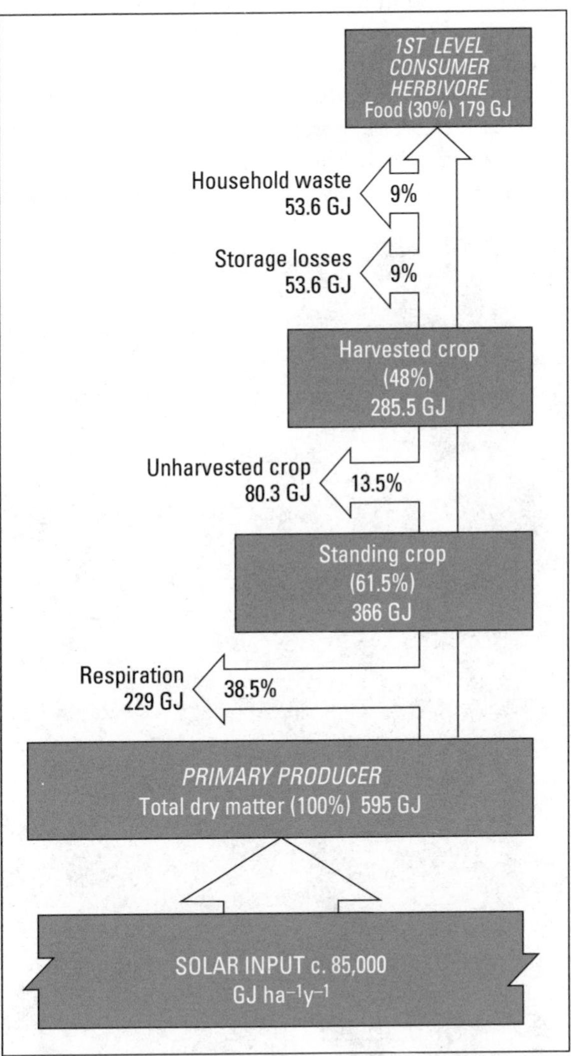


FIGURE 31.10 ▼ Energy efficiency of a typical intensive farming system (potato crop). The total dry matter produced in the system is about 595 GJ $ha^{-1}$ $y^{-1}$, which is about 0.7 percent of solar radiation inputs. The human food produced is only about 0.22 percent of solar energy received (from Duckham and Masefield, 1970)

Crop production cannot be considered alone. Many other extra-ecosystem parameters are involved. Tremendous resources are used by machinery, including hydrocarbons to make the machinery work, the steel and other materials to build it. The transportation and production of fertilizers also require large inputs of energy and materials. All of these inputs to the agrosystems represent outputs from other ecosystems. Human-managed ecosystems are energy-hungry, driven by resources that have been produced in other ecosystems. Just consider the amount of energy required to develop, build, and operate even the simplest piece of equipment.

Under some conditions, cereal cultivation can lead to a marked increase in erosion. Tillage and crop removal leave the soil unprotected and prone to the winnowing effect of wind (Figure 31.11). Also, compaction by machinery, and channelling of runoff waters by plough furrows, may encourage gullying. Serious nutrient losses through erosion can result (Table 31.3). These processes are exacerbated by the loss of organic matter from the soil. Humus binds soil particles into stable aggregates, but when it is lost (e.g., through increased oxidation and reduced organic matter inputs) the structural stability of the soil declines. Damage by machinery increases, as does erosion.

FIGURE 31.11 ▼ Topsoil removed from the adjacent field by wind has filled a ditch at the roadside. In southwestern Alberta, drought conditions can promote soil erosion, particularly when inappropriate farming methods are used (photo: G.P. Kershaw)

**Table 31.3**

*Losses of nutrients by soil erosion*

| | Nutrient losses ($kg\ ha^{-1}\ y^{-1}$) | | | |
|---|---|---|---|---|
| | N | P | K | Ca |
| Erosion from continuous corn | 74 | 20 | 678 | 246 |
| Erosion from rotational crops | 29 | 9 | 240 | 95 |
| Erosion following crop removal | 134 | 25 | 112 | 45 |

*Data* N.C. Brady, *Nature and Properties of Soils* (1973), Macmillan.

Leaching losses in cereal-growing systems can be considerable. Although they may not appear significant in relation to the large inputs of nutrients in fertilizers, they can represent important contributions of nutrients to streams. Nitrogen, in particular, is lost from cultivated land in this way, for it is highly soluble, especially when it is not retained in organic form in the humus. As we will see later, one of the consequences of the washing away of fertilizers and pesticides is pollution of aquatic environments. Large quantities of these chemicals may be applied to increase production and reduce competition from plant and animal pests, but much of this is commonly lost to the streams and groundwater.

# Grazing ecosystems

## GENERAL NATURE OF GRAZING ECOSYSTEMS

Human use of grasslands is widespread. As much as 30 percent of the world's land surface is agricultural grassland, ranging in character from extensively grazed range and moorland, to intensively stocked, artificially sown pasture.

In extensive grassland ecosystems, the role of livestock is often limited, representing little more than a controlled form of the natural grazing that would occur by wildlife of the area. Often, however, this grazing is accompanied by rangeland management practices such as burning to encourage new plant growth, and these can have severe effects on the ecosystem. Burning kills or drives out much of the natural biota, and can lead to marked changes in plant and animal species composition. It also affects the soil (Figure 31.12).

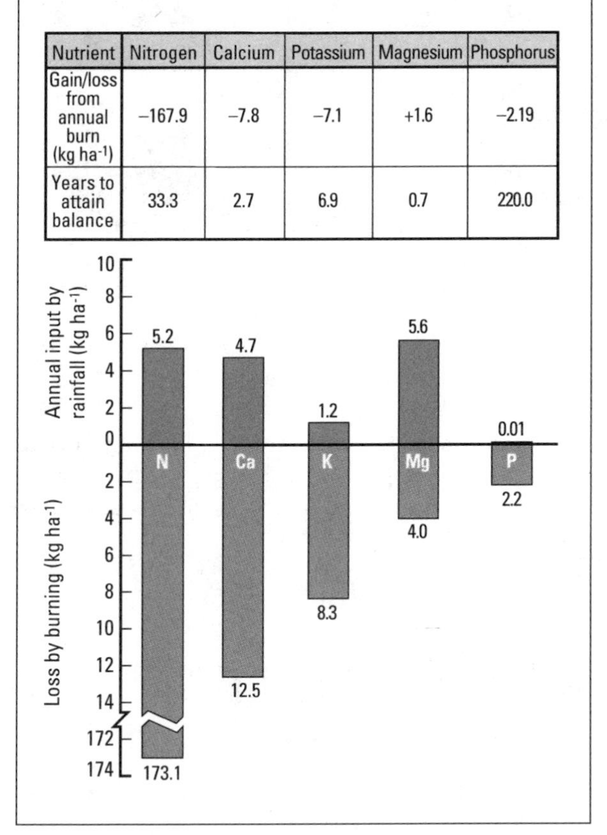

| Nutrient | Nitrogen | Calcium | Potassium | Magnesium | Phosphorus |
|---|---|---|---|---|---|
| Gain/loss from annual burn (kg ha$^{-1}$) | −167.9 | −7.8 | −7.1 | +1.6 | −2.19 |
| Years to attain balance | 33.3 | 2.7 | 6.9 | 0.7 | 220.0 |



FIGURE 31.12 ▼ Changes in soil nutrient contents of burned moorland soils, and the minimum interval between burning necessary to maintain nutrient levels (from Chapman, 1967)

Of course, in addition to the character of the soils and vegetation, animals are one of the most important attributes of grazing ecosystems. Sheep, cattle, and goats are the main animals raised in many parts of the world. Cattle include beef and dairy cattle; sheep and goats may be reared for either wool, meat, or milk. In all cases, specific varieties of animals are used. Merino sheep, for example, are the basis for much of the world's wool production. Hereford cattle are common as beef animals, while dairy farms raise other breeds, including Holstein-Friesian and Jersey cattle. The importance of crossbreeding to produce ideal animals for the conditions is illustrated by the experience in Israel. Initially, using local cattle, milk yields were in the order of 300 l yr$^{-1}$. However, with the introduction of Lebanese and Syrian crossbred cattle, this increased to 1000–2000 l yr$^{-1}$. Finally, crossbreeding with Friesian cows resulted in yields of up to 6000–8000 l yr$^{-1}$.

## NUTRIENT CYCLING IN GRAZING ECOSYSTEMS

In cereal-growing ecosystems, most nutrient cycling is carried out by decomposition of plant debris by soil organisms and chemical oxidation. In grazing systems, animals become an important part of the nutrient cycle (Figure 31.13). Plants are eaten by a grazing animal, and the nutrients are either retained in the animal's body or returned to the soil in the form of dung or urine. Relatively small proportions of the total nutrient reserves are normally retained in animals (Table 31.4), and even smaller quantities are lost through the harvest of their meat or milk. Most of the nutrients taken up by animals are returned to the soil.

The nature of the return varies from one nutrient to the next. Phosphorus, magnesium, and calcium

### Table 31.4

*Nutrient retention and returns by cattle*

| | *Nutrient* | | | | |
|---|---|---|---|---|---|
| | *Ca* | *Mg* | *P* | *K* | *Na* |
| Total intake (kg) | 120 | 95 | 85 | 135 | 30 |
| Retained in body (%) | 10 | 6 | 9 | 3 | 7 |
| Retained in milk (%) | 12 | 5 | 29 | 6 | 6 |
| Returned in dung (%) | 75 | 79 | 62 | 11 | 30 |
| Returned in urine (%) | 3 | 10 | 0 | 80 | 57 |

*Data* G.B. Davies, D.E. Hogg, and H.C. Hopwell, 'Extent of return of nutrient elements by dairy cattle: possible leaching losses,' Joint Meeting of the International Soil Science Soc., Comm. IV and V, New Zealand (1962), pp. 715–20.

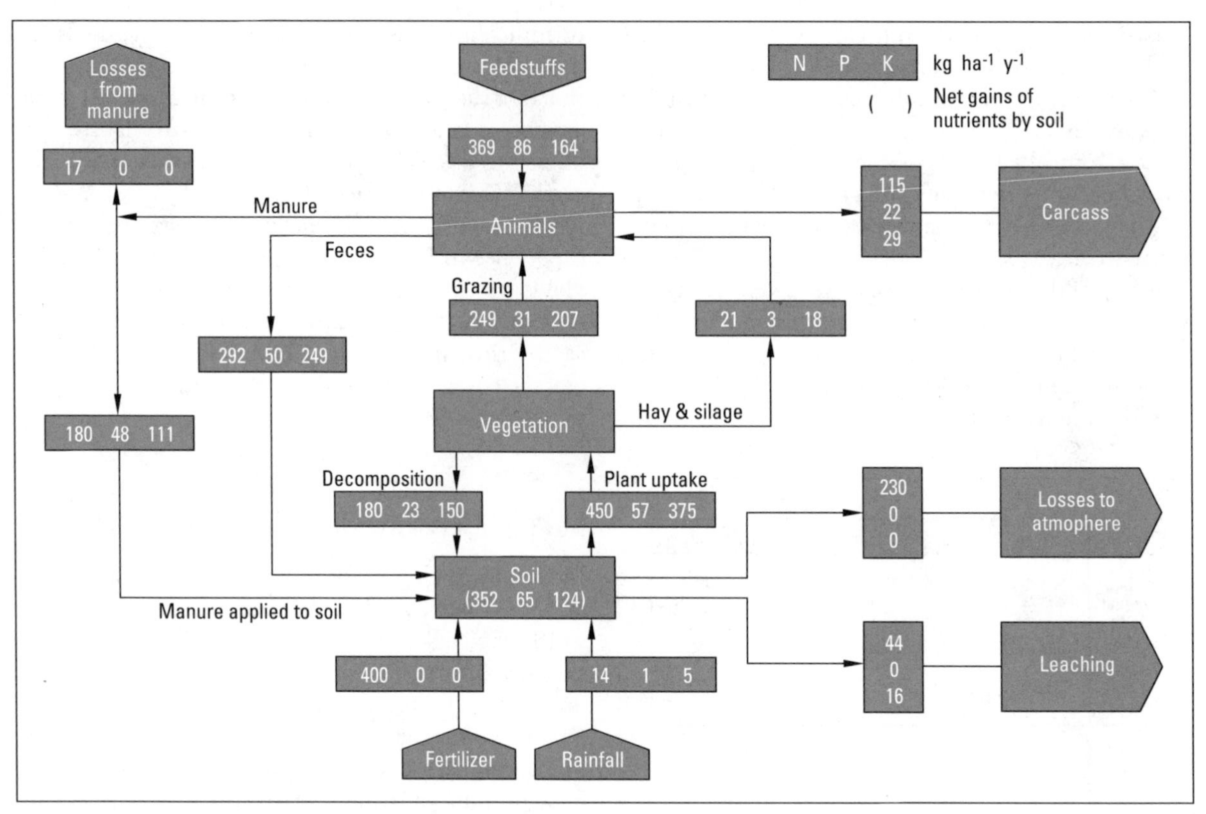


FIGURE 31.13 ▼ The nutrient cycle of an intensive grazing system: a Dutch dairy farm on a clay soil (from Henkens, in Frissel, 1978). Figures show flows of nitrogen, phosphorus, and potassium in kg $ha^{-1}$ $y^{-1}$

are returned mainly in the dung; potassium and nitrogen in the urine (Table 31.4). In both cases, but particularly in urine, the nutrients in these waste products are soluble and readily available to plants. This tends to accelerate nutrient cycling, and increase the quantity of nutrients in circulation. The abundant supply of readily digestible organic matter, together with the greater stability associated with less regular ploughing to disturb the soil fauna, results in a rise of the populations of earthworms and other soil organisms (Table 31.2).

The solubility of nutrients in animal wastes has further implications, because, like fertilizers, these components are susceptible to leaching. Pollution of water bodies may be a problem where grazing is very intensive. The nutrient cycle illustrated in Figure 31.13 shows that the efficiency of the system in terms of the energy supply to humans is relatively low, with about 0.02 percent of the initial energy inputs available to the consumer. In less intensive rangeland ecosystems, this figure may be as low as 0.004 percent. On the other hand, intensive grass-growing can be very productive. In many cases, grass crops come far closer to achieving potential yields than do cereal crops. The losses occur mainly in the steps from the grass to the animal, and from the animal to the human consumer (Figure 31.14).

# Shifting cultivation ecosystems

## GENERAL NATURE OF SHIFTING CULTIVATION

In much of the tropical and subtropical world, farmers carry out a system of cultivation that has changed little from the methods used by prehistoric peoples. Inputs to the system are few, and soil fertility relies on natural replenishment by rainwater and weathering to maintain the nutrient levels tapped by crop removal. Natural or seminatural woodland and scrub is cleared, cultivated, and cropped for a few years, then abandoned. Crops vary from one area to another, but typically include sorghum, millet, corn, and cassava.

The duration of plot rotation in many of these systems is related to population density. When pop-

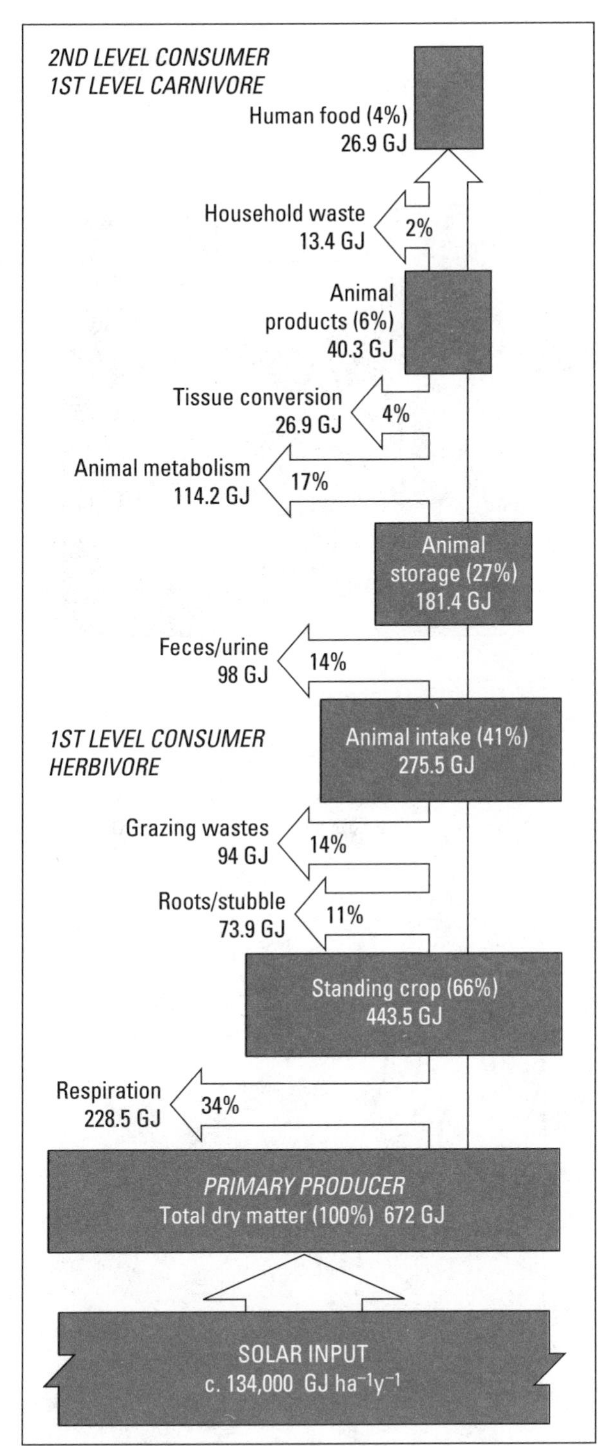


FIGURE 31.14 ▼ Energy efficiency of a typical intensive grazing system. The total dry matter produced in the system is about 672 GJ $ha^{-1}$ $y^{-1}$, which is only 0.5 percent of solar radiation inputs. The human food produced is only about 0.02 percent of solar energy received (from Duckham and Masefield, 1970)

ulation is low and ample land is available, the period of rotation may be long. Each plot may be cropped for no more than two or three years, and then abandoned for twenty or more years to regenerate. Under these conditions, regeneration is effective: the soil experiences no major decline in fertility, the ecosystem experiences no major loss of nutrients. On the other hand, as population increases, and less land is available, the rotation becomes shorter (Figure 31.4). Often, less than seven years are left for regeneration, and plots may be cropped continuously for almost as long. Soil exhaustion, erosion, and an overall decline in ecosystem nutrient levels then occur.

## NUTRIENT CYCLING UNDER SHIFTING CULTIVATION

Nutrient cycles under shifting cultivation cannot be considered on a short-term basis, as can those of cereal or grassland systems. Relatively long periods of exploitation and net nutrient loss alternate with even longer periods of rest and nutrient accumulation. The longterm nutrient budget characteristics depend on the balance between these two processes (Figure 31.15).

The period of loss is generally initiated by felling trees and burning of the residues, a practice known as slash-and-burn. This has a severe effect on nutrient reserves, as considerable quantities of nutrients are lost in smoke. Potassium, in particular, is removed in this way, because it is released mainly in the fine ash particles and gases that are swept into the atmosphere. Other nutrients are returned to the soil in the ash that collects on the surface. These are readily soluble, however, and may be leached rapidly from the soil.

Leaching is encouraged during this stage because the removal of the forest vegetation reduces the interception of rainfall and increases direct rainfall inputs to the soil. The combined effects of burning and subsequent leaching deplete the total nutrient reserves of the ecosystem. Since most of the nutrients of many tropical forests are stored in the vegetation, this can represent a serious loss (Figure 31.15). The exposed soil surface is more prone to erosion, and additional losses occur with runoff. Consequently, gullying and soil exhaustion may be initiated even before the land has been cropped.

During the following years of cropping, depletion of nutrient reserves continues through the com-

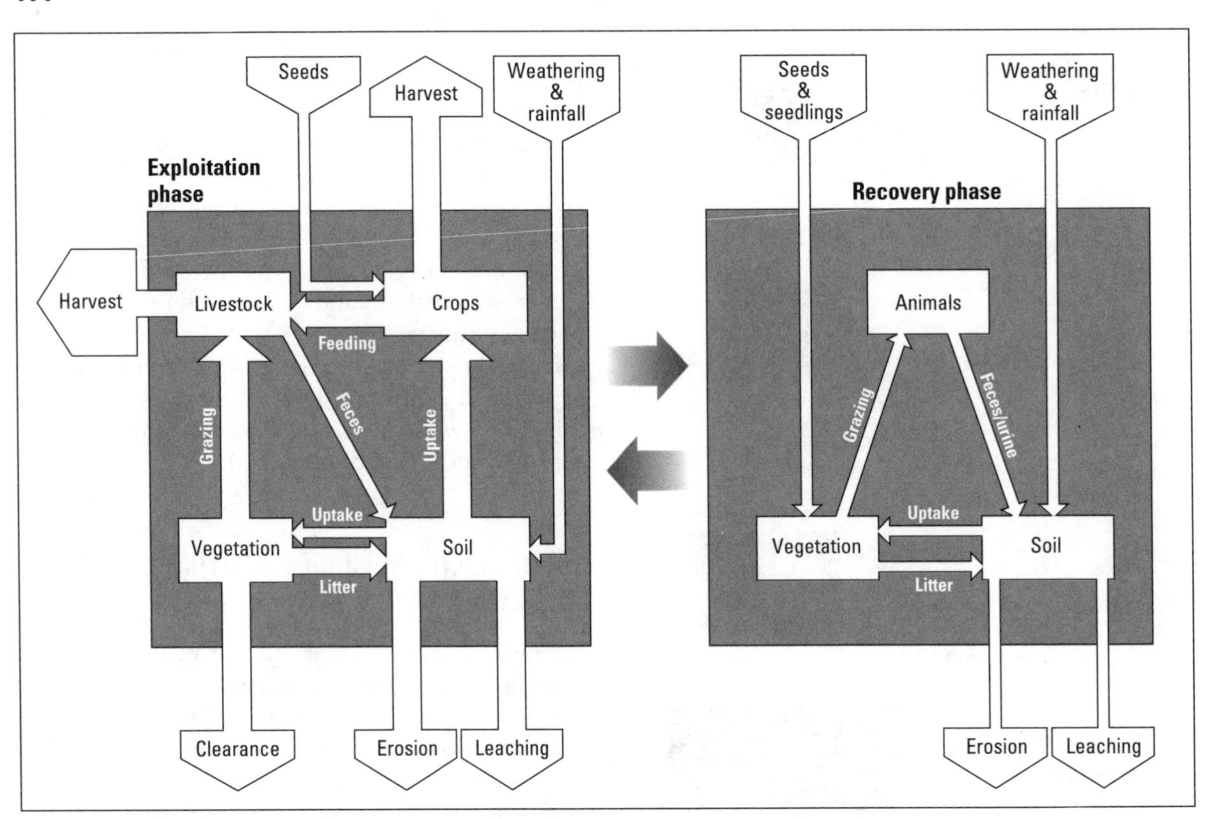


FIGURE 31.15 ▼ Nutrient cycles during exploitation and recovery phases under shifting cultivation. The width of the arrows reflects the relative magnitude of the nutrient flows

bined effects of erosion, leaching, and crop removal. Leaching and erosion are reduced to some extent, as crops often provide a reasonably continuous cover. However, the nutrients retained in the soil following burning are taken up by the plants, and considerable losses may occur with harvesting. The use of fertilizers is almost unknown, and only small quantities of animal wastes are supplied to the soil, so yields decline rather quickly as fertility falls. Diminishing yields are often due to increased competition as much as to soil exhaustion.

When small plots surrounded by lush vegetation are abandoned, regeneration occurs quite rapidly. As nutrients are taken up from the soil and held within the plants, total soil reserves may decline for a time. However, reduced leaching and erosion help to stabilize the system. Nutrient inputs from precipitation and weathering are stored, and in time the overall nutrient reserves recover.

When larger plots are abandoned, regeneration may be so slow that the bare surface becomes eroded and fertility is reduced even further. If this happens, the opportunity for revegetation may be lost and a spiral of decline may commence (Figure 31.4).

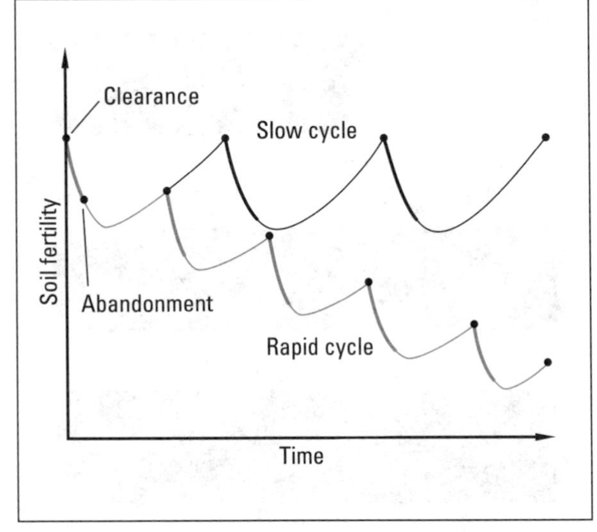


FIGURE 31.16 ▼ The effects of shifting cultivation on soil fertility. Heavy lines represent clearance-exploitation phases; light lines represent abandonment-recovery phases. Note that the soil fertility continues to fall for a short time after abandonment until the vegetation cover is reestablished

The character of these ecosystems therefore depends upon the subtle balances between nutrient removal and soil erosion during periods of cultivation, and the rates of recovery during the rest periods. Nutrient reserves of the system may be maintained over time, with marked short-term fluctuations, as long as the farmers do not return to the area too soon. If the rotation is short, however, in comparison to the rate of recovery, a gradual reduction of the nutrient reserves may occur (Figure 31.16). Moreover, if the slash-and-burn process is too widespread and effective, rapid declines due to erosion and leaching may prevent regeneration.

# Managed forest ecosystems

## GENERAL NATURE OF MANAGED FORESTS

Increasingly, forests are being harvested as a resource in their own right. Timber products include paper, building materials, fuel, and chemicals. In 1962, J. D. Ovington estimated that 42 percent of the world's harvested woodlands were used for fuel, 37 percent for building, 11 percent for pulp and paper, and 4 percent for pit props in mining.

On a global basis, timber products come from a wide range of forests managed at very different levels of intensity. At one extreme, natural and seminatural forests are harvested and left to regenerate naturally. At the other extreme, forests are planted specifically for timber production, creating monocultures that are, in some ways, similar to the ecosystems of cereal crops. These stands may be fertilized and managed until they are mature, and then harvested by clearcutting.

Management procedures in these stands include site preparation, sowing, thinning, and felling. Site preparation involves clearing the original vegetation or the previous forest crop. Large ploughs and tractors are often used for this, and disruption of the soil can be considerable. Areas may also be drained, and often the soil is ridged to allow planting on the raised ground between the wetter furrows. Young trees are sown close together to give mutual support and shelter, and sometimes fertilizers are added. As the trees grow, they start to compete for light and nutrients, so the stand is thinned out to reduce competition and encourage growth of the better trees. Wind-throw may be a problem at this stage. As the tree canopy is opened, wind may create greater turbulence at the treetops or be channelled through open areas, causing considerable pressures during gales. The roots of these trees are often relatively shallow, and are unable to withstand such pressures.

When the trees reach maturity, the stand is **clearcut**, and the procedure starts again. The age at which harvesting takes place depends on the tree species, its proposed use, and the environmental conditions. For example, Sitka spruce (*Picea sitchensis*) trees are felled about every 50 years in Britain and every 30 to 35 years in New Zealand, but in Canada, spruce trees may take 100 years or more to reach maturity. In Europe, most woodlots are intensively managed, but in Canada much of the logging is conducted in virgin forest. There are concerns that the rate of cutting is outpacing the rate of reforestation, and that the present harvest of sec-

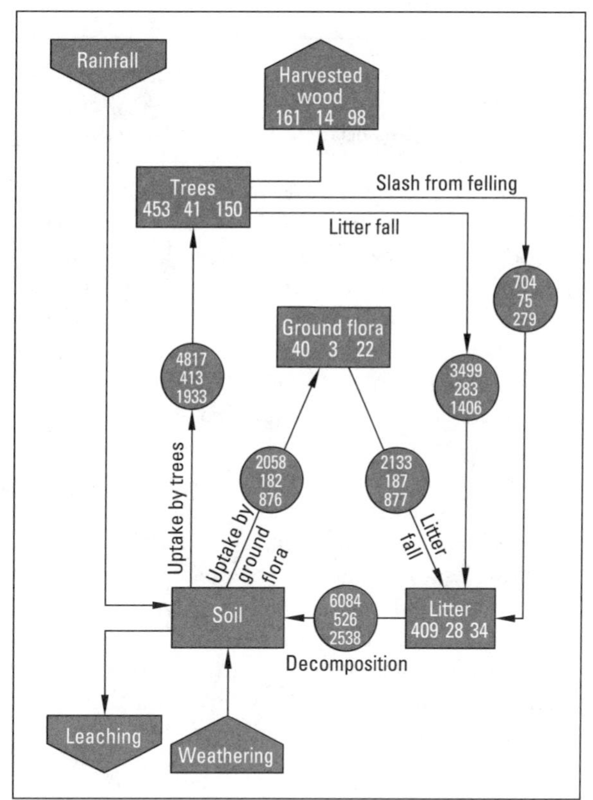


FIGURE 31.17 ▼ The nutrient cycle of a 55-year-old managed pine forest (*Pinus sylvestris*). Figures in circles show flows of nitrogen, phosphorus, and potassium totalled over the 55 years of growth; figures in boxes show the amounts retained at the end of this period. Note that inputs by rainfall and weathering, losses by leaching, and nutrients retained in the soil have not been measured (from Ovington, 1962)

## Table 31.5

*An estimate of the number of years required to replace, through input, nutrients lost in a single crop removal (left) and single crop removal plus 2 years of leaching (right)*

| | *Single crop removal* | | | | | *Crop removal plus leaching* | |
|---|---|---|---|---|---|---|---|
| | | | *Total* | | | *Total* | |
| | *N* | *P* | *K* | *Ca* | *Mg* | *N* | *Ca* |
| Red spruce—fresh till | 23 | 23 | 45 | 28 | 20 | 31 | 34 |
| Black spruce—peat | 22 | 18 | 31 | 22 | 20 | 28 | 27 |
| Black spruce—outwash sand | 24 | 21 | 42 | 22 | 22 | 31 | 27 |
| Mixed wood—fresh till | 20 | 16 | 19 | 17 | 14 | 28 | 21 |
| Mixed wood—silt, fine sand | 19 | 21 | 23 | 20 | 16 | 25 | 23 |

*Data* A.G. Gordon, 'Nutrient cycling dynamics in differing spruce and mixed wood ecosystems in Ontario and the effects of nutrient removals through harvesting.' In R.W. Wein, R.R. Riewe, and I.R. Methven (eds.), *Resources and Dynamics of the Boreal Zone* (1983), Association of Canadian Universities for Northern Studies.

ondary growth cannot be maintained without depleting the resource. In this case, the **rotation period**, the length of time between harvestings, is too short to permit regrowth of the forest resources on a **sustained yield** basis.

In many parts of the world, forests are not intensively managed, partly because much larger areas of forest are available to be exploited. Felling is more selective, and regeneration either occurs naturally, or through the planting of small, irregular plots of trees. In these cases, the trees in a stand are not of a single age, and reach maturity at different times. Nevertheless, clearcutting has occurred in many such forests, and the lack of attention to reforestation has resulted a dramatic reduction in the world's forest reserves in recent years. This has occurred in many tropical forests, but it is also a feature of woodlands in Canada and the United States.

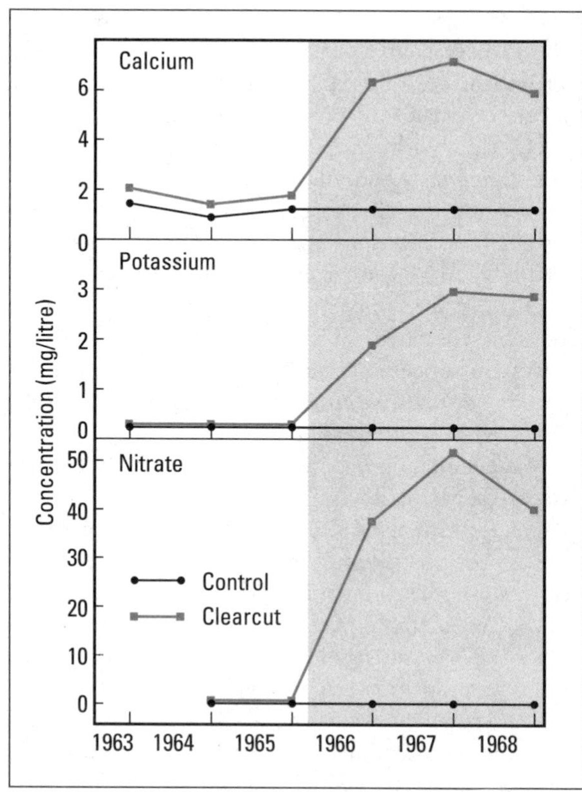


FIGURE 31.18 ▼ Nurient losses from cleared and undisturbed forests at Hubbard Brook. The shaded area represents the period when the cleared forest was devegetated (from Bormann and Likens, 1979)

## NUTRIENT CYCLING IN MANAGED FORESTS

Prior to logging, nutrient cycles in managed forests are similar to those of natural forest ecosystems. Under **traditional logging** practices, where only the tree boles are removed, nutrients are lost in the tree trunks, but the nutrient-rich leaves and branches are normally left on the ground (Figure 31.17). If these are burned, there are significant losses in smoke and by leaching, but if they are allowed to rot in situ, decomposition is slow, and nutrients are released gradually with little loss to leaching.

Modern practices include **full-tree logging,** removal of all above-ground tree biomass, and substantial lengths of time are required to replace the nutrients that are removed from the site (Table 31.5). Further nutrient losses occur with **whole-tree logging,** which rips trees out of the soil with the roots attached. Newer fibre-harvesting techniques may remove all plants in **biomass harvesting**. Biomass harvesting is advocated by consumers wanting to convert plant matter to energy (motor fuel), or to convert forest ecosystems into tree farms, where potential plant competitors are excluded and intensive management tools (e.g., fertilizers, selective herbicides) are used.

Some idea of the effects of biomass harvesting on nutrient cycles was shown by Likens and Bormann (1972) in the United States. In one experiment in Hubbard Brook, they clearcut an experimental catchment and prevented regeneration by applying herbicides. They then compared nutrient losses from the cleared catchment with those from a nearby forested catchment. The results showed increased losses by leaching following clearcutting (Figure 31.18).

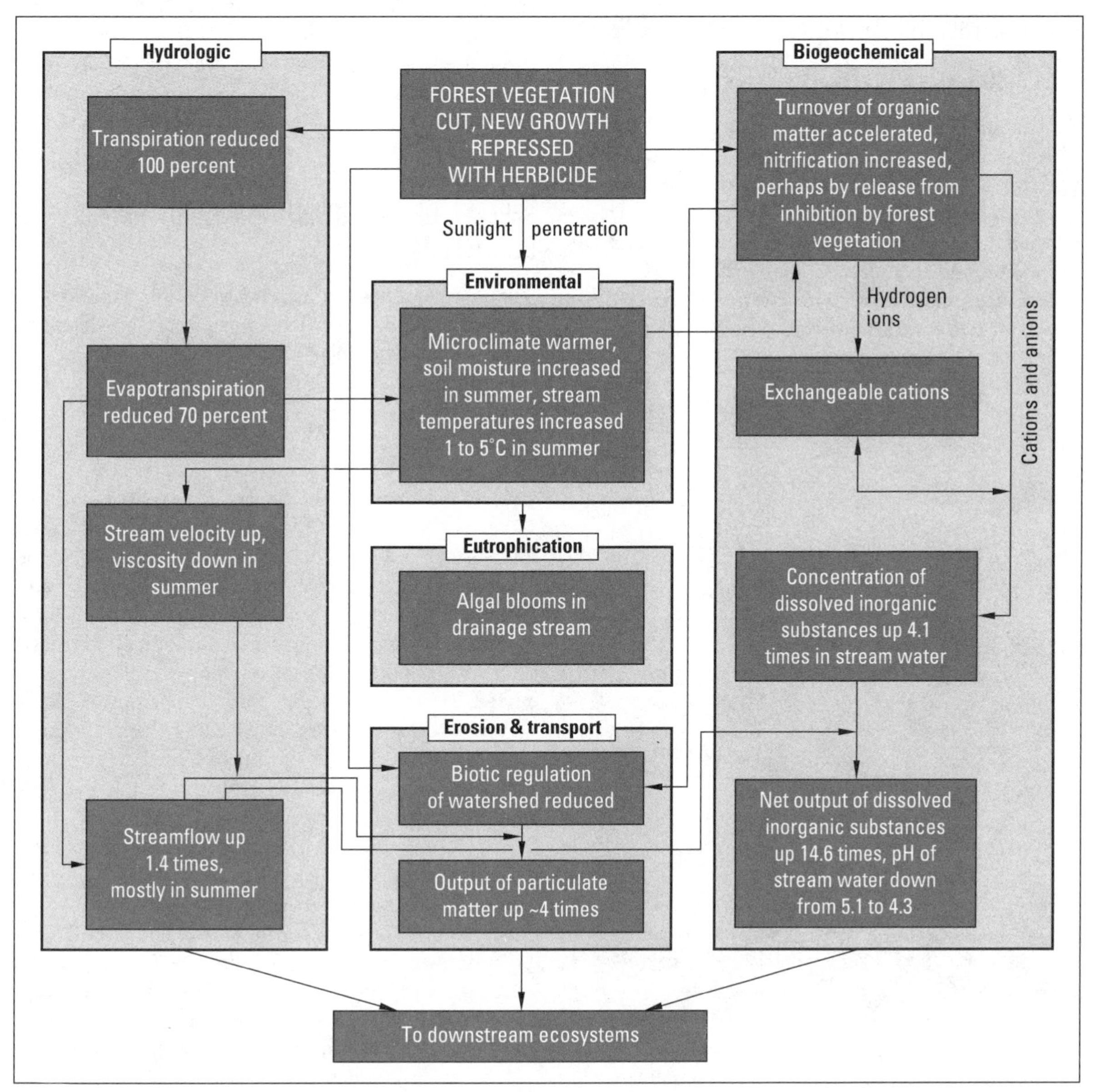


FIGURE 31.19 ▼ The effects of forest clearance on the ecosystem at Hubbard Brook (from Bormann and Likens, 1979)

These losses are a product of several related processes. Decomposition of the organic debris left after clearcutting released nutrients into the soil. Removal of the trees resulted in less rainfall interception and transpiration, greater inputs of water to the soil, and increased leaching. Erosion also increased, causing sedimentation in the rivers. A complex series of responses was initiated (Figure 31.19) that had implications, not just for the immediate area, but also for nearby water bodies and the reservoirs downstream.

# Human impact on aquatic ecosystems

## POLLUTION OF AQUATIC ECOSYSTEMS

As this last example shows, land use often has significant effects on aquatic ecosystems. This is true of agricultural and silvicultural systems, and also of many industrial activities. Waste products are washed into streams and become concentrated in lake water and in aquatic organisms.

Pollutants in these ecosystems are varied. In much of the western world, pesticides related to agriculture are important pollutants. Nitrogen and phosphorus from fertilizers are also significant. However, a wide range of industrial materials, including heavy metals and radioactive substances, may also accumulate in aquatic systems.

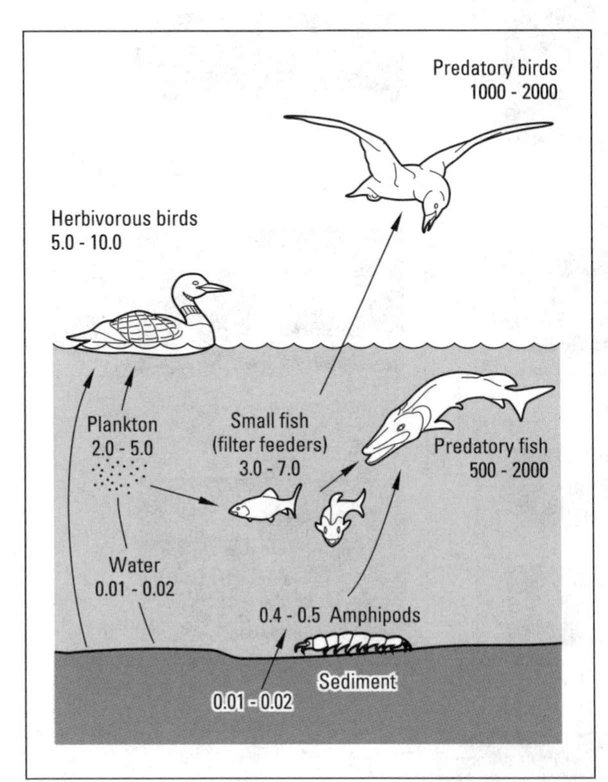


FIGURE 31.20 ▼ DDT concentation (parts per million) along a food chain: Clear Lake, California

The impact of these pollutants may be magnified by the distinctive structure and operation of aquatic ecosystems. When pollutants accumulate in the basal muds of rivers or lakes, bottom-dwelling plants may absorb large quantities of these compounds. Fish pass large quantities of water through their bodies, and in the process, may absorb many compounds. Also, predators need to eat large numbers of prey and at each step up the food chain, considerable concentration of pollutants can occur (Figure 31.20). The highest concentrations are usually found in seabirds that feed on predatory aquatic animals.

## EFFECTS OF POLLUTANTS ON LACUSTRINE ECOSYSTEMS

The effects of pollutants tend to be most marked in lacustrine environments, for these act as sinks for many of the substances carried by rivers. The lack of mixing and of dilution result in high concentrations of pollutants, with serious impacts on the local biota.

One of the most dramatic examples of this process is shown by the Great Lakes, particularly in Lake Erie and Lake Ontario. Industrial pollutants and human sewage from urban complexes along the lakeshore, together with agricultural pollutants and livestock sewage from surrounding farmland, have altered both the chemical and biological composition of the lakes. Concentrations of calcium, sulphate, chloride, sodium, and potassium all have increased markedly (Figure 31.21). In Lake Ontario, most of these changes are related to the growth of cities such as Toronto, Hamilton, and Rochester, and to industrialization along the upper Niagara River in the Buffalo area. Developments around Detroit also affected Lake Erie.

Chemical changes in these lakes have led to significant modifications in the fauna. The populations of important fish species in Lake Erie (e.g., lake herring, walleye, blue pike, and whitefish) have been drastically reduced. For example, commercial catches of lake herring were over 20 000 t $y^{-1}$ prior to 1925. By 1925, this had fallen to 7600 t, and in 1962 the catch was only 32 t. Catches of blue pike

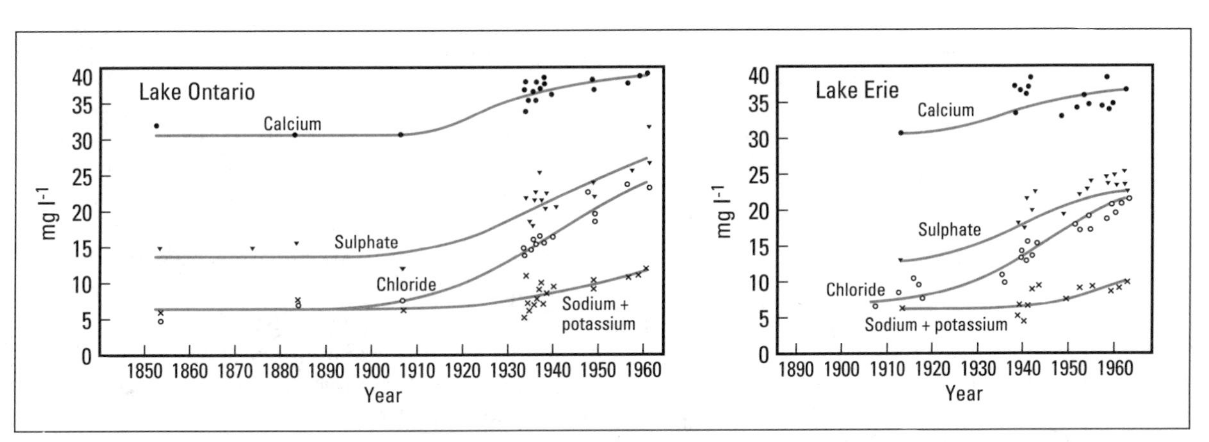


FIGURE 31.21 ▼ Changes in concentration of selected nutrient elements in Lake Ontario and Lake Erie (after Beeton, 1971)

showed a similar trend. Early in the century they averaged about 6800 t $y^{-1}$, but by 1962 they were only 0.45 t. Not all species have declined. Some, such as the sea lamprey, have found conditions more favourable and have increased in number. Lack of competition for food and breeding grounds and increased supplies of nutrients in the water have contributed to these changes.

The effects of chemical pollution in these lacustrine ecosystems also have grown over time. Nitrogen levels in the River Stour in England indicate a gradual rise in pollution levels (Figure 31.22), and when such waters enter lake basins, the effect is magnified. The Lake of the Woods, a popular tourist and cottage area in western Ontario, has also been affected by pollution. Phosphates from fertilizers and household wastes are a major problem, leading to increased eutrophication of the water and the destruction of wildlife.

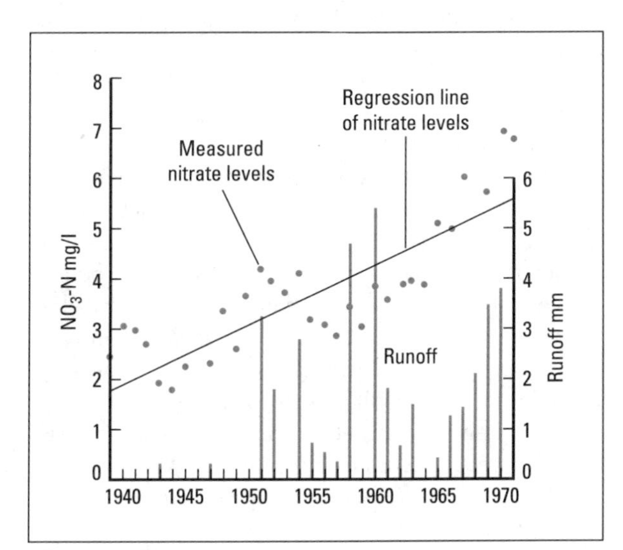


FIGURE 31.22 ▼ Nitrate levels in the River Stour, Essex, England, 1939–71 (after Edwards and Thornes, 1973)

## ACID RAIN

Recently, the effects of another form of pollution have been recognized: acid rain. Pollutants given off by the combustion of fossil fuels and by various industrial processes break down in the air and combine with water in the atmosphere to produce acids. For example, sulphur dioxide ($SO_2$), is released in large quantities from power stations, heavy industry, and vehicles. New York alone produces an estimated 2 million tonnes of $SO_2$ each year, and since 1950, emissions in Europe have risen rapidly (Figure 31.23).

In the atmosphere, $SO_2$ is gradually broken down, and over a period of about 43 days is converted to $SO_3$. During this time, it may be carried over considerable distances, especially under turbulent conditions, and consequently the pollutants may be deposited hundreds or even several thousands of kilometres from their source. For example, Scotland suffers from acid rain derived from pollutants produced in the USA, while $SO_2$ produced in Britain is said to affect large areas of Scandinavia and Germany.

Deposition occurs when the $SO_3$ is dissolved in rainwater to form a dilute solution of sulphuric acid ($H_2SO_4$). The effects of this acid are wide-ranging. It can accumulate on plant leaves and inhibit photosynthesis, and it can also accumulate in water and in soil, reducing the pH. In Germany, acid rain appears to be inhibiting the regeneration of coniferous forest, by lowering the soil pH and encouraging

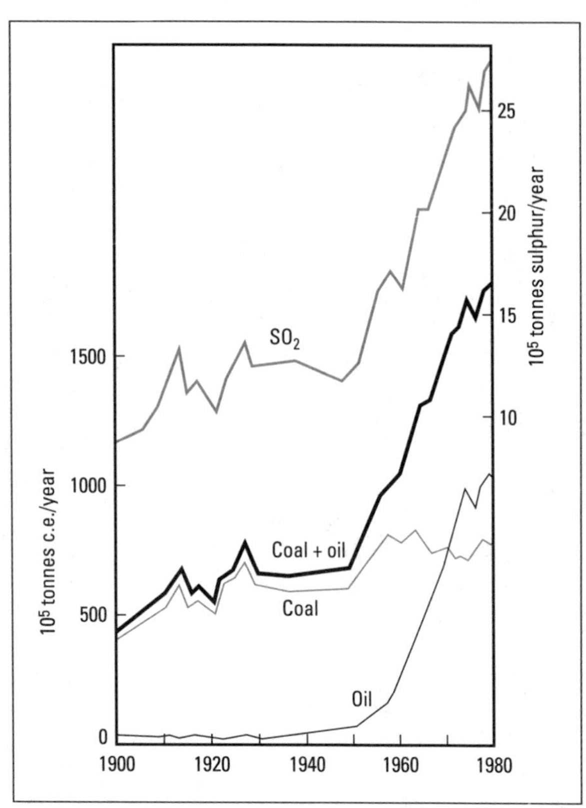


FIGURE 31.23 ▼ The consumption of coal and oil, and estimated emissions of sulphur dioxide in Europe, 1900–80 (from Semb, 1978)

the leaching of soil nutrients such as calcium and aluminum. In Scandinavia, many streams and lakes have been affected. A survey of 21 lakes in Sweden between 1966 and 1979 showed that all had experienced declines in pH of between 0.9 and 2.2 units, while sulphate levels had increased by 6 to 13 mg $l^{-1}$. Increased acidity also helps to mobilize iron, aluminum, and heavy metals such as cadmium, zinc, and copper, and these can then accumulate in waters to toxic levels, seriously affecting fish populations. One study in southern Norway suggested that 1771 lakes had lost all fish as a result of acid rain, while a further 941 had suffered a major decline in fish populations.

Scientists still are not certain that all of these effects are due solely to acid rain. Other factors, such as the natural aging of lakes, soil acidification by fertilizers and conifers, and marginal climatic changes, could also be significant factors. Nonetheless, the problem is serious enough to be a major political and research topic, and much public money is being devoted to investigating the causes and effects of acid rain.

# Conclusion

Humans are, of course, a part of the earth's ecosystems. We also tend to be much more adept and omniscient members of these ecosystems than any other species. Consequently, we try to modify natural processes, and manipulate ecosystems for our own benefit. In the process, we deliberately create and maintain artificial ecosystems. We may also inadvertently affect many natural ecosystems. These effects can have serious repercussions for the ecosystems concerned, and for ourselves. For example, forestry operations in the Maritime provinces of Canada depend heavily on red spruce (*Picea rubens*) and white spruce (*Picea glauca*) trees. Reforestation has replaced large tracts of clearcut forests with monocultures of these species. The spruce budworm (*Choristoneura fumiferana*) thrives there, as its preferred host, the spruce tree, occurs in extensive pure stands. Before widespread logging began, the forest was composed of a mosaic of stands of trees that grew to maturity after natural fires; stands were thus of varied ages and different species. However, as clearcutting occurred, this diverse forest was replaced by stands with few tree species. Under these artificial conditions, the spruce budworms reached large numbers and migrated rapidly through these controlled forests. Pesticides were then used to control spruce budworm outbreaks, but the application of pesticides was implicated in health problems in human populations. The manipulation of these forest ecosystems has produced an artificial system that will collapse without substantial efforts to maintain it, and human health is also being affected.

Unfortunately, the effects of human activity on ecosystems are not always clear, nor are they quickly evident. Consequences may be gradual and incremental. An inappropriate response to the first, subtle consequences may ultimately increase the problem. For example, farmers with soil exhaustion and low yields may try to counteract these effects by intensifying their agriculture practices, and this could lead to even more rapid soil exhaustion.

Ecosystems throughout the world have experienced impacts from human activity. Indeed, in many areas it is inappropriate to talk of natural ecosystems. Clearly, our ability to understand and manage these ecosystems effectively is critical to the survival of many species on this planet, including *Homo sapiens*.

# Appendix I

## Psychrometric Table

*Depression (Dry Bulb T°C – Wet Bulb T°C)*

| *Air Temperature (°C)* | | *0.4* | *0.8* | *1.2* | *1.6* | *2.0* | *2.4* | *2.8* | *3.2* | *3.6* | *4.0* | *4.4* | *4.8* |
|---|---|---|---|---|---|---|---|---|---|---|---|---|---|
| –6 | DP | –8.0 | –9.5 | –11.0 | –13.0 | –14.8 | –17.0 | –19.7 | –23.0 | –28.0 | –35.0 | – | – |
| | RH | 85% | 76% | 67% | 59% | 50% | 41% | 33% | 24% | 16% | 8% | – | – |
| –4 | DP | –5.7 | –7.0 | –8.3 | –10.0 | –11.5 | –13.4 | –15.5 | –18.0 | –21.0 | –25.0 | –31.0 | –43.0 |
| | RH | 88% | 80% | 72% | 64% | 56% | 48% | 40% | 33% | 25% | 18% | 10% | 3% |
| –2 | DP | –3.3 | –4.5 | –5.7 | –7.0 | –8.5 | –10.0 | –11.8 | –13.8 | –16.0 | –18.8 | –22.0 | –27.0 |
| | RH | 90% | 83% | 76% | 68% | 61% | 54% | 47% | 40% | 33% | 26% | 20% | 13% |
| 0 | DP | –1.0 | –2.0 | –3.2 | –4.3 | –5.6 | –7.0 | –8.5 | –10.0 | –12.0 | –14.0 | –16.4 | –19.0 |
| | RH | 93% | 86% | 79% | 73% | 66% | 59% | 53% | 47% | 40% | 34% | 28% | 22% |
| 2 | DP | 1.1 | 0.1 | –1.0 | –2.1 | –2.8 | –4.0 | –5.3 | –6.7 | –8.3 | –9.9 | –11.8 | –14.0 |
| | RH | 93% | 87% | 81% | 74% | 70% | 64% | 58% | 52% | 47% | 41% | 35% | 30% |
| 4 | DP | 3.1 | 2.2 | 1.3 | 0.3 | –0.8 | –1.9 | –3.2 | –4.5 | –6.0 | –6.3 | –8.0 | –9.6 |
| | RH | 94% | 88% | 82% | 77% | 71% | 65% | 60% | 54% | 49% | 47% | 42% | 36% |
| 6 | RP | 5.2 | 4.3 | 3.4 | 2.5 | 1.5 | 0.5 | –0.6 | –1.7 | –3.0 | –4.3 | –5.7 | –7.3 |
| | RH | 94% | 89% | 84% | 78% | 73% | 68% | 63% | 58% | 53% | 48% | 43% | 38% |
| 8 | RP | 7.2 | 6.4 | 5.6 | 4.8 | 3.9 | 2.9 | 1.9 | 0.9 | –0.2 | –1.4 | –2.6 | –3.9 |
| | RH | 95% | 90% | 85% | 80% | 75% | 70% | 65% | 61% | 56% | 52% | 47% | 43% |
| 10 | DP | 9.3 | 8.5 | 7.7 | 6.9 | 6.1 | 5.3 | 4.4 | 3.4 | 2.4 | 1.4 | 0.3 | –0.9 |
| | RH | 95% | 91% | 86% | 81% | 77% | 72% | 68% | 64% | 59% | 55% | 51% | 47% |
| 12 | RP | 11.3 | 10.6 | 9.9 | 9.1 | 8.3 | 7.5 | 6.7 | 5.8 | 4.9 | 4.0 | 3.0 | 2.0 |
| | RH | 96% | 91% | 87% | 83% | 78% | 74% | 70% | 66% | 62% | 58% | 54% | 50% |
| 14 | DP | 13.3 | 12.7 | 12.0 | 11.3 | 10.5 | 9.8 | 9.0 | 8.2 | 7.4 | 6.5 | 5.6 | 4.7 |
| | RH | 96% | 92% | 88% | 84% | 80% | 76% | 72% | 68% | 64% | 61% | 57% | 53% |
| 16 | DP | 15.4 | 14.7 | 14.1 | 13.4 | 12.7 | 12.0 | 11.3 | 10.5 | 9.7 | 8.9 | 8.1 | 7.2 |
| | RH | 96% | 92% | 88% | 85% | 81% | 77% | 73% | 70% | 66% | 63% | 59% | 56% |
| 18 | DP | 17.5 | 16.9 | 16.3 | 15.6 | 14.9 | 14.2 | 13.5 | 12.8 | 12.1 | 11.3 | 10.5 | 9.7 |
| | RH | 96% | 93% | 89% | 85% | 82% | 78% | 75% | 71% | 68% | 65% | 62% | 58% |
| 20 | DP | 19.5 | 19.0 | 18.5 | 17.9 | 17.3 | 16.7 | 16.0 | 15.2 | 14.5 | 13.7 | 13.0 | 12.2 |
| | RH | 96% | 93% | 89% | 86% | 83% | 79% | 76% | 73% | 70% | 67% | 64% | 61% |
| 22 | DP | 21.6 | 21.1 | 20.7 | 20.2 | 20.0 | 19.4 | 18.8 | 18.0 | 17.3 | 16.5 | 15.7 | 14.9 |
| | RH | 97% | 93% | 90% | 87% | 85% | 80% | 77% | 74% | 71% | 68% | 65% | 62% |
| 24 | DP | 23.6 | 23.2 | 22.8 | 22.3 | 21.9 | 21.5 | 21.0 | 20.5 | 20.0 | 19.4 | 18.8 | 18.1 |
| | RH | 97% | 94% | 90% | 87% | 84% | 81% | 78% | 75% | 72% | 70% | 67% | 64% |
| 26 | DP | 25.6 | 25.2 | 24.8 | 24.4 | 24.0 | 23.6 | 23.1 | 22.7 | 22.3 | 21.8 | 21.3 | 20.8 |
| | RH | 97% | 94% | 91% | 88% | 85% | 82% | 79% | 76% | 74% | 71% | 68% | 66% |
| 28 | DP | 27.6 | 27.0 | 26.8 | 26.4 | 26.0 | 25.6 | 25.2 | 24.8 | 24.4 | 23.9 | 23.5 | 23.1 |
| | RH | 97% | 97% | 91% | 88% | 85% | 83% | 80% | 77% | 75% | 72% | 69% | 67% |
| 30 | DP | 29.6 | 29.2 | 28.8 | 28.4 | 28.0 | 27.6 | 27.2 | 26.8 | 26.4 | 26.0 | 25.6 | 25.2 |
| | RH | 97% | 94% | 91% | 89% | 86% | 83% | 81% | 78% | 76% | 73% | 71% | 68% |
| 32 | DP | 31.6 | 31.2 | 30.8 | 30.4 | 30.0 | 29.6 | 29.2 | 28.8 | 28.4 | 28.0 | 27.6 | 27.2 |
| | RH | 97% | 94% | 92% | 89% | 86% | 84% | 81% | 79% | 76% | 74% | 72% | 69% |
| 34 | DP | 33.6 | 33.2 | 32.8 | 32.4 | 32.0 | 31.6 | 31.2 | 30.8 | 30.4 | 30.0 | 29.6 | 29.2 |
| | RH | 97% | 95% | 92% | 89% | 87% | 84% | 82% | 80% | 77% | 75% | 73% | 70% |
| 36 | DP | 35.6 | 35.0 | 34.8 | 34.4 | 34.0 | 33.6 | 33.2 | 32.8 | 32.4 | 32.0 | 31.6 | 31.2 |
| | RH | 97% | 95% | 92% | 90% | 87% | 85% | 83% | 80% | 78% | 76% | 73% | 71% |

| *Air Temperature (°C)* | | *5.2* | *5.6* | *6.0* | *6.4* | *6.8* | *7.2* | *7.6* | *8.0* | *8.4* | *8.8* | *9.2* | *9.6* |
|---|---|---|---|---|---|---|---|---|---|---|---|---|---|
| –6 | DP | – | – | – | – | – | – | – | – | – | – | – | – |
| | RH | – | – | – | – | – | – | – | – | – | – | – | – |
| –4 | DP | – | – | – | – | – | – | – | – | – | – | – | – |
| | RH | – | – | – | – | – | – | – | – | – | – | – | – |
| –2 | DP | –34.0 | – | – | – | – | – | – | – | – | – | – | – |
| | RH | 6% | – | – | – | – | – | – | – | – | – | – | – |
| 0 | DP | –23.0 | –28.0 | – | – | – | – | – | – | – | – | – | – |
| | RH | 16% | 10% | – | – | – | – | – | – | – | – | – | – |
| 2 | DP | –16.4 | –19.5 | –23.4 | –29.0 | –40.0 | – | – | – | – | – | – | – |
| | RH | 24% | 19% | 13% | 8% | 3% | – | – | – | – | – | – | – |
| 4 | DP | –11.5 | –13.6 | –16.2 | –19.3 | –23.3 | –29.0 | –41.0 | – | – | – | – | – |
| | RH | 31% | 26% | 21% | 16% | 12% | 7% | 2% | – | – | – | – | – |
| 6 | RP | –9.0 | –11.0 | –12.0 | –13.0 | –15.6 | –18.7 | –23.0 | –33.0 | – | – | – | – |
| | RH | 33% | 28% | 26% | 24% | 19% | 15% | 11% | 4% | – | – | – | – |
| 8 | RP | –5.4 | –7.0 | –8.8 | –10.7 | –13.0 | –16.0 | –18.0 | –19.0 | –21.6 | –27.0 | –37.0 | – |
| | RH | 38% | 34% | 31% | 27% | 21% | 17% | 15% | 13% | 10% | 6% | 2% | – |
| 10 | DP | –2.1 | –3.5 | –4.9 | –6.5 | –8.3 | –10.3 | –12.0 | –15.3 | –19.0 | –23.0 | –30.0 | – |
| | RH | 43% | 38% | 34% | 31% | 27% | 23% | 19% | 15% | 11% | 8% | 4% | – |
| 12 | RP | 0.4 | –0.3 | –1.6 | –2.9 | –4.4 | –5.9 | –7.7 | –9.7 | –12.0 | –14.7 | –18.0 | –22.0 |
| | RH | 46% | 43% | 39% | 35% | 32% | 28% | 24% | 21% | 17% | 14% | 11% | 7% |
| 14 | DP | 3.7 | 2.6 | 1.5 | 0.4 | –0.9 | –2.2 | –3.6 | –5.2 | –7.0 | –9.0 | –11.0 | –14.0 |
| | RH | 50% | 46% | 43% | 39% | 36% | 33% | 29% | 26% | 23% | 20% | 16% | 13% |
| 16 | DP | 6.3 | 5.4 | 4.4 | 3.4 | 2.3 | 1.2 | –0.1 | –1.4 | –3.0 | –4.3 | –6.0 | –8.0 |
| | RH | 53% | 49% | 46% | 43% | 40% | 37% | 33% | 30% | 27% | 24% | 21% | 19% |
| 18 | DP | 8.9 | 8.0 | 7.2 | 6.2 | 5.3 | 4.2 | 3.2 | 2.0 | 0.8 | –0.4 | –1.8 | –3.3 |
| | RH | 55% | 52% | 49% | 46% | 43% | 40% | 37% | 34% | 31% | 29% | 26% | 23% |
| 20 | DP | 11.4 | 10.6 | 9.8 | 8.9 | 8.0 | 7.1 | 6.2 | 5.2 | 4.1 | 3.0 | 1.8 | 0.6 |
| | RH | 58% | 55% | 52% | 49% | 46% | 43% | 40% | 38% | 35% | 32% | 30% | 27% |
| 22 | DP | 14.1 | 13.2 | 12.4 | 11.6 | 10.7 | 9.9 | 9.0 | 8.1 | 7.1 | 6.2 | 5.1 | 4.0 |
| | RH | 60% | 57% | 54% | 51% | 49% | 46% | 43% | 41% | 38% | 36% | 33% | 31% |
| 24 | DP | 17.3 | 16.5 | 15.6 | 14.6 | 13.7 | 12.7 | 11.8 | 11.0 | 10.0 | 9.1 | 8.2 | 7.2 |
| | RH | 61% | 59% | 56% | 53% | 51% | 48% | 46% | 43% | 41% | 39% | 36% | 34% |
| 26 | DP | 20.3 | 19.7 | 19.0 | 18.3 | 17.4 | 16.5 | 15.5 | 14.4 | 13.3 | 12.3 | 11.3 | 10.3 |
| | RH | 63% | 60% | 58% | 55% | 53% | 51% | 48% | 46% | 44% | 41% | 39% | 37% |
| 28 | DP | 22.6 | 22.2 | 21.7 | 21.2 | 20.6 | 20.0 | 19.4 | 18.6 | 17.7 | 16.7 | 15.6 | 14.4 |
| | RH | 64% | 62% | 60% | 57% | 55% | 53% | 51% | 48% | 46% | 44% | 42% | 39% |
| 30 | DP | 24.8 | 24.3 | 23.9 | 23.5 | 23.0 | 22.6 | 22.1 | 21.6 | 21.0 | 20.4 | 19.7 | 19.0 |
| | RH | 66% | 63% | 61% | 59% | 57% | 54% | 52% | 50% | 48% | 46% | 44% | 42% |
| 32 | DP | 26.8 | 26.4 | 26.0 | 25.6 | 25.2 | 24.7 | 24.3 | 23.9 | 23.4 | 23.0 | 22.5 | 22.0 |
| | RH | 67% | 65% | 62% | 60% | 58% | 56% | 54% | 52% | 50% | 48% | 46% | 44% |
| 34 | DP | 28.8 | 28.4 | 28.0 | 27.6 | 27.2 | 26.8 | 26.4 | 26.0 | 25.6 | 25.1 | 24.7 | 24.0 |
| | RH | 68% | 66% | 64% | 62% | 59% | 57% | 55% | 53% | 51% | 49% | 48% | 46% |
| 36 | DP | 30.8 | 30.4 | 30.0 | 29.6 | 29.2 | 28.8 | 28.4 | 28.0 | 27.6 | 27.2 | 27.0 | 26.4 |
| | RH | 69% | 67% | 65% | 63% | 61% | 59% | 57% | 55% | 53% | 51% | 49% | 47% |

# Appendix II

## KÖPPEN-GEIGER SYSTEM OF CLIMATIC CLASSIFICATION

Köppen's attempts at classifying climates into recognizable groups began in 1884, but his first classification system to achieve widespread use was not published until 1918. Several modifications were made to the 1918 system by Köppen and subsequently by his co-workers Geiger and Pohl. This has led to some confusion, as climatologists have used each modification, claiming them to be based on Köppen, without always specifying which classification. The version presented here was published in 1953 following revision by Geiger and Pohl.

The five major climate groups are identified by capital letters, as follows:

A Tropical rainy climates. The average monthly temperature always exceeds 18°C. There is no winter season. Annual rainfall is large, exceeding annual evaporation to give a water surplus.

B Dry climates. On average, potential evaporation always exceeds precipitation throughout the year, hence there is a net water deficit.

C Warm temperate climates. The mean temperature of the coldest month lies between 18°C and –3°C. At least one month has a mean temperature above 10°C. Seasonal differences from winter to summer are clear.

D Cold boreal forest climates. The mean temperature of the coldest month is below –3°C, but the mean temperature of the warmest month is above 10°C. This isotherm was used as it appears to coincide with the poleward limit of forest growth.

E Polar climates. Mean temperature of the warmest month is below 10°C, so even summers are cool.

Four of the groups, A, C, D, and E, are defined in terms of temperature; the fifth, B, is defined by the deficiency of precipitation in relation to potential evaporation.

Subgroups are identified by a second letter:

f Moist climates have adequate precipitation throughout the year for vegetative growth; used in types A, C, and D.

w Winter is the main dry season.

s Summer is the main dry season.

m Used only with A types, indicates rain forest climate with a brief dry season in a monsoon-type precipitation cycle.

Two capital letters, S and W, refer only to B climates:

BS Steppe or semi-arid climate, with mean annual precipitation between 380 mm and 760 mm, depending upon mean annual temperature.

BW Desert or arid climate. Mean annual precipitation is less than 250 mm.

E climates are separated:

ET Tundra climate, where mean temperature of the warmest month is between 0°C and 10°C.

EF Ice climate, where mean monthly temperatures are all below 0°C. Unless these areas are arid, they are normally ice-covered.

To denote further details of climatic characteristics, Köppen added a third letter. It is used to identify particular temperature variations:

a Hot summers, in which the warmest month has a mean temperature above 22°C. Used in C and D climates.

b Warm summers, in which the warmest month has a mean temperature below 22°C. Used in C and D climates.

c Cool, short summers, with fewer than four months having a mean temperature above 10°C. Used in C and D climates.

d Very cold winters, in which mean temperature of the coldest month is below –38°C. Used in D climate only.

h Dry and hot, where mean annual temperature is above 18°C. Used in B climate only.

k Dry and cool, where mean annual temperature is below 18°C. Used in B climate only.

H Highlands

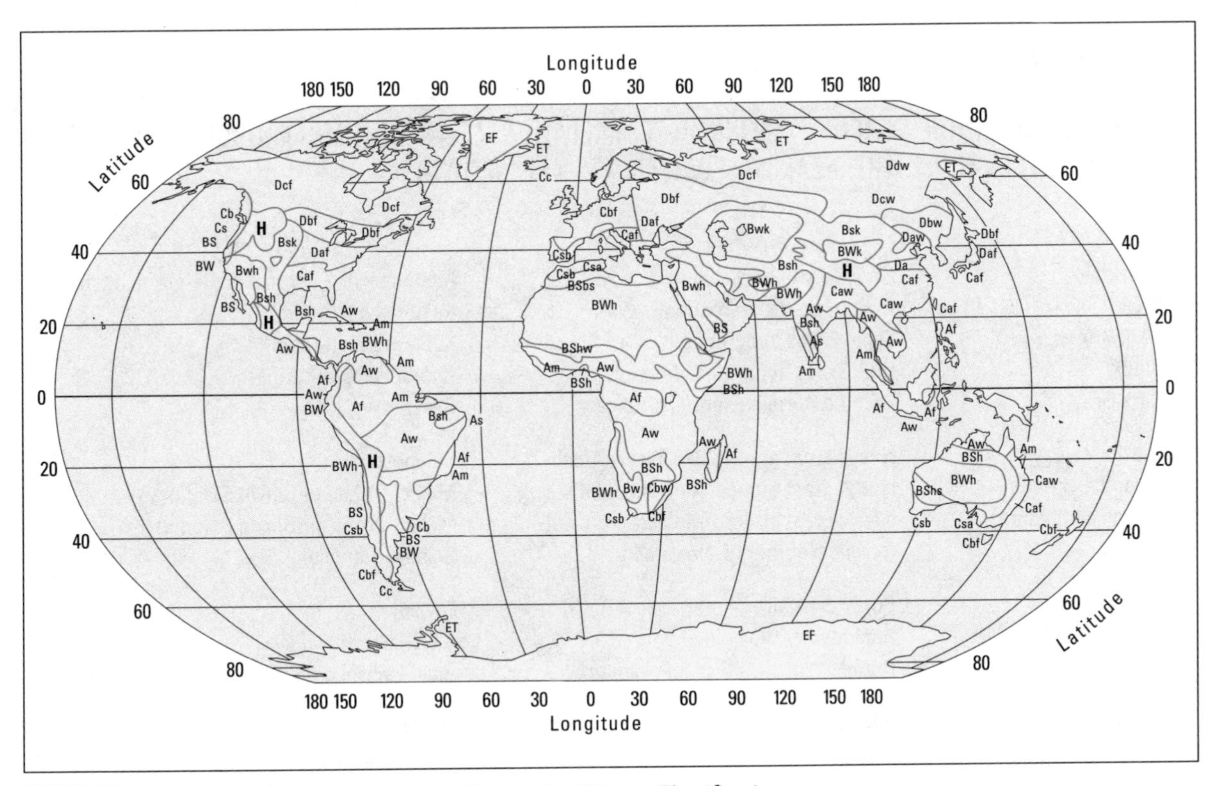


FIGURE A2 ▼ Climates of the continents according to the Köppen Classification

# Appendix III

## I. SOIL CLASSIFICATION SYSTEMS. TAXONOMIC CORRELATION AT THE CANADIAN ORDER AND GREAT GROUP LEVELS (ONLY THE NEAREST EQUIVALENTS ARE SHOWN)

| *Canadian* | *US* | *FAO* |
|---|---|---|
| Chernozemic | Boroll, some Vertisols | Kastanozem, Chernozem, Rendzina, Phaeozem |
| Brown | Aridic Boroll subgroups | Kastanozem (aridic) |
| Dark Brown | Typic Boroll subgroups | Kastanozem (aridic) |
| Black | Udic Boroll subgroups, Rendoll | Chernozem, Rendzina |
| Dark Grey | Boralfic Boroll subgroups, Alboll | Greyzem |
| Solonetzic | Natric great groups, Mollisol & Alfisol | Solonetz |
| Solonetz | Natric great groups, Mollisol & Alfisol | Mollic, Orthic, or Gleyic Solonetz |
| Solodized Solonetz | Natric great groups, Mollisol & Alfisol | Mollic, Orthic, or Gleyic Solonetz |
| Solod | Glossic Natriboroll, Natralboll | Solodic Planosol |
| Luvisolic | Boralf & Udalf | Luvisol |
| Grey Brown Luvisol | Hapludalf or Glossudalf | Albic Luvisol |
| Grey Luvisol | Boralf | Albic Luvisol, Podzoluvisol |
| Podzolic | Spodosol, some Inceptisols | Podzol |
| Humic Podzol | Cryaquod, Humod | Humic Podzol |
| Ferro-Humic Podzol | Humic Cryorthod, Humic Haplorthod | Orthic Podzol |
| Humo-Ferric Podzol | Cryorthod, Haplorthod | Orthic Podzol |
| Brunisolic | Inceptisol, some Psamments | Cambisol |
| Melanic Brunisol | Cryochrept, Eutrochrept, Hapludoll | Cambisol, Eutric Cambisol |
| Eutric Brunisol | Cryochrept, Eutrochrept | Eutric Cambisol |
| Sombric Brunisol | Umbric Dystrochrept | Dystric Cambisol |
| Dystric Brunisol | Dystrochrept, Cryochrept | Dystric Cambisol |
| Regosolic | Entisol | Fluvisol, Regosol |
| Regosol | Entisol | Regosol |
| Humic Regosol | Entisol | Fluvisol, Regosol |
| Gleysolic | Aqu-suborders | Gleysol, Planosol |
| Humic Gleysol | Aquoll, Humaquept | Mollic, Humic, Calcaric Gleysol |
| Gleysol | Aquent, Fluvent, Aquept | Eutric, Dystric Gleysol |
| Luvic Gleysol | Argialboll, Argiaquoll, Aqualf | Planosol |
| Organic | Histosol | Histosol |
| Fibrisol | Fibrist | Histosol |
| Mesisol | Hemist | Histosol |
| Humisol | Saprist | Histosol |
| Folisol | Folist | Histosol |
| Cryosolic | Pergelic subgroups | Gelic |
| Turbic Cryosol | Pergelic Ruptic subgroups | Cambisol, Regosol, Fluvisol, etc. |
| Static Cryosol | Pergelic subgroups | |
| Organic Cryosol | Pergelic Histosol or Pergelic Histic subgroups of other orders | Gelic Histosol |

## 2. THE CANADIAN SYSTEM OF SOIL CLASSIFICATION

| Order | Great Group |
|---|---|
| Brunisolic | Melanic Brunisol |
| | Eutric Brunisol |
| | Sombric Brunisol |
| | Dystric Brunisol |
| Chernozemic | Brown |
| | Dark Brown |
| | Black |
| | Dark Grey |
| Cryosolic | Turbic Cryosol |
| | Static Cryosol |
| | Organic Cryosol |
| Gleysolic | Humic Gleysol |
| | Gleysol |
| | Luvic Gleysol |
| Luvisolic | Grey Brown Luvisol |
| | Grey Luvisol |
| Organic | Fibrisol |
| | Mesisol |
| | Humisol |
| | Folisol |
| Podzolic | Humic Podzol |
| | Ferro-Humic Podzol |
| | Humo-Ferric Podzol |
| Regosolic | Regosol |
| | Humic Regosol |
| Solonetzic | Solonetz |
| | Solodized Solonetz |
| | Solod |

## 3. THE CANADIAN SYSTEM OF SOIL CLASSIFICATION CRITERIA USED FOR HORIZON DESIGNATIONS

Mineral horizons and layers are designated by capital letters followed by descriptive lowercase suffixes. Mineral horizons contain 17% or less organic carbon (about 30% organic matter) by weight.

A  A mineral horizon formed at or near the surface in the zone of leaching or eluviation of materials in solution or suspension, or of maximum accumulation or organic matter, or both. The accumulation of organic matter is seen as a darkening of the surface soil (Ah), and conversely the removal of organic matter is usually expressed by a lightening of the soil colour in the upper part of the solum (Ae). The removal of clay from the upper part of the solum (Ae) is expressed by a coarser soil texture relative to the underlying subsoil layers. The removal of iron is indicated by a paler or less red soil colour in the upper solum (Ae) relative to the lower subsoil.

B  A mineral horizon characterized by enrichment in organic matter, sesquioxides, or clay; or by the development of soil structure; or by a change of colour denoting hydrolysis, reduction, or oxidation. The accumulation in B horizons of organic matter (Bh) is shown by dark colours relative to the C horizon. Clay accumulation is indicated by finer soil textures and by clay cutans coating peds and lining pores (Bt). Soil structure in B horizons includes prismatic or columnar units with coatings or stainings and significant amounts of exchangeable sodium (Bn) and other changes of structure (Bm) from the parent material. Colour changes include browning due to oxidation of iron (Bm), and mottling and gleying of structurally altered material associated with periodic reduction (Bg).

C  A mineral horizon comparatively unaffected by the pedogenic processes operative in A and B, (C), except the process of gleying (Cg), and the accumulation of calcium and magnesium carbonates (Cca) and more soluble salts (Cs, Csa). Marl, diatomaceous earth, and rock no harder than 3 on Mohs' scale are considered C horizons.

R  A consolidated bedrock layer that is too hard to break with the hands (>3 on Mohs' scale) or to dig with a spade when moist and does not meet the requirements of a C horizon. The boundary between the R layer and any overlying unconsolidated material is called a lithic contact.

W  This is a layer of water in Gleysolic, Organic, or Cryosolic soils. Hydric layers in Organic soils are a kind of W layer.

**Descriptive suffixes**

b  A buried soil horizon.

c  A cemented (irreversible) pedogenic horizon. Ortstein, placic, and duric horizons of Podzolic soils, and a layer cemented by $CaCO_3$ are examples.

ca  A horizon of secondary carbonate enrichment in which the concentration of lime exceeds that in the unenriched parent material. It is more than 10 cm thick, and its $CaCO_3$ equivalent exceeds that of the parent material by at least 5% if the $CaCO_3$ equivalent is less than 15% (13% vs 8%), or by at least 1/3 if the $CaCO_3$ equivalent of the horizon is 15% or more (28% vs 21%). If no IC is present, this horizon is more than 10 cm thick and contains more than 5% by volume of secondary carbonates in concretions or in soft, powdery forms.

cc  Cemented (irreversible) pedogenic concretions.

e  A horizon characterized by the eluviation of clay, iron, aluminum, or organic matter alone or in combination. When dry, it is usually higher in colour value by one or more units than an underlying B horizon. It is used with A (Ae).

f  A horizon enriched with amorphous material, principally aluminum and iron combined with organic matter. It usually has a hue of 7.5YR or redder, or its hue is 10YR near the upper boundary and becomes yellower with depth. When moist, the chroma is higher than 3 or the value is 3 or less. It contains at least 0.6% pyrophosphate-extractable Al + Fe in textures finer than sand and 0.4% in sands (coarse sand, sand, fine sand, and very fine sand). The ratio of pyrophosphate-extractable Al + Fe to clay (<0.002 mm) is more than 0.05 and organic carbon exceeds 0.5%. Pyrophosphate-extractable iron is at least 0.3%, or the ratio of organic carbon to pyrophosphate-extractable iron is less than 20, or both are true. It is used with B alone (Bf), with B and h (Bhf), with B and g (Bfg), and with other suffixes. These criteria do not apply to Bgf horizons. The following f horizons are differentiated on the basis of the organic carbon content:

- Bf—0.5–5% organic carbon
- Bhf—more than 5% organic carbon

No minimum thickness is specified for a Bf or a Bhf horizon. Thin Bf and Bhf horizons do not qualify as podzolic B horizons. Some Ah and Ap horizons contain sufficient pyrophosphate-extractable Al + Fe to satisfy this criterion of f, but are designated Ah or Ap.

g  A horizon characterized by grey colours or prominent mottling, or both, indicating permanent or periodic intense reduction. Chromas of the matrix are generally 1 or less. It is used with A and e (Aeg); B alone (Bg); B and f (Bfg, Bgf); B, h, and f (Bhfg); B and t (Btg); C alone (Cg); C and k (Ckg); and several others. In some reddish parent materials, matrix colours of reddish hues and high chromas may persist despite long periods of reduction. In these soils, horizons are designated as g if there is grey mottling or marked bleaching on ped faces or along cracks.

- Aeg—This horizon must meet the definitions of A, e, and g.
- Bg—This horizon is analogous to a Bm horizon but has colours indicating poor drainage and periodic reduction. It includes horizons occurring between A and C horizons in which the main features are: (i) Colours of chromas of 1 or less, without mottles on ped surfaces or in the matrix if peds are lacking; or chromas of 2 or less in hues of 10YR or redder, on ped surfaces or in the matrix if peds are lacking, accompanied by more prominent mottles than those in the C horizon; or hues bluer than 10Y, with or without mottles on ped surfaces or in the matrix if peds are lacking. (ii) Colours indicated in (i) and a change in structure from that of the C horizon. (iii) Colours indicated in (i) and illuviation of clay too slight to meet the requirements of Bt, or an accumulation of iron oxide too slight to meet the limits of Bgf. (iv) Colours indicated in (i) and the removal of carbonates. Bg horizons occur in some Orthic Humic Gleysols and some Orthic Gleysols.
- Bfg, Bhfg, and others—When used in any of these combinations, the limits set for f, hf, t, and others must be met.
- Bgf—The dithionite-extractable iron of this horizon exceeds that of the IC by 1% or more. Pyrophosphate-extractable Al + Fe is less than the minimum limit specified for f horizons. This horizon occurs in Fera Gleysols and Fera Humic Gleysols and possibly below the Bfg of gleyed Podzols. It is distinguished from the Bfg of gleyed

Podzols on the basis of the extractability of the iron and aluminum. The iron in the Bgf horizon is thought to have accumulated as a result of the oxidation of ferrous iron. The iron oxide formed is not associated intimately with organic matter or with aluminum and is sometimes crystalline. The Bgf horizons are usually prominently mottled; more than half the soil material occurs as mottles of high chroma.

- Cg, Ckg, Ccag, Csg, Csag—When g is used with C alone, or with C and one of the lowercase suffixes k, ca, s, or sa, the horizon must meet the definition for C and for the particular suffix, as well as for g.

h A horizon enriched with organic matter. It is used with A alone (Ah), or with A and e (Ahe), or with B alone (Bh), or with B and f (Bhf).

- Ah—A horizon enriched with organic matter, it has a colour value at least one unit lower than the underlying horizon or 0.5% more organic carbon than the IC or both. It contains less than 17% organic carbon by weight.
- Ahe—An Ah horizon that has undergone eluviation as shown by streaks and splotches of different shades of grey and often by platy structure. It may be overlain by a dark-coloured Ah and underlain by a light-coloured Ae.
- Bh—This horizon contains more than 1% organic carbon, less than 0.3% pyrophosphate-extractable iron, and has a ratio of organic carbon to pyrophosphate-extractable iron of 20 or more. Generally, the colour value and chroma are less than 3 when moist.
- Bhf—Defined under f.

j This is used as a modifier of suffixes e, f, g, n, and t to denote an expression of, but failure to meet, the specified limits of the suffix it modifies. It must be placed to the right and adjacent to the suffix it modifies. For example, Bfgj means a Bf horizon with a weak expression of gleying; Bfjgj means a B horizon with weak expression of both f and g features.

- Aej—An eluvial horizon that is thin, discontinuous, or slightly discernible.
- Btj—A horizon with some illuviation of clay but not enough to meet the limits of Bt.
- Btgj, Bmgj—Horizons that are mottled but do not meet the criteria of Bg.
- Bfj—A horizon with some accumulation of pyrophosphate-extractable Al + Fe but not enough to meet the limits of Bf.
- Btnj or Bnj—Horizons in which the development of solonetzic B properties is evident but insufficient to meet the limits for Bn or Bnt.

k Denotes the presence of carbonate as indicated by visible effervescence when dilute HCl is added. It is used mostly with B and m (Bmk) or C (Ck), and occasionally with Ah or Ap (Ahk, Apk), or organic horizons (Ofk, Omk).

m A horizon slightly altered by hydrolysis, oxidation, or solution, or all three to give a change in colour or structure, or both. It has:

1. Evidence of alteration in one of the following forms:
   a. Higher chromas and redder hues than the underlying horizons.
   b. Removal of carbonates either partially (Bmk) or completely (Bm).
   c. A change in structure from the original material.
2. Illuviation, if evident, too slight to meet the requirements of a Bt or a podzolic B.
3. Some weatherable minerals.
4. No cementation or induration and lacks a brittle consistency when moist.

This suffix can be used as Bm, Bmgj, Bmk, and Bms.

n A horizon in which the ratio of exchangeable calcium to sodium is 10 or less. It must also have distinctive morphological characteristics: prismatic or columnar structure, dark coatings on ped surfaces, and hard to very hard consistency when dry. It is used with B as Bn or Bnt.

p A horizon disturbed by activities such as cultivation, logging, and habitation. It is used with A and O.

s A horizon with salts, including gypsum, which may be detected as crystals or veins, as surface crusts of salt crystals, by depressed crop growth, or by the presence of salt-tolerant plants. It is commonly used with C and k (Csk), but can be used with any horizon or combination of horizon and lowercase suffix.

sa A horizon with secondary enrichment of salts more soluble than calcium and magnesium carbonates; the concentration of salts exceeds that in the unenriched parent material. The horizon is at least 10 cm thick. The conductivity of the saturation extract must be at least 4 mS/cm and exceed that of the C horizon by at least one-third. (The unit mho has been replaced by siemens [S].)

t An illuvial horizon enriched with silicate clay. It is used with B alone (Bt), with B and g (Btg), with B and n (Bnt), etc.

- Bt—A Bt horizon contains illuvial layer-lattice clays. It forms below an eluvial horizon but may occur at the surface of a soil that has been partially truncated. It usually has a higher ratio of fine clay to total clay than the IC. It has the following properties:

1. If any part of an eluvial horizon remains and there is no lithologic discontinuity between it and the Bt horizon, the Bt horizon contains more total clay than the eluvial horizon as follows:
    a. If any part of the eluvial horizon has less than 15% total clay in the fine earth fraction (<2 mm), the Bt horizon must contain at least 3% more clay, e.g., Ae 10% clay; Bt minimum 13% clay.
    b. If the eluvial horizon has more than 15% and less than 40% total clay in the fine earth fraction, the ratio of the clay in the Bt horizon to that in the eluvial horizon must be 1.2 or more, e.g., Ae 25% clay; Bt at least 30% clay.
    c. If the eluvial horizon has more than 40% total clay in the fine earth fraction, the Bt horizon must contain at least 8% more clay, e.g., Ae 50% clay; Bt at least 58% clay.
2. A Bt horizon must be at least 5 cm thick. In some sandy soils where clay accumulation occurs, the total thickness of the lamellae should be more than 10 cm in the upper 150 cm of the profile.
3. In massive soils, the Bt horizon should have oriented clay in some pores and also as bridges between the sand grains.
4. If peds are present, a Bt horizon has clay skins on some of the vertical and horizontal ped surfaces and in the fine pores, or has illuvial oriented clays in 1% or more of the cross section as viewed in thin section.
5. If a soil shows a lithologic discontinuity between the eluvial horizon and the Bt horizon, or if only a plow layer overlies the Bt horizon, the Bt horizon need show only clay skins in some part, either in some fine pores or on some vertical and horizontal ped surfaces. Thin sections should show that the horizon has about 1% or more of oriented clay bodies.

Btj and Btg are defined under j and g.

u A horizon that is markedly disrupted by physical or faunal processes other than cryoturbation. Evidence of marked disruption such as the inclusion of material from other horizons or the absence of the horizon must be evident in at least half of the cross section of the pedon. Such turbation can result from a blowdown of trees, mass movement of soil on slopes, and burrowing animals. The u can be used with any horizon or subhorizon with the exception of A or B alone; e.g., Aeu, Bfu, Bcu.

x A horizon of fragipan character. A fragipan is a loamy subsurface horizon of high bulk density and very low organic matter content. When dry, it has a hard consistency and seems to be cemented. When moist, it has moderate to weak brittleness. It frequently has bleached fracture planes and is overlain by a friable B horizon. Air-dry clods of fragic horizons slake in water.

y A horizon affected by cryoturbation indicated by disrupted and broken horizons, incorporation of materials from other horizons, and mechanical sorting in at least half of the cross section of the pedon. It is used with A, B, and C alone or in combination with other subscripts, e.g., Ahy, Ahgy, Bmy, Cy, Cgy, Cygj.

z A frozen layer. It may be used with any horizon or layer, e.g., Ohz, Bmz, Cz, Wz.

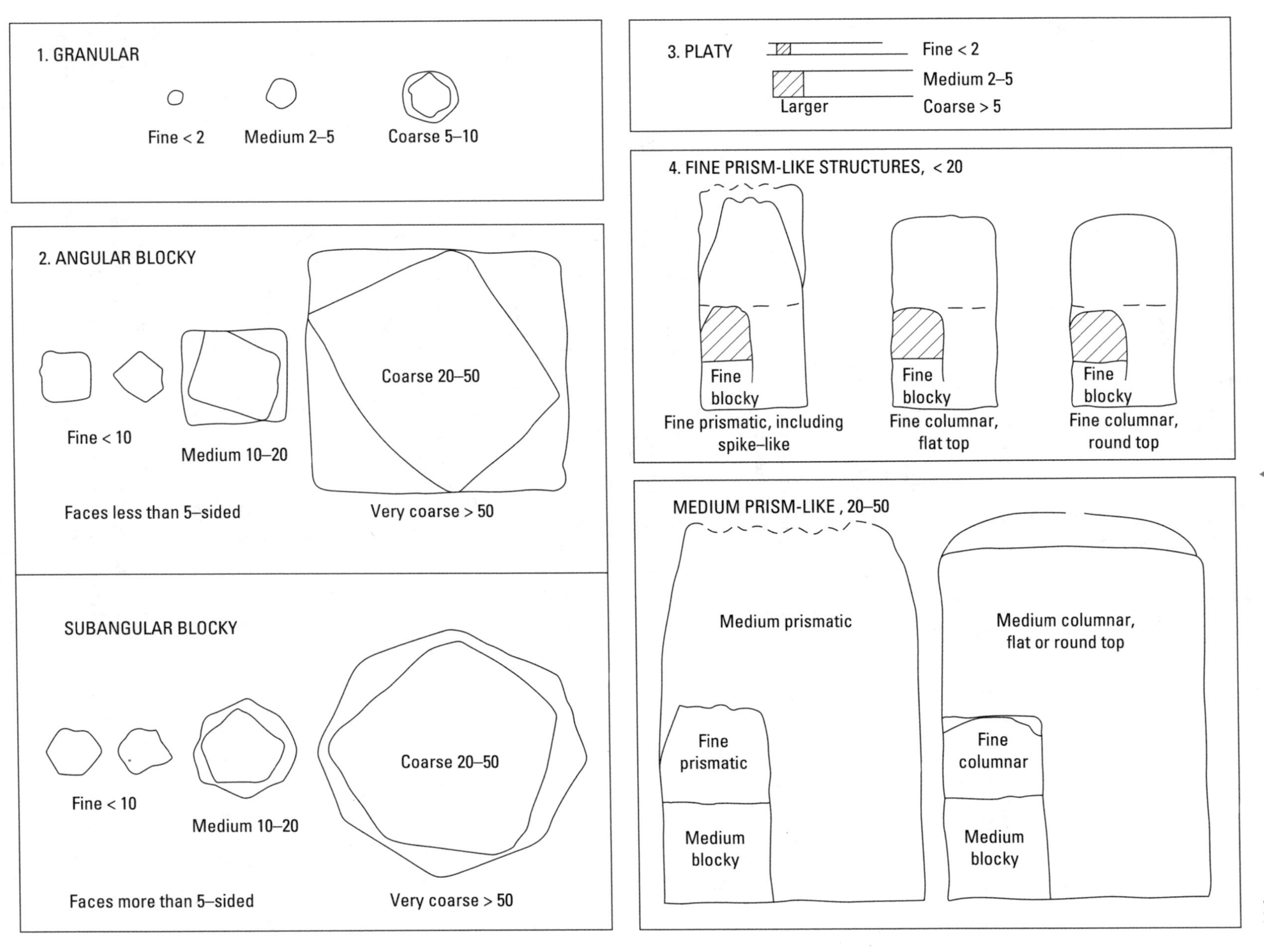


FIGURE A3 ▼ Types, kinds, and classes of soil structure (all dimensions mm)

## Table A3

*Types and Classes of Soil Structure (see Figure A3)*

| *Type* | *Kind* | *Class* | *Size (mm)* |
|---|---|---|---|
| 1. Structureless: no observable aggregation or no definite orderly arrangement around natural lines of weakness. | A. Single grain structure: loose, incoherent mass of individual particles as in sands. | | |
| | B. Amorphous (massive) structure: a coherent mass showing no evidence of any distinct arrangement of soil particles. | | |
| 2. Blocklike: soil particles are arranged around a point and bounded by flat or rounded surfaces. | A. Blocky (angular blocky): faces rectangular and flattened, vertices sharply angular. | Fine blocky | <10 |
| | | Medium blocky | 10–20 |
| | | Coarse blocky | 20–50 |
| | | Very coarse blocky | >50 |
| | B. Subangular blocky: faces subrectangular, vertices mostly oblique, or subrounded. | Fine subangular blocky | <10 |
| | | Medium subangular blocky | 10–20 |
| | | Coarse subangular blocky | 20–50 |
| | | Very coarse subangular blocky | >50 |
| | C. Granular: spheroidal, with rounded vertices. | Fine granular | <2 |
| | | Medium granular | 2–5 |
| | | Coarse granular | 5–10 |
| 3. Platelike: soil particles are arranged around a horizontal plane and generally bounded by relatively flat horizontal surfaces. | A. Platy structure: horizontal planes more or less developed. | Fine platy | <2 |
| | | Medium platy | 2–5 |
| | | Coarse platy | >5 |
| 4. Prismlike: soil particles are arranged around a vertical axis and bounded by relatively flat vertical surfaces. | A. Prismatic structure: vertical faces well-defined, and edges sharp. | Fine prismatic | <20 |
| | | Medium prismatic | 20–50 |
| | | Coarse prismatic | 50–100 |
| | | Very coarse prismatic | >100 |
| | B. Columnar structure: vertical edges near top of columns are not sharp. Columns may be flat-topped, round-topped, or irregular. | Fine columnar | <20 |
| | | Medium columnar | 20–50 |
| | | Coarse columnar | 50–100 |
| | | Very coarse columnar | >100 |

Source: Canada Soil Survey Committee, Research Branch, Agriculture Canada.

## 4. THE US COMPREHENSIVE SOIL CLASSIFICATION SYSTEM

### *Soil orders*

Entisols — Soils lacking horizons.

Inceptisols — Soils with weakly developed horizons containing weatherable materials.

Histosols — Soils with thick surface organic horizon.

Oxisols — Very old, intensely weathered soils, with an oxic horizon, and low cation exchange capacity.

Alfisols — Soils of humid and subhumid climates, with high base status and a clay-enriched B horizon.

Ultisols — Soils of warm climates, with a low base status and a clay-enriched horizon.

Vertisols — Soils of tropical and subtropical zones, with high clay content, cracking when dry.

Spodosols — Soils with a bleached, illuviated horizon, with low cation exchange capacity and lacking carbonates.

Mollisols — Midlatitude soils, with a thick, dark-coloured surface horizon of high organic matter content and base status.

Aridisols — Soils of dry climates, with accumulations of carbonates or soluble salts.

### *Soil suborders*

Aquents — Entisols of wet places; permanently saturated.

Arents — Entisols mixed by ploughing.

Fluvents — Entisols formed from recent alluvium.

Orthents — Entisols on recent erosional surfaces; shallow; stony, loamy, or clayey.

Psamments — Entisols with sandy texture throughout; usually on dune or beach sand.

Andepts — Inceptisols with a dark surface horizon, chiefly derived from recent volcanic ash.

Aquepts — Inceptisols of wet places; seasonally saturated; dark organic topsoil.

Ochrepts — Inceptisols with a pale brown topsoil of low organic matter content.

Plaggepts — Inceptisols with artificial topsoil.

Tropepts — Inceptisols of low latitudes with brownish or reddish topsoil.

Umbrepts — Inceptisols with dark, acid topsoils, in moist midlatitude areas.

Fibrists — Histosols composed mainly of sphagnum moss; permanently saturated or artificially drained.

Folists — Histosols composed of forest litter on bedrock; freely drained.

Hemists — Histosols that are almost permanently saturated and contain partly decomposed plant debris.

Saprists — Histosols saturated with water most of the year or artificially drained, containing decomposed plant debris.

Aquox — Oxisols of wet places; seasonally saturated; mottled.

Humox — Oxisols of cool moist regions with high organic carbon contents.

Orthox — Oxisols of warm, humid regions with short dry seasons; moist most of the year.

Torrox — Oxisols of arid areas, dry for more than six months, lacking organic matter.

Ustox — Oxisols with marked seasonal water regime.

Aqualfs — Alfisols of wet places; grey; seasonally saturated.

Boralfs — Alfisols of boreal forests and high mountains; grey topsoils.

Udalfs — Alfisols in areas with short dry season; formed under deciduous forest.

Ustalfs — Alfisols in areas with marked seasonal moisture regime; brownish or reddish throughout.

Xeralfs — Alfisols of Mediterranean areas; brownish-reddish throughout.

Aquults — Ultisols of wet places; seasonally saturated; dominantly grey throughout.

Humults — Ultisols with organic-rich topsoils; in mid and low latitudes.

Udults — Ultisols with marked, short dry season; reddish or yellowish B horizons.

Ustults — Ultisols with marked seasonal moisture regime; brownish to reddish throughout.

Xerults — Ultisols of Mediterranean areas; brownish to reddish throughout.

Torrerts — Vertisols of dry areas; cracks remain open through year.

Uderts — Vertisols with short dry seasons; cracks are open only briefly.

Usterts — Vertisols with marked seasonal water regime; cracks remain open for long periods.

Xererts — Vertisols of Mediterranean areas; cracks close in winter.

| | |
|---|---|
| Aquods | Spodosols of wet places; seasonally saturated; topsoil lacks free iron; B horizon often cemented. |
| Ferrods | Freely drained Spodosols with iron-enriched B horizons. |
| Humods | Spodosols in which the B horizon is enriched with organic matter and aluminum, but not iron. |
| Orthods | Freely drained Spodosols with B horizons enriched with organic matter, iron, and aluminum. |
| Albolls | Mollisols with bleached horizon over slowly permeable clay; seasonally saturated. |
| Aquolls | Mollisols of wet places; black organic-rich topsoil on mottled B horizon; seasonally saturated. |
| Borolls | Mollisols of cold-winter arid plains (steppe) or high mountains. |
| Rendolls | Mollisols on calcareous parent material; high carbonate contents in subsoil; formed in moist climates under forest. |
| Udolls | Mollisols with short dry season; no carbonate accumulation; brownish throughout. |
| Ustolls | Mollisols with marked seasonal moisture regime; marked carbonate accumulation in subsoil. |
| Xerolls | Mollisols of Mediterranean areas. |
| Argids | Aridisols with clay enriched horizon. |
| Orthids | Aridisols with no clay enrichment, but evidence of horizonation to at least 25 cm. |

# Glossary

**ablation** Reduction of glacial ice by melting, evaporation, and sublimation; occurs primarily in the zone below the equilibrium line of a glacier.

**ablation till** Rock debris lying within or on stagnant ice. As the ice melts, meltwater washes away the finer rock particles until particles coarser than those in lodgement till remain.

**ablation zone** The area of a glacier where there is a net loss of mass during the year through melting, sublimation, or iceberg calfing. This is predominantly the lower altitude areas.

**abrasion** Erosion of bedrock by particles carried by wind, water, or ice.

**absolute instability** The environmental lapse rate is cooling more rapidly than the dry adiabatic lapse rate, thus the adiabat remains warmer than the surrounding air, and it will remain buoyant and rise.

**absolute stability** The environmental lapse rate is cooling more slowly than the dry adiabatic lapse rate, thus the adiabat remains cooler than the surrounding air, and it would require an external force to make it rise.

**absolute zero** 273.15 Celsius degrees below the freezing temperature of water, the point at which all molecular movement ceases; the lowest possible temperature.

**absorption** Electromagnetic energy absorbed and changed into heat energy by the substance, usually gas or liquid, through which it is passing.

**abyssal plain** The largest flat surface on the earth's crust, making up large areas of the ocean floor at depths below 4000 m.

**accumulation zone** The area of a glacier where there is a net gain of mass through precipitation and avalanching.

**acidophile** An organism adapted to life in an acid medium.

**active layer** The top layer of ground subject to annual thawing and freezing in areas underlain by permafrost.

**active-layer failure** A shallow landslide in a permafrost area characterized by detachment of a thawed layer from the underlying frozen material.

**actual evapotranspiration** Actual amount of water evaporated and transpired at a given site.

**adiabat** A parcel of air that usually has different temperature and humidity levels than the surrounding air; from the Greek; 'without exchange.'

**adiabatic process** As an adiabat rises or falls in the atmosphere the sensible temperature within the parcel of air changes because of compression or expansion; this occurs without gain or loss of heat from the surrounding air.

**advection fog** Fog formed when a moist layer of air moves over a cool land or water surface and is cooled below the dew point temperature.

**aerosol** A general term for any particle in the air.

**aestivate** Of plants; to enter a dormant state during the growing season because of water shortage and heat.

**aftershock** One of a number of small shocks following an earthquake, located near or at the original seismic focus.

**air mass** An extensive body of air that has quasi-uniform temperature and moisture content.

**aklé** Complex sand dunes formed of sinuous ridges, aligned transverse to the wind, and composed of crescentic sections, pointing alternately upwind and downwind.

**albedo** The percentage of shortwave radiation that is reflected from a surface.

**allelopath** Direct inhibition of one species by another using noxious or toxic chemicals.

**alluvial fan** A cone-shaped feature composed of alluvium, found at the base of a slope created by a distributary stream that loses velocity and deposits its sediment; a prominent feature in arid regions.

**alluvium** Sediments deposited by a stream.

**altitude** The height of a point above mean sea level, measured vertically. Or, measured in degrees from the plane of the observer's horizon, the angular height of a heavenly body. Or, measured from base to summit, the vertical height of mountain or hill peak.

**ana-front** The air in the warm sector between fronts rising and creating a distinctive pattern of cloud development; often associated with severe storms and tornadoes.

**anabatic winds** Occur during the day when a mountain slope is heated and the air rises, creating an upslope wind.

**anchor ice** Submerged ice attached to an ocean or lake bed.

**ancillary minerals** Small quantities of minerals, including heavy minerals, that are contained in sedimentary rocks.

**angle of incidence** The angle at which incoming solar radiation strikes a surface. Depending on the nature of the surface, it determines the amount of energy reflected.

**angle of initial yield** The angle at which movement of materials downslope is initiated.

**angle of repose** The slope angle that material tends to adopt naturally; varies according to the nature of the material.

**angle of residual shear** Another term for the angle of repose.

**angle of sliding friction** The maximum frictional strength of particles, expressed as an angle; beyond that angle sliding occurs.

**anion** An electrically charged atom, group of atoms, or compound that has gained electrons, and thus has a negative charge.

**antecedent moisture** The existing moisture content of the soil.

**anticline** An upward arch-like fold of rock; normally used in reference to sedimentary rocks.

**anticyclone** A region of high atmospheric pressure.

**antidune** A sand wave in a highly-loaded, swiftly-flowing stream, which moves progressively upstream.

**aphelion** The point at which the earth is farthest from the sun in its annual elliptical orbit; roughly $15.2 \times 107$ km at present.

**aquiclude** Rock or other material that hinders or prevents the natural movement of water in the ground.

**aquifer** Porous and permeable rock that allows the easy passage of water; because of these properties it has the capacity to store water.

**arboreal** Living in or adapted to living in trees.

**Archaeocyatha** Rare, coral-like organism of the Cambrian period.

**Arctic Front** A largely inactive frontal zone lying north of the polar zone.

**arctic sea smoke** A fog caused by moisture evaporating from open water into extremely cold air, where it condenses very rapidly.

**areal scour** Glacial scoring or erosion over an extensive area under a piedmont glacier.

**arête** A sharp ridge formed between two cirques that have expanded by erosion toward each other.

**arroyo** Another term for a gully; used in Mexico and the American west. Distinctive in form, being wide and flat-bottomed with steep sides.

**artesian basin** A region into which groundwater has flowed, creating natural hydraulic pressure.

**artesian well** A well drilled into an artesian basin. The water rises above the surrounding water table due to the hydraulic pressure.

**asthenosphere** The layer of rock in the upper mantle, immediately below the lithosphere, that is close to its melting point and has low strength.

**atoll** A circular coral reef with no island in the centre.

**aureole** see metamorphic aureole

**aurora australis** Spectacular coloured lights, electro-magnetic in origin, in the ionosphere, seen on the horizon in the night sky in the southern hemisphere.

**aurora borealis** Coloured lights seen in the night sky in the northern hemisphere.

**autotrophs** Organisms, mostly green plants, that manufacture their own food using energy from the sun, carbon dioxide from the air, water and minerals from the soil.

**available nutrients** The portion of any element or compound in the soil that can be readily absorbed and assimilated by growing plants.

**avulsion** Occurs when a stream abandons its channel and creates a new route; can occur quite suddenly.

**azimuth** Direction determined by starting at 0° and moving clockwise through 360°.

**azonal soil** A soil not completely in equilibrium with its climate because of something limiting its formation.

**backshore zone** Land lying inland from the average high-water line.

**backwash** Water that runs back down a beach after a wave has broken; it collides with the next incoming wave.

**ballistic ripples** Ripples in sand spaced at intervals and created by the impact of saltating (jumping) particles; also called impact ripples.

**banding** Layering in rock caused by segregation of minerals into discrete zones; can result in strong, massive rock.

**bank seepage** Water that enters a stream through the banks, increasing the downstream flow.

**bank-caving** Occurs when the water of a stream undermines the bank, causing a collapse.

**bankfull** The point at which all the water is contained within the river channel, but any further increase would cause flooding.

**bar** A deposit of sediment extending out from a shoreline.

**barchan dune** Crescentic dunes oriented with their 'horns' pointing downwind.

**barrier reef** A coral reef created as rising sea levels drown old reefs and new ones form offshore, forming a barrier. The best-known is the Great Barrier Reef on the eastern coast of Australia.

**basal erosion** Erosion of the bottom surface of a glacier by abrasion with the underlying rock.

**basal sliding** Slippage of a glacier due partly to a thin layer of water between the base of the ice and the bedrock.

**basal till** Debris that is carried under or deposited by a moving glacier. It often has a high clay content.

**base level** The lowest level to which a land surface can be eroded by running water.

**baseflow** The portion of a stream's flow contributed by groundwater seepage or throughflow.

**basophile** An organism adapted to life in an alkaline medium.

**batholith** Large-scale upwelling of magma with an exposed area greater than 100 $km^2$, typically formed of granite rocks.

**bathythermograph** A map showing relief of the land and depth of the oceans.

**beach cusp** A distinctive pattern formed by the fan-like intrusion of waves onto a beach.

**beach drift** Transportation of sand by landward and seaward water movements in the foreshore zone parallel to the shoreline.

**beach** Loose matter, such as mud, sand, shingle, or pebbles, which accumulates on a shoreline between the low and high tide levels.

**beach rock** Rock formed when coralline sand accumulates and is cemented together.

**bedding planes** The surface between strata of sedimentary rocks, apparently indicating an environmental change.

**Bergeron-Findeisen process** The method by which raindrops are formed in cold clouds where supercooled water and supersaturated air exist; also know as the three-phase process.

**bergy bit** A heavy, compact piece of ice, or a floating glacier, usually under 10 m wide and 5 m above sea level.

**berm** Low ridge created by wave action marking the high-water point toward the back of a beach.

**bioclimate** The meteorological characteristics affecting living organisms.

**biogeography** The study of the spatial distribution of animate nature (excluding humans) and the processes that influenced this distribution.

**biomass** Dry weight of living organic matter per unit area in an ecosystem.

**biome** An area that is ecologically integrated and uniform. It represents the soil, the vegetation, the animals, and their physical environment.

**biosphere** The life zone or the environment that contains all living organisms of the earth.

**biota** The animal and plant life of a specific area and/or time period.

**biotic** Pertaining to living organisms.

**black body** Something that is able to absorb all incoming radiation; also, an object that is a perfect radiator of energy.

**black box** A system where the boundaries, the inputs, and the outputs are known, but there is little knowledge of what goes on inside.

**blocking anticyclones** High pressure regions that build up in depression tracks, diverting cyclones and giving rise to extreme weather patterns such as the droughts of North America.

**blowout** A sudden, violent release of steam or gas. In coastal sand dune areas, a hollow made by eddying wind in light or sandy soil.

**bombs** Pieces of lava thrown out by a volcano which tend to spin and cool before hitting the earth.

**boundary layer** The lowest layer of the atmosphere, roughly 300 m deep, where airflow is influenced by the earth's surface.

**Bowen Reaction Series** The two separate sequences, called continuous and discontinuous, in which minerals form during the cooling of magma.

**braided channel** The separation of the main stream channel into a number of smaller interlocking channels.

**breaker zone** The portion of the nearshore zone on a beach where the waves break.

**bulk density** The mass of dry soil per unit bulk volume; the volume is determined before the soil is dried.

**buried ice** Ice formed at the surface and later covered by sediment and preserved within permafrost.

**butte** A small, isolated hill with steep sides. It consists of resistant rock overlying weaker layers.

**calcicoles** Plants tolerant of alkaline conditions.

**caldera** A feature formed when a strato-volcano explodes and the magma chamber collapses, leaving a vast steep-sided crater.

**caliche** A soil layer near the earth's surface, more or less cemented by secondary carbonates of calcium or magnesium precipitated from the soil solution.

**calving** The process of large blocks of ice breaking from a glacier extending into water; the floating blocks are called icebergs.

**cambering** A slow continuous downslope movement of solid rock.

**capillary movement** Movement of water in the soil that is completely independent of gravity in any direction. Water films around soil particles become thin when soil is dry, and water migrates from wet areas where the films are thick.

**capillary water** One of three main types of water in the soil; moves by capillary processes.

**carbon dioxide** One of the gases that is variable by volume in the atmosphere; it makes up roughly 0.033 percent of the earth's atmosphere.

**carbonic acid** Acid formed by the solution of carbon dioxide from the atmosphere; rainwater is a mild carbonic acid.

**carnivores** A subgroup, with herbivores, of the heterotrophs, that is, organisms that derive their energy from eating the autotrophs.

**cascading system** A system in which energy and matter move through the environment from one component to another.

**catchment area** The gathering ground for the water that flows in a stream.

**catena** A nontaxonomic grouping of a sequence of soils of about the same age, derived from similar parent material, and occurring under similar climatic conditions, but having unlike characteristics because of variations in relief and in drainage.

**cation** A positively charged ion.

**cation exchange** Because of the ratio law, a change in the absolute concentrations of two cations of different valency upsets the chemical equilibrium of the soil solution and causes cations to be released from the colloids.

**cation exchange capacity** A given amount of a soil's ability to hold and exchange cations.

**cavitation** A theory that the impact of water on rock causes the water molecules to break. The implosion of minute airless bubbles exerts considerable force, eroding the rock surface.

**cementation** Lithification or formation of rocks from materials collected by water percolating through another rock or deposit; two of the most common cements are calcium carbonate and silica.

**chalk** Soft, white, friable, fine-grained pure limestone, composed mainly of calcium carbonate; formed in shallow water in the Cretaceous period.

**channel precipitation** The portion of rain that falls directly onto the stream.

**chatter mark** A mark created by the removal of a thin flake or chip of bedrock embedded in overlying, moving glacier ice; often several occur together.

**chelation** A chemical weathering solution process: ions leave the mineral and combine within organic compounds in water.

**cheluviation** The conversion of rainwater to a weak organic acid by picking up organic compounds from vegetation and surface soils.

**chemical energy** A form of electrical energy bound up in the chemical structure of any substance.

**chemical weathering** The breakdown of soil or rock into smaller particles by chemical action.

**chert** A silicate rock composed of chalcedony; like flint, it is opaque and dark, but with a splintering fracture.

**cinder cone** A distinctive cone of volcanic ash piled up around a gas vent; it has a relatively large crater and steep slopes.

**cirque** A basin-shaped hollow or depression in which snow can collect and survive summer melting; more frequently found on the poleward side of a mountain.

**cirque lake** A lake formed in the basin eroded by a cirque glacier; often referred to as a tarn.

**cirque-lip** Erosion within a cirque results in deepening, so that a lip is formed which hinders ice when it starts to flow out.

**clast** A rock fragment or mineral grain that is a sedimentary particle.

**clastic rocks** Rocks formed from the cementing together of fragments of older rocks.

**clay mineral** The smallest particles to which rock can be reduced under natural weathering; comprised of particles less than 0.002 mm in diameter.

**clay translocation** The detachment of individual clay particles or the dissolution of clay minerals in upper layers of soil and their movement downward; also know as lessivation.

**clear cut** The forestry practice of cutting all trees in an area, rather than selective cutting of individual trees.

**cleavage** Lines of weakness formed in rocks during cooling, drying, or compaction.

**cliff** A vertical or near-vertical wall of rock, usually associated with a coastline that is rising.

**climatic climax vegetation** The ultimate and characteristic form of vegetation that evolves in a climatic region.

**Climatic Optimum** A period between 5000 and 7000 years ago when temperatures were much warmer than at present.

**climatic terraces** River terraces created by an increase in precipitation and thereby streamflow; the increased erosive capacity causes the river to cut downward.

**climatology** The pattern of weather over time, at a location or in a region.

**clints** Flat areas between grikes (eroded joints) in a karst pavement.

**closed-system pingo** A pingo formed due to freezing of injected water supplied during aggradation of permafrost around a closed talik.

**coalescence** The mechanism by which things come together, grow together, and unite to become one thing.

**coefficient of variation** A statistical expression of rainfall variability.

**cold air drainage** The downslope movement of cold air as a result of its higher density.

**cold front** A front separates two masses of air and is defined as a cold front if the temperature decreases at a station when it passes; warm air is pushed up over an advancing cold front.

**collision-coalescence** The processes of raindrop formation when water droplets collide and combine to form larger drops.

**colloidal material** A finely divided substance dispersed in a gas, liquid, or solid medium.

**colloids** Microscopic mineral particles (less than 0.001 mm diameter) that remain in suspension in water almost indefinitely.

**columnar lava** Magma emerging in the ocean is cooled very quickly and forms distinctive columns whose structure reflects their mineral content.

**community** A group of populations of plants and animals in a given place, utilizing the resources of a common habitat that they either maintain or modify.

**compaction** The combining of fine rock particles by pressure of earth movements or weight of overlying deposits. Or, the process when soil particles are pressed together by torrential rain or heavy machinery.

**competitive exclusion principle** The hypothesis, based on theoretical considerations and laboratory experiments, that two or more species cannot coexist and use a single resource that is scarce relative to the demand for it.

**compressive flow** The flow regime of a glacier where ice velocity decreases along a section of the valley, hence the ice is under compressional stress.

**conceptual model** A general reconstruction of a situation that enables better understanding of the interrelationships of the observed data.

**conditional instability** Occurs when the environmental lapse rate is between the dry and the saturated adiabatic lapse rates. A parcel of air, initially stable, becomes unstable following the addition of heat from condensation.

**conduction** The transfer of sensible heat from one molecule to another by collision; the transfer is from the warmer to the colder object.

**cone of depression** The depression of the water table around a well, caused when water is withdrawn faster than it can be replaced through the surrounding rock.

**consequent stream** A stream that flows straight down a newly formed slope.

**contact metamorphism** Changes that occur during the metamorphic processes in rock closest to the source of heat and pressure.

**continental drift** The theory proposed by Alfred Wegener in 1912, now generally accepted, that the continents float on plates that make up the crust of the earth.

**continental shelf** The extension of the continents from the coast out under the oceans to the continental slope; the transition from shelf to slope usually occurs at a depth of 200 m.

**continental shield** An area of continental crust under which there are deformed felsic igneous and metamorphic rocks.

**continental slope** The steep slope at the edge of the continent where it drops down to the deep ocean.

**continentality** Land heats and cools more rapidly than water, thus locations in the interior of continents exhibit a large annual temperature range.

**continuous series** As magma cools, minerals are formed in sequence. N.L. Bowen identified one of these as continuous because the formation of minerals is continuous rather than discrete.

**contraction crack** A crack in soil caused by shrinkage as moisture is removed; the cracking is usually polygonal and reflects the structure of the soil minerals involved.

**convection** Upward, circular movement of liquid or gas, a major process in the transfer of heat energy in the atmosphere and oceans.

**convection cell** The total circulation of liquid or gas, including upward and downward flow.

**convection current** A stream of fluid in the atmosphere or ocean, produced by convection.

**convectional storms** Convective cells in the atmosphere that generate sufficient energy to be classified as storms.

**convergence** In continental drift, the process of two plates coming together. In climate, the zone at the equator where major air currents come together.

**coral reef** A coastal feature in shallow tropical waters: carbonate material created by marine organisms.

**Coriolis force** The effect of the force produced by the earth's rotation on a body moving on its surface. The body will be deflected to the right in the northern hemisphere, to the left in the southern hemisphere.

**country rock** The original rock that has been altered by heat and pressure into a new rock.

**crest** The peak of an ocean wave; the peak of the wave in a cyclone in a frontal system.

**crevasse** A crack in the brittle layer of a glacier caused by pressure.

**crevasse fillings** Ridges produced by filling of crevasses with sediment squeezed in from the glacier bed or falling in from the glacier surface.

**critical angle of sliding** The angle at which material on a slope begins to slide due to gravity. It is usually greater than the angle of repose.

**critical entrainment velocity** The speed at which water overcomes the forces that keep a particle on the bed of a stream.

**cryosphere** The section of the earth's surface that is permanently frozen.

**cryotic** At a temperature equal to or below 0°C.

**cryoturbation** Heaving of the soil by frost action; frost churning.

**cumuliform** Clouds formed by rapid vertical movement of air, also referred to as 'heap' type clouds.

**cumulonimbus** The largest form of cumuliform clouds, commonly called the 'thunderstorm' cloud; the only cloud in which hail can form.

**cutans** Clay washed down through soil which forms a veneer around larger soil particles.

**cyclone** A region of low barometric pressure (also called a depression) characterized by rising air and extensive cloud formation. Wind circles in toward the cyclone, counterclockwise in the northern hemisphere and clockwise in the southern hemisphere.

**cyclonic rainfall** Widespread rains associated with the rising air in a cyclone or low pressure system.

**dead-ice topography** A distinctive glacial landform of small depressions, small mounds, and moraine ridges; also known as hummocky moraine or kettle and kame topography.

**debris cascade** A general term for the movement of rock material through the landscape following uplift. The process of the rock from formation through weathering, transportation, and deposition.

**debris slope** The part of a slope immediately above the pediment in which sliding and avalanches are the major processes of downslope movement.

**decomposer** An organism whose niche is part of the decomposition of organic material; it derives its food from dead organisms.

**deep-water zone** The offshore zone beyond the point where the orbital wave motion intersects the seabed.

**deflation hollow** In coastal sand dune areas, a hollow made by eddying wind in light or sandy soil.

**deflation** The removal of sand by wind in a desert environment.

**deGeer moraine** see washboard

**delta** The deposit of coarse materials at the point where a river or stream enters a lake or the sea; named for the characteristic shape of the Greek letter Δ.

**delta kame** A distinctive type of delta formed where a stream flows from a glacier into a proglacial lake.

**dendritic drainage** A river drainage system whose tributaries have a pattern that resembles the branches of a tree; from the Greek *dendros*: tree. Found usually on a gentle, almost uniform slope where the underlying structure exercises no control.

**dendrochronology** The use of tree rings to determine the age of a tree.

**depositional landforms** Landforms of any type created by deposition of material.

**depression** A cyclone or low pressure area.

**depression storage** Water stored in small hollows on the surface after the infiltration capacity has been exceeded during rainfall.

**deranged drainage** A confused drainage network that forms on haphazard distribution of glacial drift; characterisitic surface features may be small lakes, streams, marshes, and islands.

**desert varnish** A smooth, dark mineral layer on desert pebbles and boulders comprised of iron, manganese, and various trace elements.

**desiccation** The extensive removal of moisture; dehydration.

**detachment** The breaking of bonds that hold a material together; the first step in the weathering of rock.

**detrital material** The product of rock weathering.

**detritus food chain** A main component of the food chain, dominated by the organisms that convert dead material into nutrients.

**dilatation** The expansion of rock after the removal of overlying material; also known as pressure release.

**dip angle** Acute angle between an inclined natural rock plane or surface and a horizontal plane of reference; it is measured perpendicular to the strike.

**dirt cone** A small mound of sediment with an ice core.

**discontinuous series** A series of discrete steps in the sequence of magma cooling; as the temperature drops there are abrupt changes in the nature of the minerals being formed.

**diurnal temperature range** The range of temperature over a 24-hour period.

**divergence** In a high pressure system, air descends and spreads outward; usually associated with clear skies.

**divide** The height of land that separates water draining into one ocean from water draining into another; also known as continental divide.

**doldrums** The region associated with the Intertropical Convergence Zone on the equator side of the Hadley cell. Rising air means little wind, which, with the heat, made it a bad area for sailing ships.

**draa** A large sand dune, having a characteristic wavelength between 0.5 and 5.0 km.

**drainage basin** The form that rivers take in a drainage basin in response to factors such as geology, climate, and relief.

**drawdown** The lowering of a water table when water is pumped out faster than it is replaced by groundwater.

**dreikanter** The characteristic shape of pebbles polished into three facets by abrasion from consistent winds in a desert environment; from the German, 'three sided'.

**drizzle** Rain droplets less than 0.5 mm in diameter.

**drumlin** A smoothed mound of glacial debris with the longitudinal axis parallel to the direction of ice movement; often described as shaped like the underside of a teaspoon.

**drumlinoid landform** A landform that resembles a drumlin.

**dry adiabatic lapse rate** The rate of change of temperature with height before condensation occurs; a fixed rate of 10°C for each 1000 m.

**dune** An accumulation of wind-blown sand piled into distinctive forms, having a characteristic wavelength up to 0.5 km.

**dust veil index** A measure of the amount of dust in the atmosphere usually used to determine the effect of volcanic dust in blocking shortwave energy from the sun.

**dyke** Thin vertical veins of magma, which cool quickly into rock that is more resistant than the surrounding rock. Also, a retaining wall, usually of earth, built parallel to a river or stream to prevent flooding.

**dynamic equilibrium** The situation in an environmental system when there is a distinct and consistent longterm trend.

**dynamic metamorphism** Structural change in rocks with little or no alteration in composition of the rocks.

**easterly wave** A low pressure trough that forms in the tropics close to the equator. It has clear skies on the west side and a line of heavy cumulus and thunderstorms on the east.

**eccentricity of orbit** Variations in the orbit of the earth from almost circular to strongly elliptical every 100 years. The fluctuations are caused primarily by the gravitational effects of Jupiter.

**ecology** The study of interaction between living organisms and their environment.

**ecosystem** A general term for the interactions between living organisms.

**eddies** Disturbances in the flow of a gas or liquid causes by irregularities in a surface.

**eddy diffusion** The spread of energy, often in the form of heat, throughout a liquid or gas, caused by turbulence.

**elbows of capture** A distinctive turn in the direction of a river created when one river captures the flow of the tributary of another; also known as river piracy or river capture.

**element** Any of more than 100 fundamental metallic and nonmetallic substances that consist of atoms of only one kind, and that singly or in combination make up all matter.

**eluviated** A pedogenic process: the transportation of soil material in suspension or in solution within the soil by the downward or lateral movement of water.

**emmer** A hybrid wheat apparently formed naturally when wild wheat crossed with goat grass.

**endemic species** Native to and confined to a certain region.

**endogenetic energy** Energy that arises from within. For example, the energy applied to geological processes that comes from within the earth.

**englacial** Objects or substances within a glacier; usually refers to transport of sediment within the glacier.

**englacial sediment** Sediment embedded within a glacier.

**enhanced plastic flow** The increase in the rate of deformation of ice in zones of a glacier where there are local increases in stress.

**entrainment** The uptake or absorption of debris or material by water or ice by a variety of processes.

**entrenched meanders** River meanders that cut deeply into a landscape created by uplift of the land: e.g., the canyons on the Nahanni River in Yukon and British Columbia.

**environmental lapse rate** The rate at which temperature falls with increasing altitude; on average 6.4°C for each kilometre.

**epeirogenic movement** Mass earth movement; tilting or warping that influences the formation of continents, such as uplifts or depressions.

**epicentre** The point on the surface of the earth directly over the focus or point where an earthquake occurred.

**epigenetic** Refers to ground ice that formed some time after the deposition of the earth material in which it occurs.

**epigenetic wedge** A wedge-shaped body of ice that develops in preexisting permafrost (see wedge ice).

**epiphyte** A plant that uses another plant for a substratum, but which derives no sustenance from the relationship.

**equatorial trough** The almost continuous region of low pressure associated with the heat equator.

**equifinality** The result of the interaction of a number of systems.

**equilibrium line** On a glacier, the line separating the zone of accumulation from the zone of ablation; the point of zero balance.

**erg** The zones of accumulated sand in a desert.

**erosional landform** A landform shaped as bedrock or regolith is removed by erosion; for example, a canyon, glacial cirque, or marine cliff.

**esker** A long sinuous ridge of sand and gravel that was deposited on the bottom of a stream flowing under a glacier.

**eustatic change** A change in sea level caused by the increase or decrease of water volume influenced by the movement of glacial ice sheets, not by land shifts.

**eutrophication** The process of gradual decay and infilling of a lake accelerated by excessive quantities of nitrogen and phosphorus compounds, and human and animal wastes entering the lake.

**evaporation** The change of phase of water from a liquid to a gas; energy is absorbed in this process as the latent heat of evaporation.

**evaporite** After evaporation, a remaining sedimentary rock consisting of minerals precipitated from a solution.

**evapotranspiration** The combined loss of moisture from a vegetated surface by evaporation and transpiration (the loss of moisture through the leaves of plants).

**excess ice content** The volume of ground ice that exceeds the pore space that would be present if the soil were unfrozen.

**exchangeable nutrient** An atom, group of atoms, or compound that can be exchanged with ions or anions and thereby assimilated by growing plants.

**exfoliation** The peeling away of layers of weathered rock; also known as onion weathering.

**exogenetic energy** Energy that arises from an external source.

**extending flow** The flow regime of a glacier where ice velocity increases along a section of the valley, hence the ice is under tensional stress.

**extrusive igneous rock** Lava that cools above the surface to form igneous rock.

**F layer** The transitional zone between the inner and outer core of the earth's interior.

**failure plane** A plane that tends to develop where stresses within a rock are greatest, and along which earthquakes occur.

**fair weather cumulus** Cumulus clouds that form from small adiabats within an air mass; they do not evolve into heavy cumulus or thunderstorms.

**fault** A failure plane created when the rigid rocks of the crust are subjected to stress.

**fault scarp** A steep wall of rock or escarpment created when one side of a fault moves vertically relative to the other.

**faulting** The process of displacement or fracturing between two portions of the earth's crust.

**feedback** The process whereby an event causes a reaction that either retards or accelerates the original action.

**felsic minerals** Pale-coloured, low-density silicate minerals; for example, feldspars and quartz

**fermentation layer** The layer in acid-environment soils between the plant litter and the fully decomposed humus.

**ferrel cell** The middle cell of the three-cell theory of the motion of the atmosphere. It appears intermittently between the Polar and Hadley cells.

**fetch** The stretch of water over which the wind blows to create wave action.

**field capacity** The water remaining in the soil after water has drained away by gravity.

**fine earth fraction** The total of sand, silt, and clay in a soil.

**firn** Denser snow left in the zone of accumulation of a glacier after summer melting has occurred.

**first-year ice** New sea ice formed during one winter: if the ice survives the following summer and continues to develop in a second winter, it becomes second-year ice.

**fjord** A deep valley, created by glaciers flowing into the sea, that is subsequently partly filled by sea water.

**flint** A heavy, hard, grey or grey-black silica.

**flocculation** A chemical process that causes fine muds and clays to agglomerate and sink rather than remaining in suspension in water.

**flood basalts** Large areas of magma that have flooded onto the surface to form extensive plateaux.

**floristic classification** A means of grouping vegetation types based on the species composition.

**flow till** Till that has accumulated in depressions in stagnant ice and has been subject to some remodification by debris slumping and flow.

**flume** An artificial channel to simulate and study the flow of water and the action of sediment in a stream.

**flutes** Long parallel ridges and valleys formed in till; they can vary from from metres to kilometres in length.

**focus** The point in the earth at which an earthquake occurs and maximum energy is released.

**fog** Appears when air at the surface is cooled below the dew point temperature and condensation occurs.

**fold** A bending of rock strata by heat and pressure.

**foliation** Layering of rock caused by the reorientation of its particles by pressure and shearing.

**foliations** Vertical or sub-vertical bands of sediment and air bubbles found within an ice wedge.

**food chain** The one-way flow of energy through an ecosystem, in which organisms at each level consume energy found in the bodies of organisms of the immediate lower level. A food chain ends with the decomposers.

**foredune** Sand blown inland from a beach and trapped by vegetation to form a dune field.

**foreland** Land lying in front.

**foreshock** Initial shock wave that precedes the main shock waves of an earthquake.

**fossorial** An animal that lives and forages below ground.

**frazil ice** In the first stages of freezing, the small ice slivers or needles suspended in water.

**free-face** The steep face of a cliff from which material falls by gravity.

**freeze–thaw** The alternate freezing and thawing of rock, a major physical weathering process.

**freezing front** The advancing boundary between frozen and unfrozen ground.

**fringing reef** The distinctive offshore reef created by coral building outward into the surrounding sea.

**front** The boundary between two air masses.

**frost action** The process of alternating freezing and thawing of moisture in soil, rock, and other materials, and the effects that result.

**frost heave** Upward or outward movement of the ground surface (and objects in or on the ground) as a result of the formation of ice within the soil.

**frost hollow** Area of lower elevation where cold air can accumulate and frost is more likely to occur.

**frost polygon** Distinctive polygonal shapes formed in the ground in cold, dry Arctic environments when water freezes in cracks created by shrinkage.

**full-tree logging** In this logging practice, all the above-ground parts of the tree (the trunk and branches) are removed.

**galeria** A series of interconnected tunnels dug at the water table to enhance the flow of water.

**garigue** A low and open scrub with many evergreen speçies and bunchgrasses in the Mediterranean area.

**geologic epochs** Subdivisions of geologic periods into specific chapters.

**geologic eras** The four major divisions of the geologic column: the Precambrian, Paleozoic, Mesozoic, and Cenozoic.

**geologic period** The subdivision of geologic eras.

**geological column** The hierarchy or sequence of time periods of the earth's history revealed by its rocks.

**geostrophic wind** An upper level wind that blows parallel to the isobars as a result of the balance between pressure gradient force and Coriolis force.

**geosynclinal sea** A geosyncline can be very large in area and great in depth and when filled with water can be called a geosynclinal sea. This will very gradually fill with sediment. Such a situation existed between India and Asia as the subcontinent moved towards Asia after detaching from Africa. The sediments of that sea now form the Himalaya Mountains.

**geosyncline** A very large depression in the crust of the earth probably caused by downwarping or settling over a long period of time.

**glacial groove** A groove ground into bedrock by material trapped in the bottom of a glacier.

**glacial stairway** A distinctive step-like profile created by difference in the hardness and therefore the resistance of bedrock to glacial erosion.

**glacier surge** Rapid and unexpected glacial movement.

**glacier trough** The distinctive U-shaped valley created by a glacier.

**glacio-fluvial** Material or deposits that are the product of the action of water derived from ice.

**glacio-lacustrine** Deposits of sediments into a lake formed from glacial meltwater.

**glaciology** The scientific study of natural ice and its actions.

**gleying** The result of oxidation and reduction when waterlogged soil prevents oxygen acting as a sink for electrons.

**Gondwana** The southern of the two continents created by the initial breakup of Pangaea; named after the Gond region of India.

**gorge** A deep steep-sided valley created by the subsidence of a block of land between two parallel faults.

**graded profile** The smoothly descending logitudinal profile displayed by a stream.

**gradient wind** Wind that results from the effects of centripetal and centrifugal forces on a geostrophic wind when the isobars are curved.

**granular disintegration** The breakdown of the matrix that holds individual minerals together to form a rock.

**gravitational water** Water that moves down through the soil as a result of the force of gravity.

**grazing food chain** The link between plants, the herbivores that feed upon them, and the carnivores that feed upon the herbivores.

**greenhouse gases** There are three main greenhouse gases, water vapour, carbon dioxide, and methane. The term refers to their properties that allow shortwave energy to pass down to the earth's surface while restricting the upward escape of longwave or heat energy from earth.

**grey box** A system in which we understand the inputs and outputs and the internal subsystems.

**grikes** Deeply eroded joints in karst pavement.

**ground ice** Ice present within soil or rock, irrespective of its origin.

**groundwater flow** Movement of water beneath the earth's surface under the influence of gravity.

**groundwater recharge** The process that occurs when water moves downward, through the unsaturated zone, stream channels, or recharge wells, to replenish the groundwater.

**groundwater** Water beneath the water table that is interstitial within the rock.

**groyne** A timber, concrete, or stone construction jutting into lake or ocean, usually perpendicular to the coast, intended to maintain a level beach and prevent erosion.

**guano** Organic material built from the droppings of sea birds.

**gully** A V-shaped erosional channel, deeper and larger than a rill, but smaller than a stream.

**gyre** The circulation of water in an ocean basin created by global wind patterns and Coriolis force.

**habitat** The physical location in which the existing organisms are biologically suited to live.

**Hadley cell** The convective circulation cells created on both sides of the heat equator where air rises and the subtropical region where it descends.

**hail** A distinctive form of precipitation composed of spheres of ice that accumulate in the updrafts and downdrafts of a cumulonimbus cloud.

**hamadas** A coarse boulder pavement created in a hot desert when the fine material is blown away.

**hanging valley** A U-shaped valley on the side of a deeper U-shaped valley created where a tributary glacier flowed into the main glacier.

**hardware model** A model that is simply a scaled-down replica of a geologic feature, e.g., a laboratory flume.

**heat capacity** The amount of energy required to raise the temperature of a substance. Water has a high capacity.

**heavy mineral** A mineral of high density, usually with magnetic properties.

**helical flow** A type of water flow, prominent in sinuous rivers, that is important for erosion and deposition of sediment.

**herbaceous** The non-woody portion of plants that is shed during dormant phases; herbs.

**herbivore** A plant-eating organism.

**heterotroph** An organism that derives its energy from the autotrophs (plants).

**high pressure zone** A region where air is more dense because of cooler temperatures; the air descends and spirals outward.

**hoar frost** A white ice deposit with a crystalline appearance.

**Holocene** The most recent geologic epoch.

**horn** A distinctive mountain peak created by the headward erosion of cirques.

**horn peak** The resulting pyramidal peak that is created when several cirques are formed back-to-back.

**horse latitudes** The latitudes associated with the subtropical highs. Historically, Spanish sailing vessels were becalmed in these zones of descending air; the crews were obliged to throw horses overboard because of lack of feed.

**horst** An area of land thrust up between two parallel faults.

**Hortonian overland flow** The concept, proposed by R.E. Horton, that overland flow occurred when rain was more intense than the infiltration capacity of the soil.

**hot spots** Distinctly hotter regions in the mantle which 'burn' through the crust, creating highly active volcanic areas.

**humic acid** Acid created by water passing through decaying vegetation.

**humidity mixing ratio** The ratio of the mass of water vapour to the mass of dry air.

**hummocky moraine** Hilly landforms created by the deposition of till in a region of stagnant ice (also called dead-ice topography).

**humus** The layer of organic matter that forms the surface layer of most soils.

**hurricane** A large spiralling weather system of low pressure, characterized by intense winds and rain, surrounding a small, high pressure area known as the 'eye.'

**hydration** A type of chemical weathering that occurs when the whole water molecule combines with a mineral.

**hydrograph** The chart record displaying the flow of a river or stream.

**hydrologic cycle** The complete circulation of water in all phases through the earth's systems.

**hydrolysis** A chemical weathering process that occurs when free hydrogen or hydroxyl ions in water enter a mineral structure to create a new compound.

**hydrosere** The succession of vegetation that occurs in a wet or flooded surface such as a lake.

**hydrosphere** The water component of the earth's systems.

**hygroscopic particles** Particles of dust with an affinity for water that are important in the process of condensation.

**hygroscopic water** Water surrounding particles in the soil; it is immobile and unavailable to plants.

**ice crystal** A frozen water droplet.

**ice floe** Floating sea ice.

**ice-cored moraine** A moraine that forms with glacier ice enclosed within the sediments at the glacier terminus or margin.

**iceberg** A large floating ice block that has broken away from a glacier.

**icecap** A large ice sheet or continental glacier.

**igneous rock** A general term for rock formed from the cooling of magma or lava.

**illuviated** A pedogenic process: deposition of soil material removed from one horizon to another, usually from an upper to a lower horizon, including silicate clay, hydrous oxides of iron and aluminum, and organic matter.

**index cycle** The sequence of wave patterns in the jet stream, lasting four to six weeks on average.

**infiltration capacity** The maximum rate at which water flows or seeps down through soil.

**inner core** The sold centre of the earth; comprised primarily of iron and nickel.

**inselberg** A small island-like hill or mountain rising above a surrounding pediment or alluvial fan.

**insolation** An abbreviation of the words 'incoming solar radiation,' the shortwave energy of the sun arriving at the top of the atmosphere.

**interception** The process during rainfall when water falls on leaves and other surfaces and does not reach the ground.

**interflow** The process that occurs when soil water moves downslope through a permeable layer, and parallel with the surface.

**interfluves** Literally areas between flows: in glaciated areas they lie between areas of glacial advance, and though covered with ice, show little or no evidence of glaciation.

**interglacial** The overall period between the formation of continental ice sheets.

**interspecies** Between different species.

**interspecific competition** Competition between species, or between individuals of different species.

**interstadial** The period within a glacial period marked by distinct advances and retreats of the ice sheet.

**Intertropical Convergence Zone (ITCZ)** The zone between the tropics where the trade winds converge and rise at the centre of the Hadley cell.

**intraspecies** Within a species.

**intraspecific competition** Competition between individuals of the same species.

**intrazonal soils** A major group of soils associated with poor drainage, salinity, or calcareous parent materials.

**intrusive ice** A layer or mass of water injected under pressure into sediments and then frozen.

**intrusive igneous rock** Rock formed from magma cooling within the crust of the earth.

**inversion** A situation where air temperature increases with height.

**ironpan** In a soil, a layer of iron that has been precipitated from above, mainly by cheluviation.

**iso** Prefix: having equal properties or qualities.

**isobar** A line joining points of equal barometric pressure.

**isogonal** A line joining points of equal magnetic declination.

**isohyet** A line joining points of equal precipitation.

**isomorphous replacement** A basic process when ions in water replace similar ions held within clay.

**isostatic change** Adjustments in the elevation of a landmass due to the load imposed by sedimentation or glaciers or the load removed by deglaciation or erosion.

**isotherm** A line joining points of equal temperature.

**isovel** A line joining points of equal velocity.

**jet stream** A rapidly flowing river of air in the upper westerlies.

**joint** A natural line of separation within rock.

**jokulhlaup** A catastrophic flood of water created by volcanic action under a glacier.

***K*-selection** A *K*-selected organism is usually long-lived, matures slowly, has limited reproductive potential, and is viewed as living a more constant and predictable life with strong competitive ability.

**kame** A Scottish word for a small hill of glacial debris.

**kame terrace** More or less continuous terrace on a hillside formed by meltwater flowing along a glacier margin.

**karst** A general term for the processes and landforms of limestone regions.

**karst pavement** Fissures and grooves formed by dissolution on a limestone surface.

**kata-front** A cyclonic front where the uplift is decreasing, the temperature contrast across the front is weak, and the clouds are thinning.

**katabatic wind** A wind created by downslope cold air drainage.

**kettle** A Scottish word used to describe small hollows left when small buried blocks of ice melted.

**kettle hole** Originally, a pothole in a river. Now, a pothole formed by whirling stones in a stream under a glacier; a circular hollow in a stretch of glacial sands, gravels, and clays.

**kinetic energy** The motion energy within a body, derived from the body's vibration and stated as a temperature.

**Köppen system** The original climate classification system, developed by Wladimir Köppen, based on temperature, precipitation, and natural vegetation. See Appendix II.

**L wave** The slowest wave generated by an earthquake; it has the longest wavelength and can travel right around the earth.

**labile nutrients** Nutrients that are exchangeable.

**laccolith** A relatively small dome of magma that has pushed up under the surface rocks and cooled to form a distinctive intrusive feature; differentiated from a batholith primarily by size.

**lag** The time between the start of rain and the peak in flow as measured and recorded on a hydrograph.

**lag deposit** Either on a stream bed or a hot desert surface, the coarse material that remains after sorting by water or wind.

**lagoon** The shallow body of water enclosed between an active coral reef and the mainland.

**laminar flow** Horizontal flow of water in streamlined layers.

**land breeze** The wind created by the cooling of the land at night; a high pressure area forms and the air moves to the lower pressure area over the water.

**landfast ice** Ice formed during freeze-up, when water in the coastal zone freezes onto the sediments and becomes firmly attached to the land.

**landscape denudation** The gradual, continuous erosion of the landscape ultimately to a level surface, provided there is no interruption.

**latent heat** Heat used in the process of evaporation, melting, or sublimation (gas to solid) and held in storage, to be released in the process of condensation, freezing, and sublimation (solid to gas).

**latent heat of condensation** The release of heat energy as sensible heat when water changes from a gas to a liquid.

**latent heat of evaporation** Energy absorbed and stored when water changes from a liquid to a gas.

**latent heat of fusion** Energy absorbed in the change of state of a solid to a liquid.

**latent heat of vaporization** Another term for the latent heat of evaporation; the energy used when a liquid changes to a gas.

**lateral moraine** Deposits of glacial debris formed along the sides of a glacier.

**latitude** A line, parallel to the equator, encircling the globe; defined by the angle between that line, the centre of the earth and the plane of the equator.

**Laurasia** The northern of two continents created when Pangaea split apart some 200 million years ago; it comprised the present land masses of North America, Greenland, and Eurasia.

**Law of the Minimum** The most limiting factor that controls the response of the individual organism.

**leaching** The removal of soil materials downward in the soil profile by water.

**leaf area index** A measure of the amount of sunlight reaching the ground; a function of crop height, type, and the number of leaves.

**leaf drip** Water that falls to the ground from leaves after they have become saturated.

**lee-side** The side of an object that is away from the direction of movement; the east side of the Rocky Mountains are on the lee-side of the westerly winds.

**lessivage** Leaching, the removal from soil of materials in solution.

**levées** Natural dykes built up along the banks of rivers by material deposited during flood stages; they contain the river within its natural channel.

**lifting condensation level** The height above the ground at which condensation begins in air that is being lifted; a visible line defined by the flat base of cumulus clouds.

**lightning** Flashes of light from electrical discharge occurring between or within clouds, or between clouds and the earth.

**limestone pavement** The upper surface of rock subjected to the processes of chemical weathering; more properly called a karst pavement, as the processes affect several sedimentary rock types.

**lithification** The conversion of raw sediment into rock.

**lithosphere** The rigid portion of the earth's upper layer that includes the plates.

**litter** Layer of organic material, leaves, and twigs lying on the soil surface, which may decompose to form humus.

**Little Ice Age** A distinctly cold period of climate that lasted variously from A.D. 1450 to 1850.

**lodgement till** Typically compact, unbedded, and poorly sorted glacial debris formed from saturated material and usually squeezed into cavities in the ice.

**loess** Windblown silts derived from glacial deposits; extensive deposits formed during the Pleistocene occur in Kansas, southern Poland, and north central China.

**long profile** The longitudinal profile of a river showing the overall gradient.

**longitudinal crevasse** A crack in the brittle upper surface of a glacier formed by differential flow rates.

**longitudinal dunes** Deposits of sand that form parallel to the wind and have steep sides and sharp crests; also called seif dunes.

**longshore current** A lateral flow created by a buildup of water along a coastline.

**longshore drift** The movement of material along a shoreline primarily caused by the angular difference between the wave washing up on the beach and the water running back.

**longwave radiation** Radiation emitted by the earth. Many authorities define longwave radiation as energy at wavelengths above 4 microns.

**lysimeter** An installation designed to directly measure potential or actual evapotranspiration from a vegetated surface.

**mafic minerals** Dense, dark-coloured ferromagnesian minerals, largely silicates.

**magma** A general term for the molten material within the mantle of the earth.

**magmatic fractionation** The creation of different minerals in a body of cooling magma because the constituents tend to crystalize at different temperatures.

**magnetotail** The trailing part of the earth's magnetic field, formed by the pressure of the solar wind.

**mantle plume** Site in the upper mantle where magma rises to the surface of the earth; plumes occur where permanent 'hot spots' occur in the mantle.

**mantle** That portion of the earth that lies immediately below the crust and above the liquid outer core.

**maquis** A shrub formation found in Mediterranean areas, usually evergreen, often including abundant spiny species.

**marine cliff** Cliff formed in areas exposed to large waves created over stretches of open ocean.

**marine limit** The maximum position of the sea level on the landmass as a result of the rise in sea level during deglaciation, before the majority of isostatic rebound has taken place.

**mass balance** The balance between accumulation of ice in the upper portion of a glacier and melting in the lower portion.

**mass movement** A collective term for the processes by which material is moved downslope.

**mathematical model** A conceptual tool that replaces natural phenomena with symbols. A formula is a classic mathematical model.

**matric force** The forces of adhesion and cohesion that counteract gravity and retain water within the soil.

**Maunder Minimum** The period from A.D. 1645 to 1715 when there were apparently far fewer sunspots than at present; named for E. Walter Maunder.

**meandering** A regular, sinuous pattern of side-to-side movement in graded streams, in which the outer portion of each curve is eroded, while the inner portion receives sediments.

**meanders** Bends in streams which have a sinuosity above 1.5.

**mechanical weathering** The breakdown of rock by physical processes.

**medial moraine** Glacial debris within a glacier, formed when two glaciers join and the lateral moraines combine.

**meltout till** Moraine or till, often crudely sorted, that is laid down as glacier ice melts.

**mesa** A high, extensive tableland which consists of horizontal strata capped by a more resistant layer or layers. One or all sides are steep cliffs.

**mesopause** The boundary in the atmosphere between the thermosphere and the mesosphere.

**mesosphere** The highest zone of our atmosphere, between 50 and 80 km above the surface; temperature increases with height in this region.

**metamorphic aureole** A zone of country rock surrounding an igneous intrusion.

**metamorphic rock** Rock that has been changed by heat and/or pressure to a new rock type.

**metasomatic metamorphism** The replacement of elements in the minerals of rocks by active gases and liquids permeating the rocks.

**metastable equilibrium** A condition of stability or equilibrium which if forced becomes temporarily unstable until a new equilibrium is established.

**meteorology** The study of the physics of the atmosphere.

**microclimate** The climate of a small region near or at the ground.

**mid-ocean ridge** Volcanic ridge formed by lava emerging through the rift where the oceanic crust is being pulled apart; usually refers to the mid-Atlantic ridge.

**midlatitude cyclone** A large area of low atmospheric pressure occurring along a front in the middle latitudes. Circulation is clockwise in the southern hemisphere and counterclockwise in the north. They are linked to disturbances in the upper planetary or Rossby waves.

**Milankovitch** Milan Milankovitch; Yugoslavian scientist who calculated the variation in energy balance and thereby temperature that is the result of changes in the tilt and orbit of the earth.

**mineral** A homogeneous, naturally-occurring, inorganic substance that has a constant chemical composition and characteristic physical structure.

**Mohorovicic boundary** A discontinuity that marks that boundary between the mantle and the crust; associated with a zone of low velocity in the seismic wave.

**monadnock** A small island-like hill or mountain rising above a surrounding pediment or alluvial fan.

**monsoon** A tropical climatic region that is identified by a very distinct wet and dry seasonal precipitation pattern.

**mor** A distinctive humus in moorland soils with a clear sequence of plant litter, partly decomposed organic matter, and fully decomposed humus.

**morphological system** A system defined by the physical properties of its components.

**mottles** Varied soil colours created by alternate phases of reduction and oxidation due to fluctuations in the water table.

**mudflat** Accumulations of fine sediments in sheltered inlets or estuaries, or behind spits.

**mudflow** A thick slurry of saturated material that loses stability and flows downslope.

**mull** A humus associated with soils where organisms are very active; the organic matter is mixed with the soil and intimately bound up with the mineral particles.

**mycorrhizal** The association, usually symbiotic, of specific fungi with the roots of higher plants.

**neap tide** Lower levels of high tides that are created when the sun, the earth, and the moon form a right-angle.

**nearshore zone** The area of the coastal zone landward of the position where wave motion intersects the sea bed.

**negative feedback** A situation in which an initial change in a system is damped down by countering processes.

**net radiation budget** Also known as the radiation balance, the difference in energy flux between the incoming solar radiation and the outgoing terrestrial radiation on a horizontal surface. In the daytime it is usually positive with more incoming than outgoing while at night it is usually negative with more outgoing than incoming.

**net radiation** The difference between incoming and outgoing radiation.

**niche** The specific part of a habitat that can support an organism.

**nival** A runoff regime found in a periglacial area characterized by the dominance of the snowmelt period.

**nivation hollow** The overdeepening of a cirque by freeze–thaw action.

**non-clastic rocks** Rocks formed from material precipitated from other rocks by chemical or biological processes.

**noncryotic** At a temperature greater than 0°C.

**non-hygroscopic particles** Particles without an affinity for water; humidities above 100 percent are required before they can act as condensation nuclei.

**non-vascular plant** Bryophytes, plants that have poorly developed systems for transferring nutrients and water; includes mosses, liverworts, and lichens.

**normal fault** A break in rock along a steeply inclined plant where one side slips vertically relative to the other; the movement is caused by tensional forces.

**nuée ardente** A glowing cloud of dust and gas that travels rapidly down the steepest slopes of a volcano, with devastating effect; also referred to as a volcanic avalanche.

**nunatak** That portion of a mountain that rises up through a surrounding glacier; Mt. Erebus is a well-known volcanic nunatak in Antarctica.

**obliquity of the ecliptic** Variation in the tilt of the earth's axis between 21.8° and 24.4°. A complete cycle from one extreme to the other occurs over a period of about 40 000 years.

**obsequent stream** A small stream draining a scarp slope, according to the Davis nomenclature for trellis drainage.

**occluded front** A weather front created when the cold front catches up with the warm front and lifts it aloft.

**ocean basin** The deep portion of the ocean, with depth varying from 3 to 6 km.

**ocean trench** The deepest portion of the ocean; most trenches are located close to continents, where the oceanic crust is subducted under the continental crust.

**offshore zone** Coastal water that is seaward of the foreshore zone.

**open system** A system in which there is a free flow of inputs and outputs across the boundaries.

**open-system pingo** A pingo formed due to freezing of injected water supplied by groundwater moving downslope through taliks.

**orogenesis** Mountain-building process caused by tectonic activity.

**orogeny** A major episode of tectonic deformation.

**orographic precipitation** Rain or snow created when air is lifted up the side of a mountain and clouds are formed.

**outer core** The liquid portion of the earth, lying between the mantle and the solid inner core.

**outwash plain** The area of glacial deposits in front of an icecap that is partly sorted by meltwater.

**overbank flow** Water that leaves the river channel and inundates the floodplain.

**overflow** Overland flow

**overland flow** Water that, during intense rainfall, is unable to infiltrate the soil and therefore flows across the surface.

**overthrust** A fold or fault in which the compressional pressure pushes one side of the fault over the other.

**overthrust fault** A landform caused when one crustal mass overrides another along an inclined fault plane; a result of crustal compression during orogeny.

**oxbow lake** A lake created when a loop of a river is cut off by a diversion of the main channel; named for its shape, which resembles the neck harness for an ox.

**oxidation** The process by which compounds lose an electron.

**ozone** A form of oxygen molecule having three atoms, instead of the normal two.

**P wave** The first seismic wave that emanates from an earthquake; it is the fastest and moves in a push-pull fashion through the earth.

**paleosol** A soil formed on a landscape during the past and subsequently buried by sedimentation.

**Palmer Drought Index** Measurement developed by Wayne Palmer to determine a crop moisture index. Weekly total precipitation and mean temperature are used to establish evapotranspiration deficits and surpluses which are compared to data of previous weeks and long-term records. Maps are produced indicating extreme drought (-4), normal (0), and extreme moist (+4) conditions.

**palsa** A peaty permafrost mound containing layers of segregated ice.

**Pangaea** The name given by Alfred Wegener to the single 'supercontinent' that existed 300 million years ago.

**parallel drainage** Drainage patterns often associated with areas of steep relief or non-cohesive materials.

**parent material** The unconsolidated and more or less chemically weathered mineral or organic matter from which soil is developed by pedogenic processes.

**patterned ground** Ground in regions of permafrost that has distinct geometric shapes created by cryogenic processes.

**ped** An aggregate of soil formed from smaller particles brought together by various forces.

**pediment** The gently concave portion at the bottom of a slope.

**pedogenesis** The soil-forming process, especially from unconsolidated parent material.

**pedon** A three-dimensional soil column extending to the base of the soil.

**peneplain** A land surface that is low and undulating, representing the final stage of landmass denudation; usually used to refer to ancient surfaces in humid environments.

**peneplanation** The final stage of the Davis cycle of landscape evolution, when the surface has been reduced to an almost level plane.

**Penman formula** A method for calculating potential evapotranspiration.

**penumbra** The outer, less dark area surrounding the dark complete shadow produced when an opaque object blocks light from an external source.

**perched aquifer** Groundwater contained in a porous rock above the larger regional water table, often resulting in springs along the sides of valleys.

**percolation** The drainage of water downward through the soil.

**periglacial** The conditions, processes, and landforms associated with cold, nonglacial environments.

**perihelion** The point at which the earth is closest to the sun in its annual orbit.

**permafrost creep** The slow, continuous deformation of permafrost on slopes due to gravity.

**permafrost** Perennially cryotic ground.

**permafrost table** The upper boundary of permafrost, generally coinciding with the bottom of the active layer.

**permeability** The extent to which something can by penetrated by, or allow the passage of, a fluid.

**pH scale** Potential hydrogen; the standard measure of acidity (low pH) or alkalinity (high pH) of a substance. Neutral pH is 7.2, the pH of pure water.

**photosphere** The luminous envelope of the sun, from which its light and heat radiate.

**photosynthesis** The conversion of energy from the sun into chemical energy by plants.

**physical model** A scaled-down model of reality, similar to a hardware model.

**physical weathering** The breakdown of rock by a physical or mechanical process that weakens the matrix holding the particles of rock together.

**physiognomic classification** A means of categorizing vegetation types based on the architecture or growth form structure of the plants.

**piedmont glacier** A glacier that spreads out onto the lowlands at the end of a valley.

**pillow lava** Rapidly cooled lave formed on the bottom of the ocean, where magma breaks through the thin crust of the sea floor.

**pingo** A large perennial ice-cored mound found in permafrost areas.
**plant community** An assemblage of plants.
**plate tectonics** The movement of the plates that make up the crust of the earth on convective currents in the mantle.
**playa lake** A lake found in a desert-environment valley; periodically dries up, leaving extensive areas of mineral deposits.
**Pleistocene** The period from about three million years ago to 12 000 years ago, during which large glaciers formed and retreated several times.
**plucking** The removal of material from bedrock by glacial ice.
**plunge pool** The deep, smooth hollow gorged out by water at the bottom of a waterfall.
**pluvial** A period of higher rainfall in desert latitudes coincident with periods of glaciation in higher latitudes.
**poikilohydric** Drought-tolerant; plants, for example, having the ability to become dormant in dry periods, but which upon wetting will quickly begin normal physiological processes.
**point of zero energy balance** The latitudinal point at which there is an equal amount of incoming solar radiation and outgoing longwave radiation resulting in an energy balance.
**polar cell** The convective cell of atmospheric circulation completed by air rising in the middle latitudes and descending over the poles.
**polar easterlies** Winds that blow out from the polar high pressure area and are deflected in an easterly direction by the rotation of the earth.
**polar jet stream** The strong circumpolar river of wind that blows from east to west in the high middle latitudes.
**polder** A large dry area of land reclaimed from the sea.
**polycyclic landform** A land feature that passes through several erosion cycles or part-cycles.
**polygenetic landform** A feature created by several different processes.
**polypedon** A group of pedons that are contiguous.
**polysaccharide** A complex organic polymer.
**population** A group of individuals of the same species.
**pore ice** Found in permafrost, it is subsurface water frozen in place, that is, in the pore space of sediments.
**porosity** The total volume of pore space within a rock or soil, expressed as a percentage.
**positive feedback** A situation in which an initial change in a system is amplified.
**post-glacial uplift** The spontaneous rise of a continental crust after the melting of an ice sheet, restoring isostatic equilibrium.
**potential energy** The energy, related to gravity, that could be generated if an object were allowed to fall.
**potential evaporation** A measure of the ability of the atmosphere to remove water, assuming unlimited supply.
**potential evapotranspiration** The amount of water that plants in a given area could evaporate or transpire, given an unlimited water supply.
**prairie** A continuous area of tall grassland; usually refers to North America.
**precession of the equinoxes** The progressive change in the date on which the earth is closest to the sun.
**pressure gradient force** The difference in pressure between two points as a function of the distance between them.
**pressure melting point** The point at which the overlying weight of glacier ice begins to increase the melting temperature of the ice.
**pressure melting** Situation where ice at the glacier bed may melt at temperatures below 0°C due to high stresses.
**pressure release** The expansion of rock after it has been released from the burden of overlying material.
**pressure ridge** A wall of hummocked ice, formed where two ice floes press against each other.
**principal shock** The second, most severe shock, often lasting only a few seconds, after an earthquake has occurred.
**principal stress** The stress vector in the ice as a result of friction at the glacier bed and along the glacier margin.
**process-response system** Formed when a cascading system and morphological system become linked.
**psammosere** The sequence of development of vegetation on sand dunes.
**push moraine** A moraine created when a glacier overrides its own deposits and bulldozes them into a mound.
**qanat** A series of tunnels dug at the water table on a slope to increase the flow of water in arid regions.
**quarrying** The detachment of particles by the action of the erosive agent itself.

**r-selection** An *r*-selected organism usually develops rapidly, is capable of reproduction early in its life history, is generally small in body size, and relatively short-lived. These organisms generally have high numbers of offspring.

**radiant energy** The energy transmitted from the sun in wave form.

**radiation balance** The difference between incoming and outgoing radiation.

**radiation fog** Fog formed by cooling of air near the ground below the dew point temperature, through the loss of heat energy by radiation.

**rain** Precipitation in which drops are greater than 0.5 mm in diameter, and usually between 1 and 2 mm.

**rain shadow** The area of decreased precipitation on the lee or downwind side of mountains. The interior of British Columbia is in the rain shadow of the coastal range of the Rocky Mountains.

**raindrop impact** The amount of energy imparted by a falling raindrop to loosen soil particles.

**recessional moraine** The mound of glacial debris created when a glacier pauses in retreat.

**rectangular drainage** The drainage network that forms on a landscape with a rectilinear joint pattern, where larger streams run at right angles to each other, and each stream has sections of roughly equal length. The characteristic 'grid' pattern is usually imposed by the structure of the bedrock.

**recumbent fold** An anticline that has been pushed by compressional tectonic forces so that it folds over on itself.

**recurrence interval** The time between repeating physical events, for example, the time between floodings in a river basin.

**reduction** The process of reducing by chemical or electrochemical means; deoxidation

**reef limestone** Calcareous rock formed by reef-building organisms.

**regelation** The process that occurs when ice that has been melting under pressure refreezes as the pressure is released.

**regelation layer** A debris-laden layer in the bottom of a glacier created when pressure-melted water refreezes.

**regional metamorphism** A large area over which heat, pressure, and volatile material escaping from the underlying magma have changed the existing rock.

**regolith** The transition layer between bedrock and the soil horizons represented by weathered rock particles and minerals. It is often referred to as the weathering front.

**regs** Areas of gravel found in deserts.

**relative humidity** The ratio of the amount of water vapour in the air against the amount that it could hold at a given temperature. It is expressed as a percentage.

**relict** A feature that was formed during a previous period under different conditions than those prevailing at the present. For example, buried permafrost that remains because of the insulating effects of the overlying soils, but which would not return if thawed out.

**residual uplift** The final slow stages of isostatic adjustment to deglaciation at the end of the Pleistocene.

**restrained rebound** The first stages of the isostatic adjustment to deglaciation which is primarily elastic in nature.

**retrogressive thaw slump** A thermokarst landform developed on slopes; consists of an ice-rich headwall, thawed overburden, and a fluid mudflow.

**reverse fault** Compressional forces cause one side of a fault to be upthrust relative to the other.

**riffles** A buildup of regularly spaced sedimentary material along the channel of a stream.

**rift valley** A long, steep-side valley, created by parallel faults separated by tensional tectonic forces, allowing the centre section to drop down.

**rill** The smallest form of channel flow; also called gully flow.

**rillenkarren** Grooves in limestone created by chemical weathering in the natural joint structure.

**rime** Ice formed when supercooled water droplets freeze to a surface that is below freezing temperature.

**rinnenkarren** Grooves in limestone joint structures; larger than rillenkarren.

**rip current** In coastal water, a current formed when strongly outflowing surface water meets the incoming surf.

**ripple index** The ratio of wavelength to height for a ripple field.

**ripples** Small-scale dunes or ridges of sediment, parallel to the flow of water in a stream, usually only a few centimetres in height and spacing.

**river capture** Occurs when a tributary of one stream erodes backward into the slope until it cuts into the drainage basin of another stream; a tributary of this stream is then redirected to flow down the new stream.

**river piracy** Another term for river capture.

**river regime** The pattern of the discharge of a river through time; for example, the changes in discharge that occur from winter low flow, to spring snow melt, to summer and fall responses to precipitation, and returning to winter low flow is the annual regime.

**river terrace** Flat area along the side of a river valley created by changes in the terrain or flow of the river.

**roche moutonnée** A small mound smoothed on the upslope side and plucked on the downslope slide by a glacier; literally, 'rock sheep,' describing the shape they take lying in a field.

**rock flute** Linear ridges in the grooves in the bedrock, made by glacial erosion, that are aligned in the direction of the flow of the ice.

**rockfall** Material breaking from a steep surface and falling to accumulate at the bottom of the cliff.

**rockslide** Movement of an individual rock mass sliding down a gentle slope.

**rogen moraine** Long, low ridge of moraine formed at right angles to the direction of flow of the ice.

**Rossby waves** Wave-like patterns in the circumpolar flow of the upper atmosphere.

**rotation period** An interval of time when land is left uncultivated before it is planted again with a particular crop.

**rotational slip** The downslope sweeping movement of material along clearly defined planes of weakness.

**runnel** A small stream.

**runoff** The flow of water over the ground in distinct channels.

**S wave** The second seismic wave emanating from an earthquake; characterized by an up-and-down motion.

**salinity** The quantity of salt (sodium chloride) in water.

**salinization** Usually used to refer to the buildup of salt in soil.

**salt crystallization** The formation of salt crystals between rock particles during evaporation. The pressure created leads to physical weathering in arid regions.

**salt marsh** Vegetated mudflats in sheltered ocean estuaries.

**saltation** The leaping motion of sand particles as they are lifted from the bed of the stream and carried a short distance downstream; from the Latin, meaning 'to jump.'

**sand bar** A general term for various deposits of sand in coastal environments.

**sandur** A large, flat plain of glacial sediment in front of a glacier, usually subjected to fluvial sorting by meltwater; also called an outwash plain.

**saturated adiabatic lapse rate** The rate of change of temperature with height in an adiabat after condensation has begun.

**saturated overland flow** After the surface layers of soil become saturated, subsequent rainfall runs downslope across the surface.

**saturation equilibrium** The condition when the number of molecules evaporating from the surface water equals the number entering.

**saturation vapour pressure** The portion of atmospheric pressure created by the water vapour molecules.

**savanna** A vegetation biome that ranges from grassland to woodland with trees widely spaced.

**saxicolous lichen** A lichen growing on a rock substrate.

**scarp** An abbreviated form of escarpment, used in such terms as 'marine scarp' or 'fault scarp.' Refers to a steep slope between two relatively level surfaces.

**scattering** The redirection of shortwave energy after it has interacted with gas molecules and particles in the atmosphere without change of wavelength.

**scavenger** An animal whose food is derived from already-dead organisms.

**sclerophilous** Plants with thick, firm-textured leaves that are usually evergreen and resistant to water loss.

**scour** The grinding, eroding, and polishing of bedrock by a glacier.

**scour and fill** The continuous process of erosion, deposition, and erosion at the stream bed.

**sea breeze** An onshore wind created by the heating of land during the daytime to create a low pressure region relative to the cooler ocean.

**sea floor spreading** The splitting and separation of the sea floor by convective flow in the underlying mantle.

**second-year ice** see first-year ice

**sediment** Rock particles deposited by ice, water, or wind.

**sedimentary rock** Rock formed from deposits of sediment, usually on the bottom of an ocean.

**segregated ice** Ground ice formed from water attracted by suction forces to the freezing front, where it forms discrete layers or lenses.

**seif dune** A sand ridge that has formed parallel to the prevailing winds; may run for many kilometres.

**seismic waves** The shock waves that emanate from an earthquake and pass through the crust of the earth.

**sensible heat** Heat that can measured by a thermometer and therefore sensed by humans. It is a measure of the rate of molecular movement, molecules moving more rapidly as the temperature increases.

**seral succession** The sequence of change in soil and microclimatic conditions that are brought about by vegetation.

**serotinous cone** Seed cones that remain on the tree and retain viable seeds, often for many growing seasons.

**Seventh Approximation** The final and successful attempt to create a soil classification system.

**shear planes** Clearly defined planes in ice along which fractures may occur under internal stress.

**shear strength** The capacity of a material to resist shear.

**sheetflow** Broad expanse of water derived from precipitation, flowing downslope when existing channels cannot carry it.

**sheetwash** A thin sheet of water flowing over the surface, especially in areas of smooth soil.

**shield volcano** A gently sloping volcano formed from low-viscosity lava, created when magma passes up through oceanic crust.

**shortwave radiation** Energy from the sun that is radiated in wave form, with wavelengths less than 4 microns.

**silicate magma** Magma formed at depths of 15 to 25 km in the earth, from which most igneous rocks are created.

**sill** An intrusive igneous feature formed when lava flows into horizontal planes of weakness within rock.

**sinuous** The adjective describing the tendency for gas or water to meander from side to side when passing through a uniform medium.

**slab failure** The downslope movement of a large, flat piece of rock due to failure along an inclined bedding plane.

**slaking** The breakdown of soil aggregates resulting from alternating saturation with water and drying-out.

**slope decline** The concept developed by W.M. Davis, that slopes gradually decline to a peneplanation or flat surface.

**snow line** The level on a mountain above which the snow never completely melts.

**snow** Precipitation in which clusters of ice crystals form snowflakes.

**soft hail** Opaque pellets of ice 2–5 mm in diameter falling in showers; also know as snow pellets.

**soil** The mixture of rock debris and organic material that develops at the earth's surface.

**soil creep** The gradual and inexorable downslope movement of soil due to gravity.

**soil erosion** The processes that wear away the land surface by natural agents (running water, ice, wind, wave action), and the resulting transport of the rock debris. Erosion does not include the weathering of rocks in situ.

**soil profile** The vertical cross section of the soil in which the layers or horizons are visible.

**soil reaction** Also know as the pH, it is a measure of the number of free hydrogen ions in the soil; the more ions, the more acid the soil.

**soil suction/soil tension** The terms used to describe the force in the soil by adhesion and cohesion which causes water to be drawn upward against the pull of gravity.

**solar constant** The average amount of solar radiation received at the top of the atmosphere. It has a value of 1370 W $m^{-2}$.

**solar energy** Energy with electromagnetic radiation from the sun as its source.

**solar wind** The outflow of particles, notably electrons and protons, from the sun; varies according to the amount of activity on the surface of the sun, seen as flares, prominences, and sunspots.

**solifluction** Slow downslope movement of thawed soil over permafrost or seasonally frozen ground.

**solution** Substances dissolved in water transported as individual ions.

**solution basin** Small bowl-shaped hollow in exposed bedrock produced by solutional weathering processes.

**Southern Oscillation** The periodic oscillation between high and low pressures on each side of the Pacific Ocean that creates changes in climate.

**species** The fundamental biological classification, comprising a subdivision of a genus and consisting of a number of plants or animals all of which have a high degree of similarity, can generally interbreed only among themselves, and show persistent differences from members of allied species.

**spit** An accumulation of sediment outward from the shore.

**splay** Material that has become spread over a floodplain area after being carried through breaks in low sections of a river course.

**spring tide** The highest tides, created when the moon, sun, and earth are aligned.

**stable** The condition when an adiabat is cooler than the surrounding air, or when warm air lies over cool air.

**stadial** A shorter cold period within the overall glacial epoch.

**stalactites** Columns formed from accumulations of precipitated material built down from the ceiling of a cave as water percolates downward.

**stalagmites** Columns built up from the floor of a cave by accumulations of precipitated material as water percolates through the earth and drips to the ground.

**standard atmospheric pressure** The pressure exerted by the weight of the air above the earth's surface; at sea level, 1 atmosphere = 101.325 kiloPascals, or 1013.25 millibars.

**standing wave** The wave form produced by two wave motions moving simultaneously through a medium in opposite directions with identical amplitude, frequency, and velocity.

**steady state equilibrium** The tendency of a system to maintain its general structure and character over long periods of time.

**stemflow** That portion of precipitation that flows down the trunk of trees.

**steppe** A grassland with few shrubs and lacking trees; usually refers to Asia.

**stomata** The small pores in the surface of a leaf through which water is transpired.

**stone net** A distinctive pattern of stone formations in regions of permafrost.

**stone polygons** Distinctive geometric shapes of stones in regions of permafrost.

**stone stripes** Lines of stones on a slope created when stone polygons are stretched by solifluction.

**storm hydrograph** The chart record of a specific storm event; the change in streamflow recorded during the storm.

**stoss-side** The upslope side of a glacial feature such as a roche moutonnée or drumlin.

**stratiform** A general term for layer-type clouds.

**strato-volcano** A volcano alternately made up of layers of ash and lava formed from magma that has pushed up through continental crust.

**stratopause** The boundary between the stratosphere and the mesosphere.

**stratosphere** The layer in the atmosphere above the troposphere, between 18 and 45 km above the surface of the earth.

**stream discharge** The total volume of water flowing in a stream channel.

**stream metamorphosis** Changes in the form of a river channel due to human activity.

**striation** A scratch or groove in bedrock gouged by rock embedded in the bottom of a glacier.

**strike** The direction of a horizontal line on an inclined rock layer, at right angles to the direction or the true dip of the rocks.

**structural landforms** Large-scale landforms created by tectonic movements.

**structure** The disposition and arrangement of rocks making up the earth's crust; as well, the morphological features of rocks.

**subarctic** The zone of open-canopy coniferous forest that lies south of the Arctic tundra and north of the taiga.

**subduction** The movement of one global plate down under the edge of another; from the Latin, 'to lead under.'

**subglacial** Processes and events occurring under the ice.

**subglacial sediment** Sediment located on the underside of a glacier.

**sublimation** The change of phase directly from solid to gas or gas to solid; the same term is used for both processes.

**submarine canyon** A deep canyon incised into the continental shelf.

**subnivean animal** An animal that must live beneath the snow because of poor adaptations to cold and moisture loss in winter.

**subsequent stream** A stream running at right angles to the main or consequent stream, according to the stream nomenclature of W.M. Davis.

**subsidence** The downward movement of a large mass of air, usually in the centre of an anticyclone or high pressure area.

**subtropical high** Regions of high pressure on the poleward side of the Hadley cell and generally located between 15° and 30° of latitude.

**subtropical jet stream** The river of strong winds in the upper atmosphere on the poleward side of the Hadley cell.

**supercooled** Water that exists in liquid form below freezing temperature.

**superimposed ice** Ice forming during snowmelt on a glacier when meltwater percolates through the previous winter's snowpack and refreezes at the base of the layer.

**supersaturated** Air in which water vapour exists even though the dew point temperature has been reached.

**supraglacial** Processes or features that occur on the surface of the ice.

**supranivean animal** An animal that is adapted to winter conditions above the snowpack; waders wallow through the snow with long legs, floaters are light enough to stay on the surface of the snow.

**surf zone** The zone above the point on a beach where the wave breaks, but below the swash zone.

**surge** A sudden and rapid advance of a glacier.

**suspension** Fine material such as clay carried in a stream, requires very calm water in order to settle out.

**sustained yield** In forestry, a practice aimed at ensuring the forest remains a productive resource.

**swash zone** The zone in which water runs up the beach as an increasingly shallow, turbulent film.

**sweep zone** Over time a beach undergoes periods of erosion and deposition. The sweep zone is defined by the extremes of elevation changes at each point on a profile down a beach.

**syncline** A downward fold in sedimentary rock.

**syngenetic** Refers to ground ice that formed at roughly the same time as the deposition of the earth material in which it occurs.

**syngenetic wedge** A wedge-shaped body of ice that develops at the same time as permafrost forms within newly depositied sediments (see wedge ice).

**synusia** A group of plants of similar form, each filling the same ecological niche.

**tabular iceberg** A large flat-topped ice mass floating in the ocean.

**talik** A body of unfrozen ground in a permafrost area.

**talus cone** A talus accumulation on a slope, which forms a cone with the apex at top.

**talus slope** The surface of a talus cone.

**tarn** A small lake in a rock basin in a cirque or glacial trough.

**tephra** General term for particles of solidified magma ejected from the vent of a volcano by gas pressure; includes all particle sizes.

**terminal moraine** Glacial debris accumulated at the front of a glacier that has reached the point of maximum advance.

**terminal velocity** The limiting uniform velocity attained by a falling or sliding body when resistance (of air or friction) is equal to the force of gravity.

**texture** Refers to the feel of soil rubbed between the fingers.

**thalassostatic terrace** A terrace created in a river valley when the sea level is lowered; from the Greek word thalassa: 'sea.'

**thalweg** The deepest point in the cross section of a river and the line of maximum velocity; also referred to as the navigable channel.

**thermal energy** Heat energy; the energy involved in the motion of extremely small components of matter.

**thermal metamorphism** The process that changes rock from one form to another by heat alone.

**thermo-erosional niche** A recess under a river bank or coastal cliff formed by thermal and mechanical erosion of ice-rich permafrost.

**thermokarst** Thaw of ground ice in a permafrost area, and the irregular hummocky topography that results.

**thermokarst lake** A lake formed in a depression resulting from the melting of ground ice in a permafrost area.

**thermosphere** The region in our atmosphere above the mesosphere. It is the topmost layer, in which particles from the sun give high temperatures but low heat.

**Thornthwaite system** A climate classification system based on the concept of potential evapotranspiration and soil moisture budget. A departure from vegetation-based systems.

**throughfall** That portion of rainfall that passes through the canopy of a tree.

**throughflow** Movement of water downslope through the regolith when it cannot percolate downward because the flow is too great.

**thufur** Small hummocks in periglacial environments.

**tidal current** Ocean currents created by the rise and fall of the tides. Usually their force is limited, but can be very powerful in restricted areas such as the Bay of Fundy.

**tidal range** That portion of a coastline between the highest and lowest tides affected by the erosional and depositional forces of the water.

**till** Glacial debris formed beneath the active ice as a result of downwasting, but deposited directly from the ice.

**till fabric** The predominant orientation of the long axis of the stones in till.

**tillplain** A rock formed from glacial till deposited in earlier glaciations. Tillites provide evidence of the extent of those glaciations.

**timber line** The altitudinal limit of trees, beyond the closed forest at lower elevations.

**tombolo** An accumulation of sand in the form of a bar between the mainland and an island.

**tor** A prominent, isolated mass of jointed and weathered rock, usually granite.

**tornado** An area of extremely intense low pressure that produces a rapidly spinning funnel that descends from a very unstable cumulonimbus cloud.

**traction** The hydraulic movement of large rocks and boulders along the bottom of a stream.

**trade winds** The surface winds that blow from the subtropical high to the Intertropical Convergence Zone and that are deflected eastward on both sides of the equator.

**traditional logging** In North America, this practice would entail removal of all trees from a cut block of woodland.

**transform fault** A fault in the crust of the earth created by two sections of the crust moving horizontally in opposite directions to each other.

**transpiration** Moisture that is passed from the ground through the roots of a plant, and then through the leaves to the atmosphere.

**transport** The general term in the earth sciences for the movement of sediment from land to sea.

**transverse crevasse** A crack in the brittle layer of the ice surface, formed where the ice is subject to tension from obstructions in the floor or changes in the bedrock slope.

**transverse dune** Long, relatively straight or somewhat serated ridge of sand oriented at right angles to the wind direction.

**tree line** The latitudinal limit of trees, beyond the closed forest at lower latitudes.

**tree ring** A concentric ring of light and dark material created during one growth year, visible in the cross section of a tree trunk.

**tree throw** Uprooted trees retain soil in their root ball when overturned by wind or snow loading, resulting in irregular surface topography and a mixing of horizons.

**trellis drainage** A rectilinear drainage network, especially in areas with folded sedimentary rock.

**trophic levels** The major and distinctive levels of organisms through which energy is cycled in the ecosystem.

**trophic pyramid** The pyramid created by the increasing numbers of organisms necessary at one level of the food chain to support organisms at the next higher level.

**tropopause** The boundary between the troposphere and the stratosphere. It is marked by two distinct breaks that correlate with the locations of the polar and subtropical jet streams.

**troposphere** The area of the atmosphere from the surface to the tropopause. It varies in thickness between the poles (8–10 km) and the equator (17–18 km).

**trough** A long valley of low pressure; an easterly wave is a typical example.

**truncated spur** During glaciation a valley tends to be straightened by glacial erosion. This removes the ends of higher elevation areas which existed in the sinuous course of the river, leaving ridges with sharply eroded ends.

**truncation** A spur or ridge, extending out into a glacial valley, that has been eroded or cut off by a glacier moving down the valley.

**tsunami** Large, long-wavelength waves generated in the ocean by an earthquake or displacement caused by an earthquake.

**tundra** Arid, treeless regions of intense cold in which permafrost occurs.

**turbidite** Sediment carried by a turbidity current.

**turbidity current** Water mixed with a dense mass of mud and silt moving over the ocean floor under the effect of gravity.

**turbulent flow** Interference with the laminar or smooth flow of water by an object so that energy is dissipated.

**typhoon** The name given in the past to a hurricane in the Pacific Ocean. Typhoons tend to be more intense and much larger than their Atlantic Ocean counterparts. Now called a hurricane.

**umbra** The dark, complete shadow produced when an opaque object blocks light from an external source.

**underfit stream** A stream that occupies a valley larger than it was capable of forming.

**underflow** Water entering a lake or ocean with a sediment content that makes it denser than the surrounding water.

**uniformitarianism** The philosophical view in science that arose from Darwin's concept that things evolved very slowly over long periods of time, and that present processes were the same in the past.

**unstable** The condition when an adiabat or parcel of air is warmer than the surrounding air.

**upfreezing** The freezing of the ground that causes larger materials to be pushed up to the surface.

**upwelling** Cold water that rises to the surface along coastal regions to replace warmer water blown away by offshore winds.

**urban heat island** The distinct dome of warm air, formed over a city, created by physical changes and human activity.

**utilidor** An insulated tunnel enclosing water and sewage services, built above the ground to prevent thawing of permafrost.

**valley glacier** A glacier that has formed and flows within a mountain valley.

**vapour pressure** That part of the total atmospheric pressure created by water vapour.

**varve** A layer of coarse and fine sediment that represents one year's deposit in glacial lake sedimentation. The number and thickness of the layers can indicate the age of the sediment and general climatic conditions at the time of formation.

**vegetation** The total plant cover of an area, consisting of one or more communities.

**virga** Precipitation that evaporates before if reaches the ground.

**volatiles** Larger pieces of material thrown out by a volcano.

**wadi** Deep gully or trench cut by intense storm and flash flood events in desert regions.

**warm front** The retreat of cold air that allows warm air to advance over a region.

**washboard or deGeer moraine** A series of nearly parallel, relatively low, sandy moraine ridges that occur in an area that has been subjected to the advance, stagnation, and melting of an ice sheet or glacier.

**water table** The upper surface of the groundwater.

**water vapour** Water in the vapour state.

**watershed** The height of land separating one drainage basin from another.

**wave height** The vertical distance, in a wave, from the trough to the crest of a wave.

**wave of translation** A water wave after its circular motion has been broken at the bottom by contact with the shore and its energy translated to a foreward motion.

**wave refraction** Along a coastal area, the change of direction of a wave as it touches the bottom.

**wave-cut notch** A notch cut by wave action in a cliff composed of resistant rock.

**wavelength** The distance, in a wave, from one trough to another or from one crest to another.

**weathering** Processes at the earth's surface that physically disrupt and/or chemically decompose rock and regolith.

**weathering front** The boundary separating weathered and unweathered rock.

**weathering zone** An area where a specific set of weathering processes are active.

**wedge ice** Ground ice formed by precipitation repeatedly entering an open thermal contraction and freezing as an ice vein.

**wedge** The portion of the earth's crust that rides over the suducted crust or slab, and through which magma rises to create volcanoes.

**weight-load-on-track** The mass of an animal divided by the total foot area in contact with the surface.

**westerlies** The winds of the middle latitudes region in which the prevailing direction is from west to east.

**white box** The total system within which are enclosed the grey and black boxes.

**whole-tree logging** see full-tree logging

**wilting point** The point at which a plant is unable to extract any water from the ground.

**windgap** A valley, cut through a scarp, that has been abandoned because the stream that formed it has been captured by another stream.

**xerosere** A succession of vegetation in a dryland region.

**zero-tillage** The practice of agriculture without plowing the land.

**zonal index** The difference in barometric pressure between two parallels of latitude, normally 5°.

**zone of recurvature** The latitudinal zone within which hurricanes and tropical storms begin to turn away from their westward track toward the north and then the east.

# Bibliography

Arkley, R.J. (1963), "Calculation of Carbonate and Water Movement in Soil from Climatic Data," *Soil Science*, 96, 239–48

Atkinson, B.W. and Smithson, P.A. (1972), "An Investigation Into Meso-scale Precipitation Distributions in a Warm Sector Depression," *Quart. Jl. Roy. Meteor. Soc.*, 98, 353–68

Atkinson, B.W. and Smithson, P.A. (1976), "Precipitation," in T.J. Chandler and S. Gregory (eds.), *The Climate of the British Isles*, Longman, 129–82

Atkinson, T.C. (1971) "Hydrology and Erosion in a Limestone Terrain" (Unpublished PhD thesis, University of Bristol)

Austin, P.M. and Houze, R.A. (1972), "Analysis of the Structure of Precipitation Patterns in New England," *J. Appl. Meteor.*, 11, 926–35

Bagnold, R.A. (1954), *The Physics of Blown Sand and Desert Dunes*, Methuen

Balme, E.O. (1953), "Edaphic and Vegetational Zoning on the Carboniferous Limestone of the Derbyshire Dales," *J. Ecology*, 41, 331–44

Barry, R.G. (1969), "Evaporation and Transpiration," in R.J. Chorley (ed.), *Water, Earth and Man*, Methuen, 169–84

Barry, R.G. (1979), "Recent Advances in Climate Theory Based on Simple Climate Models," *Progr. in Phys. Geography*, 3, 119–31

Barry, R.G. and Chorley, R.J. (1982), *Atmosphere, Weather and Climate* (4th ed.), Methuen

Beeton, A.M. (1971), "Eutrophication of the St. Lawrence Great Lakes," in T.R. Detwyler (ed.), *Man's Impact on the Environment*, New York: McGraw-Hill, 233–45

Bennett, H.H. (1939), *Soil Conservation*, New York: McGraw-Hill

Bennett, R.J. and Chorley, R.J. (1978), *Environmental Systems*, Methuen

Bliss, L.C. (1977), *Truelove Lowland, Devon Island, Canada: A High Arctic Ecosystem*, Edmonton: University of Alberta Press

Bonython, C.W. and Mason, B. (1953), "The Filling and Drying of Lake Eyre," *Geogr. J.*, 119, 321–30

Bormann, F.H. and Likens, G.E. (1979), *Pattern and Process in a Forested Ecosystem*, New York: Springer-Verlag

Boulton, G.S., Jones, A.S., Clayton, K.M. and Kenning, M.J. (1977), "A British Ice-sheet Model and Pattern of Glacial Erosion and Deposition in Britain," in F.W. Shotton (ed.), *British Quaternary Studies: Recent Advances*, Clarendon Press, 231–46

Bourcher, K. (1975), *Global Climate*, English Universities Press

Bowen, A.J. (convenor) (1981), *Basic Nearshore Processes*, Burlington, Ont.: Associate Committee for Research on Shoreline Erosion and Sedimentation, National Research Council Canada

Brady, N.C. (1973), *Nature and Properties of Soils*, Macmillan

Bridgman, H.A. (1969), "The Radiation Balance of the Southern Hemisphere," *Arch. Meteor. Geophys. u. Bioklim.*, Ser. B, 17, 325–44

Briggs, D.J. (1977), *Soils*, Butterworth

Brown, L. (1976), *British Birds of Prey*, Collins

Bruce, J.P. and Clark, R.H. (1966), *Introduction to Hydrometeorology*, Pergamon Press

Budyko, M.I., Yefimova, N.A., Aubenok, L.I. and Strokhina, L.A. (1962), "The Heat Balance of the Surface of the Earth," *Soviet Geogr.*, 3, 3–16

Bullen, K.E. (1955), "The Interior of the Earth," *Scientific Amer.*, report no. 804, San Francisco: Freeman

Burnham, C.P. (1970), "The Regional Pattern of Soil Formation in Great Britain," *Scott. Geogr. Mag.*, 86, 25–34

CRM Books (1973), *Geology Today*

Carson, M.A. and Petley, D.J. (1970), "The Existence of Threshold Slopes in the Denudation of the Landscape," *Institut. Brit. Geogr., Trans.*, 49, 71–95

Chandler, T.J. (1965), *The Climate of London*, Hutchinson

Changnon, S.A. (1971), "A Note on Hailstone Size Distributions," *J. Appl. Meteorol.*, 10, 168–70

Changnon, S.A. (1976), "Inadvertent Weather Modification," *Water Res. Bull.*, 12, 695–718

Chapman, S.B. (1967), "Nutrient Budgets for a Dry Heath Ecosystem in the south of England," *J. Ecol.*, 55, 677–89

CLIMAP Project Members (1976), "The Surface of the Ice Age Earth," *Science*, 191, 1131–7

Collinson, A.S. (1977), *Introduction to World Vegetation*, George Allen and Unwin

Cooke, G.W. (1983), *Fertilizing for Maximum Yield*, Granada

Coope, G.R. (1975), "Climatic Fluctuations in Northwest Europe Since the Last Interglacial, Indicated by Fossil Assemblages of Coleoptera," in A.E. Wright and F. Moseley (eds.), *Ice Ages:*

*Ancient and Modern, Geological Journal*, spec. issue no. 6, Liverpool, 153–68

Critchfield, H.J. (1983), *General Climatology* (4th ed.), Englewood Cliffs, NJ: Prentice-Hall Inc.

Crowe, P.R. (1951), "Wind and Weather in the Equatorial Zone," *Instit. Brit. Geogr., Trans.*, 17, 23–76

Crowe, P.R. (1971), *Concepts in Climatology*, Longman

Dansereau, P.M. (1957), *Biogeography: An Ecological Perspective*, New York: Ronald Press

Davenport, A.G. (1965), "Relationship of Wind Structure to Wind Loading," *Proc. Conf. Wind Effects on Structures*, Symp. 16, vol. 1, HMSO, 53–102

Davies, E.B., Hogg, D.E. and Hopewell, H.G. (1962), "Extent of Return of Nutrient Elements by Dairy Cattle: Possible Leaching Losses," Jt Meeting of International Soil Science Soc., Commission IV & V, New Zealand, 715–20

Decker, R. and Decker, B. (1982), "Volcanoes and the Earth's Interior," *Readings from Scientific American*, San Francisco: Freeman

Department of Environment (Water Data Unit) (1983), *Surface Water: United Kingdom, 1977–1980*, HMSO

Department of Navigation and Ocean Development (1977), *Assessment and Atlas of Shoreline Erosion Along the Californian Coast*, State of California Dept of Navigation and Ocean Development

Derbyshire, E., Gregory, J.J. and Hails, J.R. (1979), *Geomorphological Processes*, Butterworth

DeWall, A.E., Pritchett, P.C., Galvin, C.J. (1977), *Beach changes caused by the Atlantic Coast storm of 17 Dec 1970*, Coastal Engineering Research Center Technical Paper 77-1

Dietz, R.S. and Holden, J.C. (1970), "The Breakup of Pangaea," *Scientific American*, 223, 30–41

Duckham, A.N. and Masefield, G.B. (1970), *Farming Systems of the World*, Chatto & Windus

Duvigneaud, P. and Denaeyer-De Smet, S. (1970), "Biological Cycling of Minerals in Temperate Deciduous Forests," in D.E. Reichle (ed.), *Analysis of Temperate Forest Ecosystems*, Berlin: Springer-Verlag, 199–225

Eastwood, T. (1964), *Stanford's Geological Atlas of Great Britain*, Edward Stanford Ltd

Edwards, A.M.C. and Thornes, J.B. (1973), "Annual Cycle of River Water Quality: A Time Series Approach," *Water Res. Research*, 9, 1286–95

Edwards, C.A. (1969), "Soil Pollutants and Soil Animals," *Scientific American*, 220, 88–89

Eichenlaub, V.L. (1970), "Lake Effect Snowfall to the Lee of the Great Lakes: Its Role in Michigan," *Bull. Amer. Meteorol. Soc.*, 51, 403–12

Embleton, C.E. and Thornes, J.B. (1979), *Process in Geomorphology*, Edward Arnold

Ericson, D.B. and Wollin, G. (1968), "Pleistocene Climates and Chronology in Deep-sea Sediments," *Science*, 162, 1227–34

Evans, G.H. (1970), "Pollen and Diaton Analysis of Late-glacial Quaternary Deposits in the Blelham Basin, North Lancashire," *New Phytologist*, 69, 821–74

Eyre, S.R. (1968), *Vegetation and Soils: A World Picture*, Arnold

Fairbridge, R.W. (1961), *Eustatic Changes in Sea-level. Phys. & Chem of the Earth*, IV: New York: Pergamon Press, 99–185

Finlayson, B. and Statham, I. (1980), *Hillslopes*, Butterworth

Fleagle, R.G. and Businger, J.A. (1963), *An Introduction to Atmospheric Physics*, Int. Geophysics Ser., vol. 5, New York: Academic Press

Flint, R.F. (1957), *Glacial and Pleistocene Geology*, New York: Wiley

Flint, R.F. (1971), *Glacial and Quaternary Geology*, New York: Wiley

Frissel, M.J. (ed.) (1978), "Cycling of Mineral Nutrients in Agricultural Ecosystems," Proc. First Environmental Symp., Royal Netherlands Land Development Society, Amsterdam. 31 May–4 June 1976. Amsterdam: Elsevier. Reprinted from *Agroecosystems*, 4, 1–354

Fritts, H.C., Smith, D.G. and Stokes, M.A. (1965), "The Biological Model for Paleoclimatic Interpretation of Mesa Verde Tree-ring Series," *Amer. Antiquity*, 31, pt 2, 101–21

Fuggle, R.G. and Oke, T.R. (1970), "Infra-red Flux Divergence and the Urban Heat Island," in *Urban Climates*, WMO technical note no. 108, Geneva: World Meteorological Organization, 70–8

Fuh, Baw-puh (1962), "The Influence of Slope Orientation on Micro-climate," *Acta Meteorologica Sinica*, 32, 71–86

Fulton, R.J. (ed.)(1989), "Quaternary Geology of Canada and Greenland," in *Geology of North America*, vol. K1, Ottawa: Geological Survey of Canada

Gabites, J.F. (1943), "Flying Conditions in the Tropical South Pacific," New Zealand Meteorological Office, series A, no. 3

Gardiner, V. (1983), "The Relevance of Geomorphology to Studies of Quaternary Morphogenesis," in D.J. Briggs and R.S. Waters (eds.), *Studies in Quaternary Geomorphology*, Geo Books, 1–18

Gay, L.W., Knoerr, K.N. and Braaten, M.O. (1971), "Solar Radiation Variability on the Floor of a Pine Plantation," *Agricultural Meteorol.*, 8, 39–50

Gerard, R.D. (1966), "Salinity in the Ocean," in R. W. Fairbridge (ed.), *Encyclopedia of Oceanography*, New York: Van Nostrand-Reinhold, 758–63

Gimingham, C.H. (1972), *Ecology of Heathlands*, Chapman & Hall

Godwin, H. (1975), *The History of the British Flora* (2nd ed.), Cambridge University Press

Goldthwait, R.P. (1976), "Frost Sorted Patterned Ground: A Review," *Quaternary Res.*, 6, 27–35

Golley, F. (1978), "Decomposition and Biogeochemical Cycles," in *Tropical Forest Ecosystems*, Pairs: UNESCO, 270–85

Goltsberg, I.A. (1969), *Microclimate of the USSR*, Jerusalem: Israel Program for Scientific Translation

Gordon, A.G. (1983) "Nutrient Cycling Dynamics in Different Spruce and Mixed Wood Ecosystems in Ontario and the Effects of Nutrient Removals through Harvesting;" in Wein, R.W., Rieve, R.R., and Methven, I.R. (eds.) (1983), *Resources and Dynamics of the Boreal Zone*, Ottawa: Association of Canadian Universities for Northern Studies

Goudie, A.S. (1983), *Environmental Change* (2nd ed.), Oxford University Press

Goudie, A.S., Cooke, R.U. and Evans, I.S. (1970), "Experimental Investigation of Rock Weathering by Salts," *Area*, 4, 42–8

Gray, W.M. (1978), "Hurricanes: Their Formation, Structure and Likely Role in the Tropical Circulation," in D.B. Shaw (ed.), *Meteorology Over the Tropical Oceans*, Royal Meteorological Society, 155–218

Gregory, K.J. and Walling, D.E. (1973), *Drainage Basin Form and Process*, Edward Arnold

Gregory, S. (1968), *Statistical Methods and the Geographer* (2nd ed.), Longman

Griggs, D.T. (1936), "The Factor of Fatigue in Rock Weathering," *J. Geology*, 44, 781–96

Grindley, J. (1972), "Estimation and Mapping of Evaporation," in *Symposium on World Water Balance*, publ. no. 92, IASH-UNESCO, 200–13

Hare, F.K. and Thomas, M.K. (1979), *Climate Canada* (2nd ed.), Toronto: Wiley

Harris, S.A. (1986), *The Permafrost Environment*, Barnes and Noble

Harvey, J. (1976), *Atmosphere and Ocean: Our Fluid Environments*, Artemis

Hastenrath, S. and Lamb, P.J. (1977), *Climatic Atlas of the Tropical Atlantic and Eastern Pacific Ocean*, Madison: University of Wisconsin Press

Hastenrath, S. and Lamb, P.J. (1978), *Heat Budget Atlas of the Tropical Atlantic and Eastern Pacific Ocean*, Madison: University of Wisconsin Press

Hebert, P.J. (1980), "A 'normal' year for hurricanes," *Weatherwise*, 33, 26–30

Hebert, P.J. and Taylor, G. (1979), "Hurricanes," *Weatherwise*, 32, 60–7, 100–7

Hilder, E.J. (1964), "The Distribution of Plant Nutrients by Sheep at Pasture," *Proc. Australian Soc. of Animal Production*, 5, 241–8

Hjulstrom, F. (1935), "Studies of the Morphological Activity of Rivers as Illustrated by the River Fyris," *Bull. Geol. Instit., Uppsala*, 25, 227–527

Holmes, A. (1965), *Principles of Physical Geology*, Nelson

Huff, F.A. and Shipp, W.L. (1969), "Spatial Correlations of Storm, Monthly and Seasonal Precipitation," *J. Appl. Meteorol*, 8, 542–50

Hurley, P.M. (1968), "The Confirmation of Continental Drift," *Scientific American*, 218, 52–64

Huss, E. and Stranz, D. (1970), "Die Wind-verhältnisse am Bodensee," *Pure and Appl. Geo-phys.*, 81, 323–56

Imbrie, J. and Imbrie, K.P. (1979), *Ice Ages: Solving the Mystery*, Macmillan

Ingmanson, D.E. and Wallace, W.J. (1973), *Oceanology: An Introduction*, Belmont, California: Wadsworth

Israelsen, O.W. and Hansen, V.E. (1962), *Irrigation Principles and Practices* (3rd ed.), New York: Wiley

Jackson, C.I. (1963), "Some Climatological Grumbles," Weather 18, 278–82

Jackson, I.J. (1977), *Climate, Water and Agriculture in the Tropics*, Longman

James, D.E. (1973), "The Evolution of the Andes," *Scientific American*, 229, 60–9

Jarvis, R.A. (1968), *Soils of the Reading District. Memoir of the Soil Survey of Great Britain*, Harpenden: Rothampsted Experimental Station

Johnston, D.W. and Odum, E.P. (1956), "Breeding Bird Populations in Relation to Plant Succession on the Piedmont of Georgia," *Ecology*, 37, 50–62

Jones, P.D. and Kelly, P.M. (1983), "The Spatial and Temporal Characteristics of Northern Hemisphere Surface Air Temperature Variations," *J. Climatol.*, 3, 243–52

Kellogg, W.W and Schneider, S.H. (1974), "Climate Stabilization For Better or For Worse," *Science*, 186, 1163–72

King, C.A.M. (1962), *Oceanography for Geographers*, Edward Arnold

Kirkby, A. (1969), "Primitive Irrigation," in R.J. Chorley (ed.), *Water, Earth and Man*, Methuen, 209–12

Kittredge, J. (1973), *Forest Influences*, New York: Dover

Knighton, A.D. (1981), "Longitudinal Changes in the Size and Shape of Stream Bed Material: Evidence of Variable Transport Conditions," *Catena*, 9, 25–34

Knighton, A.D. (1984), *Fluvial Forms and Processes*, Edward Arnold

Kononova, M.M. (1966), *Soil Organic Matter*, Pergamon Press

Koteswaram, P. and George, C.A. (1958), "On the Formation of Monsoon Depressions in the Bay of Bengal," *Indian Jl Meteorology and Geophysics*, 9, 9–22

Krajina, V.J. and Brooke, R.C. (1969), "Ecology of Western North America," vol. 2, nos. 1 and 2, Department of Botany, University of British Columbia, pp. 282, 288

Lachenbruch, A. (1962), "Mechanisms of Thermal Contraction Cracks and Ice Wedge Polygons in Permafrost," *Geological Society of America, Special Paper* 70

Ladochy, S., Scott, G.A.J. and Sweet, G. (1987), *A Laboratory Manual in Physical Geography* (2nd ed.), Kendall Hunt Publishing Co.

Lagrula, J. (1966), "Hypsographic curve," in R.W. Fairbridge (ed.), *Encyclopedia of Oceanography*, New York: Van Nostrand Reinhold, 364–6

Lamb, H.H. (1977), *Climate: Present, Past and Future*, vol. 2, *Climatic History and the Future*, Methuen

Lamb, H.H. (1982), *Climate, History and the Modern World*, Methuen

Lauenroth, W.K. (1979), "Grassland Primary Production: North American Grasslands in Perspective," in N. French (ed.), *Perspectives in Grassland Ecology*, New York: Springer-Verlag, 32, 3–24

Leopold, L.B. (1973), "River Channel Change with Time: An Example," *Bull., Geol. Soc. America.*, 84, 1845–60

Leopold, L.B. and Langbein, W.B. (1979), "River Meanders," in *The Physics of Everyday Phenomena*, Readings from *Scientific American*, 28–35, New York: Freeman

Leopold, L.B. and Wolman, M.G. (1957), "River Channel Patterns: Braided, Meandering and Straight," U.S. Geological Survey Professional Paper 282–B

Leopold, L.B., Wolman, M.G. and Miller, J.P. (1964), *Fluvial Processes in Geomorphology*, San Francisco: Freeman

Letréguilly, A. and Reynaud, L. (1989), "Spatial Patterns of Mass-balance Fluctuations of North American Glaciers," *J. Glaciology*, 35(120), 163

Lieth, H. (1964–5), "Versuch Einer Kartographischen Darstellung der Productivat der Pflanzendecke auf die Erde," *Geographisches Taschenbuch*, 72–80

Lorius, C., Jouzel, J., Ritz, C. *et al* (1985), "A 150,000 Year Climatic Record from Antarctic Ice," *Nature*, 316 (6029), 591–596

Ludlam, F.H. (1961), "The Hailstorm," *Weather*, 16, 152–62

Lundberg, J. (1977), "Karren of the Littoral Zone, Burren District, Co. Clare, Ireland," in *Proceedings 7th Int. Speleological Congress*, Sheffield 1977, 2, Karst morphology, British Cave Research Association, 291–3

Lvovitch, M.I. (1972), "World Water Balance (general report)," in *Symposium on World Water Balance*, publ. no. 92, IASH-UNESCO, 401–15

MacArthur, R.H., and Wilson, E.O. (1967), *The Theory of Island Biogeography*, Princeton: Princeton University Press

Mackay, J.R. (1970), "Disturbances to the Tundra and Forest Tundra Environment of the Western Arctic," *Canadian Geotechnical Journal*, 7, 420–432

Mackay, J.R. (1979), "Pingos of the Tuktoyaktuk Peninsula area, Northwest Territories," *Géographie Physique et Quaternaire*, 33, 3–61

Malkus, J.S. (1958), "Tropical Weather Disturbances—Why Do So Few Become Hurricanes?," *Weather*, 13, 75–89

Marshall, J.S. and Palmer, W.McK. (1948), "The Distribution of Raindrops with Size," *J. Meteorol.*, 5, 165–6

Mason, B.J. (1970), "Future Developments in Meteorology: An Outlook to the Year 2000," *Quart. Jl Roy. Meteorol. Soc.*, 96, 349–68

Mason, B.J. (1976), "Towards the Understanding and Prediction of Climatic Variations," *Quart. Jl Roy. Meteorol. Soc.*, 102, 473–98

Mather, J.R. (1974),*Climatology: Fundamentals and Applications*, New York: McGraw-Hill

Mather, K.B. and Miller G.S. (1967), "Notes on Topographic Factors Affecting the Surface Wind in Antarctica, with special reference to katabatic winds," Univ. Alaska, Tech. Rep. UAG–K–189

Mattingly, G.E.G., Chater, M. and Johnston, A.E. (1975), "Experiments Made on Stackyard Field, Woburn, 1867–1974. III. Effects of NPK Fertilizers and Farmyard Manure on Soil Carbon, Nitrogen and Organic Phosphorus," Report Rothamsted Experimental Station for 1974, pt 2, 61–77

McGowan, A. and Derbyshire, E. (1977), "Genetic Influences on the Properties of Tills," *Quarterly Jl Engineering Geol.*, 10, 391

Meyer, L.D. and Wischmeier, W.H. (1969), "Mathematical Simulation of the Process of Soil Erosion by Water," *Trans. Amer. Soc. Agricult. Engnr.*, 12, 754–8; 762

Michel, B. (1978), *Ice Mechanisms*, Montreal: Presses de l'Université Laval

Miller, A. (1966), *Meteorology*, Columbus, Ohio: Merrill Physical Science Series

Miller, D.H. (1965), "The Heat and Water Budget of the Earth's Surface," *Advances in Geophysics*, 11, 175–302

Miller, D.H. (1977a), *Water at the Surface of the Earth*, New York: Academic Press

Miller, D.H. (1977b), *International Geophysics*, no. 21, New York: Academic Press

Mitchell, J.M. (1968), Concluding remarks, in "Causes of Climatic Change," Amer. Meteorological Soc., *Meteorol. Monog.*, 8 no. 30, 155–9

Moffitt, B.J. and Ratcliffe, R.A.S. (1972), "Northern Hemisphere Monthly Mean 500 mb and 1000–500mb Thickness Charts and Some Derived Statistics," *Geophysical Memoir*, 117

Monkhouse, F.J. (1975), *Principles of Physical Geography* (8th Ed.), Hodder & Stoughton

Moore, P. and Hunt, D. (1987), *Atlas of the Solar System*, Rand McNally

More, R.J. (1967), "Hydrological Models and Geography," in R.J. Chorley and P. Haggett (eds.), *Physical and Information Models in Geography*, Methuen, 145–85

Muller, F. (1968), "Pingos, Modern," in R.W. Fairbridge (ed.), *The Encyclopedia of Geomorphology*, Reinhold, 845–847

Munn, R.E. (1966), "Descriptive Micrometeorology," *Advances in Geophysics*, supplement 1, New York: Academic Press

National Academy of Sciences (1975), *Understanding Climatic Change: A Program for Action*, Washington: US National Academy of Sciences

National Environment Research Council (1975), Flood Studies Report, vol. 2, Meteorological Studies, NERC

Neiburger, M., Edinger, J.G. and Bonner, W.D. (1982), *Understanding Our Atmospheric Environment* (2nd ed.), San Francisco: Freeman

Newell, R.E., Kidson, J.W., Vincent, D.G. and Boer, G.J. (1974), *The General Circulation of the Tropical Atmosphere*, vol. 2, Cambridge, Mass: MIT Press

Newell, R.E., Vincent, D.G., Dopplick, T.G., Ferruza, D. and Kidson, J.W. (1969), "The Energy Balance of the Global Atmosphere," in G.A. Corby (ed.), *The Global Circulation of the Atmosphere*, Royal Meteorological Society, 42–90

Nichols, H. (1967), "The Postglacial History of Vegetation and Climate at Ennedai Lake, Keewatin and Lynn Lake, Manitoba (Canada)," *Eiszeitalter und Segenwant*, 18, 176–97

Nicholson, S.E. (1983), "Sub-Saharan Rainfall in the Years 1976–1980," *Monthly Weather Review*, 111, 1646–54

Nieuwolt, S. (1968), "Uniformity and Variation in an Equatorial Climate," *J. Tropical Geogr.*, 27, 23–39

Nieuwolt, S. (1977), *Tropical Climatology: An Introduction to the Climates of Low Latitudes*, Wiley

Nkemdirim, L.C. (1976), "Dynamics of An Urban Temperature field—A Case Study," *Jl Appl. Meteorol.*, 15, 818–28

Nye, D.H. and Greenland, D.J. (1960), *The Soil Under Shifting Cultivation*, Commonwealth Bureau of Soils, Technical Communication 51, Harpenden: Commonwealth Agricultural Bureaux

Oke, T.R. (1978), *Boundary-layer Climates*, Methuen

Oke, T.R. and Maxwell, G.B. (1975), "Urban Heat Island Dynamics in Montreal and Vancouver," *Atmospheric Environment*, 9, 191–200

Olson, R., Hastings R., Geddes, F., eds. (nd), *Northern Ecology and Resource Management*, University of Alberta Press

Ostrem, G., Haakensen, N., Kjollmoen, B., Laumann, T. and Wold, B. (1991), "Massebalansemalinger pa Norske breer 1988–89," Norwegian Hydroelectric Board publication no. 11

Ovington, J.D. (1962), "Quantitative Ecology and the Woodland Ecosystems Concept," in J.B. Craggs (ed.), *Advances in Ecological Research*, New York: Academic Press, 103–91

Oxburgh, E.R. (1974), "The Plain Man's Guide to Plate Tectonics," *Proc. Geol. Assoc.*, 85, 299–357

Palmén, E. and Newton, C.W. (1969), "Atmospheric Circulation Systems," Int. Geophys. Ser. 15, New York: Academic Press

Palmer, R.C. (1976), "Soils in Herefordshire IV," Sheet SO 74 (Malvern), Soil Survey Record no. 36., Rothamsted Experimental Station

Paterson, J.H. (1975), *North America*, New York: Oxford University Press

Paterson, W.S.B. (1981), *The Physics of Glaciers*, Oxford: Pergamon Press

Paterson, W.S.B. (1987) *The Physics of Glaciers* (2nd ed.), Oxford: Pergamon Press

Peltier, L.C. (1950), "The Geographic Cycle in Periglacial Regions as it is Related to Climatic Geomorphology," *Annals Ass. Amer. Geographers*, 40, 214–36

Penman, H.L. (1948), "Natural Evaporation from Open Water, Bare Soil and Grass," *Proc. Roy. Soc.* 193A, 120–48

Petterssen, S. (1950), "Some Aspects of the General Circulation of the Atmosphere," *Centenary Proc. Roy. Meteorol. Soc.*, Royal Meteorological Society, 120–55

Péwé, T.L. (1983), "The Periglacial Environment in North America During Wisconsin Time," in S. C. Porter (ed.), *The Late Pleistocene*, University of Minnesota Press, Minneapolis, 157–189

Phillips, D. (1990), *The Climates of Canada*, Ottawa: Ministry of Supply and Services Canada

Phillpot, H.R. and Zillman, J.W. (1969), *The Surface Temperature Over the Antarctic Continent*, Commonwealth Bureau of Meteorology, Melbourne

Pihlainen, J.A. and Johnston, G.H. (1963), *Guide to a Field Description of Permafrost*, National Research Council of Canada, Technical Memorandum No. 79, 21 p

Pirie, R.G. (1973), "Oceanography," *Contemporary Readings in Ocean Science*, New York: Oxford University Press

Pittock, A.B., Frakes, L.A., Jenssen, D., Peterson, J.A. and Zellman, J.W. (1978), *Climatic Change and Variability: A Southern Perspective*, Cambridge University Press

Plumley, W. (1948), "Black Hills Terrace Gravels: A Study in Sediment Transport," *Jl Geology*, 56, 526–77

Pruitt, W.O. (1957), "Observations on the Bioclimate of some Taiga Mammals," *Arctic*, 10

Pruitt, W.O. (1978), *Boreal Ecology*, Edward Arnold

Richey, J.E. (1964), *Scotland: The Tertiary Volcanic Districts* (3rd ed.), Geological Survey of GB

Richter, D.A. and Dahl, R.A. (1958), "Relationship of Heavy Precipitation to the Jet Maximum in the Eastern U.S.," *Monthly Weather Review*, 86, 368–76

Riehl, H. (1979), *Climate and Weather in the Tropics*, New York: Academic Press

Rodda, J.C. (1969), "An Assessment of Precipitation," in R.J. Chorley (ed.), *Water, Earth and Man*, Methuen, 130–4

Rodin, L.E. and Bazilevich, N.I. (1965), *Production and Mineral Cycling in Terrestrial Vegetation*, Oliver & Boyd

Rona, P.A. (1973), "Plate Tectonics and Mineral Resources," *Scientific American*, 229, 86–95

Rorison, I. (1969), "Ecological Inferences from Laboratory Experiments of Mineral Nutrition," in I. Rorison (ed.), *Ecological Aspects of the Mineral Nutrition of Plants*, Symposia of British Ecological Society, 9, 155–75, Blackwell

Rubin, J. and Bielorai, H. (1957), "A Study of the Irrigation Requirements and Consumptive Water Use by Sugar Beet in the Northern Negev," in *Final Report of the Ford Foundation Israel Project*, A5, spec. bull. Volcani Inst. Agric. Res. no. 6, 20–4

Rusin, N.P. (1964), *Meteorological and Radiational Regime of Antarctica*, Jerusalem: Israel Program for Scientific Translations

Russell, E.W. (1974), *Soil Conditions and Plant Growth*, Longman

Saugier, B. (1977), "Micrometeorology on Crops and Grassland," in J.J. Landsberg and C.V. Cutting (eds.), *Environmental Effects on Crop Physiology*, Academic Press

Savage, J.C. and Paterson, W.S.B. (1963), "Borehole Measurements in the Athabasca Glacier," *Jl Geophysical Res.*, 68, 4521–36

Sawyer, J.S. (1956), "Rainfall of Depressions Which Pass Eastward Over or Near the British Isles," *Professional Notes*, 7 (118), HMSO

Schmidt, W. (1930), "Der Tiefsten Minimum Temperatur in Mitteleuropa," *Naturwissenschaft*, 18, 367–9

Schumm, S.A. (1977), *The Fluvial System*, New York: Wiley

Sellers, W.D. (1965), *Physical Climatology*, Chicago: University of Chicago Press

Semb, A. (1978), "Sulphur Emissions in Europe," *Atmospheric Environment*, 12, 455–60

Shellard H.C. (1976), "Wind," in T.J. Chandler and S. Gregory (eds.), *The Climate of the British Isles*, Longman, 39–73

Short, S.K. and Jacobs, J.D. (1982), "A 1 100 Year Paleoclimatic Record from Burton Bay-Tarr Inlet, Baffin Island," *Canadian Journal of Earth Sciences*, 19(3), 398-409

Smith, D.I. (1981), "Actual and Potential Flood Damage: A Cast Study for Urban Lismore, NSW, Australia," *Applied Geography*, 1, 31–9

Smith, K. (1975), *Principles of Applied Climatology*, McGraw-Hill

Smith, R.L. (1966), *Ecology and Field Biology*, New York: Harper & Row

Sparks, B.W. (1971), *Rocks and Relief*, Longman

Stearns, C.R. (1969), "Surface Heat Budget of the Pampa de La Joya, Peru," *Monthly Weather Review*, 97, 860–6

Steiner, J.T. (1980), "The Climate of the South-West Pacific Region," New Zealand Meteorological Service, misc. publ., 166

Stewart, J.H. and Lamarche, V.C. (1967), "Erosion and Deposition Produced by the Floods of December 1964 on Coffee Creek, Trinity County, California," *US Geol. Survey Professional Paper*, 422–K

Strahler, A.N. and Strahler, A. H. (1979), *Elements of Physical Geography* (2nd ed.), New York: Wiley

Strakhov, N.M. (1967), *Principles of Lithogenesis*, vol. I. Oliver & Boyd

Stride, A.H. (1982), "Sand Transport," in A.H. Stride (ed.), *Offshore Tidal Sands—Processes and Deposits*, Chapman Hall, 58–94

Sugden, D.E. and John, B.S. (1977), *Glaciers and Landscape*, Edward Arnold

Sverdrup, H.U., Johnson, M.W. and Fleming, R.H. (1970), *The Oceans: Their Physics, Chemistry and General Biology* (2nd ed.), Englewood Cliffs, NJ: Prentice-Hall Inc.

Thompson, B.W. (1965), *The Climate of Africa*, Nairobi: Oxford University Press

Thornbury, W.D. (1954), *Principles of Geomorphology*, New York: Wiley

Toksoz, M.N. (1975), "The Subduction of the Lithosphere," *Scientific American*, 233, 88–98

Trafford, B.D. (1970), "Field Drainage," *Jl Roy. Agricult. Soc. England*, 131, 129–52

Tsung-lien Chou (1976), "The Yellow River: Its Unique Features and Serious Problems," Rivers '76, American Society of Civil Engineers, 507–26

Turner, C. (1970), "The Middle Pleistocene Deposits of Marks Tey, Essex," *Phil. Trans. Roy. Soc.*, 257B, 373–437

Van Bavel, C.H.M., Fritschen, L.J. and Reeves, W.E. (1963), "Transpiration by Sudangrass As an Externally Controlled Process," *Science*, 141, 269–70

Visher, S.S. (1945), "Climatic Maps of Geologic Interest," *Bull. Geol. Soc. America*, 56, 713–76

Walling, D.E. (1974), "Suspended Sediment and Solute Yields from a Small Catchment Prior to Urbanisation," in K.J. Gregory and D.E. Walling (eds.), *Fluvial Processes in Instrumented Watersheds*, Institute of British Geographers, spec. publication no. 6, 169–92

Walling, D.E. (1978), "Suspended Sediment and Solute Response Characteristics of the River Exe, Devon, England," in R. Davidson-Arnott and W. Nickling (eds.), *Research in Fluvial Systems*, Geobooks, 169–97

Ward, R.C. (1975), *Floods: A Geographical Perspective*, Macmillan

Wein, R.W. and MacLean, D.A. (1983), *The Role of Fire in Northern Circumpolar Ecosystems*, published on behalf of the Scientific Committee on Problems of the Environment of the International Council of Scientific Unions; Toronto: Wiley

Weyl, P.K. (1970), *Oceanography: An Introduction*, New York: Wiley

Whittaker, R.H. (1975), *Communities and Ecosystems*, New York: Macmillan

Whittaker, R.H. and Likens, G.E. (1975), "The Biosphere and Man," in H. Lieth and R.H. Whittaker (eds.), *Primary Productivity of the Biosphere, Ecological Studies*, 14, 305–28, New York: Springer-Verlag

Wiman, S. (1963), "A Preliminary Study of Experimental Frost Weathering," *Geografiska Annaler*, 45A, 113–21

Wise, A.F.E. (1971), "Effects Due to Groups of Buildings," *Phil. Trans. Roy. Soc.*, 269A, 469–85

Wolman, M.G. (1955), "The Natural Channel of Brandywine Creek, Pennsylvania," US Geol. Survey, professional paper no. 271

Wolman, M.G. (1967), "A Cycle of Sedimentation and Erosion in Urban River Channels," *Geografiska Annaler*, 49A, 385–95

Woodwell, G.M. (1963), "The Ecological Effects of Radiation," *Scientific American*, 208, 40–9

Yoshino, M.M. (1975), *Climate in a Small Area*, Tokyo: University of Tokyo Press

Young, C.P. and Gray, E.M. (1978), *Nitrate in Groundwater*, Technical Report TR69, Medmenham: Water Research Centre

# Index

(*Italic* page numbers are illustrations)